Handbook of Molecular Descriptors

Roberto Todeschini and Viviana Consonni

WILEY-VCH

Methods and Principles in Medicinal Chemistry

Edited by
R. Mannhold
H. Kubinyi
H. Timmerman

Handbook of Molecular Descriptors

Roberto Todeschini and Viviana Consonni

WILEY-VCH

Weinheim · New York · Chichester · Brisbane · Singapore · Toronto

Series Editors:
Prof. Dr. Raimund Mannhold
Biomedical Research Center
Molecular Drug Research Group
Heinrich-Heine-Universität
Universitätsstraße 1
40225 Düsseldorf
Germany

Prof. Dr. Hugo Kubinyi
Combinatorial Chemistry
and Molecular Modelling
ZHF/G, A 30
BASF AG
67056 Ludwigshafen
Germany

Prof. Dr. Hendrik Timmerman
Faculty of Chemistry
Dept. of Pharmacochemistry
Free University of Amsterdam
De Boelelaan 1083
1081 HV Amsterdam
The Netherlands

Authors:
Prof. Dr. R. Todeschini
Dr. V. Consonni
Milano Chemometrics and QSAR Research Group
Dip. di Scienze dell'Ambiente e del Territorio
Universita degli Studi Di Milano-Bicocca
Piazza della Scienza 1
20126 Milano
Italy

Cover: Staatliche Museen zu Berlin – Preußischer Kulturbesitz
Vorderasiatisches Museum

Library of Congress Card No. applied for

British Library Cataloguing-in-Publication Data: A catalogue record for this book is available from the British Library

Die Deutsche Bibliothek – CIP-Cataloguing-in-Publication Data:
A catalogue record for this book is available from Die Deutsche Bibliothek

ISBN 3-52-29913-0

Printed on acid-free paper.

Composition: Kühn & Weyh, D-79111 Freiburg
Printing: Betzdruck GmbH, D-63291 Darmstadt
Bookbinding: Osswald & Co., D-67433 Neustadt (Weinstraße)

Printed in the Federal Republic of Germany.

To our loved ones
R.T. and V.C.

Any alternative viewpoint with a different emphasis leads to an inequivalent description. There is only one reality but there are many viewpoints. It would be very narrow-minded to use only one: we have to learn to be able to imagine several.

Hans Primas

In: Chemistry, Quantum Mechanics and Reductionism
(Springer-Verlag, 1981)

Cover

Dragon
Babylon Ištar Gate (600 – 500 B.C.)
(Pergamon Museum of Berlin)

A molecular descriptor can be thought of as a mythological animal having several different meanings which depend on one's point of view.

Preface

In the late 1930s, the Hammett equation marked a breakthrough in the understanding of organic chemistry. It describes rate and equilibrium constants of the reactions of aromatic acids, phenols and anilines, as well as other compounds, in a quantitative manner, by using reaction and substituent parameters, ϱ and σ. In the same manner, the lipophilicity parameter π, derived from *n*-octanol/water partition coefficients by Corwin Hansch, led to a breakthrough in quantitative structure-activity relationships (QSAR) in biology. Like σ, π also is an additive constitutive molecular property. Their combination, later also with molar refractivity values MR, Taft's steric E_s values or the Verloop steric parameters, allowed the derivation of quantitative models for many biological *in vitro* activity values. Nonlinear lipophilicity models describe *in vivo* biological activities, where substance transport through membranes and distribution within the biological systems play an important role.

Practical problems in the estimation of the lipophilicity of araliphatic and aliphatic compounds led to the *f* hydrophobicity scales of Rekker and Leo/Hansch. However, all such descriptor scales depend on experimental determinations. New molecular descriptors were developed from scratch, starting with the work of Randic, Kier and Hall, i.e. the various molecular connectivity parameters χ. Later the electrotopological state parameters and the Todeschini WHIM parameters were added. Whereas topological descriptors are mathematical constructs that have no unique chemical meaning, they are clearly related to some physicochemical properties and are suited to the description of compound similarities in a quantitative manner. Thus, despite several critical comments in the past, they are now relatively widely used in QSAR studies. Only a meaningless and excessive application in quantitative models, as far as the number of tested and included variables is concerned, still deserves criticism.

This book is a long-awaited monograph on the various properties and molecular descriptors that are of importance in studies of chemical, physicochemical, and biological properties. It is a must for every research worker who is active in this field because it provides an encyclopedic overview of all known descriptors, whether they are physicochemical or topological in their nature. An exhaustive list of references points the way to the original literature.

The series editors wish this book a wide distribution. It is up to the reader to find out which of the properties and descriptors might be most suitable for describing the data. However, the early warnings by Corwin Hansch, John Topliss, and others should not be forgotten: make your model as simple as possible; test and include only a few parameters; try to achieve an understanding of your model; use a test set to check the external predictivity of your model. Molecular descriptors are powerful tools in QSAR studies – but their abuse may lead nowhere. May this book further contribute to their selective and proper use!

August 2000

Raimund Mannhold, Düsseldorf
Hugo Kubinyi, Ludwigshafen
Hendrik Timmerman, Amsterdam

Contents

Introduction

The effort being made today to organize Knowledge is also a way of participating in the evolution of Knowledge itself. In fact, the significance of attempting such organization can be looked for in its ability not only to give information but also to create know-how: it provides not only a collection of facts – a store of information – but also a contribution to the evolution of Knowledge. The fact is that splitting the organization of Knowledge from its production is completely arbitrary: actually, Knowledge organization is itself one way of doing research.

The true end of an encyclopedic guide is to contribute to the growth of knowledge, but not to knowledge given once and for all, based on some final basic theories, but as a *network of models* in progress. A network primarily consists of knots, i.e. objects, facts, theories, statements, and models, the links between the knots being relationships, comparisons, differences, and analogies: such a network of models is something more than a collection of facts, resulting in a powerful engine for analogical reasoning.

With these purposes in mind, the Authors conceived this *Handbook of Molecular Descriptors* as an encyclopedic guide to molecular descriptors.

First, let us look at the definition of molecular descriptor.

The molecular descriptor is the final result of a logical and mathematical procedure which transforms chemical information encoded within a symbolic representation of a molecule into an useful number or the result of some standardized experiment.

Attention is paid to the term "useful" with its double meaning: it means that the number can give more insight into the *interpretation* of the molecular properties *and/or* is able to take part in a model for the *prediction* of some interesting property of other molecules.

Why must we also accept "or"?

A fundamental task is how to predict and understand experimental facts, i.e. physico-chemical properties, biological activities, and environmental behaviour, from symbolic representations of real objects such as molecules by molecular descriptors.

Because of the huge complexity of this problem, it must be understood that during the development of new descriptors, their interpretation can be weak, provisional, or completely lacking, but their predictive ability or usefulness in application to actual problems can be a strong motive for their use. On the other hand, descriptors with poor predictive power can be usefully retained in models when they are well theoretically founded and interpretable because of their ability to encode structural chemical information.

The incompletely realized comprehension of the chemical information provided by molecular descriptors cannot be systematically ascribed to weakness in the descriptors. Actually, our inability to reduce descriptor meanings to well-established chemical concepts is often because newly emergent concepts need new terms in the language and new hierarchically connected levels for scientific explanation. Thus, what is often considered as scientific failure is sometimes the key to new useful knowledge.

In any case, all the molecular descriptors must contain, to varying extents, chemical information, must satisfy some basic invariance properties and general requirements, and must be derived from well-established procedures which enable molecular descriptors to be calculated for any set of molecules. It is obvious – almost trivial – that a single descriptor or a small number of numbers cannot wholly represent the molecular complexity or model all the physico-chemical responses and biological interactions. As a consequence, although we must get used to living with approximate models (*nothing is perfect*!), we have to keep in mind that "approximate" is not a synonym of "useless".

The field of molecular descriptors is strongly interdisciplinary and involves a mass of different theories. For the definition of molecular descriptors, a knowledge of algebra, graph theory, information theory, computational chemistry, and theories of organic reactivity and physical chemistry is usually required, although at different levels. For the use of the molecular descriptors, a knowledge of statistics, chemometrics, and the principles of the QSAR/QSPR approaches is necessary in addition to the specific knowledge of the problem. Moreover, programming and sophisticated software and hardware are often inseparable fellow-travellers of the researcher in this field.

The historical development of molecular descriptors reflects some of the distinctive characteristics of the most creative scientists, i.e. their capability of being at the same time engaged and/or detached, rational and/or quirky, serious and/or not so serious. Science is a game and the best players appreciate not only the beauty of a discovery by precise and logical reasoning, but also the taste of making a guess, of proposing eccentric hypotheses, of being doubtful and uncertain when confronted by new and complex problems. Molecular descriptors constitute a field where the most diverse strategies for scientific discovery can be found.

Molecular descriptors will probably play an increasing role in scientific growth. In fact, the availability of large numbers of theoretical descriptors containing diverse sources of chemical information would be useful to better understand relationships between molecular structure and experimental evidence, also taking advantage of more and more powerful methods, computational algorithms and fast computers. However, as before, deductive reasoning and analogy, theoretical statements and hazardous hypotheses, determination and perplexity still remain fundamental tools.

The *Handbook of Molecular Descriptors* tries to meet the great interest that the scientific community is showing in this topic. In fact, as well as the solid interest in the quantitative modelling of biological activity, physico-chemical properties, and environmental behaviour of compounds, an increasing interest has been shown by the scientific community in recent years in the fields of combinatorial chemistry, high-throughput screening, substructural analysis, and similarity searching, for which several approaches which are particularly suitable for informatic treatment have been proposed. Thus, several disciplines such as chemistry, pharmacology, environmental sciences, drug design, toxicology, and quality control for health and safety derive great advantages from these methodologies in their scientific and technological development.

Although experimental measurements would be the most direct way of obtaining safe and high-quality information, the time, costs and hazards involved in the experimentation are relevant limiting factors if we wish to entrust our knowledge growth to fully experiment-based strategies. Moreover, and most importanly, experiments are always suggested by theories, and theoretical models are the way we try to understand reality. Therefore, both general and local models play a fundamental role in scientific growth, leading to a better understanding of the properties of studied phenomena.

The *Handbook of Molecular Descriptors* collects the definitions, formulas and short comments of molecular descriptors known in chemical literature, our intention being to consider all the known molecular descriptors. The definitions of technical terms, around 1800 in all, are organized in alphabetical order. The importance of a molecular descriptor definition is not related to its length. Only a few old descriptors, abandoned or demonstrated as wrong, have been intentionally left out to avoid confusion. An effort was also made to collect bibliographic information appropriate for the work proposed in this Handbook. We are sorry if any relevant descriptor and/or work has been missed out; this has not be done deliberately and we take full responsibility for any omissions.

Many molecular descriptors have been grouped into classes using a mixed taxonomy based on different points of view, in keeping with the leading idea of the Handbook to promote learning by comparison. Descriptors have been often distinguished by their *physico-chemical meaning* or the specific *mathematical tool* used for their calculation.

Some basic concepts and definitions of statistics, chemometrics, algebra, graph theory, similarity/diversity, which are fundamental tools in the development and application of molecular descriptors, are also presented in the Handbook in some detail. More attention has been paid to information content, multivariate correlation, model complexity, variable selection, and parameters for model quality estimation, as these are the characteristic components of modern QSAR/QSPR modelling.

The Handbook contains nothing about the combinatorial algorithms for the generation and enumeration of chemical graphs, the basic principles of statistics, algorithms for descriptor calculations, or experimental techniques for measuring physico-chemical and biological responses. Moreover, relevant chemometric methods such as Partial Least Squares regression (PLS) and other regression methods, classification methods, cluster analysis, and artificial neural networks, which are also widely applied on molecular descriptors, are simply cited; references are given, but no theoretical aspect is presented. Analogously, computational chemistry methods are only quoted as important tools for calculations, but no claim is made here to their detailed explanation.

Information exchange

The Authors would be grateful to all researchers who would like to send their observations and comments on the Handbook contents, information about new descriptors, and bibliographic references. For e-mail submissions address to: *moldes@disat.unimib. it.*

The Authors are activating a website (*http://disat.unimib.it/chm/*) where recent progress in the molecular descriptor field will be reported, tables of calculated descriptor values for standard data sets collected, and links to other related websites proposed. All information received from interested and collaborative researchers will be useful to develop this Internet tool for sharing ideas and experiences on molecular descriptors and QSAR.

Bibliographic references

The reference list covers a period between 1858 and 2000, and includes 3300 references, about 3000 authors and 250 periodicals. The symbol [R] at the end of a reference denotes a publication with a significant list of references.

Acknowledgements

The idea of producing a *Handbook of Molecular Descriptors* was welcomed by several colleagues whom we warmly thank for their suggestions, revisions, bibliographic information, and moral support. We particularly thank Alexander Balaban,

Subhash Basak, Laura Belvisi, Pierre-Alain Carrupt, Claudio Chiorboli, Johann Gasteiger, Paola Gramatica, Peter Jurs, Lemont Kier, Douglas Klein, Hugo Kubinyi, Silvia Lanteri, Alessandro Maiocchi, Marjana Novic, Demetrio Pitea, Lionello Pogliani, Milan Randic, Gianfranco Tantardini, Bernard Testa, Giuseppe Triacchini, and Jure Zupan.

We are also grateful to Bracco SPA of Milan, EniRicerche of Milan, and the National Institute of Chemistry of Ljiubliana for support in the bibliographic search.

Roberto Todeschini and Viviana Consonni

Milano, June, 2000

User's Guide

This handbook consists of definitions of technical terms in alphabetical order, each technical term being an *entry* of the Handbook.

Each topic is organized in a hierarchical fashion. By following cross-references (→ and typeset in italics) one can easily find all the entries pertaining to a topic even if they are not located together. Starting from the topic name itself, one is referred to more and more specific entries in a top-down manner.

Each entry begins with an entry line.

There are three different kinds of entries: *regular*, *referenced*, and *synonym*.

A **regular entry** has its definition immediately after the entry line. A regular entry is typeset in bold face; it is followed by its (ACRONYM and/or SYMBOL), if any, and by its (*: synonyms*), if any. For example:

Wiener index (*W*) (*: Wiener number*)

A **referenced entry** has its definition in the text of another entry indicated by the symbol → and typeset in bold face. For example:

Wiener orthogonal operator → **algebraic operators**

A **synonym entry** is followed by the symbol " *:* " and its synonym typeset in italics. To find the definition of a synonym entry, if the synonym is a regular entry, one goes directly to the text under the entry line of the synonym word; otherwise, if the synonym is a referenced entry, one goes to the text of the entry indicated by →, typeset in bold face letters. For example:

Wiener number *: Wiener index*

walk number *: molecular walk count* → **walk counts**

The text of a regular entry may include the definition of one or more referenced entries highlighted in bold face. When there are many referenced entries collected under one regular entry, called a "mega" entry, they are often organized in hierarchical fashion, denoting them by the symbol •. The sub-entries can be in either alphabetic or logical order. For example, in the mega entry "steric descriptors", the first sub-entries, each followed by the corresponding text, are:

- **gravitational indices**
- **Kier steric descriptor**
- **Austel branching index**

Finally, a referenced entry within a sub-entry has its definition in the text of the sub-entry denoted by the symbol (⊙ ...). For example:

WHIM shape → **WHIM descriptors** (⊙ global WHIM descriptors)

indicates that the index "WHIM shape" is defined in the sub-entry "global WHIM descriptors" of the main entry "WHIM descriptors".

In the text of a regular entry one is referred to other relevant terms by words in italics indicated by →. In order to reach a complete view of the studied topic, we highly recommend reading also the definitions of these words in conjunction with the original entry. For example:

count descriptors

These are simple molecular descriptors based on counting the defined elements of a compound. The most common chemical count descriptors are → *atom number A*, → *bond number B*, → *cyclomatic number C*, → *H-bond acceptor index* and → *H-bond donor index* counts, → *distance-counting descriptors*, → *path counts*, → *walk counts*. …

Finally, words in italics not indicated by → in the text of a main entry (or sub-entry) denote relevant terms for the topic which are not further explained or whose definition is reported in a successive part of the same entry.

The symbol 🕮 at the end of each entry denotes a list of suggested bibliographic references.

We have made a special effort to keep mathematical notation simple and uniform. A collection of the most often appearing symbols are in the next paragraph Notations and Symbols. Moreover, a list of acronyms helps to decipher and locate the full terminologies given in the book.

Notations and Symbols

The notations and symbols used in the Handbook are listed below. In some cases, notations slightly different from those proposed by the Authors are used to avoid confusion with other descriptors and quantities.

Objects

$\mathcal{X}$	molecular descriptor
$\mathcal{M}$	molecule, compound
$\mathcal{M}$	experimental measure
$\mathcal{P}$	experimental property
$\mathcal{G}$	graph, molecular graph
$\mathcal{MG}$	multigraph, molecular multigraph

Sets

$\mathcal{V}$	set of vertices of a graph
$\mathcal{E}$	set of edges of a graph
$\mathcal{F}$	set of fragments of a molecule partition
$\mathcal{G}$	set of points in a 3D grid
${}^{m}P_{ij}$	set of atoms of the path of order m from the ith to the jth atoms
${}^{m}\mathcal{P}$	set of paths of order m

Counts

A	number of atoms of a molecule
B	number of bonds of a molecule
C	number of cycles of a molecule
C^{+}	number of cycles with overlapping of a molecule
G	number of equivalence classes
h_a	number of hydrogens bonded to the atom a
L	principal quantum numbers
n	number of objects, data, molecules
n_x	number of elements with an x-value
N	generic number of elements
$\mathrm{N_X}$	number of atoms, groups, fragments of X-type
M	number of significant principal components or latent variables
p	number of variables
${}^{m}P$	number of paths of length m
P	total number of paths of a graph
Z_a	atomic number of the atom a

Matrix operators

$\mathcal{C}$	column sum operator
$\mathcal{D}$	diagonal operator
$\mathcal{R}$	row sum operator
$\mathcal{S}$	total sum operator
$\mathcal{W}$	Wiener operator

Indices and characteristic symbols

a	index on the atoms of a molecule
b	index on the bonds of a molecule
g	index on the equivalence classes
i,j,k,l,f,m	generic indices
x,y,z	geometric coordinates
d	data distances
d	topological distances
r	geometric distances
δ	vertex degree
δ^{b}	bond vertex degree
δ^{v}	valence vertex degree
ε	edge degree
η	atom eccentricity
π	bond order
σ	vertex distance degree
m	order of a descriptor, exponent
w	weights, atom properties
p	probability
q	atomic charge
ℓ_{jm}	PCA loadings of the mth component for the jth variable
λ	eigenvalue
$\lambda_j(\mathbf{M})$	jth eigenvalue from the matrix **M**
$[\mathbf{M}]_{ij}$, m_{ij}	i-j element of the matrix **M**
t_{im}	ith score of the mth component from PCA or PLS
$\mathbf{t}_m$	mth vector score
$\mathbf{v}$	generic column vector
$\mathbf{p}_i$	ith grid point of coordinates (x,y,z)
$\mathbf{a}$	vector of atoms in a path
D	dimension (0,1,2,3)
D	diameter
R	radius
I	binary or indicator variable
I	information content

Acronyms

The most well-known acronyms used to define research fields, methods, and molecular descriptors are listed below, in alphabetical order.

AAA	Active Analog Approach
AAC	Augmented Atom Codes
AID	Atomic ID number
ANN	Artificial Neural Networks
ATS	Autocorrelation of a Topological Structure
AWC	Atomic Walk Count
BP-ANN	Back-Propagation Artificial Neural Networks
BIC	Bonding Information Content
BID	Balaban ID number
BLOGP	Bodor LOG*P*
CADD	Computer-Aided Drug Design
CAMD	Computer-Aided Molecular Design
CAMM	Computer-Aided Molecular Modelling
CASE	Computer-Automated Structure Evaluation
CHEMICALC	Combined Handling of Estimation Methods Intended for Completely Automated Log*P* Calculation
CIC	Complementary Information Content
CID	Connectivity ID number
CLOGP	Calculated LOG*P*
CoMFA	Comparative Molecular Field Analysis
CoMMA	Comparative Molecular Moment Analysis
CoMSIA	Comparative Molecular Similarity Indices Analysis
COSV	Common Overlap Steric Volume
CP-ANN	Counter-Propagation Kohonen Artificial Neural Networks
CPK	Corey-Pauling-Koltun volume
CPSA	Charged Partial Surface Areas
CR	Continuum Regression
CSA	Cluster Significance Analysis
DARC	Description, Acquisition, Retrieval Computer system
DD	Drug Design
DFT	Density Functional Theory
DG	Distance Geometry
EAID	Extended Adjacency ID number
EC	Extended Connectivity
ECA	Extended Connectivity Algorithm
ECI	Electronic Charge Index
EEVA	Electronic EigenVAlue descriptors
EVA	EigenVAlue descriptors
FEVA	First EigenValue Algorithm
FW	Free-Wilson analysis
GA	Genetic Algorithms
GAI	General a_N-Index
GA-VSS	Genetic Algorithms – Variable Subset Selection
GCSA	Generalized Cluster Significance Analysis
GERM	Genetically Evolved Receptor Models

GFA	Genetic Function Approximation
GIPF	General Interaction Properties Function
GOLPE	Generating Optimal Linear PLS Estimations
G-WHIM	Grid-Weighted Holistic Invariant Molecular descriptors
HASL	Hypothetical Active Site Lattice
HBA	Hydrogen Bond Acceptor
HBD	Hydrogen Bond Donor
HQSAR	Hologram QSAR
HFED	Hydration Free Energy Density
HINT	Hydrophatic INTeractions
HOC	Hierarchically Ordered extended Connectivity
HOMO	Highest Occupied Molecular Orbital
HSA	Hydrated Surface Area
HXID	Hu-Xu ID number
IC	neighbourhood Information Content
ILGS	Iterated Line Graph Sequence
ILS	Intermediate Least Squares regression
ISA	Isotropic Surface Area
IVEC	Iterative Vertex and Edge Centricity algorithm
IVS-PLS	Interactive Variable Selection – Partial Least Squares
K-ANN	Kohonen Artificial Neural Networks
KLOGP	Klopman LOG*P*
LASRR	Linear Aromatic Substituent Reactivity Relationships
LFER	Linear Free Energy Relationships
LSER	Linear Solvation Energy Relationships
LOMO	Lowest Occupied Molecular Orbital
LOVI	LOcal Vertex Invariant
LSER	Linear Solvation Energy Relationships
LUMO	Lowest Unoccupied Molecular Orbital
MCD	MonteCarlo version of MTD
MCIs	Molecular Connectivity Indices
MCS	Maximum Common Substructure
MEP	Molecular Electrostatic Potential
MFTA	Molecular Field Topology Analysis
MID	Molecular ID number
MLP	Molecular Lipophilicity Potential
MQSI	Molecular Quantum Similarity Indices
MQSM	Molecular Quantum Similarity Measures
MSA	Molecular Shape Analysis
MSD	Minimal Steric Difference
MSG	Molecular SuperGraph
MTD	Minimal Topological Difference
MTI	Molecular Topological Index
MUSEUM	MUtation and SElection Uncover Models
MWC	Molecular Walk Count
NN	Neural Networks
OASIS	Optimized Approach based on Structural Indices Set
OLS	Ordinary Least Squares regression
PAR	Property-Activity Relationships
PCA	Principal Component Analysis

PCR	Principal Component Regression
PELCO	Pérturbation d'un Environnement Limité Concentrique Ordonné
PID	Prime ID number
PLS	Partial Least Squares regression
PSA	Polar Surface Area
QSAR	Quantitative Structure-Activity Relationships
QShAR	Quantitative Shape-Activity Relationships
QSiAR	Quantitative Similarity-Activity Relationships
QSPR	Quantitative Structure-Property Relationships
QSRC	Quantitative Structure/Response Correlations
QSRR	Quantitative Structure-Reactivity Relationships
RID	Ring ID number
RR	Ridge Regression
RBSM	Receptor Binding Site Model
RSM	Receptor Surface Model
SA	Surface Area
SAR	Structure-Activity Relationships
SASA	Solvent-Accessible Surface Area
SAVOL	Solvent-Accessible VOLume
SBL	Smallest Binary Label
SIBIS	Steric Interactions in BIological Systems
SIC	Structural Information Content
SID	Self-returning ID number
SPP	Submolecular Polarity Parameter
SPR	Structure-Property Relationships
SRC	Structure/Response Correlations
SRW	Self-Returning Walk
SWIM	Spectral Weighted Invariant Molecular descriptors
SWM	Spectral Weighted Molecular signals
SWR	StepWise Regression
TI	Topological Index
TIC	neighbourhood Total Information Content
TLP	Topological Lipophilicity Potential
TLSER	Theoretical Linear Solvation Energy Relationships
TMSA	Total Molecular Surface Area
TOSS-MODE	TOpological SubStructure MOlecular DEsign
VFA	Voronoi Field Analysis
VR	Variable Reduction
VS	Variable Selection
VSS	Variable Subset Selection
WHIM	Weighted Holistic Invariant Molecular descriptors
WID	Weighted ID number
WLN	Wiswesser Line-formula Notation
3D-MoRSE	3D-Molecule Representation of Structures based on Electron diffraction descriptors

A

A_{x1}, A_{x2}, A_{x3} eigenvalue indices → **eigenvalue-based descriptors**

absolute hardness → **quantum-chemical descriptors** (⊙ hardness indices)

acceptor superdelocalizability : *electrophilic superdelocalizability* → **quantum-chemical descriptors**

ACC transforms → **autocorrelation descriptors**

***ACGD* index** → **charged partial surface area descriptors**

acid dissociation constant → **hydrogen-bonding descriptors**

activation hardness → **quantum-chemical descriptors** (⊙ hardness indices)

acyclic graph : *tree* → **graph**

ADAPT approach

A QSAR approach [Jurs *et al.*, 1979; Jurs *et al.*, 1988], implemented in the homonimous software ADAPT (Automated Data Analysis and Pattern Recognition Toolkit), based on the following steps: a) molecular descriptor generation; b) objective feature selection to discard descriptors which contain redundant or minimal information; c) multiple regression analysis by genetic algorithm or simulated annealing variable selection, or computational → *artificial neural networks.*

ADAPT descriptors fall into three general categories: → *topological indices*, → *geometrical descriptors* (including → *principal moments of inertia*, → *volume descriptors* and → *shadow indices*), and → *electronic descriptors* (including partial atomic charges and the → *dipole moment*); moreover, → *molecular weight*, → *count descriptors*, and a large number of → *substructure descriptors* are also generated. In addition, the → *charged partial surface area descriptors* constitute a fourth class of descriptors derived by combining electronic and geometrical information.

Several molecular properties have been modeled by the ADAPT approach, such as biological activities [Henry *et al.*, 1982; Jurs *et al.*, 1983; Jurs *et al.*, 1985; Walsh and Claxton, 1987; Wessel *et al.*, 1998; Eldred and Jurs, 1999a; Eldred *et al.*, 1999b], boiling points [Smeeks and Jurs, 1990; Stanton *et al.*, 1991; Stanton *et al.*, 1992; Egolf and Jurs, 1993a; Egolf *et al.*, 1994; Wessel and Jurs, 1995a; Wessel and Jurs, 1995b; Goll and Jurs, 1999a], chromatographic indices [Anker *et al.*, 1990; Sutter *et al.*, 1997], aqueous solubilities [Dunnivant *et al.*, 1992; Nelson and Jurs, 1994; Sutter and Jurs, 1996; Mitchell and Jurs, 1998a], critical temperature and pressures [Turner *et al.*, 1998], ion mobility constants [Wessel and Jurs, 1994; Wessel *et al.*, 1996], reaction rate constants [Bakken and Jurs, 1999b; Bakken and Jurs, 1999a].

📖 [Stanton and Jurs, 1992] [Egolf and Jurs, 1992] [Russell *et al.*, 1992] [Egolf and Jurs, 1993b] [Engelhardt and Jurs, 1997] [Mitchell and Jurs, 1997] [Johnson and Jurs, 1999] [Goll and Jurs, 1999b]

additivity model : *Free-Wilson model* → **Free-Wilson analysis**

additive-constitutive models → **group contribution methods**

additive model of inductive effect → **electronic substituent constants** (⊙ inductive electronic constants)

adjacency matrix (A) *(: vertex adjacency matrix)*

Derived from the → *molecular graph* $\mathcal{G}$, the adjacency matrix **A** represents the whole set of connections between adjacent pairs of atoms [Trinajstic, 1992]. The entries a_{ij} of the matrix equal one if vertices v_i and v_j are adjacent (i.e. the atoms i and j are bonded) and zero otherwise. The adjacency matrix is symmetric with dimension $A \times A$, where A is the number of atoms and it is usually derived from an → *H-depleted molecular graph.*

The ith row sum of the adjacency matrix is called → *vertex degree* δ_i, defined as:

$$\delta_i = \mathcal{R}_i(\mathbf{A}) = \sum_{j=1}^{A} a_{ij}$$

where $\mathcal{R}_i$ is the → *row sum operator*; it represents the number of σ-bonds of the ith atom.

The **total adjacency index** A_V is the sum of all the entries of the adjacency matrix of a molecular graph, and is twice the → *bond number B* [Harary, 1969a]:

$$A_V = S(\mathbf{A}) = \sum_{i=1}^{A}\sum_{j=1}^{A} a_{ij} = \sum_{i=1}^{A} \delta_i = 2B$$

where S is the → *total sum operator*. Therefore, the number of entries equal to one in the adjacency matrix is $2B$, while the number of entries equal to zero is $A^2 - 2B$; in particular, for acyclic graphs the total number of entries equal to one is $2\,(A-1)$ and the number of entries equal to zero is $A^2 - 2\,(A-1)$; for monocyclic graphs they are $2A$ and $A^2 - 2A$, respectively. The total adjacency index is sometimes calculated as the half-sum of the adjacency matrix elements.

Simple → *topological information indices* can be calculated on both the equality and magnitude of adjacency matrix elements. Moreover, → *walk counts* and → *self-returning walk counts* which coincide with the spectral moments of the adjacency matrix are calculated by the increasing powers of the adjacency matrix [McKay, 1977; Jiang *et al.*, 1984; Kiang and Tang, 1986; Hall, 1986; Jiang and Zhang, 1989; Jiang and Zhang, 1990; Markovic and Gutman, 1991; Jiang *et al.*, 1995; Markovic and Stajkovic, 1997; Markovic, 1999].

In order to take into account the heteroatoms in the molecule, the **augmented adjacency matrix** was proposed by Randic [Randic, 1991c; Randic, 1991d; Randic and Dobrowolski, 1998b] replacing the zero diagonal entries of the "normal" adjacency matrix with specific values empirically obtained and characterizing different atoms in the molecule. The row sums of this adjacency matrix are → *local vertex invariants* encoding the connectivity of each atom and its atom type; therefore they can be viewed as augmented vertex degrees. The inverse of the square root of the product of the augmented degrees of the vertices incident with a bond is used as bond weight in calculating the → *weighted path counts.*

Other topological matrices are derived from the adjacency matrix, such as → *atom connectivity matrices*, → *Laplacian matrix* and the powers of the adjacency matrix used to obtain walk counts and the corresponding → *molecular descriptors.*

Moreover, the adjacency matrix can be transformed into a **decimal adjacency vector** $\mathbf{a}^{10}$ of A elements each being a local vertex invariant obtained by the following expression [Schultz and Schultz, 1991]:

$$a_i^{10} = (2 \cdot a_{i1})^{A-1} + (2 \cdot a_{i2})^{A-2} + \ldots + (2 \cdot a_{iA})^{0}$$

where a_{ij} is the jth column element of the ith row of the adjacency matrix **A** (zero or one). In this way, the information contained in the adjacency matrix is compressed into an A-dimensional vector. For example, a row of the adjacency matrix [0 1 1 1 0] gives a value of 14, obtained as

$$a_i^{10} = (2 \cdot 0)^{5-1} + (2 \cdot 1)^{5-2} + (2 \cdot 1)^{5-3} + (2 \cdot 1)^{5-4} + (2 \cdot 0)^{0} = 14$$

The elements of the decimal adjacency vector are integers which were used for → *canonical numbering* of molecular graphs [Randic, 1974].

From the decimal adjacency vector, three different indices were proposed as molecular descriptors:

a) the sum of the elements of the $\mathbf{a}^{10}$ vector, i.e.

$$A1 = \sum_{i=1}^{A} a_i^{10}$$

b) the sum of the linear combination of vertex degrees δ_i each weighted by the corresponding decimal adjacency vector elements a_i^{10}, i.e.

$$A2 = \sum_{i=1}^{A} \delta_i \cdot a_i^{10}$$

c) the sum of the elements of the A-dimensional vector **d** obtained by multiplying the topological → *distance matrix* **D** by the decimal adjacency vector, i.e.

$$A3 = \sum_{i=1}^{A} [\mathbf{d}]_i$$

where the vector **d** is calculated as:

$$\mathbf{d} = \mathbf{D} \cdot \mathbf{a}^{10}$$

Example : 2-methylpentane

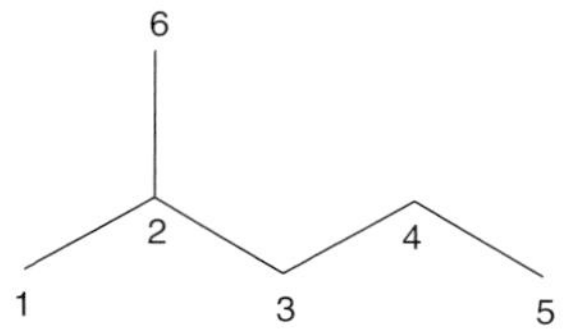

adjacency matrix **A**

Atom	1	2	3	4	5	6	δ_i
1	0	1	0	0	0	0	1
2	1	0	1	0	0	1	3
3	0	1	0	1	0	0	2
4	0	0	1	0	1	0	2
5	0	0	0	1	0	0	1
6	0	1	0	0	0	0	1

Atom	$\mathbf{a}^{10}$
1	16
2	41
3	20
4	10
5	4
6	16

$$A_V = \sum_{i=1}^{6}\sum_{j=1}^{6} a_{ij} = \sum_{i=1}^{6} \delta_i = 2 \cdot B = 10$$

$A_1 = 107$ $A_2 = 219$ $A_3 = 1028$

Box A-1.

admittance matrix : *Laplacian matrix*

adsorbability index (AI)

An empirical molecular descriptor proposed by Abe *et. al.* (1986) derived from a group contribution method based on molecular refractivity to predict the activated carbon adsorption of 157 compounds from aqueous solutions [Okouchi and Saegusa, 1989]. The adsorbability index for a molecule is calculated by the expression:

$$\mathrm{AI} = \sum_i A_i + \sum_i I_i$$

where the sums run over the atoms or functional groups; A indicates the atomic or group factors of increasing or decreasing adsorbability in the molecule and I represents special correction factors accounting for functional group effects.

For example, for benzene: $\mathrm{AI} = 6 \cdot A_{\mathrm{C}} + 6 \cdot A_{\mathrm{H}} + 3 \cdot A_{\mathrm{C=C}} = 6 \cdot 0.26 + 6 \cdot 0.12 + 3 \cdot 0.19 = 2.85$; for 1,1,2-trichloroethane: $\mathrm{AI} = 2 \cdot A_{\mathrm{C}} + 3 \cdot A_{\mathrm{H}} + 3 \cdot A_{\mathrm{Cl}} = 2 \cdot 0.26 + 3 \cdot 0.12 + 3 \cdot 0.59 = 2.65$

Table A-1. Values of A and I factors proposed by Abe *et al.*

Atom / Group	A	*Group*	I
C	0.26	Aliphatic	
H	0.12	–OH (alcohols)	–0.53
N	0.26	–O– (ethers)	–0.36
O	0.17	–CHO (aldehydes)	–0.25
S	0.54	N (amines)	–0.58
Cl	0.59	–COOR (esters)	–0.28
Br	0.86	>C=O (ketones)	–0.30
NO_2	0.21	–COOH (fatty acids)	–0.03
–C = C–	0.19		
Iso	–0.12	α-Amino acids	–1.55
Tert	–0.32		
Cyclo	–0.28	All groups in aromatics	0

Aihara resonance energy → **resonance indices**

a_N-index → **determinant-based descriptors** (⊙ general a_N-index)

algebraic operators

Algebraic operators play a meaningful role in the framework of → *molecular descriptors*, both in deriving molecular descriptors and directly as molecular descriptors.

Let **M** be a generic matrix with n rows and p columns, denoted as:

$$\mathbf{M} \equiv [m_{ij}] = \begin{vmatrix} m_{11} & m_{12} \ldots\ldots m_{1p} \\ \vdots & \vdots \\ m_{n1} & m_{n2} \ldots\ldots m_{np} \end{vmatrix}$$

The corresponding matrix elements m_{ij} are denoted as:

$$m_{ij} \equiv [\mathbf{M}]_{ij} \equiv (i,j)$$

A column vector **v** is a special case of a matrix having n rows and one column; the row vector $\mathbf{v}^{\mathrm{T}}$ is a special case of a matrix having one row and n columns.

Some definitions of matrix algebra [Golub and van Loan, 1983; Mardia *et al.*, 1988], algebraic operators and set theory are given below.

- **characteristic polynomial**

Let **M** be a square matrix ($n \times n$) and x a scalar variable, the characteristic polynomial $\mathcal{P}(\mathbf{M}, x)$ is defined as:

$$\mathcal{P}(\mathbf{M}, x) = \det(x\mathbf{I} - \mathbf{M}) = \sum_{k=0}^{n} a_k x^{n-k}$$

where **I** is the identity matrix, i.e. a matrix having the diagonal elements equal to one and all the off-diagonal elements equal to zero, and a_k the polynomial coefficients. Therefore, the characteristic polynomial is obtained by expanding the determinant and then collecting terms with equal powers of x.

The **eigenvalues** λ of the matrix **M** are the n roots of its characteristic polynomial and the set of the eigenvalues is called **spectrum of a matrix** $\Lambda(\mathbf{M})$. Determinant and trace of **M** are given by the following expressions:

$$\det(\mathbf{M}) = \prod_{k=1}^{n} \lambda_k \qquad tr(\mathbf{M}) = \sum_{k=1}^{n} \lambda_k$$

respectively.

For each eigenvalue λ_k, there exists a non-zero vector $\mathbf{t}_k$ satisfying the following relationship:

$$\mathbf{M}\mathbf{t}_k = \lambda_k \mathbf{t}_k$$

The n-dimensional vectors $\mathbf{t}_k$ are called **eigenvectors** of **M**.

- **cardinality of a set**

The cardinality of a set S is the number of elements in S and is indicated as $|S|$.

- **column sum operator**

This operator C_j performs the sum of the elements of the jth matrix column:

$$C_j(\mathbf{M}) \equiv \sum_{i=1}^{n} m_{ij}$$

The **column sum vector**, denoted by **cs**, is a p-dimensional vector collecting the results obtained by applying C_j operator on all the p columns of the matrix.

- **determinant**

The determinant of an $n \times n$ square matrix **M**, denoted by $\det(\mathbf{M})$, is scalar and is defined as:

$$\det(\mathbf{M}) = \sum\nolimits_{\pi} s(\pi) \cdot m_{1,i_1} \cdot m_{2,i_2} \cdot \ldots \cdot m_{n,i_n}$$

where the summation ranges over all $n!$ permutations π of the symbols 1, 2, ..., n. Each permutation π of degree n is given by

$$\pi = \begin{pmatrix} 1\,2\,\ldots n \\ i_1\,i_2\ldots i_n \end{pmatrix}$$

where i_1, i_2, ..., i_n are the symbols 1, 2, ..., n in some order. The sign function $s(\pi)$ is defined as:

$$s(\pi) = \begin{cases} +1 & \text{if } \pi \text{ is even} \\ -1 & \text{if } \pi \text{ is odd} \end{cases}$$

Related to the definition of determinant are permanent, pfaffian and hafnian.

The **permanent**, denoted by per(**M**), also referred to as positive determinant, is defined by omitting the sign function $s(\pi)$ [Schultz *et al.*, 1992; Yang *et al.*, 1994; Cash, 1995a; Cash, 1998] as:

$$\mathrm{per}(\mathbf{M}) = \sum\nolimits_{\pi} m_{1,i_1} \cdot m_{2,i_2} \cdot \ldots \cdot m_{n,i_n}$$

where π runs over the $n!$ permutations.

The **immanant**, denoted by $d_\lambda(\mathbf{M})$, is defined as:

$$d_\lambda(\mathbf{M}) = \sum\nolimits_{\pi} \chi_\lambda(\pi) \cdot m_{1,i_1} \cdot m_{2,i_2} \cdot \ldots \cdot m_{n,i_n}$$

where π runs over the $n!$ permutations. $\chi_\lambda(\pi)$ is an irreducible character of the symmetric group indexed by a partition λ of n.

The **pfaffian**, denoted by pfa(**M**), is analogous to the determinant where the summation over all the permutations π $(i_1, i_2, ..., i_n)$ must also satisfy the limitations

$$i_1 < i_2,\ i_3 < i_4,\ \ldots,\ i_{n-1} < i_n; \qquad i_1 < i_3 < i_5 < \ \ldots\ < i_{n-1}$$

The entries of the main diagonal are excluded from the calculation of the pfaffian [Caianiello, 1953; Caianiello, 1956].

The **hafnian**, denoted by haf(**M**), is analogous to the permanent where the summation over all the permutations π $(i_1, i_2, ..., i_n)$ must also satisfy the limitations

$$i_1 < i_2,\ i_3 < i_4,\ \ldots,\ i_{n-1} < i_n; \qquad i_1 < i_3 < i_5 < \ \ldots\ < i_{n-1}$$

The entries of the main diagonal are excluded from the calculation of the hafnian.

The hafnian calculated considering only the entries above the main diagonal is called the **short-hafnian**, shaf(**M**), while the hafnian calculated considering both entries above and below the main diagonal can also be referred to as the **long-hafnian**, lhaf(**M**) [Schultz and Schultz, 1992]. Hafnians and pfaffians differ only in the sign function $s(\pi)$ included in the definition of pfaffian.

For example, for a matrix **M** of order 4, pfaffian, long-hafnian and short-hafnian are the following:

$$\mathrm{pfa} = m_{12} \cdot m_{34} - m_{13} \cdot m_{24} + m_{14} \cdot m_{23}$$

$$\mathrm{shaf} = m_{12} \cdot m_{34} + m_{13} \cdot m_{24} + m_{14} \cdot m_{23}$$

$$\mathrm{lhaf} = m_{12} \cdot m_{21} \cdot m_{34} \cdot m_{43} + m_{13} \cdot m_{31} \cdot m_{24} \cdot m_{24} + m_{14} \cdot m_{41} \cdot m_{23} \cdot m_{32}$$

Some molecular descriptors, called → *determinant-based descriptors*, are calculated as the determinant of a → *matrix representation of a molecular structure*. Moreover, permanents, short- and long-hafnians, calculated on the topological → *distance matrix* **D**, were used as graph invariants by Schultz and called **per(D) index**, **shaf(D) index**, **lhaf(D) index** [Schultz *et al.*, 1992; Schultz and Schultz, 1992].

📖 [Schultz *et al.*, 1993; Schultz and Schultz, 1993; Schultz *et al.*, 1994; Schultz *et al.*, 1994; Schultz *et al.*, 1995; Schultz *et al.*, 1996; Chan *et al.*, 1997]

- **diagonal matrix**

A diagonal matrix **M** is a square matrix whose diagonal terms m_{ii} are the only nonzero elements. The **diagonal operator** $\mathcal{D}(\mathbf{M})$ is an operator which transforms a generic square matrix **M** into a diagonal matrix:

$$\mathcal{D}(\mathbf{M}) = \begin{vmatrix} m_{11} & \ldots & 0 & \ldots & 0 \\ 0 & \ldots & m_{ii} & \ldots & 0 \\ 0 & \ldots & 0 & \ldots & m_{nn} \end{vmatrix}$$

• **Hadamard matrix product**

The Hadamard product of two matrices **A** and **B** of same dimension is defined as:

$$[\mathbf{A} \bullet \mathbf{B}]_{ij} = [\mathbf{A}]_{ij}[\mathbf{B}]_{ij}$$

i.e. the elements of the resulting matrix are obtained by the scalar product of the corresponding elements of **A** and **B** matrices.

• **hyper-Wiener operator**

For any square symmetric $A \times A$ matrix **M** representing a → *molecular graph* with A vertices, the hyper-Wiener operator $\mathcal{HyW}$ is defined as [Ivanciuc *et al.*, 1997a]:

$$\mathcal{HyW}(\mathbf{M}) = \frac{1}{2}\sum_i \sum_j \left(m_{ij}^2 + m_{ij}\right) \qquad i \neq j$$

It takes its name from the definition of the → *hyper-Wiener index.*

• **Ivanciuc-Balaban operator**

For any square symmetric $A \times A$ matrix **M** representing a → *molecular graph* with A vertices, the Ivanciuc-Balaban operator $\mathcal{IB}$ is defined as a modified Randic operator [Ivanciuc *et al.*, 1997a]:

$$\mathcal{IB}(\mathbf{M}) = \frac{B}{C+1} \cdot \sum_{\{i,j\}} \left[\mathcal{R}_i(\mathbf{M}) \cdot \mathcal{R}_j(\mathbf{M})\right]^{-1/2}$$

where the sum runs over all the B pairs of adjacent vertices of the graph, C is the number of independent rings in the graph (→ *cyclomatic number*), and $\mathcal{R}$ is the row sum operator. It takes its name from the definition of the → *Balaban distance connectivity index*.

• **polynomial**

A polynomial $\mathcal{P}(x)$ of the x variable is a linear combination of its powers, usually written as:

$$\mathcal{P}(x) = a_0 x^n + a_1 x^{n-1} + \ldots + a_{n-1} x + a_n$$

where n is the order of the polynomial. The values of x for which $\mathcal{P}(x)$ is zero are the *roots* of the polynomial.

• **product of matrices**

Let **A** (n, m) and **B** (m, p) be two matrices. The product of the two matrices is defined as:

$$\mathbf{A} \cdot \mathbf{B} = \mathbf{C}$$

where the resulting product matrix **C** has n rows and p columns. Each scalar element c_{ij} of the **C** matrix is obtained by the scalar product between the ith row of the **A** matrix, i.e. the row vector $\mathbf{a}_i^{\mathrm{T}}$, and the jth column of the **B** matrix, i.e. the column vector $\mathbf{b}_j$:

$$\mathbf{a}_i^{\mathrm{T}} \cdot \mathbf{b}_j \equiv c_{ij} = \sum_{k=1}^{m} a_{ik} \cdot b_{kj}$$

A basic condition for the product of two matrices is that the number of columns of the left matrix and rows of the right matrix are equal (m).

The kth power matrix is a special case of the matrix product.

The main properties of the product of two matrices are:

a) $\mathbf{A} \cdot \mathbf{B} \neq \mathbf{B} \cdot \mathbf{A}$ b) $(\mathbf{A} \cdot \mathbf{B}) \cdot \mathbf{C} = \mathbf{A} \cdot (\mathbf{B} \cdot \mathbf{C})$ c) $(\mathbf{A} \cdot \mathbf{B})^{\mathrm{T}} = \mathbf{B}^{\mathrm{T}} \cdot \mathbf{A}^{\mathrm{T}}$

- **Randic operator**

For any square symmetric $A \times A$ matrix **M** representing a → *molecular graph* with A vertices, the Randic operator χ is defined as:

$$\chi(\mathbf{M}) = \sum_{\{i,j\}} \left[\mathcal{R}_i(\mathbf{M}) \cdot \mathcal{R}_j(\mathbf{M})\right]^{-1/2}$$

where the sum runs over all the pairs of adjacent vertices of the graph and $\mathcal{R}$ is the row sum operator. It takes its name from the definition of the → *Randic connectivity index*.

- **row sum operator**

This operator $\mathcal{R}_i$ performs the sum of the elements of the ith matrix row:

$$\mathcal{R}_i(\mathbf{M}) \equiv \sum_{j=1}^{p} m_{ij}$$

The **row sum vector**, **rs**, is a n-dimensional vector collecting the results obtained by applying $\mathcal{R}_i$ operator on all the n rows of the matrix.

- **scalar product of vectors**

Let **a** and **b** be two column vectors with the same dimension n. The scalar product between the two vectors is defined as:

$$\mathbf{a}^{\mathrm{T}} \cdot \mathbf{b} = \mathbf{b}^{\mathrm{T}} \cdot \mathbf{a} = \sum_{k=1}^{n} a_k \cdot b_k$$

i.e. as the sum of the products of the corresponding elements of the row vector $\mathbf{a}^{\mathrm{T}}$ with the column vector **b** or of the row vector $\mathbf{b}^{\mathrm{T}}$ with the column vector **a**.

- **sparse matrices**

A **sparse matrix** is a matrix with relatively few nonzero entries. A **binary sparse matrix B** is a matrix whose all nonzero elements are equal to one.

Let **M** be a matrix $A \times A$ representing a → *molecular graph* $\mathcal{G}$ where A is the number of vertices. To obtain an ***m*th order sparse matrix** $^{\mathbf{m}}\mathbf{M}$ from any matrix **M** the Hadamard matrix product is performed as the following:

$$^{\mathbf{m}}\mathbf{M} = \mathbf{M} \bullet {}^{\mathbf{m}}\mathbf{B}$$

where the superscript "m" of the sparse matrix $^{\mathbf{m}}\mathbf{M}$ means that all of the **M** matrix elements are taken as zero but the entries corresponding to vertices v_i and v_j whose → *topological distance* is equal to m and $^{\mathbf{m}}\mathbf{B}$ matrix is a binary sparse matrix whose elements equal one only for vertices v_i and v_j at distance m.

The → *adjacency matrix* **A** of a molecular graph $\mathcal{G}$ is an example of binary sparse matrix, only the off-diagonal entries *i-j*, where v_i and v_j are adjacent vertices, i.e. vertices connected by a bond, being equal to one. Using the adjacency matrix as multiplier in the Hadamard product it follows:

$$^{\mathbf{1}}\mathbf{M} = \mathbf{M} \bullet \mathbf{A}$$

where $^{\mathbf{1}}\mathbf{M}$ is a **1st-order sparse matrix**.

Opposite to sparse matrices, are **dense matrices**, i.e. matrices with several nonzero entries [Randic and DeAlba, 1997].

- **sum of matrices**

Let **A** (n, p) and **B** (n, p) be two matrices. The sum of the two matrices is defined as:

$$\mathbf{A} + \mathbf{B} = \mathbf{C}$$

Each scalar element c_{ij} of the **C** matrix is obtained by summing up the corresponding elements of the two matrices, i.e.

$$c_{ij} = a_{ij} + b_{ij}$$

A basic condition for the sum of two matrices is that the two matrices have the same dimension.

The main properties of the sum of two matrices are:

a) $\mathbf{A} + \mathbf{B} = \mathbf{B} + \mathbf{A}$ b) $(\mathbf{A} + \mathbf{B}) + \mathbf{C} = \mathbf{A} + (\mathbf{B} + \mathbf{C})$ c) $\alpha\,(\mathbf{A} + \mathbf{B}) = \alpha\,\mathbf{A} + \alpha\,\mathbf{B}$

where α is a scalar value.

- **total sum operator**

This operator S performs the sum of all of the elements of the matrix:

$$S(\mathbf{M}) \equiv \sum_{i=1}^{n}\sum_{j=1}^{p} m_{ij} = \sum_{i=1}^{n} \mathcal{R}_i(\mathbf{M}) = \sum_{j=1}^{p} C_j(\mathbf{M})$$

- **trace**

The trace of a square matrix **M** (i.e. $n = p$), denoted by $tr(\mathbf{M})$, is the sum of the diagonal elements:

$$tr(\mathbf{M}) \equiv \sum_{i=1}^{n} m_{ii}$$

- **transposition of a matrix**

The matrix $\mathbf{M}^{\mathrm{T}}$ is the transposed matrix of **M** if its elements are:

$$\left[\mathbf{M}^{\mathrm{T}}\right]_{ij} = [\mathbf{M}]_{ji}$$

If the dimension of **M** is $n \times p$, the transposed matrix $\mathbf{M}^{\mathrm{T}}$ has dimensions $p \times n$.

- **Wiener operator**

For any square symmetric matrix **M**, the Wiener operator $\mathcal{W}$ is defined as half the sum of the off-diagonal entries of the matrix **M** [Ivanciuc *et al.*, 1997a], i.e. as:

$$\mathcal{W}(\mathbf{M}) = \frac{1}{2}\sum_i \sum_j m_{ij} \qquad i \neq j$$

It takes its name from the definition of the → *Wiener index*.

- **Wiener orthogonal operator** (: *orthogonal operator*)

For any unsymmetric matrix **M**, the orthogonal Wiener operator $\mathcal{W}'$ is defined as:

$$\mathcal{W}'(\mathbf{M}) = \sum_{\{i,j\}} m_{ij} \cdot m_{ji} = \sum_i \sum_{j=i+1} m_{ij} \cdot m_{ji}$$

where $\{i, j\}$ indicates that all the pairs i-j are considered only once. The Wiener orthogonal operator applied on the unsymmetric matrix **M** gives the same result as the application of the Wiener operator on the symmetric matrix $\mathbf{M}'$ obtained as:

$$\mathbf{M}' = \mathbf{M} \bullet \mathbf{M}^{\mathrm{T}}$$

where the symbol • indicates the Hadamard matrix product.

algebraic structure count → Kekulé number

alignment rules

In → *grid-based QSAR techniques*, the energy value at each grid point **p** constitutes a molecular descriptor. For the selected → *molecular interaction fields* (steric, hydrophobic, coulombic, etc.), the calculated value at each grid point **p** depends on the relative orientation of the compound with respect to the grid. As a consequence, the use of the grid points as molecular descriptors requires the mandatory step of aligning the molecules of the considered → *data set* in such a way that each of the thousands of grid points represents, for all the molecules, the same kind of information, and not spurious information due to lack of invariance in the rotation of the molecules in the grid.

Therefore, in applying grid-based QSAR techniques there are, in most cases, two closely related problems: the selection of the active molecular conformations of the studied compounds and the relative alignment of the conformations, either among themselves or with respect to any → *receptor*, if its structure is known.

The ideal choice of conformers for QSAR would be the bioactive one. Wherever experimental structural data (e.g. X-ray data) exist concerning ligands bound to targets, the bioactive ligand conformation is available and should be used to derive an alignment rule.

When no structural data are available for the receptor, methods that explore conformational space may find the best relative match among the different ligands. During this process, low energy conformations are selected to obtain the best match from all the different conformations. The solution is usually not unique because other conformations may bind to the unknown receptor and multiple alignment rules, based on different starting hypotheses, should be considered when no structural information and no rigid compounds are available.

The success or failure of the grid-based methods in finding acceptable → *quantitative structure-activity relationships* is strongly dependent on how the molecules are aligned in the grid on which the molecular interaction fields are sampled. In fact, problems may be mainly due to a) an alignment where the resulting common structure, i.e. the pharmacophore, is not reliable, and b) the same grid points in different molecules representing chance variation in model geometry, thus introducing a noise in the grid points which are the variables of the grid-based techniques.

The most used alignment techniques are listed below.

- **point-by-point alignment**

For a set of congeneric compounds, the atoms of each compound are superimposed on their common backbone, aligning as much of each structure as possible.

For structurally diverse compounds, hypotheses on the → *pharmacophore* can provide an approach to overcoming ambiguities in atom superimposition and identifying a suitable alignment.

- **field-fitting alignment**

In this approach the molecules are aligned by maximizing the degree of similarity between their molecular interaction fields. Different types of probes result in different fields as well as different molecular alignments. Therefore, the selection of suitable fields (and how to weight them) depends on external considerations.

Moreover, a difficulty in field-based alignments is that irrelevant molecular regions, i.e. those not involved with the receptor sites, may distort the alignment.

📖 [Kato *et al.*, 1987] [Mayer *et al.*, 1987] [Kearsley and Smith, 1990] [Manaut *et al.*, 1991] [Cramer III *et al.*, 1993] [Dean, 1993] [Klebe, 1993] [Waller *et al.*, 1993] [Waller and Marshall, 1993] [Klebe *et al.*, 1994b] [Cramer III *et al.*, 1996] [Petitjean, 1996] [Greco *et al.*, 1997] [Langer and Hoffmann, 1998b] [Handschuh *et al.*, 1998] [Klebe, 1998] [Norinder, 1998] [Robinson *et al.*, 1999]

all-path matrix → **path counts**

all-path Wiener index → **path counts**

all possible models → **variable selection**

ALOGP method → **lipophilicity descriptors** (⊙ Ghose-Crippen hydrophobic atomic constants)

Altenburg polynomial

This is a function of the → *H-depleted molecular graph* $\mathcal{G}$ defined as:

$$\alpha(\mathcal{G}, a) = \sum_{k=1}^{D} {}^{k}f \cdot a_k$$

where the sum runs over all the distances in the graph, D being the → *topological diameter*, i.e. the maximum distance in the graph, ${}^{k}f$ the → *graph distance count* of kth order, i.e. the number of distances equal to k in the graph, and a_k the independent variables [Altenburg, 1961]. The Altenburg polynomial is closely related to the → *Wiener index* of a graph: graphs with the same Altenburg polynomials always have just the same Wiener numbers (the contrary does not always hold).

amino acid side chain descriptors

Due to the relevance and the complexity of the amino acids chains and macromolecules, some descriptors were defined to represent amino acid side chains and sequences of amino acids (e.g. peptides).

Amino acid properties were modelled, for example, by → *connectivity indices* [Gardner, 1980; Pogliani, 1992b; Pogliani, 1992a; Pogliani, 1993b; Pogliani, 1993a; Pogliani, 1994a; Pogliani, 1994c; Lucic *et al.*, 1995b; Pogliani, 1995a; Pogliani, 1996a; Pogliani, 1997c; Pogliani, 1997a; Pogliani, 1999a], → *substituent descriptors* [Charton and Charton, 1982; Charton, 1990], → *charge descriptors* [Collantes and Dunn III, 1995], → *principal properties* [Kidera *et al.*, 1985a; Kidera *et al.*, 1985b; Norinder, 1991; Cocchi and Johansson, 1993], and → *G-WHIM descriptors* [Zaliani and Gancia, 1999].

Other specific descriptors for amino acids are listed below.

A general index based on the → *total information content* of biological compounds, such as amino acid sequences, is the **information index on amino acid composition**, defined as:

$$I_{\text{AAC}} = k \cdot \left(\ln \text{N}! - \sum_{g=1}^{G} \ln n_g! \right)$$

where k is the Boltzmann constant, N is the total number of amino acid residues, G the number of → *equivalence classes*, i.e. the number of different amino acid residues, and n_g the number of the same amino acid residues, i.e. belonging to the gth equivalence class.

Unlike other information indices, factorials are used in the expression to take into account combinations of amino acid residues.

To characterize size and shape of side chains in amino acids, a topological descriptor was proposed [Raychaudhury *et al.*, 1999] based on a graph-theoretical approach applied to rooted weighted molecular graphs (hydrogen included) representing the side chains. Each vertex of the chain other than the link vertex (C_α carbon atom) is weighted and all shortest weighted paths between the link vertex C_α (assumed at zero position) and terminal vertices are taken into account; the weight of each path is given by the sum of the atomic weights of all involved atoms. Moreover, if there is more than one shortest path between two vertices, then the shortest path is considered to be that for which the sum of the weights of its vertices is a minimum.

A probability value p_i is assigned to the directed path connecting the link vertex to each ith terminal vertex, calculated as the following:

$$p_i = (\delta'_1 \cdot \ldots \cdot \delta'_{i-1})^{-1}$$

where δ' is the number of incident bonds of each atom involved in the path without considering those already counted at the previous step. Bonds in the rings not involved in any path should be deleted to get probability values.

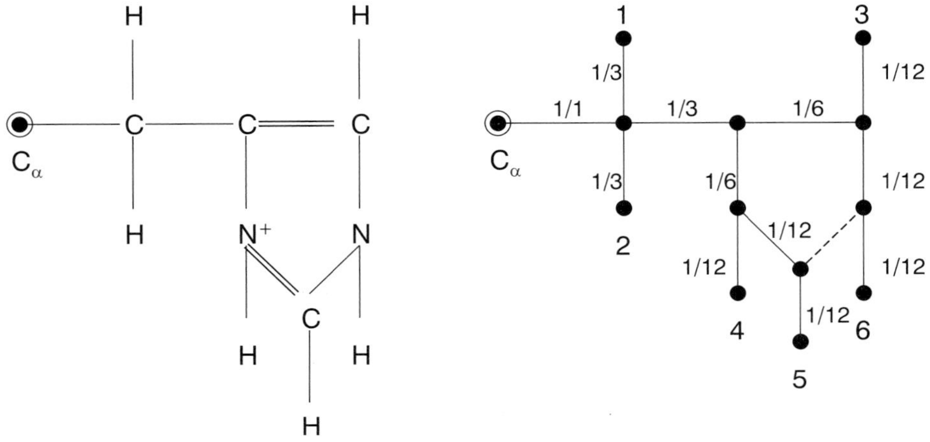

Figure A-1. Histidine molecule and corresponding molecular graph. The number associated to each bond is the probability corresponding to the incident vertex, calculated starting from C_α.

By using the calculated probability values, the path value $P_{i,p}$ is calculated as:

$$P_{i,p} = p_i \cdot \sum_k w_k$$

where the sum runs over all the vertices between the link atom and the ith vertex; w represents the weights of the atoms involved in the path. A molecular shape and size related index M_S^S, here called the **side chain topological index**, is calculated for the link vertex C_α as the sum of all the path values:

$$M_S^S = \sum_{i=1}^{N_T} P_{i,p}$$

where N_T is the number of terminal vertices in the side chain.

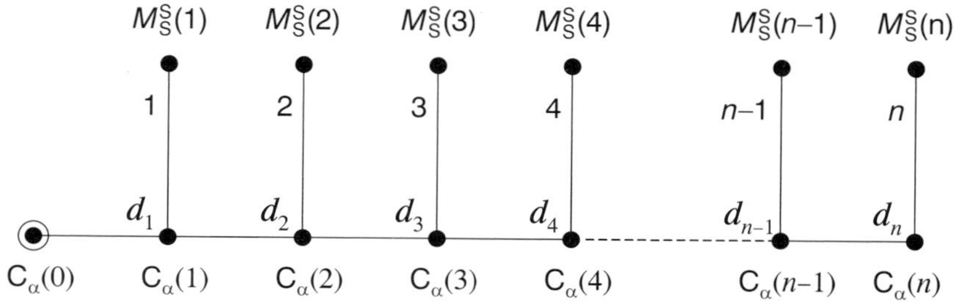

Figure A-2. Amino acid sequence starting from C_α.

From this index, a descriptor for a sequence of amino acids, called **distance exponent index** D^x was also proposed, defined as:

$$D^x = \sum_k (M_S^S)_k \cdot d_k^x$$

where the sum runs over the considered sequence, being each term at topological distance d_k from the C_α representing the origin of the sequence. Each side chain topological index M_S^S is calculated independently of its link atom C_α. The exponent x may take any real values; values $x = -3$ and -4 were usefully proved in modelling side chain properties [Raychaudhury and Klopman, 1990].

📖 [Fauchère and Pliska, 1983] [Nakayama *et al.*, 1988] [Klopman and Raychaudhury, 1988] [Klopman and Raychaudhury, 1990] [Naray-Szabo and Balogh, 1993] [Stein *et al.*, 1999] [Raychaudhury and Nandy, 1999] [Testa *et al.*, 1999]

Amoore shape indices → **shape descriptors**

Andrews descriptors → **count descriptors**

a_N-index → **determinant-based descriptors** (⊙ general a_N-index)

anisometry → **shape descriptors**

anisotropy of the polarizability → **electric polarization descriptors**

arcs → **graph**

arithmetic topological index → **vertex degree**

aromatic-bond count → **multiple bond descriptors**

aromaticity → **resonance indices**

aromaticity indices → **resonance indices**

artificial neural networks → **chemometrics**

aryl electronic constants → **electronic substituent constants**

ASIIg index → **charge descriptors** (⊙ charge-related indices)

asphericity → **shape descriptors**

association coefficients → **similarity / diversity**

atom-atom polarizability → **electric polarization descriptors**

atom-centred fragment codes : *augmented atom codes* → **substructure descriptors**

atom connectivity matrices → **weighted matrices**

atom connectivity matrix → **weighted matrices**

atom count : *atom number*

atom detour eccentricity → **detour matrix**

atom eccentricity → **distance matrix**

atom electronegativity → **quantum-chemical descriptors**

atomic charge → **charge descriptors**

atomic charge weighted negative surface area → **charged partial surface area descriptors**

atomic charge weighted positive surface area → **charged partial surface area descriptors**

atomic composition indices (: *composition indices)*

Molecular → *0D-descriptors* with high degeneracy, derived from the chemical formula of compounds and defined as → *information indices* of the elemental composition of the molecule, can be considered → *molecular complexity indices* that take account of molecular diversity in terms of different atom types.

→ *Average molecular weight* and → *relative atom-type count* are simple molecular descriptors that encode information on atomic composition. Other important descriptors of the atomic composition are based on the → *total information content* and the → *mean information content*, defined as:

- **total information index on atomic composition** (I_{AC})

The total information content on atomic composition of the molecule is calculated from the complete molecular formula, hydrogen included:

$$I_{AC} = A^h \cdot \log_2 A^h - \sum_g A_g \cdot \log_2 A_g$$

where A^h is the total number of atoms (hydrogen included) and A_g is the number of atoms belonging to the same gth chemical element [Dancoff and Quastler, 1953].

For example, benzene has 6 carbon and 6 hydrogen atoms; then, as $A^h = 12$ and $A_g = 6$ for both equivalence classes, $I_{AC} = 12$.

- **mean information index on atomic composition** ($\bar{I}_{AC}$)

The mean information content on atomic composition is the mean value of the total information content and is calculated as:

$$\bar{I}_{AC} = -\sum_g \frac{A_g}{A^h} \cdot \log_2 \frac{A_g}{A^h} = -\sum_g p_g \cdot \log_2 p_g$$

where A^h is the total number of atoms (hydrogen included), A_g is the number of equal-type atoms in the gth equivalence class, and p_g is the probability of randomly selecting a gth type atom [Dancoff and Quastler, 1953]. For example, for benzene $\bar{I}_{AC} = 1$.

atomic connectivity indices : *local connectivity indices* → **connectivity indices**

atomic dispersion coefficient → **hydration free energy density**

atomic ID numbers → **ID numbers**

atomic information content → **atomic information indices**

atomic information indices

Atomic descriptors related to the internal composition of atoms [Bonchev, 1983].

The **atomic information content** I_{at} is the → *total information content* of an atom viewed as a system whose structural elements, i.e. protons p, neutrons n, and electrons z, are partitioned into nucleons, p + n, and electrons z:

$$I_{at} = (N_n + N_p + N_{el}) \cdot \log_2(N_n + N_p + N_{el}) - N_{el} \cdot \log_2 N_{el} - (N_n + N_p) \cdot \log_2(N_n + N_p)$$

where N_n, N_p, and N_{el} are the numbers of neutrons, protons, and electrons, respectively [Bonchev and Peev, 1973].

In order to account for the different isotopes of a given chemical element the **information index on isotopic composition** I_{IC} was defined as:

$$I_{IC} = \sum_k (I_{at})_k \cdot f_k$$

where the sum runs over all isotopes of the considered chemical element, $(I_{at})_k$ is the atomic information content of the kth isotope and f_k its relative amount [Bonchev and Peev, 1973].

The **information index on proton-neutron composition** $I^{n,p}$ is an atomic descriptor defined as total information content of the atomic nucleus:

$$I^{n,p} = (N_n + N_p) \cdot \log_2(N_n + N_p) - N_n \cdot \log_2 N_n - N_p \cdot \log_2 N_p$$

where N_n and N_p are the numbers of neutrons and protons, respectively [Bonchev *et al.*, 1976c]

The **nuclear information content** I_{NUCL} is a molecular descriptor calculated as the sum of information on the proton-neutron composition of all the nuclei of the molecule:

$$I_{NUCL} = \sum_{i=1}^{A} I_i^{n,p}$$

where A is the number of atoms and $I_i^{n,p}$ is the information index on proton-neutron composition of the nucleus of the ith atom. This index also accounts for molecular size by means of the number of atomic nuclei.

atomic moments of energy → **self-returning walk counts**

atomic multigraph factor → **bond order indices** (⊙ conventional bond order)

atomic path count → **path counts**

atomic path count sum → **path counts**

atomic path number : *atomic path count* → **path counts**

atomic path/walk indices → **shape descriptors** (⊙ path/walk shape indices)

atomic polarizability → **electric polarization descriptors**

atomic polarization → **electric polarization descriptors**

atomic properties → **molecular descriptors**

atomic self-returning walk count → **self-returning walk counts**

atomic sequence count → **sequence matrices**

atomic solvation parameter ($\Delta\sigma$)

An empirical atomic descriptor $\Delta\sigma$ proposed to calculate solvation free energy of a molecular group X in terms of atomic contributions by the equation:

$$\Delta G_X = \sum_i \Delta\sigma_i \cdot SA_i$$

where the sum runs over all the non-hydrogen atoms of the X group, SA is the → *solvent accessible surface area* of the ith atom and $\Delta\sigma$ the corresponding atomic contribution to solvation energy [Eisenberg and McLachlan, 1986]. The atomic contributions $\Delta\sigma_i \cdot SA_i$ are the free energy of transfer of each atom to the solution; note that the areas SA_i depend on molecule conformation. Proposed to study protein folding and binding, the estimated values for the atomic solvation parameters $\Delta\sigma$ (in cal Å^{-2} mol^{-1}) are: $\Delta\sigma_C = 16$, $\Delta\sigma_N = -6$, $\Delta\sigma_O = -6$, $\Delta\sigma_{N^+} = -50$, $\Delta\sigma_{O^-} = -24$, $\Delta\sigma_S = 21$, for carbon, nitrogen, oxygen, nitrogen cation, oxygen anion and sulfur, respectively.

atomic walk count → **walk counts**

atomic walk count sum → **walk counts**

atom-in-structure invariant index → **charge descriptors** (⊙ charge-related indices)

atomistic topological indices → **count descriptors**

atom leverage centre → **centre of a graph**

atom number (*A*) *(: atom count)*

This is the simplest measure with regard to molecular size, defined as the total number of atoms in a molecule. It is a global, zero-dimensional, highly degenerate descriptor. In several applications for the calculation of → *molecular descriptors*, the atom number A refers only to non-hydrogen atoms.

The **information index on size** is the → *total information content* on the atom number, defined as:

$$I_{SIZE} = A^h \cdot \log_2 A^h$$

where the atom number A^h also takes hydrogen atoms into account [Bertz, 1981]. This index can also be calculated without considering hydrogen atoms.

Other related molecular descriptors are → *atomic composition indices*, several → *information indices* and → *graph invariants*.

atom-pair matching function → **molecular shape analysis**

atom pairs → **substructure descriptors**

atom polarizability → **electric polarization descriptors**

atom-type count → **count descriptors**

atom-type *E*-state indices → **electrotopological state indices**

atom-type *HE*-state indices → **electrotopological state indices**

***ATS* descriptor** → **autocorrelation descriptors** (⊙ Moreau-Broto autocorrelation)

attractive steric effects → **minimal topological difference**

augmented adjacency matrix → **adjacency matrix**

augmented atom codes → **substructure descriptors**

augmented fragments *: augmented atom codes* → **substructure descriptors**

Austel branching index → **steric descriptors**

autocorrelation descriptors

→ *Molecular descriptors* based on the autocorrelation function AC_l, defined as:

$$AC_l = \int_a^b f(x) \cdot f(x+l) \cdot dx$$

where $f(x)$ is any function of the variable x and l is the *lag* representing an interval of x; a and b define the total studied interval of the function. The function $f(x)$ is usually a time-dependent function such as a time-dependent electrical signal, or a spatial-dependent function such as the population density in space.

Function AC_l is a summation of function value products calculated at x and $x + l$. This function expresses how numerical values of the function at intervals equal to the *lag* are correlated.

Autocorrelation functions AC_l can also be calculated for any ordered sequence of n values $f(x_i)$ by summing the products of the ith value with the $(i + l)$th value as:

$$AC_l = \sum_{i=1}^{n-l} f(x_i) \cdot f(x_{i+l})$$

where l is the *lag*. The *lag* assumes values between 1 and L, where the maximum value L can be $n - 1$. However, in several applications, L is a small number ($L < 8$); a *lag* value of zero corresponds to the sum of the square values of the function.

A property of the autocorrelation function is that it does not change when the origin of the x variable is shifted.

To obtain spatial autocorrelation molecular descriptors, function $f(x_i)$ is any physico-chemical property calculated for each atom of the molecule, such as atomic mass, polarizability, etc., and → *local vertex invariants* such as → *vertex degree*. Therefore, the molecule atoms represent the set of discrete points in space and the atomic property the function evaluated at those points.

For spatial autocorrelation molecular descriptors calculated on a → *molecular graph*, *lag* l is defined as the → *topological distance d*.

Autocorrelation descriptors can also be calculated for 3D-spatial molecular geometry. In this case, the distribution of a molecular property can be a mathematical function $f(x, y, z)$, x, y, and z being the spatial coordinates, defined either for each point of molecular space (i.e. a continuous property such as electronic density or molecular interaction energy) or only for points occupied by atoms (i.e. atomic properties). An example of 3D autocorrelation descriptors are the → *spectral weighted invariant molecular descriptors* defined by → *SWM signals*.

The plot of ordered sequence of autocorrelation descriptors from *lag* zero to *lag* L is called the **autocorrelogram**, usually used in similarity analysis.

The most well-known autocorrelation descriptors are listed below.

• **Moreau-Broto autocorrelation** (: *Autocorrelation of a Topological Structure, ATS*)
This is a spatial autocorrelation defined on a molecular graph $\mathcal{G}$ as:

$$ATS_d = \sum_{i=1}^{A}\sum_{j=1}^{A} \delta_{ij} \cdot (w_i \cdot w_j)_d = \mathbf{w}^{\mathbf{T}} \cdot {}^{\mathbf{m}}\mathbf{B} \cdot \mathbf{w}$$

where w is any atomic property, A is the → *atom number*, d is the considered topological distance (i.e. the *lag* in autocorrelation terms), δ_{ij} is Kronecker delta ($\delta_{ij} = 1$ if $d_{ij} = d$, zero otherwise). ${}^{\mathbf{m}}\mathbf{B}$ is the mth order → *binary sparse matrix* and $\mathbf{w}$ the A-dimensional vector of atomic properties [Broto *et al.*, 1984a]. The autocorrelation ATS_0 defined for path of length zero is calculated as:

$$ATS_0 = \sum_{i=1}^{A} w_i^2$$

i.e. the sum of the squares of the atomic properties. Typical atomic properties are atomic masses, polarizabilities, charges, electronegativities.

For each atomic property w, the set of the autocorrelation terms defined for all existing distances in the graph is the ***ATS* descriptor** defined as:

$$\langle ATS_0, ATS_1, ATS_2, \ldots, ATS_D \rangle_w$$

where D is the → *topological diameter*, i.e. the maximum distance in the graph. The plot of the *ATS* descriptor is the corresponding autocorrelogram.

Average spatial autocorrelation descriptors are obtained by dividing each term by the corresponding number of contributions, thus excluding any dependence on molecular size:

$$\overline{ATS_d} = \frac{1}{\Delta} \cdot \sum_{i=1}^{A} \sum_{j=1}^{A} \delta_{ij} \cdot (w_i \cdot w_j)_d$$

where Δ is the sum of the Kronecker deltas, i.e. the number of vertex pairs at distance equal to d [Wagener *et al.*, 1995].

Example : 4-hydroxy-2-butanone

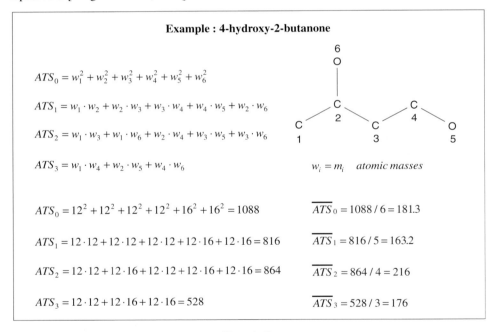

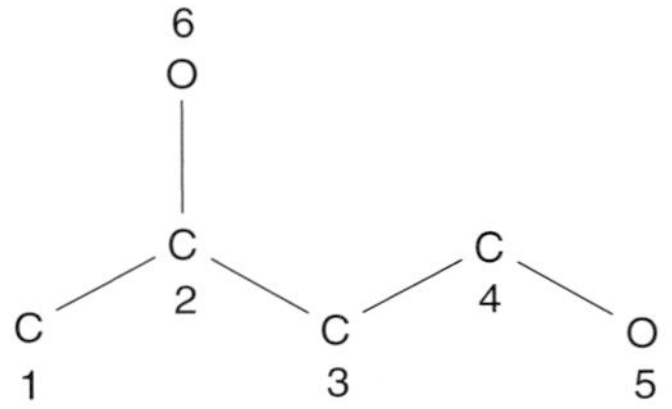

$ATS_0 = w_1^2 + w_2^2 + w_3^2 + w_4^2 + w_5^2 + w_6^2$

$ATS_1 = w_1 \cdot w_2 + w_2 \cdot w_3 + w_3 \cdot w_4 + w_4 \cdot w_5 + w_2 \cdot w_6$

$ATS_2 = w_1 \cdot w_3 + w_1 \cdot w_6 + w_2 \cdot w_4 + w_3 \cdot w_5 + w_3 \cdot w_6$

$ATS_3 = w_1 \cdot w_4 + w_2 \cdot w_5 + w_4 \cdot w_6$

$w_i = m_i$ *atomic masses*

$ATS_0 = 12^2 + 12^2 + 12^2 + 12^2 + 16^2 + 16^2 = 1088$ $\quad \overline{ATS_0} = 1088/6 = 181.3$

$ATS_1 = 12 \cdot 12 + 12 \cdot 12 + 12 \cdot 12 + 12 \cdot 16 + 12 \cdot 16 = 816$ $\quad \overline{ATS_1} = 816/5 = 163.2$

$ATS_2 = 12 \cdot 12 + 12 \cdot 16 + 12 \cdot 12 + 12 \cdot 16 + 12 \cdot 16 = 864$ $\quad \overline{ATS_2} = 864/4 = 216$

$ATS_3 = 12 \cdot 12 + 12 \cdot 16 + 12 \cdot 16 = 528$ $\quad \overline{ATS_3} = 528/3 = 176$

Box A-2.

The *ATS* descriptor is a graph invariant describing how the considered property is distributed along the topological structure. In fact, assuming an additive scheme, the *ATS* descriptor corresponds to a decomposition of the square molecular property Φ in different atomic contributions:

$$\Phi^2 = \left(\sum_{i=1}^{A} w_i \right)^2 = \sum_{i=1}^{A} w_i^2 + \sum\nolimits_{i \neq j} 2 \cdot w_i \cdot w_j = ATS_0 + 2 \cdot \sum_{d=1}^{D} ATS_d$$

where ATS_0 contains all atomic contributions to the square molecular property and ATS_d the interactions between each pair of atoms.

Autocorrelation descriptors calculated for 3D-spatial molecular geometry are based on interatomic distances collected in the → *geometry matrix* and the property function is defined by the set of atomic properties.

The variable x, which is an interatomic distance r, is divided into elementary distance intervals of 0.5 Å. All interatomic distances falling in the same interval are considered identical. The autocorrelation function values AC_l are obtained by summing all the products $w_i\ w_j$ of all pairs of atoms i and j, for which the interatomic distance r_{ij} falls within the considered interval $[x\ , x + 0.5]_l$.

📖 [Moreau and Broto, 1980a; Moreau and Broto, 1980b; Broto *et al.*, 1984c; Broto and Devillers, 1990; Zakarya *et al.*, 1993a]

• **Moran coefficient** ($I(d)$)

General index of spatial autocorrelation that, if applied to a molecular graph, can be defined as:

$$I(d) = \frac{\frac{1}{\Delta} \cdot \sum_{i=1}^{A} \sum_{j=1}^{A} \delta_{ij} \cdot (w_i - \bar{w}) \cdot (w_j - \bar{w})}{\frac{1}{A} \cdot \sum_{i=1}^{A} (w_i - \bar{w})^2}$$

where w_i is any atomic property, $\bar{w}$ is its average value on the molecule, A is the atom number, d is the considered topological distance (i.e. the *lag* in autocorrelation terms), δ_{ij} is a Kronecker delta ($\delta_{ij} = 1$ if $d_{ij} = d$, zero otherwise). Δ is the sum of the Kronecker deltas, i.e. the number of vertex pairs at distance equal to d [Moran, 1950].

The Moran coefficient usually takes a value in the interval [–1,+1]. Positive autocorrelation corresponds to positive values of the coefficient whereas negative autocorrelation produces negative values.

• **Geary coefficient** ($c(d)$)

General index of spatial autocorrelation that, if applied to a molecular graph, can be defined as:

$$c(d) = \frac{\frac{1}{2\Delta} \cdot \sum_{i=1}^{A} \sum_{j=1}^{A} \delta_{ij} \cdot (w_i - w_j)^2}{\frac{1}{(A-1)} \cdot \sum_{i=1}^{A} (w_i - \bar{w})^2}$$

where w_i is any atomic property, $\bar{w}$ is its average value on the molecule, A is the atom number, d is the considered topological distance (i.e. the lag in the autocorrelation terms), δ_{ij} is a Kronecker delta ($\delta_{ij} = 1$ if $d_{ij} = d$, zero otherwise). Δ is the sum of the Kronecker deltas, i.e. the number of vertex pairs at distance equal to d [Geary, 1954].

The Geary coefficient is a distance-type function varying from zero to infinity. Strong autocorrelation produces low values of this index; moreover, positive autocorrelation translates into values between 0 and 1 whereas negative autocorrelation produces values larger than 1; therefore, the reference "no correlation" is $c = 1$.

Crosscorrelation descriptors can be calculated in a way analogous to that for the autocorrelation descriptors, i.e. for any pairs of ordered sequences of values $f(x_i)$ and $f(y_i)$ by summing the products of the ith value of the first sequence x with the $(i + l)$th value of the second sequence y, as:

$$ACC_l = \sum_{i=1}^{n-l} f(x_i) \cdot f(y_{i+l})$$

where n is the lowest cardinality of the two sets.

An example of auto- and crosscorrelation descriptors are **ACC transforms**, derived from → *CoMFA fields* (steric and electrostatic fields) using as *lags* the distances between grid points and their neighbours along each coordinate axis, along the diagonal or along any intermediate direction. The crosscorrelation terms are obtained by the products of the → *interaction energy values* for steric and electrostatic fields in grid points at a distance equal to the *lag*. Different kinds of interactions, namely positive-positive, negative-negative, and positive-negative, are kept separated, thus resulting in 10 ACC terms for each *lag*. The major drawback of ACC transforms is that their values depend on molecule orientation along the axes [Clementi *et al.*, 1993b; van de Waterbeemd *et al.*, 1993b].

📖 [Chastrette *et al.*, 1986] [Devillers *et al.*, 1986] [Grassy and Lahana, 1993] [Zakarya *et al.*, 1993b] [Blin *et al.*, 1995] [Sadowski *et al.*, 1995] [Bauknecht *et al.*, 1996] [Patterson *et al.*, 1996] [Huang *et al.*, 1997] [Anzali *et al.*, 1998] [Legendre and Legendre, 1998]

autocorrelation of a topological structure : *Moreau-Broto autocorrelation* → **autocorrelation descriptors**

autocorrelogram → **autocorrelation descriptors**

automorphism group → **graph**

average atom charge density → **quantum-chemical descriptors** (⊙ electron density)

average atom eccentricity → **distance matrix**

average bond charge density → **quantum-chemical descriptors** (⊙ electron density)

average cyclicity index → **detour matrix**

average distance degree → **distance matrix**

average distance/distance degree → **distance/distance matrix**

average distance sum connectivity : *Balaban distance connectivity index* → **Balaban distance connectivity indices**

average electrophilic superdelocalizability → **quantum-chemical descriptors** (⊙ electrophilic superdelocalizability)

average Fukui function → **quantum-chemical descriptors** (⊙ Fukui function)

average geometric distance degree → **molecular geometry**

average information content based on centre → **centric indices**

average local ionization energy → **quantum-chemical descriptors** (⊙ electron density)

average molecular weight → **molecular weight**

average nucleophilic superdelocalizability → **quantum-chemical descriptors** (⊙ nucleophilic superdelocalizability)

average span → **size descriptors** (⊙ span)

average vertex distance degree → **Balaban distance connectivity indices**

average writhe → **mean overcrossing number**

AZV descriptors → **MPR descriptors**

B

backward Fukui function → **quantum-chemical descriptors** (⊙ Fukui function)

Balaban centric indices → **centric indices**

Balaban distance connectivity index → **Balaban distance connectivity indices**

Balaban distance connectivity indices

The formerly proposed and the most important of this series of topological indices is the **Balaban distance connectivity index** J (also called **distance connectivity index** or **average distance sum connectivity**). It is one of the most discriminating → *molecular descriptors* and its values do not increase substantially with molecule size or number of rings; it is defined in terms of sums over each ith row of the → *distance matrix* **D**, i.e. the → *vertex distance degree* σ [Balaban, 1982; Balaban, 1983a]. It is defined as:

$$\mathrm{J} = \frac{B}{C+1} \cdot \sum\nolimits_b \left(\sigma_i \cdot \sigma_j\right)_b^{-1/2} = \frac{1}{C+1} \cdot \sum\nolimits_b \left(\bar{\sigma}_i \cdot \bar{\sigma}_j\right)_b^{-1/2}$$

where σ_i and σ_j are the vertex distance degrees of two adjacent atoms, and the sum runs over all the molecular bonds b; B is the number of bonds in the molecular graph $\mathcal{G}$, and C, called the → *cyclomatic number*, the number of rings. The denominator $C + 1$ is a normalization factor against the number of rings in the molecule. $\bar{\sigma}_i = \sigma_i / B$ is the **average vertex distance degree**; it was observed that within an isomeric series the average distance degrees are low in the more branched isomers.

To further improve the discriminant power of the Balaban index J, a set of new LOVIs was defined as:

$$t_i = \frac{\sigma_i}{\delta_i}$$

where δ_i is the ith → *vertex degree*. Therefore, the $\mathbf{J_t}$ **index** was defined as:

$$\mathrm{J_t} = \frac{B}{C+1} \cdot \sum\nolimits_b \left(t_i \cdot t_j\right)_b^{-1/2}$$

The idea behind these LOVIs is that usually the vertices with the highest distance sums have the lowest vertex degrees, thus enhancing the intramolecular differences [Balaban, 1994a].

The J index for multigraphs is calculated by the distance sums of the → *multigraph distance matrix* where the distances are obtained by weighting each edge with the inverse of its → *conventional bond order* (→ *relative topological distance*); the sum runs over all pairs of adjacent vertices and B is the number of edges in the graph without accounting for their multiplicity.

Example : 2-methylpentane

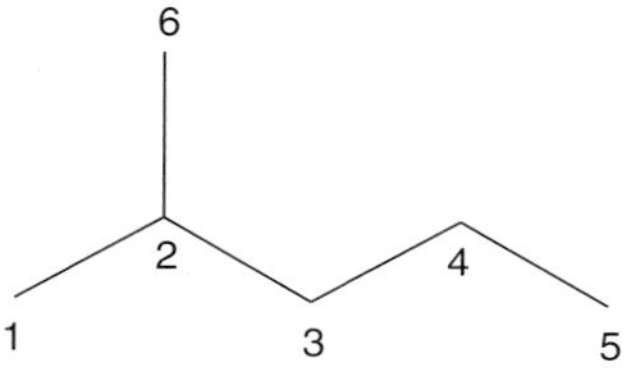

Distance matrix **D**

Atom	**1**	**2**	**3**	**4**	**5**	**6**	σ_i
1	0	1	2	3	4	2	12
2	1	0	1	2	3	1	8
3	2	1	0	1	2	2	8
4	3	2	1	0	1	3	10
5	4	3	2	1	0	4	14
6	2	1	2	3	4	0	12

Vertex degrees

Atom	δ_i	t_i
1	1	12
2	3	2.667
3	2	4
4	2	5
5	1	14
6	1	12

$$\mathrm{J} = \frac{B}{C+1} \cdot \left[(\sigma_1 \cdot \sigma_2)^{-1/2} + (\sigma_6 \cdot \sigma_2)^{-1/2} + (\sigma_2 \cdot \sigma_3)^{-1/2} + (\sigma_3 \cdot \sigma_4)^{-1/2} + (\sigma_4 \cdot \sigma_5)^{-1/2} \right] =$$
$$= 5 \cdot \left[(12 \cdot 8)^{-1/2} + (12 \cdot 8)^{-1/2} + (8 \cdot 8)^{-1/2} + (8 \cdot 10)^{-1/2} + (10 \cdot 14)^{-1/2} \right] = 2.6272$$

$$\mathrm{J}_t = \frac{B}{C+1} \cdot \left[(t_1 \cdot t_2)^{-1/2} + (t_6 \cdot t_2)^{-1/2} + (t_2 \cdot t_3)^{-1/2} + (t_3 \cdot t_4)^{-1/2} + (t_4 \cdot t_5)^{-1/2} \right] =$$
$$= 5 \cdot \left[(12 \cdot 2.667)^{-1/2} + (12 \cdot 2.667)^{-1/2} + (2.667 \cdot 4)^{-1/2} + (4 \cdot 5)^{-1/2} + (5 \cdot 14)^{-1/2} \right] = 5.0141$$

Box B-1.

In order to account for both bond multiplicity and heteroatoms, **Balaban modified distance connectivity indices** J^{X} and J^{Y} were proposed [Balaban, 1986a; Balaban *et al.*, 1990a]. These are defined in the same way as the Balaban distance connectivity index but derived from the → *multigraph distance matrix* $^*\mathbf{D}$ instead of the original distance matrix **D**:

$$\mathrm{J}^{\mathrm{X}} = \frac{B}{C+1} \cdot \sum\nolimits_b \left(^*\sigma_i^X \cdot {}^*\sigma_j^X \right)^{-1/2}$$
$$\mathrm{J}^{\mathrm{Y}} = \frac{B}{C+1} \cdot \sum\nolimits_b \left(^*\sigma_i^Y \cdot {}^*\sigma_j^Y \right)^{-1/2}$$

where B is the bond number, C is the cyclomatic number, and the sum runs over all bonds b in the graph, each being weighted by the inverse square root of the product of the → *multigraph distance degree* of the incident vertices. The distance degrees are calculated as:

$$ {}^{*}\sigma_i^X = X_i \cdot {}^{*}\sigma_i = X_i \cdot \sum_{j=1}^{A} [{}^{*}\mathbf{D}]_{ij} \quad \text{and} \quad X_i = 0.4196 - 0.0078 \cdot Z_i + 0.1567 \cdot L_i $$

$$ {}^{*}\sigma_i^Y = Y_i \cdot {}^{*}\sigma_i = Y_i \cdot \sum_{j=1}^{A} [{}^{*}\mathbf{D}]_{ij} \quad \text{and} \quad Y_i = 1.1191 + 0.0160 \cdot Z_i - 0.0537 \cdot L_i $$

where ${}^{*}\sigma$ is the vertex distance degree calculated on multigraph distance matrix ${}^{*}\mathbf{D}$, the quantities X and Y are recalculated atomic Sanderson electronegativities and covalent radii relative to carbon atom, obtained as a function of the atomic number Z_i and the principal quantum number L_i of the atom; for atoms different from B, C, N, O, F, Si, P, S, Cl, As, Se, Br, Te, and I the X and Y values are set at one. X and Y indices account for the presence of heteroatoms in the molecule.

Another generalization of the Balaban index J, so as to account for heteroatoms in the molecule, is the → *Barysz index* calculated on the → *Barysz distance matrix*.

→ *JJ indices* derived from the → *Wiener matrix* were proposed as a generalization of the Balaban index in analogy with the Kier-Hall → *connectivity indices*.

The **3D-Balaban index** ^{3D}J was derived from the → *geometry matrix* G as:

$$ {}^{3D}J = \mathcal{IB}(G) = \frac{B}{C+1} \cdot \sum\nolimits_b \left({}^{G}\sigma_i \cdot {}^{G}\sigma_j \right)_b^{-1/2} $$

where $\mathcal{IB}$ is the → *Ivanciuc-Balaban operator*; ${}^{G}\sigma_i$ and ${}^{G}\sigma_j$ are the → *geometric distance degree* of the two vertices incident with the *b* bond [Mihalic *et al.*, 1992a].

A **Balaban-type index** *DJ* [Balaban and Diudea, 1993] was defined as:

$$ DJ = \sum_{i=1}^{A} dj_i = \sum_{i=1}^{A} \sum\nolimits_{j \in \mathcal{V}_{i1}} \left(\frac{\sigma_i}{w_i(1+f_i)} \cdot \frac{\sigma_j}{w_j(1+f_j)} \right)^{-1/2} $$

where *A* is the → *atom number*, *f* is the → *multigraph factor*, *w* is a weighting factor accounting for heteroatoms, and the inner sum runs over all vertices *j* at distance 1 from the *i*th atom, i.e. vertices bonded to the *i*th atom; *dj* are local vertex invariants accounting for heteroatoms and bond multiplicity. When the factor *w* is equal to one and the multigraph factor is equal to zero then the index *DJ* is related to the Balaban index J by the following:

$$ DJ = 2 \cdot J \cdot \frac{C+1}{B} $$

📖 [Balaban and Quintar, 1983] [Barysz *et al.*, 1983a] [Balaban and Filip, 1984] [Balaban *et al.*, 1985e] [Sabljic, 1985] [Mekenyan *et al.*, 1987] [Balaban and Ivanciuc, 1989] [Balaban *et al.*, 1990b] [Balaban *et al.*, 1992a] [Nikolic *et al.*, 1993a] [Guo and Randic, 1999]

Balaban ID number → **ID numbers**

Balaban modified distance connectivity indices → **Balaban distance connectivity indices**

Balaban-type index → **Balaban distance connectivity indices**

Bartell resonance energy → **resonance indices**

barycentre : *centre of mass* → **centre of a molecule**

Barysz index → **weighted matrices**

Barysz distance matrix → **weighted matrices**

basis descriptors

Any set of → *molecular descriptors* used to represent molecules in a fixed and ordered sequence [Randic, 1992d; Randic and Trinajstic, 1993b; Baskin *et al.*, 1995]. Basis descriptors can be defined: a) by selecting a class of homogeneous naturally ordered descriptors (→ *uniform length descriptors*, → *path counts*, → *connectivity indices*, → *3D-MoRSE descriptors*, → *EVA descriptors*, → *Burden eigenvalues*, etc.); b) by using a few *ad hoc* selected descriptors such as one for each class among → *lipophilicity descriptors*, → *steric descriptors* and → *electronic descriptors*, as used in → *classical QSAR*.

A basis of descriptors can be viewed as a representation of the → *molecular structure* suitable for use in different applications. An interesting characteristic of basis descriptors is that they can be orthogonalized, thus providing a new basis of → *orthogonalized descriptors*.

Bate-Smith-Westall retention index → **chromatographic descriptors**

BC(DEF) coordinates : *BC(DEF) parameters*

BC(DEF) parameters *(: BC(DEF) coordinates)*

Proposed by Cramer, they are → *principal properties* (i.e. significant components calculated by Principal Component Analysis) of a data matrix given by 6 physico-chemical properties for 114 diverse liquid-state compounds [Cramer III, 1980a].

The → *physico-chemical properties* used to derive BC(DEF) descriptors are: activity coefficient in water, → *octanol-water partition coefficient*, boiling point, → *molar refractivity*, liquid-state → *molar volume*, heat of vaporization. The eigenvalues and corresponding cumulative explained variance of the five significant components (denoted by B, C, D, E, F) are reported in Table B-1. It may be noted that the first two principal properties B and C already explain 95.7 % of the original variance of the six physico-chemical properties; further analysis using different compounds and properties showed B and C to be independent of the data set used in their derivation, identifying them respectively as measures of molecular bulk and cohesiveness. The other three parameters, D, E, and F, are of minor importance; however, they were retained because of their significance in the correlations involving some physico-chemical properties.

In general BC(DEF) parameters describe molecular properties related to non-specific intermolecular interactions in the liquid state and could therefore be useful in predicting biological activity or physico-chemical properties depending on such non-specific interactions; 29 linear models were calculated by multivariate regression analysis that correlates BC(DEF) parameters to 29 different physico-chemical properties.

Table B-1. Eigenvalues and cumulative variances of BC(DEF) principal properties.

Principal property	*Eigenvalue*	*Cumulative variance (%)*
B	3.870	64.4
C	1.870	95.7
D	0.168	98.5
E	0.045	99.2
F	0.029	99.7

The calculation of BC(DEF) parameters for new compounds different from the original 114 compounds can be accomplished either by their physico-chemical properties or their structure.

The property-derived BC(DEF) values are calculated from a set of known property values and the corresponding property models previously derived from the original 114 × 6 data set. A property model has the general form:

$$y = b_0 + b_1 \cdot \mathrm{B} + b_2 \cdot \mathrm{C} + b_3 \cdot \mathrm{D} + b_4 \cdot \mathrm{E} + b_5 \cdot \mathrm{F}$$

where y is the known experimental property value and b the known regression coefficients taken from the specific property model. Using a set of at least six property models, all the BC(DEF) values together with their confidence intervals can be obtained as solutions of the linear equation system [Cramer III, 1983a]. In this case the physico-chemical properties should be considered as independent variables and the BC(DEF) values as dependent variables.

Alternatively, the BC(DEF) values can be obtained by → *additive-constitutive models* based on the contributions of individual fragments and some correction factors to each parameter [Cramer III, 1980b]. A hierarchical additive-constitutive model was derived by multivariate regression between the BC(DEF) values of 112 original compounds (water and methane were excluded from the model) and 35 molecular fragments. Moreover, in the same way a linear additive-constitutive model was also proposed; the fragment contributions to BC(DEF) parameters are reported in Table B-2.

Table B-2. Fragment contributions to BC(DEF) parameters. [a]Value when attached to an aliphatic system.

Fragment	*B*	*C*	*D*	*E*	*F*
intercept	−0.506	−0.056	0.007	0.031	0.028
−H	0.066	0.018	−0.027	−0.019	−0.019
$-CH_3$	0.142	−0.020	−0.016	−0.023	−0.015
$-CH_2-$	0.076	−0.038	0.011	−0.004	0.003
>CH−	0.003	−0.058	0.053	0.018	0.015
>C<	−0.075	−0.076	0.091	0.043	0.034
−CH=CH−	0.147	−0.043	0.028	0.010	0.003
$-CH{=}CH_2$	0.212	−0.025	0.000	−0.009	−0.015
$>CH{=}CH_2$	0.147	−0.043	0.028	0.010	0.003
−C≡CH	0.171	0.074	0.027	0.002	−0.012
$-C_6H_5$	0.467	−0.007	0.012	0.007	−0.017
≈CH− (aromatic)	0.088	0.002	−0.007	0.001	−0.003
−naphthyl	0.766	0.018	−0.026	0.024	−0.028
−cyclohexyl	0.489	−0.148	0.004	−0.029	−0.009
−F [a]	0.078	0.088	0.009	−0.019	−0.020
−Cl	0.165	0.087	−0.024	−0.012	−0.021
−Br [a]	0.213	0.095	−0.033	−0.008	−0.020
−I [a]	0.302	0.103	−0.056	−0.010	−0.031
$-CF_3$	0.150	0.017	0.035	−0.037	−0.013
$-CCl_3$	0.410	0.015	−0.009	−0.017	−0.017
−OH [a]	0.202	0.324	−0.012	−0.015	0.003

Fragment	*B*	*C*	*D*	*E*	*F*
–O– [a]	0.044	0.155	0.061	0.019	–0.022
–C=O– [a]	0.135	0.246	0.061	0.023	–0.021
–CH=O [a]	0.219	0.244	0.010	–0.014	–0.027
–COO– [a]	0.167	0.170	0.062	0.015	–0.027
–COOH [a]	0.323	0.342	–0.011	–0.017	0.008
$–NH_2$ [a]	0.167	0.269	0.037	0.027	–0.014
–NH– [a]	0.082	0.251	0.095	0.056	–0.010
–N– [a]	–0.006	0.189	0.125	0.069	0.014
–CN [a]	0.241	0.269	–0.007	–0.023	–0.041
–N= [a] (pyridine)	0.102	0.183	0.031	–0.011	–0.020
$–NO_2$ [a]	0.238	0.241	–0.012	–0.027	–0.037
$–CONH_2$ [a]	0.444	0.499	–0.019	–0.039	–0.012
–S– [a]	0.136	0.130	0.028	0.032	–0.020
–SH[a]	0.231	0.155	–0.026	–0.011	–0.013

📖 [Cramer III, 1983b]

BCUT descriptors → **eigenvalue-based descriptors** (⊙ Burden eigenvalues)

benzene-likeliness index → **resonance indices**

Bertz branching index : *connection number* → **edge adjacency matrix**

Bertz complexity index → **molecular complexity**

Bertz-Herndon relative complexity index → **molecular complexity**

Betti numbers → **Mezey 3D shape analysis**

binary descriptors → **indicator variables**

binary distance measures → **similarity/diversity**

binary QSAR → **structure/response correlations**

binary sparse matrix → **algebraic operators** (⊙ sparse matrices)

binding affinity → **drug design**

binding property pairs → **substructure descriptors** (⊙ atom pairs)

binding property torsions → **substructure descriptors** (⊙ topological torsion descriptor)

binding site cavity → **drug design**

binormalized centric index → **centric indices** (⊙ Balaban centric index)

binormalized quadratic index → **Zagreb indices**

Blurock spectral descriptors

These are atomic or bond descriptors derived from a spectral representation of the molecules, such as the following: a property is associated with each atom (or bond) in such a way as to also represent its environment, the range of the property values is divided into constant intervals and the number of times the values of the molecule atoms fall within each interval is counted. The spectrum of the molecule is the distribution of these property values [Blurock, 1998].

The atomic properties considered are partial charges, electron densities and polarizabilities, calculated by → *computational chemistry* methods; moreover, bond properties have been proposed as the difference between the property values of the atoms forming the bond. The range of each property is determined by the maximum and minimum values for all the atoms in all the molecules, thus obtaining uniform spectrum length for all the molecules in the data set.

Inductive learning was suggested for the prediction of the molecular property values.

Bocek-Kopecky analysis → Free-Wilson analysis

Bocek-Kopecky model → Free-Wilson analysis

Bodor hydrophobic model → lipophilicity descriptors

Bodor LOG*P* : *Bodor hydrophobic model* → **lipophilicity descriptors**

Bonchev centric information indices → centric indices

Bonchev complexity information index → molecular complexity

Bonchev topological complexity indices → molecular complexity

bond alternation coefficient → resonance indices

bond angles → molecular geometry

bond connectivity indices → connectivity indices

bond count : *bond number*

bond dipole moment → bond ionicity indices

bond distances → molecular geometry

bond distance-weighted edge adjacency matrix → edge adjacency matrix

bond eccentricity → edge distance matrix

bond *E*-state *index* → electrotopological state indices

bond flexibility → flexibility indices

bond flexibility index → flexibility indices

bond index → quantum-chemical descriptors (⊙ electron density)

bonding information content → indices of neighbourhood symmetry

bond ionicity indices

Such indices represent bond descriptors and describe the bond character, the importance of bond character to the physical and chemical behaviour of compounds being well known. Bond character is closely related to the capacity of bonded atoms to exchange electrons, and such capacity is commonly well represented by the → *electronegativity* χ of the bonded atoms.

The difference in electronegativity between two bonded atoms was called **bond dipole moment** [Malone, 1933], defined as:

$$\mu_{ij} = \left|\chi_i - \chi_j\right|$$

but there was poor correlation between this index and bond ionicity. Therefore, starting from bond dipole moment, several empirical relationships have been proposed to define bond ionicity indices f_b . The most popular are [Barbe, 1983]:

1. $f_b = 1 - \exp\left[-0.25 \cdot \left(\chi_i - \chi_j\right)^2\right]$ 2. $f_b = 1 - \exp\left[-0.21 \cdot \left(\chi_i - \chi_j\right)^2\right]$

3. $f_b = 0.160 \cdot \left(\chi_i - \chi_j\right) + 0.035 \cdot \left(\chi_i - \chi_j\right)^2$

4. $f_b = \frac{\chi_i - \chi_j}{\chi_i + \chi_j}$ 5. $f_b = \frac{\chi_i - \chi_j}{2}$ 6. $f_b = \frac{\chi_i - \chi_j}{\chi_i}$ with $\chi_i > \chi_j$

Bondi volume → **molecular surface** (⊙ van der Waals molecular surface)

bond number (*B*) (*: edge counting; bond count*)

This is the simplest graph invariant obtained from the → *adjacency matrix* **A**, defined as the number of bonds in the → *molecular graph* $\mathcal{G}$ where multiple bonds are considered as single edges. Bond number is calculated as half the → *total adjacency index* A_V:

$$B = \frac{1}{2} \cdot A_V = \frac{1}{2} \cdot \sum_{i=1}^{A} \sum_{j=1}^{A} a_{ij}$$

where a_{ij} are the entries of the adjacency matrix.

It is related to molecular size and gives equal weight to chemically nonequivalent groups, such as CH_2–CH_2, CH_2=CH, CH_2–NH_2, CH_2–Cl.

When bond multiplicity in the molecule must be considered several → *multiple bond descriptors* can be used instead of the bond number.

The number of bonds is considered in the → *cyclomatic number* and appears in several → *molecular descriptors* such as the → *Balaban distance connectivity index*, the → *mean Randic branching index*, the → *information bond index*, and several → *topological information indices.*

bond order → **quantum-chemical descriptors** (⊙ electron density)

bond order indices

These are descriptors for molecule bonds proposed with the aim of estimating the → *bond order* defined in quantum-chemical theory or of generally defining bond weights so as to distinguish the bonds in a → *molecular graph.*

Two main definitions of bond order indices are reported below. Moreover, the term **fractional bond order** was suggested to refer to the inverse of any bond order index. Fractional bond order permits individual treatment of σ and π molecular systems; σ bonds give simple graphs, while π bonds introduce a weighted molecular framework with weights less than one [Randic *et al.*, 1980].

- **conventional bond order** (π^*)

Within the framework of the graph theory, specifically the → *multigraph*, the conventional bond order π^* is defined as being equal to 1, 2, 3, and 1.5 for single, double, triple and aromatic bonds, respectively.

In order to consider chemical information relative to multiple bonds in terms of → *topological distance* between pairs of bonded atoms, the inverse powers of the conventional bond order were proposed [Balaban, 1993c; Balaban *et al.*, 1993a]. The **relative topological distance** is defined as:

$$RTD_{ij} = \left(\pi^*_{ij}\right)^{-1}$$

and, to obtain values more related to the standardized experimental interatomic average distances (as reference is taken the distance of single bonds, Table B-3), the **chemical distance** is defined as:

$$CD_{ij} = \left(\pi_{ij}^{*}\right)^{-1/4}$$

Using relative topological distance or chemical distance as well as conventional bond order to weight each edge in the graph → *weighted matrices* can be obtained.

Table B-3. Average experimental bond distances r, carbon relative values r^*, conventional bond orders π^*, relative topological distances *RTD*, and chemical distances *CD*.

Bond type	r (Å)	r^*	π^*	*RTD*	*CD*
C – C	1.54	1.00	1	1.00	1.00
C = C (Ar)	1.40	0.91	1.5	0.67	0.90
C = C	1.33	0.86	2	0.50	0.84
C ≡ C	1.20	0.78	3	0.33	0.76
C – N	1.47	1.00	1	1.00	1.00
C = N	1.29	0.88	2	0.50	0.84
C ≡ N	1.16	0.79	3	0.33	0.76
N – N	1.45	1.00	1	1.00	1.00
N = N	1.26	0.87	2	0.50	0.90
C – O	1.41	1.00	1	1.00	1.00
C = O	1.21	0.86	2	0.50	0.84
O – O	1.45	1.00	1	1.00	1.00
N – O	1.47	1.00	1	1.00	1.00
N = O	1.15	0.78	2	0.50	0.84
C – S	1.81	1.00	1	1.00	1.00
C = S	1.61	0.89	2	0.50	0.84

From the conventional bond order, the **atomic multigraph factor** (or **multigraph factor**) of the ith atom is defined as:

$$f_i = \sum\nolimits_{a \in V_{i1}} \left(\pi_{ia}^{*} - 1\right)$$

where the sum runs over all vertices a bonded to the ith atom, i.e. of the set V_{i1}, where the subscript "1" refers to vertices located at a distance equal to one from the ith atom [Balaban and Diudea, 1993]. This factor is zero for atoms without multiple bonds and it is used, for example, in the molecular descriptors derived from the → *branching layer matrix* and in the → *Balaban-type index*.

• **graphical bond order**

The graphical bond order $(TI' \,/\, TI)_b$ of the bth bond is derived from the → *H-depleted molecular graph* of the molecule by calculating a graph invariant TI' for the subgraph $\mathcal{G}'$ obtained by erasing the considered bond from the graph and then dividing it by the corresponding graph invariant TI calculated on the whole molecular graph $\mathcal{G}$ [Randic *et al.*, 1994c]. If more than one subgraph is obtained by the erasure of each bond, the single contributions are summed to give the graphical bond order. The ratio $(TI' \,/\, TI)_b$ can be interpreted as a measure of the relative importance of the edge in the graph. The first proposed graphical bond order was that calculated from the → *Hosoya Z index* and was originally called **topological bond order**; it was shown to represent the weight of a bond in distributing π-electrons over the molecular graph [Hosoya *et al.*, 1975a; Hosoya and Murakami, 1975b]. Moreover, graphical bond orders constitute special cases of → *normalized fragment topological indices.*

Molecular descriptors are derived by the additive contributions of the graphical bond orders of all bonds in the molecule as:

$$TI'/TI = \sum_{b=1}^{B} \left(\frac{TI'}{TI} \right)_b$$

They are usually called **graphical bond order descriptors**.

The first proposed graphical bond order was calculated using the → *total path count* P as a molecular invariant and is therefore denoted by the ratio $(P' / P)_{ij}$, where i and j refer to the vertices incident to the bth bond erased from the graph [Randic, 1991b].

From the path graphical bond orders, a square symmetric matrix of dimension $A \times A$, called the **P matrix**, can be derived; its elements are defined by the following:

$$[\mathbf{P}]_{ij} = \begin{cases} (P'/P)_{ij} & \text{if the vertices } v_i \text{ and } v_j \text{ are adjacent} \\ 0 & \text{otherwise} \end{cases}$$

"Otherwise" means that either the vertices v_i and v_j are not adjacent or $i = j$.

From the **P** matrix, the corresponding molecular descriptor, called the **P′/P index**, is calculated applying the → *Wiener operator* $\mathcal{W}$ as:

$$P'/P = \mathcal{W}(\mathbf{P}) = \frac{1}{2} \cdot \sum_{i=1}^{A} \sum_{j=1}^{A} [\mathbf{P}]_{ij}$$

Other encountered graphical bond order descriptors are the **χ′/χ index**, the **W′/W index**, the **WW′/WW index**, the **J′/J index**, the **CID′/CID index**, and the → *Z′/Z index* derived, respectively, from the → *Randic connectivity index*, the → *Wiener index*, the → *hyper-Wiener index*, the → *Balaban distance connectivity index*, the → *Randic connectivity ID number*, and the Hosoya Z index.

Example : 2-methylpentane

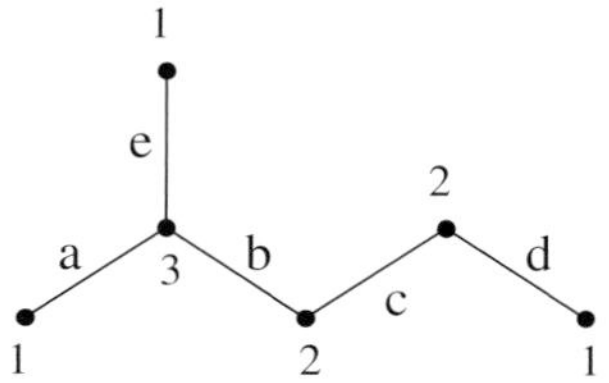

Randic connectivity index

$$^{1}\chi(G) = 2\cdot(1\cdot 3)^{-1/2} + (3\cdot 2)^{-1/2} + (2\cdot 2)^{-1/2} + (2\cdot 1)^{-1/2} = 2.7701$$

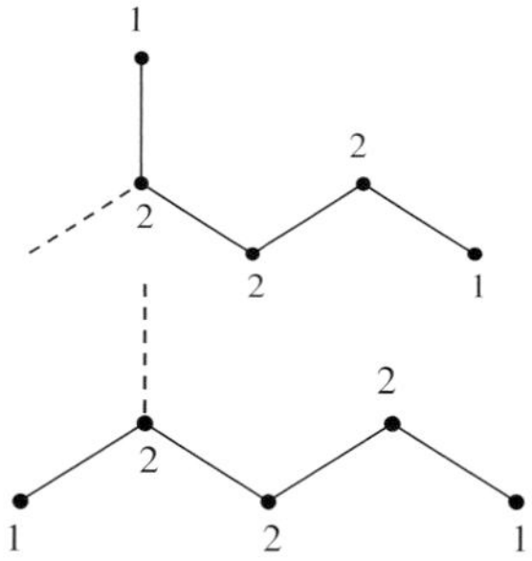

$$^{1}\chi(G-\mathrm{a}) = {}^{1}\chi(G-\mathrm{e}) = 2\cdot(2\cdot 1)^{-1/2} + 2\cdot(2\cdot 2)^{-1/2} = 2.4142$$

$$\left(\frac{\chi'}{\chi}\right)_{\mathrm{a}} = \left(\frac{\chi'}{\chi}\right)_{\mathrm{e}} = \frac{^{1}\chi(G-\mathrm{a})}{^{1}\chi(G)} = \frac{2.4142}{2.7701} = 0.8715$$

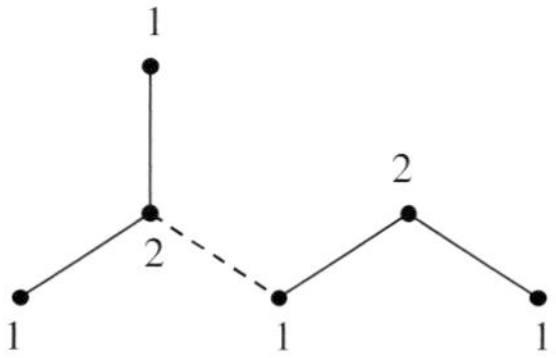

$$^{1}\chi(G-\mathrm{b}) = 4\cdot(2\cdot 1)^{-1/2} = 2.8284$$

$$\left(\frac{\chi'}{\chi}\right)_{\mathrm{b}} = \frac{^{1}\chi(G-\mathrm{b})}{^{1}\chi(G)} = \frac{2.8284}{2.7701} = 1.0210$$

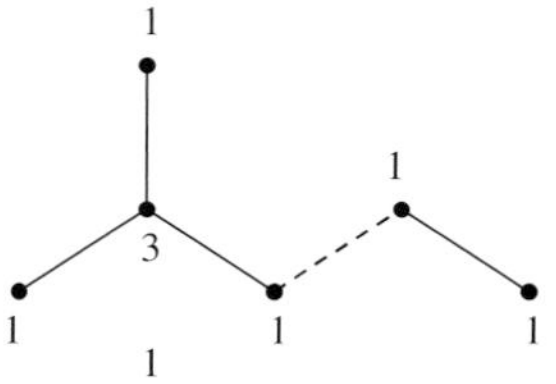

$$^{1}\chi(G-\mathrm{c}) = (1\cdot 1)^{-1/2} + 3\cdot(1\cdot 3)^{-1/2} = 2.7321$$

$$\left(\frac{\chi'}{\chi}\right)_{\mathrm{c}} = \frac{^{1}\chi(G-\mathrm{c})}{^{1}\chi(G)} = \frac{2.7321}{2.7701} = 0.9863$$

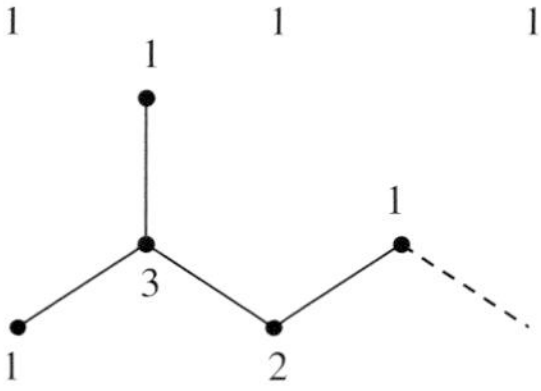

$$^{1}\chi(G-\mathrm{d}) = 2\cdot(1\cdot 3)^{-1/2} + (1\cdot 2)^{-1/2} + (2\cdot 3)^{-1/2} = 2.2701$$

$$\left(\frac{\chi'}{\chi}\right)_{\mathrm{d}} = \frac{^{1}\chi(G-\mathrm{d})}{^{1}\chi(G)} = \frac{2.2701}{2.7701} = 0.8195$$

$$\frac{\chi'}{\chi} = \left(\frac{\chi'}{\chi}\right)_{\mathrm{a}} + \left(\frac{\chi'}{\chi}\right)_{\mathrm{e}} + \left(\frac{\chi'}{\chi}\right)_{\mathrm{b}} + \left(\frac{\chi'}{\chi}\right)_{\mathrm{c}} + \left(\frac{\chi'}{\chi}\right)_{\mathrm{d}} = 2\cdot 0.8715 + 1.0210 + 0.9863 + 0.8195 = 4.5698$$

Box B-2.

An explicit formula for the direct calculation of the χ'/χ index from the molecular graph was derived as:

$$\chi'/\chi = \frac{1}{\chi(\mathcal{G})} \cdot$$

$$\sum_{\{i,j\} \in \mathcal{E}(\mathcal{G})} \frac{(A + C - \delta_i - \delta_j) \cdot (\delta_i \cdot \delta_j)^{1/2} + \delta_i \cdot [(\delta_i - 1) \cdot \delta_j]^{1/2} + \delta_j \cdot [(\delta_j - 1) \cdot \delta_i]^{1/2}}{\delta_i \cdot \delta_j}$$

where $\chi(\mathcal{G})$ is the Randic connectivity index for the whole molecular graph, the sum runs over all the edges in the graph; δ_i and δ_j are the vertex degrees of the vertices incident with the considered edge, A is the number of graph vertices and C the → *cyclomatic number*. This formula holds for every connected graph with $A > 1$ vertices [Plavsic *et al.*, 1998].

In the same way, a general formula valid for any graph based on the number A of graph vertices and the distances between pairs of vertices was derived for the calculation of the WW'/WW index as [Plavsic, 1999]:

$$WW'/WW = A + 1 - \frac{\sum_{i=1}^{A-1} \sum_{j=i+1}^{A} d_{ij} \cdot (d_{ij} + 1) \cdot (d_{ij} + 2)}{\sum_{i=1}^{A-1} \sum_{j=i+1}^{A} d_{ij} \cdot (d_{ij} + 1)}$$

Only for acyclic graphs, after special rearrangment, does the formula take the form:

$$WW'/WW = A + 1 - \frac{\sum_{m=1}^{D} \frac{(m+2)!}{(m-1)!} \cdot {}^{m}P}{\sum_{m=1}^{D} \frac{(m+1)!}{(m-1)!} \cdot {}^{m}P}$$

where m is the length of the considered paths, ${}^{m}P$ is the → *path count* of mth order and D the → *topological diameter*, i.e. the maximum path length in the graph.

📖 [Gutman *et al.*, 1978] [Randic, 1991g] [Randic, 1993c] [Randic, 1994b] [Plavsic *et al.*, 1996b]

bond order-weighted edge adjacency matrix → **edge adjacency matrix**

bond order-weighted edge connectivity index → **edge adjacency matrix**

bond profiles → **molecular profiles**

bond rigidity → **flexibility indices**

bond rigidity index → **flexibility indices** (⊙ bond flexibility index)

bond vertex degree → **vertex degree**

bootstrap → **validation techniques**

Bowden-Wooldridge steric constant → **steric descriptors** (⊙ number of atoms in substituent specific positions)

Bowden-Young steric constant → **steric descriptors** (⊙ Charton steric constant)

branching index : *Randic connectivity index* → **connectivity indices**

branching layer matrix → **layer matrices**

Brillouin redundancy index → **information content**

Broto-Moreau-Vandicke hydrophobic atomic constants → **lipophilicity descriptors**

Buckingham potential function → **molecular interaction fields** (⊙ steric interaction fields)

bulk descriptors → **steric descriptors**

bulk representation → **molecular descriptors**

Burden eigenvalues → **eigenvalue-based descriptors**

Burden matrix → **eigenvalue-based descriptors** (⊙ Burden eigenvalues)

Burden modified eigenvalues → **eigenvalue-based descriptors** (⊙ Burden eigenvalues)

C

Calculated LOG*P* → **lipophilicity descriptors** (⊙ Leo-Hansch hydrophobic fragmental constants)

Camilleri model based on surface area → **lipophilicity descriptors**

Cammarata-Yau analysis → **Free-Wilson analysis**

Cammarata-Yau model → **Free-Wilson analysis**

canonical numbering *(: unique atomic ordering; unique atomic code)*

This is a procedure which assigns unique labels to the graph vertices so that the resulting matrix representations are in canonical form. The principal aim is to find a suitable numerical code for each given graph which characterizes the graph up to isomorphism [Kvasnicka and Pospichal, 1990; Faulon, 1998].

The main canonical ordering procedures are listed below. Several different → *local vertex invariants* showing regular variation from central to terminal vertices were studied for canonical numbering of graph vertices [Filip *et al.*, 1987; Bonchev and Kier, 1992]; examples are → *vertex distance degree*, → *local connectivity indices*, → *electrotopological state indices*, → *weighted atomic self-returning walk counts*, → *Randic atomic path code*, → *MPR descriptors*, → *centric operator* and → *centrocomplexity operator*, → *vertex complexity* and → *vertex distance complexity*. The → *iterative vertex and edge centricity algorithm* (IVEC) also provides canonical ordering of vertices and edges in the graph, and an algorithm based on the eigenvalues and eigenvectors of the adjacency matrix of the graph was also proposed [Liu and Klein, 1991].

- **Morgan's Extended Connectivity Algorithm** (*: Extended Connectivity Algorithm, ECA*)

Graph vertices are ordered on the basis of their extended connectivity values obtained after a number of iterations of the Morgan method until constant atom ordering is obtained in two consecutive steps [Morgan, 1965]. The **extended connectivity** (EC) of a vertex is calculated as the iterative summation of connectivities of all first neighbours as the following:

$$\mathrm{EC}_i^k = \sum_{j=1}^{\delta_i} \mathrm{EC}_j^{k-1}$$

where δ_i is the number of first neighbours of vertex v_i and k represents the step of the iterative procedure; at the beginning ($k = 0$) the connectivity of each atom is simply the → *vertex degree* δ.

The **normalized extended connectivity** (NEC) was derived as:

$$\mathrm{NEC}_i = \lim_{n>>1} \left(\mathrm{EC}_i^k \Big/ \sum_{i=1}^{A} \mathrm{EC}_i^k \right) \cdot \sum_{i=1}^{A} \mathrm{EC}_i^1$$

where the last summation coincides with twice the number of bonds *2B* [Bonchev *et al.*, 1993a].

It must be pointed out that the extended connectivity EC^k of Morgan cannot be considered a new local invariant as it coincides with the → *atomic walk count* awc^k calculated as row sum of the kth power of the adjacency matrix **A** [Razinger, 1982; Rücker and Rücker, 1993; Figueras, 1993].

The Morgan algorithm was later improved by a better formalization and by considering stereochemical aspects [Wipke and Dyott, 1974]. The Stereochemical Extension of the Morgan Algorithm (SEMA) resulted in a higher discriminating ability of graph vertices than ECA; it is based on the iterative summation of the properties of neighboring atoms.

🕮 [Ouyang *et al.*, 1999]

• **First EigenVector Algorithm** (*FEVA*)
Vertices in a graph are ordered according to the relative magnitudes of the coefficients of the first eigenvector (corresponding to the largest eigenvalue) of the → *adjacency matrix* **A** [Randic, 1975d]. Generally nonequivalent vertices have different coefficient magnitudes, while equivalent vertices, i.e. vertices constituting the same orbits, have to be distinguished according to some alternative rule.

In order to obtain a vertex numbering similar to the Morgan algorithm, the convention to associate label 1 to the vertex with the largest coefficient and label A (the number of atoms) to that with the smallest one was established.

The largest coefficients correspond to → *central vertices* and the smallest to → *terminal vertices* and their neighbours.

• **Smallest Binary Label** (SBL)
A binary label assigned to each graph vertex which consists of the corresponding row of the adjacency matrix; this binary label can also be expressed as a decimal number, as in the → *decimal adjacency vector*. The unique numbering given by the smallest binary label of each vertex can be achieved by iteratively renumbering the vertices of the graph, i.e. iterative reordering of the adjacency matrix rows [Randic, 1974; Randic, 1975c; Mackay, 1975].

• **Jochum-Gasteiger canonical numbering**
A canonical numbering algorithm where nonterminal atoms are treated first, monovalent atoms (hydrogen and non-hydrogen atoms) then being numbered corresponding to the nonterminal atoms [Jochum and Gasteiger, 1977]. The algorithm is based on the following steps:

1. the nonterminal atoms are put into the same equivalence class on the basis of their → *vertex eccentricity*; the classes are ordered according to increasing eccentricity values and the atoms within each equivalent class are then ordered separately by the following sequential rules, beginning with the first equivalence class;
2. for each equivalent class, the atom with the highest atomic number has priority;
3. the atom with the most free electrons has priority;
4. the atom with the highest number of first neighbours (i.e. highest vertex degree) has priority;
5. the atom which has a first-neighbour atom with an atomic number higher than the others has priority;
6. the atom which has a first-neighbour atom with more free electrons than the others has priority;
7. the atom which has more bonds to first-neighbour atoms than the others has priority;
8. the atom with the highest bond order to the heavier first-neighbour atom has priority;
9. the atom which lies closer to an atom already numbered has priority;
10. the atom which has a higher bond order than that of an atom already numbered has priority;

finally, terminal atoms are then numbered according to rules 2 and 9.

• **Hierarchically Ordered extended Connectivities algorithms** (: *HOC algorithms*)
The proposed iterative procedure finds topological equivalence classes (graph orbits) and provides canonical numbering of vertices in molecular graphs. It is based on the extended connectivities such as Morgan's algorithm but also on the hierarchical ordering at each stage provided by the rank of the previous iteration [Balaban *et al.*, 1985a].

The whole procedure consists of algorithms which allow handling using graphs of different levels of complexity.

The main algorithm for ordering the vertices in graph orbits is called HOC-1 and the steps are:

Step 1. Vertices of the → *H-depleted molecular graph* are partitioned into equivalence classes according to their vertex degree δ_i, i.e. a first rank ${}^{\mathrm{I}}K_i = \delta_i$ is assigned to each vertex.

Step 2. For each vertex, the first ranks of its adjacent vertices are listed in increasing order as

$${}^{\mathrm{I}}K_i^1, {}^{\mathrm{I}}K_i^2, \ldots, {}^{\mathrm{I}}K_i^{\delta_i}$$

Step 3. An additional discrimination within each class is performed by means of the extended connectivities EC which are the sums of the vertex degrees (ranks) of the nearest neighbours, as

$${}^{\mathrm{I}}\mathrm{EC}_i = \sum_{r=1}^{\delta_i} {}^{\mathrm{I}}K_i^r$$

where the sum runs over the neighbour ranks of the ith vertex considered in increasing order. A second rank ${}^{\mathrm{II}}K_i$ is assigned to each vertex according to the increasing order of the extended connectivities. When two or more vertices are of the same rank, i.e. ${}^{\mathrm{II}}K_i = {}^{\mathrm{II}}K_j$, but the individual values of the terms contributing to the extended connectivity are different, i.e. ${}^{\mathrm{I}}K_i^r \neq {}^{\mathrm{I}}K_j^r$, then the ordering is made according to the rank of the first different addendum.

Step 4. Steps 2 and 3 are iteratively repeated, replacing first ranks by second ranks until all the ${}^{k+1}K_i$ ranks become equal to ${}^{k}K_i$ ranks of the preceding stage for all vertices.

The HOC-3 algorithm is used for the canonical numbering of graph vertices and is equal to the HOC-1 algorithm from step 1 to step 4 with an additional step based on an artificial discrimination into the largest orbits including two or more vertices. In order to make such a discrimination, one of the vertices inside the largest orbit, with the highest cardinality ${}^{k}K_i$ is arbitrarily assigned a higher rank ${}^{k}K_i + 1$, and steps 2–4 are iteratively repeated.

The final vertex numbering is the inverse of the final HOC ranks so as to assign the lowest numbers to the most central vertices.

The HOC-2 and HOC-2A algorithms were proposed for the vertex ordering of special molecular graphs with peri-condensed rings.

Based on HOC ranks, a *Unique Topological Representation* (UTR) was proposed in which topological equivalent vertices of the same rank in the graph are placed at the same level.

Moreover, two **HOC rank descriptors** based on the extended connectivity of graph vertices and the vertex rank obtained by HOC algorithms were proposed [Mekenyan *et al.*, 1984b]:

$$\mathcal{M} = \sum_{i=1}^{A} M_i \qquad \text{and} \qquad \mathcal{N} = \sum_{i=1}^{A} M_i^2$$

where the sums run over all atoms and the term M_i is a local invariant defined as

$$M_i = \sum_{j=1}^{i} S_j \quad \text{and} \quad S_j = \sum\nolimits_m K_m$$

where S_j is the sum of the ranks K of the adjacent vertices to the jth vertex restricted to those adjacent vertices of rank greater than j. The S_j values calculated for each vertex are ordered according to the increasing j index and are then summed up to i index to give the M_i local invariant. By this procedure, M_i gives a nondecreasing sequence.

Both descriptors increase with the size and cyclicity of the graph.

📖 [Mekenyan *et al.*, 1985b; Balaban *et al.*, 1985b; Bonchev *et al.*, 1985; Mekenyan *et al.*, 1984a; Mekenyan *et al.*, 1985a; Ralev *et al.*, 1985]

- **matrix method for canonical ordering**

Graph vertices are partitioned and ordered into topological equivalence classes, i.e. orbits, according to some special matrices developed for each atom [Bersohn, 1987]. These matrices give a representation of the whole molecule as seen from the considered atom.

The procedure consists first in assigning to each atom an atomic property P_i defined as:

$$P_i = 1024 \cdot \mathrm{Z}_i + 64 \cdot \mathrm{N}_i^{\mathrm{uns}} + 16 \cdot (4 - h_i)$$

where Z_i, $\mathrm{N}_i^{\mathrm{uns}}$ and h_i are the atomic number, the number of unsaturation and the number of attached hydrogen atoms of the ith atom, respectively. The unsaturation number is 8 for an atom involved in a triple bond, 4 for either of the atoms in a double bond involving a heteroatom, 6 for an allene or ketene central atom with two double bonds, 2 for vinyl carbon atoms, 1 for aromatic atoms in a six-membered ring and finally 0 for saturated atoms. The coefficients 1024, 64 and 16 have been chosen so that two atoms with the same atomic number cannot have the same property values.

The matrix representing the environment of the ith atom contains in the mth row the property values P_{mj} of the atoms located at a distance equal to m from the ith atom. The first row collects the property values of the first neighbours of the considered atom. The P_{mj} values are listed in descending order in the first entries of the matrix; the other entries are set to zero but are of no significance. The matrix dimension can be chosen for convenience.

The first partition of the vertices depends on their property values P_i; two atoms X and Y are considered topologically equivalent if: (i) they have identical matrices; (ii) if X has a neighbour I belonging to a different equivalence class; then there must exist a neighboor J of the atom Y in the same equivalence class as the atom I.

Finally, the canonical ordering is performed on the basis of P_i values and in the case of equivalence, according to the values of the matrix. Atoms with the greatest values are assigned the smallest numerical labels in the canonical ordering, such atoms being closest to the graph centre.

Property values can also be modified to take into account geometric isomerism and chirality.

- **self-returning walk ordering**

Graph vertices are ordered on the basis of their → *self-returning walk counts*, i.e. the number of walks of a given length starting and ending at the same vertex. In particular, local invariants representing the relative occurrence of the self-returning walks of the considered atom to all self-returning walks (*SRWs*) in the molecule, i.e. → *topological atomic charge* TAC_i, provide a canonical numbering of graph vertices [Bonchev *et al.*, 1993a]. The most important factors influencing the ordering, i.e. the number of *SRWs*, are vertex branching, centrality and cyclicity.

🕮 [Randic, 1977] [Hall and Kier, 1977a] [Randic, 1978a] [Randic, 1979] [Randic *et al.*, 1979] [Wilkins and Randic, 1980] [Randic, 1980b] [Randic *et al.*, 1981] [Bonchev and Balaban, 1981] [Bonchev *et al.*, 1986] [Polanski and Bonchev, 1987] [Klopman and Raychaudhury, 1988] [Herndon, 1988] [Diudea *et al.*, 1992a] [Balaban *et al.*, 1992] [Randic, 1995b] [Balasubramanian, 1995a] [Babic *et al.*, 1995] [Laidboeur *et al.*, 1997] [Agarwal, 1998] [Lukovits, 1999]

capacity factor → **chromatographic descriptors**

Carbó similarity index → **similarity/diversity**

cardinality layer matrix → **layer matrices**

cardinality of a set → **algebraic operators**

Carter resonance energy → **resonance indices**

Cartesian coordinates → **molecular geometry**

CASE approach → **lipophilicity descriptors** (⊙ Klopman hydrophobic models)

cavity term → **linear solvation energy relationships**

central edges → **graph**

centralization → **distance matrix**

central vertices → **graph**

centre distance-based criteria → **centre of a graph**

centre of a graph *(: graph centre)*

This is the set of central vertices and edges, where the definitions depend on the approach used to determine them [Bonchev, 1989]. Therefore, the graph centre can be a single vertex, a single edge, or a single group of equivalent vertices. Several graph centre definitions are derived from → *canonical numbering* approaches. Other ways to identify central vertices are the → *pruning of the graph* $\mathcal{G}$ and the application of → *centric operator* and → *centrocomplexity operator* to → *layer matrices* of the graph [Diudea *et al.*, 1992a].

According to the most popular definition, the central vertices in a graph are those vertices having the smallest → *atom eccentricity*. In acyclic graphs the centre coincides with a single vertex (i.e. a central vertex) or two adjacent vertices (i.e. a single central edge), while in cyclic graphs it usually coincides with a group of vertices.

A **generalized graph centre** concept is obtained by a hierarchy of criteria applied recursively so as to reduce the number of vertices qualifying as central vertices [Bonchev *et al.*, 1980a; Bonchev *et al.*, 1981a].

The graph **centre distance-based criteria** 1D – 4D for a vertex v_i to belong to the graph centre are:

Criterion 1D: minimum atom eccentricity η

$$\min_i(\eta_i)$$

Criterion 2D: for the vertices satisfying the first criterion, minimum → *vertex distance degree* σ

$$\min_i(\sigma_i)$$

Criterion 3D: for the vertices satisfying the previous criteria, minimum number of occurrences of the largest distance in the → *vertex distance code*

$$\min_i({}^{\eta_i}f_i)$$

where ${}^{\eta_i}f_i$ is the frequency of the maximum distance η_i from the vertex v_i to any other vertex, i.e. atom eccentricity. If the largest distance occurs the same number of times for two or more vertices, the frequency of the next largest distance $\eta_i - 1$ is considered and so on.

The graph vertices qualified as central according to the first three criteria constitute a smaller graph called **pseudocentre** (or **graph kernel**).

Criterion 4D: iterative process of the first three criteria 1D – 3D applied to the pseudocentre instead of the whole graph.

If more than one central vertex results from distance-based criteria, the graph centre is called the **polycentre**.

In order to discriminate even further among the vertices of the polycentre, the graph **centre path-based criteria** 1P – 4P can be applied to the polycentre:

Criterion 1P: minimum → *vertex path eccentricity* ${}^{\Delta}\eta$:

$$\min_i({}^{\Delta}\eta_i)$$

Criterion 2P: for vertices satisfying the first criterion, minimum → *vertex path sum* π (i.e. the sum of the lengths m of all paths starting from the considered vertex v_i)

$$\min_i(\pi_i)$$

Criterion 3P: for the vertices satisfying the previous criteria, minimum number of occurrences of the largest order path in the → *vertex path code*

$$\min_i\left({}^{\Delta\eta_i}P_i\right)$$

where ${}^{\Delta\eta_i}P_i$ is the count of the maximum order path starting from vertex v_i to any other vertex. If the longest path occurs the same number of times for two or more vertices, the path count of the next largest order is considered and so on.

Criterion 4P: iterative process of the first three criteria 1P – 3P applied to the small set of vertices selected according to the above described procedure.

The central vertices resulting from the criteria 1P – 4P are called the **oligocentre**. To further discriminate among the vertices of the oligocentre, analogous graph **centre self-returning walk-based criteria** 1W – 4W can be applied.

All the criteria defined above can also be applied to search for central edges in the graphs using the information provided from the → *edge distance matrix*. For example, the centre distance-based criteria 1D – 4D are defined as:

Criterion 1D: minimum → *bond eccentricity* ${}^{b}\eta$

$$\min_i({}^{b}\eta_i)$$

Criterion 2D: for the edges satisfying the first criterion, minimum → *edge distance degree* ${}^{E}\sigma$

$$\min_i({}^{E}\sigma_i)$$

Criterion 3D: for the edges satisfying the previous criteria, minimum number of occurrences of the longest distance in the → *edge distance code*

$$\min_i \left({}^{\mathrm{b}\eta_i} f_i \right)$$

where ${}^{\mathrm{b}\eta_i} f_i$ is the frequency of the maximum distance ${}^{\mathrm{b}}\eta_i$ from the edge e_i to any other edge, i.e. the bond eccentricity. If the longest distance occurs the same number of times for two or more edges, the frequency of the next longest distance ${}^{\mathrm{b}}\eta_i - 1$ is considered and so on.

Criterion 4D: iterative process of the first three criteria 1D – 3D applied to the pseudocentre instead of the whole graph.

Moreover, the application of the centre distance-based criteria on simultaneously both the vertex distance matrix **D** and the edge distance matrix $^{\mathbf{E}}\mathbf{D}$ resulted in a new algorithm, called the **Iterative Vertex and Edge Centricity algorithm** (IVEC), for graph centre definition and vertex → *canonical numbering* [Bonchev *et al.*, 1989]. The graph centre is selected through the sequential centric ordering of the graph vertices and edges on the basis of their metric properties and incidence.

In the initial step, the distance-based criteria 1D – 4D are applied to graph vertices to order them into equivalence classes identified by ranks 1,2,3... . Rank 1 is assigned to the polycentre and the maximum rank to the most external vertices. Then the same procedure is applied to the edges on the basis of the edge distance matrix.

Additional discrimination within the vertex equivalence classes is obtained by summing the ranks of the edges incident to each vertex of the considered class. New ranks are assigned to the vertices on this basis: lower ranks are assigned to vertices with a smaller sum. If the same rank sum is obtained for two or more vertices, the vertex where the addendum in the sum is smaller is assigned the lower rank. The same operation is performed for the graph edges by summing the new ranks of their incident vertices.

The algorithm continues iteratively until the same vertex and edge equivalence classes are obtained in two consecutive iterations. The centre of the graph then includes the vertices and edges of lowest rank.

A modified IVEC algorithm was also proposed to search for the centre of graphs where multiple bonds are present [Balaban *et al.*, 1993a].

A general definition for central vertex recognition is given by applying the concept of leverage to 3D → *molecular geometry*; called **atom leverage centre**, it is defined as the set of atoms with the minimum value of the diagonal elements h_{ii} of the → *molecular influence matrix* H ($A \times A$) derived from the atom centred spatial coordinates [Authors, This Book], i.e.

$$\{a_i : h_i = \min_i(h_i)\}$$

The molecular influence matrix is calculated as:

$$\mathrm{H} = \mathrm{M} \cdot \left(\mathrm{M}^{\mathrm{T}} \cdot \mathrm{M}\right)^{-1} \cdot \mathrm{M}^{\mathrm{T}}$$

where M is the rectangular matrix ($A \times 3$) of the atom spatial coordinates (x, y, z) of the molecule, i.e. the → *molecular matrix*. The diagonal values h_{ii} are always between 0 and 1.

📖 [Bonchev and Balaban, 1981] [Bonchev, 1983] [Barysz *et al.*, 1986a] [Bonchev and Balaban, 1993]

centre of a molecule *(: molecule centre)*

The centres of a molecule are reference points used to calculate distributional properties of the molecule and, mathematically speaking, are the first-order moments of the considered property. Arithmetic means and weighted arithmetic means are the usual way to calculate centres.

For example, the **geometric centre** of a molecule is defined as the average value of atom coordinates calculated separately for each axis:

$$\bar{x} = \frac{1}{A} \cdot \sum_{i=1}^{A} x_i \qquad \bar{y} = \frac{1}{A} \cdot \sum_{i=1}^{A} y_i \qquad \bar{z} = \frac{1}{A} \cdot \sum_{i=1}^{A} z_i$$

where A is the → *atom number.*

The weighted centre is analogously defined, by multiplying each ith atom coordinate by a corresponding weight w_i:

$$\bar{x} = \frac{1}{W} \cdot \sum_{i=1}^{A} w_i \cdot x_i \qquad \bar{y} = \frac{1}{W} \cdot \sum_{i=1}^{A} w_i \cdot y_i \qquad \bar{z} = \frac{1}{W} \cdot \sum_{i=1}^{A} w_i \cdot z_i$$

where W is the sum of the weights. For example, if the weights are the atomic masses m, the **centre of mass** (or **barycentre**) of the molecule is obtained.

Geometric and mass centres of a molecule are not molecular descriptors, but they are used to translate the molecule at the centre, thus providing a unique reference origin which allows invariance to translation and separation between translational and rotational motions.

By weighting atoms by charges, for neutral molecules, the first order moment of the charges is the → *dipole moment.* Moreover, for molecules with zero net charge and nonvanishing dipole moment, the **centre-of-dipole** was recently defined as the appropriate centre for multipolar expansions to obtain rotational invariance [Silverman and Platt, 1996].

centre-of-dipole → **centre of a molecule**

centre of mass → **centre of a molecule**

centre path-based criteria → **centre of a graph**

centre self-returning walk-based criteria → **centre of a graph**

centric indices

→ *Molecular descriptors* proposed to quantify the degree of compactness of molecules by distinguishing between molecular structures organized differently with respect to their centres. Based on the recognition of the → *graph centre* these indices are mainly defined by the information theory concepts applied to a partition of the graph vertices made according to their positions relative to the centre. Moreover, → *centric operator* and → *centrocomplexity operator* have been proposed to calculate from → *layer matrices* → *local vertex invariants* and corresponding molecular descriptors which account for molecular centricity.

The main centric indices are listed below; they have been divided into two main groups, one containing the indices proposed by Balaban and the other the indices proposed by Bonchev.

• **Balaban centric index** (B)

A topological index defined for acyclic graphs based on the **pruning of the graph**, a stepwise procedure for removing all the → *terminal vertices*, i.e. vertices with a → *vertex degree* of one ($\delta_i = 1$), and the corresponding incident edges from the → *H-depleted molecular graph* $\mathcal{G}$. The vertices and edges removed at the gth step are n_g and the total number of steps to remove all vertices is R [Balaban, 1979].

The **pruning partition** of the graph is the reversed sequence of n_g numbers provided by the pruning procedure:

$$\langle n_R, n_{R-1}, \ldots, n_1 \rangle$$

The pruning partition is related to → *molecular branching,* and the reversed order of n_g numbers is due to the fact that the number of branches cannot decrease when starting from the centre of the tree. Moreover, the first entry is always equal to one (centre) or two (bicentre) and the pruning partition is a partition of A, i.e. the number of molecule atoms.

The Balaban centric index is calculated from the pruning partition in analogy with the → *1*[st] *Zagreb index* M_1 as:

$$\mathrm{B} = \sum_{g=1}^{R} n_g^2$$

This index provides a measure of molecular branching: the higher the value of B, the more branched the tree. It is called centric index because it reflects the topology of the tree as viewed from the centre.

The **normalized centric index** C is derived by normalization, i.e. imposing the same lower bound, equal to zero for the least branched (linear) trees, on all the graphs. It is defined as:

$$C = \frac{\mathrm{B} - 2 \cdot A + U}{2}$$

where A is the number of graph vertices. The term U is defined as:

$$U = \frac{1 - (-1)^A}{2} = \begin{cases} 0 & \text{if } A \text{ is even} \\ 1 & \text{if } A \text{ is odd} \end{cases}$$

The **binormalized centric index** C' is derived by a binormalization, i.e. imposing on all the graphs the same lower bound and an upper bound equal to one for star graphs. In practice, it is obtained from the normalized centric index C divided by its corresponding value for the star graph:

$$C' = \frac{\mathrm{B} - 2 \cdot A + U}{(A-2)^2 - 2 + U}$$

The binormalized centric index provides information on the topological shape of trees in a similar way to the → *binormalized quadratic index* Q'.

• **lopping centric information index** ($\bar{I}_B$)

An index defined as the → *mean information content* derived from the pruning partition of a graph:

$$\bar{I}_B = -\sum_{g=1}^{R} \frac{n_g}{A} \cdot \log_2 \frac{n_g}{A}$$

where n_g is the number of terminal vertices removed at the gth step, A the number of graph vertices and R the number of steps to remove all graph vertices [Balaban, 1979].

Example : 2-methylpentane

Pruning partition: $\langle 1, 2, 3 \rangle$

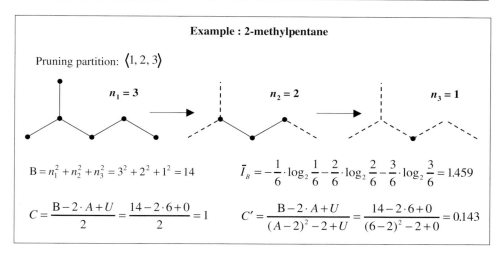

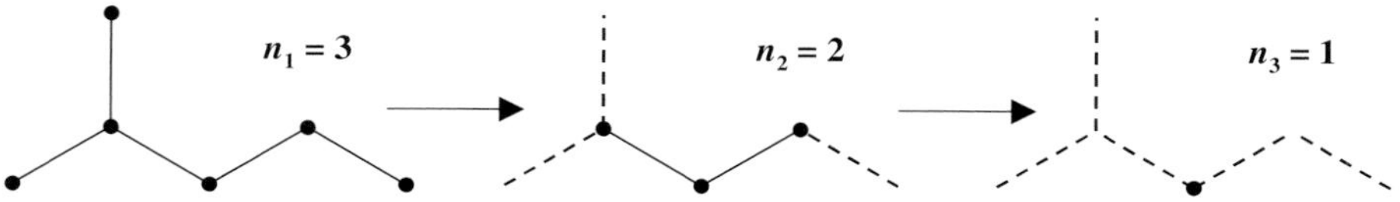

$$B = n_1^2 + n_2^2 + n_3^2 = 3^2 + 2^2 + 1^2 = 14 \qquad \bar{I}_B = -\frac{1}{6} \cdot \log_2 \frac{1}{6} - \frac{2}{6} \cdot \log_2 \frac{2}{6} - \frac{3}{6} \cdot \log_2 \frac{3}{6} = 1.459$$

$$C = \frac{B - 2 \cdot A + U}{2} = \frac{14 - 2 \cdot 6 + 0}{2} = 1 \qquad C' = \frac{B - 2 \cdot A + U}{(A-2)^2 - 2 + U} = \frac{14 - 2 \cdot 6 + 0}{(6-2)^2 - 2 + 0} = 0.143$$

Box C-1.

• **information content based on centre** (*IBC*)
Defined only for acyclic graphs and substituents, it is calculated as → *total information content* based on the shells around the centre of the graph:

$$IBC = 2W \cdot \log_2 2W - \sum\nolimits_k q_k \cdot \log_2 q_k$$

where W is the → *Wiener index* and q_k the sum of the → *vertex distance degree* of the vertices located at a → *topological distance* equal to k from the centre [Balaban *et al.*, 1994b].

• **average information content based on centre** (*AIBC*)
Defined only for acyclic graphs and substituents, it is the average of the information content based on centre *IBC*, i.e.

$$AIBC = \frac{IBC}{W}$$

IBC is divided by the Wiener index W rather than $2W$ in order to have *AIBC* values higher than one [Balaban *et al.*, 1994b].

Bonchev centric information indices are centric indices derived from the vertex → *distance matrix* **D** and the → *edge distance matrix* $^{\mathbf{E}}\mathbf{D}$, based on the concept of → *graph centre* and calculated as mean information content [Bonchev *et al.*, 1980a; Bonchev, 1983; Bonchev, 1989].

For **vertex centric indices**, the number of equivalent vertices of each equivalence class is calculated applying the → *centre distance-based criteria* 1D – 4D to the graph vertices, i.e. the subsequent application of these criteria increases the discrimination of graph vertices. Analogously, for **edge centric indices**, the number of equivalent edges of each equivalence class is calculated applying the centre distance-based criteria to the graph edges.

Once the → *polycentre* of the graph has been found, four other centric information indices, called **generalized centric information indices**, are calculated on the vertex (edge) partition from the average topological distance between each vertex (edge) and the atoms of the polycentre. An increasing discrimination of the graph vertices (edges) is obtained by subsequently applying the remaining criteria 2D – 4D.

Other centric information indices can be calculated by the same formulas on both vertex and edge graph partition based on graph → *centre path-based criteria* 1P – 4P and → *centre self-returning walk-based criteria* 1W – 4W. Moreover, **edge centric indices for multigraphs** have different values from those calculated on the parent graph, the edge distance matrix of the multigraph being different from the edge distance matrix of the parent graph.

The Bonchev centric information indices and the corresponding generalized centric information indices are listed below.

- **radial centric information index** $({}^{V}\bar{I}_{C,R})$

It is defined as:

$$ {}^{V}\bar{I}_{C,\,R} = -\sum_{g=1}^{G} \frac{n_g}{A} \log_2 \frac{n_g}{A} $$

where n_g is the number of graph vertices having the same → *atom eccentricity* η, G is the number of different vertex equivalence classes and A the number of graph vertices.

- **distance degree centric index** $({}^{V}\bar{I}_{C,\,\mathrm{deg}})$

It is defined as:

$$ {}^{V}\bar{I}_{C,\,\mathrm{deg}} = -\sum_{g=1}^{G} \frac{n_g}{A} \log_2 \frac{n_g}{A} $$

where n_g is the number of graph vertices having both the same atom eccentricity η and the same → *vertex distance degree* σ, G is the number of different vertex equivalence classes and A the atom count.

- **distance code centric index** $({}^{V}\bar{I}_{C,\,code})$

It is defined as:

$$ {}^{V}\bar{I}_{C,\,code} = -\sum_{g=1}^{G} \frac{n_g}{A} \log_2 \frac{n_g}{A} $$

where n_g is the number of graph vertices contemporarily having the same atom eccentricity η, the same vertex distance degree σ, and the same → *vertex distance code*; G is the number of different vertex equivalence classes and A the atom count.

- **complete centric index** $({}^{V}\bar{I}_{C,\,C})$

It is defined as:

$$ {}^{V}\bar{I}_{C,\,C} = -\sum_{g=1}^{G} \frac{n_g}{A} \log_2 \frac{n_g}{A} $$

where n_g is the number of graph vertices contemporarily having the same atom eccentricity η, the same vertex distance degree σ, and the same vertex distance code, but also distinguishing the atoms defining the → *pseudocentre*, i.e. removing existing degeneracy of pseudocentre atoms. G is the number of different vertex equivalence classes and A the atom count.

- **generalized radial centric information index** $({}^{V}\bar{I}^{G}_{C,\,R})$

It is defined as:

$$ {}^{V}\bar{I}^{G}_{C,\,R} = -\sum_{g=1}^{G} \frac{n_g}{A} \log_2 \frac{n_g}{A} $$

where n_g is the number of graph vertices having the same average topological distance to the polycentre, G is the number of different vertex equivalence classes and A the atom count.

• **generalized distance degree centric index** (${}^{V}\bar{I}^{G}_{C,\deg}$)
It is defined as:

$$ {}^{V}\bar{I}^{G}_{C,\deg} = -\sum_{g=1}^{G} \frac{n_g}{A} \log_2 \frac{n_g}{A} $$

where n_g is the number of graph vertices having both the same average topological distance to the polycentre and the same vertex distance degree σ, G is the number of different vertex equivalence classes and A the atom count.

• **generalized distance code centric index** (${}^{V}\bar{I}^{G}_{C,code}$)
It is defined as:

$$ {}^{V}\bar{I}^{G}_{C,code} = -\sum_{g=1}^{G} \frac{n_g}{A} \log_2 \frac{n_g}{A} $$

where n_g is the number of graph vertices having the same average topological distance to the polycentre, the same vertex distance degree σ, and the same vertex distance code; G is the number of different vertex equivalence classes and A the atom count.

• **generalized complete centric index** (${}^{V}\bar{I}^{G}_{C,C}$)
It is defined as:

$$ {}^{V}\bar{I}^{G}_{C,C} = -\sum_{g=1}^{G} \frac{n_g}{A} \log_2 \frac{n_g}{A} $$

where n_g is the number of graph vertices contemporarily having the same average topological distance to the polycentre, the same vertex distance degree σ, the same vertex distance code, but also distinguishing the atoms defining the pseudocentre, i.e. removing existing degeneracy of pseudocentre atoms. G is the number of different vertex equivalence classes and A the atom count.

• **edge radial centric information index** (${}^{E}\bar{I}_{C,R}$)
It is defined as:

$$ {}^{E}\bar{I}_{C,R} = -\sum_{g=1}^{G} \frac{n_g}{B} \log_2 \frac{n_g}{B} $$

where n_g is the number of graph edges having the same → *bond eccentricity* ${}^{b}\eta$, G is the number of different edge equivalence classes and B the number of bonds.

• **edge distance degree centric index** (${}^{E}\bar{I}_{C,\deg}$)
It is defined as:

$$ {}^{E}\bar{I}_{C,\deg} = -\sum_{g=1}^{G} \frac{n_g}{B} \log_2 \frac{n_g}{B} $$

where n_g is the number of graph edges having both the same bond eccentricity ${}^{b}\eta$ and the same → *edge distance degree* ${}^{E}\sigma$, G is the number of different edge equivalence classes and B the bond count.

• **edge distance code centric index** $({}^{E}\bar{I}_{C,code})$
It is defined as:

$${}^{E}\bar{I}_{C,\,code} = -\sum_{g=1}^{G} \frac{n_g}{B} \log_2 \frac{n_g}{B}$$

where n_g is the number of graph edges contemporarily having the same bond eccentricity ${}^{b}\eta$, the same edge distance degree ${}^{E}\sigma$, and the same → *edge distance code*; G is the number of different edge equivalence classes and B the bond count.

• **edge complete centric index** $({}^{E}\bar{I}_{C,C})$
It is defined as:

$${}^{E}\bar{I}_{C,\,C} = -\sum_{g=1}^{G} \frac{n_g}{B} \log_2 \frac{n_g}{B}$$

where n_g is the number of graph edges contemporarily having the same bond eccentricity ${}^{b}\eta$, the same edge distance degree ${}^{E}\sigma$, the same edge distance code, but also distinguishing the edges defining the pseudocentre, i.e. removing existing degeneracy of pseudocentre edges. G is the number of different edge classes and B the bond count.

• **generalized edge radial centric information index** $({}^{E}\bar{I}^{G}_{C,R})$
It is defined as:

$${}^{E}\bar{I}^{G}_{C,\,R} = -\sum_{g=1}^{G} \frac{n_g}{B} \log_2 \frac{n_g}{B}$$

where n_g is the number of graph edges having the same average topological distance to the polycentre, G is the number of different edge equivalence classes and B the bond count.

• **generalized edge distance degree centric index** $({}^{E}\bar{I}^{G}_{C,\deg})$
It is defined as:

$${}^{E}\bar{I}^{G}_{C,\,\deg} = -\sum_{g=1}^{G} \frac{n_g}{B} \log_2 \frac{n_g}{B}$$

where n_g is the number of graph edges having both the same average topological distance to the polycentre and the same edge distance degree ${}^{E}\sigma$, G is the number of different edge equivalence classes and B the bond count.

• **generalized edge distance code centric index** $({}^{E}\bar{I}^{G}_{C,code})$
It is defined as:

$${}^{E}\bar{I}^{G}_{C,\,code} = -\sum_{g=1}^{G} \frac{n_g}{B} \log_2 \frac{n_g}{B}$$

where n_g is the number of graph edges having the same average topological distance to the polycentre, the same edge distance degree ${}^{E}\sigma$, and the same edge distance code; G is the number of different edge equivalence classes and B the bond count.

• **generalized edge complete centric index** $({}^{E}\bar{I}^{G}_{C,C})$
It is defined as:

$${}^{E}\bar{I}^{G}_{C,\,C} = -\sum_{g=1}^{G} \frac{n_g}{B} \log_2 \frac{n_g}{B}$$

where n_g is the number of graph edges contemporarily having the same average topological distance to the polycentre, the same edge distance degree ${}^{E}\sigma$, the same edge distance code, but also distinguishing the edges defining the pseudocentre, i.e. removing existing degeneracy of pseudocentre edges. G is the number of different equivalence edge classes and B the bond count.

centric operator → **layer matrices**

centric topological index → **layer matrices**

centrocomplexity operator → **layer matrices**

centrocomplexity topological index → **layer matrices**

***$CHAA_1$* index** → **charged partial surface area descriptors**

***$CHAA_2$* index** → **charged partial surface area descriptors** (⊙ *$CHAA_1$* index)

chain subgraph → **molecular graph**

chance correlation → **validation techniques**

characteristic polynomial → **algebraic operators**

characteristic polynomial of the graph

Given a graph $\mathcal{G}$, this is the → *characteristic polynomial* $\mathcal{P}(\mathcal{G}; x)$ of its → *adjacency matrix* **A**, defined as:

$$\mathcal{P}(\mathcal{G}; x) = \det(x\mathbf{I} - \mathbf{A}) = \sum_{k=0}^{A} a_k x^{A-k}$$

where **I** is the identity matrix of dimension $A \times A$, x is a scalar variable, a_k are the polynomial coefficients and A is the number of graph vertices [Bonchev and Rouvray, 1991; Trinajstic, 1992]. The **graph eigenvalues** λ are the n roots of its characteristic polynomial and the set of the eigenvalues is called **spectrum of a graph** $\Lambda(\mathbf{A})$.

For any acyclic graphs two general rules are observed: i) the power of x decreases by two, and ii) the absolute values of $\mathcal{P}(\mathcal{G}; x)$ coefficients are equal to the coefficients of the → *Z-counting polynomial* $\mathcal{Q}(\mathcal{G}; x)$. Therefore, the characteristic polynomial for an acyclic graph is given by:

$$\mathcal{P}(\mathcal{G}; x) = \sum_{k=1}^{[A/2]} (-1)^k a(\mathcal{G}, k) x^{A-2k}$$

where $a(\mathcal{G}, k)$ is the → *non-adjacent number* of the → *Hosoya Z index*, i.e the coefficients of the Z-counting polynomial, and the square brackets refer to the largest integer of $A/2$.

If two graphs are → *isomorphic graphs*, their characteristic polynomials coincide; the converse is not true.

The characteristic polynomial is a molecular descriptor able to discriminate well among several graphs; however, some nonisomorphic graphs have the same characteristic polynomial (same eigenvalues), and for this reason they are called isospectral graphs [Herndon, 1974a]. Moreover, the sum of the coefficients of the characteristic polynomial can be used as a scalar molecular descriptor.

📖 [Balaban and Harary, 1971] [Zivkovic *et al.*, 1975] [Randic *et al.*, 1976] [Balasubramanian, 1982] [Balasubramanian and Randic, 1982] [Randic, 1983] [Balasubramanian, 1984b] [Balasubramanian, 1984a] [Krivka *et al.*, 1985] [Dias, 1987a] [Dias, 1987b] [Trinajstic, 1988] [Balasubramanian, 1990] [Zivkovic, 1990] [Shalabi, 1991] [Randic *et al.*, 1997a]

characteristic ratio → **shape descriptors**

characteristic root index → **eigenvalue-based descriptors**

charge density matrix → **quantum-chemical descriptors** (⊙ electron density)

charge descriptors

These are → *electronic descriptors* defined in terms of atomic charges and used to describe electronic aspects both of the whole molecule and of particular regions, such as atoms, bonds, and molecular fragments. Charge descriptors are calculated by → *computational chemistry* and therefore can be considered among → *quantum-chemical descriptors* [Lowe, 1978; Streitweiser, 1961].

Electrical charges in the molecule are the driving force of electrostatic interactions, and it is well known that local electron densities or charges play a fundamental role in many chemical reactions, physico-chemical properties and receptor-ligand → *binding affinity*.

A list of the most well-known charge descriptors is presented below.

- **atomic charge** (*: net atomic charge, q*)

The basic approach to calculating atomic charge is called **Mulliken population analysis**; this is a method for allocating electrons to atoms in order to generate atomic charges as local descriptors [Mulliken, 1955b; Mulliken, 1955a]. The partitioning scheme is based on the use of the → *charge density matrix* $\boldsymbol{P}$ and the overlap matrix $\boldsymbol{S}$ to allocate, in some fractional manner, the electrons of a molecule among its various parts (atoms, bonds, orbitals).

The net atomic charge q_a of the ath atom is defined as:

$$q_a = Z_a - \sum_{\mu=1}^{N_{AO}} \sum_{\nu=1}^{N_{AO}} P_{\mu\nu} \cdot S_{\mu\nu} \qquad \mu \in a$$

where N_{AO} is the number of atomic orbitals (the basis functions), Z_a is the effective nuclear charge of the ath atom, $P_{\mu\nu}$ the element of the charge density matrix corresponding to the atomic orbital μ centred on the ath atom and $S_{\mu\nu}$ the element corresponding to the orbitals μ and ν of the overlap matrix. The sum runs over all the atomic orbitals of the ath atom.

The charges measure the extent of electronic density localization in a molecule. Negative q_a values mean that excess of electronic charge is at centre a while positive values mean that centre a is electron-deficient.

As with other schemes of partitioning the electron density in molecules, Mulliken population analysis is arbitrary and is strongly dependent on the particular basis set employed. However, the comparison of population analyses for a series of molecules is useful for a quantitative description of intramolecular interactions, chemical reactivity and structural information. In another approach, the **Löwdin population analysis**, the atomic orbitals are first transformed to an orthogonal set, as are the molecular orbital coefficients [Löwdin, 1970].

Moreover, net atomic charges can also be calculated distinguishing σ-electron density $q_{a,\sigma}$ from π-electron density $q_{a,\pi}$.

- **net orbital charge**

It is the charge of an orbital. **Electrophilic charge** q^E and **nucleophilic charge** q^N are molecular descriptors defined as the net atomic charges calculated for the → *highest occupied molecular orbital* (HOMO) and the → *lowest unoccupied molecular orbital* (LUMO), respectively. Moreover, also of particular interest are the corresponding local invariants **HOMO electron density on the *a*th atom** q_a^{E} and **LUMO electron density on the *a*th atom** q_a^{N}.

• **maximum positive charge** (Q^+_{max})
The maximum positive charge of the atoms in a molecule:

$$Q^+_{\mathrm{max}} = \max_a(q_a^+)$$

where q^+ are the net atomic positive charges.

• **maximum negative charge** (Q^-_{max})
The maximum negative charge of the atoms in a molecule:

$$Q^-_{\mathrm{max}} = \max_a(q_a^-)$$

where q^- are the net atomic negative charges.

• **total positive charge** (Q^+)
The sum of all of the positive charges of the atoms in a molecule:

$$Q^+ = \sum\nolimits_a q_a^+$$

where q^+ are the net atomic positive charges.

• **total negative charge** (Q^-)
The sum of all of the negative charges of the atoms in a molecule:

$$Q^- = \sum\nolimits_a q_a^-$$

where q^- are the net atomic negative charges.

• **total absolute atomic charge** (Q)
The sum over all atoms in a molecule of the absolute values of the atomic charges q_a:

$$Q = \sum\nolimits_a |q_a|$$

It is a measure of molecule polarity, also called **Electronic Charge Index** (*ECI*); for example, it has been used to study amino acid side chains [Collantes and Dunn III, 1995].

The **charge polarization** is the mean absolute atomic charge in a molecule, defined as:

$$P = \frac{\sum_a |q_a|}{A} = \frac{Q}{A}$$

where A is the number of atoms.

A similar measure of molecular polarity is obtained from the **total squared atomic charge**, defined as:

$$Q^2 = \sum\nolimits_a q_a^2$$

These total charge descriptors can also be calculated restricted to a molecular fragment as well as to a functional group.

• **potential of a charge distribution** (ϕ)
The theoretical potential function of a discrete charge distribution, i.e. of atomic point charges q_i, is given at point $\mathbf{r}$ as the following:

$$\phi(\mathbf{r}) = \sum_{i=1}^{A} \frac{q_i}{r_i}$$

where r_i is the geometric distance from each atom to the point $\mathbf{r}$.

- **Submolecular Polarity Parameter** (SPP; $^1\Delta$)

An electronic descriptor defined as the maximum excess charge difference for a pair of atoms in the molecule [Kaliszan *et al.*, 1985; Osmialowski *et al.*, 1985], i.e. calculated from the difference between the atomic maximum positive charge Q^+_{max} and the atomic maximum negative charge Q^-_{max} in a molecule:

$$^1\Delta = \left| Q^+_{max} - Q^-_{max} \right|$$

The **second-order submolecular polarity parameter** $^2\Delta$ is determined analogously, and is the second largest difference of excess charges [Luco *et al.*, 1995].

The interatomic distance r between the two atoms bearing the maximum positive and negative charges is used to derive the **DP descriptor** as follows:

$$\mathrm{DP} = \frac{\left| Q^+_{max} - Q^-_{max} \right|}{r^2_\pm} = \frac{^1\Delta}{r^2_\pm}$$

where the denominator accounts for the decreasing atom interaction with increasing distance.

- **topological electronic descriptors** (T^E)

Topographic descriptors [Osmialowski *et al.*, 1986; Katritzky and Gordeeva, 1993] calculated from partial atomic charges q as the following:

$$T^E = \sum_{i=1}^{A-1} \sum_{j=i+1}^{A} \frac{\left| q_i - q_j \right|}{r^2_{ij}} \qquad {}^cT^E = \sum_{b=1}^{B} \left(\frac{\left| q_i - q_j \right|}{r^2_{ij}} \right)_b$$

where the first index considers all pairs of atoms (both connected and disconnected) and the second is restricted to all pairs *i-j* of bonded atoms; r_{ij} are → *interatomic distances*, A and B are the number of atoms and bonds, respectively. These descriptors are calculated in such a way that they reflect, to some extent, differences in size, shape and constitution, these quantities affecting the electronic charge distribution and interatomic distances of the molecules.

- **partial charge weighted topological electronic index** ($\mathrm{PCWT^E}$)

A molecular electronic descriptor defined as [Katritzky *et al.*, 1996b]:

$$\mathrm{PCWT^E} = \frac{1}{Q^-_{max}} \cdot \sum_{b=1}^{B} \left(\frac{\left| q_i - q_j \right|}{r^2_{ij}} \right)_b$$

where q are the Zefirov partial atomic charges [Zefirov *et al.*, 1987] of the bonded atoms i and j, r_{ij} is the corresponding bond distance, and Q^-_{max} is the maximum negative charge. The sum runs over all of the bonded atom pairs.

- **local dipole index** (D)

A molecular descriptor calculated as the average of the charge differences over all *i-j* bonded atom pairs:

$$D = \frac{\sum_b \left| q_i - q_j \right|_b}{B}$$

where B is the number of bonds [Clare and Supuran, 1994; Karelson *et al.*, 1996].

- **electronic-topological descriptors** (E^T)

Proposed in analogy with the → *connectivity indices*, they are calculated for a hydrogen-included molecular graph using absolute values of partial charges q_i as vertex weights instead of vertex degrees δ as [Katritzky and Gordeeva, 1993]:

$$ {}^{0}E^{\mathrm{T}} = \sum_{i=1}^{A} \left(|q_i|\right)^{-1/2} \qquad {}^{1}E^{\mathrm{T}} = \sum_{b=1}^{B} \left(|q_i \cdot q_j|\right)_b^{-1/2} \qquad {}^{2}E^{\mathrm{T}} = \sum_{k=1}^{N_2} \left(|q_i \cdot q_l \cdot q_j|\right)_k^{-1/2} $$

$$ {}^{3}E^{\mathrm{T}} = \sum_{k=1}^{{}^{3}P} \left(|q_i \cdot q_l \cdot q_h \cdot q_j|\right)_k^{-1/2} $$

where A is the number of atoms, B is the number of bonds, N_2 is the → *connection number*, and ${}^{3}P$ is the number of paths of length three. Each term in the sums is the inverse square root of the product of the absolute partial charges of the vertices involved in the considered path.

• **electron charge density connectivity index** (Ω)
Very similar to the electronic-topological descriptors, electron charge density connectivity index is defined for an → *H-depleted molecular graph* in which net atomic charges, calculated by computational chemistry, are used as weights for vertices [Estrada and Montero, 1993; Estrada, 1995d]:

$$ \Omega = \sum_{b=1}^{B} \left(\delta_i^q \cdot \delta_j^q\right)_b^{-1/2} $$

where the sum runs over all the B bonds and δ_i^q and δ_j^q are the **electron charge density weight** of the two vertices v_i and v_j incident to the bth bond defined as:

$$ \delta_i^q = q_i - h_i $$

where q_i is the net atomic charge on the ith atom and h_i is the number of hydrogen atoms bonded to it.

In order to consider the influence of the different hydrogen charges, a **corrected electron charge density weight** is calculated as:

$$ \tilde{\delta}_i^q = q_i - \sum_{j=1}^{h_i} q_j $$

where the sum is on the total charges of the h_i hydrogen atoms bonded to the ith atom.

Therefore a **corrected electron charge density connectivity index** is defined as:

$$ \tilde{\Omega} = \sum_{b=1}^{B} \left(\tilde{\delta}_i^q \cdot \tilde{\delta}_j^q\right)_b^{-1/2} $$

Both molecular descriptors are able to discriminate among groups having the same number of valence electrons but with different hybridization and different chemical environments. Moreover, they show high sensitivity to geometric isomers, conformations, and heteroatoms.

• **charge-related indices**
These are global molecular descriptors derived from an H-depleted molecular graph where each vertex is weighted by a local vertex invariant called **Atom-in-Structure Invariant Index** (*ASII*) defined as [Bangov, 1988]:

$$ ASII_i = ASII_i^0 - h_i + q_i $$

where $ASII^{0}$ is a standard value for the atom type and hybridization state of a given atom (Table C-1), h is the number of attached hydrogen atoms, and q the net atomic charge of the considered atom.

Table C-1. Standard values of the atom-in-structure invariant index for different atom types.

Atom	$ASII^0$	*Atom*	$ASII^0$
C sp_3	4	O sp_3	23
C sp_2	11	O sp_2	25
C sp_2 *ar*	13	S	28
C *sp*	7	F	32
N sp_3	15	Cl	33
N sp_2	18	Br	34
N *sp*	20	I	35

Two specific charge-related indices called ***ASIIg* index** and **Charge Topological Index** (*CTI*), respectively, were defined as the following:

$$ASIIg = \frac{10}{\sqrt{\sum_{i=1}^{A} ASII_i}} \cdot \left[\sum_{b=1}^{B} \left(ASII_i \cdot ASII_j \right)_b \right]^{1/2}$$

$$CTI = \sum_{i=1}^{A-1} \sum_{j=i+1}^{A} \frac{ASII_i \cdot ASII_j}{d_{ij}}$$

where A and B are the numbers of atoms and bonds, respectively, d_{ij} is the → *topological distance* between vertices v_i and v_j, and the sum in the *CTI* index runs over all the pairs of vertices. The *ASIIg* index was particularly useful in dealing with isomers [Bangov, 1990]; the *CTI* index was proposed as a highly discriminant index with a low degree of degeneracy [Demirev *et al.*, 1991].

📖 [Del Re, 1958] [Buydens *et al.*, 1983b] [Gasteiger *et al.*, 1988] [Abraham and Smith, 1988] [Baumer *et al.*, 1989] [Gombar and Enslein, 1990] [Dixon and Jurs, 1992] [Reynolds *et al.*, 1992b] [Palyulin *et al.*, 1995] [Hannongbua *et al.*, 1996a] [Payares *et al.*, 1997]

Charged Partial Surface Area descriptors *(: CPSA descriptors)*

A set of thirty different descriptors [Stanton and Jurs, 1990] which combine shape and electronic information to characterize molecules and therefore encode features responsible for polar interactions between molecules. The molecule representation used for deriving *CPSA* descriptors views molecule atoms as hard spheres defined by the → *van der Waals radius*. The → *solvent-accessible surface area SASA* is used as the molecular surface area; it is calculated using a sphere with a radius of 1.5 Å to approximate the contact surface formed when a water molecule interacts with the considered molecule. Moreover, the contact surface where polar interactions can take place is characterized by a specific electronic distribution obtained by mapping atomic partial charges on the solvent-accessible surface.

Let SA_a^+ and SA_a^- be the surface area contributions of the ath positive and negative atoms, respectively; q_a^+ and q_a^- the partial atomic charges for the ath positive and negative atoms; Q^+ and Q^- the total sum of partial positive and negative charges in the molecule, respectively.

The *CPSA* descriptors are defined as the following:

- **partial negative surface area** ($PNSA_1$)

It is the sum of the solvent-accessible surface areas of all negatively charged atoms, i.e.

$$PNSA_1 = \sum_{a-} SA_a^-$$

where the sum is restricted to negatively charged atoms a^-.

- **partial positive surface area** ($PPSA_1$)

It is the sum of the solvent-accessible surface areas of all positively charged atoms, i.e.

$$PPSA_1 = \sum_{a+} SA_a^+$$

where the sum is restricted to positively charged atoms a^+.

- **total charge weighted negative surface area** ($PNSA_2$)

It is the partial negative solvent-accessible surface area multiplied by the → *total negative charge* Q^-, i.e.

$$PNSA_2 = Q^- \cdot \sum_{a-} SA_a^-$$

- **total charge weighted positive surface area** ($PPSA_2$)

It is the partial positive solvent-accessible surface area multiplied by the → *total positive charge* Q^+, i.e.

$$PPSA_2 = Q^+ \cdot \sum_{a+} SA_a^+$$

- **atomic charge weighted negative surface area** ($PNSA_3$)

It is the sum of the products of atomic solvent-accessible surface areas and partial charges q_a^- over all negatively charged atoms, i.e.

$$PNSA_3 = \sum_{a-} q_a^- \cdot SA_a^-$$

- **atomic charge weighted positive surface area** ($PPSA_3$)

It is the sum of the products of atomic solvent-accessible surface areas and partial charges q_a^+ over all positively charged atoms, i.e.

$$PPSA_3 = \sum_{a+} q_a^+ \cdot SA_a^+$$

- **difference in charged partial surface area** ($DPSA_1$)

It is the partial positive solvent-accessible surface area minus the partial negative solvent-accessible surface area, i.e.

$$DPSA_1 = PPSA_1 - PNSA_1$$

- **difference in total charge weighted surface area** ($DPSA_2$)

It is the total charge weighted positive solvent-accessible surface area minus the total charge weighted negative solvent-accessible surface area, i.e.

$$DPSA_2 = PPSA_2 - PNSA_2$$

- **difference in atomic charge weighted surface area** ($DPSA_3$)

It is the atomic charge weighted positive solvent-accessible surface area minus the atomic charge weighted negative solvent-accessible surface area, i.e.

$$DPSA_3 = PPSA_3 - PNSA_3$$

- **fractional charged partial negative surface areas** ($FNSA_1$, $FNSA_2$, $FNSA_3$)

They are the partial negative surface area ($PNSA_1$), the total charge weighted negative surface area ($PNSA_2$) and the atomic charge weighted negative surface area ($PNSA_3$), divided by the total molecular solvent-accessible surface area ($SASA$), i.e.

$$FNSA_1 = \frac{PNSA_1}{SASA} \qquad FNSA_2 = \frac{PNSA_2}{SASA} \qquad FNSA_3 = \frac{PNSA_3}{SASA}$$

- **fractional charged partial positive surface areas** ($FPSA_1$, $FPSA_2$, $FPSA_3$)

They are the partial positive surface area ($PPSA_1$), the total charge weighted positive surface area ($PPSA_2$) and the atomic charge weighted positive surface area ($PPSA_3$), divided by the total molecular solvent-accessible surface area ($SASA$), i.e.

$$FPSA_1 = \frac{PPSA_1}{SASA} \qquad FPSA_2 = \frac{PPSA_2}{SASA} \qquad FPSA_3 = \frac{PPSA_3}{SASA}$$

- **surface weighted charged partial negative surface areas** ($WNSA_1$, $WNSA_2$, $WNSA_3$)

They are the partial negative surface area ($PNSA_1$), the total charge weighted negative surface area ($PNSA_2$) and the atomic charge weighted negative surface area ($PNSA_3$), multiplied by the total molecular solvent-accessible surface area ($SASA$) and divided by 1000, i.e.

$$WNSA_1 = \frac{PNSA_1 \cdot SASA}{1000} \qquad WNSA_2 = \frac{PNSA_2 \cdot SASA}{1000} \qquad WNSA_3 = \frac{PNSA_3 \cdot SASA}{1000}$$

- **surface weighted charged partial positive surface areas** ($WPSA_1$, $WPSA_2$, $WPSA_3$)

They are the partial positive surface area ($PPSA_1$), the total charge weighted positive surface area ($PPSA_2$) and the atomic charge weighted positive surface area ($PPSA_3$), multiplied by the total molecular solvent-accessible surface area ($SASA$) and divided by 1000, i.e.

$$WPSA_1 = \frac{PPSA_1 \cdot SASA}{1000} \qquad WPSA_2 = \frac{PPSA_2 \cdot SASA}{1000} \qquad WPSA_3 = \frac{PPSA_3 \cdot SASA}{1000}$$

- **relative negative charge** ($RNCG$)

It is the partial charge of the most negative atom divided by the total negative charge, i.e.

$$RNCG = \frac{Q^-_{\max}}{Q^-}$$

- **relative positive charge** ($RPCG$)

It is the partial charge of the most positive atom divided by the total positive charge, i.e.

$$RPCG = \frac{Q^+_{\max}}{Q^+}$$

- **relative negative charge surface area** ($RNCS$)

It is the solvent-accessible surface area of the most negative atom divided by the relative negative charge ($RNCG$), i.e.

$$RNCS = \frac{SA^-_{\max}}{RNCG}$$

- **relative positive charge surface area** ($RPCS$)

It is the solvent-accessible surface area of the most positive atom divided by the relative positive charge ($RPCG$), i.e.

$$RPCS = \frac{SA^{+}_{\max}}{RPCG}$$

- **total hydrophobic surface area** ($TASA$)

It is the sum of solvent-accessible surface areas of atoms with absolute value of partial charges less than 0.2, i.e.

$$TASA = \sum\nolimits_a SA_a \qquad \forall a \; : \; |q_a| < 0.2$$

- **total polar surface area** ($TPSA$)

It is the sum of solvent-accessible surface areas of atoms with absolute value of partial charges greater than or equal to 0.2.

$$TPSA = \sum\nolimits_a SA_a \qquad \forall a \; : \; |q_a| \geq 0.2$$

- **relative hydrophobic surface area** ($RASA$)

It is the total hydrophobic surface area ($TASA$) divided by the total molecular solvent-accessible surface area ($SASA$), i.e.

$$RASA = \frac{TASA}{SASA}$$

- **relative polar surface area** ($RPSA$)

It is the total polar surface area ($TPSA$) divided by the total molecular solvent-accessible surface area ($SASA$), i.e.

$$RPSA = \frac{TPSA}{SASA}$$

The set of *CPSA* descriptors was further developed to account for any particular type of polar interaction such as hydrogen bonding. **Hydrogen-Bond Charged Partial Surface Area descriptors** (or ***HB-CPSA* descriptors**) were proposed in analogy with *CPSA* descriptors [Stanton *et al.*, 1992]. Hydrogen-bond donor groups are considered to be any heteroatoms (i.e. O, S, or N) possessing a proton that can be donated. Other types of functional groups such as the alkynes were also included in the donor class. Acceptor groups include any functional group possessing sufficient electron density to participate in a hydrogen bond. In order to simplify the calculations the halogens, some double and some aromatic bonds were not included in the *HB-CPSA* descriptors.

Katritzky later enlarged this set of hydrogen-bonding descriptors [Katritzky *et al.*, 1996a]. All the H-bond descriptors are assigned to zero if no hydrogen atoms in the molecule can be donated; moreover, hydrogen-bond acceptors are usually restricted to oxygen, nitrogen, and sulfur atoms (e.g. carbonyl oxygen atoms except in –COOR, hydroxy oxygen atoms, amino nitrogen atoms, aromatic nitrogens, and mercapto sulfur atoms).

The two simplest *HB-CPSA* descriptors are the → *hydrogen-bond acceptor number HBA* and the → *hydrogen-bond donor number HBD.*

Let SA_d and SA_a be the solvent accessible surface areas of hydrogen-bonding donors (d) and acceptors (a), respectively, q_d and q_a the corresponding partial atomic charges, $SASA$ the molecular solvent-accessible surface area: the *HB-CPSA* descriptors are then defined as follows (note that the two different symbols encountered in the literature for some are considered as synonymous).

• ***RHTA* index**
It is the ratio of the number of donor groups to the number of acceptor groups, i.e.

$$RHTA = \frac{HBD}{HBA}$$

• ***SSAH* index** (: *HDSA index*)
It is the sum of the surface areas of the hydrogens which can be donated:

$$SSAH \equiv HDSA = \sum\nolimits_d SA_d$$

• ***RSAH* index**
It is the average surface area of hydrogens which can be donated.

$$RSAH = \frac{\sum_d SA_d}{HBD}$$

• ***RSHM* index** (: *FHDSA index*)
It is the fraction of the total molecular surface area associated with hydrogens which can be donated:

$$RSHM \equiv FHDSA = \frac{\sum_d SA_d}{SASA}$$

• ***SSAA* index** (: *HASA index*)
It is the sum of the surface areas of all H-bond acceptor atoms:

$$SSAA \equiv HASA = \sum\nolimits_a SA_a$$

The ***HASA*₂ index** is a variant of the *HASA* index defined as [Katritzky *et al.*, 1998d]:

$$HASA_2 = \sum\nolimits_a \sqrt{SA_a}$$

• ***RSAA* index**
It is the average surface area of H-bond acceptor groups:

$$RSAA = \frac{\sum_a SA_a}{HBA}$$

• ***RSAM* index** (: *FHASA index*)
It is the fraction of the total molecular surface area associated with H-bond acceptor groups:

$$RSAM \equiv FHASA = \frac{\sum_a SA_a}{SASA}$$

Based on the $HASA_2$ index, the **$FHASA_2$ index** is defined as [Katritzky *et al.*, 1998c]:

$$FHASA_2 = \frac{\sum_a \sqrt{SA_a}}{SASA}$$

• ***HDCA* index**
It is the sum of charged surface areas of hydrogens which can be donated:

$$HDCA = \sum\nolimits_d q_d \cdot SA_d$$

The charged surface area of hydrogens atoms, called the **$CSA2_H$ index**, and the charged surface area of chlorine atoms, called the **$CSA2_{Cl}$ index**, are two other similar H-bond descriptors defined as:

$$CSA2_H = \sum\nolimits_h q_h \cdot \sqrt{SA_h} \qquad and \qquad CSA2_{Cl} = \sum\nolimits_{Cl} q_{Cl} \cdot \sqrt{SA_{Cl}}$$

where q_h, q_{Cl} and SA_h, SA_{Cl} are partial atomic charges and solvent-accessible surface areas of hydrogen and chlorine atoms, respectively [Katritzky *et al.*, 1998d].

• ***FHDCA* index**
It is the charged surface area of hydrogens which can be donated relative to the total molecular surface area:

$$FHDCA = \frac{\sum_d q_d \cdot SA_d}{SASA}$$

• ***HDCA*₂ index**
It is a hydrogen-bonding descriptor based on solvent-accessible area of hydrogen-bond donor atoms and corresponding partial charges proposed as variant of *FHDCA* index [Katritzky *et al.*, 1996b]:

$$HDCA_2 = \frac{\sum_d q_d \cdot \sqrt{SA_d}}{\sqrt{SASA}}$$

The summation is performed over the number of simultaneously possible hydrogen bonding donor and acceptor pairs per solute molecule; also hydrogen atoms attached to carbon atoms connected directly to carbonyl or cyano groups are considered as hydrogen bonding donors.

The ***HDSA*₂ index** is another hydrogen-bonding donor descriptor with a definition similar to the $HDCA_2$ index [Katritzky *et al.*, 1996c]:

$$HDSA_2 = \frac{\sum_d q_d \cdot \sqrt{SA_d}}{SASA}$$

where the summation is performed over all possible hydrogen bonding donor sites in a molecule.

• ***HACA* index** (: $SCAA_1$ *index*)
It is the sum of charged surface areas of hydrogen-bond acceptors:

$$HACA \equiv SCAA_1 = \sum_a q_a \cdot SA_a$$

An average charged surface area called the ***SCAA*₂ index** was also calculated as:

$$SCAA_2 = \frac{\sum_a q_a \cdot SA_a}{HBA}$$

where *HBA* is the number of hydrogen-bond acceptors [Turner *et al.*, 1998; Mitchell and Jurs, 1998b].

• ***FHACA* index**
It is the charged surface area of hydrogen-bond acceptors relative to the total molecular surface area:

$$FHACA = \frac{\sum_a q_a \cdot SA_a}{SASA}$$

• ***HBSA* index**
It is the sum of the surface areas of both hydrogens which can be donated and hydrogen acceptor atoms:

$$HBSA = HDSA + HASA$$

• ***FHBSA* index**
It is the surface area of both hydrogens which can be donated and hydrogen acceptor atoms relative to the total molecular surface area:

$$FHBSA = \frac{HBSA}{SASA}$$

• ***HBCA* index**
It is the sum of charged surface areas of both hydrogens which can be donated and hydrogen acceptor atoms:

$$HBCA = HDCA + HACA$$

• ***FHBCA* index**
It is the charged surface area of both hydrogens which can be donated and hydrogen acceptor atoms relative to the total molecular surface area:

$$FHBCA = \frac{HBCA}{SASA}$$

• **$CHAA_1$ index**
It is the sum of partial charges on hydrogen-bonding acceptor atoms [Mitchell and Jurs, 1998b]:

$$CHAA_1 = \sum\nolimits_a q_a$$

The average value of $CHAA_1$ is called the **$CHAA_2$ index** and is defined as :

$$CHAA_2 = \frac{\sum_a q_a}{HBA}$$

• ***ACGD* index**
It is the average difference in charge between all pairs of H-bond donors.

• ***HRPCG* index**
It is the relative positive charge (*RPCG*) restricted to H-bond donor atoms.

• ***HRNCG* index**
It is the relative negative charge (*RNCG*) restricted to H-bond acceptor atoms.

• ***HRPCS* index**
It is the relative positively charged surface area (*RPCS*) restricted to H-bond donor atoms, i.e. the positively charged surface area corresponding to the most positively charged atom that is also a possible hydrogen donor.

• ***HRNCS* index**
It is the relative negatively charged surface area (*RNCS*) restricted to H-bond acceptor atoms, i.e. the negatively charged surface area corresponding to the most negatively charged atom that is also a possible hydrogen acceptor.

• ***CHGD* index**
It is the maximum difference in charge between a hydrogen which can be donated and its covalently bonded heteroatom.

📖 [Stanton and Jurs, 1992] [Nelson and Jurs, 1994] [Mitchell and Jurs, 1998a]

charge-matching function → **molecular shape analysis**

charge polarization → **charge descriptors** (⊙ total absolute atomic charge)

charge-related indices → **charge descriptors**

charge term matrix → **topological charge indices**

charge topological index → **charge descriptors** (⊙ charge-related indices)

charge transfer constant → **electronic substituent constants**

Charton characteristic volume : *Charton steric constant* → **steric descriptors**

Charton inductive constants → **electronic substituent constants** (⊙ inductive electronic constants)

Charton steric constant → **steric descriptors**

chemical descriptors : *molecular descriptors*

chemical distance → **bond order indices** (⊙ conventional bond order)

chemical distance matrix → **weighted matrices**

chemical formula → **molecular descriptors**

chemical graph → **graph**

chemical hardness : *absolute hardness* → **quantum-chemical descriptors** (⊙ hardness indices)

chemical invariance → **molecular descriptors** (⊙ invariance properties of molecular descriptors)

chemically intuitive molecular index → **eigenvalue-based descriptors** (⊙ Burden eigenvalues)

chemometrics

Chemometrics is a discipline that collects mathematical and statistical tools to deal with complex chemical data [Brereton, 1990; Devillers and Karcher, 1991; Frank and Todeschini, 1994; van de Waterbeemd, 1995b; Massart *et al.*, 1997; Massart *et al.*, 1998; Legendre and Legendre, 1998].

The main characterizing strategies are the multivariate approach to the problem, searching for relevant information, model validation to build models with predictive power, comparison of the results obtained by using different methods, and definition and use of indices capable of measuring the quality of extracted information and the obtained models.

Chemometrics is a most useful tool in QSAR and QSPR studies, in that it forms a firm base for data analysis and modelling and provides a battery of different methods. Moreover, a relevant aspect of the chemometric philosophy is the attention it pays to the predictive power of the models (estimated by using → *validation techniques*), → *model complexity*, and the continuous search for suitable parameters to assess the model qualities, such as → *classification parameters* and → *regression parameters.*

Chemometrics includes several fields of mathematics and statistics as listed below.

- **Artificial Neural Networks** (ANN)

A set of mathematical methods, models and algorithms designed to mimic information processing and knowledge acquisition methods of the human brain. ANNs are especially suitable for dealing with non-linear relationships and trends and are proposed for facing a large variety of mathematical problems such as data exploration, pattern recognition, modelling of continuous and categorized responses, multiple response problems, etc. [Zupan and Gasteiger, 1999; Anzali *et al.*, 1998].

Some historically important artificial neural networks are *Hopfield Networks*, *Perceptron Networks* and *Adaline Networks*, while the most well-known are *Backpropagation Artificial Neural Networks* (BP-ANN), *Kohonen Networks* (K-ANN, or *Self-Organizing Maps*, SOM), *Radial Basis Function Networks* (RBFN), *Probabilistic Neural Networks* (PNN), *Generalized Regression Neural Networks* (GRNN), *Learning Vector Quantization Networks* (LVQ), and *Adaptive Bidirectional Associative Memory* (ABAM).

📖 [Klopman, 1984; Klopman and Buyukbingol, 1988; Aoyama *et al.*, 1990a; Aoyama *et al.*, 1990b; Andrea and Kalayeh, 1991; Bodor *et al.*, 1991; Klopman and Henderson, 1991; Bodor *et al.*, 1992c; Livingstone and Salt, 1992; Salt *et al.*, 1992; So and Richards, 1992; Judson, 1992a; Liu *et al.*, 1992a; Liu *et al.*, 1992b; Maggiora *et al.*, 1992; Cambon and Devillers, 1993; Egolf and Jurs, 1993a; Lohninger, 1993; Manallack and Livingstone, 1993; Kvasnicka *et al.*, 1993; Nefati *et al.*, 1993; Campbell and Johnson, 1993; Villemin *et al.*, 1993; Ghoshal *et al.*, 1993; Balaban *et al.*, 1994a; Bienfait, 1994; Bodor *et al.*, 1994; Chastrette *et al.*, 1994; Gakh *et al.*, 1994; Gasteiger *et al.*, 1994a; Gasteiger *et al.*, 1994b; Manallack *et al.*, 1994; Villemin *et al.*, 1994; Wessel and Jurs, 1994; Xu *et al.*, 1994; Manallack and Livingstone, 1994; Tetko *et al.*, 1995; Manallack and Livingstone, 1995; Barlow, 1995; Kireev, 1995; Rorije *et al.*, 1995; Sadowski *et al.*, 1995; Wagener *et al.*, 1995; Combes and Judson, 1995; Hasegawa *et al.*, 1995a; Andrea, 1995; Anzali *et al.*, 1996; Beck *et al.*, 1996; Hall and Story, 1996; Hatrìk and Zahradnìk, 1996; Klawun and Wilkins, 1996b; Klawun and Wilkins, 1996a; Kränz *et al.*, 1996; Kyngas and Valjakka, 1996; Nefati *et al.*, 1996; Selzer *et al.*, 1996; So and Karplus, 1996a; So and Karplus, 1996b; Domine *et al.*, 1996; Ivanciuc, 1996; Wessel *et al.*, 1996; Anzali *et al.*, 1997; Baskin *et al.*, 1997; Hall and Story, 1997a; Liu *et al.*, 1997; Mitchell and Jurs, 1997; Novic *et al.*, 1997; Polanski, 1997; Pompe *et al.*, 1997; Svozil *et al.*, 1997; Vracko, 1997; Zhang *et al.*, 1997; Burden, 1998; Doucet and Panaye, 1998; Duprat *et al.*, 1998; Garcìa-Domenech and De Julián-Ortiz, 1998; Huuskonen *et al.*, 1998; Kireev *et al.*, 1998; Bernard *et al.*, 1998; Kovalishyn *et al.*, 1998; Liang and Gallagher, 1998; Mitchell and Jurs, 1998b; Mitchell and Jurs, 1998a; Polanski *et al.*, 1998; Satoh *et al.*, 1998; Tetko, 1998; Turner *et al.*, 1998; Wang and Chen, 1998; Wessel *et al.*, 1998; Burden and Winkler, 1999a; Burden and Winkler, 1999b; Gini *et al.*, 1999; Goll and Jurs, 1999b; Kaiser and Niculescu, 1999; Liu and Guo, 1999; Lucic and Trinajstic, 1999a; Pompe and Novic, 1999; Tetteh *et al.*, 1999; Vendrame *et al.*, 1999; Schweitzer and Morris, 1999]

• **classification**

Assignment of objects to one of several classes based on a classification rule. The *classes* are defined a priori by groups of objects in a training set belonging to those classes. The goal is to calculate a *classification rule* and, possibly, *class boundaries* based on the training set objects of known classes, and to apply this rule to assign a class to objects of unknown classes [Hand, 1981; Frank and Friedman, 1989; Hand, 1997]. Classification methods are suitable for modelling several QSAR responses, such as, for example, active/non-active compounds, low/medium/high toxic compounds, mutagenic/non-mutagenic compounds.

The most popular classification methods are *Linear Discriminant Analysis* (LDA), *Quadratic Discriminant Analysis* (QDA), *Regularized Discriminant Analysis* (RDA), *K*th *Nearest Neighbours* (KNN), *classification tree methods* (such as CART), *Soft-Independent Modeling of Class Analogy* (SIMCA), *potential function classifiers* (PFC), *Nearest Mean Classifier* (NMC) and *Weighted Nearest Mean Classifier* (WNMC). Moreover, several classification methods can be found among the artificial neural networks.

Classification model performance is evaluated by classification parameters, both for fitting and predictive purposes.

📖 [Henry and Block, 1979; Henry and Block, 1980b; Serrano *et al.*, 1985; Takahashi *et al.*, 1985; McFarland and Gans, 1986; McFarland and Gans, 1986; McFarland and Gans, 1990a; McFarland and Gans, 1990b; Rose *et al.*, 1991; Rose *et al.*, 1992; Moriguchi *et al.*, 1992a; Qian *et al.*, 1992; Miyashita *et al.*, 1993; Hirono *et al.*, 1994; Raevsky *et al.*, 1994; Jonathan *et al.*, 1996; Nouwen *et al.*, 1997; Garcìa-Domenech and De Julián-Ortiz, 1998; Pizarro Millán *et al.*, 1998; Dixon and Villar, 1999; Gramatica *et al.*, 1999b; Hunt, 1999; Tominaga, 1999; Worth and Cronin, 1999]

- **cluster analysis**

A special case of exploratory data analysis aimed at grouping similar objects in the same cluster and less similar objects in different clusters [Massart and Kaufman, 1983; Willett, 1987]. Cluster analysis is based on the evaluation of the → *similarity/diversity* of all the pairs of objects of a data set. This information is collected into the → *similarity matrix* or → *data distance matrix*.

Many different methods belong to cluster analysis; the most popular are the *hierarchical agglomerative methods* (i.e. *average linkage*, *complete linkage*, *single linkage*, *weighted average linkage*, etc.) which are more widely used than the *hierarchical divisive methods*. Other very popular methods are *non-hierarchical methods*, such *k-means method* and the *Jarvis-Patrick method*.

🕮 [Dunn III and Wold, 1980; Dean and Callow, 1987; Nakayama *et al.*, 1988; Willett, 1988; Lawson and Jurs, 1990; Jurs and Lawson, 1991; Good and Kuntz, 1995; Shemetulskis *et al.*, 1995; Nouwen *et al.*, 1996; Brown and Martin, 1996; Brown and Martin, 1997; Dunbar Jr, 1997; Junghans and Pretsch, 1997; McGregor and Pallai, 1997; Brown and Martin, 1998; Reynolds *et al.*, 1998; Rose and Wood, 1998]

- **experimental design**

Statistical procedures for planning an experiment, i.e. collecting appropriate data which, after analysis by statistical methods, result in valid conclusions. The design includes the selection of experimental units, the specification of the experimental conditions, i.e. the specification of factors whose effect will be studied on the outcome of the experiment, the specification of the level of the factors involved and the combination of such factors, the selection of response to be measured, and the choice of statistical model to fit the data [Box *et al.*, 1978; Carlson, 1992].

An *experiment* consists of recording the values of a set of variables from a measurement process under a given set of experimental conditions.

The most known experimental designs are *complete factorial designs*, *fractional factorial designs*, *composite designs*, and *optimal designs*.

🕮 [Borth and McKay, 1985; Pastor and Alvarez-Builla, 1991; Bonelli *et al.*, 1991; Norinder and Hogberg, 1992; Norinder, 1992; Baroni *et al.*, 1993a; Baroni *et al.*, 1993c; Marsili and Saller, 1993; Pastor and Alvarez-Builla, 1994; Rovero *et al.*, 1994; Cruciani and Clementi, 1994; van de Waterbeemd *et al.*, 1995; Sjöström and Eriksson, 1995; Austel, 1995; Borth, 1996; Eriksson and Johansson, 1996; Eriksson *et al.*, 1997].

- **exploratory data analysis**

Exploratory data analysis is a collection of techniques that search for structure in a data set before calculating any statistic model [Krzanowski, 1988]. Its purpose is to obtain information about the data distribution, about the presence of outliers and clusters, and to disclose relationships and correlations between objects and/or variables. → *Principal component analysis* and *cluster analysis* are the most well-known techniques for data exploration [Jolliffe, 1986; Jackson, 1991; Basilevsky, 1994].

🕮 [Weiner and Weiner, 1973; Stuper and Jurs, 1978; Wold, 1978; Cramer III, 1980a; Henry and Block, 1980a; Streich *et al.*, 1980; Alunni *et al.*, 1983; McCabe, 1984; Takahashi *et al.*, 1985; Dunn III and Wold, 1990; Livingstone *et al.*, 1992; Langer and Hoffmann, 1998a]

- **MultiCriteria Decision Making** (MCDM)

The multicriteria decision making (MCDM) approach is based on methods to rank the studied objects (events, molecules, cases, etc.) on the basis of multiple criteria [Hendriks *et al.*, 1992; Carlson, 1992]. In particular, *desirability functions* and *utility functions* are multicriteria decision functions able to assign a score to each object on a user-defined tuning.

Multicriteria decision functions may be used to represent chemical information by harmonizing structural information, experimental knowledge and other specific characteristics of the studied problem, such as environmental or health parameters.

📖 [Brüggemann and Bartel, 1999]

- **optimization**

Finding the optimal value (minimum or maximum) of a numerical function f, called the objective function, with respect to a set of parameters **p**, $f(p_1, p_2, ..., p_p)$. If the values that the parameters can take on are constrained, the procedure is called *constrained optimization*.

The most popular optimization techniques are *Newton-Raphson optimization*, *steepest ascent optimization*, *steepest descent optimization*, *Simplex optimization*, *Genetic Algorithm optimization*, *simulated annealing*. → *Variable reduction* and → *variable selection* are also among the optimization techniques.

📖 [Holland, 1975; Papadopoulos and Dean, 1991; Carlson, 1992; Kalivas, 1995; Hall, 1995; Sundaram and Venkatasubramanian, 1998; Handschuh *et al.*, 1998; Wehrens *et al.*, 1998]

- **regression analysis**

A set of statistical methods using a mathematical equation to model the relationship between an observed or measured response and one or more predictor variables. The goal of this analysis is twofold: modelling and predicting. The relationship is described in algebraic form as:

$$y = f(x) + e \quad \text{or} \quad \mathbf{y} = \mathbf{X} \cdot \mathbf{b}$$

where x denotes the predictor variable(s), y the response variable(s), $f(x)$ is the systematic part of the model, and e is the random error, also called model error or residual; **y** and **b** are the vectors of the responses and of regression coefficients to be estimated; the matrix **X** is usually called the *model matrix*, i.e. its columns are the independent variables used in the regression model.

The mathematical equation used to describe the relationship among response and predictor variables is called the *regression model* [Draper and Smith, 1998; Frank and Friedman, 1993; Ryan, 1997; Wold, 1994].

Regression analysis includes not only the estimation of model regression parameters, but also the calculation of → *goodness of fit* and → *goodness of prediction* statistics, *regression diagnostics*, *residual analysis*, and *influence analysis* [Atkinson, 1985].

In particular, the **leverage matrix H**, also called *influence matrix*, is an important tool in regression diagnostics, containing information on the independent variables on which the model is built.

Let **X** be a matrix with n rows and p' columns, where p' is the number of model parameters. The leverage matrix **H** is a symmetric $n \times n$ matrix defined as:

$$\mathbf{H} = \mathbf{X} \cdot \left(\mathbf{X}^{\mathrm{T}} \cdot \mathbf{X}\right)^{-1} \cdot \mathbf{X}^{\mathrm{T}}$$

where the matrix **X** is the model matrix. Moreover, a column where all the values are equal to one is added to the model matrix if the model is not constrained in the origin of the independent variables but an offset is allowed. To distinguish the two cases a parameter c is used; $c = 1$ for the former and $c = 0$ for the latter.

The main properties of the leverage matrix are:

$$\text{a)}\ \frac{c}{n} \leq h_{ii} \leq 1 \qquad \text{b)}\ \sum_{i=1}^{n} h_{ii} = p' \qquad \text{c)}\ \bar{h} = \frac{\sum_{i=1}^{n} h_{ii}}{n} = \frac{p'}{n} \qquad \text{d)}\ \sum_{j=1}^{n} h_{ij} = c$$

where $\bar{h}$ is the average value of the leverage.

The leverage matrix is related to the response vector **y** by the following relationship:

$$\hat{\mathbf{y}} = \mathbf{H} \cdot \mathbf{y}$$

where $\hat{\mathbf{y}}$ is the calculated response vector from the model.

Usually the diagonal elements h_{ii} are those used for regression diagnostics: the ith object whose diagonal element h_{ii} is greater than two or three time the average value $\bar{h}$ can be considered as having a great influence (leverage) on the model.

Besides the well-known *Ordinary Least Squares regression* (*OLS*), *biased regression*, *nonlinear regression*, and *robust regression* models are also important. Among the biased methods, the most popular regression methods are *Principal Component Regression* (*PCR*), *Partial Least Squares regression* (*PLS*), *Ridge Regression* (*RR*), *Continuum Regression* (*CR*), and *StepWise Regression* (*SWR*).

Among the nonlinear methods, there are, besides *nonlinear least squares regression*, i.e. *polynomial regression*, the *nonlinear PLS* method, *Alternating Conditional Expectations* (*ACE*), *SMART*, and *MARS*. Moreover, some Artificial Neural Networks techniques have also to be considered among nonlinear regression methods, such as the *back-propagation method*.

📖 [Hoerl and Kennard, 1970; Wold and Sjöström, 1978; Dunn III *et al.*, 1984; Wold *et al.*, 1984; Geladi and Kowalski, 1986; Myers, 1986; Cramer III *et al.*, 1988a; Friedman, 1988; Höskuldsson, 1988; Osten, 1988; Rawlings, 1988; Martens and Naes, 1989; Brekke, 1989; Geladi and Tosato, 1990; Miller, 1990; Devillers and Lipnick, 1990; Tysklind *et al.*, 1992; Miyashita *et al.*, 1992; Cramer III, 1993; Egolf and Jurs, 1993a; Randic, 1993a; Tysklind *et al.*, 1993; Wold *et al.*, 1993; Piggott and Withers, 1993; Lindgren, 1994; Meloun *et al.*, 1994; Norinder *et al.*, 1994; Eriksson *et al.*, 1994a; Mager, 1995a; Mager, 1995b; Eriksson *et al.*, 1995; Cummins and Andrews, 1995; Dunn III and Rogers, 1996; Hasegawa *et al.*, 1996a; Soskic *et al.*, 1996b; Wold *et al.*, 1996; Nguyen-Cong and Rode, 1996a; Nguyen-Cong *et al.*, 1996b; Hirst, 1996; Luco and Ferretti, 1997; Lindgren and Rännar, 1998; Lucic and Trinajstic, 1999a; Hasegawa *et al.*, 1999a]

***CHGD* index** → **charged partial surface area descriptors**

chiral factors → **weighted matrices**

chirality descriptors

An n-dimensional object is called *chiral* if it is nonsuperimposable on its mirror image by any rotation in the n-dimensional space. *Chirality* is the property of chiral objects.

If looked in an isolation, → *physico-chemical properties* (and mathematical) of a chiral molecule and its antipodal counterpart all coincide. However, when chiral structures are considered in an environment, their behaviour can be discriminated, such as occurs, for example, when a chiral molecule interacts with a receptor. Thus, chirality descriptors are useful for modeling properties related to interactions involving chiral centres.

Two general classes of chirality measures have been recognized: in the first, the degree of chirality expresses the extent to which a chiral object differs from an achiral reference object, while in the second it expresses the extent to which two enantiomorphs differ from each other [Buda *et al.*, 1992]. The **continuous chirality measure** (CCM) recently proposed [Zabrodsky and Avnir, 1995] is an example of chirality measure belonging to the first class and is based on the general definition of *continuous symmetry measure* defined as:

$$S(\mathrm{G}) = \frac{100}{A} \cdot \sum_{i=1}^{A} \|\mathbf{p}_i - \hat{\mathbf{p}}_i\|^2 \qquad 0 < S \leq 100$$

where G is a given symmetry group, $\mathbf{p}_i$ is the coordinate vector of the ith atom of the original chiral configuration, $\hat{\mathbf{p}}_i$ is the coordinate vector of the corresponding atom in the nearest G-symmetric configuration, and A is the number of atoms of the molecule. In practice, since the minimal requirement for an object to be achiral is that it should possess either a reflection mirror (σ), an inversion centre (i), or a higher order improper rotation axes (S_{2n}), the function $S(\mathrm{G})$ has to be screened over symmetry groups having these elements. This continuous chirality measure is a total (normalized) distance of the original chiral configuration from the considered G-symmetry configuration, bounded between 0 and 100.

The interest in using chirality descriptors in → *structure/response correlations* is quite recent and two of them are explained below. Weighted matrices have also been devised for obtaining the → *chiral modification number* that is added to any topological index in order to discriminate *cis/trans* isomers. Other interesting methods to quantify chirality are Kuz'min's method based on the *dissymmetry function* [Kuz'min *et al.*, 1992a; Kuz'min *et al.*, 1992b; Kutulya *et al.*, 1992] and Mezey's method [Mezey, 1997b] based on the → *Mezey 3D shape analysis.*

- **Randic chirality index**

A molecular descriptor that was proposed to discriminate between a chiral molecule and its mirror image; it is restricted to molecules embedded in 2D space, such as benzenoids, and is based on → *periphery codes* [Randic, 1998a].

Starting from a selected ith atom on the molecule periphery, the vertex degrees δ of the periphery atoms define an ordered code for the molecule; two different codes are obtained whether the clockwise or the anticlockwise direction is chosen. These codes, referring to the ith atom, are then transformed by making partial sums, adding elements of the series successively; the two codes obtained encode information on the asymmetry of the molecular periphery.

Example : benzanthracene

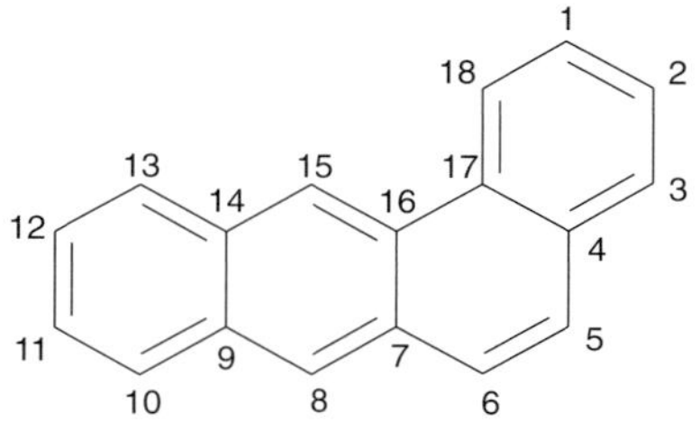

Clockwise direction

labels	**1**	**2**	**3**	**4**	**5**	**6**	**7**	**8**	**9**	**10**	**11**	**12**	**13**	**14**	**15**	**16**	**17**	**18**
atom 1	2	2	2	3	2	2	3	2	3	2	2	2	2	3	2	3	3	2

Anticlockwise direction

labels	**1**	**18**	**17**	**16**	**15**	**14**	**13**	**12**	**11**	**10**	**9**	**8**	**7**	**6**	**5**	**4**	**3**	**2**
atom 1	2	2	3	3	2	3	2	2	2	2	3	2	3	2	2	3	2	2

↓

Clockwise direction CD_j

atom 1	2	4	6	9	11	13	16	18	21	23	25	27	29	32	34	37	40	42

Anticlockwise direction AD_j

atom 1	2	4	7	10	12	15	17	19	21	23	26	28	31	33	35	38	40	42

$$D_1 = \sum_{j=1}^{18} \left(AD_j - CD_j \right) = +14$$

↓ repeating the procedure for all the atoms

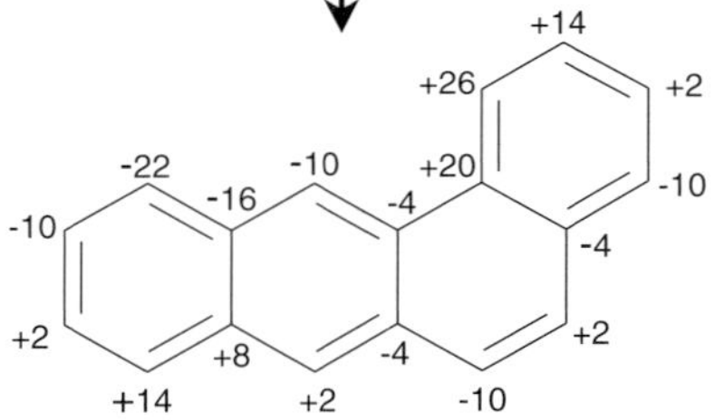

Randic chirality index $CH_R^{\,k}$

k	*3*	*5*	*7*	*9*	*11*
Benzanthracene	+ 2.17284	+ 5.07240	+ 10.97954	+ 23.70935	+ 51.05061
Mirror image	– 2.17284	– 5.07240	– 10.97954	– 23.70935	– 51.05061

Box C-2.

To obtain a single value descriptor D_i for the ith atom, the differences between the corresponding elements in the anticlockwise and clockwise codes are computed and then summed:

$$D_i = \sum_{j=1}^{A} (AD_{ij} - CD_{ij})$$

where AD_{ij} represents the jth element in the ith anticlockwise code and CD_{ij} represents the jth element in the ith clockwise code. This procedure is repeated for all the atoms on the molecule periphery. Finally, the Randic chirality index is defined as:

$$CH_{\mathrm{R}}^{k} = \frac{1}{A^k} \cdot \sum_{i=1}^{A} D_i^k$$

where k is an odd power exponent. Using different odd powers, a sequence of chirality indices can be used to better characterize the molecule and its mirror image.

For achiral molecules, all the chirality indices (and the corresponding vector elements) equal zero.

- **Moreau chirality index**

A molecular descriptor proposed with the aim of quantifying the chirality of a molecule by means of the positive or negative quantification of the chirality of the environment of the atoms in the molecule for any scalar atomic property [Moreau, 1997].

The basic idea is that asymmetrical environments are not the exclusive privilege of atoms in chiral molecules, however they are the most frequent situations. The Moreau chirality index is defined as:

$$CH_{\mathrm{M}} = \sum_{i=1}^{A} AS_i$$

where the sum runs over all atoms in the molecule, and AS_i is the measure of the chirality of the environment of the ith atom given by the following expression:

$$AS_i = 10^3 \cdot C_i \cdot \{XYZ\}_i \cdot S_i = 10^3 \cdot \frac{(\lambda_1 - \lambda_2) \cdot (\lambda_2 - \lambda_3) \cdot \lambda_3}{\left(\sum_k \lambda_k\right)^3} \cdot \frac{XYZ_i}{(e_X \cdot e_Y \cdot e_Z)} \cdot (\mathbf{v}_1, \mathbf{v}_2, \mathbf{v}_3)$$

λ_1, λ_2, and λ_3 are the eigenvalues of the square symmetric matrix **C** closely related to the covariance matrix of the cartesian coordinates (x_j, y_j, z_j) of the atoms M_j in the environment of the ith atom and defined as:

$$\mathbf{C} = \begin{vmatrix} \sum_j p_j \cdot w_j \cdot x_j^2 & \sum_j p_j \cdot w_j \cdot x_j \cdot y_j & \sum_j p_j \cdot w_j \cdot x_j \cdot z_j \\ \sum_j p_j \cdot w_j \cdot x_j \cdot y_j & \sum_j p_j \cdot w_j \cdot y_j^2 & \sum_j p_j \cdot w_j \cdot y_j \cdot z_j \\ \sum_j p_j \cdot w_j \cdot x_j \cdot z_j & \sum_j p_j \cdot w_j \cdot y_j \cdot z_j & \sum_j p_j \cdot w_j \cdot z_j^2 \end{vmatrix}$$

The origin of the coordinates is the barycentre of the environment of the considered atom and the eigenvectors $\mathbf{v}_1$, $\mathbf{v}_2$, $\mathbf{v}_3$ associated with the eigenvalues are unit vectors which define the three principal axes of this environment. The considered ith atom can be included or not in its environment. To each atom of the environment is assigned an atomic property p_j (e.g. unitary property, atomic mass, atomic electronegativity, atomic van der Waals volume) and a weight w_j which is a function of the distance of the jth environment atom from the ith atom.

In this approach, the principal planes of the environment are taken as the best approximation of the potential symmetry planes, and the asymmetry of the environment as seen from the considered ith atom is defined as proportional to the distance from i to the symmetry plane.

The coefficient C_i in the expression for AS_i accounts for the asymmetry of the environment and is equal to zero when two eigenvalues are equal, or the third eigenvalue equals zero, or the last two eigenvalues equal zero.

The term $\{XYZ\}_i$ is the product of the coordinates of the ith atom with respect to the principal axes normalized by the product of the half-thicknesses e_X, e_Y, e_Z of the "slab" which approximates the set of weighted environment atoms. This term indicates that asymmetry is positive or negative according to the octant in which the ith atom is.

The term S is the box-product of the three eigenvectors and is equal to +1 or –1 depending on the handedness of the principal axis space. It was introduced into the expression of AS in order to have an intrinsic measure of chirality.

📖 [King, 1991] [Gilat, 1994] [Serilevy *et al.*, 1994b] [Winberg and Mislow, 1995] [Franke *et al.*, 1995] [Randic and Razinger, 1996] [Randic and Mezey, 1996] [Fujita, 1996] [De Julián-Ortiz *et al.*, 1996] [Petitjean, 1996] [Balaban, 1997b] [Klein and Babic, 1997b] [Mislow, 1997] [Nemba and Balaban, 1998]

chiral modification number → **weighted matrices**

chromatic decomposition

A decomposition $A(\mathcal{V}_1, \mathcal{V}_2, \ldots, \mathcal{V}_G)$ of the set $\mathcal{V}$ of the graph $\mathcal{G}$ vertices into G → *equivalence classes* is the said chromatic decomposition of $\mathcal{G}$ (or **vertex chromatic decomposition**) if, for any pair of vertices v_i and v_j belonging to $\mathcal{V}_g$, the edge e_{ij} connecting the considered vertices does not belong to the set of edges $\mathcal{E}$ of the graph; it means that two vertices belonging to the same chromatic class $\mathcal{V}_g$ cannot be adjacent [Bonchev, 1983].

A decomposition $B(\mathcal{E}_1, \mathcal{E}_2, \ldots, \mathcal{E}_G)$ of the set $\mathcal{E}$ of the graph $\mathcal{G}$ edges into G equivalence classes is said to be **edge chromatic decomposition** of $\mathcal{G}$ if any pair of edges e_{ij} and e_{kl} belonging to $\mathcal{E}_g$ does not belong to the set ${}^2\mathcal{P}$ of the 2nd order paths of the graph (i.e. the two edges are not adjacent).

A **graph colouring** of vertices (or edges) is an assignment of a minimal number of different colours to the vertices (or edges) of $\mathcal{G}$ such that no two adjacent vertices (or edges) have the same colour. Graph colouring produces a → *chromatic graph*.

The subsets $\mathcal{V}_g$ are called **colour classes**. The simplest descriptor that can be defined by a vertex chromatic decomposition is called the **chromatic number** $k(\mathcal{G})$ [or **vertex chromatic number**, ${}^{V}k(\mathcal{G})$] and is the smallest number of colour equivalence classes (i.e. G). In general, there is not a unique chromatic decomposition of a graph with the smallest number of colours. Analogously, the descriptor obtained by an edge chromatic decomposition is called the **edge chromatic number** ${}^{E}k(\mathcal{G})$.

The **chromatic information index** (or **vertex chromatic information index**) [Mowshowitz, 1968d] is the minimum value of the → *mean information content* of all possible vertex chromatic decompositions with a number of colours equal to the vertex chromatic number $k(\mathcal{G})$ and is defined as:

$$ {}^{V}\bar{I}_{\mathrm{CHR}} = \min\left(-\sum_{g=1}^{k(\mathcal{G})} \frac{|\mathcal{V}_g|}{A} \log_2 \frac{|\mathcal{V}_g|}{A} \right) $$

where $|\mathcal{V}_g|$ is the number of vertices (i.e. the cardinality of the gth set) within the same equivalence class for the decomposition and A is the → *atom number*.

The **edge chromatic information index** is the minimum value of the mean information content of all possible edge chromatic decompositions having a number of colours equal to the edge chromatic number ${}^{E}k(\mathcal{G})$ and is defined as:

$$^{E}\bar{I}_{CHR} = \min\left(- \sum_{g=1}^{^{E}k(\mathcal{G})} \frac{|\mathcal{E}_g|}{B} \log_2 \frac{|\mathcal{E}_g|}{B} \right)$$

where $|\mathcal{E}_g|$ is the number of edges (i.e. the cardinality of gth set) within the same equivalence class for the decomposition and B is the → *bond number*.

chromatic graph → **graph**

chromatic information index → **chromatic decomposition**

chromatic number → **chromatic decomposition**

chromatographic descriptors

→ *Molecular descriptors* obtained from chromatographic techniques, i.e. from gas chromatography (GC), high-performance liquid chromatography (HPLC), thin-layer chromatogaphy (TLC) and paper chromatography (PC) [Kaliszan, 1987; Kaliszan, 1992].

The most important ones are listed below.

- **Kovats retention index** (I_i)

An index characteristic of a gas chromatographed compound on a given column at a definite temperature defined as:

$$I_i = 100 \cdot \frac{\log t'_{Ri} - \log t'_{R(Nc)}}{\log t'_{R(Nc+1)} - \log t'_{R(Nc)}} + 100 \cdot N_C$$

where $t'_{R(Nc)}$ is the adjusted retention time of a homologue standard with a carbon number N_C; $t'_{R(Nc+1)}$ is an analogous parameter for another standard with carbon number $N_C + 1$; $t'_{R_i)}$ is the adjusted retention time of the ith compound [Kováts, 1968]. The measured total retention time t_R is a sum of two factors $t_R = t_M + t'_R$, where t_M is the initial dead time and t'_R is the adjusted retention time of the compound.

- **capacity factor** (k')

Also called **phase capacity ratio**, it is a measure of the degree of retention of a solute in HPLC, as:

$$k' = \frac{t_R - t_M}{t_M} = \frac{V_R - V_M}{V_M}$$

where t_R and V_R are the retention time and retention volume, respectively, of the solute; t_M and V_M are respectively the retention time and the retention volume of the unretained solute. The quantity log k' can be considered analogous to the Bate-Smith-Westall retention index R_M and, like this, is related to → *partition coefficients*.

- **Bate-Smith-Westall retention index** (R_M)

A retention index R_M derived from thin-layer and paper chromatography [Bate-Smith and Westall, 1950], defined as:

$$R_M = \log\left(\frac{1}{R_f} - 1\right)$$

where R_f is the ratio of a distance passed by the solute to that attained by the solvent front.

The quantity R_M is proportional to the partition coefficient log$P$ and has been used in its place in many QSAR models.

📖 [Michotte and Massart, 1977] [Kaliszan and Foks, 1977] [Kaliszan, 1977] [Kaliszan and Lamparczyk, 1978] [Randic, 1978b] [Markowki *et al.*, 1978] [Millership and Woolfson, 1978] [Kaliszan, 1979] [Kier and Hall, 1979] [Baker, 1979] [Dimov and Papazova, 1979] [McGregor, 1979] [Millership and Woolfson, 1979] [Millership and Woolfson, 1980] [Kaliszan, 1981] [Buydens and Massart, 1981] [Wells *et al.*, 1981] [Calixto and Raso, 1982] [Wells *et al.*, 1982] [Bojarski and Ekiert, 1983] [Szász *et al.*, 1983] [Buydens *et al.*, 1983a] [Buydens *et al.*, 1983b] [Doherty *et al.*, 1984] [Govers *et al.*, 1984] [Bernejo *et al.*, 1984] [Sabljic and Protic, 1984] [Sabljic, 1985] [Rohrbaugh and Jurs, 1985] [Kaliszan, 1986] [Rohrbaugh and Jurs, 1986] [Robbat Jr. *et al.*, 1986a] [Robbat Jr. *et al.*, 1986b] [Lehtonen, 1987] [Klappa and Long, 1992] [Kaliszan *et al.*, 1992a] [Cabala *et al.*, 1992] [Kaliszan *et al.*, 1992b] [Camilleri *et al.*, 1993] [Dimov *et al.*, 1994] [Kaliszan *et al.*, 1994] [Corbella *et al.*, 1995] [Azzaoui and Morinallory, 1995b] [Liu and Qian, 1995] [García-March *et al.*, 1996] [Booth and Wainer, 1996a] [Booth and Wainer, 1996b] [Lee *et al.*, 1996] [Gautzsch and Zinn, 1996] [Salo *et al.*, 1996] [Dimov and Osman, 1996] [Roussel *et al.*, 1997] [Nord *et al.*, 1998] [Dross *et al.*, 1998] [Mõssner *et al.*, 1999]

CID′/CID index → **bond order indices** (⊙ graphical bond order)

CIM index : *Chemically Intuitive Molecular index* → **eigenvalue-based descriptors** (⊙ Burden eigenvalues)

circuit : *cyclic path* → **graph**

cis/trans binary factor → **cis/trans descriptors**

cis/trans descriptors

Cis/trans isomerism is usually easily distinguished by using → *geometrical descriptors*, i.e. descriptors derived from 3D molecular structures or structures embedded in a 3D space [Randic *et al.*, 1990]. Otherwise, the simplest way to distinguish *cis/trans* isomers is the **cis/trans binary factor** which takes value –1 for *cis*-isomers and +1 for *trans*-isomers [Lekishvili, 1997].

However, when molecular descriptors are derived from molecular graphs, *cis/trans* isomerism is not usually recognized and some molecular descriptors were proposed in order to discriminate between *cis/trans* isomers, such as the → *corrected electron charge density connectivity index*, and → *periphery codes*. → *Weighted matrices* were also devised for obtaining the → *geometric modification number* that is added to any topological index in order to discriminate *cis/trans* isomers.

Another topological descriptor specifically proposed for *cis/trans* isomerism is the **Pogliani cis/trans connectivity index** χ_{CT}, defined in terms of the → *Randic connectivity index* χ as:

$$\chi_{CT} = \chi - \chi_{CIS} = \chi - \sum_k \left(\delta_1^r \cdot \delta_2 \cdot \delta_3 \cdot \delta_4^r\right)_k^{-1.5}$$

where the sum runs over all the *cis*-butadienic or *cis*-2-butenic fragments in the graph; δ is the → *vertex degree* of the involved atoms and δ^r is the raised vertex degree of 1,4 vertices, i.e. the vertex degree obtained by joining the two *cis* vertices by a virtual bond, forming a virtual 4-membered ring [Pogliani, 1994b] (Figure C-1). For all-trans molecular graphs, $\chi_{CT} = \chi$.

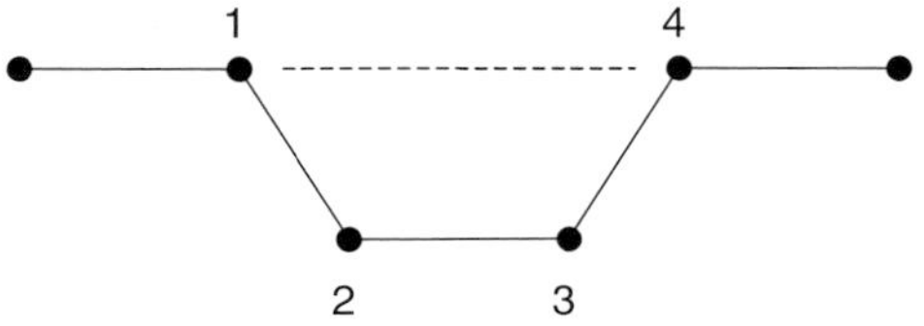

δ_1=2+1; δ_2=2; δ_3=2; δ_4=2+1

$\chi_{CIS} = 0.046$

Figure C-1.

📖 [Balaban, 1976b] [Balaban, 1998]

classical QSAR → structure/response correlations

classification → chemometrics

classification parameters

Statistical indices used to evaluate the performance of classification models [Frank and Todeschini, 1994]. They are derived from two kind of statistics, called → *goodness of fit* and → *goodness of prediction*.

All the classification parameters can be derived from the *confusion matrix*, where the rows represent the known true classes and the columns the classes assigned by the classification method. It is a non-symmetric matrix of size $G \times G$, where G is the number of classes. For example, for a three-class problem ($G = 3$):

Assigned classes

	class	A′	B′	C′	n_g
	A	c_{11}	c_{12}	c_{13}	n_a
True classes	B	c_{21}	c_{22}	c_{23}	n_b
	C	c_{31}	c_{32}	c_{33}	n_c
	$n_{g'}$	$n_{a'}$	$n_{b'}$	$n_{c'}$	n

where A,B, and C represent labels for the true classes and A′, B′, and C′ labels for the assigned classes. n_g represents the total number of objects belonging to the gth class and $n_{g'}$ the total number of objects assigned to the gth class. The diagonal elements c_{gg} represent the correctly classified objects, while the off-diagonal elements $c_{gg'}$ represent the objects erroneously classified from class g to class g'. Usually, two confusion matrices are obtained, one in the fitting and one after a validation procedure.

From the confusion matrix entries, the following parameters are defined:

Non-Error Rate (*NER*). The simplest measure of the quality of a classification model; usually expressed as a percentage, it is defined as:

$$NER\% = \frac{\sum_g c_{gg}}{n} \times 100$$

where c_{gg} are the diagonal elements of the confusion matrix and n the total number of objects. The complementary quantity is called *error rate* (*ER*) and is defined as:

$$ER\% = \frac{n - \sum_g c_{gg}}{n} \times 100 = 100 - NER\%$$

To evaluate the efficiency of a classification model the error rate can be compared with the *no-model error rate* (*NOMER*), that represents the error rate without a classification model and is calculated considering all the objects of the smaller classes as erroneously classified in the largest class n_M:

$$NOMER\% = \frac{n - n_M}{n} \times 100$$

Misclassification Risk (*MR*). This is defined as:

$$MR\% = \sum_g \frac{\left(\sum_{g'} L_{gg'} c_{gg'}\right) P_g}{n_g} \times 100$$

where P_g is the *prior class probability*, defined a priori, usually as $P_g = 1 / G$ or $P_g = n_g / n$, where G is the total number of classes and n_g the number of objects belonging to the gth class. $L_{gg'}$ are elements of the *loss matrix* **L**, which is a user-defined penalty matrix for classification error, whose diagonal elements are zero, i.e. no penalty is applied for correct classification.

Sensitivity (*Sn*). A parameter which characterizes the ability of a classifier to correctly catch objects of the gth class, defined as:

$$Sn_g = \frac{c_{gg}}{n_g} \times 100$$

Sensibility (*Sp*). A parameter which characterizes the ability of a classifier to catch only objects of the correct gth class within the calculated class, defined as:

$$Sp_g = \frac{c_{gg}}{c_{gg} + \sum_{g' \neq g} c_{gg'}} \cdot 100 = \frac{c_{gg}}{n_{g'}} \times 100$$

Cluj delta matrix → **Cluj matrices**

Cluj-detour index → **Cluj matrices**

Cluj-detour matrix → **Cluj matrices**

Cluj-distance index → **Cluj matrices**

Cluj-distance matrix → **Cluj matrices**

Cluj matrices (CJ)

These are square unsymmetric matrices $A \times A$ (A being the number of graph vertices) defined following the principle of single endpoint characterization of a path; symmetric Cluj matrices are derived from the unsymmetric Cluj matrices [Diudea, 1996b; Diudea, 1997a; Diudea, 1997b]. Several indices can be calculated on Cluj matrices, either directly by the → *Wiener orthogonal operator* from the unsymmetric matrices or as the half-sum of entries in the corresponding symmetric matrices by the → *Wiener operator*.

A first square unsymmetric $A \times A$ Cluj matrix is derived from the → *distance matrix* **D** and is called the **Cluj-distance matrix $\mathbf{CJD_U}$**. The off-diagonal entry *i-j* of the matrix is the count $N_{i,p_{ij}}$ of all vertices lying closer to the focused vertex i, but out of the shortest path p_{ij} connecting vertices v_i and v_j, i.e. the count of the external paths on the side of v_i including the path *i-j*. The focused vertex v_i is included in the set $N_{i,p_{ij}}$. In the event of more than one shortest path *i-j* being encountered, the maximal cardinality of the sets of vertices is taken. Therefore, the off-diagonal matrix elements are:

$$[\mathbf{CJD_U}]_{ij} = N_{i,p_{ij}}$$

and

$$N_{i,p_{ij}} = \max\left|\{a | a \in \mathcal{V}(\mathcal{G});\ d_{ia} < d_{ja};\ p_{ia} \cap p_{ij} = \{i\};\ |p_{ij}| = \min\}\right|$$

where $\mathcal{V}(\mathcal{G})$ is the set of molecular graph vertices; d_{ij} is the → *topological distance*. No restrictions related to the length of the path p_{ia} are imposed. The diagonal entries of the matrix are always equal to zero by definition.

The main properties of the Cluj matrix are:

- the row sums of $\mathbf{CJD_U}$ are equal to the corresponding row sums of the 1st order → *sparse Wiener matrix* $^1\mathbf{W}$, namely:

$$\mathcal{R}_i(\mathbf{CJD_U}) = \mathcal{R}_i(^1\mathbf{W})$$

where $\mathcal{R}_i$ is the → *row sum operator*.

- the column sums of $\mathbf{CJD_U}$ are equal to the corresponding column sums (and row sums) of the distance matrix **D**, namely:

$$\mathcal{C}_j(\mathbf{CJD_U}) = \mathcal{C}_j(\mathbf{D}) = \mathcal{R}_i(\mathbf{D}) \quad \text{for } i = j$$

where $\mathcal{C}_j$ is the → *column sum operator*.

- from the previous relationships it follows:

$$\sum_{i=1}^{A} \mathcal{R}_i(\mathbf{CJD_U}) = \sum_{i=1}^{A} \mathcal{R}_i(^1\mathbf{W}) = \sum_{i=1}^{A} \mathcal{R}_i(\mathbf{D}) = \sum_{j=1}^{A} \mathcal{C}_j(\mathbf{CJD_U}) = 2W$$

where W is the → *Wiener index*.

A square **symmetric Cluj-distance matrix CJD**, of dimension $A \times A$, is obtained from the unsymmetric Cluj-distance matrix $\mathbf{CJD_U}$ by the relation:

$$\mathbf{CJD} = \mathbf{CJD_U} \bullet \mathbf{CJD_U^T} \quad \text{or} \quad [\mathbf{CJD}]_{ij} = [\mathbf{CJD_U}]_{ij}[\mathbf{CJD_U}]_{ji} = N_{i,p_{ij}} \cdot N_{j,p_{ij}}$$

where the symbol • indicates the → *Hadamard matrix product*; $N_{i,p_{ij}}$ and $N_{j,p_{ij}}$ are the number of vertices lying closer to vertices v_i and v_j, respectively, according to the above definition.

It is worth pointing out that the symmetric Cluj-distance matrix **CJD** is identical to the Wiener matrix **W** for acyclic graphs; obviously, the corresponding → *graph invariants* coincide.

The **Cluj-detour matrix** $\mathbf{CJ\Delta_U}$ is analogously derived from the → *detour matrix* Δ [Diudea *et al.*, 1997e; Diudea *et al.*, 1998]. It is a square unsymmetric matrix $A \times A$ whose off-diagonal entry *i-j* is the count $N_{i,p_{ij}}$ of all vertices lying close to the focused vertex v_i, but off the longest path p_{ij}. In the event of more than one longest path *i-j* being encountered, the maximal cardinality of the sets of vertices is taken. The off-diagonal matrix elements are:

$$[\mathbf{CJ\Delta_U}]_{ij} = N_{i,p_{ij}}$$

where

$$N_{i,p_{ij}} = \max\left|\{a | a \in \mathcal{V}(\mathcal{G});\ d_{ia} < d_{ja};\ p_{ia} \cap p_{ij} = \{i\};\ |p_{ij}| = \max\}\right|$$

where $\mathcal{V}(\mathcal{G})$ is the set of molecular graph vertices. No restrictions related to the length of the path p_{ia} are imposed. The diagonal entries of the matrix are always equal to zero by definition.

A square **symmetric Cluj-detour matrix CJΔ**, of dimension $A \times A$, is obtained from the unsymmetric Cluj-detour matrix by the relation:

$$\mathbf{CJ\Delta} = \mathbf{CJ\Delta_U} \bullet \mathbf{CJ\Delta_U^T} \quad \text{or} \quad [\mathbf{CJ\Delta}]_{ij} = [\mathbf{CJ\Delta_U}]_{ij}[\mathbf{CJ\Delta_U}]_{ji} = N_{i,p_{ij}} \cdot N_{j,p_{ij}}$$

where $N_{i,p_{ij}}$ and $N_{j,p_{ij}}$ are the number of vertices lying closest to vertices v_i and v_j, respectively, according to the above definition.

For acyclic graphs, the matrix **CJΔ** is equal to the symmetric Cluj-distance matrix **CJD** and therefore to the Wiener matrix **W**, while it is different from the → *Szeged matrix* **SZ**.

For cyclic graphs, all these matrices are different.

The 1st order → *sparse matrix* of the symmetric Cluj-distance matrix $^1\mathbf{CJD}$ is calculated by the Hadamard matrix product applied to the **CJD** matrix and the → *adjacency matrix* **A**, as:

$$^1\mathbf{CJD} = \mathbf{CJD} \bullet \mathbf{A}$$

By the same operation, the unsymmetric 1st order sparse Cluj-distance matrix $^1\mathbf{CJD_U}$ is calculated from the unsymmetric $\mathbf{CJD_U}$ matrix. Analogously, from the Cluj-detour matrices the corresponding 1st order sparse matrices are defined as $^1\mathbf{CJ\Delta}$ and $^1\mathbf{CJ\Delta_U}$.

For the above-defined sparse Cluj matrices, the following relationships hold for any graph:

$$^1\mathbf{CJD} = {}^1\mathbf{CJ\Delta} = {}^1\mathbf{SZ} \quad \text{and} \quad {}^1\mathbf{CJD_U} = {}^1\mathbf{CJ\Delta_U} = {}^1\mathbf{SZ_U}$$

where $^1\mathbf{SZ}$ is the sparse → *symmetric Szeged matrix*, and $^1\mathbf{SZ_U}$ the corresponding sparse unsymmetric Szeged matrix.

From Cluj matrices molecular → *Wiener-type indices* are derived.

The **Cluj-distance index** *CJD* is obtained from the symmetric Cluj-distance matrix **CJD** as the sum of the matrix entries corresponding to all the 1st order paths (i.e. bonds) above the main diagonal, i.e. applying the Wiener operator $\mathcal{W}$ to the 1st order sparse symmetric Cluj-distance matrix $^1\mathbf{CJD}$ or from the sparse unsymmetric Cluj-distance matrix $^1\mathbf{CJD_U}$ applying the Wiener orthogonal operator $\mathcal{W}'$:

$$CJD = \mathcal{W}(^1\mathbf{CJD}) = \frac{1}{2}\sum_{i=1}^{A}\sum_{j=1}^{A}[^1\mathbf{CJD}]_{ij} = \mathcal{W}'(^1\mathbf{CJD_U}) = \sum_{i=1}^{A}\sum_{j=i}^{A}[^1\mathbf{CJD_U}]_{ij}[^1\mathbf{CJD_U}]_{ji}$$

The **Cluj-detour index** $CJ\Delta$ is analogously obtained from the 1st order sparse symmetric Cluj-detour matrix $^1\mathbf{CJ\Delta}$ applying the Wiener operator $\mathcal{W}$ or from the unsymmetric Cluj-detour matrix $\mathbf{CJ\Delta_U}$ applying the Wiener orthogonal operator $\mathcal{W}'$:

$$CJ\Delta = \mathcal{W}(^1\mathbf{CJ\Delta}) = \frac{1}{2}\sum_{i=1}^{A}\sum_{j=1}^{A}[^1\mathbf{CJ\Delta}]_{ij} = \mathcal{W}'(^1\mathbf{CJ\Delta_U}) = \sum_{i=1}^{A}\sum_{j=i}^{A}[^1\mathbf{CJ\Delta_U}]_{ij}[^1\mathbf{CJ\Delta_U}]_{ji}$$

For acyclic graph, $CJD = CJ\Delta = SZD = W$, where *W* is the Wiener index and *SZD* the → *Szeged index*. For cyclic graphs, $CJD = SZD \neq CJ\Delta \neq W$.

The **hyper-Cluj-distance index** CJD_P is defined as the following:

$$CJD_P = \mathcal{W}(\mathbf{CJD}) = \frac{1}{2}\sum_{i=1}^{A}\sum_{j=1}^{A}[\mathbf{CJD}]_{ij} = \mathcal{W}'(\mathbf{CJD_U}) = \sum_{i=1}^{A}\sum_{j=i}^{A}[\mathbf{CJD_U}]_{ij}[\mathbf{CJD_U}]_{ji}$$

and the **hyper-Cluj-detour index** $CJ\Delta_P$ as:

$$CJ\Delta_P = \mathcal{W}(\mathbf{CJ\Delta}) = \frac{1}{2}\sum_{i=1}^{A}\sum_{j=1}^{A}[\mathbf{CJ\Delta}]_{ij} = \mathcal{W}'(\mathbf{CJ\Delta_U}) = \sum_{i=1}^{A}\sum_{j=i}^{A}[\mathbf{CJ\Delta_U}]_{ij}[\mathbf{CJ\Delta_U}]_{ji}$$

For acyclic graphs, $CJD_P = CJ\Delta_P = WW = D_P \neq SZD_P$, where *WW* is the → *hyper-Wiener index*, D_P the → *hyper-distance-path index*, and SZD_P the → *hyper-Szeged index*, while for cyclic graphs all these descriptors are different.

Derived from Cluj matrices, the **reciprocal Cluj-distance matrix** $\mathbf{CJD^{-1}}$ and **reciprocal Cluj-detour matrix** $\mathbf{CJ\Delta^{-1}}$ are the matrices whose elements are the reciprocal of the corresponding symmetric Cluj matrix elements [Diudea *et al.*, 1998; Diudea *et al.*, 1998]:

$$[\mathbf{CJD^{-1}}]_{ij} = [\mathbf{CJD}]_{ij}^{-1} \quad \text{and} \quad [\mathbf{CJ\Delta^{-1}}]_{ij} = [\mathbf{CJ\Delta}]_{ij}^{-1}$$

Example : ethylbenzene

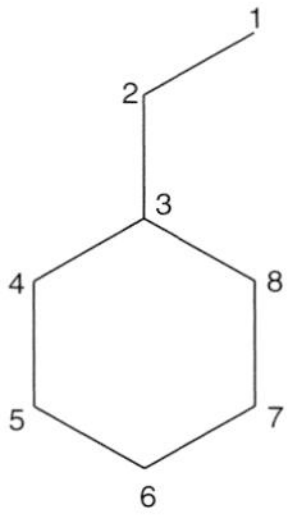

Cluj-distance matrix $\mathbf{CJD}_U$

Atom	1	2	3	4	5	6	7	8
1	0	1	1	1	1	1	1	1
2	7	0	2	2	2	2	2	2
3	6	6	0	5	4	4	4	5
4	4	4	3	0	5	4	4	2
5	3	3	2	3	0	5	2	2
6	3	2	2	2	3	0	3	2
7	3	3	2	2	2	5	0	3
8	4	3	3	2	4	4	5	0

$CJD = 109$

Symmetric Cluj-distance matrix **CJD**

Atom	1	2	3	4	5	6	7	8
1	0	7	6	4	3	3	3	4
2	7	0	12	8	6	4	6	6
3	6	12	0	15	8	8	8	15
4	4	8	15	0	15	8	8	4
5	3	6	8	15	0	15	4	8
6	3	4	8	8	15	0	15	8
7	3	6	8	8	4	15	0	15
8	4	6	15	4	8	8	15	0

$CJD_p = 226$

Cluj-detour matrix $\mathbf{CJ\Delta}_U$

Atom	1	2	3	4	5	6	7	8
1	0	1	1	1	1	1	1	1
2	7	0	2	2	2	2	2	2
3	6	6	0	3	3	4	3	3
4	1	1	1	0	1	1	4	1
5	2	2	1	1	0	1	1	2
6	3	2	2	1	1	0	1	1
7	2	2	1	2	1	1	0	1
8	1	1	1	1	4	1	1	0

$CJ\Delta = 29$

Symmetric Cluj-detour matrix $\mathbf{CJ\Delta}$

Atom	1	2	3	4	5	6	7	8
1	0	7	6	1	2	3	2	1
2	7	0	12	2	4	4	4	2
3	6	12	0	3	3	8	3	3
4	1	2	3	0	1	1	8	1
5	2	4	3	1	0	1	1	8
6	3	4	8	1	1	0	1	1
7	2	4	3	8	1	1	0	1
8	1	2	3	1	8	1	1	0

$CJ\Delta_p = 94$

Box C-3.

→ *Harary indices* and → *hyper-Harary indices* are defined for these matrices applying the Wiener operator.

The **Cluj delta matrix** $\mathbf{CJ}_\Delta$ is obtained in the same way as the symmetric Cluj matrix **CJ** (**CJD** or $\mathbf{CJ\Delta}$), but defining it on paths larger than unity [Ivanciuc *et al.*, 1997a].

📖 [Diudea, 1997d] [Gutman and Diudea, 1998a]

cluster analysis → **chemometrics**

cluster analysis feature selection → **variable reduction**

cluster expansion of chemical graphs

Given a → *molecular graph* $\mathcal{G}$, where vertices are labelled by the chemical element of the corresponding atoms, cluster expansion in the additive form is among the → *group contribution methods* expressing a molecular property Φ as a sum of contributions of all the connected subgraphs of $\mathcal{G}$, i.e.

$$\Phi(\mathcal{G}) = \sum_k \phi_k(\mathcal{G}') \cdot \mathrm{N}_k$$

where ϕ_k is the contribution to the molecular property of the kth fragment and k runs over all the connected subgraphs $\mathcal{G}'$. N_k are called **embedding frequencies** and are the number of times a given substructure (cluster) appears in a chemically isomorphic subgraph within the molecular graph. In practice, embedding frequencies are → *count descriptors* such as atom-type counts, two-atom fragment counts, etc. [Smolenskii, 1964; Gordon and Kennedy, 1973; Essam *et al.*, 1977; Klein, 1986; Schmalz *et al.*, 1992]. The property contributions of the fragments are estimated by multivariate regression analysis.

Usually this method is used on an → *H-depleted molecular graph*, truncated expansions being obtained considering only fragments up to a user-defined size. Some methods for → *log P* estimations are based on cluster expansion. Moreover, a new method for the calculation of embedding frequencies for acyclic trees based on → *spectral moments of iterated line graph sequence* was proposed recently [Estrada, 1999].

📖 [Schmalz *et al.*, 1987] [Poshusta and McHughes, 1989] [McHughes and Poshusta, 1990] [Baskin *et al.*, 1995] [Grassy *et al.*, 1995] [Kvasnicka and Pospichal, 1995] [Klein *et al.*, 1999]

cluster significance analysis → **variable selection**

cluster subgraph → **molecular graph**

CODESSA method

The CODESSA method, implemented in the homonymous software, is a QSAR approach based on the calculation of theoretical molecular descriptors and is a procedure for finding the best multivariate linear models based on → *variable selection*.

Among the several CODESSA descriptors [Katritzky and Gordeeva, 1993; Katritzky *et al.*, 1995a; Katritzky *et al.*, 1996a] are → *molecular weight*, → *molecular volume*, → *count descriptors*, → *topological indices*, → *charge descriptors*, → *shadow indices*, → *charged partial surface area descriptors*, → *quantum-chemical descriptors* and → *electric polarization descriptors*.

A stepwise selection procedure is performed to search for QSPR/QSAR models after the preliminary exclusion of → *constant and near-constant variables*. The → *pair correlation cutoff selection* of variables is then performed to avoid highly correlated descriptor variables within the model.

Several molecular properties have been modeled by the CODESSA method, such as chromatographic indices [Katritzky *et al.*, 1994; Pompe and Novic, 1999], boiling [Katritzky *et al.*, 1996c; Katritzky *et al.*, 1998d; Ivanciuc *et al.*, 1998b] and melting points [Katritzky *et al.*, 1997a], critical temperatures [Katritzky *et al.*, 1998e], gas solubilities [Katritzky *et al.*, 1996b; Huibers and Katritzky, 1998; Katritzky *et al.*, 1998f], critical micelle concentrations [Huibers *et al.*, 1996; Huibers *et al.*, 1997], → *solvent polarity scales* [Katritzky *et al.*, 1997b], and mutagenic activities [Maran *et al.*, 1999].

📖 [Murugan *et al.*, 1994] [Katritzky *et al.*, 1998c] [Katritzky *et al.*, 1998a] [Karelson and Perkson, 1999] [Ivanciuc *et al.*, 1999b]

colour classes → **chromatic decomposition**

column sum operator → **algebraic operators**

column sum vector → **algebraic operators** (⊙ column sum operator)

combined descriptors

Fixed combinations of selected descriptors accounting for molecular properties of interest. The simplest combined descriptors are the differences and average values of → *basis descriptors* such as → *connectivity indices* or → *path numbers*, and the ratios of different descriptors defined with the aim of normalization to obtain, for example, size-independent indices. Moreover, optimal linear combinations of highly correlated descriptors are combined descriptors calculated so as to reduce the number of independent variables (e.g. → *principal properties*).

Simple sums of different molecular descriptors were proposed as → *superindices* in order to obtain highly discriminant indices; particular superindices were suggested to account for → *molecular complexity*.

Examples of combined descriptors are reported below.

Differential connectivity indices are defined as differences between connectivity indices ${}^{m}\chi$ and → *valence connectivity indices* ${}^{m}\chi^{v}$ [Hall and Kier, 1986; Kier and Hall, 1991]:

$${}^{m}\Delta\chi_{q} = {}^{m}\chi_{q} - {}^{m}\chi_{q}^{v}$$

where the superscript m denotes the order of connectivity indices and the subscript q the type of → *molecular subgraph*. These are descriptors proposed to encode electronic information in terms of π and lone pair electrons on that part of the molecule defined by m and q; moreover, it was found that such descriptors are related to differences in inductive and mesomeric effects [Gálvez *et al.*, 1994a]. To account for non-dispersive force effects, the relative valence connectivity indices to non-polar compounds were defined as:

$$\Delta\chi_{np} = (\chi^{v})_{np} - \chi^{v}$$

where the non-polar connectivity index $(\chi^{v})_{np}$ is calculated substituting oxygen and nitrogen atoms in the considered molecule by carbon atoms but keeping the number of bonds to all non-hydrogen atoms constant [Bahnick and Doucette, 1988; Schramke *et al.*, 1999]. Exceptions to bond constancy were made by replacing the carbonyl group with C –C instead of C = C unless the oxygen atom was directly bonded to an unsaturated ring system (uracils), or the nitrile group with C = C.

A **topological Hammett function** σ_t was also proposed and is defined by the most significant differences between the connectivity indices as:

$$\sigma_t = b_0 + b_1 \cdot \left({}^{4}\chi_{p} - {}^{4}\chi_{p}^{v}\right) + b_2 \cdot \left({}^{4}\chi_{pc} - {}^{4}\chi_{pc}^{v}\right)$$

where ${}^{4}\chi_{p}$, ${}^{4}\chi_{p}^{v}$, ${}^{4}\chi_{pc}$ and ${}^{4}\chi_{pc}^{v}$ are the fourth-order atom and valence connectivity indices for path (p) and path-cluster (pc) graph decompositions; b_j are estimated regression coefficients.

The **L index** was proposed as the molecular descriptor defined as the simple linear combination of molecular path counts of order one ${}^1P$ (the number of bonds), order two ${}^2P$ (the → *connection number* N_2), and order three ${}^3P$:

$$L = 2 \cdot {}^1P + {}^2P - {}^3P - 2$$

It was found to correlate the sum of ^{13}C atomic chemical shifts in alkanes [Miyashita *et al.*, 1989].

Moreover, path count differences ${}^1P - {}^2P$ and ${}^2P - {}^3P$ [Randic and Trinajstic, 1988] and connectivity differences ${}^1\chi - {}^2\chi$ and ${}^2\chi - {}^3\chi$ are usually encountered in QSAR modeling; the following path count combination $P_0 + P_1 + P_2 - P_3$ was also found as the critical parameter in the correlation of carbon-13 chemical shift sums in alkanes [Miyashita *et al.*, 1989].

Other examples of combined descriptors based on connectivity indices are the following:

$${}^1C = \frac{{}^1\chi}{{}^1\chi^v + 1} \qquad {}^4C_p = \frac{{}^4\chi_p}{{}^4\chi_p^v + 1} \qquad \chi^{23} = \frac{{}^2\chi + {}^3\chi}{2}$$

where the first two were proposed by Galvez [Gálvez *et al.*, 1996b], and the last one was found to correlate well with the van der Waals area [Randic, 1991g].

Semiempirical molecular connectivity terms X are special combinations of connectivity indices that make use of empirical parameters, dielectric constants, molar masses and other *ad hoc* related parameters accounting for non-covalent interactions [Pogliani, 1997a; Pogliani, 1999a; Pogliani, 1999b; Pogliani, 1999c]; an example is:

$$X = \frac{{}^1\chi}{{}^2\chi + b \cdot {}^3\chi}$$

where b is an optimization parameter. These connectivity terms are derived by a trial-and-error procedure based on connectivity indices of lower order.

Other examples of ratios between two different descriptors are:

$$\frac{W}{Z} \qquad \frac{\mathrm{CID}}{{}^1\chi} \qquad \frac{{}^4\chi_{pc}}{MW} \qquad \frac{N_X}{MW}$$

where W is the → *Wiener index*, Z the → *Hosoya Z index*, CID the → *connectivity ID number*, N_x the number of X atom type, and MW the molecular weight [Boethling and Sabljic, 1989].

Examples of combined descriptors using products are the contributions $q_a \cdot SA_a$, largely used in → *CPSA descriptors*, where q and SA are partial charges and atomic surface areas [Bakken and Jurs, 1999a].

📖 [Stanton and Jurs, 1992] [Randic, 1993a]

CoMFA descriptors → **comparative molecular field analysis**

CoMFA fields → **molecular interaction fields**

CoMFA lattice → **comparative molecular field analysis**

CoMFA-like approaches : *grid-based QSAR techniques*

CoMFA model → **comparative molecular field analysis**

CoMMA descriptors → **comparative molecular moment analysis**

common overlap length → **molecular shape analysis** (⊙ common overlap steric volume)

common overlap surface → **molecular shape analysis** (⊙ common overlap steric volume)

common overlap steric volume → **molecular shape analysis**

compactness → **Wiener index**

Comparative Molecular Field Analysis (CoMFA)

This, the most popular QSAR approach among the → *grid-based QSAR techniques*, compares the molecular potential energy fields of a set of molecules and searches for differences and similarities that can be correlated with differences and similarities in the considered property values [Cramer III *et al.*, 1988b; Marshall and Cramer III, 1988]. Whereas CoMFA and related methods could be used to model a variety of biological and physico-chemical properties, their most common application has, by far, been focused on ligand-target binding properties [Cramer III *et al.*, 1993; Folkers *et al.*, 1993a; Folkers *et al.*, 1993b; Kim, 1995a; Martin, 1998; Norinder, 1998; Oprea and Waller, 1997].

The first step of the CoMFA approach consists in the selection of a group of compounds having a common → *pharmacophore*, in the generation of three-dimensional structures of reasonable conformation and in their alignment. Once the molecules are aligned, a rigid → *grid* of regularly spaced points (**CoMFA lattice**) representing an approximation of binding site cavity space is established around each compound; the grid point distance is arbitrarily chosen (2 Å by default), bearing in mind that even small desirable distances lead to too great a number of grid points; the walls of the grid usually extend at least 4 Å beyond the union volume of the superimposed molecules. The rigidity of receptor walls derived from the use of a rigid grid is a basic assumption and approximation in the CoMFA method.

For each molecule embedded in the grid, → *molecular interaction fields* are calculated by measuring, at every grid point, the interaction energy between hypothetical → *probe* atoms and ligand atoms. Unreasonably large positive energy values, i.e. grid points inside the molecules, are set constant at a chosen cut-off value. They mainly derive from the large values of van der Waals repulsions caused by even a small overlap of ligand atoms and probe atoms.

In the original CoMFA method, only two fields of noncovalent ligand-receptor interactions were calculated: the steric field is a → *Lennard-Jones 6–12 potential function*, and the electrostatic field is a → *Coulomb potential energy function*. Usually, the two fields are kept separate in order to facilitate the interpretation of the final results. As steric and electrostatic fields account only for enthalpic contributions to free binding energy, other fields that account for solvation and entropic terms should be added. For example, hydrophobic interactions which are entropic properties have been imported into CoMFA studies using → *HINT* and → *molecular liphophilicity potential*. Since the Lennard-Jones potential is characterized by very steep increases in energy at short distances from the molecular surface, it was recently proposed to use the van der Waals volume intersections between probe and ligand molecule for steric field calculation [Sulea *et al.*, 1997]; this molecular interaction field was called the intersection volume field (INVOL).

→ *Interaction energy values* at each grid point are the **CoMFA descriptors** and are collected into a QSAR matrix where rows represent the molecules and columns the grid points for each considered field. Grid points without variance, i.e. within the volume shared by all molecules, or with small variance, i.e. far away from the van der Waals surface of molecules, are discarded. Moreover, other parameters such as → *logP* or → *quantum-chemical descriptors* can be added as variables to the QSAR matrix after properly weighted scaling. The combination of global parameters and CoMFA fields leads to a **mixed CoMFA approach** [Kubinyi, 1993b].

Partial Least Squares regression (PLS) is usually performed on a → *data matrix* to search for a correlation between the thousands of CoMFA descriptors and biological response. However, usually after → *variable selection*, the PLS model is transformed into and presented as a multiple regression equation to allow comparison with classical QSAR models.

The **CoMFA model** is defined as:

$$\hat{y}_i = b_0 + \sum_{j=1}^{F}\sum_{x=1}^{N_x}\sum_{y=1}^{N_y}\sum_{z=1}^{N_z} b_{j,xyz} \cdot E_{ij,xyz} = \sum_{j=1}^{F}\sum_{k=1}^{N_G} b_{jk} \cdot E_{ijk}$$

where F is the number of fields used in the analysis, i.e. the number of molecular interaction fields; N_x, N_y, and N_z are the number of grid points along the X-axis, Y-axis, and Z-axis, respectively; $N_G = N_x N_y N_z$ is the total number of grid points. $E_{ij,xyz}$ is the potential interaction energy of the *i*th compound for the *j*th field at the grid coordinate (x, y, z). The k index denotes a vectorial representation of the grid points.

The **mixed CoMFA model** is defined as:

$$\hat{y}_i = b_0 + \sum_{j=1}^{F}\sum_{x=1}^{N_x}\sum_{y=1}^{N_y}\sum_{z=1}^{N_z} b_{j,xyz} \cdot E_{ij,xyz} + \sum_{j=1}^{J} b_j \cdot \Phi_{ij} = \sum_{j=1}^{F}\sum_{k=1}^{N_G} b_{jk} \cdot E_{ijk} + \sum_{j=1}^{J} b_j \cdot \Phi_{ij}$$

where Φ_{ij} is any global molecular property and J the total number of considered molecular properties.

Developments of the CoMFA approach have been also proposed based on a selection of molecule regions of interest for binding interactions [Cho and Tropsha, 1995; Tropsha and Cho, 1998; Cruciani *et al.*, 1998].

A critical review of CoMFA applications is given in [Kim *et al.*, 1998b] and a complete list of references 1993–1997 in [Kim, 1998a].

📖 [Greco *et al.*, 1991] [Kellogg *et al.*, 1991] [Kim and Martin, 1991a] [Kim and Martin, 1991b] [Kim and Martin, 1991c] [Kim, 1992a] [Kim, 1992c] [Broughton *et al.*, 1992] [Nicklaus *et al.*, 1992] [Waller and McKinney, 1992] [Waller *et al.*, 1993] [Waller and Marshall, 1993] [Debnath *et al.*, 1993] [El Tayar and Testa, 1993] [Greco *et al.*, 1993] [Kim *et al.*, 1993] [Kim, 1993c] [Kim, 1993d] [Kim, 1993e] [Kim, 1993f] [Klebe and Abraham, 1993] [Martin *et al.*, 1993] [DePriest *et al.*, 1993] [Oprea *et al.*, 1993] [Thibaut, 1993] [van de Waterbeemd *et al.*, 1993a] [van de Waterbeemd *et al.*, 1993b] [Langlois *et al.*, 1993a] [Horwitz *et al.*, 1993] [Poso *et al.*, 1993] [Agarwal *et al.*, 1993] [Poso *et al.*, 1994] [Horwitz *et al.*, 1994] [Debnath *et al.*, 1994a] [Caliendo *et al.*, 1994] [Carrieri *et al.*, 1994] [Greco *et al.*, 1994a] [Greco *et al.*, 1994b] [Langer, 1994] [Greco *et al.*, 1994] [Gantchev *et al.*, 1994] [Vansteen *et al.*, 1994] [Thibaut *et al.*, 1994] [Jiang *et al.*, 1994] [Waller *et al.*, 1994] [Caliendo *et al.*, 1995] [Kim and Kim, 1995] [Kim, 1995c] [Kroemer *et al.*, 1995] [Thull *et al.*, 1995] [Hocart *et al.*, 1995] [Fabian *et al.*, 1995] [Briens *et al.*, 1995] [Waller *et al.*, 1995] [Kireev *et al.*, 1995] [Belvisi *et al.*, 1996] [Cho *et al.*, 1996] [Cramer III *et al.*, 1996] [Elass *et al.*, 1996b] [Kellogg *et al.*, 1996] [Kroemer *et al.*, 1996] [Masuda *et al.*, 1996] [Navajas *et al.*, 1996] [Norinder, 1996] [Patterson *et al.*, 1996] [Waller *et al.*, 1996a] [Waller and Kellogg, 1996b] [Gamper *et al.*, 1996] [Hannongbua *et al.*, 1996b] [Tong *et al.*, 1996] [Cruciani *et al.*, 1997] [Kellogg, 1997] [Tominaga and Fujiwara, 1997b] [Debnath, 1998] [Demeter *et al.*, 1998] [Durst, 1998] [Kimura *et al.*, 1998] [Langer and Hoffmann, 1998b] [Swaan *et al.*, 1998] [Timofei and Fabian, 1998] [Tong *et al.*, 1998] [Coats, 1998] [Chen *et al.*, 1999] [So and Karplus, 1999] [Sulea and Purisima, 1999].

Comparative Molecular Moment Analysis (CoMMA)

The Comparative Molecular Moment Analysis method based on the 3D → *geometrical representation* of the molecule calculates different molecular moments with respect to the → *centre of mass*, centre of charge and → *centre-of-dipole* of the molecule [Silverman and Platt, 1996; Platt and Silverman, 1996].

CoMMA descriptors are the following 14 molecular descriptors:

$$\left\langle MW; \mathrm{I}_x, \mathrm{I}_y, \mathrm{I}_z; \mu; \mathrm{Q}; \mu_x, \mu_y, \mu_z; d_x, d_y, d_z; \mathrm{Q}_{xx}, \mathrm{Q}_{yy} \right\rangle$$

The first descriptor MW is the → *molecular weight*, i.e. the zero-order molecular moment with respect to the centre of mass. The three → *principal moments of inertia* I are the 2nd order moments with respect to the centre of mass. μ and Q are the magnitudes of → *dipole moment* and → *quadrupole moment* which are the 1st and the 2nd order moments with respect to the centre of charge, respectively. The dipole moment components μ_x, μ_y, and μ_z and the components of displacement d between the centre of mass and the centre of dipole are calculated with respect to the → *principal inertia axes*. Finally, the quadrupole components Q_{xx} and Q_{yy} are calculated with respect to a translated initial reference frame whose origin coincides with the centre-of-dipole.

By calculating molecular descriptors based on 3D geometry without a common orientation frame, the Comparative Molecular Moment Analysis overcomes the problems due to the molecular alignment.

In order to extend the CoMMA approach to account for the lipophilicity of the molecule, the → *Leo-Hansch hydrophobic fragmental constants* [Abraham and Leo, 1987] have been proposed as a set of atomic lipophilic weights for the calculation of lipophilic molecular multipole moments (hydropoles) [Burden and Winkler, 1999a].

Table C-2. Molecular moments of order zero, one and two.

Order of the moment	*Unit*	*Mass*	*Charge*	*Lipophilicity*
0	A	MW	$\sum_i q_i$	$\sum_i f_i$
1	0	0	μ	Lipophilic dipole moment
2	Moments of geometry	Moments of inertia	Electrostatic quadrupole moments	Lipophilic quadrupole moments

A is the number of atoms, MW the molecular weight, q the atomic charges, μ the total dipole moment, and f the hydrophobic atomic constants.

📖 [Silverman *et al.*, 1998a] [Silverman *et al.*, 1998b] [Silverman *et al.*, 1999]

Comparative Molecular Similarity Indices Analysis (CoMSIA)

Comparative Molecular Similarity Indices Analysis, among the → *grid-based QSAR techniques*, implements the steric, electrostatic, hydrophobic, and hydrogen-bonding → *similarity indices* utilized in the molecular alignment program SEAL [Klebe *et al.*, 1994a; Klebe, 1998; Klebe and Abraham, 1999].

Using the → *similarity score* A_F based on the weighted combination of steric, electrostatic and hydrophobic properties, molecule alignment is performed starting from a random orientation of two molecules relative to each other; the best alignment is achieved with the maximum similarity score.

Moreover, → *molecular interaction fields* are calculated for each molecule in terms of similarity indices instead of the usual interaction potential functions, such as Lennard-Jones and Coulomb potential functions. Similarity fields are calculated representing the similarity between molecules and different probe atoms. In particular, the similarity values at the intersections of the regularly spaced grid (1.1 and 2.0 Å) relative to the jth physico-chemical property between the ith compound and a probe atom is calculated as:

$$(A_F)_{ik,j} = \sum_t w_{\text{probe},j} \cdot w_{tj} \cdot e^{-\alpha \cdot r_{tk}^2}$$

where the sum runs over all atoms of the molecule; $w_{\text{probe},j}$ and w_{tj} are, respectively, the actual values of the jth property of the probe and the tth atom of the target molecule; α is an attenuation factor and r_{tk} is the geometric distance between the probe atom at the kth grid point and the tth atom of the molecule. Large values of α give rise to a strong distance-dependent attenuation of the similarity measures, i.e. only local similarities are considered; otherwise, for small α values, gobal molecular features are of greater importance.

The studied properties are electrostatic, steric, hydrophobic, hydrogen-acceptor and hydrogen-donor abilities; for electrostatic properties the probe assumes charge +1, for steric properties radius 1 Å, for hydrophobicity and hydrogen-bonding abilities a value of +1.

These indices replace the distance functions used in the standard Lennard-Jones and Coulomb potential functions which generate unrealistically extreme values as the surface of the considered molecule is approached.

A molecular → *similarity matrix* can be obtained both from the similarity scores between pairs of molecule and any distance function applied to similarity fields [Kubinyi *et al.*, 1998d].

Compass descriptors → **Compass method**

Compass method

A QSAR method based on the search for the best model predicting compound activity and likely bioactive conformations and alignments from a set of physical properties measured only near the surface of the molecules [Jain *et al.*, 1994a; Jain *et al.*, 1994b; Jain *et al.*, 1995]. The basic assumption is that the enthalpy of ligand-target binding depends on the interactions occurring at the ligand-target interface. Therefore, the main features characterizing the Compass method are the definition of descriptors related to surface properties, an automatic selection of the optimal molecular conformation and alignment, and the use of → *artificial neural networks* with back-propagation to take into account also nonlinear structure-activity relationships.

The method is based on three fundamental phases. The first phase consists in the generation of low-energy conformations for each molecule and in the choice of one conformer as the one most likely to be bioactive; all selected conformers are aligned, along with the identified pharmacophore or a substructure common to all molecules in the data set. A molecule *pose* is a conformation of the molecule in a particular alignment.

The second phase proceeds iteratively through three steps: a) For each molecule pose **Compass descriptors** are calculated as → *geometric distance* representing the surface shape or polar functionalities of the pose in the proximity of a given point in the space; Compass steric descriptors measure distances from sampling points to the van der Waals surface of a molecule, while donor/acceptor ability descriptors measure the distance from a sampling point to the nearest H-bond donor or acceptor group. A few sampling points are scattered on a surface 2.0 Å outside the average van der Waals envelope of the → *hypermolecule* obtained by alignment in an invariant and common reference frame. b) A neural network model is built relating the structural features (Compass descriptors) of molecule poses to biological activity. The network is trained by the back-propagation algorithm and is constituted by three layers with Gaussian input units and standard sigmoid units in the hidden layer. c) In the third step the model is used to realign the molecules in order to find better poses which are then used to give an improved model until convergence is reached.

The third step predicts the activity and bioactive pose of a new molecule.

With respect to → *CoMFA*, the Compass method effectively reduces the number of descriptors, performing a physicochemically based → *variable reduction* and overcomes the problem of guessing the best conformation and alignmet of the molecules.

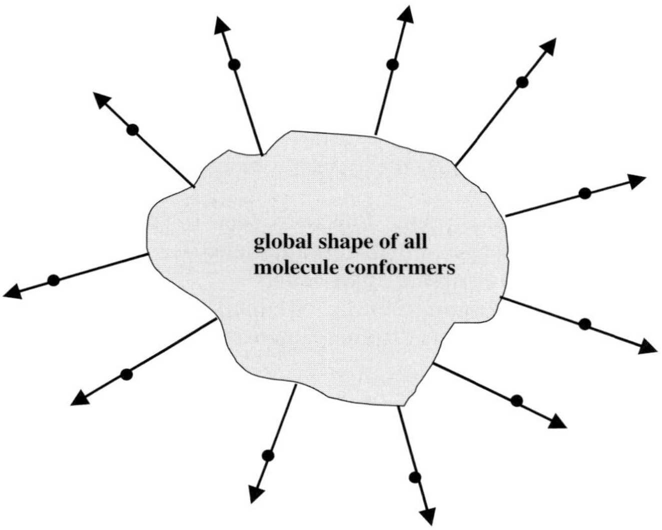

Figure C-2. Compass descriptors arising from the molecular surface.

complementary information content → **indices of neighborhood symmetry**

complete centric index → **centric indices**

complete graph → **graph**

composite nuclear potential → **quantum-chemical descriptors**

composition indices : *atomic composition indices*

computational chemistry

In a broad sense, the term computational chemistry includes several fields such as quantum chemistry, statistical molecular mechanics, molecular modelling, molecular graphics and visualization, evaluation of experimental data in X-ray crystallography,

NMR spectroscopy and related techniques; moreover, in this sense, analysis, exploration and modelling performed by → *chemometrics* on experimental data, information retrieval from chemical data banks, and expert chemical systems are also included in computational chemistry.

Theoretical chemistry and, especially, quantum chemistry constitute the basic core of computational chemistry, and their success covers the field of molecular geometries and energies, reactivity, spectroscopic properties, behaviour of electrons in atoms and molecules, and various other fundamental chemical topics [Lipkowitz and Boyd, 1990]. Therefore, the term computational chemistry is also used in a more restricted sense to denote the mathematical approaches (and their software implementations) to the calculation of molecular properties from theoretical chemistry. → *Quantum-chemical descriptors* are derived from computational chemistry in this restricted sense.

Theoretical chemistry ranges from the very simple approach, like the Hückel (HMO) and Parr-Pariser-Pople (PPP) methods, to exact calculations such as *ab initio* methods. Several methods have been proposed between these two extremes, most falling within an approach called the Self-Consistent-Field Molecular-Orbital (SCF-MO) method [Streitweiser, 1961; Dewar, 1969a]. The approach uses the concept of the molecular orbital, i.e. the wavefunction of one electron moving in the potential of the nuclei and the average effective potential of the other electrons; moreover, in most cases, molecular orbitals are defined in terms of Linear Combinations of Atomic Orbitals (LCAO-MO). Depending on the level of approximation, several different methods such as CNDO/2, MINDO/3, MNDO, AM1, PM3, etc. can be used to compute the wave function ψ_i [Leach, 1996]. Moreover, Density Functional Theory (DFT) constitutes an important development in theoretical chemistry using electron density as a basic variable instead of the wave function [Parr and Yang, 1989].

Other important and effective approaches to computational chemistry are those called Empirical Force-Field methods (EFF methods), based on a mechanistic view of the molecule in terms of force constants of bonds, bending, torsion and other special interaction terms. The set of force constants constitutes a field of empirical parameters used for the calculation of molecular geometries and energies.

Calculations based on computational chemistry methods can be performed by means of software packages, such as MOPAC, GAUSSIAN98, GAMESS, METECC, AMPAC, SPARTAN, etc.

📖 [Lewis, 1916] [Lewis, 1923] [Mulliken, 1928a] [Mulliken, 1928b] [Hückel, 1930] [Hückel, 1932] [Pauling, 1932] [Pauling and Wilson, 1935] [Pauling, 1939] [Coulson, 1939] [Eyring *et al.*, 1944] [Mulliken, 1955a] [Coulson, 1960] [Streitweiser, 1961] [Dewar, 1969a] [Murrell and Harget, 1972] [Lowe, 1978] [Löw and Saller, 1988] [Stewart, 1990]

Computer-Aided Drug Design → **drug design**

Computer-Aided Molecular Design → **drug design**

Computer-Aided Molecular Modelling → **drug design**

conformational invariance → **molecular descriptors** (⊙ invariance properties of molecular descriptors)

congenericity

Congenericity is a fuzzy concept related to the structures of the molecules of the studied dataset. With respect to some defined molecular structural characteristics, chemical analogues can be considered congeneric compounds, their structural differences being the interesting part of the study. Mono-substituted benzenes, polychlorobiphenyls, triazines, and poly-aromatic hydrocarbons are all examples of congeneric datasets.

The **congenericity principle** is the assumption that "similar compounds give similar responses" and is the basic requirement of several → *structure/response correlations*.

congenericity principle → **congenericity**

connected graph → **graph**

connectedness index → **Wiener index**

connection → **edge adjacency matrix**

connection number → **edge adjacency matrix**

connection orbital information content → **orbital information indices**

connectivity bond layer matrix → **layer matrices**

connectivity ID number : *Randic connectivity ID number* → **ID numbers**

connectivity index : *Randic connectivity index* → **connectivity indices**

connectivity indices (: *Molecular Connectivity Indices, MCIs*)

Connectivity indices are among the most popular → *topological indices* and are calculated from the → *vertex degree* δ of the atoms in the → *H-depleted molecular graph*.

The **Randic connectivity index** was the first connectivity index proposed [Randic, 1975b]; it is also called the **connectivity index** or **branching index**, and is defined as:

$$\chi_R \equiv {}^1\chi = \sum\nolimits_b (\delta_i \cdot \delta_j)_b^{-1/2}$$

where b runs over the B bonds i-j of the molecule; δ_i and δ_j are the vertex degrees of the atoms incident with the considered bond. It is closely related to the → *2nd Zagreb index M_2* and was proposed as measure of → *molecular branching*.

Each single term of the sum is called **edge connectivity** $(\delta_i \cdot \delta_j)^{-1/2}$ and can be used to characterize edges as a primitive → *bond order* accounting for bond accessibility, i.e. the accessibility of a bond to encounter another bond in intermolecular interactions, as the reciprocal of the vertex degree δ is the fraction of the total number of non-hydrogen sigma electrons contributed to each bond formed with a particular atom [Kier and Hall, 2000].

This interpretation places emphasis on the bimolecular encounter possibility among molecules, reflecting the collective influence of the bond accessibilities of each molecule with other molecules in its immediate environment. Therefore, the Randic connectivity index ${}^1\chi$ can be interpreted as the contribution of one molecule to the bimolecular interaction arising from the encounters of bonds of two identical molecules:

$${}^1\chi = \sqrt{\sum_{b=1}^{B}\sum_{b'=1}^{B} (\delta_i \cdot \delta_j)_b^{-1/2} \cdot (\delta_k \cdot \delta_l)_{b'}^{-1/2}}$$

where the two sums run over all the bonds of the molecules and δ are the vertex degrees.

→ *Information connectivity indices* based on the partition of the edges in the graph according to the equivalence and the magnitude of their edge connectivity values were derived.

The **mean Randic connectivity index** (or **mean Randic branching index**) is defined as:

$$\bar{\chi}_R = \frac{\chi_R}{B}$$

where B is the number of edges in the molecular graph.

A variant of the Randic connectivity index was also proposed as:

$$\chi'_R = A \cdot \chi_R$$

where A is the number of atoms [Mihalic *et al.*, 1992a].

Kier and Hall defined [Kier and Hall, 1986; Kier and Hall, 1977b] a general scheme based on the Randic index to calculate also zero-order and higher-order descriptors, thus obtaining **connectivity indices of *m* th order**, usually known as **Kier-Hall connectivity indices**. They are calculated by the following:

$$^0\chi = \sum_a \delta_a^{-1/2} \qquad ^1\chi = \sum_b (\delta_i \cdot \delta_j)_b^{-1/2} \qquad ^2\chi = \sum_{k=1}^{^2P} (\delta_i \cdot \delta_l \cdot \delta_j)_k^{-1/2}$$

$$^m\chi_q = \sum_{k=1}^{K} \left(\prod_{a=1}^{n} \delta_a \right)_k^{-1/2}$$

where k runs over all of the mth order subgraphs constituted by n atoms ($n = m + 1$ for acyclic subgraphs); K is the total number of mth order subgraphs present in the molecular graph and in the case of the path subgraphs equals the mth order path count ^mP. The product is over the simple vertex degrees δ of all the vertices involved in each subgraph. The subscript "q" for the connectivity indices refers to the type of → *molecular subgraph* and is ch for chain or ring, pc for path-cluster, c for cluster, and p for path (that can also be omitted).

Example : 2-methylpentane

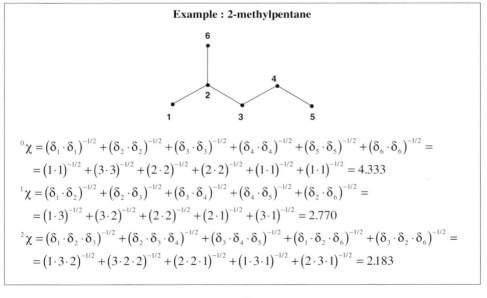

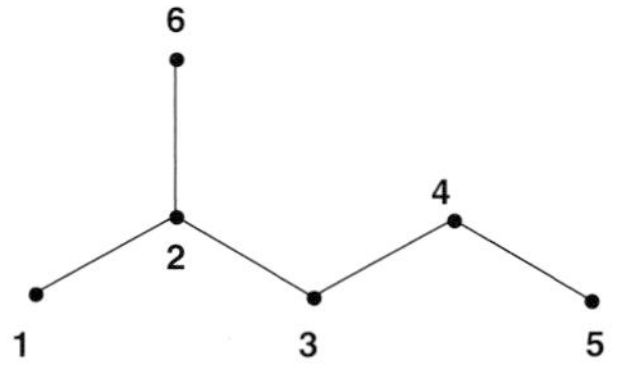

$$^0\chi = (\delta_1 \cdot \delta_1)^{-1/2} + (\delta_2 \cdot \delta_2)^{-1/2} + (\delta_3 \cdot \delta_3)^{-1/2} + (\delta_4 \cdot \delta_4)^{-1/2} + (\delta_5 \cdot \delta_5)^{-1/2} + (\delta_6 \cdot \delta_6)^{-1/2} =$$

$$= (1 \cdot 1)^{-1/2} + (3 \cdot 3)^{-1/2} + (2 \cdot 2)^{-1/2} + (2 \cdot 2)^{-1/2} + (1 \cdot 1)^{-1/2} + (1 \cdot 1)^{-1/2} = 4.333$$

$$^1\chi = (\delta_1 \cdot \delta_2)^{-1/2} + (\delta_2 \cdot \delta_3)^{-1/2} + (\delta_3 \cdot \delta_4)^{-1/2} + (\delta_4 \cdot \delta_5)^{-1/2} + (\delta_2 \cdot \delta_6)^{-1/2} =$$

$$= (1 \cdot 3)^{-1/2} + (3 \cdot 2)^{-1/2} + (2 \cdot 2)^{-1/2} + (2 \cdot 1)^{-1/2} + (3 \cdot 1)^{-1/2} = 2.770$$

$$^2\chi = (\delta_1 \cdot \delta_2 \cdot \delta_3)^{-1/2} + (\delta_2 \cdot \delta_3 \cdot \delta_4)^{-1/2} + (\delta_3 \cdot \delta_4 \cdot \delta_5)^{-1/2} + (\delta_1 \cdot \delta_2 \cdot \delta_6)^{-1/2} + (\delta_3 \cdot \delta_2 \cdot \delta_6)^{-1/2} =$$

$$= (1 \cdot 3 \cdot 2)^{-1/2} + (3 \cdot 2 \cdot 2)^{-1/2} + (2 \cdot 2 \cdot 1)^{-1/2} + (1 \cdot 3 \cdot 1)^{-1/2} + (2 \cdot 3 \cdot 1)^{-1/2} = 2.183$$

Box C-4.

By replacing the vertex degree δ by the → *valence vertex degree* δ^v in the formulas reported above, similar **valence connectivity indices** were proposed [Kier and Hall, 1981; Kier and Hall, 1983b], denoted by ${}^m\chi_q^v$, able to account for the presence of heteroatoms in the molecule as well as double and triple bonds. Analogously, **bond connectivity indices**, denoted by ${}^m\chi_q^b$, were also defined by using the → *bond vertex degree* δ^b instead of the simple vertex degree δ to specifically account for multiplicity in the molecule. Moreover, the same connectivity indices have been also calculated using the → *Z-delta number* δ^Z and therefore denoted by ${}^m\chi^Z$ [Pogliani, 1999b].

Some molecular descriptors and → *local vertex invariants* proposed as a generalization or modification of the original connectivity indices are reported below.

- **generalized connectivity indices**

Proposed as above, these are defined using different weights for molecular subgraphs:

$${}^m\chi_q = \sum_{k=1}^{K} \left(\prod_{a=1}^{n} \delta_a \right)_k^p$$

where the typical exponent –1/2 is generalized as p. If $p = 1$ and $m = 1$, the → *2nd Zagreb index* is obtained; the cases $p = -1$ and $p = 1/2$ were considered by Altenburg [Altenburg, 1980], and the cases $p = -1/3$ and $p = -1/4$ have also received attention [Randic *et al.*, 1988a; Estrada, 1995c; Amic *et al.*, 1998a].

- **total structure connectivity index**

It is a connectivity index contemporarily accounting for all the atoms in the graph as [Needham *et al.*, 1988]:

$$\chi_T = \left(\prod_{i=1}^{A} \delta_i \right)^{-1/2}$$

Note that the total structure connectivity index is the square root of the → *simple topological index* proposed by Narumi for measuring molecular branching.

- **local connectivity indices** (${}^m\bar{\chi}_i$) (: *atomic connectivity indices*)

Computed for individual atoms in a graph [Hall and Kier, 1977a]; they were developed by equally partitioning each term ${}^m w_{ij} = (\delta_i \cdot \delta_k \cdot \ldots \cdot \delta_j)^{-1/2}$ of the connectivity indices defined above among all of the atoms involved in the path i-j of mth order, i.e. as:

$${}^m\bar{\chi}_i = \frac{1}{m+1} \cdot \sum_{j=1}^{{}^mP_i} {}^m w_{ij}$$

where ${}^m w_{ij}$ is the → *path connectivity*, the index j represents the terminal vertex v_j and runs over all the mth order paths ${}^mP_i$ starting from vertex v_i.

For example, the 1st order local connectivity index for the ith vertex is defined as:

$${}^1\bar{\chi}_i = \frac{1}{2} \cdot \sum_{j=1}^{\delta_i} {}^1 w_{ij} = \frac{1}{2} \cdot \sum_{j=1}^{\delta_i} (\delta_i \cdot \delta_j)^{-1/2}$$

where δ_i is the vertex degree of the ith vertex, i.e. the number of bonds incident with the ith atom; δ_j is the vertex degree of the first neighbours of the ith vertex.

The 0-order local connectivity index is defined simply as:

$${}^0\bar{\chi}_i = (\delta_i)^{-1/2}$$

By summing these local connectivity indices over all the non-hydrogen atoms, the molecular connectivity indices previously defined are reproduced.

• **fragment molecular connectivity indices** (*FMC*)
First-order connectivity indices computed for predefined positions on molecular fragments in congeneric series [Takahashi *et al.*, 1985]. By superimposition of all congeneric compounds, a template structure is derived whose vertices define the positions for the *FMC* indices; the vertices of the common parent structure are not considered in defining the positions. For a *k* th position the corresponding fragment connectivity index is defined as:

$$FMC_k = \sum_i (\delta_k \cdot \delta_i)^{-1/2}$$

Example : Fragment Molecular Connectivity indices

TEMPLATE

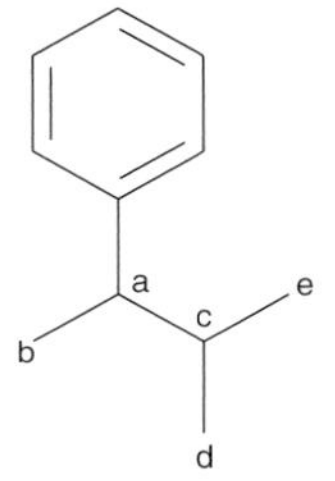

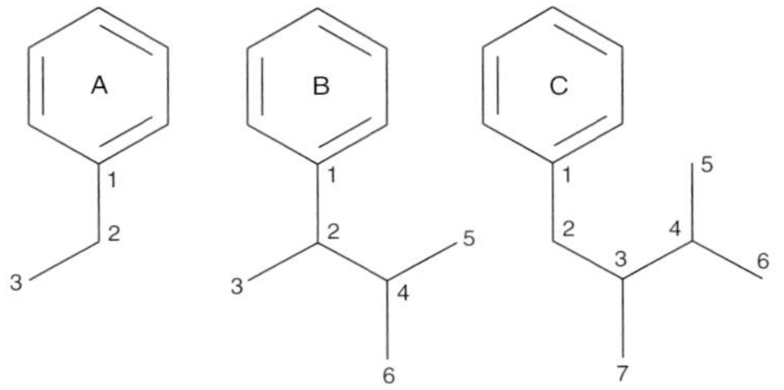

$$FMC_a(\mathrm{A}) = (\delta_2 \cdot \delta_1)^{-1/2} + (\delta_2 \cdot \delta_3)^{-1/2} = (2 \cdot 3)^{-1/2} + (2 \cdot 1)^{-1/2} = 1.115 \qquad FMC_c(\mathrm{A}) = 0$$

$$FMC_a(\mathrm{B}) = (\delta_2 \cdot \delta_1)^{-1/2} + (\delta_2 \cdot \delta_3)^{-1/2} + (\delta_2 \cdot \delta_4)^{-1/2} = (3 \cdot 3)^{-1/2} + (3 \cdot 1)^{-1/2} + (3 \cdot 3)^{-1/2} = 1.244$$

$$FMC_c(\mathrm{B}) = (\delta_4 \cdot \delta_2)^{-1/2} + (\delta_4 \cdot \delta_5)^{-1/2} + (\delta_4 \cdot \delta_6)^{-1/2} = (3 \cdot 3)^{-1/2} + (3 \cdot 1)^{-1/2} + (3 \cdot 1)^{-1/2} = 1.488$$

$$FMC_a(\mathrm{C}) = (\delta_2 \cdot \delta_1)^{-1/2} + (\delta_2 \cdot \delta_3)^{-1/2} = (2 \cdot 3)^{-1/2} + (2 \cdot 3)^{-1/2} = 0.816$$

$$FMC_c(\mathrm{C}) = (\delta_3 \cdot \delta_2)^{-1/2} + (\delta_3 \cdot \delta_4)^{-1/2} + (\delta_3 \cdot \delta_7)^{-1/2} = (3 \cdot 2)^{-1/2} + (3 \cdot 3)^{-1/2} + (3 \cdot 1)^{-1/2} = 1.319$$

Molecule	FMC_a	FMC_b	FMC_c	FMC_d	FMC_e
A	1.115	0	0	0	0
B	1.244	0.911	1.488	0.577	0.577
C	0.816	0	1.319	0.577	1.488

Box C-5.

where δ can be the simple vertex degree or the valence vertex degree, and i denotes each vertex bonded to the vertex in the kth position (the link vertex of the parent molecule is also considered). By definition, *FMC* is equal to zero if there is no atom of the substituent in the considered position. Each molecule is then described by a number of *FMC* values, corresponding to the number of predefined positions.

- **3D-connectivity indices** ($^m\chi\chi_q$)

Similar to the previously defined connectivity indices but relative to the → *geometry matrix* G, they are defined using the → *geometric distance degree* $^G\sigma$ in place of the topological vertex degree δ:

$$^m\chi\chi_q = \sum_{k=1}^{K} \left(\prod_{a=1}^{n} {}^G\sigma_a \right)_k^{-1/2}$$

where k runs over all of the mth order subgraphs constituted by n atoms ($n = m + 1$ for acyclic subgraphs); K is the total number of mth order subgraphs, and in the case of path subgraphs equals the mth order path count ^mP. The subscript "q" for the connectivity indices refers to the type of molecular subgraph [Randic, 1988a; Randic, 1988b; Randic *et al.*, 1990].

- **modified Randic index** ($^1\chi_{\text{mod}}$)

Molecular descriptor proposed as the sum of atomic properties, accounting for valence electrons and extended connectivities in the H-depleted molecular graph using a Randic connectivity index-type formula [Lohninger, 1993]:

$$^1\chi_{\text{mod}} = \frac{1}{2} \cdot \sum_{i=1}^{A} \sum_{j=1}^{\delta_i} \frac{Z_i}{(\delta_i \cdot \delta_j)^{1/2}}$$

where the first sum runs over all the atoms in the molecular graph while the second runs over the first neighbours of the considered atom; δ is the vertex degree, and Z_i the atomic number of the ith atom.

- **Kupchik modified connectivity indices**

These are modifications of the Randic connectivity index defined in such a way as to account for the presence of heteroatoms in the molecule [Kupchik, 1986; Kupchik, 1988; Kupchik, 1989]:

$$^1\chi^r = \sum\nolimits_b \left(\delta_i^{\text{het}} \cdot \delta_j^{\text{het}} \right)_b^{-1/2} \quad \text{and} \quad ^1\chi^b = \sum\nolimits_b \frac{r_{ij}}{r_{\text{CC}}} \cdot \left(\delta_i \cdot \delta_j \right)_b^{-1/2}$$

where the sums run over all the bonds in the molecule and i, j denote the vertices incident with the considered bond; r_{ij} is the carbon-heteroatom bond length and r_{CC} a standard carbon-carbon bond length; δ is the simple vertex degree, i.e. the number of first neighbors. The modified vertex degree δ^{het} is calculated as:

$$\delta_i^{\text{het}} = \frac{R_{\text{C}}}{R_i} \cdot (Z - h_i)$$

where R_i and R_{C} are the covalent radius of the ith atom and the carbon atom, respectively; Z is the atomic number and h_i the number of hydrogen atoms bonded to v_i.

These modified connectivity indices were found to be related to the → *molar refractivity* of alkanes, alkylsilanes, and alkylgermanes.

- **solvation connectivity indices** ($^m\chi_q^s$)

Molecular descriptors defined in order to model solvation entropy and describe dispersion interactions in solution. Taking into account the characteristic dimension of the molecules by atomic parameters, they are defined as:

$$^{m}\chi_q^s = \frac{1}{2^{m+1}} \cdot \sum_{k=1}^{K} \frac{\left(\prod_{a=1}^{n} \mathrm{L}_a\right)_k}{\left(\prod_{a=1}^{n} \delta_a\right)_k^{1/2}}$$

where L_a is the principal quantum number (2 for C, N, O atoms, 3 for Si, S, Cl, ...) of the ath atom in the kth subgraph and δ_a the corresponding vertex degree; K is the total number of mth order subgraphs; n is the number of vertices in the subgraph. The normalization factor $1/(2^{m+1})$ is defined in such a way that the indices $^{m}\chi$ and $^{m}\chi^s$ for compounds containing only second-row atoms coincide.

For example, the 1st order solvation connectivity index is

$$^{1}\chi^s = \frac{1}{4} \cdot \sum_{b=1}^{B} \frac{(\mathrm{L}_i \cdot \mathrm{L}_j)_b}{(\delta_i \cdot \delta_j)_b^{1/2}}$$

where b runs over all the B bonds; L_i and L_j are the principal quantum numbers of the two vertices incident to the considered bond. This index coincides with the Randic connectivity index $^{1}\chi$ for the hydrocarbons; L = 2 for all the atoms.

These molecular descriptors are defined for an H-depleted molecular graph; furthermore, fluorine atoms are not taken into account, their dimension being very close to that of the hydrogen atom.

- **perturbation connectivity indices** ($^{m}\chi_q^p$)

Modified connectivity indices based on **perturbation delta values** δ^p defined as:

$$^{m}\chi_q^p = \sum_{k=1}^{K} \left(\prod_{a=1}^{n} \delta_a^p\right)_k^{-1/2}$$

where

$$\delta_i^p = \delta_i^v + \sum_{j=1}^{\delta_i} \gamma_{ij} \cdot \delta_j^p$$

where k runs over all of the mth order subgraph of type q constituted by n atoms; K is the total number of mth order subgraphs [Gombar *et al.*, 1987b]. Perturbation delta values are obtained from valence vertex degrees δ^v modified by atomic environments. The perturbation term of the ith atom is the sum of the valence vertex degrees of its first neighbours, each weighted by parameters γ_{ij} accounting for the type of the bond i-j. γ values should be functions of the properties of the connected atoms i and j (e.g. between –0.30 and +0.30). For $\gamma = 0$, perturbation delta values coincide with the corresponding valence vertex degrees.

The widespread use of connectivity descriptors in modeling many molecular properties has been noticeable in the literature since 1975. Some examples of QSAR/QSPR studies by connectivity descriptors are reported below.

Biological activites and toxicological indices: [Kier *et al.*, 1975a; Kier *et al.*, 1975b; Murray *et al.*, 1976; Di Paolo *et al.*, 1977; Kier and Hall, 1977a; Hall and Kier, 1977b; Kier and Hall, 1978; Bonjean and Luu Duc, 1978; Di Paolo, 1978a; Di Paolo, 1978b; Glennon and Kier, 1978; Kier *et al.*, 1978; Kier and Glennon, 1978; Glennon *et al.*, 1979; Bindal *et al.*, 1980; Richard and Kier, 1980; Hall and Kier, 1981; Bojarski and Ekiert, 1982; Koch, 1982; Samata *et al.*, 1982; Schultz *et al.*, 1982; Sabljic, 1983; Gupta *et al.*, 1983; Hall and Kier, 1984; Melkova, 1984; Rouvray, 1986b; Hall *et al.*, 1989; Leegwater, 1989; Garcìa-Domenech *et al.*, 1991; Sabljic, 1991; Ivanusevic *et al.*, 1991;

Soskic and Sabljic, 1993; Soskic and Sabljic, 1995; Soskic *et al.*, 1995; Cash, 1995d; Garcìa-Domenech and De Julián-Ortiz, 1998; Shapiro and Guggenheim, 1998a; Casaban-Ros *et al.*, 1999; Gough and Hall, 1999a; Cercos-del-Pozo *et al.*, 2000].

Chromatographic indices: [Kaliszan and Foks, 1977; Kaliszan, 1977; Michotte and Massart, 1977; Kaliszan and Lamparczyk, 1978; Markowki *et al.*, 1978; Millership and Woolfson, 1978; Kaliszan, 1979; Kier and Hall, 1979; McGregor, 1979; Millership and Woolfson, 1979; Millership and Woolfson, 1980; Wells *et al.*, 1981; Calixto and Raso, 1982; Wells *et al.*, 1982; Bojarski and Ekiert, 1983; Szász *et al.*, 1983; Doherty *et al.*, 1984; Govers *et al.*, 1984; Sabljic and Protic, 1984; Sabljic, 1985; Robbat Jr. *et al.*, 1986a; Robbat Jr. *et al.*, 1986b; Lehtonen, 1987].

Solubilities: [Hall *et al.*, 1975; Cammarata, 1979; Edward, 1982a; Nirmalakhandan and Speece, 1988a; Nirmalakhandan and Speece, 1989a; Lucic *et al.*, 1995c; Nikolic and Trinajstic, 1998a]

Lipophilicity: [Murray *et al.*, 1975; Boyd *et al.*, 1982; Niemi *et al.*, 1992; Sabljic *et al.*, 1993; Luco *et al.*, 1994; Blum *et al.*, 1994; Finizio *et al.*, 1995; Gombar and Enslein, 1996; Thomsen *et al.*, 1999].

Soil sorption coefficients: [Sabljic and Protic, 1982; Sabljic, 1984; Gerstl and Helling, 1987; Sabljic, 1987; Bahnick and Doucette, 1988; Sabljic, 1989; Meylan *et al.*, 1992; Sekusak and Sabljic, 1992; Sabljic *et al.*, 1995; Hu *et al.*, 1995; Hong *et al.*, 1996; Müller and Kördel, 1996].

Bioconcentration factors: [Govers *et al.*, 1984; Sabljic and Protic, 1984; Sabljic, 1988].

Various physico-chemical properties: [Kier *et al.*, 1976a; Kier *et al.*, 1976b; Murray, 1977; Sasaki *et al.*, 1980; Edward, 1982b; Singh *et al.*, 1984; Rouvray and El-Basil, 1988; Hall and Aaserud, 1989; Moliner *et al.*, 1991; Boecklen and Niemi, 1994; Pogliani, 1996a; Pogliani, 1999c; Schramke *et al.*, 1999].

📖 [Kier and Hall, 1976a] [Kier and Hall, 1976b] [Kier *et al.*, 1977] [Hall and Kier, 1978b] [Freeland *et al.*, 1979] [Henry and Block, 1979] [Henry and Block, 1980a] [Henry and Block, 1980b] [Kier, 1980b] [Kier and Hall, 1983a] [Gupta and Sharma, 1986] [Randic *et al.*, 1988b] [Mokrosz, 1989] [Kier and Hall, 1990b] [Hall, 1990] [Kunz, 1990] [Sabljic *et al.*, 1990b] [Sabljic and Piver, 1992] [Sabljic and Horvatic, 1993] [Nirmalakhandan and Speece, 1993] [Pogliani, 1994a] [Dang and Madan, 1994] [Gálvez *et al.*, 1994a] [Saxena, 1995a] [Saxena, 1995b] [Stankevitch *et al.*, 1995] [Patil *et al.*, 1995] [Perez-Gimenez *et al.*, 1995] [Dowdy *et al.*, 1996] [Pogliani, 1996b] [Pogliani, 1997c] [Garcìa-Domenech *et al.*, 1997] [Caporossi *et al.*, 1999] [Mitchell *et al.*, 1999].

connectivity indices of *m*th order → **connectivity indices**

connectivity valence layer matrix → **layer matrices**

Connolly surface area → **molecular surface** (⊙ solvent-accessible molecular surface)

constant interval reciprocal indices → **distance matrix**

constant and near-constant variables → **variable reduction**

constitutional descriptors

These are the most simple and commonly used descriptors, reflecting the molecular composition of a compound without any information about its → *molecular geometry.*

The most common constitutional descriptors are number of atoms (→ *atom number*), number of bonds (→ *bond number*), absolute and relative numbers of specific atom types (→ *count descriptors*), absolute and relative numbers of single, double, triple, and aromatic bonds, number of rings (→ *cyclomatic number*), number of rings divided by the number of atoms or bonds, number of benzene rings, number of benzene rings divided by the number of atoms, → *molecular weight* and → *average molecular weight*, → *atomic composition indices*, → *information index on size*, etc.

These descriptors are insensitive to any conformational change, do not distinguish among isomers, and are either → *0D-descriptors* or → *1D-descriptors*.

constitutional graph : *molecular graph*

contact surface → **molecular surface** (⊙ solvent-accessible molecular surface)

continuous chirality measure → **chirality descriptors**

contour length → **size descriptors** (⊙ Kuhn length)

contour profiles → **molecular profiles**

conventional bond order → **bond order indices**

conventional bond order ID number → **ID numbers**

Corey-Pauling-Koltun volume → **volume descriptors**

corrected electron charge density connectivity index → **charge descriptors** (⊙ electron charge density connectivity index)

corrected electron charge density weight → **charge descriptors** (⊙ electron charge density connectivity index)

corrected second moments : *topological atomic valencies* → **self-returning walk counts**

corrected structure count : *algebraic structure count* → **Kekulé number**

corrected Taft steric constant → **steric descriptors** (⊙ Taft steric constant)

cospectral graphs : *isospectral graphs* → **graph**

Coulomb potential energy function → **molecular interaction fields** (⊙ electrostatic interaction fields)

count descriptors

These are simple molecular descriptors based on counting the defined elements of a compound. The most common chemical count descriptors are → *atom number A*, → *bond number B*, → *cyclomatic number C*, → *hydrogen-bond acceptor number* and → *hydrogen-bond donor number*, → *distance-counting descriptors*, → *path counts*, → *walk counts*.

When the different chemical nature of atoms is considered, the **atom-type count** is defined as the number of atoms of the same chemical element. A → *molecular graph* $\mathcal{G}$ can be characterized by a vector of atom-type counts as

$$\langle N_C; N_H; N_O; N_N; N_S; N_F; N_{Cl}; N_{Br}; N_I; \; \ldots \rangle$$

whose entries represent the number of carbon, hydrogen, oxygen, nitrogen, sulfur, fluorine, chlorine, bromine, iodine atoms, respectively. These descriptors are derived from the chemical formula, i.e. they are → *0D-descriptors*. The **relative atom-type count** is the ratio between a given atom count and the total number of atoms, therefore the following vector can be defined:

$$\langle \bar{N}_C; \bar{N}_H; \bar{N}_O; \bar{N}_N; \bar{N}_S; \bar{N}_F; \bar{N}_{Cl}; \bar{N}_{Br}; \bar{N}_I; \; \ldots \rangle$$

where

$$\bar{N}_X = \frac{N_X}{A}$$

The **atomistic topological indices** were proposed by Burden [Burden, 1996] as atom-type counts where each atom is classified by its element and the number of connections, thus also accounting for atom hybridization. In particular, N_p, N_s, N_t, N_q are the number of primary, secondary, tertiary and quaternary carbon atoms, respectively, and N_{C4}, N_{C3}, N_{C2} are the numbers of sp^3, sp^2, and *sp* carbon atoms, respectively.

Analogously, the **functional group count** can be defined considering the well-known **functional chemical groups**, i.e. groups of atoms having a characteristic and specific reactivity, such as

$$\langle N_{OH}; N_{COOH}; N_{NH_2}; N_{C=O}; N_{OCH_3}; N_{SH}; N_{H_2C=};\ N_{BENZ}; \ldots \rangle$$

whose entries represent the number of oxhydryl, carboxylic, aminic, carbonylic, methoxy, thio, methylene, and phenyl functional groups, respectively.

Andrews descriptors are particular atom and functional group counts relative to those groups found to be statistically significant in receptor binding modelling [Andrews *et al.*, 1984]: CO_2^-, PO_4^-, N^+, N, OH, C=O, ether and thioether groups, halogens, sp^3 and sp^2 carbon atoms, and the → *rotatable bond number*.

Even more general is the defintion of **fragment count** as the number of a specific kind of **molecular fragments** (or simply **fragments**), i.e. arbitrarily selected groups of adjacent atoms in a molecule. A general method for modelling → *physico-chemical properties* using fragment counts is the → *cluster expansion of chemical graphs.*

The **subgraph count set** (SCS) is a vector descriptor where each entry is the number of time-specific subgraphs obtained by cutting one bond at a time in an → *H-depleted molecular graph* [Oberrauch and Mazzanti, 1990]:

$$\langle N_{METHYL\bullet},\ N_{ETHYL\bullet},\ N_{PROPYL\bullet},\ N_{ISOPROPYL\bullet},\ \ldots\ldots \rangle$$

The order of counts is not defined *a priori* and a subset of relevant subgraph counts can be used instead of the complete SCS. In chemical terms, these subgraphs are recognized as radicals.

Both the functional group count and the fragment count can be derived from recognized substructures within the molecule, i.e. they are → *1D-descriptors*; in fact they are also considered specific → *substructure descriptors.*

Count descriptors give local chemical information, are insensitive to isomers and to conformational changes and show a high level of degeneracy. However, due to their immediate availability, they are among the most used descriptors.

📖 [Tosato *et al.*, 1992] [Chiorboli *et al.*, 1993a] [Chiorboli *et al.*, 1993b] [Chiorboli *et al.*, 1993c] [Chiorboli *et al.*, 1996] [Okey and Stensel, 1996a] [Okey *et al.*, 1996b] [Winkler *et al.*, 1998b] [Kaiser and Niculescu, 1999]

covalent hydrogen-bond acidity → **theoretical linear solvation energy relationships**

covalent hydrogen-bond basicity → **theoretical linear solvation energy relationships**

covalent radius-weighted distance matrix → **weighted matrices**

CPK volume : *Corey-Pauling-Koltun volume* → **volume descriptors**

***CPSA* descriptors** : *charged partial surface area descriptors*

crosscorrelation descriptors → **autocorrelation descriptors**

cross-validation → **validation techniques**

CSA2$_{Cl}$ **index** → **charged partial surface area descriptors** (⊙ *HDCA index*)

CSA2$_{H}$ **index** → **charged partial surface area descriptors** (⊙ *HDCA index*)

cube root molecular weight → **molecular weight**

cycle : *cyclic path* → **graph**

cycle matrices (C)

They are topological rectangular matrices derived from a → *molecular graph* $\mathcal{G}$ where each column represents a graph → *circuit*. Two main cycle matrices are defined: the *vertex cycle matrix* $\mathbf{C_V}$ whose rows are the A vertices (i.e. the atoms) and the *edge cycle matrix* $\mathbf{C_E}$ whose rows are the B edges (i.e. the bonds) of the graph [Bonchev, 1983].

Based on total and mean information content, several → *topological information indices* can be calculated both on the vertex cycle matrix (→ *information indices on the vertex cycle matrix*) and the edge cycle matrix (→ *information indices on the edge cycle matrix*).

- **vertex cycle matrix ($\mathbf{C_V}$)**

A rectangular cycle matrix whose rows are the vertices (atoms) and columns the circuits of the graph, i.e. having a dimension $A \times C^+$, where C^+ is the → *cyclicity*, i.e. the number of circuits. Derived from the → *H-depleted molecular graph*, its elements are $c_{ij} = 1$ if the vertex v_i belongs to the jth circuit, otherwise $c_{ij} = 0$.

The sum over all the entries in the ith row is called **vertex cyclic degree** γ :

$$\gamma_i = \mathcal{R}_i(\mathbf{C_V}) = \sum_{j=1}^{C^+} c_{ij}$$

where $\mathcal{R}$ is the → *row sum operator*, and C^+ is the cyclicity. The sum over all the vertex cyclic degrees is called **total vertex cyclicity** C_{VC} of the graph:

$$C_{VC} = \sum_{i=1}^{A} \gamma_i = S(\mathbf{C_V}) = \sum_{i=1}^{A} \sum_{j=1}^{C^+} c_{ij}$$

where S is the → *total sum operator* and A the number of atoms. It increases rapidly with the number of cycles and can be used as a general descriptor for molecular cyclicity.

- **edge cycle matrix ($\mathbf{C_E}$)**

A rectangular cycle matrix whose rows are the edges (bonds) and columns the circuits of the graph, i.e. having a dimension $B \times C^+$, where C^+ is the cyclicity. Derived from H-depleted molecular graph, its elements are $c_{ij} = 1$ if the ith edge belongs to the jth circuit, otherwise $c_{ij} = 0$.

The sum over all the entries in the ith row is called **edge cyclic degree** γ^e :

$$\gamma_i^e = \mathcal{R}_i(\mathbf{C_E}) = \sum_{j=1}^{C^+} c_{ij}$$

where $\mathcal{R}$ is the row sum operator and C^+ the cyclicity. The sum over all the edge cyclic degrees is called **total edge cyclicity** C_{EC} of the graph:

$$C_{EC} = \sum_{i=1}^{B} \gamma_i^e = S(\mathbf{C_E}) = \sum_{i=1}^{B} \sum_{j=1}^{C^+} c_{ij}$$

where S is the total sum operator and B the number of bonds.

Example : anthracene

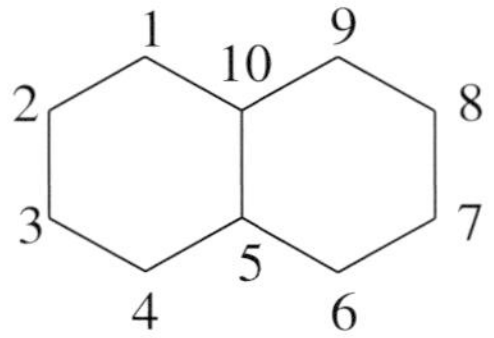

Vertex cycle matrix $\mathbf{C_V}$

atom	*circuit* 1	2	3	γ_i
1	1	0	1	2
2	1	0	1	2
3	1	0	1	2
4	1	0	1	2
5	1	1	1	3
6	0	1	1	2
7	0	1	1	2
8	0	1	1	2
9	0	1	1	2
10	1	1	1	3

Edge cycle matrix $\mathbf{C_E}$

edge	*circuit* 1	2	3	γ_i^e
(1, 2)	1	0	1	2
(2, 3)	1	0	1	2
(3, 4)	1	0	1	2
(4, 5)	1	0	1	2
(5, 6)	0	1	1	2
(6, 7)	0	1	1	2
(7, 8)	0	1	1	2
(8, 9)	0	1	1	2
(9, 10)	0	1	1	2
(1, 10)	1	0	1	2
(5, 10)	1	1	0	2

Box C-6.

cyclicity → **graph**

cyclicity index → **detour matrix**

cyclic path → **graph**

cyclomatic number *(: ring number)*

It is the number of independent cycles C (or rings) in a molecule, and, more exactly, the number of non-overlapping cycles. The cyclomatic number of a polycyclic graph is equal to the minimum number of edges that must be removed from the graph to transform it to the related acyclic graph. It is equal to zero for trees and to one for monocyclic graphs.

The cyclomatic number is the simplest descriptor which discriminates cyclic compounds from acyclic ones and is related to the number of bonds B and atoms A in a molecule as follows:

$$C = B - A + 1 = {}^{1}P - A + 1$$

where ${}^{1}P$ is the number of 1st order paths (i.e. the molecule bonds). The cyclomatic number is the usual way that a chemist counts rings in a molecular structure.

This descriptor appears as a part of → *molecular descriptors* such as the → *Balaban distance connectivity index*.

The cyclomatic number must not be confused with the graph → *cyclicity* C^{+}. Thus, for example, naphthalene has a cyclomatic number equal to two (the two benzene rings) and a cyclicity equal to three (the two benzene rings plus the more external 10-atom ring).

[Balaban, 1976d] [Hanser *et al.*, 1996]

D

DARC/PELCO analysis

DARC/PELCO analysis, a topological method dealing with structural environments, was originally proposed, and then later refined and broadened, by Dubois [Dubois *et al.*, 1966; Dubois *et al.*, 1973a; Dubois *et al.*, 1973b; Dubois *et al.*, 1976; Dubois, 1976].

The method calls for a combination of the DARC system (**Description, Acquisition, Retrieval and Computer-aided design**) and the PELCO procedure (**Perturbation of an Environment Limited Concentric and Ordered**).

DARC/PELCO analysis is based on the simultaneous representation of all the experimental compounds and the population of the compounds structurally contained in them, the whole set of experimental compounds generating an ordered multidimensional space which constitutes a → *hyperstructure*.

Each molecule is represented by an ordered → *chromatic graph* which describes the topological and chemical nature of each site, i.e. the graph vertices represent non-hydrogen atoms and the edge bonds between these atoms. Vertex chromatism corresponds to the chemical nature of the atoms, edge chromatism to the bond multiplicity.

The hyperstructure is built following the operations of focalization, organization, ordering, and chromatic evaluation of each data set molecule.

The *focus* is the → *maximum common substructure* among the experimental compounds corresponding to a molecule with a measured response, i.e. it is the parent molecule.

The *environment* is organized in concentric layers centred around the focus and is *limited* and *concentric* (*ELC*). The vertices at an odd distance from the focus belong to layer A; the vertices at an even distance belong to layer B. Each pair of successive A-B layers starting from the focus constitutes an *environment limited* to B (*EB*). The environment is *ordered* (*ELCO*) in the sense that each site (vertex) is located unambiguously by means of a topological coordinate (A_i or B_{ij}); each topological coordinate gives a development direction for the environment.

The superimposition of all the ordered graphs of the molecule data set provides the hyperstructure whose central topological vertex corresponds to the focus, and the environment is also organized in concentric layers A-B where each vertex corresponds to a site present in at least one experimental compound. From the first concentric environment of the hyperstructure, the different substitution sites give origin to the main development directions starting from the focus. Note that each hyperstructure vertex can be labeled differently, in accordance with the chemical nature of the atoms occupying the same site; this vertex therefore corresponds to multiple vertices, i.e. topochromatic sites.

The hyperstructure can be mathematically represented by the **DARC/PELCO matrix**, there being n rows, the number of experimental compounds, and N_S columns, the topochromatic sites of the hyperstructure. Each ith row of the DARC/PELCO matrix is called the **topochromatic vector $\mathbf{I}_i$** and directly accounts for the overall topology of the ith molecule. Therefore, the **DARC/PELCO descriptors** of the ith molecule are the elements of its topochromatic vector, being binary variables I_{is} where $I_{is} = 1$ if the sth topochromatic site of the hyperstructure corresponds to an atom of the ith molecule and $I_{is} = 0$ otherwise.

The **DARC/PELCO model** is defined as:

$$\hat{y}_i = y_0 + \sum_{s=1}^{N_S} b_s \cdot I_{is}$$

where y_0 is the response of the parent molecule, i.e. the focus of the hyperstructure, and b_s the regression coefficients called *perturbations*. In spite of the formal analogy with the → *Free-Wilson model*, where the contribution to the biological activity of each group in a substitution site is considered additive and independent of the structural variation in the rest of the molecule, the DARC/PELCO model considers the contribution of a substituent group to the biological activity as the sum of ordered perturbations given by all the vertices starting from the focus and characterizing that group.

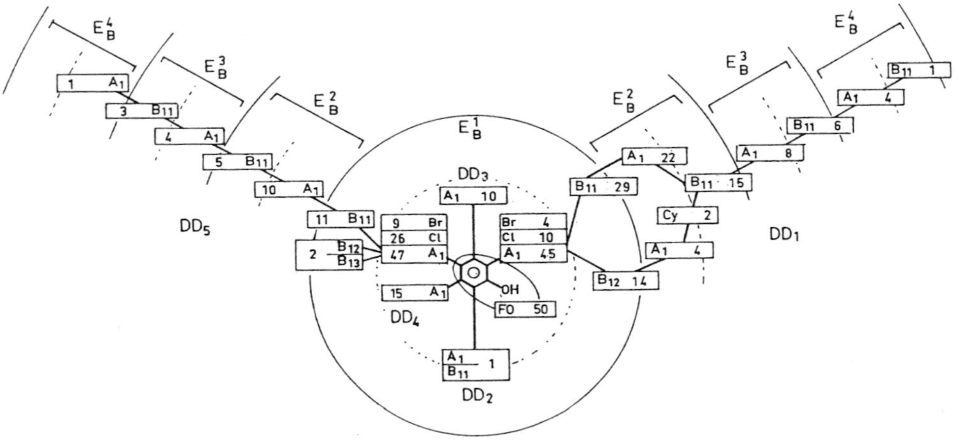

Figure D-1. Example of DARC/PELCO hyperstructure. The focus is constituted by the phenol.

A particular advantage of this method is the determination of the structural area for reliable predictions. This area, called the **proference**, consists of all the structures generated by the hyperstructure that do not belong to the population of experimental compounds (**population trace**). In other words, predictable structures are localized in the hyperstructure and their activity is predicted by interpolation, using the corresponding topochromatic vector where each present vertex ($I_{is} = 1$) refers to a site existing in at least one experimental compound. Different levels of reliability are determined depending on the extent to which the structure is surrounded by experimental compounds.

📖 [Dubois *et al.*, 1967] [Dubois *et al.*, 1975a] [Dubois *et al.*, 1975b] [Duperray *et al.*, 1976a] [Duperray *et al.*, 1976b] [Dubois *et al.*, 1979] [Mercier and Dubois, 1979] [Dubois *et al.*, 1980a] [Panaye *et al.*, 1980] [Doucet *et al.*, 1983] [Dubois *et al.*, 1984] [Dubois *et al.*, 1986] [De La Guardia *et al.*, 1988] [Bonchev, 1989] [Mercier *et al.*, 1990] [Mercier *et al.*, 1991] [Mekenyan *et al.*, 1993a] [Carabédian and Dubois, 1998]

DARC/PELCO descriptors → **DARC/PELCO analysis**

DARC/PELCO matrix → **DARC/PELCO analysis**

DARC/PELCO model → **DARC/PELCO analysis**

Dash-Behera steric density parameter : *steric density parameter* → **steric descriptors**

data → **data set**

data distance matrix → **similarity/diversity**

data matrix → **data set**

data set

A collection of *objects* described by one or more *variables*. An **object** is a basic unit in data analysis, e.g. an individual, a molecule, an experiment. Each object is described by one or more measurements, called **data**. A **variable** represents a characteristic of the objects that may take any value from a specified set, e.g. a physico-chemical property, a molecular descriptor.

A data set is often considered as a sample from a population and the sample parameters calculated from the data set as estimates of the population parameters (→ *statistical indices*). Moreover, it is usually used to calculate statistical models such as quantitative → *structure/response correlations*. In this case the data set is organized into a **data matrix X** with n rows and p columns, where each row corresponds to an object of the data set and each column to a variable; therefore each element x_{ij} represent the value of the jth variable for the ith object ($i = 1, ..., n; j = 1, ..., p$).

Data set variables can be distinguished by their role in the models as independent and dependent variables. **Independent variables** (or **explanatory variables**, **predictor variables**) are those variables assumed to be capable of taking part in a function suitable to model the response variable. **Dependent variables** (or **response variables**) are variables (often obtained from experimental measurements) for which the interest is to find a statistical dependence on one or more independent variables. Independent variables constitute the data matrix **X**, while dependent variables are collected into a matrix **Y** with n rows and r columns ($r = 1$ when only one response variable is defined) (Figure D-2).

$$\mathbf{X} = \begin{vmatrix} x_{11} & x_{12} & \cdots & \cdots & x_{1p} \\ x_{21} & x_{22} & & & x_{2p} \\ \cdots & \cdots & \cdots & \cdots & \cdots \\ \cdots & \cdots & \cdots & \cdots & \cdots \\ x_{n1} & x_{n2} & & & x_{np} \end{vmatrix} \qquad \mathbf{Y} = \begin{vmatrix} y_{11} & y_{12} & \cdots & \cdots & y_{1r} \\ y_{21} & y_{22} & & & y_{2r} \\ \cdots & \cdots & \cdots & \cdots & \cdots \\ \cdots & \cdots & \cdots & \cdots & \cdots \\ y_{n1} & y_{n2} & & & y_{nr} \end{vmatrix}$$

Figure D-2. n objects described by an **X** data matrix of p independent variables and a **Y** matrix of r responses.

In order to estimate the predictive capabilities of a model by → *validation techniques*, the data set can be split into different parts: the **training set** (or **learning set**), the set of objects used for modelling, a **test set**, the set of objects used to optimize the goodness of prediction of a model obtained from the training set, and the **external evaluation set** (or **evaluation set**), which is a new data set used to perform further external validation of the model obtained from the training set.

The use of several variables in the original data set increases the complexity of the data and therefore the → *model complexity*: noise, variable correlation, redundancy of information provided by the variables, unbalanced information and information that is not useful give the data an intrinsic complexity that must be resolved. Simple cases of these situations are spectra, each, for example, constituted by 800–1000 digitalized signals, i.e. containing 800–1000 highly correlated variables. Usually, → *variable reduction* and → *variable selection* improve the quality of models (in particular, their predictive power) and data information. → *Chemometrics* provides several useful tools able to check the different kinds of information contained in the data [Frank and Todeschini, 1994]

Similarity and diversity among the objects of a data set are encoded in the → *similarity matrix* and in the → *data distance matrix*, respectively. By transposing the data matrix, the corresponding matrices measuring similarity/diversity among variables can be obtained.

D/D index → **distance/distance matrix**

decimal adjacency vector → **adjacency matrix**

decomposition → **equivalence classes**

degeneracy of molecular descriptors → **molecular descriptors**

degree complexity : *mean information content on the vertex degree magnitude* → **topological information indices**

degree distance of the graph → **Schultz molecular topological index**

delocalized effect : *resonance effect* → **electronic substituent constants**

delta matrix ($\mathbf{D}_\Delta$)

A square symmetric matrix of dimension $A \times A$, A being the number of graph vertices, derived as the difference between the → *distance-path matrix* $\mathbf{D_P}$ and the → *distance matrix* $\mathbf{D}$ [Diudea, 1996a; Diudea *et al.*, 1997d; Ivanciuc *et al.*, 1998]:

$$\mathbf{D}_\Delta = \mathbf{D_P} - \mathbf{D}$$

The entries of the matrix represent the number of all paths larger than unity included in the shortest path between the considered vertices:

$$[\mathbf{D}_\Delta]_{ij} = \begin{cases} \binom{d_{ij}}{2} = \dfrac{d_{ij} \cdot (d_{ij} - 1)}{2} & \text{if } i \neq j \\ 0 & \text{if } i = j \end{cases}$$

where d_{ij} is the → *topological distance* between vertices v_i and v_j.

Applying the → *Wiener operator* $\mathcal{W}$ to the delta matrix, a molecular descriptor called the **delta number** D_Δ is obtained as:

$$D_\Delta = \mathcal{W}'(\mathbf{D}_\Delta) = \frac{1}{2} \cdot \sum_{i=1}^{A} \sum_{j=1}^{A} [\mathbf{D}_\Delta]_{ij} = \frac{1}{2} \cdot \sum_{i=1}^{A} \sum_{j=1}^{A} \binom{d_{ij}}{2}$$

The delta number can be also derived by:

$$D_\Delta = D_P - W$$

where W is the → *Wiener index* and D_P the → *hyper-distance-path index* which coincides with the → *hyper-Wiener index* WW for acyclic graphs. It follows that the delta number is the "non-Wiener" part of the hyper-Wiener index.

Moreover, the number D_Δ can be related to the distance matrix and the Wiener index W by:

$$D_\Delta = \frac{tr(\mathbf{D}^2) - 2 \cdot W}{4} = \frac{D_2 - W}{2}$$

where $\mathbf{D}^2$ is the distance matrix raised to the second power, and D_2 is the → *unnormalized second moment of distances*, i.e. the sum of the square distances in the graph.

Example : 2-methylpentane

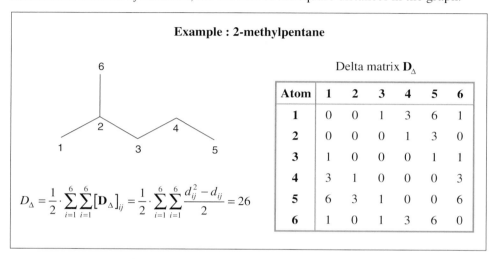

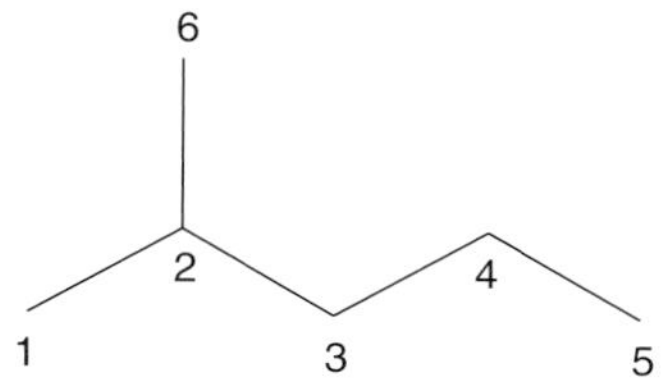

$$D_\Delta = \frac{1}{2} \cdot \sum_{i=1}^{6} \sum_{i=1}^{6} [\mathbf{D}_\Delta]_{ij} = \frac{1}{2} \cdot \sum_{i=1}^{6} \sum_{i=1}^{6} \frac{d_{ij}^2 - d_{ij}}{2} = 26$$

Delta matrix $\mathbf{D}_\Delta$

Atom	1	2	3	4	5	6
1	0	0	1	3	6	1
2	0	0	0	1	3	0
3	1	0	0	0	1	1
4	3	1	0	0	0	3
5	6	3	1	0	0	6
6	1	0	1	3	6	0

Box D-1.

delta number → **delta matrix**

dense matrices → **algebraic operators** (⊙ sparse matrices)

dependent variables → **data set**

description, acquisition, retrieval and computer-aided design → **DARC/PELCO analysis**

desolvation energy fields → **molecular interaction fields**

det|A + D| index → **determinant-based descriptors**

det|A| index → **determinant-based descriptors**

det|D| index → **determinant-based descriptors**

determinant → **algebraic operators**

determinant-based descriptors

Descriptors obtained by the calculation of the → *determinant* of a matrix representing a → *molecular graph*. Molecular descriptors similar to the determinant-based descriptors could be obtained applying → *permanent* and → *hafnian* to any matrix representing a molecular graph, such as → *per(D) index*, → *shaf(D) index*, and → *lhaf(D) index* [Schultz *et al.*, 1992; Schultz and Schultz, 1992].

The most popular determinant-based descriptors are listed below.

• **det|A| index**

The determinant of the → *adjacency matrix* **A**. It was observed that this determinant is often equal to zero and this is a necessary and sufficient condition for the presence of non-bonding molecular orbitals in Hückel theory. The actual numerical value of det|**A**| is correlated to the thermodynamic stability of the molecule [Graovac and Gutman, 1978; Trinajstic, 1992].

• **det|D| index**

The determinant of the → *distance matrix* **D**. In an isomeric series, this index takes the same values for all compounds, alternating from negative values for isomers with an even number of non-hydrogen atoms to positive values for compounds with an odd number of non-hydrogen atoms [Schultz *et al.*, 1990; Schultz *et al.*, 1993]. In fact, for acyclic alkanes with the same number A of atoms, the following relation holds [Knop *et al.*, 1991]:

$$\det|\mathbf{D}| = -(-2)^{A-2} \cdot (A - 1)$$

• **det|A + D| index**

The determinant of the matrix resulting from the sum of the adjacency matrix **A** and the distance matrix **D** of an → *H-depleted molecular graph*, this same matrix is used to calculate the → *Schultz molecular topological index* [Schultz *et al.*, 1990; Knop *et al.*, 1991; Schultz *et al.*, 1993]. Demonstrated to be more discriminant than previously described determinant-based descriptors, the absolute value of this index increases with the size of the molecules, being negative for molecules with an even number of non-hydrogen atoms and positive for those with an odd number. Moreover, it decreases with increase of branching in an isomeric series of compounds, i.e. the degree of substitution increases.

The logarithm of this index was used to model different → *physico-chemical properties* [Cash, 1995c], showing that it can suffer from the drawback that the determinant values are often equal to zero.

The determinant of ($\mathbf{A}^w$ + G) was also proposed as a molecular topographic descriptor where $\mathbf{A}^w$ is the weighted adjacency matrix (→ *weighted matrices*) where the entries corresponding to bonded atoms are → *bond distances* and G the → *geometry matrix* [Mihalic *et al.*, 1992a].

• **general a_N-index** (*GAI*)

A topological index defined as the absolute value of the determinant of the **orbital interaction matrix of linked atoms (OIMLA)** [Xu *et al.*, 1992a; Xu *et al.*, 1992b; Xu, 1992]:

$$GAI = |\det(\mathbf{OIMLA})|$$

OIMLA is a symmetric weighted adjacency matrix of dimension $2B \times 2B$ whose diagonal elements are the relative energies of the corresponding atomic hybrid orbitals (setting to zero the energy of the C_{sp3} orbital) and the off-diagonal elements represent the interaction type of hybrid orbitals, assumed to be proportional to the corresponding overlap integrals. This matrix is derived from an H-depleted molecular graph called the **orbital interaction graph of linked atoms** (OIGLA) which is a → *directed graph* where arcs (ordered pairs of vertices) are used to describe interactions between hybrid orbitals. Entries equal to zero indicate that no interaction between hybrid orbitals is considered. Both atomic hybrid orbital energies and overlap integrals are obtained by methods of → *computational chemistry* (e.g. CNDO/2).

GAI was found to be a useful index for the discrimination of *cis/trans* isomerism (→ *cis/trans descriptors*) and to model the chromatographic behaviour of phosphorus derivatives.

GAI is an extension to molecules containing heteroatoms and/or multiple bonds of the **a_N-index**, previously defined only for alkane derivatives. The a_N-index was calculated as the absolute value of the constant term of the characteristic polynomial of **OIMLA** where diagonal entries are zero and off-diagonal entries are calculated in a similar way to for *GAI* [Yang and Kiang, 1983].

detour distance → **detour matrix**

detour/distance matrix → **detour matrix**

detour index → **detour matrix**

detour matrix (Δ)

The detour matrix **Δ** of a graph $\mathcal{G}$ (or **maximum path matrix**) is a square symmetric $A \times A$ matrix, A being the number of graph vertices, whose entry *i-j* is the length of the longest path from vertex v_i to vertex v_j ($^{max}p_{ij}$), [Buckley and Harary, 1990; Ivanciuc and Balaban, 1994b]:

$$[\mathbf{\Delta}]_{ij} = \begin{cases} \Delta_{ij} = |^{max}p_{ij}| & if\ i \neq j \\ 0 & if\ i = j \end{cases}$$

The length of the longest path between the vertices v_i and v_j is the maximum number of edges which separate the two vertices and is called **detour distance** (Δ_{ij}).

This definition is exactly the “opposite” of the definition of the → *distance matrix* whose off-diagonal elements are the lengths of the shortest paths between the considered vertices. However, the distance and detour matrices coincide for acyclic graphs, there being only one path connecting any pair of vertices.

For edge-weighted graphs, the **weighted detour matrix** $^{w}\mathbf{\Delta}$ was proposed [Nikolic *et al.*, 1996a]. The off-diagonal *i-j* entry is defined as the maximum path weight, i.e. the maximum sum of edge weights along the path between the vertices v_i and v_j, which is not necessarily the longest possible path between them.

The maximum value entry in the *i*th row is called **atom detour eccentricity** $^{\Delta}\eta_i$ (also **vertex path eccentricity** or simply **path eccentricity**):

$$^{\Delta}\eta_i = \max_j\left([\mathbf{\Delta}]_{ij}\right)$$

From the distribution of the element values in the *i*th row of the detour matrix, the **maximum path degree sequence** of the *i*th vertex is derived as a local vector-descriptor defined as:

$$\langle n_{i0}, n_{i1}, \ldots, n_{im}, \ldots, n_{ik} \rangle$$

where n_{im} is the number of vertices in the molecular graph located at a detour distance equal to m from the vertex v_i, and k is the maximum detour distance in the graph; n_{i0} is equal to one by definition. Analogously, from the distribution of the element values in the upper or lower triangle of the detour matrix, the **maximum path frequency sequence** is derived as a molecular vector-descriptor defined as:

$$\langle {}^{\Delta}F_0, {}^{\Delta}F_1, \ldots, {}^{\Delta}F_m, \ldots, {}^{\Delta}F_k \rangle$$

where $^{\Delta}F_m$ is the number of detour distances equal to m in the molecular graph; obviously, $^{\Delta}F_0$ equals the number of vertices in the graph.

The **maximum path sum** of the ith vertex, denoted by $MPVS$, is a local vertex invariant defined as the sum of the length of the longest paths between vertex v_i and any other vertex in the molecular graph, i.e.

$$MPVS_i = \mathcal{R}_i(\mathbf{\Delta}) = \sum_{j=1}^{A} [\mathbf{\Delta}]_{ij}$$

where $\mathcal{R}$ is the → *row sum operator*.

A Wiener-type index, originally called **MPS topological index** [Ivanciuc and Balaban, 1994b] but usually known as **detour index** [Amic and Trinajstic, 1995c; Lukovits, 1996b; Lukovits and Razinger, 1997b], was proposed as the sum of the detour distances between any two vertices in the molecular graph. It is calculated as:

$$w = \mathcal{W}(\mathbf{\Delta}) = \frac{1}{2} \cdot \sum_{i=1}^{A} \sum_{j=1}^{A} [\mathbf{\Delta}]_{ij} = \frac{1}{2} \cdot \sum_{i=1}^{A} MPVS_i$$

where $\mathcal{W}$ is the → *Wiener operator*.

The characteristic polynomial of the detour matrix is called the **detour polynomial** and is defined as:

$$\pi(\mathcal{G}; x) = \det|x\mathbf{I} - \mathbf{\Delta}| = x^A - \sum_{i=1}^{A} b_i \cdot x^{i-1}$$

where $\mathbf{I}$ is the $A \times A$ identity matrix, x the independent variable, A the number of atoms in the molecule and b_i the polynomial coefficients. The roots of the detour polynomial are the eigenvalues of the detour matrix constituting the **detour spectrum**.

A modified detour matrix was proposed by substituting diagonal zero elements with the length of the longest path from each vertex to itself (i.e. the size of the cycle containing the considered vertex). From this modified matrix, the same molecular descriptors defined above can be calculated [Rücker and Rücker, 1998].

The **detour-path matrix $\mathbf{\Delta_P}$**, analogously defined as the → *distance-path matrix* $\mathbf{D_P}$, is a square symmetric matrix $A \times A$ whose off-diagonal entry i-j is the count of all paths of any length m ($1 \leq m \leq \Delta_{ij}$) that are included within the longest path from vertex v_i to vertex v_j (Δ_{ij}) [Diudea, 1996a]. The diagonal entries are zero.

Each entry i-j of the $\mathbf{\Delta_P}$ matrix is calculated from the detour matrix $\mathbf{\Delta}$ as the following:

$$[\Delta_\mathbf{P}]_{ij} = \binom{\Delta_{ij} + 1}{2} = \frac{\Delta_{ij}^2 + \Delta_{ij}}{2}$$

i.e. as all the possible combinations of two elements taken from $\Delta_{ij} + 1$ elements (binomial coefficient).

The **hyper-detour index** ww can be obtained applying the Wiener operator $\mathcal{W}$ to the detour-path matrix, as:

$$ww = \mathcal{W}(\mathbf{\Delta_P}) = \frac{1}{2} \cdot \sum_{i=1}^{A} \sum_{j=1}^{A} [\mathbf{\Delta_P}]_{ij}$$

or to the → *symmetric Cluj-detour matrix* as:

$$ww = \mathcal{W}(\mathbf{CJ\Delta}) = \frac{1}{2} \cdot \sum_{i=1}^{A} \sum_{j=1}^{A} [\mathbf{CJ\Delta_U}]_{ij} [\mathbf{CJ\Delta_U}]_{ji}$$

For acyclic graphs, the hyper-detour index ww is equal to the → *hyper-distance-path index* D_P obtained from the distance-path matrix $\mathbf{D_P}$ and to the → *hyper-Wiener index* WW obtained from the → *Wiener matrix* $\mathbf{W}$.

From the detour matrix and the distance matrix, a combined matrix, called **detour/distance matrix Δ/D** (or **maximum/minimum path matrix**), is defined as [Ivanciuc and Balaban, 1994b]:

$$[\mathbf{\Delta}/\mathbf{D}]_{ij} = \begin{cases} \Delta_{ij} & if\ i > j \\ 0 & if\ i = j \\ d_{ij} & if\ i < j \end{cases}$$

It is a square unsymmetric $A \times A$ matrix, where the upper triangle of the matrix contains the elements of the detour matrix (information about the longest paths) and the lower triangle contains the elements of the topological distance matrix (information about the shortest paths).

The **maximum/minimum path sum** of the ith vertex, denoted by $MmPVS$, is a local vertex invariant defined as the sum of the lengths of the longest and shortest paths between vertex v_i and any other vertex in the molecular graph. It is calculated as the sum of elements over the ith row and ith column in the **Δ**/D matrix, or, alternatively, as the sum of the → *vertex distance degree* σ_i calculated on the distance matrix **D** and the maximum path sum $MPVS_i$ of the ith vertex calculated on the detour matrix **Δ**:

$$MmPVS_i = \mathcal{R}_i(\mathbf{\Delta}/\mathbf{D}) + C_{j=i}(\mathbf{\Delta}/\mathbf{D}) = \sum_{j=1}^{A} [\mathbf{\Delta}/\mathbf{D}]_{ij} + \sum_{i=1}^{A} [\mathbf{\Delta}/\mathbf{D}]_{ij} = MPVS_i + \sigma_i$$

where $\mathcal{R}$ and C are the row sum and column sum operators, respectively.

A combined molecular index called the **detour/Wiener index** w/W (or **MmPS topological index**) is defined as the sum of the longest and shortest paths between any two vertices in the molecular graph:

$$w/W = \sum_{i=1}^{A} \sum_{j=1}^{A} [\mathbf{\Delta}/\mathbf{D}]_{ij} = \frac{1}{2} \cdot \sum_{i=1}^{A} MmPVS_i = w + W$$

where w is the detour index, i.e. the sum of the lengths of the longest paths, and W is the → *Wiener index*, i.e. the sum of the lengths of the shortest paths.

It must be noted that for acyclic graphs the following relation holds:

$$w = W = (w/W)/2$$

The **distance/detour quotient matrix D/Δ** is also derived from detour and distance matrices but it is a square symmetric matrix $A \times A$ whose off-diagonal entries are the ratios of the lengths of the shortest to the longest path between any pair of vertices [Randic, 1997c]. It is defined as:

$$[\mathbf{D}/\mathbf{\Delta}]_{ij} = \begin{cases} \dfrac{d_{ij}}{\Delta_{ij}} & if\ i \neq j \\ 0 & if\ i = j \end{cases}$$

where d_{ij} and Δ_{ij} are the topological and detour distances between vertices v_i and v_j, respectively. Some local and graph invariants have been calculated on this matrix. The row sums were proposed as local invariants showing a high discriminatory ability; branching vertices tend to have smaller row sums than bridging vertices. If the **D/Δ** matrix row sums of vertices belonging to single rings (or cycles) in the molecule are summed up, the **D/Δ** ring indices which can be considered special substructure descriptors reflecting local geometrical environments in complex cyclic systems are obtained. Moreover, the half sum of all row sums which corresponds to the half sum of all entries of the **D/Δ** matrix was proposed as an index of → *molecular cyclicity*, showing regular variation with increase in cyclicity in graphs of the same size. It is called the **D/Δ** index and is defined as:

Example : ethylbenzene

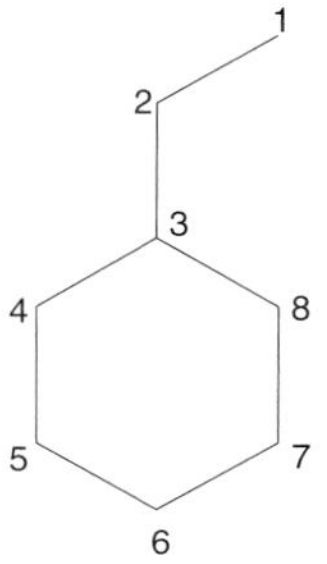

Detour matrix Δ

Atom	1	2	3	4	5	6	7	8	$^{\Delta}\eta_i$	*MPVS*$_i$
1	0	1	2	7	6	5	6	7	7	34
2	1	0	1	6	5	4	5	6	6	28
3	2	1	0	5	4	3	4	5	5	24
4	7	6	5	0	5	4	3	4	7	34
5	6	5	4	5	0	5	4	3	6	32
6	5	4	3	4	5	0	5	4	5	30
7	6	5	4	3	4	5	0	5	6	32
8	7	6	5	4	3	4	5	0	7	34

Detour-path matrix Δ_P

Atom	1	2	3	4	5	6	7	8
1	0	1	3	28	21	15	21	28
2	1	0	1	21	15	10	15	21
3	3	1	0	15	10	6	10	15
4	28	21	15	0	15	10	6	10
5	21	15	10	15	0	15	10	6
6	15	10	6	10	15	0	15	10
7	21	15	10	6	10	15	0	15
8	28	21	15	10	6	10	15	0

Detour/distance matrix **Δ/D**

Atom	1	2	3	4	5	6	7	8	*MmPVS*$_i$
1	0	1	2	7	6	5	6	7	56
2	1	0	1	6	5	4	5	6	44
3	2	1	0	5	4	3	4	5	36
4	3	2	1	0	5	4	3	4	48
5	4	3	2	1	0	5	4	3	48
6	5	4	3	2	1	0	5	4	48
7	4	3	2	3	2	1	0	5	48
8	3	2	1	2	3	2	1	0	48

Distance/detour quotient matrix **D/Δ**

Atom	1	2	3	4	5	6	7	8	*sums*
1	0	1	1	0.429	0.666	1	0.666	0.429	5.190
2	1	0	1	0.333	0.600	1	0.600	0.333	4.866
3	1	1	0	0.200	0.500	1	0.500	0.200	4.400
4	0.429	0.333	0.200	0	0.200	0.500	1	0.500	3.162
5	0.666	0.600	0.500	0.200	0	0.200	0.500	1	3.666
6	1	1	1	0.500	0.200	0	0.200	0.500	3.566
7	0.666	0.600	0.500	1	0.500	0.200	0	0.200	3.666
8	0.429	0.333	0.200	0.500	1	0.500	0.200	0	3.126

$w = 124$

$ww = 365$

$w/W = w + W =$

$= 124 + 64 = 188$

$D/\Delta = 15.821$

Box D-2.

$$D/\Delta = \mathcal{W}(\mathbf{D}/\mathbf{\Delta}) = \frac{1}{2} \cdot \sum_{i=1}^{A} \sum_{j=1}^{A} [\mathbf{D}/\mathbf{\Delta}]_{ij}$$

where $\mathcal{W}$ is the Wiener operator.

The D/Δ index decreases as the cyclicity of the molecule increases, so that it reaches the maximum value for the monocyclic graph C_A and the minimum for the → *complete graph* K_A, A being the number of vertices of the actual graph $\mathcal{G}_A$. Therefore, a more suitable measure of molecular cyclicity was proposed as a standardized D/Δ index called the **cyclicity index** γ:

$$\gamma = \frac{D/\Delta(C_A) - D/\Delta(\mathcal{G}_A)}{D/\Delta(C_A) - D/\Delta(K_A)}$$

The **average cyclicity index** is calculated simply as:

$$\bar{\gamma} = \frac{\gamma}{A}$$

where A is the number of graph vertices. Both the cyclicity and the average cyclicity index allow comparison of cyclic systems of different size; they represent the deviation of the cyclicity of the considered molecule from that of the size-corresponding monocyclic molecule C_A.

For both detour and detour/distance matrices there can also be defined the **reciprocal detour matrix** $\mathbf{\Delta}^{-1}$ and the **reciprocal detour/distance matrix** $\mathbf{\Delta}/\mathbf{D}^{-1}$, as:

$$[\mathbf{\Delta}^{-1}]_{ij} = \begin{cases} \Delta_{ij}^{-1} & if\ i \neq j \\ 0 & if\ i = j \end{cases} \quad \text{and} \quad [\mathbf{\Delta}/\mathbf{D}^{-1}]_{ij} = \begin{cases} \Delta_{ij}^{-1} & if\ i > j \\ 0 & if\ i = j \\ d_{ij}^{-1} & if\ i < j \end{cases}$$

All elements equal to zero are left unchanged in the reciprocal matrix.

Applying the Wiener operator results in → *Harary indices* and → *hyper-Harary indices* for these matrices and for the corresponding 1st order → *sparse matrix*, respectively.

📖 [Harary, 1969a] [Diudea *et al.*, 1997e] [Trinajstic *et al.*, 1997] [Randic *et al.*, 1998a]

detour-path matrix → **detour matrix**

detour polynomial → **detour matrix**

detour spectrum → **detour matrix**

detour/Wiener index → **detour matrix**

Dewar-Grisdale approach → **electronic substituent constants** (⊙ field/resonance effect separation)

Dewar-Golden-Harris approach → **electronic substituent constants** (⊙ field/resonance effect separation)

Dewar resonance energy → **resonance indices**

DFT-based descriptors → **quantum-chemical descriptors**

diagonal matrix → **algebraic operators**

diagonal operator → **algebraic operators** (⊙ diagonal matrix)

dielectric constant → **electric polarization descriptors**

dielectric susceptibility → **electric polarization descriptors**

difference in atomic charge weighted surface area → **charged partial surface area descriptors**

difference in charged partial surface area → **charged partial surface area descriptors**

difference in total charge weighted surface area → **charged partial surface area descriptors**

differential connectivity indices → **combined descriptors**

differential descriptors

They are → *molecular descriptors* or → *substituent descriptors* calculated by difference between the considered compound (or functional group, fragment) and a → *reference structure* or a → *hyperstructure*. Examples of differential descriptors are those obtained by the → *minimal topological difference* (MTD) and → *molecular shape analysis* (MSA).

digraph → **graph**

dipolarity / polarizability term → **linear solvation energy relationships**

dipole moment → **electric polarization descriptors**

dipole moment components → **electric polarization descriptors**

dipole polarization → **electric polarization descriptors**

dipole term : *dipolarity / polarizability term* → **linear solvation energy relationships**

directed graph : *digraph* → **graph**

directional WHIM density → **WHIM descriptors** (⊙ directional WHIM descriptors)

directional WHIM descriptors → **WHIM descriptors**

directional WHIM shape → **WHIM descriptors** (⊙ directional WHIM descriptors)

directional WHIM size → **WHIM descriptors** (⊙ directional WHIM descriptors)

directional WHIM symmetry → **WHIM descriptors** (⊙ directional WHIM descriptors)

disconnected graph → **graph** (⊙ connected graph)

disjoint principal properties → **principal component analysis**

dispersion → **distance matrix**

distance code centric index → **centric indices**

distance connectivity index : *Balaban distance connectivity index* → **Balaban distance connectivity indices**

distance-counting descriptors

Proposed by Clerc and Terkovics [Clerc and Terkovics, 1990], called **start-end vectors** (or **SE-vectors**), these are a sequence of descriptors based on → *path counts* of an → *H-depleted molecular graph*. For each ith atom, the atomic path counts $^{m}P_i$ of length m ($m = 0, ..., L$) are calculated; the path counts of the same mth order are summed over all atoms to give the corresponding mth order molecular path count ^{m}P, divided by 2 for lengths $m > 0$. L is the maximum considered path length and is usually set to a reasonable number depending on the → *data set* (typically set to 5). The sequence of molecular path counts of increasing length is an **SE**-vector called a **TT**-vector, i.e. the calculation takes into account all atoms of the graph. This vector descriptor encodes information about branching, size, and the cyclicity of molecules.

To account for heteroatoms in the molecule, vertices of the graph corresponding to heteroatoms are marked by a specific label and only paths starting at the marked atoms are counted. In this case, as each path is counted only once, no correction is required in the calculation of the corresponding molecular path count. Depending on the kind of heteroatoms in the molecule, different **SE**-vectors are obtained; for example, the **NT**-vector consists of the counts of paths starting from the nitrogen atoms and ending at all of the remaining atoms.

Moreover, the topological relationship between the heteroatoms is considered by counting the paths running between a pair of heteroatoms for which a new **SE**-vector is calculated as above; however, if the considered heteroatoms are of the same type then the paths of non-zero length are counted twice, therefore the molecular path count is obtained by dividing the sum of the corresponding atomic path counts by 2. For example the **NN**-vector consists of the counts of the paths starting from the nitrogen atoms and ending at the other nitrogen atoms in the molecule.

To encode multiple bonds in the molecule, vertices of the graph incident to a multiple bond are marked by a specific label, i.e. are considered as pseudo-heteroatoms, and only paths starting at the marked atoms are counted and then summed to give new **SE**-vectors. Depending on the presence of double and triple bonds, **SE**-vectors such as **2T**-vector, **3T**-vector, etc., are obtained. Resonant bonds in aromatic systems are replaced by alternating double bonds. Accounting contemporarily for heteroatoms and multiple bonds, an atom can be marked twice if it is a heteroatom involved in a multiple bond; examples of these vectors are the **N2**-vector and **O2**-vector representing the mutual position of double bonds and nitrogen or oxygen atoms in the molecule, respectively.

The first element of each **SE**-vector, i.e. the 0-order molecular path count, corresponds to the count of the considered graph elements: for example, the first entry in the **TT**-vector is the number of atoms in the molecule, in the **OT**-vector it is the number of oxygen atoms, and in the **2T**-vector the number of double bonds.

The final distance-counting descriptor is obtained by chaining in an arbitrary but constant way all of the calculated **SE**-vectors, such as

$$\langle \mathrm{TT}_0, \mathrm{TT}_1, \cdots, \mathrm{TT}_L; \mathrm{NT}_0, \mathrm{NT}_1, \cdots, \mathrm{NT}_L; \mathrm{NN}_1, \mathrm{NN}_2, \cdots, \mathrm{NN}_L; \cdots \rangle$$

where the number of entries is given by the following:

$$\text{number of entries} = \frac{N(N+1)}{2} \cdot (L+1)$$

N being the number of different atom types (and pseudo-heteroatoms) and L the maximum path length. It follows that the dimension of the **SE**-vectors increases linearly with the maximum path length and with the square of the number of heteroatoms.

SE-vectors describe the global topology of a molecule, also taking into account the presence of heteroatoms and multiple bonds as well as their numbers and relative positions. However, they do not encode information about stereochemistry of molecules.

In order to obtain **SE**-vectors not depending on molecular size a normalization scheme is introduced. The **TT**-vector is normalized with reference to the chain graph, by first subtracting the ^{CG}**TT**-vector of the chain graph and then dividing it:

$$\left\langle \frac{\mathrm{TT}_0 - {}^{CG}\mathrm{TT}_0}{{}^{CG}\mathrm{TT}_0}, \frac{\mathrm{TT}_1 - {}^{CG}\mathrm{TT}_1}{{}^{CG}\mathrm{TT}_1}, \ldots, \frac{\mathrm{TT}_L - {}^{CG}\mathrm{TT}_L}{{}^{CG}\mathrm{TT}_L} \right\rangle$$

Example : 1-ene-2-methyl-butenoic acid

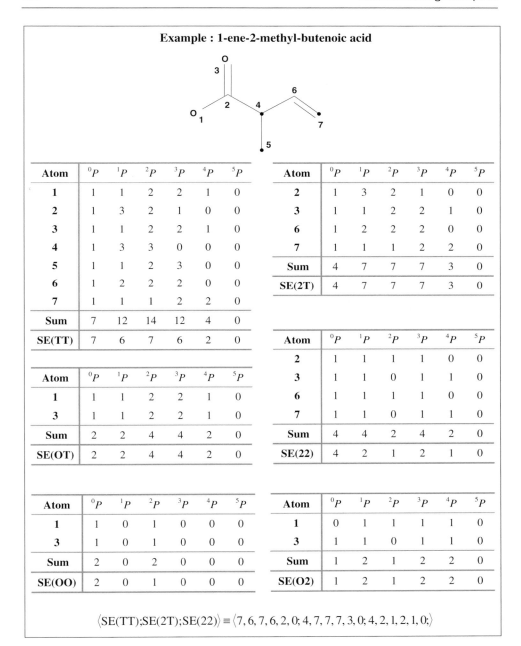

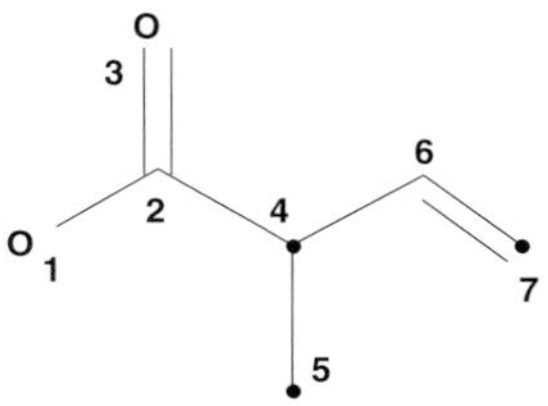

Atom	^0P	^1P	^2P	^3P	^4P	^5P
1	1	1	2	2	1	0
2	1	3	2	1	0	0
3	1	1	2	2	1	0
4	1	3	3	0	0	0
5	1	1	2	3	0	0
6	1	2	2	2	0	0
7	1	1	1	2	2	0
Sum	7	12	14	12	4	0
SE(TT)	7	6	7	6	2	0

Atom	^0P	^1P	^2P	^3P	^4P	^5P
2	1	3	2	1	0	0
3	1	1	2	2	1	0
6	1	2	2	2	0	0
7	1	1	1	2	2	0
Sum	4	7	7	7	3	0
SE(2T)	4	7	7	7	3	0

Atom	^0P	^1P	^2P	^3P	^4P	^5P
1	1	1	2	2	1	0
3	1	1	2	2	1	0
Sum	2	2	4	4	2	0
SE(OT)	2	2	4	4	2	0

Atom	^0P	^1P	^2P	^3P	^4P	^5P
2	1	1	1	1	0	0
3	1	1	0	1	1	0
6	1	1	1	1	0	0
7	1	1	0	1	1	0
Sum	4	4	2	4	2	0
SE(22)	4	2	1	2	1	0

Atom	^0P	^1P	^2P	^3P	^4P	^5P
1	1	0	1	0	0	0
3	1	0	1	0	0	0
Sum	2	0	2	0	0	0
SE(OO)	2	0	1	0	0	0

Atom	^0P	^1P	^2P	^3P	^4P	^5P
1	0	1	1	1	1	0
3	1	1	0	1	1	0
Sum	1	2	1	2	2	0
SE(O2)	1	2	1	2	2	0

$$\langle \text{SE(TT);SE(2T);SE(22)} \rangle \equiv \langle 7,6,7,6,2,0;4,7,7,7,3,0;4,2,1,2,1,0; \rangle$$

Box D-3.

The other **SE**-vectors are normalized dividing each entry by the number of considered heteroatoms or multiple bonds, i.e. the value of the first entry in the vector, as for example:

$$\left\langle \frac{NT_0}{NT_0}, \frac{NT_1}{NT_0}, \ldots, \frac{NT_L}{NT_0} \right\rangle$$

Since for the **XY**-vectors, where X and Y are two different properties (heteroatoms and/or multiple bonds), the first entry is zero by definition; thus they are normalized dividing each entry by the total sum of the vector entries.

A further development of **SE**-vectors was aimed at considering the information about molecular geometry. Both the geometry matrix and the topological distance matrix are built and the mth order path counts are calculated by summing the → *geometric distance* between all those atoms separated by m bonds divided by the corresponding topological distances, i.e. m. Then to obtain an integer each term is divided by the mean length of all paths running from the starting atom.

📖 [Baumann *et al.*, 1997a] [Baumann and Clerc, 1997b] [Affolter *et al.*, 1997] [Baumann, 1999]

distance degree centric index → **centric indices**

distance degree sequence : *vertex distance code* → **distance matrix**

distance/detour quotient matrix → **detour matrix**

distance/distance matrix (DD)

A square symmetric matrix $A \times A$, A being the number of molecule atoms, whose entries are the quotient of the corresponding elements of the molecular → *geometry matrix* G and the graph → *distance matrix* **D**; therefore, each entry is defined as:

$$[\mathbf{DD}]_{ij} = \frac{[G]_{ij}}{[\mathbf{D}]_{ij}} = \frac{r_{ij}}{d_{ij}}$$

where r_{ij} is the → *geometric distance* and d_{ij} the → *topological distance* [Randic *et al.*, 1994b; Randic and Krilov, 1999].

The row sums of this matrix contain information on the molecular folding; in fact, in highly folded structures, they tend to be relatively small as the interatomic distances are small while the topological distances increase as the size of the structure increases. Therefore, the average row sum is a molecular invariant called the **average distance/distance degree**, i.e.

$$ADDD = \frac{1}{A} \cdot \sum_{i=1}^{A} \sum_{j=1}^{A} \frac{r_{ij}}{d_{ij}}$$

while the half sum of all distance/distance matrix entries is another molecular descriptor called the **D/D index**, i.e.

$$D/D = \mathcal{W}(\mathbf{DD}) = \frac{1}{2} \cdot \sum_{i=1}^{A} \sum_{j=1}^{A} \frac{r_{ij}}{d_{ij}}$$

where $\mathcal{W}$ is the → *Wiener operator*.

From the largest eigenvalue of the distance/distance matrix a → *folding degree index* was also defined.

distance distribution moments → **distance matrix**

distance-enhanced exponential sum connectivities → **exponential sum connectivities**

distance exponent index → **amino acid side chain descriptors**

Distance Geometry (DG)

A QSAR method proposed with the aim of automatically finding the simplest receptor binding site consistent with the binding data, based on the following assumptions: 1) binding is observed to occur on a single receptor site; 2) each ligand molecule has a well-determined chemical 3D-structure, but its flexibility is also taken into account; 3) no chemical modification of the molecules occurs during the binding, although their conformations may change; 4) the free energy of such a conformational change is small compared to the free energy of the binding; 5) the experimental free energy of binding is modeled by summing the interaction energies for all contact distances between parts of the ligand molecule and receptor sites; 6) the receptor site is considered relatively rigid with respect to ligand conformational flexibility [Crippen, 1979; Crippen, 1980; Crippen, 1981; Ghose and Crippen, 1990].

The basis of the DG method is that each ligand molecule is represented as a collection of points in space, each corresponding to atoms or groups of atoms. For a chosen molecular representation, the conformation of the molecule is described in terms of distance between points, i.e. → *interatomic distances*. The matrix containing interatomic distances between all possible pairs of points is the → *geometry matrix* of the molecule. To account for molecular flexibility, a matrix of lower bounds on the considered interatomic distances and a matrix of upper bounds are also defined; fixed interatomic distances are represented by equal values in these matrices. It should be noted that the calculation of the atomic coordinates is a really difficult problem, and several mathematical techniques have been proposed [Crippen, 1977; Crippen, 1978; Crippen, 1991].

The binding sites of the receptor are represented in an analogous way, i.e. representing the interesting binding site regions by points and collecting their relative positions in a receptor geometry distance matrix. However, unlike molecule points, the site points may be called either "empty" or "filled". An empty site point is a vacant place where a ligand point might be lying when binding takes place, whereas a filled site point indicates the position of receptor steric blocking groups, precluding the presence of any ligand point during binding.

The free energy of binding is calculated in a simplified all-or-nothing fashion by adding up the contribution from each ligand point and site point "contact". The individual interaction energy contributions, taken from a reference list, are collected into an energy matrix where each row corresponds to a type of ligand point and each column to a type of receptor site point. If the fit between the calculated and the experimental binding free energy is not satisfactory, the interaction energy contributions or the number and/or geometry of the site points can be changed.

An extension of the distance geometry approach is given by the **Voronoi binding site models**, proposed with the aim of reducing excessive details in site model shape [Crippen, 1987; Srivastava *et al.*, 1993]. In this approach the receptor sites are not represented by points, but by non-overlapping regions, called → *Voronoi polyhedra*, that cover the whole space. Each atom would always lie in one and only one region, i.e. in a Voronoi polyhedron, and a binding mode would consist of a listing of the regions in which each atom is located [Boulu and Crippen, 1989; Boulu *et al.*, 1990].

📖 [Gordon, 1980] [Ghose and Crippen, 1982] [Ghose and Crippen, 1983] [Ghose and Crippen, 1984] [Ghose and Crippen, 1985a] [Ghose and Crippen, 1985b] [Sheridan *et al.*, 1986] [Ghose *et al.*, 1995] [Grdadolnik and Mierke, 1997]

distance index : *vertex distance degree* → **distance matrix**

distance matrix (D)

Derived from the → *molecular graph* G, the distance matrix (viz., **vertex distance matrix**) summarizes in matrix form the topological distance information between all the atom pairs [Trinajstic, 1992]. The **topological distance** d_{ij} is the number of edges in the shortest → *path* $^m p_{ij}$ between the vertices v_i and v_j, i.e. the length of the → *geodesic* between v_i and v_j :

$$d_{ij} = \begin{cases} |^{\min}p_{ij}| & if\ i \neq j \\ 0 & if\ i = j \end{cases}$$

The off-diagonal entries d_{ij} of the distance matrix equal one if vertices v_i and v_j are adjacent (i.e. the atoms *i* and *j* are bonded and $d_{ij} = a_{ij} = 1$ where a_{ij} are elements of the → *adjacency matrix* **A**) and are greater than one otherwise. The diagonal elements are of course equal to zero. The distance matrix is symmetric with dimension $A \times A$, where A is the number of atoms, and is usually derived from an → *H-depleted molecular graph*.

For edge-weighted graphs, the distance matrix entry *i-j* could be defined as the minimum sum of edge weights along the path between the vertices v_i and v_j, which is not necessarily the shortest possible path between them, or otherwise as the sum of the weights of the edges along the shortest path between the considered vertices (→ *weighted distance matrices*).

The maximum value entry in the *i*th row is called the **atom eccentricity** η_i (or **vertex eccentricity**), i.e. the maximum distance from the *i*th vertex to any other vertices:

$$\eta_i = \max_j(d_{ij})$$

From the eccentricity definition, a graph G can be immediately characterized by two → *molecular descriptors* known as **topological radius** R and **topological diameter** D. The topological radius of a molecule is defined as the minimum atom eccentricity and the topological diameter is defined as the maximum atom eccentricity, according to the following

$$R = \min_i(\eta_i) \quad \text{and} \quad D = \max_i(\eta_i)$$

Other simple molecular descriptors are calculated as functions of atom eccentricity [Konstantinova, 1996], viz. **eccentricity** η, **average atom eccentricity** $\bar{\eta}$, and **eccentric** $\Delta\eta$ are defined, respectively, as the following:

$$\eta = \sum_{i=1}^{A} \eta_i \qquad \bar{\eta} = \frac{1}{A} \cdot \sum_{i=1}^{A} \eta_i \qquad \Delta\eta = \frac{1}{A} \cdot \sum_{i=1}^{A} |\eta_i - \bar{\eta}|$$

From the frequencies of the row entries, the **vertex distance code** (or **distance degree sequence** of a vertex) is defined as the ordered sequence of the occurrences of the increasing distance values for the *i*th considered vertex,

$$\langle {}^1f_i, {}^2f_i, {}^3f_i, \ldots, {}^{\eta_i}f_i \rangle$$

where ${}^1f_i$, ${}^2f_i$, ${}^3f_i$, ... , called **vertex distance counts**, indicate the frequencies of distances equal to 1, 2, 3, ..., respectively, from vertex v_i to any other vertex and η_i is the *i*th atom eccentricity. The vertex distance count of first order ${}^1f_i$ coincides with the → *vertex degree* δ_i, i.e. the number of first neighbours, while ${}^2f_i$ and ${}^3f_i$ correspond to the → *connection number* (i.e. number of second neighbors) and *polarity number* (i.e. number of third neighbors) for the ith vertex, respectively.

The total number of distances in the graph equal to k is called **graph distance count** of kth order ${}^kf$; it is obtained as:

$$ {}^kf = \frac{1}{2} \cdot \sum_{i=1}^{A} {}^kf_i $$

The **graph distance code** is the ordered sequence of graph distance counts:

$$ \langle {}^1f, {}^2f, {}^3f, \ldots, {}^Df \rangle $$

where D is the topological diameter.

The **graph distance index** is defined as the squared sum of all graph distance counts:

$$ GDI = \sum_{k=1}^{D} \left({}^kf\right)^2 $$

The **distance distribution moments** D_m are the moments for the distribution of topological distances d_{ij} in a molecular graph defined as:

$$ D_m = \sum_{i=1}^{A-1} \sum_{j=i+1}^{A} d_{ij}^m $$

where m denotes the mth power of the distance matrix elements [Klein and Gutman, 1999]. Note that D_1 is the → *Wiener index W*. Normalized distance distribution moments were used to define → *molecular profiles* and the second moment D_2 takes part in defining the → *hyper-Wiener index*.

The **mean square distance index** *MSD* [Balaban, 1983a] is calculated from the second order distance distribution moment as:

$$ MSD \equiv D^{(2)} = \frac{\left(\sum_{i=1}^{A} \sum_{j=1}^{A} \left(d_{ij}^2 \right) \right)^{1/2}}{A \cdot (A-1)} = \left(\frac{\sum_{k=1}^{D} {}^kf \cdot k^2}{\sum_{k=1}^{D} {}^kf} \right)^{1/2} $$

where ${}^kf$ is the graph distance count of order k, D is the topological diameter, and A is the number of atoms. The same index restricted to the → *terminal vertices* (vertices of degree one) is called the **endpoint mean square distance index** D_1 and is applicable only on acyclic graphs.

Both the *MSD* and D_1 indices decrease with increasing → *molecular branching* in an isomeric series.

The **vertex distance degree** (or **distance number**, **distance index**, **distance rank**, **vertex distance sum**, **distance of a vertex**) is the row sum σ_i of the distance matrix **D**:

$$ \sigma_i = \mathcal{R}_i(\mathbf{D}) = \sum_{j=1}^{A} d_{ij} = \sum_{k=1}^{\eta_i} {}^kf_i \cdot k $$

where $\mathcal{R}$ is the → *row sum operator*, ${}^kf_i$ is the vertex distance count of kth order and the sum runs over the different distance values.

All these quantities are → *local vertex invariants*. High values of the vertex distance sum σ are observed for → *terminal vertices* while low values for → *central vertices*. Moreover, among the terminal vertices, the vertex distance degrees are small if the vertex is close to a branching site and larger if the terminal vertex is far away.

Topological distance, distance degrees, eccentricities, topological radius and diameter, and frequencies are used to calculate several → *topological information indices* and other molecular descriptors such as → *eccentric connectivity index*, and → *Petitjean shape indices*. The most popular topological index is the Wiener index W calculated as the half-sum of all distance matrix entries.

The **polarity number** p (or **Wiener polarity number**) was also defined by Wiener in 1947 [Wiener, 1947c] as the number of pairs of graph vertices which are separated by three edges. It is usually assumed that the polarity number accounts for the flexibility of acyclic structures, p being equal to the number of bonds around which free rotations can take place. Moreover, it relates to the steric properties of molecules.

The polarity number is usually calculated on the distance matrix as the number of pairs of vertices at a topological distance equal to three, i.e.

$$p_2 = {}^3f$$

where ${}^3f$ is the graph distance count of 3rd order, i.e. the count of entries equal to three in the upper or lower triangular submatrix **D** [Platt, 1947; Platt, 1952]. In this case the polarity number is denoted by p_2 to distinguish it from the original p. For acyclic graphs $p = p_2$, while in cycle-containing graphs the two numbers are generally different. Recently two other definitions of polarity number, based on the same original concept, were introduced to have a greater discriminating ability among cyclic structures suitable for QSAR modelling purposes [Lukovits and Linert, 1998]. The polarity number p_3 was defined as the number of ways a path of length three can be laid upon the hydrogen-depleted graph. Moreover, the polarity number p_4 was defined as the number of ways a path of length three can be laid upon the acyclic edges of the graph (including those cases, when the second edge of the considered path coincides with a cyclic edge) + $1.8 \times \mathrm{N}$, where N is the number of ways the path of length three can be laid upon the cyclic part of the graph; all edges not belonging to a cycle are acyclic edges and the product $1.8 \times \mathrm{N}$ has to be rounded to yield an integer. For acyclic structures, $p = p_2 = p_3 = p_4$ (Table D-1).

Table D-1. Values of polarity numbers p, p_2, p_3, p_4 and Wiener index W for some molecules.

Compound	p	p_2	p_3	p_4	W
cyclopropane	0	0	3	5	3
cyclobutane	4	0	4	7	8
methyl-cyclopropane	2	0	3	5	8
n-pentane	2	2	2	2	28
cyclopentane	5	0	5	9	15
i-propylcyclopentane	11	6	11	9	62
n-propylcyclopentane	11	5	10	10	67

The average row sum of the distance matrix is a molecular invariant called the **average distance degree** defined as:

$$\bar{\sigma} = \frac{1}{A} \cdot \sum_{i=1}^{A} \sigma_i = \frac{2W}{A}$$

where W is the Wiener index; the **mean distance degree deviation** $\Delta\sigma$ is defined as:

$$\Delta\sigma = \frac{1}{A} \cdot \sum_{i=1}^{A} |\sigma_i - \bar{\sigma}|$$

The minimum value of the vertex distance degrees is another molecular invariant called the **unipolarity**:

$$\sigma^* = \min_i(\sigma_i)$$

The total sum of the entries of the distance matrix is called the **Rouvray index** I_{ROUV} (or **rank distance** or **total vertex distance**), which is twice the Wiener index W:

$$I_{\mathrm{ROUV}} \equiv S(\mathbf{D}) = \sum_{i=1}^{A}\sum_{j=1}^{A} d_{ij} = \sum_{i=1}^{A} \sigma_i = 2\,W$$

where S is the → *total sum operator*.

Example : 2-methylpentane

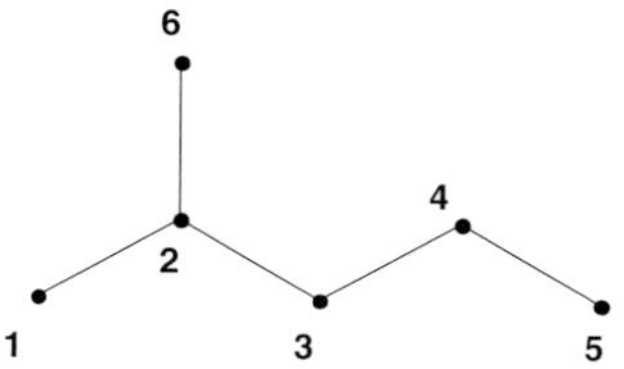

Distance matrix **D**

Atom	**1**	**2**	**3**	**4**	**5**	**6**	σ_i	η_i
1	0	1	2	3	4	2	12	4
2	1	0	1	2	3	1	8	3
3	2	1	0	1	2	2	8	2
4	3	2	1	0	1	3	10	3
5	4	3	2	1	0	4	14	4
6	2	1	2	3	4	0	12	4

Distance counts

Atom	^1f	^2f	^3f	^4f
1	1	2	1	1
2	3	1	1	0
3	2	3	0	0
4	2	1	2	0
5	1	1	1	2
6	1	2	1	1
half-sum	5	5	3	2

$$R = \min_i(\eta_i) = 2 \qquad D = \max_i(\eta_i) = 4$$

$$\eta = \sum_{i=1}^{6} \eta_i = 20 \qquad \bar{\eta} = \frac{1}{6}\cdot\sum_{i=1}^{6}\eta_i = \frac{20}{6} = 3.333 \qquad \Delta\eta = \frac{1}{6}\cdot\sum_{i=1}^{6}\left|\eta_i - 3.333\right| = 0.6667$$

$$I_{\mathrm{ROUV}} = 2W = \sum_{i=1}^{6}\sum_{j=1}^{6} d_{ij} = \sum_{i=1}^{6}\sigma_i = 64 \qquad \bar{\sigma} = \frac{1}{6}\cdot\sum_{i=1}^{6}\sigma_i = \frac{64}{6} = 10.667 \qquad \Delta\sigma = \frac{1}{6}\cdot\sum_{i=1}^{6}\left|\sigma_i - 10.667\right| = 2$$

$$\sigma^* = \min_i(\sigma_i) = 8 \qquad \Delta\sigma^* = 2W - 6\cdot\sigma^* = 64 - 6\cdot 8 = 16 \qquad \Delta\sigma^+ = \max_i(\sigma_i - 8) = 6$$

$$\log(\mathrm{PRS}) = \sum_{i=1}^{6}\log(\sigma_i) = 6.1107$$

$$GDI = \sum_{k=1}^{4}\left({}^k f\right)^2 = 63 \qquad MSD = \left(\frac{\sum_{k=1}^{4} {}^k f\cdot k^2}{\sum_{k=1}^{4} {}^k f}\right)^{1/2} = \left(\frac{5+5\cdot 4+3\cdot 9+2\cdot 16}{5+5+3+2}\right)^{1/2} = 2.3664$$

Box D-4.

Other molecular invariants immediately derived from the distance matrix are the **centralization** $\Delta\sigma^*$, **variation** $\Delta\sigma^+$, and **dispersion** σ_2^*, defined respectively as:

$$\Delta\sigma^* = 2 \cdot W - A \cdot \sigma^* \qquad \Delta\sigma^+ = \max_i(\sigma_i - \sigma^*) \qquad \sigma_2^* = \min_i\left(\frac{1}{A} \cdot \sum_{j=1}^{A} d_{ij}^2\right)$$

Another molecular descriptor is the **PRS index** (or **Product of Row Sums index**), defined as the product of the vertex distance degrees σ_i:

$$PRS = \prod_{i=1}^{A} \sigma_i \qquad \text{or} \qquad \log(PRS) = \log\left(\prod_{i=1}^{A} \sigma_i\right) = \sum_{i=1}^{A} \log(\sigma_i)$$

where the second expression can be preferred due to the large values that can be reached by the PRS index [Schultz *et al.*, 1992]. This index is related to the → *permanent* of the distance matrix.

Other important molecular descriptors obtained from the distance matrix are → *determinant-based descriptors* and → *Balaban distance connectivity indices.*

The **Molecular Distance-Edge vector** (MDE vector), denoted by λ, was proposed as a molecular ten-dimensional vector descriptor based on the geometric means of the topological distances between carbon atoms of predefined type [Liu *et al.*, 1998; Liu *et al.*, 1999]; the four types of carbon atoms were classified simply as primary carbon C_1 (3 bonded hydrogens), secondary C_2 (2 bonded hydrogens), tertiary C_3 (1 bonded hydrogens), and quaternary C_4 (no bonded hydrogens). The single elements are defined as:

$$\lambda_{st} = \frac{n_{st}}{\bar{d}_{st}^2} \qquad s = 1,\ 2,\ 3,\ 4; \quad t \geq s$$

where

$$\bar{d}_{st} = \prod\nolimits_{s \leq t} \left(d_{i(s),j(t)}\right)^{1/(2 \cdot n_{st})}$$

The geometric mean takes into account all the distances between carbon atoms i and j of types s and t; n_{st} is the number of possible atom pairs for a fixed combination of carbon types. λ_{st} is set at zero by definition if no atom pairs with types s and t are present in the molecular graph.

The **reciprocal distance matrix $\mathbf{D}^{-1}$** (or **Harary matrix**) is a square symmetric $A \times A$ matrix derived from the distance matrix whose entries are defined as:

$$\left[\mathbf{D}^{-1}\right]_{ij} = \begin{cases} 1/d_{ij} & \text{if } i \neq j \\ 0 & \text{if } i = j \end{cases}$$

i.e. each off-diagonal element is the reciprocal of the topological distance d between the considered vertices [Ivanciuc *et al.*, 1993b; Plavsic *et al.*, 1993b]; the diagonal elements are zero by definition. The original → *Harary index H* is calculated from this matrix as an analog of the Wiener index.

The **reciprocal distance sum** RDS_i of the ith vertex is a local invariant defined as the sum of the reciprocal distance matrix elements in the ith row:

$$RDS_i = \mathcal{R}_i\left(\mathbf{D}^{-1}\right) = \sum_{j=1}^{A} d_{ij}^{-1}$$

where $\mathcal{R}$ is the row sum operator. From these local vertex invariants, two molecular descriptors, called ***RDSQ* index** and ***RDCHI* index** respectively, were defined, based on a Randic-type formula [Ivanciuc *et al.*, 1993b]:

$$RDSQ = \sum_{b=1}^{B} \left(RDS_i \cdot RDS_j\right)_b^{1/2} \qquad RDCHI = \sum_{b=1}^{B} \left(RDS_i \cdot RDS_j\right)_b^{-1/2}$$

Example : 2-methylpentane

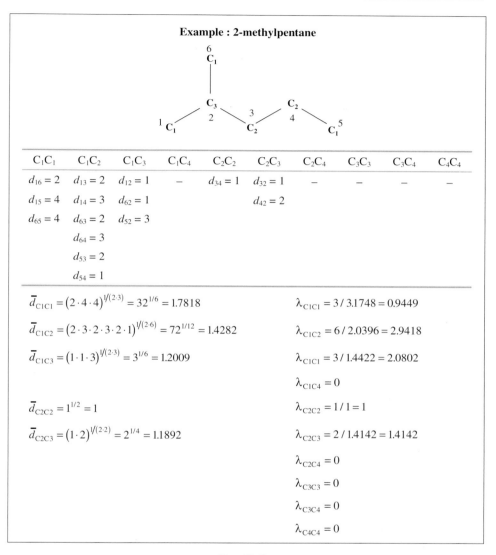

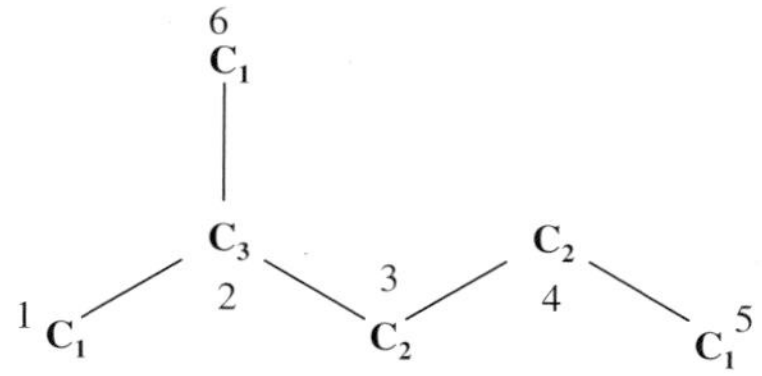

C_1C_1	C_1C_2	C_1C_3	C_1C_4	C_2C_2	C_2C_3	C_2C_4	C_3C_3	C_3C_4	C_4C_4
$d_{16} = 2$	$d_{13} = 2$	$d_{12} = 1$	–	$d_{34} = 1$	$d_{32} = 1$	–	–	–	–
$d_{15} = 4$	$d_{14} = 3$	$d_{62} = 1$			$d_{42} = 2$				
$d_{65} = 4$	$d_{63} = 2$	$d_{52} = 3$							
	$d_{64} = 3$								
	$d_{53} = 2$								
	$d_{54} = 1$								

$$\bar{d}_{\mathrm{C1C1}} = (2 \cdot 4 \cdot 4)^{1/(2\cdot 3)} = 32^{1/6} = 1.7818 \qquad \lambda_{\mathrm{C1C1}} = 3/3.1748 = 0.9449$$

$$\bar{d}_{\mathrm{C1C2}} = (2 \cdot 3 \cdot 2 \cdot 3 \cdot 2 \cdot 1)^{1/(2\cdot 6)} = 72^{1/12} = 1.4282 \qquad \lambda_{\mathrm{C1C2}} = 6/2.0396 = 2.9418$$

$$\bar{d}_{\mathrm{C1C3}} = (1 \cdot 1 \cdot 3)^{1/(2\cdot 3)} = 3^{1/6} = 1.2009 \qquad \lambda_{\mathrm{C1C1}} = 3/1.4422 = 2.0802$$

$$\lambda_{\mathrm{C1C4}} = 0$$

$$\bar{d}_{\mathrm{C2C2}} = 1^{1/2} = 1 \qquad \lambda_{\mathrm{C2C2}} = 1/1 = 1$$

$$\bar{d}_{\mathrm{C2C3}} = (1 \cdot 2)^{1/(2\cdot 2)} = 2^{1/4} = 1.1892 \qquad \lambda_{\mathrm{C2C3}} = 2/1.4142 = 1.4142$$

$$\lambda_{\mathrm{C2C4}} = 0$$

$$\lambda_{\mathrm{C3C3}} = 0$$

$$\lambda_{\mathrm{C3C4}} = 0$$

$$\lambda_{\mathrm{C4C4}} = 0$$

Box D-5.

where sums run over all the pairs *i-j* of bonded atoms. While the *RDSQ* index increases with both molecular size and → *molecular branching*; the *RDCHI* index increases with molecular size but decreases with molecular branching. Moreover, the reciprocal distance sums of the graph vertices were proposed as an alternative to vertex distance degrees to detect the → *graph centre* [Plavsic *et al.*, 1993b].

Analogously, the **reciprocal square distance matrix $\mathbf{D}^{-2}$** is derived from the distance matrix by the following:

$$[\mathbf{D}^{-2}]_{ij} = \begin{cases} 1/d_{ij}^2 & \text{if } i \neq j \\ 0 & \text{if } i = j \end{cases}$$

i.e. each off-diagonal element is the square reciprocal of the corresponding element of the distance matrix. This matrix can be useful to take into account the fact that atom interactions decrease as the square distance between atoms increases, such as in the → *topological charge indices*. A variant of the Harary index based on this matrix was also proposed [Plavsic *et al.*, 1993b].

A modified reciprocal distance matrix was proposed in order to obtain **constant interval reciprocal indices**; the idea was that reciprocal distances should be uniformly spaced as topological distances [Schultz and Schultz, 1998; Schultz and Schultz, 2000]. The matrix elements are:

$$rd_{ij} = \frac{d^{\max} - d_{ij} + 1}{d^{\max}}$$

where d_{ij} are elements of the distance matrix, $d^{\max}$ is the largest integer in the distance matrix or in the ith row or arbitrarily user-defined. If $d^{\max}$ is chosen as the largest value in each row, the resulting reciprocal matrix is unsymmetric. Using these modified reciprocal distance matrices instead of the classical reciprocal distance matrix, CIRD indices are defined in analogy with the Harary index H, and CIRMTI and CIRS indices in analogy with the → *reciprocal Schultz indices*.

Opposite to the distance matrix is the → *detour matrix*, where the entries correspond to the length of the longest path between the vertices. Other related topological matrices are the → *distance/distance matrix*, the → *detour/distance matrix* and the → *distance/detour quotient matrix*.

The distance matrix derived from a graph must not be confused with the distance matrix derived from a → *data set*, called here → *data distance matrix*.

📖 [Entringer *et al.*, 1976] [Doyle and Garver, 1977] [Bersohn, 1983] [Plesnik, 1984] [Rouvray, 1986] [Müller *et al.*, 1987] [Senn, 1988] [Mihalic *et al.*, 1992a] [Thangavel and Venuvanalingam, 1993] [Kunz, 1993] [Kunz, 1994] [Dobrynin and Gutman, 1994] [Dobrynin, 1995] [Chepoi, 1996] [Ivanciuc *et al.*, 1997b] [Ivanciuc and Ivanciuc, 2000]

distance measures → **similarity/diversity**

distance-normalized exponential sum connectivities → **exponential sum connectivities**

distance number : *vertex distance degree* → **distance matrix**

distance-path matrix (D_P)

A square symmetric matrix $A \times A$ whose off-diagonal entry *i-j* is the count of all paths of any length m ($1 \leq m \leq d_{ij}$) that are included in the shortest path from vertex v_i to vertex v_j (whose → *topological distance* is d_{ij}) [Diudea, 1996a]. The diagonal entries are zero.

Each entry *i-j* of the $\mathbf{D_P}$ matrix is calculated from the → *distance matrix* $\mathbf{D}$ as the following:

$$[\mathbf{D_P}]_{ij} = \binom{d_{ij}+1}{2} = \frac{d_{ij}^2 + d_{ij}}{2}$$

i.e. as all the possible combinations of two elements taken from $d_{ij} + 1$ elements (binomial coefficient).

The **hyper-distance-path index** D_P can be obtained applying the → *Wiener operator* $\mathcal{W}$ to the distance-path matrix:

$$D_P = \mathcal{W}(\mathbf{D_P}) = \frac{1}{2} \cdot \sum_{i=1}^{A} \sum_{j=1}^{A} [\mathbf{D_P}]_{ij} = \frac{1}{2} \cdot \sum_{i=1}^{A} \sum_{j=1}^{A} d_{ij} + \frac{1}{2} \cdot \sum_{i=1}^{A} \sum_{j=1}^{A} \binom{d_{ij}}{2} = W + D_\Delta$$

where in the expression on the right the first term is just the → *Wiener index W* and the second the → *delta number* D_Δ, which can be considered the "non-Wiener part" of the index [Diudea *et al.*, 1997d].

For acyclic graphs the hyper-distance-path index D_P coincides with the → *hyper-Wiener index WW* derived from the → *Wiener matrix* and with the → *hyper-detour index* derived from the → *detour-path matrix*. Moreover, it was proposed as an extension of the hyper-Wiener index for any graph [Klein *et al.*, 1995].

The **reciprocal distance-path matrix** $\mathbf{D_P}^{-1}$ is a matrix whose elements are the reciprocal of the corresponding distance-path matrix elements:

$$[\mathbf{D_P^{-1}}]_{ij} = [\mathbf{D_P}]_{ij}^{-1}$$

All elements equal to zero are left unchanged in the reciprocal matrix.

→ *Hyper-Harary indices* are defined for this matrix and → *Harary indices* for the corresponding 1st-order → *sparse matrix* $^1\mathbf{D_P}^{-1}$ applying the Wiener operator.

Example : ethylbenzene

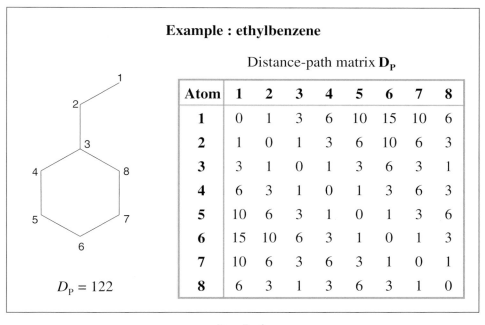

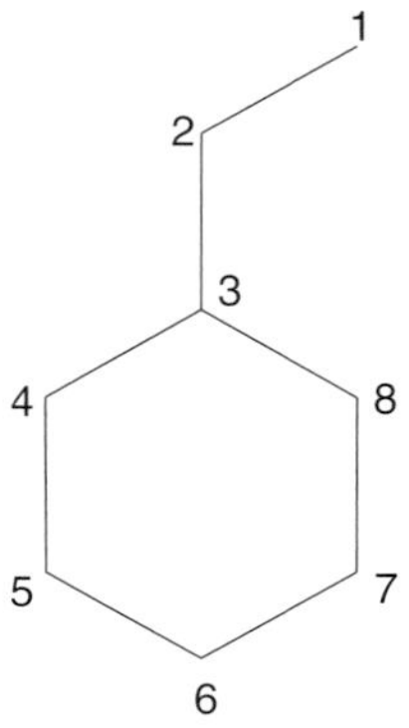

$D_P = 122$

Distance-path matrix $\mathbf{D_P}$

Atom	1	2	3	4	5	6	7	8
1	0	1	3	6	10	15	10	6
2	1	0	1	3	6	10	6	3
3	3	1	0	1	3	6	3	1
4	6	3	1	0	1	3	6	3
5	10	6	3	1	0	1	3	6
6	15	10	6	3	1	0	1	3
7	10	6	3	6	3	1	0	1
8	6	3	1	3	6	3	1	0

Box D-6.

distance of a vertex : *vertex distance degree* → **distance matrix**

distance rank : *vertex distance degree* → **distance matrix**

distance-sequence matrix → **sequence matrices**

distance sum layer matrix → **layer matrices**

diversity → **similarity/diversity**

diversity matrix : *data distance matrix* → **similarity/diversity**

Dosmorov complexity index → **molecular complexity**

double-bond count → **multiple bond descriptors**

double invariants

A general approach for the construction of molecular graph-invariants where associated with each graph $\mathcal{G}$ is a matrix whose elements are molecular subgraphs [Randic *et al.*, 1997b]. Matrices of this kind are called **Generalized Graph matrices** (or **GG matrices**); they are square symmetric matrices of dimension $A \times A$, A being the number of graph vertices, whose entry *i-j* is defined as the subgraph that contains all the shortest paths between vertices v_i and v_j. In acyclic graphs and graphs with odd-membered rings, only one shortest path exists between each pair of vertices; therefore all **GG** matrix entries are paths of different lengths; for this reason the **GG** matrix is called the **path matrix**. In cyclic graphs with even-membered rings there may be several shortest paths between a pair of vertices, resulting in cyclic subgraphs. The diagonal elements of the **GG** matrix are → *isolated vertices.*

Example : 2-methylpentane

First invariant: leading eigenvalue of the adjacency matrix **A**
Second invariant: leading eigenvalue of **GG** matrix

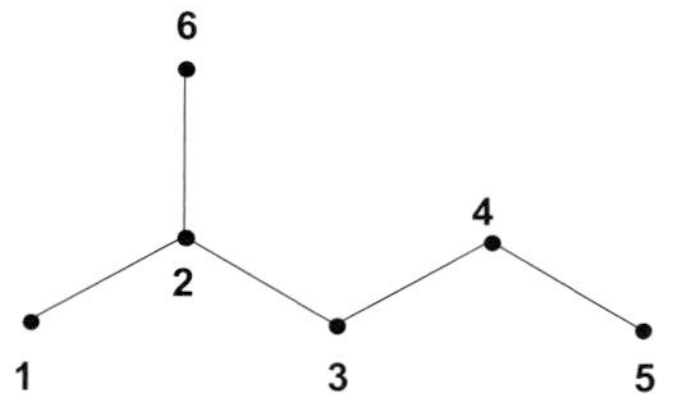

$\lambda\lambda_1 = 6.8313$

GG matrix

Atom	1	2	3	4	5	6
1	0	^1p	^2p	^3p	^4p	^2p
2	^1p	0	^1p	^2p	^3p	^1p
3	^2p	^1p	0	^1p	^2p	^2p
4	^3p	^2p	^1p	0	^1p	^3p
5	^4p	^3p	^2p	^1p	0	^4p
6	^2p	^1p	^2p	^3p	^4p	0

Leading eigenvalues of **GG** matrix subgraphs

Atom	1	2	3	4	5	6
1	0	1.0000	1.4142	1.6180	1.7321	1.4142
2	1.0000	0	1.0000	1.4142	1.6180	1.0000
3	1.4142	1.0000	0	1.0000	1.4142	1.4142
4	1.6180	1.4142	1.0000	0	1.0000	1.6180
5	1.7321	1.6180	1.4142	1.0000	0	1.7321
6	1.4142	1.0000	1.4142	1.6180	1.7321	0

Box D-7.

Double invariants are descriptors derived from **GG** matrices by the following:

Each entry (a subgraph) of the **GG** matrix can be transformed into a matrix which is a submatrix of the **GG** matrix and any topological matrix (e.g. → *adjacency matrix* **A**, → *distance matrix* **D**, → *Wiener matrix* **W**, → *distance/distance matrix* **DD**, → *distance/detour quotient matrix* **D/Δ**, etc) can be chosen to represent molecular subgraphs.

a) the entries (subgraphs) of the **GG** matrix are transformed into numerical values by any chosen graph-invariant (first invariant, I_1) applied to the submatrices representing the subgraphs;

b) a second invariant I_2, which can be different from the first invariant, is applied to the **GG** matrix whose entries are transformed as defined in a).

Therefore, double invariants are functions of the type:

$$I_2\{I_1(\mathbf{GG})\}$$

It must be noted that the two invariants have to be compatible, i.e. the domain of I_2 is in the range of I_1; moreover, the two invariant operators do not commute, i.e.

$$I_2\{I_1(\mathbf{GG})\} \neq I_1\{I_2(\mathbf{GG})\}$$

Given the large number of available graph-invariants, there are many possible combinations of compatible invariants, resulting in an explosion of new descriptors.

A molecular branching index was proposed for acyclic graphs, based on the double invariant approach and called → *$\lambda\lambda_1$ branching index*. A calculation example [Randic, 1997d; Randic, 1998a] for 2-methylpentane is given in Box D-7, where the symbol ^{m}p represents a path of length m.

This approach can be extended to 3D molecular geometry, defining **generalized geometry matrices** and corresponding **3D-double invariants**.

DP descriptor → **charge descriptors** (⊙ submolecular polarity parameter)

drug → **drug design**

drug design

Drug design is a research field where several disciplines are involved, including not only the design of drugs, but also pharmacokinetics and toxicity. Appropriate chemometric tools, such as experimental design, multivariate analysis, and artificial neural networks are usually used in the planning and evaluation of pharmacokinetics and toxicological experiments, as well as in the modelling of the biological activity of drugs. Moreover, → *similarity searching* and → *substructural analysis* are actually very useful for screening a large database of chemical compounds in the search for new drugs and pharmacophores [Purcell *et al.*, 1973; Martin, 1978; Ariëns, 1979; Franke, 1984; Hadzi and Jerman-Blazic, 1987; Dean, 1987; Tute, 1990; Ramsden, 1990; Kubinyi, 1993a; van de Waterbeemd, 1995b; van de Waterbeemd *et al.*, 1997b; Kubinyi *et al.*, 1998b; Kubinyi *et al.*, 1998c].

Drug design techniques are all implemented in software packages that run on powerful workstations and can be distinguished as three main groups: **Computer-Aided Drug Design** (CADD) involves all computer-assisted techniques used to discover, design and optimize biologically active compounds with a possible use as drugs; **Computer-Aided Molecular Design** (CAMD) involves all computer-assisted techniques used to discover, design and optimize biologically active compounds with desired structure and properties. **Computer-Aided Molecular Modelling** (CAMM, or simply **Molecular Modelling**) is the investigation of molecular structures and properties using → *computational chemistry* and graphical visualization techniques.

Some fundamental terms and concepts in drug design are briefly reviewed below [van de Waterbeemd *et al.*, 1997a; IUPAC Recommendations, 1997; IUPAC Recommendations, 1998].

A **reference compound** (or **reference structure**) is a compound assumed as the reference for some considered aspect, such as a physico-chemical or biological property, or a molecular skeleton common to a set of compounds (→ *maximum common substructure*). A **drug** is any substance for treating, curing, or preventing disease in human beings or animals. A drug may also be used to make a medical diagnosis or to restore, correct, or modify physiological functions. A **lead compound** is a compound which, because of its biological properties, is taken as a reference compound.

Lead discovery, generation and optimization are basic activities in drug design which is devoted to identifying active new chemical entities, developing new active compounds and optimizing those able to be transformed into clinically useful drugs.

Binding affinity is the tendency of a molecule to associate with another. In particular, the affinity of a drug is its ability to bind to its biological target. A **ligand** is a compound which can bind to a biological target.

A **receptor** is a molecule or a polymeric structure in a cell which acts as the biological target, recognizing and binding a compound. **Receptor mapping** is the technique used to describe geometric, electronic and other physico-chemical features of a binding site. The active site cavity of a receptor is called the **binding site cavity** (or **receptor cavity**).

A **pharmacophore** is the ensemble of steric, electronic and other physico-chemical properties that is necessary to ensure optimal supramolecular interactions with a specific biological target structure. In other words, the pharmacophore concept is based on the kinds of interactions observed in molecular recognition: hydrogen bonding, charge-charge and hydrophobic interactions characterized by defined spatial arrangements. A pharmacophore does not represent a real molecule or a real association of functional groups; it is purely an abstract concept that accounts for the common molecular interaction capacity of a set of compounds towards the target structure. A pharmacophore can be considered to be the largest common denominator shared by a set of active molecules.

The concepts of receptor and pharmacophore play a basic role in the alignment of molecules in → *grid-based QSAR techniques.*

📖 [Hansch, 1971] [Cramer III *et al.*, 1974] [Höltje and Kier, 1974] [Höltje, 1975] [Höltje, 1976] [Kier and Hall, 1976a] [Martin, 1979] [Marshall *et al.*, 1979] [Höltje and Vogelgesang, 1979] [Hübel *et al.*, 1980] [Höltje, 1982] [Austel, 1983] [Jolles and Wooldridge, 1984] [Höltje and Tintelnot, 1984] [Höltje *et al.*, 1985] [Franke and Streich, 1985a] [Franke and Streich, 1985b] [Streich and Franke, 1985] [van de Waterbeemd and Testa, 1987] [Klopman and Buyukbingol, 1988] [Ghose and Crippen, 1990] [Magee, 1990] [Seibel and Kollman, 1990] [Magee, 1991] [Doweyko, 1991] [Pastor and Alvarez-Builla, 1991] [Boudreau and Efange, 1992] [Kim, 1993b] [Höltje *et al.*, 1993] [Marshall, 1993] [Walters and Hinds, 1994] [Kaminski, 1994] [van de Waterbeemd, 1994a] [Clark and Murray, 1995] [Hahn, 1995] [Ghose *et al.*, 1995] [Kier and Testa, 1995] [Ortiz *et al.*, 1995] [Luo *et al.*, 1996] [van de Waterbeemd, 1996a] [Lipkowitz and Boyd, 1997] [Greco *et al.*, 1997] [Mekenyan and Veith, 1997] [Fujita, 1997] [Hahn and Rogers, 1998] [Ghose and Wendoloski, 1998]

DSI index → **electronegativity-based connectivity indices**

dual electronic constants σ^+ and σ^- → **electronic substituent constants** (⊙ resonance electronic constants)

Dubois steric constant → **steric descriptors** (⊙ Taft steric constant)

dummy variables : *indicator variables*

Dunn model based on surface area → **lipophilicity descriptors**

***DV* index** → **multiple bond descriptors**

d-WDEN indices : *directional WHIM density* → **WHIM descriptors** (⊙ directional WHIM descriptors)

d-WSHA indices : *directional WHIM shape* → **WHIM descriptors** (⊙ directional WHIM descriptors)

d-WSIZ indices : *directional WHIM size* → **WHIM descriptors** (⊙ directional WHIM descriptors)

d-WSYM indices : *directional WHIM symmetry* → **WHIM descriptors** (⊙ directional WHIM descriptors)

dynamic QSAR → **structure/response correlations**

dynamic reactivity indices → **reactivity indices**

D/Δ index → **detour matrix**

D/Δ ring indices → **detour matrix**

E

EA indices : *Extended Adjacency matrix indices* → **eigenvalue-based descriptors**

***EAmax* index** → **eigenvalue-based descriptors** (⊙ extended adjacency matrix indices)

***EA*Σ index** → **eigenvalue-based descriptors** (⊙ extended adjacency matrix indices)

eccentric → **distance matrix**

eccentric connectivity index (ξ^C)

A molecular descriptor based on both → *atom eccentricity* η (calculated from the → *distance matrix* **D**) and → *vertex degree* δ (calculated from the → *adjacency matrix* **A**) in the → *H-depleted molecular graph* [Sharma *et al.*, 1997]. It is defined as the sum of the products between eccentricity η and vertex degree δ over the *A* atoms of the graph:

$$\xi^c = \sum_{i=1}^{A} \eta_i \cdot \delta_i$$

eccentricity → **distance matrix**

ECN index → **electronegativity-based connectivity indices**

ECP index → **electronegativity-based connectivity indices**

edge adjacency matrix (E)

Derived from the → *molecular graph* $\mathcal{G}$, the edge adjacency matrix **E** encodes the connectivity between graph edges. It is a square symmetric matrix of dimension $B \times B$, where B is the number of bonds, and is usually derived from an → *H-depleted molecular graph* [Bonchev, 1983]. It is to be noted that the edge adjacency matrix of a graph $\mathcal{G}$ is equal to the → *adjacency matrix* of the → *line graph* of $\mathcal{G}$ [Gutman and Estrada, 1996].

The entries $[\mathbf{E}]_{ij}$ of the matrix equal one if edges e_i and e_j are adjacent (the two edges thus forming a → *path* of length two) and zero otherwise. For multigraphs, the edge adjacency matrix can be augmented by a row and a column for each multiple edge.

The **edge degree** ε_i provides the simplest information related to the considered bond and is calculated from the edge adjacency matrix as follows:

$$\varepsilon_i = \mathcal{R}_i(\mathbf{E}) = \sum_{j=1}^{B} [\mathbf{E}]_{ij}$$

where $\mathcal{R}_i$ is the → *row sum operator*.

The edge degree ε is related to the → *vertex degree* δ by the following relationship:

$$\varepsilon_i = \delta_{i(1)} + \delta_{i(2)} - 2$$

where $i(1)$ and $i(2)$ refer to the two vertices incident to the ith edge.

The number of edges whose same edge degree equals g is called the **edge degree count** ${}^gF_E$; therefore the vector

$$\langle {}^1F_E; {}^2F_E; \ldots; {}^6F_E; \rangle$$

can be associated with each graph $\mathcal{G}$, provided the maximum edge degree equals six.

Related to the previous definition is the **edge type count** $ne_{gg'}$, defined as the number of edges with the same vertex degree of the incident vertices, where g and g' are the degree values of the incident vertices.

The **total edge adjacency index** A_E (also known as **Platt number**, F [Platt, 1947; Platt, 1952]) is the sum over all entries of the edge adjacency matrix:

$$A_E \equiv F = \sum_{i=1}^{B} \sum_{j=1}^{B} [\mathbf{E}]_{ij} = \sum_{i=1}^{B} \varepsilon_i = 2 \cdot N_2$$

where N_2 is the number of graph connections.

Two adjacent edges constitute a 2nd order path and this subgraph is called **connection** (or **link**). The **connection number** N_2, also known as **Gordon-Scantlebury index** (N_{GS}) [Gordon and Scantlebury, 1964] or **Bertz branching index** (BI), is the simplest graph-invariant obtained from the edge adjacency matrix which considers both vertices and edges and is calculated as:

$$N_2 \equiv N_{GS} \equiv BI \equiv {}^2P = A_E/2$$

where A_E is the total edge adjacency index and ${}^2P$ the 2nd order path count.

The number of connections N_2 of a molecular graph is related to the → *1st Zagreb index* M_1 and the → *quadratic index* Q by the following relationships:

$$N_2 = M_1/2 - A + 1 = Q + A - 2$$

where A is the number of atoms.

When multiple bonds are considered, the connection number can be calculated on multigraphs by using more general approaches. Given an H-depleted molecular graph $\mathcal{G}$, the number of connections is equal to the number of edges in the line graph of $\mathcal{G}$. Moreover, the connection number can also be calculated as the sum of atom contributions, taking into account that each connection intersects at a particular atom and is therefore simply a function of the number of hydrogens on that atom:

$$N_2 = \frac{1}{2} \cdot \sum_{i=1}^{A} (4 - h_i) \cdot (3 - h_i) - n_{=} - 3 \cdot n_{\equiv}$$

where h_i is the number of hydrogen atoms attached to the ith atom and $n_{=}$ and $n_{\equiv}$ the number of double and triple bonds in the molecule, respectively; the sum runs over all atoms in the molecule [Hendrickson *et al.*, 1987].

As the number of connections is sensitive to different features of molecular structure such as size, branching, cyclicity and multiple bonds it was used by Bertz to define its → *molecular complexity* index.

From the edge adjacency matrix, a graph-theoretical invariant analogous to the → *Randic connectivity index* was derived by Estrada [Estrada, 1995a]; it is called **edge connectivity index**, denoted by ε, and defined as:

$$\varepsilon = \sum\nolimits_k \left(\varepsilon_i \cdot \varepsilon_j\right)_k^{-1/2}$$

where k runs over all connections N_2, and ε_i and ε_j are the edge degrees of the two edges in the connection. It coincides with the Randic connectivity index χ of the line graph of $\mathcal{G}$.

The **extended edge connectivity indices** were defined as a generalization of the edge connectivity index in analogy to the → *Kier-Hall connectivity indices*:

$${}^m\varepsilon_q = \sum_{k=1}^{K} \left(\prod\nolimits_b \varepsilon_b\right)_k^{-1/2}$$

where k runs over all of the mth order subgraphs, m being the number of edges in the subgraph; K is the total number of mth order subgraphs; the edge degrees of all the edges in the subgraph are considered. The subscript "q" for the connectivity indices refers to the type of → *molecular subgraph* and is ch for chain or ring, pc for path-cluster, c for cluster and p for path (this can also be omitted) [Estrada *et al.*, 1998a; Estrada, 2000; Estrada and Rodriguez, 2000]. Some mathematical relationships between the extended edge connectivity indices and → *line graph connectivity indices* were found.

Example : 2-methylpentane

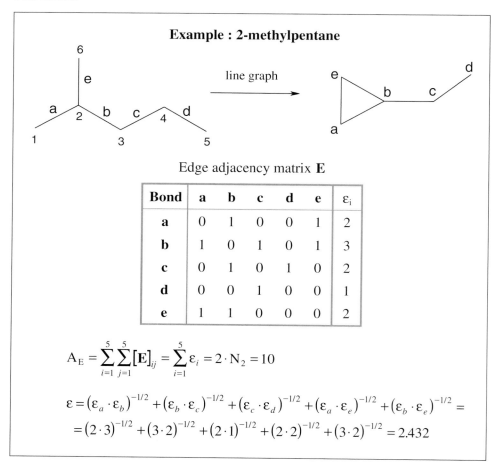

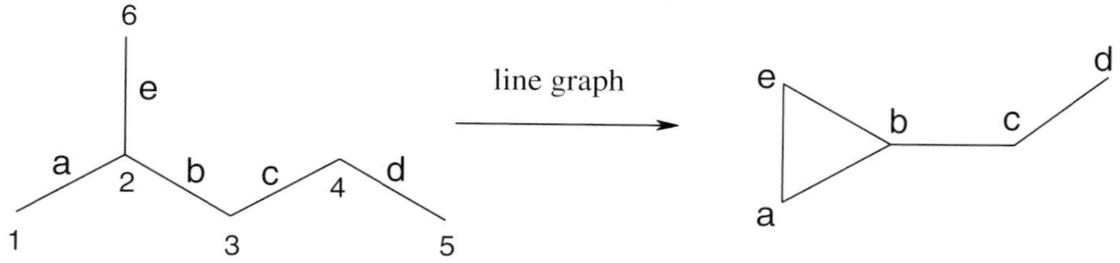

Edge adjacency matrix **E**

Bond	**a**	**b**	**c**	**d**	**e**	ε_i
a	0	1	0	0	1	2
b	1	0	1	0	1	3
c	0	1	0	1	0	2
d	0	0	1	0	0	1
e	1	1	0	0	0	2

$$A_E = \sum_{i=1}^{5}\sum_{j=1}^{5}[\mathbf{E}]_{ij} = \sum_{i=1}^{5}\varepsilon_i = 2 \cdot N_2 = 10$$

$$\varepsilon = (\varepsilon_a \cdot \varepsilon_b)^{-1/2} + (\varepsilon_b \cdot \varepsilon_c)^{-1/2} + (\varepsilon_c \cdot \varepsilon_d)^{-1/2} + (\varepsilon_a \cdot \varepsilon_e)^{-1/2} + (\varepsilon_b \cdot \varepsilon_e)^{-1/2} =$$
$$= (2 \cdot 3)^{-1/2} + (3 \cdot 2)^{-1/2} + (2 \cdot 1)^{-1/2} + (2 \cdot 2)^{-1/2} + (3 \cdot 2)^{-1/2} = 2.432$$

Box E-1.

The **spectral moments of the edge adjacency matrix E** were defined [Estrada, 1996] as:

$$\mu_k = tr(\mathbf{E}^k)$$

where k is the power of the adjacency matrix and tr the → *trace*, i.e. the sum of the diagonal elements. The zero-order spectral moment μ_0 corresponds to the number of edges in the graph, that is the trace of the resulting B-dimensional identity matrix.

Since the kth order spectral moment μ_k corresponds to the sum of all self-returning walks of length k in the line graph of the molecular graph G, it can be expressed as the linear combination of the → *embedding frequencies* of the molecular graph, i.e. the counts of different structural fragments (subgraphs) in the graph, each fragment corresponding to a specific self-returning walk. Several relations between spectral moments and embedding frequencies were derived by Estrada [Estrada, 1998b; Markovic and Gutman, 1999]. A graph-theoretical approach called **TOpological SubStructural MOlecular DEsign** (TOSS-MODE) was proposed to express physical and biological properties in terms of substructural features of molecules by the spectral moments of the edge adjacency matrix [Estrada *et al.*, 1998c]. The spectral moments are substituted in the QSAR model by their expressions in terms of structural fragments of the molecule, obtaining an equation that relates the property directly with the molecular structure and allowing → *reversible decoding*.

A **bond order-weighted edge adjacency matrix** $^{\pi}\mathbf{E}$ is obtained for weighted graphs whose edges are weighted by the → *bond order* π calculated by → *computational chemistry* methods on selected molecular geometries [Estrada and Montero, 1993]. The off-diagonal elements of the edge adjacency matrix for a weighted graph are zero for non-adjacent edges; otherwise, if two edges e_i and e_j are adjacent, the entry i-j is defined by the bond order of the jth edge, while the symmetric entry j-i is defined by the bond order of the ith edge, thus resulting in an unsymmetric matrix:

$$[^{\pi}\mathbf{E}]_{ij} = \begin{cases} 0 & \text{if } e_i \text{ and } e_j \text{ are non–adjacent} \\ \pi_j & \text{otherwise} \end{cases}$$

Therefore, the ith edge degree calculated on the weighted edge adjacency matrix is the sum of the bond orders associated with all ε_i bonds adjacent to the ith edge:

$$^{\pi}\varepsilon_i = \sum_{j=1}^{B} [^{\pi}\mathbf{E}]_{ij} = \sum_{j=1}^{\varepsilon_i} \pi_j$$

The **bond order-weighted edge connectivity index** $^{\pi}\varepsilon$ is then defined as [Estrada and Ramirez, 1996]:

$$^{\pi}\varepsilon = \sum\nolimits_k \left(^{\pi}\varepsilon_i \cdot {}^{\pi}\varepsilon_j\right)_k^{-1/2}$$

where k runs over all connections N_2.

A characteristic of this index is its sensitivity to the presence of heteroatoms, to heteroatom position in the molecule (greater values referring to central positions), and to the discriminating power of conformational isomers.

In analogy with the bond order-weighted edge adjacency matrix, a **resonance-weighted edge adjacency matrix** $^{\mathbf{k}}\mathbf{E}$ was also proposed, replacing the bond orders with parameters $k_{C\text{–}X}$ used in the Hückel matrix and related to the resonance integral $\beta_{C\text{–}X}$ of the bond between the heteroatom X and the carbon atom by the relationship:

$$\beta_{C-X} = k_{C-X} \cdot \beta_{C-C}$$

where $\beta_{C\text{–}C}$ is the resonance integral of the carbon-carbon bond [Estrada, 1995b]. The literature reports several values for $k_{C\text{–}X}$ parameters, and some proposed by Estrada are collected in Table E-1.

Table E-1. Values of the $k_{C\text{–}X}$ constants proposed by Estrada.

C – X bond	$k_{C\text{–}X}$	C – X bond	$k_{C\text{–}X}$
C – C	1.0	C – S	0.7
C – N	1.0	C – F	0.7
C – O	0.8	C – Cl	0.4
C = O	1.6	C – Br	0.3

The **resonance-weighted edge connectivity index** $^k\varepsilon$ is derived from the row sums of the above defined matrix in the same way as the edge connectivity index ε and is sensitive to the presence of heteroatoms and multiple bonds in the molecule.

A **bond distance-weighted edge adjacency matrix** $^{\mathbf{b}}\mathbf{E}$ is derived from the edge adjacency matrix using → *bond distances* calculated by computational chemistry methods as diagonal entries on selected molecular geometries [Estrada, 1997]:

$$[{}^{\mathbf{b}}\mathbf{E}]_{ij} = \begin{cases} [\mathbf{E}]_{ij} & \text{if i} \neq \text{j} \\ r_{ii} & \text{if i} = \text{j} \end{cases}$$

where $[\mathbf{E}]_{ij}$ are the usual entries of the edge adjacency matrix and r_{ii} is the bond distance, i.e. the geometric distance between the vertices incident to the ith edge. Bond distances are used as weights for edges of the molecular graph, thus allowing discrimination among the different isomers, heteroatoms, conformations, etc.

As for the edge adjacency matrix, the **spectral moments of the bond distance-weighted edge adjacency matrix** $^{\mathbf{b}}\mathbf{E}$ were defined [Estrada, 1997; Estrada, 1998a] as:

$$\mu_k = tr({}^{\mathbf{b}}\mathbf{E}^k)$$

where k is the power of the bond distance-weighted adjacency matrix and tr the trace of this matrix.

📖 [Nikolic and Trinajstic, 1998a] [Nikolic *et al.*, 1998b] [Estrada *et al.*, 1997]

edge centric indices → **centric indices**

edge centric indices for multigraphs → **centric indices**

edge chromatic decomposition → **chromatic decomposition**

edge chromatic information index → **chromatic decomposition**

edge chromatic number → **chromatic decomposition**

edge complete centric index → **centric indices**

edge connectivity → **connectivity indices**

edge connectivity index → **edge adjacency matrix**

edge counting : *bond number*

edge cyclic degree → **cycle matrices** (⊙ edge cycle matrix)

edge cycle matrix → **cycle matrices**

edge degree → **edge adjacency matrix**

edge degree count → **edge adjacency matrix**

edge distance code → **edge distance matrix**

edge distance code centric index → **centric indices**

edge distance counts → **edge distance matrix**

edge distance degree → **edge distance matrix**

edge distance degree centric index → **centric indices**

edge distance index : *edge distance degree* → **edge distance matrix**

edge distance matrix (^{E}D)

Derived from the → *graph* $\mathcal{G}$, the edge distance matrix is the edge-analog of the vertex → *distance matrix*, and summarizes in matrix form the topological distance information among all the pairs of bonds [Bonchev, 1983; Estrada and Gutman, 1996]. It is simply the distance matrix **D** of the → *line graph* of $\mathcal{G}$.

The **topological edge distance** $[^{E}\mathbf{D}]_{ij}$ between the edges e_i and e_j is defined as the length of the shortest → *path* between them, i.e. number of vertices in the shortest path connecting edges e_i and e_j not counting the terminal vertices of the path.

As each ith edge is characterized by two vertices $v_{i(1)}$ and $v_{i(2)}$ incident to the edge, the topological edge distance between the edges e_i and e_j can also be obtained from the minimum → *topological distance* d between two pairs of vertices as:

$$\left[^{E}\mathbf{D}\right]_{ij} = \min\left\{d_{i(1),j(1)}, d_{i(1),j(2)}, d_{i(2),j(1)}, d_{i(2),j(2)}\right\} + 1$$

For acyclic graphs the topological edge distance can be calculated by the following formula:

$$\left[^{E}\mathbf{D}\right]_{ij} = \frac{1}{4} \cdot \left(d_{i(1),j(1)} + d_{i(1),j(2)} + d_{i(2),j(1)} + d_{i(2),j(2)}\right)$$

The off-diagonal entries $[^{E}\mathbf{D}]_{ij}$ of the edge distance matrix equal one if edges e_i and e_j are adjacent (i.e. the edges e_i and e_j are connected and $[^{E}\mathbf{D}]_{ij} = [\mathbf{E}]_{ij} = 1$ where $[\mathbf{E}]_{ij}$ are elements of the → *edge adjacency matrix* **E**), otherwise they are greater than one. The diagonal elements equal zero. The edge distance matrix, usually derived from an → *H-depleted molecular graph*, is square symmetric with dimension $B \times B$, where B is the number of bonds.

The maximum value entry in the ith row is called **bond eccentricity** $^{b}\eta_i$ (or **edge eccentricity**):

$$^{b}\eta_i = \max_j\left(\left[^{E}\mathbf{D}\right]_{ij}\right)$$

From the eccentricity definition, a graph $\mathcal{G}$ can be immediately characterized by two molecular descriptors known as **topological radius from edge eccentricity** ^{E}R and **topological diameter from edge eccentricity** ^{E}D. The radius is defined as the minimum bond eccentricity and the diameter as the maximum bond eccentricity, according to:

$$^{E}R = \min_i\left(^{b}\eta_i\right) \quad \text{and} \quad ^{E}D = \max_i\left(^{b}\eta_i\right)$$

From the edge distance matrix several → *topological information indices* are calculated. Moreover, the atomic and molecular descriptors already defined for the vertex distance matrix are analogously defined for the edge distance matrix.

From the frequencies of the row edge distance entries, the **edge distance code** is defined as the ordered sequence of the occurrence of increasing edge distance values for the ith considered edge,

$$\left\langle {}^{1}f_i, {}^{2}f_i, {}^{3}f_i, \ldots, {}^{^{b}\eta_i}f_i \right\rangle$$

where $^{1}f_i$, $^{2}f_i$, $^{3}f_i$, ... , called **edge distance counts**, indicate the frequencies of the edge distances equal to 1, 2, 3, ..., respectively, from edge e_i to any other edge and $^{b}\eta_i$ is the ith bond eccentricity.

The **edge distance degree** (or **edge distance index**, **edge distance sum**) is the row sum $^{E}\sigma_i$ obtained by summing the ith row entries of the edge distance matrix:

$$^{E}\sigma_i = \mathcal{R}_i\left(^{E}\mathbf{D}\right) = \sum_{j=1}^{B}\left[^{E}\mathbf{D}\right]_{ij} = \sum_{k=1}^{^{b}\eta_i} {}^{k}f_i \cdot k$$

where $\mathcal{R}_i$ is the → *row sum operator* and $^{k}f_i$ is the edge distance count of kth order which runs over the different edge distance values.

The sum of the edge distance degrees, i.e. the sum of all matrix elements, is called **total edge distance** D_E and defined as:

$$D_E = S(^{\mathbf{E}}\mathbf{D}) = \sum_{i=1}^{B}\sum_{j=1}^{B}[^{\mathbf{E}}\mathbf{D}]_{ij} = \sum_{i=1}^{B}{}^{E}\sigma_i$$

where S is the → *total sum operator* and B the number of bonds.

A Wiener-type index – **edge Wiener index** ^{E}W – can be obtained from the edge distance matrix $^{\mathbf{E}}\mathbf{D}$ as:

$$^{E}W = \mathcal{W}(^{\mathbf{E}}\mathbf{D}) = \frac{1}{2}\cdot\sum_{i=1}^{B}\sum_{j=1}^{B}[^{\mathbf{E}}\mathbf{D}]_{ij}$$

where $\mathcal{W}$ is the → *Wiener operator* and B the number of bonds [Gutman and Estrada, 1996]. It was demonstrated that ^{E}W differs from the standard → *Wiener index W* only by a constant and, consequently, reflects the same molecular structure features, i.e.:

$$^{E}W = W - \frac{A\cdot(A-1)}{2}$$

where A is the number of vertices in $\mathcal{G}$, i.e. the number of atoms in the considered molecule.

More interesting is the → *edge-type Schultz index* derived from both the edge distance matrix and the edge adjacency matrix.

Bond multiplicity is taken into account by augmenting the edge distance matrix with a supplementary column and row where the elements are conventional bond orders, therefore obtaining an **edge distance matrix for multigraphs** [Bonchev, 1983]. All the local vertex invariants and molecular descriptors defined above can also be calculated on this matrix.

Moreover, if the chemical bond distances are calculated in real molecules, a **geometric edge distance matrix** ^{E}G can be obtained from which analogous geometric descriptors are derived. Each entry of this matrix is calculated from the → *interatomic distances* between the endpoint vertices of the considered pair of edges:

$$[^{\mathbf{E}}G]_{ij} = \frac{1}{4}\cdot\left(r_{i(1),j(1)} + r_{i(1),j(2)} + r_{i(2),j(1)} + r_{i(2),j(2)}\right)$$

where r is the Euclidean distance; $i(1)$ and $i(2)$ are the vertices incident to the ith edge, while $j(1)$ and $j(2)$ are the vertices incident to the jth edge.

The **reciprocal edge distance matrix** $^{\mathbf{E}}\mathbf{D}^{-1}$ is a square symmetric $B\times B$ matrix whose off-diagonal entries are reciprocal distances between considered edges:

$$\left[^{\mathbf{E}}\mathbf{D}^{-1}\right]_{ij} = \begin{cases} 1/[^{\mathbf{E}}\mathbf{D}]_{ij} & \text{if } i \neq j \\ 0 & \text{if } i = j \end{cases}$$

The diagonal entries are zero by definition. This matrix is the → *reciprocal distance matrix* of the line graph of the molecular graph $\mathcal{G}$ [Ivanciuc *et al.*, 1997a].

📖 [Estrada *et al.*, 1997] [Estrada and Rodriguez, 1997]

edge distance matrix for multigraphs → **edge distance matrix**

edge distance sum : *edge distance degree* → **edge distance matrix**

edge eccentricity : *bond eccentricity* → **edge distance matrix**

edge layer matrix → **layer matrices**

edge orbital information content → **orbital information indices**

edge radial centric information index → **centric indices**

edges → **graph**

edge type count → **edge adjacency matrix**

edge-type Gutman index → **Schultz molecular topological index**

edge-type Schultz index → **Schultz molecular topological index**

edge Wiener index → **edge distance matrix**

EEVA descriptors → **EVA descriptors**

effective resonance constant → **electronic substituent constants**
(⊙ resonance electronic constants)

effective solute hydrogen-bond acidity → **linear solvation energy relationships**
(⊙ hydrogen-bond parameters)

effective solute hydrogen-bond basicity → **linear solvation energy relationships**
(⊙ hydrogen-bond parameters)

eigenvalue-based descriptors

Descriptors calculated by the eigenvalues of a square (usually) symmetric matrix representing a → *molecular graph*.

These descriptors can be selected eigenvalues (usually the largest eigenvalue) or functions of several or all eigenvalues of the considered matrix [Hall, 1981; Lee and Yeh, 1993].

Examples of such descriptors are → *WHIM descriptors* and → *G-WHIM descriptors* of size and shape, → *EA indices*, → *quasi-Wiener index*, → *Mohar indices*, several → *shape descriptors*, → *EVA descriptors* and → *EEVA descriptors*.

Other popular eigenvalue-based descriptors are the following:

- **eigenvalues of the adjacency matrix** (λ_i^A)

The eigenvalues of the → *adjacency matrix* **A** can be used as molecular descriptors. These eigenvalues take both positive and negative values, their sum being equal to zero.

Level pattern indices (LPI) of molecules are defined in Hückel theory as the numbers of bonding N_+, non-bonding N_0 and anti-bonding N_- molecular orbitals and correspond to the numbers of positive, zero and negative eigenvalues of the adjacency matrix of the molecule [Jiang and Zhu, 1994]. The level pattern indices relate to the stability and valence state of the molecule: if $N_0 > 0$, the molecule will be either a free radical or an unstable species with a low → *HOMO-LUMO energy gap*; if $N_0 = 0$ and $N_+ \neq N_-$ the molecule in the ground state will be an anion or cation. The presence of non-bonding orbitals N_0 can easily be detected by the determinant of the adjacency matrix:

$$|\mathbf{A}| = \prod_{i=1}^{A} \lambda_i^A = 0$$

The sum of the level pattern indices corresponds to the number A of atoms in the molecule: $A = N_+ + N_0 + N_-$. Moreover, the eigenvalues may be closely related to the π-electron energy of the molecule:

$$E_\pi = \sum_{i=1}^{A} g_i \cdot \lambda_i^A$$

where A is the number of atoms, λ_i^A are the eigenvalues of the adjacency matrix and g_i is the occupation number on the ith molecular orbital which can take values 0, 1, or 2.

The largest eigenvalue of adjacency matrix **A** is among the most popular graph invariants and is known as the **Lovasz-Pelikan index** λ_1^{LP}, also called **leading eigenvalue** λ_1, [Lovasz and Pelikan, 1973]. This eigenvalue has been suggested as an index of → *molecular branching*, the smallest values corresponding to chain graphs and the highest to the most branched graphs. It is not a very discriminant index because in many cases the same value is obtained for two or more nonisomorphic graphs.

Moreover, the coefficients of the eigenvector associated with the largest eigenvalue were used by Randic for → *canonical numbering* of graph vertices.

Balaban proposed [Balaban *et al.*, 1991] the use of the coefficients of the eigenvector associated with the lowest (largest negative) eigenvalue as LOVIs able to provide discrimination among graph vertices; lower values correspond to vertices of lower degree, farther from the centre or from a vertex of high degree. Some molecular descriptors based on eigenvalues and corresponding eigenvectors of the adjacency matrix have also been proposed. The **VAA indices** were defined as:

$$\mathrm{VAA1} = \sum_{k=1}^{\mathrm{N}_+} \lambda_\mathrm{k}^+ \qquad \mathrm{VAA2} = \frac{\mathrm{VAA1}}{A} \qquad \mathrm{VAA3} = \frac{A}{10} \cdot \log(\mathrm{VAA1})$$

where λ^+ are the positive eigenvalues and N_+ is the number of positive eigenvalues; A is the number of molecular graph vertices.

The **VEA indices** are defined by the coefficients (i.e. loadings) ℓ_{iA} of the eigenvector associated with the largest negative eigenvalue (i.e. the Ath eigenvalue of the decreasing eigenvalue sequence):

$$\mathrm{VEA1} = \sum_{i=1}^{A} \ell_{iA} \qquad \mathrm{VEA2} = \frac{\mathrm{VEA1}}{A} \qquad \mathrm{VEA3} = \frac{A}{10} \cdot \log(\mathrm{VEA1})$$

The **VRA indices** are defined by applying the → *Randic operator* to the coefficients ℓ_{iA} of the eigenvector associated with the largest negative eigenvalue:

$$\mathrm{VRA1} = \sum\nolimits_b \left(\ell_{iA} \cdot \ell_{jA}\right)_b^{-1/2} \qquad \mathrm{VRA2} = \frac{\mathrm{VRA1}}{A} \qquad \mathrm{VRA3} = \frac{A}{10} \cdot \log(\mathrm{VRA1})$$

where the sum runs over all of the bonds in the molecular graph; ℓ_{iA} and ℓ_{jA} are the LOVIs of the two vertices incident to the considered bond.

→ *Schultz-type indices* that are based on the eigenvector associated with the lowest (largest negative) eigenvalue of the adjacency matrix have also been proposed.

📖 [Hall, 1992; Balaban, 1992]

• Burden eigenvalues

Molecular descriptors defined as eigenvalues of a modified connectivity matrix, which could be called **Burden matrix B** [Burden, 1989]. The **B** matrix representing an → *H-depleted molecular graph* is defined as the following: the diagonal elements B_{ii} are the atomic numbers Z_i of the atoms; the off-diagonal elements B_{ij} representing two bonded atoms i and j are equal to $\pi^* \cdot 10^{-1}$ where π^* is the → *conventional bond order*, i.e. 0.1, 0.2, 0.3 and 0.15 for a single, double, triple and aromatic bond, respectively; off-diagonal elements B_{ij} corresponding to terminal bonds are augmented by 0.01; all other matrix elements are set at 0.001.

The ordered sequence of the n smallest eigenvalues of the **B** matrix was proposed as a molecular descriptor with high discrimination power, to be used in the recognition and ordering of molecular structures. The basic assumption was that the lowest eigenvalues contain contributions from all atoms and thus reflect the topology of the whole molecule.

The diagonal elements of the **B** matrix can be set in different ways to account for different features of the molecule. It was proposed that a **B** matrix could represent a hydrogen-included molecular graph: the diagonal elements are roughly proportional to the electronegativity of the atoms, based on an electronegativity scale where the carbon atom value is assumed equal to zero, while the off-diagonal terms corresponding to pairs of bonded atoms are the square roots of conventional bond order, i.e. $B_{ij} = \sqrt{\pi^*_{ij}}$.

Including information on the electronic environment of the atoms in the matrix should relate the matrix eigenvalues to the electronic distribution of the whole molecule. This led to the proposing of the **Chemically Intuitive Molecular index** (**CIM index**): this is an ordered sequence of the n largest absolute eigenvalues of the **B** matrix and is based on atomic electronegativities as defined above (the same electronegativity value, $X = 2.3$, was used for all the halogen atoms) [Burden, 1997].

This vector descriptor was later improved [Benigni *et al.*, 1999b] by a more extended electronegativity scale which distinguished halogens and took sulfur and phosphorus atoms into consideration (Table E-2).

Table E-2. Atom electronegativity values for the **B** matrix.

C	0.00	F	2.30	S	0.50
H	0.15	Cl	0.90	P	0.50
N	0.90	Br	0.80		
O	0.90	I	0.50		

The **Burden modified eigenvalues** are the largest absolute eigenvalues of the above defined matrix:

$$\mathbf{BME} \equiv \langle \lambda_1, \lambda_2, \ldots, \lambda_L \rangle$$

where L is a user-defined maximum length (e.g., $L = 15$) resulting in → *uniform-length descriptors*.

Recently proposed to address searching for chemical → *similarity/diversity* on large data bases [Pearlman and Smith, 1998; Pearlman, 1999], **BCUT descriptors** (Burden – CAS – University of Texas eigenvalues) are based on a significant extension of the Burden approach, considering three classes of matrices whose diagonal elements correspond to 1) atomic charge-related values, 2) atomic polarizability-related values, and 3) atomic H-bond abilities. Additionally, a variety of definitions were considered for the off-diagonal terms, including functions of interatomic distance, overlaps, computed bond orders, etc. Moreover, for the off-diagonal terms not only was a 2D-approach used, but also a 3D approach, to account for geometric → *interatomic distances*.

Further types of atomic features could also be considered, together with other proximity measures and weighting schemes. Among the eigenvalues obtained from each of these matrices, the highest and the lowest have been demonstrated to reflect relevant aspects of molecular structure, and are therefore useful for similarity searching.

📖 [Menard *et al.*, 1998a; Menard *et al.*, 1998b; Burden and Winkler, 1999a; Stanton, 1999; Pearlman and Smith, 1999]

- **eigenvalues of the distance matrix** (λ_i^{D})

The largest eigenvalue of the topological → *distance matrix* **D** representing an H-depleted molecular graph was proposed as a molecular descriptor, and was called the **leading eigenvalue of the distance matrix** λ_1^{D} [Schultz *et al.*, 1990]. It was found to be a good discriminant in a series of compounds of increasing size.

Balaban proposed [Balaban *et al.*, 1991] the unique negative eigenvalue λ_-^{D} of the distance matrix as a molecular descriptor, together with two of its transformations; these descriptors were called **VAD indices**:

$$\mathrm{VAD1} = \lambda_-^{D} \qquad \mathrm{VAD2} = \frac{\mathrm{VAD1}}{A} \qquad \mathrm{VAD3} = \frac{A}{10} \cdot \log(\mathrm{VAD1})$$

The coefficients of the eigenvector associated with the unique negative eigenvalue of the distance matrix were used as LOVIs, able to provide discrimination among graph vertices; higher values correspond to vertices of lower degree, farther from the center or from a vertex of high degree. Based on the sum of these LOVIs, the **VED indices** were proposed as molecular descriptors:

$$\mathrm{VED1} = \sum_{i=1}^{A} \ell_{iA} \qquad \mathrm{VED2} = \frac{\mathrm{VED1}}{A} \qquad \mathrm{VED3} = \frac{A}{10} \cdot \log(\mathrm{VED1})$$

where A is the number of molecular graph vertices and ℓ_{iA} are the coefficients (i.e. loadings) of the eigenvector associated with the largest negative eigenvalue (i.e. the A th eigenvalue of the decreasing eigenvalue sequence).

The **VRD indices** were defined by applying the Randic operator to the coefficients ℓ_{iA} of the eigenvector associated with the largest negative eigenvalue as:

$$\mathrm{VRD1} = \sum\nolimits_{b} \left(\ell_{iA} \cdot \ell_{jA}\right)_b^{-1/2} \qquad \mathrm{VRD2} = \frac{\mathrm{VRD1}}{A} \qquad \mathrm{VRD3} = \frac{A}{10} \cdot \log(\mathrm{VRD1})$$

where the sum runs over all the bonds in the molecular graph; ℓ_{iA} and ℓ_{jA} are the LOVIs of the two vertices incident to the considered bond.

→ *Schultz-type indices* were also proposed based on the eigenvector associated with the lowest (largest negative) eigenvalue of the distance matrix.

- **leading eigenvalue of (A + D)** (λ_1^{AD})

It is the largest eigenvalue of the matrix obtained as the sum of the adjacency matrix **A** and the topological distance matrix **D** representing a molecular graph [Schultz *et al.*, 1990]. Its logarithm was used to model physico-chemical properties [Cash, 1995c].

- **$\lambda\lambda_1$ branching index** ($\lambda\lambda_1$)

One of the recently proposed → *double invariants*, this is the largest eigenvalue of the path matrix which is among the generalized graph matrices [Randic, 1997d; Randic, 1998b]. The path matrix contains the paths between each pair of vertices of the molecular graph and each path is characterized by the Lovasz-Pelikan index λ_1^{LP} calculated from the adjacency matrix representing the path itself. Therefore it is a generalization of the Lovasz-Pelikan index to characterize molecular branching of acyclic molecules; differently from λ_1^{LP}, it assumes the highest values for chain graphs and the lowest for the most branched graphs.

- **Characteristic Root Index** (*CRI*)

The sum of the positive eigenvalues λ of the → *χ matrix*, based on the path connectivities calculated by the → *valence vertex degree* δ^v of the atoms in the path:

$$CRI = \sum\nolimits_m \lambda_m \qquad \text{for } \lambda_m > 0$$

The *CRI* descriptor encodes information about all connectivities in the H-depleted molecular graph and is sensitive to the presence of heteroatoms in the molecule.

📖 [Saçan and Inel, 1993; Saçan and Inel, 1995; Saçan and Balcioglu, 1996]

- **PPP eigenvalues** (λ_i^{PPP})

The eigenvalues of the Hamiltonian matrix obtained by the *Pariser-Parr-Pople* (PPP) *method*.

A molecular descriptor λ^{PPP} is defined as [Balasubramanian, 1991; Balasubramanian, 1994]:

$$\lambda^{PPP} = \sum_{i=1}^{A} \left| \frac{\lambda_i^{PPP} - \alpha - \gamma/2}{\beta} \right|$$

where the sum runs over all the eigenvalues; α and β are the Hückel parameters and γ the $\gamma_{\mu\mu}$ PPP integral [Pariser and Parr, 1953a; Pariser and Parr, 1953b; Pople, 1953; Randic, 1991b]. The set of PPP-eigenvalues was also proposed to measure the similarity between pairs of molecules.

- **Wiener matrix eigenvalues** (λ^W)

They are the eigenvalues obtained from the → *Wiener matrix* **W**. In particular, in analogy with the Lovasz-Pelikan index, the **1st Wiener matrix eigenvalue** λ_1^W (or **Wiener matrix leading eigenvalue**) is taken as an alternative descriptor for the molecular branching [Randic *et al.*, 1994a].

- **Extended Adjacency matrix indices** (*: EA indices*)

They are two molecular descriptors calculated from → *extended adjacency matrices* **EA** [Yang *et al.*, 1994]. The first is the sum of the absolute eigenvalues of the **EA** matrix, called the ***EA*Σ index**:

$$EA\Sigma = \sum_{i=1}^{A} |\lambda_i^{EA}|$$

The second molecular descriptor is the maximum absolute eigenvalue of the **EA** matrix, called the ***EA*max index**:

$$EA\max = \max_i |\lambda_i^{EA}|$$

These descriptors account for heteroatoms and multiple bonds, possess high discriminating power, and correlate well with a number of physico-chemical properties and the biological activities of organic compounds.

- **folding degree index** (φ)

The largest eigenvalue λ_1^{DD} obtained by the diagonalization of the → *distance/distance matrix* **DD**, then normalized dividing it by the number of atoms *A* [Randic *et al.*, 1994b; Randic and Krilov, 1999]:

$$\phi = \frac{\lambda_1^{DD}}{A} \qquad 0 < \phi < 1$$

This quantity tends to one for linear molecules (of infinite length) and decreases in correspondence with the folding of the molecule. For example, φ values for transoid-molecules are always greater than the values for the corresponding cisoid-molecules.

Thus, ϕ can be thought of as a measure of the folding degree of the molecule because it indicates the degree of departure of a molecule from strict linearity. This index allows a quantitative measure of similarity between chains of the same length but with different geometries; it is sensitive to conformational changes and can also be considered among the → *shape descriptors*.

The **folding profile** of a molecule was proposed as:

$$\langle {}^1\phi, {}^2\phi, {}^3\phi, \ldots, {}^k\phi, \ldots \rangle$$

where ${}^k\phi$ is the normalized first eigenvalue of the $\mathbf{DD^k}$ matrix whose elements are raised to the kth power. Obviously, ${}^1\phi$ is the folding degree index. These vector descriptors were used to study the folding of proteins [Randic, 1997b].

The folding degree index is a measure of the conformational variability of the molecule, i.e. the capability of a flexible molecule (often macromolecules, proteins) to assume a conformation that closes over upon itself. Other descriptors related to the folding degree are the → *average span* and → *characteristic ratio*.

• **A_{x_1}, A_{x_2}, A_{x_3} eigenvalue indices**
They are molecular descriptors defined as:

$$A_{x_1} = \lambda_1^{\max}/2 \qquad A_{x_2} = \lambda_2^{\max}/2 \qquad A_{x_3} = \lambda_3^{\max}/2$$

where $\lambda^{\max}$ refers to the largest eigenvalue of three symmetrized augmented path → *sparse matrices* ($^+\mathbf{P}$) [Yao *et al.*, 1993a; Xu *et al.*, 1995]:

$$^{1+}\mathbf{P}^{\mathbf{T}} \bullet {}^{1+}\mathbf{P} \qquad ^{2+}\mathbf{P}^{\mathbf{T}} \bullet {}^{2+}\mathbf{P} \qquad ^{3+}\mathbf{P}^{\mathbf{T}} \bullet {}^{3+}\mathbf{P}$$

where the superscript "**T**" indicates the transpose matrix and the symbol • the → *Hadamard matrix product*.

The augmented matrices $^{1+}\mathbf{P}$, $^{2+}\mathbf{P}$ and $^{3+}\mathbf{P}$ are obtained by adding two columns to each matrix $^1\mathbf{P}$, $^2\mathbf{P}$ and $^3\mathbf{P}$: in the first column there is the addition of the square roots of vertex degrees δ and in the second column the square roots of the van der Waals radii of the atoms. The corresponding sparse path matrices $^1\mathbf{P}$, $^2\mathbf{P}$ and $^3\mathbf{P}$ of dimension $A \times A$ are defined as the following:

$$[^1\mathbf{P}]_{ij} = \begin{cases} 1 & \textit{if i and j are adjacent} \\ 0 & \textit{otherwise} \end{cases}$$

$$[^2\mathbf{P}]_{ij} = \begin{cases} 2 & \textit{if there is a } 2^{nd} \textit{order path between i and j} \\ 0 & \textit{otherwise} \end{cases}$$

$$[^3\mathbf{P}]_{ij} = \begin{cases} 3 & \textit{if there is a } 3^{rd} \textit{order path between i and j} \\ 0 & \textit{otherwise} \end{cases}$$

It can be observed that $^1\mathbf{P}$ is the adjacency matrix **A** and that for acyclic graphs each of the above defined path matrices is coincident with the corresponding sparse matrix of the distance matrix.

🕮 [McClelland, 1982] [Mohar, 1991a] [Hall, 1993] [Dias, 1999] [Cvetkovic and Fowler, 1999]

EigenVAlue descriptors : *EVA descriptors*

eigenvalues → **algebraic operators** (⊙ characteristic polynomial)

eigenvalues of the adjacency matrix → **eigenvalue-based descriptors**

eigenvectors → **algebraic operators** (⊙ characteristic polynomial)

electric dipole moment : *dipole moment* → **electric polarization descriptors**

electric polarization descriptors

Electric polarization, dipole moments and other related physical quantities, such as multipole moments and polarizabilities, constitute another group of both local and molecular descriptors, which can be defined either in terms of classical physics or quantum mechanics. They encode information about the charge distribution in molecules [Böttcher *et al.*, 1973]. They are particularly important in modelling solvation properties of compounds which depend on solute/solvent interactions and in fact are frequently used to represent the → *dipolarity/polarizability term* in → *linear solvation energy relationships.* Moreover, they can be used to model the polar interactions which contribute to the determination of the → *lipophilicity* of compounds.

The **dipole moment** μ (or **electric dipole moment**) is a vector quantity which encodes displacement with respect to the centre of gravity of positive and negative charges in a molecule, defined as:

$$\mu = \sum\nolimits_i q_i \cdot \mathbf{r}_i$$

where q_i are point charges located at positions $\mathbf{r}_i$. The components of the μ vector are called **dipole moment components**:

$$\mu_x = \sum\nolimits_i q_i \cdot x_i \qquad \mu_y = \sum\nolimits_i q_i \cdot y_i \qquad \mu_z = \sum\nolimits_i q_i \cdot z_i$$

where x, y, z are the coordinates of the charges. Molecules with zero dipole moments are called *nonpolar*, others *polar*; moreover, dipole moments equal to zero indicate molecules with a centre of symmetry.

In analogy to the definition of electric dipole moment, electric multipole moments are also defined. In particular, the **quadrupole moment Q** and the **octupole moment U** are defined as:

$$\mathbf{Q} = \frac{1}{2!} \cdot \sum_{i=1}^{A} q_i \cdot \mathbf{r}_i \cdot \mathbf{r}_i \qquad \mathbf{U} = \frac{1}{3!} \cdot \sum_{i=1}^{A} q_i \cdot \mathbf{r}_i \cdot \mathbf{r}_i \cdot \mathbf{r}_i$$

where **Q** is a tensor of second degree (a 3 × 3 symmetric matrix) and **U** a tensor of third degree (a 3 × 3 × 3 symmetric matrix). For example, when the charge distribution is spherically symmetrical, all the diagonal terms of the quadrupole moment are equal to zero. Therefore, the trace of the electric quadrupole moment is a measure of molecular charge distribution deviation from sphericity.

When a molecule is embedded in a uniform electric field $\mathbf{E_0}$ in a vacuum an **induced dipole moment** μ_{IND} arises, defined by the relationship:

$$\mu_{IND} = \alpha \cdot \mathbf{E}_0$$

where the scalar constant of proportionality is called the **polarizability** α (or **static polarizability**). This scalar polarizability may be regarded as the sum of the **electronic polarizability** α_E and the **atomic polarizability** α_A. A polarizable molecule shows an induced dipole moment different from zero.

For substituent groups, the **excess electron polarizability** $\Delta\alpha_E$ was also defined [Dearden *et al.*, 1991b] as the difference between the calculated electron polarizability for straight chain alkyl groups by using a model based on the → *McGowan characteristic volume* $\mathrm{V_X}$ and the effective electron polarizability of the substituent:

$$\Delta\alpha_E = \alpha_E - [0.135 \cdot \mathrm{V_X} + 0.052]$$

In general, the scalar polarizability α is not sufficient to describe the induced polarization, therefore, a **polarizability tensor** $\vec{\alpha}$ is used to better encode induced polarization and represents **molecular polarizability**. In such a general case the induced dipole moment need not have the same direction as the applied field, but the direction will depend on the position of the molecule relative to the polarizing field [Miller and Savchik, 1979; Miller, 1990b].

Each molecule, polar or nonpolar, is polarizable, i.e. its electrons can be shifted under an electric field **E** so that **polarization** P is induced in the molecule. The polarization is proportional to the electric field strength **E** and the simplest relationships between P and **E** are:

$$P = \chi \cdot \mathbf{E} = \frac{\varepsilon - 1}{4\pi} \cdot \mathbf{E}$$

where the scalar proportionality constant χ is called the **dielectric susceptibility**. In the second equation, ε is the **dielectric constant** (or **permittivity**).

Polarization can be factorized into two main contributions: **induced polarization** P_α, due to translation effects, and **dipole polarization** P_μ, due to orientation of permanent dipoles. Moreover, induced polarization can be viewed as being due to the contribution of **electronic polarization** P_E and **atomic polarization** P_A:

$$P = P_\alpha + P_\mu = P_E + P_A + P_\mu$$

Other important quantities related to polarization and dipole moments are listed below.

- **molar polarization**

The dipole moment induced per unit of volume V is called the molar polarization P_M, and is defined by the Clausius-Mossotti equation as:

$$P_M = \frac{\varepsilon - 1}{\varepsilon + 2} \cdot \frac{MW}{\varrho} = \frac{4\pi}{3} \cdot \mathrm{N_A} \cdot \alpha \cdot \mathbf{E} = \frac{n_D^2 - 1}{n_D^2 + 2} \cdot \frac{MW}{\varrho} = \mathrm{MR}$$

where MW is the molecular weight, ϱ the density, and ε the dielectric constant; in the second equation, $\mathrm{N_A}$ is the Avogadro number, α the scalar polarizability, **E** the electric field, and MR the → *molar refractivity*. For high frequency fields the relationship $\varepsilon = n^2$ holds, where n is the → *refractive index*; the subscript D indicates the value of the refractive index corresponding to the sodium D line, which is usually used.

- **atom polarizability**

This is the polarization effect at atomic level, where dipoles $\mu_{IND,i}$ are induced on each atom as

$$\mu_{IND,i} = \alpha_i \cdot \mathbf{E}_i$$

where $\mathbf{E}_i$ is the electric field at the ith atom and α_i the corresponding polarizability, assumed to be isotropic. Atom polarizabilities are linearly correlated with their → *hardness* (Politzer, 1987). Atomic contributions to polarizability (Table E-3) were estimated by several authors [Kang and Jhon, 1982; Miller, 1990a; No *et al.*, 1993] and are used to calculate the mean polarizability of a molecule by summing the atomic contributions.

Table E-3. Atomic polarizability values: 1. Kang-Jhon atomic hybrid polarizabilities α_A(ahp); 2. Miller-Savchik average atomic polarizabilities α_A^*(ahc); 3. No-Cho-Jhon-Sheraga atomic polarizabilities $\alpha_{ij,0}^*$ and linear charge coefficient α_{ij}.
[a] Data from [Miller, 1990a]

	1	2	3	4
Atom / Hybrid	α_A	α_A^*	$\alpha_{ij,0}^*$	α_{ij}
H bonded to Xsp_3	0.386	0.392	0.396	0.219
H bonded to Xsp_2	0.386	0.392	0.298	0.404
C sp_3	1.064	1.116	1.031	0.590
C sp_2 arom.	1.382	1.369	1.450	0.763
C sp_2 carbonyl			1.253	0.862
C sp_2 ethylene			1.516	0.568
C sp	1.279	1.294		
N sp_3	1.094	1.077	0.966	0.437
N sp_2 amide			0.821	0.422
N sp_2 pyrrole	1.090	0.851	0.871	0.424
N sp_2 pyridine	1.030	0.910	0.656	0.436
N sp	0.852	0.972		
O sp_3	0.664	0.780	0.623	0.281
O sp_2 carbonyl	0.460	0.739	0.720	0.347
O sp_2 aromatic	0.422	0.586	0.720	0.347
S sp_3	3.000[a]	3.056	2.688	1.319
S sp_2 thione	3.729[a]	3.661		
S sp_2 aromatic	2.700[a]	2.223		
P sp_3	1.538[a]	1.647		
F	0.296[a]	0.527	0.226	0.144
Cl	2.315[a]	2.357	2.180	1.089
Br	3.013[a]	3.541	3.114	1.402
I	5.415[a]	5.573	5.166	2.573

• **mean polarizability**

The molecular polarizability is a tensor, not a scalar quantity, when the molecule is not perfectly spherical. The mean polarizability α of a molecule is calculated by the relation:

$$\alpha = \frac{\alpha_{xx} + \alpha_{yy} + \alpha_{zz}}{3}$$

where α_{xx}, α_{yy}, and α_{zz} are the polarizabilities along each principal component axis of the molecule, obtained by diagonalization of the polarizability tensor [Cartier and Rivail, 1987]. When, for practical purposes, the effect due to anisotropy of polarizability is small, the mean polarizability is calculated simply as:

$$\alpha = \frac{tr(\vec{\alpha})}{3}$$

where $tr(\vec{\alpha})$ is the trace of the non-diagonalized polarizability tensor $\vec{\alpha}$.

Molecular polarizability can also be approximated by simply summing atomic polarizabilities over all the molecule atoms. Moreover, molecular polarizability α expressed as polarizability volume is called **polarizability volume** and is defined as:

$$\alpha' = \frac{\alpha}{4\pi\varepsilon_0}$$

where ε_0 is the dielectric constant in a vacuum.

- **atom-atom polarizability**

Index of chemical reactivity calculated from the perturbation theory as:

$$P_{ab} = 4 \cdot \sum_i \sum_j \sum_\mu \sum_\nu \frac{c_{i\mu,a} \cdot c_{j\mu,a} \cdot c_{i\nu,b} \cdot c_{j\nu,b}}{\varepsilon_i - \varepsilon_j}$$

where i and j run over the molecular orbitals, μ and ν over the atomic orbitals of the atoms a and b; c are the coefficients of the linear combination of atomic orbitals ϕ defining each molecular orbital ψ.

The **self-atom polarizability** P_{aa} is analogously defined as:

$$P_{aa} = 4 \cdot \sum_i \sum_j \sum_\mu \sum_\nu \frac{c_{i\mu,a} \cdot c_{j\mu,a} \cdot c_{i\nu,a} \cdot c_{j\nu,a}}{\varepsilon_i - \varepsilon_j}$$

An index for **total polarizability** is obtained as:

$$P_{TOT} = \sum_{a=1}^{A} P_{aa}$$

where P_{aa} is the self-atom polarizability of the ath atom.

- **anisotropy of the polarizability** (β^2)

A measure of the deviation of the molecular polarizability from a spherical shape, defined as:

$$\beta^2 = \frac{(\alpha_{xx} - \alpha_{yy})^2 + (\alpha_{yy} - \alpha_{zz})^2 + (\alpha_{zz} - \alpha_{xx})^2}{2}$$

where α_{xx}, α_{yy}, and α_{zz} are the polarizabilities along each principal component axis of the molecule, obtained by diagonalization of the polarizability tensor $\vec{\alpha}$.

Some empirical **polarity / polarizability descriptors** which were proposed to measure the ability of the compound to influence a neighbouring charge or dipole by virtue of dielectric interactions are reported below.

- **electrostatic factor**

A molecular descriptor proposed for solvent classification and defined as:

$$EF = \varepsilon \cdot \mu$$

where ε is the dielectric constant and μ the magnitude of the dipole moment of the solvent.

In general, values between 0 and 2 indicate hydrocarbon solvents, between 3 and 20 electron-donor solvents, between 20 and 50 hydroxylic solvents, and values greater than 50, dipolar aprotic solvents.

- **molecular polarizability effect index** (MPEI)

A descriptor of molecular polarizability based on the principle that molecules are polarized by electric fields and calculated by summing polarizability contributions by different atoms in the molecule [Chenzhong and Zhiliang, 1998]:

$$\text{MPEI} = \sum_{i=1}^{A} \text{PEI}_i$$

where A is the number of atoms and PEI_i the **polarizability effect index** for the ith atom.

The polarizability effect index is based on the stabilizing energy E_X caused by the polarizability effect for a substituent X interacting with a point charge q. For alkanes and aliphatic alcohol substituents, the PEI values were defined by the following:

$$PEI_X = K \cdot \sum \left[N_i \cdot \frac{1 + \cos\theta}{q - \cos\theta} - \frac{2 \cdot \cos\theta \cdot \left(1 - (\cos\theta)^{N_i}\right)}{(1 - \cos\theta)^2} \right]^{-2}$$

where $K = -2.16\; q^2/(2D \cdot r^4_{CC})$, D being the effective dielectric constant and r_{CC} the carbon-carbon bond distance of the probe from the substituent centre, the sum running over all the atoms in the molecule. For each carbon atom located at unit distance from the probe with charge q, N_i represents the distance from the probe and θ is a bond angle equal to 70.5° (180° – 109.5°) for sp_3 hybridization.

- **Kier-Hall solvent polarity index** (${}^1\chi^v_f$)

It is a solvent polarity index defined [Kier and Hall, 1986] as the first-order valence connectivity index ${}^1\chi^v$ divided by the number N_f of discrete, isolated functional groups in order to account for multiple interaction sites as:

$${}^1\chi^v_f = {}^1\chi^v / N_f$$

It was assumed that functional groups influencing solvent polarity are π-electron systems and lone pairs. Thus, for example, N_f(benzene) = 1, N_f(nitrobenzene) = 2 (one π-electron system and one lone pair), N_f(pyridine) = 2 (one π-electron system and one lone pair), N_f(nitro group) = 1. Halogen atoms are considered not to give a significant contribution to a molecule in enhancing its solvent polarity and thus for a halogen-containing molecule ${}^1\chi^v$ is used unmodified.

📖 [Sekusak and Sabljic, 1992]

- **local polarity index (Π)**

It is defined as the average deviation of the surface → *molecular electrostatic potential* $V(\mathbf{r})$ calculated as:

$$\Pi = \frac{1}{SA} \cdot \int_0^{SA} \left| V(\mathbf{r}) - \bar{V} \right| dSA = \lim_{n \to \infty} \frac{1}{n} \cdot \sum_{i=1}^{n} \left| V(\mathbf{r}_i) - \bar{V} \right|$$

where $V(\mathbf{r}_i)$ is the potential energy value at ith grid point on the molecular surface SA, $\bar{V}$ is the average potential energy value on the surface for the considered molecule and n is the number of grid points on the molecular surface [Brinck *et al.*, 1993; Murray *et al.*, 1993].

Π ranges from zero for a neutral atom to 21.6 Kcal/mol for water. It is a measure of charge separation or local polarity; it has been shown to correlate with the → *dipolarity / polarizability term* Π^* as well as with the → *dielectric constant* ε. The local polarity index is among the descriptors used in the → *GIPF approach* and was used to calculate log$P$ by the → *Politzer hydrophobic model.*

- **Q polarity index**

A topological polarity index derived by the electrotopological → *intrinsic state I* of the atoms in the molecule and defined as:

$$\mathrm{Q} = \frac{A^2 \cdot \sum_{i=1}^{A} I_i^{\mathrm{ALK}}}{\left(\sum_{i=1}^{A} I_i\right)^2}$$

where A is the number of atoms, I_i^{ALK} is the intrinsic state of the ith atom in the skeleton structure of the molecule in which each non-hydrogen atom is replaced with an sp_3 carbon atom on the corresponding isoconnective alkane as reference structure; I_i is the intrinsic state of the ith atom in the actual compound. The basic idea is that Q value for the considered molecule lies between two extremes of minimal and maximal polarity: the minimal polarity is given by a molecule only constituted of sp_3 carbon atoms and the maximal polarity is approximated by the square of the number of atoms A in the molecule [Kier and Hall, 1999].

- **polar hydrogen factor** ($\mathrm{Q_H}$)

An index of the molecular polarity due to C–H bonds restricted to halogenated hydrocarbons [Di Paolo *et al.*, 1979]. It is calculated as the sum of the contributions to the polarity of all the C–H bonds in a molecule. For each C–H bond are considered three different contributions due to halogens linked to the same carbon atom of the C–H bond, halogens in α-position and halogens in β-position with respect to the considered C–H bond:

$$\mathrm{Q_H} = \sum\nolimits_b \left[\sum\nolimits_C k_C + \sum\nolimits_\alpha k_\alpha + \sum\nolimits_\beta k_\beta\right]$$

where the external sum runs over all the C–H bonds and the three internal sums run over all halogens directly attached to the carbon atom of a C–H bond, in the α- and β-positions, respectively. When no halogen is attached to a C–H bond, its contribution to $\mathrm{Q_H}$ is taken as zero. The only exception is made for a methylene group CH_2 flanked by two halogen-substituted methyl groups: in this case C–H bonds are considered in the calculation.

The constant values k (Table E-4) are defined according to the Swain-Lupton → *field-inductive constant* $\mathcal{F}$, i.e. they represent the relative field effect of an atom or a group.

Table E-4. k_H parameters for the $\mathrm{Q_H}$ polar hydrogen factor.

halogen	k_H *contribution*	*halogen*	k_H *contribution*
F	0.43	α-Cl	0.10
Cl	0.41	α-Br	0.09
Br	0.44	β-F	0.05
I	0.40	β-Cl	0.05
α-F	0.13	β-Br	0.05

For example, $\mathrm{Q_H}$ values for $CHCl_3$, CH_2Cl_2, $CF_2Cl–CH_3$, $CF_3–CCHFCl$ and $CF_3–CH_2–CF_2Cl$ are $3 \times 0.41 = 1.23$, $4 \times 0.41 = 1.64$, 0, $0.43 + 0.41 + 3 \times 0.13 = 1.23$, $2 \times (5 \times 0.13 + 0.10) = 1.50$, respectively.

📖 [Hannay and Smyth, 1946] [McClellan, 1963] [Buckingham, 1967] [Böttcher *et al.*, 1973] [Exner, 1975] [Lien *et al.*, 1979] [Lien *et al.*, 1982] [Li *et al.*, 1984] [Topsom, 1987c] [Lewis, 1989] [Miller, 1990b] [Beck *et al.*, 1996] [Stuer-Lauridsen and Pedersen, 1997] [Norinder *et al.*, 1998] [Beck *et al.*, 1998] [Norinder *et al.*, 1999]

electromeric effect → electronic substituent constants

electron affinity → quantum-chemical descriptors

electron charge density connectivity index → charge descriptors

electron charge density weight → charge descriptors (⊙ electron charge density connectivity index)

electron density → quantum-chemical descriptors

electron donor-acceptor substituent constant → electronic substituent constants

electronegativity → quantum-chemical descriptors

electronegativity-based inductive constant → electronic substituent constants (⊙ inductive electronic constants)

electronegativity-based connectivity indices

Molecular descriptors defined in analogy with the → *Randic connectivity index* and calculated on hydrogen-included molecular graphs where heavy vertices are weighted by valence group electronegativities [Diudea *et al.*, 1996a].

The first proposed electronegativity-based connectivity index was the **DSI index** [Diudea and Silaghi-Dumitrescu, 1989] based on the → *Sanderson group electronegativity* ESG_i used as a local vertex invariant. It is defined as:

$$\mathrm{DSI} = \sum_{b=1}^{B} \left(\mathrm{EVG}_i \cdot \mathrm{EVG}_j\right)_b^{-1/2}$$

EVG_i is the valence group electronegativity of the ith vertex calculated from the Sanderson group electronegativity (geometric mean of electronegativities of the atoms belonging to the considered group G_i), accounting for the atom valence as:

$$\mathrm{EVG}_i = (\mathrm{ESG}_i)^{1/(1+\delta_i)} = \left[\left(\chi_{S,i} \cdot \left(\chi_{S,\mathrm{H}}\right)^{h_i}\right)^{1/(1+h_i)}\right]^{1/(1+\delta_i)}$$

where $\chi_{S,i}$ and $\chi_{S,\mathrm{H}}$ are the Sanderson electronegativities of the ith heavy atom and the hydrogen atom, respectively; δ_i is the → *vertex degree* of the ith atom; h_i is the number of hydrogen atoms belonging to the group G_i of the ith vertex and calculated as:

$$h_i = 8 - \mathrm{L}_i - \delta_i$$

where L is the principal quantum number. When $\delta_i > 8 - \mathrm{L}_i$, then $h_i = 0$ by definition; moreover, if multiple bonds are present the vertex degree should be replaced by the sum of the conventional bond orders (i.e. bond vertex degree δ_i^{b}).

An extension of DSI index to paths of higher order was also proposed as:

$${}^{m}\mathrm{DSI} = \sum\nolimits_{{}^{m}p_{ij}} \left(\mathrm{EVG}_i \cdot \mathrm{EVG}_k \cdot \ \ldots \ \cdot \mathrm{EVG}_j\right)^{-1/2}$$

where the sum is over all the paths of length m in the graph and the product over the valence group electronegativities of all the vertices involved in each path.

Two further electronegativity-based connectivity indices are the **ECP index** and **ECN index** based on the valence group carbon-related electronegativities EC_i proposed by Diudea [Diudea *et al.*, 1996a]:

$$\mathrm{ECP} = \sum_{i=1}^{A} \mathrm{ecp}_i = \sum_{i=1}^{A} \sum\nolimits_{j(i)} \left(\mathrm{EC}_i \cdot \mathrm{EC}_{j(i)}\right)^{1/2}$$

$$\mathrm{ECN} = \sum_{i=1}^{A} \mathrm{ecn}_i = \sum_{i=1}^{A} \sum\nolimits_{j(i)} \left(\mathrm{EC}_i \cdot \mathrm{EC}_{j(i)}\right)^{-1/2}$$

where ecp_i and ecn_i are local vertex invariants calculated by summing the products of the electronegativity of the considered ith vertex by the electronegativities of its bonded atoms $j(i)$; A is the number of heavy atoms in the molecule.

electronegativity scales → **quantum-chemical descriptors**

electronegativity-weighted distance matrix → **weighted matrices**

electronegativity-weighted walk degrees → **walk counts**

electronic charge index : *total absolute atomic charge* → **charge descriptors**

electronic chemical potential → **quantum-chemical descriptors**

electronic descriptors

Local or global molecular descriptors related to the electronic distribution in the molecule; they are fundamental to many chemical reactions, physico-chemical properties, and ligand-macromolecule interactions. The theory of electronic density is based on a quantum-mechanical approach; however, → *electronegativity* and charges, which are not physical observables, are also important quantities for the definition of several electronic descriptors.

Many → *quantum-chemical descriptors* are derived from the charge distribution in a molecule or the → *electron density* of specified atoms or molecular regions, and from conformational energy values such as the → *Joshi electronic descriptors.* Moreover, several → *charge descriptors* and → *electric polarization descriptors* are obtained from atomic charge estimations.

Electronic information is combined with shape and steric information to characterize molecules in → *charged partial surface area descriptors.* Other approaches, different from those closely related to quantum-chemistry, refer to electronic descriptors, such as → *electronic substituent constants,* → *electrotopological state indices,* → *topological charge indices.* → *Reactivity indices* and → *resonance indices* are also related to electronic descriptors.

Electronic EigenVAlue descriptors : *EEVA descriptors* → **EVA descriptors**

electronic polarizability → **electric polarization descriptors**

electronic polarization → **electric polarization descriptors**

electronic substituent constants (: *Hammett substituent constants, σ electronic constants)*

Derived from the → *Hammett equation*, σ electronic constants are calculated for different molecular substituents from the rate or equilibrium constant of specific reactions, with respect to a reference compound [Topsom, 1976; Charton, 1981; Taft and Topsom, 1987; Topsom, 1987b].

These substituent descriptors are usually used in → *Hansch analysis* as molecular electronic properties in the case of a mono-substituted series of compounds, while they are summed over all the molecule substituents to obtain global molecular descriptors in the general case of poly-substituted compounds.

The possible modes of action of substituents in modifying the electron distribution of the parent molecule can be distinguished in two main effects: a **polar effect** (or **field-inductive effect** or **localized effect**) and a **resonance effect** (or **delocalized effect**) (Figure E-1). The term polar effect is used to characterize the influence of unconjugated, sterically remote substituents on equilibrium or rate processes. The polar substituent effect is transmitted through bonds (**inductive effect** or **static inductive effect**) involving a polarization, either of the σ-bond network (**σ-inductive effect**) or the π-bond network (**π-inductive effect**). It can also be transmitted through solvent space according to classical laws of electrostatics (**field effect**) by the formation of bond dipole moments. Transmission efficiency of the localized effects is empirically measured by a **transmission coefficient** ε ($0 \leq \varepsilon \leq 1$).

Note that the separability of the two contributions of the polar effect is attained with difficulty; in fact, attempts to separate the polar effect contributions have been unsuccessful, so they are usually considered together. However, from a theoretical point of view, field effects have been studied using the Kirkwood-Westheimer [Kirkwood and Westheimer, 1938; Westheimer and Kirkwood, 1938] and Tanford models [Tanford, 1957].

Resonance effect is an energy stabilization due to delocalization of electrons in the bond network of the molecule and can be attributed to a **mesomeric effect**, i.e. the delocalization of π electrons on the π orbital network, a **hyperconjugation effect**, i.e. a delocalization of σ electrons in a π orbital aligned with the σ bond, and **secondary mesomeric effects**, such as repulsion of the π electrons by nonbonded electrons on a substituent or solvent, or by time-dependent effects due to polarizabilities (for the last, the term **electromeric effect** is sometimes used).

If the substituent is bonded to an sp^3 carbon atom not involved in the π molecular orbitals formation, its electrical effect is only local, assuming that σ electron delocalization is negligible. Substituents bonded to *sp* or sp^2 carbon atoms can exert both localized and delocalized effects.

Taking into account the basic contributions defined above, the overall electronic effect σ of the substituent can be represented by the following equation:

$$\sigma = \ell \cdot \sigma_I + d \cdot \sigma_R$$

where σ_I and σ_R are the polar and resonance contributions, respectively; ℓ and d weight their importance in determining the overall effect. Equivalent expressions with the same meaning can be represented by different symbols by other authors, such as:

$$\sigma = f \cdot \mathcal{F} + r \cdot \mathcal{R} \quad \text{or} \quad \sigma = \lambda \cdot \sigma_D + \delta \cdot \sigma_L$$

From these coefficients, the percentage of the resonance effect $D\%$ can be obtained as:

$$D\% = \frac{d}{d + \ell} \cdot 100$$

Several electronic substituent constants were defined so as to represent both global and particular electronic effects. The σ values obtained unambiguously from experimentally accessible data or from the many possible reaction series are called *primary values* and the corresponding set the *primary standard*. The σ values derived from the primary values, by rescaling with modified ϱ constants or correlation equations, are called *secondary values* and the corresponding set the *secondary standard*.

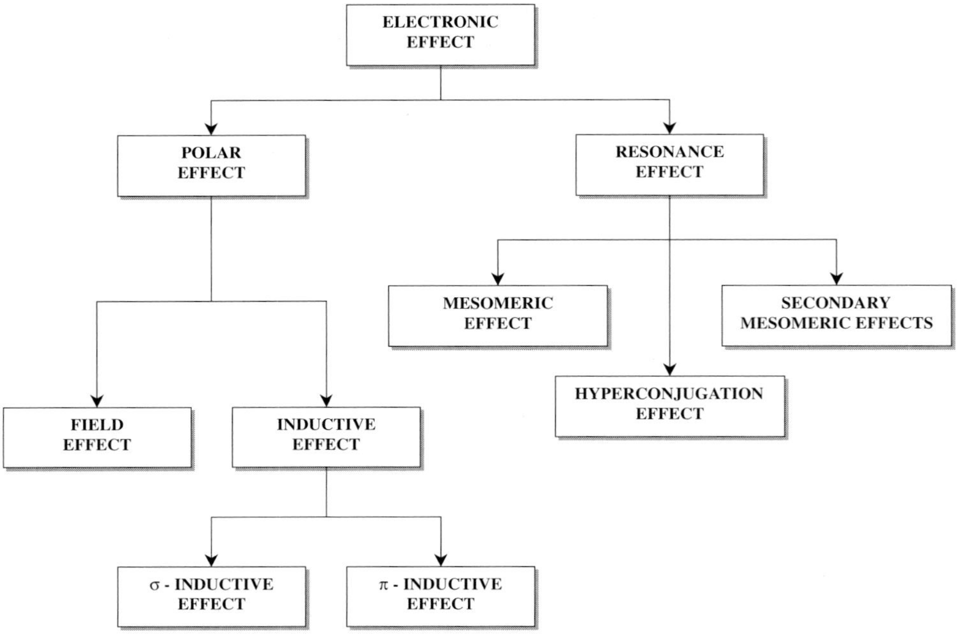

Figure E-1. Scheme of the relationships among the electronic effects.

The most popular electronic substituent descriptors are listed below; Table E-5 collects the main information concerning all the electronic substituents.

• **overall electronic constants σ_m and σ_p**

The original Hammett substituent constants [Hammett, 1937; Hammett, 1970] measuring the overall electronic effect of the *meta*- and *para*-substituents of benzene derivatives having the functional group in the side chain. They were originally calculated from the variation of the acid dissociation constant K_a of substituted benzoic acids (*m*-, *p*-XC_6H_4COOH) in water at 25°C, with respect to the unsubstituted compound (i.e. benzoic acid):

$$\sigma_{m,p} = \frac{1}{\varrho} \cdot \log\left(\frac{K^X}{K_0}\right) = \frac{1}{\varrho} \cdot \left(pK_a^0 - pK_a^X\right)$$

where the reaction constant ϱ is arbitrarily assumed equal to one. K^X and K^0 are the acid dissociation constants of the X-substituted and parent compound, respectively. The subscripts m and p are used to identify *meta*- and *para*-substituents, respectively. Other values of $\sigma_{m,\,p}$ were also obtained from the hydrolysis of benzoic esters and other reaction series, based on both equilibrium and rate constants (*K* and *k*, respectively).

σ values measure the substituent total electronic effect with respect to hydrogen and are, in principle, independent of the nature of the reaction; however, for large and/or charged substituents the σ values are estimated with less precision.

• **unbiased constants σ^0**

The unbiased constants σ^0 were defined by Taft in order to avoid overestimating the resonance effect due to direct resonance interactions between substituent and reaction site [Taft, 1960], i.e. they are a measure only of the interaction between the substituent and the molecular skeleton as felt at the reaction site.

These σ^0 constants were evaluated by the dissociation of *meta*-substituted phenylacetic acids and esters and were defined for a selected group of m-XC_6H_4– substituents which exhibit a precise linear free energy relationship. Their values are independent of whether the process is rate or equilibrium, the solvent and reaction conditions.

They represent inductive constants for the *meta*-substituted phenyl groups (m-XC_6H_4–) relative to the unsubstituted phenyl group (C_6H_5–), because substituent X in the *meta* position is not directly conjugated with the side chain reaction centre Y, thus specific –C_6H_4Y resonance effects do not contribute to the overall electronic effect. However, σ^0 values contain contributions from resonance interactions within m-XC_6H_4– groups. The same does not hold for the corresponding *para*-substituted phenyl substituents (p-XC_6H_4–). Therefore, by using the linear free energy relationships for selected *meta*-substituted groups to determine the ϱ reaction constant, the effective $\bar{\sigma}$ values were obtained for all p-XC_6H_4– :

$$\bar{\sigma} = 1/\varrho \cdot \log(k/k_0)$$

The difference $\sigma - \sigma^0$ can be considered as a measure of the resonance effect between the aromatic group and the reaction centre Y for the dissociation of *meta*-substituted benzoic acids in water. Moreover, deviations from the relationship $\log(k/k_0) = \sigma^0\varrho$ provide a useful measure of the specific polar and resonance effects dependent upon both solvent conditions and reaction type.

• **inductive electronic constants**

Electronic substituent constants representing the polar effect exerted by the substituent on the active site of a molecule. Different reaction series and data derived from various statistical procedures have led to several proposals for inductive constants; only the most popular are considered in the following.

The **Roberts-Moreland inductive constant** σ' is based on the dissociation constants K of 4-substituted bicyclo[2,2,2]octane-1-carboxylic acids in 50 % ethanol by volume at 25°C defined as:

$$\sigma' = \frac{1}{1.464} \cdot (\log K_X - \log K_H)$$

where $\varrho = 1.464$ is assumed in the Hammett equation [Roberts and Moreland, 1953].

As the X substituent in bicyclo[2,2,2]octane-1-carboxylic acids is bonded to an sp^3 carbon atom it exerts only an inductive effect. Moreover, the chosen reference compound is free from conformational effects and no steric effect is observed, as the substituent and the active site are not in close proximity to each other.

The **Holtz-Stock inductive constant** σ' was calculated from the dissociation constants K of 4-substituted bicyclo[2,2,2]octane-1-carboxylic acids in 50 % ethanol by weight at 25°C using $\varrho = 1.65$ in the Hammett equation [Holtz and Stock, 1964].

The **Taft σ* constant** (or **σ* electronic constant**, **Taft polar constant**) was proposed by Taft [Taft, 1956] to measure the inductive effect in the aliphatic series:

$$\sigma^* = \frac{1}{2.48} \cdot \left[\log\left(\frac{k_X}{k_{Me}}\right)_B - \log\left(\frac{k_X}{k_{Me}}\right)_A\right] = \frac{1}{2.48} \cdot \left[\log\left(\frac{k_X}{k_{Me}}\right)_B - E_S\right]$$

where A and B, respectively, stand for acid-catalyzed and base-catalyzed, k_X and k_{Me} are the rate constants of acid- and base-catalyzed hydrolysis or esterification of substituted and unsubstituted esters, respectively, and E_S is the → *Taft steric constant.*

The factor 2.48 corresponds to the average of the available ϱ values of alkaline hydrolysis from the Hammett equation and attempts to place σ^* on the same scale as the σ_m and σ_p electronic constants. Since base-catalyzed hydrolysis involves both inductive and steric effects and acid-catalyzed hydrolysis involves only the steric effect, removing the steric effect leaves only the inductive effect, assuming that in both reactions the steric effects are the same.

When defining the σ^* values as a measure of inductive effects, the choice in reactivity types is such that specific steric, resonance and other effects are apparently constant and a linear free energy relationship of the Hammett type holds.

An **Additive Model of Inductive Effect** was recently proposed to estimate the σ^* inductive constant of substituents on the basis of the fundamental characteristics of the constituent atoms [Cherkasov *et al.*, 1998a; Cherkasov and Jonsson, 1998b]:

$$\sigma^* = \sum_{i=1}^{n} \frac{\sigma_i^A}{r_i^2} = 7.84 \cdot \sum_{i=1}^{n} \frac{\Delta\chi_i \cdot R_i^2}{r_i^2}$$

where the sum runs over all the atoms of the substituent, r_i is the distance of the ith atom of the substituent to the reaction centre, R_i the covalent radius of the atom. σ^A is an empirical atomic parameter reflecting the ability of an atom to attract (or donate) electrons and its values were estimated by multivariate regression analysis on Taft σ^* constants for several substituents. The values were found to have good correlation with the difference in electronegativity $\Delta\chi$ between a given atom and the reaction centre, reflecting the driving force of electron density displacement and, with the square of the covalent radius of the atom, reflecting the ability to delocalize the charge.

The **Taft-Lewis inductive constant** σ_I was proposed [Taft and Lewis, 1958] to measure the inductive effect in aliphatic series on a scale for direct comparison with aromatic σ values, derived from the σ^* constant as:

$$\sigma_I = 0.45 \cdot \sigma^* = \frac{1}{5.51} \cdot \left[\log\left(\frac{k_X}{k_H}\right)_B - \log\left(\frac{k_X}{k_H}\right)_A \right]$$

The derived ϱ value of 5.51 was later modified by Taft into $\varrho = 6.23$ [Taft, 1960]. In this way, σ_I values can also be considered as a measure of the inductive effect of substituents bonded to aromatic carbons.

These σ_I values of Taft and Lewis were used as a basis set by Charton [Charton, 1963; Charton, 1964] with the aim of obtaining a large number of inductive constants. Acid dissociation constants of substituted acetic acids (XCH_2COOH) in water were correlated with σ_I constants of the basis set at temperatures from 5° to 50°C in terms of the equation:

$$\sigma_I = b \cdot pK_a^X + a$$

The regression coefficients a and b were estimated separately for each reaction series, and then additional σ_I values (**Charton inductive constants**) were estimated and a set of recommended values also suggested. In general, steric and resonance effects in the acetic acid system can be considered negligible. Moreover, unlike the Taft inductive constants, those defined by Charton require only one experiment for their determination.

The **Siegel-Kormany inductive constant** σ'' was calculated by the dissociation constants of 4-substituted cyclohexanecarboxylic acids in three different solvents (in water, water/ethanol 50 % by weight and by volume) at 25°C using $\varrho = 1$ in the Hammett equation [Siegel and Komarmy, 1960]. The expected relationship between this inductive constant and the inductive constant of Roberts-Moreland or Taft σ^* was confirmed.

Based on the acid dissociation of 4-substituted quinuclidines in water at 25°C, the **Grob inductive constant** σ_I^q was proposed [Taft and Grob, 1974; Grob and Schlageter, 1976; Grob, 1985]:

$$\sigma_I^q = pK_a^X - pK_a^H$$

assuming $\varrho = 1$ in the Hammett equation. The quinuclidine system is free of steric and conformational effects. Moreover, it is much more sensitive to electronic effects than the bicyclooctane system.

^{19}F inductive constant σ_m^F was estimated from ^{19}F-NMR for *meta*-substituents of F-benzene in very dilute CCl_4 solution [Taft, 1960; Taft *et al.*, 1963a] by equation:

$$\sigma_m^F \equiv \sigma_I = 0.084 - \frac{\delta_m^F}{7.1}$$

where δ_m^F is the ^{19}F chemical shift of *meta*-substituted fluorobenzenes. The NMR chemical shifts measure the inductive perturbation of *meta*-substituents on the charge density of the fluorine neighbour. This relationship is based on the correlation between chemical shift and the Taft-Lewis inductive constant σ_I.

The **Inamoto-Masuda inductive constant** ι (*iota*) [Inamoto and Masuda, 1977] was empirically defined by modifying the electronegativity as defined by Gordy [Gordy, 1946]. The inductive constant ι is based on the → *electronegativity* of the substituent atom directly bonded to the skeletal group and defined as:

$$\iota = \frac{Z_{eff} + 1}{L_{eff}}$$

where Z_{eff} is the effective nuclear charge in the valence shell of the considered atom and L_{eff} is the effective principal quantum number (Slater rule). If the atom belongs to the second period group, the inductive constant is estimated by the equation:

$$\iota = 0.64 \cdot \chi_X + 0.53$$

where χ_X is the substituent charge obtained from the bond dipole moment [Inamoto *et al.*, 1978].

The **electronegativity-based inductive constant** σ_χ was derived from atomic charge densities on the hydrogen atom of XH, XCH_2H and XCH_2CH_2H derivatives [Marriott *et al.*, 1984] defined as:

$$\sigma_\chi = 1 - q_H$$

where q_H is the atomic charge on the H atom and is calculated from → *computational chemistry*. The electronegativity of a substituent is primarily determined by the electronegativity of the nearest attached atom. Moreover, the electronegativity of the atom is affected by both changes in hybridization and the polarity of other atoms in the group. The values $1 - q_H$ being related to electronegativity scales, such values can be considered a measure of the inductive effect of the substituent.

- **resonance electronic constants**

Electronic substituent constants representing the resonance effect exerted by the substituent on the active site of a molecule. Several proposals of resonance constants were made on the basis of different reaction series or derived from statistical procedures; only the most popular are considered in the following.

The **Taft resonance constants** were calculated from the overall electronic effect in specific reaction series by subtracting the inductive contribution based on the Taft-Lewis σ_I values [Taft and Lewis, 1959; Taft *et al.*, 1959; Taft, 1960; Ehrenson *et al.*, 1973] as:

$$\sigma_R^0 = \sigma^0 - \sigma_I \qquad \bar{\sigma}_R = \bar{\sigma} - \sigma_I \qquad \sigma_R^+ = \sigma^+ - \sigma_I \qquad \sigma_R^- = \sigma^- - \sigma_I$$

In particular, σ_R^0 called **resonance polar effect** [Taft, 1956] is defined for any benzene derivative where there is no direct conjugation between substituent and reactive; it can be considered constant for a particular solvent, therefore expressing resonance interactions between substituent and skeletal group. $\bar{\sigma}_R$ is usually referred to as the **effective resonance constant**; σ_R^+ and σ_R^- hold for electrophilic and nucleophilic reaction series, respectively.

Moreover, Taft also tried to propose a general σ_R scale as:

$$\sigma_R = \sigma - \sigma_I$$

However, the σ_R values for *para*-substituents show great variability with reaction type, and it was therefore inferred that a widely applicable and precise σ_R-scale could not be devised [Taft and Lewis, 1958; Taft and Lewis, 1959].

Resonance constants σ_R^0 were also estimated from ^{19}F-NMR based on ^{19}F chemical shifts in *para*- and *meta*-substituted F-benzene in very dilute CCl_4 solution [Taft *et al.*, 1959] by equation:

$$\sigma_R^0 = \frac{1}{2.97} \cdot \left(\delta_p^F - \delta_m^F\right)$$

where δ^F is the ^{19}F chemical shift. The quantity $\delta_p^F - \delta_m^F$ can be considered a measure of the perturbation in electron density detected by the fluorine atom caused by the resonance interaction between the *para*-substituent and fluorobenzene system; δ_m^F is mainly related to the perturbation derived from bond polarizations, and the inductive contributions to δ_m^F and δ_p^F values are assumed to be equal.

Analogously, a significant correlation was found between δ_p^C (and $\delta_p^C - \delta_m^C$) and σ_R^0 or σ_R , δ^C being the chemical shift of the ^{13}C atom of the aromatic ring in the *meta*- or *para*-position relative to the substituent [Maciel and Natterstad, 1965].

Dual electronic constants σ^+ and σ^- were proposed to measure the "exaltation" of the resonance effect which appears when the substituent and active site bonded to a skeletal group give origin to a direct conjugation between them.

The **electrophilic substituent constant** σ^+ measures the electronic effects for electron-releasing substituents (e.g. –OMe, –Me, –OH, $-NH_2$) and for a strong electron-acceptor active site, while the **nucleophilic substituent constant** σ^- measures the electronic effects for electron-acceptor substituents (e.g. $-NO_2$, –CN, $-CO_2H$) and strong electron-releasing active site.

Standard σ^+ values were estimated by Brown-Okamoto [Brown and Okamoto, 1958a] from the rate constants k of the solvolysis of substituted *t*-cumyl chlorides ($XC_6H_4C(CH_3)_2Cl$) in 90 % acetone-water at 25°C, as:

$$\sigma^+ = \frac{1}{-4.54} \cdot \log \frac{k}{k_0}$$

where the reaction constant $\varrho = -4.54$ was estimated from a set of *meta*-substituents.

The σ^+ values correlate with ionization potentials obtained from substituted benzyl radicals [Harrison *et al.*, 1961].

Standard σ_p^- values were obtained from acid dissociation constants K_a of *para*-substituted phenols (p-XC_6H_4OH) in water or water/ethanol 50 % at 25°C [Cohen and Jones, 1963] and successively from those of *para*-substituted anilines.

When strong resonance interactions are less relevant, σ^+ and σ^- constants are equal to the normal σ values obtained from substituted benzoic acids.

The **Yukawa-Tsuno equation** (also referred to as **Linear Aromatic Substituent Reactivity relationship**, *LASR*) modifies the Hammett equation, taking into account the exaltation of the resonance effects of electron-releasing and electron-attracting substituents on the reaction centre [Yukawa *et al.*, 1972a; Yukawa *et al.*, 1972b].

For electron-releasing substituents, the Yukawa-Tsuno equation is:

$$\log\left(\frac{k_X}{k_H}\right) = \varrho \cdot \left[\sigma^0 + r \cdot (\sigma^+ - \sigma^0)\right] = \varrho \cdot \left[\sigma^0 + r \cdot \Delta\bar{\sigma}^+\right]$$

where k_X and k_H are the respective rate constants of the X-substituted and unsubstituted compounds, ϱ is the Hammett reaction constant, and r is the contribution of the enhanced resonance effect of the substituent. σ^0 is the "normal" (i.e. unbiased) substituent constant derived by Yukawa *et al.* from the rate constants of alkaline hydrolysis of ethyl phenyl-acetates ($HO^- + XC_6H_4CH_2COOEt$) in water at 25°C.

$\Delta\sigma^+$ was proposed as the substituent constant measuring the exaltation of the resonance effect of a *para*-substituent on an electrophilic reaction, while the first term $\varrho\sigma^0$ accounts primarily for the electronic effect of *meta*-substituents and *para*-substituents whose σ^0 and σ^+ are equivalent.

Analogously, for electron-attracting substituents, the Yukawa-Tsuno equation is:

$$\log\left(\frac{k_X}{k_H}\right) = \varrho \cdot \left[\sigma^0 + r \cdot (\sigma^- - \sigma^0)\right] = \varrho \cdot \left[\sigma^0 + r \cdot \Delta\bar{\sigma}^-\right]$$

Both these equations were originally proposed using σ values instead of σ^0 values [Yukawa and Tsuno, 1959]. If $r = 0$, the Yukawa-Tsuno equations reduce to the classical Hammett equation, while, if $r = 1$, they correspond to the correlations with only σ^+ or σ^- constants.

A resonance constant R was calculated from two distinct reaction series as:

$$R = \log\left(\frac{K_X}{K_0}\right)_I - \log\left(\frac{K_X}{K_0}\right)_{II}$$

where the subscript I represents the series of dissociation constants of 4-substituted pyridinium ions in water at 25°C and II the series of dissociation constants of 4-substituted quinoclidinium ions in water at 25°C [Taft and Grob, 1974].

This resonance constant is justified by the following correlation equations found separately for the two series:

$$\log\left(\frac{K_X}{K_0}\right)_I = 5.15 \cdot \sigma_I + 2.69 \cdot \sigma_R^+$$

$$\log\left(\frac{K_X}{K_0}\right)_{II} = 5.15 \cdot \sigma_I$$

Since the inductive effects are essentially the same in the two series, it follows that the difference in log (K_X/K_0) values for corresponding substituents in I and II reaction series gives a measure of the resonance effect for the dissociation of 4-substituted pyridinium ions.

- **field / resonance effect separation**

In order to give a uniform view of the different σ scales, considering both field-inductive and resonance effect, a number of proposed approaches were aimed at separating the two main contributions within a unique theoretical framework.

According to the **Swain-Lupton approach** (SL), the σ electronic constant is defined as a linear combination of the two basic electronic contributions [Swain and Lupton Jr., 1968; Swain *et al.*, 1983; Swain, 1984; Reynolds and Topsom, 1984] based on the equation:

$$\sigma^{SL} = f \cdot \mathcal{F} + r \cdot \mathcal{R}$$

where $\mathcal{F}$ is the **field-inductive constant** (or **Swain-Lupton field constant**) and $\mathcal{R}$ the **resonance constant** (or **Swain-Lupton resonance constant**), f and r being weighting factors that depend on the system used to define the particular σ-scale (analogous to the coefficients ℓ and d defined above).

Swain and Lupton defined the field constant $\mathcal{F}$ assuming that the polar effect was a component in both Hammett electronic constants σ_m and σ_p:

$$\mathcal{F} = b_0 + b_1 \cdot \sigma_m + b_2 \cdot \sigma_p$$

where the coefficients were evaluated by least square regression using pK_a values of 4-substituted bicyclo[2,2,2]octane-1-carboxylic acids of Holtz-Stock. While Swain-Lupton assumed $\varrho = 1$ so as to put the $\mathcal{F}$ values on the same scale as Hammett constants, Hansch used σ' as originally calculated by $\varrho = 1.65$, thus obtaining the following equation [Hansch *et al.*, 1973a]:

$$\mathcal{F} \equiv \sigma' = -0.009 + 1.369 \cdot \sigma_m - 0.373 \cdot \sigma_p$$

The resonance constant $\mathcal{R}$ was estimated by:

$$\mathcal{R} = \sigma_p - 0.921 \cdot \mathcal{F}$$

assuming $r = 1$ and the coefficient f was evaluated assuming that $\mathcal{R} = 0$ for $N^+(CH_3)_3$. The main assumption is that the substituents in the *para* position give the primary resonance effect.

Based on the same previous assumptions, but using a different ϱ value ($\varrho = 1.56$) and an extended data set of σ_I values, the following equation was used to define the field constant $\mathcal{F}$ and the corresponding resonance constant $\mathcal{R}$:

$$\mathcal{F} \equiv \sigma_I = 0.033 + 1.297 \cdot \sigma_m - 0.385 \cdot \sigma_p$$

Several sets of resonance constants were defined according to the different types of σ_p values.

Resonance substituent constants $\mathcal{R}^+$, $\mathcal{R}^-$, and $\mathcal{R}^0$ are derived from the corresponding σ values as:

$$\mathcal{R}^+ = \sigma_p^+ - f \cdot \mathcal{F} \qquad \mathcal{R}^- = \sigma_p^- - f \cdot \mathcal{F} \qquad \mathcal{R}^0 = \sigma_p^0 - f \cdot \mathcal{F}$$

using the appropriate field constant $\mathcal{F}$ values calculated by Swain-Lupton (the coefficient f is usually taken as equal to one).

Also in the Taft-Lewis approach, the resonance R and polar effects I can be viewed as additive contributions to the overall electronic effect, defined as:

$$\log\left(\frac{k}{k_0}\right) = I + R = \varrho_I \cdot \sigma_I + \varrho_R \cdot \sigma_R$$

where the inductive effect I is defined as in the Hammett equation.

This equation is applied separately to the effects of *meta*- and *para*-substituents, both being σ and ϱ position-dependent.

However, assuming

$$\varrho_I = \varrho_I^m = \varrho_I^p = \varrho_R^m = \varrho_R^p = \varrho_R$$

for *meta*- and *para*-substituents of benzene derivatives, the two following equations were proposed:

$$\log\left(\frac{K_p}{K_0}\right) = \varrho_I \cdot (\sigma_I + \sigma_R)$$

$$\log\left(\frac{K_m}{K_0}\right) = \varrho_I \cdot (\sigma_I + \alpha \cdot \sigma_R)$$

Moreover, the inductive constants are assumed equal in both cases while the resonance constant of the substituent in the *meta* position is considered a fraction of its resonance effect in the *para* position. The coefficient α was originally proposed as equal to 1/3 for the dissociation of benzoic acids; other values were proposed to account for enhanced resonance effects.

By combining the two expressions, a general equation to measure the inductive effect was proposed as:

$$\varrho_I \cdot \sigma_I = \left(\frac{1}{1-\alpha}\right) \cdot \left[\log\frac{K_m}{K_0} - \alpha \cdot \log\frac{K_p}{K_0}\right]$$

According to the original **Dewar-Grisdale approach** (DG), also called **FM method** (Field and Mesomeric method) [Dewar and Grisdale, 1962a; Dewar and Grisdale, 1962b], delocalized and localized long-range field effects are represented by the quantities M and F, respectively. Electronic substituent constants σ_{ij} of a given substituent X, bonded at position i of the skeletal group, acting on a reaction centre bonded at position j, are defined by the equation:

$$\sigma_{ij}^{DG} = \frac{F}{r_{ij}} + M \cdot q_{ij}$$

where q_{ij} and r_{ij} represent the charge in position j and the distance (in units of benzene bond length) between i and j atoms. F and M values are calculated from σ_m and σ_p constants by substituting appropriate values in the equation. Once F and M are known for each substituent, σ_{ij}^{DG} constants can be calculated for any structure at the i-j positions (e.g. *ortho*, *meta*, *para* positions in benzene derivatives, or other positions in naphthalene derivatives), using appropriate values of r_{ij} and q_{ij}. The term q_{ij} is taken as a measure of the transmission of mesomeric effects between position i and j of a conjugated system. In this approach, both substituent and reaction site are approximated by single point charges located at points i and j (Figure E-2).

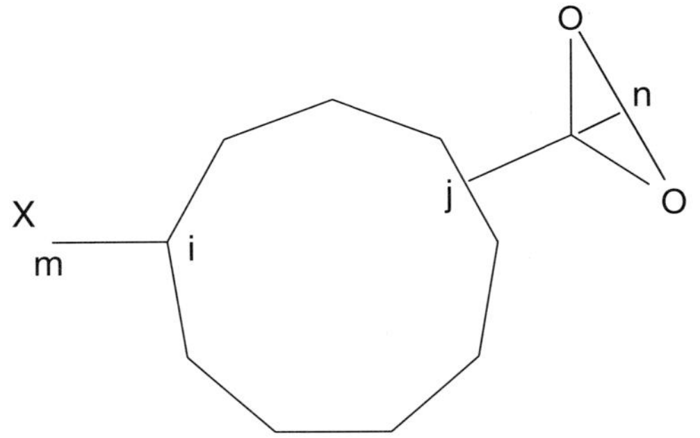

Figure E-2. Geometrical scheme for the calculation of the Dewar-Grisdale electronic constants.

Substituting the charge term q_{ij} by the atom-atom polarizability term α_{ij} results in an analogous set of F' and M' values.

The **Dewar-Golden-Harris approach** (DGH), also called **FMMF method** (i.e. Field, Mesomeric and Mesomeric-Field method) [Dewar *et al.*, 1971a] is a modification of the Dewar-Grisdale approach, where the substituent X is approximated by a finite dipole (represented by two point charges along the i-X bond) and the reaction site at position j as a single point charge. This approach is based on the equation:

$$\sigma_{ij}^{DGH} = \frac{F}{r_{ij}} + M \cdot q_{ij} + M_F \cdot \sum_{k \neq j} \frac{q_{ik}}{r_{kj}}$$

where F and M respectively represent localized and delocalized effects of the substituent X, and q_{ik} is the charge of the carbon atom at different positions k in the skeletal moiety, and r is the distance (in units of benzene bond length) between atom j and any other atom in the skeletal group. The third substituent parameter M_F is the mesomeric-field constant, describing the ability of the substituent to polarize adjacent π systems; if there is no direct resonance interaction between the substituent and the reaction centre, M_F should be proportional to M.

In the event of there being substituent effects on the dissociation of the carboxylic acids, the previous equation becomes:

$$\sigma_{ij}^{\mathrm{DGH}} = F \cdot R_{ij} + M \cdot q_{ij} + M_F \cdot \sum_{k \neq j} \frac{q_{ik}}{r_{kn}}$$

where the quantity R_{ij} is defined as:

$$R_{ij} = \frac{1}{r_{in}} - \frac{0.9}{r_{mn}}$$

where m is a point charge (with charge $q = -0.9$) at distance 1.40 Å from the point charge i (with charge $q = 1$) along the i-X bond; n is the point charge in the middle of the axis joining the two oxygens of the carbonyl group of the reaction site.

Further modifications of two previous approaches were proposed by Forsyth [Forsyth, 1973], calculating σ_{ij}^{+} constants.

Other substituent electronic constants were defined for specific different reference compounds (e.g. heterocycles) and reactions.

Aryl electronic constants σ_a are electronic constants defined for substituents on an aromatic ring different from benzene, such as pyridine [Otsuji *et al.*, 1960]. The concept may be generalized to various isocyclic and heterocyclic rings such as thiophene, furan, etc., and various types of aryl electronic constants σ_a^0, σ_a^+, σ_a^- may be also defined. A particular set of this kind is given by σ_r **electronic constants** obtained by the protonation reaction of hydrocarbons, assuming as the reference compound α-naphthyl ($\sigma_r = 0$).

Phosphorus electronic constants σ^{Φ} are electronic substituent constants derived from the dissociation constants of dialkylphosphinic acids for substituent groups directly bonded to a phosphorus atom [Mastryukova and Kabachnik, 1971]. Assuming the alkyl groups exert only an inductive effect, these electronic constants can be distinguished as σ_I^{Φ} inductive constant and σ_R^{Φ} resonance constant [Charton and Charton, 1978].

Radical electronic constants are substituent constants derived from free-radical reactions. The most popular are: the $\boldsymbol{E_R}$ **radical parameter** defined on the basis of the radical abstraction of α-hydrogens of substituted cumenes [Yamamoto and Otsu, 1967] and the $\sigma_\alpha^{\cdot}$ **radical substituent constants** defined by the benzylic α-hydrogen hyperfine coupling constants [Wayner and Arnold, 1984].

The **charge transfer constant** C_T (or **group charge transfer**) is an electronic substituent constant defined from the dissociation constant of a complex between tetracyanoethylene and an X-substituted parent compound [Hetnarski and O'Brien, 1975]:

$$C_T = \log K_{\mathrm{X}} - \log K_{\mathrm{H}}$$

This definition is analogous to that used for σ electronic substituent constants and accounts for the formation ability of a charge-transfer complex (CTC), such as the π-complex formation ability of aromatic systems.

The **electron donor-acceptor substituent constant** κ is an electronic constant proposed to measure the ability of a group to modify the stability of an electron donor-acceptor complex which is often referred to as charge transfer complex [Foster *et al.*, 1978; Livingstone *et al.*, 1979]. It is defined as:

$$\kappa = \log K_{\mathrm{X}}^{\mathrm{App}} - \log K_{\mathrm{H}}^{\mathrm{App}}$$

where K^{App} is the apparent equilibrium constant for the formation of a complex between the electron-acceptor 1,3,5-trinitrobenzene and the X-substituted benzene in CCl_4 solution at 33.4 °C. Such equilibrium constants were determined by the NMR technique. Benzene was chosen as the reference compound.

Table E-5. Summary of the electronic substituent constants.

ID	*Symbol*	*Reaction*	ϱ	*D* %	*References*
1	σ_p	*K*; 4-XC_6H_4COOH	1.00	53	[McDaniel and Brown, 1958; Lewis and Johnson, 1959]
2	σ_m	*K*; 3-XC_6H_4COOH	1.00	22	[McDaniel and Brown, 1958; Lewis and Johnson, 1959]
3	σ_p^+	*k*; 4-$XC_6H_4C(CH_3)_2Cl$	–4.54	66	[Brown and Okamoto, 1958a]
4	σ_m^+	*k*; 3-$XC_6H_4C(CH_3)_2Cl$	–4.54	33	[Brown and Okamoto, 1958a]
5	σ_p^-	*K*; *p*-XC_6H_4OH	2.23	56	[Lewis and Johnson, 1959; Hine, 1962; Cohen and Jones, 1963]
6	σ_m^0	*K* and *k*; unbiased values	1.00	23	[Taft, 1960]
7	$\bar{\sigma}_p$	*K* and *k*; effective values	1.00	42	[Taft, 1960]
8	σ^0	*k*; HO^- + $XC_6H_4CH_2COOEt$	0.98	37	[Yukawa *et al.*, 1966]
9	σ_p^n	*K* and *k*; "normal" values	1.00	47	van Bekkum
10	σ^*	*k*; XCH_2COOY	2.48	6	[Taft, 1952; Taft, 1953b; Taft and Lewis, 1958; Taft, 1956]
11	σ^*	*k*; XCOOY	2.48	37	[Taft, 1952; Taft, 1953b; Taft and Lewis, 1958; Taft, 1956]
12	σ^*	*k*; $X_2CHCOOY$	2.48	0	[Taft, 1952; Taft, 1953b; Taft and Lewis, 1958; Taft, 1956]
13	σ^*	*k*; $X_2(CH_2)_2COOY$	2.48	15	[Taft, 1952; Taft, 1953b; Taft and Lewis, 1958; Taft, 1956]
14	σ^*	*k*; $X_2(CH_2)_3COOY$	2.48	33	[Taft, 1952; Taft, 1953b; Taft and Lewis, 1958; Taft, 1956]
15	σ^*	*k*; 2-XC_6H_4COOY	1.00	53	[Taft, 1952; Taft, 1953b; Taft and Lewis, 1958; Taft, 1956]
16	σ'	*K*; 4-$XC_8H_{12}COOH$	1.464	3	[Roberts and Moreland, 1953]
17	σ'	*K*; 4-$XC_8H_{12}COOH$	1.65	1	[Holtz and Stock, 1964; Baker, 1966]
18	σ_I	$\sigma_I = 0.45\ \sigma^*$ *k*; XCH_2COOY	6.23	0	[Taft, 1952; Taft, 1953b; Taft and Lewis, 1958; Taft, 1956]
19	σ_R	$\sigma_R = \sigma - \sigma_I$	1.00	92	[Taft *et al.*, 1959; Taft and Lewis, 1959; Taft *et al.*, 1963a]
20	σ_R^0	$\sigma_R^0 = \sigma^0 - \sigma_I$	1.00	84	[Taft *et al.*, 1959; Taft and Lewis, 1959; Taft *et al.*, 1963a]
21	σ_I^q	*k*; $XC_7H_{12}N$	1.00	–	[Grob and Schlageter, 1976]
22	σ''	*K*; 4-$XC_6H_{10}COOH$	1.00	0	[Siegel and Komarmy, 1960]
23	σ_j^{DG}	*K*; j-$XC_{10}H_8COOH$	1.46	29	[Dewar and Grisdale, 1962b]
24	σ_I	$\sigma_I = b \cdot pK_a^R + a$	–	–	[Charton, 1964]
25	σ_m^I	ionization potential, $XC_6H_4CH_2^\bullet$	1.00	26	[Harrison *et al.*, 1961]
26	σ_p^I	ionization potential, $XC_6H_4CH_2^\bullet$	1.00	65	[Harrison *et al.*, 1961]

ID	*Symbol*	*Reaction*	ϱ	*D* %	*References*
27	σ^{Q}	$\sigma^{Q} = (f - 34.826) / 1.024$ ^{35}Cl quadrupole resonance	1.00	26	[Bray and Barnes, 1957]
28	σ_p^F	^{19}F nmr chemical shift, XC_6H_4F	1.00	65	[Taft, 1960]
29	σ_p^C	^{13}C nmr chemical shift, XC_6H_5	1.00	68	[Maciel and Natterstad, 1965]
30	ι (iota)	$\iota = 1.755 - \delta^{C} / 54.9$ ^{13}C nmr shift, XC_6H_4Y	–	–	[Inamoto *et al.*, 1978; Inamoto and Masuda, 1977]
31	σ_χ	theoretical calculations	–	–	[Marriott *et al.*, 1984; Marriott and Topsom, 1982]
32	*I*	*K*; acids	1.00	5	[Branch and Calvin, 1941]
33	σ_R^0	$\sigma_R^0 = 0.0079\ A^{1/2} - 0.027$ infrared spectroscopy	1.00	96	[Brownlee *et al.*, 1965; Brownlee *et al.*, 1966]
34	σ_R^0	$\sigma_R^0 = (\delta_p^F - \delta_m^F) / 2.97$ ^{19}F nmr chemical shift, XC_6H_4F	–	–	[Taft *et al.*, 1959]
35	*R*	*K*; 4-X-pyridinium ions	1.00	–	[Taft and Grob, 1974]
36	*F*	$f(\sigma_m, \sigma_p)$	1.00	22	[Dewar and Grisdale, 1962b]
37	*M*	$f(\sigma_m, \sigma_p)$	1.00	93	[Dewar and Grisdale, 1962b]
38	*F′*	$f(\sigma_m, \sigma_p)$	1.00	27	[Dewar and Grisdale, 1962b]
39	*M′*	$f(\sigma_m, \sigma_p)$	1.00	93	[Dewar and Grisdale, 1962b]
40	σ_p^+	$f(\sigma_m, \sigma_p)$	1.00	70	[Swain and Lupton Jr., 1968]
41	$\mathcal{F}$	$f(\sigma_m, \sigma_p)$	1.00	0	[Swain and Lupton Jr., 1968]
42	$\mathcal{R}$	$f(\sigma_m, \sigma_p)$	1.00	100	[Swain and Lupton Jr., 1968]
43	$\Delta\sigma$	$\sigma_p - \sigma_m$	1.00	92	[McDaniel and Brown, 1958; Lewis and Johnson, 1959]
44	C_T	X-cyanoethylenes	–	–	[Hetnarski and O'Brien, 1975]
45	σ^{Φ}	dialkylphosphinic acids	–	–	[Mastryukova and Kabachnik, 1971]
46	σ_a	heterocyclic rings	–	–	[Otsuji *et al.*, 1960]

Notes on the Table E-5. The symbols *K* and *k* in the reaction column represent equilibrium $(1/\varrho)\log(K_X / K_H)$ values and kinetic $(1/\varrho)\log(k_X / k_H)$ values, respectively; subscripts X and H refer to the X-substituted and H-substituted compounds, respectively.

For all the series the unsubstituted compounds have $\sigma = 0$ except for series 15–17, 27, 28, and 29, whose values are: 15) $\sigma = 7.760$; 16) $\sigma = 7.760$; 17) $\sigma = 34.622$; 27) $\sigma = 0.490$; 28) $\sigma = -0.100$; 29) $\sigma = -0.115$. The values of the sensitivity to resonance effect (*D*%) are taken from Swain-Lupton [Swain and Lupton Jr., 1968].

σ_j electronic constants of Dewar-Grisdale (36 – 39) represent σ values determined from *j*-$XC_{10}H_8COOH$ substituted compounds (1-naphthoic acids) with j = 3, 4, 5, 6, 7; the corresponding percentages of resonance (*D* %) are: 29, 57, 38, 43, 48.

σ^{Q} values (27) were estimated from the ^{35}Cl quadrupole resonance frequency *f* in *ortho*-substituted chlorobenzenes; it is related to the Taft σ^* polar constant.

σ_R^0 values (33) were estimated from the integrated intensity A of the ν_{16} band in IR spectra.

📖 [Gutowsky *et al.*, 1952] [Brown *et al.*, 1958b] [Okamoto *et al.*, 1958b] [Okamoto *et al.*, 1958c] [Okamoto and Brown, 1958a] [Biggs and Robinson, 1961] [Hine, 1962] [Taft *et al.*, 1963b] [Lambert, 1966] [Baker *et al.*, 1967] [Charton, 1971] [Sawada *et al.*, 1972] [De Maria *et al.*, 1973] [Hetnarski and O'Brien, 1973] [Sjöström and Wold, 1976] [Shorter, 1978] [Charton, 1978a] [Ewing, 1978] [Exner, 1978] [Godfrey, 1978] [Ford *et al.*, 1978] [Taft, 1983] [Reynolds, 1983] [Craik and Brownlee, 1983] [Bijloo and Rekker, 1984a] [Laurence *et al.*, 1984] [Charton, 1984] [Charton, 1987] [Topsom, 1987a] [Bowden, 1990] [Hansch *et al.*, 1991] [Hansch *et al.*, 1995b] [Tompe *et al.*, 1995]

electronic-topological descriptors → **charge descriptors**

electron isodensity contour surface → **molecular surface**

electron-topological matrix of congruity → **electron-topological method**

Electron-Topological method (: *ET method*)

A QSAR approach based on a matrix representation of chemical compounds involving topological, geometrical and electronic features. Comparison of matrices representing compounds with their biological activity aim to find the submatrix presenting specific molecular fragments responsible for activity [Dimoglo *et al.*, 1988; Bersuker and Dimoglo, 1991a].

For each molecule with A atoms, the square symmetric $A \times A$ matrix, called the **electron-topological matrix of congruity ETMC**, is defined as the following:

a) the diagonal elements are electronic atomic parameters such as atomic charges, polarizabilities, orbital energies, etc.;

b) the off-diagonal elements corresponding to bonded atoms are electronic parameters of the considered bonds such as bond order, bond energy or polarizability;

c) the off-diagonal elements corresponding to non-bonded atoms are the corresponding geometric → *interatomic distances* r_{ij}.

For a given conformation of a molecule the interatomic distances are fixed while the electronic parameters relative to atoms and bonds can be combined in different ways, each giving a different **ETMC** matrix. If all these matrices are taken together, a three-dimensional matrix **TDETMC** is obtained with $A \times A \times M$ elements where M is the number of all the considered combinations of electronic parameters.

By comparing the **ETMC** matrices of all the active compounds with those of the inactive ones, the **electron-topological submatrix of activity ETSA** is derived which contains matrix elements that are present in all the active compounds and absent from the inactive ones. The information contained in this matrix allows the design of new active compounds as well as the screening of several compounds with respect to their activity.

📖 [Bersuker *et al.*, 1991c] [Bersuker *et al.*, 1991b]

electron-topological submatrix of activity → **electron-topological method**

electrophilic atomic frontier electron density → **quantum-chemical descriptors**

electrophilic charge → **charge descriptors** (⊙ net orbital charge)

electrophilic frontier electron density index → **quantum-chemical descriptors** (⊙ electrophilic atomic frontier electron density)

electrophilic indices → **reactivity indices**

electrophilic substituent constant → **electronic substituent constants**
(⊙ resonance electronic constants)

electrophilic superdelocalizability → **quantum-chemical descriptors**

electropy index (ε)

It is an information index proposed to characterize the global electronic structure of molecules calculated from the molecular structure but avoiding quantum chemical approaches [Mekenyan *et al.*, 1987].

Example : methylethyl ether

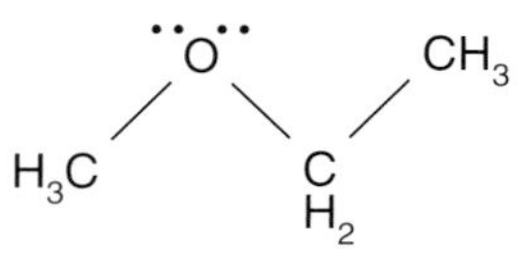

Molecular subspace types	*Subspace electrons,* n_g	*Subspaces of each type, n*
1s(C)	2	3
1s(O)	2	1
CH_3	6	2
CH_2	4	1
C – O	2	2
C – C	2	1
oxygen lone pairs	4	1
total number of electrons, N_{el}	= 34	
total number of subspaces, *G*	= 11	

$$\varepsilon = \log_2(34!) - 3\cdot\log_2(2!) - 1\cdot\log_2(2!) - 2\cdot\log_2(6!) - 1\cdot\log_2(4!) + \\ - 2\cdot\log_2(2!) - 1\cdot\log_2(2!) - 1\cdot\log_2(4!) = 92.642$$

Box E-2.

It is defined as:

$$\varepsilon = \log_2(\mathrm{N_{el}}!) - \sum_{g=1}^{G} \log_2(n_g!)$$

where N_{el} is the total number of electrons in the molecule, i.e. the sum of all electrons (inner and valence electrons) of all atoms; n_g is the number of electrons in the gth molecular subspace and G is the number of all possible subspaces.

Molecular subspaces are: a) the core parts of each atom type (e.g. C_{1S}, O_{1S}, N_{1S},); b) σ and π bond spaces, where σ space is further divided into different independent bond spaces such as C–C, C–H, CH_2, CH_3, C–O, etc.; c) the lone pairs on each atom.

The electropy index may be viewed as a measure of the degree of freedom for electrons to occupy different subspaces during the process of molecular formation.

electrostatic balance term → **GIPF approach**

electrostatic factor → **electric polarization descriptors**

electrostatic hydrogen-bond acidity → **theoretical linear solvation energy relationships**

electrostatic hydrogen-bond basicity → **theoretical linear solvation energy relationships**

electrostatic interaction fields → **molecular interaction fields**

electrotopological state index : *E-state index* → **electrotopological state indices**

electrotopological state indices

The electrotopological state S_i of the ith atom in the molecule, called the ***E*-state index** (or **electrotopological state index**) gives information related to the electronic and topological state of the atom in the molecule and is defined as [Kier and Hall, 1990a; Kier and Hall, 1999]:

$$S_i = I_i + \Delta I_i = I_i + \sum_{j=1}^{A} \frac{I_i - I_j}{(d_{ij} + 1)^k}$$

where I_i is the **intrinsic state** of the ith atom and ΔI_i is the field effect on the ith atom calculated as perturbation of the intrinsic state of the ith atom by all other atoms in the molecule; d_{ij} is the → *topological distance* between the ith and the jth atoms; A is the number of atoms. The exponent k is a parameter to modify the influence of distant or nearby atoms for particular studies. Usually it is taken as $k = 2$.

The intrinsic state of the ith atom is calculated by:

$$I_i = \frac{(2/\mathrm{L}_i)^2 \cdot \delta_i^v + 1}{\delta_i}$$

where L_i is the principal quantum number (2 for C, N, O, F atoms, 3 for Si, S, Cl, ...), δ_i^v is the number of valence electrons (→ *valence vertex degree*) and δ_i is the number of sigma electrons (→ *vertex degree*) of the ith atom in the → *H-depleted molecular graph*. (Table E-6 and Table E-7).

Table E-6. Intrinsic state values.

Group	*I*	*Group*	*I*
>C<	1.250	≡N, –OH	6.000
>CH–	1.333	=O	7.000
$-CH_2-$	1.500	–F	8.000
–C=	1.667	–SH	3.222
$-CH_3$, =CH–, >N–	2.000	–S–	1.833
≡C–, –NH–	2.500	=S	3.667
$=CH_2$, =N–	3.000	–Cl	4.111
–O–	3.500	–Br	2.750
≡CH, $-NH_2$	4.000	–I	2.120
=NH	5.000		

The intrinsic state of an atom can be simply thought of as the ratio of π and lone pair electrons to the count of the σ bonds in the molecular graph for the considered atom. Therefore, the intrinsic state reflects the possible partitioning of non-σ electrons influence along the paths starting from the considered atom; the less partitioning of the electron influence, the more available are the valence electrons for intermolecular interactions. From the intrinsic state the → *Q polarity index* was derived.

The perturbation $\Delta_{ij} = I_i - I_j$ of the ith intrinsic state by the jth atom can be viewed as an "electronegative gradient" whose sign gives the direction of influence of surrounding atom intrinsic states.

From the definition of intrinsic states and field effects, it can be seen that large positive values of *E*-states S_i relate to atoms of high electronegativity and/or terminal atoms or atoms that lie on the mantle of the molecule; small or negative *E*-state values correspond to atoms possessing only σ electrons and/or buried in the interior of the molecule or close to higher electronegative atoms. Therefore, the *E*-state index is a measure of the electronic accessibility of an atom and can be interpreted as a probability of interaction with another molecule. However, the index cannot be considered a pure electronic descriptor: it is, in fact, a descriptor of atom polarity and steric accessibility.

Note that:

$$\sum_{i=1}^{A} \Delta I_i = 0 \rightarrow \sum_{i=1}^{A} S_i = \sum_{i=1}^{A} I_i$$

the sum of the field effects over all atoms in the molecule being equal to zero.

This corresponds to an electronegativity equalization principle and means that the sum of the *E*-states in the molecule depends on only the number and type of atoms, not on their mutual interactions.

Table E-7. Valence vertex degrees δ^v, vertex degrees δ, their differences, and Kier-Hall electronegativity values for different atoms.

Atom / hybrid	δ^v	δ	$\delta^v - \delta$	*KHE*
C_{sp3}	4	4	0	0.00
C_{sp2}	4	3	1	0.25
C_{sp}	4	2	2	0.50
N_{sp3}	5	3	2	0.50

Atom / hybrid	δ^v	δ	$\delta^v - \delta$	*KHE*
N_{sp2}	5	2	3	0.75
N_{sp}	5	1	4	1.00
O_{sp3}	6	2	4	1.00
O_{sp2}	6	1	5	1.25
S_{sp3}	6	2	4	0.44
S_{sp2}	6	1	5	0.55
F	7	1	6	1.50
Cl	7	1	6	0.67
Br	7	1	6	0.38
I	7	1	6	0.24

The electrotopological states are → *local vertex invariants.* After rescaling, they are also used as atomic weighting factors for the calculation of the → *WHIM descriptors.*

Since the *E*-state values derive from an H-depleted graph, they are calculated for each → *hydride group*, i.e. they encode electronic and topological information about both heavy atoms and their bonded hydrogens. For molecules with highly polar groups, these two contributions can be treated separately by the **hydrogen electrotopological state index** HS_i (or ***HE*-state index**), which was defined to complement the *E*-state index to encode electronic and topological information about the hydrogens. It is defined as [Kier and Hall, 1999]:

$$HS_i = KHE_i + [KHE_i - KHE(H_i)] + \sum_{j \neq i} \frac{KHE_j - KHE(H_i)}{\left(d_{ij} + 1\right)^2} =$$

$$= KHE_i + [KHE_i + 0.2] + \sum_{j \neq i} \frac{KHE_j + 0.2}{\left(d_{ij} + 1\right)^2}$$

where KHE_i is the → *Kier-Hall electronegativity* of the *i*th heavy atom in the H-depleted molecular graph chosen as a measure of the intrinsic state of the attached hydrogen atom whose electronegativity $KHE(H_i)$ is taken to be –0.2; the perturbation term is given by the sum over all other heavy atoms in the molecule of the difference between their electronegativity and hydrogen electronegativity divided by the square of the topological distance *d*; note that the distance is calculated between each *j* heavy atom and the *i*th heavy atom to which the hydrogen is bonded.

Therefore, given an H-depleted molecular graph, two state values can be calculated for each vertex, the *E*-state value which measures the electron density and accessibility of the atom and the *HE*-state value which measures the reaction and interaction ability of the bonded hydrogens.

In this approach to the calculation of hydrogen *E*-states HS_i, topology is not considered a relevant factor in determining the *E*-state of a hydrogen; only the relative electronegativity is used to characterize the polarity of bonds with hydrogen in the molecule. However, in other definitions of hydrogen electrotopological states the topology is also accounted for. In particular, in the first approach to *HE*-states the state of the hydrogen atom in the X–H bond is mainly determined by the electronegativity of the X atom and, to a lesser extent, by its topology [Kellogg *et al.*, 1996]. It is defined as:

$$HS_i = I(H_i) + \sum_j \frac{\Delta I_{ij}}{(d_{ij}+1)^2} = \frac{(\delta_i^v - \delta_i)^2}{\delta_i} + \sum_j \frac{\Delta I_{ij}}{(d_{ij}+1)^2}$$

where $I(H_i)$ is the intrinsic state of the hydrogen bonded to the ith atom, defined in terms of squared electronegativity of the ith atom to accentuate the electronic influence on H in the X–H bond. ΔI_{ij} is the perturbation of the hydrogen intrinsic state by the intrinsic state of the jth atom in the molecule and d is the topological distance. Hydrogen intrinsic states seem to be a measure of the H-donor ability of X–H groups.

Another approach [Kier and Hall, 1999] is to apply the E-state definition to each vertex of the molecular graph where the hydrogens of polar groups (–OH, –NH, –COOH, etc.) are explicitly considered as independent vertices. In this case, the intrinsic state of polar hydrogens will be:

$$I_i(H) = \frac{(2/\mathrm{L}_i)^2 \cdot \delta_i^v + 1}{\delta_i} = 5$$

where L, δ^v and δ always equal one. For each polar group, the difference between the HE-state and the E-state of the bonded atoms H–X reflects the polarity of the considered bond.

Based on the same approach used to define E-state indices, a **bond E-state index** BS_b was also tentatively proposed as:

$$BS_b = BI_b + \Delta BI_{bt} = \sqrt{(I_i \cdot I_j)_b} + \sum_{t \neq b} \frac{BI_b - BI_t}{(d_{bt}+1)^2}$$

where b is the considered bond constituted by the atoms i and j, t runs over all the remaining bonds different from the b bond, BI is the bond intrinsic state defined by the intrinsic states I of the adjacent vertices, ΔBI the perturbation term, and d the → *topological edge distance* between bonds b and t [Kier and Hall, 1999].

E-state and HE-state values were also used as atomic properties to calculate → *molecular interaction fields* [Kellogg *et al.*, 1996]. The ***E*-state fields** are defined by superimposing a 3D fixed → *grid* over the molecule and calculating at each kth grid point an interaction energy value:

$$\mathrm{E}_k = \sum_i S_i \cdot f(r_{ik})$$

where S_i is the electrotopological state value of the ith atom and $f(r_{ik})$ is a function of the distance r between each ith atom of the target molecule and the considered kth grid point. This function is not defined *a priori* but has to be empirically searched for; explored functions are $1/r$, $1/r^2$, $1/r^3$, $1/r^4$, and e^{-r}. Since the $1/r^n$ functions are discontinuous at grid points close to atoms, field default values of zero are set for grid points within the van der Waals envelope of the molecule. In the same way ***HE*-state fields** are calculated using hydrogen electrotopological state values HS for each molecule atom instead of S values.

Moreover, **atom-type *E*-state indices** were proposed as molecular descriptors encoding topological and electronic information related to particular atom types in the molecule [Hall and Kier, 1995a; Hall *et al.*, 1995b]. They are calculated by summing the E-state values of all atoms of the same atom type in the molecule. Each atom type is first defined by atom identity, based on the atomic number Z, and valence state, itself identified by the **valence state indicator** (VSI) defined as:

$$VSI = \delta^v + \delta$$

where δ^v and δ are the valence vertex degree and the vertex degree of the atom; an indicator variable $\mathrm{I_{AR}}$ is also used to discriminate between atoms of an aromatic system ($\mathrm{I_{AR}} = 1$) and nonaromatic atoms ($\mathrm{I_{AR}} = 0$). In order to distinguish particular atoms, classified as the same atom type according to atom number, valence state indicator and aromatic indicator, the analysis of bonded atoms and the difference between valence δ^v and simple vertex degree δ are used. Each atom type *E*-state symbol is a composite of three parts. The first part is "S" which refers to the sum of the *E*-states of all atoms of the same type. The second part is a string representing the bond types associated with the atom ("s", "d", "t", "a" for single, double, triple and aromatic bonds, respectively). The third part is the symbol identifying the chemical element and, if any, bonded hydrogens, such as CH3, CH2, F, etc.

The atom-type *E*-state indices combine structural information about the electron accessibility associated with each atom type, an indication of the presence or absence of a given atom type and a count of the number of atoms of a given atom type.

Analogously, **atom-type *HE*-state indices** were proposed as molecular descriptors calculated by summing hydrogen electrotopological states of all atoms of the same atom type [Kier and Hall, 1999].

A Balaban-type index, called the → *E-state topological parameter* was derived from *E*-state indices.

📖 [Kier *et al.*, 1991] [Hall *et al.*, 1991a] [Hall *et al.*, 1991b] [Hall and Kier, 1992a] [Kier and Hall, 1992b] [Kier and Hall, 1992c] [Tsantilikakoulidou and Kier, 1992a] [Tsantilikakoulidou *et al.*, 1992b] [Hall *et al.*, 1993c] [Kier, 1995] [Palyulin *et al.*, 1995] [Hall and Story, 1996] [Aboushaaban *et al.*, 1996] [Kier and Hall, 1997b] [Hall and Vaughn, 1997b] [Huuskonen *et al.*, 1998] [de Gregorio *et al.*, 1998] [Gough and Hall, 1999a] [Gough and Hall, 1999b]

elongation → graph

embedded correlation → multivariate *K* correlation index

embedding frequencies → cluster expansion of chemical graphs

empirical descriptors

The class of the empirical descriptors is a fuzzy, not well-defined class. In principle, empirical descriptors are those not defined on the basis of a general theory such as, for example, quantum chemistry or graph theory. Rather they are defined by practical rules derived from chemical experience, e.g. considering specific or local structural factors present in the molecules, often sets of congeneric compounds. As a consequence, in most cases, empirical descriptors represent limited subsets of compounds and cannot be extended to classes of compounds different from those for which they were defined. Empirical descriptors have not to be confused with experimentally derived descriptors even if it is well known that several of them are empirically derived.

Empirical descriptors can be a few user-defined values for discriminating among special molecular fragments or counting local specific atom/fragment occurrences within a molecule.

Examples of empirical descriptors can be considered to be the → *Taillander index* (restricted to substituted benzenes), → *second-grade structural parameters* (restricted to alkenes), → *polar hydrogen factor* (restricted to halogenated hydrocarbons), → *hydrophobic fragmental constants*, → *six-position number*, → *Idoux steric constant*, → *hydrophilicity index*, → *adsorbability index*, → *bond flexibility index*, and → *atomic solvation parameter*.

📖 [Chiorboli *et al.*, 1993c] [Carlton, 1998]

endospectral graph → **self-returning walk counts**

endospectral vertices → **self-returning walk counts**

endpoint mean square distance index → **distance matrix**

end-to-end distance → **size descriptors**

energy-based descriptors → **quantum-chemical descriptors**

enthalpic fields → **molecular interaction fields**

enthalpic hydrophobic substituent constants → **lipophilicity descriptors**
(⊙ Hansch-Fujita hydrophobic substituent constants)

entropic fields → **molecular interaction fields**

entropic hydrophobic substituent constants → **lipophilicity descriptors**
(⊙ Hansch-Fujita hydrophobic substituent constants)

environment descriptors

Descriptors of molecular fragments defined as the → *Randic connectivity index* calculated on fragment atoms and their first neighbours [Jurs *et al.*, 1979]. The value of the Randic connectivity index for a given fragment represents the immediate surroundings of the substructure as embedded within the molecule. If the fragment is not present in the molecule, zero value is given.

Environment descriptors are closely related to → *substructure descriptors*, differing from the latter in using real values in place of binary variables or counts. The set of fragments is defined by the user depending on the data set and the specific problem.

E_K polarity scale → **linear solvation energy relationships**
(⊙ dipolarity / polarizability term)

E_T polarity scale → **linear solvation energy relationships**
(⊙ dipolarity / polarizability term)

equipotent walks → **walk counts**

equivalence classes

Subsets of equivalent elements according to a specified equivalence relation.

A **decomposition** $\mathcal{N}$ of a system containing N elements is any partition of these elements into disjoint subsets of equivalent elements by a selected equivalence relation.

A **finite probability scheme** can be associated to this decomposition as follows:

Equivalence classes	$1, 2, \ldots.., G$
Element partition	$n_1, n_2, \ldots., n_g, \ldots., n_G$
Probability distribution	$p_1, p_2, \ldots., p_g, \ldots., p_G$

where G is the total number of equivalence clsasses, $p_g = n_g / N$ is the probability of a randomly chosen element belonging to the gth subset having n_g elements and $N = \sum_{g=1}^{G} n_g$.

→ *Information content* is a fundamental measure derived from the partitioning of elements into equivalence classes; several → *molecular descriptors* are derived as → *information indices*.

E_R radical parameter → **electronic substituent constants**

***E*-state index** → **electrotopological state indices**

***E*-state fields** → **electrotopological state indices**

***E*-state topological parameter** (TI^E)

It is derived by applying the → *Ivanciuc-Balaban operator* to the → *E-state index* values used to characterize molecule atoms [Voelkel, 1994]:

$$TI^E = \frac{B}{C+1} \cdot \sum_{b=1}^{B} (S_i \cdot S_j)_b^{-1/2}$$

where B is the → *bond number*, C the → *cyclomatic number*, i.e. the number of rings in the molecule, S_i and S_j the *E*-state values for the two atoms incident to the bth bond. The sum runs over all bonds in the molecule.

ET method : *Electron-Topological method*

Eulerian walk → **graph**

EVA descriptors *(: EigenVAlue descriptors)*

EVA descriptors were recently proposed by Ferguson *et al.* [Ferguson *et al.*, 1997; Turner *et al.*, 1997] as a new approach to extract chemical structural information from mid- and near-infrared spectra. The approach is to use, as a multivariate descriptor, the vibrational frequencies of a molecule, a fundamental molecular property characterized reliably and easily from the potential energy function. The EigenVAlue (EVA) descriptor is a function of the eigenvalues obtained from the normal coordinate matrix; they correspond to the fundamental vibrational frequencies of the molecule, which can be calculated using standard quantum or molecular mechanical methods from → *computational chemistry*.

The eigenvectors, corresponding to atomic displacement, are not considered as molecular descriptors.

Since the number of vibrational normal modes varies with the number of atoms A in a molecule (actually $3A - 6$ for a molecule without axial symmetry or $3A - 5$ for a linear molecule), each set of eigenvalues will generally be of different dimensionality. Thus to obtain comparability among the molecules and → *uniform length descriptors*, the frequency range chosen is 0 and 4000 cm^{-1} to encompass the frequencies of all fundamental molecular vibrations, and the eigenvalues are projected onto a bounded frequency scale (BFS) where the vibrational frequencies are represented by points along this axis, obtaining a scale of fixed dimensionality. Then a Gaussian function of fixed standard deviation σ is centred at each eigenvalue projection over the BFS axis, resulting in a series of $3A - 6$ (or $3A - 5$) identical and overlapping Gaussians.

The value of the **EVA function** at any point x on the BFS axis is determined by summing the contributions from each and every one of the $3A - 6$ (or $3A - 5$) overlaid Gaussians at that point:

$$EVA_x = \sum_{i=1}^{3A-6} \frac{1}{\sigma \cdot \sqrt{2\pi}} \cdot e^{-(x-\lambda_i)^2/2\sigma^2}$$

where λ_i is the ith vibrational frequency (eigenvalue) for the molecular structure.

As the final step, the EVA function is sampled at fixed increments of L cm^{-1} along the BFS axis, this samping results in the $4000/L$ values that define the EVA descriptor uniform length.

The choice of σ defines the degree to which the fundamental vibrations overlap: σ values determine the number and extent to which vibrations of a particular frequency in one structure can be statistically related to those in the other structures (interstructural overlap); moreover, such values govern the extent to which vibrations within the same structure may overlap at non-negligible values (intrastructural overlap).

The frequency range being fixed, the sampling parameter L determines the total number of EVA descriptor elements; L should be maximized so as to reduce computational overhead and minimized to catch all the useful information. The optimal L value depends on the selected σ value.

Characteristic values of the Gaussian standard deviation σ is 10 cm^{-1} (range 10 – 20 cm^{-1}) and of the sampling increment L is 5 cm^{-1} (range 2 – 20 cm^{-1}), resulting in 800 descriptor variables.

The EVA descriptors are among → *3D-descriptors*, independent of any molecular alignment, giving information about molecular size, shape and electronic properties. Moreover, the EVA descriptors show only a moderate dependence on conformations.

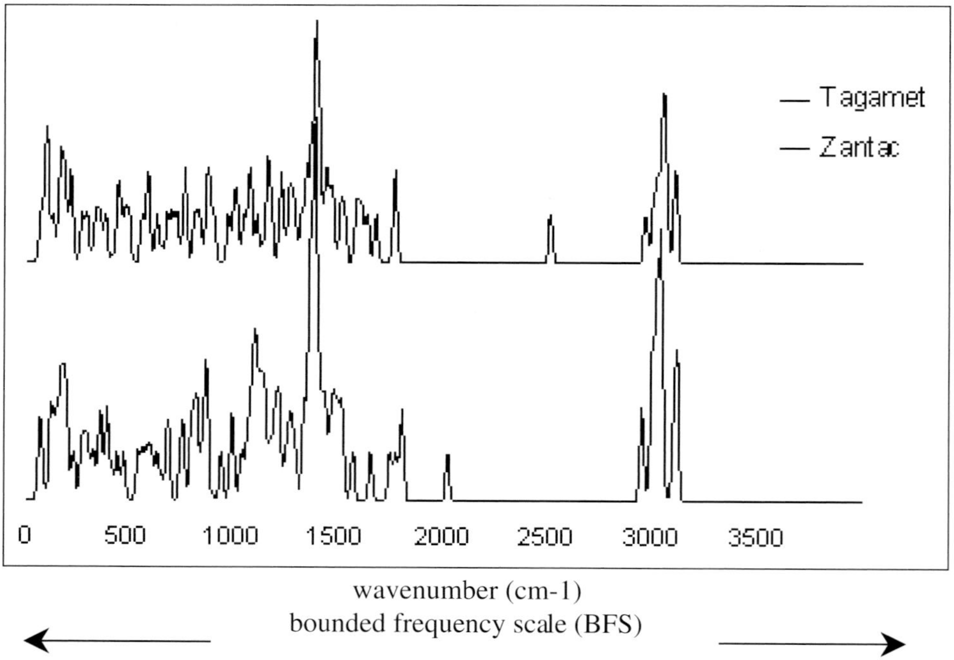

Figure E-3. Bounded Frequency Scale
with superimposed EVA descriptors for two compounds.

EEVA descriptors (or **Electronic EigenVAlue descriptors**) are vector-descriptor proposed as a modification of EVA [Tuppurainen, 1999a]. Semi-empirical molecular orbital energies, i.e. the eigenvalues of the Schrödinger equation, are used instead of the vibrational frequencies of the molecule. Each molecular orbital energy of the molecule is first projected onto a bounded energy scale (the range can be –45 to 10 eV, but this depends on the quantum-chemical method used to calculate orbital energies). Then a Gaussian function of fixed standard deviation σ (0.50 eV was proposed) is centred at each MO energy projection, resulting in a series of identical and overlapping Gaussians. Once an appropriate sampling interval L (0.25 eV was proposed) has been chosen, the whole range is sampled; the EEVA descriptors at each sampling point x are defined as:

$$EEVA_x = \sum_i \frac{1}{\sigma \cdot \sqrt{2\pi}} \cdot e^{-(x-E_i)^2/2\sigma^2}$$

where the sum runs over all Gaussian functions and E_i is the ith molecular orbital energy of the molecule. By using the parameter values defined above, this procedure provides a descriptor vector consisting of 220 (i.e. 55/0.25) elements so that dimensionality is much lower than that of the EVA descriptor vector.

📖 [Ginn *et al.*, 1997] [Heritage *et al.*, 1998] [Turner *et al.*, 1999] [Benigni *et al.*, 1999a] [Benigni *et al.*, 1999b] [Baumann, 1999]

EVA function → **EVA descriptors**

evaluation set : *external evaluation set* → **data set**

Evans extended connectivity indices

Two molecular descriptors proposed to generalize the → *Randic connectivity index*, defined as [Evans *et al.*, 1978]:

$$\chi_2^{\mathrm{ext}} = \sum_b \left[\left({}^2\delta_i n_i \cdot {}^2\delta_j n_j\right)_b \cdot \pi_b^*\right]^{-1/2} \quad \text{and} \quad \chi_3^{\mathrm{ext}} = \sum_b \left[\left({}^3\delta_i n_i \cdot {}^3\delta_j n_j\right)_b \cdot \pi_b^*\right]^{-1/2}$$

where i and j are indices for the two adjacent vertices incident to the b bond, n_i is an integer describing the ith atom type, and ${}^2\delta_i$ and ${}^3\delta_i$ are the 2nd and 3th order vertex degree of the ith atom, i.e. the sum of the vertices at topological distances 2 and 3 from the ith vertex, respectively. π_b^* is the → *conventional bond order*, and the sum runs over all bonds.

EV$_{\mathrm{TYPE}}$ descriptors → **van der Waals excluded volume method**

EV$_{\mathrm{WHOLE}}$ descriptors → **van der Waals excluded volume method**

excess electron polarizability → **electric polarization descriptors**

excess molar refractivity → **molar refractivity**

expanded distance Cluj matrices → **expanded distance matrices**

expanded distance indices → **expanded distance matrices**

expanded distance matrices

The original **expanded distance matrix $\tilde{\mathbf{D}}$**, proposed by Tratch [Tratch *et al.*, 1990b] is a square symmetric $A \times A$ matrix, representing an → *H-depleted molecular graph* with A vertices, whose diagonal entries are equal to zero and the off-diagonal entry $\tilde{d}_{ij}$ is defined as:

$$\tilde{d}_{ij} = \mu_{ij} \cdot v_{ij} \cdot d_{ij}$$

d_{ij} being the → *topological distance* between vertices v_i and v_j, v_{ij} the number of external paths including the shortest path between the considered vertices with length $m \geq d_{ij}$, and μ_{ij} the number of shortest paths between v_i and v_j that is equal to one for any pair of vertices in acyclic graphs. The number of external paths v_{ij} with respect to the shortest path i-j is just the same for each of the shortest paths connecting vertices v_i and v_j; therefore, the product $v_{ij} \cdot \mu_{ij}$ gives the total number of external paths in cyclic graphs.

Applying the → *Wiener operator* $\mathcal{W}$ to the expanded distance matrix, a molecular descriptor called **expanded Wiener number** $\tilde{W}$ is defined as:

$$\tilde{W} = \mathcal{W}(\tilde{\mathbf{D}}) = \frac{1}{2} \cdot \sum_{i=1}^{A} \sum_{j=1}^{A} \tilde{d}_{ij}$$

For acyclic graphs, the expanded distance matrix can be obtained simply as:

$$\tilde{\mathbf{D}} = \mathbf{D} \bullet \mathbf{W}$$

where **D** and **W** are the → *distance matrix* and the → *Wiener matrix*, respectively, and • indicates the → *Hadamard matrix product*. Therefore, the corresponding expanded Wiener number is calculated as:

$$\tilde{W} = \sum_{i=1}^{A-1} \sum_{j=i+1}^{A} d_{ij} \cdot \mathrm{N}_i \cdot \mathrm{N}_j$$

where N_i and N_j are the number of vertices on each side of the path *i-j*, including both vertices *i* and *j*, respectively. If only the bond contributions are considered ($d_{ij} = 1$), the expanded Wiener number coincides with the → *Wiener index*.

Based on different powers of the topological distance, **generalized expanded Wiener numbers** $\tilde{W}^n$, also called **Tratch-Stankevitch-Zefirov-type indices**, were proposed as [Klein and Gutman, 1999]:

$$\tilde{W}^n = \sum_{i=1}^{A-1} \sum_{j=i+1}^{A} d_{ij}^n \cdot \mathrm{N}_i \cdot \mathrm{N}_j$$

where n is any integer and the relation is valid only for trees.

Note that $n = 0$ and $n = 1$ result in, respectively, the → *hyper-Wiener index* and the expanded Wiener number; formally, for $n \to -\infty$, the generalized expanded Wiener number should coincide with the well-known Wiener index. For cycle-containing graphs, the generalized expanded Wiener numbers were calculated as:

$$\tilde{W}^n = \sum_{i=1}^{A-1} \sum_{j=i+1}^{A} d_{ij}^n \cdot \#_{ij}$$

where $\#_{ij}$ is the number of pairs of vertices of the graph such that there is a → *geodesic* between them containing vertices v_i and v_j.

A generalization of the expanded distance matrix was proposed by Diudea [Diudea and Gutman, 1998] in order to define new matrices derived from the Hadamard matrix product between the distance matrix **D** and a general square $A \times A$ matrix **M** as:

$$\mathbf{DM} = \mathbf{D} \bullet \mathbf{M}$$

If **M** is one among the → *Cluj matrices*, then **expanded distance Cluj matrices** are obtained. Next, if **M** is the → *Szeged matrix*, then the **expanded distance Szeged matrix** is derived; analogously, **expanded distance Szeged property matrices** [Minailiuc *et al.*, 1998] and **expanded distance walk matrices** are derived from → *Szeged property matrices* and → *walk matrices*.

From these matrices, two kinds of molecular indices, called **expanded distance indices** DM_p and **expanded square distance indices** D^2M_p, are obtained by applying the Wiener operator $\mathcal{W}$ or the → *orthogonal Wiener operator* $\mathcal{W}'$ as:

$$DM_p = \mathcal{W}(\mathbf{DM}) = \frac{1}{2} \cdot \sum_{i=1}^{A} \sum_{j=1}^{A} [\mathbf{DM}]_{ij}$$

and, only for unsymmetric **DM** matrices,

Example : 2-methylpentane

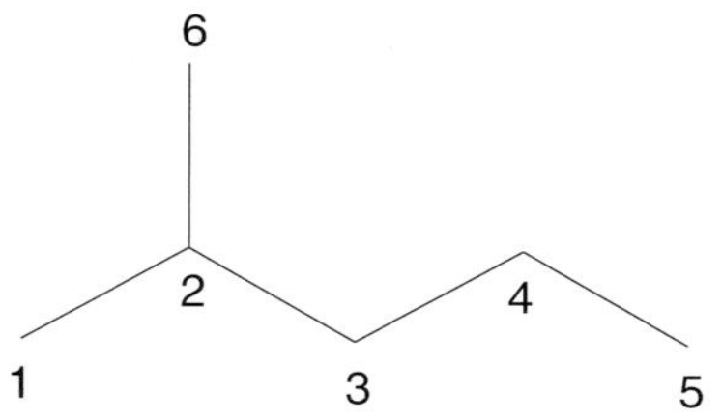

Vertex pair (i, j)	d_{ij}	N_i	N_j	$d_{ij} \cdot N_i \cdot N_j$
(1, 2)	1	1	5	5
(1, 3)	2	1	3	6
(1, 4)	3	1	2	6
(1, 5)	4	1	1	4
(1, 6)	2	1	1	2
(2, 3)	1	3	3	9
(2, 4)	2	3	2	12
(2, 5)	3	3	1	9
(2, 6)	1	5	1	5
(3, 4)	1	4	2	8
(3, 5)	2	4	1	8
(3, 6)	2	3	1	6
(4, 5)	1	5	1	5
(4, 6)	3	2	1	6
(5, 6)	4	1	1	4

Expanded Wiener number:

$$\tilde{W} = 5+6+6+4+2+9+12+9+5+8+8+6+5+6+4 = 95$$

Wiener index :

$$W = d_{12} \cdot N_1 \cdot N_2 + d_{23} \cdot N_2 \cdot N_3 + d_{26} \cdot N_2 \cdot N_6 + d_{34} \cdot N_3 \cdot N_4 + d_{45} \cdot N_4 \cdot N_5 = \\ = 5+9+5+8+5 = 32$$

Box E-3.

$$D^2M_p = \mathcal{W}'(\mathbf{DM}) = \sum_{i=1}^{A}\sum_{j=i}^{A}\left([\mathbf{DM}]_{ij}\cdot[\mathbf{DM}]_{ji}\right) = \frac{1}{2}\cdot\sum_{i=1}^{A}\sum_{j=1}^{A}\left[\mathbf{DM}\bullet\mathbf{DM}^{\mathbf{T}}\right]_{ij}$$

Note that D^2M_p indices involve squared topological distances.

Using the reciprocal distance matrix $\mathbf{D}^{-1}$ instead of the simple distance matrix in the Hadamard matrix product, new **expanded reciprocal distance matrices HM** are obtained as:

$$\mathbf{HM} = \mathbf{D}^{-1}\bullet\mathbf{M}$$

where **M** can be any square $A \times A$ matrix as defined above.

From these matrices, the **expanded reciprocal distance indices** HM_p and **expanded reciprocal square distance indices** H^2M_p are derived as:

$$HM_p = \mathcal{W}(\mathbf{HM}) = \frac{1}{2}\cdot\sum_{i=1}^{A}\sum_{j=1}^{A}[\mathbf{HM}]_{ij}$$

and, only for unsymmetric **HM** matrices,

$$H^2M_p = \mathcal{W}'(\mathbf{HM}) = \sum_{i=1}^{A}\sum_{j=i}^{A}\left([\mathbf{HM}]_{ij}\cdot[\mathbf{HM}]_{ji}\right) = \frac{1}{2}\cdot\sum_{i=1}^{A}\sum_{j=1}^{A}\left[\mathbf{HM}\bullet\mathbf{HM}^{\mathbf{T}}\right]_{ij}$$

Moreover, following the same procedure, new matrices are defined [Minailiuc *et al.*, 1998] replacing the topological distance matrix **D** by the → *geometry matrix* G as:

$$\mathbf{GM} = \mathrm{G}\bullet\mathbf{M} \qquad \text{and} \qquad \mathbf{H_GM} = \mathrm{G}^{-1}\bullet\mathbf{M}$$

where **M** can be any square $A \times A$ matrix as defined above and G^{-1} is the → *reciprocal geometry matrix*. Therefore, **GM** and $\mathbf{H_GM}$ matrices can be called **expanded geometric distance matrices** and **expanded reciprocal geometric distance matrices**, respectively. The corresponding molecular indices are defined as:

$$GM_p = \mathcal{W}(\mathbf{GM}) = \frac{1}{2}\cdot\sum_{i=1}^{A}\sum_{j=1}^{A}[\mathbf{GM}]_{ij}\ ; \qquad H_GM_p = \mathcal{W}(\mathbf{H_GM}) = \frac{1}{2}\cdot\sum_{i=1}^{A}\sum_{j=1}^{A}[\mathbf{H_GM}]_{ij}$$

and, only for unsymmetric **GM** and $\mathbf{H_GM}$ matrices,

$$G^2M_p = \mathcal{W}'(\mathbf{GM}) = \sum_{i=1}^{A}\sum_{j=i}^{A}\left([\mathbf{GM}]_{ij}\cdot[\mathbf{GM}]_{ji}\right) = \frac{1}{2}\cdot\sum_{i=1}^{A}\sum_{j=1}^{A}\left[\mathbf{GM}\bullet\mathbf{GM}^{\mathbf{T}}\right]_{ij}$$

$$H_G^2M_p = \mathcal{W}'(\mathbf{H_GM}) = \sum_{i=1}^{A}\sum_{j=i}^{A}\left([\mathbf{H_GM}]_{ij}\cdot[\mathbf{H_GM}]_{ji}\right) = \frac{1}{2}\cdot\sum_{i=1}^{A}\sum_{j=1}^{A}\left[\mathbf{H_GM}\bullet\mathbf{H_GM}^{\mathbf{T}}\right]_{ij}$$

The symbols of the molecular descriptors derived from the most common expanded distance matrices are collected in the Table E-8.

Table E-8. Molecular descriptors derived from the expanded distance matrices.

Molecular descriptor	*.... obtained from the matrix ....*
$\tilde{W}$	expanded distance matrix
HW_p	expanded reciprocal distance Wiener matrix
GW_p	expanded geometric distance Wiener matrix
H_GW_p	expanded reciprocal geometric distance Wiener matrix
D^2CJD_p	expanded distance unsymmetric Cluj-distance matrix
H^2CJD_p	expanded reciprocal distance unsymmetric Cluj-distance matrix
G^2CJD_p	expanded geometric distance unsymmetric Cluj-distance matrix

Molecular descriptor	*.... obtained from the matrix ....*
$H_G^2CJD_p$	expanded reciprocal geometric distance unsymmetric Cluj-distance matrix
$D^2CJ\Delta_p$	expanded distance unsymmetric Cluj-detour matrix
$H^2CJ\Delta_p$	expanded reciprocal distance unsymmetric Cluj-detour matrix
$G^2CJ\Delta_p$	expanded geometric distance unsymmetric Cluj-detour matrix
$H_G^2CJ\Delta_p$	expanded reciprocal geometric distance unsymmetric Cluj-detour matrix
DSZ_p	expanded distance symmetric Szeged matrix
HSZ_p	expanded reciprocal distance symmetric Szeged matrix
GSZ_p	expanded geometric distance symmetric Szeged matrix
H_GSZ_p	expanded reciprocal geometric distance symmetric Szeged matrix
D^2SZ_p	expanded distance unsymmetric Szeged matrix
H^2SZ_p	expanded reciprocal distance unsymmetric Szeged matrix
G^2SZ_p	expanded geometric distance unsymmetric Szeged matrix
$H_G^2SZ_p$	expanded reciprocal geometric distance unsymmetric Szeged matrix
D^2SZP_p	expanded distance unsymmetric Szeged property matrix
H^2SZP_p	expanded reciprocal distance unsymmetric Szeged property matrix
G^2SZP_p	expanded geometric distance unsymmetric Szeged property matrix
$H_G^2SZP_p$	expanded reciprocal geometric distance unsymmetric Szeged property matrix
D^2W_p	expanded distance unsymmetric walk matrix
H^2W_p	expanded reciprocal distance unsymmetric walk matrix
G^2W_p	expanded geometric distance unsymmetric walk matrix
$H_G^2W_p$	expanded reciprocal geometric distance unsymmetric walk matrix

If the calculation of the above defined indices is performed by summing only the matrix entries corresponding to pairs of adjacent vertices, then similar edge-defined indices can be calculated.

📖 [Diudea *et al.*, 1997f]

expanded distance matrix → **expanded distance matrices**

expanded distance Szeged matrix → **expanded distance matrices**

expanded distance Szeged property matrices → **expanded distance matrices**

expanded distance walk matrices → **expanded distance matrices**

expanded geometric distance matrices → **expanded distance matrices**

expanded reciprocal distance indices → **expanded distance matrices**

expanded reciprocal distance matrices → **expanded distance matrices**

expanded reciprocal geometric distance matrices → **expanded distance matrices**

expanded reciprocal square distance indices → **expanded distance matrices**

expanded square distance indices → **expanded distance matrices**

expanded Wiener number → **expanded distance matrices**

experimental design → **chemometrics**

experimental measurements

Experimental measurements are the basis from which numerical or graphical information can be extracted by experiments. An experiment is a well-defined operational procedure where a quantity is measured for a given sample; hence, experimental quantities are quantities measured by experimental measurement.

Physico-chemical properties constitute the most important class of experimental measurements, also playing a fundamental role as → *molecular descriptors* both for their availability as well as their interpretability. Examples of physico-chemical measurable quantities are refractive indices, molar refractivities, parachors, densities, solubilities, partition coefficients, dipole moments, chemical shifts, retention times, spectroscopic signals, rate constants, equilibrium constants, vapor pressures, boiling and melting points, acid dissociation constants, etc. [Lyman *et al.*, 1982; Reid *et al.*, 1988; Horvath, 1992; Baum, 1998].

Biological activities, toxicities, environmental parameters are other experimental quantities more often considered as responses in QSAR modelling. Examples of biological measurable quantities are effects due to a concentration of a compound, binding affinities, toxicities, inhibition constants, mutagenicity, teratogenicity. Examples of environmental measurable quantities are biochemical oxygen demand, chemical oxygen demand, biodegradability, bioconcentration factors, atmospheric residence time, volatilization from soil, rate constants of atmospheric degradation reactions.

Physico-chemical properties and biological activities $\mathcal{P}$ can be distinguished with respect to their behaviour in a system $\mathcal{S}$. A molecular property $\mathcal{P}$ may be categorized in terms of its behaviour under the hypothesis that a system $\mathcal{S} = \mathcal{A} \cup \mathcal{B}$ breaks up into two separate non-interacting subsystems $\mathcal{A}$ and $\mathcal{B}$ [Trinajstic *et al.*, 1986b].

The physical behaviour of the property has at least four mathematical possibilities of interest:

$$1.\ \mathcal{P}(\mathcal{S}) = \mathcal{P}(\mathcal{A}) + \mathcal{P}(\mathcal{B})$$
$$2.\ \mathcal{P}(\mathcal{S}) = \mathcal{P}(\mathcal{A}) \quad or \quad \mathcal{P}(\mathcal{B})$$
$$3.\ \mathcal{P}(\mathcal{S}) = \mathcal{P}(\mathcal{A}) \cdot \mathcal{P}(\mathcal{B})$$
$$4.\ \mathcal{P}(\mathcal{S}) = \partial\mathcal{P}(\mathcal{A}) \cdot \mathcal{P}(\mathcal{B}) - \mathcal{P}(\mathcal{A}) \cdot \partial\mathcal{P}(\mathcal{B})$$

where $\mathcal{A} \cap \mathcal{B} = \emptyset$. These properties are termed *additive*, *constantive*, *multiplicative*, and *derivative*, respectively. Additive and constantive properties correspond to those properties called in physical language extensive and intensive properties.

Many physico-chemical properties and biological acitivities seem to fall within the domain of additive properties. Examples of constantive properties are local molecular properties, such as dissociation energy for a localized bond or ionization potential. Characteristic multiplicative "properties" are wave functions, Kekulé structure counts and probabilities. The derivative properties are associated to the correspnding multiplicative property $\mathcal{P}$.

The choice of suitable QSAR/QSPR approaches as well as effective molecular descriptors also depends on the characteristic behaviour of the studied property.

For each experimental quantity, several estimation methods are known; moreover, for many of them, theoretical methods and/or empirical models were proposed to obtain reliable estimates avoiding experimental measurements.

In any case, experimental values or their estimated values are commonly used as molecular descriptors (i.e. as predictors in X block) or constitute the response (i.e. in Y block) which has to be modelled by other descriptors, i.e. reproduced by theoretical models.

Some experimental quantities, such as, for example, spectra, need to be transformed in some way before they are used as descriptors. For example, infrared spectra signals (IR spectra) sampled at 10 cm^{-1} in the fingerprint region (1500 – 600 cm^{-1}) were used as → *uniform length descriptors*, each spectrum being scaled in the range 0 – 100 [Benigni *et al.*, 1999a]. Moreover, **modulo-*L* descriptors** were proposed as uniform length descriptors to condense the chemical information of a molecule characterized by mass spectra [Scsibrany and Varmuza, 1992a]. For each mass spectrum, L descriptors H_k (k = 1, L) are calculated by summing the peak height h_i at each ith mass by the following modulo-L function:

$$H_k = \left(\sum\nolimits_i h_i\right)_k \qquad k = i \bmod L$$

where the sum runs over all the masses. These H descriptors are successively normalized to a constant sum equal to one for each spectrum. Modulo-14 descriptors were proposed as optimal value. In this case, for example, the first descriptor H_1 is obtained by summing the peak heights at masses 1, 15, 29, ...; and the fourteenth descriptor H_{14} by summing the peak heights at masses 14, 28, 42, ... These descriptors were used, after → *principal component analysis*, to find maximum common substructures in large mass spectrometric databases.

📖 [Horvath, 1988] [Jochum *et al.*, 1988] [Gasteiger, 1988] [Dearden, 1990] [Müller and Klein, 1991] [Horvath, 1992] [Chaumat *et al.*, 1992] [Crebelli *et al.*, 1992]

explanatory variables : *independent variables* → **data set**

exploratory data analysis → **chemometrics**

exponential product-sum connectivities → **exponential sum connectivities**

exponential similarity index → **similarity/diversity**

exponential sum connectivities

→ *Local vertex invariants* proposed by Balaban [Balaban and Catana, 1993] with the aim of obtaining high discrimination among the vertices of an → *H-depleted molecular graph*. They are denoted by c_i and are expressed in logarithmic units as:

$$\log c_i = \left(\sum_{k=1}^{\eta_i} k^z \cdot \prod\nolimits_{v_j \in V_{ik}} G_{jk}\right) \cdot \log G_i$$

where the sum runs over all the distances from vertex v_i to any other vertex in the graph up to its → *atom eccentricity* η_i, i.e. the maximum distance, and it involves the products of G values of the atoms v_j, located at distance k from the vertex v_i and constituting the vertex subset V_{ik}; G_i is the value of local invariant G for the considered vertex v_i defined as:

$$\log G_i = \left(\prod_{j=1}^{\delta_i} g_j\right) \cdot \log g_i \qquad \text{being} \qquad g_i = (\delta_i)^{-1/2} \cdot \left(\sum_{j=1}^{\delta_i} \delta_j\right)^{-1}$$

where the product involves all g values of the δ_i first neighbours of the vertex v_i, the local invariant g being defined as a function of the → *vertex degree* δ_i of the considered atom and the vertex degrees δ_j of its first neighbours. The exponent z in the formula of the c invariant can be either equal to +1, leading to **distance-enhanced exponential sum connectivities**, or –1 leading to **distance-normalized exponential sum connectivities**.

Analogously, other highly discriminating local invariants were proposed, denoted by c'_i and called **exponential product-sum connectivities**:

$$\log c'_i = \left(\sum_{k=1}^{\eta_i} k^z \cdot \prod\nolimits_{v_j \in V_{ik}} G'_{jk} \right) \cdot \log G'_i$$

where

$$\log G'_i = \left(\prod_{j=1}^{\delta_i} g'_j \right) \cdot \log g'_i \qquad \text{and} \qquad g'_i = (\delta_i)^{-1/2} \cdot \left(\sum_{j=1}^{\delta_i} \delta_j + \prod_{j=1}^{\delta_i} \delta_j \right)^{-1}$$

Exponential sum-connectivities c_i and c'_i take real values in the 0 – 1 range. Larger values are assigned to vertices that have higher degrees, are closer to the → *graph centre* and are closer to a vertex of higher degree.

By summing the exponential sum connectivities of all vertices, the corresponding molecular descriptors are derived:

$$XC = \sum_{i=1}^{A} c_i \qquad \text{and} \qquad XC' = \sum_{i=1}^{A} c'_i$$

where A is the number of atoms in the molecule. These topological indices decrease asymptotically toward zero with increasing the number of vertices in linear chain graphs, while they increase toward infinity with increasing the number of vertices in highly branched graphs.

Exponential sum connectivities c_i and c'_i can also be calculated for graph fragments or organic substituents X, considering the dimer molecule X–X (e.g. for the ethyl group n-butane is taken) [Balaban and Catana, 1994]. The values of c_i or c'_i are calculated for each half of such a dimer and assigned as LOVIs to each fragment. Moreover, the following descriptors for the whole fragment were proposed:

$$G_1 = \sum\nolimits_i e^{-d_i} \cdot c_i \qquad G'_1 = \sum\nolimits_i e^{-d_i} \cdot c'_i$$

$$G_2 = \sum\nolimits_i 2^{-d_i} \cdot c_i \qquad G'_2 = \sum\nolimits_i 2^{-d_i} \cdot c'_i$$

$$G_3 = \sum\nolimits_i 10^{1-d_i} \cdot c_i \qquad G'_3 = \sum\nolimits_i 10^{1-d_i} \cdot c'_i$$

$$G_4 = \sum\nolimits_i d_i^{-3} \cdot c_i \qquad G'_4 = \sum\nolimits_i d_i^{-3} \cdot c'_i$$

where the sum runs over all the atoms of the fragment, d is the topological distance of the considered atom from the root vertex.

For groups containing heteroatoms or multiple bonds, the LOVIs of each vertex are multiplied by the parameter f_i :

$$f_i = \frac{R_i}{R_{Csp3}}$$

where R_i is the covalent radius of the ith atom in its hybridization state and R_{Csp3} is the covalent radius of C_{sp3} (0.77 Å).

Moreover, for monocyclic substituents, the following descriptor was proposed:

$$G_1 = \sum\nolimits_i c_i \cdot f_i \cdot \exp(-d_i - d_i/N_{BR})$$

where N_{BR} is the number of ring adjacencies (i.e. 6 for cyclohexane, 9 for benzene). An analogous formula holds for G'_1 on replacing c_i by c'_i.

extended adjacency ID number → ID numbers

extended adjacency matrices

The extended adjacency matrices **EA** are → *weighted adjacency matrices* $A \times A$ whose elements are defined as a function of → *local vertex invariants* of the → *adjacency* *matrix* **A** and of some → *atomic properties* [Yang *et al.*, 1994; Hu and Xu, 1996]. The defined functions have the aim of removing the degeneracy of the entries of the adjacency matrix that is a binary matrix.

The **extended adjaceny matrix from vertex degrees** is an adjacency matrix **EA** whose entries a'_{ij} are defined as:

$$a'_{ij} = \begin{cases} a_{ij} \cdot \frac{\delta_i/\delta_j+\delta_j/\delta_i}{2} & \text{if } i \neq j \\ 0 & \text{if } i = j \end{cases}$$

where a_{ij} are the entries of the adjacency matrix and δ is the → *vertex degree*.

A correction factor can be introduced to account for heteroatoms, such as the → *atomic electronegativity* eln_i; the entries a^h_{ij} of the **heteroatom-corrected extended adjacency matrix EA**$^{\mathbf{h}}$ are the following:

$$a^h_{ij} = \begin{cases} a_{ij} \cdot \frac{\delta^h_i/\delta^h_j+\delta^h_j/\delta^h_i}{2} & \text{if } i \neq j \\ eln_i & \text{if } i = j \end{cases}$$

where $\delta^h_i = \delta_i \cdot eln_i$.

A further extension of the extended adjacency matrix, in terms of → *conventional bond order* π^*, can also be made to consider bond multiplicity; the entries of the **heteroatom/multiplicity-corrected extended adjacency matrix EA**$^{\mathbf{hb}}$ are the following:

$$a^{hb}_{ij} = \begin{cases} a_{ij} \cdot \frac{\delta^{hb}_i/\delta^{hb}_j+\delta^{hb}_j/\delta^{hb}_i}{2} & \text{if } i \neq j \\ a^{hb}_{ii} = a^h_{ii} & \text{if } i = j \end{cases}$$

where $\delta^{hb}_i = \delta^h_i + (1 - \frac{1}{\pi^*})$, the increment being due to the multiplicity:

$$\begin{cases} 0.00 & \text{if } \pi^* = 1 \\ 0.33 & \text{if } \pi^* = 1.5 \\ 0.50 & \text{if } \pi^* = 2 \\ 0.67 & \text{if } \pi^* = 3 \end{cases}$$

where π^* is the conventional bond order.

Note. In the original paper, the heteroatom/multiplicity degrees are defined as the following:

$$\begin{cases} \delta^{hb}_i = \delta^h_i & \text{if } \pi^* = 1 \\ \delta^{hb}_i = \delta^h_i + 1/2 & \text{if } \pi^* = 2 \\ \delta^{hb}_i = \delta^h_i + 1/3 & \text{if } \pi^* = 3 \end{cases}$$

However, by this definition, the ranking of the corrected degrees due to different multiplicities is doubtful, the triple bond increment being intermediate to single and double bonds.

From the extended adjacency matrix defined above, two → *eigenvalue-based descriptors*, called → *extended adjacency matrix indices*, were proposed as molecular descriptors.

extended adjacency matrix indices → **eigenvalue-based descriptors**

extended adjacency matrix from vertex degrees → **extended adjacency matrices**

extended connectivity → **canonical numbering** (⊙ Morgan's extended connectivity algorithm)

extended connectivity algorithm : *Morgan's extended connectivity algorithm* → **canonical numbering**

extended edge connectivity indices → **edge adjacency matrix**

extended local information on distances → topological information indices

external evaluation set → data set

external fragment topological indices → fragment topological indices

external validation → validation techniques

extrathermodynamic approach

Sometimes used as a synonym of → *Hansch analysis*, the extrathermodynamic approach refers to models based on empirical relationships of → *physico-chemical properties* with thermodynamic parameters such as free energies, enthalpies and entropies for various reactions.

These relationships, based on thermodynamic parameters but not requiring the formal thermodynamic theory, are therefore "extrathermodynamic".

The Hammett, Taft and Hansch-Fujita equations defining electronic, steric and hydrophobic constants are examples of extrathermodynamic relationships, being based on logarithms of rate or equilibrium constants, i.e. free-energy related quantities, of standard organic reactions of congeneric compound series. Since these correlation equations are often linear with respect to at least one variable, they are called **Linear Free Energy Relationships** (LFER) [Wells, 1968a; Mekenyan and Bonchev, 1986; Fujita, 1990].

Between 1961 and 1964, Hansch and co-workers used, for the first time, an extrathermodynamic approach to mathematically relate biological activity to the physico-chemical properties of molecules. The basic assumption is that the introduction of different substituents into a reference compound modifies its biological activity, which can be expressed to a first-order approximation by the following relationship:

$$\Delta K = f(\Delta\Phi_1, \Delta\Phi_2, \ldots, \Delta\Phi_J)$$

where Φ are molecular physico-chemical properties, usually electronic, steric and hydrophobic properties, and K (or k) is the equilibrium (or rate) constant of the biological interaction.

📖 [Hansch *et al.*, 1968] [Verloop, 1972] [Chapman and Shorter, 1978] [Charton, 1978a] [Kubinyi, 1993b]

F

feature selection : *variable reduction*

Fibonacci numbers → **symmetry descriptors** (⊙ Merrifield-Simmons index)

field effect → **electronic substituent constants**

field-fitting alignment → **alignment rules**

field-inductive constant → **electronic substituent constants** (⊙ field/resonance effect separation)

field-inductive effect : *polar effect* → **electronic substituent constants**

field / resonance effect separation → **electronic substituent constants**

^{19}F inductive constant → **electronic substituent constants** (⊙ inductive electronic constants)

finite probability scheme → **equivalence classes**

first eigenvector algorithm → **canonical numbering**

***FHACA* index** → **charged partial surface area descriptors**

***FHASA* index** : *RSAM index* → **charged partial surface area descriptors**

***FHASA*$_2$ index** → **charged partial surface area descriptors** (⊙ *RSAM index*)

***FHBCA* index** → **charged partial surface area descriptors**

***FHBSA* index** → **charged partial surface area descriptors**

***FHDCA* index** → **charged partial surface area descriptors**

***FHDSA* index** : *RSHM index* → **charged partial surface area descriptors**

flexibility index based on path length → **flexibility indices**

flexibility indices

Molecular descriptors proposed for the quantification of **molecular flexibility** (or dual **molecular rigidity**) and **bond flexibility** (or dual **bond rigidity**) [von der Lieth *et al.*, 1996].

The concept of flexibility is a widely used qualitative term in chemistry since molecular flexibility influences the chemical and biological properties of compounds as well as their interactions with other molecules.

Only a few attempts to quantify this concept can be found in the literature. Molecular dynamics and → *grid-based QSAR techniques* have been developed more recently to account for conformational flexibility in compounds [Clark *et al.*, 1992; Clark *et al.*, 1993; Hahn, 1997].

In most cases, a local description of the flexibility is also required to handle flexible and rigid parts of molecules. Therefore, both bond flexibility indices and molecular flexibility quantification were proposed.

Usually, the flexibility of a bond within a molecule is related to → *bond order*, the nature of the atoms incident to the bond, bond participation in one or more cyclic structures, and the branching of adjacent atoms. A single acyclic bond formed by two C_{sp3} atoms is regarded as freely rotatable, while polyaromatic ring systems, double and triple bonds are usually regarded as rigid. Analogously, the structural features decreasing molecular flexibility (or increasing molecular rigidity) are few atoms, the presence of rings and branching. It is usually assumed that a completely flexible molecule has an endless chain of C_{sp3} atoms, i.e. its flexibility is infinite.

All the flexibility indices are usually obtained from the → *H-depleted molecular graph*. The most popular indices are the following:

- **rotatable bond number**

The number of bonds which allow free rotation around themselves [Bath *et al.*, 1995]. For example, the number of $C_{sp3} - C_{sp3}$ bonds in the molecule. In automatic conformational analysis, methyl groups are often not included in rotatable bonds.

- **flexibility index based on path length** (F_K)

A topological index for molecular flexibility defined as a function of the length L of the longest chain in the molecule, and of the count ^3P of length 3 paths [Kier and Hall, 1983c]:

$$F_K = \frac{L}{1 - 1/^3P}$$

The F_K index increases with increasing chaining and decreases with increasing branching.
F_K is set at zero if no three-bond path is present, i.e. for all compounds with $^3P = 0$.

- **Kier molecular flexibility index** (Φ)

A measure of molecular flexibility derived from the → *Kier alpha-modified shape descriptors* $^1\kappa_\alpha$ and $^2\kappa_\alpha$:

$$\Phi = \frac{^1\kappa_\alpha \cdot {}^2\kappa_\alpha}{A}$$

where the Kier shape indices calculated from the H-depleted molecular graph depend on the heteroatoms by the parameter α [Kier, 1989]. $^1\kappa_\alpha$ encodes information about the count of atoms and relative cyclicity of molecules, while $^2\kappa_\alpha$ encodes information about branching or relative spatial density of molecules. The atom count A allows comparisons among isomers.

- **global flexibility index** (GS)

A measure of molecular flexibility derived from additive contributions of path flexibilities:

$$GS = \frac{2}{A \cdot (A-1)} \cdot \sum_{i=1}^{A-1} \sum_{j=i+1}^{A} LS_{ij}$$

where A is the → *atom number* of the H-depleted molecular graph [von der Lieth *et al.*, 1996]. LS_{ij} is the **local simple flexibility index** between atoms i and j in a molecule, defined as:

$$LS_{ij} = (d_{ij} + 1) - \left(\frac{NRB + 0.75 \cdot {}^4F + 0.50 \cdot {}^3F}{2} \right)_{ij}$$

where d_{ij} is the → *topological distance* between atoms i and j; NRB is the number of nonrotatable bonds in the path p_{ij}; ^4F and ^3F are the number of branching atoms with → *vertex degree* equal to four and three, i.e 4th and 3rd order → *vertex degree count*, respectively, in the path p_{ij}.

An acyclic nonbranched chain of C_{sp3} atoms is regarded as completely flexible and LS simply equals the number of atoms involved in the considered path.

- **bond flexibility index** (Φ_{BD})

A bond flexibility index calculated as the mean of the atom flexibility indices Φ^a :

$$\Phi_{BD} = \frac{\Phi_i^a + \Phi_j^a}{2}$$

where i and j are the atoms incident to the b bond. Φ_{BD} is between 0 (not flexible) and 10 (completely flexible) [von der Lieth *et al.*, 1996].

Flexibility indices Φ^a are defined for each molecule atom, provided that it belongs to one of the defined molecular substructures: aromatic rings, multiple bonds, conjugated systems, simple rings, chains and bridges. For all the atoms belonging to double bonds, triple bonds or aromatic rings, $\Phi^a = 0.5$. For all atoms belonging to simple rings, Φ^a is equal to the Kier molecular flexibility index Φ; for atoms in condensed ring systems, Kier molecular flexibility index Φ is calculated for both simple rings and the condensed system, the smallest value is taken as the Φ^a value for each atom in the ring. Moreover, for each substitution group in the ring or condensed ring, a value of 0.2 is subtracted from the Φ^a value. The C_{sp3} atoms in the chains and bridges are assigned as $\Phi^a = 10$, and this value is corrected by subtracting the sum of the atom masses of all the atoms in the next four adjacent shells, divided by 100.

The **bond rigidity index** ϱ_{BD}, obtained from the bond flexibility index Φ_{BD} in the same range but with opposite meaning, is defined as:

$$\varrho_{BD} = 10 - \Phi_{BD}$$

- **Kier bond rigidity index** (ϱ_{KB})

A measure of the rigidity/flexibility of a bond within a molecule, derived from the Kier molecular flexibility index Φ. This index is defined as:

$$\varrho_{KB} = \sum\nolimits_k \Phi_k^f - \Phi + 1$$

where Φ^f is the Kier flexibility index calculated for a single fragment. The sum runs over all fragments obtained by breaking the bond of interest [von der Lieth *et al.*, 1996]. It is a measure of bond rigidity because it represents the increase in flexibility of the fragments with respect to the parent molecule; this difference increases as the rigidity of the broken bond increases.

📖 [Luisi, 1977]

F matrix → **layer matrices** (⊙ cardinality layer matrix)

FM method : *Dewar-Grisdale approach* → **electronic substituent constants** (⊙ field/resonance effect separation)

FMMF method : *Dewar-Golden-Harris approach* → **electronic substituent constants** (⊙ field/resonance effect separation)

folding degree index → **eigenvalue-based descriptors**

folding profile → **eigenvalue-based descriptors** (⊙ folding degree index)

forest → **graph**

forward Fukui function → **quantum-chemical descriptors** (⊙ Fukui function)

fractals

Fractals are geometric structures of fractional dimension; their theoretical concepts and physical applications were early studied by Mandelbrot [Mandelbrot, 1982]. By definition, any structure possessing a self-similarity or a repeating motif invariant under a transformation of scale is called *fractal* and may be represented by a fractal dimension. Mathematically, the fractal dimension D_f of a set is defined through the relation:

$$\mathrm{N}(\varepsilon) \propto \varepsilon^{-D_f}$$

where $\mathrm{N}(\varepsilon)$ denotes the number of spheres of radius ε needed to cover the whole set. The fractal dimension of a set can be interpreted as the amount of information needed to fully specify a point of the set.

The concept of fractal dimension has been applied, for example, to represent protein surface geometries [Isogai and Itoh, 1984; Wagner *et al.*, 1985; Åqvist and Tapia, 1987; Wang *et al.*, 1990; Tominaga and Fujiwara, 1997a; Tominaga, 1998a] and flexibility of alkanes [Rouvray and Pandey, 1986; Rouvray and Kumazaki, 1991].

fractional bond order → **bond order indices**

fractional charged partial negative surface areas → **charged partial surface area descriptors**

fractional charged partial positive surface areas → **charged partial surface area descriptors**

fragmental constants : *group contributions* → **group contribution methods**

fragments : *molecular fragments* → **count descriptors**

fragment count → **count descriptors**

fragment ID numbers → **ID numbers**

fragment molecular connectivity indices → **connectivity indices**

fragment topological indices (FTI)

Derived from → *topological indices* calculated by graph dissections, they were proposed to reflect the interactions between the excised fragment and the remainder of the molecule [Mekenyan *et al.*, 1988a].

Let $\mathcal{G}$ an → *H-depleted molecular graph* having A vertices and $\mathcal{G}'$ a subgraph (i.e. a fragment $\mathcal{F}$) of $\mathcal{G}$ with A' vertices, with $A' < A$ by definition. Topological indices TI which consider only vertices and edges belonging to the subgraph are called **internal fragment topological indices** (IFTI). The two following requirements were proposed for IFTI:

$$0 \leq \mathrm{IFTI}(\mathcal{F}) < \mathrm{TI}(\mathcal{G}) \quad \text{and} \quad 0 \leq \mathrm{IFTI}(\mathcal{G} - \mathcal{F}) < \mathrm{TI}(\mathcal{G})$$

where TI denotes one among the common defined topological indices, restricted to those which increase with the increase in the number of graph vertices. $\mathrm{IFTI}(\mathcal{G} - \mathcal{F})$ is the topological index calculated on the complementary subgraph.

Indices which describe a fragment in connection with the remainder of the graph are called **external fragment topological indices** (EFTI). They were proposed as the difference in value between the topological index for the whole graph and the internal fragment indices for both the fragment and remainder of the molecule:

$$\mathrm{EFTI}(\mathcal{F}) = \mathrm{TI}(\mathcal{G}) - \left[\mathrm{IFTI}(\mathcal{F}) - \sum\nolimits_k \mathrm{IFTI}(\mathcal{G} - \mathcal{F})_k\right]$$

where the sum runs over all the $\mathcal{G}-\mathcal{F}$ disconnected components; there will be only one component if the subgraph $\mathcal{G}-\mathcal{F}$ is a connected graph. The following requirement was proposed for EFTI:

$$\mathrm{EFTI}(\mathcal{F}) < \mathrm{TI}(\mathcal{G})$$

where TI denotes one among the common defined topological indices, restricted to those which increase with the increase in the number of graph vertices.

Example : 2,3,4-trimethylpentane

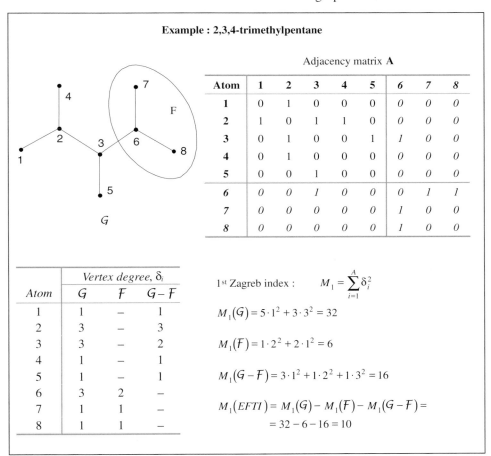

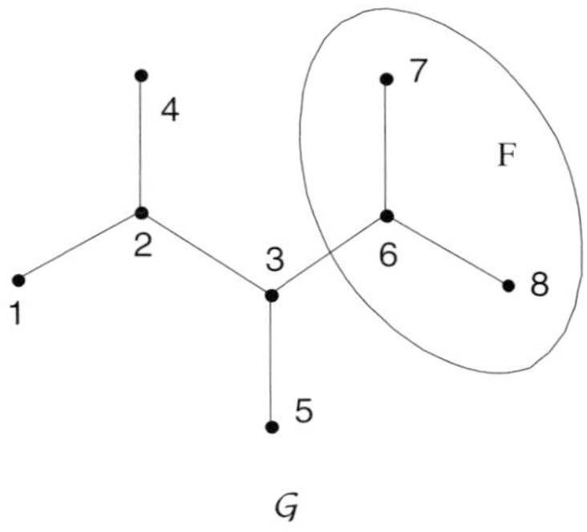

Adjacency matrix **A**

Atom	**1**	**2**	**3**	**4**	**5**	*6*	*7*	*8*
1	0	1	0	0	0	*0*	*0*	*0*
2	1	0	1	1	0	*0*	*0*	*0*
3	0	1	0	0	1	*1*	*0*	*0*
4	0	1	0	0	0	*0*	*0*	*0*
5	0	0	1	0	0	*0*	*0*	*0*
6	*0*	*0*	*1*	*0*	*0*	*0*	*1*	*1*
7	*0*	*0*	*0*	*0*	*0*	*1*	*0*	*0*
8	*0*	*0*	*0*	*0*	*0*	*1*	*0*	*0*

Atom	*Vertex degree, δ_i*		
	$\mathcal{G}$	$\mathcal{F}$	$\mathcal{G}-\mathcal{F}$
1	1	–	1
2	3	–	3
3	3	–	2
4	1	–	1
5	1	–	1
6	3	2	–
7	1	1	–
8	1	1	–

1st Zagreb index : $M_1 = \sum_{i=1}^{A} \delta_i^2$

$$M_1(\mathcal{G}) = 5 \cdot 1^2 + 3 \cdot 3^2 = 32$$

$$M_1(\mathcal{F}) = 1 \cdot 2^2 + 2 \cdot 1^2 = 6$$

$$M_1(\mathcal{G}-\mathcal{F}) = 3 \cdot 1^2 + 1 \cdot 2^2 + 1 \cdot 3^2 = 16$$

$$M_1(EFTI) = M_1(\mathcal{G}) - M_1(\mathcal{F}) - M_1(\mathcal{G}-\mathcal{F}) = $$
$$= 32 - 6 - 16 = 10$$

Box F-1.

The **normalized fragment topological indices** (NIFTI and NEFTI) may be obtained by dividing them by the topological index for the whole graph, i.e. as:

$$\mathrm{NIFTI}(\mathcal{F}) = \frac{\mathrm{IFTI}(\mathcal{F})}{\mathrm{TI}(\mathcal{G})} \quad \text{and} \quad \mathrm{NEFTI}(\mathcal{F}) = \frac{\mathrm{EFTI}(\mathcal{F})}{\mathrm{TI}(\mathcal{G})}$$

A special case of normalized fragment topological indices NIFTI is the → *graphical bond order*.

It has to be noted that IFTI($\mathcal{F}$) is a constant for a given fragment of any molecule, whereas NIFTI($\mathcal{F}$), EFTI($\mathcal{F}$) and NEFTI($\mathcal{F}$) depend upon the molecule as a whole. In particular, the following general relation holds:

$$\mathrm{TI}(\mathcal{G}_1) > \mathrm{TI}(\mathcal{G}_2) \quad \rightarrow \quad \mathrm{EFTI}(\mathcal{F} \subset \mathcal{G}_1) > \mathrm{EFTI}(\mathcal{F} \subset \mathcal{G}_2)$$

free energy of hydration density tensor → **hydration free energy density**

free molecular volume → **volume descriptors** (⊙ molar volume)

free valence index → **quantum-chemical descriptors** (⊙ electron density)

Free-Wilson analysis *(: FW analysis)*

Free-Wilson analysis is a QSAR approach searching for a relationship between a biological response and the presence/absence of substituent groups on a common molecular skeleton [Free and Wilson, 1964; Kubinyi, 1990; Kubinyi, 1993b]. The approach, called the *de novo approach* when presented in 1964, is based on the assumption that each substituent gives an additive and constant effect to the biological activity regardless of the other substituents in the rest of the molecule, i.e. the substituent effects are considered to be independent of each other. Compounds → *congenericity* is also another basic requirement.

Once a common skeleton for the chemical analogues is defined, regression analysis is performed, considering a number S of substitution sites R_s ($s = 1, S$), and for each site a number N_s of different substituents. Hydrogen atoms are also considered as substituents if present in a substitution site of some compounds.

The **Free-Wilson descriptors** of the ith compound are → *indicator variables* $I_{i,ks}$ where $I_{i,ks} = 1$ if the kth substituent is present in the sth site and $I_{i,ks} = 0$ otherwise. These descriptors are usually collected in a table called the **Free-Wilson matrix** (Table F-1), where the rows represent the data set molecules and each column represents a substituent in a specific site.

Example. Eight derivatives of toluene are considered, with two substitution sites (X and Y) (Figure F-1).

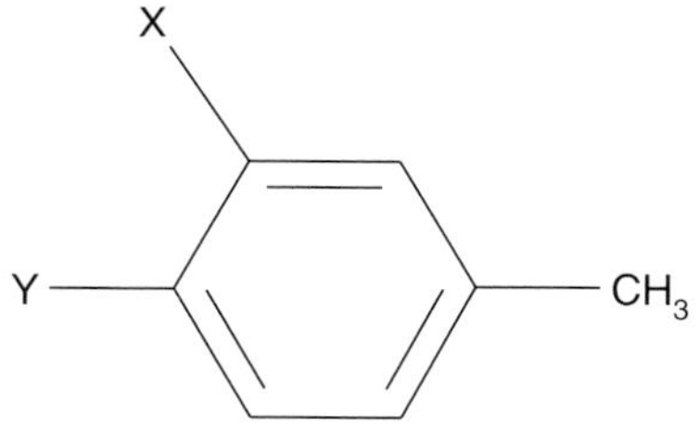

Figure F-1. Toluene parent molecule.

In the site X, ethyl, fluorine, chlorine, and bromine substituents are considered ($N_X = 4$), while in the site Y, only chlorine and bromine substituents are considered ($N_Y = 2$). The eight possible derivatives are coded in the Free-Wilson approach as shown in Table F-1.

Table F-1. Free-Wilson matrix of eight toluene derivatives.

Site:	X				Y	
Compound	C_2H_5	F	Cl	Br	Cl	Br
1	1	0	0	0	1	0
2	1	0	0	0	0	1
3	0	1	0	0	1	0
4	0	1	0	0	0	1
5	0	0	1	0	1	0
6	0	0	1	0	0	1
7	0	0	0	1	1	0
8	0	0	0	1	0	1

The total number of binary descriptors (i.e. the matrix columns) is

$$p = \sum_{s=1}^{S} \mathrm{N}_s$$

Given the number of substitution sites S and the number of substituents for each site N_s, the maximum number of compounds that can be studied, i.e. obtained by all of the substituent combinations, is calculated as:

$$n^{\max} = \prod_{s=1}^{S} \mathrm{N}_s$$

The **Free-Wilson model** (also called **additivity model**) is defined as:

$$\hat{y}_i = b_0 + \sum_{s=1}^{S} \sum_{k=1}^{\mathrm{N}_s} b_{ks} \cdot \mathrm{I}_{i,\,ks}$$

where b_0 is the intercept of the model corresponding to the average biological response calculated from the data set and b_{ks} are the regression coefficients. The biological response y is usually used in the form log(1/C), where C is the concentration achieving a fixed effect. The regression coefficients b_{ks} of the Free-Wilson model give the importance of each kth substituent in each sth site in increasing/decreasing the response with respect to the mean response, i.e. the activity contribution of the substituent.

This contribution can be calculated as the average activity $\bar{y}_{ks}$ of the compounds containing the considered group k in the specific site s with respect to the overall average activity $\bar{y}$:

$$b_{ks} = \Delta y_{ks} = \bar{y}_{ks} - \bar{y} = \frac{1}{f_{ks}} \cdot \sum_{i=1}^{n} y_i \cdot \mathrm{I}_{i,\,ks} - \frac{1}{n} \cdot \sum_{i=1}^{n} y_i$$

where

$$f_{ks} = \sum_{i=1}^{n} \mathrm{I}_{i,\,ks}$$

represents the number of occurrences of the kth substituent in the sth site in the data set, i.e. the number of entries equal to one in the ksth column of Free-Wilson matrix. n is the number of molecules in the data set, y_i the observed response.

Taking into account all of the substituents for each site, the sum of descriptor values (1 or 0) over all available substituents in the site is obviously equal to one, i.e. for each compound there is only one substituent in that site. Therefore this constraint for each site produces S internal relationships:

$$\sum_{k=1}^{N_s} I_{i,ks} = 1 \qquad \sum_{k=1}^{N_s} f_{ks} = n \qquad \text{and} \qquad \sum_{k=1}^{N_s} f_{ks} \cdot b_{ks} = 0 \qquad s = 1, 2, \ldots, S$$

where b_{ks} are the group contributions on the response. As a consequence, the minimum number of compounds to obtain the regression coefficients of the model is:

$$n^{\min} = p - S + 1$$

i.e. S variables are completely correlated to the others.

The → *embedded correlation* K_{EMB} in the Free-Wilson matrix, i.e. the intrinsic multivariate correlation of the data set, is:

$$K_{\mathrm{EMB}} = \frac{S}{p-1} \text{ for } n \geq p \qquad or \qquad K_{\mathrm{EMB}} = \frac{S + (p-n)}{p-1} \text{ for } p > n$$

where n is the number of compounds effectively used in the model. For example, for three substituent sites (S = 3), each with four substituent groups (N_s = 4), the total number of variables is p = 12 and the minimum number of required compounds is $n^{\min}$ = 12 – 3 + 1 = 10. For n = $n^{\min}$ = 10, i.e. $p > n$, K_{EMB} = [3 + (12 – 10)] / 11 = 0.45; for n = 15, K_{EMB} = 3 / 11 = 0.27. For the embedded correlation in the Free-Wilson matrix the following relation holds:

$$\frac{S}{p-1} \leq K_{\mathrm{EMB}} \leq \frac{2 \cdot S - 1}{p-1}$$

Fujita-Ban analysis is a **modified Free-Wilson analysis** where the activity contribution of each substituent is relative to the activity of a → *reference compound* [Fujita and Ban, 1971]. Any compound can be chosen as the reference, but usually the H-substituted compound (all R_s = H) is adopted. The Free-Wilson matrix in the Fujita-Ban analysis does not contain the descriptors corresponding to the substituents of the reference compound, i.e. the number of indicator variables is diminished by the number of sites S with respect to the corresponding original Free-Wilson approach. Moreover the row vector corresponding to the reference compound is characterized by all the descriptor binary values equal to zero.

The **Fujita-Ban model** is defined as:

$$\hat{y}_i = b_o + \sum_{s=1}^{S} \sum_{k=1}^{N_s} b_{ks} \cdot I_{i,ks}$$

which differs from the Free-Wilson model because the intercept b_0 corresponds to the estimated biological activity of the reference compound, i.e. $b_0 = \hat{y}_{\mathrm{REF}}$, while in the Free-Wilson model it corresponds to the theoretical biological activity of a "naked" compound, i.e. without any substituent.

The Fujita-Ban model is a linear transformation of the classical Free-Wilson model: indeed, group contributions of the Free-Wilson model can be transformed to Fujita-Ban group contributions by subtracting the group contributions of the corresponding substituents of the reference compound.

The **Cammarata-Yau analysis** is similar to the Fujita-Ban analysis, the only difference being that the intercept is maintained constant and equal to the response of the reference compound [Cammarata and Yau, 1970]; this means that the observed responses are first transformed:

$$y_i' = y_i - y_{\text{REF}}$$

accordingly, the **Cammarata-Yau model** is a regression through the origin, defined as:

$$\hat{y}_i = \sum_{s=1}^{S} \sum_{k=1}^{N_s} b_{ks} \cdot \text{I}_{i,\,ks}$$

The **Bocek-Kopecky analysis** is another modified Free-Wilson approach proposed to take into account interaction terms, i.e. nonlinear effects [Bocek *et al.*, 1964; Kopecky *et al.*, 1965]. The **Bocek-Kopecky model** is defined as:

$$\hat{y}_i = b_o + \sum_{s=1}^{S} \sum_{k=1}^{N_s} b_{ks} \cdot \text{I}_{i,\,ks} + \sum_{s=1}^{S} \sum_{s'=s+1}^{S} \sum_{k=1}^{N_s} \sum_{k'=1}^{N_{s'}} b_{kk',\,ss'} \cdot \text{I}_{i,\,ks} \cdot \text{I}_{i,\,k's'}$$

The Fujita-Ban model having the hydrogen substituted compound as reference compound is related to the → *Hansch linear model* by the following relationship:

$$b_{ks} \approx \sum_{j=1}^{J} b_j \cdot \Phi_{ks,\,j}$$

where b_j are the Hansch regression coefficients, J the number of considered substituent properties (e.g., lipophilic, electronic, steric properties), and $\phi_{ks,\,j}$ the jth substituent group constant for the kth substituent in the sth site. This relationship means that the group contribution b_{ks} in the Fujita-Ban model of the kth substituent in the sth site is numerically equivalent to the weighted sum of all the → *physico-chemical properties* of that substituent [Singer and Purcell, 1967; Kubinyi and Kehrhahn, 1976; Kubinyi, 1988b].

A great advantage of these approaches is the possibility of a complete → *reversible decoding*, i.e. the possibility to interpret by the model *how* and *where* the response is increased/decreased.

The main shortcomings of the Free-Wilson approaches are that 1) structural variation is necessary in at least two different sites; 2) a relatively large number of variables is necessary to describe a relatively small number of compounds; 3) the models can be used to predict a maximum number of compounds equal to $n^{\max} - n$, where n is the number of compounds effectively used in the model; 4) predictions of substituents not included in the analysis are usually not reliable.

Related to the Free-Wilson analysis is the → *DARC/PELCO analysis*, which is an extension of the former to the → *hyperstructure* concept [Duperray *et al.*, 1976a].

📖 [Cammarata, 1972] [Cammarata and Bustard, 1974] [Kubinyi, 1976a] [Hall and Kier, 1978a] [Schaad *et al.*, 1981] [Duewer, 1990] [Liwo *et al.*, 1992] [Franke and Buschauer, 1992] [Simmons *et al.*, 1992] [Franke and Buschauer, 1993] [Henrie II *et al.*, 1993] [Norinder, 1993] [Singh *et al.*, 1993] [De Castro and Reissmann, 1995] [Hasegawa *et al.*, 1995b] [Hatrìk and Zahradnìk, 1996] [Hasegawa *et al.*, 1996b]

Free-Wilson descriptors → **Free-Wilson analysis**

Free-Wilson matrix → **Free-Wilson analysis**

Free-Wilson model → **Free-Wilson analysis**

frontier orbitals → **quantum-chemical descriptors**

frontier orbital electron densities → **quantum-chemical descriptors**

F strain : *substituent front strain* → **steric descriptors**

Fujita-Ban analysis → **Free-Wilson analysis**

Fujita-Ban model → **Free-Wilson analysis**

Fujita steric constant → **steric descriptors** (⊙ Taft steric constant)

Fukui function → **quantum-chemical descriptors**

functional group count → **count descriptors**

functional chemical groups → **count descriptors**

FW analysis : *Free-Wilson analysis*

G

Galvez matrix → **topological charge indices**

ganglia-augmented atom codes → **substructure descriptors** (⊙ augmented atom codes)

GAP : *HOMO-LUMO energy gap* → **quantum-chemical descriptors**

Geary coefficient → **autocorrelation descriptors**

general a_N-index → **determinant-based descriptors**

general free valence index → **quantum-chemical descriptors** (⊙ electron density)

general interaction properties function approach : *GIPF approach*

generalized centric information indices → **centric indices**

generalized cluster significance analysis → **variable selection** (⊙ cluster significance analysis)

generalized complete centric index → **centric indices**

generalized connectivity indices → **connectivity indices**

generalized edge complete centric index → **centric indices**

generalized edge distance code centric index → **centric indices**

generalized edge distance degree centric index → **centric indices**

generalized edge radial centric information index → **centric indices**

generalized expanded Wiener numbers → **expanded distance matrices**

generalized distance code centric index → **centric indices**

generalized distance degree centric index → **centric indices**

generalized geometry matrices → **double invariants**

generalized graph centre → **centre of a graph**

generalized graph matrices → **double invariants**

generalized Hosoya indices → **Hosoya Z index**

generalized Hosoya Z matrix → **Hosoya Z matrix**

generalized radial centric information index → **centric indices**

generating optimal linear PLS estimations → **variable selection**

genetic algorithm – variable subset selection → **variable selection**

genetic function approximation → **variable selection**

geodesic → **graph**

geometrical descriptors

→ *Molecular descriptors* defined in several different ways but always derived from the three-dimensional structure of the molecule. Generally, geometrical descriptors are calculated either on some optimized → *molecular geometry* obtained by the methods of → *computational chemistry* or on crystallographic coordinates.

→ *Topographic indices* constitute a special subset of geometrical descriptors, being calculated on the graph representation of molecules but using the geometric distances between atoms instead of the topological distances.

Examples of geometrical descriptors are the → *quantum-chemical descriptors*, → *moment of inertia*, → *length-to-breadth ratio*, → *surface areas*, → *volume descriptors*, → *CPSA descriptors*, → *EVA descriptors*, → *WHIM descriptors*, → *3D-MoRSE descriptors*, → *interaction energy values*, → *spectrum-like descriptors.*

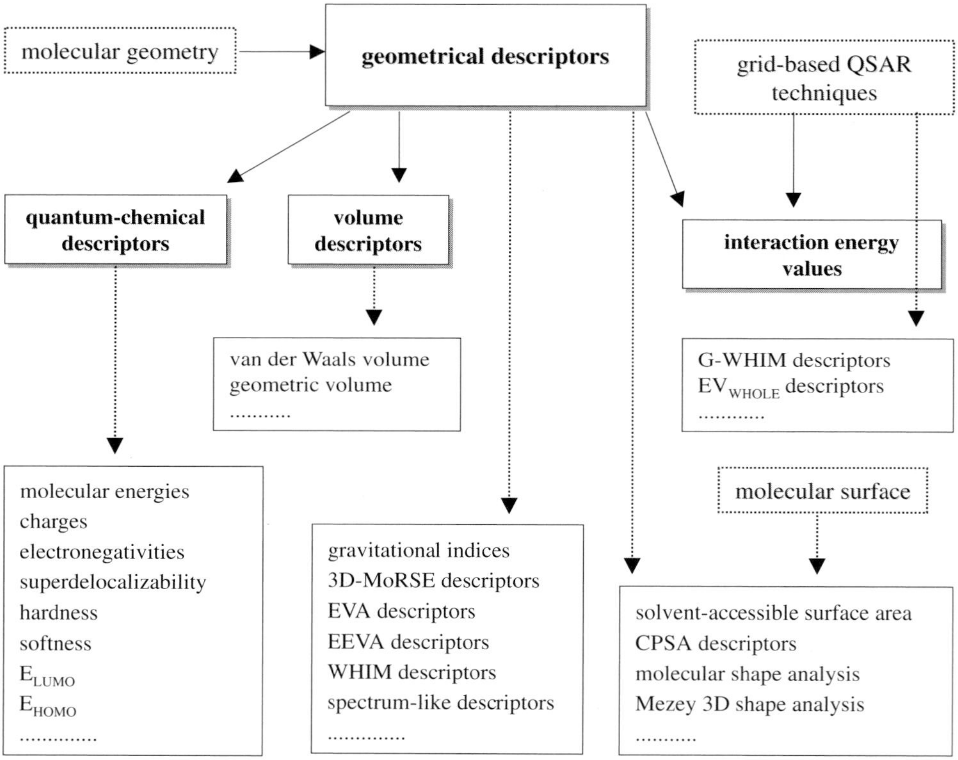

Figure G-1. Scheme of the molecular descriptors classified as geometrical descriptors.

geometrical representation → **molecular descriptors**

geometrical shape coefficient → **shape descriptors** (⊙ Petitjean shape indices)

geometric atom pairs → **substructure descriptors** (⊙ atom pairs)

geometric binding property pairs → **substructure descriptors** (⊙ atom pairs)

geometric centre → **centre of a molecule**

geometric diameter → **molecular geometry**

geometric distance → **molecular geometry**

geometric distance degree → **molecular geometry**

geometric distance matrix : *geometry matrix* → **molecular geometry**

geometric eccentricity → **molecular geometry**

geometric edge distance matrix → **edge distance matrix**

geometric factors → **weighted matrices**

geometric modification number → **weighted matrices**

geometric radius → **molecular geometry**

geometric sum layer matrix → **layer matrices**

geometric topological index → **vertex degree**

geometric volume → **volume descriptors**

geometry matrix → **molecular geometry**

GG matrices : *generalized graph matrices* → **double invariants**

Ghose-Crippen hydrophobic atomic constants → **lipophilicity descriptors**

Gini index → **information content**

GIPF approach (: *General Interaction Properties Function approach*)

A general method, proposed by Politzer and coworkers, to estimate physico-chemical properties depending on noncovalent interactions [Brinck *et al.*, 1993; Murray *et al.*, 1993; Politzer *et al.*, 1993; Murray *et al.*, 1994]. This is based on molecular surface area in conjunction with some statistically-based quantities related to the → *molecular electrostatic potential* (MEP) at the → *molecular surface*. The → *electron isodensity contour surface* [0.001 a.u. contour of $\varrho(\mathbf{r})$] is taken as the molecular surface model.

The general GIPF model for a physico-chemical property Φ is:

$$\Phi = f\left(SA,\ \Pi,\ \sigma^2_{\text{tot}},\ \nu\right)$$

where SA is the surface area and Π is the → *local polarity index*. The other two molecular surface indices are defined as the following:

$$\sigma^2_{\text{tot}} = \sigma^2_+ + \sigma^2_- = \frac{1}{n^+} \cdot \sum_{i=1}^{n^+} \left[V^+(\mathbf{r}_i) - \bar{V}^+\right]^2 + \frac{1}{n^-} \cdot \sum_{i=1}^{n^-} \left[V^-(\mathbf{r}_i) - \bar{V}^-\right]^2$$

$$\nu = \frac{\sigma^2_+ \cdot \sigma^2_-}{\left(\sigma^2_{\text{tot}}\right)^2}$$

where σ^2_{tot} is the **surface electrostatic potential variance**, which measures the electrostatic interaction tendency of the molecule; σ^2_+ and σ^2_- are the variances of positive and negative regions of the molecular surface potential; V^+ and V^- are the positive and negative values of the MEP at a grid point $\mathbf{r}$ on the molecular surface, $\bar{V}^+$ and $\bar{V}^-$ are their average values, n^+ and n^- are the numbers of grid points with positive and negative values. ν is the **electrostatic balance term** which reaches a maximum value of 0.25 when σ^2_+ and σ^2_- are equal.

Site-specific molecular quantities can be added to the global molecular descriptors in the GIPF model depending on the physico-chemical property to be estimated. Some of these site-specific descriptors are:

$\bar{I}_{S,\min}$ is the lowest value of the → *average local ionization energy* found on the molecular surface; this reflects the tendency for charge transfer and polarization at any particular molecular site [Haeberlein and Brinck, 1997].

$V_{S,\min}$ and $V_{S,\max}$ are the most negative and positive values of the molecular electrostatic potential on the molecular surface; the maximum reflects the tendency for long-range attraction of nucleophiles at a specific site, while the minimum reflects the tendency for long-range attraction of electrophiles at a specific site. $V_{S,\min}$ and $V_{S,\max}$ for a large variety of molecules correlate with hydrogen-bond basicity and acidity, respectively [Murray and Politzer, 1998].

SA^+ and SA^- are the portions of the surface area over which $V(\mathbf{r})$ is positive and negative, respectively.

Several properties have been estimated by the GIPF approach such as heat of vaporization, sublimation [Politzer *et al.*, 1997] and fusion [Murray *et al.*, 1996], boiling points and critical constants [Murray *et al.*, 1993a], surface tension, liquid and solid densities [Murray *et al.*, 1996], crystal lattice energies [Politzer and Murray, 1998], impact sensitivities [Murray *et al.*, 1998], diffusion coefficients [Politzer *et al.*, 1996], solubilities [Politzer *et al.*, 1992; Murray *et al.*, 1995], aqueous solvation free energies [Murray *et al.*, 1999; Politzer *et al.*, 2000], → *hydrogen-bonding parameters* [Lowrey *et al.*, 1995a] and → *lipophilicity*.

🕮 [Murray *et al.*, 1991] [Brinck *et al.*, 1993] [Murray *et al.*, 1993b] [Beck *et al.*, 1998]

global flexibility index → **flexibility indices**

global site-property analysis → **Hansch analysis**

global synthetic invariant → **iterated line graph sequence**

global topological charge index → **topological charge indices**

global weighted walk numbers → **walk matrices**

global WHIM descriptors → **WHIM descriptors**

globularity factor → **shape descriptors** (⊙ ovality index)

Gombar hydrophobic model → **lipophilicity descriptors**

goodness of fit → **regression parameters**

goodness of prediction → **regression parameters**

Gordon-Scantlebury index : *connection number* → **edge adjacency matrix**

graph

A graph is a mathematical object defined within the *graph theory* [Harary, 1969a; Harary, 1969b; Balaban and Harary, 1976; Rouvray and Balaban, 1979; Rouvray, 1990a; Bonchev and Rouvray, 1991; Trinajstic, 1992].

A graph is usually denoted as $\mathcal{G} = (\mathcal{V}, \mathcal{E})$, where $\mathcal{V}$ is a set of **vertices** and $\mathcal{E}$ is a set of elements representing the binary relationship between pairs of vertices; unordered vertex pairs are called **edges**, ordered vertex pairs are called **arcs**, and elements of $\mathcal{E}$ which relate a vertex to itself are called **loops**. If two vertices occur as an unordered pair more than once, they define a **multiple edge**; if two vertices occur as an ordered pair more than once, they define a **multiple arc**. Two edges in a graph $\mathcal{G}$ are said to be **independent edges** if they have no common vertex. A collection of k mutually (i.e. pairwise) independent edges in a graph $\mathcal{G}$ ($k \geq 2$) is called a **k-matching** of $\mathcal{G}$.

$\mathcal{G}$ and $\mathcal{G}'$ are called **isomorphic graphs** if a bijective mapping of the vertex and edge sets exists, i.e.

$$\mathcal{V}(\mathcal{G}) \leftrightarrow \mathcal{V}(\mathcal{G}) \qquad \text{and} \qquad \mathcal{E}(\mathcal{G}) \leftrightarrow \mathcal{E}(\mathcal{G})$$

or, in other words, if there exists a one-to-one correspondence between the vertices and edges, such that adjacency is preserved. A **graph automorphism** is an isomorphic mapping of a graph $\mathcal{G}$ onto itself, i.e. is a bijective mapping of the vertex and edge sets onto themselves which preserves the number of links joining any two vertices:

$$\mathcal{V}(\mathcal{G}) \leftrightarrow \mathcal{V}(\mathcal{G}') \qquad \text{and} \qquad \mathcal{E}(\mathcal{G}) \leftrightarrow \mathcal{E}(\mathcal{G}')$$

The set of all automorphisms of a graph forms an **automorphism group**. The occurrence of automorphisms depends on the symmetry of the graph; in particular, it depends on the presence of equivalent vertices that can be mapped automorphically onto each other, i.e. can interchange preserving the adjacency of the graph. The cardinality of the automorphism group of a graph is called **symmetry number** and is considered among the → *symmetry descriptors.*

Topologically equivalent vertices constitute disjoint subsets of vertices called **orbits**.

A graph $\mathcal{G}'$ is a **subgraph** of the graph $\mathcal{G}$ if the following relationships hold:

$$\mathcal{V}(\mathcal{G}') \subseteq \mathcal{V}(\mathcal{G}) \qquad \text{and} \qquad \mathcal{E}(\mathcal{G}') \subseteq \mathcal{E}(\mathcal{G})$$

The → *vertex degree* is the number of edges incident to a given vertex. If two vertices are connected by an arc, two degrees are assigned to each vertex; the **indegree** counts the arcs ending on the vertex, the **outdegree** counts the arcs starting from the vertex. **Terminal vertices** are the vertices of a graph with degree equal to one; **terminal edges** are the edges incident to terminal vertices. **Central vertices** and **central edges** are the vertices (edges) belonging to the → *graph centre.* All vertices with vertex degree equal to zero are called **isolated vertices**.

Branching of a graph is a fuzzy concept which can be based on the presence in the graph of vertices with degrees equal to three or higher.

A **walk** (or **random walk**) in $\mathcal{G}$ is a sequence of vertices $w = (v_1, v_2, \ldots, v_k)$ such that $(v_i, v_{i+1}) \in \mathcal{E}$ for each $i = 1$, to k–1, i.e. is a sequence of pairwise adjacent edges leading from vertex v_1 to vertex v_k; any vertex or edge can be traversed several times. The **walk length** is the number of edges traversed by the walk.

A **path** (or **self-avoiding walk**) is a walk without any repeated vertices. The **path length** is the number of edges associated with the path. The smallest path between two considered vertices is called **geodesic** and its length corresponds to the → *topological distance*; **elongation** is the longest path between two considered vertices and its length corresponds to the → *detour distance* between the vertices.

A walk closed in itself is called a **self-returning walk**, i.e. a walk starting and ending on the same vertex.

A self-returning path is called a **cyclic path** (or **cycle** or **circuit**), i.e. a cycle is a walk with no repeated vertices (i.e. a path) other than its first and last ones ($v_1 = v_k$). The number of independent cycles (or rings) in a graph is the → *cyclomatic number.* **Cyclicity** C^+ is the number of all possible cycles in a graph.

A **trail** is a walk in which vertices can be revisited but edges can be traversed only once; an **Eulerian walk** is a trail in which all vertices of the graph must be encountered.

A **Hamiltonian path** is a path in which all vertices of the graph must be visited once and the beginning and the end are different. A **Hamiltonian circuit** is a path in which all vertices of the graph must be visited once, starting and ending on the same vertex.

A list of graphs of practical interest follows.

- **simple graph** (*: graph*)

Graph having no arcs, no multiple edges or loops.

• **digraph** (*: directed graph*)
Graph in which all vertex pairs are arcs. If any vertex pair is associated with only one arc, the graph is called **oriented graph**.

• **multigraph**
Graph having no arcs or loops, but including multiple edges.

• **pseudograph**
Graph having no arcs or multiple edges, but containing loops.

• **connected graph**
Graph in which for each pair of vertices $\{i,j\} \in \mathcal{V}(\mathcal{G})$ at least one path exists. Otherwise $\mathcal{G}$ is called a **disconnected graph**. The simplest disconnected graph is a graph with an isolated vertex and the vertices not joined by a path belong to different components of the graph. The number of components of a graph is denoted $k(\mathcal{G})$.

• **regular graph**
Graph having all vertices with the same degree.

• **tree** (*: acyclic graph*)
Connected graph without cycles. The number of edges B and the number of vertices A are related by the condition $B = A - 1$. A **rooted tree** is a tree having one vertex distinguished from the others; if this vertex is an endpoint, the graph is called **planted tree**. In chemistry, rooted and planted trees can be used to represent molecular substituents.

• **linear graph** (*: path graph*)
A tree without branching; there are exactly two terminal vertices of degree 1 and $A - 2$ vertices of degree two.

• **star graph**
A maximally branched tree; there are $A - 1$ terminal vertices of degree 1 and one vertex of degree $A - 1$.

• **complete graph**
Graph in which all vertices and edges are mutually adjacent, i.e. all vertices have degree of $A - 1$. The maximal number of edges is:

$$B = \binom{A}{2}$$

A complete graph contains the maximal number of cycles and is denoted as $K_A(\mathcal{G})$, A being the number of vertices.

• **forest**
A set of disjoint trees $\mathcal{F} = \{(\mathcal{V}_1, \mathcal{E}_1), \ldots, (\mathcal{V}_k, \mathcal{E}_k)\}$; a forest does not contain cycles.

• **spanning tree**
A connected acyclic subgraph containing all vertices of $\mathcal{G}$.

• **minimal spanning tree**
A spanning tree in which the number of edges is minimal.

• **indexed graph**
A graph $\mathcal{G}$ associated to a mapping ϕ such as:

$$\phi : \mathcal{V}(\mathcal{G}) \rightarrow \{1, 2, \ldots, n\}$$

where $\mathcal{V}(\mathcal{G})$ is the set of n vertices of a graph and the indexing function ϕ assigns an integer number to each vertex of the graph. Uniquivocally defined indexed graphs are obtained by → *canonical numbering*.

• **weighted graph**
A graph $\mathcal{G}$ in which a weight $w_{ij} \geq 0$ is assigned to each edge $\{i,j\} \in \mathcal{E}(\mathcal{G})$ or a weight w_i is assigned to each vertex $i \in \mathcal{V}(\mathcal{G})$.

• **line graph**
A line graph $L(\mathcal{G})$ is a graph obtained by representing the edges of the graph $\mathcal{G}$ by points and then by joining two such points with a line, if the edges they represent were adjacent in the original graph; the following relation holds:

$$|\mathcal{V}(L(\mathcal{G}))| = |\mathcal{E}(\mathcal{G})|$$

Multiple edges are considered as independent vertices in the line graph.

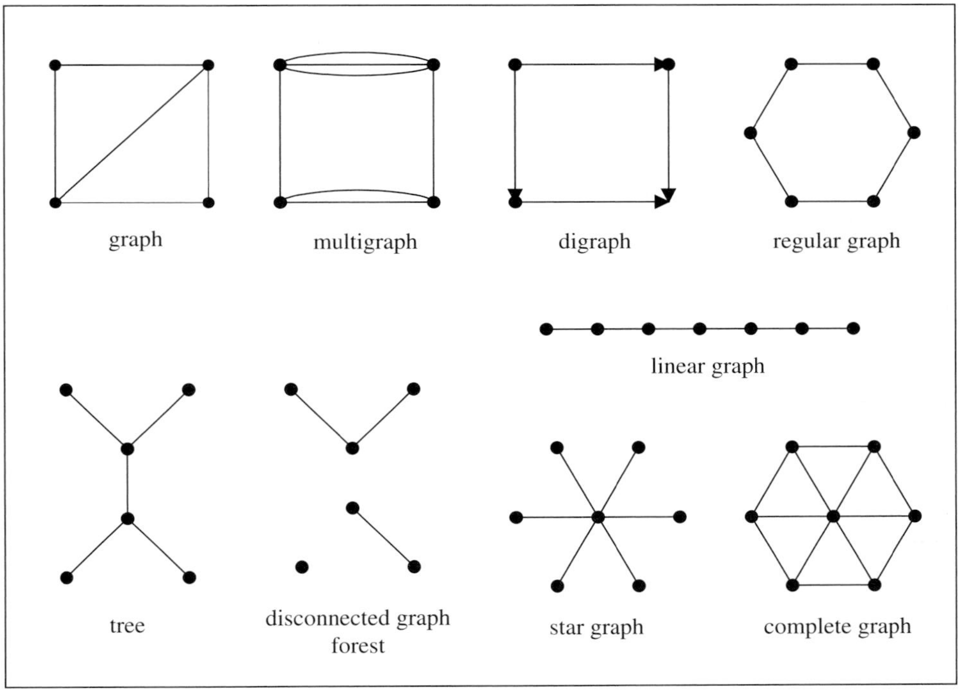

Figure G-2. Examples of graphs.

• **chromatic graph**
Graphs whose vertices or edges are symbolically differentiated by assigning a minimal number of different colours to the vertices (or edges) of $\mathcal{G}$ such that no two adjacent vertices (or edges) have the same colour (→ *chromatic decomposition*).

• **isocodal graphs**
Graphs with identical atomic codes for all the vertices, i.e. there exists a one-to-one correspondence between the atomic codes of all vertices. The most well-known atomic codes are → *walk count atomic code*, → *self-returning walk count atomic code*, → *atomic path code*.

• **isospectral graphs** (: *cospectral graphs*)
Non-isomorphic graphs having the same → *characteristic polynomial*.

• **subspectral graphs**
Graphs whose characteristic polynomial eigenvalues contained in the spectrum of another graph.

• **homeomorphic graphs**
Graphs obtained from the same graph by a sequence of line subdivisions.

A chemically interpreted graph is called **chemical graph**, i.e. graph representing a chemical system such as molecules, reactions, crystals, polymers, and orbitals. The common feature of chemical systems is the presence of sites (atoms, electrons, molecules, molecular fragments, etc.) and connections (bonds, reaction steps, van der Waals forces, etc.) between them. In the graph representation of a chemical system, sites are replaced by vertices and connections by edges. The most common chemical graphs are molecular graphs and reaction graphs. The former correspond to specific chemical structures, the latter to sets of chemical reactions. A topological representation of a molecule is given by a → *molecular graph*.

Note. **Graph theory** is a branch of mathematics which studies the structure of graphs and networks. Graph theory started in 1736, when Euler solved the problem known as the Königsberg bridges problem, which was reduced by him in a graph-theoretical problem. Although the term "graph" was first introduced into literature by the mathematician J.J. Sylvester in 1878, it was derived by him from the contemporary chemical term "graphical notation", which denoted the chemical structure of a molecule.

📖 [Cayley, 1874] [Hosoya, 1972a] [Balaban, 1976d] [Read and Corneil, 1977] [Knop *et al.*, 1981] [Quintas and Slater, 1981] [Randic *et al.*, 1983b] [King, 1983] [Grossman, 1985] [Balaban *et al.*, 1988] [Hansen and Jurs, 1988a] [Balaban and Tomescu, 1988] [Rücker and Rücker, 1991a] [Gutman, 1991] [Liu and Klein, 1991] [Rouvray, 1991] [Polansky, 1991] [Bonchev and Rouvray, 1992] [Figueras, 1992] [Ivanciuc and Balaban, 1992a] [Bangov, 1992] [Gautzsch and Zinn, 1992a] [Gautzsch and Zinn, 1992b] [Rücker and Rücker, 1992] [Balaban, 1993e] [Gautzsch and Zinn, 1994] [Müller *et al.*, 1995] [Lukovits, 1996a] [Lepovic and Gutman, 1998] [Ivanciuc and Balaban, 2000]

graph automorphism → **graph**

graph centre : *centre of a graph*

graph colouring → **chromatic decomposition**

graph distance code → **distance matrix**

graph distance complexity → **topological information indices**

graph distance count → **distance matrix**

graph distance index → **distance matrix**

graph eigenvalues → **characteristic polynomial of the graph**

graphical bond order → **bond order indices**

graphical bond order descriptors → **bond order indices** (⊙ graphical bond order)

graph invariants *(: graph-theoretical invariants)*

→ *Molecular descriptors* based on a graph representation of the molecule and representing graph-theoretical properties that are preserved by isomorphism, i.e. properties with identical values for → *isomorphic graphs*. A graph invariant may be a → *characteristic polynomial*, a sequence of numbers, or a single numerical index obtained by the application of → *algebraic operators* to matrices representing molecular graphs and whose values are independent of vertex numbering or labelling [Kier and Hall, 1976a; Bonchev and Trinajstic, 1977; Bonchev *et al.*, 1979b; Balaban *et al.*, 1983; Kier and Hall, 1986; Basak *et al.*, 1987b; Hansen and Jurs, 1988a; Rouvray, 1989; Basak *et al.*, 1990c; Trinajstic, 1992; Randic, 1993a; Balaban, 1997a; Basak *et al.*, 1997a; Diudea and Gutman, 1998; Balaban, 1998; Randic, 1998c; Balaban and Ivanciuc, 2000; Devillers and Balaban, 2000].

Single numerical indices derived from the topological characteristics of a → *molecular graph* are usually called **topological indices** (TIs) or **molecular topological indices** (MTIs). These are numerical quantifiers of molecular topology that are mathematically derived in a direct and unambiguous manner from the structural graph of a molecule, usually an → *H-depleted molecular graph*. They can be sensitive to one or more structural features of the molecule such as size, shape, symmetry, branching and cyclicity and can also encode chemical information concerning atom type and bond multiplicity. In fact, topological indices were proposed divided into two categories: topostructural and topochemical indices [Basak *et al.*, 1997c; Gute *et al.*, 1999]. **Topostructural indices** encode only information on the adjacency and distance of atoms in the molecular structure; **topochemical indices** quantify information on topology but also specific chemical properties of atoms such as their chemical identity and hybridization state.

→ *Topological information indices* are graph invariants, based on information theory and calculated as → *information content* of specified equivalence relationships on the molecular graph.

Topological indices are mainly based on distances between atoms calculated by the number of intervening bonds and are thus considered *through-bond* indices; they differ from → *topographic indices* which are, instead, considered *through-space* indices because they are based on geometric distances [Diudea *et al.*, 1995b; Balaban, 1997a].

In general, TIs do not uniquely characterize molecular topology, different structures may have some of the same TIs. A consequence of topological indices non-uniqueness is that they do not, in general, allow reconstruction of the molecule. Therefore, suitably defined ordered sequences of TIs can be used to characterize molecules with higher discrimination.

There are several ways to obtain topological descriptors. Simple topological indices consist in the counting of some specific graph elements; examples are the → *Hosoya Z index*, → *path counts*, → *walk counts*, → *self-returning walk counts*, → *Kier shape descriptors*, → *path/walk shape indices*. However, the most common TIs are derived by applying some algebraic operators (e.g. the → *Wiener operator*) to a → *matrix representation of the molecular structure*, such as → *adjacency matrix* **A**, → *distance matrix* **D**, → *detour matrix* Δ, → *Szeged matrix* **SZ**, → *Cluj matrices* **CJ**, → *layer matrices* **LM**, → *walk matrices* **W**; among them there are the → *Wiener index*, → *Randic connectivity index* and related indices, → *Balaban distance connectivity indices*, → *Schultz molecular topological index*, → *hyper-Wiener index*, → *quasi-Wiener index*, → *eigenvalue-based descriptors*, → *determinant-based descriptors*, → *Harary indices*.

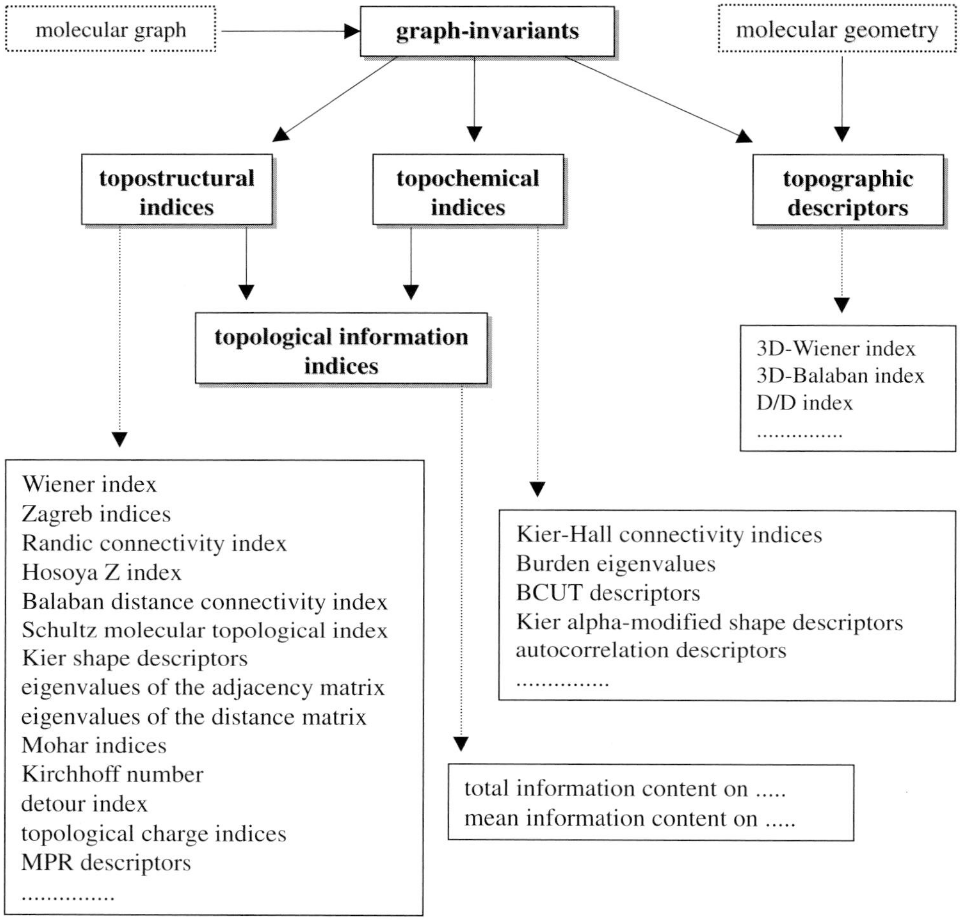

Figure G-3. Scheme of the molecular descriptors belonging to different classes of graph invariants.

Other topological molecular descriptors can be obtained by using suitable functions applied to → *local vertex invariants*; the most common functions are atom and/or bond additives, resulting in descriptors which give correlation of physico-chemical properties that are atom and/or bond additives themselves. → *Zagreb indices* and → *ID numbers* are derived according to this approach.

Another way to derive topological indices is by generalizing the existing indices or topological matrices. → *Kier-Hall connectivity indices*, → *higher-order Wiener numbers*, and → *generalized Hosoya indices* are all examples of the generalization of existing indices, while → *expanded distance matrices* and → *generalized graph matrices* (double invariants) are examples of generalized matrices. Moreover, → *MPR descriptors* are proposed based on a general matrix-vector multiplication approach and several → *combined descriptors* are linear combinations of existing descriptors.

Several → *fragment topological indices* can be derived by any topological index calculated for molecular subgraphs.

Particular topological indices are derived from weighted molecular graphs where vertices and/or edges are weighted by quantities representing some 3D features of the molecule, such as those obtained by → *computational chemistry*. The graph invariants obtained in this way encode both information on molecular topology and → *molecular geometry*. Examples of these topological descriptors are → *BCUT descriptors*, → *electronic-topological descriptors*, → *electron charge density connectivity index*, and several descriptors obtained from → *weighted matrices*.

Graph-invariants have been successfully applied in characterizing the structural similarity/dissimilarity of molecules and in QSAR / QSPR modelling.

Due to the large proliferation of graph invariants, the result of many authors following the procedures outlined above and other general schemes, some rules are needed to critically analyze such invariants, paying particular attention to their effective role in correlating physico-chemical properties and other experimental responses as well as their chemical meaning. In this respect, a list of desirable attributes for topological indices was suggested by Randic [Randic, 1991b].

📖 [Rouvray, 1973] [Gutman, 1974] [Rouvray, 1975] [Rouvray, 1976] [Balaban, 1979] [Balaban and Motoc, 1979] [Bonchev *et al.*, 1980b] [Lall, 1981a] [Lall, 1981c] [Motoc and Balaban, 1981] [Sabljic and Trinajstic, 1981] [Basak *et al.*, 1982] [Motoc *et al.*, 1982b] [Balaban, 1983b] [Basak *et al.*, 1983] [Basak *et al.*, 1984a] [Lall, 1984] [Razinger *et al.*, 1985] [Grossman *et al.*, 1985] [Barysz *et al.*, 1986b] [Basak *et al.*, 1986b] [Gutman and Polansky, 1986a] [Rouvray, 1986a] [Basak *et al.*, 1987a] [Narumi, 1987] [Rouvray, 1987] [Balaban, 1988c] [Basak, 1988] [Gao and Hosoya, 1988] [Randic *et al.*, 1988a] [Rouvray, 1988] [Gargas and Seybold, 1988] [Stankevitch *et al.*, 1988] [Sabljic, 1988] [Baskin *et al.*, 1989] [Rouvray, 1989] [Balaban and Feroiu, 1990] [Petitjean and Dubois, 1990] [Basak *et al.*, 1990a] [Randic, 1990a] [Randic, 1990b] [Sabljic, 1990] [Schultz *et al.*, 1990] [Fabic-Petrac *et al.*, 1991] [Labanowski *et al.*, 1991] [Osmialowski and Kaliszan, 1991] [Randic, 1991b] [Balaban *et al.*, 1992] [Bonchev *et al.*, 1992] [Dubois *et al.*, 1992] [Hall and Kier, 1992b] [Mihalic and Trinajstic, 1992] [Mihalic *et al.*, 1992a] [Randic, 1992b] [Todeschini *et al.*, 1992] [Romanowska, 1992] [Balaban, 1993b] [Balaban, 1993c] [Gordeeva *et al.*, 1993] [Morales and Araujo, 1993] [Randic and Trinajstic, 1993a] [Sekusak and Sabljic, 1993] [Hosoya, 1994] [Ivanciuc and Balaban, 1994a] [Ivanov *et al.*, 1994] [Kirby, 1994] [Kuz'min *et al.*, 1994] [Mekenyan and Basak, 1994] [Salo *et al.*, 1994] [Amic *et al.*, 1995a] [Balaban *et al.*, 1995] [Balaban, 1995b] [Rastelli *et al.*, 1995] [Basak *et al.*, 1995] [Basak *et al.*, 1996] [Chepoi, 1996] [Garcìa-Domenech *et al.*, 1996] [Heinzer and Yunes, 1996] [van Aalten *et al.*, 1996] [Diudea *et al.*, 1997c] [Gawlik *et al.*, 1997] [Gute and Basak, 1997] [Liu *et al.*, 1997] [Lucic and Trinajstic, 1997] [Randic, 1997b] [Randic and DeAlba, 1997] [Randic *et al.*, 1997b] [Chan *et al.*, 1998] [Merrifield and Simmons, 1998] [Meurice *et al.*, 1998] [Rios-Santamarina *et al.*, 1998] [Shankar Raman and Maranas, 1998] [Balaban *et al.*, 1999] [Luco, 1999] [Pompe and Novic, 1999] [Rücker and Rücker, 1999] [Tao and Lu, 1999a]

graph kernel : *pseudocentre* → **centre of a graph**

graph potentials → **MPR descriptors**

graph-theoretical invariants : *graph invariants*

graph-theoretical shape coefficient → **shape descriptors** (⊙ Petitjean shape indices)

graph theory → **graph**

graph vertex complexity → **topological information indices**

graph walk count : *molecular walk count* → **walk counts**

gravitational indices → **steric descriptors**

Green resonance energy → **resonance indices**

grid → **grid-based QSAR techniques**

GRID method

Among the → *grid-based QSAR techniques*, this method is a computational procedure for detecting favourable binding sites on a molecule of known structure [Goodford, 1985; Goodford, 1995]. An orthogonal grid of points is created round and through this target, and another small molecule, such as water (the → *probe*) is placed at the first of this points; → *interaction energy values* are then calculated. The probe is rotated at the grid point until it makes the most favourable energetic interactions with the target. The procedure is repeated at each grid point and the final 3D array of energies constitute the *GRID Map* for that particular probe. These energies are calculated by a sophisticated Empirical Force-Field method (*GRID Force Field*). Typically grid spacing of 0.5 Å are used.

The main advantage of the GRID method is the great variety of available probes, represented by several functional groups such as water, methyl, ammonium, carboxylate, and benzene; in particular, among them there are probes which can both accept and donate a hydrogen bond (e.g. water), probes which cannot turn around (e.g. carbonyl probe), and a recently proposed hydrophobic probe, named DRY.

The GRID method, unlike other grid-based techniques, explicitly takes the flexibility of the molecule into account in ligand-receptor interactions. The conformational flexibility of amino acid side chains is studied, allowing them to be attracted or repelled by the probe as the probe is moving around [Liljefors, 1998]. This algorithm works by dividing the target molecule into a flexible core and flexible side chain on an atom basis.

📖 [Boobbyer *et al.*, 1989] [Wade *et al.*, 1993b] [Wade and Goodford, 1993c] [Wade, 1993a] [Davis *et al.*, 1994] [Goodford, 1996] [Langer and Hoffmann, 1998b]

grid-based QSAR model → **grid-based QSAR techniques**

grid-based QSAR techniques

These are QSAR techniques, sometimes also referred to as **CoMFA-like approaches**, based on descriptors defined as molecular interaction energy values representing → *molecular interaction fields* or, in other words, the interaction energy between a → *probe* and a target compound embedded in a grid.

Interaction fields of a set of congeneric molecules are compared to search for differences and similarities that can be correlated with differences and similarities in the considered property values. Usually biological activities are studied by grid-based QSAR techniques, ligand-target binding properties being of particular interest. Grid-based descriptors mainly characterize molecular shape and charge distribution in the 3D-space responsible for nonbonding interactions involved in ligand-receptor binding.

These approaches give the possibility of representing the molecular interaction regions of interest graphically, a big advantage in pharmacological studies.

The basic starting steps are the generation of three-dimensional structures, conformation search and compound alignment with respect to a compound itself or a → *pharmacophore*.

Once the molecules are aligned, a rigid grid of regularly spaced points representing an approximation of binding site cavity space is established around each compound. A **grid** is a regular 3D array of $N_xN_yN_z$ points (N_G), a lattice of grid points with N_x points along the X-axis, N_y points along the Y-axis, and N_z points along the Z-axis, each point **p** being characterized by the grid coordinates (x, y, z). The grid can be chosen to cover all the atoms of all compounds or else to cover common particular regions of interest in the compounds. The density of the grid must be such as to sample the potential energies of the theoretically continuous scalar field reliably. A density sampling of about 0.25 – 0.50 Å for sharp fields such as molecular electrostatic potential seems to preserve field invariance [Todeschini *et al.*, 1997a]. Steps of 2 Å are the most commonly used; however, a finer grid was suggested to obtain better predictive models [Liljefors, 1998].

For each molecule embedded in the grid, molecular interaction fields are calculated; **interaction energy values** at each grid point are the molecular descriptors within the framework of grid-based QSAR techniques. Usually a reasonable selection of interaction energy values is performed based on energy cut-off values, selected molecular regions or other specific criteria, depending on the method.

The **grid-based QSAR model** has the form:

$$\hat{y}_i = b_0 + \sum_{j=1}^{F}\sum_{x=1}^{N_x}\sum_{y=1}^{N_y}\sum_{z=1}^{N_z} b_{j,\,xyz} \cdot E_{ij,\,xyz} = \sum_{j=1}^{F}\sum_{k=1}^{N_G} b_{jk} \cdot E_{ijk}$$

where F is the number of fields used in the analysis, i.e. the number of molecular interaction fields; N_x, N_y, and N_z are the number of grid points along the X-axis, Y-axis, and Z-axis, respectively; $N_G = N_x\ N_y\ N_z$ is the total number of grid points. $E_{ij,\,xyz}$ is the potential interaction energy of the ith compound for the jth field in the grid coordinate (x, y, z). The k index runs over the grid points in a vectorial representation.

Due to the large number of descriptors (commonly 15,000 – 20,000 for each field), the multivariate regression analysis is usually performed by partial least squares regression (PLS), with or without → *variable selection*. Moreover, a → *similarity matrix* can be calculated from distance functions based on interaction fields between pairs of molecules.

The most popular grid-based QSAR techniques are → *GRID method* and → *CoMFA*; a number of related techniques has also been proposed such as → *CoMSIA*, → *Voronoi Molecular Field Analysis*, → *hydration free energy density*, and the approach based on → *G-WHIM descriptors*.

→ *Dynamic QSAR* also takes into account conformational variability of each molecule, selecting the conformation of each molecule providing the best QSAR model. A specific approach to the conformational variability was recently proposed [Albuquerque *et al.*, 1998], where each molecule conformation is placed in the grid and the frequency of the occupation of each grid cell is computed for each of seven predefined pharmacophoric classes of atoms. These frequencies are called **grid cell occupancy descriptors** (GCODs) and used as independent variables in constructing QSAR models.

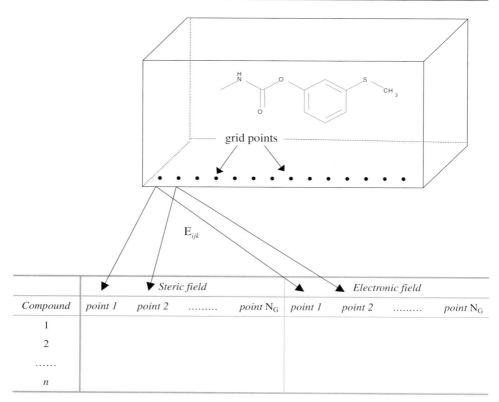

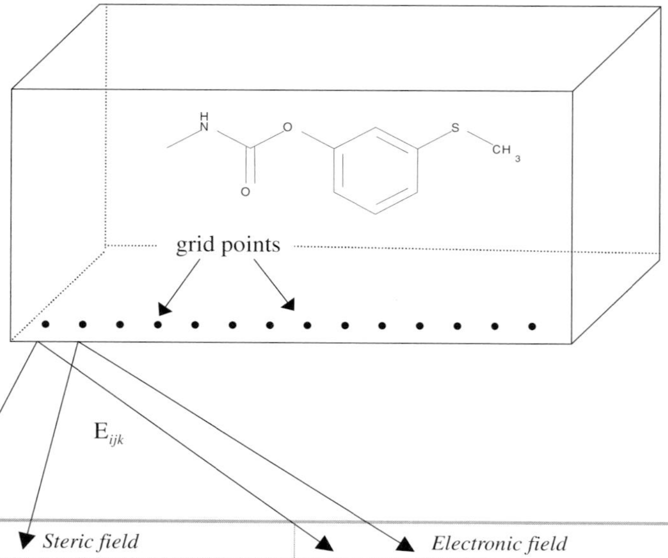

	Steric field				Electronic field			
Compound	*point 1*	*point 2*	....	*point* N_G	*point 1*	*point 2*	....	*point* N_G
1								
2								
......								
n								

Figure G-4. Construction of the data set in the grid-based QSAR techniques.

grid cell occupancy descriptors → **grid-based QSAR techniques**

grid region selection methods → **variable selection**

GRID electrostatic energy function → **molecular interaction fields**
(⊙ electrostatic interaction fields)

Grid-Weighted Holistic Invariant Molecular descriptors : *G-WHIM descriptors*

Grob inductive constant → **electronic substituent constants** (⊙ inductive electronic constants)

group charge transfer : *charge transfer constant* → **electronic substituent constants**

group contribution methods (GCM)

Group contribution methods search for relationships between structural properties and a physico-chemical or biological response based on the following models:

$$y_i = f(G_1, G_2, \ldots, G_m)$$

where the experimental property y_i for the ith compound is a function of m group contributions G_j [Reinhard and Drefahl, 1999]. The **group contributions**, also known as **fragmental constants**, are numerical quantities associated with substructures of the molecule, such as single atoms, atom pairs, atom-centred substructures, molecular fragments, functional groups, etc. For example, atom contribution models exhibit a one-to-one correspondence between atoms and property contributions, i.e. the molecular property is a function of all the single atomic properties. The specification of the structural groups depends on the particular GCM scheme adopted. → *Cluster expansion of chemical graphs* is an example of a group contribution method based on all the connected subgraphs of the molecular graph.

Generally, the application of GCM to a molecule requires the following steps:

1) Identification of all groups in the molecule applicable to the particular GCM scheme. An automated search for substructures of interest for a given property is performed by the → *CASE approach*.

2) Calculation of fragmental constants measuring contributions to the molecular property of the considered fragments by employing the function associated with the particular GCM.

3) Evaluation of some correction factors which should account for interactions among molecular groups.

Linear GCM models are defined as the following:

$$y_i = k_0 + \sum_{j=1}^{m} G_j \cdot \mathrm{I}_{ij} \qquad \text{or} \qquad y_i = k_0 + \sum_{j=1}^{m} G_j \cdot \mathrm{N}_{ij}$$

where k_0 is a model-specified constant, j runs over the m group contributions defined within the GCM scheme, G_j is the contribution of the jth group. I_{ij} and N_{ij} are → *substructure descriptors*, and in particular, I_{ij} is a binary variable taking a value equal to one if the jth group is present in the ith molecule, zero otherwise, while N_{ij} is the number of times the jth group occurs in the ith molecule.

Nonlinear GCM models are usually defined as:

$$y_i = k_0 + \sum_{j=1}^{m} G_j \cdot \mathrm{N}_{ij} + \left(\sum_{j=1}^{m} G_j \cdot \mathrm{N}_{ij} \right)^2$$

Moreover, mixed GCM models are defined by adding, usually, one or more molecular descriptors to the group contributions:

$$y_i = k_0 + \sum_{j=1}^{m} G_j \cdot \mathrm{N}_{ij} + \sum_{j'=1}^{J} \Phi_{ij'}$$

where the second summation runs over the J molecular descriptors defined in the GCM scheme and $\Phi_{ij'}$ is the j'th descriptor value for the ith molecule.

The group contributions G are usually estimated by multivariate regression analysis, but they can also be experimental, theoretical or user-defined quantities. For example, in the latter case, the molecular weight can be viewed as a simple linear atom contribution model, where the group contributions are atomic masses. In the first case, large training sets are used to obtain reliable estimates of the group contributions. Usually a battery of group contributions (a field of scalar parameters) is defined taking into account several structural characteristics of the molecules, also sometimes adding extra terms (correction factors) referring to special substructures. If correction factors are considered, the GCM models are usually called **additive-constitutive models**.

Group contribution models were proposed for several molecular property estimations, such as boiling and melting points [Wang *et al.*, 1994; Krzyzaniak *et al.*, 1995; Le and Weers, 1995], → *molar refractivity* [Huggins, 1956; Ghose and Crippen, 1987], $\mathrm{p}K_\mathrm{a}$ [Perrin *et al.*, 1981; Hilal *et al.*, 1995], critical temperatures, solubilities [Hine and Mookerjee, 1975; Klopman *et al.*, 1992; Myrdal *et al.*, 1993; Thomsen *et al.*, 1999], soil sorption coefficients [Tao *et al.*, 1999b], and several thermodynamic properties [Thinh and Trong, 1976; Yoneda, 1979; Reid *et al.*, 1988]. The Rekker method [Nys and Rekker, 1973; Rekker, 1977a] is a group contribution method largely applied to the estimation of → *log P*. Another well-known group contribution model is that proposed by Atkinson for the evaluation of reaction rate constants with hydroxyl radicals of organic compounds [Atkinson, 1987; Atkinson, 1988].

📖 [Smolenskii, 1964] [Essam *et al.*, 1977] [Ghose and Crippen, 1986] [Klein, 1986] [Elbro *et al.*, 1991] [Gao *et al.*, 1992] [Drefahl and Reinhard, 1993] [Klopman *et al.*, 1994] [Yalkowsky *et al.*, 1994a] [Yalkowsky *et al.*, 1994b] [Bhattacharjee, 1994] [Meylan and Howard, 1995] [Klein *et al.*, 1999] [Wildman and Crippen, 1999] [Viswanadhan *et al.*, 1999] [Platts *et al.*, 1999] [Platts *et al.*, 2000]

group contributions → **group contribution methods**

group electronegativity → **quantum-chemical descriptors**

group molar refractivity → **molar refractivity**

Gutman index *: 1st Zagreb index* → **Zagreb indices**

Gutman molecular topological index → **Schultz molecular topological index**

Gutmann's acceptor number → **linear solvation energy relationships** (⊙ hydrogen- bond parameters)

Gutmann's donor number → **linear solvation energy relationships** (⊙ hydrogen-bond parameters)

G-WHIM descriptors *(: Grid-Weighted Holistic Invariant Molecular descriptors)*

Based on a similar approach to that used to define → *WHIM descriptors*, G-WHIM descriptors [Todeschini *et al.*, 1997a; Todeschini and Gramatica, 1998] are global molecular descriptors of → *molecular interaction fields*.

Once the optimal choices for the grid have been made, the G-WHIM descriptors are used to condense the whole information contained in the scalar field constituted by the calculated → *interaction energy values* into a few global parameters, whose values are independent of the molecular orientation within the grid.

For each molecule, the G-WHIM descriptors are calculated by the following steps:

a. The molecule is freely and separately embedded in the centre of the grid.

b. The interaction energy values are calculated at all the grid points by using the selected probe.

c. The interaction energy values are used as weights for the grid points. This is the main difference between G-WHIM and WHIM descriptors, where the defined atomic properties are used to weight molecule atoms.

d. Finally, the G-WHIM descriptors are calculated in the same way as for the WHIM descriptors, i.e. by the calculation of a weighted covariance matrix, principal component analysis and the calculation of statistical parameters on the projected points along each principal component (i.e., on the score values).

It should be noted that only points with non-zero interaction energy are effective in the computation of the descriptors. Secondly, when the calculated interactions give both positive and negative values, interaction energy values cannot be used directly in this form as statistical weights, which must always be defined semi-positively. In this case, the scalar field values are divided in two blocks: a grid-negative (positive) block containing the grid coordinates associated with negative (positive) interaction values, giving their absolute values and setting the positive (negative) values to zero.

This assumption leads to two sets of G-WHIM descriptors: one describing the positive part of the molecular field, the other describing the negative part.

Thus, for each region [positive (+) and negative (–)], the G-WHIM descriptors consist of 8 directional plus 5 non-directional molecular descriptors (26 for a complete description of each interaction field), calculated from each molecule:

$$\lambda_1(\pm) \quad \lambda_2(\pm) \quad \lambda_3(\pm) \quad \vartheta_1(\nu) \quad \vartheta_2(\pm) \quad \eta_1(\pm) \quad \eta_2(\pm) \quad \eta_3(\pm)$$

and

$$\mathrm{T}(\pm), \mathrm{A}(\pm), \mathrm{V}(\pm), \mathrm{K}(\pm), \mathrm{D}(\pm)$$

The directional γ and the non-directional G parameters, defined for WHIM descriptors and containing information about the molecular symmetry, are not considered in the frame of the G-WHIM approach as their meaning becomes doubtful, depending heavily upon the point sampling. However, information regarding molecular symmetry can be obtained by directly using the WHIM symmetry parameters.

The meaning of the G-WHIM descriptors is that previously defined for WHIM descriptors, but now the descriptors refer to the interaction molecular field instead of the molecule. For example, the eigenvalues λ_1, λ_2 and λ_3, relate to the interaction field size; the eigenvalue proportions ϑ_1 and ϑ_2 relate to the interaction field shape; the group of descriptors constituted by the inverse function of the kurtosis (κ), i.e., $\eta_m = 1/\kappa_m$ relate to the interaction field density along each axis. Moreover, global information about the interaction field is obtained as for → *global WHIM descriptors* (T, A, V, K, D), with the same meaning.

First [Todeschini *et al.*, 1997a], the acentric factor was used as the shape descriptor, defined as $\omega = \vartheta_1 - \vartheta_3$, ranging between zero (spherical interaction field) and one (linear interaction field); this shape descriptor was substituted by the global K shape descriptor; moreover, among the global G-WHIM, only the eigenvalue sum T had previously been used.

It must be noted that the invariance to rotation of G-WHIM descriptors, i.e. the independence of any molecular alignment rule, is obtained if the grid points are dense enough. In fact, a too sparse distribution of grid points represents an inadequate sampling of the ideal scalar field and is not able to guarantee that the calculated scalar field is representative of the ideal scalar field in such a way as to preserve rotational invariance.

In fact, if a molecule is placed in an infinite, isotropic and even, very dense grid, the scalar field $\mathcal{F}$ calculated at the grid points must contain the same information, independent of the molecule's orientation and depending only on the potential energy of the selected probe and the mathematical functions representing the interaction. Thus $\mathcal{F}$ contains the whole information about the interaction properties of the molecule.

In practice, this ideal situation cannot be achieved, but it can be simulated by plunging the molecule into a finite grid $\mathcal{N}$: the aim is to represent the theoretical $\mathcal{F}$ scalar field by a finite sampling of this field.

G-WHIM descriptors can be calculated for any selected region of the field. To avoid irrelevant or unreliable chemical information due to long-range interactions, use is made of an energy cut-off criterion, which takes into account only interaction energy values relevant to the considered interaction (e.g., long-range or chemical interaction). In this way, points far from the molecule and not contributing to the interaction are not included in the calculations, e.g., the field values inside the van der Waals surface of the molecule are not considered. Moreover, specific chemical information is gained by using different energy cut-off values to select the regions, keeping in mind that the higher the cut-off value, the smaller the considered region around the molecule (i.e. the total number of non-zero weighted grid points).

The G-WHIM approach integrates the information contained in WHIM descriptors and overcomes any problems due to the alignment of the different molecules and the explosion of variables arising from traditional → *grid-based QSAR techniques.* In particular, the G-WHIM approach can take into account either all the points within the cut-off values, excluding only positive interactions within the inner part of the molecule, or the surface points at a cut-off value, i.e. points on an iso-potential-energy surface. G-WHIM descriptors calculated on the → *Connolly surface area* are also called MS-WHIM [Bravi *et al.*, 1997].

The ability to take the individual parameters provided by different cut-off values into account when defining different molecular interaction regions could possibly lead to a deeper chemical insight into molecular interactions and properties.

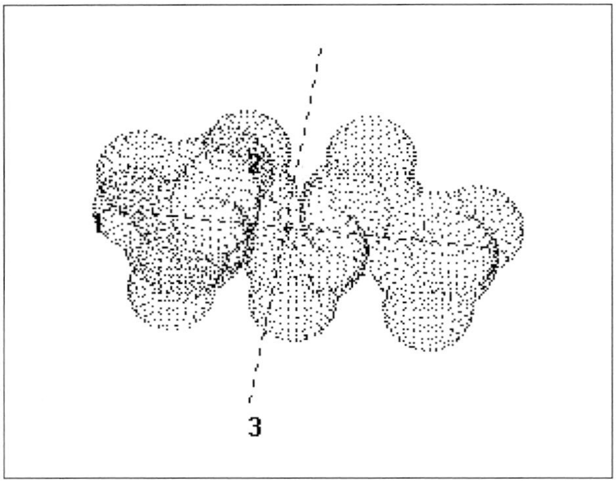

Figure G-5. Principal axes of the grid interaction energy values of the 2-methylpentane.

📖 [Zaliani and Gancia, 1999] [Ekins *et al.*, 1999]

H

***HACA* index** → **charged partial surface area descriptors**

Hadamard matrix product → **algebraic operators**

hafnian → **algebraic operators** (⊙ determinant)

Hamiltonian circuit → **graph**

Hamiltonian path → **graph**

Hammett electronic constant → **Hammett equation**

Hammett equation

Proposed by Hammett in 1937 [Hammett, 1937; Hammett, 1938], the Hammett equation is defined for the rate constants k and equilibrium constants K of reactions of *meta*- and *para*-substituted benzoic acid derivatives:

$$\varrho \cdot \sigma = \log\left(\frac{k_X}{k_H}\right) \quad \text{and} \quad \varrho \cdot \sigma = \log\left(\frac{K_X}{K_H}\right)$$

where the constants k_H and K_H refer to an unsubstituted compound (i.e. with hydrogen in the substitution site), while k_X and K_X refer to a *meta*- or *para*-X-substituted compound. *Ortho*- substituents are less used as the electronic effect could be complicated by steric interactions.

The substituent constant σ, called the **Hammett electronic constant**, depends only on the nature and position of the substituent and is related to the electronic effect the substituent has on the rate or equilibrium of the reaction, relative to the unsubstituted compound.

The **reaction constant** ϱ depends upon the reaction, the conditions under which it is studied and the nature of the reaction series. The magnitude of ϱ gives the susceptibility of a given reaction to polar substituents. Large positive values are obtained from all base-catalyzed reactions, while for acid-catalyzed reactions ϱ values are of variable sign but in all cases quite small.

Therefore, σ is an electronic descriptor of the substituent estimated by measured rate or equilibrium constants of a reaction, under the control parameter ϱ.

From the Hammett equation, several σ → *electronic substituent constants* are derived from different reactions and different experimental conditions; a modification of the Hammett equation was defined as the → *Yukawa-Tsuno equation*.

📖 [Hammett, 1935] [Jaffé, 1953] [Yamamoto and Otsu, 1967] [Shorter, 1978]

Hammett substituent constants : *electronic substituent constants*

Hancock steric constant : *corrected Taft steric constant* → **steric descriptors** (⊙ Taft steric constant)

Hansch analysis

Derived from physical organic chemistry and the → *Hammett equation*, this can be considered to be the first approach to modern QSAR studies. Proposed by Hansch and co-workers in the early 1960s [Hansch *et al.*, 1962; Hansch *et al.*, 1963; Hansch and Fujita, 1964; Hansch *et al.*, 1965; Hansch and Anderson, 1967; Hansch, 1969; Hansch, 1971; Hansch, 1978], it is the investigation of the quantitative relationships between the biological activity of a series of compounds and their physico-chemical parameters representing hydrophobic, electronic, steric and other effects using multivariate regression methods [Kubinyi, 1993b].

Hansch analysis assumes that variations in the magnitude of a certain biological activity exhibited by a series of bioactive compounds can be correlated to variations in different physico-chemical factors associated with their structure. Therefore, the basic QSAR equation in the Hansch analysis is defined as:

$$\text{biological activity} = f(\Phi_1, \Phi_2, \ldots, \Phi_J)$$

where Φ are → *physico-chemical properties* of congeneric compounds having a common skeleton but varying substituents. Together with the most significant parameters, factors for hydrogen bonding, van der Waals and charge-transfer forces, etc. can be used, depending on the situation.

The biological activity is usually defined as log (1/*C*), where *C* is the molar concentration of a compound producing a fixed effect.

Hansch analysis tries to correlate biological activity with physico-chemical properties by linear and nonlinear regression analysis, finding property-activity relationship models.

The simplest Hansch analysis is based on the **Hansch linear model** [Kubinyi, 1988b], defined as:

$$\hat{y}_i = b_0 + \sum_{j=1}^{J} b_j \cdot \Phi_{ij}$$

where Φ_{ij} represents the jth physico-chemical property of the ith compound, b_j are the regression coefficients, and J is the total number of considered properties. The intercept b_0 corresponds to a theoretical biological acitivity of a compound all of whose property values are zero. This condition is approximately fullfilled for a hydrogen-substituted reference compound as several property values are normalized to zero. Depending on the regression coefficient significance, some factors that are not relevant can result.

For example, a typical Hansch linear equation for mono-substituted derivatives is the following:

$$\log(1/C) = b_0 + b_1 \cdot \pi + b_2 \cdot \sigma + b_3 \cdot \mathrm{E_s}$$

where π are the → *Hansch-Fujita hydrophobic substituent constants*, σ are the → *electronic substituent constants*, and $\mathrm{E_s}$ is the → *Taft steric constant*.

The jth molecular property of the ith compound Φ_{ij} can be defined as the sum of the jth substituent constant values ϕ of all substituents of that compound:

$$\Phi_{ij} = \sum_{s=1}^{S} \sum_{k=1}^{\mathrm{N_s}} \phi_{ks,\,j} \cdot \mathrm{I}_{i,\,ks}$$

where $\mathrm{I}_{i,ks}$ are → *indicator variables* (such as in → *Free-Wilson analysis*) indicating the presence, i.e. $\mathrm{I}_{i,ks} = 1$, of the kth substituent in the sth site for the ith compound; otherwise $\mathrm{I}_{i,ks} = 0$; $\phi_{j,ks}$ is the jth property of the kth substituent in the sth site; S is the number of substitution sites and $\mathrm{N_s}$ is the number of group substituents in the sth site. As in each site only one substituent is present for a given compound, S is the total number of contributions to the considered molecular property. Alternatively, the jth molecular property Φ_{ij} of the ith compound is defined as the jth global molecular property such as → *logP* or some global → *steric descriptors*.

Both the molecular physico-chemical properties Φ obtained by the above relationship and the → *substituent constants* ϕ are usually known as **Hansch descriptors** [Hansch *et al.*, 1995b].

In order to also take into account some nonlinear contributions of the properties [Hansch and Clayton, 1973; Kubinyi, 1993b], a **Hansch parabolic model** is defined as:

$$\hat{y}_i = b_0 + \sum_{j=1}^{J} b_j \cdot \Phi_{ij} + \sum_{j=1}^{J} b''_j \cdot \Phi_{ij}^2$$

Among the quadratic terms, the most used is usually $(\log P)^2$, in order to mimic the nonlinear behaviour of the interchange between a two-phase system (e.g. aqueous/ organic system), i.e. too low or too high lipophilicity values act as a limiting factor. The most common parabolic model is specifically defined as:

$$\log(1/C) = b_0 - b_1 \cdot (\log P)^2 + b_2 \cdot \log P + b_3 \cdot \sigma$$

The general **Hansch nonlinear model** is defined as:

$$\hat{y}_i = b_0 + \sum_{j=1}^{J} b_j \cdot \Phi_{ij} + \sum_{j=1}^{J} \sum_{j'=j}^{J} b''_{jj'} \cdot \Phi_{ij} \cdot \Phi_{ij'}$$

also taking into account the combined effects of the properties, even if not usually considered.

Besides the nonlinear models and, specifically, the parabolic model, other models were proposed for nonlinear dependence of the biological response from hydrophobic interactions. Among them, the most important are the **Hansch bilinear models** [Kubinyi, 1977; Kubinyi, 1979] such as:

$$\log(1/C) = b_0 + b_1 \cdot \log P - b_2 \cdot \log(\beta \cdot P + 1)$$

Special cases of such bilinear models are the **McFarland model** [McFarland, 1970], where $b_2 = 2b_1$ and $\beta = 1$ and the **Higuchi-Davis model** [Higuchi and Davis, 1970], where $b_2 = 1$ and $\beta = V_{lip} / V_{aq}$, which is the ratio between the volume of the lipid phase V_{lip} and the volume of the aqueous phase V_{aq}.

The Hansch linear model is related to the → *Fujita-Ban model* when, in both models, the hydrogen substituted compound is taken as the reference compound; each Fujita-Ban regression coefficient b_{ks} corresponds to the Hansch equation for a single substituent:

$$b_{ks} \approx \sum_{j=1}^{J} b_j \cdot \phi_{ks,j}$$

where J is the number of considered properties (e.g., lipophilic, electronic, steric properties), and $\phi_{ks,j}$ the jth substituent group property for the kth substituent in the sth site. This relationship means that the group contribution b_{ks} in the Fujita-Ban model of the kth substituent in the sth site is numerically equivalent to the weighted sum of all the physico-chemical properties of that substituent [Kubinyi and Kehrhahn, 1976]. Substituting the previous relationship in the Fujita-Ban model, it can be observed that the two models are closely related:

$$\hat{y}_i = b_0 + \sum_{s=1}^{S} \sum_{k=1}^{N_s} b_{ks} \cdot I_{i,ks} = b_0 + \sum_{s=1}^{S} \sum_{k=1}^{N_s} \sum_{j=1}^{J} b_j \cdot \phi_{ks,j} \cdot I_{i,ks} =$$

$$= b_0 + \sum_{j=1}^{J} b_j \cdot \sum_{s=1}^{S} \sum_{k=1}^{N_s} \phi_{ks,j} \cdot I_{i,ks} = b_0 + \sum_{j=1}^{J} b_j \cdot \Phi_{ij}$$

In particular, the Fujita-Ban group contributions implicitly contain all the possible physico-chemical contributions of a substituent; as a consequence, the Fujita-Ban models always give an upper limit of correlation which can be achieved by Hansch linear models.

Hansch-Free-Wilson mixed models were also proposed [Kubinyi, 1976a] by combining the two approaches in a single model. A quadratic term accounting for hydrophobic interactions (usually log$P$ or π Hansch-Fujita constant) can be added to the Free-Wilson (or Fujita-Ban) model as:

$$\hat{y}_i = b_0 + \sum_{s=1}^{S} \sum_{k=1}^{N_s} b_{ks} \cdot I_{i,\,ks} + b' \cdot (\log P_i)^2$$

where the first part is the Free-Wilson model; S and N_s are, respectively, the number of substitution sites and substituent groups in each sth site; $I_{i,ks}$ is an indicator variable for the ith compound denoting the presence (1) or absence (0) of the kth group in the sth site.

Mixed models can also be obtained by mixing the two approaches, each describing a different group of substituents:

$$\hat{y}_i = b_o + \sum_{s=1}^{S} \sum_{k=1}^{N_s} b_{ks} \cdot I_{i,\,ks} + \sum_{j=1}^{J} b'_j \cdot \Phi_{ij} + b'' \cdot (\log P_i)^2$$

where the Free-Wilson part accounts for a set of substituents and the Hansch part for another set.

Another mixed model, called here **Site-Property analysis** (*SP analysis*), can be obtained [Authors, this book]; it represents information regarding the presence of each substituent group in each site by the corresponding physico-chemical properties, i.e. the information of the indicator variables $I_{i,ks}$ of the Fujita-Ban analysis is preserved in each site but is represented by the set of selected properties:

$$\hat{y}_i = b_0 + \sum_{s=1}^{S} \sum_{j=1}^{J} b_{sj} \cdot \phi_{is,\,j}$$

where S is the number of substitution sites, J is the number of properties; $\phi_{is,j}$ is the jth substituent group property in the sth site for the ith compound, i.e.

$$\phi_{is,\,j} = \sum_{k=1}^{N_s} \phi_{ks,\,j} \cdot I_{i,\,ks}$$

where

$$\sum_{k=1}^{N_s} I_{i,\,ks} = 1$$

Therefore, *SP analysis* can be performed only if all substituent group constants are available for all the substituents in the data set. The total number of variables is $S \times J$. This approach allows complete → *reversible decoding*, i.e. the possibility to interpret by the model *how* and *where* the response is increased/decreased.

By assuming that a response would depend on both the holistic properties of molecules and on the local specific group contributions a mixed **Global-Site-Property analysis** (*GSP analysis*) can be achieved by a generalized model such as:

$$\hat{y}_i = b_0 + \sum_{l=1}^{p'} b'_l \cdot \Phi_{il} + \sum_{s=1}^{S} \sum_{j=1}^{J} b_{sj} \cdot \phi_{is,\,j}$$

where Φ are generic global properties, i.e. global descriptors obtained by any method, and p' the number of selected descriptors; $\phi_{is,j}$ is the jth substituent group property in the sth site for the ith compound. The total number of variables is $S \times J + p'$.

In Hansch analysis and related approaches, the statistical problems due to the relatively high number of variables with respect to the number of compounds have to be faced using → *variable selection* procedures.

Although the predictive power of a model is considered to be a criterion for the relevance of QSAR models, the main purpose of Hansch analysis and related approaches such as Free-Wilson analysis concerns not prediction, but a better understanding of the chemical problem.

[Singer and Purcell, 1967] [Hansch *et al.*, 1968] [Hansch and Lien, 1968] [Mouvier and Dubois, 1968] [Hansch *et al.*, 1969] [Kutter and Hansch, 1969] [Hansch, 1970] [Hansch and Lien, 1971] [Hansch and Dunn III, 1972] [Hansch *et al.*, 1973a] [Unger and Hansch, 1973] [Hansch and Yoshimoto, 1974] [Silipo and Hansch, 1975] [Hansch and Calef, 1976] [Duperray *et al.*, 1976a] [Duperray *et al.*, 1976b] [Fukunaga *et al.*, 1976] [Kubinyi, 1976b] [Schaad and Hess Jr., 1977] [Hall and Kier, 1978a] [Schaad *et al.*, 1981] [Denny *et al.*, 1982] [Silipo and Vittoria, 1990] [Fujita, 1990] [Schultz *et al.*, 1990] [El Tayar *et al.*, 1991a] [Breyer *et al.*, 1991] [Debnath *et al.*, 1992a] [Debnath *et al.*, 1992b] [Debnath *et al.*, 1992c] [Debnath and Hansch, 1992d] [Hansch and Zhang, 1992] [van de Waterbeemd, 1992] [Smith *et al.*, 1992] [Debnath *et al.*, 1993] [Peijnenburg *et al.*, 1993] [Carrieri *et al.*, 1994] [Hadjipavloulitina and Hansch, 1994] [Hansch *et al.*, 1995a] [Hansch, 1995b] [Hansch *et al.*, 1995c] [Hansch and Leo, 1995] [Fujita, 1995] [Kawashima *et al.*, 1995] [Pavlikova *et al.*, 1995] [Susarla *et al.*, 1996] [Hansch *et al.*, 1996] [Puri *et al.*, 1996] [Singh *et al.*, 1996] [Hansch and Gao, 1997] [Hansch *et al.*, 1998] [Tuppurainen, 1999b]

Hansch bilinear models → **Hansch analysis**

Hansch descriptors → **Hansch analysis**

Hansch-Free-Wilson mixed models → **Hansch analysis**

Hansch-Fujita hydrophobic substituent constants → **lipophilicity descriptors**

Hansch linear model → **Hansch analysis**

Hansch nonlinear model → **Hansch analysis**

Hansch parabolic model → **Hansch analysis**

Harary Cluj-detour indices → **Harary indices**

Harary Cluj-distance indices → **Harary indices**

Harary detour/distance indices → **Harary indices**

Harary detour indices → **Harary indices**

Harary index → **Harary indices**

Harary indices *(: Harary numbers)*

The **Harary index** H (also called **Harary number**) is a molecular topological index [Ivanciuc *et al.*, 1993b; Plavsic *et al.*, 1993b] derived from the → *reciprocal distance matrix* $\mathbf{D}^{-1}$ by the → *Wiener operator* $\mathcal{W}$:

$$H = \mathcal{W}(\mathbf{D}^{-1}) = \frac{1}{2} \cdot \sum_{i=1}^{A} \sum_{j=1}^{A} d_{ij}^{-1}$$

The Harary index increases with both molecular size and → *molecular branching*; it is therefore a measure of molecular compactness like the → *Wiener index*. However, the Harary index seems to be a more discriminating index than the Wiener index.

By generalization, Harary indices and **hyper-Harary indices** (or **hyper-Harary numbers**) are all → *molecular descriptors* derived from the application of the Wiener operator to reciprocal topological matrices; Harary indices are obtained from the 1st-order → *sparse matrix*, i.e. considering only the graph edges, while the hyper-Harary indices are from the whole matrices, i.e. considering the paths [Diudea, 1997c].

Another set of topological indices based on a modified reciprocal distance matrix are the → *constant interval reciprocal indices.*

Example : 2-methylpentane

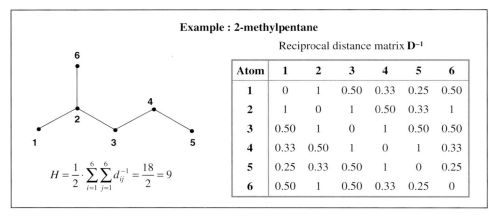

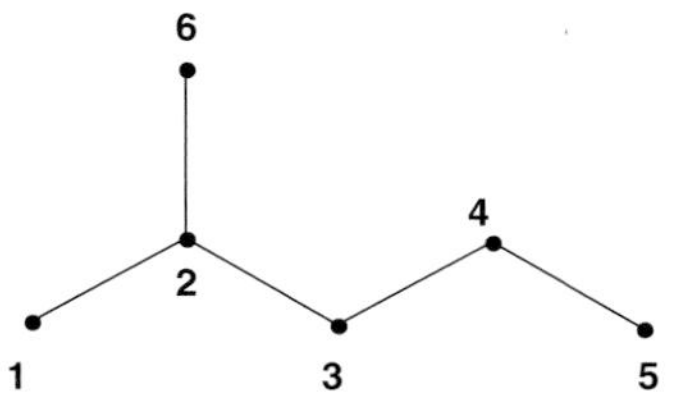

$$H = \frac{1}{2} \cdot \sum_{i=1}^{6} \sum_{j=1}^{6} d_{ij}^{-1} = \frac{18}{2} = 9$$

Reciprocal distance matrix $\mathbf{D}^{-1}$

Atom	1	2	3	4	5	6
1	0	1	0.50	0.33	0.25	0.50
2	1	0	1	0.50	0.33	1
3	0.50	1	0	1	0.50	0.50
4	0.33	0.50	1	0	1	0.33
5	0.25	0.33	0.50	1	0	0.25
6	0.50	1	0.50	0.33	0.25	0

Box H-1.

The most important Harary indices are listed below.

- **Harary Wiener indices**

The Harary index and hyper-Harary index, defined only for acyclic graphs, are obtained from, respectively, the 1st-order sparse → *reciprocal Wiener matrix* $^{\mathbf{1}}\mathbf{W}^{-\mathbf{1}}$:

$$^{1}H_{\mathrm{W}} = \mathcal{W}(^{\mathbf{1}}\mathbf{W}^{-\mathbf{1}}) = \frac{1}{2} \cdot \sum_{i=1}^{A} \sum_{j=1}^{A} [^{\mathbf{1}}\mathbf{W}^{-\mathbf{1}}]_{ij}$$

and the reciprocal Wiener matrix $\mathbf{W}^{-\mathbf{1}}$:

$$H_{\mathrm{WW}} = \mathcal{W}(\mathbf{W}^{-\mathbf{1}}) = \frac{1}{2} \cdot \sum_{i=1}^{A} \sum_{j=1}^{A} [\mathbf{W}^{-\mathbf{1}}]_{ij}$$

It must be noted that, while the indices obtained by applying the Wiener operator to the distance matrix **D** and to the 1st-order → *sparse Wiener matrix* $^{\mathbf{1}}\mathbf{W}$ are equal (i.e. the Wiener index), the corresponding Harary indices are not, i.e. $H \neq {}^{1}H_{\mathrm{W}}$.

- **hyper-Harary distance index**

This is obtained from the → *reciprocal distance-path matrix* $\mathbf{D}_{\mathbf{P}}^{-\mathbf{1}}$:

$$H_{\mathrm{DP}} = \mathcal{W}(\mathbf{D}_{\mathbf{P}}^{-\mathbf{1}}) = \frac{1}{2} \cdot \sum_{i=1}^{A} \sum_{j=1}^{A} [\mathbf{D}_{\mathbf{P}}^{-\mathbf{1}}]_{ij}$$

For acyclic graphs, the equality between the → *hyper-distance-path index* D_P and the → *hyper-Wiener index WW* is not true for the corresponding hyper-Harary indices, i.e. $H_{DP} \neq H_{WW}$.

• **Harary detour indices**

The Harary index and hyper-Harary index [Diudea *et al.*, 1998] are obtained from, respectively, the 1st-order sparse → *reciprocal detour matrix* $^{\mathbf{1}}\mathbf{\Delta}^{-\mathbf{1}}$:

$$^{1}H_{\Delta} = \mathcal{W}(^{\mathbf{1}}\mathbf{\Delta}^{-\mathbf{1}}) = \frac{1}{2} \cdot \sum_{i=1}^{A} \sum_{j=1}^{A} [^{\mathbf{1}}\mathbf{\Delta}^{-\mathbf{1}}]_{ij}$$

and the reciprocal detour matrix $\mathbf{\Delta}^{-\mathbf{1}}$:

$$H_{\Delta} = \mathcal{W}(\mathbf{\Delta}^{-\mathbf{1}}) = \frac{1}{2} \cdot \sum_{i=1}^{A} \sum_{j=1}^{A} [\mathbf{\Delta}^{-\mathbf{1}}]_{ij}$$

• **Harary detour/distance indices**

The Harary index and hyper-Harary index are obtained from, respectively, the 1st-order sparse → *reciprocal detour/distance matrix* $^{\mathbf{1}}\mathbf{\Delta}/\mathrm{D}^{-\mathbf{1}}$:

$$^{1}H_{\Delta/\mathrm{D}} = \mathcal{W}(^{1}\mathbf{\Delta}/\mathbf{D}^{-\mathbf{1}}) = \frac{1}{2} \cdot \sum_{i=1}^{A} \sum_{j=1}^{A} [^{1}\mathbf{\Delta}/\mathbf{D}^{-\mathbf{1}}]_{ij}$$

and the reciprocal detour/distance matrix $\mathbf{\Delta}/\mathbf{D}^{-\mathbf{1}}$:

$$H_{\Delta/\mathrm{D}} = \mathcal{W}(\mathbf{\Delta}/\mathbf{D}^{-\mathbf{1}}) = \frac{1}{2} \cdot \sum_{i=1}^{A} \sum_{j=1}^{A} [\mathbf{\Delta}/\mathbf{D}^{-\mathbf{1}}]_{ij}$$

• **Harary Cluj-distance indices**

The Harary index and hyper-Harary index are obtained from, respectively, the 1st-order sparse → *reciprocal Cluj-distance matrix* $^{\mathbf{1}}\mathbf{CJD}^{-\mathbf{1}}$:

$$^{1}H_{\mathrm{CJD}} = \mathcal{W}(^{\mathbf{1}}\mathbf{CJD}^{-\mathbf{1}}) = \frac{1}{2} \cdot \sum_{i=1}^{A} \sum_{j=1}^{A} [^{\mathbf{1}}\mathbf{CJD}^{-\mathbf{1}}]_{ij}$$

and the reciprocal Cluj-distance matrix $\mathbf{CJD}^{-\mathbf{1}}$:

$$H_{\mathrm{CJD}} = \mathcal{W}(\mathbf{CJD}^{-\mathbf{1}}) = \frac{1}{2} \cdot \sum_{i=1}^{A} \sum_{j=1}^{A} [\mathbf{CJD}^{-\mathbf{1}}]_{ij}$$

In acyclic graphs, the Harary Cluj-distance index $^{1}H_{\mathrm{CJD}}$ coincides with the Harary Wiener index $^{1}H_{\mathrm{W}}$ and the Harary Szeged index $^{1}H_{\mathrm{SZ}}$ ($^{1}H_{\mathrm{CJD}} = {}^{1}H_{\mathrm{W}} = {}^{1}H_{\mathrm{SZ}}$); for the corresponding hyper-Harary indices the following relationships hold: $H_{\mathrm{CJD}} = H_{\mathrm{W}} \neq H_{\mathrm{SZ}}$. In cyclic graphs, only $^{1}H_{\mathrm{CJD}} = {}^{1}H_{\mathrm{SZ}}$, while the other indices give distinct values [Diudea *et al.*, 1997f].

• **Harary Cluj-detour indices**

The Harary index and hyper-Harary index [Diudea *et al.*, 1998] are obtained from, respectively, the 1st-order sparse → *reciprocal Cluj-detour matrix* $^{\mathbf{1}}\mathbf{CJ\Delta}^{-\mathbf{1}}$:

$$^{1}H_{\mathrm{CJ}\Delta} = \mathcal{W}\left(^{\mathbf{1}}\mathbf{CJ\Delta}^{-\mathbf{1}}\right) = \frac{1}{2} \cdot \sum_{i=1}^{A} \sum_{j=1}^{A} \left[^{\mathbf{1}}\mathbf{CJ\Delta}^{-\mathbf{1}}\right]_{ij}$$

and the reciprocal Cluj-detour matrix $\mathbf{CJ\Delta}^{-\mathbf{1}}$:

$$H_{CJ\Delta} = \mathcal{W}(\mathbf{CJ\Delta^{-1}}) = \frac{1}{2} \cdot \sum_{i=1}^{A} \sum_{j=1}^{A} [\mathbf{CJ\Delta^{-1}}]_{ij}$$

- **Harary Szeged indices**

The Harary index and hyper-Harary index are obtained from, respectively, the 1st-order sparse → *reciprocal Szeged matrix* $\mathbf{^1SZ^{-1}}$:

$$^1H_{SZ} = \mathcal{W}(\mathbf{^1SZ^{-1}}) = \frac{1}{2} \cdot \sum_{i=1}^{A} \sum_{j=1}^{A} [\mathbf{^1SZ^{-1}}]_{ij}$$

and the reciprocal Szeged matrix $\mathbf{SZ^{-1}}$:

$$H_{SZ} = \mathcal{W}(\mathbf{SZ^{-1}}) = \frac{1}{2} \cdot \sum_{i=1}^{A} \sum_{j=1}^{A} [\mathbf{SZ^{-1}}]_{ij}$$

In acyclic graphs, the Harary Szeged index $^1H_{SZ}$ coincides with the Harary Wiener index ^1H_W ($^1H_{SZ} = {}^1H_W$) while the corresponding hyper-Harary indices are different ($H_{SZ} \neq H_W$); in cyclic graphs all these indices differ [Diudea *et al.*, 1997f].

- **Harary walk indices**

The Harary walk indices are obtained from the → *reciprocal walk matrix* $\mathbf{W}^{-1}_{(\mathbf{M_1},\mathbf{M_2},\mathbf{M_3})}$:

$$H_{W(\mathbf{M_1},\mathbf{M_2},\mathbf{M_3})} = \mathcal{W}\left(\mathbf{W}^{-1}_{(\mathbf{M_1},\mathbf{M_2},\mathbf{M_3})}\right) = \frac{1}{2} \cdot \sum_{i=1}^{A} \sum_{j=1}^{A} \left[\mathbf{W}^{-1}_{(\mathbf{M_1},\mathbf{M_2},\mathbf{M_3})}\right]_{ij}$$

where $\mathbf{M_1}$, $\mathbf{M_2}$, and $\mathbf{M_3}$ are square $A \times A$ matrices.

📖 [Estrada and Rodriguez, 1997]

Harary matrix : *reciprocal distance matrix* → **distance matrix**

Harary number : *Harary index* → **Harary indices**

Harary numbers : *Harary indices*

Harary Szeged indices → **Harary indices**

Harary walk indices → **Harary indices**

Harary Wiener indices → **Harary indices**

hardness density : *local hardness* → **quantum-chemical descriptors** (⊙ hardness indices)

hardness indices → **quantum-chemical descriptors**

harmonic oscillator model of aromaticity index : *HOMA index* → **resonance indices**

harmonic topological index → **vertex degree**

Hartley information → **information content**

***HASA* index** : *SSAA index* → **charged partial surface area descriptors**

***HASA*$_2$ index** → **charged partial surface area descriptors** (⊙ *SSAA index*)

hashed fingerprints → **substructure descriptors**

HB$_1$ and HB$_2$ parameters → **hydrogen-bonding descriptors**

***HBCA* index** → **charged partial surface area descriptors**

***HB-CPSA* descriptors** : *hydrogen-bond charged partial surface area descriptors* → **charged partial surface area descriptors**

H-bonding descriptors : *hydrogen-bonding descriptors*

HB parameter → **hydrogen-bonding descriptors**

***HBSA* index** → **charged partial surface area descriptors**

HCD-descriptors → **molecular descriptors** (⊙ invariance properties of molecular descriptors)

***HDCA* index** → **charged partial surface area descriptors**

***$HDCA_2$* index** → **charged partial surface area descriptors**

H-depleted molecular graph → **molecular graph**

***HDSA* index** : *SSAH index* → **charged partial surface area descriptors**

***$HDSA_2$* index** → **charged partial surface area descriptors** (⊙ *$HDCA_2$ index*)

Herndon resonance energy → **resonance indices**

***HE*-state fields** → **electrotopological state indices**

***HE*-state index** : *hydrogen electrotopological state index* → **electrotopological state indices**

heteroatom-corrected extended adjacency matrix → **extended adjacency matrices**

heteroatom/multiplicity-corrected extended adjacency matrix → **extended adjacency matrices**

hierarchically ordered extended connectivities algorithms → **canonical numbering**

higher-order Wiener numbers → **Wiener matrix**

highest occupied molecular orbital → **quantum-chemical descriptors**

highest occupied molecular orbital energy → **quantum-chemical descriptors**

highest scoring common substructure → **maximum common substructure**

Higuchi-Davis model → **Hansch analysis**

Hildebrand solubility parameter (δ_H)

A measure of the intermolecular interactions between solute molecules and their environment, defined as:

$$\delta_H = \sqrt{\frac{-E_c}{\bar{V}}} = \sqrt{\frac{\Delta H_v - RT}{\bar{V}}}$$

where E_c is the cohesion energy between liquid molecules defined as a function of polarizability, ionization potential and dipole moment; $\bar{V}$ is the → *molar volume* of the compound, ΔH_v the vaporisation enthalpy of the liquid at 298 K, T the absolute temperature and R the universal gas constant [Hildebrand and Scott, 1950]. For apolar or moderately polar compounds the vaporization enthalpy can be estimated by their boiling points bp (K) as:

$$\Delta H_v(cal/mole) = -2950 + 23.7 \cdot \mathrm{bp} + 0.02 \cdot \mathrm{bp}^2$$

Often referred to as the **solvent cohesive energy density**, the Hildebrand solubility parameter is considered to be a measure of the solvent contribution to the → *cavity term*, and is used as a correction factor in the → *solvatochromic equation*. It is related to the general definition of **London cohesive energy** between two interacting species:

$$\varepsilon_{\mathrm{L}} = \frac{3 \cdot \alpha_i \cdot \alpha_j}{2 \cdot r_{ij}^6} \cdot \frac{\mathrm{IP}_i \cdot \mathrm{IP}_j}{\mathrm{IP}_i + \mathrm{IP}_j}$$

where α and IP are the → *polarizability* and the → *ionization potential* of the two species, and r their → *geometric distance*.

Moreover, the solubility parameter can also be calculated as [Small, 1953]:

$$\delta_{\mathrm{S}} = \frac{\sum_k \mathrm{F}_k}{\bar{\mathrm{V}}}$$

where F_k is the molar attraction constant of the kth substituent of the compound and the sum runs over all substituents; $\bar{\mathrm{V}}$ is the molar volume.

📖 [Kamlet *et al.*, 1981b] [Pussemier *et al.*, 1989]

Hill potential function → **molecular interaction fields** (⊙ steric interaction fields)

HOC algorithms : *hierarchically ordered extended connectivities algorithms* → **canonical numbering**

HOC rank descriptors → **canonical numbering** (⊙ hierarchically ordered extended connectivities algorithms)

Hodes statistical-heuristic method

This is a QSAR method for selecting active compounds proposed by taking → *substructure descriptors* as the predictors of the biological activity [Hodes *et al.*, 1977; Hodes, 1981a; Hodes, 1981b]. For each present substructure in a large data set, a weight for activity and a weight for inactivity are derived by calculating separately the incidence of each substructure in active and inactive compounds. The binomial distribution is assumed with mean and standard deviation given by:

$$mean = n \cdot \mathrm{p} \qquad std.dev. = \sqrt{n \cdot \mathrm{p} \cdot (1 - \mathrm{p})}$$

where n is the number of active (or inactive) compounds and p the incidence of the substructure in the whole set of compounds. Therefore, each weight is calculated as the inverse of the probability that the number of standard deviations away from the mean can occur by chance.

For example, a substructure occurs 17.7 % in the whole data set, i.e. p = 0.177. In the subset of 33 active compounds ($n = 33$), 5.84 compounds (0.177 × 33) are expected to have this substructure, assuming that it has nothing to do with activity. This number is assumed as the mean of the binomial distribution, with a standard deviation equal to 2.19. If the actual number of compounds containing the considered substructure is 11, the number of standard deviations away from the mean is:

$$\frac{11 - 5.84}{2.19} = 2.36$$

The one-tailed probability obtained using the normal approximation to the binomial probability is 0.0091, and the logarithm of the inverse of this probability value is used as the activity weight w^{AC} for the considered substructure, i.e. log(109.89) = 2.04, the smaller the probability, the larger being the significance. The inactivity weight w^{IN} is analogously calculated by using the subset of inactive compounds.

The score s_i of the activity for a new compound is calculated by adding the weights for each substructural feature present in the considered compound, i.e.:

$$s_i = \sum_f \log\left(w_{if}^{AC}\right) = \log\left(\prod_f w_{if}^{AC}\right)$$

where the logarithm is used to give convenient magnitudes to the scores. The score is an estimate of the probability that the compound belongs to the active set.

Note. The two-tailed probability values originally proposed have here been substituted by the one-tailed probability values because a number of active compounds in the data set significantly smaller than the expected mean value would give a significantly small two-tailed probability value, i.e. an incorrect large activity weight. In the previous numerical example, if no active compounds instead of 11 contained the considered substructure, the active weight would be 2.12; with 6 active compounds, the active weight would be 0.02 !

Hodgkin similarity index → **similarity/diversity**

Hologram QSAR → **substructure descriptors**

Holtz-Stock inductive constant → **electronic substituent constants**
(⊙ inductive electronic constants)

HOMA index → **resonance indices**

HOMO electron density on the *a*th atom → **charge descriptors** (⊙ net orbital charge)

homeomorphic graphs → **graph**

HOMO-LUMO energy gap → **quantum-chemical descriptors**

HOMO/LUMO energy fraction → **quantum-chemical descriptors**

Hosoya graph decomposition → **Hosoya Z index**

Hosoya ID number → **Hosoya Z matrix**

Hosoya non-adjacent number : *non-adjacent number* → **Hosoya Z index**

Hosoya mean information index → **Hosoya Z index**

Hosoya resonance energy → **resonance indices**

Hosoya total information index → **Hosoya Z index**

Hosoya Z index (: *Z index*)

The Hosoya Z index of a graph $\mathcal{G}$ is derived by a combinatorial algorithm and is defined as [Hosoya, 1971]:

$$Z = \sum_{k=0}^{[A/2]} a(\mathcal{G}, k)$$

where $a(\mathcal{G}, k)$, called the **non-adjacent number** (or **Hosoya non-adjacent number**), is the number $a(\mathcal{G}; k)$ of ways k edges may be selected from all of the B edges of graph $\mathcal{G}$ so that no two of them are adjacent. A is the → *atom number*, and the Gaussian brackets [ ] represent the greatest integer of $A/2$. The Z index is calculated by summing the $a(\mathcal{G}; k)$ coefficients over all different k values. For any graph, $a(\mathcal{G}, 0) = 1$ and $a(\mathcal{G}, 1) = B$. For $k = A/2$, the non-adjacent number $a(\mathcal{G}, A/2)$ corresponds to the → *Kekulé number*, i.e. to the number of Kekulé molecular structures.

The Hosoya Z index depends on the molecular size as well as on branching and ring closure. It was also found to correlate well with the boiling point.

Hosoya graph decomposition is a graph edge decomposition defined as:

$$N_k(\mathcal{E}_1, \mathcal{E}_2, \ldots, \mathcal{E}_j, \ldots, \mathcal{E}_G)$$

where G is the non-adjacent number $a(\mathcal{G}, k)$, $\mathcal{E}_j$ is the jth → *k-matching* of the graph, i.e. a subset of k non-adjacent bonds; N_k is the cardinality of the union of all subsets.

Hosoya found that for linear graphs the Z indices are the → *Fibonacci numbers*, i.e. $Z = F_A$, where A is the number of atoms in the molecular graph; therefore, for a linear graph, it is closely related to the → *Merrifield-Simmons index* [Gutman *et al.*, 1992; Randic *et al.*, 1996b].

Using non-adjacent numbers $a(\mathcal{G}, k)$ as coefficients, the **Z-counting polynomial** $\mathcal{Q}$ of $\mathcal{G}$ is derived as:

$$\mathcal{Q}(\mathcal{G}; x) = \sum_{k=0}^{[A/2]} a(\mathcal{G}, k) \cdot x^k$$

Therefore, the Hosoya Z index can be obtained from the $\mathcal{Q}$ polynomial for $x = 1$. For acyclic graphs the Z-counting polynomial coefficients $a(\mathcal{G}, k)$ coincide with the absolute values of the coefficients of the graph → *characteristic polynomial* $\mathcal{P}$ [Nikolic *et al.*, 1992].

In order to calculate the Z index for large graphs a composition principle for Z was developed [Hosoya, 1971]: the Z index value of the graph $\mathcal{G}$ is obtained as the product of the Z values of graphs $\mathcal{G}'$ and $\mathcal{G}''$, derived from $\mathcal{G}$ by cutting a b bond, plus the product of the Z values of all graphs derived from $\mathcal{G}'$ and $\mathcal{G}''$ by cutting all bonds incident to the b bond in the original graph $\mathcal{G}$. The Z value for an empty graph is set at one. It was demonstrated that the Z value for the graph $\mathcal{G}$ is uniquely obtained independently of the choice of the bond b to cut in the first step.

The Hosoya Z index for an edge-weighted graph $\mathcal{G}^*$ takes into account the edge weights w_b defining the non-adjacent numbers $a(\mathcal{G}^*, k)$ as:

$$a(\mathcal{G}^*, k) = \sum_{j=1}^{G} \left(\prod_{b=1}^{k} w_b \right)_{\mathcal{E}_j}$$

where the product runs over all bonds b of a set $\mathcal{E}$ constituted by k non-adjacent bonds and the sum runs over all G subsets of the Hosoya graph decomposition. Obviously, it follows that

$$a(\mathcal{G}^*, 1) = \textstyle\sum_b w_b \text{ and, by definition, } a(\mathcal{G}^*, 0) = 1$$

The **Hosoya total information index** is calculated on the Hosoya graph decomposition and is based on the distribution of $a(\mathcal{G}; k)$ coefficients. It is defined as → *total information content* as the following:

$$I_Z = Z \cdot \log_2 Z - \sum_{k=0}^{[A/2]} a(\mathcal{G}, k) \cdot \log_2 a(\mathcal{G}, k)$$

where Z is the Hosoya index. Analogously, the **Hosoya mean information index** is defined as → *mean information content*:

$$\bar{I}_Z = - \sum_{k=0}^{[A/2]} \frac{a(\mathcal{G}, k)}{Z} \cdot \log_2 \frac{a(\mathcal{G}, k)}{Z}$$

It is noteworthy that I_Z and $\bar{I}_Z$ indices coincide with → *information indices on polynomial coefficients* for acyclic graphs.

Generalized Hosoya indices Z_m were proposed [Hermann and Zinn, 1995] as counts of non-adjacent molecular paths in the graph $\mathcal{G}$:

Example : 2-methylpentane

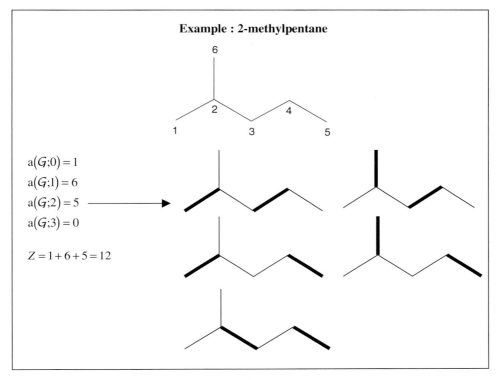

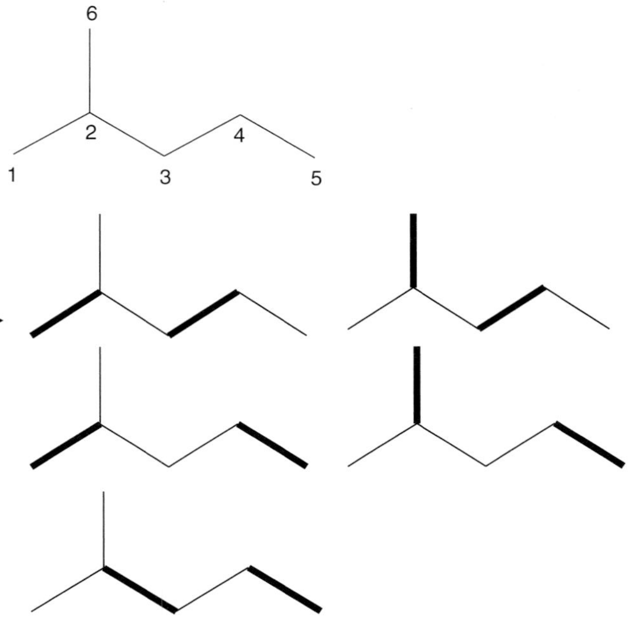

$a(\mathcal{G};0) = 1$
$a(\mathcal{G};1) = 6$
$a(\mathcal{G};2) = 5$
$a(\mathcal{G};3) = 0$

$Z = 1 + 6 + 5 = 12$

Box H-2.

$$Z_m = \sum_{k=0}^{G} a(\mathcal{G}, k)_m$$

where the subscript m refers to the order of the index and indicates the path length, $a(\mathcal{G}, k)_m$ is the number of all possible combinations of k non-adjacent paths of length m, $a(\mathcal{G}, 0) = 1$ by definition, and G is the maximum possible number of k, dependent on the selected path length and the molecule size. The Z_1 index coincides with the original Hosoya Z index and the Z_0 index counts all possible combinations of non-adjacent atoms. These indices must not be confused with the → *^{m}Z numbers* based on the sequential erasure of each path from the original graph.

Hosoya Z index was also used to define → *resonance indices* and → *graphical bond order*.

📖 [Hosoya, 1972b] [Hosoya, 1972c] [Hosoya and Murakami, 1975b] [Narumi and Hosoya, 1980] [Hosoya and Ohkami, 1983] [Narumi and Hosoya, 1985] [Hosoya, 1988] [Gutman, 1988] [Hosoya, 1991] [Gutman, 1992b] [Salvador *et al.*, 1998] [Hosoya *et al.*, 1999]

Hosoya Z matrix

A square symmetric matrix of dimension $A \times A$, A being the number of vertices in the → *H-depleted molecular graph* $\mathcal{G}$. The original Hosoya **Z** matrix is defined only for acyclic graphs; its entry z_{ij} is equal to the → *Hosoya Z index* of the subgraph $\mathcal{G}'$

obtained from the graph $\mathcal{G}$ by erasing all edges along the path connecting the vertices v_i and v_j as:

$$z_{ij} = \begin{cases} Z(\mathcal{G}') & \text{if } i \neq j \\ 0 & \text{if } i = j \end{cases}$$

the diagonal entries z_{ii} are zero by definition [Randic, 1994b]. If more than one subgraph is obtained by the erasure procedure, the entry value z_{ij} is calculated by summing all of the Hosoya Z indices of the subgraphs.

A general definition of the Hosoya **Z** matrix (**generalized Hosoya Z matrix**) able to represent both acyclic and cyclic graphs is the following [Plavsic *et al.*, 1997]:

$$z_{ij} = \begin{cases} \dfrac{\sum_{^{\min}p_{ij}} Z(\mathcal{G}')}{^{\min}P_{ij}} & \text{if } i \neq j \\ Z(\mathcal{G}) & \text{if } i = j \end{cases}$$

where $Z(\mathcal{G}')$ is the Z index of the graph $\mathcal{G}'$ obtained from the graph $\mathcal{G}$ by erasing all edges along the shortest path connecting the vertices v_i and v_j, i.e. the → *geodesic* $^{\min}p_{ij}$, and the sum runs over all $^{\min}P_{ij}$ geodesics between the considered vertices. The diagonal entries z_{ii} are simply equal to $Z(\mathcal{G})$, i.e. the Hosoya Z index of the original graph.

It is interesting to observe that the magnitude of the entries in the **Z** matrix decreases as the separation between the vertices increases; it can therefore be expected to simulate the interactions between the pairs of vertices well.

The **Z′/Z index** is among the → *graphical bond order descriptors* and can be obtained from the Hosoya **Z** matrix only by considering the entries relative to adjacent vertices (i.e. bonds):

$$Z'/Z = \frac{1}{Z} \cdot \mathcal{W}(\mathbf{Z} \bullet \mathbf{A}) = \frac{1}{2} \cdot \sum_{i=1}^{A} \sum_{j=1}^{A} \left(\frac{z_{ij}}{Z} \right) = \sum_{b=1}^{B} \left(\frac{Z(\mathcal{G}')}{Z} \right)_b$$

where Z is the Hosoya Z index of the whole graph, $\mathcal{W}$ is the → *Wiener operator* applied to the Hosoya → *sparse matrix* where the non-zero elements correspond to pairs of adjacent vertices, **A** being the → *adjacency matrix* and • the → *Hadamard matrix product*; B is the total number of bonds. Each term in the sum is a → *graphical bond order*.

Other graph invariants derived from the Hosoya **Z** matrix were the → *eigenvalues* and the coefficients of the → *characteristic polynomial*. Moreover, sequences of weighted paths and the weighted path counts were defined using as path weights the magnitude of the **Z** matrix elements. **^{m}Z numbers** are calculated as the sum of the magnitude of the entries corresponding to pairs of vertices separated by the shortest path of length m:

$$^{m}Z = \sum_{^{m}p_{ij}} z_{ij} = \sum_{^{m}p_{ij}} Z(\mathcal{G} - {}^{m}p_{ij})$$

where the term in the second summation is the Hosoya Z number of the graph $\mathcal{G}$ from which the path $^{m}p_{ij}$ is erased. ^{m}Z numbers must not be confused with → *generalized Hosoya indices* Z_m.

Summing up all ^{m}Z numbers, a global molecular descriptor called the **Hosoya ID number** ZID is obtained:

$$ZID = A + \sum_{m} {}^{m}Z$$

where A is the number of atoms corresponding to ^{0}Z and the sum runs over all path lengths.

From row sums of the **Z** matrix using → *Ivanciuc-Balaban operator* $\mathcal{IB}$, the $\mathbf{L_Z}$ **index** is calculated as:

$$L_Z = \mathcal{IB}(\mathbf{Z}) = \frac{B}{C+1} \cdot \sum_b (rs_i \cdot rs_j)_b^{-1/2}$$

where B is the number of bonds, C the → *cyclomatic number*, and rs_i and rs_j are the matrix row sums corresponding to the vertices v_i and v_j incident to the bth bond.

Analogously, applying the Wiener operator $\mathcal{W}$ to the **Z** matrix, the $\mathbf{K_Z}$ **index** is obtained:

$$K_Z = \mathcal{W}(\mathbf{Z}) = \frac{1}{2} \cdot \sum_{i=1}^{A} \sum_{j=1}^{A} z_{ij}$$

📖 [Plavsic *et al.*, 1996a]

***HRNCG* index** → **charged partial surface area descriptors**

***HRNCS* index** → **charged partial surface area descriptors**

***HRPCG* index** → **charged partial surface area descriptors**

***HRPCS* index** → **charged partial surface area descriptors**

$\mathbf{H_1}$ topological index

According to the → *Randic connectivity index*, this is defined as [Li and You, 1993a; Li and You, 1993b; Li *et al.*, 1995]:

$$\mathrm{H}_1 = \left(\sum_{b=1}^{B} \frac{1}{(1+\Delta_b)\sqrt{\delta_i \cdot \delta_j}} \right)^2 =$$

$$= \sum_{b=1}^{B} \left(\frac{1}{(1+\Delta_b)\sqrt{\delta_i \cdot \delta_j}} \right)_b^2 + \sum_{b=1}^{B-1} \sum_{b'=b+1}^{B} \left(\frac{1}{(1+\Delta_b)\sqrt{\delta_i \cdot \delta_j}} \right)_b \cdot \left(\frac{1}{(1+\Delta_{b'})\sqrt{\delta_i \cdot \delta_j}} \right)_{b'}$$

where the sums run over all B bonds in the molecular graph where hydrogen atoms are not suppressed; δ_i and δ_j are the → *vertex degree* of the two vertices incident to the considered bth bond. Δ_b is a bond parameter representing the interaction between the two bonded vertices i and j and calculated as the following:

$$\Delta_b = \alpha \cdot (\mathrm{IP}_i - \mathrm{EA}_j)_b + (1-\alpha) \cdot (\mathrm{IP}_j - \mathrm{EA}_i)_b$$

where IP and EA are the → *ionization potential* and → *electron affinity*, respectively. The first term in the equation represents the electron transfer interaction from the HOAO (Highest Occupied Atomic Orbital) of the ith atom to the LUAO (Lowest Unoccupied Atomic Orbital) of the jth atom, the second term represents the feedback interaction from the HOAO of the jth atom to LUAO of the ith atom. The parameter α is used to modulate the importance of the two kinds of interaction; it is generally taken equal to 0.5.

Hückel resonance energy → **resonance indices**

Hu-Xu ID number → **ID numbers**

hydrated surface area → **molecular surface** (⊙ solvent-accessible molecular surface)

Hydration Free Energy Density (HFED)

The empirical hydration free energy density is expressed by a linear combination of some physical properties calculated around the molecule with net atomic charges, polarizabilities, dispersion coefficients of the atoms in the molecule, and solvent accessible surface [Son *et al.*, 1999]. These physical properties are the result of the interaction of the molecule with its environment. To calculate the HFED of a molecule a grid model was proposed; a shell of critical thickness r_C was defined around the solvent-accessible surface with a number of grid points inside (e.g. 8 points / A^3).

The hydration free energy density, denoted by g_k, is calculated at each kth grid point of the shell as:

$$g_k = \frac{b_0}{N_G} + \frac{b_1}{N_G} \cdot \sum_{k=1}^{N_G} R_k + b_2 \cdot \left| \sum_{i=1}^{A} \frac{q_i}{r_{ik}} \right| + b_3 \cdot \sum_{i=1}^{A} \frac{q_i^2}{r_{ik}} + b_4 \cdot \sum_{i=1}^{A} \frac{\alpha_i}{r_{ik}^3} + b_5 \cdot \sum_{i=1}^{A} \frac{D_i}{r_{ik}^6}$$

where N_G and A are the numbers of grid points and molecule atoms, respectively; R_k is the distance between the centre of mass of the molecule and the kth grid point, r_{ik} the distance between the ith atom and the kth grid point; q_i is the net atomic charge of the ith atom; α and D are atomic polarizabilities and dispersion coefficients. $b_0,\ldots,$ b_5 are regression coefficients to be determined by multivariate regression analysis.

The atomic polarizability is the charge-dependent effective atomic polarizabilty (CDEAP) calculated by an empirical method as a linear function of the net atomic charge q_i:

$$\alpha_i = \alpha_i^0 - a_i \cdot q_i$$

where α^0 and a are the effective atomic polarizability of a neutral atom and the charge coefficient, respectively [No *et al.*, 1993]. The **atomic dispersion coefficient** D is calculated by the Slater-Kirkwood formula [Slater and Kirkwood, 1931]:

$$D_i = \frac{3}{4} \cdot \left(\frac{e \cdot h}{m \cdot \sqrt{e}} \right) \cdot \frac{\alpha_i^2}{\sqrt{\alpha_i / N_{el}}}$$

where h, m and e are the Planck constant, the mass and charge of the electrons; N_{el} is the number of effective electrons of the ith atom.

The hydration free energy ΔG_{HYD} is obtained by summing over the HFED within a threshold distance r_t :

$$\Delta G_{HYD} = \sum_{k=1}^{N_G} g_k$$

where N_G is the number of grid points. The quantity ΔG_{HYD} is the scalar representation of the field around the molecule given by the hydration free energy density; to encode information on the spatial distribution of this physical property the **free energy of hydration density tensor** was also proposed. The elements of the tensor in cartesian coordinates are defined as:

$$\vec{g}_{xx} = \frac{1}{2} \cdot \sum_{k=1}^{N_G} g_k \cdot \left(2x_k^2 - y_k^2 - z_k^2\right)$$

$$\vec{g}_{xy} = \frac{3}{2} \cdot \sum_{k=1}^{N_G} g_k \cdot x_k \cdot y_k$$

where $\vec{g}_{xx}$ and $\vec{g}_{xy}$ are xx and xy components of the tensor, and x, y, and z the coordinates of the grid point.

hydride group → **molecular graph**

hydrogen bond acceptors → **hydrogen-bonding descriptors**

hydrogen-bond acceptor number → **hydrogen-bonding descriptors**

hydrogen-bond charged partial surface area descriptors → **charged partial surface area descriptors**

hydrogen bond acidity → **linear solvation energy relationships** (⊙ hydrogen-bond parameters)

hydrogen bond basicity → **linear solvation energy relationships** (⊙ hydrogen-bond parameters)

hydrogen bond donors → **hydrogen-bonding descriptors**

hydrogen-bond donor number → **hydrogen-bonding descriptors**

hydrogen-bond electron-acceptor power : *hydrogen bond basicity* → **linear solvation energy relationships** (⊙ hydrogen-bond parameters)

hydrogen-bond electron-drawing power : *hydrogen bond acidity* → **linear solvation energy relationships** (⊙ hydrogen-bond parameters)

hydrogen-bonding ability constants → **hydrogen-bonding descriptors**

hydrogen-bonding descriptors *(: H-bonding descriptors)*

The hydrogen bond is the bond arising from the interaction between a hydrogen and an electron donor atom, such as oxygen and nitrogen; hydrogen-bonding modifies the electron distribution of the neighbour of the electron-donor atom, thus influencing reactivity. Hydrogen bonding causes an association of molecules, i.e. large aggregates of single molecules. This association influences several → *physico-chemical properties*, such as the compressibility factor, vaporization energy, density, surface tension, parachor, conductivity, dielectric constant, molar refractivity, boiling and melting points. Moreover, the hydrogen-bonding ability of molecules has long been recognized as being very important in biological reactions, including drug actions.

The theory of hydrogen bonding was fully discussed by Pimentel and McClellan [Pimentel and McClellan, 1960] and Vinogradov and Linnell [Vinogradov and Linnell, 1971].

Intramolecular and intermolecular hydrogen bonds can occur in biological and chemical systems. Moreover, functional groups in the molecule can be divided into **Hydrogen Bond Donors** (HBD) and **Hydrogen Bond Acceptors** (HBA), the former group having strong electron-withdrawing substituents such as –OH, –NH, –SH, and –CH and the latter different groups such as –PO, –SO, –CO, –N, –O, –S and –F; even a π-electron rich system can be considered an H-bond acceptor. Some groups are amphiprotic, i.e. with both acceptor and donor ability, such as –OH and –NH [Gutmann, 1978].

Hydrogen-bonding ability within a congeneric series of compounds having a common H-bond acceptor or donor can be correlated with the electronic effects of substituents using either → *electronic substituent constants* or other physico-chemical properties such as pK_a values. In particular, $\mathbf{p}\boldsymbol{K}_\mathbf{a}$ is the **acid dissociation constant**, i.e. a measure of the extent of ionization of weakly acid organic compounds. Most approaches for estimating pK_a are based on → *group contribution methods*; other approachs are based on quantum-chemical calculations [Grüber and Buss, 1989; Dixon and Jurs, 1993; Sixt *et al.*, 1996; Schüürmann *et al.*, 1997; Duboc, 1978; Amic *et al.*, 2000].

Hydrogen-bonding descriptors were introduced in → *Hansch analysis* as well as in → *grid-based QSAR techniques* in the form of → *hydrogen-bonding fields*; moreover, → *hydrogen bond acidity* and → *hydrogen bond basicity* scales, which are among the → *solvatochromic parameters*, were derived both for solutes and solvents by an empirical approach. → *Hydrogen-bond charged partial surface area descriptors* were derived from → *computational chemistry* based on surface areas and partial charges of HBA and HBD atoms or groups.

In analogy with solvatochromic parameters but based on quantum theoretical chemistry, a set of → *quantum-chemical descriptors* intended to describe the hydrogen bonding effects of molecules by → *theroretical linear solvation energy relationships* (TLSER) were proposed.

Moreover, H-bond donor ability was estimated by using atomic charge on the most positively charged hydrogen atom in the molecule (Q_H) in conjunction with the → *lowest unoccupied molecular orbital energy* ε_{LUMO}; in an analogous way, H-bond acceptor ability was estimated by using the charge of the most negatively charged atom which is also capable of hydrogen bonding (Q_{MN}) in conjunction with the → *highest occupied molecular orbital energy* ε_{HOMO} [Dearden and Ghafourian, 1995; Urrestarazu Ramos *et al.*, 1998]. H-bond donor ability was also estimated by using electron → *donor superdelocalizability* S^+ and → *self-atom polarizability* P [Dearden *et al.*, 1997].

The other most popular hydrogen-bonding descriptors are listed below. Reviews on the subject of hydrogen-bonding are [Hadzi *et al.*, 1990; Dearden and Ghafourian, 1999].

- **HB parameter**

This is the most simple hydrogen-bonding molecular descriptor, defined as a binary variable and accounting for the general ability of the molecule to give hydrogen bonds [Fujita *et al.*, 1977]. A modified HB parameter was proposed [Charton and Charton, 1982] on the basis of the number of hydrogen bonds that a molecule or substituent is capable of forming; for example, $HB(-NH_2) = 2$ as a proton donor and $HB(-NH_2) = 1$ as a proton acceptor, and HB(–OH) = 1 as a proton-donor and HB(–OH) = 2 as a proton-acceptor.

- **I_{HA} and I_{HD} parameters**

These are the simplest hydrogen-bonding substituent descriptors defined as binary variables accounting for the ability of the substituent to give hydrogen bonds [Hansch and Leo, 1979]. I_{HA} is equal to one if the substituent includes at least one H-bond acceptor, otherwise, zero. In the same way, I_{HD} is equal to one if the substituent includes at least one H-bond donor, otherwise, zero.

- **Hydrogen-Bond Acceptor number** (*HBA*)

A measure of the hydrogen-bonding ability of a molecule expressed in terms of number of possible hydrogen-bond acceptors. In particular it is calculated as the count of lone pairs on oxygen and nitrogen atoms in the molecule.

- **Hydrogen-Bond Donor number** (*HBD*)

A measure of the hydrogen-bonding ability of a molecule expressed in terms of number of possible hydrogen-bond donors. In particular it is calculated as the count of hydrogen atoms bonded to oxygen and nitrogen atoms in the molecule.

📖 [Winiwarter *et al.*, 1998]

- **HB_1 and HB_2 parameters**

Substituent descriptors of the hydrogen-bonding ability of functional groups, defined by a set of rules [Yang *et al.*, 1986].

The HB_1 parameter is a count descriptor based on atoms in a group which possess the ability to form hydrogen bonds. This includes both H-bond acceptors and H-bond donors. The rules for HB_1 are:

a) an oxygen atom is counted as 1, but as zero in $-OCF_3$;

b) a nitrogen atom is counted as 1, but as zero if it is bonded to an oxygen atom; moreover, the fragment N–N is counted as 1 and the fragment $-N_3$ as zero.

c) a hydrogen atom when bonded to oxygen or nitrogen atoms is counted as 1; moreover, it is also counted as 1 in the –C≡C–H fragment.

For example, HB_1 values for –NO, $-NO_2$, $-SO_2NHCH_3$, and $-CONH_2$ are 1, 2, 4, and 4, respectively.

The HB_2 parameter is defined as the number of atoms in a group able to form hydrogen bonds multiplied by the value of the strength of the hydrogen bond, then divided by 10. The total number of H-bond acceptors and H-bond donors is calculated by following the rules defined for HB_1. The multiplicative parameters accounting for H-bond strength are 6.05 for oxygen atoms, 5.5 for nitrogen atoms, 2.5 for hydrogen atoms.

For example, HB_2 values for –NO, $-NO_2$, $-SO_2NHCH_3$, and $-CONH_2$ are 0.61, 1.21, 2.01, and 1.66, respectively.

📖 [Basak *et al.*, 1990b; Basak, 1990]

• **hydrogen-bonding ability constants** (I_H)

Substituent descriptors defined by measuring the additive contributions of molecular fragments to the hydrogen bonding ability of a molecule [Seiler, 1974]. Such descriptors are calculated from the difference in log*P* value in two solvent/water systems:

$$\Delta \log P = \log P_{\text{octanol}} - \log P_{\text{solvent}} = b_0 + \sum_k (I_H)_k \cdot N_k$$

where $(I_H)_k$ and N_k are the hydrogen-bonding ability constant and the number of occurences of the *k*th fragment in the molecule, respectively. b_0 and $(I_H)_k$ are regression coefficients estimated by multivariate regression analysis. These substituent descriptors reflect both H-bond donor and H-bond acceptor ability, and are also a function of molecule polarity [Dearden and Ghafourian, 1999].

A set of 23 hydrogen-bonding constants was determined using octanol/water and cyclohexane/water systems; the calculated model was derived from 195 compounds, with intercept b_0 = –0.16, gives r^2 = 0.935 and s = 0.333. Some I_H substituent values are reported in Table H-1.

Table H-1. Hydrogen-bonding ability constants for some substituent groups.

Substituent	I_H
–N=N–NH– (triazole)	4.24
aliphatic –COOH	2.88
aromatic –COOH	2.87
aromatic –OH	2.60
–CONH–	2.56
$-SO_2NH-$	1.93
aliphatic –OH	1.82
aliphatic $-NH_2$	1.33
aromatic $-NH_2$	1.18
=N–	1.01

Substituent	I_H
$-CO-CH_2-CO$	0.59
$-NR_1R_2$ (R_1, $R_2 \neq H$)	0.55
$-NO_2$	0.45
$>C=O$	0.31
$-C\equiv N$	0.23
$-O-$	0.11
ortho-substitution to $-OH$, $-COOH$, $-N R_1R_2$	-0.62

- **Raevsky H-bond indices**

A set of descriptors characterizing relative H-bond donor and H-bond acceptor abilities of compounds calculated to reproduce the free energy ΔG and enthalpy ΔH of the hydrogen bond complex formation as defined in the thermodynamic equation:

$$\Delta G = \Delta H - T \cdot \Delta S$$

where ΔS is the entropy of complexation, T the temperature in Kelvin. ΔG was thought of as a multiplicative function of H-bond donor and H-bond acceptor ability as:

$$\Delta G = b_0 + b_1 \cdot C_{HD} \cdot C_{HA}$$

where b_0 and b_1 are regression coefficients, C_{HD} and C_{HA} are the free energy H-bond donor and H-bond acceptor factors, respectively [Raevsky *et al.*, 1992a; Raevsky, 1997]. Based on known experimental ΔG values, C_{HD} and C_{HA} values were estimated for 414 and 1298 compounds, respectively, by using the HYBOT program. A value of $C_{HA} = 4.00$ was selected for the standard H-bond acceptor (hexamethylphosphoramide), and a value of $C_{HD} = -2.50$ was selected for the standard H-bond donor (phenol).

The enthalpy contributions were also estimated and H-bond donor E_{HD} and H-bond acceptor E_{HA} enthalpy factors were also calculated.

Based on these four H-bond factors, Raevsky H-bond indices were therefore proposed as:

$$C_{HD}^{max} \quad C_{HA}^{max} \quad E_{HD}^{max} \quad E_{HA}^{max}$$

$$\sum C_{HD} \quad \sum C_{HA} \quad \sum E_{HD} \quad \sum E_{HA}$$

$$\frac{\sum C_{HD}}{MW} \quad \frac{\sum C_{HA}}{MW} \quad \frac{\sum E_{HD}}{MW} \quad \frac{\sum E_{HA}}{MW}$$

where the first four descriptors are free energy and enthalpy factors for the strongest H-bond donor atom and H-bond acceptor atom in the molecule; the second set is based on the sums of the free energy and enthalpy factors for all H-bond donor atoms and H-bond acceptor atoms in the molecule; the third set is constituted by the second set normalized on the molecular weight *MW*.

📖 [Raevsky *et al.*, 1992b; Raevsky *et al.*, 1993; Schneider *et al.*, 1993; Raevsky *et al.*, 1995]

Other reliable indicators for hydrogen-bonding are $\Delta \log P$ determined using octanol/water – heptane/water systems and octanol/water – chloroform/water systems [El Tayar *et al.*, 1991b].

📖 [Allerhand and Schleyer, 1963a] [Allerhand and Schleyer, 1963b] [Grob and Schlageter, 1976] [Gutman, 1976] [Di Paolo *et al.*, 1979] [Taft *et al.*, 1984] [Kamlet *et al.*, 1986c] [Maria *et al.*, 1987] [Schultz, 1987a] [Schultz and Cajina-Quezada, 1987b] [Abraham *et al.*, 1989a] [Abraham *et al.*, 1989b] [Boobbyer *et al.*, 1989] [Hadzi *et al.*, 1990] [Fan *et al.*, 1990]

[Wilson and Famini, 1991] [Abraham *et al.*, 1991] [Murray and Politzer, 1991] [El Tayar *et al.*, 1992a] [Tsai *et al.*, 1993] [Abraham, 1993a] [Abraham, 1993b] [Cramer *et al.*, 1993] [Kim, 1993f] [Testa *et al.*, 1993] [Wade and Goodford, 1993c] [Wade *et al.*, 1993b] [Zakarya *et al.*, 1993a] [Fanelli *et al.*, 1993] [Dziembowska, 1994] [ter Laak *et al.*, 1994] [Fan *et al.*, 1994] [Luco *et al.*, 1995] [Headley *et al.*, 1995] [Potts and Guy, 1995] [van de Waterbeemd *et al.*, 1996] [Berthelot *et al.*, 1996] [Elass *et al.*, 1996a] [Norinder *et al.*, 1997] [Sjöberg, 1997] [Schüürmann *et al.*, 1997] [Mitchell and Jurs, 1998b] [Turner *et al.*, 1998] [Klebe and Abraham, 1999] [Yamagami *et al.*, 1999]

hydrogen bonding fields → **molecular interaction fields**

hydrogen-bond parameters → **linear solvation energy relationships**

hydrogen-depleted molecular graph : *H-depleted molecular graph* → **molecular graph**

hydrogen electrotopological state index → **electrotopological state indices**

hydrophatic atom constants → **molecular interaction fields** (⊙ hydrophobic fields)

Hydrophatic INTeractions : *Kellogg and Abraham interaction field* → **molecular interaction fields** (⊙ hydrophobic fields)

hydrophilic effect : *Moriguchi polar parameter* → **lipophilicity descriptors**

hydrophilicity index (*Hy*)

A simple empirical index related to hydrophilicity of compounds based on → *count descriptors* [Todeschini *et al.*, 1997b]. It is defined as:

$$Hy = \frac{(1+N_{Hy}) \cdot \log_2(1+N_{Hy}) + N_C \cdot \left(\frac{1}{A} \cdot \log_2 \frac{1}{A}\right) + \sqrt{\frac{N_{Hy}}{A^2}}}{\log_2(1+A)}$$

where N_{Hy} is the number of hydrophilic groups (–OH, –SH, –NH), N_C the number of carbon atoms and A the number of atoms (hydrogen excluded). *Hy* index is between –1 and 3.64 (Table H-2).

Table H-2. Hydrophilicity values *Hy* for some compounds.

Compound	N_{Hy}	N_C	A	*Hy*
H_2O_2	2	0	2	3.64
H_2O	1	0	1	3.44
4-OH	4	4	8	3.30
Triols	3	3	6	2.54
Carbonic acid	2	3	6	2.05
Diols	2	2	4	1.84
Methanol	1	1	2	1.40
Ethanol	1	2	3	0.71
Decanediols	2	10	12	0.52
Propanol	1	3	4	0.37
Butanol	1	4	5	0.17
Pentanol	1	5	6	0.03
Methane	0	1	1	0.00
$N_{Hy} = 0$ and $N_C = 0$	0	0	2	0.00

Compound	N_{Hy}	N_C	*A*	*Hy*
Decanols	1	10	11	–0.28
Ethane	0	2	2	–0.63
Pentane	0	5	5	–0.90
Decane	0	10	10	–0.96
Alkane with N_C = 1000	0	1000	1000	–1.00

hydrophobic fields → **molecular interaction fields**

hydrophobic fragmental constants : *Nys-Rekker hydrophobic fragmental constants* → **lipophilicity descriptors**

hydrophobicity → **lipophilicity descriptors**

hydrophobic substituent constants : *Hansch-Fujita hydrophobic substituent constants* → **lipophilicity descriptors**

hyper-Cluj-detour index → **Cluj matrices**

hyper-Cluj-distance index → **Cluj matrices**

hyperconjugation effect → **electronic substituent constants**

hyper-detour index → **detour matrix**

hyper-distance-path index → **distance-path matrix**

hyper-Harary distance index → **Harary indices**

hyper-Harary indices → **Harary indices**

hyper-Harary numbers : *hyper-Harary indices* → **Harary indices**

hypermolecule → **hyperstructure-based QSAR techniques**

hyperstructure → **hyperstructure-based QSAR techniques**

hyperstructure-based QSAR techniques

QSAR techniques based on the construction of a **hyperstructure** defined as a virtual structure built by overlapping the training set structures such that some atoms and bonds of different structures coincide.

A hyperstructure built by overlapping molecular graphs is called a **molecular supergraph** (MSG), and it can be considered as a certain graph such that each training set structure can be represented as its subgraph. A 3D hyperstructure based on the → *molecular geometry* of the training set compounds is called a **hypermolecule**.

The most important QSAR techniques based on a hyperstructure are → *minimal topological difference*, → *DARC/PELCO analysis*, → *molecular field topology analysis.*

hyper-Szeged index → **Szeged matrix**

hyper-Wiener index → **Wiener matrix**

hyper-Wiener operator → **algebraic operators**

I

I_{HA} and I_{HD} parameters → **hydrogen-bonding descriptors**

I_R aromaticity indices → **resonance indices**

ICD-descriptors → **molecular descriptors** (⊙ invariance properties of molecular descriptors)

IDentification numbers : *ID numbers*

ID numbers *(: IDentification numbers)*

Molecular ID numbers (MID) are molecular descriptors defined as → *weighted path counts* or weighted → *walk counts*, mainly proposed to uniquivocally identify a molecule by a single real number, the aim being to obtain highly discriminatory power [Randic, 1984b; Szymanski *et al.*, 1986a].

These indices are usually calculated by assigning a weighting factor w_{ij} to each → *path* or → *walk* of the → *molecular graph* with v_i and v_j as endpoints. By summing all weighted paths (or walks) starting from the ith vertex, **atomic ID numbers** (AID) are obtained as → *local vertex invariants* characterizing the atomic environment.

In most the cases, molecular identification numbers are obtained as the half-sum of atomic identification numbers over all graph vertices. Otherwise, by summing atomic ID numbers only over the atoms belonging to molecular fragments such as functional groups, **fragment ID numbers** (or **subgraph ID numbers**) are obtained.

A list of molecular identification numbers is given below. Other important ones are the → *Hosoya ID number*, the → *hyper-Wiener index*, the → *restricted walk ID number*, and the → *total topological state*, defined elsewhere.

Note. To avoid confusion among different acronyms, some ID number names have been changed with respect to the original author definition. In particular, the Randic Connectivity ID number has been changed from ID into CID, the ID numbers proposed by Balaban from SID to BID, from MINID to MINCID, and from MINSID to MINBID.

- **Randic Connectivity ID number** (CID) *(: Connectivity ID number)*

It is the molecular identification number proposed first [Randic, 1984b; Szymanski *et al.*, 1985; Randic and Trinajstic, 1993a] and is defined as a weighted molecular path count:

$$\mathrm{CID} = A + \sum\nolimits_{^m p_{ij}} w_{ij}$$

where A is the number of atoms, $^m p_{ij}$ denotes a path of length m (i.e. a sequence of m edges) from the vertex v_i to vertex v_j, and w_{ij} is the path weight. The sum runs over all paths of the graph; each path of length zero is given a unit weight.

The weight w_{ij} is calculated by multiplying the → *edge connectivity* (Table I-1) of all m edges (bonds) of the path $^m p_{ij}$ as:

$$w_{ij} = \prod_{b=1}^{m} \left(\delta_{b(1)} \cdot \delta_{b(2)}\right)_b^{-1/2}$$

where $\delta_{b(1)}$ and $\delta_{b(2)}$ are the → *vertex degree* of the two atoms incident to the bth edge and b runs over all of the m edges of the path. The use of edge weights smaller than one results in a gradual attenuation of the role of paths of longer lengths; therefore the Randic ID number, unlike the molecular path count, is a graph invariant in which local features are more pronounced.

• **Prime ID number** (PID)
A modification of the Randic connectivity ID number, aimed at improving the discriminating power [Randic, 1986b]. The Prime ID number is analogously defined as:

$$\mathrm{PID} = A + \sum\nolimits_{{}^m p_{ij}} w_{ij}$$

where A is the number of atoms, ${}^m p_{ij}$ denotes a path of length m (i.e. a sequence of m edges) from the vertex v_i to vertex v_j, and w_{ij} is the path weight. The sum runs over all paths of the graph. Unlike the Randic ID number, the weight w_{ij} is calculated by multiplying the edge weights of all m edges (bonds) of the path ${}^m p_{ij}$ defined in a different way:

$$w_{ij} = \prod_{b=1}^{m} pn_b^{-1/2}$$

where pn_b is a prime number associated with the b bond according to the scheme of Table I-1, i.e. the prime number is chosen according to the vertex degrees of the atoms incident to the considered bond, removing the degeneracy between the bond type (1, 4) and (2, 2).

Table I-1. Bond edge connectivities and weights.

(δ_i, δ_j)	*Edge connectivity*	*Prime number*	*Edge weight*
(1, 2)	0.7071	2	0.7071
(1, 3)	0.5774	3	0.5774
(1, 4)	0.5000	5	0.4472
(2, 2)	0.5000	7	0.3780
(2, 3)	0.4082	11	0.3015
(2, 4)	0.3536	13	0.2774
(3, 3)	0.3333	17	0.2425
(3, 4)	0.2887	19	0.2294
(4, 4)	0.2500	23	0.2085

In practice, prime ID numbers are calculated by substituting the edge connectivity of the CID numbers with a different edge weight based on the first nine prime numbers.

• **conventional bond order ID number** (πID)
A molecular weighted path number obtained by weighting graph edges with → *conventional bond order* [Randic and Jurs, 1989]:

$$\pi\mathrm{ID} = A + \sum\nolimits_{{}^m p_{ij}} w_{ij} = A + \frac{1}{2} \cdot \sum_{i=1}^{A} \left(P_i^w - 1\right)$$

where A is the number of atoms, ${}^m p_{ij}$ denotes a path of length m (i.e. a sequence of m edges) from the vertex v_i to vertex v_j, and w_{ij} is the path weight. The sum runs over all paths of the graph; each path of length zero is given a unit weight; therefore the weighted path count of length zero coincides with the number of atoms. P_i^w is the atomic ID number for the ith atom, i.e. the sum of all weighted paths starting from vertex v_i of any length, zero included.

The weight w_{ij} is calculated by multiplying the conventional bond order π^* of all m edges (bonds) of the path ${}^m p_{ij}$:

$$w_{ij} = \prod_{b=1}^{m} \pi_b^*$$

This ID number accounts for multiple bonds in the molecule; for saturated molecules each bond weight is equal to one, therefore the ID number coincides with the → *total path count*.

An alternative ID number is calculated in the same way using → *fractional bond order* instead of conventional bond order to accomplish a gradual attenuation of the role of paths of longer lengths.

• **Ring ID number** (RID)
The Randic ID number restricted to path contributions from atoms of a ring in the molecule, i.e. it is calculated by summing up the atomic ID numbers of the atoms belonging to the selected ring [Randic, 1988c]. The ring ID number is a particular case of fragment ID numbers.

• **Balaban ID number** (BID)
A molecular identification number [Balaban, 1987a] defined as:

$$\text{BID} = A + \sum_{^m p_{ij}} w_{ij}$$

where A is the number of atoms, $^m p_{ij}$ denotes a path of length m (i.e. a sequence of m edges) from the vertex v_i to vertex v_j, and w_{ij} is the path weight. The sum runs over all paths of the graph. The weight w_{ij} is calculated by multiplying the edge weights of all m edges (bonds) of the path $^m p_{ij}$ by the following:

$$w_{ij} = \prod_{b=1}^{m} \left(\sigma_{b(1)} \cdot \sigma_{b(2)}\right)_b^{-1/2}$$

where $\sigma_{b(1)}$ and $\sigma_{b(2)}$ are the → *vertex distance degree* of the two atoms incident to the b edge and b runs over all of the m edges of the path.

• **MINCID**
A molecular ID number of a graph with A vertices derived from the Randic connectivity ID number with the aim of obtaining much faster calculations, also for large polycyclic graphs [Ivanciuc and Balaban, 1996b].

It is analogously defined as a weighted molecular path count:

$$\text{MINCID} = A + \sum_{^{\min} p_{ij}} w_{ij}$$

where A is the number of atoms, $^{\min} p_{ij}$ denotes the shortest path (i.e. → *geodesic*) of length d_{ij} from the vertex v_i to vertex v_j, and w_{ij} is the path weight. The sum runs over all shortest paths of the graph.

The weight w_{ij} is calculated by multiplying the edge connectivity of all edges (bonds) of the shortest path $^{\min} p_{ij}$:

$$w_{ij} = \prod_{b=1}^{d_{ij}} \left(\delta_{b(1)} \cdot \delta_{b(2)}\right)_b^{-1/2}$$

where $\delta_{b(1)}$ and $\delta_{b(2)}$ are the vertex degrees of the two atoms incident to the bth edge and b runs over all of the d_{ij} edges of the path.

For cyclic molecular graphs, the set of geodesics represents a subset of the set of paths, while, for acyclic graphs, the two sets are identical; therefore, for a graph $\mathcal{G}$:

$$\text{CID}(\mathcal{G}) > \text{MINCID}(\mathcal{G}) \qquad \text{for a cyclic graph}$$

$$\text{CID}(\mathcal{G}) = \text{MINCID}(\mathcal{G}) \qquad \text{for an acyclic graph}$$

• **MINBID**

A molecular ID number of a graph with A vertices derived from the Balaban ID number with the aim of obtaining much faster calculations, also for large polycyclic graphs [Ivanciuc and Balaban, 1996b].

It is analogously defined as a weighted molecular path count:

$$\text{MINBID} = A + \sum\nolimits_{^{\min}p_{ij}} w_{ij}$$

where A is the number of atoms, $^{\min}p_{ij}$ denotes the shortest path (i.e. → *geodesic*) of length d_{ij} from the vertex v_i to vertex v_j, and w_{ij} is the path weight. The sum runs over all the shortest paths of the graph.

The weight w_{ij} is calculated by multiplying the edge weights of all edges (bonds) of the path $^{\min}p_{ij}$:

$$w_{ij} = \prod_{b=1}^{d_{ij}} \left(\sigma_{b(1)} \cdot \sigma_{b(2)}\right)_b^{-1/2}$$

where $\sigma_{b(1)}$ and $\sigma_{b(2)}$ are the vertex distance degrees of the two atoms incident to the b edge and b runs over all of the d_{ij} edges of the path.

For cyclic molecular graphs, the set of geodesics represents a subset of the set of paths, while, for acyclic graphs, the two sets are identical; therefore, for a graph $\mathcal{G}$:

$$\text{BID}(\mathcal{G}) > \text{MINBID}(\mathcal{G}) \qquad \text{for a cyclic graph}$$

$$\text{BID}(\mathcal{G}) = \text{MINBID}(\mathcal{G}) \qquad \text{for an acyclic graph}$$

• **Weighted ID number** (WID)

A molecular ID number of a graph with A vertices defined as a function of the sum of weighted walks [Szymanski *et al.*, 1986b]. The weight of an edge is the reciprocal of the square root of the product of the vertex distance degrees σ of the vertices incident to the edge.

Let **W** be the weighted adjacency matrix defined as:

$$w_{ij} = \begin{cases} \left(\sigma_i \cdot \sigma_j\right)^{-1/2} & \textit{if } a_{ij} = 1 \\ 0 & \textit{otherwise} \end{cases}$$

where a_{ij} are the entries of the → *adjacency matrix* **A**.

An auxiliary identification number ID* is obtained as:

$$\text{ID}^* = \sum_{i=1}^{A}\sum_{j=1}^{A} w_{ij}^* \qquad A \leq \text{ID}^* \leq A^2$$

where A is the number of vertices and w_{ij}^* are the entries of the $\mathbf{W}^*$ matrix which is obtained by summing the matrix powers $\mathbf{W^m}$ of the edge-weighted matrix:

$$\mathbf{W}^* = \sum_{m=0}^{A-1} \mathbf{W^m} = \mathbf{I} + \sum_{m=1}^{A-1} \mathbf{W^m}$$

where $\mathbf{W^0} = \mathbf{I}$, **I** being the identity matrix.

Each $\mathbf{W^m}$ matrix contains, at position *i-j*, the sum of the weights of all walks of length m from vertex v_i to vertex v_j; therefore:

$$[\mathbf{W^m}]_{ij} = \sum\nolimits_{w} \prod_{b=1}^{m} \left(\sigma_{b(1)} \cdot \sigma_{b(2)}\right)_b^{-1/2}$$

where $\sigma_{b(1)}$ and $\sigma_{b(2)}$ are the vertex distance degrees of the two atoms incident to the b edge and b runs over all of the m bonds of the walk and the sum runs over all the walks of length m between the vertices v_i and v_j. The diagonal entry *i-i* is the sum of all weighted self-returning walks for vertex v_i of length m. Thus, the matrix $\mathbf{W}^*$ contains, at position i-j, the sum of the weights of all walks from vertex v_i to vertex v_j of length less than A; the entries at position i-i on the diagonal correspond to the sum of the weights of all self-returning walks for vertex v_i of length less than A.

The weighted walk ID number is defined as:

$$\mathrm{WID} = A - \frac{1}{A} + \frac{\mathrm{ID}^*}{A^2} \qquad A \leq \mathrm{WID} < A + 1$$

where ID* = A^2 for complete graphs and ID* = A for a null graph.

The **Self-returning ID number** (SID) is a molecular ID number [Müller *et al.*, 1993] of a graph defined as the sum of weighted self-returning walks of any length:

$$\mathrm{SID} = \sum_{i=1}^{A} w_{ii}^*$$

where w_{ii}^* are the diagonal entries of the $\mathbf{W}^*$ matrix.

- **Extended Adjacency ID number** (EAID)

A complex ID number calculated by the matrix powers of an extended adjacency matrix, derived by weighting the vertices of the graph [Hu and Xu, 1996].

Let **W** be the edge-weighted matrix defined as:

$$w_{ij} = \begin{cases} \sqrt{\frac{S_i}{S_j}} + \sqrt{\frac{S_j}{S_i}} & if\ i \neq j \\ 0 & if\ i = j \end{cases}$$

where the local vertex invariant S_i accounting for heteroatoms is defined as:

$$S_i = lcv_{i0} + \sum_{k=1}^{D} lcv_{ik} \cdot lcb_{ik} \cdot 10^{-k}$$

where lcv_{ik} and lcb_{ik} are the entries of the → *connectivity valence layer matrix* **LCV** and the → *connectivity bond layer matrix* **LCB**, and D is the number of layers around the focused ith vertex.

An extended adjacency matrix **EA** is defined here by the following:

$$[\mathbf{EA}]_{ij} = \begin{cases} \sqrt{R_i}/6 & if\ i = j \\ \left(\sqrt{\pi_{ij}^*} \cdot w_{ij}\right)/6 & if\ i \neq j \end{cases}$$

where R_i is the → *covalent radius* of the ith atom and π_{ij}^* is the → *conventional bond order* of the i-j bond; w_{ij} are the elements of the edge-weighted matrix defined above. For all the non-adjacent vertex pairs the matrix elements are equal to zero.

The extended adjacency ID number EAID is calculated as:

$$\mathrm{EAID} = \sum_{i=1}^{A} [\mathbf{EA}^*]_{ii}$$

where $[\mathbf{EA}^*]_{ii}$ are the diagonal entries of the $\mathbf{EA}^*$ matrix which is obtained by summing the A power matrices $\mathbf{EA^m}$:

$$\mathbf{EA}^* = \sum_{m=0}^{A-1} \mathbf{EA^m} = \mathbf{I} + \sum_{m=1}^{A-1} \mathbf{EA^m}$$

where $\mathbf{EA^0} = \mathbf{I}$, the identity matrix.

• **Hu-Xu ID number** (HXID)

A molecular ID number where each path ${}^m p_{ij}$ of length m is weighted by the following:

$$w_{ij} = \prod_{a=2}^{m+1} \left(\frac{\pi^*_{a-1,a}}{a} \cdot \frac{1}{\delta'_{a-1} \cdot \delta'_a} \right)^{1/2}$$

where π^* is the → *conventional bond order*, a is the sequence number of the vertices along the path between vertices v_i and v_j, and δ' is a modified valence vertex degree, defined as:

$$\delta'_a = \delta_a \cdot \sqrt{Z_a}$$

where Z_a is the atomic number of the considered atom and δ_a the vertex degree [Hu and Xu, 1997].

Example : 2-methylpentene

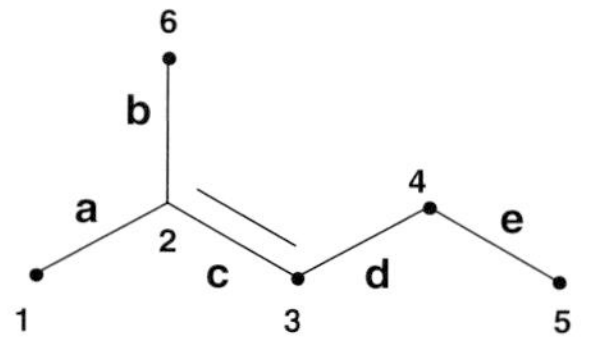

Connectivity ID numbers

Atom	${}^0P_i^w$	${}^1P_i^w$	${}^2P_i^w$	${}^3P_i^w$	${}^4P_i^w$	P_i^w
1	1	0.5774	1.3905	0.1178	0.0680	3.1537
2	1	1.5630	0.2041	0.1178		2.8849
3	1	0.9082	0.7601			2.6683
4	1	1.0774	0.2041	0.2357		2.5172
5	1	0.5774	0.2887	0.1178	0.1361	2.1200
6	1	0.5774	1.3905	0.1178	0.0680	3.1537
	${}^0P^w$	${}^1P^w$	${}^2P^w$	${}^3P^w$	${}^4P^w$	CID
total	6	2.6404	2.1190	0.3535	0.1361	11.2489

Prime ID numbers

Atom	${}^0P_i^w$	${}^1P_i^w$	${}^2P_i^w$	${}^3P_i^w$	${}^4P_i^w$	P_i^w
1	1	0.5774	1.3289	0.0658	0.0380	3.0101
2	1	1.4563	0.1140	0.0658		2.6361
3	1	0.6795	0.5664			2.2459
4	1	0.9554	0.1140	0.1316		2.2010
5	1	0.5774	0.2183	0.0658	0.0760	1.9375
6	1	0.5774	1.3289	0.0658	0.0380	3.0101
	${}^0P^w$	${}^1P^w$	${}^2P^w$	${}^3P^w$	${}^4P^w$	PID
total	6	2.4117	1.8353	0.1974	0.0760	10.5204

Conventional bond order ID numbers

Atom	${}^0P_i^w$	${}^1P_i^w$	${}^2P_i^w$	${}^3P_i^w$	${}^4P_i^w$	P_i^w
1	1	1	2.5	2.25	3.375	10.125
2	1	3.5	2.25	3.375		10.125
3	1	3	5.25			9.250
4	1	3	2.25	4.5		10.750
5	1	1.5	2.25	3.375	6.75	14.875
6	1	1	2.5	2.25	3.375	10.125
	${}^0P^w$	${}^1P^w$	${}^2P^w$	${}^3P^w$	${}^4P^w$	πID
total	6	6.5	8.5	7.875	6.750	35.625

Balaban ID numbers

Atom	${}^0P_i^w$	${}^1P_i^w$	${}^2P_i^w$	${}^3P_i^w$	${}^4P_i^w$	P_i^w
1	1	0.1021	0.0232	0.0014	0.0001	1.1268
2	1	0.3292	0.0140	0.0012		1.3444
3	1	0.2368	0.0350			1.2718
4	1	0.1963	0.0140	0.0029		1.2132
5	1	0.0845	0.0094	0.0012	0.0002	1.0953
6	1	0.1021	0.0232	0.0014	0.0001	1.1268
	${}^0P^w$	${}^1P^w$	${}^2P^w$	${}^3P^w$	${}^4P^w$	BID
total	6	0.5255	0.0594	0.0041	0.0002	6.5892

Box I-1.

Then, the molecular ID number is defined as:

$$\mathrm{HXID} = \sum_{i=1}^{A} \mathrm{AID}_i^2$$

AID being the atomic ID number obtained by summing the weights of all paths starting from the ith vertex,

$$\mathrm{AID}_i = \sum_{j=1}^{A} w_{ij}$$

where w_{ij} are the path weights and j runs over all vertices different from vertex v_i.

Idoux steric constant → **steric descriptors** (⊙ number of atoms in substituent specific positions)

immanant → **algebraic operators** (⊙ determinant)

Inamoto-Masuda inductive constant → **electronic substituent constants** (⊙ inductive electronic constants)

incidence matrix (I)

Derived from the → *H-depleted molecular graph*, the incidence matrix [Bonchev and Trinajstic, 1977] is a rectangular matrix representation of a graph whose rows are the vertices (atoms, A) and columns are the edges (bonds, B), i.e. having a dimension $A \times B$. Their elements are $\mathrm{i}_{ij} = 1$ if the edge e_j is incident to the vertex v_i, otherwise $\mathrm{i}_{ij} = 0$.

Example : 2-methylpentane

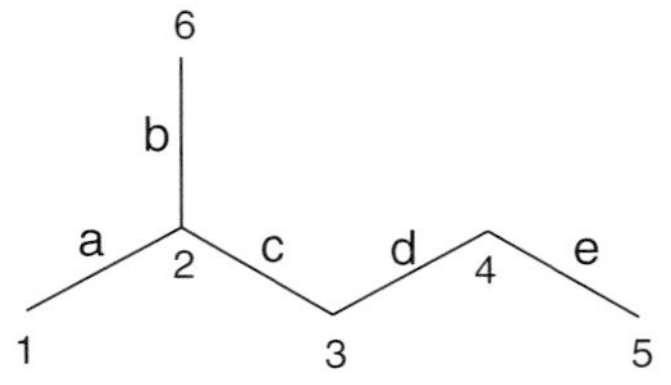

Atoms/Bonds	a	b	c	d	e
1	1	0	0	0	0
2	1	1	1	0	0
3	0	0	1	1	0
4	0	0	0	1	1
5	0	0	0	0	1
6	0	1	0	0	0

Box I-2.

The **total information content on the incidence matrix** I_{INC} and the **mean information content on the incidence matrix** $\bar{I}_{\mathrm{INC}}$ are respectively defined as [Bonchev, 1983]:

$$I_{\mathrm{INC}} = A \cdot B \cdot \log_2 A - B \cdot (A-2) \cdot \log_2(A-2) - 2 \cdot B$$

$$\bar{I}_{\mathrm{INC}} = \log_2 A - \frac{A-2}{A} \cdot \log_2(A-2) - \frac{2}{A}$$

It must be noted that the mean information content does not depend on the number and type of bonds; therefore it will be the same for the → *multigraph* $\mathcal{MG}$ and the corresponding graph $\mathcal{G}$.

However, the total information content of the multigraph is always greater than that in the corresponding graph by the quantity:

$$I_{\text{INC}}(\mathcal{MG}) - I_{\text{INC}}(\mathcal{G}) = A \cdot \log_2 A - (A-2) \cdot \log_2(A-2) - 2$$

indegree → **graph**

independent edges → **graph**

independent variables → **data set**

indexed graph → **graph**

indicator variables (I)

Also known as **dummy variables**, indicator variables are one of the most simple descriptors and are used when the problem of interest cannot be represented by real numerical values. They usually take positive, negative or zero integer values, indicating the states of some quantity. For example, the presence and discrimination of *cis/trans* isomers can be represented by an indicator variable such as:

$$\mathrm{I} = \begin{cases} 1 & \textit{trans isomer} \\ 0 & \textit{no cis/trans isomer} \\ -1 & \textit{cis isomer} \end{cases}$$

This indicator variable takes the value 1 when a *cis* isomer is present, –1 when a *trans* isomer is present and 0 when the characteristic is not applicable to the molecule, i.e. when no *cis/trans* isomerism is present. Indicator variables are in general multi-valued variables, and are thus able to discriminate among multi-state quantities.

The most common subcases of indicator variables are the **binary descriptors**, which are bi-valued variables taking the value of 1 when the considered characteristic is present in the molecule and the value of 0 when the characteristic is absent; these descriptors are usually indicated by the symbol I_{char}, where *char* is the considered characteristic.

For example, a typical use of binary descriptors is to represent:

- presence/absence of a molecular fragment
- presence/absence of an aromatic substructure
- presence/absence of a specified functional group
- *cis/trans* isomer discrimination
- chirality discrimination

Binary descriptors should be used when the considered characteristic is really a dual characteristic of the molecule or when the considered quantity cannot be represented in a more informative numerical form. In any case, the → *mean information content* of a binary descriptor $\bar{I}_{\text{char}}$ is low (the maximum value is 1 when the proportions of 0 and 1 are equal), thus the → *standardized Shannon's entropy* $H^*_{char} = \bar{I}_{\text{char}}/\log_2 n$, where n is the number of elements, gives a measure of the efficiency of the collected information.

For specific purposes, single binary variables can also be combined using the logical operators "and" ($\wedge$) and "or" ($\vee$) [Streich and Franke, 1985].

Binary descriptors are used in → *Free-Wilson analysis* and → *DARC/PELCO analysis.*

indices of neighbourhood symmetry *(: multigraph information content indices)*

→ *Topological information indices* of a graph based on neighbour degrees and edge multiplicity. The index based on first neighbour degrees and edge multiplicity was originally proposed by Sarkar [Sarkar *et al.*, 1978] and called "*information content for a multigraph*". They are calculated by partitioning graph vertices into → *equivalence classes* defined as the following: two vertices v_i and v_j of a → *multigraph* $\mathcal{MG}$, being of the same chemical element and having the same → *vertex degree*, are said to be topologically equivalent with respect to rth order neighbourhood if, and only if, to each rth order path $^r p_i$ starting from the vertex v_i there corresponds a distinct rth order path $^r p_j$ starting from the vertex v_j characterized by the same → *conventional bond order* of the edges in the path and the same chemical element and vertex degree of the involved vertices [Magnuson *et al.*, 1983; Roy *et al.*, 1984].

In other words, a basic requirement for the topological equivalence of two vertices is that the corresponding neighbourhoods of the rth order are the same, where the rth order **neighbourhood of a vertex** v_i in the graph $\mathcal{MG}$ is the subset $V_{ir}(\mathcal{MG})$ of vertices defined as:

$$V_{ir}(\mathcal{MG}) = \{a | a \in V(\mathcal{MG}); d_{ia} < r\}$$

where d_{ia} is the → *topological distance* of the vertex a from the focused vertex v_i and r is any non-negative real number. The vertex neighbourhood can be thought of as an open sphere $S(v_i\ , r)$ constituted by all the vertices a in the graph, such that their distance from the vertex v_i is less than r. Obviously, $S(v_i\ , 0) = \emptyset$, $S(v_i\ , 1) = v_i$ for $0 < r < 1$, and if $1 < r < 2$, then $S(v_i\ , r)$ is the set consisting of v_i together with all its adjacent vertices.

In practice, an unordered sequence called a *coordinate* is assigned to each graph vertex v_i:

$$\left\{(a_1, \delta_1, \pi_1^*; \ ....; a_r, \delta_r, \pi_r^*)_1; (a_1, \delta_1, \pi_1^*; \ ....; a_r, \delta_r, \pi_r^*)_2; .... ; (a_1, \delta_1, \pi_1^*; \ ....; a_r, \delta_r, \pi_r^*)_{^rP_i}\right\}$$

where the coordinate is composed of ordered sequences each representing a distinct path of length r starting from the v_i vertex, ^rP_i is the total number of rth order paths starting from v_i (i.e. the rth order → *atomic path count*), a and δ are the chemical element and the vertex degree of the vertices involved in the considered path, and π^* is the conventional bond order of the edge connecting the considered neighbour of v_i and the previous one. The ordered subscripts 1, 2, ... , r of the chemical element, vertex degree and bond order in each path refer to 1st, 2nd, ... , r^{th} neighbours of the v_i vertex along with the considered path. Therefore, two vertices are topologically equivalent if their coordinates are the same.

From the obtained equivalence classes in the hydrogen-filled multigraph, for each rth order (usually $r = 0 - 6$), the rth order **neighbourhood Information Content** IC_r is calculated as defined by → *Shannon's entropy*:

$$IC_r = -\sum_{g=1}^{G} \frac{A_g}{A} \cdot \log_2 \frac{A_g}{A} = -\sum_{g=1}^{G} p_g \cdot \log_2 p_g$$

where g runs over the G equivalence classes, A_g is the cardinality of the gth equivalence class, A is the total number of atoms, and p_g is the probability of randomly selecting a vertex of the gth class. It represents a measure of structural complexity per vertex.

This descriptor calculated for the → *H-depleted molecular graph* coincides with the → *vertex orbital information content* in the case of atoms of the same chemical element and maximal order of neighbourhood:

$$\bar{I}_{ORB} = IC_r \quad \text{for } r = \max(r)$$

Example : 2-methyl-2-butene

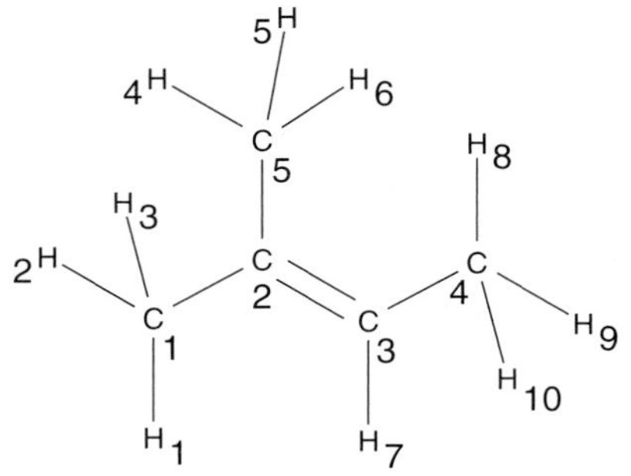

Order	Equivalent vertices	Probability	Descriptors
0	(C_1, C_2, C_3, C_4, C_5)	5/15	$IC_0 = 0.9183$
	(H_1, H_2, H_3, H_4, ... , H_{10})	10/15	$CIC_0 = 2.9886$
			$TIC_0 = 13.7744$
			$SIC_0 = 0.2350$
			$BIC_0 = 0.2350$
1	(C_1, C_4, C_5)	3/15	$IC_1 = 1.3753$
	(C_2)	1/15	$CIC_1 = 2.5316$
	(C_3)	1/15	$TIC_1 = 20.6292$
	(H_1, H_2, H_3, H_4, ... , H_{10})	10/15	$SIC_1 = 0.3520$
			$BIC_1 = 0.3520$
2	(C_1, C_5)	2/15	$IC_2 = 1.8716$
	(C_2)	1/15	$CIC_2 = 2.0353$
	(C_3)	1/15	$TIC_2 = 28.0740$
	(C_4)	1/15	$SIC_2 = 0.4790$
	(H_1, H_2, H_3, H_4, H_5, H_6, H_8, H_9, H_{10})	9/15	$BIC_2 = 0.4790$
	(H_7)	1/15	
3	(C_1, C_5)	2/15	$IC_3 = 2.4226$
	(C_2)	1/15	$CIC_3 = 1.4843$
	(C_3)	1/15	$TIC_3 = 36.3387$
	(C_4)	1/15	$SIC_3 = 0.6201$
	(H_1, H_2, H_3, H_4, H_5, H_6)	6/15	$BIC_3 = 0.6201$
	(H_7)	1/15	
	(H_8, H_9, H_{10})	3/15	

Box I-3.

From the rth order neighbourhood information content, the following information indices were also derived:

- **neighbourhood Total Information Content** (TIC_r)

The rth order TIC_r, is defined as A times IC_r:

$$TIC_r = A \cdot IC_r$$

where A is the atom number. This descriptor represents a measure of the graph complexity.

- **Structural Information Content** (SIC_r)

The rth order SIC_r is defined in a normalized form of the information content to delete the influence of graph size:

$$SIC_r = \frac{IC_r}{\log_2 A}$$

where A is the atom number.

- **Bonding Information Content** (BIC_r)

The rth order BIC_r is defined in a normalized form as the SIC_r index, but taking into account the number of bonds and their multiplicity:

$$BIC_r = \frac{IC_r}{\log_2\left(\sum_{b=1}^{B} \pi_b^*\right)}$$

where B is the number of bonds and π_b^* is the conventional bond order of the b bond. In the original definition the denominator was simply considered to be the bond number B.

- **Complementary Information Content** (CIC_r)

The rth order CIC_r measures the deviation of IC_r from its maximum value, which corresponds to the vertex partition into equivalence classes containing one element each:

$$CIC_r = \log_2 A - IC_r$$

where A is the atom number.

- **redundant information content** (R_r)

A measure of relative redundancy of a graph obtained by normalizing the complementary information content, defined as [Roy *et al.*, 1984]:

$$R_r = \frac{CIC_r}{\log_2 A} = 1 - SIC_r$$

- **order of neighbourhood** (O)

The order of neighbourhood O is a molecular descriptor defined as the order r of the IC_r index when it reaches the maximum value:

$$O = r \quad \text{where} \quad r : \max_r(IC_r)$$

To account for steric effects in molecule-receptor interactions, the **weighted information indices by volume** have been proposed [Ray *et al.*, 1985]. These molecular descriptors are calculated in the same way as the indices of neighbourhood symmetry defined above using the atomic van der Waals volumes to get the probabilities of the equivalence classes. In other words, the van der Waals volumes of the atoms belonging to each equivalent class are summed to give a molecule subvolume, then divided by the total molecule volume. For example, the weighted information content by volume is defined as:

$$IC_r^{\mathrm{V}} = -\sum_{g=1}^{G} \frac{\mathrm{V}_g}{\mathrm{V}_{\mathrm{VDW}}} \cdot \log_2 \frac{\mathrm{V}_g}{\mathrm{V}_{\mathrm{VDW}}}$$

where $\mathrm{V}_{\mathrm{VDW}}$ is the → *van der Waals volume* and V_g is the sum of the effective van der Waals volumes of the atoms in the gth class. The effective van der Waals volume of an atom is defined as the van der Waals volume of the atom minus half the sphere overlapping of the atom due to covalent bonding of the adjacent atoms in the molecule.

📖 [Bonchev *et al.*, 1976b] [Ray *et al.*, 1981] [Ray *et al.*, 1982] [Basak and Magnuson, 1983] [Roy *et al.*, 1983] [Ray *et al.*, 1983] [Basak *et al.*, 1984a] [Basak *et al.*, 1984b] [Basak *et al.*, 1986a] [Basak, 1987] [Basak *et al.*, 1987b] [Basak *et al.*, 1988b] [Basak *et al.*, 1990b] [Basak, 1990] [Basak *et al.*, 1991] [Niemi *et al.*, 1992] [Boecklen and Niemi, 1994] [Basak *et al.*, 1994] [Basak *et al.*, 1995] [Basak *et al.*, 1996a] [Basak and Gute, 1997b] [Basak *et al.*, 1999] [Basak, 2000]

induced dipole moment → **electric polarization descriptors**

induced polarization → **electric polarization descriptors**

induction parameter → **multiple bond descriptors**

inductive effect → **electronic substituent constants**

inductive electronic constants → **electronic substituent constants**

inertia matrix → **principal moments of inertia**

inertia principal moments : *principal moments of inertia*

inertial shape factor → **shape descriptors**

informational energy content → **information content**

information bond index (I_B)

Proposed by Dosmorov [Dosmorov, 1982] in analogy with the → *total information index on atomic composition* of a molecule as:

$$I_B = B \cdot \log_2 B - \sum_{g=1}^{G} B_g \cdot \log_2 B_g$$

where B is the → *bond number* and B_g the number of bonds of type g; the sum runs over all G different types of bonds in the molecule. The first simple partition of molecule bonds is performed according to the → *conventional bond order*, i.e. single, double, triple and aromatic bonds [Bonchev, 1983].

information connectivity indices

→ *Topological information indices* based on the partition of the edges in the graph according to the equivalence and the magnitude of their → *edge connectivity* values [Bonchev *et al.*, 1981b].

The **mean information content on the edge equality** ${}^{E}\bar{I}_{\chi}^{E}$ is based on the partition of edges according to edge connectivity equivalence and is defined as:

$${}^{E}\bar{I}_{\chi}^{E} = -\sum_{g=1}^{G} \frac{B_g}{B} \log_2 \frac{B_g}{B}$$

where B_g is the number of edges having the same edge connectivity, G is the number of different connectivity values and B the → *bond number*.

The **mean information content on the edge magnitude** ${}^{E}\bar{I}_{\chi}^{M}$ is based on the magnitude of edge connectivities and is defined as:

$$ {}^{E}\bar{I}_{\chi}^{M} = -\sum_{b=1}^{B} \frac{(\delta_i \cdot \delta_j)_b^{-1/2}}{{}^{1}\chi} \log_2 \frac{(\delta_i \cdot \delta_j)_b^{-1/2}}{{}^{1}\chi} $$

where ${}^{1}\chi$ is the → *Randic connectivity index*, $(\delta_i \cdot \delta_j)_b^{-1/2}$ is the edge connectivity of the b bond, and B the bond number.

information content

The information content of a system having n elements is a measure of the degree of diversity of the elements in the set [Klir and Folger, 1988]; it is defined as:

$$ I_C = \sum_{g=1}^{G} n_g \log_2 n_g $$

where G is the number of different → *equivalence classes* and n_g is the number of elements in the gth class and

$$ n = \sum_{g=1}^{G} n_g $$

Each gth equivalence class is built by the definition of some relationships among the elements of the system. The logarithm is taken at base 2 for measuring the information content in bits.

The information content is zero, i.e. no equivalence relationships are known if all the elements are different from each other, i.e. there are $G = n$ different equivalence classes. On the contrary, the information content is maximal if all the elements of the set are recognized as belonging to the same class ($G = 1$). This quantity is called **maximal information content** ${}^{\max}I_C$ and represents the information content needed to characterize all of the n alternatives, i.e. the elements of the considered set:

$$ {}^{\max}I_C = n \log_2 n $$

The **total information content** (or **negentropy**) of a system having n elements is defined by the following:

$$ I = {}^{\max}I_C - I_C = n \log_2 n - \sum_{g=1}^{G} n_g \log_2 n_g = n \cdot H $$

The term H is *Shannon's entropy*, defined below.

The total information content represents the residual information contained in the system after G relationships are defined among the n elements.

The **mean information content** $\bar{I}$, also called **Shannon's entropy** H [Shannon and Weaver, 1949], is defined as:

$$ \bar{I} \equiv H = \frac{I}{n} = -\sum_{g=1}^{G} \frac{n_g}{n} \log_2 \frac{n_g}{n} = -\sum_{g=1}^{G} \mathrm{p}_g \log_2 \mathrm{p}_g $$

where p_g is the probability of randomly selecting an element of the gth class, and I is the total information content (Table I-2).

The maximum value of the entropy is $\log_2 n$, obtained when $n_g = 1$ for all G classes. This term, called **Hartley information**, is defined as:

$$ I_n = \log_2 n $$

where n can be interpreted as the number of alternatives regardless of whether they are realized by one selection from a set or by a sequence of selections [Hartley, 1928].

The Hartley information is based on the uncertainty associated with the choice among a certain number n of alternatives and is a simple measure of non-specificity: it represents the information content needed to characterize one of the n alternatives.

The **standardized Shannon's entropy** (or **standardized information content**) is the ratio between the actual mean information content and the maximum available information content (i.e. the Hartley information):

$$H^* = \frac{H}{I_n} = \frac{H}{\log_2 n} = \frac{I}{n \log_2 n} \qquad 0 \leq H^* \leq 1$$

The standardized Shannon's entropy is a measure of the relative efficiency of the collected information, i.e. the mean information per unit.

From the mean information content, Brillouin [Brillouin, 1962] defined a complementary quantity, called **Brillouin redundancy index** R (or **redundancy index**), to measure the information redundancy of the system:

$$R = 1 - \frac{H}{\log_2 n} = 1 - H^*$$

Both total and mean information content are widely used as molecular descriptors and are called → *information indices.*

Table I-2. Elemental contributions to the information content functions (n: number of elements; p: probability).

n	$log_2\, n$	$n\; log_2\, n$	p	$-log_2$ p	$-$ p log_2 p
0	$-\infty$	0.000	0.0	∞	0.000
1	0.000	0.000	0.1	3.322	0.332
2	1.000	2.000	0.2	2.322	0.464
3	1.585	4.755	0.3	1.737	0.521
4	2.000	8.000	0.4	1.322	0.529
5	2.322	11.610	0.5	1.000	0.500
6	2.585	15.510	0.6	0.737	0.442
7	2.807	19.651	0.7	0.515	0.360
8	3.000	24.000	0.8	0.322	0.258
9	3.170	28.529	0.9	0.152	0.137
10	3.322	33.219	1.0	0.000	0.000

Another measure of entropy is given by the **Gini index** G defined as:

$$\mathrm{G} = \sum\nolimits_{g \neq g'} \mathrm{p}_g \cdot \mathrm{p}_{g'} \qquad 0 \leq \mathrm{G} \leq \frac{n-1}{2 \cdot n}$$

where g and g' are two different equivalence classes. The Gini index increases as the diversity of the system elements increases. A complementary quantity to the Gini index is the **informational energy content** defined as [Onicescu, 1966]:

$$I_E = \sum\nolimits_g \mathrm{p}_g^2 \qquad \frac{1}{n} \leq I_E \leq 1$$

It corresponds to a redundancy measure whose maximum and minimum values are 1 and $1/n$, respectively.

Example: given a set of 7 elements (n = 7): {3, 3, 4, 4, 4, 5, 8}.

Equivalence class	n_g	*Probability*	$n_g \log_2 n_g$	$-\log_2 p_g$	$-p_g \log_2 p_g$
1	2	$p_1(3) = 2 / 7 = 0.286$	2.000	1.806	0.516
2	3	$p_2(4) = 3 / 7 = 0.429$	4.755	1.221	0.524
3	1	$p_3(5) = 1 / 7 = 0.143$	0.000	2.806	0.401
4	1	$p_4(8) = 1 / 7 = 0.143$	0.000	2.806	0.401

$$^{max}I_C = n \cdot \log_2 n = 19.651 \qquad I_C = 6.755 \qquad I = 11.896 \qquad H = 1.842$$
$$H^* = 0.656 \qquad R = 0.344 \qquad G = 0.327 \qquad I_E = 0.307$$

📖 [Shannon, 1948] [Dancoff and Quastler, 1953] [Rashevsky, 1960] [Bonchev and Kamenska, 1978] [Kier, 1980c] [Rouvray, 1997] [Agrafiotis, 1997]

information content based on centre → **centric indices**

information content ratio → **model complexity**

information energy content → **information content**

information index on amino acid composition → **amino acid side chain descriptors**

information index on isotopic composition → **atomic information indices**

information index on molecular conformations (I_{CONF})

Molecular index defined as → *total information content* based on the number of conformations of a molecule (usually below a cut-off energy value) [Bonchev, 1983]:

$$I_{CONF} = N_{CONF} \cdot \log_2 N_{CONF}$$

where N_{CONF} is the number of molecular conformations. The corresponding **mean information index on molecular conformations** is defined as:

$$\bar{I}_{CONF} = \log_2 N_{CONF}$$

For example, rigid molecules (a unique conformation) have $I_{CONF} = 0$ and $\bar{I}_{CONF} = 0$; molecules with two conformations (e.g. chair/boat) have $I_{CONF} = 1$ and $\bar{I}_{CONF} = 1$ [Dosmorov, 1982].

information index on molecular symmetry → **symmetry descriptors**

information indices on polynomial coefficients

Information indices defined as → *total information content* and → *mean information content* based on the partition of the coefficients of the → *characteristic polynomial of the graph* $\mathcal{P}(\mathcal{G}; x)$.

For acyclic molecules they coincide with the → *Hosoya total information index* and → *Hosoya mean information index*, respectively.

information index on proton-neutron composition → **atomic information indices**

information index on size → **atom number**

information indices

Molecular descriptors calculated as → *information content* of molecules. Different criteria are used for defining → *equivalence classes*, i.e. equivalency of atoms in a molecule such as chemical identity, ways of bonding through space, molecular topology and symmetry, → *local vertex invariants* [Bonchev, 1983].

Among these molecular descriptors, the most important are → *topological information indices.*

Other information indices are → *atomic composition indices*, → *information bond index*, → *Morowitz information index*, → *information index on size*, → *information index on molecular symmetry*, → *information index on amino acid composition*, → *information index on molecular conformations*, → *Bertz complexity index*, → *Dosmorov complexity index*, → *Bonchev complexity index*, → *atomic information indices*, and → *electropy index.*

📖 [Bonchev *et al.*, 1976a]

information indices on the adjacency matrix → **topological information indices**

information indices on the distance matrix → **topological information indices**

information indices on the edge adjacency matrix → **topological information indices**

information indices on the edge cycle matrix → **topological information indices**

information indices on the edge distance matrix → **topological information indices**

information indices on the vertex cycle matrix → **topological information indices**

information layer index (H_{LC})

A molecular descriptor derived from the → *cardinality layer matrix* **LC** of a molecular graph $\mathcal{G}$ defined as → *mean information content* [Konstantinova, 1996]:

$$H_{LC} = -\sum_{i=1}^{A}\sum_{k=0}^{\eta_i}\frac{n_{ik}}{A}\cdot\log_2\frac{n_{ik}}{A} = \sum_{i=1}^{A} H_{LC(i)}$$

where n_{ik} is the cardinality of the set V_{ik} constituted by all the vertices at distance k from ith vertex; η_i is the → *atom eccentricity*, and $H_{LC(i)}$ is the **vertex information layer index** of the ith vertex.

information on the possible valence bonds → **Morowitz information index**

information Wiener index : *mean information content on the distance magnitude* → **topological information indices**

integrated spatial difference in field potential → **molecular shape analysis**

interatomic distances → **molecular geometry**

interaction energy values → **grid-based QSAR techniques**

interaction fields : *molecular interaction fields*

interactive polar parameter → **lipophilicity descriptors**

interactive variable selection for PLS → **variable selection**

intermediate least squares regression → **variable selection**

intermolecular interatomic distances → **molecular geometry**

internal coordinates → **molecular geometry**

internal fragment topological indices → **fragment topological indices**

interstitial volume → **molecular surface** (⊙ solvent-accessible molecular surface)

intramolecular interatomic distances → **molecular geometry**

intricacy numbers → **Schultz molecular topological index**

intrinsic state → **electrotopological state indices**

intrinsic molecular volume : *van der Waals volume* → **molecular surface** (⊙ van der Waals molecular surface)

invariance properties of molecular descriptors → **molecular descriptors**

inverse QSAR : *reversible decoding* → **structure/response correlations**

ionization energy : *ionization potential* → **quantum-chemical descriptors**

ionization potential → **quantum-chemical descriptors**

isocodal graphs → **graph**

isolated vertices → **graph**

isomorphic graphs → **graph**

isospectral graphs → **graph**

isotropic surface area → **molecular surface** (⊙ solvent-accessible molecular surface)

Iterated Line Graph Sequence (*ILGS*)

An ordered sequence of line graphs $L(\mathcal{G})$ obtained by an iterative procedure starting from the → *molecular graph* $\mathcal{G}$ [Diudea *et al.*, 1992a]:

$$\langle L_0, L_1, L_2, \ldots, L_m \rangle$$

where L_0 is the → *line graph* of zero-order coinciding with the original molecular graph $\mathcal{G}$; L_1 is the line graph of $\mathcal{G}$; L_2 is the line graph of the first line graph L_1; L_m is the mth line graph of $\mathcal{G}$, i.e. $L_m(\mathcal{G}) = L(L_{m-1}(\mathcal{G}))$.

The numbers of vertices A_m and edges B_m in the graph L_m are given by the following relations:

$$A_m = B_{m-1}$$

$$B_m = \sum_{i=1}^{A_{m-1}} \binom{\delta_i}{2} = \frac{1}{2} \cdot \sum_{i=1}^{A_{m-1}} \delta_i^2 - B_{m-1}$$

where A_{m-1} and B_{m-1} represent the number of vertices and edges in the line graph of $(m-1)$th order, respectively; δ_i is the → *vertex degree*; it may be noted that the number of edges in the mth line graph L_m coincides with the → *connection number* of the $(m-1)$th line graph L_{m-1}.

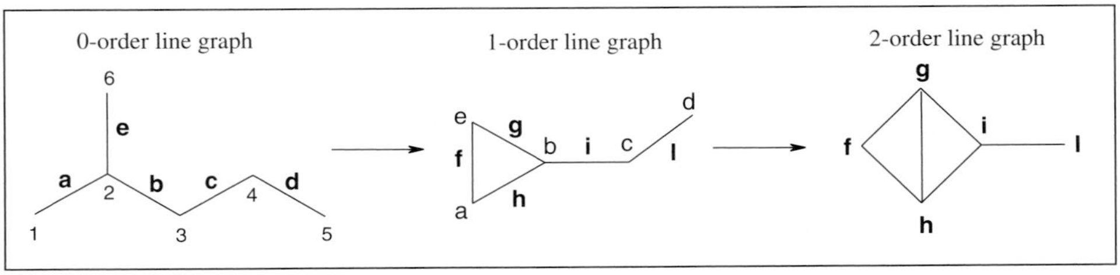

Figure I-1. First and second line graphs of 2-methylpentane.

Each vertex i_m of the current line graph L_m denotes a pair of vertices of the lower-order graph L_{m-1}:

$$i_m = (j_{m-1},\ k_{m-1})$$

where the two vertices j and k in L_{m-1} are necessarily connected by an edge of the graph and are themselves pairs of vertices of graph L_{m-2}. This relation between the ith edge and the corresponding vertices j and k can be represented as:

$$j_{m-1} \in i_m \quad \text{and} \quad k_{m-1} \in i_m$$

The relatedness of vertices in the line derivative process can be represented by the Kronecker delta as:

$$\delta(i_m,\ i_{m+1}) = \begin{cases} 1 & \text{if } (i_m \in i_{m+1}) \\ 0 & \text{otherwise} \end{cases}$$

This relation can be extended to any arbitrary rank of derivative m and n ($m \geq n$), stating that $\delta(i_m, i_n) = 1$ only if the vertex i_m appears in at least one of the subsets defining the vertex i_n.

Based on iterated line graph sequence, local and global centricities of molecular graphs were derived on the basis of the → *branching layer matrix* **LB** by the MOLCEN algorithm [Diudea *et al.*, 1992a]. This algorithm was proposed to obtain → *canonical numbering* of subgraphs of various length in molecular graphs.

MOLORD algorithm, proposed by Diudea [Diudea *et al.*, 1995a], calculates local and global graph invariants from a series of line graphs $L_0, L_1, ..., L_m$ derived from a molecular graph $\mathcal{G}$. Let ${}^mI_i$ be any local invariant of the vertex i and ${}^mGI$ the corresponding global invariant on each L_m within the set of derivative graphs $L_0, L_1, ..., L_n$, i.e.

$${}^mGI = \sum_{i=1}^{A_m} {}^mI_i$$

A **partial local invariant** ${}^mPI(i_n)$ of a vertex i_n with respect to the mth order line graph L_m is defined as:

$${}^mPI(i_n) = \frac{{}^nGI}{{}^mGI} \cdot \sum_{i=1}^{A_m} {}^mI_i \cdot \delta(i_n, i)$$

where ${}^mI_i$ is the local vertex invariant of the vertex i_m calculated from the topology of the graph L_m. The partial invariant of the vertex i_n of the graph L_n with respect to L_m is calculated by summing all local invariants ${}^mI_i$ of those vertices in L_m which are "related" to i_n according to the $m - n$ successive derivatives, L_n. ..., L_m. The ratio ${}^nGI\ /\ {}^mGI$ is used as a scaling factor allowing a comparison of PI values irrespective of the current L_m for which they are evaluated.

For a series of successive derivative graphs L_n, ..., L_m, a **local synthetic invariant** ${}^mSI(i_n)$ of the vertex i_n in the nth order line graph is calculated as:

$${}^mSI(i_n) = \sum_{k=n}^{m} {}^kPI(i_n) \cdot f^{n-k}$$

where the superscript m in ${}^mSI(i_n)$ means that the last line graph L_m has been taken into account; f is an empirical factor used to give a different weighting to the contributions arising from derivatives of various ranks. It must be observed that in the case of $n = m$, the synthetic local invariant ${}^mSI(i_n)$ reduces to the classical invariant ${}^nI_i$.

Finally, a **global synthetic invariant** ${}^mGSI(L_n)$ of a graph L_n is defined as:

$${}^mGSI(L_n) = \sum_{i_n=1}^{A_n} {}^mSI(i_n)$$

Studies of these descriptors were performed using, as local invariants, the → *centric operator* c_i and the → *centrocomplexity operator* x_i calculated on → *layer matrices.*

A vector of molecular topological descriptors can be calculated for the whole iterated line graph sequence. For example, a **line graph Randic connectivity index** was calculated as:

$$\chi(\mathrm{L}_n) = \sum\nolimits_b \left[\delta_i(\mathrm{L_n}) \cdot \delta_j(\mathrm{L}_n)\right]_b^{-1/2}$$

where the sum runs over all edges in the nth order line graph and δ_i and δ_j are → *the vertex degree* of the two vertices incident to the b bond [Estrada *et al.*, 1998b]. Note that the χ index of the first line graph L_1 coincides with the → *edge connectivity index* ε. Moreover, the **line graph connectivity indices** were proposed in analogy with the Kier-Hall connectivity indices as:

$${}^m\chi_q(\mathrm{L}_n) = \sum_{k=1}^{K} \left(\prod\nolimits_a \delta_a(\mathrm{L}_n)\right)_k^{-1/2}$$

where k runs over all of the mth order subgraphs, m being the number of edges in the subgraph; K is the total number of mth order subgraphs; the product is over all the vertex degrees of the vertices involved in the subgraph. The subscript "q" for the connectivity indices refers to the type of → *molecular subgraph* and is ch for chain or ring, pc for path-cluster, c for cluster, and p for path (that can also be omitted) [Estrada *et al.*, 1998a]. It was shown that line graph connectivity indices are linear combinations of → *extended edge connectivity indices* for some molecular graphs.

Spectral moments of iterated line graph sequence were derived from the → *adjacency matrix* **A** of the line graph L_n of any order n as:

$$\mu_k(\mathrm{L}_n) = tr\left[\mathbf{A^k}(\mathrm{L}_n)\right]$$

where μ_k is the kth order spectral moment and tr is the → *trace* of the kth power of the adjacency matrix $\mathbf{A^k}$ of the considered line graph L_n [Estrada, 1999]. They were also expressed as linear combinations of some → *embedding frequencies* of the molecular graph, i.e. number of occurrences of specified subgraphs in the original molecular graph $\mathcal{G}$.

Obviously, spectral moments μ_k of the 1st order line graph L_1 are the → *spectral moments of the edge adjacency matrix*, and 0- order spectral moments μ_0 of any line graph L_n coincide with the number of vertices in the considered line graph.

📖 [Gutman and Estrada, 1996]

iterative vertex and edge centricity algorithm → **centre of a graph**

Ivanciuc-Balaban operator → **algebraic operators**

J

Jenkins steric parameter → **steric descriptors**

J_t index → **Balaban distance connectivity indices**

***J'/J* index** → **bond order indices** (⊙ graphical bond order)

JJ indices → **Wiener matrix**

Jochum-Gasteiger canonical numbering → **canonical numbering**

Joshi electronic descriptors

Molecular → *electronic descriptors* assuming that the minimum energy conformation of a molecule represents the optimal picture of the electronic charge distribution in the whole molecule [Joshi *et al.*, 1993; Joshi *et al.*, 1994].

The Joshi electronic descriptors JSn (JS1 – JS5) are defined as:

$$\mathrm{JS1} = \frac{\mathrm{E_R}}{\mathrm{E_H}} \qquad \mathrm{JS2} = \frac{\mathrm{E_R} - \mathrm{E_H}}{\mathrm{E_H}} \qquad \mathrm{JS3} = \frac{\mathrm{E_R} - \mathrm{E_{HS}}}{\mathrm{E_H}}$$

$$\mathrm{JS4} = \frac{\mathrm{E_R} - \sum_j \mathrm{E}_{\mathrm{R}j}}{\mathrm{E_H}} \qquad \mathrm{JS5} = \frac{\mathrm{E_R} - \sum_j \mathrm{E}_{\mathrm{R}j} - \mathrm{E_{HS}}}{\mathrm{E_H}}$$

where E is the ΔH_f conformational energy value of the global minimim energy conformer calculated by → *computational chemistry* (AM1 method). The subscripts R, H, and HS refer to an R-substituted compound, the unsubstituted compound, and a compound where the aromatic moiety is unsubstituted but the side chain is substituted in a similar way. $\mathrm{E}_{\mathrm{R}j}$ is the energy contribution due to formation of the *j*th substituent group calculated by subtracting the energy value of methane from that of the correspondingly substituted methane. The summation in JS4 and JS5 depends on the series of studied compounds.

Joshi steric descriptor → **steric descriptors**

Jurs shape indices : *shadow indices*

K

Kaliszan shape parameter → **shape descriptors**

Kantola – Villar – Loew hydrophobic models → **lipophilicity descriptors**

***K* correlation analysis** → **variable reduction**

***K* correlation index** : *multivariate K correlation index*

Kekulé number (K) (: *Kekulé structure count, SC)*

The number of Kekulé structures in an aromatic system [Trinajstic, 1992]. It can be calculated by extensive enumeration of the structures or by using appropriate algorithms.

For benzenoid systems, the Kekulé number K is obtained from the positive eigenvalues λ_i of the → *adjacency matrix* as:

$$\mathrm{K} = \prod_{i=1}^{A/2} \lambda_i$$

where A is the number of atoms. The logarithm of the Kekulé number is related to the resonance energy of the compound and used among the → *resonance indices.*

The Kekulé number of alternant hydrocarbons is equal to the sum of "even" K^+ and "odd" K^- Kekulé structures:

$$\mathrm{K} = \mathrm{K}^+ + \mathrm{K}^-$$

The even or odd parity is determined in the Dewar-Longuet-Higgins scheme [Dewar and Longuet-Higgins, 1952] according to whether the number of transpositions of double bonds required to transform one Kekulé structure into another one is even or odd.

The difference between the even and odd Kekulé structures is called **algebraic structure count** ASC [Wilcox Jr, 1968; Wilcox Jr, 1969] or **corrected structure count** CSC [Herndon, 1973a; Herndon, 1974b]:

$$\mathrm{ASC} = \mathrm{K}^+ - \mathrm{K}^- \qquad \mathrm{K}^+ \geq \mathrm{K}^-$$

ASC represents a structure count exclusive of structures which do not contribute to stabilizing resonance interactions. However, the concept of parity and the derived ASC descriptor do not work in non-alternant systems with three odd-membered rings [Randic and Trinajstic, 1993a].

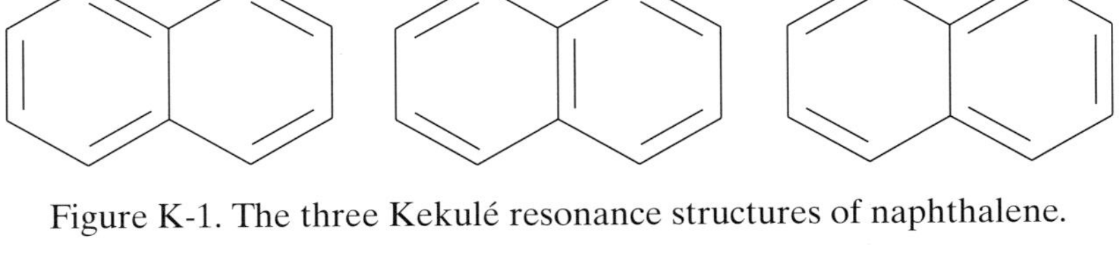

Figure K-1. The three Kekulé resonance structures of naphthalene.

📖 [Kekulé, 1865] [Herndon, 1973b] [Gutman and Trinajstic, 1973b] [Balaban and Tomescu, 1985] [Dias, 1992] [Ivanciuc and Balaban, 1992b] [Balaban *et al.*, 1993b] [Guo and Zhang, 1993] [Guo *et al.*, 1996] [Mishra and Patra, 1998] [Cash, 1998]

Kekulé structure count : *Kekulé number*

Kellog and Abraham interaction field → **molecular interaction fields**
(⊙ hydrophobic fields)

Kier alpha-modified shape descriptors → **Kier shape descriptors**

Kier bond rigidity index → **flexibility indices**

Kier-Hall connectivity indices : *connectivity indices of mth order* → **connectivity indices**

Kier-Hall connectivity matrix → **weighted matrices**

Kier-Hall electronegativity → **vertex degree**

Kier-Hall solvent polarity index → **electric polarization descriptors**

Kier molecular flexibility index → **flexibility indices**

Kier steric descriptor → **steric descriptors**

Kier shape descriptors (κ)

Topological shape descriptors $^{m}\kappa$ defined in terms of the number of graph vertices A and the number of paths ^{m}P with length m (m = 1,2,3) in the → *H-depleted molecular graph*, according to the following:

$$^{1}\kappa = 2 \cdot \frac{^{1}P_{\max} \cdot {}^{1}P_{\min}}{(^{1}P)^2} = \frac{A(A-1)^2}{(^{1}P)^2} \qquad ^{2}\kappa = 2 \cdot \frac{^{2}P_{\max} \cdot {}^{2}P_{\min}}{(^{2}P)^2} = \frac{(A-1)(A-2)^2}{(^{2}P)^2}$$

$$^{3}\kappa = 4 \cdot \frac{^{3}P_{\max} \cdot {}^{3}P_{\min}}{(^{3}P)^2} = \begin{cases} \dfrac{(A-3)(A-2)^2}{(^{3}P)^2} & \text{for even } A \ (A > 3) \\[2ex] \dfrac{(A-1)(A-3)^2}{(^{3}P)^2} & \text{for odd } A \ \ (A > 3) \end{cases}$$

where $^{m}P_{\min}$ and $^{m}P_{\max}$ are the minimum and maximum mth order → *path count* in the molecular graphs of molecules with the same → *atom number A* [Kier, 1985; Kier, 1986b]. These extremes are obtained from two reference structures chosen in an isomeric series and, for the ith molecule, is therefore:

$$^{m}P_{\min} \leq {}^{m}P_i \leq {}^{m}P_{\max}$$

The reference structure for $^{1}P_{\min}$ is the → *linear graph* while for $^{1}P_{\max}$ it is the → *complete graph* in which all atoms are bonded to each other; their numerical values are calculated as follows:

$$^{1}P_{\min} = A - 1 \qquad ^{1}P_{\max} = \frac{A(A-1)}{2}$$

The scaling factor of 2 in the numerator of $^{1}\kappa$ index formula makes the value $^{1}\kappa = A$ when there are no cycles in the graph of the molecule. Monocyclic molecules have a lower value and bicyclic structures have an even lower value. The structural information encoded in $^{1}\kappa$ is related to the complexity, or more precisely, the number of cycles of a molecule.

The reference structure for $^{2}P_{\min}$ is the linear graph, while for $^{2}P_{\max}$ it is the → *star graph,* in which all atoms but one are adjacent to a central atom; their numerical values are calculated as follows;

$$^{2}P_{\min} = A - 2 \qquad ^{2}P_{\max} = \frac{(A-1)(A-2)}{2}$$

where A is the total number of vertices in the graph. The scaling factor of 2 in the numerator of $^2\kappa$ index formula makes the value $^2\kappa = A - 1$ for all linear graphs. The information encoded by $^2\kappa$ index is related to the degree of star graph-likeness and linear graph-likeness, i. e. $^2\kappa$ encodes information about the spatial density of atoms in a molecule.

The reference structure for $^3P_{\min}$ is the linear graph while for $^3P_{\max}$ it is the *twin star graph*; their numerical values are calculated as follows:

$$^3P_{\min} = A - 3 \qquad ^3P_{\max} = \begin{cases} \dfrac{(A-2)^2}{4} & \text{for even } A \\ \dfrac{(A-1)(A-3)}{4} & \text{for odd } A \end{cases}$$

The scaling factor of 4 is used in the numerator of $^3\kappa$ index to bring $^3\kappa$ onto approximately the same numerical scale as the other kappa indices. The $^3\kappa$ values are larger when → *molecular branching* is nonexistent or when it is located at the extremities of a graph; $^3\kappa$ encodes information about the centrality of branching.

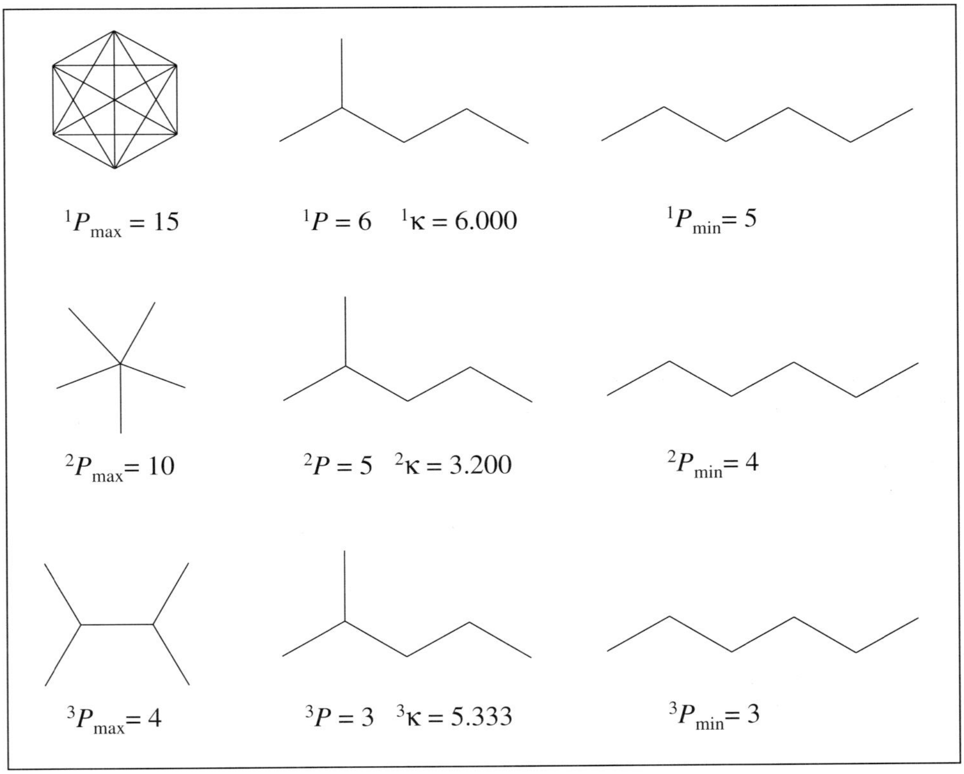

Box K-1.

To take into account the different shape contribution of heteroatoms and hybridization states, **Kier alpha-modified shape descriptors** $^m\kappa_\alpha$ (m = 1,2,3) were proposed [Kier, 1986a] by the following:

$$^{1}\kappa_{\alpha} = \frac{(A+\alpha)(A+\alpha-1)^2}{(^{1}P+\alpha)^2} \qquad ^{2}\kappa_{\alpha} = \frac{(A+\alpha-1)(A+\alpha-2)^2}{(^{2}P+\alpha)^2}$$

$$^{3}\kappa_{\alpha} = \begin{cases} \dfrac{(A+\alpha-3)(A+\alpha-2)^2}{(^{3}P+\alpha)^2} & \text{for even } A \ (A>3) \\ \dfrac{(A+\alpha-1)(A+\alpha-3)^2}{(^{3}P+\alpha)^2} & \text{for odd } A \ (A>3) \end{cases}$$

where α is a parameter derived from the ratio of the → *covalent radius* R_i of the ith atom relative to the sp^3 carbon atom (R_{Csp3}):

$$\alpha = \sum_{i=1}^{A} \left(\frac{R_i}{R_{Csp^3}} - 1 \right)$$

The only non-zero contributions to α are given by heteroatoms or carbon atoms with a valence state different from sp^3 (Table K-1).

Table K-1. Covalent radius R and α parameter values.

Atom / Hybrid	R (Å)	α	*Atom / Hybrid*	R (Å)	α
C_{sp3}	0.77	0	P_{sp3}	1.10	0.43
C_{sp2}	0.67	–0.13	P_{sp2}	1.00	0.30
C_{sp}	0.60	–0.22	S_{sp3}	1.04	0.35
N_{sp3}	0.74	–0.04	S_{sp2}	0.94	0.22
N_{sp2}	0.62	–0.20	F	0.72	–0.07
N_{sp}	0.55	–0.29	Cl	0.99	0.29
O_{sp3}	0.74	–0.04	Br	1.14	0.48
O_{sp2}	0.62	–0.20	I	1.33	0.73

Kappa indices can also be calculated for molecular fragments and functional groups X. The calculation of these indices for groups was performed using a "pseudo-molecule" X–X: two fragments X of the same kind are linked together, kappa values are calculated for the pseudo-molecule and this is then divided by two.

In order to quantify the shape of the whole molecule, Kier proposed a linear combination of the above defined κ indices, each representing a particular shape attribute of the molecule:

$$shape = b_0 \cdot {}^{0}\kappa + b_1 \cdot {}^{1}\kappa + b_2 \cdot {}^{2}\kappa + b_3 \cdot {}^{3}\kappa$$

where $^{0}\kappa$ is the → *Kier symmetry index* used to encode the shape contributions due to symmetry.

Specific combinations of κ indices were also proposed as indices of molecular flexibility (→ *Kier molecular flexibility index*) and steric effects (→ *Kier steric descriptor*).

📖 [Kier, 1986c] [Kier, 1987a] [Kier, 1987b] [Kier, 1987c] [Gombar and Jain, 1987a] [Mokrosz, 1989] [Kier, 1990] [Hall and Kier, 1991] [Skvortsova *et al.*, 1993] [Kier, 1997] [Hall and Vaughn, 1997b]

Kier symmetry index → **symmetry descriptors**

K_Z index → **Hosoya Z matrix**

K inflation factor → **variable reduction** (⊙ *K correlation analysis)*

Kirchhoff matrix : *Laplacian matrix*

Kirchhoff number → **resistance matrix**

Klopman-Henderson cumulative substructure count

A QSAR graph theory-based method involving a heuristic processing of H-depleted molecular graphs represented by common substructure counts [Klopman and Henderson, 1991]. The method consists in extracting substructure descriptors from the → *data set* and then searching for their significance in correlating the biological activity.

From the frequencies of the row entries of the → *distance matrix* of each molecule, the → *vertex distance code* is defined as:

$$\langle {}^1f_i, {}^2f_i, {}^3f_i, \ldots, {}^{\eta_i}f_i \rangle$$

where ${}^1f_i$, ${}^2f_i$, ${}^3f_i$, ... are the → *vertex distance counts* indicating the frequencies of distances equal to 1, 2, 3, ..., respectively, from vertex v_i to any other vertex and η_i is the ith → *atom eccentricity.*

A cumulative path matrix of dimension $p \times D$ is calculated, where each row represents one among the p derived fragment descriptors relative to the considered molecule, D being the → *topological diameter* of the molecule. Each fragment descriptor is represented by a vertex distance code, called a **structural environment vector (SEV)**, which identifies a particular atom-centred fragment that can occur more than once in the molecule. In other words, the different vertex distance codes in the molecule are collected in the cumulative path matrix, each identifying a particular atom-centred fragment. Then, all the different structural environment vectors in the data set are selected as fragment descriptors related to the studied biological response through a function of likeliness instead of the classical multivariate regression analysis.

Therefore, the potential influence for the biological activity of **SEV** is evaluated by using a **likeliness function** defined as:

$$L_j(k) = 1 - 4 \cdot \frac{\left(\mathrm{N}_j^{\mathrm{AC}} \cdot k\right) \cdot \left(\mathrm{N}_j^{\mathrm{IN}} \cdot k\right)}{\left(\mathrm{N}_j^{\mathrm{AC}} + \mathrm{N}_j^{\mathrm{IN}} + 2 \cdot k\right)^2}$$

where $\mathrm{N}_j^{\mathrm{AC}}$ and $\mathrm{N}_j^{\mathrm{IN}}$ are the number of active and inactive compounds, containing the jth fragment represented by the jth SEV, respectively, in the data set; k is an adjustable parameter whose optimal value was determined to be in the 0.01 to 10 range. The larger the $L(k)$ value of a descriptor, the stronger its association with biological activity.

To establish the statistical significance of the likeliness function of each jth SEV, the binomial probability p_j is calculated as:

$$\mathrm{p}_j\left(m > \mathrm{N}_j^{\mathrm{AC}}\right) = \sum_{m=\mathrm{N}_j^{\mathrm{AC}}}^{\mathrm{N}_j} \frac{\mathrm{N}_j!}{m! \cdot (\mathrm{N}_j - m)!} \cdot \mathrm{p}^m \cdot \mathrm{q}^{\mathrm{N}_j - m}$$

where $\mathrm{N}_j^{\mathrm{AC}}$ is the number of active compounds in which the jth fragment occurs, N_j is the total number of compounds containing the jth fragment; the probabilities p and q are defined as:

$$\mathrm{p} = \frac{\mathrm{N_{AC}}}{n} \qquad \mathrm{q} = 1 - \mathrm{p}$$

Once the significant **SEV**s most related to the biological activity have been found, → *reversible decoding* can be easily peformed.

Example : 2-methylpentane

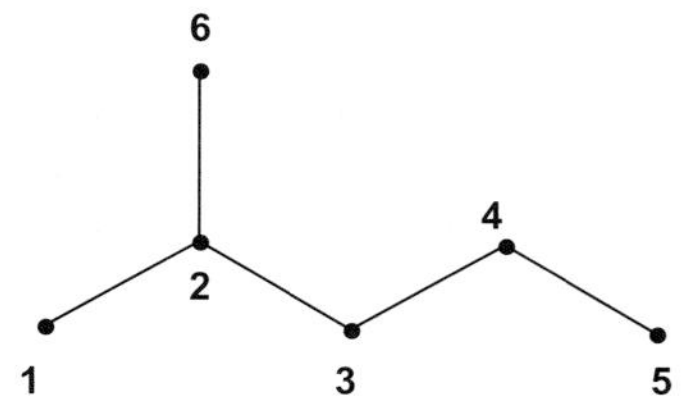

Atom	^1f	^2f	^3f	^4f
1	1	2	1	1
2	3	1	1	0
3	2	3	0	0
4	2	1	2	0
5	1	1	1	2
6	1	2	1	1

2-methylpentane SEVs

Fragment	^1f	^2f	^3f	^4f
1	1	2	1	1
2	3	1	1	0
3	2	3	0	0
4	2	1	2	0
5	1	1	1	2

Box K-2.

Klopman hydrophobic atomic contants → **lipophilicity descriptors** (⊙ Klopman hydrophobic models)

Klopman hydrophobic models → **lipophilicity descriptors**

Klopman LOG*P* → **lipophilicity descriptors** (⊙ Klopman hydrophobic models)

***k*-matching** → **graph**

KOKOS descriptors → **principal component analysis**

Koppel-Paju *B* scale → **linear solvation energy relationships** (⊙ hydrogen-bond parameters)

Kovats retention index → **chromatographic descriptors**

Kuhn length → **size descriptors**

Kupchik modified connectivity indices → **connectivity indices**

L

Laplacian matrix (L) *(: admittance matrix, Kirchhoff matrix)*

A square $A \times A$ symmetric matrix, A being the number of vertices in the → *molecular graph*, obtained as the difference between the **vertex degree matrix V** and the → *adjacency matrix* **A** [Mohar, 1989a; Mohar, 1989b]:

$$\mathbf{L} = \mathbf{V} - \mathbf{A}$$

where **V** is a diagonal matrix of dimension $A \times A$ whose diagonal entries are the vertex degrees of molecule atoms:

$$[\mathbf{V}]_{ij} = \begin{cases} \delta_i & \text{if } i = j \\ 0 & \text{if } i \neq j \end{cases}$$

δ_i being the → *vertex degree* of the ith atom. Therefore, the entries of the Laplacian matrix are:

$$[\mathbf{L}]_{ij} = \begin{cases} \delta_i & \text{if } i = j \\ -1 & \text{if } \{i,j\} \in \mathcal{E}(\mathcal{G}) \\ 0 & \text{if } \{i,j\} \notin \mathcal{E}(\mathcal{G}) \end{cases}$$

where $\mathcal{E}(\mathcal{G})$ is the set of edges of the molecular graph $\mathcal{G}$.

The diagonalization of the Laplacian matrix gives A real eigenvalues λ_i which constitute the **Laplacian spectrum** [Mohar, 1991b; Trinajstic *et al.*, 1994] and are conventionally labelled so that

$$\lambda_1 \geq \lambda_2 \geq \ldots \geq \lambda_A$$

Among the several properties of the Laplacian eigenvalues, three important ones are:

(a) the Laplacian eigenvalues are non-negative numbers;

(b) the last eigenvalue λ_A is always equal to zero;

(c) the eigenvalue λ_{A-1} is greater than zero if, and only if, the graph $\mathcal{G}$ is connected; therefore, for a molecular graph all the Laplacian eigenvalues except the last are positive numbers.

Moreover, the sum of the positive eigenvalues is equal to twice the → *bond number* B, i.e.

$$\sum_{i=1}^{A-1} \lambda_i \equiv tr(\mathbf{L}) = 2 \cdot B$$

The sum of the reciprocal $A - 1$ positive eigenvalues was proposed as a molecular descriptor [Mohar *et al.*, 1993; Gutman *et al.*, 1993b] and called the **quasi-Wiener index** W^* [Markovic *et al.*, 1995]; it is defined as:

$$W^* = A \cdot \sum_{i=1}^{A-1} \frac{1}{\lambda_i}$$

For acyclic graphs, the quasi-Wiener index W^* coincides with the → *Wiener index* W, i.e. $W^* = W$, while for cycle-containing graphs the two descriptors differ. Moreover, it has been demonstrated that the quasi-Wiener index coincides with the → *Kirchhoff number* for any graph [Gutman and Mohar, 1996].

The product of the positive $A - 1$ eigenvalues of the Laplacian matrix gives the **spanning tree number** T^* of the molecular graph $\mathcal{G}$ as:

$$T^* = \frac{1}{A} \cdot \prod_{i=1}^{A-1} \lambda_i = \frac{|a|}{A}$$

where the → *spanning tree* is a connected acyclic subgraph containing all the vertices of $\mathcal{G}$ [Trinajstic *et al.*, 1994]. The term a in the second equality is the coefficient of the linear term in the → *characteristic polynomial* of the Laplacian matrix [Nikolic *et al.*, 1996b]. The number of spanning trees of a graph is used as a measure of → *molecular complexity* for polycyclic graphs; it increases with the complexity of the molecular structure. It should be noted that some algorithmic methods have been proposed to calculate the number of spanning trees in molecular graphs of cata-condensed systems [John *et al.*, 1998].

Example : 2-methylpentane

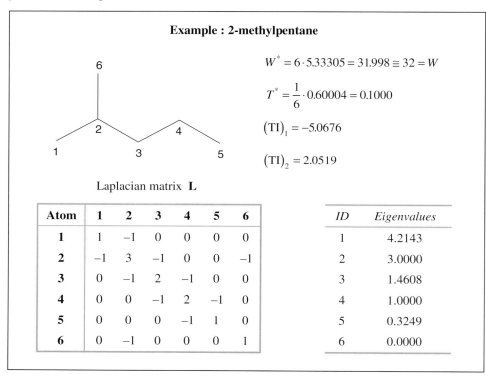

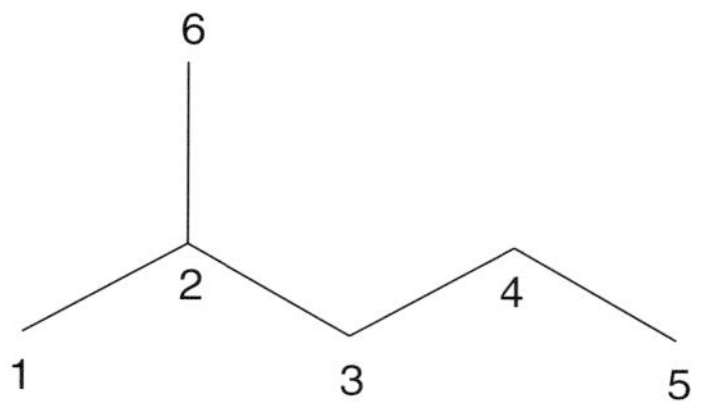

$$W^* = 6 \cdot 5.33305 = 31.998 \cong 32 = W$$

$$T^* = \frac{1}{6} \cdot 0.60004 = 0.1000$$

$$(\mathrm{TI})_1 = -5.0676$$

$$(\mathrm{TI})_2 = 2.0519$$

Laplacian matrix **L**

Atom	**1**	**2**	**3**	**4**	**5**	**6**
1	1	−1	0	0	0	0
2	−1	3	−1	0	0	−1
3	0	−1	2	−1	0	0
4	0	0	−1	2	−1	0
5	0	0	0	−1	1	0
6	0	−1	0	0	0	1

ID	*Eigenvalues*
1	4.2143
2	3.0000
3	1.4608
4	1.0000
5	0.3249
6	0.0000

Box L-1

Also derived from the Laplacian matrix are the **Mohar indices** $(\mathrm{TI})_1$ and $(\mathrm{TI})_2$, defined as:

$$(\mathrm{TI})_1 = 2 \cdot A \cdot \log\left(\frac{B}{A}\right) \cdot \sum_{i=1}^{A-1} \frac{1}{\lambda_i} = 2 \cdot \log\left(\frac{B}{A}\right) \cdot W^* \qquad (\mathrm{TI})_2 = \frac{4}{A \cdot \lambda_{A-1}}$$

where λ_{A-1} is the first non-zero eigenvalue and W^* the quasi-Wiener index [Trinajstic *et al.*, 1994]. Being $W^* = W$ for acyclic graphs, the first Mohar index $(\mathrm{TI})_1$ is closely related to the Wiener index for acyclic graphs.

📖 [Ivanciuc, 1993] [Gutman *et al.*, 1994a] [Nikolic *et al.*, 1996b] [Chan *et al.*, 1997]

Laplacian spectrum → **Laplacian matrix**

lateral validation → **validation techniques**

lattice representation : *stereoelectronic representation* → **molecular descriptors**

layer matrices (LM) (: *shell matrices*)

A layer matrix **LM** of a → *molecular graph* $\mathcal{G}$ is a rectangular unsymmetric matrix $A \times (D+1)$, A being the number of molecule atoms and D the → *topological diameter*. The entry *i-k* (lm_{ik}) is the sum of the weights of the vertices located in the concentric shell (layer) at → *topological distance* k around the vertex v_i [Diudea *et al.*, 1991; Diudea, 1994; Skorobogatov and Dobrynin, 1988]. The kth layer of the vertex v_i is the set $\mathcal{V}_{ik}(\mathcal{G})$ of vertices defined as follows:

$$\mathcal{V}_{ik}(\mathcal{G}) = \{a | a \in \mathcal{V}(\mathcal{G}); d_{ia} = k\}$$

where d_{ia} is the topological distance of the ath vertex from v_i.

The entries of the layer matrix are:

$$lm_{ik} = \sum\nolimits_{a \in \mathcal{V}_{ik}} w_a$$

where w_a represents the considered weight (i.e. atomic property) of the atom a.

The columns of the layer matrix are $D + 1$, D being the topological diameter, and the case $k = 0$ is also considered, meaning that the property of the focused ith atom is also considered.

The weights w_a can be any chemical or topological atomic properties. Examples of chemical → *atomic properties* are → *van der Waals volume*, atomic mass, → *polarizability*; examples of → *local vertex invariants* are → *vertex degree*, → *path degree*, → *walk degree*.

Example : 2-methylpentane

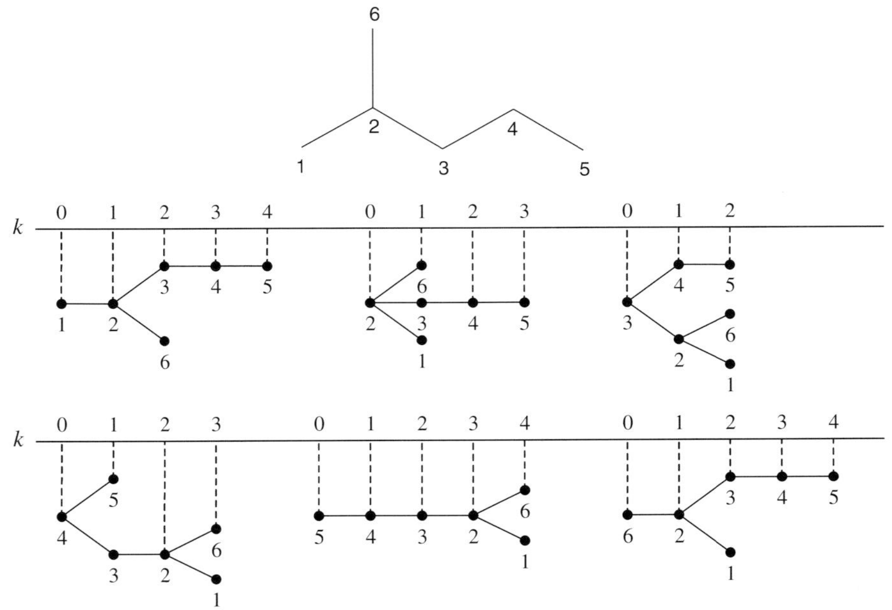

Box L-2.

Based on the different atomic properties, several layer matrices can be obtained; the most common are defined below.

- **cardinality layer matrix (LC)**

The simplest layer matrix obtained weighting all atoms by a weight equal to one. Therefore, the entry *i-k* of the matrix is the number lc_{ik} of atoms located at distance k from the focused ith vertex. For the cardinality layer matrix the following relations hold:

$$\sum_{k=0}^{D} lc_{ik} = \sum_{i=1}^{A} lc_{i0} = A \quad \text{and} \quad \sum_{i=1}^{A} lc_{i1} = 2 \cdot B$$

where A is the number of graph vertices and B the number of graph edges.

The cardinality layer matrix was originally called the **λ matrix** [Skorobogatov and Dobrynin, 1988] and **F matrix** [Diudea and Pârv, 1988].

By this matrix, the → *information layer index* is calculated. Moreover, four different local vertex invariants derived from the cardinality layer matrix have been proposed [Wang *et al.*, 1994] as follows:

$$\gamma_i = \log\left(\sum_{k=1}^{D} lc_{ik} \cdot 2^{-k}\right) \qquad \gamma_i = \log\left(\sum_{k=1}^{D} lc_{ik} \cdot 4^{-k}\right)$$

$$\gamma_i = \log\left(\sum_{k=1}^{D} lc_{ik}/(k+1)\right) \qquad \gamma_i = \log\left(\sum_{k=1}^{D} lc_{ik}/(k+1)^{3/2}\right)$$

where D is the maximum topological distance in the graph.

- **branching layer matrix (LB)**

A layer matrix obtained weighting all the atoms by their → *vertex degrees* δ [Diudea *et al.*, 1991]. Therefore, the entry *i-k* of the matrix is the sum of the vertex degrees over all vertices in the kth layer around the focused ith vertex.

Each ith row of the **LB** matrix expresses the global state of vertex degrees from the viewpoint of vertex i, i.e. the distribution of sums of vertex degrees in shells around the ith vertex. The sum of each ith row element is a constant equal to $2B$, twice the number of bonds. Moreover, more branched and → *central vertices* show higher values in the first layers within the corresponding rows, while less branched and → *terminal vertices* have higher values in the far layers. The vertex degrees of the atoms are in the first column ($k = 0$).

- **connectivity valence layer matrix (LCV)**

Similar to the branching layer matrix, this matrix is obtained weighting the atoms by their → *valence vertex degree* δ^v instead of the simple vertex degree δ [Hu and Xu, 1996].

- **edge layer matrix (LE)**

Analogously to the branching layer matrix, the edge layer matrix **LE** is obtained weighting all vertices by the number of distinct edges incident to the vertices of the kth layer around the vertex v_i, without counting any edge already counted in a preceding layer [Diudea *et al.*, 1991]. The sum of the elements of each ith row is a constant equal to the number of bonds B, while the sum of the elements of each kth column is a constant equal to $2B$. The vertex degrees of the atoms are in the first column ($k = 0$).

- **connectivity bond layer matrix (LCB)**

A layer matrix whose entry *i-k* is defined as the sum of the → *conventional bond order* π^* of the bonds connecting the vertices situated in the kth layer with the vertices of the $(k-1)$th layer with respect to the focused ith vertex [Hu and Xu, 1996].

• **sum layer matrix (LS)**
To increase the discriminating power of atomic and molecular descriptors derived from the layer matrices the sum layer matrix **LS** was also defined, and its entries are the sums of the corresponding entries of the branching layer matrix **LB** (lb_{ik}) and edge layer matrix **LE** (le_{ik}) [Diudea *et al.*, 1991]:

$$ls_{ik} = lb_{ik} + le_{ik}$$

The sum of the elements of each ith row is a constant equal to $3B$, B being the number of graph edges.

• **distance sum layer matrix (LDS)**
A layer matrix obtained weighting the atoms by their → *vertex distance degree* σ, i.e. the row sum of the → *distance matrix* **D** [Balaban and Diudea, 1993]. Therefore, the entry i-k of the layer matrix is the sum of the vertex distance degrees of the atoms located at distance k from the focused ith vertex. It is obvious that the entries of the first column ($k = 0$) are only vertex distance degrees. Moreover, the sums over each row in **LDS** are all equal to twice the → *Wiener index*, i.e. the following relation holds:

$$\sum_{k=0}^{D} lds_{ik} = \sum_{i=1}^{A} lds_{i0} = 2 \cdot W$$

where A is the number of graph vertices and W the Wiener index.

• **geometric sum layer matrix (LGS)**
A layer matrix defined in analogy with the distance sum layer matrix deriving the atom weights from the → *geometry matrix* G instead of the distance matrix **D** [Diudea *et al.*, 1995b]. Therefore, the entry i-k of the layer matrix is the sum of the → *geometric distance degree* $^{G}\sigma$ of the atoms located at distance k from the focused ith vertex, where the geometric distances are obtained from → *computational chemistry.* The sums over each row in **LGS** as well as the zero-column sum are all equal to twice the → *3D-Wiener index*.

• **path degree layer matrix (LPD)**
A layer matrix obtained weighting all atoms by the → *path degree* ξ. Therefore, the entry i-k of the matrix is the sum of the path degrees of all atoms located at distance k from the focused ith vertex. The half-sum of the elements in the first column ($k = 0$) is equal to the half-sum of the elements in each row of **LPD** matrix and corresponds to a molecular descriptor recently reproposed as the → *all-path Wiener index* W^{AP}, i.e. the following relations hold:

$$\sum_{k=0}^{D} lpd_{ik} = \sum_{i=1}^{A} lpd_{i0} = 2 \cdot W^{AP}$$

where A is the number of graph vertices.

Note that for acyclic graphs, the path degree layer matrix **LPD** coincides with the distance sum layer matrix **LDS**.

• **walk degree layer matrix ($LW^{(m)}$)**
A layer matrix obtained weighting the atoms by their → *walk degree* (i.e. the → *atomic walk count* of length m, $awc_i^{(m)}$). Therefore, the entry i-k of the layer matrix (lw_{ik}) is the sum of the walk degrees of the atoms located at distance k from the focused ith vertex [Diudea *et al.*, 1994]. Different $\mathbf{LW}^{(m)}$ matrices can be calculated according to the chosen order m of the walk degrees $awc_i^{(m)}$. The walk degree layer matrix $\mathbf{LW}^{(1)}$ coincides with the branching layer matrix **LB**. The elements of the first column ($k = 0$) in the $\mathbf{LW}^{(m)}$ matrix represent only the walk degrees of order m.

Moreover, the half-sum of both the entries in the first column ($k = 0$) and in each row is the total number of walks of length m in the graph, as can be seen from the following relationships:

$$\frac{1}{2}\cdot\sum_{k=0}^{D} lw_{ik}^{(m)} = \frac{1}{2}\cdot\sum_{i=1}^{A} lw_{i0}^{(m)} = \frac{1}{2}\cdot\sum_{i=1}^{A} awc_i^{(m)} = mwc^{(m)}$$

where A is the number of graph vertices, $awc_i^{(m)}$ the atomic walk count, and $mwc^{(m)}$ is the mth order → *molecular walk count*.

Analogous layer matrices can be obtained using, as atom weights, → *weighted walk degrees* instead of the simple vertex walk degrees.

Example : 2-methylpentane

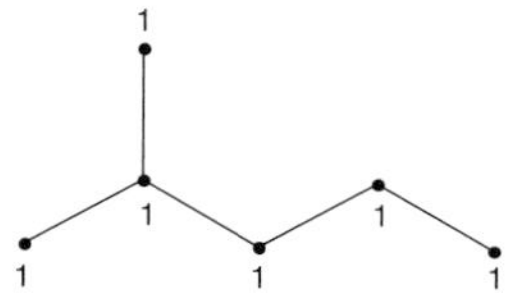

cardinality layer matrix **LC**

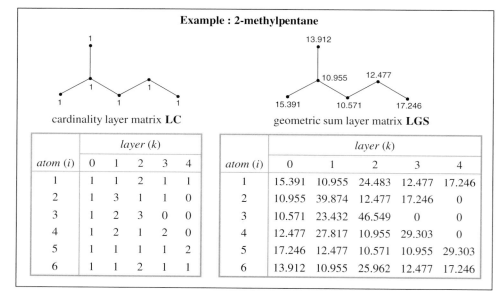

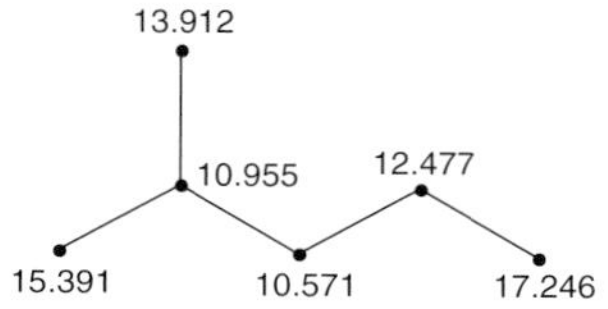

geometric sum layer matrix **LGS**

atom (i)	*layer* (k) 0	1	2	3	4
1	1	1	2	1	1
2	1	3	1	1	0
3	1	2	3	0	0
4	1	2	1	2	0
5	1	1	1	1	2
6	1	1	2	1	1

atom (i)	*layer* (k) 0	1	2	3	4
1	15.391	10.955	24.483	12.477	17.246
2	10.955	39.874	12.477	17.246	0
3	10.571	23.432	46.549	0	0
4	12.477	27.817	10.955	29.303	0
5	17.246	12.477	10.571	10.955	29.303
6	13.912	10.955	25.962	12.477	17.246

Box L-3.

Derived from layer matrices, two main types of → *local vertex invariants* were defined on the basis of two types of operators, the **centric operator** c_i and the **centro-complexity operator** x_i [Diudea, 1994; Diudea *et al.*, 1995b]:

$$c_i(\mathbf{LM}) = \left[\sum_{k=1}^{D} (lm_{ik})^{k/d}\right]^{-1} \quad \text{and} \quad x_i(\mathbf{LM}) = \left[\frac{1}{w_i}\cdot\sum_{k=0}^{D} lm_{ik}\cdot 10^{-zk} \pm l_i\right]^{\pm 1}\cdot t_i$$

where

$$l_i = f_i\cdot\left(\frac{lm_{i0}}{10}+\frac{lm_{i1}}{100}\right) \quad \text{and} \quad f_i = \sum\nolimits_{a\in V_{i1}} (\pi_{ia}^{*} - 1)$$

where d is a specified topological distance larger than the topological diameter D (for example, 10); w_i is the atomic property of the ith atom, z is the number of digits of the maximum entry lm_{ik} value; l_i is a local parameter accounting for multiple bonds, f_i being the → *atomic multigraph factor* obtained by summing the conventional bond orders π_{ia}^{*} of the atoms a bonded to the ith vertex; t_i is a weighting factor accounting for heteroatoms by means of atomic numbers, electronegativities, covalent radii, and so forth.

A third type of operator generating local invariants from layer matrices is defined as follows [Balaban and Diudea, 1993]:

$$xj_i(\mathbf{LM}) = \sum_{j \in V_{i1}} \left[\frac{\sum_{k=0}^{D} lm_{ik} \cdot 10^{-zk}}{t_i \cdot (1+f_i)} \cdot \frac{\sum_{k=0}^{D} lm_{jk} \cdot 10^{-zk}}{t_j \cdot (1+f_j)} \right]^{-1/2}$$

where the sum runs over all atoms j bonded to the ith atom, i.e. located at distance one from v_i.

Centrocomplexity invariant values should measure the location of the considered vertex with respect to a vertex "of importance", i.e. a vertex of highest branching degree, electronegativity, etc., while centric invariant values should measure the centricity of a vertex, i.e. its location with respect to the → *graph centre.*

The local indices obtained by applying centrocomplexity operators to the branching layer matrix **LB** are called **regressive vertex degrees** [Diudea *et al.*, 1991]. In fact, such indices are an extension of the concept of vertex degree, taking into account the contributions of distant vertices from the focused ith vertex; these contributions decrease with increasing distance, slightly augmenting the value of the classical vertex degree. High values of regressive vertex degrees should correspond to vertices of highest degree, closest to branching sites and to the graph centre. The two types of originally proposed centrocomplexity operators were defined as:

$$RVD_i \equiv x_i^{R1}(\mathbf{LB}) = \sum_{k=0}^{D} lb_{ik} \cdot k^{-3} = \delta_i + \sum_{k=1}^{D} lb_{ik} \cdot k^{-3}$$

$$RVD_i \equiv x_i^{R2}(\mathbf{LB}) = \sum_{k=0}^{D} lb_{ik} \cdot 10^{-k} = \delta_i + \sum_{k=1}^{D} lb_{ik} \cdot 10^{-k}$$

where **LB** is the branching layer matrix and δ the vertex degree.

In analogy with the regressive vertex degrees, the **regressive distance sums** (or **regressive incremental distance sums**) are local invariants obtained by applying the centrocomplexity operator x_i to the distance sum layer matrix **LDS** [Balaban and Diudea, 1993]. Also in this case, a simplified form of the centrocomplexity operator has been proposed:

$$RDS_i \equiv x_i(\mathbf{LDS}) = \sum_{k=0}^{D} \left(lds_{ik} \cdot 10^{-zk}\right) = \sigma_i + \sum_{k=1}^{D} \left(lds_{ik} \cdot 10^{-zk}\right)$$

where σ_i is the distance sum (i.e. vertex distance sum) of the ith vertex and z is the number of digits of the maximum entry lds_{ik} value.

Example : 2-methylpentane

Branching layer matrix **LB** ≡
walk degree layer matrix $\mathbf{LW}^{(1)}$

Atom (*i*)	*Layer* (*k*)				
	0	1	2	3	4
1	1	3	3	2	1
2	3	4	2	1	0
3	2	5	3	0	0
4	2	3	3	2	0
5	1	2	2	3	2
6	1	3	3	2	1

vertex degrees, δ_i

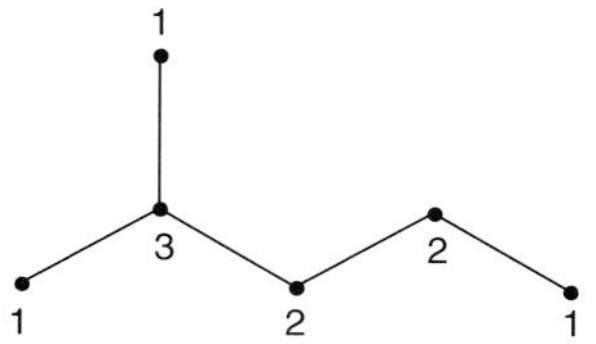

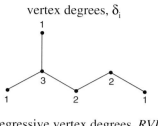

regressive vertex degrees, RVD_i

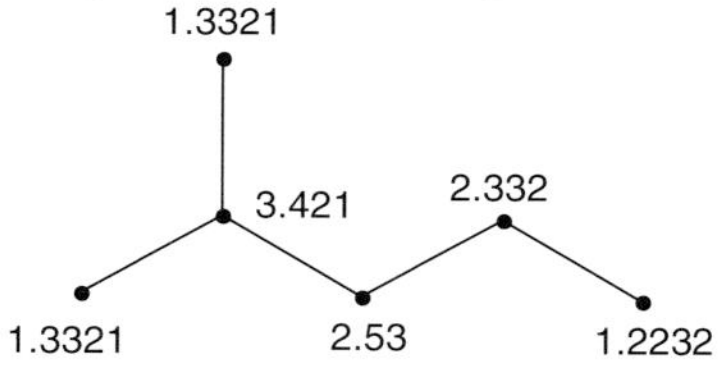

$$RVD_1 \equiv RVD_6 = \sum_{k=0}^{4} lb_{1k} \cdot 10^{-k} = 1 + 0.3 + 0.03 + 0.002 + 0.0001 = 1.3321$$

$$RVD_2 = \sum_{k=0}^{4} lb_{2k} \cdot 10^{-k} = 3 + 0.4 + 0.02 + 0.001 = 3.421$$

$$RVD_3 = \sum_{k=0}^{4} lb_{3k} \cdot 10^{-k} = 2 + 0.5 + 0.03 = 2.53$$

$$RVD_4 = \sum_{k=0}^{4} lb_{4k} \cdot 10^{-k} = 2 + 0.3 + 0.03 + 0.002 = 2.332$$

$$RVD_5 = \sum_{k=0}^{4} lb_{5k} \cdot 10^{-k} = 1 + 0.2 + 0.02 + 0.003 + 0.0002 = 1.2232$$

Box L-4.

Example : 2-methylpentane

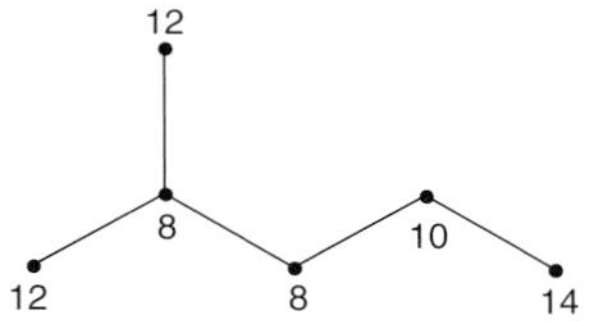

Distance sum layer matrix **LDS** ≡
path degree layer matrix **LPD**

Atom (i)	*Layer* (k)				
	0	1	2	3	4
1	12	8	20	10	14
2	8	32	10	14	0
3	8	18	38	0	0
4	10	22	8	24	0
5	14	10	8	8	24
6	12	8	20	10	14

regressive distance sums, RDS_i

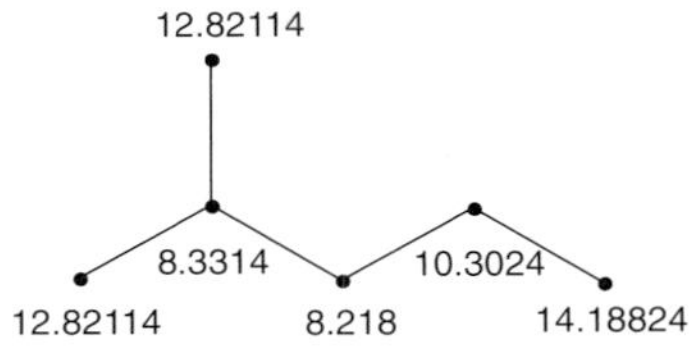

regressive decremental distance sums, $RDDS_i$

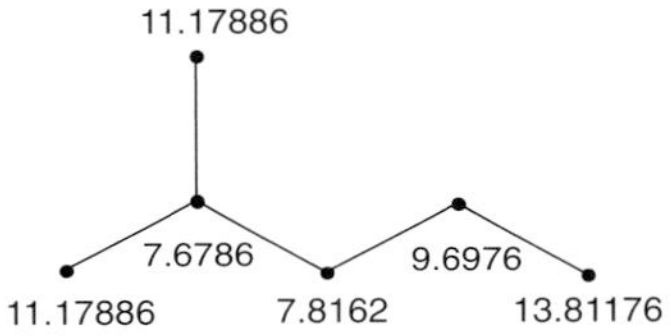

$$RDS_1 \equiv RDS_6 = \sum_{k=0}^{4}\left(lds_{1k}\cdot 10^{-k}\right) = 12 + 0.8 + 0.020 + 0.0010 + 0.00014 = 12.82114$$

$$RDS_2 = \sum_{k=0}^{4}\left(lds_{2k}\cdot 10^{-k}\right) = 8 + 0.32 + 0.010 + 0.0014 = 8.3314$$

$$RDS_3 = \sum_{k=0}^{4}\left(lds_{3k}\cdot 10^{-k}\right) = 8 + 0.18 + 0.038 = 8.218$$

$$RDS_4 = \sum_{k=0}^{4}\left(lds_{4k}\cdot 10^{-k}\right) = 10 + 0.22 + 0.08 + 0.0024 = 10.3024$$

$$RDS_5 = \sum_{k=0}^{4}\left(lds_{5k}\cdot 10^{-k}\right) = 14 + 0.10 + 0.08 + 0.008 + 0.00024 = 14.18824$$

$$RDDS_1 \equiv RDDS_6 = \sigma_1 - \sum_{k=1}^{4}\left(lds_{1k}\cdot 10^{-k}\right) = 12 - 0.8 - 0.020 - 0.0010 - 0.00014 = 11.17886$$

$$RDDS_2 = \sigma_2 - \sum_{k=1}^{4}\left(lds_{2k}\cdot 10^{-k}\right) = 8 - 0.32 - 0.010 - 0.0014 = 7.6786$$

$$RDDS_3 = \sigma_3 - \sum_{k=1}^{4}\left(lds_{3k}\cdot 10^{-k}\right) = 8 - 0.18 - 0.038 = 7.8162$$

$$RDDS_4 = \sigma_4 - \sum_{k=1}^{4}\left(lds_{4k}\cdot 10^{-k}\right) = 10 - 0.22 - 0.08 - 0.0024 = 9.6976$$

$$RDDS_5 = \sigma_5 - \sum_{k=1}^{4}\left(lds_{5k}\cdot 10^{-k}\right) = 14 - 0.10 - 0.08 - 0.008 - 0.00024 = 13.81176$$

Box L-5.

To obtain greater discrimination between terminal and central vertices the **regressive decremental distance sums** were proposed [Balaban, 1995b]. They are calculated from the distance sum layer matrix **LDS** by the following:

$$RDDS_i = \sigma_i - \sum_{k=1}^{D} \left(lds_{ik} \cdot 10^{-zk} \right)$$

where σ_i is the distance sum of the ith vertex. In this way, the progressively attenuated contributions due to more distant vertices are subtracted from the distance degree of the focused vertex.

The sums of the local vertex invariants x_i and c_i over all of the atoms give the corresponding molecular descriptors, called **centrocomplexity topological index** X and **centric topological index** C, respectively:

$$X(\mathbf{LM}) = \sum_{i=1}^{A} x_i(\mathbf{LM}) \quad \text{and} \quad C(\mathbf{LM}) = \sum_{i=1}^{A} c_i(\mathbf{LM})$$

where **LM** represents any layer matrix. These descriptors are related to → *molecular complexity*. Normalized centrocomplexity and centric local invariants x'_i and c'_i are obtained dividing each local invariant by the corresponding global topological index.

📖 [Diudea and Bal, 1990] [Diudea and Kacso, 1991] [Diudea *et al.*, 1992b] [Ivanciuc *et al.*, 1992] [Ivanciuc *et al.*, 1993b] [Dobrynin, 1993] [Wang *et al.*, 1994]

LCD-descriptors → **molecular descriptors** (⊙ invariance properties of molecular descriptors)

lead compound → **drug design**

leading eigenvalue : *Lovasz-Pelikan index* → **eigenvalue-based descriptors** (⊙ eigenvalues of the adjacency matrix)

leading eigenvalue of the distance matrix → **eigenvalue-based descriptors** (⊙ eigenvalues of the distance matrix)

leading eigenvalue of (A + D) → **eigenvalue-based descriptors**

learning set : *training set* → **data set**

leave-more-out technique → **validation techniques** (⊙ cross-validation)

leave-one-out technique → **validation techniques** (⊙ cross-validation)

length-to-breadth ratio → **shape descriptors** (⊙ Kaliszan shape parameter)

Lennard-Jones 6–12 potential function → **molecular interaction fields** (⊙ steric interaction fields)

Leo-Hansch hydrophobic fragmental constants → **lipophilicity descriptors**

level pattern indices → **eigenvalue-based descriptors** (⊙ eigenvalues of the adjacency matrix)

leverage matrix → **chemometrics** (⊙ regression analysis)

lhaf(D) index → **algebraic operators** (⊙ determinant)

ligand → **drug design**

likeliness function → **Klopman-Henderson cumulative substructure count**

L index → **combined descriptors**

L_Z index → **Hosoya Z matrix**

linear aromatic substituent reactivity relationships : *Yukawa-Tsuno equation* → **electronic substituent constants** (⊙ resonance electronic constants)

linear free energy relationships → **extrathermodynamic approach**

linear graph → **graph**

linear notation systems → **molecular descriptors**

linear similarity index → **similarity/diversity**

Linear Solvation Energy Relationships (LSERs)

Linear solvation energy relationships constitute the basis on which effects of solvent-solute interactions on physico-chemical properties and reactivity parameters are studied. In general, a property ϕ of a species A in a solvent S can be expressed as:

$$\phi_{A,S} = \sum_j \varphi_j(A, S)$$

where φ are complex functions of both solvents and solutes [Kamlet *et al.*, 1981a]. By assuming that these functions can be factorized in two contributions separately dependent on solute and solvent, the property can be represented as:

$$\phi_{A,S} = \sum_j f_j(A) \cdot g_j(S)$$

where f are functions of the solute and g functions of the solvent.

The underlying philosophy of the linear solvation energy relationships is based on the possibility of studying these two functions, after a proper choice of the reference systems and properties. Moreover, it has been recognized that solution properties ϕ mainly depend on three factors: a cavity term, a polar term, and hydrogen-bond term:

ϕ = *intercept* + *cavity term* + *dipolarity/polarizability term* + *hydrogen-bond term*

Therefore, a typical linear solvation energy relationship is expressed as [Kamlet *et al.*, 1987a]:

$$\phi_{A,S} = b_0 + b_1 \cdot \left(\delta_H^2\right)_1 \cdot V_2 + b_2 \cdot \Pi_1^* \cdot \Pi_2^* + b_3 \cdot \alpha_1 \cdot \beta_2 + b_4 \cdot \beta_1 \cdot \alpha_2$$

where b are estimated regression coefficients, and the subscripts 1 and 2 in the solvent/solute property parameters refer to the solvent S and the solute A, respectively. This equation is usually known as **solvatochromic equation**, even if it is extended to cover some nonspectroscopic properties, and the parameters of polarity/dipolarizability and hydrogen-bonding as **solvatochromic parameters**. The term *solvatochromic* is derived from the origin of this approach referring to the effect solvent has on the colour of an indicator which is used for quantitative determination of some molecular attributes (*solvatochromic parameters*).

From the general solvatochromic equation, two special cases can be encountered. When dealing with effects of different solvents on properties of a specific solute, the general equation is explicitly on solvent parameters:

$$\phi_{A,S_i} = b_0 + b_1 \cdot \left(\delta_H^2\right)_i + b_2 \cdot \Pi_{1,i}^* + b_3 \cdot \alpha_{1,i} + b_4 \cdot \beta_{1,i}$$

This equation has been used in several correlations of solvent effects on solute properties such as reaction rates and equilibrium constants of solvolyses, energy of electronic transitions, solvent-induced shifts in UV/visible, IR, and NMR spectroscopy, fluorescence lifetimes, and formation constants of hydrogen-bonded and Lewis acid/base complexes [Kamlet *et al.*, 1986b].

Conversely, when dealing with solubilities, lipophilicity, or other properties of a set of different solutes in a specific solvent, the general equation is explicitly on the solute parameters:

$$\phi_{A_i,S} = b_0 + b_1 \cdot V_i + b_2 \cdot \Pi^*_{2,i} + b_3 \cdot \alpha_{2,i} + b_4 \cdot \beta_{2,i}$$

This equation has been mainly used in correlations of aqueous solubility of compounds, octanol/water partition coefficients and some other partition parameters together with some biological properties [Kamlet *et al.*, 1984; Kamlet *et al.*, 1986a; Kamlet *et al.*, 1987a; Kamlet *et al.*, 1987c; Kamlet *et al.*, 1988c].

Two other general linear solvation energy relationships for solute physico-chemical properties in a fixed phase [Abraham *et al.*, 1990b; Abraham *et al.*, 1991a; Abraham *et al.*, 1991b; Abraham, 1993b; Abraham *et al.*, 1994a] are:

$$\log\left(\phi_{A_i,S}\right) = b_0 + b_1 \cdot V_{X,i} + b_2 \cdot R_{2,i} + b_3 \cdot \pi^H_{2,i} + b_4 \cdot \alpha^H_{2,i} + b_5 \cdot \beta^H_{2,i}$$

$$\log\left(\phi_{A_i,S}\right) = b_0 + b_1 \cdot L^{16}_i + b_2 \cdot R_{2,i} + b_3 \cdot \pi^H_{2,i} + b_4 \cdot \alpha^H_{2,i} + b_5 \cdot \beta^H_{2,i}$$

where the first can be applied to processes within condensed phases and the second to processes involving gas-condensed phase transfer.

The descriptors of the solvatochromic equation are specified below.

- **cavity term**

The cavity term is a measure of the endoergic cavity-forming process, i.e. the free energy necessary to separate the solvent molecules, overcoming solvent-solvent cohesive interactions, and provides a suitably size cavity for the solute. The magnitude of the cavity term depends on the → *Hildebrand solubility parameter* δ_H and → *volume descriptors* of the solute. The solute volume can be measured in different ways, such as by → *van der Waals volume* V_{VDW} [Leahy, 1986], → *molar volume* $\bar{V}$ or → *Mc Gowan's characteristic volume* V_X; in some cases, also → *molecular weight MW* has been used. Usually the volumes are divided by 100 $(\bar{V}/100)$ to obtain a more homogeneous scale with respect to the other parameters.

The **Ostwald solubility coefficient** L is the gas-liquid partition coefficient defined as:

$$L = \frac{C(\text{solution})}{C(\text{gas})}$$

where C(solution) and C(gas) are the solute concentrations in solution and gas phases, respectively. The parameter L^{16} is the Ostwald solubility coefficient on n-hexadecane at 298 K; it includes both general dispersion interactions and the endoergic cavity term and was proposed for modelling properties of solutes in processes involving gas-condensed phase transfer such as gas-liquid chromatographic parameters [Abraham*et al.*, 1987].

- **dipolarity / polarizability term** (: *dipole term*)

This term is a measure of the exoergic balance (i.e. release of energy) of solute-solvent and solute-solute dipolarity / polarizability interactions. This term, denoted by Π^*, describes the ability of the compound to stabilize a neighbouring charge or dipole by virtue of nonspecific dielectric interactions and is in general given by → *electric polarization descriptors* such as → *dipole moment* or other empirical → *polarity / polarizability descriptors* [Abraham *et al.*, 1988]. Other specific polarity parameters empirically derived for linear solvation energy relationships are reported below.

Several **solvent polarity scales** were proposed to quantify the polar effects of solvents on physical properties and reactivity parameters in solution, such as rate of solvolyses, energy of electronic transitions, and solvent-induced shifts in IR or NMR spectroscopy. Most of the polarity scales were derived by an empirical approach based on the principles of the → *linear free energies relationships* applied to a chosen reference property and system where hydrogen-bonding effects are assumed negligible [Reichardt, 1965; Kamlet *et al.*, 1981a; Kamlet *et al.*, 1983; Reichardt, 1990].

The most important scales of solvent polarity are:

π^* polarity scale. A solvent polarity parameter (also denoted as π_1^*) proposed by Kamlet, Abboud and Taft [Kamlet *et al.*, 1977; Kamlet *et al.*, 1981b], based on the solvatochromic shifts on the frequency maxima of the $\pi \rightarrow \pi^*$ transitions of seven different benzene derivatives. The π^* values are averaged on the seven compounds to prevent the inclusion of specific effects or spectral anomalies and are normalized so that π^* equals zero for cyclohexane and one for dimethylsulfoxide. This scale is one of the most comprehensive for the number of considered solvents and is widely used. Moreover, for solution properties ϕ involving different relative contributions of polarity and polarizability, π^* values can be corrected as $(\pi^* - d\delta_H)$, where δ_H is the Hildebrand solubility parameter. The term d is calculated by dividing the difference in ϕ at $\pi^* = 0.7$, as obtained separately for nonpolychlorinated aliphatic and aromatic solvents, by the average of the slopes of the solvatochromic equations for aliphatic and aromatic solvents. The Hildebrand parameter is assumed to be $\delta_H = 0.0$ for nonpolychlorinated aliphatic solutes, $\delta_H = 0.5$ for polychlorinated aliphatics, and $\delta_H = 1.0$ for aromatic solutes; the term d ranges between zero, for maximal polarizability contributions to the studied property, and –0.40 for minimal contributions. The → *excess molar refractivity* R_2 has also been used as the polarizability correction term instead of the Hildebrand parameter δ_H [Abraham *et al.*, 1991].

Y polarity scale. A solvent polarity scale proposed by Grunwald and Winstein [Grunwald and Winstein, 1948] based on solvolytic rate k_0 of *t*-butyl chloride in 80 % aqueous ethanol at 25 °C. The Y polarity value for a given solvent is calculated by:

$$Y = \log k_S - \log k_0$$

where k_S is the solvolytic rate of *t*-butyl chloride in the considered solvent. Y scale was proposed as measure of an empirical "ionizing power" of solvents.

E_T polarity scale. A solvent polarity scale proposed by Dimroth, Reichardt and coworkers [Dimroth *et al.*, 1963; Reichardt, 1965] based on the solvatochromic band shifts of the 4-(2,4,6-triphenylpyridinium)-2,6-diphenylphenoxide and its trimethyl derivative. This scale is one of the most comprehensive for the number of considered solvents and is widely used.

E_K polarity scale. A solvent polarity scale proposed by Walther [Walther, 1974] based on the hypsochromic shift of the longest wavelength absorption of a molybdenum complex.

***Z* polarity scale.** A solvent polarity scale proposed by Kosower [Kosower, 1958a; Kosower, 1958b] based on the energy of the electronic transition of the 1-ethyl-4-carbomethoxypyridinium iodide, which is strongly solvent dependent. This is a measure of an internal charge transfer process. The original set of Z values being quite small, it was successively extended by means of other indicators.

Table L-1. Empirical parameters of the solvent polarity from different sources.

Solvent	Y	Z	E_T	E_K	π^*
H_2O	3.493	94.6	63.1		1.09
HCOOH	2.054				0.65
$HCONH_2$	0.604	83.3	56.6		
CH_3COOH	–1.64	79.2	51.1	55.0	0.62
CH_3OH	–1.09	83.6	55.5	56.3	0.60
CH_3NO_2			46.3		0.85
CH_3CN		71.3	46.0		0.75
C_2H_5OH	–2.03	79.6	51.9	55.3	0.54
C_5H_5N		64.0	40.2	57.0	0.87
CCl_4			32.5	49.9	0.294
CH_2Cl_2		64.2	41.1	53.9	0.802
C_6H_6		54.0	34.5	53.4	0.588
C_6H_5Cl			37.5	53.9	0.709
C_6H_5Br			37.5	53.9	0.794
C_6H_5CN			42.0		0.933
$C_6H_5NO_2$			42.0		1.006
cyclo-C_6H_{12}			31.2	49.0	0.000

A wide variety of correlations among solvent polarity scales were studied [Reichardt and Dimroth, 1968]; however, because of the different reference compounds used to define them, direct comparison should be done with caution [Bentley and von Schleyer, 1977].

The **solute polarity parameter** π_2^* originally was taken as identical with the solvent polarity parameter π^* for non-associated liquids only [Taft *et al.*, 1985b]. Then an alternative solute polarity parameter π_2^H (or $\sum \pi_2^H$) was proposed based on experimental procedures that include, at least in principle, all types of solute molecules [Abraham *et al.*, 1991a; Abraham and Whiting, 1992]. Values of this solute parameter were determined by back-calculation solving "inverse" solvation equation systems based on 30–70 stationary phases for each solute. π_2^H values refer to a situation in which a solute molecule is surrounded by an excess of solvent molecules, and so they are effective values, more correctly denoted as $\sum \pi_2^H$, accounting for combined effects due to polyfunctional groups in the molecule.

- **hydrogen-bond parameters**

These are measures of the exoergic effects (i.e. release of energy) of the complexation between solutes and solvents.

The hydrogen bond donor (HBD) power of a compound is called the **hydrogen bond acidity** (or **hydrogen-bond electron-drawing power**) and is denoted by α_1 and α_2 for solvents and solutes, respectively.

The most important scales for **solvent HBD acidity** are reported here.

The **α scale** proposed to measure solvent hydrogen bond acidity, i.e. the ability of a bulk solvent to act as hydrogen bond donor toward a solute, was derived from 16 diverse properties involving 13 solutes as averaged values [Taft and Kamlet, 1976; Kamlet *et al.*, 1983].

The **Gutmann's Acceptor Number** (AN) was proposed [Gutmann, 1978] as a quantitative empirical parameter of solvent hydrogen bond acidity based on ^{31}P-nmr shifts of thiethylphosphine oxide at infinite dilution, calculated as $AN = -\delta_{\infty}^{corr} \cdot 2.349$.

For **solute HBD acidity** different scales were proposed mainly based on complexation constants and enthalpies of complexation. The most important are reported here.

The **α_m scale** was proposed for solute HBD acidity of "monomer" amphihydrogen-bonding compounds acting as non-self-associated solutes [Taft *et al.*, 1985b; Kamlet *et al.*, 1986a]. In particular, α_m values were derived from log$K$ values for complexation with pyridine *N*-oxide in cyclohexane; this set of values was successively extended through various back-calculations using the solvatochromic equation.

The **α_2^H scale** was proposed for solute HBD acidity based on log$K$ values for 1:1 complexation of series of acids against a given base in dilute solution of CCl_4 [Abraham *et al.*, 1989b]. Forty-five linear equations have been solved for each considered base by a series of acids:

$$\log K_i = b_0^B + b_1^B \cdot \log K_A^{H_i}$$

where b^B are the regression coefficients characterizing each reference base, and $\log K_A^H$ values are characteristics of hydrogen-bonding acids, and hence represent the solute hydrogen bond acidities. All the equations intersect at a "magic" point where log$K$ =–1.1 (K measured on molar scale). The general $\log K_A^H$ values were then transformed into α_2^H values suitable for multivariate regression analysis by the following:

$$\alpha_2^H = \frac{\log K_A^H + 1.1}{4.636}$$

A fairly good correlation was found between the α_2^H scale and the α_m scale. Moreover, the set of original α_2^H values was then enlarged by solving a system of solvatochromic equations on partition coefficients, thus including several new compounds and molecular fragments [Abraham *et al.*, 1994c]. The **effective solute hydrogen-bond acidity** $\sum \alpha_2^H$ was back-calculated by a number of multiple linear regression equations for solutes surrounded by a large excess of solvent and hence undergoing multiple hydrogen-bonding. This hydrogen bond descriptor agrees with α_2^H values for monofunctional compounds, while for polyfunctional compounds it differs significantly [Abraham, 1993b].

The hydrogen bond acceptor (HBA) power of a compound is called **hydrogen bond basicity** (or **hydrogen-bond electron-acceptor power**) and is denoted by β_1 and β_2 for solvents and solutes, respectively.

The most important scales for **solvent HBA basicity** are reported here.

The **β scale** was proposed to measure solvent hydrogen bond basicity, i.e. the ability of a bulk solvent to act as hydrogen bond acceptor. This scale was derived by systematic application of the solvatochromic comparison method; the final β values were calculated by averaging 13 β parameters for each solvent obtained with different solutes and different physicochemical properties [Kamlet *et al.*, 1981a; Kamlet *et al.*, 1983].

The **Koppel-Paju B scale** was proposed to measure solvent hydrogen bond basicity, based on solvent shifts of the IR stretching frequencies of the free and hydrogen bonded OH group of phenol in CCl_4 [Koppel and Paju, 1974].

The **Gutmann's Donor Number** (DN) was proposed [Gutmann, 1978] as a quantitative empirical parameter for solvent nucleophilicity. For most solvents it was found to correlate well with the β scale.

The most important scales for **solute HBA basicity** are reported here.

The **β_m scale** was proposed for solute HBA basicity of "monomer" amphihydrogen-bonding compounds acting as non-self-associated solutes. In particular, β_m values are taken equal to β values for non-self-associating compounds.

The **β_2^H scale** was proposed for solute HBA basicity based on log$K$ values for 1:1 complexation of series of bases against a number of reference acids in dilute solution of CCl_4 [Abraham *et al.*, 1990]. Thirty-four linear equations have been solved for each considered reference acid by a series of bases:

$$\log K_i = b_0^A + b_1^A \cdot \log K_B^{H_i}$$

where b^A are the regression coefficients characterizing each reference acid, and $\log K_B^H$ values are characteristics of the bases representing the solute hydrogen bond basicities. All the equations intersect at a "magic" point where log$K$ =–1.1 (K measured on molar scale). The general $\log K_B^H$ values were then transformed into β_2^H values suitable for multivariate regression analysis by the following:

$$\beta_2^H = \frac{\log K_B^H + 1.1}{4.636}$$

This transformation was proposed to obtain a basicity scale with a zero-point corresponding to all non-hydrogen-bonding bases, such as alkanes and cycloalkanes. Moreover, on this scale, hexamethylphosphoric triamide basicity is equal to one. The set of original β_2^H values was then enlarged by solving a system of solvatochromic equations on partition coefficients, thus including several new compounds and molecular fragments [Abraham *et al.*, 1994c]. The **effective solute hydrogen-bond basicity** $\sum\beta_2^H$ was back-calculated by a number of multiple linear regression equations for solutes surrounded by a large excess of solvent and hence undergoing multiple hydrogen-bonding. This hydrogen bond descriptor agrees with β_2^H values for monofunctional compounds, while for polyfunctional compounds it significantly differs [Abraham *et al.*, 1991a; Abraham and Whiting, 1992; Abraham, 1993b].

For most solutes, the effective hydrogen-bond basicity is constant over all the solvent systems; however, in the case of some specific solutes, including anilines and pyridines, the effective solute hydrogen-bond basicity varies with the solvent system. Therefore, the descriptor $\sum\beta_2^H$ is preferably used for partition between water and nonaqueous solvent systems, while an alternative $\sum\beta_2^0$ can be used for partition between water and aqueous solvent systems [Abraham and Rafols, 1995].

Table L-2. LSER parameter values for some solutes. Symbols defined in the text and data taken from [Kamlet *et al.*, 1987a; Abraham *et al.*, 1994a].

Solute	$\bar{V}/100$	V_X	*log* L^{16}	R_2	$\sum\pi_2^H$	α_m	β_m	$\sum\alpha_2^H$	$\sum\beta_2^H$
diethyl ether	1.046	0.7309	2.015	0.041	0.25	0.00	0.47	0.00	0.45
di-*n*-butyl ether	1.693	1.2945	3.924	0.000	0.25	0.00	0.46	0.00	0.45
1-propanol	0.757	0.5900	2.031	0.236	0.42	0.33	0.45	0.37	0.48
2-propanol	0.765	0.5900	1.764	0.212	0.36	0.33	0.51	0.33	0.56
1-butanol	0.915	0.7309	2.601	0.224	0.42	0.33	0.45	0.37	0.48
2-butanol	0.917	0.7309	2.338	0.217	0.36	0.33	0.51	0.33	0.56
1-pentanol	1.086	0.8718	3.106	0.219	0.42	0.33	0.45	0.37	0.48
cyclopentanol	1.009	0.7630	3.241	0.427	0.54	0.33	0.51	0.32	0.56
trichloroethene	0.897	0.7146	2.997	0.524	0.40	0.00	0.10	0.08	0.03
1,1,1-trichloroethane	0.989	0.7576	2.733	0.369	0.41	0.00	0.10	0.00	0.09
tetrachloromethane	0.968	0.7391	2.823	0.458	0.38	0.00	0.10	0.00	0.00

Solute	$\bar{V}/100$	V_X	*log* L^{16}	R_2	$\sum\pi_2^H$	α_m	β_m	$\sum\alpha_2^H$	$\sum\beta_2^H$
1,2-dichloroethane	0.787	0.6352	2.573	0.416	0.64	0.00	0.10	0.10	0.11
butanone	0.895	0.6879	2.287	0.166	0.70	0.00	0.48	0.00	0.51
cyclopentanone	0.986	0.7202	3.221	0.373	0.86	0.00	0.52	0.00	0.52
cyclohexanone	1.136	0.8611	3.792	0.403	0.86	0.00	0.53	0.00	0.56
acetonitrile	0.521	0.4042	1.739	0.237	0.90	0.15	0.35	0.04	0.33
benzene	0.989	0.7164	2.786	0.610	0.52	0.00	0.10	0.00	0.14
phenol	0.989	0.7751	3.766	0.805	0.89	0.61	0.33	0.60	0.31
m-cresol	1.163	0.9160	4.329	0.822	0.88	0.55	0.33	0.57	0.34
nitrobenzene	1.127	0.8906	4.557	0.871	1.11	0.00	0.30	0.00	0.28
2-nitrotoluene	1.217	1.0315	4.878	0.866	1.11	0.00	0.30	0.00	0.28
benzonitrile	1.120	0.8711	4.039	0.742	1.11	0.00	0.35	0.00	0.33
tetrahydrofuran	0.911	0.6223	2.636	0.289	0.52	0.55	0.00	0.00	0.48

📖 [Berson *et al.*, 1962] [Brooker *et al.*, 1965] [Lassau and Jungers, 1968] [Kamlet and Taft, 1979b] [Kamlet *et al.*, 1979a] [Kamlet and Taft, 1979c] [Taft and Kamlet, 1979] [Taft *et al.*, 1984] [Krygowski *et al.*, 1985] [Taft *et al.*, 1985a] [Kamlet *et al.*, 1987b] [Kamlet *et al.*, 1988a] [Kamlet *et al.*, 1988b] [Abraham *et al.*, 1989a] [Abraham *et al.*, 1990a] [Schüürmann, 1990] [Murray and Politzer, 1991] [Vallat *et al.*, 1992] [Leahy *et al.*, 1992a] [Leahy *et al.*, 1992b] [Abraham, 1993a] [Altomare *et al.*, 1993] [Abraham *et al.*, 1994b] [Leahy *et al.*, 1994] [Pagliara *et al.*, 1995] [Lowrey *et al.*, 1995a] [Cronin and Dearden, 1995d] [Abraham *et al.*, 1996] [Yang *et al.*, 1996] [Pagliara *et al.*, 1997a] [Steyaert *et al.*, 1997] [Svozil *et al.*, 1997] [Katritzky *et al.*, 1999b] [Katritzky *et al.*, 1999a]

linear subfragment descriptors → **substructure descriptors**

line graph → **graph**

line graph connectivity indices → **iterated line graph sequence**

line graph Randic connectivity index → **iterated line graph sequence**

link : *connection* → **edge adjacency matrix**

linking number → **mean overcrossing number**

lipole → **lipophilicity descriptors**

lipophilicity → **lipophilicity descriptors**

lipophilicity descriptors

Lipophilicity is the measure of the partitioning of a compound between a lipidic and an aqueous phase [Taylor, 1990]. It is usually represented by **partition coefficients** P defined as the ratio of the concentrations of a compound in organic and aqueous phases of a two-compartment system under equilibrium conditions:

$$P = \frac{[C]_{org}}{[C]_{aq}}$$

where $[C]_{org}$ and $[C]_{aq}$ are the concentrations of the solute in organic and aqueous phases, respectively. The partition coefficients are usually transformed into a logarithmic form as:

$$\log P = \log \frac{[C]_{org}}{[C]_{aq}} = \log[C]_{org} - \log[C]_{aq}$$

In order to avoid possible associations of the solute in the organic phase, partition coefficients should be measured at low concentrations or extrapolated to infinite dilution of the solute. They are dimensionless measures of the relative affinity of a molecule with respect to the two phases and depend on absorption, transport, and partitioning phenomena. Compounds for which $P > 1$ or $\log P > 0$ are *lipophilic*, and compounds for which $P < 1$ or $\log P < 0$ are *hydrophilic*. In particular, lipophilicity depends on solute bulk, polar and hydrogen-bonding effects.

The most widely used molecular descriptor encoding this property is the **octanol-water partition coefficient** K_{ow} (and **log K_{ow}**, or also **log$P$** when no further specifications are given), i.e. the partition coefficient between 1-octanol and water:

$$\log P \equiv \log K_{ow} = \log \frac{[C]_{1-octanol}}{[C]_{water}} = \log[C]_{1-octanol} - \log[C]_{water}$$

Other commonly used partition coefficients are defined for n-alkane systems, such as n-heptane-water.

Lipophilicity can be factorized in two main terms as [Carrupt *et al.*, 1997]:

lipophilicity = hydrophobicity – polarity

where **hydrophobicity** refers to nonpolar interactions (such as dispersion forces, hydrophobic interactions, ...) of the solute with organic and aqueous phases and *polarity* to polar interactions (such as ion-dipole interactions, hydrogen bonds, induction and orientation forces, ...). As can be observed, in this case the term hydrophobicity is not synonymous with lipophilicity, but is a component of it.

Usually hydrophobicity is encoded by → *steric descriptors* such as molar or → *molecular volume* which account satisfactorily for nonpolar interactions; polarity can be described by polar terms which are negatively related to lipophilicity. An important factorization of lipophilicity is provided by the → *solvatochromic parameters*. Moreover, a measure of the global polarity of a given solute was proposed by Testa and coworkers [Testa and Seiler, 1981; El Tayar and Testa, 1993; Vallat *et al.*, 1995] and called the **interactive polar parameter** Λ (or the **Testa lipophobic constant**). It is defined as the difference between the experimentally measured lipophilicity and that estimated for a hypothetical n-alkane of the same molecular volume V as:

$$\Lambda = (\log P)_{exp} - (\log P)_{calc}$$

where the calculated log$P$ is obtained from the following equation:

$$(\log P)_{calc} = 0.0309 \cdot \mathrm{V} + 0.346$$

The interactive parameter Λ should by definition encode the same information as the solvatochromic polar parameters; it takes negative values for lipophobic fragments [van de Waterbeemd and Testa, 1987; El Tayar *et al.*, 1992a; Testa *et al.*, 1993].

For apolar compounds an analogous linear relationship should be expected between log$P$ and → *van der Waals volume* V_{VDW} which accounts for steric contributions to log$P$ [Moriguchi, 1975; Moriguchi *et al.*, 1976; Moriguchi and Kanada, 1977]. An improved relation is obtained by incorporating into V_{VDW} a correction accounting for → *molecular branching*. Moreover, to extend the relation to polar compounds, a correction factor called the **Moriguchi polar parameter** (or **hydrophilic effect**) V_H was proposed as in the following equation:

$$\log P = 2.71 \cdot (V_{VDW} - V_H) + 0.12$$

Thus, the hydrophilic effect V_H is calculated as:

$$V_H = V_{VDW} - \frac{\log P - 0.12}{2.71}$$

and accounts for intramolecular hydrophobic bonding; moreover, it was found to correlate well with the interactive polar parameter Λ.

There are several methods developed for the calculation of log*P* from molecular structure, based on → *substituent constants*, → *fragmental constants*, → *electronic descriptors*, → *steric descriptors*, → *connectivity indices*, surface areas, → *volume descriptors*, and → *chromatographic descriptors* [Leo, 1990; Hansch *et al.*, 1995b; Carrupt *et al.*, 1997; Reinhard and Drefahl, 1999].

The most popular approaches to log*P* calculation are listed below.

- **Hansch-Fujita hydrophobic substituent constants**
  (: *hydrophobic substituent constants*, π)

The lipophilicity is calculated by analogy with the → *Hammett equation* as:

$$\log \frac{P_X}{P_H} = \varrho \cdot \pi_X$$

where P_X and P_H are the partition coefficients of an X-substituted and unsubstituted compound, respectively; π_X is the hydrophobic constant of the substituent X; the ϱ constant reflects the characteristics of the solvent system and is assumed equal to one for the octanol/water solvent system [Fujita *et al.*, 1964].

These hydrophobic substituent constants are commonly used in → *Hansch analysis* to encode the lipophilic behavior of the substituents; the lipophilicity of the whole molecule is obtained by adding to the lipophilicity of the unsubstituted parent compound ($\log P_H$) the lipophilic contributions of the substituents:

$$\log P(\text{molecule}) = \log P_H + \sum_{s=1}^{S} \pi_{X_s}$$

where S is the number of substitution sites and π_{Xs} are the hydrophobic constants of the substituents in the molecule. Distinct values of the π constants were defined for aromatic and aliphatic compounds. The Hansch-Fujita hydrophobic constants are still widely used in QSAR studies, but not for calculating log*P* values.

The major drawback of this approach is that π values depend on their electronic environment. When electronic interactions of the substituent X with other substituents in the compound are possible, more realistic π values have to be used. In particular, the π^- lipophilic constant, also known as the **Norrington lipophilic constant** [Norrington *et al.*, 1975], measures the lipophilic contribution of strong electron-releasing groups such as $-OH$, $-NH_2$, $-NHR$ or $-NR_1R_2$ when they are attached to a conjugated system (usually phenol or aniline); the π^+ lipophilic constant measures the lipophilic contribution of strong electron-attracting groups such as cyano or nitro groups, conjugated with the functional group. The use of this last constant is very rare.

Based on the decomposition of log*P* into enthalpic P_h and entropic P_s contributions:

$$\log P = \frac{-\Delta G_p^0}{2.303 \cdot RT} = \frac{-\Delta H_p^0}{2.303 \cdot RT} + \frac{\Delta S_p^0}{2.303 \cdot R} = P_h + P_s$$

the hydrophobic substituent constant π was decomposed [Da *et al.*, 1992; Da *et al.*, 1993] into the **enthalpic hydrophobic substituent constant** π_h and the **entropic hydrophobic substituent constant** π_s, respectively:

$$\pi = \pi_h + \pi_s$$

$$\pi_h = (P_h)_X - (P_h)_H \qquad P_h = \frac{-\Delta H_p^0}{2.303 \cdot RT}$$

$$\pi_s = (P_s)_X - (P_s)_H \qquad P_s = \frac{+\Delta S_p^0}{2.303 \cdot R}$$

where ΔG_p^0, ΔH_p^0, and ΔS_p^0 are the Gibbs free energy, enthalpy, and entropy of transfer for partition, respectively; R is the gas constant and T the absolute temperature; the subscripts X and H denote the substituted and unsubstituted compound, respectively. The enthalpic contribution P_h can be interpreted as a new hydrophobic parameter that reflects the heat evolved when a solute is transferred from water to a non-aqueous phase. Similarly, the entropic contribution P_s can be interpreted to reflect the change of randomness induced in the solution when a solute is transferred from water to a non-aqueous phase.

📖 [Hansch *et al.*, 1962; Hansch *et al.*, 1963; Hansch and Anderson, 1967; Leo *et al.*, 1971b; Martin and Lynn, 1971; Hansch and Dunn III, 1972; Hansch *et al.*, 1972; Hansch *et al.*, 1973a; Fujita and Nishioka, 1976; Fujita, 1983; Leo, 1993; Gago *et al.*, 1994; Amic *et al.*, 1998b]

- **Nys-Rekker hydrophobic fragmental constants** (f)

Also simply called **hydrophobic fragmental constants**, these are measures of the absolute lipophilicity contribution of specific molecular fragments to the lipophilicity of the molecule [Nys and Rekker, 1973; Nys and Rekker, 1974; Rekker, 1977a; Rekker, 1977b; Rekker and De Kort, 1979].

The log$P$ of a molecule is calculated by summing the fragmental contributions and applying the appropriate correction factors as:

$$\log P = \sum_{k=1}^{K} f_k \cdot \mathrm{N}_k + \sum_{j=1}^{C} F_j \cdot \mathrm{N}_j + c$$

where f_k and N_k are the hydrophobic constant and the number of occurrences of the kth fragment in the considered compound, K is the number of different fragments, N_j is the number of occurrences of the jth correction factor, and C is the number of applied correction factors. F is the value of the considered correction factor describing some special structural features (proximity effects, hydrogen atoms attached to polar groups, aryl-aryl conjugation, etc); in practice, it can be calculated as

$$F_j = k_j \cdot 0.219$$

where 0.219 is the so-called "magic constant" and k_j is an integer value characterizing the jth correction factor.

Different sets of fragmental constants were derived by multiple regression analysis for fragments depending on their attachment to an aliphatic or aromatic carbon atom. In this approach the effects of intramolecular interactions on lipophilicity are taken into account by the correction factors and implicitly in the definition of the molecular fragments. However, group interactions are evaluated more by the topological distance between the groups rather than by the chemical nature of the groups and their geometry.

A drawback of this approach is that, for the same compound, different selections of fragments give different log*P* values. Moreover, for complex compounds, the decomposition of the molecular structure into appropriate fragments is not unique and is a difficult task.

📖 [Mayer *et al.*, 1982; Takeuchi *et al.*, 1990; Rekker and Mannhold, 1992; Rekker, 1992; Rekker and de Vries, 1993; Mannhold *et al.*, 1998b; Rekker *et al.*, 1998]

- **Leo-Hansch hydrophobic fragmental constants** (f')

Hydrophobic constants are calculated for molecular fragments by a "constructionist approach" which consists of determining very accurately the log*P* values of simple compounds usually having a single functional group and then calculating fundamental hydrophobic fragmental constants from these values [Hansch and Leo, 1979; Leo, 1987; Leo, 1993]. Log*P* of the studied compounds are calculated by using Rekker's additive scheme, based on different fragmental constants f'' and correction factors F'. The decomposition of the molecular structure into fragments is performed by using a unique and simple set of rules, thus obtaining a unique solution; the fragments are either atoms or polyatomic groups. Correction factors were derived from compounds with more than one substituent to better approximate experimental log*P* values. They take into account proximity effects due to multiple halogenation and groups giving hydrogen bonds, intramolecular hydrogen bonds involving oxygen and nitrogen atoms, electronic effects in aromatic systems, unsaturation, branching, chains, and rings. Over 200 fragmental constants and 14 correction factors have been determined.

The software version of the Leo-Hansch fragmental method is known as **Calculated LOG*P*** (*CLOGP*) [Chou and Jurs, 1979].

📖 [Leo and Hansch, 1971a; Leo *et al.*, 1975; Lyman *et al.*, 1982; Mayer *et al.*, 1982; Abraham and Leo, 1987; Leo, 1991; Leo, 1993]

- **Klopman hydrophobic models**

The first model for the prediction of log*P* proposed by Klopman and Iroff [Klopman and Iroff, 1981] was based on the assumption that partition coefficients of molecules depend on the charge densities on each atom of the molecule. The following equation was proposed including both atom and group counting descriptors and charge density descriptors:

$$\log P = b_0 + \sum_i b_i \cdot \mathrm{N}_i + \sum_i b'_i \cdot q_i + \sum_i b''_i \cdot q_i^2 + \sum_j F_j \cdot \mathrm{N}_j$$

where the first three summations run over the different types of atoms, b are the estimated regression coefficients, N_i the number of occurrences of the ith atom, and q_i the charge density on the ith atom; N_j are the occurrences or the presence/absence of some specific functional groups (acid/ester, nitrile, amide groups) whose influence on molecule lipophilicity is described by selected correction factors F_j.

Another model was proposed based on atomic composition of the molecule only, ignoring the influence of the charge density descriptors on the log*P* calculation [Klopman *et al.*, 1985]. The best fitted proposed model is:

$$\log P = -0.206 + 0.332 \cdot \mathrm{N_C} + 0.071 \cdot \mathrm{N_H} - 0.860 \cdot \mathrm{N_O} - 1.124 \cdot \mathrm{N_N} + 0.688 \cdot \mathrm{N_{Cl}}$$
$$+ 0.981 \cdot (\mathrm{N_{ac}} + \mathrm{N_{est}}) - 0.138 \cdot \mathrm{N_{Ph}} + 2.969 \cdot \mathrm{N_{NO2}} + 1.053 \cdot \mathrm{I_{aliph}}$$

$$n = 195; \quad r^2 = 0.949; \quad s = 0.293; \quad F = 33$$

where $\mathrm{N_C}$, $\mathrm{N_H}$, $\mathrm{N_O}$, $\mathrm{N_N}$, $\mathrm{N_{Cl}}$ are the numbers of carbon, hydrogen, oxygen, nitrogen, and chlorine atoms, respectively; $\mathrm{N_{ac}}$ and $\mathrm{N_{est}}$ are the numbers of acid and ester groups, respectively; N_{Ph}, and N_{NO_2} are the numbers of phenyl rings and nitrile groups; I_{aliph} is an indicator variable for aliphatic hydrocarbons.

The regression coefficients of the model relative to the atomic counting descriptors can be viewed as individual log*P* contributions of each atom, **Klopman hydrophobic atomic constants**. A better evaluation of these atomic contributions was proposed by classifying the atoms also accounting for their environment represented by the first atom neighbours [Klopman and Wang, 1991]. Moreover, this method uses for the evaluation of log*P* of the molecules only those atom-centred groups that are identified by stepwise regression as the most significant groups determining log*P*.

A further developed model (**Klopman LOG*P***, *KLOGP*), was proposed as:

$$\log P = b_0 + \sum_i b_i \cdot N_i + \sum_j F_j \cdot N_j$$

where N_i is the occurrence of the ith atom-centred fragment in the molecule and N_j are the occurrences of particular fragments accounting for the interactions between groups whose influence on molecule lipophilicity is described by calculated correction factors F_j [Klopman *et al.*, 1994]. Basic atom-centred groups are of two types: a) atomic groups defined by their chemical element, hybridization and number of attached hydrogen atoms; b) functional groups containing at least one heteroatom. Correction factors are supplementary hydrophobic constants relative to specific substructures with more than two non-hydrogen atoms.

The regression coefficients b_i are the Klopman hydrophobic atomic constants measuring the hydrophobic contributions of atom types in the same way as the hydrophobic fragmental constants f_i defined in Rekker and Leo-Hansch approaches. The best evaluation of 64 atomic constants plus 30 correction factors was obtained by a training set of 1663 compounds, $r^2 = 0.93$, $s = 0.38$, $F = 218$.

The automated recognition of fragments and correction factors is performed by the **CASE approach** (*Computer Automated Structure Evaluation*). Basically, CASE is an artificial intelligence system capable of identifying structural fragments that may be associated with the properties of the training molecules, such as biological activity and physico-chemical properties [Klopman, 1984]. MULTICASE is the most recent upgraded software version [Klopman, 1992; Klopman, 1998].

- **Suzuki-Kudo hydrophobic fragmental constants**

The contribution method of Suzuki and Kudo [Suzuki and Kudo, 1990; Suzuki, 1991] is based on hydrophobic fragmental constants f_k and is defined as:

$$\log P = b_0 + \sum_k f_k \cdot N_k$$

where N_k is the occurrence of the kth fragment in the molecule. A first set of 415 basic hydrophobic constants was proposed, representing the lipophilic contributions of groups, each described by its structural environment. Several basic groups were first defined as CH_3, CH_2, CO, SO_2, etc., which were further distinguished according to their neighbouring atoms with their connectivities. Groups of atoms such as cyano and nitro are considered as univalent heteroatoms. In addition, extended fragments based on the basic fragments plus some other functional groups were selected together with some user-defined fragments. A training set of 1465 compounds plus a test set of 221 compounds were used to evaluate the hydrophobic constants by multivariate regression analysis. The software version of the Suzuki-Kudo method is **CHEMICALC** (*Combined Handling of Estimation Methods Intended for Completely Automated LogP Calculation*).

- **Broto-Moreau-Vandycke hydrophobic atomic constants**

The Broto-Moreau-Vandicke contribution method is based on hydrophobic atomic constants a_k measuring the lipophilic contributions of atoms, each described by its nature, neighbouring atoms and associated connectivities, thus implicitly considering some proximity effects and interactions in conjugated systems [Broto *et al.*, 1984b]. Hydrogen atoms and correction factors are not explicitly considered. The model is defined as:

$$\log P = b_0 + \sum_k a_k \cdot \mathrm{N}_k$$

where N_k is the occurrence of the kth atom type in the molecule. The atomic constants a_k were estimated by multivariate regression analysis.

Atom types are classified according to their structural environment; carbon atom-types are differentiated by their bonds to non-hydrogen atoms; heteroatoms are differentiated by their bonds to non-hydrogen first neighbours and the nature of the neighbours, moreover if the neighbour atom is a carbon atom its bond environment is also accounted for. A conjugation contribution is considered as correction factor for sp_2 carbon atoms in conjugated systems.

Using a training set of 1868 compounds, a set of 222 atomic constants was proposed and the best fitted model gave a standard error about 0.4 log unit.

The software program SMILOGP for the calculation of log$P$ is based on the Broto-Moreau-Vandycke hydrophobic constants and the SMILES notation for the recognition of molecule atom-centred fragments [Convard *et al.*, 1994].

- **Ghose-Crippen hydrophobic atomic constants**

The Ghose-Crippen contribution method is based on hydrophobic atomic constants a_k measuring the lipophilic contributions of atoms in the molecule, each described by its neighbouring atoms [Ghose and Crippen, 1986; Ghose *et al.*, 1988; Viswanadhan *et al.*, 1989].

The model is defined as:

$$\log P = \sum_k a_k \cdot \mathrm{N}_k$$

where N_k is the occurrence of the kth atom type. The hydrophobic constants have been evaluated for hydrogen atoms, carbon atoms and heteroatoms. Hydrogen and halogen atoms are classified by the hybridization and oxidation state of the carbon atom to which they are bonded; for hydrogens, heteroatoms attached to a carbon atom in α-position are further considered. Carbon atoms are classified by their hybridization state and depending on whether their neighbours are carbon or heteroatoms.

A set of 120 atom types was proposed. The corresponding hydrophobic constants were evaluated by multivariate regression analysis using a training set of 893 compounds, $r^2 = 0.856$ and $s = 0.496$. This approach is actually called the **ALOGP method** [Viswanadhan *et al.*, 1993], and a similar approach was also proposed for → *molar refractivity*. As in the Broto-Moreau-Vandycke approach, correction factors are avoided, while hydrogen atoms are considered instead.

- **Moriguchi model based on structural parameters**

A model described by a regression equation based on 13 structural parameters and defined as [Moriguchi *et al.*, 1992b; Moriguchi *et al.*, 1994]:

$$\log P = -1.014 + 1.244 \cdot (F_{\mathrm{CX}})^{0.6} - 1.017 \cdot (\mathrm{N_O} + \mathrm{N_N})^{0.9} + 0.406 \cdot F_{\mathrm{PRX}} - 0.145 \cdot \mathrm{N}_{\mathrm{UNS}}^{0.8}$$
$$+0.511 \cdot \mathrm{I_{HB}} + 0.268 \cdot \mathrm{N_{POL}} - 2.215 \cdot F_{\mathrm{AMP}} + 0.912 \cdot \mathrm{I_{ALK}} - 0.392 \cdot \mathrm{I_{RNG}} - 3.684 \cdot F_{\mathrm{QN}}$$
$$+0.474 \cdot \mathrm{N_{NO2}} + 1.582 \cdot F_{\mathrm{NCS}} + 0.773 \cdot \mathrm{I}_{\beta\mathrm{L}}$$

$$n = 1230; \quad r^2 = 0.91; \quad s = 0.411; \quad F = 900.4$$

The meanings of the structural parameters and corresponding regression coefficients are shown in Table L-3.

Table L-3. Regression coefficients of the Moriguchi model.

Symbol	*Descriptor*	b_i
b_0	intercept	–1.014
F_{CX}	summation of number of carbon and halogen atoms weighted by C = 1.0; F = 0.5; Cl = 1.0; Br = 1.5; I = 2.0	1.244
$N_O + N_N$	total number of nitrogen and oxygen atoms	–1.017
F_{PRX}	proximity effect of N/O: X-Y = 2; X-A-Y = 1 (X, Y: N/O; A: C, S, or P) with correction –1 for carbox-amide/sulfonamide	0.406
N_{UNS}	total number of unsaturated bonds (not those in NO_2)	–0.145
I_{HB}	dummy variable for the presence of intramolecular H-bonds	0.511
N_{POL}	number of polar substituents	0.268
F_{AMP}	amphoteric property: α-amino = 1.0; aminobenzoic acid or pyridinecarboxylic acid = 0.5	–2.215
I_{ALK}	dummy variable for alkane, alkene, cycloalkane, cycloalkene (hydrocarbons with 0 or 1 double bond)	0.912
I_{RNG}	dummy variable for the presence of ring structures (not benzene and its condensed rings)	–0.392
F_{QN}	quaternary nitrogen = 1.0; N-oxide = 0.5	–3.684
N_{NO2}	number of nitro groups	0.474
F_{NCS}	–N=C=S group = 1.0; –S–CN group = 0.5	1.582
$I_{\beta L}$	dummy variable for the presence of β-lactam	0.773

• **Moriguchi model based on surface area**

Model for predicting lipophilicity of compounds based on the → *solvent-accessible surface area SASA* generated by a solvent probe of radius 1.4 Å and a set of parameters encoding hydrophilic effects of polar groups [Iwase*et al.*, 1985]:

$$\log P = -1.06 + 1.90 \cdot SASA - 1.00 \cdot \sum_k S_{H_k}$$

$$n = 138; \quad r^2 = 0.99; \quad s = 0.13; \quad F = 7284$$

where S_H are measures of surface area of polar groups contributing negatively to log$P$ of the compounds. The latter parameters can be considered as fragmental correction factors whose values are derived separately for polar groups in aliphatic and aromatic systems.

• **Dunn model based on surface area**

A model for predicting log$P$ values in different solvent systems [Dunn III *et al.*, 1987; Koehler *et al.*, 1988], defined by the equation:

$$\log P_{solv} = a_{solv} \cdot ISA - b_{solv} \cdot f(HSA)$$

where *ISA* is the → *isotropic surface area*, related to the solute surface accessible to nonspecific solvent interactions, and *HSA* the solvent-accessible → *hydrated surface area*, associated with hydration of polar functional groups. *f*(*HSA*) is the *hydrated fraction surface area*, i.e. *HSA* / *SASA*., encoding the polar component of the lipophilicity as the S_H parameter in the Moriguchi model based on surface area.

• **Camilleri model based on surface area**

A model based on the factorization of the solvent-accessible surface area in 12 contributions relative to 12 molecular fragments [Camilleri *et al.*, 1988]:

$$\log P = b_0 + \sum_{k=1}^{12} b_k \cdot \mathrm{N}_k$$

where b are the regression coefficients associated with the surface area contributions and $\mathrm{N_k}$ the number of occurrences of the kth molecular fragment (Table L-4).

Table L-4. Regression coefficients of the Camilleri model.

Type	*Fragment*	b_i
A_0	intercept	–23.9
A_1	aromatic hydrocarbon	2.49
A_2	saturated hydrocarbon chains not A_3, A_6, A_{10}, A_{12}	2.731
A_3	single saturated carbon atom attached to a nonhydrocarbon group plus hydrogens	–2.237
A_4	OH group	–1.809
A_5	oxygen atom of OR group not A_{11}	–0.042
A_6	hydrocarbon part of OR group not A_{12}	0.963
A_7	Cl atom	3.634
A_8	NH_2 or NH group	–3.197
A_9	C(=O)R group	–0.712
A_{10}	hydrocarbon chain part in C(=O)R group	0.697
A_{11}	oxygen atom of OR group in C(=O)OR group	–8.54
A_{12}	hydrocarbon part of OR group in C(=O)OR group	3.526

• **Politzer hydrophobic model**

A model obtained by applying the → *GIPF approach* (*General Interaction Properties Function approach*) proposed by Politzer and co-workers [Brinck *et al.*, 1993; Murray *et al.*, 1993; Murray *et al.*, 1994] as a general method to estimate physico-chemical properties in terms of → *molecular electrostatic potential* (MEP) properties calculated at the → *molecular surface.*

The Politzer hydrophobic model was proposed as the following:

$$\log P = -0.504 + 0.0300 \cdot SA - 0.00472 \cdot (\mathrm{N_N} + 2\mathrm{N_O}) \cdot \sigma_-^2 - 0.000963 \cdot SA \cdot \Pi$$

$$n = 70; \quad r^2 = 0.97; \quad s = 0.277$$

where SA is the molecular surface area, $\mathrm{N_N}$ and $\mathrm{N_O}$ are the numbers of nitrogen and oxygen atoms, respectively, σ_-^2 is the variance of the negative regions of the molecular surface potential, Π is the → *local polarity index.*

Improvements of the Politzer hydrophobic model were later proposed using additional quantum-chemical descriptors derived from the molecular electrostatic potential, dipole moment, and ionization energies. These descriptors were searched for to give the best estimations of the cavity term, polarity/dipolarizability term, and hydrogen-bond parameters defined in → *linear solvation energy relationships* [Haeberlein and Brinck, 1997].

• **Meylan-Howard hydrophobic model**

This hydrophobic model is derived from an atom/fragment contribution method providing hydrophobic atomic and fragmental constants f_k measuring the lipophilic contributions of atoms and fragments in the molecule [Meylan *et al.*, 1992; Meylan and Howard, 1995].

The model is defined as:

$$\log P = 0.229 + \sum_k f_k \cdot \mathrm{N}_k + \sum_j F_j \cdot \mathrm{N}_j$$

$$n = 2351; \quad r^2 = 0.982; \quad s = 0.216$$

where N_k is the occurrence of the kth atom type or fragment, and N_j is the occurrence of the jth correction factor F_j.

The hydrophobic constants f_k have been evaluated by a first linear regression analysis of 1120 compounds, without considering correction factors. The correction factors were then derived from a linear regression on additional 1231 compounds, correlating the differences between experimental log$P$ and the log$P$ estimated by the first regression model.

A set of 130 atom/fragment hydrophobic constants was evaluated, together with 235 correction factors.

- **Gombar hydrophobic model** (: *VLOGP model*)

A model for the assessment of log$P$ based on 363 molecular descriptors derived from the molecular topology and obtained from 6675 diverse chemicals, with r^2 = 0.986 and s = 0.20 [Gombar and Enslein, 1996; Gombar, 1999]. Among the molecular descriptors considered are → *molecular weight*, → *electrotopological state indices*, → *Kier shape descriptors* of order 1 – 7, and some → *symmetry descriptors*. In particular, several descriptors are defined as the sum of the E-states of the atoms involved in the whole molecule or in predefined molecular fragments.

- **Bodor hydrophobic model** (: *Bodor LOGP, BLOGP*)

A nonlinear 18-parameter model based on 10 molecular descriptors calculated by semi-empirical quantum-chemistry methods, starting from optimized 3D geometries [Bodor *et al.*, 1989; Bodor and Huang, 1992b; Huang and Bodor, 1994]:

$$\log P = 9.552 + 0.005286 \cdot MW + 0.08325 \cdot \mathrm{N_C} + 1.0392 \cdot \mathrm{I_{alk}} - 0.05726 \cdot \mu$$
$$- 7.6661 \cdot O - 5.5961 \cdot O^2 + 2.1059 \cdot O^4 + 0.05984 \cdot SA - 0.0001141 \cdot SA^2$$
$$- 0.2741 \cdot Q - 8.5144 \cdot Q_\mathrm{N} + 31.243 \cdot Q_\mathrm{N}^2 - 17.377 \cdot Q_\mathrm{N}^4$$
$$- 4.6249 \cdot Q_\mathrm{O} + 20.346 \cdot Q_\mathrm{O}^2 - 5.4195 \cdot Q_\mathrm{O}^4 - 5.0040 \cdot Q_\mathrm{ON}$$

$$n = 302; \quad r^2 = 0.96; \quad s = 0.306; \quad F = 368$$

The molecular descriptors are shown in Table L-5:

Table L-5. Regression coefficients of the Bodor model.

Symbol	*Descriptor*	b_i
b_0	intercept	9.5524
MW	→ *molecular weight*	0.005286
$\mathrm{N_C}$	number of carbon atoms	0.08325
$\mathrm{I_{ALK}}$	indicator variable for the presence of alkanes	1.0392
μ	→ *dipole moment*	–0.05726
O	→ *ovality index*	–7.6661
O^2	second power of the ovality index	–5.5961
O^4	fourth power of the ovality index	2.1059
SA	→ *van der Waals surface area*	0.05984

Symbol	*Descriptor*	b_i
SA^2	second power of the van der Waals surface area	−0.0001141
Q	→ *total absolute atomic charge*	−0.2741
Q_N	square root of the sum of the squared charges of nitrogen atoms	−8.5144
Q_N^2	second power of Q_N	31.243
Q_N^4	fourth power of Q_N	−17.377
Q_O	square root of the sum of the squared charges of oxigen atoms	−4.6249
Q_O^2	second power of Q_O	20.346
Q_O^4	second power of Q_O	−5.4195
Q_{ON}	sum of the absolute charges of oxygen and nitrogen atoms	−5.0040

The model shows high correlation between the independent variables and some not-significant regression coefficients.

A model based on the same set of quantum-chemical descriptors was also proposed for aqueous solubility by a neural network approach [Bodor *et al.*, 1991; Bodor *et al.*, 1992c; Bodor *et al.*, 1994].

- **Kantola – Villar – Loew hydrophobic models**

A lipophilicity model based on atomic charges, surface areas, dipole moments, and a set of adjustable parameters depending only on the atomic number [Kantola *et al.*, 1991]. The parameter values are determined in order to reproduce experimental log*P* values using the following general model:

$$\log P = \sum_{i=1}^{A} \left[\alpha_i \cdot SA_i + \beta_i \cdot SA_i \cdot q_i^2 + \gamma_i \cdot q_i\right] + \delta \cdot \mu$$

where SA are atomic contributions to the solvent-accessible surface area, q are atomic charges, μ the dipole moment. Setting to zero some of the parameters α, β, γ, and δ, different submodels are obtained. The model descriptors are calculated by → *computational chemistry* methods, thus resulting in conformationally dependent hydrophobicity values.

- **molecular lipophilicity potential model**

A log*P* model based on the → *molecular lipophilicity potential* (*MLP*) defined as:

$$\log P = -0.10 + 0.00286 \cdot \sum MLP^+ + 0.00152 \cdot \sum MLP^-$$

$$n = 114; \quad r^2 = 0.94; \quad s = 0.37; \quad F = 926$$

where $\Sigma\ MLP^+$ and $\Sigma\ MLP^-$ descriptors are the sum of the positive and negative *MLP* values, respectively [Gaillard *et al.*, 1994b]. They represent the hydrophobic and polar contributions of the molecule.

The specific expression for *MLP* used in this model is the following:

$$MLP_i = \sum_k a_k \cdot \frac{1 + \exp(b \cdot c)}{1 + \exp[b \cdot (r_{ik} - c)]}$$

where MLP_i is the molecular lipohilicity potential at the ith grid point, a_k are the Broto-Moreau-Vandycke hydrophobic atomic constants, r_{ik} the → *geometric distance* between the kth fragment and the ith grid point; b and c are the two parameters defining the shape of the Fermi-type function used to calculate *MLP* values (b = 1.33 and c = 3.25). *MLP* values are calculated using the → *solvent-accessible surface area* as integration space.

• **lipole**
The lipole of a molecule is a measure of the lipophilic distribution and is calculated from atomic lipophilicity values l_i as:

$$L = \sum_{i=1}^{A} r_i \cdot l_i$$

where r_i and l_i are the distance from the → *centre of mass* and the lipophilicity of the ith atom, respectively; A is the number of atoms in the molecule [Oxford Molecular Ltd., 1999].

• **topological index of hydrophobicity**
A topological index based on molecular connectivity defined as:

$$\chi_H = {}^1\chi + \sum_k \delta\chi_k$$

where ${}^1\chi$ is the → *Randic connectivity index* and $\delta\chi$ are correction factors determined experimentally. [Sakhartova and Shatz, 1984; Shatz *et al.*, 1984]. The corrections are $\delta\chi$(alkanes) = 0, $\delta\chi$(alkylbenzenes) = –1.597, $\delta\chi$(ketones) = –3.076.

📖 [Tanford, 1973] [Smith *et al.*, 1975] [Hopfinger and Battershell, 1976] [Kubinyi, 1976b] [Pleiss and Grunewald, 1983] [Ray *et al.*, 1983] [Sakhartova and Shatz, 1984] [Shatz *et al.*, 1984] [Dearden, 1985] [Unger *et al.*, 1986] [van de Waterbeemd, 1986] [Kamlet *et al.*, 1987b] [Kamlet *et al.*, 1988a] [Kasai *et al.*, 1988] [Schüürmann, 1990] [Still *et al.*, 1990] [Sasaki *et al.*, 1991] [Tsai *et al.*, 1991] [El Tayar *et al.*, 1992b] [Niemi *et al.*, 1992] [Garrone *et al.*, 1992] [Ikemoto *et al.*, 1992] [Rekker *et al.*, 1993] [Abraham and Kellogg, 1993] [Altomare *et al.*, 1993] [Dubost, 1993] [Floersheim *et al.*, 1993] [Jaworska and Schultz, 1993] [Kaliszan, 1993] [Kaliszan *et al.*, 1993] [Mekenyan and Veith, 1993] [ter Laak *et al.*, 1994] [Waller, 1994] [Debnath *et al.*, 1994b] [Tanaka *et al.*, 1994] [Hansch, 1995a] [Abraham *et al.*, 1995] [Basak and Grunwald, 1995a] [Cramer, 1995] [Grunenberg and Herges, 1995] [Reynolds, 1995] [Kim, 1995c] [Pagliara *et al.*, 1995] [Tsai *et al.*, 1995] [Salter and Kell, 1995] [Feng *et al.*, 1996a] [Tanaka and Fujiwara, 1996] [Basak *et al.*, 1996a] [Caron *et al.*, 1996] [Caron *et al.*, 1997] [Pagliara *et al.*, 1997a] [Pagliara *et al.*, 1997b] [Dross *et al.*, 1998] [Duprat *et al.*, 1998] [Schultz and Cronin, 1999] [Violon, 1999]

LMO technique : *leave-more-out technique* → **validation techniques** (⊙ cross-validation)

loading matrix → **principal component analysis**

local Balaban index → **local vertex invariants**

local connectivity indices → **connectivity indices**

local dipole index → **charge descriptors**

local hardness → **quantum-chemical descriptors** (⊙ hardness indices)

local information on distances : *relative vertex distance complexity* → **topological information indices**

localized effect : *polar effect* → **electronic substituent constants**

local polarity index → **electric polarization descriptors**

local profiles → **molecular profiles**

local quantum-chemical properties → **quantum-chemical descriptors**

local simple flexibility index → **flexibility indices** (⊙ global flexibility index)

local softness → **quantum-chemical descriptors** (⊙ softness indices)

local synthetic invariant → **iterated line graph sequence**

LOcal Vertex Invariants (LOVIs) *(: vertex topological indices, VTIs)*

Descriptors of the graph vertices used to characterize local properties in a molecule; they are numbers associated with graph vertices independently of any arbitrary vertex numbering. They can be either purely topological if heteroatoms are not distinguished from carbon atoms, or chemical if the heteroatoms are assigned distinct values from carbon atoms, even when these are topologically equivalent [Filip *et al.*, 1987; Balaban, 1987a; Ivanciuc *et al.*, 1993b; Balaban, 1994a]. An ideal set of LOVIs is such that distinct LOVIs are relative to nonequivalent vertices in any graph.

Local vertex invariants are used to calculate several molecular → *topological indices* by applying different operators such as addition of LOVIs, addition of squares of LOVIs, addition of reciprocal geometric means for any pair of adjacent vertices. Moreover, they can be used to obtain → *canonical numbering* of molecular graphs and compare molecules in order to study → *molecular branching* and centricity.

The most well-known LOVIs are → *vertex degree*, → *valence vertex degree* and → *bond vertex degree*, → *atom eccentricity*, → *vertex distance degree*, → *walk degree*, → *atomic path counts*, → *atomic ID numbers*, → *path degree*, → *extended connectivity*, → *exponential sum connectivities*, → *MPR descriptors*, → *graph potentials*, LOVIs calculated by applying → *centric operator* and → *centrocomplexity operator* to → *layer matrices*.

A set of local vertex invariants $EFTI_i$ was derived from → *fragment topological indices*, when one non-hydrogen atom at a time is considered:

$$EFTI_i = TI(\mathcal{G}) - IFTI(\mathcal{G}') - IFTI(i)$$

where TI is any topological index, increasing with increase in the number of graph vertices, $\mathcal{G}'$ is the subgraph obtained by erasing the ith vertex with its incident edges, IFTI(i) is the corresponding topological index calculated for the ith vertex, which is often equal to zero or constant [e.g. IFTI(i) = 1 for the → *Hosoya Z index*] [Mekenyan *et al.*, 1988a].

Another set of local vertex invariants was proposed by Ivanciuc [Ivanciuc, 1989] as combinations of topological distances d_{ij} and vertex degrees δ by the following general expression:

$$VTI - N_i = \sum_{j=1}^{A} d_{ij}^{\alpha} \cdot \delta_i^{\beta} \cdot \delta_j^{\gamma}$$

where $\alpha = 0, 1$; $\beta = 0, 1, -1$; $\gamma = 0, 1, -1$, giving a total of 18 LOVIs (N = 1, 2, ...,18 in VTI–N).

The vertex distance degree σ_i is a particular case where $\alpha = 1$, $\beta = 0$ and $\gamma = 0$.

Another set of local vertex invariants was proposed by Diudea *et al.* [Diudea *et al.*, 1996a] using a Randic-type formula as:

$$conn(P)_i = \sum\nolimits_{j(i)} \left(P_i \cdot P_{j(i)}\right)^{\alpha}$$

where P_i and $P_{j(i)}$ are → *atomic properties* of the ith vertex and its first neighbour vertices, respectively; the sum runs over all vertices bonded to the ith vertex; α is usually equal to –1/2 and sometimes to +1/2. By summing all LOVIs over all atoms, the corresponding graph invariant is obtained.

If the atomic property is the vertex degree, the obtained LOVIs correspond to twice the first order → *local connectivity indices*. Moreover, the **Randic-Razinger index** $\chi_i^{^kW}$ is defined as:

$$\chi_i^{^kW} = \sum\nolimits_{j(i)} \left({}^kW_i \cdot {}^kW_{j(i)}\right)^{-1/2}$$

where ${}^kW$ is the walk degree of the kth order and the **local Balaban index** J_i is defined as:

$$\mathrm{J}_i = \sum\nolimits_{j(i)} \left(\sigma_i \cdot \sigma_{j(i)}\right)^{-1/2}$$

where σ is the vertex distance degree [Diudea *et al.*, 1997a]. It can be noted that the sums of these LOVIs over all the atoms correspond to specific functions of the related graph invariants:

$$2 \cdot \chi^{^kW} = \sum_{i=1}^{A} \chi_i^{^kW} \qquad \frac{2 \cdot \mathrm{J} \cdot (C+1)}{B} = \sum_{i=1}^{A} \mathrm{J}_i$$

where $\chi^{^kW}$ is one of the → *walk connectivity indices* and J the → *Balaban distance connectivity index*; C and B are the → *cyclomatic number* and the number of bonds, respectively.

📖 [Klopman *et al.*, 1988] [Klopman and Raychaudhury, 1990] [Balaban and Balaban, 1991] [Balaban *et al.*, 1991] [Balaban and Balaban, 1992] [Balaban, 1992] [Balaban *et al.*, 1992] [Bonchev and Kier, 1992] [Ivanciuc *et al.*, 1992] [Kier and Hall, 1992a] [Balaban and Diudea, 1993] [Bonchev *et al.*, 1993a] [Balaban, 1994c] [Balaban, 1995b] [Diudea *et al.*, 1995a] [Medeleanu and Balaban, 1998]

log K_{ow} : *octanol-water partition coefficient* → **lipophilicity descriptors**

log$P$: *octanol-water partition coefficient* → **lipophilicity descriptors**

London cohesive energy → **Hildebrand solubility parameter**

long hafnian → **algebraic operators** (⊙ determinant)

LOO technique : *leave-one-out technique* → **validation techniques** (⊙ cross-validation)

loops → **graph**

lopping centric information index → **centric indices**

Lovasz-Pelikan index → **eigenvalue-based descriptors**
(⊙ eigenvalues of the adjacency matrix)

Löwdin population analysis → **charge descriptors** (⊙ atomic charge)

lowest unoccupied molecular orbital → **quantum-chemical descriptors**

lowest unoccupied molecular orbital energy → **quantum-chemical descriptors**

LUMO electron density on the *a*th atom → **charge descriptors** (⊙ net orbital charge)

M

mathematical representation of molecular descriptors → molecular descriptors

matrix method for canonical ordering → canonical numbering

matrix representation of a molecular structure

Matrices are the most common mathematical tool to encode structural information of molecules. They usually are the starting point for the calculation of many → *molecular descriptors* and → *local vertex invariants*; moreover, they constitute the mathematical form used as molecule input in the majority of software packages. In most cases, rows and columns of the matrix represent the molecule atoms and the matrix entries information about pairs of vertices such as their connectivities, topological or geometric distances, sums of the weights of the atoms along the connecting paths; the diagonal entries can encode chemical information about the atoms.

The most important matrices or classes of matrices are listed in Table M-1.

Table M-1. Matrix representation of the molecular structure: symbol, current name, number of rows and columns (*A*: number of atoms, *B*: number of bonds, C^+*: cyclicity, D: topological diameter, K: maximum walk length), symmetry (Y: symmetric matrix, N: unsymmetric matrix; D: diagonal matrix).*

Symbol	*Matrix*	*Rows*	*Col.*	*Symm.*
$\mathbf{A}$	adjacency matrix	*A*	*A*	Y
$\mathbf{\Omega}^{\mathbf{AP}}$	all-path matrix	*A*	*A*	Y
$\mathbf{C}$	atom connectivity matrix	*A*	*A*	Y
$\mathbf{D}^{\mathbf{Z}}$	Barysz distance matrix	*A*	*A*	Y
${}^{b}\mathbf{E}$	bond distance-weighted edge adjacency matrix	*B*	*B*	Y
${}^{\pi}\mathbf{E}$	bond order-weighted edge adjacency matrix	*B*	*B*	Y
$\mathbf{CT}$	charge term matrix	*A*	*A*	N
$\mathbf{CJ}$	Cluj matrices	*A*	*A*	
$\mathbf{CJ\Delta}$	Cluj-detour matrix	*A*	*A*	
$\mathbf{CJD}$	Cluj-distance matrix	*A*	*A*	
$\mathbf{D}_{\Delta}$	delta matrix	*A*	*A*	Y
$\mathbf{\Delta}$	detour matrix	*A*	*A*	Y
$\mathbf{\Delta/D}$	detour/distance matrix	*A*	*A*	N
$\mathbf{\Delta}_{\mathbf{P}}$	detour-path matrix	*A*	*A*	Y
$\mathbf{D}$	distance matrix	*A*	*A*	Y
$\mathbf{DD}$	distance/distance matrix	*A*	*A*	Y
$\mathbf{D/\Delta}$	distance/detour quotient matrix	*A*	*A*	Y
$\mathbf{D}_{\mathbf{P}}$	distance-path matrix	*A*	*A*	Y
$\mathbf{E}$	edge adjacency matrix	*B*	*B*	Y
$\mathbf{C}_{\mathbf{E}}$	edge cycle matrix	*B*	C^+	
${}^{E}\mathbf{D}$	edge distance matrix	*B*	*B*	Y
$\tilde{\mathbf{D}}$	expanded distance matrix	*A*	*A*	Y
$\mathbf{DM}$	expanded distance matrices	*A*	*A*	Y

Symbol	*Matrix*	*Rows*	*Col.*	*Symm.*
GM	expanded geometric distance matrices	*A*	*A*	Y
HM	expanded reciprocal distance matrices	*A*	*A*	Y
$\mathbf{H_GM}$	expanded reciprocal geometric distance matrices	*A*	*A*	Y
EA	extended adjacency matrix	*A*	*A*	Y
M	Galvez matrix	*A*	*A*	N
^{E}G	geometric edge distance matrix	*B*	*B*	Y
G	geometry matrix	*A*	*A*	Y
I	incidence matrix	*A*	*B*	N
Z	Hosoya matrix	*A*	*A*	Y
L	Laplacian matrix	*A*	*A*	Y
LM	layer matrices	*A*	*D* + 1	N
M	molecular matrix	*A*	3	N
***D**	multigraph distance matrix	*A*	*A*	Y
P	P matrix	*A*	*A*	Y
$\mathbf{\Delta^{-1}}$	reciprocal detour matrix	*A*	*A*	Y
$\mathbf{CJ\Delta^{-1}}$	reciprocal Cluj-detour matrix	*A*	*A*	
$\mathbf{CJD^{-1}}$	reciprocal Cluj-distance matrix	*A*	*A*	
$\mathbf{\Delta/D^{-1}}$	reciprocal detour/distance matrix	*A*	*A*	Y
$\mathbf{D^{-1}}$	reciprocal distance matrix	*A*	*A*	Y
G^{-1}	reciprocal geometry matrix	*A*	*A*	Y
$\mathbf{D^{-2}}$	reciprocal square distance matrix	*A*	*A*	
$\mathbf{SZ^{-1}}$	reciprocal Szeged matrix	*A*	*A*	
$\mathbf{W^{-1}}$	reciprocal Wiener matrix	*A*	*A*	Y
$\mathbf{\Omega}$	resistance distance matrix	*A*	*A*	Y
RRW	restricted random walk matrix	*A*	*A*	N
SM	sequence matrices	*A*	*K*	N
SZ	Szeged matrix	*A*	*A*	
$\mathbf{SZ_UP}$	Szeged property matrices	*A*	*A*	N
T	topological state matrix	*A*	*A*	Y
$\mathbf{C_V}$	vertex cycle matrix	*A*	C^{+}	N
V	vertex degree matrix	*A*	*A*	D
$\mathbf{^{k}W_M}$	walk diagonal matrix	*A*	*A*	D
$\mathbf{W_{(M1,\,M2,\,M3)}}$	walk matrix	*A*	*A*	N
W	Wiener matrix	*A*	*A*	Y
$\boldsymbol{\chi}$	χ matrix	*A*	*A*	Y

📖 [Rouvray, 1976] [Kunz, 1989] [Randic *et al.*, 1993a] [Ivanciuc *et al.*, 1997a]

maximal information content → **information content**

maximum bond length → **resonance descriptors** (⊙ RC index)

maximum common substructure (MCS)

The maximum common substructure (often "maximal common substructure") of two compounds is the largest possible substructure that is present in both structures. The recognition of a maximum common substructure depends on the defined matching conditions; for example, two substructures are considered to be identical if all atoms and all bonds (single, double, triple, aromatic) can be matched. A further restriction can be applied concerning the number of hydrogen atoms: two non-hydrogen atoms are considered to be identical only if the number of hydrogens bonded to them is equal.

The MCS is a measure and a description of the similarity of two structures whose numerical value *MCS* is the number of common elements provided by the matching conditions, i.e. a measure of the size of the maximum common substructure. It is commonly used in → *similarity searching* [Scsibrany and Varmuza, 1992b].

The MCS of a set of N compounds, however, may be very small or may not even exist if an exotic structure is contained in the set. Therefore, the common structural characteristics of a set of structures are better described by a set of MCS, each of them being the MCS of a pair of structures. Such a set is obtained by determining the MCS for all the N(N-1)/2 pairs of compounds; then the number N_i of occurrences of each different MCS is counted in the set. Finally, an ordered set of MCS is obtained by a ranking function which consideres both frequency and size of the MCS:

$$R_i = (1-k)\cdot\frac{N_i}{N} + k\cdot\frac{A_i}{A^{\max}}$$

where A_i is the number of non-hydrogen atoms in MCS_i and $A^{\max}$ is the maximum number of non-hydrogen atoms in all MCS. k is a user-adjustable parameter (ranging between 0 and 1) which determines the different influence of the frequency and size of MCS; for $k = 1$, only size is considered in the ranking, while for $k = 0$, only frequencies [Varmuza *et al.*, 1998]. The ordered set of MCS characterizes common and typical structural properties in the investigated set of compounds.

A measure of similarity obtained by the maximum common substructure between two compounds s and t is given by:

$$SI_{st} = \frac{(A+B)_{MCS}}{(A+B)_s}\cdot\frac{(A+B)_{MCS}}{(A+B)_t}$$

where $(A + B)$ is the sum of atoms and bonds in the maximum common substructure MCS, in the sth compound and tth compound, respectively [Durand *et al.*, 1999]. A topological distance between the sth and tth compounds is usually defined as:

$$d_{st} = (A+B)_s + (A+B)_t - 2\cdot(A+B)_{MCS}$$

Usually the considered MCSs are connected graphs, i.e. continuous bonded substructures, but disconnected substructures can also be allowed, using a corrected *MCS*, as for example:

$$MCS_{st} = A_{MCS} - k\cdot(N_{FRAG} - 1)$$

where N_{FRAG} is the number of disconnected fragments, k a penalty function between 0 and 1, and A_{MCS} the atom number of the MCS between the sth and tth compounds.

Therefore, the **highest scoring common substructure** (*HSCS*) value is a standardized variable (Z-score) proposed to measure the similarity between the two molecules s and t [Sheridan and Miller, 1998], defined as:

$$HSCS_{st} = \frac{MCS_{st} - Mean(A_s, A_t)}{Std(A_s, A_t)}$$

where MCS_{st} is the score of the actual MCS, *Mean* and *Std* are the mean expected score and standard deviation of the *MCS* within a large sample of randomly selected molecules of the same size; they are calculated by regression analysis as:

$$Mean(A_s, A_t) = b_0^{\text{mean}} + b_1^{\text{mean}} \cdot \min(A_s, A_t)$$

$$Std(A_s, A_t) = b_0^{\text{std}} + b_1^{\text{std}} \cdot \min(A_s, A_t)$$

where $\min(A_s, A_t)$ is the number of atoms in the smallest molecule for a pair of randomly selected molecules of the same size as molecules *s* and *t*. *HSCS* values greater than 4.0 can be considered highly significant.

📖 [Cone *et al.*, 1977] [Brint and Willett, 1987b] [Brint and Willett, 1987a]

maximum electrophilic superdelocalizability → **quantum-chemical descriptors**
(⊙ electrophilic superdelocalizability)

maximum/minimum path matrix : *detour/distance matrix* → **detour matrix**

maximum/minimum path sum → **detour matrix**

maximum negative charge → **charge descriptors**

maximum nuclear repulsion for C–H bond index

A molecular descriptor accounting for nuclear repulsion energy between bonded carbon and hydrogen nuclei. It is defined as:

$$\mathrm{E}_{nm}^{\mathrm{CH}} = \max_k \left(\frac{\mathrm{Z_C} \cdot \mathrm{Z_H}}{r_{\mathrm{CH}}} \right)_k$$

where Z are the atomic numbers, r_{CH} the C–H bond length, and *k* refers to a pair of bonded carbon and hydrogen atoms [Katritzky *et al.*, 1998a]. It possibly encodes the information about the hybridization state of carbon atoms because the C–H bond length depends on the carbon hybridization state.

maximum nucleophilic superdelocalizability → **quantum-chemical descriptors**
(⊙ nucleophilic superdelocalizability)

maximum path degree sequence → **detour matrix**

maximum path frequency sequence → **detour matrix**

maximum path matrix : *detour matrix*

maximum path sum → **detour matrix**

maximum positive charge → **charge descriptors**

MCB index → **multiple bond descriptors**

McClelland resonance energy → **resonance indices**

McFarland model → **Hansch analysis**

McGowan's characteristic volume → **volume descriptors**

mean distance degree deviation → **distance matrix**

mean extended local information on distances → **topological information indices**

mean information content → **information content**

mean information content on the adjacency equality → **topological information indices**

mean information content on the vertex cyclic degree equality
→ **topological information indices**

mean information content on the vertex cyclic degree magnitude
→ **topological information indices**

mean information content on the vertex degree equality
→ **topological information indices**

mean information content on the vertex degree magnitude
→ **topological information indices**

mean information index on atomic composition → **atomic composition indices**

mean information index on molecular conformations
→ **information index on molecular conformations**

mean local information on distances : *vertex distance complexity*
→ **topological information indices**

mean overcrossing number

A descriptor for polymer chains accounting for the occurrence of entanglements caused by polymer chains interpenetrating each other. The mean overcrossing number $\bar{N}$ is a → *geometrical descriptor* defined as the number of bond-bond crossings in a regular 2D-projection of the chain, averaged over all possible projections and calculated on the → *molecular geometry* [Arteca, 1999]. It is a suitable descriptor of DNA chains, polymer geometrical shape, the rheological and dynamic properties of polymer melts and concentrated solutions, being explained by the occurrence of entanglements that cause geometrically constrained chain motion.

Moreover, the **average writhe** $\bar{W}_r$ was also defined as the observed overcrossings sum for each given 2D-projection, by distinguishing right-handed crossing (+1) from left-handed crossing (–1). By definition, $\bar{N} \geq \bar{W}_r$. Both $\bar{N}$ and $\bar{W}_r$ provide useful information: in a compact random configuration a large value of $\bar{N}$ and a vanishing value of $\bar{W}_r$ are expected, whereas in a configuration with regular dihedral angles (e.g. a compact helix) both $\bar{N}$ and $\bar{W}_r$ are expected to be large.

These two descriptors can be combined to produce an effective polymer shape parameter, called **order parameter** ς, such as:

$$\varsigma = \frac{\bar{N} - \bar{W}_r}{\bar{N}}$$

which exhibits two regular trends: $\varsigma \to 0$ in a nonentangled regular configuration and $\varsigma \to 1$ in an entangled random configuration.

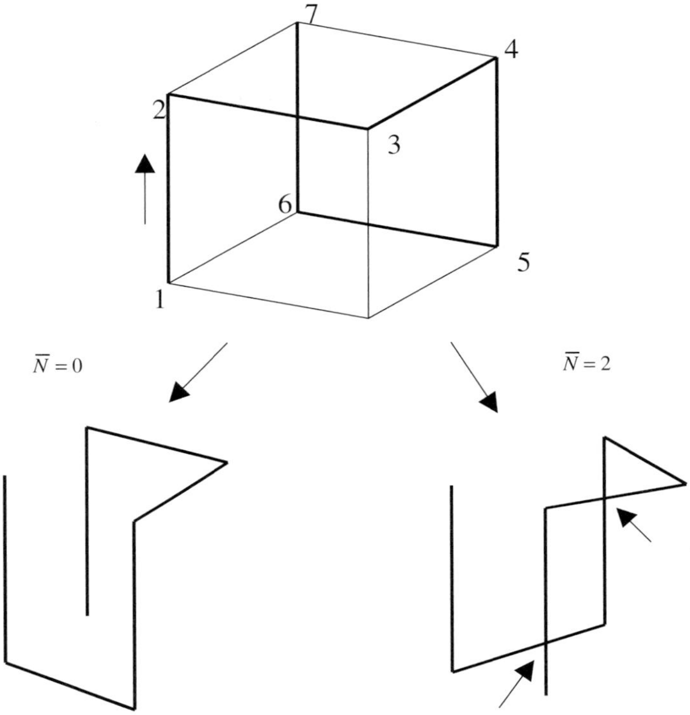

Figure M-1. Example of calculation of the mean overcrossing number.

Another descriptor of the macromolecular topology is the **linking number** $\mathcal{L}$, which characterizes the entanglements of molecules having at least two molecular loops [White, 1969]. For two disjoint curves C_1 and C_2, viewed along a direction in space, the linking number is computed as the sum of the handedness indices of only overcrossings for which curve C_1 is underneath C_2, ignoring the overcrossings of each curve with itself. Two separate, nonentangled curves yield $\mathcal{L} = 0$; the simplest nontrivial link of two loops, $\mathcal{L} = 1$.

📖 [Fuller, 1971] [Arteca and Mezey, 1990]

mean polarizability → **electric polarization descriptors**

mean Randic branching index : *mean Randic connectivity index* → **connectivity indices**

mean Randic connectivity index → **connectivity indices**

mean square distance index → **distance matrix**

mean topological charge index → **topological charge indices**

mean Wiener index → **Wiener index**

Merrifield-Simmons bond order → **symmetry descriptors** (⊙ Merrifield-Simmons index)

Merrifield-Simmons index → **symmetry descriptors**

mesomeric effect → **electronic substituent constants**

Meyer anchor sphere volume → **size descriptors**

Meyer-Richards similarity index → **similarity/diversity**

Meyer visual descriptor of globularity → **shape descriptors**

Meylan-Howard hydrophobic model → **lipophilicity descriptors**

Mezey 3D shape analysis

This is an approach to shape analysis and shape comparison of molecules based on algebraic topological methods suitable for algorithmic, nonvisual analysis and coding of molecular shapes by → *computational chemistry* [Mezey, 1985; Mezey, 1993c]. Several topological methods were proposed for the analysis and coding of molecular shapes, most of them based on the concept of → *molecular surface.* In particular, the **Shape Group Method** (SGM) is a topological shape analysis technique of any almost everywhere twice continuously differentiable 3D functions (e.g. → *electron density*). A detailed description of these methods is given in [Mezey, 1990c].

In general, topological methods are based on subdividing the molecular surface into domains, according to physical or geometrical conditions [Mezey, 1991b].

For example, if two molecular surfaces of the same molecule are considered to be based on two different physical properties such as the → *electron isodensity contour surface* $G_1(m_1)$ and the → *molecular electrostatic potential contour surface* $G_2(m_2)$, then the former can be subdivided into domains according to electrostatic potential values. The interpenetration of the two surfaces provides several closed loops on the isodensity contour surface; these loops are sets of points of G_1 with equal value m_2 of → *molecular electrostatic potential* (MEP) and define the boundaries of the surface G_1 regions that are characterized by MEP values either greater than the threshold value m_2 or lower for all the points in the region. Using different threshold values of MEP, several different electrostatic potential ranges can be mapped on the isodensity surface; these ranges define a subdivision of $G_1(m_1)$ into domains whose topological interralations can be characterized by a numerical code which provides a shape characterization of the molecule.

Applying the same approach to several molecules, the similarity of their shape can be searched for by comparing the topological relations among the corresponding domains on the molecular surfaces.

An alternative approach to domain subdivisions of the molecular surface is based on local curvature properties. It is applicable only to differentiable molecular surfaces such as contour surfaces, e.g. the electron isodensity contour surface $G(m)$, m being the threshold value defining the contour surface.

The local curvature properties of the surface $G(m)$ in each point $\mathbf{r}$ of the surface are given by the eigenvalues of the local Hessian matrix. Moreover, for a defined reference curvature b, the number $\mu(\mathbf{r}, b)$ is defined as the number of local canonical curvatures (Hessian matrix eigenvalues) that are less than b. Usually b is chosen equal to zero and therefore the number $\mu(\mathbf{r}, 0)$ can take values 0, 1, or 2 indicating that at the point $\mathbf{r}$ the molecular surface is locally concave, saddle-type, or convex, respectively. The three disjoint subsets A_0, A_1, and A_2 are the collections of the surface points at which the molecular surface is locally concave, saddle-type, or convex, respectively; the maximum connected components of these subsets A_0, A_1, and A_2 are the surface domains denoted by $D_{0,k}$, $D_{1,k}$ and $D_{2,k}$ where the index k refers to an ordering of these domains, usually according to decreasing surface size.

The mutual arrangement of the domains along the molecular surface can be represented by the topological neighbour relation, two domains being neighbours if they have a common boundary line. Therefore, a symmetric square *shape matrix* can be built up where the rows and columns are given by the surface domains; the off-diagonal entries of the shape matrix can be equal to 1 if the considered domains are adjacent and 0 otherwise; the diagonal entries are equal to the number μ (usually 0, 1, or 2) depending on the domain type. If the surface domains are listed in increasing order according to the size of their surface areas, then the shape matrix encodes both shape and size information. Moreover, from the shape matrix the corresponding *shape graph* can be derived as an additional tool for shape characterization of the molecular surface with respect to the reference curvature *b*; in fact, shape matrix and graph matrix are topological descriptors of molecular surface shape regarded as 3D topological shape codes. They are particularly useful in quantifying the similarity of shapes of the different molecules; the comparison of molecular shape is reduced to a comparison of shape matrices or graphs.

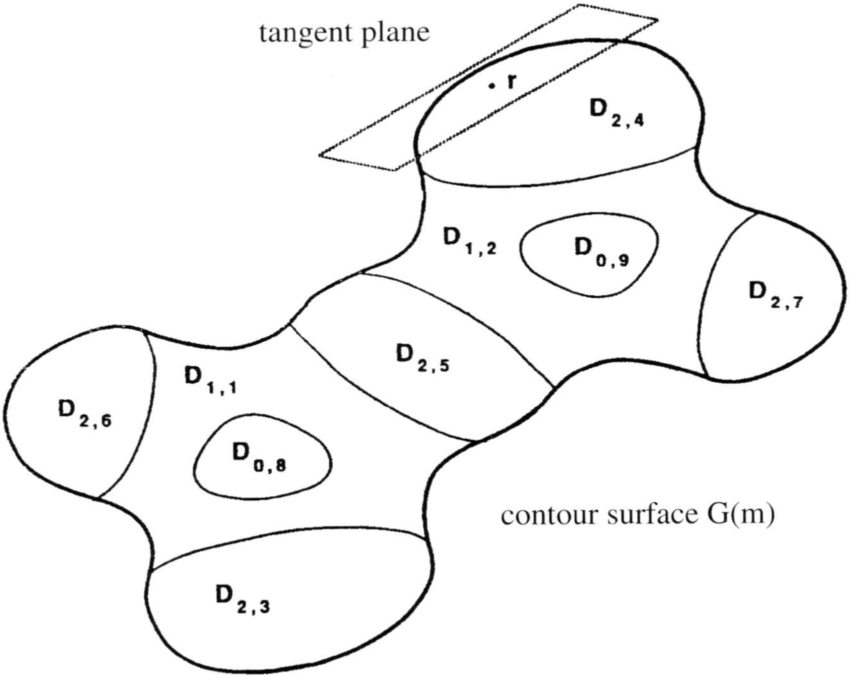

Figure M-2. A subdivision of a molecular contour surface $G(m)$, based on local curvature properties. The contour surface is subdivided into locally concave ($D_{0,k}$), saddle-type ($D_{1,k}$), and locally convex ($D_{2,k}$) domains with respect to the local tangent plane in each point **r**. The second index *k* refers to an ordering of these domains according to decreasing surface area.

It is worth noting that the topological relations among the domains are invariant within a given range of different molecular conformations. In fact, change in molecular conformation can lead to change in size, location and even in domain existence, but for certain conformational changes the existence and the mutual neighbour relations of the domains remain invariant.

Considering a finite number of threshold values m, a set of contour surfaces $G(m)$ is studied for each molecule combined with a set of reference curvature values b. Therefore, for each pair (m, b) of parameters, the curvature domains $D_0(m, b)$, $D_1(m, b)$ and $D_2(m, b)$ are computed and the truncation of contour surfaces $G(m)$ is performed by removing all curvature domains D_μ of specified type μ (in most applications $\mu = 2$) from the contour surface, thus obtaining a truncated surface $G(m, \mu)$ for each (m, b) pair. For most small changes of the parameter values, the truncated surfaces remain topologically equivalent, and only a finite number of equivalence classes is obtained for the entire range of m and b values.

In the next step, the *shape groups* of the molecular surface, that is, the zero-, one-, and two-dimensional algebraic homology groups of the truncated surfaces, are computed. The zero-, one-, and two-dimensional **Betti numbers** $b_\mu^0(m, b)$, $b_\mu^1(m, b)$, and $b_\mu^2(m, b)$ are the ranks of these zero-, one-, and two-dimensional homology groups, i.e. the shape groups. They are a list of topologically invariant numbers, represent a numerical shape code of the molecule, providing a detailed shape characterization of the distribution of the property used to define the molecule surface. In practice, the Betti numbers of 1D shape groups give the most important chemical information, and the analysis is often restricted to this class of Betti numbers.

📖 [Mezey, 1987a] [Mezey, 1987b] [Arteca and Mezey, 1987] [Mezey, 1988a] [Mezey, 1988b] [Mezey, 1988c] [Arteca *et al.*, 1988a] [Arteca *et al.*, 1988b] [Arteca and Mezey, 1988a] [Arteca and Mezey, 1988b] [Mezey, 1989] [Arteca and Mezey, 1989] [Mezey, 1990b] [Mezey, 1991c] [Mezey, 1992] [Mezey, 1993a] [Mezey, 1993b] [Mezey, 1993d] [Mezey, 1994] [Walker *et al.*, 1995] [Mezey, 1996] [Mezey, 1997a] [Mezey, 1999]

MIM complexity indices → **molecular complexity**

MINBID → **ID numbers**

MINCID → **ID numbers**

minimal spanning tree → **graph**

minimal steric difference → **minimal topological difference**

Minimal Topological Difference (*MTD*)

Among the → *hyperstructure-based QSAR techniques*, the *MTD* method is based on the approximate atom-per-atom superimposition of the n molecules of a → *data set* in order to build a → *hypermolecule* (i.e. 3D → *hyperstructure*): hydrogen atoms, small differences in atomic positions, bond lengths and bond angles are neglected. The S vertices of the hypermolecule correspond to the positions of the data set molecule atoms [Simon *et al.*, 1973; Simon and Szabadai, 1973; Simon *et al.*, 1984].

The basic idea is that the geometry of the hypermolecule is related to the geometry of the receptor binding site, and molecule steric affinity to the binding site is obtained by comparing between geometry of the molecule and that of the hypermolecule. Moreover, in order to represent the active regions of the hypermolecule, vertices within the → *binding site cavity* (*cavity vertices*) are labelled with a parameter $\varepsilon = -1$, vertices in the cavity walls (*wall vertices*) with $\varepsilon = +1$, and vertices in the external part of the cavity, i.e. in the aqueous solution, with $\varepsilon = 0$.

In this way, ***MTD* descriptors** measuring the → *steric misfit* between the binding site cavity and the considered molecules are calculated as:

$$MTD_i = c + \sum_{s=1}^{S} \varepsilon_s \cdot \mathrm{I}_{is}$$

where the subscript i refers to the considered molecule; c is the total number of cavity vertices of the hypermolecule and should be a measure of the volume of the binding site cavity; S is the total number of hypermolecule vertices; I_{is} is a binary variable for the sth hypermolecule vertex equal to 1 if the ith molecule occupies the sth vertex with an atom, otherwise equal to zero.

The MTD_i descriptor is a measure of steric misfit of the ith molecule with respect to the receptor cavity and is equal to the number of unoccupied cavity vertices plus the number of occupied wall vertices; it can be considered among both → *steric descriptors* and → *differential descriptors.*

For a molecule coincident with the active region of the hypermolecule, it can be observed that c atoms are located at the hypermolecule cavity vertices ($\varepsilon = -1$), and no atoms coincide with the hypermolecule wall vertices, i.e. there is no steric misfit and $MTD = 0$.

The ***MTD* model** is obtained by including *MTD* descriptors in the → *Hansch model*:

$$\hat{y}_i = b_0 + \sum_{j=1}^{J} b_j \cdot \Phi_{ij} - b' \cdot MTD_i$$

where Φ are J selected → *physico-chemical properties* of a Hansch-type model and the sign minus of the *MTD* coefficient indicates the detrimental contribution to the activity due to steric misfit.

The final optimal set of ε_1, ε_2, ..., ε_s values (*receptor map*) is searched for according to an optimization procedure, starting from random or arbitrary assignment of hypermolecule vertices to cavity, wall or external regions. To each set of ε_s values corresponds a set of *MTD* descriptors and a set of calculated responses. The optimization procedure is based on the maximization of the correlation of the biological responses calculated by the *MTD* model with the experimental responses.

The ***MTD-MC* method** is a modified version of the *MTD* method, accounting for the existence of several low-energy conformations of molecules used to derive the hypermolecule by overlap. Each molecule is described by a vector of binary variables $\mathrm{I}_{i(k)s}$ equal to one if the sth vertex of the hypermolecule is occupied by the ith molecule in the kth conformation. If more than one low-energy conformation is allowed for a molecule, the conformations considered will be the one which best fits the binding site cavity, i.e. the one with the lowest *MTD* value:

$$MTD_i = \min_k \left(c + \sum_{s=1}^{S} \varepsilon_s \cdot \mathrm{I}_{i(k)s} \right)$$

where k runs over all of the considered conformations of the ith molecule.

A modification of the *MTD* approach that also acts on biological responses, called the ***MTD-ADJ* method**, was proposed to improve the performance of the modelling power of the method, accounting for the relative contribution of the active conformation to the activity of each compound [Sulea *et al.*, 1998].

Let C_A be the concentration of active conformation A; the following relationships hold:

$$C_\mathrm{A} = \alpha_\mathrm{A} \cdot C \qquad y^{\mathrm{exp}} = -\log C \qquad y^{\mathrm{adj}} = -\log(\alpha_\mathrm{A} C) = y^{\mathrm{exp}} - \log\alpha_\mathrm{A}$$

where C is the total concentration, y^{exp} and y^{adj} the experimental and adjusted biological responses. The factor α_A for each conformation is calculated by the Boltzmann distribution:

$$\alpha_A = \frac{g_A \cdot \exp^{-E_A/RT}}{\sum_{k=1}^{N} g_k \cdot \exp^{-E_k/RT}}$$

where E are the calculated total conformational energies, g the degeneration degree of the conformational energy levels; R and T the gas constant and the absolute temperature, respectively.

For each conformation of all compounds, the corresponding *MTD* value and adjusted biological response are calculated. By using the optimization and → *validation techniques* of the *MTD* model, each compound's conformation which best fits the adjusted response is retained and should be considered the active conformation of the compound. If more than one conformation is selected for the same compound, all these conformers have the same *MTD* values, while the adjusted response is calculated considering the conformer contributions to the total population, i.e. $\Sigma\alpha$.

The *MTD-ADJ* method provides additional information concerning the active conformations of the compounds.

The **Minimal Steric Difference** (*MSD*) method is the first version of the *MTD* QSAR approach, where each molecule is compared with the molecule with the highest biological acitivity in the data set, taken as the → *reference structure* instead of the hypermolecule. The assumption is that the most active molecule fits into the binding site best [Simon and Szabadai, 1973].

MSD_i is a descriptor of steric misfit defined as the number of unsuperposable non-hydrogen atoms for the maximal superimposition of the ith molecule to the reference molecule. *MSD* coincides with *MTD* when simple minimal steric differences are calculated with respect to the most active compound, no external atoms are considered in the hypermolecule, and the number of hypermolecule cavity vertices correspond to the atoms of the reference molecule.

The **Monte Carlo version of MTD** (*MCD*) approach is a modification of the *MSD* approach, where the MCD_i descriptor is calculated as the nonoverlapping volume (NOV_i) of the ith molecule with respect to the reference molecule and the reference molecule with respect to the ith molecule [Motoc *et al.*, 1977].

The two superimposed molecules are included within a cube with volume V, and a large number of points N are randomly dispersed within the cube. The **nonoverlapping volume** NOV_i of the ith molecule is calculated as:

$$NOV_i = \mathrm{V} \cdot \left(\frac{\mathrm{N_{REF}} + \mathrm{N}_i}{\mathrm{N}} \right)$$

where $\mathrm{N_{REF}}$ is the number of points falling into the → *van der Waals molecular surface* of the reference molecule but not into that of the ith molecule, and N_i is the number of points falling into the van der Waals envelope of the ith molecule but not into that of the reference molecule; N is the total number of points randomly distributed throughout volume V of the cube.

A further modification of the *MTD* approach is called **Steric Interactions in BIolo-gical Systems** (*SIBIS*), where **attractive steric effects** (*SMDC*) and **repulsive steric effects** (*SMDW*) are considered separately [Motoc and Dragomir, 1981; Motoc, 1984a; Motoc, 1984b]. Moreover, the optimization procedure searching for the receptor map, i.e. the optimal set of ε_s values, is modified by the introduction of connectivity restrictions, where all the cavity vertices have to form a single topological connected network, i.e. the receptor cavity is not fragmented into several subcavities.

The two steric contributions are defined as:

$$SMDC_i = \sum_{s=1}^{S_{cav}} b'_{is} \cdot \mathrm{I}_{is} \qquad SMDW_i = \sum_{s=1}^{S_{wall}} b''_{is} \cdot \mathrm{I}_{is}$$

where b_{is} are correction factors accounting for the size of the atom of the ith molecule in the sth position of the hypermolecule; I_{is} is a binary variable for the sth hypermolecule vertex equal to 1 if the ith molecule occupies the sth vertex with an atom, and equal to zero otherwise. The first descriptor *SMDC* is a sum running over all hypermolecule cavity positions S_{cav} ($\varepsilon_s = -1$) and the second descriptor *SMDW* a sum over all the hypermolecule cavity wall positions S_{wall} ($\varepsilon_s = +1$).

Therefore, the **SIBIS model** is defined as:

$$\hat{y}_i = b_0 + \sum_{j=1}^{J} b_j \cdot \Phi_{ij} + b' \cdot SMDC_i - b'' \cdot SMDW_i$$

where Φ are selected J physico-chemical properties of the Hansch model; the plus sign of the *SMDC* coefficient indicates a favourable contribution (attractive steric effects) to activity, and the minus sign of the *SMDW* coefficient a detrimental contribution (repulsive steric effects).

📖 [Simon, 1974] [Simon and Szabadai, 1975] [Simon *et al.*, 1976a] [Simon *et al.*, 1976b] [Simon *et al.*, 1977] [Popoviciu *et al.*, 1978] [Balaban *et al.*, 1980c] [Motoc, 1983] [Simon *et al.*, 1985a] [Balaban *et al.*, 1987b] [Magee, 1991] [Simon and Bohl, 1992] [Simon, 1993] [Ciubotariu *et al.*, 1993] [Oprea *et al.*, 1993] [Muresan *et al.*, 1995] [Sulea *et al.*, 1995] [Fabian *et al.*, 1995] [Mracec *et al.*, 1996] [Timofei *et al.*, 1996] [Polanski, 1997] [Hadaruga *et al.*, 1999]

Minoli complexity index → **molecular complexity**

mixed CoMFA approach → **comparative molecular field analysis**

mixed CoMFA model → **comparative molecular field analysis**

MmPS topological index : *detour/Wiener index* → **detour matrix**

model complexity (*cpx*)

Model complexity is an important parameter to compare different QSAR/QSPR models. Moreover, the prediction power of a model is inversely related to its complexity.

In general, model complexity is related to the number of variables selected for modelling purposes. Let **I** be the vector of length p, where p is the total number of variables, constituted of p binary variables. Each variable takes a value equal to zero ($\mathrm{I}_j = 0$) if the jth variable is excluded and a value equal to one ($\mathrm{I}_j = 1$) if the jth variable remains included within the variables.

The general problem of excluding variables from data, i.e. of estimating the best $\mathbf{I}^*$ vector, can be divided in two main blocks: methods for → *variable reduction* and methods for → *variable selection*. The first group of methods evaluates the variable exclusion by inner relationships among the p descriptor variables, i.e.

$$\mathbf{I} = f(\mathbf{x}_1, \mathbf{x}_2, \ldots, \mathbf{x}_p)$$

while the second group evaluates the variable exclusion by considering the relationships among the $\mathbf{x}_j$ variables and a response variable $\mathbf{y}$, i.e.

$$\mathbf{I} = f(\mathbf{x}_1, \mathbf{x}_2, \ldots, \mathbf{x}_p; \mathbf{y})$$

In the first case, attention is paid to excluding variables carrying low or redundant information, in the second, to excluding variables which are not functionally related to the studied response. In the latter, besides the exclusion of specific variables, one can condense the information from all the original variables into a few significant latent variables (linear combinations) by methods such as *Principal Component Regression* and *Partial Least Squares regression*.

The main measures of model complexity are reported below.

- **number of terms in the model**

The simplest definition of model complexity is based on the number of terms in the model or, in other words, the model complexity is made up by the number of model variables from *Ordinary Least Squares regression* ($cpx = p$), the number M of significant principal components from Principal Component Regression ($cpx = M$), and the number of significant latent variables from Partial Least Squares regression ($cpx = M$)

- **standardized regression coefficients sum**

Model complexity is defined as the sum of standardized regression coefficients:

$$cpx = \sum_{j=1}^{p} \left| b'_j \right| = \sum_{j=1}^{p} \left| \frac{b_j \cdot s_j}{s_y} \right|$$

where b'_j is the jth standardized regression coefficient, b_j the ordinary regression coefficient, s_y and s_j the standard errors of the response and jth variable, respectively, and p is the total number of model variables.

- **information content ratio**

Model complexity is defined as the ratio between the **multivariate entropy** S_X of the X-block (n objects and p variables) of the model and → *Shannon's entropy* H_Y of the **y** response vector, thus also accounting for the information content of the **y** response [Authors, This Book]:

$$cpx = \frac{S_X}{H_Y} \qquad 0 \leq cpx \leq \frac{p \cdot \log_2 n}{H_Y} \leq p$$

where H_Y and S_X are defined as

$$H_Y = -\sum\nolimits_k \frac{n_Y}{n} \log_2 \frac{n_Y}{n}$$

$$S_X = [1 + (p-1)(1-K)] \cdot \frac{\sum_{j=1}^{p} H_j}{p}$$

where k runs on the different equivalence classes for **y** and n_Y is the number of equal y values; H_j is the Shannon entropy of the jth variable; p is the total number of variables and n is the total number of objects. K is the → *multivariate K correlation index*.

When all the y and x values (for each jth variable) are different (i.e. the Shannon's entropy of each variable is $\log_2 n$), the model complexity depends only on the total number p of model variables and the K correlation in the X-block as:

$$cpx = \frac{[1 + (p-1)(1-K)] \cdot \dfrac{p \log_2 n}{p}}{\log_2 n} = 1 + (p-1)(1-K)$$

Then, $cpx = p$ indicates the presence in the model of perfectly p uncorrelated x-variables, while cpx values lower than 1 indicate insufficient information in the X-block to completely model the **y** response.

model of the frontier steric effect → **steric descriptors** (⊙ Taft steric constant)

modified Free-Wilson analysis → **Free-Wilson analysis**

modified Hosoya index → **resonance indices** (⊙ Hosoya resonance energy)

modified Randic index → **connectivity indices**

modified spectrum-like descriptors → **spectrum-like descriptors**

modified valence vertex degree → **vertex degree**

modulo-*L* descriptors → **experimental measurements**

Mohar indices → **Laplacian matrix**

molar polarization → **electric polarization descriptors**

molar refractivity (MR)

Molecular descriptor of a liquid which contains both information about molecular volume and polarizability, usually defined by the Lorenz-Lorentz equation [Lorentz, 1880; Lorentz, 1880]:

$$\mathrm{MR} = \frac{\mathrm{n}^2 - 1}{\mathrm{n}^2 + 2} \cdot \frac{MW}{\varrho}$$

where MW is the → *molecular weight*, ϱ the liquid density (the ratio MW / ϱ corresponds to the → *molar volume* $\bar{\mathrm{V}}$), and n the **refractive index** of the liquid. The refractive index is defined as a ratio of the velocity of light in vacuum to the velocity of light in the substance of interest. Used as an indicator of the purity of organic compounds, it is related to several electric and magnetic properties such as polarizability as well as to critical temperature, surface tension, density, and boiling point. Usually, the refractive index is measured at the sodium D-line and indicated as n_D^2. Moreover, the **refractive index function** $f(\mathrm{n})$ defined as:

$$f(\mathrm{n}) = \frac{\mathrm{n}^2 - 1}{\mathrm{n}^2 + 2}$$

was proposed as a molecular descriptor, accounting for composite solute interactions [Fuchs *et al.*, 1982].

Alternative definitions of molar refractivity were proposed by Gladstone and Dale ($\mathrm{MR_{GD}}$) [Gladstone and Dale, 1858] and by Vogel ($\mathrm{MR_V}$) [Vogel, 1948] as:

$$\mathrm{MR_{GD}} = (\mathrm{n} - 1) \cdot \frac{MW}{\varrho} \qquad \mathrm{MR_V} = \mathrm{n} \cdot MW$$

Molar refractivity values estimated by substituting the molar volume by → *Mc Gowan's characteristic volume* $\mathrm{V_X}$ was proposed by Abraham et al. [Abraham *et al.*, 1990b] as:

$$\mathrm{MR_A} = 10 \cdot f(\mathrm{n}) \cdot \mathrm{V_X}$$

where $f(\mathrm{n})$ is the refractive index function. Moreover, in order to remove cohesive dispersion interactions, it was proposed to subtract the molar refractivity of the *n*-alkane with the same characteristic volume $\mathrm{V_X}$:

$$R_2 = \mathrm{MR_A} - \mathrm{MR_A^*} = \mathrm{MR_A} - (2.83195 \cdot \mathrm{V_X} - 0.52553)$$

where $\mathrm{MR_A}$ is the molar refractivity of the considered compound and $\mathrm{MR_A^*}$ the molar refractivity of the *n*-alkane with the same characteristic volume $\mathrm{V_X}$. The parameter R_2 can be considered a polarizability descriptor and is called **excess molar refractivity**. By definition, $R_2 = 0$ for all *n*-alkanes, and the same holds for branched alkanes.

When molar refractivity is determined using the sodium D-line, it coincides with the → *electron polarization* P_E. Therefore, it can be considered contemporarily as being among → *electronic descriptors* as well as → *steric descriptors*.

As molar refractivity is essentially an additive property, **group molar refractivity** is calculated as the difference between the molar refractivity of an X-substituted compound and the reference compound:

$$\mathrm{MR_X} = \mathrm{MR_{X+REF}} - \mathrm{MR_{REF}}$$

This parameter is often used as a substituent steric constant in → *Hansch analysis*. In order to put the molar refractivities of the substituents on approximately the same scale as the → *hydrophobic substituent constants* π, the substituent MR values are often scaled down by the factor 0.1.

The difference between the molar refractivity of a substituent X and hydrogen was used to estimate the difference in the interaction energy of a hydrogen-substituted parent compound and a X-substituted analogue compound:

$$\Delta E_{\mathrm{INT}} = \frac{-1673.6}{r_{\mathrm{XB}}^6} \cdot (\mathrm{MR_X} - \mathrm{MR_H}) \quad KJ/mol$$

where r_{XB} is the distance in angstroms between the group and the binding site [Pauling and Pressman, 1945].

Values for the atomic molar refractivity were also estimated by → *group contribution methods* [Ghose and Crippen, 1987].

📖 [Vogel, 1948] [Vogel *et al.*, 1951] [Topliss and Shapiro, 1975] [Dunn III, 1977] [Wilkerson *et al.*, 1994] [Arnaud *et al.*, 1994] [Estrada and Ramirez, 1996] [Katritzky *et al.*, 1998b]

molar volume → volume descriptors

molecular branching → molecular complexity

molecular cyclicity → molecular complexity

molecular complexity

The concept of *molecular complexity* was introduced into chemistry only quite recently and is mainly based on the → *information content* of molecules. Several different measures of complexity can be obtained according to the diversity of the considered structural elements such as atom types, bonds, connections, cycles, etc. The first attempts to quantify molecular complexity were based on the elemental composition of molecules; later other molecular characteristics were considered such as the symmetry of molecular graphs, molecular branching, molecular cyclicity and centricity [Bonchev and Seitz, 1996].

A composite hierarchical concept of molecular complexity was proposed by Bonchev-Polansky [Bonchev and Polansky, 1987] according to their *general complexity scheme*, which begins with molecule size and proceeds through topology; molecules of the same size and topology are distinguished by their atom and bond types; moreover, a further discrimination is provided by geometric → *interatomic distances* and molecular symmetry.

In particular, topological complexity is hierarchically defined, and the main features are *molecular branching*, *molecular centricity* and *molecular cyclicity*.

Molecular branching is a molecule property comprising several structural variables such as number of branching, valence, distances apart, distances from the → *graph centre* and length of branches [Kirby, 1994]. Given this multifaced definition of branching, its quantification is not an easy task. However, operational definitions of branching can be given by selected molecular indices which, to some extent, reflect the branching of molecules as intended in an intuitive way [Randic, 1975b; Gutman and Randic, 1977; Bonchev *et al.*, 1979a; Barysz *et al.*, 1985; Bertz, 1988; Rouvray, 1988; Bonchev, 1995; Randic, 1997d; Klein and Babic, 1997b].

The → *Wiener index* was the first proposed index of molecular branching [Bonchev and Trinajstic, 1977]; it is a function, inversely related to branching, of the number, length and position of branches as well as of the number of atoms. For an isomeric series it can be considered to be mainly dependent on molecular branching. Other specific molecular descriptors proposed as measures of branching are the → *Lovazs-Pelikan index*, → *ramification index*, → *ramification pair indices*, → *Zagreb indices*, and → $\lambda\lambda_1$ *branching index*. The → *Balaban centric index*, the → *Randic connectivity index*, the → *mean information content of the distance equality* [Bonchev and Trinajstic, 1978], and the → *Merrifield-Simmons index* can also be used to discriminate among isomeric molecules with different branching patterns.

Molecular cyclicity is another important feature of molecular complexity, defined in terms of number of molecule cycles and the manner in which the cycles are connected. It was first characterized by Bonchev [Bonchev *et al.*, 1980c; Bonchev and Mekenyan, 1983] by a system of rules based on the number of atoms, the number of cycles, the number of atoms in a cycle, the number of cycles having a common edge, etc. The → *cyclomatic number* is a simple measure of independent molecule cycles. The Wiener index was initially chosen as a single-valued molecular descriptor related to the cyclicity degree of isomeric molecules; it decreases for molecules whose cyclic structures are more complex than those of molecules of the same size but with fewer rings. The → *Harary index* and → *Kirchhoff number* were also proposed as discriminating cyclicity indices [Bonchev *et al.*, 1994a]. Recently, Randic [Randic, 1997c] defined the molecular cyclicity in terms of the cyclicity of the corresponding molecular graph *as the departure of the cyclic character of the graph from that of the monocyclic graph relative to the departure of the complete graph from the monocyclic graph.* The cyclic character of the graph is given by the → *D/Δ index* calculated on the → *distance/detour quotient matrix*. In practice, the degree of cyclicity of a molecule with A atoms is calculated by the comparison of the corresponding molecular graph with the two extreme graphs of the same size (monocycle and → *complete graph*). The → *cyclicity index* is a quantitative measure of molecular cyclicity as defined by Randic.

Molecular centricity is considered less important than branching and cyclicity, but it contributes to the quantification of molecular complexity by distinguishing between molecular structures organized differently with respect to their centres [Bonchev, 1997]. → *Centric indices* are topological descriptors related to molecular centricity. Moreover, the → *centric topological index* and → *centrocomplexity topological index* calculated on the → *branching layer matrix* of an → *iterated line graph sequence* were proposed to measure molecular centricity [Diudea *et al.*, 1992a].

Molecular complexity indices are mainly based on → *information indices* defined to account for molecule complexity. These indices may be broadly divided into topological complexity indices and chemical complexity indices [Basak, 1987]. The former are calculated as the information content of molecular graphs where atoms are not distinguished; among these only those able to account for multiplicity in the graph are used to measure molecular complexity. The latter account for the chemical nature of the individual atoms in terms of bonding topology of weighted graphs or through the use of the physico-chemical characteristics of the atoms in the molecule.

The most well-known complexity indices are listed below. Other molecular descriptors which give information about molecular complexity are the → *indices of neighbourhood symmetry* and the → *total adjacency index*. The latter was proposed by Bonchev and Polansky [Bonchev and Polansky, 1987] as a simple measure of topological complexity, being a measure of the degree of connectedness of molecular graph.

- **Bertz complexity index** (I_{CPX})

The most popular complexity index was introduced by Bertz [Bertz, 1981; Bertz, 1983a; Bertz, 1983b], taking into account both the variety of kinds of bond connectivities and atom types of a → *H-depleted molecular graph*.

A general form of a molecular complexity index I_{CPX} is:

$$I_{CPX} = I_{CPB} + I_{CPA}$$

where I_{CPB} and I_{CPA} are the information contents related to the bond connectivity and the atom type diversity, respectively. The term I_{CPB} was originally defined as:

$$I_{CPB} \equiv C(TI) = 2 \cdot TI \cdot \log_2 TI - \sum\nolimits_g TI_g \cdot \log_2 TI_g$$

where TI is any graph invariant, and TI_g is the number of equivalent elements forming the graph invariant TI. The choice of the graph invariant should be based on the assumption that molecular complexity increases with size, branching, vertex and edge weights, etc. The → *connection number* N_2, i.e. the number of bond pairs, was proposed by Bertz as a good choice for evaluating molecular complexity as it measures both the size and symmetry of the graph. Therefore, the two terms of the Bertz complexity index are defined as:

$$I_{CPB} = 2 \cdot N_2 \log_2 N_2 - \sum\nolimits_g (N_2)_g \log_2 (N_2)_g = N_2 \log_2 N_2 + {}^{CONN}I_{ORB}$$

$$I_{CPA} \equiv I_{AC} = A \log_2 A - \sum\nolimits_g A_g \log_2 A_g$$

where $(N_2)_g$ is the number of symmetrically identical connections of type g; A is the total number of atoms (hydrogen excluded); A_g is the number of atoms of the gth element, ${}^{CONN}I_{ORB}$ is the → *total connection orbital information content*. The term I_{CPB} measures the complexity of a molecule given by the partition of equivalent connections sensitive to branching, rings and multiple bonds of the molecule; when all the connections are the same, the bond complexity term is equal to $N_2 \log_2 N_2$ to take into account the size of the molecule, together with its symmetry in terms of bond connectivity. The atom complexity term I_{CPA} accounts for the presence of heteroatoms in the molecule and corresponds to the → *total information index on atomic composition* calculated for H-depleted molecular graphs.

- **Rashevsky complexity index** ($\bar{I}_{RASH}$)

A quantitative measure of graph complexity per vertex based on the sum of a chemical and a topological term:

$$\bar{I}_{RASH} = \bar{I}_{AC} + \bar{I}_{TOP}$$

where the two terms are the → *mean information index on atomic composition* $\bar{I}_{AC}$ and the → *topological information content* $\bar{I}_{TOP}$, respectively [Rashevsky, 1955]. Note that the topological information content proposed by Rashevsky is not based on graph orbits as is the most general topological information content later proposed by Trucco [Trucco, 1956a; Trucco, 1956b]. In fact, two vertices v_i and v_j are considered topologically equivalent if for each kth neighbouring vertex (k ranging between 1 and the → *atom eccentricity*) of vertex v_i, there exists a kth neighbouring vertex of the same degree for vertex v_j.

The Rashevsky complexity index was further developed by Mowshowitz in order to obtain a measure of relative complexity of undirected and directed graphs [Mowshowitz, 1968a; Mowshowitz, 1968b; Mowshowitz, 1968c; Mowshowitz, 1968d].

- **Dosmorov complexity index** (I_{DOSM})

A molecular complexity index defined as a linear combination of five single information indices:

$$I_{DOSM} = I_{AC} + I_{at} + I_{B} + I_{SYM} + I_{CONF}$$

where I_{AC} is the → *total information index on atomic composition* calculated for H-depleted molecular graphs; I_{at} is the → *atomic information content*; I_B is the → *information bond index*; I_{SYM} is the → *information index on molecular symmetry*; I_{CONF} is the → *information index on molecular conformations*. This combination of indices was proposed as a general index of molecular complexity accounting for the chemical nature of atoms, molecular size, number and kind of molecular bonds, symmetry and conformations. Moreover, by incorporating the atomic information content the Dosmorov index becomes more discriminating than the Bertz index in the case of different substituent atoms of the same valence [Bonchev, 1983].

- **Bonchev complexity information index** (I_{BONC})

A molecular descriptor defined in analogy with the Dosmorov complexity index, also accounting for electronic properties of molecules [Bonchev, 1983]:

$$I_{BONC} = I_{IC} + I_{NUCL} + I_{EL} + I_{Topology} + I_{SYM} + I_{CONF}$$

where I_{IC} is the → *information index on isotopic composition*; I_{EL} is one of the electronic information indices; $I_{Topology}$ can be any topological information index accounting for structural complexity; I_{SYM} is the → *information index on molecular symmetry*; I_{CONF} is the → *information index on molecular conformations*; I_{NUCL} is the → *nuclear information content*.

- **Minoli complexity index**

A measure of complexity of a molecular graph monotonically increasing on the number of vertices and edges, and reflecting the degree of connectedness of the graph [Minoli, 1976]. It is defined for undirect graphs, with no loops and multiple edges as:

$$\chi = \frac{A \cdot B}{A + B} \cdot \sum_{i=1}^{A-1} \sum_{j=i+1}^{A} P_{ij}$$

where A is the atom number, B the bond number, and P_{ij} the number of paths of any length between vertices v_i and v_j.

- **Bertz-Herndon relative complexity index** (C_{BH})

A simple measure of structural complexity of a molecule based on its graph representation $\mathcal{G}$ compared with the parent → *complete graph* $K(\mathcal{G})$, i.e. the complete graph with the same number of vertices. It is defined as:

$$C_{BH} = \frac{K_{\mathcal{G}}}{K_{K(\mathcal{G})}}$$

where K is the total number of connected subgraphs in $\mathcal{G}$ and $K(\mathcal{G})$, respectively [Bertz and Herndon, 1986; Bonchev, 1997].

- **Bonchev topological complexity indices** (: *topological complexity indices, TC*)

Complexity indices derived from the graph representation of molecules and mainly based on branching; the basic idea is that "The higher the connectivity of a molecular graph *and its connected subgraphs*, the more complex is the molecule" [Bonchev, 1997].

The *total topological complexity TC* was defined as:

$$TC = \sum_{m=0}^{B} {}^{m}TC = \sum_{k=1}^{K} \mathrm{A_V}(\mathcal{G}_k) = \sum_{k=1}^{K} \sum_{i=1}^{\mathrm{N}_k} \sum_{j=1}^{\mathrm{N}_k} a_{ij}$$

where the first sum runs over all orders of the molecular subgraphs (i.e. the number of edges in the considered subgraphs); ${}^{m}TC$ is the mth order *partial topological complexity* defined by all connected subgraphs having the same number m of edges (e.g. all isolated vertices if $m = 0$, all edges if $m = 1$, etc.); the second sum runs over all K connected subgraphs and $\mathrm{A_V}(\mathcal{G}_k)$ is the total adjacency index of the kth subgraph $\mathcal{G}_k$; N_k is the number of vertices in the kth connected subgraph; a_{ij} are the elements of the → *adjacency matrix* of the → *H-depleted molecular graph* $\mathcal{G}$. In practice, the → *vertex degree* is assigned to each vertex in the molecular graph, then for each subgraph the sum over the degrees of all involved vertices is calculated, and to obtain the partial topological complexity this quantity is summed over all subgraphs of the same order.

A vector of partial complexities was defined as:

$$\langle {}^{0}TC;\ {}^{1}TC;\ {}^{2}TC;\ \ldots;\ {}^{B}TC \rangle$$

where ${}^{B}TC$ is simply the total adjacency index of the molecular graph.

A different definition of topological complexities $TC1$ was given by considering the vertex degrees as they are in isolated subgraphs. The inequality $TC > TC1$ always holds.

- **MIM complexity indices**

A study of a → *leverage matrix* obtained from the → *molecular matrix* M of the atomic coordinates (hydrogen atoms included), i.e. a matrix with A rows (the atoms) and three columns (the x, y, z geometric coordinates), is in progress [Authors, This Book]. This symmetric $A \times A$ matrix H, called the **molecular influence matrix** (MIM), is calculated on the atom coordinates after their centring:

$$\mathrm{H} = \mathrm{M} \cdot \left(\mathrm{M}^{\mathbf{T}} \cdot \mathrm{M}\right)^{-1} \cdot \mathrm{M}^{\mathbf{T}}$$

In this case, the sum of the diagonal elements corresponds to molecule dimensionality, i.e. 1, 2, or 3 for linear, planar, or 3D-molecules, respectively. The atoms of the molecule can be viewed as the objects of a → *data set* projected onto the space of the independent variables.

The diagonal elements h_{ii} of the molecular influence matrix represent the "influence" of each molecule atom in determining the whole shape of the molecule; in fact mantle atoms always have higher h_{ii} values than atoms near the molecule centre. They can be interpreted as the atom accessibility to outer interactions and their magnitudes depend on the size and shape of the molecule.

As the values of the leverage matrix are sensitive to the whole molecule structure, they automatically contain information about the molecular complexity, which is a function of the size, symmetry, elemental molecular composition, molecular branching, and centricity. Thus, 3D complexity descriptors can be obtained from this matrix as total and standardized information content:

$$I_{\mathrm{TC}} = A \cdot \log_2 A - \sum_{g=1}^{G} \mathrm{N}_g \cdot \log_2 \mathrm{N}_g \qquad \text{and} \qquad I_{\mathrm{TC}}^{*} = 1 - \frac{\sum_{g=1}^{G} \mathrm{N}_g \cdot \log_2 \mathrm{N}_g}{A \cdot \log_2 A}$$

where the A atoms are partitioned into G equivalence classes according to the magnitude of their h_{ii} values; N_g is the number of vertices with equal diagonal elements h_{ii} of the molecular influence matrix.

If all the atoms have different h_{ii} values, i.e. the molecule does not show any element of symmetry, $I_{TC} = A \cdot \log A$ and $I^*_{TC} = 1$; otherwise, if all the atoms have equal h_{ii} values (a perfectly symmetric theoretical case), $I_{TC} = 0$ and $I^*_{TC} = 0$.

These indices encode information on the molecule entropy (thermodynamic entropy) and are useful in modelling physico-chemical properties related to entropy and symmetry.

molecular complexity indices → **molecular complexity**

molecular connectivity indices *: connectivity indices*

molecular descriptor properties → **molecular descriptors**

molecular descriptors *(: chemical descriptors)*

Molecular descriptors play a fundamental role in chemistry, pharmaceutical sciences, environmental protection policy, health research, and quality control, being the way molecules, thought of as real bodies, are *transformed* into numbers, allowing some mathematical treatment of the chemical information contained in the molecule. Therefore molecular descriptors allow us to find → *structure/response correlations* and perform → *similarity searching*, → *substructure searching*, etc.

Molecular descriptors are formally mathematical representations of a molecule obtained by a well-specified algorithm applied to a defined *molecular representation* or a well-specified experimental procedure: *the molecular descriptor is the final result of a logical and mathematical procedure which transforms chemical information encoded within a symbolic representation of a molecule into a useful number or the result of some standardized experiment.*

The term "useful" has a double meaning: it means that the number can give more insight into the interpretation of the molecular properties and/or is able to take part in a model for the prediction of some interesting molecular properties. Even if the interpretation of a molecular descriptor may be weak, provisional, or even completely lacking, it could be strongly correlated to some molecular properties to give models with high predictive power. On the other hand, descriptors with poor predictive power can be usefully retained in models when they are well theoretically well-founded and interpretable because of their ability to encode structural chemical information.

Although several molecular quantities were defined from the beginnning of quantum chemistry and graph theory, the term "molecular descriptor" has become popular with the development of structure-property correlation models. The → *Platt number* [Platt, 1947] and → *Wiener index* [Wiener, 1947c], defined in 1947, are sometimes referred to as the first molecular descriptors.

By the definition given above, molecular descriptors are divided into two main classes: → *experimental measurements*, such as → *logP*, → *molar refractivity*, → *dipole moment*, → *polarizability*, and **theoretical molecular descriptors**, which are derived from a symbolic representation of the molecule and can be further classified according to the different types of *molecular representation.*

The **molecular representation** is the way a molecule, i.e. a phenomenological real body, is symbolically represented by a specific formal procedure and conventional rules. The quantity of chemical information which is transferred to the symbolic representation of the molecule depends on the kind of representation [Testa and Kier, 1991; Jurs *et al.*, 1995].

The simplest molecular representation is the **chemical formula**, which is the list of the different atom types, each accompanied by a subscript representing the number of occurrences of the atoms in the molecule. For example, the chemical formula of *p*-chlorotoluene is C_7H_7Cl, indicating the presence in the molecule of A = 8 (number of atoms, hydrogen excluded), N_C = 7, N_H = 7, and N_{Cl} = 1 (the subscript "1" is usually omitted in the chemical formula). This representation is independent of any knowledge concerning the molecular structure, and hence molecular descriptors obtained from the chemical formula can be called **0D-descriptors**. Examples are the → *atom number A*, → *molecular weight MW*, → *atom type count* N_X, and, in general, → *constitutional descriptors* and any function of the *atomic properties*.

The **atomic properties** constitute the weights used to characterize molecule atoms; the most common atomic properties are atomic mass, → *atomic charge*, → *van der Waals radius*, → *atomic polarizability*, and hydrophobic atomic constants. Atomic properties can also be defined by the → *local vertex invariants* (LOVIs) derived from graph therory.

The **substructure list representation** can be considered as a one-dimensional representation of a molecule and consists of a list of structural fragments of a molecule; the list can only be a partial list of fragments, functional groups, or substituents of interest present in the molecule, thus not requiring a complete knowledge of the molecule structure. The descriptors derived by this representation can be called **1D-descriptors** and are typically used in → *substructural analysis* and → *substructure searching*.

The two-dimensional representation of a molecule considers how the atoms are connected, i.e. it defines the connectivity of atoms in the molecule in terms of the presence and nature of chemical bonds. Approaches based on the → *molecular graph* allow a two-dimensional representation of a molecule, usually known as the **topological representation**. Molecular descriptors derived from the algorithms applied to a topological representation are called **2D-descriptors**, i.e. they are the so-called → *graph invariants*.

Two-dimensional representations alternative to the molecular graph are the **linear notation systems**, like, for example, *Wiswesser Line Notation system* (WLN) [Smith and Baker, 1975] and *SMILES notation* [Weininger, 1988; Weininger *et al.*, 1989; Weininger, 1990; Convard *et al.*, 1994; Hinze and Welz, 1996].

The three-dimensional representation views a molecule as a rigid geometrical object in space and allows a representation not only of the nature and connectivity of the atoms, but also the overall spatial configuration of the molecule. This representation of a molecule is called **geometrical representation**, and molecular descriptors derived from this representation are called **3D-descriptors**. Examples of 3D-descriptors are → *geometrical descriptors*, several → *steric descriptors*, and → *size descriptors*.

Several molecular descriptors derive from multiple molecular representations and can then be classified with difficulty. For example, graph invariants derived from a molecular graph weighted by properties obtained by → *computational chemistry* are both 2D- and 3D-descriptors.

The **bulk representation** of a molecule describes the molecule in terms of a physical object with 3D attributes such as bulk and steric properties, surface area and volume.

The **stereoelectronic representation** (or **lattice representation**) of a molecule is a molecular description related to those molecular properties arising from electron distribution – interaction of the molecule with probes characterizing the space surrounding them (e.g. → *molecular interaction fields*). This representation is typical of → *grid-based QSAR techniques*. Descriptors at this level can be considered **4D-descriptors**, being characterized by a scalar field, i.e. a lattice of scalar numbers associated with the 3D → *molecular geometry*.

Finally, the **stereodynamic representation** of a molecule is a time-dependent representation which adds structural properties to the 3D representations, such as flexibility, conformational behaviour, transport properties, etc. → *Dynamic QSAR* is an example of a multi-conformational approach.

Mathematics and statistics, graph theory, computational chemistry and molecular modelling techniques enable the definition of a large number of theoretical descriptors characterizing physico-chemical and biological properties, reactivity, shape, steric hindrance, etc. of the whole molecule, molecular fragments and substituents.

The fundamental difference between theoretical descriptors and experimentally measured ones is that theoretical descriptors contain no statistical error due to experimental noise, which is not the case for experimental measurements.

However, the assumptions needed to facilitate calculation and numerical approximation are themselves associated with an inherent error, although in most cases the direction, but not the magnitude, of the error is known. Moreover, within a series of related compounds the error term is usually considered to be approximately constant. All kinds of error are absent only for the most simple theoretical descriptors such as count descriptors or for descriptors directly derived from exact mathematical theories such as graph invariants.

Theoretical descriptors derived from physical and physico-chemical theories show some natural overlap with experimental measurements. Several → *quantum-chemical descriptors*, surface areas and → *volume descriptors* are examples of such descriptors also having an experimental counterpart.

With respect to the experimental measurements, the greatest recognized advantages of the theoretical descriptors are usually (but not always) in terms of cost, time and availability.

Within the two main classes of descriptors, experimental measurements and theoretical descriptors, several other subclasses of molecular descriptors can be recognized on the basis of a rational analysis of **molecular descriptor properties**.

The main properties of the descriptors can be represented by a four-level taxonomy. Together with the first-level classification based on the molecular representation, as defined above, the other three levels are summarized below.

- **mathematical representation of molecular descriptors**

The descriptors can be represented by a scalar value, a vector, a two-way matrix, a tensor, or a scalar field, which can be discretized into a lattice of grid points.

- **invariance properties of molecular descriptors**

The invariance properties of molecular descriptors can be defined as the ability of the algorithm for their calculation to give a descriptor value that is independent of the particular characteristics of the molecular representation, such as atom numbering or labelling, spatial reference frame, molecular conformations, etc. Invariance to molecular numbering or labelling is assumed as a minimal basic requirement for any descriptor. **Chemical invariance** of a molecular descriptor means that its values are independent of the atom types and multiple bonds, i.e., the descriptor is not able to account for heteroatoms and bond orders in the molecules. Such invariance is considered explicitly in classifying topological indices as → *topostructural indices* and → *topochemical indices*.

Two other important invariance properties, **translational invariance** and **rotational invariance**, are the invariance of a descriptor value to any translation or rotation of the molecules in the chosen reference frame. These invariance properties have to be considered when dealing with descriptors derived from → *molecular geometry* G. For all descriptors based on → *internal coordinates*, roto-translational invariance is naturally guaranteed. For descriptors based on spatial atomic coordinates, translational invariance is usually easily attained by centring the atomic coordinates; rotational invariance may be satisfied by using, as the reference frame, an uniquivocally defined frame such as the principal axes of each molecule. In some QSAR methods, such as grid-based QSAR techniques, the problem of invariance to rotation is overcome by adopting → *alignment rules*.

Conformational invariance means that molecular descriptor values are independent of the conformational changes in molecules. *Conformations* of molecules are the different atom dispositions in the 3D space, i.e. configurations that flexible molecules can assume without any change to their connectivity. Usually interest in different conformations of a molecule is related to those conformations where the total energies are relatively close to the minimum energy, i.e. within a cut-off energy value of some kcal/mol.

Molecular descriptors can be devided into four classes according to their conformational invariance degree as suggested by Charton [Charton, 1983]:

a) *No Conformational Dependence* (**NCD-descriptors**); this is typical of all descriptors which do not depend on 3D molecular geometry, such as molecular weight, → *count descriptors* and → *topological indices*.

b) *Low Conformational Dependence* (**LCD-descriptors**); this is the case of molecular descriptors whose values show small variations only in the presence of relevant conformational changes, such as, for example, *cis/trans* configurations. Examples are → *cis/trans descriptors* and usually → *charge descriptors*.

c) *Intermediate Conformational Dependence* (**ICD-descriptors**); these are molecular descriptors whose values show small variations in the presence of any conformational changes. Typical descriptors of this class are → *EVA descriptors*, and descriptors based on mass distribution, e.g., → *radius of gyration*.

d) *High Conformational Dependence* (**HCD-descriptors**); this is the case of descriptors with values very sensitive to any conformational change in the molecule. Typical descriptors of this class are → *interaction energy values* obtained from molecular interaction fields, → *3D-MoRSE descriptors*, → *WHIM descriptors*, → *G-WHIM descriptors*, → *spectrum-like descriptors*, → *shape descriptors* based on molecular geometries, etc.

Among the → *quantum-chemical descriptors*, descriptors of different kinds of conformational dependence can be found; → *ionization potential*, → *electron affinity*, molecular orbital energies are often LCD- or ICD-descriptors, while molecular energies are usually HCD-descriptors.

It should be noted that some invariance properties such as invariance to atom numbering and roto-translations are mandatory for molecular descriptors used in QSAR/QSPR modelling; in several cases, chemical invariance is required, particularly when dealing with a series of compounds with different substituents; moreover, conformational invariance is closely dependent on the considered problem.

- **degeneracy of molecular descriptors**

This property refers to the ability of a descriptor to avoid equal values for different molecules. In this sense, descriptors can show no degeneracy at all (N), low (L), intermediate (I), or high (H) degeneracy. The degree of degeneracy of a descriptor can naturally be measured by → *Shannon's entropy*. Moreover, the degree of degeneracy depends on the molecules present in the considered data set. Suitable measures of molecular descriptor degeneracy can be provided by using a data set consisting of an extended hydrocarbon series also containing heteroatoms and cycles.

Degeneracy is considered an undesirable characteristic for all molecular descriptors that are used for the characterization of molecules in store and retrieval database systems; however, in QSAR modelling, degenerate properties are better modelled by molecular descriptors showing analogous degeneracy [Todeschini *et al.*, 1998].

Based on the previous criteria, examples of an indicative classification of molecular descriptors are shown in Table M-2.

Table M-2. Mathematical properties of some molecular descriptors. *Non-specified properties closely depend on the original descriptors used for the calculation of the similarity matrices.

Descriptor	*Molecular representation*	*Mathematical representation*	*Invariance properties*	*Degeneracy*
Molecular weight	0D	Scalar	NCD	H
Atom-type counts	0D	Scalar	NCD	H
Fragment counts	1D	Scalar	NCD	H
Topological information indices	2D	Scalar	NCD	L/I
Molecular profiles	2D	Vector	NCD	N
Substituent constants	3D	Scalar	NCD/LCD	L/I
WHIM descriptors	3D	Vector	HCD	N
3D-MoRSE descriptors	3D	Vector	LCD/MCD	N
Similarity indices *	–	Matrix	–	–
Surface/volume descriptors	3D	Scalar	HCD/MCD	L
Quantum-chemical descriptors	3D	Scalar	MCD/HCD	N/L
Compass descriptors	3D	Vector	HCD	N
Interaction energy values	4D	Lattice	HCD/RD	N

Suitable molecular descriptors, besides the trivial invariance properties, should satisfy some basic requirements. The list of desirable requirements of chemical descriptors suggested by Randic [Randic, 1996a] is shown in Table M-3.

Table M-3. List of desirable requirements for molecular descriptors.

#	*Descriptors*
1	Should have structural interpretation
2	Should have good correlation with at least one property
3	Should preferably discriminate among isomer
4	Should be possible to apply to local structure
5	Should be possible to generalize to "higher" descriptors
6	Descriptors should be preferably independent
7	Should be simple
8	Should not be based on properties
9	Should not be trivially related to other descriptors
10	Should be possible to construct efficiently
11	Should use familiar structural concepts
12	Should have the correct size dependence
13	Should change gradually with gradual change in structures

Molecular descriptors are usually classified into several classes by a mixed taxonomy based on different points of view. For example, descriptors are often distinguished by their physico-chemical meaning such as → *electronic descriptors*, → *steric descriptors*, → *lipophilicity descriptors*, → *hydrogen-bonding descriptors*, → *shape descriptors*, → *charge descriptors*, → *electric polarization descriptors*, → *reactivity descriptors*; moreover, on the basis of the specific mathematical tool used for the calculation of the molecular descriptors, → *autocorrelation descriptors*, → *eigenvalue-based descriptors*, → *determinant-based descriptors*, → *Wiener-type indices*, → *Schultz-type indices* can be distinguished.

📖 [Duperray *et al.*, 1976a] [Jurs *et al.*, 1979] [Jurs *et al.*, 1985] [Lavenhar and Maczka, 1985] [Mekenyan and Bonchev, 1986] [Jurs *et al.*, 1988] [Dearden, 1990] [Govers, 1990] [Randic, 1990b] [Silipo and Vittoria, 1990] [Weininger and Weininger, 1990] [Randic, 1991b] [Ash *et al.*, 1991] [Randic, 1991g] [Cronin, 1992] [Horvath, 1992] [Bonchev *et al.*, 1993c] [Katritzky and Gordeeva, 1993] [Randic and Trinajstic, 1993a] [Rücker and Rücker, 1993] [Dearden *et al.*, 1995b] [Basak *et al.*, 1996a] [Karelson *et al.*, 1996] [Balaban, 1997a] [Basak *et al.*, 1997a] [Basak *et al.*, 1997c] [Klein and Babic, 1997b] [Matter, 1997] [Basak *et al.*, 1998] [Gasteiger, 1998] [Lee *et al.*, 1998] [Baumann, 1999] [Jalali-Heravi and Parastar, 1999]

molecular distance-edge vector → **distance matrix**

molecular eccentricity → **shape descriptors**

molecular electronegativity → **quantum-chemical descriptors**

molecular electrostatic potential → **quantum-chemical descriptors**

molecular electrostatic potential contour surface → **molecular surface**

molecular energies → **quantum-chemical descriptors**

Molecular Field Topology Analysis (MFTA)

The method of Molecular Field Topology Analysis is one of the → *hyperstucture-based QSAR techniques*. It was proposed as a "topological analogue" of the → *CoMFA* method because it is based on topological rather than spatial alignment of structures [Zefirov *et al.*, 1997a; Palyulin *et al.*, 2000].

The quantitative description of structural features is provided by local physico-chemical parameters. First, for a set of structures of known activity (a training set) the so-called → *molecular supergraph* (MSG) is automatically constructed. The MSG is a certain graph such that each training set structure can be represented as its subgraph. It enables the construction of → *uniform length descriptors* for all structures in the set. To build each descriptor vector, the MSG vertices and edges corresponding, respectively, to the atoms and bonds of a given structure are assigned the values of local descriptors (e.g., atomic charge q and van der Waals radius R) for these atoms and bonds, and other vertices and edges are labelled with neutral descriptor values that provide a reasonable simulation of properties in an unoccupied region of space. The descriptor vector formation is illustrated in Fig. M-3.

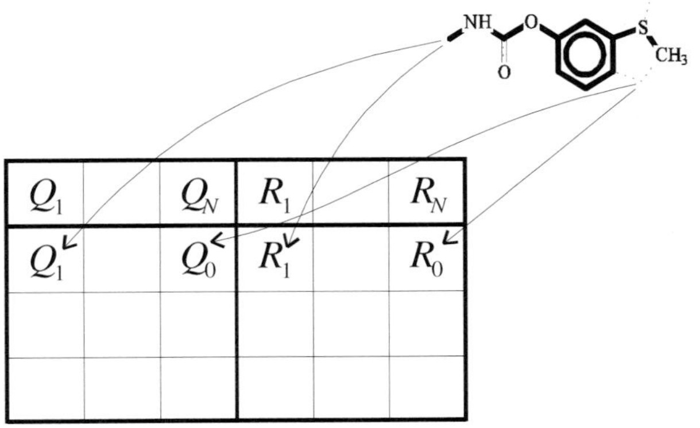

Figure M-3. Descriptor vector formation in MFTA.

The following local descriptors are currently calculated: Gasteiger's → *atomic charge* q estimated with the electronegativity equalization approach, Sanderson's → *electronegativity* χ_{SA}, Bondi's → *van der Waals radius* R, atomic contribution to the → *van der Waals molecular surface* SA_{VDW}, relative steric accessibility defined as $Ac = SA_{VDW}/ SA_{FREE}$ (where SA_{FREE} is the van der Waals surface of the "free" (isolated) atom of the same type), → *electrotopological state* S, atomic lipophilicity contribution l taking into account the environment of an atom, and group lipophilicity L_g defined as a sum of contributions for both a non-hydrogen atom and attached hydrogens, the ability of an atom in a given environment to be a donor (H_d) and acceptor (H_a) of a hydrogen bond characterized by the binding constants, local stereochemical indicator variables, and the site occupancy factors for atoms I_a and bonds I_b (which have the value 1 if a given feature is present in the structure and 0 otherwise). This set of local descriptors provides sufficient coverage of major interaction types that are important for the interaction of a ligand with a biological target. However, the set is open and can be easily extended to account for the specific features of the problem.

Since the number of descriptors is rather large (though much smaller than in CoMFA), partial least squares (PLS) regression is used to analyze the descriptor-activity relationships. As a result, the quantitative characteristic of the influence on activity of each descriptor in each position, including common structural fragments, can be determined (Figure M-4). Such characteristics provide a basis for designing new, potentially more active structures as well as being anchor points for spatial structure alignment.

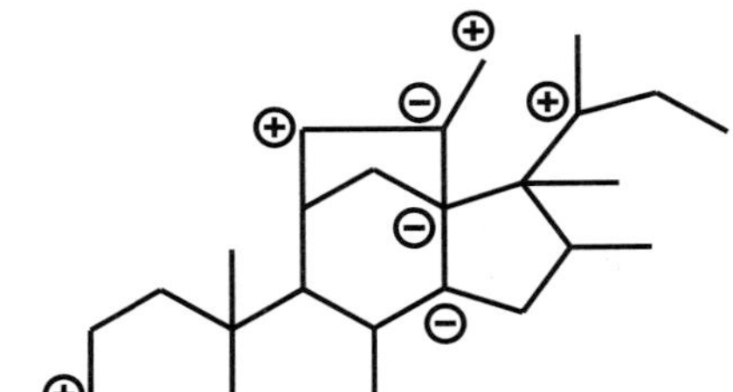

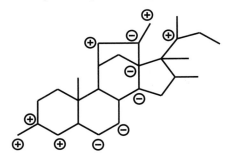

Figure M-4. Molecular supergraph of steroid data set with major atomic charge contributions into the CBG affinity.

MFTA often gives models that are comparable in quality of description and prediction to models based on the widely used classical QSAR methods and 3D approaches.

molecular fingerprints → **substructure descriptors**

molecular flexibility → **flexibility indices**

molecular fragments → **count descriptors**

molecular geometry

A molecule is the smallest fundamental group of atoms of a chemical compound that can take part in a chemical reaction. The atoms of the molecule are organized in a 3D structure; the **molecular matrix** M is a rectangular matrix $A \times 3$ whose rows represent the molecule atoms and the columns the atom **Cartesian coordinates** (x, y, z) with respect to any rectangular coordinate system with axes X, Y, Z. The cartesian coordinates of a molecule usually correspond to some optimized molecular geometry obtained by the methods of → *computational chemistry.* The molecular geometry can also be obtained from crystallographic coordinates or from 2D-3D automatic converters.

In practice, the molecular matrix is an augmented matrix M′, where the first column denotes the atom type (e.g. carbon, hydrogen, chlorine atoms) and the last four columns contain the labels of the atoms connected with the ith atom:

$$\mathrm{M} = \begin{vmatrix} x_1 & y_1 & z_1 \\ x_2 & y_2 & z_2 \\ \ldots & \ldots & \ldots \\ x_A & y_A & z_A \end{vmatrix} \qquad \mathrm{M}' = \begin{vmatrix} at.1 & x_1 & y_1 & z_1 & c_{11} & c_{12} & c_{13} & c_{14} \\ at.2 & x_2 & y_2 & z_2 & c_{21} & c_{22} & c_{23} & c_{24} \\ \ldots & \ldots & \ldots & \ldots & \ldots & \ldots & \ldots & \ldots \\ at.A & x_A & y_A & z_A & c_{A1} & c_{A2} & c_{A3} & c_{A4} \end{vmatrix}$$

An alternative to the molecular matrix representation of a molecule is that of **internal coordinates**, where the relative position of each atom to the other atoms in the molecule is given: these coordinates are bond distances, bond angles, and torsion angles. **Bond distances** r_{st} are the interatomic distances between bonded atoms (usually expressed in Angström Å); **bond angles** ϑ_{stv} are plane angles among triples of connected atoms (s, t, v) within the molecule; **torsion angles** ω_{stvz} are dihedral angles among 4-ples of connected atoms (s, t, v, z) (Figure M-5). Note that bond distances and bond angles are less sensitive to conformational change than interatomic distances and torsion angles.

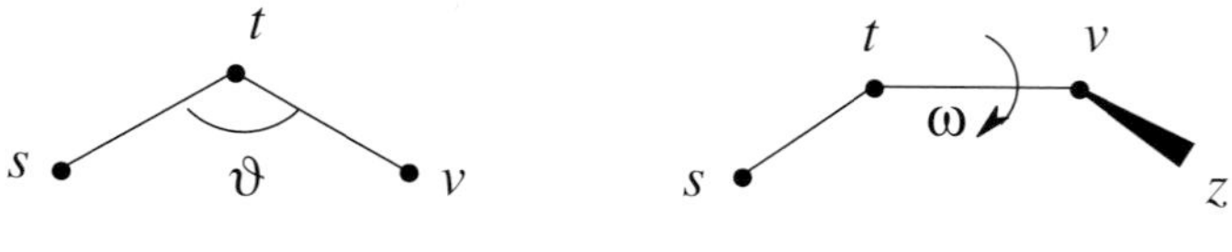

Figure M-5. Bond and torsion angles.

Internal coordinates are collected in the so-called **Z-matrix**, which is a rectangular matrix, whose rows are the atoms, defined as:

$$\mathbf{Z} = \begin{vmatrix} at.1 & & & & 0 & 0 & 0 \\ at.2 & r_{12} & & & 1 & 0 & 0 \\ at.3 & r_{23} & \vartheta_{321} & & 2 & 1 & 0 \\ at.4 & r_{34} & \vartheta_{432} & \omega_{4321} & 3 & 2 & 1 \\ \dots & \dots & \dots & \dots & \dots & \dots & \dots \\ \dots & \dots & \dots & \dots & \dots & \dots & \dots \\ at.A & r_{As} & \vartheta_{Ast} & \omega_{Astv} & s & t & v \end{vmatrix}$$

where r, ϑ, and ω are the molecular internal coordinates considered among the → *geometrical descriptors*. The last three columns show the labels of atoms involved in bonds, bond angles and torsion angles.

Other simple geometrical descriptors are **interatomic distances** r_{st} between pairs of atoms s and t. Interatomic distances are devided into **intramolecular interatomic distances**, i.e. distances between any pair of atoms (s, t) within the molecule, and **intermolecular interatomic distances**, i.e. distances between atoms of a molecule and atoms of a receptor structure, a reference compound or another molecule. While classical computational chemistry describes molecular geometry in terms of three-dimensional Cartesian coordinates or internal coordinates, the → *distance geometry* (DG) method takes the interatomic distances as the fundamental coordinates of molecules, exploiting their close relationship to experimental quantities and molecular energies.

In general, geometrical descriptors can be derived from the molecular matrix, Z-matrix or geometry matrix.

The **geometry matrix** G (or **geometric distance matrix**) of a molecule, obtained from the molecular matrix M, is a square symmetric matrix $A \times A$ whose entry r_{st} is the **geometric distance** calculated as the Euclidean distance between the atoms s and t:

$$\mathbf{G} = \begin{vmatrix} 0 & r_{12} & \dots & r_{1A} \\ r_{21} & 0 & \dots & r_{2A} \\ \dots & \dots & \dots & \dots \\ r_{A1} & r_{A2} & \dots & 0 \end{vmatrix}$$

Diagonal entries are always zero. Geometric distances are intramolecular interatomic distances.

Like the molecular matrix, the geometry matrix contains information about molecular configurations and conformations; however, the geometry matrix does not contain information about atom connectivity. Thus, for several applications, it is accompanied by a connectivity table where, for each atom, there are listed the identification numbers of the atoms bonded to it. The geometry matrix can also be calculated on geometry-based standardized bond lengths and bond angles and derived by embedding a graph on a regular two-dimensional or three-dimensional grid; in these cases, the geometry matrix is often referred to as the **topographic matrix** [Balaban, 1997a].

From the geometry matrix used to represent a → *molecular graph*, a number of → *local vertex invariants* and related → *molecular descriptors*, called **topographic indices**, can be derived [Randic and Wilkins, 1979d; Randic, 1988a; Randic *et al.*, 1990; Diudea *et al.*, 1995b; Randic and Razinger, 1995a; Balaban, 1997b].

Analogously to the → *vertex distance degree*, the ith row sum of the geometry matrix is called the **geometric distance degree** ${}^{G}\sigma_i$ that is a local vertex invariant used, for example, in the development of the → *3D-connectivity indices* $\chi\chi$. In general, the row sum of this matrix represents a measure of the centrality of an atom; atoms that are close to the → *centre of the molecule* have smaller atomic sums, while those far from the centre have large atomic sums. The smallest and the largest row sums give the extreme values of the first eigenvalue of the G matrix; therefore, when all the atoms are equivalent, i.e. the distance degrees are all the same, the geometric distance degree yields exactly the first eigenvalue. The average sum of all geometric distance degrees is a molecular invariant called the **average geometric distance degree**, i.e.

$$^G\bar{\sigma} = \frac{1}{A} \cdot \sum_{i=1}^{A} {}^G\sigma_i = \frac{1}{A} \cdot \sum_{i=1}^{A} \sum_{j=1}^{A} r_{ij}$$

while the half sum of all geometric distance degrees is another molecular descriptor called the → *3D-Wiener index* in analogy with the → *Wiener index* calculated from the topological distance matrix.

The maximum value entry in the ith row is a local descriptor called the **geometric eccentricity** $^G\eta_i$ representing the longest geometric distance from the ith vertex to any other vertex in the molecule:

$$^G\eta_i = \max_j (r_{ij})$$

From the eccentricity definition, the **geometric radius** ^GR and **geometric diameter** ^GD can immediately characterize a molecule. The radius of a molecule is defined as the minimum geometric eccentricity, and the diameter is defined as the maximum geometric eccentricity in the molecule, according to the following:

$$^GR = \min_i (^G\eta_i) \quad \text{and} \quad ^GD = \max_i (^G\eta_i)$$

These parameters are → *size descriptors* also depending on the molecular shape (→ *Petitjean shape indices*), such as their topological counterpart, i.e. → *topological radius* and → *topological diameter*.

The **reciprocal geometry matrix** $\mathbf{G}^{-1}$ is obtained from the geometry matrix as the following:

$$[\mathbf{G}^{-1}]_{ij} = \begin{cases} (r_{ij})^{-1} & i \neq j \\ 0 & i = j \end{cases}$$

From the geometry matrix, the usual → *graph invariants* can be calculated such as → *characteristic polynomial*, → *eigenvalue-based descriptors*, → *path counts*, → *ID numbers*, → *3D-Balaban index*, → *3D-Schultz index* and so forth [Randic, 1988b; Nikolic *et al.*, 1991]. It is noteworthy that all these indices are sensitive to molecular geometry. Moreover, the geometry matrix is used for the calculation of size descriptors and → *3D-MoRSE descriptors*.

Important derived matrices are the powers of the geometry matrix, used to define → *molecular profiles* descriptors, and the → *distance/distance matrix* **DD**, which unifies the topological and geometric molecular information.

The molecular matrix $\mathbf{M}$ and the $\mathbf{Z}$-matrix are the natural starting points for the calculation of several 3D atomic and molecular descriptors, such as → *quantum-chemical descriptors*, → *molecular interaction fields*, → *EVA descriptors*, → *WHIM descriptors*, → *CoMMA descriptors*, → *Compass descriptors*, and → *molecular surface* descriptors.

📖 [Turro, 1986] [Mihalic and Trinajstic, 1991] [Mihalic *et al.*, 1992a] [Warthen *et al.*, 1993] [Kunz, 1993] [Kunz, 1994] [Balasubramanian, 1995b] [Estrada and Ramirez, 1996] [Zhu and Klein, 1996a] [Randic and Razinger, 1997] [Laidboeur *et al.*, 1997] [Ivanciuc *et al.*, 1998a] [Tao and Lu, 1999a] [Ivanciuc and Ivanciuc, 2000]

Example : 2-methylpentane

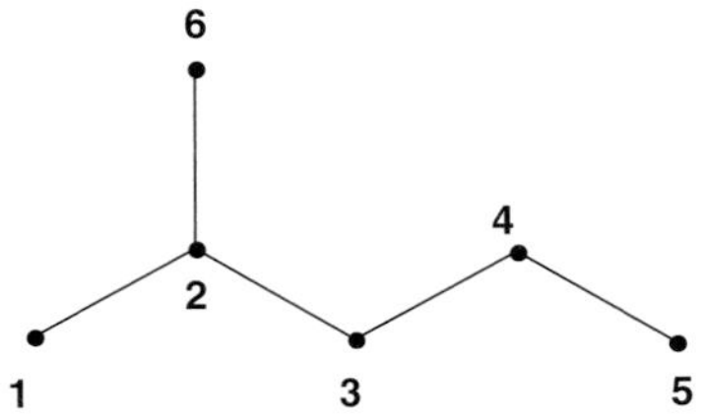

Geometry matrix G

Atom	1	2	3	4	5	6	${}^{G}\sigma_i$	${}^{G}\eta_i$
1	0	1.519	2.504	3.856	5.014	2.498	15.391	5.014
2	1.519	0	1.530	2.521	3.864	1.521	10.955	3.864
3	2.504	1.530	0	1.521	2.509	2.507	10.571	2.509
4	3.856	2.521	1.521	0	1.511	3.038	12.447	3.856
5	5.014	3.864	2.509	1.511	0	4.348	17.246	5.014
6	2.498	1.521	2.507	3.038	4.348	0	13.912	4.348

$${}^{G}R = \min_i\left({}^{G}\eta_i\right) = 2.509 \qquad {}^{G}\bar{\sigma} = \frac{1}{6}\cdot\sum_{i=1}^{6} {}^{G}\sigma_i = \frac{80.522}{6} = 13.4203$$

$${}^{G}D = \max_i\left({}^{G}\eta_i\right) = 5.014 \qquad {}^{3D}W = \frac{1}{2}\cdot\sum_{i=1}^{6} {}^{G}\sigma_i = 40.261$$

Box M-1.

molecular graph *(: structural graph, constitutional graph)*

A → *graph G* where vertices and edges are chemically interpreted as atoms and bonds [Harary, 1969a; Harary, 1969b; Balaban and Harary, 1976; Rouvray and Balaban, 1979; Rouvray, 1990; Bonchev and Rouvray, 1991; Trinajstic, 1992]. A molecular graph obtained excluding all the hydrogen atoms is called an **H-depleted molecular graph** (or **hydrogen-depleted molecular graph**). Such a graph depicts the connectivity of atoms in a molecule irrespective of the metric parameters such as equilibrium → *interatomic distances* between nuclei, → *bond angles* and → *torsion angles*, representing the 3D → *molecular geometry*. Thus, a molecular graph is a → *topological representation* of the molecule, and it is from this that several → *molecular descriptors* are derived.

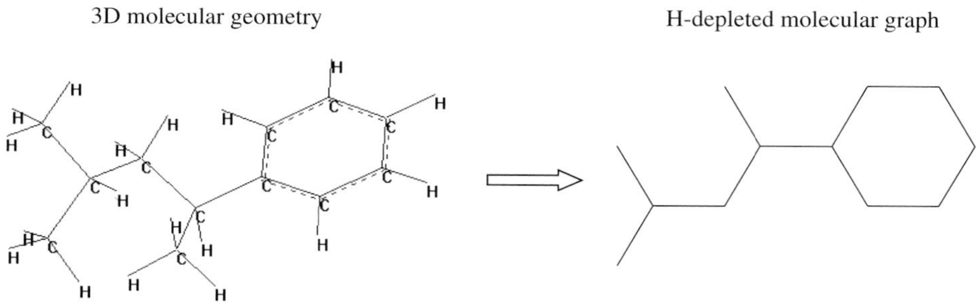

Figure M-6. The transition from 3D-geometry to 2D-topology.

Some vertices of the H-depleted molecular graph can be more precisely defined as the **hydride group**, which is a heavy atom plus its bonded hydrogens. For example, hydride groups are $-CH_3$, $-CH_2-$, $=NH$, $-NH_2$, $-OH$.

A **molecular subgraph** is a subset of atoms and related bonds which is itself a valid graph usually representing molecular fragments and functional groups.

There are four commonly used subgraph types: **path subgraph**, **cluster subgraph**, **path-cluster subgraph**, and **chain subgraph** (or *Ring*), emphasizing different aspects of atom connectivity within the molecule. They are defined according to the following rules:

1) if the subgraph contains a cycle it is of type Chain (CH); otherwise, 2) if all vertex degrees in the subgraph (not in the whole graph) are either greater than 2 or equal to 1, the subgraph is of type Cluster (C); otherwise, 3) if all vertex degrees in the subgraph are either equal to 2 or 1, the subgraph is of type Path (P); otherwise, 4) the subgraph is of type Path-Cluster (PC).

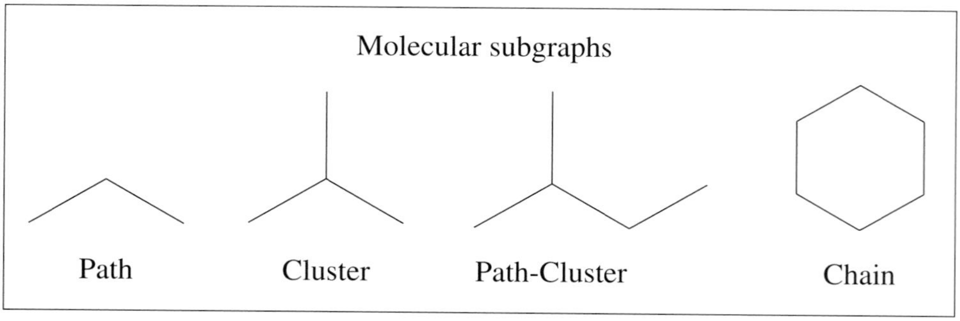

Figure M-7. Elementary molecular subgraphs.

The **order of a subgraph** is the number of edges within it. Note that subgraphs of order 0 are considered of type Path, subgraphs of order 1 and 2 are only of type Path, and subgraphs of order 3 can be of type Path, Cluster, or Chain only.

Referring to the subgraph order, rth order indices can be defined as → *count descriptors*, i.e. the number of rth order subgraphs in the graph $\mathcal{G}$. The zero order index is simply the → *atom number A*, i.e. the number of graph vertices; 1st order index is the → *bond number B*, i.e. the number of graph edges; 2nd order index is the

→ *connection number*, i.e. the 2nd order → *molecular path count*; 3rd order indices are the number of paths of length 3, i.e. the number of 3-edge clusters, or the number of 3-edges cycles.

The total number K of connected subgraphs of a molecular graph $\mathcal{G}$ is a very simple measure of → *molecular complexity*, obviously referring only to the structural complexity of the molecule; it can be called the **total subgraph count**.

The simplest form to represent the chemical information contained in a molecular graph is by a → *matrix representation of a molecular structure*. Examples are → *adjacency matrix* **A**, → *edge adjacency matrix* **E**, vertex → *distance matrix* **D**, → *edge distance matrix* $^{\mathbf{E}}\mathbf{D}$, → *incidence matrix* I, → *Wiener matrix* **W**, → *Hosoya Z-matrix* **Z**, → *Cluj matrices* **CJ**, → *detour matrix* **Δ**, → *Szeged matrix* **SZ**, → *distance/distance matrix* **DD**, and → *detour/distance matrix* **Δ/D**.

📖 [Gutman and Trinajstic, 1972] [Gutman *et al.*, 1975] [Randic, 1975a] [Polanski and Rouvray, 1976a] [Polanski and Rouvray, 1976b] [Balaban, 1976a] [Balaban, 1978] [Balaban and Rouvray, 1980] [Mekenyan *et al.*, 1980] [Mekenyan *et al.*, 1981] [Trinajstic *et al.*, 1983] [Balaban, 1985a] [Balaban, 1985b] [Hosoya, 1986] [Trinajstic *et al.*, 1986a] [Balaban, 1987a] [Hansen and Jurs, 1988b] [Balaban, 1991] [Balaban, 1993e] [Dias, 1993] [Balaban, 1995a] [Klein, 1997a] [John *et al.*, 1998] [Bytautas and Klein, 1998] [Bytautas and Klein, 1999]

molecular holograms → **substructure descriptors**

molecular ID numbers → **ID numbers**

molecular influence matrix → **molecular complexity** (⊙ MIM complexity indices)

molecular interaction fields (: *interaction fields*)

A molecular interaction field is a scalar field of → *interaction energy values* between a molecule, whose → *molecular geometry* is known, and a → *probe* [Wade, 1993a; Andrews, 1993; Leach, 1996]. For QSAR studies, molecular interaction fields are calculated using one or more probes for a number of compounds previously aligned by specific → *alignment rules* and embedded in the same fixed → *grid*, i.e. a regular 3D array of N_G points, each point **p** being characterized by grid coordinates (x, y, z). The interaction energy values are calculated by moving the probe in each grid point.

Depending on the selected probe and the defined potential energy function, several molecular interaction fields can be calculated. The most common are *steric fields* and *electrostatic fields*, sometimes referred as **CoMFA fields** because originally implemented in → *CoMFA*. Several interaction fields are actually calculated in the → *GRID method*.

Recently, derived from a topological approach, → *E-state fields* and → *HE-fields* were defined.

The **enthalpic fields** are all of the molecular interaction fields accounting for enthalpic contributions to the free energy of ligand-receptor binding, such as *steric fields* and *electrostatic fields*. On the other hand, the **entropic fields** are all of the molecular interaction fields accounting for entropic contributions to the free energy of ligand-receptor binding. The entropy of binding is related to hydrophobic interactions between nonpolar ligand and receptor lipophilic chemical groups after the release of water molecules formerly structured around the receptor groups, and to the loss of conformational freedom due to ligand immobilization at the binding site. The entropy of binding is mainly modelled by *hydrophobic fields* or *hydrogen bonding fields*; however sometimes also the degrees of torsional freedom in the molecule were considered to account for the entropy change resulting from the reduced conformational freedom of the ligand in the receptor complex [Greco *et al.*, 1997].

Some molecular interaction fields are listed below.

- **steric interaction fields** (: *van der Waals interaction fields*)

A steric interaction field is obtained by calculating the van der Waals interaction energy E_{vdw} between probe and target in each grid point [Kim, 1992b]. Different potential energy functions were proposed to model van der Waals interactions between atoms. The most common are *Lennard-Jones potential*, *Buckingham potential*, and *Hill potential* [Leach, 1996].

The general formula of the **Lennard–Jones 6–12 potential function** [Lennard-Jones, 1924; Lennard-Jones, 1929] is:

$$E_{vdw} = \sum_s \sum_t \left(\frac{A}{r_{st}^{12}} - \frac{C}{r_{st}^{6}} \right)$$

where the first sum runs over all probe atoms and the second over all atoms of the target molecule; r_{st} is the interatomic distance between the sth atom of the probe and the tth atom of the target; A and C are two functions defined as:

$$A = \sqrt{\varepsilon_s \cdot \varepsilon_t} \cdot (R_s + R_t)^{12} \text{ and } C = 2 \cdot \sqrt{\varepsilon_s \cdot \varepsilon_t} \cdot (R_s + R_t)^{6}$$

where ε is the well depth and R one half the separation at which the energy passes through a minimum (i.e. the → *van der Waals radius*). The Lennard-Jones potential is characterized by an attractive component that varies as r^{-6} and a repulsive component that varies as r^{-12}. The energy function modelling the steric repulsion between pairs of atoms becomes large and positive at interatomic distances r less than the sum of the van der Waals radii of the probe atom and the target atom.

The **Buckingham potential function** [Buckingham, 1938] is defined in an exponential form as:

$$E_{vdw} = \sum_s \sum_t \left(A \cdot \exp^{-B \cdot r_{st}} - \frac{C}{r_{st}^{6}} \right)$$

where the first sum runs over all probe atoms and the second over all atoms of the target molecule; r_{st} is the interatomic distance between the sth atom of the probe and the tth atom of the target; A, B and C are functions of the well depth ε, the van der Waals radius R, and an adjustable parameter α. The exponential energy function is commonly used for small molecules, the Lennard-Jones 12–6 function for macromolecules.

The **Hill potential function** is an exponential function defined as:

$$E_{vdw} = \sum_s \sum_t \left[-2.25 \cdot \sqrt{\varepsilon_s \cdot \varepsilon_t} \cdot \left(\frac{R_s + R_t}{r_{st}} \right)^{6} + 8.28 \cdot 10^{5} \cdot \sqrt{\varepsilon_s \cdot \varepsilon_t} \cdot \exp\left(-\frac{r_{st}}{0.073 \cdot (R_s + R_t)} \right) \right]$$

where ε is the well depth and R van der Waals radius. The coefficients were determined by fitting them to data for the rare gases [Hill, 1948].

- **electrostatic interaction fields**

Molecular interaction fields obtained by calculating electrostatic interaction energy E_{el} between probe and target in each grid point. Besides the → *molecular electrostatic potential* (MEP), the most common energy function for electrostatic interactions is the **Coulomb potential energy function** defined as:

$$E_{el} = \sum_s \sum_t \frac{q_s \cdot q_t}{4\pi \cdot \varepsilon_0 \cdot \varepsilon_m \cdot r_{st}}$$

where the first sum runs over all probe atoms and the second over all atoms of the target molecule; r_{st} is the interatomic distance between the sth atom of the probe and the tth atom of the target; q is the → *partial atomic charge*; ε_0 is the permittivity of the free space and ε_m is the relative dielectric constant of the surrounding medium.

The **GRID electrostatic energy function** was proposed to account for the dielectric discontinuity between a solute and the solvent as [Goodford, 1985]:

$$E_{el} = \sum_s \sum_t \frac{q_s \cdot q_t}{K \cdot \zeta} \cdot \left(\frac{1}{r_{st}} + \frac{(\zeta - \varepsilon)/(\zeta + \varepsilon)}{\sqrt{(r_{st}^2 + 4 \cdot s_s \cdot s_t)}} \right)$$

where the first sum runs over all probe atoms and the second over all atoms of the target molecule; r_{st} is the interatomic distance between the sth atom of the probe and the tth atom of the target; q is the partial atomic charge; K is a constant; ζ and ε are the relative dielectric constants of the protein and the target solution phases, respectively; s_s and s_t are the nominal depths at which the probe atom and the target atom are respectively buried in the target phase. These depths are calculated by counting the number of neighbouring target atoms within a distance of 4 Å and translating this into an equivalent depth using a calibrated scale.

- **molecular orbital fields**

Fields restricted to the regions occupied by selected molecular orbitals; of particular interest are fields related to the → *highest occupied molecular orbital* (HOMO) and to the → *lowest unoccupied molecular orbital* (LUMO) [Oprea and Waller, 1997; Navajas *et al.*, 1996; Durst, 1998].

Molecular orbital fields are descriptors particularly useful when an ionic or charge transfer reaction is part of the ligand-receptor interaction; in this case, electrostatic fields are not able to fully represent the electronic characteristics of molecules.

In order to calculate a molecular orbital field, semiempirical single-point calculations are performed on the molecule optimized geometry and the electron density at each grid point in the region of the selected orbital is determined.

- **hydrophobic fields**

Molecular descriptors based on hydrophobic interaction energy between nonpolar surfaces of ligand and receptor. The energy of hydrophobic interactions derives from the disruption of the water structure around nonpolar surfaces resulting in a gain of entropy [Abraham and Kellogg, 1993].

Kellogg and Abraham interaction field, also called **Hydropatic INTeractions** (HINT), is a hydrophobic field calculated by → *Leo-Hansch hydrophobic fragmental constants* scaled by surface area and a distance-dependent function [Kellogg *et al.*, 1991; Kellogg and Abraham, 1992a; Abraham and Kellogg, 1993]. *Hydropathy* is a term used in structural molecular biology to represent the hydrophobicity of amino acid side chains.

The hydropathic field in each grid point is calculated as:

$$E_{hy} = \sum_s \sum_t (SA_s \cdot h_s \cdot SA_t \cdot h_t \cdot R_{st} + R'_{st})$$

where the first sum runs over all probe atoms and the second over all atoms of the target molecule; SA is the atomic → *solvent-accessible surface area*, h the hydropathic atom constant, and R_{st} and R'_{st} are functions of the interatomic distance r_{st} between the sth atom of the probe and the tth atom. The function R_{st} scales the product between solvent-accessible surface area and hydropathic constant with a distance usually defined as:

$$R_{st} = I_{st} \cdot \exp^{-r_{st}}$$

where I_{kt} is a sign-flip function recognizing acid-base interactions. The function R'_{st} is a Lennard-Jones-type potential accounting for close contacts of atoms by the van der Waals radius term:

$$R'_{st} = \mathrm{A} \cdot \varepsilon_{st} \cdot \left[\left(\frac{R_s + R_t}{r_{st}} \right)^{12} - 2 \cdot \left(\frac{R_s + R_t}{r_{st}} \right)^{6} \right]$$

where A is a scaling factor, ε_{st} is the depth of the Lennard-Jones potential well, and R is the van der Waals radius of the considered atoms. The probe is usually taken as a single atom and its parameters are set to unity.

Hydrophatic atom constants h are derived from Leo-Hansch hydrophobic fragmental constants in such a way that:

- the sum of hydropathic atom constants in a group is consistent with the group fragmental constant;
- frontier atoms in a group are more important than shielded atoms;
- bond, chain, branch and proximity factors are applied in an additive scheme, the first three to all eligible atoms, the last to the central atoms of polar groups.

Positive hydropathic constants indicate hydrophobic atoms, while negative constants indicate hydrophilic atoms.

The **Molecular Lipophilicity Potential** (*MLP*) describes the combined lipophilic effect of all fragments in a molecule on its environment and can be calculated at any point in space around the molecule [Audry *et al.*, 1986; Fauchère *et al.*, 1988; Furet *et al.*, 1988; Audry *et al.*, 1989; Audry *et al.*, 1992]. It is defined by considering a molecule surrounded by non-polar or low polarity organic solvent molecules, and assuming that the solvent molecule distribution around the considered molecule depends on the fragmental or atomic contributions to log$P$ and the distances at which the solvent molecules are from the target molecule. Therefore, the molecular lipophilicity potential at each kth grid point is calculated as:

$$\mathrm{MLP}_k = \sum_{i=1}^{A} \frac{a_i}{1 + r_{ki}}$$

where the sum runs over all atoms (or fragments) of the target molecule; a_i are the → *Ghose-Crippen hydrophobic atomic constants* for the ith atom (or fragments) in the target molecule, and r_{ki} is the distance between the considered atom (or fragments) and the kth grid point. Only non-hydrogen atoms A of the molecule are usually considered.

In contrast to other potentials, the lipophilicity potential is not obtained by calculating the interactions between a probe and the molecule.

Different MLP functions can be obtained according to the selection of the fragmental constant values and the distance function [Croizet *et al.*, 1990; Gaillard *et al.*, 1994a; Gaillard *et al.*, 1994b; Testa *et al.*, 1996; Carrupt *et al.*, 1997]. The MLP has later been adapted to a new atomic hydrophobic parameter called the **Topological Lipophilicity Potential** (TLP) defined for each jth atom of the molecule as [Langlois *et al.*, 1993b; Dubost, 1993]:

$$\mathrm{TLP}_j = \sum_{i=1}^{A} \frac{a_i}{1 + d_{ij}}$$

where d_{ij} is the → *topological distance* between atoms i and j of the molecule.

📖 [Masuda *et al.*, 1996; Gussio *et al.*, 1996; Testa *et al.*, 1999]

• **hydrogen bonding fields**

Descriptors accounting for hydrogen-bonding interactions between ligand and receptor. Hydrogen bonding fields are obtained by calculating the energy E_{hb} due to the formation of hydrogen-bonds between probe and target in each grid point [Leach, 1996; Oprea and Waller, 1997].

The hydrogen bonding potential energy is calculated as [Wade, 1993a]:

$$E_{hb} = \sum_s \sum_t E_r(r_{st}) \cdot E_s \cdot E_t$$

where the first sum runs over probe atoms and the second over atoms of the target molecule; E_r is an energy component dependent on the interatomic distance r_{st} between probe and target atoms involved in the hydrogen-bond; E_s and E_t are energy components dependent on the angle made by the hydrogen bond at the probe and target atoms, respectively. E_s and E_t values are between 0 and 1. The component E_r is usually defined by a Lennard-Jones function as:

$$E_r = \frac{A}{r_{st}^m} - \frac{C}{r_{st}^n}$$

where A and C are constants dependent on the chemical type of the hydrogen-bonding atoms; m and n are parameters taking different values: for example, $m = 12$ and $n = 10$ are commonly used values, $m = 8$ and $n = 6$ were used in the GRID hydrogen bonding energy function [Boobbyer *et al.*, 1989].

A more sophisticated hydrogen bonding potential energy based on the geometry of the hydrogen-bonding systems was proposed by Kim [Kim, 1993a; Kim, 1993f; Kim *et al.*, 1993] and implemented in the GRID program:

$$E_{hb} = \left(\frac{C}{r_{st}^6} - \frac{D}{r_{st}^4} \right) \cdot \cos(m \cdot \theta)$$

where the energy is evaluated in each grid point; r_{st} is the interatomic distance between probe and target atoms involved in the hydrogen bond; C and D are parameters taken from tables; m is usually equal to one; θ is the angle made by donor, hydrogen and acceptor atoms. The probe used is a neutral H_2O molecule with an effective radius of 1.7 Å, free to rotate around the grid point.

• **total interaction energy fields**

Potential energy descriptors accounting for the total non-covalent interaction potential energy which determines the binding affinity of a molecule to the considered receptor. They are generally calculated as the pairwise sum of the interaction energies between each probe atom and each target atom as [Wade, 1993a]:

$$E = \sum_s \sum_t (E_{vdw} + E_{el} + E_{hb})_{st}$$

where the first sum runs over probe atoms and the second over atoms of the target molecule; E_{vdw} is the van der Waals interaction energy, E_{el} the electrostatic energy and E_{hb} the hydrogen-bonding energy. Other noncovalent energy contributions can be included.

• **desolvation energy fields**

Potential energy descriptors proposed as an indicator of hydrophobicity [Oprea and Waller, 1997]. Originally, they were calculated using the finite difference approximation method; the linearized Poisson-Boltzmann equation was solved numerically to compute the electrostatic contribution to solvation at each grid point. Desolvation energy field values were calculated as the difference between solvated (grid dielectric = 80) and in vacuo (grid dielectric = 1).

📖 [Richard, 1991] [Kim, 1993b] [Balogh and Naray-Szabo, 1993] [Nusser *et al.*, 1993] [Naray-Szabo and Balogh, 1993] [van de Waterbeemd *et al.*, 1996] [Liljefors, 1998]

molecular lipophilicity potential → **molecular interaction fields** (⊙ hydrophobic fields)

molecular lipophilicity potential model → **lipophilicity descriptors**

molecular matrix → **molecular geometry**

molecular modelling : *computer-aided molecular modelling* → **drug design**

molecular moment of energy → **self-returning walk counts**

molecular orbital contour surface → **molecular surface**

molecular orbital energies → **quantum-chemical descriptors**

molecular orbital fields → **molecular interaction fields**

molecular path code → **path counts**

molecular path count → **path counts**

molecular path number : *molecular path count* → **path counts**

molecular path/walk indices → **shape descriptors** (⊙ path/walk shape indices)

molecular polarizability → **electric polarization descriptors**

molecular polarizability effect index → **electric polarization descriptors**

molecular profiles

Molecular descriptors ${}^kD$ derived from the → *distance distribution moments* of the → *geometry matrix* G, defined as the average row sum of its entries raised to the kth power, normalized by the factor $k!$:

$${}^kD = \frac{1}{k!} \cdot \frac{\sum_{i=1}^{A}\sum_{j=1}^{A} r_{ij}^k}{A}$$

where r_{ij}^k is the kth power of the i-j entry of the geometry matrix and A the number of atoms [Randic, 1995a; Randic, 1995b; Randic and Razinger, 1995b].

Using several increasing k values, a sequence of molecular invariants called the *molecular profile* is obtained as:

$$\langle {}^1D, {}^2D, {}^3D, {}^4D, {}^5D, {}^6D, \ldots \rangle$$

As the exponent k increases, the contributions of the most distant pairs of atoms become the most important.

The maximum non-zero value of ${}^kD$ is for the power corresponding to the number of atoms of the molecule ($k = A$); to obtain → *uniform length descriptors*, values for $k > A$ are set to zero.

For large k values, ${}^kD$ values tend to zero, due to the effect of the factorial normalization factor.

Another set of theoretical invariants can be obtained by averaging the row sums as:

$${}^kd = \frac{1}{k!} \cdot \frac{\sum_{i=1}^{A}\sum_{j=1}^{A} r_{ij}^k}{A^2}$$

obtaining the vector

$$\langle {}^1d, {}^2d, {}^3d, {}^4d, {}^5d, {}^6d, \ldots \rangle$$

For characterization of 2D structures, molecular profiles are computed in the same way by the distance distribution moments of the topological → *distance matrix* **D**.

If one is interested in the characterization of molecular local features, i.e. **local profiles**, the calculation of the ${}^kD$ values can be restricted to the local environment of interest, i.e. only the row sums corresponding to atoms of interest are considered, obtaining a vector of local theoretical invariants:

$$\langle {}^1L, {}^2L, {}^3L, {}^4L, {}^5L, {}^6L, \ldots \rangle$$

In this way, different types of profiles can be derived, such as **shape profiles**, which are local profiles taking into account only atoms on the molecular periphery:

$$\langle {}^1S, {}^2S, {}^3S, {}^4S, {}^5S, {}^6S, \ldots \rangle$$

In this case, the row sums of the geometry matrix are obtained by summing only the geometric distance powers of the atoms belonging to the periphery, and the average is found from the number of the contributing atoms only. Each atomic distance sum is considered as a local indicator of molecular shape, and each molecular invariant ${}^kS$ is considered a global shape descriptor.

In the case of 3D space-filled molecular models, one can represent the molecule by **contour profiles**, which are shape profiles calculated for all individual contours used to map the molecule. Each contour profile is then defined by a sequence:

$$\langle {}^1C, {}^2C, {}^3C, {}^4C, {}^5C, {}^6C, \ldots \rangle$$

where each element of the profile is the normalized average row sum of an augmented geometry matrix, where additional points defining the contour are also considered.

Particular contour profiles are obtained by randomly distributed points over the surface of the molecule.

Moreover, an arbitrary number of points can be considered along the molecule bonds, thus deriving **bond profiles**:

$$\langle {}^1B, {}^2B, {}^3B, {}^4B, {}^5B, {}^6B, \ldots \rangle$$

Bond profiles constitute a generalization of atomic molecular profiles since they provide a characterization of molecular connectivity, which is not explicitly contained in the geometry matrix [Randic, 1996a; Randic and Krilov, 1996].

Volume profiles ${}^kV$ of a molecule can be calculated by distributing random points throughout the molecular interior defined by the → *van der Waals molecular surface* and then constructing the corresponding augmented geometry matrices whose elements are raised to the kth power [Randic and Krilov, 1997b]. Moreover, to characterize the molecular surface, random points are restricted to the surface, thus obtaining **surface profiles** ${}^kSA$. Based on the same principles of the → *surface-volume ratio* $G' = SA / V$, the **volume-to-surface profiles** ${}^kV / {}^kSA$ have been proposed as → *shape descriptors* defined as the ordered sequence of the ratios of the volume to surface profile elements of corresponding order.

📖 [Randic, 1996c] [Randic and Razinger, 1997] [Randic and Krilov, 1997a] [Zefirov and Tratch, 1997b]

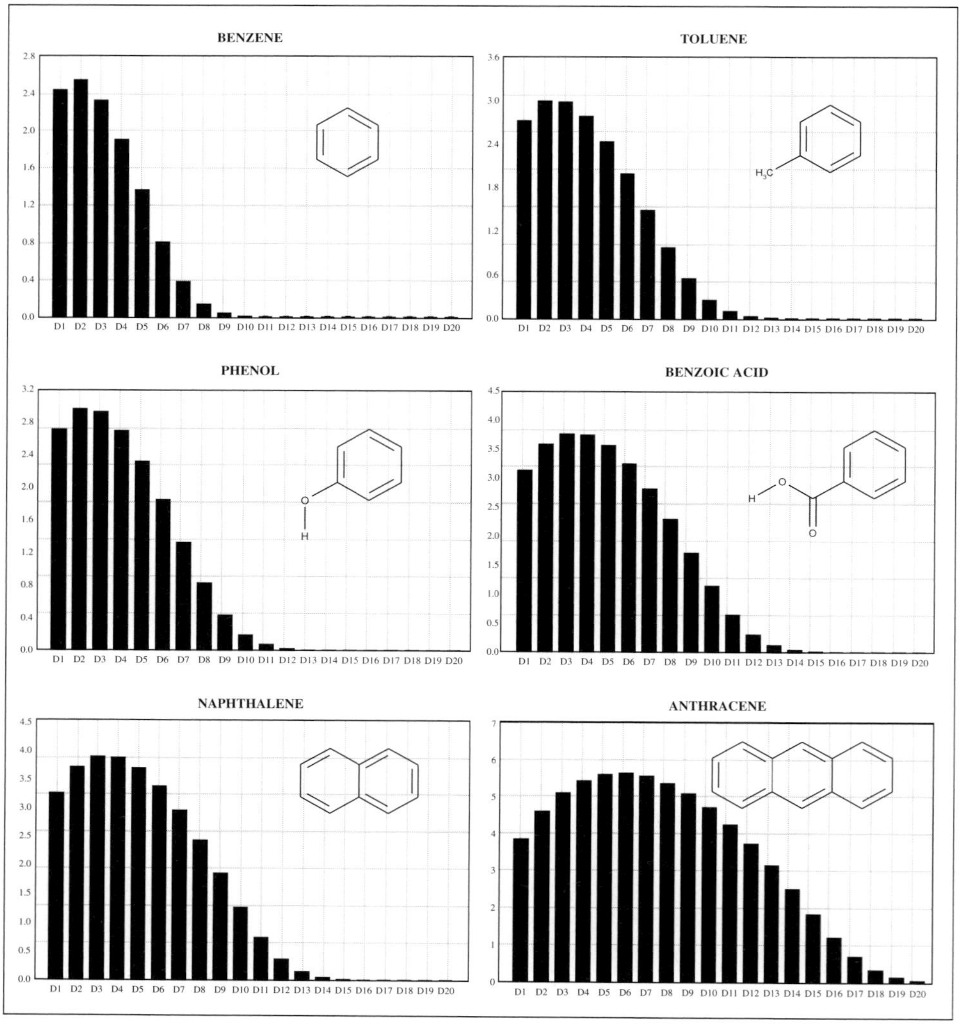

Figure M-8. Molecular profiles for some compounds.

molecular quantum similarity indices → **similarity/diversity**

molecular quantum similarity measures → **similarity/diversity**

molecular representation → **molecular descriptors**

molecular rigidity → **flexibility indices**

molecular self-returning walk count → **self-returning walk counts**

molecular sequence code → **sequence matrices**

molecular sequence count → **sequence matrices**

Molecular Shape Analysis (MSA)

A QSAR approach based on a set of methods that combines molecular shape similarity and commonality measures with other → *molecular descriptors* both to search for similarities among molecules and to build QSAR models [Hopfinger, 1980; Burke and Hopfinger, 1993]. The term *molecular shape similarity* refers to molecular similarity on the basis of a comparison of three-dimensional molecular shapes represented by some property of the atoms composing the molecule, such as the van der Waals spheres. The *molecular shape commonality* is the measure of molecular similarity when conformational energy and molecular shape are simultaneously considered [Hopfinger and Burke, 1990].

The main assumption of this approach is that the shape of the molecule is closely related to the shape of the → *binding site cavity* and, as a consequence, to the biological activity. Therefore, a shape reference compound is chosen which represents the binding site cavity, and the similarity (or commonality) measured between the reference shape and the shape of other compounds is used to determine the biological activity of these compounds. As well as the shape similarity measures, other molecular descriptors such as those in → *Hansch analysis* can be used to evaluate the biological response. The MSA model is thus defined as:

$$\hat{y}_i = b_0 + \sum_k b_k \cdot \Phi_{ik} + \left[\mathrm{f}_0(M(i,j)) + \varrho\left(\mathrm{n}_j - \mathrm{n}_i\right) - \beta \cdot \Delta\mathrm{E}_i\right]$$

where i refers to any compound of the data set and j the reference compound; y_i is the estimated biological response of the ith compound, usually expressed as a logarithm of the ligand concentration inverse; b_0 is a constant characteristic of the reference compound (j); Φ_k is any molecular descriptor representing → *physico-chemical properties* such as → *Hansch descriptors*, topological, geometrical, electronic or thermodynamic features of the molecules. The last term (in square brackets) is a 3D molecular structure term involving molecular shape and conformational thermodynamics; $\mathrm{f}_0(M(i, j))$ is a molecular shape similarity function, i.e. a function of the measure of the relative shape similarity between i and j, $\varrho \cdot (\mathrm{n}_j - \mathrm{n}_i)$ is the difference in intramolecular conformational entropy (flexibility) between i and j, and $\beta \cdot \Delta\mathrm{E}_i$ is a measure of the relative stability of the bioactive conformation of compound i with respect to its global intramolecular energy minimum. The quantity $\mathrm{I_c}(i, j) = \mathrm{f}_0(M(i, j)) - \beta \cdot \Delta\mathrm{E}_i$ is the shape commonality index which takes into account the balance between a gain in molecular shape similarity at the expense of loss in conformational stability.

There are seven operations involved in the MSA approach:

a) conformational analysis;
b) active conformation hypothesis;
c) shape reference compound selection;
d) pair-wise molecular superimposition;
e) molecular shape similarity (or commonality) measure calculation;
f) other molecular descriptor calculation;
g) trial QSAR model development.

For each MSA operation there exists a set of choices which are experimented in the trial QSAR model; the final selection of the requirements for each MSA operation is based on optimizing the fitting ability of the QSAR model.

The most active compound is usually assumed as the reference structure, but a set of overlapped structures can also be assumed to define a reference shape.

Some **molecular shape similarity descriptors** (or **MSA descriptors**) are reported below. They represent a measure of the matching between the shapes of two molecules i and j, one of them being by definition the reference structure; the representation of molecular shape is given in different ways.

- **Common Overlap Steric Volume** (*COSV*)

Defined as the volume shared by two superimposed molecules, i.e.

$$M_0(i,j) \equiv \mathrm{V}_0(i,j) = \mathrm{V}_i \cap \mathrm{V}_j$$

where V_i and V_j are the → *van der Waals volume* of the ith and jth molecules, respectively.

Two arbitrary functions of the common overlap steric volume were also introduced as alternative molecular shape descriptors:

$$S_0 = \mathrm{V}_0^{2/3} \quad \text{and} \quad L_0 = \mathrm{V}_0^{1/3}$$

where S_0 is the **common overlap surface** (or **overlap surface**) and L_0 the **common overlap length**. Despite the terms, S_0 has the dimensions of area but is not a physical measure of the common surface area between two molecules, and the same holds for L_0. Therefore, if the shape of the reference molecule is a good approximation for the acceptor site cavity, V_0 should measure the part of the cavity volume occupied by the considered ligand, while S_0 should be an approximation for the contact surface area of the ligand with receptor.

The **nonoverlap steric volume** $\mathrm{V}_{\mathrm{non}}$ is another MSA descriptor defined as [Tokarski and Hopfinger, 1994]:

$$\mathrm{V}_{\mathrm{non}}(i,j) = \mathrm{V}_{ij} - \mathrm{V}_j$$

where V_{ij} is the composite steric volume of the two aligned molecules i and j. In practice, the non-overlap volume measures the regions of the ith molecule volume not shared by the reference compound, i.e. it represents the → *steric misfit*.

- **atom-pair matching function**

Defined as:

$$M_r(i,j) = \sum_{a=1}^{A_i} \sum_{a'=1}^{A_j} K_{aa'} \cdot r_{aa'}$$

where the sums run over all pairs of atoms of the two considered molecules i and j, $r_{aa'}$ is the interatomic distance between each pair of atoms from the ith and jth molecules, $K_{aa'}$ is a user-defined constant providing the relative importance of the considered distance. For $M_r \to 0$, the superposition between i and j becomes better.

- **charge-matching function**

Defined as:

$$M_c(i,j) = \sum_{a=1}^{A_i} \sum_{a=1}^{A_j} \frac{q_a \cdot q_{a'}}{\mathrm{Q_T}} \cdot r_{aa'}$$

where the sums run over all pairs of atoms of the two considered molecules i and j, q are atomic partial charges and r the → *interatomic distances* between atoms from molecules i and j, Q_{T} is a normalization term calculated as:

$$\mathrm{Q_T} = \sum_{a=1}^{A_i} \sum_{a'=1}^{A_j} q_a \cdot q_{a'}$$

The partial charges q_a and $q_{a'}$ are assumed to always have the same sign; otherwise they would not be matched.

- **Integrated Spatial Difference in Field Potential** (*ISDFP*)

A field-based shape descriptor derived from the representation of molecular body by → *molecular interaction fields* and defined as:

$$M_p(i,j) = \frac{1}{\Phi} \cdot \left[\int_\Phi \left[\mathrm{E}_i(R,\Theta,\phi) - \mathrm{E}_j(R,\Theta,\phi) \right]^2 \cdot d\Phi \right]^{1/2}$$

where E are the → *interaction energy values*, as measured by a probe, at the spherical coordinate position (R, Θ, ϕ), and Φ is the considered integration volume. To calculate this descriptor, it is assumed that molecules i and j are superimposed.

- **weighted combination of *COSV* and *ISDFP***

A combination of two complemetary measures of shape similarity defined as:

$$M_w(i,j) = w \cdot \left[(\mathrm{V}_j \cap \mathrm{V}_j) - M_0(i,j) \right] + (1 - w) \cdot M_p(i,j)$$

where $M_0(i, j)$ and $M_p(i, j)$ are the common overlap steric volume and the integrated spatial difference in field potential, and w is a weighting factor between zero and one. The two descriptors are considered complementary in the sense that the overlap volume measures the shape within the van der Waals surface formed by superimposition of i and j, while *ISDFP* measures the shape outside the van der Waals surface.

📖 [Hopfinger, 1981] [Battershell *et al.*, 1981] [Hopfinger and Potenzone Jr., 1982] [Hopfinger, 1983] [Hopfinger, 1984] [Mabilia *et al.*, 1985] [Walters and Hopfinger, 1986] [Hopfinger *et al.*, 1987] [Nagy *et al.*, 1994] [Rowberg *et al.*, 1994] [Rhyu *et al.*, 1995] [Holzgrabe and Hopfinger, 1996]

molecular shape similarity descriptors → **molecular shape analysis**

molecular similarity matrices → **similarity/diversity**

molecular structure

" ... *the term molecular structure represents a set of nonequivalent conceptual entities. There is no reason to believe that when we discuss different topics (e.g., organic synthesis, reaction rate theories, spectroscopic transitions, reaction mechanisms, ab-initio calculations) using the concept of molecular structure, the different meanings we attach to the term molecular structure ultimately flow from the same concept.*" [Basak and Gute, 1997b].

Together with the concepts of synthesis and chemical composition, the concept of molecular structure is one of the most fruitful of 20th century scientific research. This concept is conveniently studied by considering several levels of description, i.e. the molecular structure is a part of a hierarchical system organized in different levels; to each level correspond characteristic language, properties, and relationships within its constitutional elements at that level as well as relationships between higher and lower levels. Thus, particles, atoms, molecules, compounds, cells, bodies, etc. are hierarchically organized levels of a complex system. At each level emergent properties arise from the organization of elements characterizing that level, i.e. the presence of organizing relationships gives birth to new properties and constraints that influence the complexity of the system.

The above considerations also hold for different hierarchical descriptions of a system at a given level, i.e. the same level is traversed by an inner hierarchical organization because of different descriptions of the same elements. The molecular representations are hierarchical descriptions of the molecular system, therefore, derived from the different representations of the molecular structure; several → *molecular descriptors* are calculated with different chemical information contents.

Each molecular representation reflects hypotheses and ideas; a theory of unknown but supposedly relevant relationships between molecules and their behaviour. Much chemical research makes efforts to accurately predict properties, or to accurately classify chemical structures according to their properties, on the basis of chemical structure alone.

Each → *molecular representation* is a model which highlights only a part of the reality, or, in other words, which explains a part of the experimental evidence. Also a simple chemical formula such as C_6H_5Cl already gives chemical information, at least about chemical composition and stoichiometric atom type relationships.

Although chemical theories are the framework within which molecular structure has been developed, experimental properties define the reference framework in which the concept or, better still, the concepts of molecular structure have been continuously verified, evaluated and modified.

📖 [Woolley, 1976] [Woolley, 1978a] [Primas, 1981] [Weininger, 1984] [Turro, 1986] [Wirth, 1986] [Weininger and Weininger, 1990] [Ash *et al.*, 1991] [Randic, 1992a] [Wentang *et al.*, 1993] [Dietz, 1995] [Bauerschmidt and Gasteiger, 1997] [Testa *et al.*, 1997]

molecular subgraph → **molecular graph**

molecular supergraph → **hyperstructure-based QSAR techniques**

molecular surface

The term molecular surface usually refers to any surface surrounding some or all of the nuclei of the molecule. In the strict quantum mechanical sense, molecules do not have precisely defined surfaces; however, in analogy to macroscopic objects, the electron distribution may be regarded as a 3D *molecular body* whose boundary is the molecular surface. In other words, the molecular surface can be viewed as the formal boundary that separates the 3D space into two parts: within the surface one is expected to find the whole molecule and beyond the rest of the universe [Meyer, 1986c; Mezey, 1991b].

Different physical properties and molecular models have been used to define the molecular surface; the most common are reported below together with the descriptors proposed as measures of **surface areas** and molecular volume (→ *volume descriptors*). Molecular surface area and volume are parameters of molecules that are very important in understanding their structure and chemical behaviour such as their ability to bind ligands and other molecules. An analysis of molecular surface shape is also an important tool in QSAR and → *drug design*; in particular, both → *molecular shape analysis* and → *Mezey 3D shape analysis* were developed to search for similarities among molecules, based on their molecular shape.

- **van der Waals molecular surface**

This is the surface that envelops fused hard spheres centred at the atom coordinates (atomic nuclei) and having radii equal to some of the recommended values of the van der Waals radii. The spheres interpenetrate one another in such a way that the distance between the centres of two spheres equals the formal bond length.

The **van der Waals surface area** SA_{VDW} (also known as **Total molecular Surface Area**, *TSA*) is then defined as the exterior surface of the union of all such spheres in the molecule, i.e. the area of the van der Waals molecular surface.

Analogously, the **van der Waals volume** (V_{VDW}), also called **intrinsic molecular volume** V_I, is the volume of the space within the van der Waals molecular surface. The

van der Waals radius R is the distance at which the attractive and repulsive forces between two non-bonded atoms are balanced; thus the van der Waals volume may be regarded as an impenetrable volume for other molecules (Table M-4).

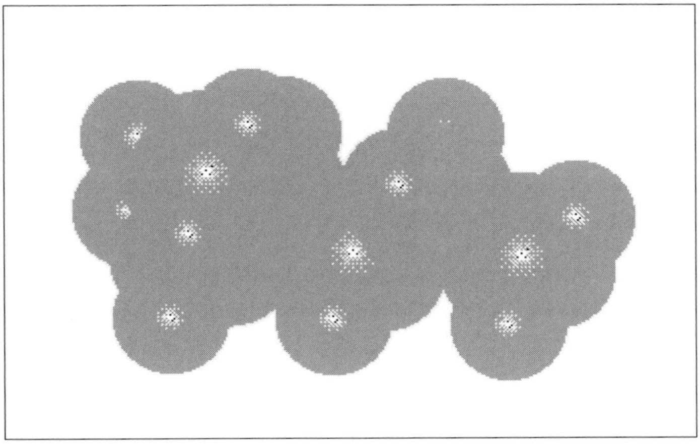

Figure M-9. Van der Waals molecular surface of 2-methylpentane.

The accurate calculation of van der Waals volume and surface area is quite complex, and different approaches have been proposed, most being based on a numerical integration method [Leo *et al.*, 1976; Testa and Purcell, 1978; Pearlman, 1980; Pavani and Ranghino, 1982; Gavezzotti, 1983; Motoc and Marshall, 1985; Meyer, 1985a; Meyer, 1985b; Meyer, 1986a; Stouch and Jurs, 1986; Meyer, 1988b; Meyer, 1988a; Bodor *et al.*, 1989; Meyer, 1989]. For example, a molecule or fragment can be described by a set of vectors $\mathbf{S}_i$, connecting the atomic nuclei to a chosen origin, and a set of radii, R_i (one for each atomic species), each of which defines a sphere around each nucleus i [Gavezzotti, 1983]. An envelope space V_{ENV}, containing all the vectors $\mathbf{S}_i + R_i$ for the molecule, is defined and filled with a very large number (N) of probe points; then, the number of points (N_{OCC}) inside at least one of the atomic spheres is counted and the molecule volume V is calculated as:

$$V = V_{ENV} \cdot \frac{N_{OCC}}{N}$$

Bondi developed a method based on covalent bond distances and van der Waals radii to calculate van der Waals volume [Bondi, 1964]. The volume calculated in this way is sometimes called the **Bondi volume**. It is obtained easily by summing appropriate volume contributions of atoms and functional groups, as proposed by Bondi; note that the Bondi volume does not account for the overlaps which are possible whenever three or more atomic spheres intersect; it is roughly 60 – 70 % of the → *molecular volume.*

Table M-4. Van der Waals radii for some atoms:
[a] [Bondi, 1964]; [b] [Rohrbaugh and Jurs, 1987a];
[c] [Gavezzotti, 1983].

Atom	*Bondi*[a]	*Rohrbaugh-Jurs*[b]	*Gavezzotti*[c]
H	1.20	1.20	1.17
C	1.70	1.70	1.75
N	1.55	1.55	1.55
O	1.52	1.52	1.40
S	1.80	1.80	–
F	1.47	1.50	1.30
Cl	1.75	1.75	1.77
Br	1.85	1.85	1.95
I	1.98	1.97	2.10

• **solvent-accessible molecular surface**
In the case of large complex and folded molecular structures, a part of the van der Waals surface is buried in the interior and is thus inaccessible to solvent interactions which mainly govern the chemical behaviour of molecules in solution. Therefore, in order to obtain the best representation of the outer surface and overall shape of the molecule the solvent-accessible molecular surface was proposed. It was originally defined [Lee and Richards, 1971] as the surface across which the centre of a spherical approximation of the solvent is passed when the solvent sphere is rolled over the van der Waals surface of the molecule. The radius (1.5 Å) of the solvent sphere is usually chosen to approximate the contact surface formed when a water molecule interacts with the considered molecule. If there are several grooves or minor cavities on the van der Waals surface where the rolling sphere cannot enter, then the solvent-accessible surface will be significantly different from the van der Waals surface.

A few years later, Richards [Richards, 1977] gave a new definition of solvent-accessible surface, dividing it into two parts: the *contact surface* and the *reentrant surface*. The **contact surface** is that part of the van der Waals surface that is accessible to the probe sphere representing the solvent molecule. The **reentrant surface** comes from the inward-facing surface of the probe sphere when it is simultaneously in contact with more than one atom.

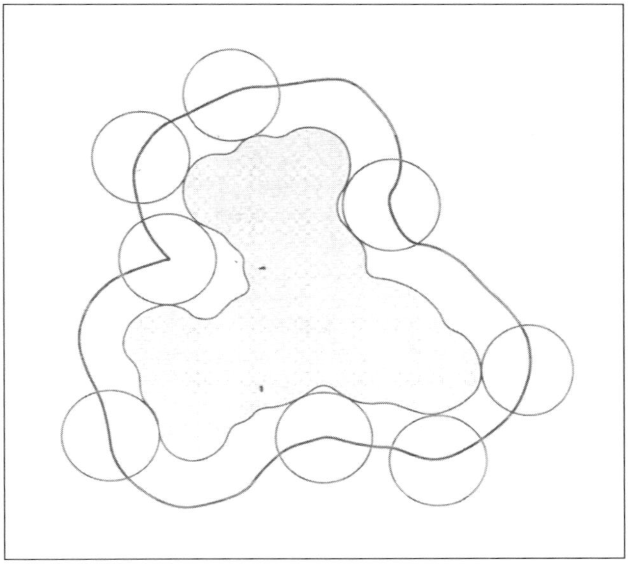

Figure M-10. Solvent-accessible molecular surface defined by the centres of spheres rolled along the molecular contour surface. The radius of the sphere is chosen according to the size of the solvent molecule.

The area of the solvent-accessible surface is called the **Solvent-Accessible Surface Area** *SASA* (or **Total Solvent-Accessible Surface Area**, *TSASA*). Several algorithms were proposed that implement both the first original definition of *SASA* and that of Richards. One of the most popular algorithms that implements Richards' solvent-accessible surface was proposed by Connolly [Connolly, 1983a]. It is an analytical method for computing molecular surface, and is based on surface decomposition into a set of curved regions of spheres and tori that join at circular arcs; spheres, tori and arcs are defined by analytical expressions in terms of atomic coordinates, van der Waals radii and the probe radius. The molecular surface calculated in such a way is sometimes referred to as **Connolly surface area**. This algorithm also allows the calculation of solvent-accessible atomic areas.

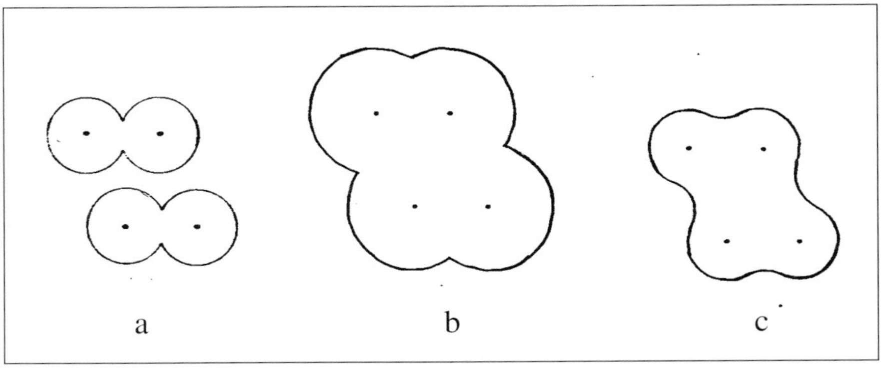

Figure M-11. Comparison among (a) van der Waals surface, (b) solvent-accessible surface, and (c) contact surface.

An alternative to the hard sphere model is a recently proposed method for *SASA* calculations, based on atomic Gaussian functions describing the exposure of atoms and molecular fragments to solvent. A simple integral function of these atomic Gaussians is used to define a Gaussian neighbourhood which behaves in a complementary fashion to the conventional definition of solvent accessibility, i.e. the smaller the Gaussian neighbourhood, the more exposed is the atom and hence the greater its accessibility [Grant and Pickup, 1995; Grant *et al.*, 1996].

Several → *charged partial surface area descriptors* (CPSA) and → *hydrogen-bond charged partial surface area descriptors* (HB-CPSA) are based on portions of the solvent-accessible surface area relative to polar or hydrophobic regions of the molecule, in some cases weighted by the corresponding local charges. Moreover, the **Hydrated Surface Area** (*HSA*) is the portion of the solvent-accessible surface area associated with hydration of polar functional groups. The **Isotropic Surface Area** (*ISA*) is the surface of the molecule accessible to nonspecific interactions with the solvent, i.e. the surface of the molecule involved in specific hydrogen-bonding with water is not considered [Collantes and Dunn III, 1995; Koehler *et al.*, 1988]. A hydration complex model needs to estimate the isotropic surface area. The **Polar Surface Area** (*PSA*) is defined as the part of the surface area of the molecule associated with oxygens, nitrogens, sulfurs and the hydrogens bonded to any of these atoms. This surface descriptor is related to the hydrogen-bonding ability of compounds [Palm *et al.*, 1998; Winiwarter *et al.*, 1998].

The volume of space bounded by the solvent-accessible molecular surface is called the **solvent-excluded volume** because it is the volume of space from which solvent is excluded by the presence of the molecule when the solvent molecule is also modelled as a hard sphere. Moreover, the **interstitial volume** is the volume consisting of packing defects between the atoms that are too small to admit a probe sphere of a given radius; in practice, it is calculated as the difference between the solvent-excluded volume and the van der Waals volume. An analytical method was developed by Connolly able to calculate the solvent-excluded volume [Connolly, 1983b]; several other numerical and analytical approaches have been proposed.

📖 [Silla *et al.*, 1991; Hirono *et al.*, 1991]

- **electron isodensity contour surface**

The collection of all those points **r** of the space where the value of the → *electron density* $\varrho(\mathbf{r})$ is equal to a threshold value m [Mezey, 1991b], i.e.

$$G(m) = \{\mathbf{r} \; : \; \varrho(\mathbf{r}) = m\}$$

Any positive value as threshold m can be chosen, although a relatively small value is usually used to define a suitable molecular surface because the electron density converges rapidly to zero at short distances from the nuclei. The positive values of the threshold are due to the usual convention that a large negative charge means a large positive value of the electron density. For large values of m the molecular surface is composed of several disconnected surfaces each surrounding one nucleus, while for too small values of m the surface is an essentially spherical surface surrounding all of the nuclei and containing no information on the shape of the molecule.

- **molecular electrostatic potential contour surface**

The collection of all those points **r** of the space where the value of the → *molecular electrostatic potential* (MEP) $V(\mathbf{r})$ is equal to a threshold value m [Mezey, 1991b], i.e.

$$G(m) = \{\mathbf{r} \; : \; V(\mathbf{r}) = m\}$$

The contour parameter m as well as the electrostatic potential can take both positive and negative values. An analysis of the shape of MEP surfaces is of particular interest in drug design as the electrostatic potential has a marked influence on the binding interactions between ligand and receptor.

- **molecular orbital contour surface**

A molecular surface defined as the contour surface of individual molecular orbitals such as HOMO and LUMO, other frontier orbitals, or localized and delocalized orbitals [Mezey, 1991b]. In practice, it is the collection of all those points $\mathbf{r}$ of the space where the value of the electronic wavefunction $\Psi(\mathbf{r})$ of the considered molecular orbital is equal to a threshold value m, i.e.

$$G(m) = \{\mathbf{r} : \Psi(\mathbf{r}) = m\}$$

The contour parameter m can take both positive and negative values.

[Hermann, 1972] [Amidon *et al.*, 1975] [Arteca *et al.*, 1988b] [Leicester *et al.*, 1988] [Marsili, 1988] [Pascual-Ahuir and Silla, 1990] [Valko and Slegel, 1992] [Brusseau, 1993] [Leicester *et al.*, 1994a] [Leicester *et al.*, 1994b] [Schüürmann, 1995] [Palm *et al.*, 1996] [Lee *et al.*, 1996] [Randic and Krilov, 1997b] [Brickmann, 1997] [Zweerszeilmaker *et al.*, 1997] [Whitley, 1998]

molecular surface interaction terms (MSI)

A set of → *molecular descriptors* including the molecular surface area and empirically derived descriptors accounting for dispersion, polar and hydrogen bonding interactions [Grigoras, 1990]. They were proposed to empirically express the molecular surface energy using atomic contributions to the total molecular surface. The molecular surface interaction terms are:

$$\mathbf{A} = \sum_i SA_i \qquad \mathbf{A}_- = \sum_i SA_i \cdot b_i \cdot q_i^- \qquad \mathbf{A}_+ = \sum_i SA_i \cdot b_i \cdot q_i^+$$

$$\mathbf{A}_{\mathrm{HB}} = \sum_i SA_i \cdot b_i \cdot q_i^{\mathrm{H}}$$

where $\mathbf{A}$ is the → *total molecular surface area* calculated as the sum of all the atomic surface areas SA_i; this is a dispersion molecular surface interaction term. $\mathbf{A}_-$ is the electrostatic negative molecular surface interaction term calculated as the sum of surface areas of negatively charged atoms multiplied by their corresponding scaled → *net atomic charge* q_i^-. $\mathbf{A}_+$ is the electrostatic positive molecular surface interaction term calculated as the sum of surface areas of positively charged atoms multiplied by their corresponding scaled net atomic charge q_i^+. $\mathbf{A}_{\mathrm{HB}}$ is the hydrogen-bonding molecular surface interaction term calculated as the sum of the surface areas of hydrogen-bonding hydrogen atoms multiplied by their corresponding scaled net atomic charge q_i^{H}. The coefficient b_i is an empirical charge scaling factor which is the same for the atoms of the same chemical type (Table M-5).

Table M-5. Charge scaling factors b for different atom chemical types.

Atom/Hybrid	*Coefficient b*	*Atom/Hybrid*	*Coefficient b*
H (at C_{sp3})	3.29	N_{sp3}	0.155
H (at C_{sp2})	7.77	N_{sp}	1.59
H_{HB}	1.00	N_{AROM}	2.79
C_{sp3}	1.00	O_{sp3}	1.32
C_{sp2}	0.00	O_{sp2}	1.51
C_{sp}	2.33	F	0.00
C_{AROM}	9.25	Cl	1.78

The hydrogen atom considered for the calculation of the hydrogen bonding term are not taken into account in the $\mathbf{A}_+$ term.

molecular topological index : *Schultz molecular topological index*

molecular topological indices : *topological indices* → **graph invariants**

molecular transform → **3D-MoRSE descriptors**

molecular volume → **volume descriptors** (⊙ molar volume)

molecular walk count → **walk counts**

molecular weight (MW)

Among the → *size descriptors*, this is the simplest and most used molecular → *0D-descriptor*, calculated as the sum of the atomic weights. It is related to molecular size and is atom-type sensitive. It is defined as:

$$MW = \sum_{i=1}^{A} m_i$$

where m is the atomic mass and i runs over the A number of atoms of the molecule. The **average molecular weight** defined as:

$$\overline{MW} = \frac{1}{A} \cdot \sum_{i=1}^{A} m_i = \frac{MW}{A}$$

is also used as a molecular descriptor and is related to → *atomic composition indices.*

Square root molecular weight ($MW2$), defined as $MW2 = MW^{1/2}$, and **cube root molecular weight** ($MW3$), defined as $MW3 = MW^{1/3}$ and corresponding to a linear dimension of size, are also used as molecular size descriptors.

molecule centre : *centre of a molecule*

MOLORD algorithm → **iterated line graph sequence**

moment of inertia → **principal moments of inertia**

Monte Carlo version of MTD → **minimal topological difference**

Moran coefficient → **autocorrelation descriptors**

Moreau-Broto autocorrelation → **autocorrelation descriptors**

Moreau chirality index → **chirality descriptors**

Morgan's extended connectivity algorithm → **canonical numbering**

Moriguchi model based on structural parameters → **lipophilicity descriptors**

Moriguchi model based on surface area → **lipophilicity descriptors**

Moriguchi polar parameter → **lipophilicity descriptors**

Morovitz information index (I_{MOR})

An information index accounting for the structural features of a molecule [Morovitz, 1955]. It is defined as:

$$I_{\mathrm{MOR}} = I_{\mathrm{AC}} + I_{\mathrm{PB}}$$

where the first term is → *total information index on atomic composition* and the second term is the **information on the possible valence bonds** I_{PB} defined as:

$$I_{PB} = \sum_{g=1}^{G} A_g \cdot \log_2 V_g$$

where g runs over all the different atom types, A_g is the number of atoms of the gth type, and V_g the number of possible bonds which can be formed by an atom of the gth type, calculated as:

$$V_{g,r} = \binom{6+r-1}{r} = \frac{(6+r-1)!}{r! \cdot 5!}$$

where 6 is assumed as the maximum possible valence and r is the actual valence of the gth type atom. For example, $V_H = 6$, $V_O = 21$, $V_{N,3} = 56$, $V_{C,4} = 126$ [Bonchev, 1983].

MPR descriptors

→ *Local vertex invariants* (LOVIs) obtained as the solutions of a linear equation system defined as:

$$\mathbf{M_p} \cdot \mathbf{s} = \mathbf{r} \qquad \mathbf{M_p} = \mathbf{M} + \mathbf{p} \cdot \mathbf{I}$$

where $\mathbf{M_p}$ is a square $A \times A$ matrix representing the → *molecular graph*, **p** is an A-dimensional column vector containing weights for graph vertices which are used as diagonal elements of the **M** matrix, **r** the A-dimensional column vector of atomic properties, **I** is the identity matrix and **s** the A-dimensional column vector which is the solution of the system [Filip *et al.*, 1987; Ivanciuc *et al.*, 1992].

MPR (*Matrix-Property-Response*) descriptors are thus the elements of the **MPR** vector calculated as:

$$\mathbf{MPR} \equiv \mathbf{s} = \left(\mathbf{M_p^T} \cdot \mathbf{M_p}\right)^{-1} \cdot \mathbf{M_p^T} \cdot \mathbf{r}$$

The vertex properties encoded in the column vectors **p** and **r** can be either topological, e.g. → *vertex degree*, → *vertex distance degree*, or chemical, e.g. atomic number, → *electronegativity*, → *ionization potential*.

Among the LOVIs obtained by this general approach, the most known are **AZV descriptors** derived from the → *adjacency matrix* **A** whose diagonal elements are substituted by the atomic numbers Z_i and the A-dimensional vector **r** contanining the vertex degrees δ_i.

Different sets of LOVIs can be obtained by different choices of matrices and vectors defining the linear equation system; several combinations were studied on linear alkanes (Table M-6).

Table M-6. A: adjacency matrix; D: distance matrix; V: vector of vertex degrees δ; S: vector of distance sums σ; Z: vector of atomic numbers; N: vector of numbers of graph vertices; 1: unit vector. LOVI range values for linear alkanes.

ID	*MPR*	*LOVIs range*	*ID*	*MPR*	*LOVIs range*
1	AZV	0.1 – 1	11	DSN	0.05 – 0.7
2	ASV	0.01 – 0.2	12	DN^2N	0.06 – 0.2
3	DSV	–0.02 – 0.12	13	ANS	1 – 4
4	AZS	2 – 9	14	ANV	0.08 – 0.5
5	ASZ	0.1 – 1	15	AZN	0.3 – 1.5
6	DN^2S	0.1 – 0.3	16	ANZ	0.5 – 1.7

ID	*MPR*	*LOVIs range*	*ID*	*MPR*	*LOVIs range*
7	DN^21	0 – 0.09	17	AN1	0.1 – 0.3
8	AS1	0.02 – 0.1	18	DSZ	0.06 – 0.6
9	DS1	0 – 0.3	19	ANN	0.7 – 0.9
10	ASN	0.2 – 0.7	20	DN^2Z	0.03 – 0.5

Example : 1,3 - butandiol

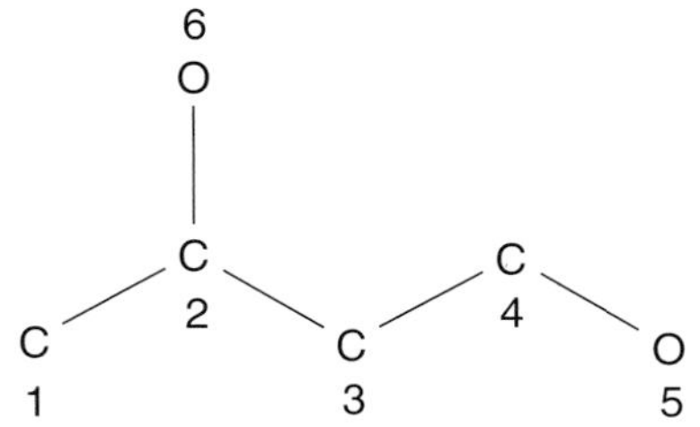

Adjacency matrix **A** augmented by atomic numbers Z

Atom	**1**	**2**	**3**	**4**	**5**	**6**
1	6	1	0	0	0	0
2	1	6	1	0	0	1
3	0	1	6	1	0	0
4	0	0	1	6	1	0
5	0	0	0	1	8	0
6	0	1	0	0	0	8

•

AZV_i

Atom	s_i
1	s_1
2	s_2
3	s_3
4	s_4
5	s_5
6	s_6

=

Vertex degrees

Atom	δ_i
1	1
2	3
3	2
4	2
5	1
6	1

$$\begin{cases} 6\cdot s_1 + s_2 = 1 \\ s_1 + 6\cdot s_2 + s_3 + s_6 = 3 \\ s_2 + 6\cdot s_3 + s_4 = 2 \\ s_3 + 6\cdot s_4 + s_5 = 2 \\ s_4 + 8\cdot s_5 = 1 \\ s_2 + 8\cdot s_6 = 1 \end{cases} \longrightarrow \begin{array}{l} s_1 = 0.09382 \\ s_2 = 0.43708 \\ s_3 = 0.21334 \\ s_4 = 0.28284 \\ s_5 = 0.08996 \\ s_6 = 0.07036 \end{array}$$

Box M-2.

Closely related to MPR descriptors are local vertex invariants called **graph potentials** denoted by U_i [Golender *et al.*, 1981; Ivanciuc *et al.*, 1992]. They are calculated as the solutions of a linear equation system defined as:

$$\mathbf{W} \cdot \mathbf{s} = \mathbf{r}$$

where **W** is a weighted matrix, **r** the A-dimensional column vector of atomic properties, and **s** the A-dimensional column vector of solutions of the system which are local invariants U_i.

The weighted matrix **W** is defined as:

$$[\mathbf{W}]_{ij} = \begin{cases} w_{ii} + \sum_k w_{ik} & \text{if } i = j \\ -w_{ij} & \text{if } \{i,j\} \in \mathcal{E}(\mathcal{G}) \\ 0 & \text{if } \{i,j\} \notin \mathcal{E}(\mathcal{G}) \end{cases}$$

where w_{ii} is any topological or chemical semi-positive definite atomic property and the sum runs over the first neighbours of the ith atom; w_{ij} is any topological or chemical semi-positive definite bond weight and $\mathcal{E}(\mathcal{G})$ the set of edges of the molecular graph $\mathcal{G}$. If the weights w are all set equal to one, then the **W** matrix is:

$$\mathbf{W} = \mathbf{V} - \mathbf{A} + \mathbf{I} = \mathbf{L} + \mathbf{I}$$

where **V**, **A**, **I**, and **L** are the diagonal → *vertex degree matrix*, the adjacency matrix, the identity matrix and the → *Laplacian matrix*.

📖 [Balaban, 1993c] [Balaban, 1994a]

MoRSE descriptors : *3D-MoRSE descriptors*

MPS topological index : *detour index* → **detour matrix**

MSA descriptors : *molecular shape analysis descriptors* → **molecular shape analysis**

***MTD-ADJ* method** → **minimal topological difference**

***MTD* descriptors** → **minimal topological difference**

***MTD-MC* method** → **minimal topological difference**

***MTD* model** → **minimal topological difference**

***MTI′* index** : *S index* → **Schultz molecular topological index**

***m*th order sparse matrix** → **algebraic operators** (⊙ sparse matrices)

Mulliken electronegativity → **quantum-chemical descriptors**

Mulliken population analysis → **charge descriptors** (⊙ atomic charge)

multicriteria decision making → **chemometrics**

multigraph → **graph**

multigraph distance degree → **weighted matrices**

multigraph distance matrix → **weighted matrices**

multigraph factor : *atomic multigraph factor* → **bond order indices** (⊙ conventional bond order)

multigraph information content indices : *indices of neighbourhood symmetry*

multiple arc → **graph**

multiple edge → **graph**

multiple bond count → **multiple bond descriptors**

multiple bond descriptors

The simplest descriptors of the degree of unsaturation of a molecule are → *count descriptors* based on the presence of double bonds, triple bonds, and aromatic bonds; they are the **double-bond count** $n_=$, the **triple-bond count** $n_\equiv$ and the **aromatic-bond count** $n_\approx$. The **MCB index** was proposed as the number of multiple C–C bonds in the molecule accounting for double, triple, and aromatic bonds [Bakken and Jurs, 1999b].

Another simple global **multiple bond count** is defined as:

$$b^* = \sum\nolimits_b \pi^*_{ij} - B$$

where π^* is the → *conventional bond order* and the summation runs over all B bonds. For saturated compounds, $b^* = 0$.

The **unsaturation index** *UI* was also defined as:

$$UI = \log_2(1 + b)$$

where b is calculated as:

$$b = \left[2\mathrm{N_C} + 2 - \mathrm{N_H} - \mathrm{N_X} + \mathrm{N_N} + \mathrm{N_P} + 2\left(\mathrm{N_{O\text{-}S}} - \mathrm{N_{SO_3}}\right)\right]/2 - C$$

N_C, N_H, N_X, N_N, N_P, and C are the number of carbon atoms, hydrogen, halogen, nitrogen, phosphoros, and independent cycles, respectively. $N_{O\text{–}S}$ and N_{SO3} are the number of oxygen atoms bonded to sulfur and the number of SO_3 groups, respectively. When no sulfur atoms are present, this index can be easily calculated from the 0D-representation of the molecule; otherwise, it is coincident with the index calculated replacing b with b^*.

A multiple bond descriptor was proposed in terms of → *valence vertex degree* δ^v, called the **DV index**, defined as:

$$DV = \sum\nolimits_b \left[\left(\delta^v_i\right)^{-1/2} + \left(\delta^v_j\right)^{-1/2}\right]_b$$

where the sum runs over all the multiple bonds and i and j denote the atoms involved in the considered bond [Millership and Woolfson, 1980].

The **induction parameter** was proposed to estimate the interaction ability of polar and nonpolar groups in the molecule [Thomas and Eckert, 1984]; it is based on the degree of unsaturation and defined as:

$$q_{ind} = 1 - \frac{n_=}{A}$$

where $n_=$ is the number of double bonds in the molecule. For saturated molecules, $q_{ind} = 1$.

→ *Multigraph information content indices* are → *information indices* encoding the bond multiplicity in the molecules.

To take into account the absolute contribution that a single double-bond makes to the whole size and shape of alkene molecules, **second-grade structural parameters** were derived from a → *molecular graph* [Zhang *et al.*, 1997]. The topological descriptors representing the size w and the shape $P^=_W$ related to the presence of a double bond are, respectively:

$$w = \frac{\sum_{i=1}^{A} (d_{ik} + d_{il})}{2W} \qquad P^=_W = {}^3f_k + {}^3f_l$$

where W is the → *Wiener index*, i.e. the total sum of distances in the molecular graph, and d represents the → *topological distance* between two vertices; k and l denote the vertices incident with the considered double bond and the sum runs over all A vertices of the molecule. In the shape descriptor $P_W^{=}$, ${}^3f_k$ and ${}^3f_l$ are the number of vertices at a distance 3 from vertices v_k and v_l, respectively, i.e. their → *vertex distance count*. The size descriptor w is derived from the Wiener index while $P_W^{=}$ is derived from the → *polarity number*. An extension giving information about the presence of several double bonds can be the sum of the w and $P_W^{=}$ values defined above over all double bonds.

multivariate *K* correlation index (: *K correlation index*)

This is the total quantity of correlation contained in a → *data set* $\mathbf{X}(n, p)$ estimated from the eigenvalue distribution obtained by diagonalization of the corresponding *correlation matrix* $\mathbf{R}(p, p)$ [Todeschini, 1997; Todeschini *et al.*, 1998].

Let $\lambda_1, \lambda_2, ..., \lambda_p$ be the set of the p → *eigenvalues*. The K index is defined as the following:

$$K = \frac{\sum_{j=1}^{p} \left| EV_j - \frac{1}{p} \right|}{\frac{2 \cdot (p-1)}{p}} \qquad 0 \leq K \leq 1$$

where:

$$EV_j = \frac{\lambda_j}{\sum_{j=1}^{p} \lambda_j}$$

is the explained variance from the jth eigenvalue; the denominator corresponds to the maximum value reached by the numerator and is used to scale K values between 0 and 1. It must be observed that all of the zero eigenvalues are also considered in the summation, each giving a contribution of $1/p$.

The K correlation index is a → *redundancy index*, taking the value of 1 when all of the variables are correlated and 0 when they are uncorrelated. From a geometrical point of view, $K = 0$ corresponds to a spherical p-dimensional space and $K = 1$ corresponds to a straight line in a p-dimensional space. In the case where p, the number of variables, is greater than n, the number of objects, the rank of the correlation matrix is n, this rank being less than or equal to $\min(n, p)$.

In this case at least $p - n$ eigenvalues are zero; the previous formula holds true, i.e. the summation runs on the total number of variables p, but can be viewed as the contribution of two parts:

$$K = \frac{\sum_{j=1}^{n} \left| \frac{\lambda_j}{\sum \lambda_j} - \frac{1}{p} \right| + (p-n)\frac{1}{p}}{\frac{2 \cdot (p-1)}{p}}$$

The first part runs from the first to the nth eigenvalue and the second takes into account the contribution to the correlation due to p–n zero eigenvalues. Assuming the first part to be uncorrelated (i.e. the first n eigenvalues are p/n), an estimate of the minimum correlation within the data can be obtained from the following expression:

$$K_{\text{EMB}} = \min K = \left[n \cdot \left(\frac{1}{n} - \frac{1}{p} \right) + (p-n)\frac{1}{p} \right] \cdot \frac{p}{2 \cdot (p-1)} = \frac{p-n}{p-1}$$

This quantity is called **embedded correlation** and represents the minimum correlation of data when $p > n$. The embedded correlation can be viewed as the correlation due to the presence of N_R relationships among the p variables, i.e. as:

$$K_{EMB} = \frac{N_R}{p-1}$$

The multivariate correlation index calculated by diagonalization of the correlation matrix obtained from the → *molecular matrix* $\mathbb{M}$ is also among the → *WHIM shape* descriptors.

multivariate entropy → **model complexity** (⊙ information content ratio)

mutation and selection uncover models → **variable selection**

N

NCD-descriptors → **molecular descriptors** (⊙ invariance properties of molecular descriptors)

negentropy : *total information content* → **information content**

neighbourhood information content → **indices of neighbourhood symmetry**

neighbourhood of a vertex → **indices of neighbourhood symmetry**

neighbourhood total information content → **indices of neighbourhood symmetry**

net atomic charge : *atomic charge* → **charge descriptors**

net orbital charge → **charge descriptors**

non-adjacent number → **Hosoya Z index**

nonoverlapping volume → **minimal topological difference**

nonoverlap steric volume → **molecular shape analysis** (⊙ common overlap steric volume)

normalized atomic walk count → **walk counts**

normalized centric index → **centric indices** (⊙ Balaban centric index)

normalized extended connectivity → **canonical numbering** (⊙ Morgan's extended connectivity algorithm)

normalized fragment topological indices → **fragment topological indices**

normalized quadratic index : *quadratic index* → **Zagreb indices**

normalized Szeged property matrices → **Szeged matrix**

normalized vertex distance complexity → **topological information indices**

normalized Wiener index → **Wiener index**

Norrington lipophilic constant → **lipophilicity descriptors** (⊙ Hansch-Fujita hydrophobic substituent constants)

nuclear information content → **atomic information indices**

nucleophilic atomic frontier electron density → **quantum-chemical descriptors**

nucleophilic charge → **charge descriptors** (⊙ net orbital charge)

nucleophilic frontier electron density index → **quantum-chemical descriptors** (⊙ nucleophilic atomic frontier electron density)

nucleophilic indices → **reactivity indices**

nucleophilic substituent constant → **electronic substituent constants** (⊙ resonance electronic constants)

nucleophilic superdelocalizability → **quantum-chemical descriptors**

number of atoms in substituent specific positions → **steric descriptors**

number of terms in the model → **model complexity**

Nys-Rekker hydrophobic fragmental constants → **lipophilicity descriptors**

O

OASIS method *(: Optimized Approach based on Structural Indices Set)*

The OASIS method is based on the same assumption as → *Hansch analysis* and can be regarded as an extended and optimized version of the Hansch approach [Mekenyan and Bonchev, 1986; Mekenyan *et al.*, 1990a].

Besides → *substituent constants*, the descriptors used in the OASIS approach are → *topological indices*, → *geometrical descriptors*, → *steric descriptors*, and → *electronic descriptors*, both for molecules and their fragments.

The **OASIS model** is defined as:

$$\textit{biological activity} = f(\{\mathrm{TI}\}, \{\Phi_1\}, \{\Phi_2\}, \{\Phi_3\})$$

where $\{\Phi_i\}$ are different sets of → *physico-chemical properties*, each set representing steric, electronic and hydrophobic descriptors, respectively, and $\{\mathrm{TI}\}$ is a set of topological descriptors.

This equation allows for the possibility that more than one descriptor of the different factors contribute to the overall biological activity. Moreover, these model components can be either local, i.e. referring to atoms or fragments, or global, i.e. describing the molecule as a whole.

📖 [Mekenyan *et al.*, 1986] [Mekenyan *et al.*, 1986a] [Mekenyan *et al.*, 1986b] [Mekenyan *et al.*, 1988b] [Mekenyan *et al.*, 1990b] [Mercier *et al.*, 1991] [Bonchev *et al.*, 1993c] [Mekenyan *et al.*, 1993a] [Bonchev *et al.*, 1994b] [Kamenska *et al.*, 1996]

OASIS model → **OASIS method**

object → **data set**

octanol-water partition coefficient → **lipophilicity descriptors**

octupole moment → **electric polarization descriptors**

oligocentre → **centre of a graph**

optimization → **chemometrics**

Optimized Approach based on Structural Indices Set *: OASIS method*

orbital electronegativity → **quantum-chemical descriptors**

orbital information indices (I_{ORB})

The first information index proposed as a measure of complexity of an → *H-depleted molecular graph* is the **vertex orbital information content** $\bar{I}_{ORB}$, originally simply called **topological information content** $\bar{I}_{TOP}$ [Rashevsky, 1955; Trucco, 1956a; Trucco, 1956b], defined as → *mean information content*:

$$\bar{I}_{ORB} \equiv \bar{I}_{TOP} = -\sum_{g=1}^{G} \frac{A_g}{A} \cdot \log_2 \frac{A_g}{A}$$

where A is the total number of atoms and A_g is the number of topologically equivalent atoms of the gth type; the sum runs over all G different classes of topological equivalence. The atoms are distinguished not on the basis of their chemical nature but on their topological relationships to each other. Vertices are topologically equivalent if they belong to the same → *orbits* of the vertex → *automorphism group* of the graph.

Note that in the original definition of the topological information content proposed by Rashevsky the equivalence classes were defined by the → *vertex degree* of the neighbouring vertices instead of the graph orbits.

The **total topological information content** was defined as:

$$I_{TOP} = A \cdot \bar{I}_{TOP} = A \cdot \log_2 A - \sum_{g=1}^{G} A_g \cdot \log_2 A_g$$

where A is the number of atoms.

The **edge orbital information content** was defined by analogy as [Trucco, 1956a; Trucco, 1956b]:

$$^{E}\bar{I}_{ORB} = -\sum_{g=1}^{G} \frac{B_g}{B} \cdot \log_2 \frac{B_g}{B}$$

where B is the total number of edges in the graph and B_g is the number of edges belonging to the gth edge orbit of the graph; the sum runs over all G different edge orbits. The corresponding **total edge orbital information content** was defined as:

$$^{E}I_{ORB} = B \cdot {}^{E}\bar{I}_{ORB} = B \cdot \log_2 B - \sum_{g=1}^{G} B_g \cdot \log_2 B_g$$

Moreover, based on the connection orbits of the graph, the **connection orbital information content** was also defined [Bonchev, 1983]:

$$^{CONN}\bar{I}_{ORB} = -\sum_{g=1}^{G} \frac{(N_2)_g}{N_2} \cdot \log_2 \frac{(N_2)_g}{N_2}$$

where N_2 is the → *connection number* of the graph and $(N_2)_g$ is the number of connections belonging to the gth orbit of the graph; the sum runs over all G different connection orbits. The corresponding **total connection orbital information content** defined as:

$$^{CONN}I_{ORB} = N_2 \cdot {}^{CONN}\bar{I}_{ORB} = N_2 \cdot \log_2 N_2 - \sum_{g=1}^{G} (N_2)_g \cdot \log_2 (N_2)_g$$

is a component of the → *Bertz complexity index* when calculated for multigraphs.

The orbital information indices can also be calculated for multigraphs, giving a higher measure of graph complexity as the multiplicity of the edges provides more graph orbits than the simple graph. In the case of graph vertices of the same chemical element, the orbital information content for multigraph coincides with the → *neighbourhood information content* of maximal order calculated for the H-depleted molecular graph:

$$\bar{I}_{ORB} = IC_{max}$$

orbital interaction graph of linked atoms → **determinant-based descriptors**
(⊙ general a_N-index)

orbital interaction matrix of linked atoms → **determinant-based descriptors**
(⊙ general a_N-index)

orbits → **graph**

ordered structural code → **self-returning walk counts**

ordered walk count molecular code → **walk counts**

order of a subgraph → **molecular graph**

order of neighbourhood → **indices of neighbourhood symmetry**

order parameter → **mean overcrossing number**

oriented graph → **graph** (⊙ digraph)

orthogonal descriptors : *orthogonalized descriptors*

orthogonalized descriptors (: *orthogonal descriptors)*

These are obtained by applying an orthogonalization procedure to a selected set of → *molecular descriptors*. A descriptor X_j is made orthogonal to another descriptor X_i simply by regressing it against X_i and using as the new orthogonal descriptor the residual $X_j - \hat{X}_j$, where $\hat{X}_j$ is the value of X_j calculated by the regression model. The residual, i.e. the orthogonalized descriptor, represents the part of descriptor X_j not explained by the descriptor X_i; the symbol ${}^i\Omega^j$ is usually used to indicate that descriptor X_j is made orthogonal to descriptor X_i. The orthogonalization process is an iterative procedure until the last considered descriptor is orthogonalized against the preceding orthogonalized descriptors [Randic, 1991b; Randic, 1991e; Randic, 1991f].

The orthogonalization procedure requires a prior ordering of the descriptors to which other descriptors are subsequently made orthogonal, i.e. it requires → *basis descriptors*. Thus it is evident that different orthogonalized descriptor bases derive from different ordering. However, in the case of path counts, connectivity indices, → *uniform length descriptors*, which are naturally ordered descriptors, the orthogonalization procedure can be applied easily and the symbol Ω_j used simply for orthogonalized descriptors.

The most well-known orthogonalization technique is the Gram-Schmidt orthogonalization scheme [Golub and van Loan, 1983]. When a subset of descriptors has the same partial ordering they are simultaneously and mutually orthogonalized within the set itself after sequential orthogonalization to the preceding descriptors [Klein *et al.*, 1997]. Canonical orthogonalization, symmetrical orthogonalization and optimal orthogonalization are techniques to perform this task.

In order to better explain this procedure, a sequence of → *path counts* ${}^mP$ is considered. **Orthogonalized path counts** are calculated through the following steps: a) The first path count ${}^1P$ in the sequence is chosen as the first orthogonalized descriptor Ω_1; it can also be decomposed into the mean contribution represented by Ω_0 and the deviation from the mean represented by Ω_1. b) The second orthogonalized descriptor Ω_2 is calculated as a residual of the regression between the second path count ${}^2P$ and the first one ${}^1P$; the residual contains information independent of ${}^2P$. c) The orthogonalization continues by regressing the path number ${}^3P$ against the first path count ${}^1P$, and then the residual of this regression against the orthogonalized descriptor Ω_2; the residual of the last regression is the third orthogonalized descriptor Ω_3. In general, the pth orthogonalized descriptor Ω_p is defined as the residual in the multiple regression of the pth descriptor of the sequence against the $p-1$ previously orthogonalized descriptors [Soskic *et al.*, 1996b].

Orthogonalized descriptors are used in → *similarity/diversity* analysis and quantitative → *structure/response correlations* with the aim of eliminating the bias provided by the interdependence of common molecular descriptors. Moreover, the interpretation of regression models should be facilitated, as the information encoded in each descriptor is unique.

The similarity/diversity analysis based on previously orthogonalized descriptors is usually called **orthosimilarity** [Randic, 1996b].

📖 [Randic, 1991a] [Randic, 1991g] [Randic and Seybold, 1993] [Randic, 1993b] [Randic, 1994a] [Randic and Trinajstic, 1994] [Pogliani, 1994a] [Pogliani, 1995b] [Amic *et al.*, 1995b] [Lucic *et al.*, 1995a] [Lucic *et al.*, 1995b] [Lucic *et al.*, 1995c] [Araujo and Morales, 1996a] [Araujo and Morales, 1996b] [Soskic *et al.*, 1996a] [Soskic *et al.*, 1996b] [Amic *et al.*, 1997] [Mracec *et al.*, 1997] [Araujo and Morales, 1998] [Nikolic and Trinajstic, 1998a] [Ivanciuc *et al.*, 2000c]

orthogonalized path counts → **orthogonalized descriptors**

orthogonal operator : *Wiener orthogonal operator* → **algebraic operators**

orthosimilarity → **orthogonalized descriptors**

Ostwald solubility parameter → **linear solvation energy relationships** (⊙ cavity term)

outdegree → **graph**

ovality index → **shape descriptors**

overall electronic constants σ_m and σ_p → **electronic substituent constants**

overlap surface : *common overlap surface* → **molecular shape analysis**
(⊙ common overlap steric volume)

P

pair correlation cut-off selection → **variable reduction**

Palm steric constant → **steric descriptors** (⊙ Taft steric constant)

parachor (PA)

The parachor is defined by the Sudgen equation [Sudgen, 1924] as:

$$\mathrm{PA} = \gamma^{1/4} \cdot \frac{MW}{\varrho_\mathrm{L} - \varrho_\mathrm{V}} \approx \gamma^{1/4} \cdot \frac{MW}{\varrho_\mathrm{L}} = \gamma^{1/4} \cdot \bar{\mathrm{V}}$$

where MW is the → *molecular weight*, γ the liquid surface tension, and ϱ_L and ϱ_V the density at a given temperature of liquid and vapor, respectively. The second relationship holds when the vapor density is negligible with respect to the liquid density, $\bar{\mathrm{V}}$ being the → *molar volume.*

The parachor is related to → *physico-chemical properties* depending on the molecule volume, e.g., boiling point.

📖 [Vogel, 1948] [Quayle, 1953] [Briggs, 1981]

partial atomic charge → **quantum-chemical descriptors** (⊙ electron density)

partial charge weighted topological electronic index → **charge descriptors**

partial local invariant → **iterated line graph sequence**

partial negative surface area → **charged partial surface area descriptors**

partial positive surface area → **charged partial surface area descriptors**

partial Wiener indices → **Wiener index**

partition coefficients → **lipophilicity descriptors**

path → **graph**

path-cluster subgraph → **molecular graph**

path connectivity → **weighted matrices**

path count : *molecular path count* → **path counts**

path counts (: *path numbers*)

Path counts are atomic and molecular descriptors obtained from an → *H-depleted molecular graph* $\mathcal{G}$, based on the graph → *path.* Analogous to the → *atomic walk count awc*, the **atomic path count** (or **atomic path number**) ${}^{m}P_i$ is a local vertex invariant encoding the atomic environment, defined as the number of paths of length m starting from the ith vertex to any other vertex in the graph. The length m of the path, i.e. the number of edges involved in the path, is called **path order** [Randic *et al.*, 1979; Randic and Wilkins, 1979d; Randic, 1979].

The **vertex path code** (or **Randic atomic path code**) of the ith vertex is the ordered sequence of atomic path counts, with respect to the path length:

$$\langle {}^{1}P_i, {}^{2}P_i, \ldots, {}^{L}P_i \rangle$$

where $L = {}^{\Delta}\eta_i$ is the → *atom detour eccentricity* of the ith vertex, i.e. the length of the longest path starting from the vertex v_i; it can be derived from the → *detour matrix* as the maximum value entry in the ith row. The atomic path count of first order ${}^{1}P_i$ is the

→ *vertex degree* δ_i, while the atomic path count of zero order ^0P_i is always equal to one. Vertex path codes for all non-hydrogen atoms in the molecule can be collected into a rectangular matrix which has been called → *path-sequence matrix* **SP**. The sum of all the elements in the vertex path code is the total number of paths of any length starting from the considered vertex and can be called the **atomic path count sum** P_i:

$$P_i = \sum_{m=1}^{L} {}^mP_i$$

The **molecular path count**, also called **path count, molecular path number** or **topological bond index** with the symbol K_m, is the total number of paths of length m in the graph and is denoted by ^mP (m = 0,1,..., L), where L is the length of the longest path in the graph. ^0P coincides with the → *atom number* A , ^1P with the → *bond number* B, ^2P with the → *connection number* N_2, i.e. the number of two contiguous edges.

The **molecular path code** of the graph is the ordered sequence of molecular path counts:

$$\langle {}^0P, {}^1P, {}^2P, \ldots, {}^LP \rangle$$

it is obtained by summing the corresponding elements in the atomic path codes of all vertices, then dividing by 2 since each path has been counted twice:

$$^mP = \frac{1}{2} \cdot \sum_{i=1}^{A} {}^mP_i \qquad m \neq 0$$

The path count ^0P is simply equal to A.

Molecular path codes can be used as → *uniform length descriptors*, for example to search for similarities among molecules, by choosing a suitable value for the maximum length L with respect to the set of studied molecules.

By summing all the elements of the molecular path code, the **total path count** P (also called **total path number**) is obtained:

$$P = \sum_{m=0}^{L} {}^mP = A + \frac{1}{2} \cdot \sum_{m=1}^{L} \sum_{i=1}^{A} {}^mP_i = A + \frac{1}{2} \cdot \sum_{i=1}^{A} P_i$$

This descriptor is considered a useful quantitative measure of → *molecular complexity.*

For acyclic graphs, the total path count is simply calculated from the number A of molecular vertices as:

$$P = \frac{A^2 + A}{2}$$

For simple structures, the path counts can be derived directly from the molecular graphs; otherwise specific algorithms are needed. For example, Randic's algorithm results in path counts for nonequivalent vertices from the → *adjacency matrix* [Randic *et al.*, 1979].

Atom ID	*Path length m*					*Atomic path count sums*
	0	*1*	*2*	.....	*L*	
1	1	$^1P_1$	$^2P_1$	.....	$^LP_1$	P_1
2	1	$^1P_2$	$^2P_2$	.....	$^LP_2$	P_2
.....	.....	.....	.....	.....	.....	.....
.....	.....	.....	.....	.....	.....	.....
A	1	^1P_A	^2P_A	.....	^LP_A	P_A
Molecular path counts	A	^1P	^2P	.....	^LP	P

The → *P matrix* is a graph representation of molecules based on the total path count.

In order to take into account multiple bonds and heteroatoms, **weighted path counts** can be calculated, either by introducing the weighting factors after the paths have been enumerated or by computing the weighted paths directly [Randic and Basak, 1999c]. The sums of path weights obtained by applying several weighting schemes to the graph edges are known as the → *ID numbers*; the most common weighting schemes are based on → *bond order indices*. Moreover, the **WTPT index** was proposed as the sum of the weights of all the paths starting from heteroatoms in the molecule [Bakken and Jurs, 1999b]; it is fairly closely related to path counts of heteroatoms used in → *start-end vectors*.

Based on the length of the paths in the molecular graph, other local vertex invariants and molecular descriptors have been proposed.

The **path degree** ξ_i, or **vertex path sum**, is a local invariant defined as the sum of the lengths m of all paths starting from vertex v_i:

$$\xi_i = \sum_{m=1}^{L} {}^{m}P_i \cdot m$$

where $L = {}^{\Delta}\eta_i$ is the atom detour eccentricity of the ith vertex, i.e. the length of the longest path starting from v_i and ${}^{m}P_i$ is the number of paths of length m from v_i. For acyclic graphs, the path degree ξ_i coincides with the → *vertex distance degree* σ_i. Moreover, the path degrees have been proposed as vertex weights to obtain the → *path degree layer matrix* **LPD**.

By summing path degrees over all vertices in the graph, the **all-path Wiener index** W^{AP} is derived. A molecular descriptor proposed as a variant of the → *Wiener index* but with more discriminating power among cycle-containing structures is defined as:

$$W^{AP} = \frac{1}{2} \cdot \sum_{i=1}^{A} \xi_i = \sum_{i=1}^{A-1} \sum_{j=i+1}^{A} \sum\nolimits_{p_{ij}} |p_{ij}|$$

where the two outer sums on the right side run over all pairs of vertices in the graph and the inner sum runs over all paths p_{ij} between the vertice v_i and v_j; $|\,p_{ij}\,|$ denotes the length of the considered path [Lukovits, 1998a]. Its maximum value is equal to $A^2\ (A-1)\ 2^{A-4}$ for a → *complete graph* with A vertices.

It has to be noted that the all-path Wiener index coincides with a previously proposed global index obtained as the half-sum of any row of the path degree layer matrix **LPD**.

The all-path Wiener index can be calculated more easily on the **all-path matrix** $\mathbf{\Omega^{AP}}$, which is a square symmetric $A \times A$ matrix, A being the number of graph vertices, whose i-j entry is the sum of the lengths of all the paths p_{ij} connecting vertices v_i and v_j:

$$\left[\mathbf{\Omega^{AP}}\right]_{ij} = \begin{cases} \sum |p_{ij}| & i \neq j \\ 0 & i = j \end{cases}$$

where $|p_{ij}|$ denotes the length of a path between v_i and v_j. Diagonal elements are equal to zero by definition. → *Distance matrix* **D**, → *detour matrix* **Δ** and → *detour/distance matrix* **Δ/D** are closely related to the all-path matrix as they are based on the length of the shortest, longest, and longest plus shortest paths between any two vertices in the graph, respectively. It must be noted that for acyclic graphs all these matrices coincide, there being a unique path between two vertices.

Example : ethylbenzene

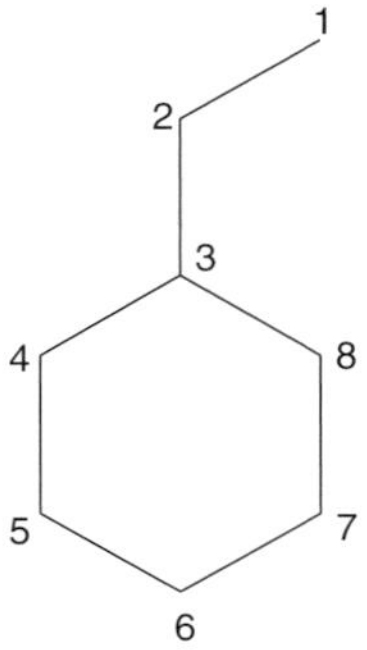

Atom (*i*)	Path length (*m*) 0	1	2	3	4	5	6	7	P_i	ξ_i
1	1	1	1	2	2	2	2	2	12	53
2	1	2	2	2	2	2	2	0	12	42
3	1	3	3	2	2	2	0	0	12	33
4	1	2	3	3	2	2	1	1	14	48
5	1	2	2	3	3	3	1	0	14	48
6	1	2	2	2	4	4	0	0	14	48
7	1	2	2	3	3	3	1	0	14	48
8	1	2	3	3	2	2	1	1	14	48
^mP	8	8	9	10	10	10	4	2		

$$P = \sum_{m=0}^{7} {}^mP = 8 + \frac{1}{2} \cdot \sum_{i=1}^{8} P_i = 61$$

$$\xi_1 = 1 \cdot 0 + 1 \cdot 1 + 1 \cdot 2 + 2 \cdot 3 + 2 \cdot 4 + 2 \cdot 5 + 2 \cdot 6 + 2 \cdot 7 = 53$$
$$\xi_2 = 1 \cdot 0 + 2 \cdot 1 + 2 \cdot 2 + 2 \cdot 3 + 2 \cdot 4 + 2 \cdot 5 + 2 \cdot 6 + 0 \cdot 7 = 42$$
$$\xi_3 = 1 \cdot 0 + 3 \cdot 1 + 3 \cdot 2 + 2 \cdot 3 + 2 \cdot 4 + 2 \cdot 5 + 0 \cdot 6 + 0 \cdot 7 = 33$$
$$\xi_4 = 1 \cdot 0 + 2 \cdot 1 + 3 \cdot 2 + 3 \cdot 3 + 2 \cdot 4 + 2 \cdot 5 + 1 \cdot 6 + 1 \cdot 7 = 48$$

....

$$\xi_8 = 1 \cdot 0 + 2 \cdot 1 + 3 \cdot 2 + 3 \cdot 3 + 2 \cdot 4 + 2 \cdot 5 + 1 \cdot 6 + 1 \cdot 7 = 48$$

$$W^{AP} = \frac{1}{2} \cdot \sum_{i=1}^{8} \xi_i = 184$$

$$W^{AP} = \frac{1}{2} \cdot \sum_{i=1}^{8} \sum_{j=1}^{8} \left[\Omega^{AP}\right]_{ij} = 184$$

All-path matrix Ω^{AP}

Atom	**1**	**2**	**3**	**4**	**5**	**6**	**7**	**8**
1	0	1	2	10	10	10	10	10
2	1	0	1	8	8	8	8	8
3	2	1	0	6	6	6	6	6
4	10	8	6	0	6	6	6	6
5	10	8	6	6	0	6	6	6
6	10	8	6	6	6	0	6	6
7	10	8	6	6	6	6	0	6
8	10	8	6	6	6	6	6	0

Box P-1.

The all-path Wiener index is derived from the all-path matrix as:

$$W^{AP} = \mathcal{W}(\mathbf{\Omega^{AP}}) = \frac{1}{2} \cdot \sum_{i=1}^{A} \sum_{j=1}^{A} [\mathbf{\Omega^{AP}}]_{ij}$$

where $\mathcal{W}$ is the → *Wiener operator*.

Because the all-path Wiener index increases exponentially with the number A of graph vertices, it was proposed to divide it by the average number k of paths between vertices:

$$\bar{W}^{AP} = W^{AP}/k$$

where k is obtained by dividing the total number of paths between vertices in the graph by $A\ (A-1)\ /\ 2$. For simple cycles, $k = 2$.

📖 [Randic and Wilkins, 1979c] [Randic and Wilkins, 1979e] [Randic, 1980a] [Randic *et al.*, 1980] [Wilkins *et al.*, 1981] [Quintas and Slater, 1981] [Randic *et al.*, 1983a] [Randic, 1984a] [Randic *et al.*, 1987] [Kunz, 1989] [Randic and Jurs, 1989] [Clerc and Terkovics, 1990] [Randic, 1990a] [Randic, 1991c] [Randic, 1992c] [Hall *et al.*, 1993a] [Hall *et al.*, 1993b] [Kier *et al.*, 1993] [Pisanski and Zerovnik, 1994] [Plavsic *et al.*, 1996b] [Randic, 1996b] [Randic, 1997a]

path degree → **path counts**

path degree layer matrix → **layer matrices**

path eccentricity : *atom detour eccentricity* → **detour matrix**

path graph : *linear graph* → **graph**

path length → **graph**

path matrix → **double invariants**

path numbers : *path counts*

path order → **path counts**

path-sequence matrix → **sequence matrices**

path subgraph → **molecular graph**

path/walk shape indices → **shape descriptors**

pendent matrix → **superpendentic index**

per(D) index → **algebraic operators** (⊙ determinant)

periphery codes

These are binary molecular codes proposed to characterize the periphery shape of molecules embedded on a 2D hexagonal lattice [Balaban and Harary, 1968; Balaban, 1976b]. They are suitable for the shape characterization of planar benzenoids and annulenes. "Inside" and "Outside" regions of closed curves are indicated by binary labels 1 and 0, respectively, associated with the graph vertices [Randic and Razinger, 1995a; Randic and Razinger, 1995b; Randic and Razinger, 1997]. In other words, digit 1 is associated with movement towards Inside and digit 0 with movement Outside of each ring; a clockwise direction is adopted and the starting point on the periphery is the vertex satisfying the convention of lexicographic minimum. Other different canonical rules can be chosen to define periphery codes [Jerman-Blazic Dzonova and Trinajstic, 1982; Müller *et al.*, 1990a].

Periphery codes can be used to evaluate → *similarity/diversity* based on molecular shape among several compounds [Randic and Razinger, 1995b]. Moreover, periphery codes can also be used to distinguish between *cis-* and *trans-* isomers [Oth and Gilles, 1968; Balaban, 1969; Balaban, 1997a] and to recognize whether a shape is chiral or not [Randic, 1998a]. In particular, for 2D-embedded molecules, the → *Randic chirality index* was proposed by calculating a particular periphery code from left to right and from right to left: if different results are obtained, then the shape is chiral.

📖 [Balaban, 1971] [Balaban, 1988a] [Randic and Mezey, 1996]

permanent → **algebraic operators** (⊙ determinant)

permittivity : *dielectric constant* → **electric polarization descriptors**

persistence length → **size descriptors**

perturbation connectivity indices → **connectivity indices**

perturbation delta values → **connectivity indices** (⊙ perturbation connectivity indices)

perturbation of an environment limited concentric and ordered → **DARC/PELCO analysis**

Petitjean shape indices → **shape descriptors**

pfaffian → **algebraic operators** (⊙ determinant)

pharmacophore → **drug design**

phase capacity ratio : *capacity factor* → **chromatographic descriptors**

physico-chemical properties → **experimental measurements**

Pisanski-Zerovnik index → **Wiener index**

pK_a : *acid dissociation constant* → **hydrogen-bonding descriptors**

planted tree → **graph** (⊙ tree)

Platt number : *total edge adjacency index* → **edge adjacency matrix**

PLS-based variable selection → **variable selection**

P matrix → **bond order indices** (⊙ graphical bond order)

Pogliani cis/trans connectivity index → **cis/trans descriptors**

Pogliani index → **vertex degree**

point-by-point alignment → **alignment rules**

polar effect → **electronic substituent constants**

polar hydrogen factor → **electric polarization descriptors**

polarity / polarizability descriptors → **electric polarization descriptors**

polarity number → **distance matrix**

polarizability → **electric polarization descriptors**

polarizability effect index → **electric polarization descriptors** (⊙ molecular polarizability effect index)

polarizability tensor → **electric polarization descriptors**

polarizability volume → **electric polarization descriptors** (⊙ mean polarizability)

polarization → **electric polarization descriptors**

polar surface area → **molecular surface** (⊙ solvent-accessible molecular surface)

Politzer hydrophobic model → **lipophilicity descriptors**

polycentre → **centre of a graph**

polynomial → **algebraic operators**

population trace → **DARC/PELCO analysis**

potential of a charge distribution → **charge descriptors**

power matrix → **algebraic operators** (⊙ product of matrices)

***P′/P* index** → **bond order indices** (⊙ graphical bond order)

PPP eigenvalues → **eigenvalue-based descriptors**

PPP pairs → **substructure descriptors**

PPP triangles → **substructure descriptors**

predictor variables : *independent variables* → **data set**

prime ID number → **ID numbers**

principal axes of a molecule → **principal moments of inertia**

principal components → **principal component analysis**

Principal Component Analysis (PCA)

A basic chemometric technique for → *exploratory data analysis*, modelling the p variables in the data matrix $\mathbf{X}$ ($n \times p$), where n is the number of objects, as linear combinations of the common factors $\mathbf{T}$ ($n \times M$), called **principal components $\mathbf{t_m}$**:

$$\mathbf{X} = \mathbf{T} \cdot \mathbf{L}^{\mathrm{T}}$$

where $\mathbf{T}$ is the **score matrix**, $\mathbf{L}$ ($p \times M$) is the **loading matrix** and M is the number of significant principal components ($M \leq p$). The columns of the loading matrix are the eigenvectors $\mathbf{l_m}$; the eigenvector coefficients ℓ_{jm}, called *loadings*, represent the importance of each original variable in the considered eigenvector [Jolliffe, 1986; Jackson, 1991; Basilevsky, 1994].

The components are calculated according to the maximum variance criterion, i.e. each successive component is an orthogonal linear combination of the original variables such that it covers the maximum of the variance not accounted for by the previous components. The eigenvalue λ_m associated with each mth component represents the variance explained by the considered component. Moreover, the sum of the variances of all the p components equals the variance of the original variables.

The principal components can also be viewed as linear combinations of the p original variables:

$$\mathbf{T} = \mathbf{X} \cdot \mathbf{L} \qquad \text{i.e.} \qquad t_{im} = x_{i1} \cdot \ell_{1m} + x_{i2} \cdot \ell_{2m} + \ \ldots \ + x_{ip} \cdot \ell_{pm} = \sum_{j=1}^{p} x_{ij} \cdot \ell_{jm}$$

where ℓ_{jm} are the coefficients of the linear combinations (i.e. the *loadings*) and t_{im} is the PCA score of the ith object (e.g., a molecule) in the mth principal component.

Mathematically, PCA consists in the diagonalization of the correlation matrix (or covariance matrix) of the data matrix $\mathbf{X}$ with size $p \times p$ (the number of variables).

Thus, the M significant components having been chosen, each ith object is represented by the vector:

$$\langle t_{i1};\ t_{i2};\ \ldots;\ t_{iM} \rangle$$

The main advantages of principal components are that:

1) each component is orthogonal to all the remaining components, i.e. the information carried by this component is unique;

2) each component represents a *macrovariable* of the data;

3) components associated with the lowest eigenvalues do not usually contain useful information (noise, spurious information, etc.).

When PCA is performed on a set of compounds characterized by → *molecular descriptors* (→ *physico-chemical properties*, structural variables, etc.) the significant principal components are called **principal properties** PP because they summarize the main information of the original molecular descriptors:

$$\langle PP_1, PP_2, \ldots, PP_M \rangle$$

Therefore, principal properties PPs are PCA scores, which are empirical scales describing physico-chemical properties of the training set objects [Alunni *et al.*, 1983; Carlson, 1992; Clementi *et al.*, 1993a]. The number of significant PPs and their meaning depend closely on the original variables used to perform PCA. Since the PPs derived from PCA are orthogonal to each other and they are small in number, they are suitable for design problems [Skagerberg *et al.*, 1989; Eriksson *et al.*, 1997].

Principal properties can be calculated for both whole molecules and substituent groups, fragments, amino acids, etc. For example, the *i*th substituent can be represented by four PPs, each having a different meaning such as PP_1 = steric, PP_2 = lipophilic, PP_3 = electrostatic, PP_4 = H-bonding properties of the substituent, respectively.

→ *BC(DEF) parameters* are principal properties of a data matrix given by 6 physico-chemical properties for 114 diverse liquid-state compounds. Moreover, ten principal properties were calculated by statistical analyses on 188 physico-chemical properties for the 20 naturally occurring amino acids [Kidera *et al.*, 1985a; Kidera *et al.*, 1985b]. These 10 properties were called **KOKOS descriptors** by Pogliani on the basis of the authors' names [Pogliani, 1994a]; they describe most of the conformational, bulk, hydrophobicity, α-helix and β-structure properties of amino acids.

Principal properties calculated on molecular → *interaction energy values* obtained by → *grid-based QSAR techniques* are usually referred to as **3D principal properties** (3D-PP) [van de Waterbeemd *et al.*, 1993b]. They were originally proposed for a theoretical description of the amino acids [Norinder, 1991; Cocchi and Johansson, 1993]. 3D-PP were also calculated from → *ACC transforms.*

When different data sets of descriptors are used separately to derive the principal properties of the same compounds, **disjoint principal properties** DPP are obtained as the whole set of significant PPs derived from each block of descriptors:

$$\left\langle PP_1^A, PP_2^A, \ldots, PP_{M_A}^A; PP_1^B, PP_2^B, \ldots, PP_{M_B}^B; PP_1^C, PP_2^C, \ldots, PP_{M_C}^C \right\rangle$$

where A, B, C represent three different blocks of variables on which PCA was performed [van de Waterbeemd *et al.*, 1995].

📖 [Weiner and Weiner, 1973] [Wold, 1978] [Dunn III *et al.*, 1978] [Dunn III and Wold, 1978] [Dunn III and Wold, 1980] [Lukovits and Lopata, 1980] [Streich *et al.*, 1980] [Lukovits, 1983] [McCabe, 1984] [Maria *et al.*, 1987] [Eriksson *et al.*, 1988] [van de Waterbeemd *et al.*, 1989] [Jonsson *et al.*, 1989] [Eriksson *et al.*, 1990] [Ridings *et al.*, 1992] [Hemken and Lehmann, 1992] [Suzuki *et al.*, 1992a] [Tysklind *et al.*, 1992] [Ordorica *et al.*, 1993] [Caruso *et al.*, 1993] [Cristante *et al.*, 1993] [Rodriguez Delgado *et al.*, 1993] [Norinder, 1994] [Franke *et al.*, 1994] [Bazylak, 1994] [Cocchi *et al.*, 1995] [Azzaoui and Morinallory, 1995b] [Bjorsvik and Priebe, 1995] [Kimura *et al.*, 1996] [Gibson *et al.*, 1996] [Clementi *et al.*, 1996] [Bjorsvik *et al.*, 1997] [Young *et al.*, 1997]

[Balasubramanian and Basak, 1998] [Langer and Hoffmann, 1998a] [Xue *et al.*, 1999] [Vendrame *et al.*, 1999] [Kuanar *et al.*, 1999]

principal component analysis feature selection → **variable reduction**

principal inertia axes : *principal axes of a molecule* → **principal moments of inertia**

principal moments of inertia (: *inertia principal moments)*

These are physical quantities related to the rotational dynamics of a molecule. The **moment of inertia** about any axis is defined as:

$$\mathrm{I} = \sum_{i=1}^{A} m_i \cdot r_i^2$$

where A is the atom number, and m_i and r_i are the atomic mass and the perpendicular distance from the chosen axis of the ith atom of the molecule, respectively. For any rectangular coordinate system, with axes X, Y, Z, three moments of inertia are defined as:

$$\mathrm{I_{XX}} = \sum_{i=1}^{A} m_i \cdot \left(y_i^2 + z_i^2\right) \qquad \mathrm{I_{YY}} = \sum_{i=1}^{A} m_i \cdot \left(x_i^2 + z_i^2\right) \qquad \mathrm{I_{ZZ}} = \sum_{i=1}^{A} m_i \cdot \left(x_i^2 + y_i^2\right)$$

where (x, y, z) are the coordinates of the atoms.

The corresponding cross-terms are called **products of inertia** and are defined as:

$$\mathrm{I_{XY}} = \mathrm{Y_{YX}} = \sum_{i=1}^{A} m_i \cdot x_i \cdot y_i \qquad \mathrm{I_{XZ}} = \mathrm{Y_{ZX}} = \sum_{i=1}^{A} m_i \cdot x_i \cdot z_i \qquad \mathrm{I_{YZ}} = \mathrm{Y_{ZY}} = \sum_{i=1}^{A} m_i \cdot y_i \cdot z_i$$

Therefore the **inertia matrix** I is a square symmetric matrix 3×3.

Principal moments of inertia are the moments of inertia corresponding to that particular and unique orientation of the axes for which one of the three moments has a maximum value, another a minimum value, and the third is either equal to one or the other or is intermediate in value between the other two. The corresponding axes are called **principal axes of a molecule** (or **principal inertia axes**). Moreover, the products of inertia all reduce to zero and the corresponding inertia matrix is diagonal. Conventionally, principal moments of inertia are labeled as:

$$\mathrm{I_A} \leq \mathrm{I_B} \leq \mathrm{I_C}$$

In general, the three principal moments of inertia have different values, but, depending on the molecular symmetry, they show characteristic equalities such as those shown in Table P-1.

A number of → *shape descriptors* are defined in terms of principal moments of inertia. Moreover, principal moments of inertia are used to provide a unique reference framework for the calculation of the → *shadow indices*, and in general are used as → *alignment rules* of the molecules. They are encountered among → *WHIM descriptors* and used by the → *CoMMA method*.

Table P-1. Principal moments for some selected symmetries.

Symmetry	*Principal moments*	*Example*
Spherical top	$\mathrm{I_A = I_B = I_C}$	CCl_4
Symmetric top	$\mathrm{I_A = I_B \neq I_C}$	NH_3
Asymmetric top	$\mathrm{I_A \neq I_B \neq I_C}$	CH_2FCl
Linear symmetry	$\mathrm{0 = I_A \neq I_B = I_C}$	HC≡CH
Planar symmetry	$\mathrm{I_A + I_B = I_C}$	C_6H_6

principal properties → **principal component analysis**

probe → **grid-based QSAR techniques**

products of inertia → **principal moments of inertia**

product of matrices → **algebraic operators**

product of row sums index : *PRS index* → **distance matrix**

proference → **DARC/PELCO analysis**

properties matrix → **topoelectric matrices**

PRS index → **distance matrix**

pruning of the graph → **centric indices** (⊙ Balaban centric index)

pruning partition → **centric indices** (⊙ Balaban centric index)

pseudocentre → **centre of a graph**

pseudograph → **graph**

Q

Q polarity index → **electric polarization descriptors**

quadratic index → **Zagreb indices**

quadrupole moment → **electric polarization descriptors**

quantitative information analysis → **structure/response correlations**

quantitative shape-activity relationships → **structure/response correlations**

quantitative similarity-activity relationships → **similarity/diversity**

quantitative structure-activity relationships → **structure/response correlations**

quantitative structure-chromatographic relationships → **structure/response correlations**

quantitative structure-enantioselective retention relationships → **structure/response correlations**

quantitative structure-property relationships → **structure/response correlations**

quantitative structure-reactivity relationships → **structure/response correlations**

quantitative structure/response correlations → **structure/response correlations**

quantitative structure-toxicity relationships → **structure/response correlations**

quantum-chemical descriptors

Quantum-chemical descriptors are obtained from quantum-mechanical calculations (→ *computational chemistry*), in which the energy levels E_i and the wave functions ψ_i of a molecule are computed by solving the *Schrödinger equation*:

$$\mathcal{H}\psi_i = E_i\psi_i$$

where $\mathcal{H}$ is the many-electron Hamiltonian [Streitweiser, 1961; Salem, 1966; Dewar, 1969a; Kier, 1971; Leach, 1996]. The spectrum of the energy levels is itself a molecular descriptor. Moreover, from the wave function ψ_i several important physical properties are obtained.

In general, without loss of generality, it can be assumed that the wave function ψ_i is approximated by a linear combination of atomic functions ϕ_μ (i.e. atomic orbitals), i.e.

$$\psi_i = \sum_\mu c_{\mu i} \cdot \phi_\mu$$

where c are the coefficients of the linear combination of atomic orbitals defining each ith molecular orbital ψ_i.

Several different kinds of quantum-chemical descriptors have been defined, and these can be broadly divided into energy-based descriptors, local quantum-chemical properties, descriptors based on the Density Functional Theory, molecular orbital energies, superdelocalizability indices, frontier orbital electron densities, and polarizabilities [Cartier and Rivail, 1987; Bergmann and Hinze, 1996; Karelson *et al.*, 1996].

The most common quantum-chemical descriptors are listed below.

Molecular energies calculated by → *computational chemistry* methods are fundamental descriptors commonly used in QSAR models; moreover, energies are used as cutoff values for the selection of the most important conformation(s). Besides the energy levels calculated from quantum-mechanical calculations and the molecular total energy, several other kinds of energy are defined such as heat of formation, heat of solvation, heat of vaporization, steric energy, etc.

→ *Substituent front strain*, → *steric energy difference*, → *Joshi steric descriptor* and → *Joshi electronic descriptors* are examples of molecular descriptors calculated from the standard enthalpy of formation and the steric energy.

Two other important and useful molecular energy measures are ionization potential and electron affinity. The **ionization potential** IP (or **ionization energy**) is defined as the energy needed to extract one electron from a chemical system, i.e.

$$\mathrm{IP} = \mathrm{E}(\mathrm{N}_{\mathrm{el}}) - \mathrm{E}(\mathrm{N}_{\mathrm{el}} - 1)$$

where N_{el} is the number of electrons of the system.

Ionization potential is a measure of the capability of a molecule to give the corresponding positive ion.

The **electron affinity** EA is defined as the gain in energy of the chemical system when an electron is captured from the system, i.e.

$$\mathrm{EA} = \mathrm{E}(\mathrm{N}_{\mathrm{el}}) - \mathrm{E}(\mathrm{N}_{\mathrm{el}} + 1)$$

Electron affinity is a measure of the capability of a molecule to give the corresponding negative ion.

Other fundamental **energy-based descriptors** have been defined by traditional computational chemistry. The concept of **electronegativity** was recognized as a useful basic principle in chemistry more than 150 years ago [Pritchard and Skinner, 1955]. Originally defined by Pauling [Pauling, 1939], electronegativity is "the power of an atom in a molecule to attract electrons to itself".

The classical definition of **atom electronegativity** is due to Mulliken [Mulliken, 1934; Mulliken, 1955a; Mulliken, 1955b] – **Mulliken electronegativity** – :

$$\chi_{\mathrm{MU}} = \frac{\mathrm{IP} + \mathrm{EA}}{2}$$

i.e. the arithmetic mean of the ionization potential IP and the electronic affinity EA. In this definition, the use of ionization potentials and electron affinities of valence states was proposed.

The Mulliken scale of electronegativity (in volts) can be converted into the Pauling scale χ_{PA} (*Pauling units*) by the empirical relation:

$$\chi_{\mathrm{PA}} = 0.303 \cdot \chi_{\mathrm{MU}}$$

Electronegativity scales other than Pauling and Mulliken scales are the Allred-Rochow scale χ_{AR} [Allred and Rochow, 1958; Allred and Rochow, 1961] based on estimated effective nuclear potentials and covalent radii, the Gordy scale χ_{G} [Gordy, 1946; Gordy, 1951] based on the number of electrons in the valence shell of the atom and the covalent radius, the Sanderson scales χ_{S} [Sanderson, 1952; Sanderson, 1954; Sanderson, 1955; Sanderson, 1971] based on covalent radii, the Hinze-Jaffé scale χ_{HJ} [Hinze and Jaffé, 1962; Hinze *et al.*, 1963; Hinze and Jaffé, 1963a; Hinze and Jaffé, 1963b] based on orbital energies and effective charges, and the Zhang scale χ_{Z} based on ionization energies and covalent radii [Zhang, 1982a; Zhang, 1982b]. Atom electronegativity can also be estimated by a topological approach, as proposed by Kier-Hall (→ *Kier-Hall electronegativity*). An extended review of electronegativity scales has been made by Luo and Benson [Luo and Benson, 1990], and some of these are collected in Table Q-1.

Table Q-1. Electronegativity values from different sources (Pauling Units).

	H	B	C	N	O	F	Si	P	S	Cl	Br	I
Pauling 1	2.1	2.0	2.5	3.0	3.5	4.0	1.8	2.1	2.5	3.0	2.8	2.5
Pauling 2	2.20	–	2.55	3.04	3.44	3.98	1.90	2.19	2.58	3.16	2.76	2.66
Mulliken	2.28	2.01	2.63	2.33	3.17	3.91	2.44	1.81	2.41	3.00	2.76	2.56
Allred-Rochow	2.20	2.01	2.50	3.07	3.50	4.10	1.74	2.06	2.44	2.83	2.74	2.21
Sanderson 1	2.31	1.88	2.47	2.93	3.46	3.92	1.74	2.16	2.66	3.28	2.96	2.50
Sanderson 2	2.592	2.275	2.746	3.194	3.654	4.000	2.138	2.515	2.957	3.475	3.219	2.778
Mullay	2.08	1.85	2.47	3.41	3.15	4.00	1.91	1.99	2.49	3.07	2.81	2.47
Gordy	2.17		2.52	2.98	3.45	3.95			2.58	3.50	2.75	2.50
Wells	2.28		2.30	3.35	3.70	3.95			2.80	3.03	2.80	2.47
Boyd-Markus	1.94	1.95	2.53	3.23	3.53	4.00	1.81	2.34	2.65	3.14	2.78	2.48
Inamoto-Masuda	2.00		2.21	2.71	3.02	3.05	1.72	1.93	2.15	2.37	2.32	2.15
Diudea	1.680	1.501	1.831	2.240	2.680	3.024	1.424	1.646	2.026	2.512	2.279	1.879
Zhang	2.271	1.966	2.536	3.062	3.642	4.188	1.769	2.131	2.479	2.835	2.529	2.142

Pauling 1 [Pauling, 1939]; Pauling 2 [Allred and Rochow, 1961]; Mulliken [Mulliken, 1934; Mulliken, 1935]; Allred-Rochow [Allred and Rochow, 1958]; Sanderson 1 [Sanderson, 1952; Sanderson, 1955]; Sanderson 2 [Sanderson, 1988]; Mullay [Mullay, 1984]; Gordy [Gordy, 1946]; Wells [Wells, 1968b]; Boyd-Markus [Boyd and Markus, 1981]; Inamoto-Masuda [Inamoto and Masuda, 1982]; Diudea [Diudea *et al.*, 1996a]; Zhang [Zhang, 1982a]

The concept of electronegativity has become increasingly general (perhaps even ambiguous) during its revision in the different quantum-chemistry frameworks, ranging from atomic electronegativity to orbital and functional group electronegativity to molecular electronegativity. Electronegativity is a property of the state of the system; electrons tend to flow from a region of low electronegativity to a region of high electronegativity. With the formation of a molecule, the electronegativities of the constituent atoms or fragments equalize, all becoming equal to the electronegativity of the final state of the molecule (*Sanderson principle of electronegativity equalization* [Sanderson, 1951; Sanderson, 1971; Zefirov *et al.*, 1987]).

The concept of electronegativity equalization led to the calculation of dipole moments [Malone, 1933; Ferreira, 1963a], bond dissociation energies [Ferreira, 1963b], atomic charges [Stoklosa, 1973] and force constants [Polansky and Derflinger, 1963; Gordy, 1946]. Moreover, based on electronegativity, several → *bond ionicity indices* were proposed. In order to avoid chemically unacceptable results due to the total equalization of the electronegativity in a molecule [Gasteiger and Marsili, 1980], a partial equalization principle as a result of changes in orbital overlap was also taken into account [Pritchard, 1963]. In particular, *partial equalization of orbital electronegativity* (PEOE) was proposed using a topological iterative approach [Gasteiger and Marsili, 1980; Marsili and Gasteiger, 1980]; a *modified partial equalization of orbital electronegativity* (M-PEOE) was also successively proposed [No *et al.*, 1990a; No *et al.*, 1990b].

Group electronegativity was proposed for molecular substituents, functional groups and fragments, where the centre atom is the atom connected to the body of the molecule.

Group electronegativity is defined as the electronegativity of the centre atom of the substituent and is affected by the neighbours of the central atom. Several methods have been proposed to estimate group electronegativities [Clifford, 1959; McDaniel and Yingst, 1964; Huheey, 1965; Huheey, 1966; Inamoto and Masuda, 1982; Mullay, 1984; Mullay, 1985; Bratsch, 1985; Xie *et al.*, 1995]. For example, **Sanderson group electronegativity** ESG is calculated as the geometric mean of the electronegativities of atoms belonging to the considered group [Sanderson, 1983]:

$$\mathrm{ESG}_i = \left(\chi_{\mathrm{S},1}, \chi_{\mathrm{S},2}, \ldots, \chi_{\mathrm{S},m}\right)^{1/m}$$

where χ_S is the Sanderson electronegativity and m the number of atoms of the ith molecular group. Valence group carbon-related electronegativities EC were derived from Sanderson's ESG values by taking into account the covalent radii (as mean bond lengths) relative to the sp^3 carbon atom [Diudea *et al.*, 1996a]. Using group electronegativities as local invariants, → *electronegativity-based connectivity indices* were proposed as molecular descriptors.

Modern definitions of electronegativity were proposed on the basis of the density functional theory (DFT).

For a molecule of N_{el} electrons, the **electronic chemical potential** μ is defined as:

$$\mu = \left(\frac{\partial E}{\partial N_{el}}\right)_{v(\mathbf{r})}$$

where E is the ground-state energy and $v(\mathbf{r})$ is the composite nuclear potential at the point **r**. It is a measure of the escaping tendency of an electronic cloud and corresponds to the negative of the electronegativity χ. In fact, as defined by Iczkowski and Margrave [Iczkowski and Margrave, 1961], **molecular electronegativity** χ is:

$$\chi = -\left(\frac{\partial E}{\partial N_{el}}\right)_{v(\mathbf{r})} = -\mu$$

where μ is the electronic chemical potential. This definition is restricted to the ground state and results in the same value of electronegativity (and electronic chemical potential) everywhere, for molecule, atoms, solid or molecular regions.

Considering a finite difference approximation and a quadratic dependence of energy on the number of electrons, the electronegativity approximates the loss or gain in charge by the ionization potential IP and electron affinity EA, i.e.

$$\chi_{MU} = \frac{\mathrm{IP} + \mathrm{EA}}{2} = -\mu$$

which corresponds to the common definition of Mulliken electronegativity.

The **orbital electronegativity** χ_v of the μth atomic orbital is an atomic descriptor defined as:

$$\chi_\mu = -\left(\frac{\partial E}{\partial n_\mu}\right) = \left(\frac{\partial E}{\partial q_\mu}\right)$$

where n_μ and q_μ are the occupation number and the partial atomic charge of the μth atomic orbital.

From quantum theory, a number of **local quantum-chemical properties** are defined at each point **r** of the molecule space. The most important functions are listed below.

- **electron density**

A local electronic descriptor obtained by computational chemistry methods.

The electron density $\varrho(\mathbf{r})$ at a point **r** can be calculated as the sum of squares of the molecular orbitals ψ at point **r** for all occupied molecular orbitals N_{OCC}; for a closed-shell system of N_{el} electrons occupying N/2 orbitals it is defined as:

$$\varrho(\mathbf{r}) = 2 \cdot \sum_{i=1}^{N_{OCC}} |\psi_i(\mathbf{r})|^2$$

It can be seen that the integral of $\varrho(\mathbf{r})$ over all space equals the number of electrons in the system N_{el} :

$$N_{el} = \int \varrho(\mathbf{r})\, d\mathbf{r}$$

The **average local ionization energy** was introduced [Sjöberg *et al.*, 1990] to estimate the average energy required to remove an electron located at the point **r** from the molecule and is defined as:

$$\bar{I}(\mathbf{r}) = \frac{\sum_i \varrho_i(\mathbf{r}) \cdot |\varepsilon_i|}{\varrho(\mathbf{r})}$$

where the sum runs over the molecular orbitals, $\varrho_i(\mathbf{r})$ is the electronic density of the ith molecular orbital at the point **r**, and ε_i the ith orbital energy.

When molecular orbitals are expressed in terms of a basis set of K single electronic functions ϕ (often atomic orbitals), the electron density is expressed as:

$$\varrho(\mathbf{r}) = \sum_\mu \sum_\nu P_{\mu\nu} \cdot \phi_\mu(\mathbf{r}) \cdot \phi_\nu(\mathbf{r})$$

where $P_{\mu\nu}$ are the elements of the **charge density matrix *P*,** defined as:

$$P_{\mu\nu} = \sum_{i=1}^{N_{OCC}} n_i \cdot c_{\mu i} \cdot c_{\nu i} \quad \text{and} \quad P_{\mu\mu} = \sum_{i=1}^{N_{OCC}} n_i \cdot c_{\mu i}^2 = q_\mu$$

where the element $P_{\mu\nu}$ is called **bond order** [Coulson, 1939], n_i is the occupation number (1 or 2), q_μ is the **partial atomic charge** of the μth atomic orbital and the summation runs over all the occupied molecular orbitals N_{OCC} . The c are the coefficients of the linear combination of atomic orbitals defining each ith molecular orbital ψ_i.

For closed shell systems, the charge density matrix elements are simply defined as:

$$P_{\mu\nu} = 2 \cdot \sum_{i=1}^{N_{OCC}} c_{\mu i} \cdot c_{\nu i}$$

where the occupancy number is 2 and the summation runs over the occupied molecular orbitals.

Different definitions of bond order have been suggested by other authors: based on the valence bond theory [Pauling, 1939], corrected by the overlap integral [Mulliken, 1955a], and corrected by the eigenvalues of the secular equation [Ham and Ruedenberg, 1958c; Ham, 1958a; Ham and Ruedenberg, 1958b; Ruedenberg, 1958].

Some molecular descriptors related to the charge density matrix have been proposed [Salem, 1966]. **Bond index** B_{ij} was proposed to measure the multiplicity of bonds between two atoms [Wiberg, 1968]. It is defined between two atoms i and j as the square of the off-diagonal elements $P_{\mu\nu}$ of the charge density matrix between atomic orbitals μ on the ith atom and ν on the jth atom, summed over all such distinct orbitals:

$$B_{ij} = \sum_{\mu \in i} \sum_{\nu \in j} P_{\mu\nu}^2$$

The bond index is a function of the square of the charge density matrix elements, whereas the bond order is defined in terms of the charge density matrix elements themselves. It is a measure of the extent of electron sharing between two atoms, but it has the disadvantage that it is always positive and hence cannot describe antibonding situations.

The **valency index** V_i represents the valency of the ith atom as the sum of the valencies of its atomic orbitals [Armstrong *et al.*, 1973; Gopinathan and Jug, 1983a]. Thus the valency V_i of the ith atom is given by:

$$V_i = \sum_{j \neq i} \sum_{\mu \in i} \sum_{\nu \in j} P_{\mu\nu}^2 = \sum_{\mu \in i} 2 \cdot P_{\mu\mu} - \sum_{\mu \in i} \sum_{\nu \in i} P_{\mu\nu}^2 = \sum_{j \neq i} \mathrm{B}_{ij}$$

where the first summation in the first term and the last one run over all the A atoms different from the ith atom, $P_{\mu\nu}$ are the squares of the off-diagonal charge density matrix elements, and B_{ij} is the bond index.

For closed shell systems, the previous expression of atom valency is the following:

$$V_i = \sum_{\mu \in i} \left(2 \cdot q_\mu - q_\mu^2\right)$$

This quantity has a value zero when q_μ , the occupancy of the orbital ϕ_μ, is either 2 or 0; it has the maximum value of 1 when $q_\mu = 1$.

The **free valence index** was proposed as a measure of the residual valency of the ith atom in π-electron molecular orbitals [Coulson, 1946]; it is defined as:

$$F_i = \pi_i^{\max} - \pi_i^*$$

where $\pi_i^{\max}$ is the maximum bond order of the ith atom and π_i^* is the sum of the bond orders of the bonds connecting the ith atom to all its neighbours. $\pi_i^{\max}$ is usually taken as $\sqrt{3}$, the value for the central carbon atom in trimethylene methane. Topological formulas for the free-valence index were proposed by Gutman [Gutman, 1978]. Moreover, a generalization of the original free valence index accounting for σ electrons is the **general free valence index** defined as [Gopinathan and Jug, 1983b]:

$$F'_i = V'_i - V_i$$

where V_i is the valency of the ith atom and V'_i is the "reference valency" of the ith atom chosen as the integer value around which the computed valency of the atom is distributed in a large number of compounds. This general free valence index can also be defined as a percentage

$$F'_i\ \% = \frac{V'_i - V_i}{V'_i} \cdot 100$$

representing the residual covalent binding capacity of the ith atom.

Other molecular descriptors derived from the charge density matrix are the **average atom charge density** P_1 and the **average bond charge density** P_2, defined as:

$$P_1 = \frac{\sum_{\mu,\nu} P_{\mu\nu}}{A} \qquad \text{and} \qquad P_2 = \frac{\sum'_{\mu,\nu} P_{\mu\nu}}{B}$$

where A and B are the number of atoms and bonds, respectively, and the second summation is restricted to bonds [Balasubramanian, 1994].

- **composite nuclear potential**

For a given configuration of the nuclei of a molecule, the composite nuclear potential is defined as:

$$v(\mathbf{r}) = \sum_{a=1}^{A} \frac{Z_a}{|\mathbf{r} - \mathbf{R}_a|}$$

where Z_a are the nuclear charges at positions $\mathbf{R}_a$. Assuming that the electronic density is removed without changing the nuclear configuration, the composite nuclear potential is the potential experienced by a unit charge at location **r**.

- **Somoyai function**

This is a special representation of the difference between the electronic density $\varrho(\mathbf{r})$ at a point **r** and the composite nuclear potential $v(\mathbf{r})$ at the same point, defined as:

$$S(\mathbf{r}, s) = \varrho(\mathbf{r}) - s \cdot v(\mathbf{r})$$

The Somoyai parameter s has the physical dimension of bohr^{-2}. Since a notable part of the electronic density is mimicked by the composite nuclear potential, it can be assumed that only their difference provides a description of the chemical bonding. Then, for any fixed value of s, the Somoyai function gives information about the role of the electronic density in the chemical bonding.

- **Molecular Electrostatic Potential** (MEP)

An important reactivity descriptor [Bonaccorsi *et al.*, 1970], giving the interaction energy of a molecule with a unit positive charge at position **r**, and defined as:

$$v(\mathbf{r}) = \sum_{a=1}^{A} \frac{Z_a}{|\mathbf{r} - \mathbf{R}_a|} - \int \frac{\varrho(\mathbf{r}')}{|\mathbf{r} - \mathbf{r}'|} \cdot \mathbf{dr}'$$

where Z_a is the nuclear charge of the ath atom in position $\mathbf{R}_a$. MEP is among the most used → *molecular interaction fields*.

- **Fukui function**

The Fukui function $f(\mathbf{r})$ is a local electronic descriptor of reactivity which finds its origin within density-functional theory (DFT) and is defined as [Parr and Yang, 1989]:

$$f(\mathbf{r}) = \left(\frac{\partial \varrho(\mathbf{r})}{\partial N_{el}}\right)_{v(\mathbf{r})} = \left(\frac{\partial \mu}{\partial v(\mathbf{r})}\right)_{N_{el}}$$

i.e. as the first derivative of the electronic density $\varrho(\mathbf{r})$ with respect to the number of electrons N_{el} of the system, at a given external potential $v(\mathbf{r})$. The Fukui function corresponds to the first derivative of the electronic chemical potential μ with respect to the external potential $v(\mathbf{r})$ for a given number of electrons. The Fukui function indicates the regions in a molecule where the charge density changes during a reaction, and measures how sensitive a chemical potential is to external perturbation at a specific point.

Because of the discontinuity of this derivative, a **backward Fukui function** $f^-(\mathbf{r})$ and a **forward Fukui function** $f^+(\mathbf{r})$ are defined, corresponding to local descriptors for electrophilic and nucleophilic attack, respectively. In terms of the finite difference approximation, both functions can be written as:

$$f^-(\mathbf{r}) = \varrho_N(\mathbf{r}) - \varrho_{N-1}(\mathbf{r}) \qquad f^+(\mathbf{r}) = \varrho_{N+1}(\mathbf{r}) - \varrho_N(\mathbf{r})$$

where $\varrho_N(\mathbf{r})$, $\varrho_{N+1}(\mathbf{r})$, and $\varrho_{N-1}(\mathbf{r})$ are the electron densities of the N_{el}, $N_{el}+1$, and $N_{el}-1$ electron systems, respectively, all calculated at the same external potential $v(\mathbf{r})$ of the N_{el} electron system.

In order to define a reactivity index for radical attack, an **average Fukui function** $f^0(\mathbf{r})$ is also defined as:

$$f^0(\mathbf{r}) = \frac{f^+(\mathbf{r}) + f^-(\mathbf{r})}{2} = \frac{\varrho_{N+1}(\mathbf{r}) + \varrho_{N-1}(\mathbf{r})}{2}$$

In practice, of two different sites with similar disposition for reacting with a given reagent, the reagent prefers the one which is associated with the maximum response of the electronic chemical potential of the system.

Using a population analysis method, an atom-localized version of the Fukui functions is defined as [Yang and Mortier, 1986]:

$$f_a^- = q_a(N_{el}) - q_a(N_{el} - 1)$$

$$f_a^+ = q_a(N_{el} + 1) - q_a(N_{el})$$

$$f_a^0 = \frac{q_a(N_{el} + 1) - q_a(N_{el} - 1)}{2}$$

where $q_a(N_{el})$, $q_a(N_{el}+1)$, and $q_a(N_{el}-1)$ are the charges on the *a*th atom in the N_{el}, $N_{el}+1$, and $N_{el}-1$ electron systems.

DFT-based descriptors are those derived from the *Density Functional Theory* [Parr and Yang, 1989; Geerlings *et al.*, 1996a; Geerlings *et al.*, 1996b]. The most important are hardness and softness indices.

- **hardness indices**

The second derivative of the energy with respect to the number of electrons is called **absolute hardness** η (or **chemical hardness**) [Parr and Pearson, 1983], which for a molecule with N_{el} electrons is defined as:

$$\eta = \frac{1}{2}\left(\frac{\partial^2 E}{\partial N_{el}^2}\right)_{v(\mathbf{r})} = \frac{1}{2}\left(\frac{\partial \mu}{\partial N_{el}}\right)_{v(\mathbf{r})} = \int h(\mathbf{r})\, d\mathbf{r} = \frac{1}{2 \cdot S}$$

where μ is the electronic chemical potential, $h(\mathbf{r})$ is called **local hardness** (or **hardness density**) and S is the *total softness* (defined below).

From the frontier orbital energies an approximated absolute hardness is also obtained. In fact, under a finite difference approximation and a quadratic dependence of the energy on the number of electrons, absolute hardness is defined as

$$\eta = \frac{\mathrm{IP} - \mathrm{EA}}{2} = \frac{GAP}{2} \equiv \frac{\varepsilon_{\mathrm{LUMO}} - \varepsilon_{\mathrm{HOMO}}}{2}$$

where IP and AE are the ionization potential and the electron affinity, respectively, and *GAP* denotes the HOMO-LUMO energy difference. $\varepsilon_{\mathrm{LUMO}}$ and $\varepsilon_{\mathrm{HOMO}}$ are, respectively, the energies of the lowest unoccupied molecular orbital and the highest occupied molecular obital. Such hardness is that within the framework of Koopmans' theorem; high values of hardness are related to the stability of a molecule as well as to the HOMO-LUMO energy gap (see below).

The **activation hardness** is a measure of → *dynamic reactivity* obtained as the difference between absolute hardness of reactant (R) and transition states (T):

$$\Delta\eta = \eta_R - \eta_T$$

The activation hardness is sensitive to the reactivity at different molecule sites and to orientation effects.

- **softness indices**

The **total softness** S is defined as:

$$S = \left(\frac{\partial N_{el}}{\partial \mu}\right)_{v(\mathbf{r})} = \int s(\mathbf{r})\, d\mathbf{r} = \frac{1}{2\eta}$$

where η is the absolute hardness, μ the electronic chemical potential, and $s(\mathbf{r})$ the **local softness** (or **softness density**), which is related to the Fukui function $f(\mathbf{r})$, via total softness, by the relationship

$$s(\mathbf{r}) = S \cdot f(\mathbf{r})$$

indicating that the Fukui function distributes the total softness among different regions of space.

Under a finite difference approximation and a quadratic dependence of energy on the number of electrons, the total softness can also be calculated as:

$$S = \frac{1}{IP - EA}$$

where IP and EA are the ionization potential and the electron affinity.

Molecular orbital energies give information about reactivity/stability of specific regions of the molecule. Among the molecular orbitals, a fundamental role is played by the **frontier orbitals** [Fleming, 1990; Clare, 1994; Huang *et al.*, 1996], which are the orbitals involved in the formation of a transition state, i.e. the **highest occupied molecular orbital** (HOMO) and the **lowest unoccupied molecular orbital** (LUMO).

These orbitals play a major role in governing many chemical reactions and are responsible for the formation of many charge-transfer complexes.

The main descriptors based on molecular orbital energies are the following.

- **highest occupied molecular orbital energy** (ε_{HOMO})

The energy of the highest energy level containing electrons in the molecule. Molecules with high HOMO energy values can donate their electrons more easily compared to molecules with low HOMO energy values, and hence are more reactive. Therefore, the ε_{HOMO} descriptor is related to the ionizaton potential IP, is a measure of the nucleophilicity of a molecule and is important in modelling molecular properties and reactivity, in particular for radical reactions.

- **lowest unoccupied molecular orbital energy** (ε_{LUMO})

The energy of the lowest energy level containing no electrons in the molecule. Molecules with low LUMO energy values are more able to accept electrons than molecules with high LUMO energy values. Therefore, the ε_{LUMO} descriptor is related to the electron affinity EA, is a measure of the electrophilicity of a molecule and is important in the modelling of molecular properties and reactivity, in particular for radical reactions.

📖 [Debnath *et al.*, 1991]

- **HOMO-LUMO energy gap** (: *GAP*)

The difference between the HOMO and LUMO energies:

$$GAP = \varepsilon_{LUMO} - \varepsilon_{HOMO}$$

where ε_{LUMO} and ε_{HOMO} are the energies of the lowest unoccupied molecular orbital and the highest occupied molecular orbital, respectively.

GAP is an important stability index, a large *GAP* being related to the high stability of a molecule with its low reactivity in chemical reactions. It is an approximation of the lowest excitation energy of the molecule and can be used for the definition of absolute and activation hardness.

- **HOMO/LUMO energy fraction** ($f_{H/L}$)

A stability index defined as the ratio between the HOMO and LUMO energies:

$$f_{H/L} = \frac{\varepsilon_{HOMO}}{\varepsilon_{LUMO}}$$

where ε_{LUMO} and ε_{HOMO} are the energies of the lowest unoccupied molecular orbital and the highest occupied molecular orbital, respectively. Low values of $f_{H/L}$ are related to high stability of the molecule.

Frontier orbital electron densities also involve the highest occupied molecular orbital (HOMO) and the lowest unoccupied molecular orbital (LUMO), providing useful measures of donor-acceptor interactions at the point **r** of the molecular space.

The main descriptors based on molecular orbital electron densities are the following.

- **electrophilic atomic frontier electron density** (f_a^-)

A molecular descriptor defined as the electron density of the HOMO orbital as:

$$f_a^- = \sum_\mu (c_{HOMO,\,\mu})^2$$

where the sum runs over all the HOMO atomic orbitals.

To compare reactivity of atoms of different molecules, the **electrophilic frontier electron density index** is defined as:

$$F_a^- = \frac{f_a^-}{|\varepsilon_{HOMO}|}$$

where the electrophilic frontier electron density is normalized by the energy of the corresponding frontier molecular orbital.

- **nucleophilic atomic frontier electron density** (f_a^+)

A molecular descriptor defined as the electron density of the LUMO orbital as:

$$f_a^+ = \sum_\mu (c_{LUMO,\,\mu})^2$$

where the sum runs over all the LUMO atomic orbitals.

To compare the reactivity of atoms of different molecules, the **nucleophilic frontier electron density index** is defined as:

$$F_a^+ = \frac{f_a^+}{|\varepsilon_{LUMO}|}$$

where the nucleophilic frontier electron density is normalized by the energy of the corresponding frontier molecular orbital.

Superdelocalizability indices are → *dynamic reactivity indices* of occupied and unoccupied molecular orbitals, provide information about molecular interactions and allow comparisons between corresponding atoms in different molecules.

The **superdelocalizability** of an atom is related to the contribution of the atom to the stabilization energy in the formation of a charge-transfer complex with a second molecule or to the ability of a reactant to form bonds through charge transfer.

- **electrophilic superdelocalizability** (S_a^-)

Also known as the **acceptor superdelocalizability**, this is defined as the sum over all occupied molecular orbitals (N_{OCC}) and over the atomic orbitals μ of the ath atom, divided by the energy of the corresponding molecular orbital i.e.

$$S_a^- = 2 \cdot \sum_{i=1}^{N_{OCC}} \frac{\sum_\mu c_{i\mu,\,a}^2}{|\varepsilon_i|}$$

It is a measure of the availability of electrons in the ath atom. If the transition states are mainly controlled by the frontier orbital, the electrophilic superdelocalizability is calculated on the highest occupied molecular orbital (HOMO).

The **maximum electrophilic superdelocalizability** among all the atoms in a molecule is a molecular descriptor defined as:

$$S^-_{max} = \max_a(S^-_a)$$

The **total electrophilic superdelocalizability** is the sum over all of the atoms of the electrophilic superdelocalizability, i.e.

$$S^-_{TOT} = \sum\nolimits_a S^-_a$$

The **average electrophilic superdelocalizability** is derived from the total electrophilic superdelocalizability as:

$$\bar{S}^- = \frac{S^-_{TOT}}{A}$$

where A is the number of atoms.

📖 [Bearden and Schultz, 1997]

- **nucleophilic superdelocalizability** (S_a^+)

Also known as the **donor superdelocalizability**, this is defined as the sum over all unoccupied molecular orbitals ($N_{MO} - N_{OCC}$), N_{MO} being the total number of molecular orbitals, and over the atomic orbitals μ of the ath atom, divided by the energy of the corresponding molecular orbital i.e.

$$S^+_a = 2 \cdot \sum_{i=N_{OCC}+1}^{N_{MO}} \frac{\sum_\mu c^2_{i\mu,\,a}}{|\varepsilon_i|}$$

It is a measure of the availability of room on the ath atom for additional electron density.

If the transition states are mainly controlled by the frontier orbital, the nucleophilic superdelocalizability is calculated on the lowest unoccupied molecular orbital (LUMO).

The **maximum nucleophilic superdelocalizability** among all the atoms in a molecule is a molecular descriptor defined as:

$$S^+_{max} = \max_a(S^+_a)$$

The **total nucleophilic superdelocalizability** is the sum over all of the atoms of the nucleophilic superdelocalizability, i.e.

$$S^+_{TOT} = \sum\nolimits_a S^+_a$$

The **average nucleophilic superdelocalizability** is derived from the total nucleophilic superdelocalizability as:

$$\bar{S}^+ = \frac{S^+_{TOT}}{A}$$

where A is the number of atoms.

Quantum-chemical descriptors are used in several QSAR approaches, such as, for example, → *theoretical linear solvation energy relationships* (TLSERs), → *Mezey 3D shape analysis*, → *GIPF approach*, → *CODESSA method*, → *ADAPT approach*.

📖 [Mulliken, 1928a] [Mulliken, 1928b] [Schomaker and Stevenson, 1941] [Hannay and Smyth, 1946] [Fukui *et al.*, 1954] [Platt, 1954] [Hall, 1955] [Ketelaar, 1958] [Del Re, 1958] [Löwdin, 1970] [Dewar *et al.*, 1970] [McClelland, 1974] [Gelius, 1974] [Woolley, 1978a] [Miertus *et al.*, 1981] [Lukovits, 1983] [Buydens *et al.*, 1985] [Kaliszan *et al.*, 1985] [Yang and Mortier, 1986] [Zefirov *et al.*, 1987] [Grüber and Buss, 1989]

[Han, 1990] [Loew and Burt, 1990] [Bodor *et al.*, 1991] [Rabinowitz and Little, 1991] [Herdan *et al.*, 1991] [Bodor *et al.*, 1992c] [Carbó and Calabuig, 1992d] [Cocchi *et al.*, 1992] [Menziani and De Benedetti, 1992] [De Benedetti, 1992] [Venturelli *et al.*, 1992] [Cardozo *et al.*, 1992] [Shusterman, 1992] [Tuppurainen *et al.*, 1992] [Tuppurainen and Lotjonen, 1993] [Cramer *et al.*, 1993] [Hall *et al.*, 1993c] [Mekenyan *et al.*, 1993b] [Veith and Mekenyan, 1993] [Trapani *et al.*, 1993] [Nakayama *et al.*, 1993] [Nevalainen and Kolehmainen, 1994] [Gough *et al.*, 1994] [Verhaar *et al.*, 1994] [Eriksson *et al.*, 1994] [Cnubben *et al.*, 1994] [Reddy and Locke, 1994a] [Reddy and Locke, 1994b] [Doichinova *et al.*, 1994] [Kireev *et al.*, 1994] [Slater and Paynter, 1994] [Hansson and Ahnoff, 1994] [Basak and Grunwald, 1995c] [Brunoblanch and Estiu, 1995] [Lowrey *et al.*, 1995a] [Clare, 1995b] [Kourounakis and Bodor, 1995] [Luco *et al.*, 1995] [Sixt *et al.*, 1995] [Stankevitch *et al.*, 1995] [Hermens and Verhaar, 1995b] [Makovskaya *et al.*, 1995] [Lewis *et al.*, 1995a] [Lewis and Parke, 1995b] [Akagi *et al.*, 1995] [Clare, 1995a] [Clare, 1995c] [Galanakis *et al.*, 1995] [De Benedetti *et al.*, 1995] [Grassy *et al.*, 1995] [Hamerton *et al.*, 1995] [Andersson *et al.*, 1996] [Beck *et al.*, 1996] [Chen *et al.*, 1996] [Medven *et al.*, 1996] [Reddy and Locke, 1996] [Rorije and Peijnenburg, 1996] [Shpilkin *et al.*, 1996] [Sixt *et al.*, 1996] [Verhaar *et al.*, 1996] [Klamt, 1996] [Mracec *et al.*, 1996] [Purdy, 1996] [Richard and Hunter, 1996] [Hatch *et al.*, 1996] [Hatch *et al.*, 1996] [Karabunarliev *et al.*, 1996a] [Karabunarliev *et al.*, 1996b] [Vanvlaardingen *et al.*, 1996] [Schüürmann *et al.*, 1996] [Schüürmann, 1996] [Tang *et al.*, 1996] [Bjorsvik *et al.*, 1997] [Lukovits *et al.*, 1997a] [Bögel *et al.*, 1997] [Dearden *et al.*, 1997] [Ertl, 1997] [Müller, 1997] [Novic *et al.*, 1997] [Schüürmann *et al.*, 1997] [Migliavacca *et al.*, 1997] [Basak *et al.*, 1998] [Beck *et al.*, 1998] [Migliavacca *et al.*, 1998] [Schultz and Cronin, 1999] [Tuppurainen, 1999b] [Warne *et al.*, 1999a] [Warne *et al.*, 1999b]

quasi-Wiener index → **Laplacian matrix**

QUIK rule → **validation techniques**

R

radial centric information index → **centric indices**

radial distribution function descriptors : *RDF descriptors*

radius of gyration → **size descriptors**

radius-diameter diagram → **shape descriptors** (⊙ Petitjean shape indices)

Raevski H-bond indices → **hydrogen-bonding descriptors**

ramification index → **vertex degree**

ramification pair indices → **vertex degree**

Randic atomic path code : *vertex path code* → **path counts**

Randic chirality index → **chirality descriptors**

Randic connectivity ID number → **ID numbers**

Randic connectivity index → **connectivity indices**

Randic operator → **algebraic operators**

Randic-Razinger index → **local vertex invariants**

random walk : *walk* → **graph**

random walk matrices : *walk matrices*

rank distance : *Rouvray index* → **distance matrix**

Rashevsky complexity index → **molecular complexity**

RC index → **resonance indices**

***RDCHI* index** → **distance matrix**

RDF code → **RDF descriptors**

RDF descriptors (: *Radial Distribution Function descriptors*)

These descriptors are based on the distance distribution in the → *geometrical representation* of a molecule and constitute a radial distribution function code (**RDF code**) that shows certain characteristics in common with the → *3D-MoRSE code*.

Formally, the radial distribution function of an ensemble of A atoms can be interpreted as the probability distribution of finding an atom in a spherical volume of radius R. The general form of the radial distribution function code (RDF code) is represented by:

$$g(R) = f \cdot \sum_{i=1}^{A-1} \sum_{j=i+1}^{A} w_i \cdot w_j \cdot e^{-\beta \cdot (R - r_{ij})^2}$$

where f is a scaling factor, w are characteristic → *atomic properties* of the atoms i and j, r_{ij} are the → *interatomic distances* between the ith and jth atoms, respectively, and A is the number of atoms [Hemmer *et al.*, 1999]. The exponential term contains the distance r_{ij} between the atoms i and j and the smoothing parameter β, which defines the probability distribution of the individual interatomic distance; β can be interpreted as a temperature factor that defines the movement of atoms. $g(R)$ is generally calculated at a number of discrete points with defined intervals. An RDF code of 128 values was proposed, obtained by setting the β parameter in the range of 100 and 200 $Å^{-2}$ and a step size for R about 0.1 – 0.2 Å.

By including characteristic atomic properties w of the atoms i and j, the RDF code can be used in different tasks to fit the requirements of the information to be represented. These atomic properties enable the discrimination of the atoms of a molecule for almost any property that can be attributed to an atom.

The radial distribution function in this form meets all the requirements for a 3D structure descriptor: it is independent of the number of atoms, i.e., the size of a molecule, it is unique regarding the three-dimensional arrangement of the atoms, and it is invariant against translation and rotation of the entire molecule. Additionally, the RDF code can be restricted to specific atom types or distance ranges to represent specific information in a certain three-dimensional structure space, e.g., to describe steric hindrance or structure/activity properties of a molecule.

The RDF descriptor is interpretable by using simple rule sets, and thus it provides a possibility for conversion of the code back into the corresponding 3D structure. Besides information about interatomic distances in the entire molecule, the RDF code provides further valuable information, e.g., about bond distances, ring types, planar and non-planar systems and atom types. This fact is a most valuable consideration for a computer-assisted code elucidation.

***RDSQ* index** → **distance matrix**

reaction constant → **Hammett equation**

reactivity indices

Reactivity indices are usually categorized as either **electrophilic indices** or **nucleophilic indices**, depending on whether the reaction of interest involves electrophilic or nucleophilic attack. Moreover, **static reactivity indices**, such as charges, describe isolated molecules in their ground state, while **dynamic reactivity indices** refer to molecules in their transition states during a reaction.

→ *Charge descriptors* and several → *quantum-chemical descriptors* such as → *molecular orbital energies* and → *superdelocalizability indices* are examples of reactivity indices.

🕮 [Lall, 1981c] [Jug, 1984] [Simon *et al.*, 1985b] [Gasteiger *et al.*, 1987] [Gasteiger *et al.*, 1988] [Jiang and Zhang, 1990] [Rabinowitz and Little, 1991] [Balaban, 1994b] [Mekenyan and Basak, 1994] [Mekenyan and Veith, 1994] [Geerlings *et al.*, 1996b] [Bakken and Jurs, 1999a]

receptor → **drug design**

receptor cavity : *binding site caity* → **drug design**

receptor mapping → **drug design**

receptor mapping techniques → **structure/response correlations**

reciprocal Cluj-detour matrix → **Cluj matrices**

reciprocal Cluj-distance matrix → **Cluj matrices**

reciprocal detour/distance matrix → **detour matrix**

reciprocal detour matrix → **detour matrix**

reciprocal distance matrix → **distance matrix**

reciprocal distance sum → **distance matrix**

reciprocal edge distance matrix → **edge distance matrix**

reciprocal geometry matrix → **molecular geometry**

reciprocal distance-path matrix → **distance-path matrix**

reciprocal Schultz indices → **Schultz molecular topological index**

reciprocal square distance matrix → **distance matrix**

reciprocal Szeged matrix → **Szeged matrix**

reciprocal walk matrix → **walk matrices**

reciprocal Wiener matrix → **Wiener matrix**

redundancy index : *Brillouin redundancy index* → **information content**

redundant information content → **indices of neighbourhood symmetry**

reentrant surface → **molecular surface** (⊙ solvent-accessible molecular surface)

reference compound → **drug design**

reference structure : *reference compound* → **drug design**

refractive index → **molar refractivity**

refractive index function → **molar refractivity**

regression analysis → **chemometrics**

regression parameters

→ *Statistical indices* used for the evaluation of the performance of regression models. They are derived from two kind of statistics, called *goodness of fit* and *goodness of prediction*, where more attention is paid to the fitting properties of the models in the first approach, while, in the second, attention is paid to the predictive power of the models, using → *validation techniques*.

The **goodness of fit** statistic measures how well a model fits the data of the → *training set*, e.g. how well a regression model (or a classification model) accounts for the variance of the response variable.

The most important indices are listed below. The quantity df_E refers to the degrees of freedom of the error, i.e. to $n - p'$, where n is the number of objects (samples), p' the number of model parameters (for example, $n - p - 1$ for a regression model with p variables and the intercept). df_M and df_T refer to the degrees of freedom of the model and the total degrees of freedom, respectively.

Residual Sum of Squares, RSS (: *error sum of squares*). The sum of squared differences between the observed (y) and estimated response ($\hat{y}$):

$$RSS = \sum_{i=1}^{n} (y_i - \hat{y}_i)^2$$

This quantity is minimized by the least square estimator. A complementary quantity is the *Model Sum of Squares, MSS*, defined as the sum of the squared differences between the estimated responses and the average response:

$$MSS = \sum_{i=1}^{n} (\hat{y}_i - \bar{y})^2$$

This is a part of the total variance explained by the regression model as opposed to the residual sum of squares *RSS*. Moreover, a reference quantity is the *Total Sum of Squares, TSS*, defined as the sum of the squared differences between the experimental responses and the average response:

$$TSS = \sum_{i=1}^{n} (y_i - \bar{y})^2$$

This is the total variance that a regression model has to explain and is used as a reference quantity to calculate standard quality parameters such as the coefficient of determination.

Coefficient of determination, R^2. The squared multiple correlation coefficient that is the percent of total variance of the response explained by a regression model. It can be calculated from the model sum of squares *MSS* or from the residual sum of squares *RSS*:

$$R^2 = \frac{MSS}{TSS} = 1 - \frac{RSS}{TSS} = 1 - \frac{\sum_{i=1}^{n} (y_i - \hat{y}_i)^2}{\sum_{i=1}^{n} (y_i - \bar{y})^2}$$

where *TSS* is the total sum of squares around the mean. A value of one indicates perfect fit, i.e. a model with zero error term.

A related quantity is the *multiple correlation coefficient* R defined as the square root of R^2. It is a measure of linear association between the observed response and the estimated response, i.e. the response obtained by a linear combination of the predictor variables in a linear regression model. A quantity complementary to R^2 is the *coefficient of nondetermination* defined as:

$$cnd = 1 - R^2$$

Residual Mean Square, RMS or s^2 (*: mean square error, expected squared error*). The estimate s^2 of the error variance σ^2, defined as:

$$s^2 = \frac{RSS}{df_E}$$

where *RSS* is the residual sum of squares and df_E is the error degrees of freedom. The square root of the residual mean square is called *Residual Standard Deviation, RSD* or s (*: error standard deviation, residual standard error, Root Mean Square Error RMSE*); it is an estimate of the model error σ.

Standard Deviation Error in Calculation, SDEC (*: standard error in calculation, SEC*). A function of the residual sum of squares, defined as:

$$SDEC \equiv SEC = \sqrt{\frac{\sum_{i=1}^{n} (y_i - \hat{y}_i)^2}{n}} = \sqrt{\frac{RSS}{n}}$$

F-ratio test. Among the most well-known statistical tests, this is defined as the ratio between the model sum of squares *MSS* and the residual sum of squares *RSS*:

$$F = \frac{MSS/df_M}{RSS/df_E}$$

The calculated value is compared with the critical value F_{crit} for the corresponding degrees of freedom. It is a comparison between the model explained variance and the residual variance: high values of the F-ratio test indicate reliable models.

Adjusted R^2, R^2_{adj}. A parameter adjusted for the degrees of freedom, so that it can be used for comparing models with different numbers of predictor variables:

$$R^2_{adj} = 1 - \frac{RSS/df_E}{TSS/df_T} = 1 - (1 - R^2) \cdot \left(\frac{n-1}{n-p'}\right)$$

where RSS and TSS are the residual sum of squares and the total sum of squares, respectively; R^2 is the coefficient of determination.

Exner statistic, ψ^2. A parameter adjusted for the degrees of freedom, so that it can be used for comparing models with different numbers of predictor variables:

$$\psi^2 = \frac{\sum_{i=1}^{n} (y_i - \hat{y}_i)^2}{\sum_{i=1}^{n} (y_i - \bar{y})^2} \cdot \frac{n}{n-p} = \frac{RSS}{TSS} \cdot \frac{n}{n-p}$$

Fitness function. A model selection criterion used to compare models with a different number of p variables and n objects:

$$FIT = \frac{R^2 \cdot (n-p-1)}{(n+p^2) \cdot (1-R^2)}$$

where R^2 is the coefficient of determination.

J_p *statistic*. A function of the residual mean square s^2:

$$J_p = \frac{(n+p') \cdot s^2}{n}$$

where n is the total number of observations and p' the number of parameters in the model.

S_p *statistic*. It is a function of the residual mean square, defined as:

$$S_p = \frac{s^2}{df_E}$$

where df_E is the error degrees of freedom.

Mallows C_p. Model selection criterion used to compare biased regression models with the full least squares regression model:

$$C_p = \frac{RSS_p}{s^2} + 2p - n$$

where RSS_p is the residual sum of squares of the biased model with p parameters, and s^2 is the residual mean square of the full least squares model, i.e. the squared residual standard deviation. In OLS, $C_p = p$; in biased models, $C_p < p$. The best models are those that have a minimum number of parameters and are fairly close to the $C_p = p$ line, i.e.

$$\min_p(|C_p - p|)$$

Note. A *quality factor* (Q) was also proposed [Pogliani, 1994a] for measuring the quality of the regression models and defined as the ratio between the multiple correlation coefficient R and the standard deviation s. However, s being dependent on the measurement unit used for the response, it should be not used to measure the quality of the regression models.

The **goodness of prediction** statistic measures how well a model can be used to estimate future (test) data, e.g. how well a regression model (or a classification model) estimates the response variable given a set of values for predictor variables. These quantities are obtained using → *validation techniques* and are also used as criteria for model selection.

Besides some special parameters adopted in → *variable selection*, such as *Friedman's lack-of-fit function* (LOF) and *fitness function* (FIT), the most important indices are listed below.

Predictive Residual Sum of Squares, PRESS. The sum of squared differences between the observed and estimated response by validation techniques:

$$PRESS = \sum_{i=1}^{n} \left(y_i - \hat{y}_{i/i}\right)^2$$

where $\hat{y}_{i/i}$ denotes the response of the ith object estimated by using a model obtained without using the ith object. Use of validation techniques minimizes this quantity.

Cross-validated R^2, R^2_{cv} (or Q^2). The explained variance in prediction:

$$R^2_{\text{cv}} \equiv Q^2 = 1 - \frac{PRESS}{TSS} = 1 - \frac{\sum_{i=1}^{n} \left(y_i - \hat{y}_{i/i}\right)^2}{\sum_{i=1}^{n} (y_i - \bar{y})^2}$$

where $PRESS$ is the predictive error sum of squares and TSS the total sum of squares.

Generalized cross-validation, gcv. The ratio of the residual sum of squares to the squared residual degrees of freedom:

$$gcv = \frac{RSS}{(df_{\text{E}})^2}$$

Predictive Squared Error, PSE. The average value of the predictive error sum of squares:

$$PSE = \frac{PRESS}{n} = \frac{\sum_{i=1}^{n} \left(y_i - \hat{y}_{i/i}\right)^2}{n}$$

Standard Deviation Error of Prediction, SDEP (: *standard error in prediction, SEP*). A function of the predictive residual sum of squares, defined as:

$$SDEP \equiv SEP = \sqrt{\frac{\sum_{i=1}^{n} \left(y_i - \hat{y}_{i/i}\right)^2}{n}} = \sqrt{\frac{PRESS}{n}} = \sqrt{PSE}$$

Akaike Information Criterion, AIC_p. A model selection criterion for choosing between models with different parameters and defined as:

$$AIC_p = -2 \cdot \mathrm{L}_p + 2 \cdot p$$

where p is the number of parameters and L_p is the maximized log-likelihood. For regression models, the optimal complexity according to the Akaike criterion is obtained by minimizing the following:

$$\min\left(AIC_p\right) = \frac{n + p + 1}{n - p - 1} \cdot s^2$$

where s^2 is the residual mean square.

[Myers, 1986] [Rawlings, 1988] [Ryan, 1997] [Draper and Smith, 1998]

regressive decremental distance sums → layer matrices

regressive distance sums → layer matrices

regressive incremental distance sums : *regressive distance sums* → **layer matrices**

regressive vertex degrees → **layer matrices**

regular graph → **graph**

relative atom-type count → **count descriptors**

relative atomic moments → **self-returning walk counts**

relative hydrophobic surface area → **charged partial surface area descriptors**

relative negative charge → **charged partial surface area descriptors**

relative negative charge surface area → **charged partial surface area descriptors**

relative polar surface area → **charged partial surface area descriptors**

relative positive charge → **charged partial surface area descriptors**

relative positive charge surface area → **charged partial surface area descriptors**

relative topological distance → **bond order indices** (⊙ conventional bond order)

relative vertex distance complexity → **topological information indices**

repeated evaluation set technique → **validation techniques**
(⊙ training/evaluation set splitting)

repulsive steric effects → **minimal topological difference**

resistance distance → **resistance matrix**

resistance distance hyper-Wiener index → **Wiener matrix**

resistance matrix (Ω)

A square $A \times A$ symmetric matrix, A being the number of non-hydrogen atoms, whose off-diagonal entries are given by the resistance distance Ω_{ij} between any pair of vertices in the → *H-depleted molecular graph* G as [Klein and Randic, 1993]:

$$[\mathbf{\Omega}]_{ij} = \begin{cases} 0 & if\ i = j \\ \Omega_{ij} & if\ i \neq j \end{cases}$$

The **resistance distance** is based on electrical network theory and is defined as the effective electrical resistance between two vertices (nodes) when a battery is connected across them and each graph edge is considered as a resistor taking a value of 1 ohm.

The resistance distance Ω_{ij} between two adjacent vertices v_i and v_j is obviously equal to one if the edge defined by the two vertices does not belong to a cycle, i.e. there is a single path between them. Otherwise, it is calculated as the following:

$$\frac{1}{\Omega_{ij}} = \sum\nolimits_{p_{ij}} |p_{ij}|^{-1}$$

where the sum runs over all the paths p_{ij} connecting the two considered vertices and $|p_{ij}|$ is the length of the considered path p_{ij}. Multiple connections between two vertices decrease the distance between them because the difficulty of transport decreases when there is more than one possibility for their communication.

For any pair of non-adjacent vertices, the resistance distance is the effective resistance calculated according to the two classical Kirchhoff laws for series and parallel electrical circuits; some examples of calculations are given in Box R-1 for ethylbenzene.

The resistance distance between any pair of vertices can also be calculated by the → *Laplacian matrix* as the following:

$$\Omega_{ij} = \left(\mathbf{d}^{ij}\right)^{\mathbf{T}} \cdot \mathbf{L}^{+} \cdot \mathbf{d}^{ij} = \mathbf{u}_i^{\mathbf{T}} \cdot \mathbf{L}^{+} \cdot \mathbf{u}_i - \mathbf{u}_i^{\mathbf{T}} \cdot \mathbf{L}^{+} \cdot \mathbf{u}_j - \mathbf{u}_j^{\mathbf{T}} \cdot \mathbf{L}^{+} \cdot \mathbf{u}_i + \mathbf{u}_j^{\mathbf{T}} \cdot \mathbf{L}^{+} \cdot \mathbf{u}_j$$

where $\mathbf{d}^{ij}$ is the vector obtained as the difference between $\mathbf{u}_i$ and $\mathbf{u}_j$, which are column vectors with every element equal to zero except the ith and jth element, respectively; $\mathbf{L}^+$ is the Moore-Penrose generalized inverse of the Laplacian matrix $\mathbf{L}$.

Example : ethylbenzene

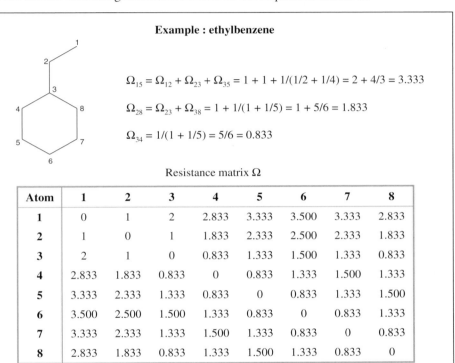

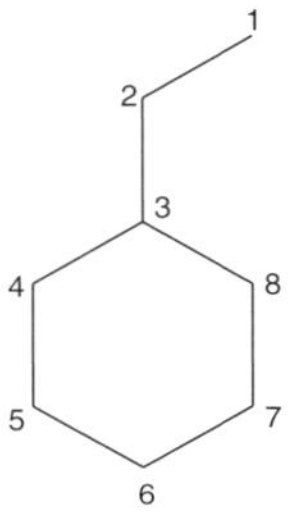

$\Omega_{15} = \Omega_{12} + \Omega_{23} + \Omega_{35} = 1 + 1 + 1/(1/2 + 1/4) = 2 + 4/3 = 3.333$

$\Omega_{28} = \Omega_{23} + \Omega_{38} = 1 + 1/(1 + 1/5) = 1 + 5/6 = 1.833$

$\Omega_{34} = 1/(1 + 1/5) = 5/6 = 0.833$

Resistance matrix Ω

Atom	1	2	3	4	5	6	7	8
1	0	1	2	2.833	3.333	3.500	3.333	2.833
2	1	0	1	1.833	2.333	2.500	2.333	1.833
3	2	1	0	0.833	1.333	1.500	1.333	0.833
4	2.833	1.833	0.833	0	0.833	1.333	1.500	1.333
5	3.333	2.333	1.333	0.833	0	0.833	1.333	1.500
6	3.500	2.500	1.500	1.333	0.833	0	0.833	1.333
7	3.333	2.333	1.333	1.500	1.333	0.833	0	0.833
8	2.833	1.833	0.833	1.333	1.500	1.333	0.833	0

Box R-1.

The sum of the resistance distances between all pairs of vertices in the graph was proposed [Klein and Randic, 1993] as a molecular descriptor analogous to the → *Wiener index* and subsequently called **Kirchhoff number** *Kf* [Bonchev *et al.*, 1994a]; it is defined as:

$$Kf = \mathcal{W}(\mathbf{\Omega}) = \frac{1}{2} \cdot \sum_{i=1}^{A} \sum_{j=1}^{A} \Omega_{ij}$$

where $\mathcal{W}$ is the → *Wiener operator* and A the number of vertices in the graph. For acyclic graphs the Kirchhoff number coincides with the Wiener index, as the resistance distances coincide with the classical topological distances for any pair of vertices. Therefore, it is mainly related to molecular size, like the Wiener index. Moreover, among the isomeric series it was proposed as a better index of → *molecular cyclicity* than the Wiener index due to its low degeneracy; it decreases when the number of cycles and their centricity decrease.

The Kirchhoff number can also be directly calculated from the Laplacian matrix by the following:

$$Kf = A \cdot tr(\mathbf{L}^{+})$$

where $tr(\mathbf{L}^{+})$ is the trace of the Moore-Penrose generalized inverse of the Laplacian matrix [Mardia *et al.*, 1988]. It was demonstrated that the Kirchhoff number and the → *quasi-Wiener index* coincide [Gutman and Mohar, 1996].

📖 [Ivanciuc *et al.*, 1997a]

resonance constant → **electronic substituent constants** (⊙ field/resonance effect separation)

resonance effect → **electronic substituent constants**

resonance electronic constants → **electronic substituent constants**

resonance energy → **resonance indices**

resonance indices

The **resonance energy** RE is a theoretical quantity introduced to explain the stability of benzene and is used for predicting the degree of electron delocalization of conjugated systems [Wheland, 1955; Salem, 1966; Randic, 1989; Trinajstic, 1991; Trinajstic, 1992]. The general definition of resonance energy is:

$$\mathrm{RE} = \mathrm{E}_{\pi}(\textit{conjugated molecule}) - \mathrm{E}_{\pi}(\textit{reference structure})$$

where E_{π} is the π-electron energy.

The resonance energy per electron (REPE) is a size-independent quantity obtained by dividing the total resonance energy for the number of π-electrons.

Resonance indices are → *molecular descriptors* based on the calculation of the resonance energy.

A number of resonance energy definitions are based on theoretical quantum-chemistry approaches. It was also recognized that the main difference between the proposed approaches lies in the definition of the nonconjugated reference structure, not in the use of different MO theories.

- **Hückel resonance energy** (HRE)

The classical definition of resonance energy obtained from a reference structure containing carbon-carbon isolated double bonds with π-electron energy of ethylene [Streitweiser, 1961]:

$$\mathrm{HRE} = \mathrm{E}_{\pi}(\textit{conjugated molecule}) - 2 \cdot \mathrm{N}_{\mathrm{C=C}}$$

where $\mathrm{N}_{\mathrm{C=C}}$ is the number of double bonds in a Kekulé structure of the molecule.

This criterion to define the resonance energy fails in many cases, overestimating the aromaticity of rather unstable compounds.

📖 [Gutman and Trinajstic, 1972]

- **Dewar resonance energy** (DRE)

Resonance energy defined as the difference between the π-energy E_{π} of the compound and the reference energy estimated by the bond contributions of the corresponding nonconjugated structure, in the framework of SCF π-MO approximation:

$$\mathrm{DRE} = \mathrm{E}_{\pi}(\textit{conjugated molecule}) - \sum\nolimits_{b} n_b \cdot \mathrm{E}_b$$

where n_b is the number of bonds having energy E_b [Dewar and Longuet-Higgins, 1952; Dewar, 1969a; Dewar and de Llano, 1969b; Dewar *et al.*, 1971b; Dewar and Gleicher, 1965].

This definition of resonance energy makes a clear distinction between aromatic (positive DRE), antiaromatic (negative DRE) and nonaromatic (near zero DRE) conjugated molecules. Extensive tables of resonance energies were also obtained in the framework of the HMO approximation by Hess and Schaad [Hess Jr. and Schaad, 1971a; Hess Jr. and Schaad, 1971b; Hess Jr. *et al.*, 1972; Hess Jr. and Schaad, 1973]. Moreover, extensions and modifications of the calculations of the reference structure energy were proposed by other authors [Baird, 1969; Baird, 1971].

📖 [Dewar *et al.*, 1969; Dewar and Harget, 1970a; Dewar and Trinajstic, 1970; Schaad and Hess Jr., 1972; Hess Jr. *et al.*, 1975]

Several resonance energy indices (RE) are derived directly from the molecular topology. The most popular of them are listed below.

- **Green resonance energy**

This is defined for benzenoid systems as:

$$\mathrm{RE} = \frac{B}{3} + \frac{B^*}{10}$$

where B is the total number of aromatic bonds and B^* is the number of bonds contained in one benzene ring and linking two others [Green, 1956].

- **Bartell resonance energy**

A definition of resonance energy which relates π-energies to the Pauling → *bond order* π by the following expression:

$$\mathrm{BRE} = \frac{4}{3} \cdot \beta \cdot \left(\mathrm{N}_{\mathrm{C=C}} - \sum\nolimits_b \pi_b^2\right)$$

where β is the resonance integral and $\mathrm{N}_{\mathrm{C=C}}$ the number of formal double bonds [Bartell, 1963; Bartell, 1964]. The Pauling bond orders are obtained from the analysis of the Kekulé resonance structures.

- **Carter resonance energy**

This is defined for benzenoid systems as:

$$\mathrm{RE} = 0.6 \cdot \mathrm{N}_{\mathrm{C=C}} + 1.5 \cdot \ln \mathrm{K} - 1$$

where 0.6 and 1.5 are empirical parameters, $\mathrm{N}_{\mathrm{C=C}}$ the number of double bonds in one Kekulé structure and K the → *Kekulé number* of the molecule [Carter, 1949].

- **Herndon resonance energy**

This is defined for benzenoid systems as:

$$\mathrm{RE} = 1.185 \cdot \ln \mathrm{K}$$

where 1.185 was obtained by fitting the Dewar-deLlano SCF π-MO resonance energy values and K is the Kekulé number of the molecule [Herndon, 1973b; Herndon, 1974b; Herndon and Ellzey Jr, 1974].

- **Wilcox resonance energy**

This is defined for general aromatic systems including alternant 4-membered rings as:

$$\mathrm{RE} = 0.445 \cdot \ln \mathrm{CSC} - 0.17 \cdot \mathrm{N}_4$$

where CSC is the → *corrected structure count* and N_4 the number of 4-membered rings [Wilcox Jr, 1968; Wilcox Jr, 1969]. Parameter values were obtained by fitting the Hess-Schaad resonance energy values [Hess Jr. *et al.*, 1975].

• **McClelland resonance energy**
A measure of resonance energy defined in terms of the number of atoms A_π and bonds B_π involved in a π-system as [McClelland, 1971]:

$$\mathrm{RE} = 0.92 \cdot \sqrt{2 \cdot A_\pi \cdot B_\pi}$$

• **Hosoya resonance energy**
This is a measure of aromatic stability of conjugated systems defined as [Hosoya *et al.*, 1975a]:

$$\mathrm{RE} = \tilde{\mathrm{Z}} - \mathrm{Z}$$

where Z is the → *Hosoya Z index* and $\tilde{Z}$ is the **modified Hosoya index** defined as the sum of the coefficients of the → *characteristic polynomial of the graph*:

$$\tilde{\mathrm{Z}} = \sum_{k=0}^{[A/2]} (-1)^k \cdot a_{2k}$$

In acyclic graphs, $\tilde{Z}$ and Z are equal, the → *Z-counting polynomial* being in this case coincident with the characteristic polynomial.

• **Aihara resonance energy**
This is a measure of the aromatic stability of conjugated systems defined as:

$$\mathrm{RE} = 6.0846 \cdot \log\left(\frac{\mathrm{Z}^*}{\mathrm{Z}}\right)$$

where Z is the Hosoya Z index and Z* is defined as:

$$\mathrm{Z}^* = \prod_{i=1}^{A} \left(1 + \lambda_i^2\right)^{1/2}$$

where λ_i are the eigenvalues of the characteristic polynomial [Aihara, 1976; Aihara, 1977a; Aihara, 1977b; Aihara, 1978].

• **Topological Resonance Energy** (TRE)
A topological resonance index defined as [Gutman *et al.*, 1977; Trinajstic, 1992]:

$$\mathrm{TRE} = \mathrm{E}_\pi - \mathrm{E}_{\mathrm{REF}} = \sum_{i=1}^{A} g_i \cdot \left(\lambda_i - \lambda_i^{ac}\right)$$

where E_π is the Hückel π-electronic energy and $\mathrm{E}_{\mathrm{REF}}$ the reference energy obtained from the matching polynomial of the corresponding molecular graph; λ_i are the eigenvalues of the characteristic polynomial of the molecule and λ_i^{ac} the eigenvalues of the corresponding reference structure (i.e. eigenvalues of the matching polynomial), g_i being the occupation number on the ith molecular orbital which can take values 0, 1, or 2.

The average TRE index is defined as TRE / N_{el}, where N_{el} is the number of π electrons in the molecule.

📖 [Babic *et al.*, 1997; Juric *et al.*, 1997]

Table R-1. Resonance indices for some benzenoid compounds.
K is the Kekulé number. Data from: [Swinborne-Sheldrake *et al.*, 1975], [Aihara, 1977b], [Trinajstic, 1992].

Compound	K	*Dewar*	*Green*	*Aihara*	*Herndon*	*TRE*
Benzene	2	0.869	2.00	0.273	0.821	0.276
Naphthalene	3	1.323	3.67	0.389	1.302	0.390
Anthracene	4	1.600	5.33	0.475	1.643	0.476
Phenanthrene	5	1.933	5.43	0.546	1.907	0.546
Pyrene	6	2.098	6.43	0.562	2.123	0.592
Naphthacene	5	1.822	7.00	0.553	1.907	0.558
3,4-Benzophenanthrene	8	2.478	–	0.687	2.464	–
1,2-Benzanthracene	7	2.291	7.10	0.643	2.306	–
Chrysene	8	2.483	7.20	0.688	2.464	0.684
Triphenylene	9	2.654	7.30	0.739	2.604	–
Perylene	9	2.619	8.20	0.598	2.604	–

Another view of the degree of electron delocalization is given by the concept of **aromaticity** [Lloyd, 1996]. This concept is one of the most important general concepts for an understanding of organic chemistry and physico-chemical properties. In spite of this, the concept of aromaticity has never been defined unequivocally; the commonly accepted description of aromaticity is as a characterictic delocalization of the π-electrons giving a stabilization of cyclic and polycyclic conjugated molecules, i.e. the increased stability of conjugated rings compared to its classical localized structure. The corresponding molecular descriptors are called **aromaticity indices** and are often calculated from the bond lengths and bond orders of the compounds [Krygowski *et al.*, 1995b].

The most well-known aromaticity indices are listed below, and some values are collected in Table R-4.

- **bond alternation coefficient** (*BAC*)

A purely geometric aromaticity index defined as:

$$BAC = \sum_b (r_{b+1} - r_b)^2$$

where r_{b+1} and r_b are consecutive bond lengths in the rings and summation runs over all π-bonds of the molecule (or fragment) under study [Binsch and Heilbronner, 1968].

- **I_R aromaticity indices**

General aromaticity indices based on the statistical degree of uniformity on the bond orders of the ring periphery and represented as I_5, I_6, and $I_{5,6}$ for 5- and 6-membered rings and 5,6-fused rings, respectively [Bird, 1985; Bird, 1986]. These indices are defined as:

$$I_R = 100 \cdot \left(1 - \frac{V}{V_R}\right)$$

where V_R is a constant depending on the considered ring (e.g. $V_R = 35$ for a five-membered ring heterocycle and $V_R = 33.3$ for a six-membered ring heterocycle; for systems consisting of a five-membered and a six-membered ring fused together, $V_R = 35$).

The term V is defined as:

$$V = \frac{100}{\bar{\pi}} \cdot \sqrt{\frac{\sum_b (\pi_b - \bar{\pi})^2}{B_\pi}}$$

where B_π is the number of π-bonds and π_b the bond orders obtained from the Gordy relationship:

$$\pi_b = \frac{a}{r_b^2} - b$$

where a and b are constants depending on the π-bond type and r_b the bond distances; $\bar{\pi}$ is the average bond order (Table R-2).

Table R-2. Values of a and b constants used in the calculation of bond orders.

Bond	*a*	*b*	*Bond*	*a*	*b*
C ≈ C	6.80	1.71	N ≈ N	5.28	1.41
C ≈ N	6.48	2.00	N ≈ O	4.98	1.45
C ≈ O	5.75	1.85	N ≈ S	10.53	2.50
C ≈ S	11.9	2.59	O ≈ S	17.05	5.58
C ≈ P	13.54	3.02	S ≈ S	19.30	3.46

- **RC index**

This is an aromaticity index based on the idea of *ring current* whose magnitude is determined by its weakest link in the ring [Jug, 1983; Jug, 1984]. The weakest link is considered as the bond with the minimum total bond order:

$$RC = \min_b(\pi_b)$$

where b runs over all the π-bond system and π_b are the bond orders.

By analogy, a complementary aromatic index, called **maximum bond length** (LB) was defined as the longest bond length in the π-electron system under consideration [Krygowski and Ciesielski, 1995a; Krygowski *et al.*, 1995b]:

$$LB = \max_b(r_b)$$

where b runs over all the π-bond system and r_b are the bond lengths.

- **HOMA index** (: *Harmonic Oscillator Model of Aromaticity index*)

This index is based on the degree of alternation of single/double bonds, measuring the bond length deviations from the optimal lengths attributed to the typical aromatic state [Kruszewski and Krygowski, 1972; Krygowski, 1993; Krygowski and Ciesielski, 1995a; Krygowski *et al.*, 1995b; Krygowski *et al.*, 1996]. The HOMA index is defined as:

$$HOMA = 1 - \frac{\sum_k \alpha_k \cdot \sum_{b=1}^{B_{\pi k}} \left(r_k^{opt} - r_b\right)^2}{B_\pi}$$

where the first sum runs over each kth aromatic bond type, $B_{\pi k}$ is the number of considered π-bond contributions of the kth aromatic bond type, r_b is the actual bond length, α_k and r_k^{opt} are a numerical constant and the typical aromatic bond length referring to the kth aromatic bond type (Table R-3). B_π is the total number of aromatic bonds.

Table R-3. Values of the parameter α and of the optimal distances r^{opt}. [a]1,3-butadiene.

Bond	α	r^{opt}	*Bond*	α	r^{opt}
$C \approx C_a$	257.7	1.388	$C \approx P$	118.91	1.698
$C \approx C$	98.89	1.397	$C \approx S$	94.09	1.677
$C \approx N$	93.52	1.334	$N \approx N$	130.33	1.309
$C \approx O$	157.38	1.265	$N \approx O$	57.21	1.248

Table R-4. Values of aromaticity indices for some compounds. Data from [Krygowski *et al.*, 1995b] and then elaborated.

Compound	HOMA	*BAC*	I_6	LB
Benzene	1.000	0.000	100.0	1.397
Naphthalene	0.802	0.088	81.3	1.424
Anthracene	0.696	0.098	79.2	1.446
Phenanthrene	0.727	0.101	77.1	1.465
Pyrene	0.728	0.094	80.1	1.438
3,4-Benzophenanthrene	0.671	0.107	73.8	1.460
Triphenylene	0.722	0.070	85.1	1.478

• **benzene-likeliness index**

It is an aromaticity index based on the first-order valence connectivity index $^1\chi^v$ divided by the number B of bonds of the molecule and normalized on the benzene molecule [Kier and Hall, 1986]:

$$B_L = \frac{^1\chi^v / B}{2/6} = \frac{3 \cdot {^1\chi^v}}{B}$$

where $^1\chi^v = 2$ and $B = 6$ for benzene. Some typical values of this index are B_L (benzene) = 1, B_L(thiophene) = 0.97, B_L (pyrrole) = 0.95, B_L (pyridine) = 0.93, B_L (imidazole) = 0.86.

📖 [Hall, 1957] [Figeys, 1970] [Randic, 1975a] [Graovac *et al.*, 1977] [Randic, 1980c] [Fratev *et al.*, 1980] [Singh *et al.*, 1984] [Gutman, 1986] [Gutman, 1987a] [Babic *et al.*, 1991] [Gutman, 1992a] [Cyranski and Krygowski, 1996] [Randic *et al.*, 1996a] [Randic, 1997e] [Randic *et al.*, 1998c] [Randic and Guo, 1999]

resonance polar effect → **electronic substituent constants** (⊙ resonance electronic constants)

resonance-weighted edge adjacency matrix → **edge adjacency matrix**

resonance-weighted edge connectivity index → **edge adjacency matrix**

response variables : *dependent variables* → **data set**

response-variable correlation cutoff → **variable selection**

restricted random walk matrix → **walk matrices**

restricted walk ID number → **walk matrices**

reversible decoding → **structure/response correlations**

REX descriptors → **substructure descriptors**

RHTA index → **charged partial surface area descriptors**

ring ID number → **ID numbers**

ring number : *cyclomatic number*

Roberts-Moreland inductive constant → **electronic substituent constants** (⊙ inductive electronic constants)

rooted tree → **graph** (⊙ tree)

root mean square Wiener index → **Wiener index**

rotamer factors → **weighted matrices**

rotamer modification number → **weighted matrices**

rotatable bond number → **flexibility indices**

rotational invariance → **molecular descriptors** (⊙ invariance properties of molecular descriptors)

Rouvray index → **distance matrix**

row sum operator → **algebraic operators**

row sum vector → **algebraic operators** (⊙ row sum operator)

RSAA index → **charged partial surface area descriptors**

RSAH index → **charged partial surface area descriptors**

RSAM index → **charged partial surface area descriptors**

RSHM index → **charged partial surface area descriptors**

rule of six : *six-position number* → **steric descriptors** (⊙ number of atoms in substituent specific positions)

S

Sanderson group electronegativity → **quantum-chemical descriptors**

***SCAA*$_1$ index** : *HACA index* → **charged partial surface area descriptors**

***SCAA*$_2$ index** → **charged partial surface area descriptors** (⊙ *HACA* index)

scalar product of vectors → **algebraic operators**

Schrödinger equation → **quantum-chemical descriptors**

Schultz molecular topological index (*MTI*)

A topological index, originally simply called by the author **Molecular Topological Index**, derived from the → *adjacency matrix* **A**, the → *distance matrix* **D** and the A-dimensional column vector **v** constituted by the → *vertex degree* δ of the A atoms in the → *H-depleted molecular graph* [Schultz, 1989]. The Schultz index is defined as:

$$MTI = \sum_{i=1}^{A} [(\mathbf{A}+\mathbf{D})\mathbf{v}]_i = \sum_{i=1}^{A} t_i$$

where t_i are the elements, called **intricacy numbers**, of the A-dimensional column vector **t** obtained as the following:

$$\mathbf{t} = (\mathbf{A} + \mathbf{D})\,\mathbf{v}$$

i.e. the **D** and **A** matrices are summed and then multiplied by the **v** vector. Intricacy numbers measure the combined influence of valence, adjacency and distance for each comparable set of vertices; the lower the intricacy number, the more intricate or complex the vertex [Schultz and Schultz, 1993].

The Schultz index can be decomposed in two distinct parts corresponding to two independent descriptors:

$$MTI = \mathbf{u}^{\mathbf{T}} \cdot [\mathbf{A} \cdot (\mathbf{A}+\mathbf{D})] \cdot \mathbf{u} = \mathbf{u}^{\mathbf{T}} \cdot \left(\mathbf{A}^2 + \mathbf{AD}\right) \cdot \mathbf{u} = \mathbf{u}^{\mathbf{T}} \cdot \mathbf{A}^2 \cdot \mathbf{u} + \mathbf{u}^{\mathbf{T}} \cdot \mathbf{AD} \cdot \mathbf{u} = M_2 + S$$

$$M_2 = \mathbf{u}^{\mathbf{T}} \cdot \mathbf{A}^2 \cdot \mathbf{u} = \sum_{i=1}^{A}\sum_{j=1}^{A} \left[\mathbf{A}^2\right]_{ij} = \sum_{i=1}^{A} \delta_i^2$$

$$S \equiv MTI' = \mathbf{u}^{\mathbf{T}} \cdot \mathbf{AD} \cdot \mathbf{u} = \sum_{i=1}^{A}\sum_{j=1}^{A} [\mathbf{AD}]_{ij}$$

where **u** is an A-dimensional unit column vector, M_2 is just the → *2nd Zagreb index* and S is the nontrivial part of MTI, originally denoted by MTI' and called **MTI′ index** [Müller *et al.*, 1990b; Mihalic *et al.*, 1992a].

The **S index** can also be written as:

$$S = \sum_{i=1}^{A}\sum_{j=1}^{A} [\mathbf{AD}]_{ij} = \frac{1}{2}\sum_{i=1}^{A}\sum_{j=1}^{A} \left(\delta_i + \delta_j\right) \cdot d_{ij} = \sum_{i=1}^{A} \delta_i \sum_{j=1}^{A} d_{ij} = \sum_{i=1}^{A} \delta_i \cdot \sigma_i \equiv D'$$

where d_{ij} is the → *topological distance* and σ_i is the ith → *vertex distance degree* [Gutman, 1994b; Schultz *et al.*, 1995]. From the above relationships follows the coincidence between the S index and the **degree distance of the graph** D' later proposed by Dobrynin [Dobrynin and Kochetova, 1994] simply summing over all graph vertices the local invariants called **vertex degree distance** defined as:

$$D'_i = \delta_i \cdot \sigma_i$$

Since the S index is a sum of δ-weighted vertex distance degrees, it can be considered a vertex-valency-weighted analogue of the → *Wiener index*.

Reciprocal Schultz indices were defined as [Schultz and Schultz, 1998]:

$$RMTI = \sum_{i=1}^{A} \left[\left(\mathbf{A} + \mathbf{D}^{-1}\right) \cdot \mathbf{v}\right]_i \quad \text{and} \quad RS = \sum_{i=1}^{A} \left[\mathbf{D}^{-1} \cdot \mathbf{v}\right]_i$$

where $\mathbf{D}^{-1}$ is the → *reciprocal distance matrix*.

A Schultz-type topological index – **Gutman molecular topological index** S_G – was also defined by Gutman [Gutman, 1994b], as:

$$S_G = \sum_{i=1}^{A} \sum_{j=1}^{A} \delta_i \delta_j \cdot d_{ij}$$

where $\delta_i \delta_j \cdot d_{ij}$ is the topological distance between the vertices v_i and v_j weighted by the product of the endpoint vertex degrees. Like the S index, the S_G index is a vertex-valency-weighted analogue of the Wiener index, whereas the weighting factor is multiplicative instead of additive.

The Schultz *MTI* was demonstrated [Klein *et al.*, 1992; Plavsic *et al.*, 1993a] to be strongly correlated to the Wiener index W according to the following formal relation:

$$MTI = 4W + 2 \cdot N_2 - (A-1) \cdot (A-2)$$

where N_2 is the → *connection number* and A is the number of molecule atoms. This relation is only true if the molecular graph $\mathcal{G}$ is a tree.

An **edge-type Schultz index** has been derived from the → *edge adjacency matrix* **E** and the → *edge distance matrix* $^{\mathbf{E}}\mathbf{D}$ [Estrada and Gutman, 1996; Estrada and Rodriguez, 1997]:

$$^{E}MTI = \sum_{b=1}^{B} \left[\left(\mathbf{E} + {}^{\mathbf{E}}\mathbf{D}\right)\mathbf{v}\right]_b$$

where $\mathbf{v}$ in this case is a B-dimensional column vector whose elements are the edge degrees ε.

In the same way an **edge-type Gutman index** can be defined as:

$$^{E}S_G = \sum_{i=1}^{B} \sum_{j=1}^{B} \varepsilon_i \varepsilon_j \cdot \left[{}^{\mathbf{E}}\mathbf{D}\right]_{ij}$$

where ε_i and ε_j are the edge degrees of the two considered edges and $[{}^{\mathbf{E}}\mathbf{D}]_{ij}$ is the topological distance between them.

A generalization of the Schultz molecular topological index to account for heteroatoms and multiple bonds was proposed based on the → *Barysz distance matrix*.

The **3D-Schultz index** ^{3D}MTI is derived from the → *geometry matrix* G as:

$$^{3D}MTI = \sum_{i=1}^{A} \left[\mathbf{A}^{w} + \mathrm{G})\mathbf{v}\right]_i$$

where $\mathbf{A}^{w}$ is a weighted adjacency matrix (→ *weighted matrices*) whose entries corresponding to bonded atoms are → *bond distances* and $\mathbf{v}$ is an A-dimensional column vector constituted by the → *vertex degree* δ of the A atoms in the H-depleted molecular graph [Mihalic *et al.*, 1992a]. Analogously, the **3D-MTI′ index** is defined as:

$$^{3D}MTI' = \sum_{i=1}^{A} \sum_{j=1}^{A} [\mathbf{A} \cdot \mathrm{G}]_{ij} = \sum_{i=1}^{A} \left[\mathbf{v}^{\mathrm{T}} \cdot \mathrm{G}\right]_i$$

where $\mathbf{v}^{\mathrm{T}}$ is the transposed vector $\mathbf{v}$ defined above.

By analogy with the Schultz molecular topological index, → *Schultz-type indices* were also proposed.

[Juric *et al.*, 1992] [Klavzar and Gutman, 1996] [Gutman and Klavzar, 1997]

Schultz-type indices

Defined in analogy with the → *Schultz molecular topological index MTI* are → *molecular descriptors* based on a product of square $A \times A$ matrices, the → *adjacency matrix* **A** being obligatory [Diudea *et al.*, 1997b; Diudea and Randic, 1998; Diudea and Gutman, 1998].

The general formula of Schultz-type indices is the following:

$$MTI_{(\mathbf{M_1},\mathbf{A},\mathbf{M_2})} = \mathbf{u}^{\mathrm{T}} \cdot [\mathbf{M_1} \cdot (\mathbf{A} + \mathbf{M_2})] \cdot \mathbf{u} = \mathbf{u}^{\mathrm{T}} \cdot [\mathbf{M_1 A} + \mathbf{M_1 M_2}] \cdot \mathbf{u} = S(\mathbf{M_1 A}) + S(\mathbf{M_1 M_2})$$

where **u** is a unit column vector of size A, A being the number of atoms; $\mathbf{M_1}$ and $\mathbf{M_2}$ are two generic square $A \times A$ matrices, and S is the → *total sum operator*.

Selecting different combinations of $\mathbf{M_1}$ and $\mathbf{M_2}$ matrices leads to the derivation of several Schultz-type indices. The original Schultz molecular topological index MTI is obtained for $\mathbf{M_1} = \mathbf{A}$ and $\mathbf{M_2} = \mathbf{D}$, where **D** is the topological → *distance matrix*. Typical Schultz indices are derived from (**D**, **A**, **D**), ($\mathbf{D}^{-1}$, **A**, $\mathbf{D}^{-1}$), (**W**, **A**, **D**), ($\mathbf{W}^{-1}$, **A**, $\mathbf{D}^{-1}$), (**W**, **A**, **W**), ($\mathbf{CJ_u}$, **A**, $\mathbf{CJ_u}$), ($\mathbf{SZ_u}$, **A**, $\mathbf{SZ_u}$), where $\mathbf{D}^{-1}$ is the → *reciprocal distance matrix*, **W** is one among → *walk matrices*, $\mathbf{W}^{-1}$ is the → *reciprocal walk matrix*, and $\mathbf{CJ_u}$ and $\mathbf{SZ_u}$ the unsymmetric Cluj and Szeged matrices, respectively.

Schultz-type indices based on eigenvectors were recently proposed [Medeleanu and Balaban, 1998] as → *local vertex invariants* and molecular descriptors, on the basis of the eigenvector of adjacency and distance matrices associated with the lowest (largest negative) eigenvalue. The LOVIs are derived from the following A-dimensional column vectors:

$$\mathbf{V1} = (\mathbf{A} + \mathbf{D}) \cdot \mathbf{VA} \qquad \mathbf{V2} = (\mathbf{A} + \mathbf{D}) \cdot \mathbf{VD}$$

$$\mathbf{V3} = \mathbf{A} \cdot \mathbf{VA} \qquad \mathbf{V4} = \mathbf{A} \cdot \mathbf{VD} \qquad \mathbf{V5} = \mathbf{D} \cdot \mathbf{VA} \qquad \mathbf{V6} = \mathbf{D} \cdot \mathbf{VD}$$

where **A** is the adjacency matrix, **D** is the distance matrix, **VA** is the eigenvector of the adjacency matrix and **VD** the eigenvector of the distance matrix. **V1** – **V6** are vectors containing different local vertex invariants; the highest values of **V1**, **V2**, **V5**, and **V6** LOVIs correspond to vertices of lower degree and further from the graph centre, **V3** LOVIs show the opposite trend as do the coefficients of the eigenvector **VA**, the variation of **V4** LOVIs is the least regular.

From these sets of LOVIs two different types of topological index were derived:

$$\mathrm{XMTn} = \sum_{i=1}^{A} [\mathbf{Vn}]_i$$

$$\mathrm{XMTnR} = \sum_{b=1}^{B} \left([\mathbf{Vn}]_i \cdot [\mathbf{Vn}]_j \right)_b^{-1/2}$$

where the first sum is over all of the LOVIs in each **Vn** vector, **Vn** being equal to **V1** – **V6**. The second index is based on a Randic-type formula, where the sum runs over the bonds in the graph, B being the total number of bonds, and the product is between the LOVIs of the two vertices incident to the considered bond.

[Diudea, 1995a] [Diudea and Pop, 1996] [Diudea and Randic, 1997]

Schultz-type indices based on eigenvectors → **Schultz-type indices**

Schultz weighted distance matrices → **weighted matrices**

score matrix → **principal component analysis**

secondary mesomeric effects → **electronic substituent constants**

second-grade structural parameters → **multiple bond descriptors**

second-order submolecular polarity parameter → **charge descriptors** (⊙ submolecular polarity parameter)

selective PLS → **variable selection** (⊙ intermediate least squares regression)

self-atom polarizability → **electric polarization descriptors** (⊙ atom-atom polarizability)

self-avoiding walk : *path* → **graph**

self-returning ID number → **ID numbers** (⊙ weighted ID number)

self-returning walk atomic code → **self-returning walk counts**

self-returning walk → **graph**

self-returning walk counts

A particular case of → *walk counts*, self-returning walk counts are atomic and molecular descriptors obtained from an → *H-depleted molecular graph* $\mathcal{G}$, based on graph walks starting and ending at the same vertex, i.e. self-returning walks (*SRWs*) [Harary, 1969a].

The length k of a walk is the total number of edges that are traversed, repeated use of the same edge or edges being allowed.

The **atomic self-returning walk count** of kth order, denoted by $srw_i^{(k)}$, is the number of walks of length k starting and ending at the ith vertex. It is easily obtained by the kth power of the → *adjacency matrix* $\mathbf{A}$; in fact each diagonal element of $\mathbf{A}^k$ can be interpreted as the sum of all self-returning walks of length k for a given vertex:

$$srw_i^{(k)} = \left[\mathbf{A}^k\right]_{ii}$$

The **molecular self-returning walk count** of kth order is the total number of self-returning walks of length k in the graph and is simply calculated by summing all of the atomic self-returning walk counts of the same order:

$$srw^{(k)} = \sum_{i=1}^{A} srw_i^{(k)} = tr\left(\mathbf{A}^k\right)$$

where A is the number of vertices in the graph and tr is the → *trace* of the kth power of the adjacency matrix, i.e. the sum of the diagonal elements. From the above relation it can be derived that molecular self-returning walk counts are the **spectral moments of the adjacency matrix**, which were also expressed as linear combinations of counts of certain fragments contained in the molecular graph, i.e. → *embedding frequencies* [Barysz and Trinajstic, 1984; Jiang *et al.*, 1984; Kiang and Tang, 1986; Jiang and Zhang, 1989; Jiang and Zhang, 1990; Markovic and Gutman, 1991; Jiang *et al.*, 1995; Markovic and Stajkovic, 1997; Markovic, 1999].

It should be noted that the self-returning walk count of order 2 for the ith atom coincides with its → *vertex degree* δ_i, i.e. the number of bonded atoms; moreover, the molecular self-returning walk counts of 1st and 2nd order coincide with the number of atoms A and twice the number of bonds B in the molecule, respectively.

For the ith atom of a molecule, the sequence of atomic self-returning walk counts of increasing length up to A defines the **self-returning walk atomic code** ($SRWAC$):

$$\left\langle srw_i^{(1)},\ srw_i^{(2)},\ srw_i^{(3)},\ \ldots,\ srw_i^{(A)} \right\rangle$$

These vector descriptors characterize the atoms in the molecule [Randic, 1980c]. It can be noted that for acyclic graphs only the counts of SRWs of even length are different from zero; for any graph, the first term in the code is equal to one.

Vertices in a molecular graph having the same $SRWAC$ are called **endospectral vertices** and the corresponding graph **endospectral graph**. Attaching any subgraph to each endospectral vertex one at a time generates → *isospectral graphs*, i.e. graphs with the same → *characteristic polynomial*.

Graphs with identical $SRWAC$s for all the atoms are → *isocodal graphs* [Ivanciuc and Balaban, 1996a].

By summing all the entries of the self-returning walk atomic code, a different local invariant called **vertex structural code** (SC, or **structural code**) is obtained:

$$SC_i = \sum_{k=1}^{A} srw_i^{(k)}$$

where k is the increasing length of the walks.

Based on these local invariants, the **ordered structural code** (OSC) is a molecular descriptor defined as the ascending ordered sequence of SC_i in the molecule [Barysz and Trinajstic, 1984]:

$$OSC = \left\langle SC_{i(1)},\ SC_{i(2)},\ SC_{i(3)},\ \ldots,\ SC_{i(A)} \right\rangle$$

where A is the number of atoms in the molecule and the numbers in parenthesis represent the ordered sequence. The OSC vectors of different sized molecules are transformed into → *uniform length descriptors* by selecting a dimension L equal to the maximum number of the atoms in the largest molecule in the considered data set and by adding zero to the empty positions.

Weighted atomic self-returning walk counts of any order k, denoted by ${}^{w}srw_i^{(k)}$, were proposed [Bonchev and Kier, 1992] by summing the weights of all self-returning walks SRW^k of a given vertex:

$${}^{w}srw_i^{(k)} = \sum\nolimits_j w\left(SRW_i^k\right)_j$$

where w is the weight of the jth self-returning walk of length k for the ith vertex, which can be calculated as:

$$w\left(SRW_i^k\right)_j = 2 \cdot \sum\nolimits_{b \in j} \left(\delta_{b(1)} \cdot \delta_{b(2)}\right)^{1/2} \qquad \text{or} \qquad w\left(SRW_i^k\right)_j = \prod\nolimits_{b \in j} \left(\delta_{b(1)} \cdot \delta_{b(2)}\right)$$

The first weight is obtained by summing the product of the vertex degrees δ of all adjacent vertices involved in the considered walk; the sum is multiplied by 2, each edge being traversed twice. The second self-returning walk weight is calculated by multiplying the weights of all edges involved in the walk, each edge weight being the product of the vertex degrees of the incident vertices. These local vertex invariants were used for the → *canonical numbering* of graph vertices. Other weighting schemes for self-returning walks have also been proposed to improve QSAR relationships [Bonchev *et al.*, 1993b].

Self-returning walk counts are indices derived from molecular topology that are closely related to the moments of energy derived from quantum chemistry [Bonchev and Kier, 1992; Bonchev *et al.*, 1993b; Bonchev and Gordeeva, 1995; Gutman *et al.*, 1995], as is explained below.

Molecular moment of energy of the kth order is defined as:

$$\mu^k = \sum_i \mathrm{E}_i^k = tr\left(\mathcal{H}^k\right)$$

where E_i are the energy levels constituting the discrete spectrum of a molecule, $\mathcal{H}$ is the Hamiltonian matrix and tr the → *trace* of the matrix. Moreover, the trace of the kth power of the Hamiltonian matrix is equal to the count of the weighted molecular self-returning walks $SRW^{(k)}$ of order k, beginning and ending at the same orbital. The weights associated with the walk are the interaction integrals $\mathrm{H}_{ia,ib}$ involving the overlapping orbitals ia and ib. The simplest weighting results from the one-electron Hückel method in which all $\mathrm{H}_{ia,ib} = \beta$ (the resonance integral) if ia and ib are p-orbitals located on atoms of the π-bonded network. For such systems the **atomic moments of energy** μ_i^k are defined as the kth molecular moment of the ith orbital, i.e. the number of walks that start at this orbital and return to it in k steps, traversing one bond in each step. From these local vertex invariants, the corresponding molecular descriptor is derived:

$$\mu^k = \sum_{i=1}^{A} \mu_i^k = \sum_{i=1}^{A} srw_i^{(k)} \cdot \beta^k = srw^{(k)} \cdot \beta^k$$

where $srw^{(k)}$ is the total number of self-returning walks of length k in the molecule.

To avoid rapidly increasing numbers with molecule dimension and branching, **relative atomic moments** (RAMs) are defined, normalizing each atomic moment as:

$$f_i^k = \frac{\mu_i^k}{\mu^k} = \frac{srw_i^{(k)}}{srw^{(k)}}$$

i.e. dividing each atomic term by the corresponding kth order molecular term.

It was demonstrated that *RAMs* always reach a limit with increasing power k, and this limit is numerically equal to the partial charge of the *Lowest Occupied Molecular Orbital* (LOMO), which is the most stable orbital:

$$TAC_i \equiv f_i = \lim_{k\to\infty}\left(f_i^k\right) = \lim_{k\to\infty}\left(\frac{srw_i^{(k)}}{srw^{(k)}}\right) = c_{i,\mathrm{LOMO}}^2$$

where $c_{i,\mathrm{LOMO}}^2$ is the squared coefficient of the first molecular orbital of the ith atom. The second equality has not been demonstrated, but is believed to have general validity.

This limit is a fractional topological charge f_i, called the **topological atomic charge** (TAC), and represents the relative occurrence of the SRWs of the ith atom to all SRWs in the molecule; it can be considered as the fractional topological charge of an atom, assuming that each SRW is associated with the movement of an electron near the nucleus of the considered atom; the larger the number of SRWs of an atom the larger the topological charge ascribable to that atom, i.e. the measure of the time an electron moves near the atom.

Topological atomic charges provide a canonical numbering of graph vertices, each vertex being ordered according to its branching, centrality and cyclicity.

Moreover, **topological atomic valencies** (TAVs, or **corrected second moments**) were derived from fractional atomic charges f_i by rescaling the values with respect to the 2nd-order molecular self-returning walk count srw^2:

$$TAV_i = f_i \cdot srw^2$$

where srw^2 represents the total number of σ-electrons taking part in the molecular graph.

Topological valence is a real local invariant close to the chemical valence of the atom. TAVs can be interpreted as atomic valencies corrected by accounting for all higher order atom-atom connectivities; atoms with low TAV values are regarded as possessing large free valence.

Analogous to the topological atomic charge definition is that of **topological bond orders**, defined as the limit of

$$p_{ij}^{(k)} = \frac{\mu_i^k + \mu_j^k}{\mu^k} = \frac{srw_i^{(k)} + srw_j^{(k)}}{srw^{(k)}}$$

i.e. as

$$p_{ij} = \lim_{k \to \infty} \left(\frac{srw_i^{(k)} + srw_j^{(k)}}{srw^{(k)}} \right)$$

This definition of → *bond order* can be regarded as an extension of the occurrence of the double bonds in all the Kekulé structures to the occurrence of a bond in all self-returning walks [Bonchev and Gordeeva, 1995].

The topological bond order thus defined can be interpreted as the one-electron distribution over the bonds in a molecule, analogous to the one-electron distribution over all atoms given by the topological atomic charges:

$$\sum\nolimits_a f_a = \sum\nolimits_b (p_{ij})_b = 1$$

where the first sum runs over the A atoms and the second over the B bonds.

📖 [Barysz *et al.*, 1986a] [Shalabi, 1991] [Bonchev and Seitz, 1995]

self-returning walk ordering → **canonical numbering**

self-returning walk-sequence matrix → **sequence matrices**

self-returning ID number → **ID numbers**

semiempirical molecular connectivity terms → **combined descriptors**

sequence matrices (SM)

A sequence matrix **SM** of a → *graph* $\mathcal{G}$ is a rectangular unsymmetric matrix $A \times K$, whose entry ik (sm_{ik}) is the number of walks of increasing length k (k = 0, 1, ..., K) starting from the ith vertex to all the other $A - 1$ vertices. K is the maximum length of the walk and depends on the type of considered → *walk* [Diudea, 1994]. The label m denotes the type of walk, d being for the shortest path (topological distance), p for path, w for random walk, and srw for self-returning walk. The maximum length K of the specified walk is the → *topological diameter D* if shortest paths between vertices are considered, the maximum → *path eccentricity* in the molecule if the paths are considered, and an arbitrary chosen number between one and infinite (however usually limited to $A - 1$ according to the Cayley-Hamilton theorem) if walks or self-returning walks are considered.

Therefore, different sequence matrices can be obtained using different types of walks. Note that the sequence matrix based on topological distances d, here called **distance-sequence matrix SD**, coincides with the → *cardinality layer matrix* **LC** and has also been proposed by other authors as the → *λ matrix* [Skorobogatov and Dobrynin, 1988] and → *F matrix* [Diudea and Pârv, 1988]. Moreover, the **path-sequence matrix SP** based on paths p was previously proposed as the **τ matrix** [Skorobogatov and Dobrynin, 1988].

The entries of the distance-sequence matrix **SD** are simply the → *vertex distance counts* ${}^1f_i$, ${}^2f_i$, ${}^3f_i$, ... ${}^Df_i$, i.e. the frequencies of distances equal to 1, 2, 3, ..., respectively, from vertex v_i to any other vertex, while each entry in the path-sequence matrix **SP** is the → *atomic path count* ${}^kP_i$, the → *atomic walk count* $awc_i^{(k)}$ in the **walk-**

sequence matrix SW, and the → *atomic self-returning walk count* $swc_i^{(k)}$ in the **self-returning walk-sequence matrix SSRW**. Note that for $k = 0$, the entries sm_{i0} are all equal to one independently of the kind of walk.

Applying the → *row sum operator* to the sequence matrices, local vertex invariants, **atomic sequence count** asc_i, are obtained as:

$$asc_i = \sum_{k=1}^{K} sm_{ik}$$

where asc_i coincides with the → *atomic path count sum* P_i in the path-sequence matrix **SP**, the → *atomic walk count sum* $awcs_i$ i.e. the total number of walks of any length starting from v_i, in the walk-sequence matrix **SW**, and the → *vertex structural code* SC_i in the self-returning walk-sequence matrix **SSRW**.

Example : 2-methylpentane

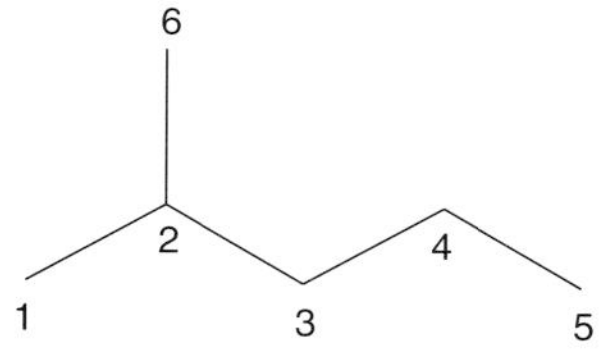

$$TSC_D \equiv TSC_P = \sum_{k=0}^{4} msc_k = 6 + \frac{1}{2} \cdot \sum_{i=1}^{6} asc_i = 21$$

Distance-sequence matrix **SD** ≡ path-sequence matrix **SP**

	Length (k)					
Atom (i)	0	1	2	3	4	asc_i
1	1	1	2	1	1	5
2	1	3	1	1	0	5
3	1	2	3	0	0	5
4	1	2	1	2	0	5
5	1	1	1	1	2	5
6	1	1	2	1	1	5
msc_k	6	5	5	3	2	21

Walk-sequence matrix **SW**

	Length (k)						
Atom (i)	0	1	2	3	4	5	asc_i
1	1	1	3	4	11	15	34
2	1	3	4	11	15	40	73
3	1	2	5	7	18	25	57
4	1	2	3	7	10	25	47
5	1	1	2	3	7	10	23
6	1	1	3	4	11	15	34
msc_k	6	5	10	18	36	65	140

$$TSC_W = \sum_{k=0}^{5} msc_k = 6 + \frac{1}{2} \cdot \sum_{i=1}^{6} asc_i = 140$$

Self-returning walk-sequence matrix **SSRW**

	Length (k)						
Atom (i)	0	1	2	3	4	5	asc_i
1	1	0	1	0	3	0	4
2	1	0	3	0	10	0	13
3	1	0	2	0	7	0	9
4	1	0	2	0	5	0	7
5	1	0	1	0	2	0	3
6	1	0	1	0	3	0	4
msc_k	6	0	5	0	15	0	26

$$TSC_{SRW} = \sum_{k=0}^{5} msc_k = 6 + \frac{1}{2} \cdot \sum_{i=1}^{6} asc_i = 26$$

Box S-1.

Applying the → *column sum operator* (divided by two) to any sequence matrix **SM**, a global index for each k value different from zero, called the **molecular sequence count** msc_k, is obtained:

$$msc_k = \frac{1}{2} \cdot C_k(\mathbf{SM}) = \frac{1}{2} \cdot \sum_{i=1}^{A} sm_{ik}$$

where A is the → *atom number*. In particular, the msc_k is the → *graph distance count* $^k f$ for the distance-sequence matrix **SD**, the → *molecular path count* $^k P$ for the path-sequence matrix **SP**, the → *molecular walk count* $mwc^{(k)}$ for the walk-sequence matrix **SW**, and the → *molecular self-returning walk count* $srw^{(k)}$ for the self-returning walk-sequence matrix **SSRW**.

The set of molecular sequence counts constitutes a molecular vector descriptor, called a **molecular sequence code**:

$$\langle msc_0, msc_1, msc_2, \ldots, msc_K \rangle$$

where msc_0 is simply the number A of vertices in the graph.

The global index measuring the total number of walks of a specified type and of any length in the graph is the **total sequence count** TSC_M defined as:

$$TSC_M = \sum_{k=0}^{K} msc_k = A + \frac{1}{2} \cdot \sum_{i=1}^{A} asc_i$$

where the subscript M denotes the type of sequence matrix. This index coincides with the → *total path count* P for the path-sequence matrix **SP** and the → *total walk count* TWC for the walk-sequence matrix **SW**.

sequential search → **variable selection**

SE-vectors : *distance-counting descriptors*

shadow indices (: *Jurs shape indices*)

A set of 3D → *geometrical descriptors*, similar to the → *Amoore shape indices*, related to size and shape of molecules. They are calculated by projecting the molecular surface on three mutually perpendicular planes XY, XZ and YZ, assuming van der Waals radii for atoms [Rohrbaugh and Jurs, 1987a; Rohrbaugh and Jurs, 1987b; Jurs *et al.*, 1988].

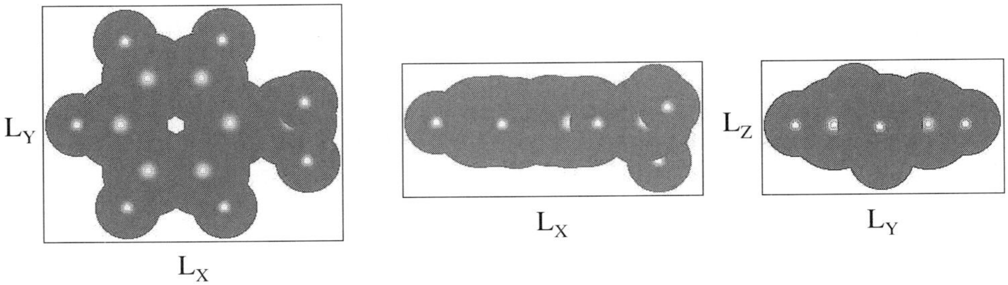

Figure S-1. Projections and embedding rectangles of toluene molecule in the three principal planes.

Basically, a molecule is flattened into a plane by disregarding the third dimension; the area of the molecule which is projected onto the remaining two dimensions defines the shadow area of interest. In order to obtain invariance to rotation of the calculated projections, the X, Y, Z molecule axes are previously aligned along the three → *principal inertia axes.*

The six shadow indices are defined as the following:

- SHDW1: area of the molecular shadow in the XY plane;
- SHDW2: area of the molecular shadow in the XZ plane;
- SHDW3: area of the molecular shadow in the YZ plane;
- SHDW4: standardized molecular shadow area in the XY plane calculated as:

$$\mathrm{SHDW4} = \frac{\mathrm{SHDW1}}{\mathrm{L_X} \cdot \mathrm{L_Y}}$$

where the denominator is the area of the embedding rectangle;

- SHDW5: standardized molecular shadow area in the XZ plane calculated as:

$$\mathrm{SHDW5} = \frac{\mathrm{SHDW2}}{\mathrm{L_X} \cdot \mathrm{L_Z}}$$

where the denominator is the area of the embedding rectangle;

- SHDW6: standardized molecular shadow area in the YZ plane calculated as:

$$\mathrm{SHDW6} = \frac{\mathrm{SHDW3}}{\mathrm{L_Y} \cdot \mathrm{L_Z}}$$

where the denominator is the area of the embedding rectangle.

The lengths $\mathrm{L_X}$, $\mathrm{L_Y}$, and $\mathrm{L_Z}$ are the maximum dimensions of the molecular surface projections.

The ratio of the largest to the smallest dimension of the box built on each molecule shadow can be considered as a shape descriptor very similar to the → *length-to-breadth ratio.*

shaf(D) index → **algebraic operators** (⊙ determinant)

Shannon's entropy : *mean information content* → **information content**

shape descriptors

Molecular shape is related to several physico-chemical processes, such as transport phenomena as well as entropy contributions, and interaction capability between ligand and receptor.

The degree of deviation from the spherical top is called the **anisometry**.

Several shape descriptors are defined within more general approaches to → *molecular descriptors.* This is the case of → *Kier shape descriptors,* → *shape profiles,* → *shadow indices,* → *WHIM shape* descriptors, → *Sterimol shape parameters* $\mathrm{L/B_1}$ and $\mathrm{B_1/B_5}$, molecular → *periphery codes,* and → *centric indices.* Other approaches to the study of molecular surface and shape are → *Mezey 3D shape analysis* and Hopfinger → *molecular shape analysis.* → *Triangular descriptors* have also been used to characterize molecular shape to search for similarities among molecules.

Other specific shape descriptors are listed below.

- **Petitjean shape indices**

A first Petitjean shape index is a topological anisometry descriptor [Petitjean, 1992], also called a **graph-theoretical shape coefficient** I_2, defined as:

$$I_2 = \frac{D - R}{R} \qquad 0 \leq I_2 \leq 1$$

where R and D are the → *topological radius* and the → *topological diameter*, respectively, obtained from the → *distance matrix* representing the considered → *molecular graph*. For strictly cyclic graphs, $D = R$ and $I_2 = 0$.

The **geometrical shape coefficient** I_3 is calculated in the same way but using the information of the → *geometry matrix* G [Bath *et al.*, 1995]:

$$I_3 = \frac{{}^{G}D - {}^{G}R}{{}^{G}R}$$

where ${}^{G}R$ and ${}^{G}D$ are the → *geometric radius* and → *geometric diameter*, respectively.

A **radius-diameter diagram** is defined as a bivariate distribution of the → *data set* compounds in the space defined by the molecular radius and diameter; it provides a summary of the similarities among the molecule chemical shapes in the topological or geometrical space.

- **Kaliszan shape parameter** (η)

Defined as the ratio of the longest to the shortest side of a rectangle having the minimum area that can envelop a molecular structure drawn assuming → *van der Waals radius* for atoms and standard bond lengths [Kaliszan *et al.*, 1979; Radecki *et al.*, 1979; Kaliszan, 1987]. It was used to model GC retention indices. A slightly different shape parameter is the **length-to-breadth ratio** L/B, which is defined as the ratio of the longest to the shortest side of the rectangle that envelops a molecular structure and at the same time maximizes the L/B ratio [Janini *et al.*, 1975; Wise *et al.*, 1981].

In general, the length-to-breadth ratio is the ratio of the longest L to the shortest B side of a rectangle containing some molecular projection, having uniquivocally defined a specific molecular orientation. For example, length-to-breadth ratio for molecular substituents is found among → *STERIMOL parameters.* Moreover, length-to-breadth ratio was calculated from the dimensions of rectangles that envelop the molecule oriented along with the → *principal inertia axes* [Collantes *et al.*, 1996]. In this case, the dimensions of the rectangle enclosing the molecule are calculated using the atomic coordinates on the principal inertia axes.

The ratio between the first and second eigenvalues derived from → *WHIM descriptors* can be used as a shape descriptor related to the length-to-breadth ratio L/B. It is defined as:

$$\mathrm{L/B}_w = \frac{\lambda_{1w}}{\lambda_{2w}}$$

where w is one among the weights defined in the WHIM approach. It may be noted that this shape parameter not only accounts for the distance between extreme atoms along the principal axes but also for the distribution of all atoms around the molecule centre [Authors, This Book].

- **Amoore shape indices**

Defined as the cross-sectional areas of the molecular surface in the inertial planes, i.e. the planes obtained by the → *principal moments of inertia* of the molecule [Amoore, 1964; Meyer, 1986a]. When the three cross-sectional areas are equal, the molecule is a spherical top.

- **inertial shape factor** (S_I)

This is a shape factor based on the → *principal moments of inertia* and defined as:

$$S_I = \frac{I_B}{I_A \cdot I_C}$$

where I are the principal moments of inertia [Lister *et al.*, 1978].

• **molecular eccentricity** (ε)
Among the → *eigenvalue-based descriptors*, molecular eccentricity is a shape descriptor obtained from the eigenvalues λ_i of the → *inertia matrix* I defined as [Arteca, 1991]:

$$\varepsilon = \frac{\left(\lambda_1^2 - \lambda_3^2\right)^{1/2}}{\lambda_1} \qquad 0 \leq \varepsilon \leq 1$$

where $\varepsilon = 0$ corresponds to spherical top molecules and $\varepsilon = 1$ to linear molecules.

It is a shape descriptor defined by analogy with the eccentricity of an ellipse, which is defined:

$$\varepsilon = \frac{(l_M^2 - l_m^2)^{1/2}}{l_M}$$

where l_M and l_m are the lengths of the major and minor elliptical axes, respectively.

• **asphericity** (Ω_A)
An anisometry descriptor which measures the deviation from the spherical shape [Arteca, 1991], calculated from the eigenvalues λ_i of the → *inertia matrix* I and defined as

$$\Omega_A = \frac{1}{2} \cdot \frac{(\lambda_1 - \lambda_2)^2 + (\lambda_1 - \lambda_3)^2 + (\lambda_2 - \lambda_3)^2}{\lambda_1^2 + \lambda_2^2 + \lambda_3^2} \qquad 0 \leq \Omega_A \leq 1$$

where $\Omega_A = 0$ corresponds to spherical top molecules and $\Omega_A = 1$ to linear molecules. For prolate molecules (cigar-shaped), $\lambda_1 \approx \lambda_2 > \lambda_3$ and $\Omega_A \approx 0.25$, whereas for oblate molecules (disk-shaped), $\lambda_1 > \lambda_2 \approx \lambda_3$ and $\Omega_A \approx 1$.

• **spherosity index** (Ω_S)
An anisometry descriptor defined as a function of the eigenvalues, obtained by → *Principal Component Analysis* applied to the correlation matrix calculated from the → *molecular matrix* M:

$$\Omega_S = \frac{3 \cdot \lambda_3}{\lambda_1 + \lambda_2 + \lambda_3} \qquad 0 \leq \Omega_S \leq 1$$

Spherosity index varies from zero for flat molecules, such as benzene, to one for totally spherical molecules [Robinson *et al.*, 1997a].

• **ovality index** (O)
The ovality index O is an anisometry descriptor, i.e. a measure of the departure of a molecule from the spherical shape, based on the property that, for a fixed volume, the spherical shape presents the minimum surface [Bodor *et al.*, 1989; Bodor *et al.*, 1998]. It is calculated from the ratio between the actual molecular surface area SA and the minimum surface area SA_0 corresponding to the actual molecule → *van der Waals volume* V_{VDW}:

$$O = \frac{SA}{SA_0} = \frac{SA}{4\pi r^2} = \frac{SA}{4\pi \cdot \left(\frac{3 \cdot V_{VDW}}{4\pi}\right)^{2/3}} \qquad O \geq 1$$

where r is the molecule radius. The ovality index is equal to 1 for spherical top molecules and increases with increasing linearity of the molecule.

The inverse of the ovality index is the **globularity factor** G ($0 < G \leq 1$) which is between zero and one [Meyer, 1986c]. For molecules with the same volume, the most spherical species have G values approximating to one, and for molecules of non-equal volume, G reflects the relative compactness. When both the effective surface area and volume of the molecule are available, the **surface-volume ratio** $G' = SA / V$ can be used as a descriptor of molecular congestion. More specifically, it was interpreted as a measure of the capability of a compound to adapt its shape to the requirements of an approaching reagent [Meyer, 1988b].

- **Meyer visual descriptor of globularity**

A geometrical shape descriptor based on the radii of three spheres centred at the barycentre of the molecule [Meyer, 1986a]. The first sphere of radius R_1 has volume equal to the → *van der Waals volume*; the second sphere of radius R_2 has volume equal to the → *molecular volume*, and the third sphere of radius R_3 is defined as the sphere embedding the whole molecule. The shape descriptor is then defined as:

$$R_M = \frac{R_3 - R_2}{R_2 - R_1}$$

Apherical top corresponds to small values of the R_M parameter.

- **characteristic ratio** (C_∞)

A descriptor of average shape features of macromolecules, polymers, and proteins, i.e. it can be considered as a measure of the degree of folding, defined as [Arteca, 1991]:

$$C_\infty = \lim_{B \to \infty} C = \frac{\langle R_G^2 \rangle}{B \cdot l^2}$$

where $\langle R_G^2 \rangle$ is the mean square → *radius of gyration* averaged on all of the conformations (or configurations), B the number of bonds, and l the → *Kuhn length*.

The characteristic ratio is also defined for the → *end-to-end distance* r_{ee} as:

$$C'_\infty = \lim_{B \to \infty} C' = \frac{\langle r_{ee}^2 \rangle}{B \cdot l^2}$$

- **path/walk shape indices**

Similar to the invariants derived from the → *distance/distance matrix* **D/D**, **atomic path/walk indices** are defined [Randic, 1999a] for each atom as the ratio between → *atomic path count* ${}^mP_i$ and → *atomic walk count* $awc_i^{(m)}$ of the same length m, i.e.

$$(p/w)_i^m = {}^mP_i / awc_i^{(m)}$$

Whereas the number of paths in a molecule is bounded and determined by the molecule diameter, the number of walks is unbounded. However, being interested only in quotients, the walk count is terminated when it exceeds the maximum allowed length of the corresponding path.

Molecular path/walk indices are defined as the average sum of atomic path/walk indices of equal length:

$$(p/w)^m = \frac{1}{A} \cdot \sum_{i=1}^{A} (p/w)_i^m$$

Alternatively, they are obtained by separately summing all the paths and walks of the same length, and then calculating the ratio between their counts:

$$(P/W)^m = \frac{1}{A} \cdot \frac{\sum_{i=1}^{A} {}^mP_i}{\sum_{i=1}^{A} awc_i^{(m)}} = \frac{1}{A} \cdot \frac{{}^mP}{mwc^{(m)}}$$

where ${}^mP$ and $mwc^{(m)}$ are the → *molecular path count* and → *molecular walk count* of the mth order, respectively.

It should be noted that the counts of the paths and walks of length one are equal and, therefore, the corresponding molecular indices always equal one for all molecules.

As the path/walk count ratio is independent of molecular size, these descriptors can be considered as shape descriptors. Both the proposed indices can be transformed into → *uniform length descriptors* by fixing the maximum length (for example, $m = 5$):

$$\left\langle (p/w)^2,\ (p/w)^3,\ (p/w)^4,\ (p/w)^5 \right\rangle \qquad \left\langle (P/W)^2,\ (P/W)^3,\ (P/W)^4,\ (P/W)^5 \right\rangle$$

• **shape factor** (E_T)

This index is defined as:

$$E_T = \frac{\sum_k n_k \cdot (k+1)}{L}$$

where n_k is the number of vertices at a topological distance equal to k from any vertex belonging to the longest path in the → *H-depleted molecular graph*; L is the length, i.e. the number of edges, of the longest path [Gálvez *et al.*, 1995b; Gozalbes *et al.*, 1999]. The lower the E_T value the more elongated the graph; the E_T represents the eccentricity of the graph if it is compared to an ellipse.

The **surface factor** S is calculated for cyclic compounds as:

$$S = \sum_f \left(E_T \cdot L^2\right)_f$$

where the sum runs over all the rings and aliphatic fragments in the molecule.

Note that the double bonds are considered simple with respect to their contribution to the surface factor.

Table S-1. Shape indices for some selected compounds.

Compound	I_2	I_3	$(p/w)^2$	$(p/w)^3$	$(p/w)^4$	$(p/w)^5$	Ω_S	Ω_A	$^1\kappa_\alpha$	$^2\kappa_\alpha$	*Ku*	*Km*
Ethane	0.000	0.424	0.000	0.000	0.000	0.000	0.341	0.435	2.000	1.000	0.384	0.659
Propane	1.000	0.989	0.333	0.000	0.000	0.000	0.202	0.411	3.000	2.000	0.426	0.634
n-Butane	0.500	0.610	0.417	0.167	0.000	0.000	0.127	0.617	4.000	3.000	0.589	0.784
n-Pentame	1.000	0.968	0.433	0.200	0.067	0.000	0.087	0.702	5.000	4.000	0.670	0.837
n-Hexane	0.667	0.731	0.444	0.214	0.089	0.033	0.063	0.785	6.000	5.000	0.747	0.886
iso-Butane	1.000	0.987	0.500	0.000	0.000	0.000	0.230	0.148	4.000	1.333	0.280	0.385
Neopentane	1.000	0.985	0.600	0.000	0.000	0.000	1.000	0.000	5.000	1.000	0.000	0.000
cyclo-Propane	0.000	0.402	0.500	0.000	0.000	0.000	0.348	0.106	1.333	1.000	0.115	0.326
cyclo-Butane	0.000	0.310	0.500	0.250	0.000	0.000	0.226	0.150	2.250	0.750	0.215	0.387
cyclo-Pentane	0.000	0.269	0.500	0.250	0.125	0.000	0.227	0.149	3.200	1.440	0.249	0.387
cyclo-Hexane	0.000	0.266	0.500	0.250	0.125	0.063	0.207	0.157	4.167	2.222	0.274	0.396
Benzene	0.000	0.283	0.500	0.250	0.125	0.063	0.000	0.250	3.412	1.606	0.500	0.500
Toluene	0.333	0.530	0.538	0.274	0.126	0.062	0.017	0.353	4.382	1.784	0.483	0.535
Phenol	0.333	0.497	0.538	0.274	0.126	0.062	0.000	0.361	4.344	1.757	0.500	0.538
Benzoic acid	0.667	0.785	0.556	0.303	0.139	0.066	0.000	0.463	5.984	2.421	0.531	0.650
Naphthalene	0.667	0.847	0.554	0.326	0.182	0.095	0.000	0.392	5.483	2.144	0.500	0.576
Anthracene	0.750	0.798	0.573	0.340	0.195	0.114	0.000	0.571	7.573	2.846	0.646	0.741

I_2 and I_3: Petijean shape indices; $(p/w)^m$: molecular path/walk indices; Ω_S: spherosity index; Ω_A: asphericity; $^1\kappa_\alpha$ and $^2\kappa_\alpha$: Kier alpha-modified shape descriptors; *Ku* and *Km*: WHIM shape indices.

📖 [Pitzer, 1955a] [Pitzer *et al.*, 1955b] [Woolley, 1978b] [Arteca, 1991] [Mezey, 1993c] [Kuz'min *et al.*, 1995] [Mezey, 1997a] [Randic and Krilov, 1997b] [Mõssner *et al.*, 1999] [Randic and Krilov, 1999]

shape factor → **shape descriptors**

shape group method → **Mezey 3D shape analysis**

shape profiles → **molecular profiles**

shell matrices : *layer matrices*

short hafnian → **algebraic operators** (⊙ determinant)

SIBIS model → **minimal topological difference**

side chain topological index → **amino acid side chain descriptors**

Siegel-Kormany inductive constant → **electronic substituent constants** (⊙ inductive electronic constants)

similarity → **similarity/diversity**

similarity/diversity

The concept of **similarity** and its dual concept of **diversity** play a fundamental role in several QSAR strategies and chemometric methods [Willett, 1987]. By definition, similarity is a binary relationship, i.e. a relationship between two objects.

Similarity searching is a standard tool for → *drug design*, based on the idea that, given a target structure with interesting properties, similar compounds chosen in large databases should have similar properties. Similarity searching involves the specification of the target structure and its characterization by one or more structural descriptors; then, this set of reference structural descriptors is compared with the corresponding sets of descriptors for each of the molecules in the database. A measure of similarity between the target structure and each of the database structures allows a ranking of decreasing similarity with the target for all the molecules. The numerical value of a similarity/diversity measure depends on three main components: a) the description of the objects (e.g. molecular descriptors), b) the weighting scheme of the description elements, and c) the selected similarity index or distance [Rouvray, 1990b; Bath *et al.*, 1994; Klein, 1995; Willett, 1998; Downs and Willett, 1999].

→ *Cluster analysis* methods, → *principal component analysis* and related techniques, and different → *artificial neural networks* (such as Kohonen neural networks) are usually used to search for clusters of similar compounds, where a cluster is constituted by distinct objects that are more similar to each other than to objects outside the group.

Different methods have been developed for similarity searches, some based on topological features of the molecule and others on three-dimensional structures; the latter are also able to account for different conformations for each molecule [Johnson and Maggiora, 1990b]. Moreover, two distinct similarity approaches have been recognized. The direct-comparison methods [Dean and Chau, 1987; Dean *et al.*, 1988] search for molecule similarity on the basis of counts of substructures common to a pair of molecules or, in general, of the best superimposition between two molecules obtained by some → *alignment rules*, such as → *molecular shape similarity descriptors* or the → *maximum common substructure*. The descriptor-based similarity methods search for similarity among molecules using any → *molecular descriptors* calculated independently for each molecule, being often → *substructure descriptors*, → *molecular profiles*, → *BCUT descriptors*, → *SWM signals*, → *EVA descriptors*, → *MoRSE descriptors*, → *autocorrelation descriptors*, and any set of → *uniform length descriptors*; as these methods are very fast they are particularly suitable for searches in large databases [Sheridan *et al.*, 1996; Brown, 1997; Lewis *et al.*, 1997].

Distance measures d_{st} between the objects s and t are scalar values representing the basic measurements of diversity, i.e. $d_{st} = 0$ for identical objects [Sneath and Sokal, 1973; Cuadras, 1989; Frank and Todeschini, 1994; Willett, 1998]. Note that the variables considered in the distance measures have to be in scale comparable, otherwise preliminary scaling procedures must be performed.

For a distance measure to be defined as a metric it must have the following properties:

$$d_{st} \geq 0$$
$$d_{ss} = 0$$
$$d_{st} = d_{ts}$$
$$d_{st} \leq d_{sz} + d_{zt}$$

The most important distance measures are the Euclidean distance and the average Euclidean distance. However, depending on the considered problem, other distance measures can be legitimately used. Some are listed below, where p is the number of real variables and x_{sj} and x_{tj} are the values of the jth element (variable, attribute, descriptor) representing s and t objects, respectively; $\mathbf{x}_s$ and $\mathbf{x}_t$ are the descriptor p-dimensional vectors of the two objects. If the objects are chemical compounds, x_{ij} are the values of the molecular descriptors chosen for their representation, such as → *topological indices*, → *physico-chemical properties*, → *molecular fingerprints*.

– *Euclidean distance*:

$$d_{st} = \sqrt{\sum_{j=1}^{p} \left(x_{sj} - x_{tj}\right)^2}$$

– *average Euclidean distance*:

$$d_{st} = \sqrt{\frac{\sum_{j=1}^{p} \left(x_{sj} - x_{tj}\right)^2}{p}}$$

– *Camberra distance*:

$$d_{st} = \sum_{j=1}^{p} \frac{\left|x_{sj} - x_{tj}\right|}{\left(x_{sj} + x_{tj}\right)} \qquad x > 0$$

– *average Camberra distance*:

$$d_{st} = \frac{1}{p} \cdot \sum_{j=1}^{p} \frac{\left|x_{sj} - x_{tj}\right|}{\left(x_{sj} + x_{tj}\right)} \qquad x > 0; \quad 0 \leq d_{st} \leq 1$$

– *Lagrange distance* (*or Chebyshev distance*):

$$d_{st} = \max_j \left|x_{sj} - x_{tj}\right|$$

– *Lance-Williams distance*: $$d_{st} = \frac{\sum_{j=1}^{p} |x_{sj} - x_{tj}|}{\sum_{j=1}^{p} (x_{sj} + x_{tj})} \qquad x > 0$$

– *Mahalanobis distance*: $$\mathrm{d}_{st} = \sqrt{(\mathbf{x}_s - \mathbf{x}_t)^{\mathrm{T}} \mathbf{S}^{-1} (\mathbf{x}_s - \mathbf{x}_t)}$$ where **S** is the covariance matrix

– *Manhattan distance (or taxi distance)*: $$\mathrm{d}_{st} = \sum_{j=1}^{p} |x_{sj} - x_{tj}| \qquad j = 1, p$$

– *average Manhattan distance*: $$\mathrm{d}_{st} = \frac{1}{p} \sum_{j=1}^{p} |x_{sj} - x_{tj}|$$

– *Minkowski distances*: $$\mathrm{d}_{st} = \left(\sum_{j=1}^{p} |x_{sj} - x_{tj}|^r \right)^{1/r}$$

– *distance between probabilities*:

$$d_{st} = \sum_{j=1}^{p} (x_{sj} - x_{tj}) \cdot \log_2 \left(\frac{x_{sj}}{x_{tj}} \right) \qquad x_{sj}, x_{tj} > 0; \quad \sum_{j=1}^{p} x_{sj} = \sum_{j=1}^{p} x_{tj} = 1$$

Special distance measures must be used for data whose variables are represented by → *binary descriptors*, i.e. variables represented by values either zero or one.

Given two objects s and t, represented by p binary values 0/1, the **binary distance measures** are based on the frequencies arising from the following table:

	$t = 1$	$t = 0$	
$s = 1$	a	b	$a + b$
$s = 0$	c	d	$c + d$
	$a + c$	$b + d$	p

where a, b, c, and d are the frequencies of the events ($s = 1$ and $t = 1$), ($s = 1$ and $t = 0$), ($s = 0$ and $t = 1$), and ($s = 0$ and $t = 0$), respectively, in the pair of binary vectors describing the objects s and t; p is the total number of variables, equal to $a + b + c + d$, which is also the vector length. In other words, a is the number of bits equal to one in both objects (common "presences") and d the number of bits equal to zero in both objects (common "absences"), $a + b$ is the number of bits equal to one in the sth object and $a + c$ the number of bits equal to one in the tth object. Therefore, the diagonal entries a and d give information about the similarity between the two vectors, while the entries b and c give information about their dissimilarity.

Several measures on binary variables provide **association coefficients** a_{st} (i.e. similarity measures between two binary representations), from which the corresponding distance measures are easily calculated by applying the transformation:

$$\mathrm{d}_{st} = \text{constant} - \mathrm{a}_{st}$$

The most popular measures are Hamming and Tanimoto coefficients, which are listed below, together with other important distance measures on binary variables:

– *Hamming coefficient*: $\mathrm{d}_{st} = b + c$ $\quad 0 \leq \mathrm{d}_{st} \leq p$

– *squared Hamming coefficient*: $\mathrm{d}_{st} = \sqrt{b + c}$ $\quad 0 \leq \mathrm{d}_{st} \leq p$

– *Tanimoto coefficient*: $\mathrm{d}_{st} = \dfrac{b + c}{a + b + c + d}$ $\quad 0 \leq \mathrm{d}_{st} \leq 1$

– *squared Tanimoto coefficient*: $\mathrm{d}_{st} = \sqrt{\dfrac{b + c}{a + b + c + d}}$ $\quad 0 \leq \mathrm{d}_{st} \leq 1$

– *simple matching coefficient*: $\mathrm{a}_{st} = \dfrac{a + d}{a + b + c + d}$ $\quad 0 \leq \mathrm{a}_{st} \leq 1$

– *Jaccard coefficient*: $\mathrm{a}_{st} = \dfrac{a}{a + b + c}$

– *Dice coefficient*: $\mathrm{a}_{st} = \dfrac{2a}{2a + b + c}$

– *Cosine coefficient*: $\mathrm{a}_{st} = \dfrac{a}{\sqrt{(a + b) \cdot (a + c)}}$

A general expression of association measure can be given as [Tversky, 1977]:

$$\mathrm{a}_{st} = \frac{a}{\alpha \cdot b + \beta \cdot c + a}$$

where α and β are user-defined parameters. In particular, equal values of α and β provide symmetrical similarity indices, such as the Jaccard coefficient when $\alpha = \beta = 1$ and the Dice coefficient when $\alpha = \beta = 1/2$; different values of α and β provide asymmetrical similarity indices, i.e. when $\alpha = 1$ and $\beta = 0$ yields $\mathrm{a}_{st} = a \,/\, (a + b)$, which can be interpreted as the fraction of object *s* which is in common with object *t*.

When the objects of the data set are ranked, measures of distance can also be applied on the ranks r_{sj} and r_{tj}, representing the ranks of the *j*th variable of the object *s* and *t*, respectively. The most important distance measures on ranked data are listed below:

– *Mahalanobis-like distance*: $\mathrm{d}_{st} = 2 \cdot \sum_{j=1}^{p} \dfrac{\left(\mathrm{r}_{sj} - \mathrm{r}_{tj}\right)^2}{\left(\mathrm{r}_{sj} + \mathrm{r}_{tj}\right)}$

– *rank distance*: $\mathrm{d}_{st} = \sum_{j=1}^{p} \dfrac{\left(\mathrm{r}_{sj} - \mathrm{r}_{tj}\right)^2}{s_j^2}$ where s_j^2 is the variance of the variable *j*.

Distance measures between two sets $\mathcal{K}_s$ and $\mathcal{K}_t$ can be defined by the following expressions [Skvortsova *et al.*, 1998]:

$$d_{st} = |\mathcal{K}_s| + |\mathcal{K}_t| - 2 \cdot |\mathcal{K}_s \cap \mathcal{K}_t| \quad \text{or} \quad \mathrm{d}_{st} = \left(1 - \frac{|\mathcal{K}_s \cap \mathcal{K}_t|^2}{|\mathcal{K}_s| \cdot |\mathcal{K}_t|}\right) \Big/ \left(|\mathcal{K}_s| + |\mathcal{K}_t|\right)$$

where $|\mathcal{K}|$ denotes the total number of elements in the set $\mathcal{K}$, i.e. its cardinality, and $|\mathcal{K}_s \cap \mathcal{K}_t|$ is the cardinality of the intersection of the two sets, i.e. the number of common elements. For example, if the sets $\mathcal{K}$ contained structural fragments of molecular graph, then their intersection would be their → *maximum common substructure* MCS. Note that the first measure corresponds to the Hamming distance.

The pairwise distances calculated on the → *data matrix* $\mathbf{X}(n \times p)$ can be arranged into a matrix, called **data distance matrix** or **diversity matrix** $(n \times n)$, in which both rows and columns correspond to objects.

Note. This matrix is usually called the *distance matrix*. In this book, the matrix is called the *data distance matrix* to avoid confusion with the topological vertex → *distance matrix*, which contains the topological distances d_{ij} between the vertices of a molecular graph. The distances between pairs of objects of the data distance matrix will be indicated by d_{st}.

Similarity indices are quantitative measures of the comparison of two objects. In the case of chemical compounds similarity indices quantify the degree of structural resemblance between their structural representations. A similarity index s_{st} calculated for objects s and t has the following properties:

$$0 \le s_{st} \le 1$$

$$s_{ss} = 1$$

$$s_{st} = s_{ts}$$

where $s_{st} = 1$ represents identical objects and $s_{st} = 0$ represents the maximum dissimilarity.

The simplest similarity measures s_{st} between two objects s and t are obtained from the distance measures, based on the two following similarity functions:

$$s_{st} = \frac{1}{1 + \mathrm{d}_{st}} \qquad s_{st} = 1 - \frac{\mathrm{d}_{st}}{\mathrm{d}^{\max}}$$

where d_{st} is any distance measure between s and t and $\mathrm{d}^{\max}$ is the maximum distance among the pairs of data set objects. Both measures range between zero and one, but the first is independent of the objects of the data set different from s and t, while the latter measures a relative similarity with respect to the pair of objects having the maximum distance, i.e. having a similarity equal to zero. A special case of similarity matrix is the *correlation matrix*. Moreover, other direct similarity indices on continuous variables are the following:

– *Jaccard coefficient*:

$$s_{st} = \frac{\sum_{j=1}^{p} x_{sj} \cdot x_{tj}}{\sum_{j=1}^{p} x_{sj}^2 + \sum_{j=1}^{p} x_{tj}^2 - \sum_{j=1}^{p} x_{sj} \cdot x_{tj}}$$

– *Dice coefficient*:

$$s_{st} = \frac{2 \cdot \sum_{j=1}^{p} x_{sj} \cdot x_{tj}}{\sum_{j=1}^{p} x_{sj}^2 + \sum_{j=1}^{p} x_{tj}^2}$$

– *Cosine coefficient*:

$$s_{st} = \frac{\sum_{j=1}^{p} x_{sj} \cdot x_{tj}}{\sqrt{\sum_{j=1}^{p} x_{sj}^2 \cdot \sum_{j=1}^{p} x_{tj}^2}}$$

For binary variables, the similarities are given by the association coefficients reported above.

Simple similarity measures between two sets $\mathcal{K}_s$ and $\mathcal{K}_t$ can be defined by the following expressions:

$$s_{st} = \frac{2 \cdot |\mathcal{K}_s \cap \mathcal{K}_t|}{|\mathcal{K}_s| + |\mathcal{K}_t|} \quad \text{and} \quad s_{st} = \frac{|\mathcal{K}_s \cap \mathcal{K}_t|^2}{|\mathcal{K}_s| \cdot |\mathcal{K}_t|}$$

where $|\mathcal{K}|$ denotes the total number of elements in the set $\mathcal{K}$, i.e. its cardinality, and $|\mathcal{K}_s \cap \mathcal{K}_t|$ is the cardinality of the intersection of the two sets, i.e. the number of common elements [Skvortsova *et al.*, 1998; Basak *et al.*, 1994]. Note that the first measure corresponds to the Dice coefficient and the second to the square of the Cosine coefficient.

The pairwise similarity indices calculated from the data matrix $\mathbf{X}(n \times p)$ can be arranged into a matrix, called **similarity matrix** $(n \times n)$, in which rows and columns correspond to objects.

For molecules described by properties distributed in the molecular space such as a lattice, **Molecular Quantum Similarity Measures** (MQSM) were proposed to quantify the comparison between the fields representing two previously superimposed molecules [Carbó and Calabuig, 1992a; Carbó and Calabuig, 1992b; Carbó and Calabuig, 1992c; Besalù *et al.*, 1995; Carbó *et al.*, 1995; Carbó-Dorca and Besalù, 1996; Lobato *et al.*, 1997; Amat *et al.*, 1998; Robert and Carbó-Dorca, 1998a; Robert *et al.*, 1999].

In general, an MQSM between the two molecules s and t is defined as:

$$s_{st} = \iint \varrho_s(\mathbf{r}_1)\, \Omega(\mathbf{r}_1, \mathbf{r}_2)\, \varrho_t(\mathbf{r}_2)\, d\mathbf{r}_1 d\mathbf{r}_2$$

where ϱ_s and ϱ_t are density functions for molecules s and t, $\mathbf{r}_1$ and $\mathbf{r}_2$ are two points in the space, and $\Omega\ (\mathbf{r}_1, \mathbf{r}_2)$ is a definite positive operator. The most commonly encountered operators are:

a) $\Omega\ (\mathbf{r}_1, \mathbf{r}_2) = \delta\ (\mathbf{r}_1 - \mathbf{r}_2)$ — *overlap-like MQSM*;
b) $\Omega\ (\mathbf{r}_1, \mathbf{r}_2) = |\mathbf{r}_1 - \mathbf{r}_2|^{-1}$ — *Coloumb-like MQSM*;
c) $\Omega\ (\mathbf{r}_1, \mathbf{r}_2) = |\mathbf{r}_1 - \mathbf{r}_2|^{-2}$ — *gravitational-like MQSM*;
d) $\Omega\ (\mathbf{r}_1, \mathbf{r}_2) = \varrho_C(\mathbf{r}_1)\ \delta\ (\mathbf{r}_1 - \mathbf{r}_2)$ — *triple-density MQSM*;

where δ is the delta Dirac function.

By different mathematical transformations, **Molecular Quantum Similarity Indices** (MQSI) are derived from molecular quantum similarity measures. They are divided into two main classes: C-class indices, referred to as *correlation-like indices* ranging from 0 (maximum dissimilarity) to 1 (maximum similarity), and D-class indices, referred to as *distance-like indices* ranging from 0 (maximum similarity) to infinity (maximum dissimilarity). C-class indices s_{st} can be transformed into D-class indices d_{st} by the following:

$$\mathrm{d}_{st} = \sqrt{1 - s_{st}^2}$$

It can be noted that distances calculated in this way range from zero to one.

Based on the molecular quantum similarity measures, **Molecular Quantum Self-Similarity Measures** (MQS-SM) were proposed as molecular descriptors where each molecule is compared with itself and all the others, and appropriate Hermitian operators Ω are associated to each molecular property [Ponec *et al.*, 1999].

The most popular MQSI based on the overlap-like MQSM operator are reported below, together with other proposed similarity indices of distributed properties [Carbó and Calabuig, 1990; Good *et al.*, 1993a; Good, 1995a; Constans and Carbó, 1995; Carbó *et al.*, 1996; Good and Richards, 1998]. They are expressed in a form suitable for a discrete molecular space represented by a → *grid* of scalar values; N is the total number of grid points, P_{sk} is the property value (density function ϱ) for the sth molecule in the kth grid point.

- **Carbó similarity index** (C)

Initially proposed to compare molecules in terms of their → *electron density*, it can be applied to compare any structural property between molecules *s* and *t* [Carbó *et al.*, 1980; Carbó and Domingo, 1987]. It is defined as:

$$C_{st} = \frac{\sum_{k=1}^{N} P_{sk} \cdot P_{tk}}{\left(\sum_{k=1}^{N} P_{sk}^2\right)^{1/2} \cdot \left(\sum_{k=1}^{N} P_{tk}^2\right)^{1/2}}$$

The Carbó index is sensitive to the shape of the property distributions, but not to their magnitude.

- **Hodgkin similarity index** (H)

With respect to the Carbó index, it is less sensitive to the shape of the property distribution but more sensitive to its magnitude [Hodgkin and Richards, 1987; Richards and Hodgkin, 1988]. It is defined as:

$$H_{st} = \frac{2 \cdot \sum_{k=1}^{N} P_{sk} \cdot P_{tk}}{\sum_{k=1}^{N} P_{sk}^2 + \sum_{k=1}^{N} P_{tk}^2} = 1 - \frac{d_{st}^2}{\sum_{k=1}^{N} P_{sk}^2 + \sum_{k=1}^{N} P_{tk}^2}$$

where d_{st}^2 is the squared Euclidean distance between the molecules *s* and *t*. This index is particularly important for calculating the → *molecular electrostatic potential* (MEP) and the molecular electric field (MEF) similarity because these properties may be of similar shape for a pair of molecules while their absolute values are significantly different.

- **linear similarity index** (L)

A measure of the similarity between molecules *s* and *t* defined as [Good, 1992]:

$$L_{st} = \frac{1}{N} \cdot \sum_{k=1}^{N} \left(1 - \frac{|P_{sk} - P_{tk}|}{\max(|P_{sk}|, |P_{tk}|)}\right)$$

- **exponential similarity index** (E)

A measure of the similarity between molecules *s* and *t* defined as [Good, 1992]:

$$E_{st} = \frac{1}{N} \cdot \sum_{k=1}^{N} e^{-d_{st,k}}$$

where $d_{st,k}$ is a distance measure between the molecules *s* and *t* at the *k*th grid point, defined as:

$$d_{st,k} = \frac{|P_{sk} - P_{tk}|}{\max(|P_{sk}|, |P_{tk}|)}$$

- **Meyer-Richards similarity index** (S)

A modified version of the Carbó similarity index [Meyer and Richards, 1991], defined as:

$$S_{st} = \frac{N_{st}}{(N_s \cdot N_t)^{1/2}}$$

where N_{st} is the number of grid points falling inside both molecules and N_s and N_t are the total number of grid points falling inside each individual molecule.

- **similarity score** (A_F)

A similarity measure between two molecules *s* and *t* in any relative orientation to each other, used in CoMSIA (→ *comparative molecular similarity analysis*) and defined as [Kearsley and Smith, 1990]:

$$A_F = -\sum_{i=1}^{A_s}\sum_{j=1}^{A_t} w_{ij} \cdot e^{-a \cdot r_{ij}^2}$$

where A_s and A_t are the number of atoms of the two molecules, the parameter a defines the distance dependence, r_{ij} is the interatomic distance between atoms i and j, and w_{ij} is the total contribution due to the property values of the two considered atoms, defined as:

$$w_{ij} = w_E \cdot q_i \cdot q_j + w_S \cdot v_i \cdot v_j + \ldots$$

where w_E , w_S , ... are user-attributed values to give different weights to electrostatic (E), steric (S), hydrophobic and hydrogen-bonding properties; q and v are partial charges and van der Waals volumes of the atoms [Kubinyi *et al.*, 1998d].

- **Tanimoto coefficient** (T)

Derived from the Jaccard coefficient applied to 3D distributed properties in the form:

$$T_{st} = \frac{\sum_{k=1}^{N} P_{sk} \cdot P_{tk}}{\sum_{k=1}^{N} P_{sk}^2 + \sum_{k=1}^{N} P_{tk}^2 - \sum_{k=1}^{N} P_{sk} \cdot P_{tk}}$$

A modified Tanimoto coefficient was also proposed to measure the degree of relatedness of the shapes of two structures [Hahn, 1997].

- **Spearman rank correlation coefficient** (R)

The Spearmann rank correlation coefficient is a widely used statistical index that can also be taken as a similarity measure of 3D distributed properties [Dean, 1990; Manaut *et al.*, 1991]; it is defined as:

$$R_{st} = 1 - \frac{6 \cdot \sum_{k=1}^{N} d_{st,k}^2}{N^3 - N}$$

where $d_{st,k}$ is a distance measure between the objects s and t at the kth grid point.

Similarity matrices can be used to describe a set of molecules in order to search for QSAR models, correlating the biological activity of the molecules with their similarity to each other [Rum and Herndon, 1991; Good *et al.*, 1993a; Kubinyi, 1997]; these matrices are usually known as **molecular similarity matrices** and the corresponding QSAR method as **Quantitative Similarity-Activity Relationships** (QSiAR) [Kubinyi *et al.*, 1998d]. In other words, similarity and diversity measures can be used as input variables for modelling molecular properties or biological activities [Horwell *et al.*, 1995; Benigni *et al.*, 1995; So and Karplus, 1997a; So and Karplus, 1997b].

The procedure consists in transforming the initial **X** block of descriptors, with n compounds and p molecular descriptors, into a similarity or diversity matrix obtaining a $n \times n$ squared symmetric matrix, after the appropriate scaling of the original variables. A regression model is performed using as molecular descriptors the columns $\mathbf{d}_j$ of the distance matrix (*diversity descriptors*), where the column elements d_{ij} represent the distances between each ith object and the jth object. Analogously, molecular descriptors can be defined as the columns $\mathbf{s}_j$ of the similarity matrix (*similarity descriptors*). Regression models are formally in the form:

$$\hat{y}_i = b_0 + \sum_{j=1}^{n} b_j \cdot d_{ij} \quad \text{or} \quad \hat{y}_i = b_0 + \sum_{j=1}^{n} b_j \cdot s_{ij}$$

where b are the regression coefficients.

As the number of variables equals the number of molecules, regression models could be better obtained by using variable selection techniques or methods such as partial least squares or principal component regression. Once the relevant distance descriptors are obtained, the model can be interpreted easily by considering the sign of each model descriptor: for a jth variable with a positive sign, the response increases when a compound is more dissimilar to the jth compound, while, for a jth variable with a negative sign, the response increases when a compound is more similar to the jth compound. The opposite interpretation holds if similarity descriptors are used. In practice, such models give easily interpretable information in terms of similarity (or diversity) of each compound with respect to a set of reference compounds corresponding to the variables which are selected as relevant.

Differential descriptors such as → *steric misfit* can also be considered as similarity/diversity descriptors.

📖 [Varkony *et al.*, 1979] [Barysz *et al.*, 1983b] [Willett and Winterman, 1986] [Herndon and Bertz, 1987] [Dean and Callow, 1987] [Jerman-Blazic *et al.*, 1989] [Johnson, 1989] [Johnson *et al.*, 1990a] [Maggiora and Johnson, 1990] [Rosen, 1990] [Burt *et al.*, 1990b] [Burt and Richards, 1990a] [Lajiness, 1990] [Ponec and Strand, 1990] [Bawden, 1990] [Ugi *et al.*, 1990] [Willett, 1990] [Willett, 1991] [Good *et al.*, 1992] [Carbó and Calabuig, 1992d] [Reynolds *et al.*, 1992a] [Takahashi *et al.*, 1992] [Good *et al.*, 1993b] [Good and Richards, 1993] [Richards, 1993] [Dean and Perkins, 1993] [Stanton *et al.*, 1993] [Serilevy *et al.*, 1994a] [Serilevy *et al.*, 1994b] [Basak and Grunwald, 1994] [Basak and Grunwald, 1995b] [Basak and Grunwald, 1995d] [Richards, 1995] [Martin *et al.*, 1995] [Young and Hawkins, 1995] [Turner *et al.*, 1995] [Walker *et al.*, 1995] [Kearsley *et al.*, 1996] [Basak *et al.*, 1996b] [Montanari *et al.*, 1996] [Chapman, 1996] [Briem and Kuntz, 1996] [Cheng *et al.*, 1996] [Hassan *et al.*, 1996] [Willett, 1997] [Robinson *et al.*, 1997b] [Mason and Pickett, 1997] [Kunz and Radl, 1998] [Dixon and Villar, 1998] [Liu *et al.*, 1998] [Cramer III *et al.*, 1998] [Balaban, 1998] [Johnson *et al.*, 1998] [Flower, 1998] [Kubinyi, 1998a] [Menard *et al.*, 1998a] [Menard *et al.*, 1998b] [Clark and Langton, 1998] [Rarey and Dixon, 1998] [Cho *et al.*, 1998] [Sello, 1998] [Zheng *et al.*, 1998] [Robert and Carbó-Dorca, 1998b] [Tominaga, 1998a] [Young and Hawkins, 1998] [Agrafiotis and Lobanov, 1999] [Filimonov *et al.*, 1999] [Schnur, 1999] [Stanton *et al.*, 1999] [Gillet *et al.*, 1999] [Bayada *et al.*, 1999] [Godden *et al.*, 2000]

similarity indices → **similarity/diversity**

similarity matrix → **similarity/diversity**

similarity score → **similarity/diversity**

similarity searching → **similarity/diversity**

simple graph → **graph**

simple topological index → **vertex degree**

***S* index** → **Schultz molecular topological index**

single evaluation set technique → **validation techniques** (⊙ training/evaluation set splitting)

site-property analysis → **Hansch analysis**

six-position number → **steric descriptors** (⊙ number of atoms in substituent specific positions)

size descriptors

→ *Molecular descriptors* related to the dimensions of the molecule and often calculated from the → *molecular geometry.* Combined with molecular shape information, they are closely related to → *steric descriptors.* The simplest size descriptors are → *atom count*, → *bond count*, → *molecular weight*, and some among the → *volume descriptors* such as → *van der Waals volume.* Other size descriptors are → *Sterimol parameters* and → *WHIM size* descriptors.

Moreover, several → *topological indices* are explicitly or implicitly related to the molecular size, such as, for example, → *Wiener index*, → *Zagreb indices*, → *Lovasz-Pelikan index*, → *connectivity indices.*

Other size descriptors are listed below. Several of them are often used to represent geometrical characteristics of long-chain molecules such as polymers and macromolecules [Volkenstein, 1963; Flory, 1969; Arteca, 1991].

- **span** (R)

A size descriptor defined as the radius of the smallest sphere, centred on the → *centre of mass*, completely enclosing all atoms of a molecule [Volkenstein, 1963]:

$$\mathrm{R} = \max_i(r_i)$$

where r_i is the distance of the ith atom from the centre of mass.

The **average span** descriptor, calculated as the average value of conformational changes and denoted by $\bar{\mathrm{R}}$, is used to describe long chain molecules, such as macromolecules, polymers, and proteins, and is related to the Kuhn length (see below).

- **Kuhn length** (l)

For long chain molecules, the Kuhn length is the mean of the → *bond distances*, i.e.

$$l = \frac{\sum_{b=1}^{B} (r_{ij})_b}{B}$$

where B is the number of bonds and r_{ij} is the bond distance between i and j bonded atoms [Flory, 1969].

It is a size descriptor used for macromolecules, polymers, and proteins. In this case, a useful parameter is also the **contour length** L_C defined as:

$$L_C = B \cdot l$$

Moreover, for large B values, Kuhn length is related to the average span $\bar{\mathrm{R}}$ by the following relationships:

$$\bar{\mathrm{R}} \cong B \cdot l \qquad \bar{\mathrm{R}} = B^{1/2} \cdot l \qquad \bar{\mathrm{R}} = B^{1/3} \cdot l$$

where the first relationship holds when almost all linear conformations are retained, the second when accessible conformations correspond to randomly folded chains, and the third when the most compact (folded) conformations are retained.

- **end-to-end distance** (r_{ee})

A simple size descriptor for long-chain molecules defined as [Flory, 1969]:

$$r_{ee} = r_{1A} = \|\mathbf{r}_1 - \mathbf{r}_A\|$$

where $r_{1\mathrm{A}}$ is the interatomic distance between the first and the last atoms of the chain and $\mathbf{r}$ is the vector of the atom coordinates with respect to the centre of mass.

• **persistence length** (L_P)
A size descriptor adopted for long-chain molecules such as polymers, determined by both geometrical and topological information [Flory, 1969]. Let a linear chain be defined as a sequence of straight line segments (the bonds): the ith bond is a vector with direction $\mathbf{r}_{i+1} - \mathbf{r}_i$ and the positions of the successive bonds relative to the ith bond are projected along the $\mathbf{r}_{i+1} - \mathbf{r}_i$ direction. Then, the persistence length L_P is defined as the conformational (or configurational) average of the sum of these projections, for any ith bond.

Under the assumption that $L_C / L_P >> 1$, persistence length is related to the → *characteristic ratio* C'_∞, the end-to-end distance r_{ee} and to contour length L_C, by the following relationships:

$$L_P = \frac{l \cdot C'_\infty}{2} = \frac{\langle r_{ee}^2 \rangle}{2 \cdot B \cdot l} = \frac{\langle r_{ee}^2 \rangle}{2 \cdot L_C}$$

where B and l are the number of bonds and the Kuhn length, respectively.

• **radius of gyration** (R_G)
A size descriptor for the distribution of atomic masses in a molecule [Tanford, 1961; Volkenstein, 1963], defined as:

$$R_G = \sqrt{\frac{\sum_{i=1}^{A} m_i \cdot r_i^2}{MW}}$$

where r_i is the distance of the ith atom from the centre of mass of the molecule, m_i is the corresponding atomic mass, A the atom number and MW the → *molecular weight*.

The radius of gyration can also be calculated from the → *principal moments of inertia* I; for planar molecules ($I_C = 0$) it is defined as:

$$R_G = \sqrt{\frac{(I_A \cdot I_B)^{1/2}}{MW}}$$

and for non-planar molecules as:

$$R_G = \sqrt{\frac{2\pi \cdot (I_A \cdot I_B \cdot I_C)^{1/3}}{MW}}$$

The radius of gyration is a measure of molecular compactness for long-chain molecules such as polymers, i.e. small values are obtained when most of the atoms are close to the centre of mass. It is also related to the → *characteristic ratio*.

A **size-shape geometrical constant** α_G is derived from the radius of gyration as [Wilding and Rowley, 1986]:

$$\alpha_G = -7.706 \cdot 10^{-4} + 0.033 \cdot R_G + 0.01506 \cdot R_G^2 - 9.997 \cdot 10^{-4} \cdot R_G^3$$

📖 [Arteca, 1996]

• **Meyer anchor sphere volume**
A substituent size descriptor defined as the volume V^a of the portion of the substituent within a sphere centred at the link atom [Meyer, 1986b]. The radius of the sphere was chosen equal to 0.3 nm to comprise the substituent portion responsible for the steric effect of the substituent. It was used, together with the → *ovality index* calculated on the substituent, to estimate substituent steric effects; for substituents with equal volume V^a, much larger steric effects are observed for globular substituents.

size-shape geometrical constant → **size descriptors** (⊙ radius of gyration)

smallest binary label → **canonical numbering**

softness density : *local softness* → **quantum-chemical descriptors** (⊙ softness indices)

softness indices → **quantum-chemical descriptors**

solute HBA basicity → **linear solvation energy relationships** (⊙ hydrogen-bond parameters)

solute HBD acidity → **linear solvation energy relationships** (⊙ hydrogen-bond parameters)

solvation connectivity indices → **connectivity indices**

solvatochromic equation → **linear solvation energy relationships**

solvatochromic parameters → **linear solvation energy relationships**

solvent-accessible molecular surface → **molecular surface**

solvent-accessible surface area → **molecular surface** (⊙ solvent-accessible molecular surface)

solvent cohesive energy density : *Hildebrand solubility parameter*

solvent-excluded volume → **molecular surface** (⊙ solvent-accessible molecular surface)

solvent HBA basicity → **linear solvation energy relationships** (⊙ hydrogen-bond parameters)

solvent HBD acidity → **linear solvation energy relationships** (⊙ hydrogen-bond parameters)

solvent polarity scales → **linear solvation energy relationships** (⊙ dipolarity/polarizability term)

Somoyai function → **quantum-chemical descriptors**

SP indices (: *subgraph property indices*)

Bond-additive molecular descriptors derived from the → *H-depleted molecular graph* defined as:

$$SP = \sum_{b=1}^{B} SP_b$$

where the sum runs over all the B bonds [Diudea *et al.*, 1996b; Diudea *et al.*, 1997a].

The bond descriptor SP_b is obtained by erasing the b bond from the molecular graph and evaluating the properties of the two remaining subgraphs $\mathcal{G}_1$ and $\mathcal{G}_2$ as the following:

$$SP_b = \mathrm{P}^*(\mathcal{G}_1) \cdot \mathrm{P}^*(\mathcal{G}_2) = \frac{\sum_{a \in \mathcal{V}_{1,b}} \mathrm{P}_a}{\sum_a \mathrm{P}_a} \cdot \frac{\sum_{a \in \mathcal{V}_{2,b}} \mathrm{P}_a}{\sum_a \mathrm{P}_a} = \mathrm{P}^*(\mathcal{G}_1) \cdot [1 - \mathrm{P}^*(\mathcal{G}_1)]$$

where P_a is the contribution to the molecular property P due to the ath vertex. P^* of the two subgraphs are the normalized properties obtained by dividing the sum of subgraph vertex contributions by the molecular property calculated on the whole graph. $\mathcal{V}_1$ and $\mathcal{V}_2$ are the subsets of vertices relative to $\mathcal{G}_1$ and $\mathcal{G}_2$ subgraphs, respectively.

Examples of vertex properties are any → *local vertex invariants*.

Example : 2-methylpentane

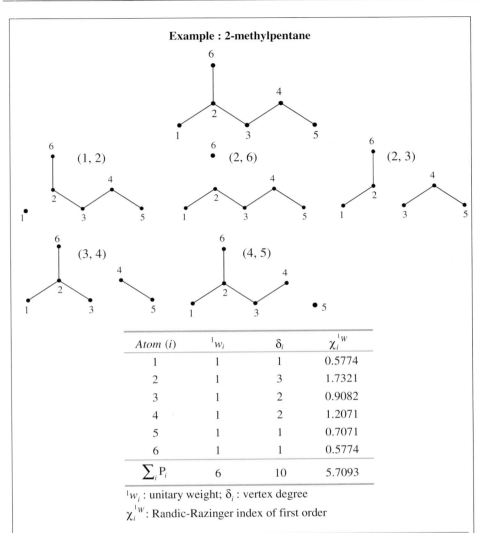

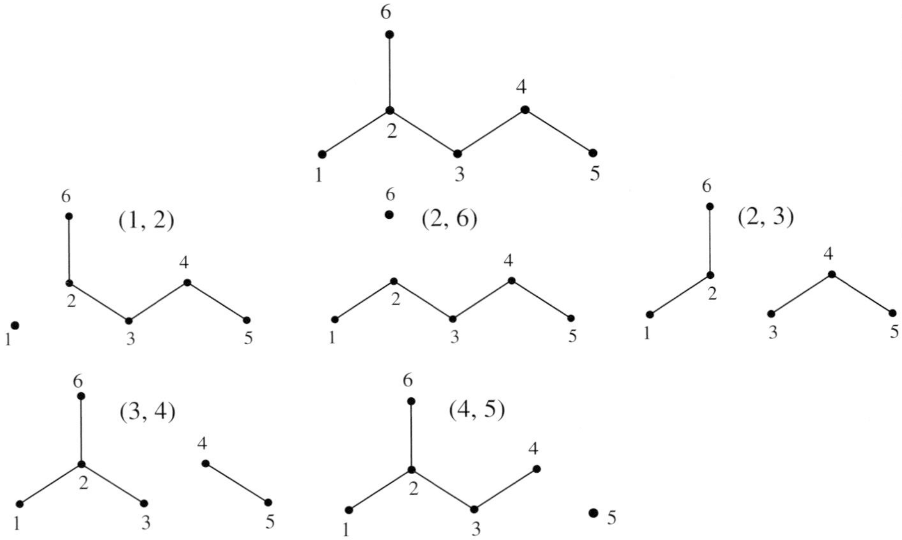

Atom (*i*)	^1w_i	δ_i	$\chi_i^{^1W}$
1	1	1	0.5774
2	1	3	1.7321
3	1	2	0.9082
4	1	2	1.2071
5	1	1	0.7071
6	1	1	0.5774
$\sum_i P_i$	6	10	5.7093

^1w_i : unitary weight; δ_i : vertex degree

$\chi_i^{^1W}$: Randic-Razinger index of first order

edge (*i*, *j*)	$S1_b$	$S\delta_b$	$S\chi^W_b$
(1, 2)	$1/6 \cdot 5/6 = 0.1389$	$1/10 \cdot 9/10 = 0.09$	$0.5774/5.7093 \cdot 5.1319/5.7093 = 0.0909$
(2, 6)	$1/6 \cdot 5/6 = 0.1389$	$1/10 \cdot 9/10 = 0.09$	$0.5774/5.7093 \cdot 5.1319/5.7093 = 0.0909$
(2, 3)	$3/6 \cdot 3/6 = 0.25$	$5/10 \cdot 5/10 = 0.25$	$2.8869/5.7093 \cdot 2.8224/5.7093 = 0.2499$
(3, 4)	$4/6 \cdot 2/6 = 0.2222$	$7/10 \cdot 3/10 = 0.21$	$3.7951/5.7093 \cdot 1.9142/5.7093 = 0.2229$
(4, 5)	$1/6 \cdot 5/6 = 0.1389$	$1/10 \cdot 9/10 = 0.09$	$5.0022/5.7093 \cdot 0.7071/5.7093 = 0.1085$
SP	0.8889	0.73	0.7631

Box S-2.

span → **size descriptors**

spanning tree → **graph**

spanning tree number → **Laplacian matrix**

sparse matrices → **algebraic operators**

sparse matrix → **algebraic operators** (⊙ sparse matrices)

sparse Wiener matrix → **Wiener matrix**

sparse χ**-E matrix** → **weighted matrices**

sparse χ **matrix** : *Kier-Hall connectivity matrix* → **weighted matrices**

Spearman rank correlation coefficient → **similarity/diversity**

spectral moments of iterated line graph sequence → **iterated line graph sequence**

spectral moments of the adjacency matrix → **self-returning walk counts**

spectral moments of the bond distance-weighted edge adjacency matrix → **edge adjacency matrix**

spectral moments of the edge adjacency matrix → **edge adjacency matrix**

spectral weighted invariant molecular descriptors → **SWM signals**

spectral weighted molecular signals : *SWM signals*

spectrum-like descriptors (S)

Recently proposed by the Zupan group [Novic and Zupan, 1996], these are → *3D-descriptors* for a spectrum-like representation of the molecular structure, defined by its → *molecular geometry.* They essentially describe the local geometry of molecules, the structure representation being obtained by the projection of the A atoms of the molecule to $2N$ points on two mutually perpendicular circles of a sphere with arbitrary radius.

In order to calculate spectrum-like descriptors, the molecules are aligned and located in a common centre within the sphere. Each circle is spanned with a fixed angle ϕ which defines the resolution of the structure representation, i.e. $N = 360 / \phi$ values are considered. For example, by fixing ϕ value at 36°, 18°, or 10°, a number of descriptors equal to 10, 20 and 36, respectively, is obtained for each circle.

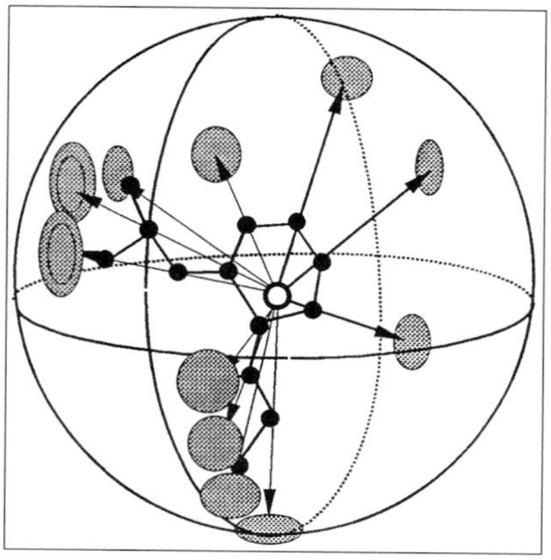

Figure S-2. Spectrum-like projections.

At each jth projection point, the ath atom contribution to the jth descriptor is obtained by a Lorentzian function L (but can be any "bell-shaped" function):

$$L_{ja} \equiv L_j\left(\varrho_a, \phi_a, \phi_j, \sigma_a\right) = \frac{\varrho_a}{\left(\phi_j - \phi_a\right)^2 + \sigma_a^2}$$

where ϱ_a is the distance of the ath atom from the centre of the projection plane, ϕ_a the angle of the atom, ϕ_j the jth spanned angle. The parameter σ_a can be used as an atom type dependent parameter. If only atom positions are considered σ_a is set to 1, otherwise the Mulliken charge of the atom can be used. The suggested σ values are 0.2, 1.0, 1.2, 1.4 for hydrogen, carbon, nitrogen, and oxygen atoms, respectively.

As can be observed, the function takes the maximum value when the ath atom is exactly along the spanned direction angle ϕ_j.

The 2N molecular descriptors S_j are obtained by summing, over all the A atoms, the Lorentzian function values of the two orthogonal circles (N for each circle):

$$S_j = \sum_{a=1}^{A} L_j\left(\varrho_a, \phi_a, \phi_j, \sigma_a\right) = \sum_{a=1}^{A} \frac{\varrho_a}{\left(\phi_j - \phi_a\right)^2 + \sigma_a^2} \qquad j = 1, \ldots, 2N$$

A modification of the Lorentzian function was proposed by Forina and co-workers [Forina *et al.*, 1997] as:

$$L_{ja} \equiv L_j\left(\varrho_a, \phi_a, \phi_j, h_a, k\right) = \frac{\varrho_a \cdot h_a}{k \cdot \left(\phi_j - \phi_a\right)^2 + 1}$$

where k is a constant, whose value controls the width of the function, i.e. the function has a width, measured in degrees on the projection circle, which increases with decreasing k (the optimal k value is chosen by cross-validation between 1 and 10). The atom type dependent parameter h_a substitutes the original σ_a and controls the height of the contribution of each atom to the Lorentzian signal.

Therefore, the **modified spectrum-like descriptors** are calculated as:

$$S_j = \sum_{a=1}^{A} L_j\left(\varrho_a, \phi_a, \phi_j, h_a, k\right) = \sum_{a=1}^{A} \frac{\varrho_a \cdot h_a}{k \cdot \left(\phi_j - \phi_a\right)^2 + 1} \qquad j = 1, \ldots, 2N$$

Spectrum-like descriptors give information about molecular size and are dependent on molecule alignment, i.e. on the two perpendicular circles onto which the atom projections fall. From → *variable selection*, partial → *reversible decoding* can also be performed.

🕮 [Zupan and Novic, 1997] [Vracko, 1997] [Baumann, 1999] [Zupan *et al.*, 2000]

spectrum of a graph → **characteristic polynomial of the graph**

spectrum of a matrix → **algebraic operators** (⊙ characteristic polynomial)

spherosity index → **shape descriptors**

square root molecular weight → **molecular weight**

***SSAA* index** → **charged partial surface area descriptors**

***SSAH* index** → **charged partial surface area descriptors**

standardized information content : *standardized Shannon's entropy*
→ **information content**

standardized regression coefficients sum → **model complexity**

standardized Shannon's entropy → **information content**

star graph → **graph**

start-end vectors : *distance-counting descriptors*

static inductive effect : *inductive effect* → **electronic substituent constants**

static polarizability : *polarizability* → **electric polarization descriptors**

static reactivity indices → **reactivity indices**

statistical indices

Statistical indices are fundamental quantities involved in → *chemometrics* and several QSAR methods as well as in several algorithmic procedures for the calculation of → *molecular descriptors.*

A short list of definitions of the most used statistical indices is given below [Zar, 1984; Arnold, 1990; Johnson and Wichern, 1992]. Symbols and indices are those reported in the definition of → *data set*, and sample statistics is adopted.

Arithmetic mean. The most important measure of the centre of an x variable, defined as:

$$\bar{x} = \frac{1}{n} \cdot \sum_{i=1}^{n} x_i$$

The generalization of the arithmetic mean is the *weighted mean*, defined as:

$$\bar{x} = \frac{\sum_{i=1}^{n} w_i \cdot x_i}{\sum_{i=1}^{n} w_i}$$

where w_i is the weight associated with each ith object.

Geometric mean. A measure of the centre, appropriate only when all the data are positive, defined as the nth root of the product of the n data:

$$\bar{x}_G = \sqrt[n]{\prod_{i=1}^{n} x_i}$$

Variance. The most important measure of the x variable dispersion with respect to its centre, defined as:

$$s^2 = \frac{\sum_{i=1}^{n} (x_i - \bar{x})^2}{n-1}$$

where $\bar{x}$ is the arithmetic mean of the x variable. The square root of the variance is called the *standard deviation* (s).

Pearson's first index. A measure of the asymmetry of the distribution of the x variable, defined as:

$$\kappa_3 = \frac{\sum_{i=1}^{n} (x_i - \bar{x})^3}{s^3 \cdot n}$$

where $\bar{x}$ is the arithmetic mean of the x variable and s^3 the third power of its standard deviation.

Negative values show a right-tailed distribution, positive values a left-tailed distribution. Other measures of distribution asymmetry are *Pearson's second index*, *skewness*, and the *Bonferroni index*.

Kurtosis. A measure of the degree of bimodality of the distribution of the x variable, defined as:

$$\kappa_4 = \frac{\sum_{i=1}^{n} (x_i - \bar{x})^4}{s^4 \cdot n}$$

where $\bar{x}$ is the arithmetic mean of the x variable and s^4 the fourth power of its standard deviation. For a peak distribution, $\kappa = \infty$, while for a complete bimodal distribution, $\kappa = 1$.

Uniform and normal distributions have $\kappa = 1.8$ and $\kappa = 3$, respectively.

Correlation. A measure of the degree of association between two variables j and k, defined as:

$$r_{jk} = \frac{\sum_{i=1}^{n} (x_{ij} - \bar{x}_j) \cdot (x_{ik} - \bar{x}_k)}{\sqrt{\sum_{i=1}^{n} (x_{ij} - \bar{x}_j)^2 \cdot \sum_{i=1}^{n} (x_{ik} - \bar{x}_k)^2}} = \frac{s_{jk}^2}{s_j \cdot s_k}$$

where $\bar{x}$ is the arithmetic mean of the variables, s_{jk}^2 is the covariance between the two variables, and s the corresponding standard deviations of the two variables j and k. The matrix whose elements are the pairwise correlations between all the variables is called the *correlation matrix*. The term at the numerator, divided by the degrees of freedom $n - 1$, is the covariance between variables j and k:

$$s_{jk}^2 = \frac{\sum_{i=1}^{n} (x_{ij} - \bar{x}_j) \cdot (x_{ik} - \bar{x}_k)}{n - 1}$$

The matrix whose elements are the pairwise covariances of all the variables is called *covariance matrix*.

stepwise regression methods → **variable selection**

stereodynamic representation → **molecular descriptors**

stereoelectronic representation → **molecular descriptors**

steric density parameter → **steric descriptors**

steric descriptors

Steric effects are among the most relevant in modelling → *physico-chemical properties* and biological activities, thus playing a fundamental role in QSAR / QSPR modelling.

For this reason, a huge number of steric parameters were defined and used from the beginning to represent the steric properties of a molecule. Steric properties influence molecule energy, reaction and conformational paths, reaction rates and equilibria, binding affinity between ligand and receptor and other thermodynamic properties.

Steric descriptors account for both the size and shape of molecules and substituents, and are thus contemporarily related to → *size descriptors* and → *shape descriptors*. However, in several QSAR models, size descriptors are encountered as estimates of steric molecular/fragment properties.

The term **bulk descriptors** can be used to denote steric descriptors of the whole molecule, referring to the measure of the hindrance of the molecule when it is considered as a part of a system and dense assembly, each molecule being constrained by its neighbours to a limited region in space.

Steric descriptors were obtained from → *experimental measurements* of equilibrium and rate constants, → *computational chemistry*, geometrical and structural characteristics, and the → *topological representation* of a molecule.

The most common steric descriptors are → *molar refractivity*, → *surface areas* and several → *volume descriptors* such as → *molecular volume*. Other steric descriptors are → *steric interaction fields*, → *MTD descriptors*, → *common overlap steric volume*, and several topological descriptors accounting for both size and → *molecular branching*.

Other popular molecular steric descriptors are listed below.

- **gravitational indices**

→ *Geometrical descriptors* reflecting the mass distribution in a molecule, defined as [Katritzky *et al.*, 1996c]:

$$G_1 = \sum_{i=1}^{A-1} \sum_{j=i+1}^{A} \frac{m_i \cdot m_j}{r_{ij}^2} \qquad G_2 = \sum_{b=1}^{B} \left(\frac{m_i \cdot m_j}{r_{ij}^2} \right)_b$$

where m_i and m_j are the atomic masses of the considered atoms, r_{ij} the corresponding → *interatomic distances*, and A and B the number of atoms and bonds of the molecule, respectively. The G_1 index takes into account all atom pairs in the molecule while the G_2 index is restricted to pairs of bonded atoms. These indices are related to the bulk cohesiveness of the molecules, accounting, simultaneously, for both atomic masses (volumes) and their distribution within the molecular space. For modelling purposes the square root and cube root of the gravitational indices were also proposed [Wessel *et al.*, 1998].

Both indices can be extended to any other atomic property different from atomic mass, such as → *atomic polarizability*, atomic → *van der Waals volume*, etc.

- **Kier steric descriptor** (Ξ)

A linear combination of → *Kier shape descriptors* obtained from an empirical relationship between Taft steric constant E_s and group κ indices [Kier, 1986b] and defined by the following [Kier, 1987d]:

$$\Xi = 2 \cdot {}^1\kappa_\alpha - {}^0\kappa - {}^3\kappa_\alpha$$

where ${}^1\kappa_\alpha$ and ${}^3\kappa_\alpha$ are the alpha-modified 1st and 3rd order Kier shape descriptors and ${}^0\kappa$ the → *Kier symmetry descriptor*. This index can be calculated for both the whole molecule and the substituent groups; it is somewhat related to the radii of the atoms involved in the substituent, and is a measure of the spatial influence of the group operating through the attached atom in the group.

- **Austel branching index** (S_X)

A geometric steric parameter for the X substituent based on the non-hydrogen atoms of the substituent and their topological distance from the atom to which the substituent is linked [Austel *et al.*, 1979]; it is calculated as:

$$S_X = \sum_i b_i \cdot \sum_j n_{ij} \cdot k_j$$

where the first sum runs over all shells of the substituent around the link point on the parent structure, where the first shell ($i = 0$) includes the link point, the second shell

($i = 1$) all the substituent atoms at a topological distance equal to 1 from the link point, and so on; b_i are coefficients accounting for the branching contribution in the ith shell to the steric effect of the substituent. n_{ij} is the number of atoms of jth type bonded to atoms in the ith shell (atoms in the previous shell $i - 1$ are not counted); k_j is a factor that accounts for the size of the jth atom type.

Both the b and k values were estimated by regression analysis from the Taft steric constant E_s and the Charton steric constant in an iterative procedure using 96 substituents. The k values are equal to 1.0 for the second period elements, except for the fluorine atom ($k = 0.8$); $k = 1.2$, 1.3, and 1.7 for the third, fourth, and fifth period elements, respectively.

In calculating n_{ij}, a correction is proposed for cyclic structures: the number n_{ij} for the atom in the ith shell is reduced by the number of ring-bonds leading from the $(i - 1)$th shell to the current ith shell.

Assuming that the relative distance contributions to the steric effect of a substituent are not significantly different, all the regression coefficients b were settled equal to one and a simplified expression was proposed as:

$$S_X = \sum_j N_j \cdot k_j$$

where the sum runs over all the different atom types and N_j is the count of the jth atom types.

For example, over this approximation, S_X values for $-CH_3$, $-C{=}OF$, $-SF_5$ substituent groups are 1 (1×1), 2.8 ($1 \times 1 + 1 \times 1 \times 1 \times 0.8$), and 5.2 ($1 \times 1.2 + 5 \times 0.8$), respectively.

- **Jenkins steric parameter** (S_{AFF})

A steric parameter based on the proton affinity A_H and methyl cation affinity A_{CH3}, calculated by computational chemistry methods, for reaction at the nitrogen atom series of compounds in which the nitrogen atom was unhindered, e.g. pyridines in which the 2- and 6-positions are unsubstituted [Jenkins *et al.*, 1994; Jenkins *et al.*, 1995; Baxter *et al.*, 1996]. Using these affinity values, a reference regression model was found which correlates the methyl cation affinity A_{CH3} to the proton affinity A_H for unhindered compounds:

$$\hat{A}_{CH3} = -446.78 + 0.971 \cdot A_H$$

For nitrogen compounds where the nucleophilic nitrogen atom is hindered, the methyl cation affinity values calculated are less than those estimated by the reference model. The difference between the two values is assumed as the steric parameter, i.e:

$$S_{AFF} = A_{CH3} - \hat{A}_{CH3}$$

- **sum of steric substituent constants**

Among molecular steric descriptors there can also be considered those obtained by summing the steric substituent constants of all the substituents present in the molecule, such as, for example

$$\Sigma = \sum_k (E_S)_k$$

where E_S are the Taft steric constants (see below).

Steric substituent constants (or **steric substituent parameters**) are descriptors of substituent groups that measure the substituent steric effects on the reactivity centres of a molecule, based on differences in the rate and equilibrium constants of selected chemical reactions.

The most popular steric substituent descriptors are listed below.

• **Taft steric constant** (E_S)
The Taft steric constant E_S was proposed as a measure of steric effects that a substituent X exerts on the acid-catalyzed hydrolytic rate of esters of substituted acetic acids XCH_2COOR [Taft, 1952]. The basic assumption is that the effect of X on acid hydrolysis is purely steric, as the reaction constant ϱ for acid hydrolysis of substituted esters is close to zero.

Therefore, the Taft steric constant is calculated as the average value from four series of kinetic data (hydrolysis of ethyl esters in 79 % aqueous acetone by volume at 25°C, esterification of carboxylic acids in methanol at 25°C, esterification of carboxylic acids in ethanol at 25°C, hydrolysis of ethyl esters in 69 % aqueous acetone by volume at 25°C):

$$E_S = \log(k_X)_A - \log(k_H)_A = \log\left(\frac{k_X}{k_H}\right)_A$$

where k_X and k_H are the rate constants for the substituted and unsubstituted esters or acids, respectively, and the subscript A denotes hydrolysis in acid solution. The bulkier the substituent, the more negative the E_S constant values.

A rescaled set of the Taft steric constant was defined for the series of esters of substituted formic acids XCOOR as:

$$E_S'' = E_S - 1.24$$

where 1.24 corresponds to the E_S value of the formic acid or ester [Motoc and Balaban, 1982a].

To correct the hyperconjugation effect due to the α-hydrogens, a **corrected Taft steric constant** E_S^C **(Hancock steric constant)** was proposed:

$$E_S^C = E_S + 0.306 \cdot (h_\alpha - 3)$$

where h_α represents the number of α-hydrogens [Hancock *et al.*, 1961b].

In order to account for both C–C and C–H hyperconjugation effects a different correction for Taft steric constant was proposed [Palm, 1972], obtaining the **Palm steric constant**:

$$E_S^0 = E_S + 0.33 \cdot (h_\alpha - 3) + 0.13 \cdot N_C$$

where h_α represents the number of α-hydrogens and N_C the number of α-carbon atoms in the substituent.

The **Dubois steric constant** E_S' is a revised Taft steric constant defined using the acid-catalyzed esterification of carboxylic acids (at 40°C) as reference reaction [MacPhee *et al.*, 1978a; MacPhee *et al.*, 1978b].

The **Taft-Kutter-Hansch steric constants** (TKH E_S) constitute a combined set of parameters of the original Taft steric constants and those extended by Kutter and Hansch [Kutter and Hansch, 1969; Hansch, 1970; Sotomatsu and Fujita, 1989] by using a correlation with the average of the minimum and maximum van der Waals radii as defined by the equation:

$$\hat{E}_S = 3.484 - 1.839 \cdot \bar{R}_{VDW}$$

The values of the van der Waals radii are taken from Charton. By using this equation, steric constants were also calculated for ortho-substituted and non-alkyl groups.

The **Fujita steric constant** E_S^F [Fujita *et al.*, 1973; Fujita and Iwamura, 1983] was defined to evaluate the global steric effect for branched alkyl substituents of the type $CR_1R_2R_3$ by the following correlation equation:

$$E_S^F = -2.104 + 3.429 \cdot E_S^C(R_1) + 1.978 \cdot E_S^C(R_2) + 0.649 \cdot E_S^C(R_3)$$

Values of the Taft steric constants are collected in Table S-2 for some substituents.

A **Model of the Frontier Steric Effect** is a theoretical approach recently proposed to estimate the Taft steric constant of substituents on the basis of the fundamental characteristics of the constituent atoms [Cherkasov and Jonsson, 1998b]:

$$R_S = -30 \cdot \log\left(1 - \sum_{i=1}^{n} \frac{R_i^2}{4 \cdot r_i^2}\right) \quad \text{and} \quad R'_S = \sum_{i=1}^{n} \frac{R_i^2}{r_i^2}$$

where the sums run over all the atoms of the substituent, r_i is the distance of the ith atom of the substituent to the reaction centre, and R_i the radius of the atom. The second definition is an approximated solution of the first one for small values of the ratio.

📖 [Taft, 1953a; Taft, 1953c; Lambert, 1966; Hopkinson, 1969; Unger and Hansch, 1976; Murray, 1977; Fujita, 1978; Panaye *et al.*, 1980; Gallo, 1983]

- **Charton steric constant** (ν_X)

Charton steric constant ν_X (or **Charton characteristic volume**) is defined as:

$$\nu_X = R_{VDW(X)} - R_{VDW(H)} = R_{VDW(X)} - 1.20$$

where $R_{VDW(X)}$ is the → *van der Waals radius* of the substituent X and $R_{VDW(H)}$ the hydrogen atom radius [Charton, 1969; Charton, 1975; Charton, 1983]. For symmetric top substituents of the type CX_3, Charton assumed that the axis of the group is the extension of the C–G_i bond, where G_i is the skeletal atom to which the group is bonded; then, he defined the minimum and maximum van der Waals radii perpendicular to the group axis as minimum and maximum widths of the group, and the van der Waals radius parallel to the axis as the length of the group. Thus, depending on the van der Waals radius, different sets of ν_X constants can be calculated to completely characterize most substituents. However, in order to take into account that groups of atoms assume conformations which minimize their repulsive effects, it is suggested to select the minimum van der Waals radius.

The average of the minimum and maximum van der Waals radii as defined by Charton was correlated to the Taft steric constant (see above).

Directly obtained from the van der Waals radii is also the **Bowden-Young steric constant** R [Bowden and Young, 1970]. This measures the steric hindrance of the substituent, calculated as the distance (in Å) from the aromatic carbon atom to which the substituent is bonded to the periphery of the van der Waals radius of the substituent, using known distances and van der Waals radii, and referred to that of the unsubstituted compound, i.e. $R_H = 0$.

Values of the Charton steric constants are collected in Table S-2 for some atoms.

📖 [Charton, 1971; Charton, 1976; Charton and Charton, 1978; Charton, 1978b]

- **number of atoms in substituent specific positions**

The number of atoms in specific positions of a substituent are among the simplest substituent steric descriptors and can be used as steric "correction site" parameters. The **six-position number** N_6 (or **rule of six**) was suggested as the major factor in steric effects by Newman [Newman, 1950; Taft, 1956], summing carbon and hydrogen atoms in the sixth position from the carbonyl oxygen considered as atom one. This empirical rule is assumed for reactions involving addition to an unsaturated function: the greater the number of atoms in position six, the greater the steric effect.

The **Idoux steric constant** $\Delta 6$ is defined as the change in the six-position number, i.e. the difference between the six-position number of the X substituent of the ester XCOOX′ and the six-position number of the same substituent X in the part X′ of the ester [Idoux *et al.*, 1973; Idoux *et al.*, 1977].

Distinguishing the contributions of carbon and hydrogen atoms, the **Bowden-Wooldridge steric constant** E_S^{BW} is defined by a correlation equation with the Taft steric constant:

$$E_S^{BW} = 0.119 - 0.347 \cdot N_{C6} - 0.075 \cdot N_{H6}$$

where N_{C6} and N_{H6} are the carbon and hydrogen atoms, respectively, in sixth position with respect to the carbonyl oxygen atom in the ester used to define E_S [Bowden and Wooldridge, 1973].

Always based on the number of substituent carbon atoms in different positions, Charton proposed a general correlation equation with its steric constant v defined as

$$v = b_0 + b_1 \cdot n_\alpha + b_2 \cdot n_\beta + b_3 \cdot n_\gamma + b_4 \cdot n_\delta$$

where n_α, n_β, etc. represent the number of atoms in α-, β-, γ-, and δ-position, respectively [Charton, 1978b].

- **steric density parameter** (SD_X) (: *Dash-Behera steric density parameter*)

A substituent descriptor which combines molecular weight and van der Waals volume according to equation:

$$SD_X = \left(\frac{MW}{V_{VDW}}\right)_X - \left(\frac{MW}{V_{VDW}}\right)_H = \left(\frac{MW}{V_{VDW}}\right)_X - 0.29$$

where MW is the → *molecular weight* and V_{VDW} the → *van der Waals volume*. X denotes the X-substituted compound and H the unsubstituted compound, i.e. with the hydrogen atom in the substitution site [Dash and Behera, 1980]. Namely, the meaning of this parameter is related to the density of the substituent. Values of SD for several substituents are derived from substituted organic acids.

- **steric vertex topological index** ($SVTI$)

A substituent steric descriptor for alkyl groups only defined in terms of the → *topological distance d* from an → *H-depleted molecular graph* [Ivanciuc and Balaban, 1996c]:

$$SVTI = \sum_{j=1}^{A_X} d_{ij} \quad \forall d_{ij} \leq 3$$

where A_X is the number of non-hydrogen atoms of the substituent X, d_{ij} is the topological distance between any vertex v_j of the substituent group and v_i, which is the atom of the attachment site of a common reference skeleton. This index was proposed to approximate the steric effects of substituents. Therefore, only distances not exceeding 3 are considered, as the steric effect of a residue is related to the volume of that portion of its body closer to the link site.

- **substituent front strain** (S_f) (: *F strain*)

Substituent steric descriptor calculated from standard enthalpy of formation obtained by computational chemistry using empirical force field methods. It is defined by the following relationship [Beckaus, 1978; Giese and Beckaus, 1978]:

$$S_f = \Delta H_f^0[XC(CH_3)_3] - \Delta H_f^0[XCH_3] + 8.87 \quad [10^4 \ J/mol]$$

where $\Delta H_f^0[XC(CH_3)_3]$ and $\Delta H_f^0[XCH_3]$ are the standard enthalpies of formation of the X-substituted *t*-butyl and methyl derivatives, respectively, methyl and *t*-butyl groups being chosen for their high symmetry. The additive term 8.87 appears on normalization of $S_f(CH_3) = 0$.

F strain constants reflect only the steric repulsion of the attacking or leaving group of a reaction and contain no additional conformational effects.

The **steric energy difference** ΔSE, quite well related to the difference in enthalpy of formation $\Delta\Delta H_f$, has also been used as a steric descriptor [DeTar and Tenpas, 1976; DeTar and Delahunty, 1983]; it was defined as the difference between the steric energy (calculated by molecular mechanic methods) of the X-substituted ortho acid $XC(OH)_3$ and the steric energy of the corresponding X-substituted carboxylic acid XCOOH.

- **Joshi steric descriptor**

A quantum-chemical descriptor proposed to measure the steric effects of one or more substituents in a congeneric series of compounds [Joshi *et al.*, 1993; Joshi *et al.*, 1994]. The Joshi steric parameter JM1 (or log (JM1)) is defined as:

$$\mathrm{JM1} = \frac{\Delta E_X}{\Delta E_H} \qquad \log(\mathrm{JM1}) = \log(\Delta E_X) - \log(\Delta E_H)$$

where ΔE_H and ΔE_X are differences in conformational energy values of the unsubstituted and X-substituted compound, respectively:

$$\Delta E_H = E_H(global) - E_H(strained) \qquad \Delta E_X = E_X(global) - E_X(strained)$$

E(*global*) is the global energy minimum of the most favourable conformation associated with the least steric interactions, E(*strained*) is the energy of a randomly chosen reference conformation which is fixed for all compounds of the series.

For the unsubstituted compound, log (JM1) = 0; if $\Delta E_H = 0$, the reference conformation is changed. The steric parameter JM1 values are calculated by computational chemistry using both the AM1 method and PCILO approximation.

The steric component is achieved by subtracting the energy E(*strained*) of a constrained or sterically hindered molecule conformation from the energy values E(*global*) of the sterically most favourable conformation (ΔE_X). The steric influences of all substituents in the compound is determined by referencing ΔE_X to ΔE_H.

📖 [Dubois *et al.*, 1980b] [Verloop, 1985] [Kim, 1992b]

steric energy difference → **steric descriptors** (⊙ substituent front strain)

steric interaction fields → **molecular interaction fields**

steric interactions in biological systems → **minimal topological difference**

steric misfit

The difference between the steric hindrance of the → *leading compound* (assumed as the → *reference compound*) and the considered molecule. It is a measure of the similarity where low values of steric misfit correspond to high similarity between molecule and reference compound, i.e. a favourable condition for the considered molecule.

Methods for evaluating the steric misfit are → *minimal topological difference* and → *molecular shape analysis*.

steric substituent constants → **steric descriptors**

steric substituent parameters : *steric substituent constants* → **steric descriptors**

steric vertex topological index → **steric descriptors**

Sterimol B parameters → **Sterimol parameters**

Sterimol length parameter → **Sterimol parameters**

Sterimol parameters (: *Verloop Sterimol parameters; Verloop parameters*)

A set of parameters proposed by Verloop [Verloop *et al.*, 1976; Verloop, 1987] to describe the size and shape of substituents in a congeneric series. These parameters were evaluated by measuring the dimensions of substituents in a restricted number of directions by a computer program (STERIMOL) which simulates 3D model building of substituent groups, using the → *Corey-Pauling-Koltun volume* (CPK atomic models). For flexible substituents minimum energy conformations are considered.

The substituent attachment atom G_1 on the parent structure (e.g. benzene) is placed at the origin of the Cartesian coordinates (x, y, z), assuming that the bond connecting the substituent to G_1 defines the X axis. The **Sterimol length parameter** L is defined as the maximum length along the X axis, i.e. it is the x coordinate of the intersection point on the X axis of the tangential plane to the substituent, perpendicular to the X axis (Figure S-3a). For example, for the substituent H of the benzene molecule, L = 2.06, obtained as sum of the C–H bond (1.06 Å) and the van der Waals radius of H (1.00 Å).

The **Sterimol B parameters** were proposed to characterize the widths of the substituent along directions perpendicular to the X axis. The width parameters B_1, B_2, B_3, B_4 are, in ascending order, taken as the distances to the X axis of tangential planes to the substituent, perpendicular to the Z and Y axes. B_1 and B_4 are defined as the smallest and largest width along the Z axis, while B_2 and B_3 are defined as the smallest and largest width along the Y axis (Figure S-3b). In other words, these parameters describe the positions, relative to the origin and the axes, of the five side planes of the box in which the substituent is embedded, made in such a way that the distance of one of its sides to the axis has the smallest possible value. In most cases, the B_4 value is almost equal to the maximum length L of the substituent.

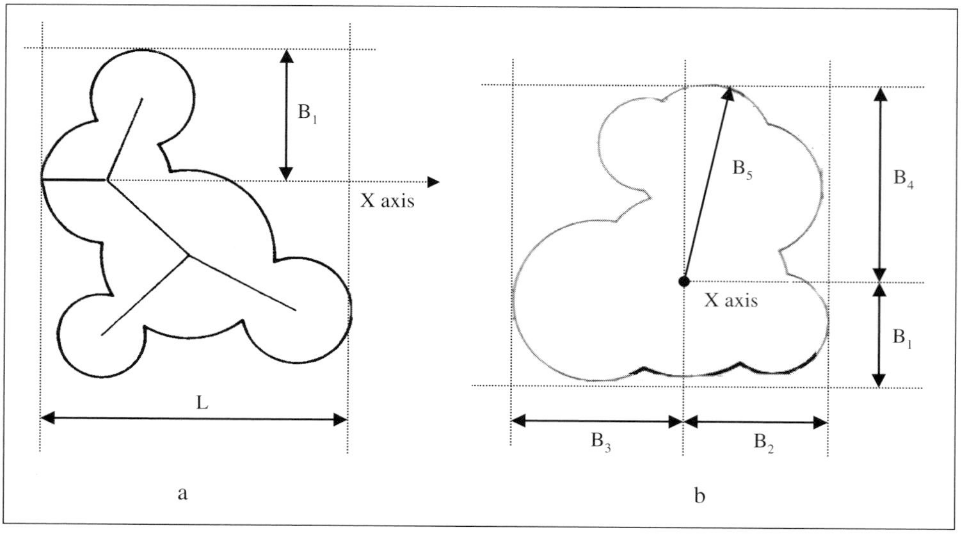

Figure S-3. a) Projection of a substituent along the X axis showing the parameters L and B_1; b) projection of a substituent perpendicular to the X axis showing B parameters.

Later, a new Sterimol width parameter B_5 was introduced to replace the B_4 parameter, defined as the maximum width (i.e. the maxmum distance from X axis) of the substituent in the Z-Y plane (perpendicular to the X axis).

Among the → *shape descriptors*, **Sterimol shape parameters** were proposed as the → *length-to-breadth ratio*, defined as L/B_1 and B_5/B_1 (or previously, B_4/B_1), giving information about the deviations of a substituent from a spherical shape.

Verloop *et al.* determined the Sterimol parameters for over 1000 substituent groups.

📖 [Arnaud *et al.*, 1994] [Kawashima *et al.*, 1994] [Draber, 1996] [Singh *et al.*, 1996]

Sterimol shape parameters → **Sterimol parameters**

structural code : *vertex structural code* → **self-returning walk counts**

structural environment vector → **Klopman-Henderson cumulative substructure count**

structural graph : *molecular graph*

structural information content → **indices of neighbourhood symmetry**

structural keys → **substructure descriptors**

structure-activity relationships → **structure/response correlations**

structure-property relationships → **structure/response correlations**

structure-reactivity relationships → **structure/response correlations**

Structure/Response Correlations (SRC)

The term *Structure/Response Correlations* and its acronym *SRC*, is proposed in this Handbook to collect under a unique framework all the approaches aimed at finding relationships between molecular structure and measured (or calculated) molecular response.

The proposed term "*structure/response correlations*" is a further generalization of the proposal "*Structure Property Correlation* (SPC)" suggested by van de Waterbeemd [van de Waterbeemd, 1992]. In fact, the Authors think that the term *property* actually has a too specific meaning and suggest the very general term *response* to encompass both physico-chemical properties and biological activities. *Response* is semantically related to the mathematical concept of *y response* in modelling, i.e. the dependent variable in any correlation equation that can easily be intended here as any experimental measured quantity, such as, for example, → *physico-chemical properties*, i.e. **Structure-Property Relationships** (SPR), or biological activities, i.e. **Structure-Activity Relationships** (SAR), or reactivity measures, i.e. **Structure-Reactivity Relationships** (SRR).

In quantitative approaches to SRC studies, i.e. **Quantitative Structure/Response Correlations** (QSRC), the aim is to obtain quantitative relationships like those represented by regression models and classification models. Thus, SRC studies involve both quantitative correlation studies and qualitative comparisons among → *molecular descriptors* and responses.

The class of the quantitative approaches to SRC studies includes all the well-known approaches called **Quantitative Structure-Activity Relationships** (QSAR), **Quantitative Structure-Property Relationships** (QSPR), **Quantitative Structure-Reactivity Relationships** (QSRR), **Quantitative Shape-Activity Relationships** (QShAR), the molecular shape being considered as a component of the molecular structure, **Quantitative Struc-**

ture-Chromatographic Relationships (QSCR), **Quantitative Structure-Toxicity Relationships** (QSTR), → *Quantitative Similarity-Activity Relationships* (QSiAR), **Quantitative Structure-Enantioselective Retention Relationships** (QSERR), and so on.

The proposed slash between the two terms "structure" and "response" denotes both "*and*" and "*or*", thus also involving property-property relationships as well as → *similarity/diversity* correlations. Therefore, quantitative property-property relationships (QPPR), property-activity relationships (QPAR), activity-activity relationships (QAAR), and similarity/diversity correlations fall, in a broader sense, within SRC studies.

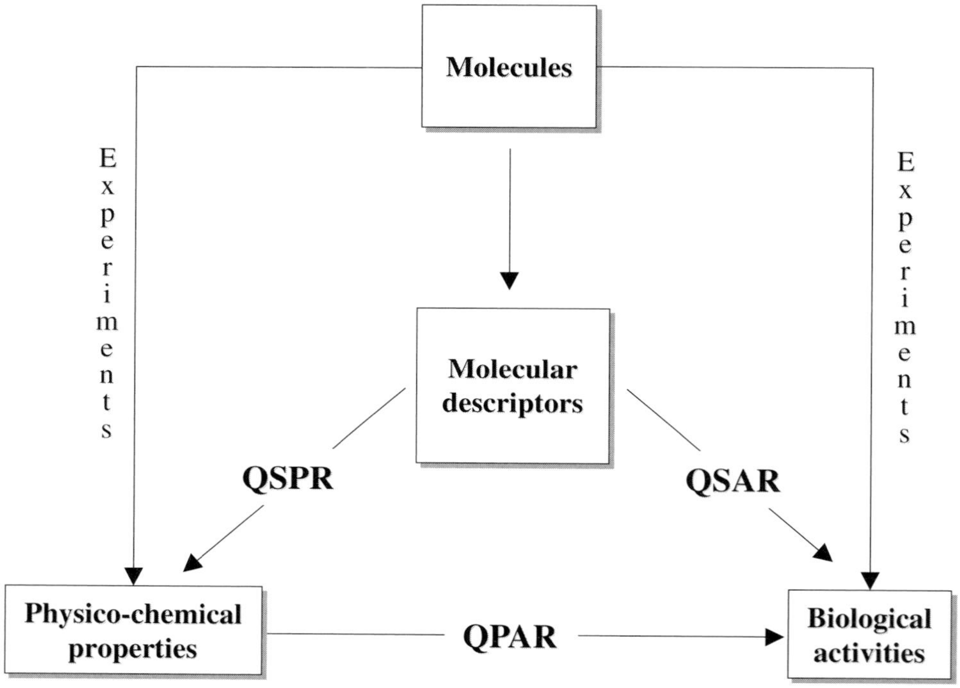

Figure S-4. Scheme of the relationships of the molecular descriptors with molecules and structure/response correlation approaches.

Quantitative Information Analysis is the term proposed by Kier to denote structure/response correlations, where the word "analysis" is chosen to avoid any restriction to QSAR/QSPR models, but naturally includes similarity/diversity analysis as well as any explorative analysis or model which refers not only to relationships with the molecular structure.

The term **classical QSAR** is often used to denote the → *Hansch analysis*, → *Free-Wilson analysis*, → *Linear Free Energy Relationships* (LFER) and → *Linear Solvation Energy Relationships* (LSER), i.e. those SRC approaches developed between 1960 and 1980 that can be considered the beginning of the modern QSAR/QSPR methods.

QSAR approaches based on the → *topological representation* of the molecules are often called **2D-QSAR**. The term **3D-QSAR** is often, though improperly, used to denote only → *grid-based QSAR techniques* and *receptor mapping techniques* (see below), but the present relevance of other kinds of descriptors obtained from 3D → *molecular geometry* naturally leads to the enlargement of this term to include all the SRC methods based on the → *geometrical representation* of the molecules.

Dynamic QSAR (also called **4D-QSAR**) denotes those recently developed SRC techniques that take conformation variability of the molecules into account [Mekenyan *et al.*, 1994; Dimitrov and Mekenyan, 1997]. **Binary QSAR** refers to those techniques where attention is paid to modelling binary responses such as active/inactive compounds [Gao *et al.*, 1999].

Because of the importance of the design of new active compounds in contemporary → *drug design*, several methods have been proposed to measure the binding affinity of a set of ligand compounds without a deep knowledge of the three-dimensional structure of the receptor site. A number of methods, called **receptor mapping techniques**, attempt to provide insight into the receptor active site and characterize receptor binding requirements for the design of a desirable ligand. Usually, in these methods, the most used molecular descriptors are → *interatomic distances* and different kinds of interaction energies.

→ *Distance Geometry* (DG) and → *minimal topological difference* (MTD) are the earliest examples of such an approach. Among them, other popular methods are *Active Analog Approach* (AAP) [Marshall *et al.*, 1979], *LOCON* and *LOGANA methods* [Streich and Franke, 1985; Franke and Streich, 1985a; Franke and Streich, 1985b; Franke *et al.*, 1985], *Receptor Surface Model* (RSM) [Hahn, 1995; Hahn and Rogers, 1995; Hahn, 1997; Hahn and Rogers, 1998], *Receptor Binding Site Model* (RBSM) [Höltje *et al.*, 1985; Höltje *et al.*, 1993], *Hypothetical Active-Site Lattice* (HASL) [Wiese, 1993; Doweyko, 1991; Doweyko and Mattes, 1992], *Genetically Evolved Receptor Models* (GERM) [Walters and Hinds, 1994; Walters, 1998; Chen *et al.*, 1998].

The years between 1860 and 1880 were characterized by a strong dispute about the concept of → *molecular structure*, arising from the studies on substances showing optical isomerism and the studies of Kekulé (1861–1867) on the structure of benzene. The concept of the molecule as a three-dimensional body was first proposed by Butlerov (1861–1865), Wislicenus (1869–1873), Van't Hoff (1874–1875) and Le Bel (1874). The publication in French of the revised edition of "La chimie dans l'éspace" by Van't Hoff in 1875 is considered a milestone in the three-dimensional conception of the chemical structures.

Crum-Brown and Fraser (1868–1869) [Crum-Brown and Fraser, 1868a; Crum-Brown and Fraser, 1868b] proposed the existence of a correlation between biological activity of different alkaloids and molecular constitution; their equation can be considered the first general formulation of a quantitative structure-activity relationship. A few years later, a hypothesis on the existence of correlations between molecular structure and physico-chemical properties was reported in the work of Körner (1874), that dealt with the synthesis of disubstituted benzenes and his discovery of *ortho*, *meta*, and *para* derivatives: the different colours of disubstituted benzenes were thought to be related to their differences in molecular structure [Körner, 1874].

The periodic table proposed by Mendeleev (1870) gives relationships between atomic structure and properties; the concept of an internal structure of atoms and molecules becomes ever more relevant, and important studies are made such as those of G. N. Lewis [Lewis, 1916; Lewis, 1923].

The quantitative property-activity models, commonly referred to as marking the beginning of QSAR/QSPR studies [Richet, 1893], have come from the search for relationships between the potency of local anesthetics and the oil/water partition coefficient [Meyer, 1899], between narcosis and chain length [Overton, 1901; Overton, 1991], and between narcosis and surface tension (Traube, 1904).

Based on the graph theory, the → *Wiener index* [Wiener, 1947c; Wiener, 1947a; Wiener, 1947b] and the → *Platt number* [Platt, 1947] are the first → *theoretical molecular descriptors* proposed in 1947 to model the boiling point of hydrocarbons.

The use of quantum-chemical descriptors in SRC studies dates back to early in 1970 [Kier, 1971], although → *quantum-chemical descriptors* have to be regarded as belonging to the development of quantum chemistry; during the period 1930–1960, the milestones are the work of Pauling [Pauling, 1932; Pauling and Wilson, 1935; Pauling, 1939] and Coulson [Coulson, 1939] on the chemical bond, Sanderson on electronegativity [Sanderson, 1952], Fukui [Fukui *et al.*, 1954] and Mulliken on electronic distribution [Mulliken, 1955a].

In the mid-1960s, led by the pioneering works of Hansch [Hansch *et al.*, 1962; Hansch *et al.*, 1963; Fujita *et al.*, 1964], the QSAR/QSPR approach began to assume its modern look. Based on the → *Hammett equation* [Hammett, 1935; Hammett, 1937; Hammett, 1938; Hammett, 1940], steric, electronic and hydrophobic constants were defined, becoming a basic tool for QSAR modelling. In the same years, Free and Wilson developed a model of additive group contributions to biological activities [Free and Wilson, 1964], giving a further push to the take-off of SRC strategies.

Since the development of SRC, methods have made important progress, particularly in the direction of the 3D- and 4D-QSAR approaches. The first formulation of a lattice model to compare molecules by aligning them in 3D space and by extracting chemical information from → *molecular interaction fields* was proposed by Cramer, Patterson, and Bunce [Cramer III *et al.*, 1988b] and Goodford [Goodford, 1985].

During the last decade, a great explosion of molecular descriptors and SRC methods has been observed; besides all the methods for pharmacophore identification, surface areas, and volume descriptors, charges and other quantum-chemical quantities have been extensively enhanced and used as descriptors of the whole molecule. Moreover, there is a tendency to extend traditional → *topological indices* to a three-dimensional representation of the molecule by including geometrical information and/or physico-chemical atomic properties. The first 3D theoretical graph invariant is the 3D-Wiener index [Bogdanov *et al.*, 1989; Mekenyan *et al.*, 1986a].

Attention to the combination of the molecular geometry with chemical atomic information has led to the development of spectral molecular representations given, for example, by → *3D-MoRSE descriptors* [Schuur and Gasteiger, 1996], → *RDF descriptors* [Hemmer *et al.*, 1999], → *spectrum-like descriptors* [Novic and Zupan, 1996], → *EVA descriptors* [Ferguson *et al.*, 1997], → *SWM signals* [Todeschini *et al.*, 1999], and → *Blurock spectral descriptors* [Blurock, 1998].

An increasing interest of the scientific community in recent years has been shown in the fields of combinatorial chemistry, high-throughput screening, → *substructural analysis*, and → *similarity searching*, for which several → *similarity/diversity* approaches have been proposed mainly based on → *substructure descriptors* such as → *molecular fingerprints*, which are particularly suitable for informatic treatment.

A relevant aspect in structure/response correlations is the ability to obtain information about molecular structure from QSAR/QSPR models. In particular, the term **reversible decoding** (or **inverse QSAR**) denotes any procedure capable of re-constructing the molecular structure or fragment starting from molecular descriptor values, i.e. once molecular codes from a structure representation are obtained, reversibility would lead to structures from codes [Baskin *et al.*, 1989; Gordeeva *et al.*, 1990; Zefirov *et al.*, 1991; Skvortsova *et al.*, 1993; Kier *et al.*, 1993; Hall *et al.*, 1993a; Hall *et al.*, 1993b; Kier and Hall, 1993b; Zefirov *et al.*, 1995; Cho *et al.*, 1998; Lukovits, 1998b].

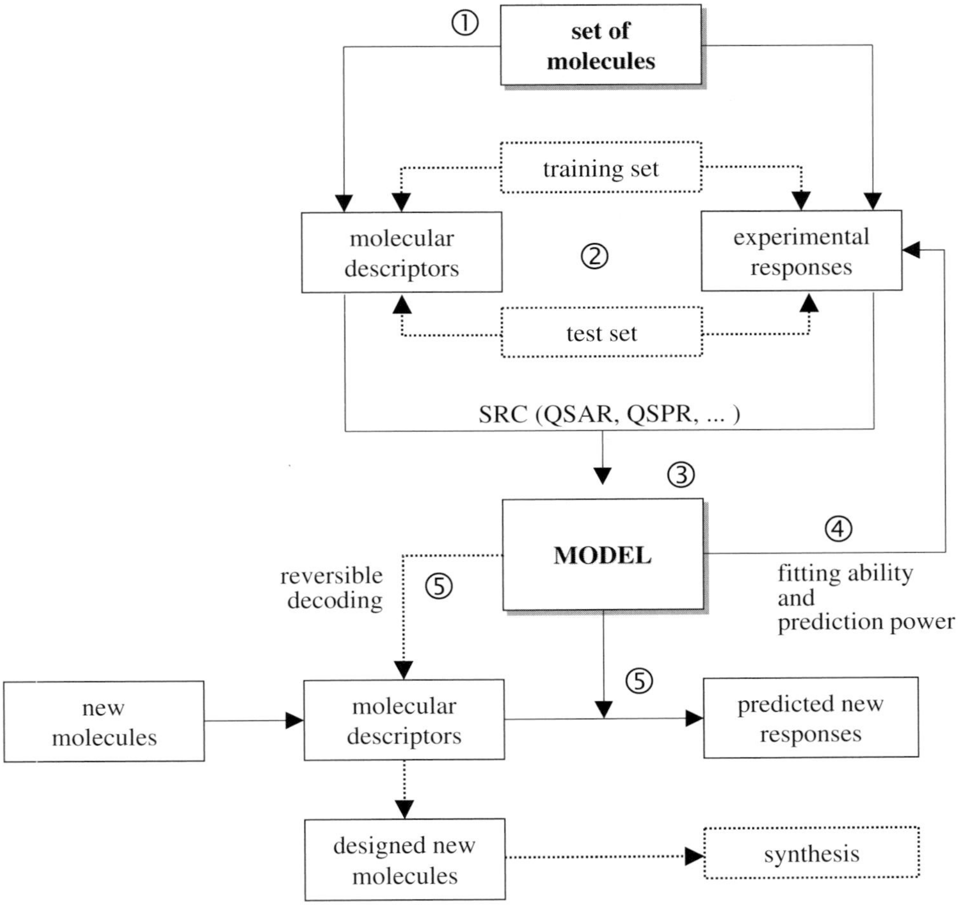

Figure S-5. Role of a model in structure/response correlations.

Reversible decoding is of great importance, since once a SRC model is established optimal values of the response can be chosen and values of the model molecular descriptors calculated by using the estimated SRC model. Then the possible molecular structures corresponding to the optimized descriptor values can be designed (and synthesized). This last operation is a troublesome task as the model molecular descriptors are not simple and easily interpretable.

Reversibility is a highly desired property of a descriptor, but is not strictly essential for structure-response studies; it is closely related to the uniqueness of the descriptor, i.e. to its degree of degeneration.

📖 [Ferguson, 1939] [Taft, 1956] [Wells, 1968a] [Hansch, 1971] [Purcell *et al.*, 1973] [Chapman and Shorter, 1978] [Exner, 1978] [Hall and Kier, 1978a] [Hansch, 1978] [Martin, 1979] [Dearden, 1983] [Golberg, 1983] [Topliss, 1983] [van de Waterbeemd and Testa, 1983] [Jolles and Wooldridge, 1984] [Simon *et al.*, 1984] [Seydel, 1985] [Trinajstic *et al.*, 1986b] [Hadzi and Jerman-Blazic, 1987] [Kier, 1987a] [Seybold *et al.*, 1987] [Hyde and Livingstone, 1988] [Kubinyi, 1988a] [Nirmalakhandan and Speece,

1988b] [Balàz *et al.*, 1990] [Brüggemann *et al.*, 1990] [Karcher and Devillers, 1990a] [Ramsden, 1990] [Stanley and Ostrowsky, 1990] [Voiculetz *et al.*, 1990] [Basak *et al.*, 1991] [Benigni and Giuliani, 1991] [Mezey, 1991a] [Silipo and Vittoria, 1991] [Ariëns, 1992] [Coccini *et al.*, 1992] [Kier and Hall, 1992b] [Rekker, 1992] [Balaban, 1993d] [Hansch, 1993] [Kim, 1993b] [Kubinyi, 1993a] [Kubinyi, 1993b] [McKone, 1993] [Randic *et al.*, 1993b] [Topliss, 1993] [van de Waterbeemd, 1993] [Wermuth, 1993] [Green and Marshall, 1995] [Hansch *et al.*, 1995b] [Hansch and Fujita, 1995] [Hansch and Leo, 1995] [Kubinyi, 1995] [Plummer, 1995] [Sanz *et al.*, 1995] [Katritzky *et al.*, 1995b] [Karelson *et al.*, 1996] [Trinajstic *et al.*, 1996] [Balaban, 1997c] [Basak *et al.*, 1997c] [Kier and Hall, 1997a] [Lipkowitz and Boyd, 1997] [Milne, 1997] [Compadre *et al.*, 1998] [Devillers *et al.*, 1998] [Devillers, 1998] [Hansch *et al.*, 1998] [Kubinyi *et al.*, 1998b] [Kubinyi *et al.*, 1998c] [Martin, 1998] [Reddy *et al.*, 1998] [Schultz *et al.*, 1998] [Selassie and Klein, 1998] [Zakarya and Chastrette, 1998] [Devillers and Balaban, 2000]

subgraph → **graph**

subgraph count set → **count descriptors**

subgraph ID numbers : *fragment ID numbers* → **ID numbers**

subgraph property indices : *SP indices*

submolecular polarity parameter → **charge descriptors**

subspectral graphs → **graph**

substituent constants

Experimentally determined descriptors of molecular substituents in congeneric series representing the variation in a measured molecular property Φ when the considered substituent X replaces a reference group or atom (usually hydrogen) on the skeletal structure.

Substituent constants ϕ_X are estimated as:

$$\phi_X = \log \Phi_X - \log \Phi_0$$

where Φ_X is the property value of the X-substituted compound and Φ_0 is the property value of the reference (parent) compound. The commonly measured → *physico-chemical properties* Φ are the equilibrium and rate constants of specific reactions.

The most popular substituent constants are → *electronic substituent constants*, → *Hansch-Fujita hydrophobic substituent constants* and → *steric substituent constants* such as → *Taft steric constant*, → *Charton steric constant*, → *substituent front strain*, and → *steric density parameter*.

📖 [McDaniel and Brown, 1955] [Hancock and Falls, 1961a] [Charton, 1969] [Charton, 1971] [Hansch *et al.*, 1973b] [Unger and Hansch, 1976] [Fujita and Nishioka, 1976] [Hansch *et al.*, 1977] [Sasaki *et al.*, 1981] [Fujita, 1981] [Fujita, 1983] [Alunni *et al.*, 1983] [Buydens *et al.*, 1983b] [Schultz and Moulton, 1985] [van de Waterbeemd *et al.*, 1989] [Livingstone *et al.*, 1992] [Sasaki *et al.*, 1992] [Stegeman *et al.*, 1993] [Sasaki *et al.*, 1993] [Hansch *et al.*, 1995b] [Hasegawa *et al.*, 1996a] [Amic *et al.*, 1998b] [Hansch *et al.*, 1998]

substituent descriptors

Substituent descriptors represent → *physico-chemical properties* of substituents or functional groups or simply their presence in specific positions of a parent molecule.

All → *structure/response correlations* based on the substituent descriptors involve the hypothesis that the modelled property closely depends more on the characteristics of the molecule substituents than on those of the whole molecule; thus the validity of the → *congenericity principle* and response additivity are assumed.

Well-known substituent descriptors are the → *substituent constants* which are experimentally determined descriptors; among them, → *electronic substituent constants*, → *steric substituent descriptors*, and lipophilicity substituent descriptors such as → *Hansch-Fujita hydrophobic constants* are the most commonly used in QSAR/QSPR modelling.

Moreover, in spite of their holistic character, several molecular physico-chemical properties have also been calculated for substituents; examples are → *molar refractivity*, → *lipohilicity*, and → *surface areas*; size properties of the substituents are often represented by → *Sterimol parameters*.

Among the → *graph-invariants*, several substituent descriptors have been defined such as, for example, → *Kier steric descriptor*, → *steric vertex topological index* and → *fragment molecular connectivity indices*.

A great advantage of the substituent descriptors is that they are calculated just once and then can be used in a congeneric series whenever the substituent is present in a site of the parent molecule, independently of the different molecules to which they are attached, without further calculations. Approaches based on the substituent descriptors usually allow us to perform a → *reversible decoding*; however, they need the congenericity of the training set compounds and in most cases do not account for substituent-substituent and substituent-molecule skeleton interactions.

The most well-known approaches based on substituent descriptors are the → *Hansch analysis* and → *Free-Wilson analysis*; in the latter technique, the substituents are defined by → *indicator variables* representing their presence/absence in the substitution sites of the parent molecule.

Table S-2. Values of some substituent descriptors.

Substituent	MR	π	σ_m	σ_p	$\mathcal{R}$	$\mathcal{F}$	L	B1	B5	*SD*
H	1.03	0.00	0.00	0.00	0.00	0.00	2.06	1.00	1.00	0.000
CH_3	5.65	0.56	−0.07	−0.17	−0.13	−0.04	2.87	1.52	2.04	0.807
C_2H_5	10.30	1.02	−0.07	−0.15	−0.10	−0.05	4.11	1.52	3.17	0.923
n-C_3H_7	14.96	1.55	−0.07	−0.13	−0.08	−0.06	4.92	1.52	3.49	
$CH(CH_3)_2$	14.96	1.53	−0.07	−0.15	−0.10	−0.05	4.11	1.90	3.17	0.970
F	0.92	0.14	0.34	0.06	−0.34	0.43	2.65	1.35	1.35	2.986
Cl	6.03	0.71	0.37	0.23	−0.15	0.41	3.52	1.80	1.80	2.664
Br	8.88	0.86	0.39	0.23	−0.17	0.44	3.82	1.95	1.95	4.994
I	13.94	1.12	0.35	0.18	−0.24	0.40	4.23	2.15	2.15	6.171
NO_2	7.36	−0.28	0.71	0.78	0.16	0.67	3.44	1.70	2.44	2.448
OH	2.85	−0.67	0.12	−0.37	−0.64	0.29	2.74	1.35	1.93	2.270
SH	9.22	0.39	0.25	0.15	−0.11	0.28	3.47	1.70	2.33	
NH_2	5.42	−1.23	−0.16	−0.66	−0.68	0.02	2.78	1.35	1.97	1.228
$NHCH_3$	10.33	−0.47	−0.30	−0.84	−0.74	−0.11	3.53	1.35	3.08	
CHO	6.88	−0.65	0.35	0.42	0.13	0.31	3.53	1.60	2.36	1.507
COOH	6.93	−0.32	0.37	0.45	0.15	0.33	3.91	1.60	2.66	2.037

Substituent	MR	π	σ_m	σ_p	$\mathcal{R}$	$\mathcal{F}$	L	B1	B5	*SD*
OCH_3	7.87	–0.02	0.12	–0.27	–0.51	0.26	3.98	1.35	3.07	
CH_2OH	7.19	–1.03	0.00	0.00	0.00	0.00	3.97	1.52	2.7	1.549
CN	6.33	–0.57	0.56	0.66	0.19	0.51	4.23	1.60	1.60	3.589
SCN	13.40	0.41	0.41	0.52	0.19	0.36	4.08	1.70	4.45	
NCS	17.24	1.15	0.48	0.38	–0.09	0.51	4.29	1.50	4.24	
$COCH_3$	11.18	–0.55	0.38	0.50	0.20	0.32	4.06	1.60	3.13	1.341
$CH{=}CH_2$	10.99	0.82	0.05	–0.02	–0.08	0.07	4.29	1.60	3.09	

MR: molar refractivity; π: hydrophobic substituent constant; σ_m and σ_p: overall electronic constants for *meta-* and *para-*position; $\mathcal{R}$ and $\mathcal{F}$: Swain-Lupton resonance and field constants; L: Sterimol length parameter; B1 and B5: Sterimol B parameters; *SD*: Dash-Behera steric density parameter.

📖 [Wootton *et al.*, 1975] [Borth and McKay, 1985] [van de Waterbeemd *et al.*, 1989] [Breyer *et al.*, 1991] [Peijnenburg *et al.*, 1992a] [Peijnenburg *et al.*, 1992b] [Harada *et al.*, 1992] [Hansch *et al.*, 1995b] [Jurs *et al.*, 1995]

substituent front strain → **steric descriptors**

substituent steric constant sum → **steric descriptors**

substructural analysis → **substructure descriptors**

substructure descriptors

In general, substructure descriptors are the counts of predefined structural features in the molecules or binary variables specifying their presence/absence.

This kind of descriptor is usually used for → *similarity/diversity* studies, i.e. to evaluate the similarity between a → *reference compound* and the large data set compounds, or as independent variables in → *group contribution methods* for the evaluation of molecular properties of interest.

Substructure descriptors can be divided into two main classes, 2D substructure descriptors which are based on a → *topological representation* of the molecules and 3D substructure descriptors based on a 3D → *geometrical representation*. They are usually collected into arrays of different lengths depending on the amount of structural information to be encoded, giving molecular descriptors which are particularly useful in the screening of large databases of chemical compounds and substructure or pharmacophore searching. **Substructure searching** is the most common molecule retrieval mechanism involving the retrieval of all molecules in a database that contains a user-defined substructure query [Hagadone, 1992; Barnard, 1993; Pearlman, 1993; Pepperrell, 1994; Clark and Murray, 1995; Brown and Martin, 1997; Lipkus, 1997; Willett, 1998; Pickett *et al.*, 1998; Wang and Zhou, 1998; Gillet *et al.*, 1998; Gao *et al.*, 1999; Lipkus, 1999].

Substructural analysis is substructure searching where weights are calculated, relating the presence of a specific substructure moiety in a molecule to the probability that the molecule is active in some biological test system [Cramer III *et al.*, 1974; Craig, 1990; Klopman, 1992]. Examples of substructural analysis are the → *Klopman-Henderson cumulative substructure count* and the → *Hodes statistical-heuristic method*.

Molecular fingerprints are string representations of chemical structures designed to enhance the efficiency of chemical database searching and analysis. They can encode the 2D and/or 3D features of molecules as an array of binary values or counts. Therefore, molecular fingerprints consist of *bins*, each bin being a substructure descriptor associated with a specific molecular feature.

3D-fingerprints encoding information on 3D → *molecular geometry* are binary string representations where each bin specifies the count or presence/absence of geometric distances between two atoms or functional groups, as well as torsion angles [Sheridan *et al.*, 1989; Pepperrell and Willett, 1991; Bath *et al.*, 1994; Allen *et al.*, 1995]. Moreover, **triangular descriptors** (or **triplet descriptors**) were proposed which describe the relative positions of three atoms or group centroids in the molecule. Each possible triplet of non-hydrogen atoms is taken as a triangle, and different triangle measures have been proposed such as individual triangle side lengths, triangular perimeter and area; these measures are integerized and transformed into single bit integers of defined length by different procedures, and their distribution is used to describe the molecule. They are used both to characterize molecular shape and for 3D pharmacophore database searching [Bemis and Kuntz, 1992; Nilakantan *et al.*, 1993; Bath *et al.*, 1994; Good and Kuntz, 1995; Good *et al.*, 1995b]. Potential pharmacophore point pairs (**PPP pairs**) and potential pharmacophore point triangles (**PPP triangles**) are 3D fingerprints encoding, respectively, the distance information between all possible combinations of two and three potential pharmacophore points [Brown and Martin, 1996]. Potential pharmacophore points usually considered are hydrogen bond donors and acceptors, sites of potential negative and positive charge, and hydrophobic atoms. All atoms of the molecule are analyzed to see whether they can be classed potentially as one of these point types. Moreover, there are other 3D fingerprints similar to PPP-triangles that are based on extended sets of potential pharmacophore points containing also polar hydrogen, nonpolar hydrogen, nitrogen, and oxygen together with aromatic ring centres and triple and double bond centres [Chen *et al.*, 1998; McGregor and Muskal, 1999].

Rigid 3D-fingerprints consider only a single conformation of the molecule, while flexible 3D-fingerprints account for all the geometric features that can be achieved, based on a conformational exploration of the considered molecule, usually based on the incremental rotation of all the rotatable bonds.

Structural keys are molecular fingerprints which define the presence or absence of predefined structural fragments, encoding this information in an array of binary values; moreover, with structural keys, it is possible to record also the number of occurrences of each fragment. Structural keys rely on the use of a predefined fragment dictionary, which is a list of fragments selected to be of importance. For an efficient use of the structural keys in library searching, the set of selected fragments should obey two principles:

1) independence of occurrence, avoiding two fragments occuring together and thus resulting in redundant information;

2) fragments should be of approximately equifrequent distribution in order to avoid lack of generality.

Hashed fingerprints are molecular fingerprints obtained by hashing single structural fragments, with the aim of reducing the length of the fingerprint, i.e. the number of fingerprint bins [Ihlenfeldt and Gasteiger, 1994]. Therefore, whereas in a traditional fingerprint a single bin represents the presence of a single fragment or the counting of the number of times each single fragment occurs, in hashed fingerprints different fragments may be counted into the same bin because of the restriction on fingerprint length (number of bins).

Molecular holograms are a special form of 2D hashed fingerprints consisting of counts of predefined molecular fragments and substructures occurring in the molecule, defined by a set of rules [Hurst and Heritage, 1997; Tripos Inc., 1997; Seel *et al.*, 1999]. First, a molecule is decomposed into a number of predefined fragments that are assigned fragment integer identification numbers; then, each fragment is counted in a particular bin based on its fragment integer ID corresponding to the hologram bin ID. Therefore, the values of the bins are the hologram descriptors, i.e. the count of all the fragments in each bin. Typically, as the number of unique fragments contained in the molecule is much larger than the number of holographic bins, the hashing procedure is used to map the different single fragments to the same bin, causing *fragment collision*. The hashing process occurs when the value of the fragment integer ID is larger than the hologram length; thus the value of the remainder when the integer is divided by the hologram length identifies the holographic bin whose value has to be incremented.

In hologram generation, some fragment parameters have to be settled: two fragment size parameters determine the maximum and minimum number of atoms in any one fragment (the default values ranging from 4 to 7 atoms in length), and five parameters allow the distinction of atoms, bonds, connectivity, hydrogens, and fragment chirality. Typical hologram lengths range from 50 to 500, even if the number of distinct fragments generated according to the above cited parameters is around 1000.

Molecular holograms do not require molecular alignment and are generated very rapidly, providing an efficient approach for screening a large number of compounds from chemical databases. The QSAR approach based on molecular holograms was called **Hologram QSAR** (HQSAR) [Winkler *et al.*, 1998b; Winkler and Burden, 1998a; Tong *et al.*, 1998; So and Karplus, 1999; Lowis, 1997; Burden and Winkler, 1999a; Viswanadhan *et al.*, 1999].

Some examples of codes for substructure identification are listed below.

- **atom pairs**

Arbitrary codes defined to describe pairs of atoms and bond types connecting them. For example, an atom pair can be defined as a substructure composed of two non-hydrogen atoms and an interatomic separation:

$$AP = \{ [i\text{th atom description}]\ [\text{separation}]\ [j\text{th atom description}] \}$$

The two considered atoms need not be directly connected and the *separation* is the → *topological distance* between them [Carhart *et al.*, 1985]; these descriptors are usually called **topological atom pairs** being based on the topological representation of the molecules. Atom type is defined by the element itself and the number of neighbours and π electrons. An extension of these atom pairs is given by **binding property pairs** that were analogously proposed, even if the atoms are assigned to seven binding classes: cations, anions, hydrogen-bond donors and acceptors, polar atoms, hydrophobic atoms and other [Kearsley *et al.*, 1996; Sheridan and Miller, 1998].

Moreover, **geometric atom pairs** and **geometric binding property pairs** were proposed, substituting the topological distance between two atom types with the → *geometric distance* [Sheridan *et al.*, 1996]. PPP-pairs are a particular case of geometric atom pairs where only pairs of the five potential pharmacophore points are considered [Brown and Martin, 1996].

Atom pairs are sensitive to long-range correlations between atoms in molecules and therefore to small changes even in large molecules. Although they can be used as substructure descriptors, they are a too generic representation of the molecule.

• **REX descriptors**
Similar to atom pairs descriptors, REX descriptors are codes defined by a pair of "terminators" and the link between them [Judson, 1992b]:

$$\text{REX} = \{\ [i\text{th terminator}]\ [j\text{th terminator}]\ [\text{link length}]\ \}$$

A terminator may be an atom, a lone pair or a bond; the link is derived from a topological representation of the molecules as the length of the path between the considered terminators. For each pair of terminators, different REX descriptors are defined according to each different link between them, i.e. all paths and not only the shortest path are considered.

As terminators, all the atoms of interest (i.e. C, H, N, P, O, S, halogens) can be considered; terminators of the molecule not explicitly considered are classed together as a single dummy terminator (M). Links having two carbon terminators are usually excluded from REX analysis, considerably reducing the total number of links associated with each structure. However, for an analysis of saturated hydrocarbons the links between carbon atoms are included.

Individual hydrogen atoms are included as terminators only if they are attached to specific atom types such as, for example, oxygen, nitrogen, or sulfur atoms. When distinguishing REX descriptors the type of atoms and bonds connecting the terminators is not taken into account.

With respect to the atom pairs descriptors, REX descriptors use more generalized linear fragments, allowing more complex substructures to be dealt with.

• **Augmented Atom Codes** (AAC) (*: atom-centred fragment codes, augmented fragments*)
The AAC is a short-range atom-centred code that describes each atom by its own atom type and the bond types and atom types of its first neighbours. Functionalities in a molecule can be represented by two to five atoms (corresponding to one bond to 4 bonds), which consist of a central atom and its neighbouring bonded atoms. Hydrogen atoms are not considered. Each molecule is described completely by A fragment codes, A being the number of non-hydrogen atoms [Adamson *et al.*, 1970; Hodes *et al.*, 1977; Hodes, 1981a; Hodes, 1981b; Baumann and Clerc, 1997b].

Each fragment is represented by a single-value AAC descriptor. All AAC descriptors representing all fragments in the data set molecules are recorded in an arbitrary but fixed way in a uniform-length multidimensional vector.

The use of these substructure descriptors greatly increases the specific chemical information regarding different functional groups, but cannot discriminate between different arrangements of functional groups within a molecule.

An extension of AAC descriptors is given by the **ganglia-augmented atom codes** (gAA), where "triplet ganglia" are defined as sets of three attached non-hydrogen atoms, together with their pendant bonds [Hodes *et al.*, 1977; Hodes, 1981a; Hodes, 1981b; Tinker, 1981].

📖 [Mekenyan *et al.*, 1987; Meylan *et al.*, 1992; Lohninger, 1994; Baumann and Clerc, 1997b]

• **linear subfragment descriptors**
Fragment codes defined by a linear sequence of three interconnected non-hydrogen atoms, each described by its bonded hydrogen atoms; moreover, two sets of labels are associated with the sequence, one indicating the multiplicity of the bonds that join the atoms of the sequence, the other indicating the presence of a side chain consisting of a terminal functionality such as a halogen, NH_2, COOH, etc. [Klopman, 1984].

• **topological torsion descriptor** (TT)

The topological torsion is defined as a linear sequence of four consecutively bonded non-hydrogen atoms *k-i-j-l*, each described by its atom type (TYPE), the number of π electrons (NPI) and the number of non-hydrogen atoms (NBR) bonded to it [Nilakantan *et al.*, 1987]. Usually NBR does not include *k-i-j-l* atoms that go to make the torsion itself; therefore it is minus 1 for *k* and *l* atoms, and minus 2 for the two central atoms *i* and *j*. The torsion around the *i-j* bond and defined by the four indices *k-i-j-l* is represented by the following TT descriptor:

$$\mathrm{TT} = \{\, [\mathrm{NPI{-}TYPE{-}NBR}]_k\, [\mathrm{NPI{-}TYPE{-}NBR}]_i\, [\mathrm{NPI{-}TYPE{-}NBR}]_j\, [\mathrm{NPI{-}TYPE{-}NBR}]_l \,\}$$

The TT descriptor is a topological analogue of the 3D torsion angle, defined by four consecutively bonded atoms. Moreover, it is an extension of the triplet ganglia-augmented atom codes of Hodes. The topological torsion is a short-range descriptor, i.e. it is sensitive only to local changes in the molecule and is independent of the total number of atoms in the molecule.

An extension of topological torsions is given by **binding property torsions** that were proposed in an analogous way, but the atoms were assigned to seven different types: cations, anions, hydrogen-bond donors and acceptors, polar atoms, hydrophobic atoms and others [Kearsley *et al.*, 1996].

📖 [Sheridan and Venkataraghavan, 1987] [Martin *et al.*, 1988] [Sheridan *et al.*, 1989] [Hicks and Jochum, 1990] [Randic, 1992c] [Tratch *et al.*, 1992] [Attias and Petitjean, 1993] [Zhou *et al.*, 1993] [Jackel and Nendza, 1994] [Merschsundermann *et al.*, 1994] [Takihi *et al.*, 1994] [Jordan *et al.*, 1995] [Brown, 1997] [Matter, 1997] [Xiao *et al.*, 1997] [Brown and Martin, 1998] [Flower, 1998] [Molchanova and Zefirov, 1998] [Matter and Pötter, 1999] [Varmuza and Scsibrany, 2000]

substructure list representation → **molecular descriptors**

substructure searching → **substructure descriptors**

sum layer matrix → **layer matrices**

sum of matrices → **algebraic operators**

sum of steric substituent constants → **steric descriptors**

supercode → **superindices**

superdelocalizability → **quantum-chemical descriptors**

superdelocalizability indices → **quantum-chemical descriptors**

superindices

Superindices were proposed as ordered sequences of → *molecular descriptors* providing different chemical information. Thus, in order to obtain indices with higher discrimination power among isomers and molecular structures, different superindices were proposed [Balaban, 1979].

The most popular superindices are → *uniform length descriptors* of → *topological information indices* such as [Bonchev *et al.*, 1981b]:

$$SI = \left\langle {}^{V}\bar{I}_D^M; {}^{E}\bar{I}_\chi^E; \bar{I}_Z; {}^{V}\bar{I}_{C,R}; \bar{I}_{ORB}; {}^{E}\bar{I}_{CHR} \right\rangle$$

where the sequence terms are → *mean information content on the distance magnitude*, → *mean information content on the edge equality*, → *Hosoya mean information index*,

→ *radial centric information index*, → *vertex orbital information content*, → *edge chromatic information index*, respectively.

In order to preserve the information contained in the element partition of a graph, a **supercode** is obtained by replacing each information index of the superindex by the cardinalities of the equivalence classes used to calculate it.

Simple sums of superindex elements can also be considered superindices belonging to the class of → *combined descriptors*; some of them were proposed as measures of → *molecular complexity*.

📖 [Motoc and Balaban, 1981] [Bonchev, 1983]

superpendentic index

A molecular descriptor derived from the → *H-depleted molecular graph* and proposed to enhance the role of terminal vertices in QSAR and QSPR studies [Gupta *et al.*, 1999]. It is calculated from the **pendent matrix,** which is a submatrix of the → *distance matrix* **D** with A rows and a number m of columns corresponding to the number of terminal vertices. The superpendentic index is calculated as the square root of the sum of the products of the nonzero row elements (the topological distances d) in the pendent matrix:

$$\int^{p} = \left(\sum_{i=1}^{A} \prod\nolimits_{m} d_{im} \right)^{1/2}$$

where m is the number of terminal vertices, i.e. the columns of the pendent matrix.

Example : 2-methylpentane

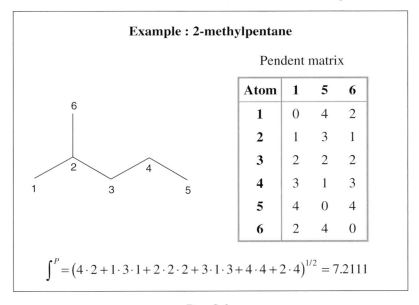

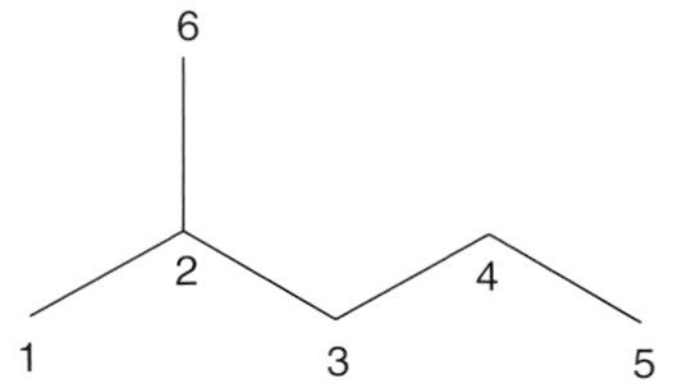

Pendent matrix

Atom	1	5	6
1	0	4	2
2	1	3	1
3	2	2	2
4	3	1	3
5	4	0	4
6	2	4	0

$$\int^{P} = (4 \cdot 2 + 1 \cdot 3 \cdot 1 + 2 \cdot 2 \cdot 2 + 3 \cdot 1 \cdot 3 + 4 \cdot 4 + 2 \cdot 4)^{1/2} = 7.2111$$

Box S-3.

📖 [Bakken and Jurs, 1999b]

surface areas → **molecular surface**

surface electrostatic potential variance → **GIPF approach**

surface factor → **shape descriptors** (⊙ shape factor)

surface profiles → **molecular profiles**

surface-volume ratio → **shape descriptors** (⊙ ovality index)

surface weighted charged partial negative surface areas → **charged partial surface area descriptors**

surface weighted charged partial positive surface areas → **charged partial surface area descriptors**

Suzuki-Kudo hydrophobic fragmental constants → **lipophilicity descriptors**

Swain-Lupton approach → **electronic substituent constants** (⊙ field/resonance effect separation)

Swain-Lupton field constant : *field-inductive constant* → **electronic substituent constants** (⊙ field/resonance effect separation)

Swain-Lupton resonance constant : *resonance constant* → **electronic substituent constants** (⊙ field/resonance effect separation)

SWIM descriptors : *spectral weighted invariant molecular descriptors* → **SWM signals**

SWM signals (: *Spectral Weighted Molecular signals*)

Spectral weighted molecular signals are calculated within the framework on which the → *WHIM descriptors* are defined. For each molecule, the ith atom score of each mth principal axis t_{im} represents the atom projection along the axis, and the weight w_{ij} of the atom is taken as the signal intensity in that position [Todeschini *et al.*, 1999]. The weights w are atomic properties, e.g. the weights used in the WHIM approach.

When more than one atom falls in the same position, i.e. has the same projection, the signal intensity is the sum of the weights of the overlapping atoms having the same score. To avoid score values which differ only in numerical approximation due to the calculation procedure, such values are approximated to the first decimal digit (0.1). Therefore, a weighted spectral representation of a molecule is given by a set of ordered signals defined by score-intensity pairs

$$\langle t_{s(i),m},\ w_{s(i),m} \rangle \qquad m = 1,\ 2,\ 3$$

where m represents the mth principal axis and $s(i)$ the ordered sequence of the scores for each principal axis, from the most negative to the most positive values. Each score in the $s(i)$ position corresponds to the ith set of atoms having the same score. The complete spectral representation is given by fixing the range of $\mathbf{t}$ vector scores (for example, between –20 and +20) and then juxtaposing the three axial spectral representations, from the first to the third principal axis.

The maximum total number of signals is $A \times 3$, where A is the number of molecule atoms; the actual total number of signals depends on the molecular symmetry.

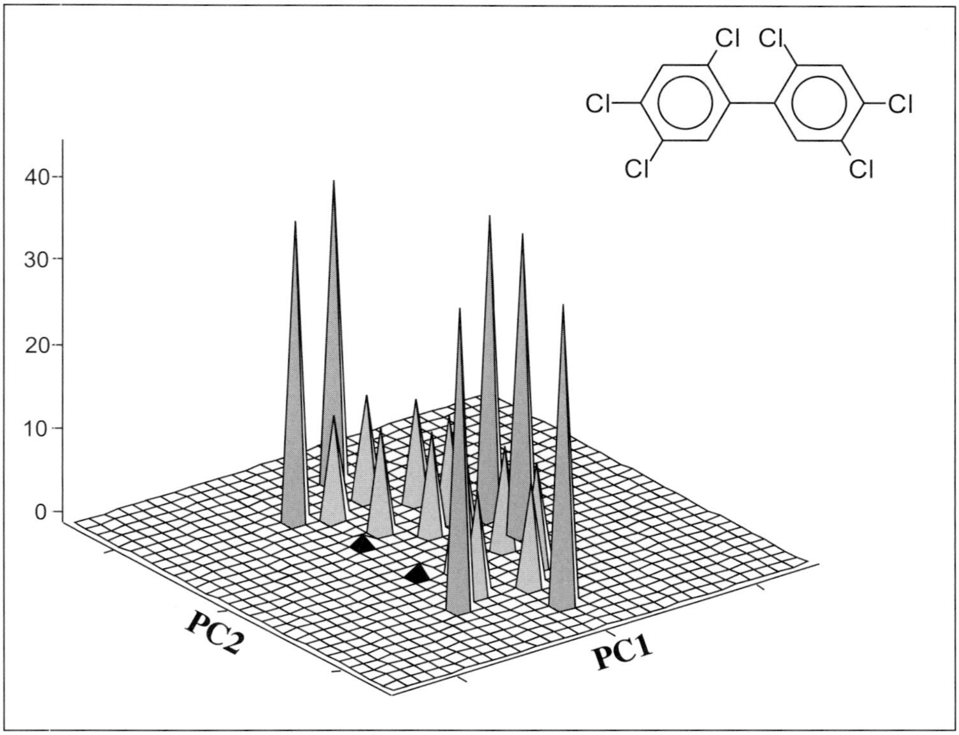

Figure S-6. SWM signals of a hexachloro-biphenyl derivative.

Because of the properties of PCA, the projection along each principal axis is not totally invariant as only direction is uniquely defined: thus, changing the signs of the scores $\boldsymbol{t} \rightarrow -\boldsymbol{t}$, the reversed axial spectrum is equally valid. The principal axes being three, 2^3 different spectra of a molecule can be obtained. Therefore, to compare different spectra there are three possibilities:

a) to take the two reversed spectra for each principal axis into account.
b) to derive invariant molecular descriptors from their spectra;
c) to first adopt → *alignment rules.*

Procedure a) was adopted for the similarity analysis of compounds that are each represented by a complete spectral representation.

Given two molecules *a* and *b*, each axial spectrum is compared by considering the two orientation possibilities for one of the two molecules; then, for each orientation, a rigid shift of spectral signals of one molecule with respect to the other one is performed by a defined step (0.1). The maximum value of similarity found by this procedure is taken as the measure of similarity between the two molecules relative to the considered axial spectrum.

The most appropriate measure of distance was taken to be the average *Camberra distance* $\mathrm{d}_{ab,m}$ between the molecules *a* and *b* along the *m* th principal axis:

$$\mathrm{d}_{ab,m} = \frac{1}{n_m} \cdot \sum\nolimits_i \frac{|w_{ai,m} - w_{bi,m}|}{(w_{ai,m} + w_{bi,m})}$$

where $w_{ai,m}$ and $w_{bi,m}$ are, respectively, the ith intensities of the molecules a and b along the mth principal axis; n_m is the number of non-overlapping signals along the mth axis; the sum runs over all the n_m non-overlapping signals which are:

$$n_m = n_{am} + n_{bm} - n_{m(\text{overlap})}$$

where n_{am} and n_{bm} are the number of signals along the mth axis for the two molecules, respectively, and $n_{m(\text{overlap})}$ the number of overlapping signals of the two molecular axial spectra, given the actual relative position due to the shift procedure.

If, in correspondence to a signal of a molecule the other molecule does not present a signal, the denominator reaches the maximum value, i.e. d = 1. The Camberra distance is transformed into the corresponding similarity measure:

$$s_{ab,m} = 1 - \mathrm{d}_{ab,m}$$

the average Camberra distance being normalized between zero and one.

The total similarity is calculated as the geometric mean of the axial similarities as:

$$S_{ab}(w_j) = \sqrt[3]{s_{ab,1} \cdot s_{ab,2} \cdot s_{ab,3}}$$

If more than one weight is considered, the final measure of similarity is given as the geometric mean of the k total similarities obtained separately for each weight:

$$S_{ab} = \sqrt[k]{S_{ab}(w_1) \cdot \ \ldots \ \cdot S_{ab}(w_k)}$$

Procedure b) was adopted to obtain a set of molecular → *autocorrelation descriptors* calculated separately from each axial spectral representation. These descriptors were called **Spectral Weighted Invariant Molecular descriptors** (or **SWIM descriptors**).

As along each component the SWM signals of a molecule are naturally ordered by the scores, the autocorrelation of *lag l* for each mth axis AC_{lm} is calculated as:

$$AC_{lm} = \sum_{i=1}^{\mathrm{N}_m - l} w_{i,m} \cdot w_{i+l,m}$$

where N_m is the number of signals of the molecule on the mth axis; $w_{i,m}$ and $w_{i+l,m}$ are the weights of the molecule at positions i and $i + l$.

For each considered weight, a molecule is represented by a vector of autocorrelation descriptors:

$$\langle AC_{01},\ AC_{02},\ AC_{03};\ AC_{11},\ \ldots;\ AC_{L1}, AC_{L2}, AC_{L3}\rangle_w$$

where L is the maximum user-defined lag (e.g. L = 5). The autocorrelations for *lag* zero (l = 0) correspond to the sums of the squared weights, i.e. the square intensity of each signal.

These descriptors are invariant to the direction of the axial spectral representations and constitute → *uniform length descriptors.*

symmetric Cluj-detour matrix → Cluj matrices

symmetric Cluj-distance matrix → Cluj matrices

symmetric Szeged matrix → Szeged matrix

symmetry descriptors

Symmetry is an important but often elusive molecular property when numerical values must be assigned. However, symmetry plays an important role in the quantum mechanical interpretation of atomic and molecular states, NMR spectra and several physico-chemical properties. Symmetry is closely related to those molecular properties which also depend on entropy contributions, such as, for example, melting point, vapour pressure, surface tension, and → *dipole moment*. Moreover, the nature of overall molecular shape depends on molecular symmetry.

Each molecule (or conformation) belongs to a definite point group of symmetry and each point group of symmetry includes a set of symmetry operations which are transformations leaving the whole system in a position equivalent to the initial one: identity, rotation, mirror reflection, inversion, mirror rotation. The various groups of symmetry are:

$$C_1, C_s, C_i, C_n, C_{nV}, C_{nh}, D_{nh}, T_d, \text{etc.}$$

Some → *shape descriptors* and → *centric indices* contain information about symmetry, while → *WHIM symmetry*, → *Bertz complexity index*, → *indices of neighbourhood symmetry*, and → *symmetry number* obtained by the → *automorphism group* of a graph are explicitly related to the symmetry.

Other specific symmetry descriptors are listed below.

- **information index on molecular symmetry** (I_{SYM})

A molecular symmetry descriptor calculated as → *total information content*:

$$I_{SYM} = A \cdot \log_2 A - \sum_{g=1}^{G} A_g \log_2 A_g$$

where A is the number of atoms, A_g the number of atoms belonging to the gth class of symmetry, and G the number of different classes of symmetry in the molecule [Bonchev *et al.*, 1976b]. Each class of symmetry includes atoms able to exchange position through operations of the symmetry point group to which the molecule belongs.

Both the number of atoms and the degree of symmetry influence the increase in information content. In practice, for molecules with the same number of atoms, the information content increases with decrease in symmetry; for example, molecules with symmetry groups C_1 or C_s have the largest information content.

In general, molecular symmetry is closely related to the 3D geometry of the molecules, and can be determined only from atomic coordinates; moreover, topological equivalence does not necessarily reflect all the symmetry of the 3D molecular geometry; however, it may be in some cases a useful approximation. In fact the information index I_{SYM}, in principle based on knowledge of the molecular geometry, is related to the → *total information index on atomic composition* I_{AC} and the → *topological information content* I_{TOP}; I_{SYM} is usually greater than these two indices. However, I_{SYM} coincides with I_{AC} when the symmetry group contains a number of symmetry operations for which all the atoms of the same chemical elements are equivalent; I_{SYM} is coincident with I_{TOP} when each different valency in the molecule is realized by atoms of only one chemical element.

I_{SYM} was used in combination with other information indices to define general measures of molecular complexity, i.e. → *Dosmorov complexity index* and → *Bonchev complexity index*.

- **Kier symmetry index** (${}^0\kappa$)

Proposed as an extension of the → *Kier shape descriptors* to account for zero order paths, i.e. the atoms A, it is defined as the total information content of the molecule [Kier, 1987b]:

$${}^0\kappa = -A \cdot \sum_{g=1}^{G} \frac{A_g}{A} \cdot \log_2 \frac{A_g}{A}$$

where A_g is the number of topologically equivalent atoms in the gth class. Each equivalence class is constituted by all atoms having the same → *valence topological state* S^v.

It was proposed with the aim of measuring molecular symmetry in terms of atom topological uniquenes, the lower the $^0\kappa$ value the greater being the topological symmetry. The same index with the opposite trend can also be calculated as the → *redundancy index*.

• **Merrifield-Simmons index** (σ)
It is defined as the number of open sets in the topology $\mathcal{T}$ of the graph $\mathcal{G}$, namely $\sigma(\mathcal{G})$ [Merrifield and Simmons, 1980; Merrifield and Simmons, 1998].

The *topology* $\mathcal{T}$ is the collection of all unions of *basis sets*, remembering that to each vertex v_i there can be associated a unique irreducible basis set $\mathcal{B}_i$. Thus, to produce only distinct unions it is required that no two elements entering the union be comparable. Since comparable basis elements are those corresponding to the adjacent vertices of $\mathcal{G}$, distinct open sets are guaranteed only if basis element unions, corresponding to sets of vertices of $\mathcal{G}$ no two of which are adjacent, are formed. In graph theory, such a set of vertices is called *stable set* of $\mathcal{G}$. Thus, $\sigma(\mathcal{G})$ is the cardinality of a graph topology and is equal to the number of stable sets of the graph $\mathcal{G}$. In other words, it is defined as:

$$\sigma(\mathcal{G}) = \sum\nolimits_k q(\mathcal{G}, k)$$

where $q(\mathcal{G}, k)$ is the number of ways of choosing k disjoint vertices from $\mathcal{G}$. It was found to be inversely correlated to the → *Hosoya Z index* [Gutman *et al.*, 1992; Hosoya *et al.*, 1999].

The method of calculating $\sigma(\mathcal{G})$ is to examine every subset S of $\mathcal{V}(\mathcal{G})$, but this is impractical even for small graphs. In practice, $\sigma(\mathcal{G})$ is usually recursively calculated, considering the relation between the stable sets of $\mathcal{G}$ and those of $\mathcal{G}$ with a vertex v_i removed, i.e. $\mathcal{G}' \equiv \mathcal{G} - v_i$. Since, each stable set of $\mathcal{G}'$ is also a stable set of $\mathcal{G}$, the recursion formula is defined as:

$$\sigma(\mathcal{G}) = \sigma(\mathcal{G}') + \sigma(\mathcal{G}' - S_{\delta_i})$$

where $\mathcal{G}' - S_{\delta i}$ is the subgraph obtained from $\mathcal{G}$ by deleting the vertex v_i together with the set $S_{\delta i}$ of all the vertices adjacent to v_i, δ_i being the → *vertex degree* of the ith atom.

The calculation is performed bearing in mind that the recursion formula:

$$\sigma(\mathcal{G}_{n+1}) = \sigma(\mathcal{G}_n) + \sigma(\mathcal{G}_{n-1})$$

holds for a linear graph of n vertices with a vertex added to one end of the graph.

This recursion formula is identical to the definition of **Fibonacci numbers**:

$$F_{n+1} = F_n + F_{n-1} \qquad \text{where} \qquad F_0 = F_1 = 1$$

which are integers with a simple combinatorial meaning: F_{n+1} is the number of subsets of the set $\{1, ..., n\}$ such that no two elements are adjacent [Gutman and El-Basil, 1986; Randic *et al.*, 1996b].

Hence, for linear graphs L_n, the Merrifield-Simmons index is given by:

$$\sigma(L_n) = F_{n+1}$$

Moreover, again for linear graphs (i.e. n-alkanes), the Hosoya Z index coincides with the Fibonacci number F_n, and thus Hosoya and Merrifield-Simmons indices are both closely and directly related. For monocyclic graphs C_n and iso-path graphs $i - L_n$ the following relationships between Fibonacci numbers and the Merrifield-Simmons index hold:

$$\sigma(C_n) = F_n + F_{n-2} \qquad \sigma(i - L_n) = F_{n+1} + F_{n-4}$$

Table S-3 collects Fibonacci numbers and the corresponding Merrifield-Simmons indices for number n of vertices between 0 – 16.

Table S-3. Fibonacci numbers (F_n) and corresponding Merrifield-Simmons indices for linear (L_n), cyclic (C), and iso-path graphs (i-L_n).

n	F_n	$\sigma(L_n)$	$\sigma(C_n)$	$\sigma(i\text{-}L_n)$
0	1	1		
1	1	2		
2	2	3	3	
3	3	5	4	
4	5	8	7	9
5	8	13	11	14
6	13	21	18	23
7	21	34	29	37
8	34	55	47	60
9	55	89	76	97
10	89	144	123	157
11	144	233	199	254
12	233	377	322	411
13	377	610	521	665
14	610	987	843	1076
15	987	1597	1364	1741
16	1597	2584	2207	2817

The Merrifield-Simmons index $\sigma(\mathcal{G})$ is a molecular descriptor quite sensitive to the molecular topology, in particular to symmetry, branching and cyclicity; $\sigma(\mathcal{G})$ increases with branching and decreases with cyclicity.

From the graph topology, quantitative measures of the contribution of each bond in the molecule to the topological space can be obtained. Such a measure is the total number of open sets containing the bond i-j, namely n_{ij}, and the sum of their cardinalities, namely c_{ij}. In particular, the quantity n_{ij} was found to be related to the → *bond order* by normalizing n_{ij} with respect to $\sigma(\mathcal{G})$. The **Merryfield-Simmons bond order** is defined as the best found nonlinear relationship between bond orders and the quantity n_{ij}/σ :

$$\pi_{ij} = \exp\left[3.81 \cdot \left(\frac{n_{ij}}{\sigma} - 0.67\right)\right]$$

• **symmetry factor** (σ_{SYM})

The symmetry factor was used in the expression of the entropy of acyclic saturated hydrocarbons proposed by Pitzer in terms of molecular constants by enumerating the partition functions for the molecular motions [Pitzer, 1940; Pitzer and Scott, 1941]. It is defined as:

$$\sigma_{\text{SYM}} = \frac{\sigma_e \cdot \sigma_i}{\text{I}_\text{R}}$$

where σ_e is the symmetry number for external rotations, σ_i the symmetry number for internal rotations, and I_R the number of racemic isomers ($\text{I}_\text{R} = 2$ for racemic mixture, $\text{I}_\text{R} = 1$ otherwise) (Table S-4).

Table S-4. Values of the symmetry numbers for some compounds.

Compound	σ_e	σ_i	I_R	σ_{SYM}
n-Butane	2	1	1	2
2-Methyl-propane	3	1	1	3
n-Pentane	2	1	1	2
1-Methyl-butane	1	1	1	1
2,2-Dimethyl-propane	12	1	1	12
n-Hexane	2	1	1	2
2-Methyl-pentane	1	1	1	1
3-Methyl-pentane	1	1	1	1
2,3-Dimethyl-butane	2	1	1	2
2,2-Dimethyl-butane	1	3	1	3
3-Methyl-hexane	1	1	2	0.5

📖 [Kitaigorodsky, 1973] [Cohen *et al.*, 1974] [Randic *et al.*, 1981] [Narumi and Hosoya, 1985] [Balaban, 1986b] [Randic *et al.*, 1986] [Mezey, 1990a] [Rücker and Rücker, 1990] [Rücker and Rücker, 1991b] [Figueras, 1992] [Dannenfelser *et al.*, 1993] [Mezey, 1997b] [Caporossi and Hansen, 1998] [Ivanov and Schüürmann, 1999]

symmetry factor → **symmetry descriptors**

symmetry number → **graph**

Szeged delta matrix → **Szeged matrix**

Szeged fragmental indices → **Szeged matrix**

Szeged index → **Szeged matrix**

Szeged matrix (SZ)

The Szeged matrix $\mathbf{SZ_U}$ of an → *H-depleted molecular graph* $\mathcal{G}$ is a square unsymmetric $A \times A$ matrix, A being the number of graph vertices, whose off-diagonal entry *i*-*j* is the number of vertices $N_{i,(ij)}$ lying closer to the focused vertex v_i [Khadikar *et al.*, 1995; Diudea, 1997a; Diudea, 1997c]:

$$[\mathbf{SZ_U}]_{ij} = N_{i,(ij)}$$

where

$$N_{i,(ij)} = \left|\{a | a \in \mathcal{V}(\mathcal{G});\ d_{ia} < d_{ja}\}\right|$$

d being the → *topological distance* and $\mathcal{V}(\mathcal{G})$ the set of graph vertices; note that the vertices equidistant to v_i and v_j are not counted and the value of $N_{i,(ij)}$ depends on both v_i and v_j. The diagonal elements are equal to zero by definition. The Szeged matrix is similar to the → *Cluj-distance matrix* $\mathbf{CJD_U}$ except for the supplementary condition in the Cluj matrix:

$$p_{ia} \cap p_{ij} = \{i\}$$

where p_{ij} is the shortest path between the considered vertices.

The square **symmetric Szeged matrix SZ**, of dimension $A \times A$, is obtained from the unsymmetric Szeged matrix $\mathbf{SZ_u}$ by the relation

$$\mathbf{SZ} = \mathbf{SZ_U} \bullet SZ_U^T \quad \text{or} \quad [\mathbf{SZ}]_{ij} = [\mathbf{SZ_U}]_{ij}[\mathbf{SZ_U}]_{ji}$$

where the symbol • indicates the → *Hadamard matrix product*.

In other words, the off-diagonal entry *i-j* of the symmetric Szeged matrix **SZ** is the product of the numbers $N_{i,(ij)}$ and $N_{j,(ij)}$ of the vertices lying closer to the vertices v_i and v_j, respectively (the equidistant vertices not being included):

$$[\mathbf{SZ}]_{ij} = N_{i,(ij)} \cdot N_{j,(ij)}$$

where

$$N_i = \left|\{a | a \in \mathcal{V}(\mathcal{G});\ d_{ia} < d_{ja}\}\right|$$
$$N_j = \left|\{a | a \in \mathcal{V}(\mathcal{G});\ d_{ja} < d_{ia}\}\right|$$

The diagonal entries of the matrix are always equal to zero by definition.

The 1st order sparse symmetric Szeged matrix $^1\mathbf{SZ}$ is obtained as:

$$^1\mathbf{SZ} = \mathbf{SZ} \bullet A$$

where the symbol • is the Hadamard matrix product and **A** the → *adjacency matrix*; the 1st order sparse unsymmetric Szeged matrix $^1\mathbf{SZ_U}$ is analogously obtained. For any graph, both sparse symmetric and unsymmetric Szeged matrices coincide with the corresponding 1st-order sparse Cluj-distance matrix and → *Cluj-detour matrix*.

The **Szeged index** *SZD* [Khadikar *et al.*, 1995; Gutman and Klavzar, 1995; Zerovnik, 1996; Zerovnik, 1999] is obtained from the symmetric Szeged matrix **SZ** as the sum of the matrix entries corresponding to all 1st-order paths above the main diagonal, i.e. applying the → *Wiener operator* $\mathcal{W}$ to the 1st-order sparse symmetric Szeged matrix $^1\mathbf{SZ}$ or from the unsymmetric Szeged matrix $\mathbf{SZ_U}$ applying the → *Wiener orthogonal operator* $\mathcal{W}'$:

$$SZD = \mathcal{W}(^1\mathbf{SZ}) = \frac{1}{2}\sum_{i=1}^{A}\sum_{j=1}^{A}[^1\mathbf{SZ}]_{ij} = \mathcal{W}'(^1\mathbf{SZ_U}) = \sum_{i=1}^{A-1}\sum_{j=i+1}^{A}[^1\mathbf{SZ_U}]_{ij}[^1\mathbf{SZ_U}]_{ji}$$

For acyclic graphs the Szeged index *SZD* is equal to the → *Wiener index W*, the → *Cluj-distance index CJD* and the → *Cluj-detour index CJΔ: SZD = W = CJD = CJΔ. For cyclic graphs,* $SZD = CJD \neq W \neq CJ\Delta$.

The **hyper-Szeged index** SZD_P is defined as:

$$SZD_P = \mathcal{W}(\mathbf{SZ}) = \frac{1}{2}\sum_{i=1}^{A}\sum_{j=1}^{A}[\mathbf{SZ}]_{ij} = \mathcal{W}'(\mathbf{SZ_U}) = \sum_{i=1}^{A-1}\sum_{j=i+1}^{A}[\mathbf{SZ_U}]_{ij}[\mathbf{SZ_U}]_{ji}$$

For any graphs, the hyper-Szeged index SZD_P is different from the other hyper-indices.

The **reciprocal Szeged matrix** $\mathbf{SZ^{-1}}$ is a matrix whose off-diagonal elements are the reciprocal of the corresponding symmetric Szeged matrix elements:

$$[\mathbf{SZ}^{-1}]_{ij} = [\mathbf{SZ}]_{ij}^{-1}$$

All elements equal to zero are left unchanged in the reciprocal matrix.

→ *Hyper-Harary indices* are defined for this matrix and → *Harary indices* for the corresponding 1st-order → *sparse matrix* $^1\mathbf{SZ}^{-1}$ by applying the Wiener operator.

The **Szeged delta matrix** $\mathbf{SZ_\Delta}$ was proposed as the difference between the Szeged matrix and its 1st order sparse matrix [Ivanciuc *et al.*, 1997a]:

$$\mathbf{SZ_D} = \mathbf{SZ} - {}^1\mathbf{SZ}$$

Example : ethylbenzene

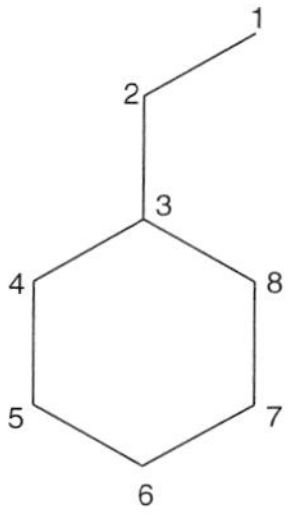

Szeged matrix $\mathbf{SZ}_U$

Atom	1	2	3	4	5	6	7	8
1	0	1	1	2	2	3	2	2
2	7	0	2	2	4	3	4	2
3	6	6	0	5	4	5	4	5
4	6	3	3	0	5	4	5	2
5	4	4	2	3	0	5	2	3
6	5	3	3	2	3	0	3	2
7	4	4	2	3	2	5	0	3
8	6	3	3	2	5	4	5	0

Symmetric Szeged matrix **SZ**

Atom	1	2	3	4	5	6	7	8
1	0	7	6	12	8	15	8	12
2	7	0	12	6	16	9	16	6
3	6	12	0	15	8	15	8	15
4	12	6	15	0	15	8	15	4
5	8	16	8	15	0	15	4	15
6	15	9	15	8	15	0	15	8
7	8	16	8	15	4	15	0	15
8	12	6	15	4	15	8	15	0

$SZD = 109$ $\quad SZD_P = 308$

Box S-4.

The **Szeged property matrices $\mathbf{SZ_U}\boldsymbol{P}$** of a graph $\mathcal{G}$ are square unsymmetric $A \times A$ matrices where each off-diagonal entry i-j is a function of a selected property of vertices $P_{i,(ij)}$ lying closer to the focused vertex v_i [Minailiuc *et al.*, 1998]:

$$[\mathbf{SZ_U}\boldsymbol{P}]_{ij} = P_{i,(ij)}$$

where

$$P_{i,(ij)} = f(P_a) \quad where \quad a | a \in \mathcal{V}(\mathcal{G});\ d_{ia} < d_{ja}$$

$$f(P_a) = m \cdot \sum_a P_a \qquad \text{or} \qquad f(P_a) = \left(\prod_a P_a\right)^{1/N_{i,(ij)}}$$

where the properties are evaluated on $N_{i,(ij)}$ vertices a which satisfy the Szeged condition. $\mathcal{V}(\mathcal{G})$ is the set of graph vertices and d is the topological distance. The two proposed functions are an additive function of the vertex properties and the geometric mean; the term m in the former is a weighting factor. $P_{i,(ij)}$ is considered a fragmental property since it collects the properties of the molecular fragment linked to the considered vertex.

The classical Szeged matrix $\mathbf{SZ_U}$ is derived using a unit vertex property and the factor $m = 1$ in the additive function. Moreover, an interesting Szeged property matrix is obtained considering $m = 1$ with a generic vertex property P_a. Thus, for example, Szeged walk-property matrices $\mathbf{SZ_U}\boldsymbol{W^k}$ can be calculated using an → *atomic walk count* of different orders; the Szeged mass-property matrix $\mathbf{SZ_UA}$ is obtained by using the atomic masses, and the Szeged electronegativity-property matrix $\mathbf{SZ_UX}$ by using the → *Sanderson group electronegativity* calculated for each atom.

Normalized Szeged property matrices $\mathbf{SZ_U}\boldsymbol{NP}$ are particular Szeged property matrices, defined using as the weighting factor m in the additive function, the reciprocal of the global molecular property $P(\mathcal{G})$:

$$f'(P_a) = \frac{\sum_a P_a}{P(\mathcal{G})}$$

Wiener-type indices, called **Szeged fragmental indices**, are derived from these matrices applying the orthogonal operator $\mathcal{W}'$ to both dense and the sparse matrices:

$$SZP = \mathcal{W}'(\mathbf{SZ_U}\boldsymbol{P}\bullet\mathbf{A}) = \sum_{i=1}^{A}\sum_{j=i}^{A}[^1\mathbf{SZ_U}\boldsymbol{P}]_{ij}\cdot[^1\mathbf{SZ_U}\boldsymbol{P}]_{ji}$$

and

$$SZP_p = \mathcal{W}'(\mathbf{SZ_U}\boldsymbol{P}) = \sum_{i=1}^{A}\sum_{j=i}^{A}[\mathbf{SZ_U}\boldsymbol{P}]_{ij}\cdot[\mathbf{SZ_U}\boldsymbol{P}]_{ji}$$

where the indices SZP are derived from the 1st-order sparse matrices $^1\mathbf{SZ_U}\boldsymbol{P}$ calculated by the Hadamard matrix product between Szeged property matrices and the adjacency matrix **A**.

📖 [Dobrynin and Gutman, 1996] [Klavzar *et al.*, 1996] [Diudea *et al.*, 1997f] [Diudea *et al.*, 1997b] [Chepoi and Klavzar, 1997] [Gutman and Dobrynin, 1998b]

Szeged property matrices → **Szeged matrix**

T

Taft-Kutter-Hansch steric constants → **steric descriptors** (⊙ Taft steric constant)

Taft-Lewis inductive constant → **electronic substituent constants** (⊙ inductive electronic constants)

Taft polar constant : *Taft σ* constant* → **electronic substituent constants** (⊙ inductive electronic constants)

Taft resonance constants → **electronic substituent constants** (⊙ resonance electronic constants)

Taft steric constant → **steric descriptors**

Taft σ* constant → **electronic substituent constants** (⊙ inductive electronic constants)

Taillander index

An empirical steric index, denoted as ΣD, for substituted benzene rings and defined as the 6-term sum of the distances, given by the L → *Sterimol length parameter*, between the six atoms bonded to the benzene carbon atoms, i.e. the value of the external perimeter of the benzene ring [Taillander *et al.*, 1983; Ravanel *et al.*, 1985]. It represents the perimeter of the efficacious section and describes the steric properties of aromatic compounds.

For example, for the unsubstituted benzene ring, the Taillander index is the sum of the six H...H distances; the chlorobenzene perimeter is shown in Figure T-1.

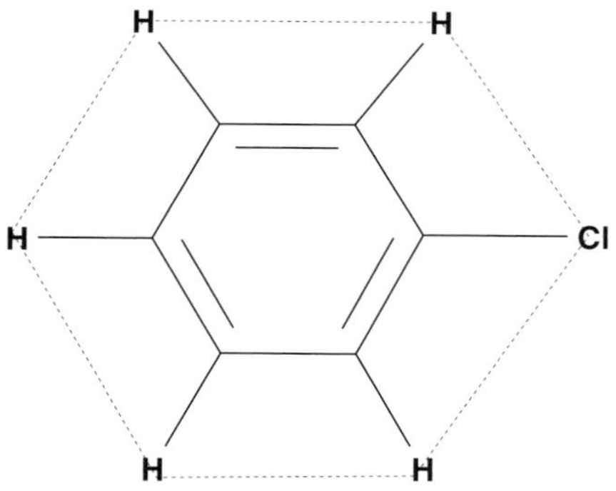

Figure T-1. Chlorobenzene perimeter.

📖 [Argese *et al.*, 1999]

Tanimoto coefficient → **similarity/diversity**

terminal edges → **graph**

terminal vertices → **graph**

Testa lipophobic constant : *interactive polar parameter* → **lipophilicity descriptors**

test set → **data set**

Theoretical Linear Solvation Energy Relationships (TLSERs)

QSAR method based on the philosophy of the → *Linear Solvation Energy Relationships*, whose empirically derived molecular descriptors are substituted by descriptors defined in the framework of → *computational chemistry* [Famini *et al.*, 1991; Famini *et al.*, 1992; Famini and Wilson, 1994a]. The TLSER descriptors were developed with the aim of optimally correlating with LSER descriptors, thereby being as generally applicable to solute/solvent interactions as are the LSER descriptors.

The general TLSER equation for a given physico-chemical property ϕ is expressed as:

$$\log \phi = b_0 + b_1 \cdot V_{VDW} + b_2 \cdot \pi_I + b_3 \cdot \varepsilon_A + b_4 \cdot q^+ + b_5 \cdot \varepsilon_B + b_6 \cdot q^-$$

where b_0 to b_6 are regression coefficients estimated by multivariate regression analysis. V_{VDW} is the → *van der Waals volume*, in units of cubic Å, representing the → *cavity term*. π_I is the **TLSER polarizability index** chosen to represent the → *dipolarity/polarizability term* and defined by the → *polarizability volume* divided by the molecular volume in order to get a size-independent parameter; it indicates the ease with which the electron cloud may be moved or polarized. ε_B and ε_A are, respectively, the **Covalent Hydrogen-Bond Basicity** (CHBB) and the **Covalent Hydrogen-Bond Acidity** (CHBA), defined as:

$$\varepsilon_B = 0.30 - \frac{|\varepsilon_{HOMO}(\text{solute}) - \varepsilon_{LUMO}(H_2O)|}{100} \quad \text{and}$$

$$\varepsilon_A = 0.30 - \frac{|\varepsilon_{LUMO}(\text{solute}) - \varepsilon_{HOMO}(H_2O)|}{100}$$

where ε_{LUMO} and ε_{HOMO} are the → *lowest unoccupied molecular orbital energy* and the → *highest occupied molecular orbital energy*, respectively.

q^- is the **Electrostatic Hydrogen-Bond Basicity** (EHBB) defined as the magnitude of the charge on the most negatively charged solute atom, and q^+ is the **Electrostatic Hydrogen-Bond Acidity** (EHBA) defined as the charge on the most positively charged solute hydrogen.

Note that the LSER hydrogen-bond parameters are split into covalent and electrostatic hydrogen-bond contributions.

The TLSER descriptors have been used to estimate a range of properties including retention indices [Donovan and Famini, 1996], gas phase acidity [Famini *et al.*, 1993], and toxicological indices and biological activities [Wilson and Famini, 1991; Sixt *et al.*, 1995; Famini *et al.*, 1998].

📖 [Famini and Penski, 1992] [Cramer *et al.*, 1993] [Famini and Wilson, 1993] [Famini and Wilson, 1994b] [Headley *et al.*, 1994] [Lowrey *et al.*, 1995a] [Lowrey and Famini, 1995b] [Chester *et al.*, 1996]

theoretical molecular descriptors → **molecular descriptors**

TLSER polarizability index → **theoretical linear solvation energy relationships**

topochemical indices → **graph invariants**

topochromatic vector → **DARC/PELCO analysis**

topoelectric indices → **topoelectric matrices**

topoelectric matrices (TEM)

Square symmetric matrices 3×3 obtained from the powers of the → *adjacency matrix* **A** and a standardized *properties matrix* $\hat{\mathbf{P}}$ defined in terms of → *atomic properties* derived from the → *ionization potential* IP and → *electron affinity* EA of the atoms, i.e. the → *Mulliken electronegativity* and the equilibrium charge [Borodina *et al.*, 1998].

They are calculated as:

$$\mathbf{TEM^m} = \frac{\hat{\mathbf{P}}^{\mathrm{T}} \cdot \mathbf{A}^m \cdot \hat{\mathbf{P}}}{A} \qquad m = 1, 2, \ldots, K$$

where A is the number of molecule atoms and K is the maximum considered power of adjacency matrix.

The **properties matrix** $\hat{\mathbf{P}}$ used to define **TEM** is an $A \times 3$ matrix, where each ith row represents the vector $\langle 1; \hat{p}_i; \hat{q}_i \rangle$, where $\hat{p}_i$ and $\hat{q}_i$ are the autoscaled Mulliken electronegativity and the autoscaled equilibrium charge (approximated by a quadratic function) of the ith atom, respectively:

$$\hat{p}_i = \frac{p_i - \bar{p}}{\sigma_p} \qquad \bar{p} = \frac{\sum_{i=1}^{A} p_i}{A} \qquad \sigma_p^2 = \frac{\sum_{i=1}^{A} (p_i - \bar{p})^2}{A}$$

$$\hat{q}_i = \frac{q_i - \bar{q}}{\sigma_q} \qquad \bar{q} = \frac{\sum_{i=1}^{A} q_i}{A} \qquad \sigma_q^2 = \frac{\sum_{i=1}^{A} (q_i - \bar{q})^2}{A}$$

where p_i and q_i are defined as:

$$p_i \equiv \chi_{\mathrm{MU},i} = \frac{\mathrm{IP}_i + \mathrm{EA}_i}{2} \qquad q_i = -\frac{p_i}{\mathrm{IP}_i - \mathrm{EA}_i} = -\frac{1}{2} \cdot \frac{\mathrm{IP}_i + \mathrm{EA}_i}{\mathrm{IP}_i - \mathrm{EA}_i}$$

Each molecule is characterized by the ordered sequence of topoelectric matrices defined for increasing powers of the adjacency matrix:

$$\langle \mathbf{TEM^1}; \mathbf{TEM^2}; \ldots; \mathbf{TEM}^K \rangle$$

For each topoelectric matrix $\mathbf{TEM^m}$, a set of **topoelectric indices** is derived, the topoelectric indices being all elements of the considered matrix; the first element $[\mathbf{TEM^m}]_{11}$ and the elements below the main diagonal are not considered, thus resulting in 5 descriptors for each matrix, TEM_1^m, ..., TEM_5^m. If K matrices are calculated for each molecule, a total $5 \times K$ topoelectric indices are used to characterize the molecules.

From topoelectric indices, a measure of similarity between two compounds s and t was also proposed:

$$s_{st} = \frac{1}{1 + \frac{1}{5 \cdot K} \cdot \sum_{m=1}^{K} \sum_{j=1}^{5} \left[TEM_j^m(s) - TEM_j^m(t) \right]^2}$$

where K is the number of considered matrices.

topographic indices → **molecular geometry**

topographic matrix → **molecular geometry**

topological atomic charge → **self-returning walk counts**

topological atomic valencies → **self-returning walk counts**

topological atom pairs → **substructure descriptors** (⊙ atom pairs)

topological bond index : *molecular path count* → **path counts**

topological bond order → **bond order indices** (⊙ graphical bond order)

topological bond orders → **self-returning walk counts**

topological charge index → **topological charge indices**

topological charge indices

Topological charge indices were proposed to evaluate the charge transfer between pairs of atoms, and therefore the global charge transfer in the molecule [Gálvez *et al.*, 1994b; Gálvez *et al.*, 1995b].

Let **M** be the matrix obtained by multiplying the → *adjacency matrix* **A** by the → *reciprocal square distance matrix* $\mathbf{D}^{-2}$, i.e.

$$\mathbf{M} = \mathbf{A} \cdot \mathbf{D}^{-2}$$

In order to avoid division by zero, the diagonal entries of the distance matrix remain the same; the obtained matrix **M**, called the **Galvez matrix**, is a square unsymmetric matrix $A \times A$, A being the number of atoms in the molecule.

An unsymmetric **charge term matrix CT** is derived from the matrix **M**; its terms CT_{ij} for each pair of vertices v_i and v_j are defined as:

$$CT_{ij} = \begin{cases} \delta_i & if\ i = j \\ m_{ij} - m_{ji} & if\ i \neq j \end{cases} \quad \text{or} \quad \mathbf{CT} = \mathcal{D}(\mathbf{M}) + \left[\mathbf{M} - \mathbf{M}^{\mathrm{T}}\right]$$

where m_{ij} are the elements of the matrix **M**, δ_i is the → *vertex degree* of the ith atom and $\mathcal{D}$ is the → *diagonal operator*. The charge terms CT_{ij} are → *graph invariants* that are related to the charge transfer between the pair of considered vertices. The diagonal entries of the **CT** matrix represent the topological valence of the atoms; the off-diagonal entries CT_{ij} represent a measure of the net charge transferred from the atom j to the atom i. To take into account also the heteroatoms, the diagonal entries of the adjacency matrix can be substituted by the Pauling's → *atom electronegativity* (taking as the reference value 2 for the chlorine atom) or simply the → *valence vertex degree* or any other atomic property.

For each path of length k, a **topological charge index** G_k is defined as:

$$G_k = \frac{1}{2} \cdot \sum_{i=1}^{A} \sum_{j=1}^{A} \left|CT_{ij}\right| \cdot \delta\left(k; d_{ij}\right)$$

where $\delta(k; d_{ij})$ indicates the Kronecker delta, namely:

$$\delta\left(k; d_{ij}\right) = \begin{cases} 1 & if\ d_{ij} = k \\ 0 & if\ d_{ij} \neq k \end{cases}$$

where d_{ij} are the elements of the → *distance matrix*.

Therefore, G_k is the half-sum of all charge terms CT_{ij} corresponding to pair of vertices with topological distance $d_{ij} = k$ and would represent the total charge transfer between atoms placed at topological distance k. The maximum number of G_k terms is equal to the → *topological diameter* D; the maximum number of summation terms in the G_k terms is equal to $A - 1$ for a linear molecule.

A **mean topological charge index** J_k is defined as:

$$J_k = \frac{G_k}{A - 1}$$

The G_1 is closely related to the → *molecular branching*. Moreover, a **global topological charge index** J is defined as:

$$J = \sum_{k=1}^{5} J_k$$

where the superior limit equal to 5 was proposed by the authors to obtain → *uniform length descriptors* such as $\langle G_1, G_2, \ldots, G_L; J_1, J_2, \ldots, J_L; J \rangle$, where L is equal to 5 or to any other selected integer. In any case, G_k (and consequently J_k and J) values are set to zero for k values greater than the diameter of the molecule.

📖 [Gálvez *et al.*, 1995a] [Gálvez *et al.*, 1996a] [Rios-Santamarina *et al.*, 1998]

Example : 2-methylpentane

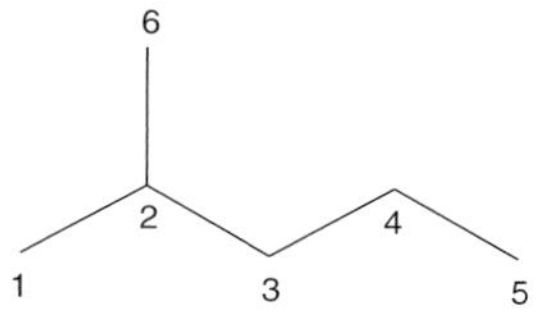

Adjacency matrix **A**

Atom	**1**	**2**	**3**	**4**	**5**	**6**
1	0	1	0	0	0	0
2	1	0	1	0	0	1
3	0	1	0	1	0	0
4	0	0	1	0	1	0
5	0	0	0	1	0	0
6	0	1	0	0	0	0

•

Reciprocal square distance matrix $\mathbf{D}^{-2}$

Atom	**1**	**2**	**3**	**4**	**5**	**6**
1	0	1	0.2500	0.1111	0.0625	0.2500
2	1	0	1	0.2500	0.1111	1
3	0.2500	1	0	1	0.2500	0.2500
4	0.1111	0.2500	1	0	1	0.1111
5	0.0625	0.1111	0.2500	1	0	0.0625
6	0.2500	1	0.2500	0.1111	0.0625	0

=

Galvez matrix **M**

=

Atom	**1**	**2**	**3**	**4**	**5**	**6**
1	1	0	1	0.25	0.1111	1
2	0.5	3	0.5	1.2222	0.3750	0.5
3	1.1111	0.25	2	0.25	1.1111	1.1111
4	0.3125	1.1111	0.25	2	0.25	0.3125
5	0.1111	0.25	1	0	1	0.1111
6	1	0	1	0.25	0.1111	1

Charge term matrix **CT**

Atom	**1**	**2**	**3**	**4**	**5**	**6**
1	1	-0.5	-0.1111	-0.0625	0	0
2	0.5	3	0.25	0.1111	0.1225	0.5
3	0.1111	-0.25	2	0	0.1111	0.1111
4	0.0625	-0.1111	0	2	0.25	0.0625
5	0	-0.1225	-0.1111	-0.25	1	0
6	0	-0.5	-0.1111	-0.0625	0	1

$G_1 = |CT_{12}| + |CT_{26}| + |CT_{23}| + |CT_{34}| + |CT_{45}| = 0.5 + 0.5 + 0.25 + 0 + 0.25 = 1.5$

$G_2 = |CT_{16}| + |CT_{13}| + |CT_{63}| + |CT_{24}| + |CT_{35}| = 0 + 0.1111 + 0.1111 + 0.1111 + 0.1111 = 0.4444$

$G_3 = |CT_{14}| + |CT_{64}| + |CT_{25}| = 0.0625 + 0.0625 + 0.1225 = 0.2475$

$G_4 = |CT_{15}| + |CT_{65}| = 0$

$J = J_1 + J_2 + J_3 + J_4 = 1.5/5 + 0.4444/5 + 0.2475/5 + 0 = 0.43838$

Box T-1.

topological complexity indices : *Bonchev topological complexity indices* → **molecular complexity**

topological diameter → **distance matrix**

topological diameter from edge eccentricity → **edge distance matrix**

topological distance → **distance matrix**

topological edge distance → **edge distance matrix**

topological electronic descriptors → **charge descriptors**

topological Hammett function → **combined descriptors**

topological index of hydrophobicity → **lipophilicity descriptors**

topological indices → **graph invariants**

topological information content : *vertex orbital information content* → **orbital information indices**

topological information indices

These are → *graph-theoretical invariants* that view the → *molecular graph* as a source of different probability distributions to which information theory definitions can be applied. They can be considered a quantitative measure of the lack of structural homogeneity or the diversity of a graph, in this way being related to symmetry associated with structure. The information content of a graph is not unique, depending on the equivalence relation defined on the graph.

Several information indices, usually calculated as → *total information content* and → *mean information content*, are based on partitioning graph elements or matrix elements in → *equivalence classes* according to two basic criteria:
- *equality criterion*: elements are considered equivalent if their values are equal;
- *magnitude criterion*: each element is considered as an equivalence class whose cardinality, i.e. number of elements, is equal to the magnitude of the element.

The symbol G is usually used to denote the number of equivalence classes and the symbol n_g to denote the cardinality of the set of elements in the gth class.

Collected below are the information indices calculated on the most important topological matrices such as → *adjacency matrix* **A**, → *distance matrix* **D**, → *edge distance matrix* $^{\mathbf{E}}\mathbf{D}$, → *edge adjacency matrix* **E**, → *vertex cycle matrix* $\mathbf{C_V}$, and → *edge cycle matrix* $\mathbf{C_E}$ [Bonchev *et al.*, 1981b; Bonchev, 1983].

Other specific information indices are → *orbital information indices*, → *information connectivity indices*, → *indices of neighbourhood symmetry*, → *information layer index*, → *chromatic information index*, information indices on the → *incidence matrix*, Hosoya information indices, and → *Bonchev centric information indices.*

Information indices on the adjacency matrix A are listed below.

- **total information content on the adjacency equality** ($^{V}I^{E}_{adj}$)

Derived from the adjacency matrix **A**, this is defined as:

$$^{V}I^{E}_{adj} = A^2 \log_2 A^2 - 2B \log_2 2B - \left(A^2 - 2B\right) \log_2\left(A^2 - 2B\right)$$

where A is the → *atom number* and B is the → *bond number*. Moreover, the entries of the adjacency matrix equal to one are $2B$, thus the entries equal to zero are $A^2 - 2B$; in particular, for acyclic graphs the total number of entries equal to one is $2\ (A - 1)$

and the number of entries equal to zero is $A^2 - 2\,(A - 1)$; for cyclic graphs they are $2A$ and $A^2 - 2A$, respectively.

This index is constant for a given number of atoms A, is insensitive to any kind of branching and can distinguish only between trees and different cyclic structures.

- **mean information content on the adjacency equality** (${}^{V}\bar{I}^{E}_{adj}$)

This is calculated by dividing the total information content on the adjacency matrix elements equality by the total number of adjacency matrix elements A^2 as:

$$ {}^{V}\bar{I}^{E}_{adj} = -\frac{2B}{A^2}\log_2\frac{2B}{A^2} - \left(1 - \frac{2B}{A^2}\right)\log_2\left(1 - \frac{2B}{A^2}\right) $$

where A is the atom number and B the bond number. It is based on the probability that a randomly selected entry will signify or not signify the adjacency.

- **total information content on the adjacency magnitude** (${}^{V}I^{M}_{adj}$)

This is derived from the partition of the adjacency matrix **A** elements according to their magnitude and is a trivial quantity, as zero entries do not contribute to the → *total adjacency index* A_V, i.e. the sum of all vertex degrees, corresponding to twice the bond number B:

$$ {}^{V}I^{M}_{adj} = A_V \cdot \log_2 A_V - 2B \cdot 1 \cdot \log_2 1 = A_V \cdot \log_2 A_V $$

where $2B$ is the number of adjacency matrix elements equal to one.

- **mean information content on the adjacency magnitude** (${}^{V}\bar{I}^{M}_{adj}$)

This is calculated by dividing the total information content on the adjacency matrix elements magnitude by the total adjacency index A_V:

$$ {}^{V}\bar{I}^{M}_{adj} = -2B \cdot \frac{1}{A_V} \cdot \log_2\frac{1}{A_V} = 1 + \log_2 B $$

where B is the bond number and $2B$ is the number of adjacency matrix elements equal to one.

- **mean information content on the vertex degree equality** (${}^{V}\bar{I}^{E}_{adj,\,\deg}$)

Derived from the adjacency matrix **A** and based on the partition of vertices according to → *vertex degree* equality, this is defined as:

$$ {}^{V}\bar{I}^{E}_{adj,\,\deg} = -\sum_{g=1}^{G}\frac{{}^{g}F}{A} \cdot \log_2\frac{{}^{g}F}{A} $$

where ${}^{g}F$ is the → *vertex degree count*, i.e. the number of vertices with the same vertex degree δ, A is the atom number, and G is the maximum vertex degree value.

- **mean information content on the vertex degree magnitude** (${}^{V}\bar{I}^{M}_{adj,\,\deg}$) (*: degree complexity*)

Based on the partition of vertices according to the vertex degree magnitude, this is defined as:

$$ {}^{V}\bar{I}^{M}_{adj,\,\deg} \equiv I^{d} = -\sum_{a=1}^{A}\frac{\delta_a}{A_V} \cdot \log_2\frac{\delta_a}{A_V} = -\sum_{g=1}^{G}{}^{g}F \cdot \frac{g}{A_V} \cdot \log_2\frac{g}{A_V} $$

where A is the atom number, δ_a is the vertex degree of the ath atom, ${}^{g}F$ is the vertex degree count, i.e. the number of vertices with vertex degree equal to g, A_V is the total adjacency index, and G is the maximum vertex degree value.

This index was proposed as a measure of → *molecular complexity* together with some other information indices derived from the distance matrix [Raychaudhury *et al.*, 1984].

Information indices on the edge adjacency matrix E are listed below.

• **total information content on the edge adjacency equality** (${}^{E}I^{E}_{adj}$)
Derived from the → *edge adjacency matrix* **E**, this is defined as:

$${}^{E}I^{E}_{adj} = B^2 \log_2 B^2 - 2N_2 \log_2 2N_2 - \left(B^2 - 2N_2\right) \log_2 \left(B^2 - 2N_2\right)$$

where B is the bond number and N_2 the → *connection number*, i.e. the number of 2nd order path counts in the → *molecular graph*. Moreover, the entries of the edge adjacency matrix equal to one are $2N_2$; thus the entries equal to zero are $B^2 - 2N_2$.

• **mean information content on the edge adjacency equality** (${}^{E}\bar{I}^{E}_{adj}$)
This is calculated by dividing the total information content of the edge adjacency matrix elements equality by the total number of edge adjacency matrix elements B^2:

$${}^{E}\bar{I}^{E}_{adj} = -\frac{2N_2}{B^2} \log_2 \frac{2N_2}{B^2} - \left(1 - \frac{2N_2}{B^2}\right) \log_2 \left(1 - \frac{2N_2}{B^2}\right)$$

where B is the bond number and N_2 the connection number. It is based on the probability that a randomly selected entry will signify or not signify the adjacency.

• **total information content on the edge adjacency magnitude** (${}^{E}I^{M}_{adj}$)
This is derived from the partition of edge adiacency matrix elements according to their magnitude and is a trivial quantity because the zero entries do not contribute to the → *total edge adjacency index* A_E (→ *Platt number*), i.e. the sum of all edge degrees, corresponding to twice the connection number N_2:

$${}^{E}I^{M}_{adj} = A_E \cdot \log_2 A_E - 2N_2 \cdot 1 \cdot \log_2 1 = A_E \cdot \log_2 A_E$$

where $2N_2$ is the number of edge adjacency matrix elements equal to one.

• **mean information content on the edge adjacency magnitude** (${}^{E}\bar{I}^{M}_{adj}$)
This is calculated by dividing the total information content on the edge adjacency matrix elements magnitude by the total adjacency index A_E:

$${}^{E}\bar{I}^{M}_{adj} = -2N_2 \cdot \frac{1}{A_E} \cdot \log_2 \frac{1}{A_E} = 1 + \log_2 N_2$$

where N_2 is the connection number and $2N_2$ is the number of edge adjacency matrix elements equal to one.

• **mean information content on the edge degree equality** (${}^{E}\bar{I}^{E}_{adj,\,\deg}$)
Derived from the edge adjacency matrix **E** and based on the partition of bonds according to the equality of their edge degrees and defined as:

$${}^{E}\bar{I}^{E}_{adj,\,\deg} = -\sum_{g=1}^{G} \frac{{}^{g}F_E}{B} \cdot \log_2 \frac{{}^{g}F_E}{B}$$

where ${}^{g}F_E$ is the → *edge degree count*, i.e. the number of edges with → *edge degree* ε equal to g, B is the bond number, and G is the maximum edge degree value.

• **mean information content on the edge degree magnitude** (${}^{E}\bar{I}^{M}_{adj,\,\deg}$)
Derived from the adjacency matrix **E** and based on the partition of bonds according to the magnitude of their edge degrees and defined as:

$${}^{E}\bar{I}^{M}_{adj,\,\deg} = -\sum_{b=1}^{B} \frac{\varepsilon_b}{A_E} \cdot \log_2 \frac{\varepsilon_b}{A_E} = -\sum_{g=1}^{G} {}^{g}F_E \cdot \frac{g}{A_E} \cdot \log_2 \frac{g}{A_E}$$

where ε_b is the degree of the bth edge, A_E is the total edge adjacency index, B is the bond number, and ${}^{g}F_E$ is the edge degree count, i.e. the number of edges with degree ε equal to g.

Information indices on the distance matrix D are listed below.

- **total information content on the distance equality** (${}^{V}I_D^E$)

Based on the equality of distances in the graph, this is defined as [Bonchev and Trinajstic, 1978]:

$$ {}^{V}I_D^E \equiv I_D^E = \frac{A(A-1)}{2} \cdot \log_2 \frac{A(A-1)}{2} - \sum_{g=1}^{G} {}^{g}f \cdot \log_2^g f $$

where A is the number of molecule atoms, ${}^{g}f$ is the number of distances with equal g value in the triangular **D** submatrix (i.e. the → *graph distance count*), and G is the maximum distance value, i.e. the → *topological diameter D*.

- **mean information content on the distance equality** (${}^{V}\bar{I}_D^E$)

Obtained by dividing the total information content of distance equality by the total number $A(A–1)/2$ of distances in the graph, this is calculated as:

$$ {}^{V}\bar{I}_D^E \equiv \bar{I}_D^E = -\sum_{g=1}^{G} \frac{2 \cdot {}^{g}f}{A(A-1)} \cdot \log_2 \frac{2 \cdot {}^{g}f}{A(A-1)} $$

where A is the number of molecule atoms, ${}^{g}f$ is the number of distances with equal g value in the triangular **D** submatrix, and G is the maximum distance value, i.e. the topological diameter D.

- **total information content on the distance magnitude** (${}^{V}I_D^M$)

The information content on the distribution of distances according to their magnitude is defined as:

$$ {}^{V}I_D^M \equiv I_D^M = W \cdot \log_2 W - \sum_{g=1}^{G} {}^{g}f \cdot g \cdot \log_2 g $$

where W is the → *Wiener index*, i.e. the total sum of distances in the graph, ${}^{g}f$ is the number of distances of equal g value in the triangular **D** submatrix, and G is the maximum distance value, i.e. the topological diameter D.

- **mean information content on the distance magnitude** (${}^{V}\bar{I}_D^M$)

This is calculated by dividing the total information content of the distance magnitude by the Wiener index W; it is therefore also called the **information Wiener index**:

$$ {}^{V}\bar{I}_D^M \equiv \bar{I}_D^M = -\sum_{g=1}^{G} {}^{g}f \cdot \frac{g}{W} \cdot \log_2 \frac{g}{W} $$

where ${}^{g}f$ is the number of distances of equal g value in the triangular **D** submatrix, and G is the maximum distance value, i.e. the topological diameter D.

- **mean information content on the distance degree equality** (${}^{V}\bar{I}_{D,\,\mathrm{deg}}^E$)

The mean information content of the partition of vertex distance degrees according to their equality is defined as:

$$ {}^{V}\bar{I}_{D,\,\mathrm{deg}}^E \equiv \bar{I}_{D,\,\mathrm{deg}}^E = -\sum_{g=1}^{G} \frac{n_g}{A} \cdot \log_2 \frac{n_g}{A} $$

where n_g is the cardinality of the gth set of vertices having equal → *vertex distance degree* σ, G is the number of equivalence classes and A the atom number.

• **mean information content on the distance degree magnitude** (${}^{V}\bar{I}^{M}_{D,\,\mathrm{deg}}$)
The mean information content of the partition of vertex distance degrees according to their magnitude is defined as:

$$ {}^{V}\bar{I}^{M}_{D,\,\mathrm{deg}} \equiv \bar{I}^{M}_{D,\,\mathrm{deg}} = -\sum_{a=1}^{A} \frac{\sigma_a}{2W} \cdot \log_2 \frac{\sigma_a}{2W} = -\sum_{g=1}^{G} n_g \cdot \frac{\sigma_g}{2W} \cdot \log_2 \frac{\sigma_g}{2W} $$

where W is the Wiener index (and $2W$ the → *Rouvray index*), σ_a is the vertex distance degree of the ath atom, n_g is the number of vertices having equal vertex distance degrees in the gth class, σ_g is the vertex distance degree of the vertices in the gth class, and G is the number of equivalence classes.

A group of local vertex invariants and corresponding molecular graph invariants derived from the distance matrix were proposed as quantities related to the → *molecular complexity* [Raychaudhury *et al.*, 1984; Balaban, 1992; Balaban, 1993c]:

• **vertex complexity** (v_i^c)
Derived from the distance matrix **D**, this is a local vertex invariant defined as [Raychaudhury *et al.*, 1984]:

$$ v_i^c = -\sum_{g=0}^{\eta_i} \frac{{}^{g}f_i}{A} \cdot \log_2 \frac{{}^{g}f_i}{A} = -\sum_{g=0}^{\eta_i} \mathrm{p}_g \cdot \log_2 \mathrm{p}_g = \frac{1}{A} \cdot \log_2 A - \sum_{g=1}^{\eta_i} \mathrm{p}_g \cdot \log_2 \mathrm{p}_g $$

where ${}^{g}f_i$ is the number of distances from the vertex v_i equal to g, i.e. the → *vertex distance counts*. ${}^{0}f_i$ is always equal to one, i.e. there is only one distance equal to zero; η_i is the → *atom eccentricity* and A the atom number; p_g is the probability of randomly selecting a distance equal to g from vertex v_i.

• **vertex distance complexity** ($\bar{v}_i^d$) (: *mean local information on distances*, u_i)
Derived from the distance matrix **D**, it is a local vertex invariant calculated on the complete molecular graph, hydrogens included, and defined as mean information content of a vertex [Raychaudhury *et al.*, 1984; Klopman *et al.*, 1988; Klopman and Raychaudhury, 1988]:

$$ \bar{v}_i^d \equiv u_i = -\sum_{j=1}^{A} \frac{d_{ij}}{\sigma_i} \cdot \log_2 \frac{d_{ij}}{\sigma_i} = -\sum_{g=1}^{\eta_i} {}^{g}f_i \cdot \frac{g}{\sigma_i} \cdot \log_2 \frac{g}{\sigma_i} $$

where g runs over all of the different distances from the vertex v_i , ${}^{g}f_i$ is the number of distances from the vertex v_i equal to g, σ_i is the ith vertex distance degree, and η_i is the ith atom eccentricity. The mean local information on distances u_i was originally proposed by Balaban [Balaban and Balaban, 1991; Ivanciuc *et al.*, 1993a] for hydrogen-depleted molecular graphs. Moreover, a molecular topological index derived from these local invariants was defined by applying the → *Ivanciuc-Balaban operator* and simply called the ***U* index**:

$$ U = \frac{B}{C+1} \cdot \sum\nolimits_b \left(u_i \cdot u_j\right)_b^{-1/2} $$

where the sum runs over all the edges in the graph and u_i and u_j are the LOVIs relative to the vertices incident to the considered bond. B and C are the number of bonds and the → *cyclomatic number*, respectively.

• **normalized vertex distance complexity** ($\tilde{v}_i^d$)
Derived from the vertex distance complexity by dividing it by its maximum value [Klopman and Raychaudhury, 1990]:

$$ \tilde{v}_i^d = \frac{\bar{v}_i^d}{\log_2 \sigma_i} $$

where σ_i is the vertex distance degree. This index is independent of the molecular size.

• **relative vertex distance complexity** (v_i) (: *local information on distances*)
Derived from the vertex distance complexity u_i and defined as the difference between the maximum information content of a vertex and its mean information content [Balaban and Balaban, 1991; Ivanciuc *et al.*, 1993a]:

$$v_i = \sigma_i \cdot \log_2 \sigma_i - u_i$$

where σ_i is the vertex distance degree. The corresponding molecular topological index was calculated applying the Ivanciuc-Balaban operator; it is simply called the ***V* index**:

$$V = \frac{B}{C+1} \cdot \sum\nolimits_b \left(v_i \cdot v_j\right)_b^{-1/2}$$

where the sum runs over all edges in the graph and v_i and v_j are the LOVIs relative to the vertices incident to the considered bond. B and C are the number of bonds and the cyclomatic number, respectively.

• **mean extended local information on distances** (y_i)
Derived from the partition of the distances from a vertex v_i according to their magnitude, defined as total information content [Balaban and Balaban, 1991; Ivanciuc *et al.*, 1993a]:

$$y_i = \sum_{g=1}^{\eta_i} {}^g f_i \cdot g \cdot \log_2 g$$

where g runs over all of the different distances from the vertex v_i, ${}^g f_i$ is the number of distances from the vertex v_i equal to g, and η_i is the ith atom eccentricity. The corresponding molecular topological index was calculated applying the Ivanciuc-Balaban operator; it is simply called the ***Y* index**:

$$Y = \frac{B}{C+1} \cdot \sum\nolimits_b \left(y_i \cdot y_j\right)_b^{-1/2}$$

where the sum runs over all edges in the graph and y_i and y_j are the LOVIs relative to the vertices incident to the considered bond. B and C are the number of bonds and the cyclomatic number, respectively.

• **extended local information on distances** (x_i)
Derived from the mean extended local information on distances y_i and defined as total information content of a vertex content [Balaban and Balaban, 1991; Ivanciuc *et al.*, 1993a]:

$$x_i = \sigma_i \cdot \log_2 \sigma_i - y_i$$

where σ_i is the vertex distance degree and y_i is the vertex information content of the distance magnitude. The corresponding molecular topological index was calculated applying the Ivanciuc-Balaban operator; it is simply called the ***X* index**:

$$X = \frac{B}{C+1} \cdot \sum\nolimits_b \left(x_i \cdot x_j\right)_b^{-1/2}$$

where the sum runs over all edges in the graph and x_i and x_j are the LOVIs relative to the vertices incident to the considered bond. B and C are the number of bonds and the cyclomatic number, respectively.

• **graph vertex complexity** (H_V)
Derived from the distance matrix **D**, it is defined as the molecular average vertex complexity [Raychaudhury *et al.*, 1984]:

$$H_V = \frac{1}{A} \cdot \sum_{i=1}^{A} v_i^c$$

where v_i^c is the vertex complexity and A is the number of atoms.

• **graph distance complexity** (H_D)
Derived from the distance matrix **D**, it is a graph invariant defined as [Raychaudhury *et al.*, 1984]:

$$H_D = \sum_{i=1}^{A} \frac{\sigma_i}{I_{\mathrm{ROUV}}} \cdot v_i^d = \sum_{i=1}^{A} \frac{\sigma_i}{2W} \cdot v_i^d$$

where v_i^d is the vertex distance complexity, σ_i is the ith vertex distance degree, I_{ROUV} is the → *Rouvray index* and W the → *Wiener index*.

Information indices on the edge distance matrix $^{\mathrm{E}}\mathbf{D}$ are listed below.

• **total information content on the edge distance equality** ($^{E}I_D^E$)
Based on the equality or inequality of edge distances in the graph, this is defined as:

$$^{E}I_D^E = \frac{B(B-1)}{2} \cdot \log_2 \frac{B(B-1)}{2} - \sum_{g=1}^{G} {}^{g}f \cdot \log_2^g f$$

where B is the bond count, ^{g}f is the number of edge distances with equal g value in the triangular $^{\mathrm{E}}\mathbf{D}$ submatrix, and G is the maximum edge distance value.

• **mean information content on the edge distance equality** ($^{E}\bar{I}_D^E$)
Obtained dividing the total information content of the edge distance equality by the total number $B(B$–$1)/2$ of edge distances in the graph, this is calculated as:

$$^{E}\bar{I}_D^E = -\sum_{g=1}^{G} \frac{2 \cdot {}^{g}f}{B(B-1)} \cdot \log_2 \frac{2 \cdot {}^{g}f}{B(B-1)}$$

where B is the bond count, ^{g}f is the number of edge distances with equal g value in the triangular $^{\mathrm{E}}\mathbf{D}$ submatrix, and G is the maximum edge distance value.

• **total information content on the edge distance magnitude** ($^{E}I_D^M$)
The information content of the distribution of edge distances according to their magnitude is defined as:

$$^{E}I_D^M = {}^{\mathrm{E}}W \cdot \log_2 {}^{\mathrm{E}}W - \sum_{g=1}^{G} {}^{g}f \cdot g \cdot \log_2 g$$

where $^{\mathrm{E}}W$ is the → *edge Wiener index*, i.e. the half of the → *total edge distance* $\mathrm{D_E}$, ^{g}f is the number of edge distances with equal g value in the triangular $^{\mathrm{E}}\mathbf{D}$ submatrix, and G is the maximum edge distance value.

• **mean information content on the edge distance magnitude** ($^{E}\bar{I}_D^M$)
This is calculated by dividing the total information content of the edge distance magnitude by the edge Wiener index $^{\mathrm{E}}W$:

$$^{E}\bar{I}_D^M = -\sum_{g=1}^{G} {}^{g}f \cdot \frac{g}{{}^{\mathrm{E}}W} \cdot \log_2 \frac{g}{{}^{\mathrm{E}}W}$$

where ^{g}f is the number of edge distances with equal g value in the triangular $^{\mathrm{E}}\mathbf{D}$ submatrix and G is the maximum edge distance value.

• **mean information content on the edge distance degree equality** ($^{E}\bar{I}_{D,\,\mathrm{deg}}^E$)
The mean information content on the partition of edge distance degrees according to their equality is defined as:

$$^{E}\bar{I}^{E}_{D,\,\mathrm{deg}} = -\sum_{g=1}^{G} \frac{n_g}{B} \cdot \log_2 \frac{n_g}{B}$$

where n_g is the cardinality of the gth set of vertices having an equal → *edge distance degree* $^{E}\sigma$, G is maximum edge distance degree value, and B is the bond number.

- **mean information content on the edge distance degree magnitude** ($^{E}\bar{I}^{M}_{D,\,\mathrm{deg}}$)

The mean information content of the partition of edge distance degrees according to their magnitude is defined as:

$$^{E}\bar{I}^{M}_{D,\,\mathrm{deg}} = -\sum_{b=1}^{B} \frac{^{E}\sigma_b}{\mathrm{D_E}} \cdot \log_2 \frac{^{E}\sigma_b}{\mathrm{D_E}} = -\sum_{g=1}^{G} n_g \cdot \frac{^{E}\sigma_g}{\mathrm{D_E}} \cdot \log_2 \frac{^{E}\sigma_g}{\mathrm{D_E}}$$

where $\mathrm{D_E}$ is the total edge distance and $^{E}\sigma_b$ is the edge degree of the bth bond, n_g is the number of bonds having equal distance degree in the gth class, $^{E}\sigma_g$ is the distance degree of the edges in the gth class, and G is the number of equivalence classes.

Information indices on the vertex cycle matrix $\mathbf{C_V}$ are listed below.

- **total information content on the vertex cycle matrix elements equality** ($^{V}I^{E}_{cyc}$)

This is derived from the → *vertex cycle matrix* $\mathbf{C_V}$ and is based on the partition of matrix elements according to their equalities:

$$^{V}I^{E}_{cyc} = A \cdot C^{+} \log_2 A \cdot C^{+} - n_1 \log_2 n_1 - n_0 \log_2 n_0$$

where A and C^{+} are the atom number and the → *cyclicity*, i.e. the total number of rings in the graph, respectively; n_1 and n_0 are the number of matrix entries equal to one and the number of entries equal to zero, respectively.

- **mean information content on the vertex cycle matrix elements equality** ($^{V}\bar{I}^{E}_{cyc}$)

This is calculated by dividing the total information content on the vertex cycle matrix elements equality by the total number of matrix elements $A \cdot C^{+}$ as:

$$^{V}\bar{I}^{E}_{cyc} = -\frac{n_1}{A \cdot C^{+}} \log_2 \frac{n_1}{A \cdot C^{+}} - \frac{n_0}{A \cdot C^{+}} \log_2 \frac{n_0}{A \cdot C^{+}}$$

where A is the atom number, C^{+} the number of rings, and n_1 and n_0 are the number of matrix entries equal to one and the number of entries equal to zero, respectively. It is based on the probability that a randomly selected entry will signify or not signify that a vertex belongs to a given cycle.

- **total information content on the vertex cycle matrix elements magnitude** ($^{V}I^{M}_{cyc}$)

This is derived from the vertex cycle matrix $\mathbf{C_V}$ and is based on the magnitude of matrix elements:

$$^{V}I^{M}_{cyc} = C_{\mathrm{VC}} \log_2 C_{\mathrm{VC}} - 1 \cdot \log_2 1 = C_{\mathrm{VC}} \log_2 C_{\mathrm{VC}}$$

where C_{VC} is the → *total vertex cyclicity*, i.e the total sum of matrix elements. Since the zero-entries do not contribute to the total vertex cyclicity, the information index $^{V}I^{M}_{cyc}$ is simply a logarithmic function of the number of matrix entries equal to one.

- **mean information content on the vertex cycle matrix elements magnitude** ($^{V}\bar{I}^{M}_{cyc}$)

This is calculated by dividing the total information content of the vertex cycle matrix elements magnitude by the total vertex cyclicity C_{VC} as:

$$^{V}\bar{I}^{M}_{cyc} = -C_{\mathrm{VC}} \cdot \frac{1}{C_{\mathrm{VC}}} \log_2 \frac{1}{C_{\mathrm{VC}}} = \log_2 C_{\mathrm{VC}}$$

This information index is simply the total vertex cyclicity expressed in bits, i.e. information units.

• **mean information content on the vertex cyclic degree equality** (${}^{V}\bar{I}^{E}_{cyc,\,deg}$)
Derived from the vertex cycle matrix $\mathbf{C_V}$ and based on the partition of vertices according to the equality of their cyclic degrees, it is defined as:

$$ {}^{V}\bar{I}^{E}_{cyc,\,deg} = -\sum_{g=1}^{G} \frac{n_g}{A} \cdot \log_2 \frac{n_g}{A} $$

where n_g is the number of vertices having the same → *vertex cyclic degree* γ^v, G is the number of different degree values, and A is the atom number.

• **mean information content on the vertex cyclic degree magnitude** (${}^{V}\bar{I}^{M}_{cyc,\,deg}$)
Derived from the vertex cycle matrix $\mathbf{C_V}$ and based on the magnitude of cyclic degrees of the vertices, this is defined as:

$$ {}^{V}\bar{I}^{M}_{cyc,\,deg} = -\sum_{i=1}^{A} \frac{\gamma_i^v}{C_{VC}} \cdot \log_2 \frac{\gamma_i^v}{C_{VC}} $$

where γ_i^v is the vertex cyclic degree of the ith vertex, C_{VC} is the total vertex cyclicity, i.e. the sum of all cyclic degrees, and A is the atom number.

Information indices on the edge cycle matrix $\mathbf{C_E}$ are listed below.

• **total information content on the edge cycle matrix elements equality** (${}^{E}I^{E}_{cyc}$)
This is derived from the → *edge cycle matrix* $\mathbf{C_E}$ and is based on the partition of matrix elements according to their equalities:

$$ {}^{E}I^{E}_{cyc} = B \cdot C^{+} \log_2 B \cdot C^{+} - n_1 \log_2 n_1 - n_0 \log_2 n_0 $$

where B and C^+ are the bond number and the cyclicity, i.e. the total number of rings in the graph, respectively; n_1 and n_0 are the number of matrix entries equal to one and the number of entries equal to zero, respectively.

• **mean information content on the edge cycle matrix elements equality** (${}^{E}\bar{I}^{E}_{cyc}$)
This is calculated by dividing the total information content on the edge cycle matrix elements equality by the total number of matrix elements $B{\cdot}C^+$:

$$ {}^{E}\bar{I}^{E}_{cyc} = -\frac{n_1}{B \cdot C^{+}} \log_2 \frac{n_1}{B \cdot C^{+}} - \frac{n_0}{B \cdot C^{+}} \log_2 \frac{n_0}{B \cdot C^{+}} $$

where B is the atom number, C^+ the number of rings, n_1 and n_0 are the number of matrix entries equal to one and the number of entries equal to zero, respectively. It is based on the probability that a randomly selected entry will signify or not signify that an edge belongs to a given cycle.

• **total information content on the edge cycle matrix elements magnitude** (${}^{E}I^{M}_{cyc}$)
This is derived from the edge cycle matrix $\mathbf{C_E}$ and is based on the magnitude of matrix elements:

$$ {}^{E}I^{M}_{cyc} = C_{EC} \log_2 C_{EC} - 1 \cdot \log_2 1 = C_{EC} \log_2 C_{EC} $$

where C_{EC} is the → *total edge cyclicity*, i.e the total sum of matrix elements. Since the zero-entries do not contribute to the total edge cyclicity, the information index ${}^{E}I^{M}_{cyc}$ is simply a logarithmic function of the number of matrix entries equal to one.

• **mean information content on the edge cycle matrix elements magnitude** (${}^{E}\bar{I}^{M}_{cyc}$)
This is calculated by dividing the total information content on the edge cycle matrix elements magnitude by the total edge cyclicity C_{EC}:

$$ {}^{E}\bar{I}^{M}_{cyc} = -C_{EC} \cdot \frac{1}{C_{EC}} \log_2 \frac{1}{C_{EC}} = \log_2 C_{EC} $$

This information index is simply the total edge cyclicity expressed in bits, i.e. information units.

• **mean information content on the edge cyclic degree equality** (${}^{E}\bar{I}^{E}_{cyc,\,\deg}$)
Derived from the edge cycle matrix $\mathbf{C_E}$ and based on the partition of edges according to the equality of their cyclic degrees, it is defined as:

$$ {}^{E}\bar{I}^{E}_{cyc,\,\deg} = -\sum_{g=1}^{G} \frac{n_g}{B} \cdot \log_2 \frac{n_g}{B} $$

where n_g is the number of edges having the same → *edge cyclic degree* γ^e, G is the number of different degree values, and B is the bond number.

• **mean information content on the edge cyclic degree magnitude** (${}^{E}\bar{I}^{M}_{cyc,\,\deg}$)
Derived from the edge cycle matrix $\mathbf{C_E}$ and based on the magnitude of cyclic degrees of the edges, this is defined as:

$$ {}^{E}\bar{I}^{M}_{cyc,\,\deg} = -\sum_{i=1}^{B} \frac{\gamma_i^e}{C_{\mathrm{EC}}} \cdot \log_2 \frac{\gamma_i^e}{C_{\mathrm{EC}}} $$

where γ_i^e is the edge cyclic degree of the ith edge, C_{EC} is the total edge cyclicity, i.e. the sum of all cyclic degrees, and B is the bond number.

📖 [Gutman and Trinajstic, 1973a] [Bonchev and Trinajstic, 1977] [Bonchev *et al.*, 1980a] [Mekenyan *et al.*, 1980] [Cvetkovic and Gutman, 1985] [Kunz, 1986] [Nikolic *et al.*, 1993a] [Ivanciuc *et al.*, 2000b]

topological lipophilicity potential → **molecular interaction fields** (⊙ hydrophobic fields)

topological radius → **distance matrix**

topological radius from edge eccentricity → **edge distance matrix**

topological representation → **molecular descriptors**

topological resonance energy → **resonance indices**

topological state → **topological state matrix**

topological state matrix (T)

The topological state matrix **T** [Hall and Kier, 1990] is a square symmetric matrix $A \times A$, A being the number of → *molecular graph* vertices, whose entries are defined as:

$$ t_{ij} = \left[GM_{ij}\right]^{b} \cdot f\left(n_{ij}\right)^{c} $$

where the powers can be $b = \pm 1$ and $c = 0, \pm 1, \pm 2, \pm 3$; GM_{ij} is the geometric mean of the → *vertex degree* δ of the atoms involved in the path i–j of length m and defined as:

$$ GM_{ij} = \left(\prod_{a=1}^{n_{ij}} \delta_a \right)^{1/n_{ij}} \qquad a \in {}^{m}p_{ij} $$

where n_{ij} is the number of atoms in the path ${}^{m}p_{ij}$ of mth order (equal to $m+1$ for acyclic paths). $f(n_{ij})$ is a function of the separation between the vertices v_i and v_j. The simplest function (as proposed by Hall-Kier) is the reciprocal of the number of vertices in the path:

$$ f\left(n_{ij}\right) = \frac{1}{n_{ij}} $$

and the simplest topological state matrix is defined with $b = 1$ and $c = 1$.

The diagonal entries t_{ii} correspond to paths of length zero, thus they are simply equal to the vertex degrees. Moreover, for cyclic graphs, there can be more than one path between the vertices v_i and v_j; hence each topological state matrix element t_{ij} is a sum of the contributions of all paths between the considered vertices.

From the topological state matrix, a local vertex invariant S_i, called the **topological state** (or **vertex topological state**), can be calculated as:

$$S_i = \mathcal{R}_i(\mathbf{T}) = \sum_{j=1}^{A} t_{ij}$$

where $\mathcal{R}_i$ is the → *row sum operator*.

A **total topological index** τ having a high discriminating power can also be derived as the following:

$$\tau = \sum_{i=1}^{A} t_{ii} + \sum_{i=1}^{A-1} \sum_{j=i+1}^{A} t_{ij}$$

As the diagonal entries t_{ii} are just the vertex degrees δ, the first term on the right is the → *total adjacency index* A_V. The topological state index is one of the → *molecular ID numbers*.

Valence topological state matrix $\mathbf{T}^v$, **valence topological state** S^v and **total valence topological index** τ^v can be obtained using → *valence vertex degree* δ^v instead of the simple vertex degrees to encode the atom identity. Valence topological states were used to search for topological equivalence among molecule atoms to define the → *Kier symmetry index*.

🕮 [Hu and Xu, 1994]

topological substructural molecular design → **edge adjacency matrix**

topological torsion descriptor → **substructure descriptors**

topostructural indices → **graph invariants**

torsion angles → **molecular geometry**

total absolute atomic charge → **charge descriptors**

total adjacency index → **adjacency matrix**

total charge weighted negative surface area → **charged partial surface area descriptors**

total charge weighted positive surface area → **charged partial surface area descriptors**

total connection orbital information content → **orbital information indices**

total edge adjacency index → **edge adjacency matrix**

total edge cyclicity → **cycle matrices** (⊙ edge cycle matrix)

total edge distance → **edge distance matrix**

total edge orbital information content → **orbital information indices**

total electrophilic superdelocalizability → **quantum-chemical descriptors** (⊙ electrophilic superdelocalizability)

total hydrophobic surface area → **charged partial surface area descriptors**

total information content → **information content**

total information content on the adjacency equality → **topological information indices**

total information content on the adjacency magnitude → **topological information indices**

total information content on the distance equality → **topological information indices**

total information content on the distance magnitude → **topological information indices**

total information content on the edge adjacency equality → **topological information indices**

total information content on the edge adjacency magnitude → **topological information indices**

total information content on the edge cycle matrix elements equality → **topological information indices**

total information content on the edge cycle matrix elements magnitude → **topological information indices**

total information content on the edge distance equality → **topological information indices**

total information content on the edge distance magnitude → **topological information indices**

total information content on the incidence matrix → **incidence matrix**

total information content on the vertex cycle matrix elements equality → **topological information indices**

total information content on the vertex cycle matrix elements magnitude → **topological information indices**

total information index on atomic composition → **atomic composition indices**

total interaction energy fields → **molecular interaction fields**

total molecular surface area : *van der Waals surface area* → **molecular surface** (⊙ van der Waals molecular surface)

total negative charge → **charge descriptors**

total nucleophilic superdelocalizability → **quantum-chemical descriptors** (⊙ nucleophilic superdelocalizability)

total path count → **path counts**

total path number : *total path count* → **path counts**

total polarizability → **electric polarization descriptors** (⊙ atom-atom polarizability)

total polar surface area → **charged partial surface area descriptors**

total positive charge → **charge descriptors**

total sequence count → **sequence matrices**

total softness → **quantum-chemical descriptors** (⊙ softness indices)

total solvent-accessible surface area : *solvent-accessible surface area* → **molecular surface** (⊙ solvent-accessible molecular surface)

total squared atomic charge → **charge descriptors** (⊙ total absolute atomic charge)

total structure connectivity index → **connectivity indices**

total subgraph count → **molecular graph**

total sum operator → **algebraic operators**

total topological index → **topological state matrix**

total topological information content → **orbital information indices**

total valence topological index → **topological state matrix**

total vertex cyclicity → **cycle matrices** (⊙ vertex cycle matrix)

total vertex distance : *Rouvray index* → **distance matrix**

total walk count → **walk counts**

trace → **algebraic operators**

trail → **graph**

training/evaluation splitting → **validation techniques**

training set → **data set**

translational invariance → **molecular descriptors** (⊙ invariance properties of molecular descriptors)

transmission coefficient → **electronic substituent constants**

transposition of a matrix → **algebraic operators**

Tratch-Stankevitch-Zefirov-type indices : *generalized expanded Wiener numbers* → **expanded distance matrices**

tree → **graph**

triangular descriptors → **substructure descriptors**

triple bond count → **multiple bond descriptors**

triplet descriptors : *triangular descriptors* → **substructure descriptors**

U

***U* index** → **topological information indices** (⊙ vertex distance complexity)

unbiased constants → **electronic substituent constants**

uniform length descriptors

Uniform length descriptors are vectors having the same cardinality independently of the considered molecule and constituted by homogeneous descriptors, i.e. an L-length vector of ordered descriptors $\mathcal{X}_k$ ($k = 1, 2, ..., L$):

$$\langle \mathcal{X}_1, \mathcal{X}_2, \ldots, \mathcal{X}_k, \ldots \mathcal{X}_L \rangle$$

Each entry of the vector represents a single descriptor defined by a function depending on an index k; in order to capture relevant information, the upper limit L is arbitrarily defined according to the specific problem [Junghans and Pretsch, 1997; Baumann, 1999].

Several algorithms to obtain numbers from a → *molecular representation* produce different numbers of descriptors, depending on the number of atoms or bonds of the molecule or, more generally, on some molecular features related to the algorithm. For example, the number of nonzero spectroscopic signals of a molecule closely depends on the molecule itself. In these cases, the algorithm must be implemented by a suitable transformation capable of obtaining uniform length descriptors, i.e. a rule to define a fixed number L of elements of the vectors and a rule to fill the vector elements when the values are not defined (for example, filling with zero values).

Examples of applications of the rules to obtain uniform length descriptors are → *EVA descriptors*, → *topological charge indices*, → *atomic walk count sequence*, → *SE-vectors*, → *molecular profiles*, → *3D-MoRSE descriptors*, → *autocorrelation descriptors*.

uninformative variable elimination by PLS → **variable subset selection**

unipolarity → **distance matrix**

unique atomic code : *canonical numbering*

unique atomic ordering : *canonical numbering*

unnormalized second moment of distances → **Wiener matrix**

unusual vertices → **walk counts**

unusual walks → **walk counts**

unsaturation index → **multiple bond descriptors**

V

VAA indices → **eigenvalue-based descriptors** (⊙ eigenvalues of the adjacency matrix)

VAD indices → **eigenvalue-based descriptors** (⊙ eigenvalues of the distance matrix)

valence connectivity indices → **connectivity indices**

valence electron descriptor → **vertex degree**

valence state indicator → **electrotopological state indices**

valence topological state matrix → **topological state matrix**

valence topological state → **topological state matrix**

valence vertex degree → **vertex degree**

valency index → **quantum-chemical descriptors** (⊙ electron density)

validation techniques

Validation techniques constitute a fundamental tool for the assessment of the validity of models obtained from a → *data set* by multivariate regression and classification methods. Validation techniques are used to check the predictive power of the models, i.e. to give a measure of their capability to perform reliable predictions of the modelled response for new cases where the response is unknown [Diaconis and Efron, 1983; Myers, 1986; Cramer III *et al.*, 1988a; Rawlings, 1988].

A necessary condition for the validity of a regression model is that the multiple correlation coefficient R^2 is as close as possible to one and the standard error of the estimate s small. However, this condition (fitting ability) is not sufficient for model validity as the models give a closer fit (smaller s and larger R^2) the larger the number of parameters and variables in the models. Moreover, unfortunately, these parameters are not related to the capability of the model to make reliable predictions on future data.

Other problems for the validity of the models arise when models, often with only a few variables, are obtained by using procedures based on → *variable selection* [Allen, 1971]. In fact, when a set with a large number of descriptors to select from is available, simple models can be found with apparently good fitting properties due to **chance correlation**, i.e. collinearity without predictive ability [Topliss and Edwards, 1979; Wold and Dunn III, 1983; Clark and Cramer III, 1993].

To avoid models with chance correlation, a check with different validation procedures must be adopted, such as, for example, cross-validation, y-scrambling and QUICK rule. A general validation procedure [Wold, 1991] would be the deletion of some objects before the selection of the variables: applying the variable selection procedure and then predicting the responses for excluded objects. The whole procedure, including variable selection, is then repeated a number of times, depending on the adopted specific validation technique.

In the best situation, a fairly representative validation set of compounds (external validation set), for which predicted response values can be compared with actual ones, should be available. However, for the obvious reasons of time and cost, adequate validation sets are rarely available. As Swante Wold said "Without a real validation set, a simulated one may be better than nothing." [Wold, 1991].

With the aim of achieving a better understanding of the relationships between response and predictors, the interpretability, simplicity and comparability of a model can always add useful information about its validity.

A number of statistical techniques have been proposed to simulate the predictive ability of a model. The most popular validation techniques are listed below.

- **cross-validation**

This is the most common validation technique, where a number of modified data sets are created by deleting, in each case, one or a small group of objects from the data in such a way that each object is taken away once and only once [Efron, 1983; Osten, 1988].

For each reduced data set, the model is calculated, and responses for the deleted objects are predicted from the model. The squared differences between the true response and the predicted response for each object left out are added to *PRESS* (*predictive residual sum of squares*). From the final *PRESS*, the Q^2 (or R^2_{CV}) and *SDEP* (*standard deviation error of prediction*) values are usually calculated [Cruciani *et al.*, 1992].

The simplest and most general cross-validation procedure is the **leave-one-out technique** (**LOO technique**), where each object is taken away, one at a time. In this case, given n objects, n reduced models have to be calculated. This technique is particularly important as this deletion scheme is unique, and the predictive ability of the different models can be compared accurately. However, in several cases, the predictive ability obtained is too optimistic, particularly when the number of objects is quite large. This is because of a too small perturbation of the data when only one object is left out.

When the number of objects is not too small, more realistic predictive abilities are obtained by deleting more than one object at each step. To apply this cross-validation procedure, called the **leave-more-out technique** (**LMO technique**), the number of cancellation groups is defined by the user, i.e. the number of blocks the data are divided into, and, at each step, all the objects belonging to a block are left out from the calculation of the model.

The cancellation groups G range from 2 to n (in this case, leave-more-out coincides with the leave-one-out technique). For example, given 60 objects (n = 60), for 2, 3, 5, 10 cancellation groups G, at each time n/G objects are left in the evaluation sets, i.e. 30, 20, 12, and 10 objects, respectively.

Rules for selecting the group of objects for the evaluation set at each step must be adopted in such a way that each object is left out only once.

- **training/evaluation set splitting**

This is a validation technique based on the splitting of the data set into a training set and an evaluation set. The model is calculated from the training set, and the predictive power is checked on the evaluation set. The splitting is performed by randomly selecting the objects belonging to the two sets. As the results are strongly dependent on the splitting of the data, this technique is better used by repeating the splitting several hundreds of times and averaging the predictive capabilities, i.e. using the **repeated evaluation set technique** [Boggia *et al.*, 1997].

The **single evaluation set technique** can be used reliably only if the splitting is performed by partitioning the objects by a well-stated criterion, such as a criterion based on experimental design or cluster analysis.

• **bootstrap**
By this validation technique, the original size of the data set (n) is preserved for the training set, by the selection of n objects with repetition; in this way the training set usually consists of repeated objects and the evaluation set of the objects left out [Efron, 1982; Efron, 1987]. The model is calculated on the training set and responses are predicted on the evaluation set. All the squared differences between the true response and the predicted response of the objects of the evaluation set are collected in *PRESS*. This procedure of building training sets and evaluation sets is repeated thousands of time, *PRESS* are summed and the average predictive power is calculated.

• **external validation**
A validation technique where a test set is retained to perform a further check on the predictive capabilities of a model obtained from a training set and with predictive power optimized by an evaluation set.

• **y-scrambling** (: *y-randomization test*)
This validation technique is adopted to check models with chance correlation, i.e. models where the independent variables are randomly correlated to the response variables. The test is performed by calculating the quality of the model (usually R^2 or, better, Q^2) randomly modifying the sequence of the response vector $\mathbf{y}$, i.e. by assigning to each object a response randomly selected from the true responses [Lindgren *et al.*, 1996]. If the original model has no chance correlation, there is a significant difference in the quality of the original model and that associated with a model obtained with random responses. The procedure is repeated several hundreds of times.

• **lateral validation**
This is a technique which refers to the method of validating a new model, i.e. obtained from a new data set, by comparing it with other models previously obtained for the same response [Kim, 1995b].

The similarity of the regression coefficients and the equality of their signs support the reliability of the models. The new model can also be based on different descriptors, but with the same physical meaning.

• **QUIK rule**
A rule based on the → *multivariate K correlation index*, which compares the multivariate correlation index K_X of the X-block of the predictor variables with the multivariate correlation index K_{XY} obtained by the augmented X-block matrix by adding the column of the response variable [Todeschini *et al.*, 1998]. Only regression models having multivariate correlation K_{XY} greater than multivariate correlation K_X can fulfill the QUIK rule, a necessary condition for the model validity, i.e.

$$K_{XY} > K_X$$

This constraint is included in the maximization (or minimization) of some goodness of prediction statistic and prevents models with collinearity but without predictive power, i.e. chance correlation, from being taken into account.

📖 [Stone, 1974] [Wold, 1978] [Hodes, 1981a] [Lanteri, 1992] [Eriksson *et al.*, 1993] [Wold and Eriksson, 1995] [Mager, 1996] [Burden *et al.*, 1997]

van der Waals excluded volume method

This is 3D-QSAR technique proposed to overcome the drawbacks of the → *grid-based QSAR techniques* such as → *CoMFA* related to the superimposition of the molecules. The descriptors produced by this method do not require → *alignment rules* and are not significantly affected by the orientation of the molecules [Tominaga and Fujiwara, 1997b; Tominaga, 1998b].

The method is based on the use of a → *probe* given by the excluded volume of two spheres with different radii and an identical centre corresponding to the → *barycentre* of the molecule. The probe is layered like an onion, each layer being the excluded volume. The first sphere, i.e. the component of the probe, is an atom (e.g. an iodine atom with van der Waals radius of 2.05 Å, a carbon atom with van der Waals radius of 1.52 Å, or a hydrogen atom with van der Waals radius of 1.08 Å) whose volume defines the first layer, then 60 atoms of the same type (e.g. iodine atoms) construct the second sphere whose surface is like a fullerene and which shares the same centre as the first sphere. The excluded volume between the first and second spheres defines the second layer. In the same way the subsequent spheres and layers are also defined.

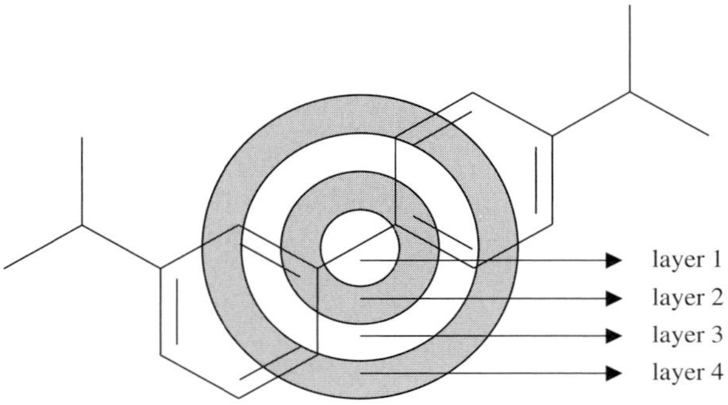

Figure V-1. Layers defining the sequence of excluded volumes.

Therefore, the molecular descriptors, called **EV_{WHOLE} descriptors**, are calculated as the excluded van der Waals volume between each molecule and probe layer. They represent the expansion of molecular volume in 3D space. In addition, molecular descriptors, called **EV_{TYPE} descriptors**, were also proposed as the excluded volume between a specific type of atom of the molecule and probe layer; 15 atom types were defined for classifying molecule atoms. Using 21 layers in the probe, there is a total of 336 molecular descriptors considering both categories.

van der Waals interaction fields : *steric interaction fields* → **molecular interaction fields**

van der Waals molecular surface → **molecular surface**

van der Waals radius → **molecular surface** (⊙ van der Waals molecular surface)

van der Waals surface area → **molecular surface** (⊙ van der Waals molecular surface)

van der Waals volume → **molecular surface** (⊙ van der Waals molecular surface)

variable → **data set**

variable reduction (: *feature selection*)

Variable reduction consists in the selection of a subset of variables able to preserve the essential information contained in the whole → *data set*, but eliminating redundancy, too highly intercorrelated variables, etc.

Variable reduction differs from → *variable selection* in the fact that the subset of variables is selected independently of the response of interest.

The most common methods for variable reduction are listed below.

• **constant and near-constant variables**
A preliminary approach to variable reduction consiting in the elimination of all the variables that take the same value for all the objects in the data set. Near-constant variables, i.e. variables that assume the same value except in one or very few cases, would also be excluded. A good measure for evaluating near-constant variables is the → *standardized Shannon's entropy*: the entropy of a variable with one different value over 10 objects is 0.141, over 20 objects is 0.066, and with two different values over 100 objects is 0.024.

• **pair correlation cut-off selection**
This method is based on the *correlation matrix* and on a selected correlation cut-off value (e.g. $r^* = 0.95$). For each pair of variables with a correlation value greater than the cut-off value, one of the two correlated variables is arbitrarily eliminated.

• **cluster analysis feature selection**
Methods of → *cluster analysis* are applied to the variables, on the so-called Q-mode data matrix, i.e. on the transposed data matrix. Once cluster analysis has been performed, one variable for each cluster is retained as representative of all the variables within that cluster. Which and how many are the retained/excluded variables depends on the chosen cluster analysis method.

• **principal component analysis feature selection**
Visual inspection of the significant loading plots obtained by → *Principal Component Analysis* can be a non-quantitative but useful tool to select the most relevant variables to preserve the most important information contained in the original data [Jolliffe, 1986; Jackson, 1991].

More quantitative approaches based on PCA have also been defined, based on the largest loadings in the absolute value of the eigenvectors, usually obtained from the correlation matrix. From among the *Jolliffe techniques of variable reduction* [Jolliffe, 1972; Jolliffe, 1973], the B_2 technique deletes those variables showing the maximum loading (in absolute value) for all the q eigenvectors having the smallest eigenvalues, where the number q of eigenvectors is selected by some arbitrary rule (for example, for eigenvalues less than 0.7). The B_1 technique is similar to the B_2, but after each variable deletion (on the last eigenvector), a PCA is newly performed on the remaining variables until the last eigenvalue assumes some significant value. The B_4 technique is similar to the B_2, but the procedure is applied starting from the eigenvectors corresponding to the largest eigenvalues and retaining the variables showing the largest loadings. Finally, the B_3 technique eliminates $p - k$ variables $\mathbf{x}_j$ satisfying the following condition:

$$\min_j \left(\sum_{m=k+1}^{p} \lambda_m \cdot \ell_{jm}^2 \right)$$

McCabe techniques of variable reduction [McCabe, 1984] are based on the calculation of the residual correlation (or covariance) matrix $\mathbf{S_M}$ of the deleted variables where the effect of the retained variable is removed. This matrix is a square symmetric matrix of order $p - k$, where k is the number of retained variables, obtained from the correlation (covariance) matrix of the retained variables $\mathbf{S_R}$ (of size $k \times k$) and the correlation (or covariance) matrix of the deleted variables $\mathbf{S_D}$ (of size $q \times q$). The retained variables by McCabe techniques are called *principal variables*. This terminology can be extended to all the sets of retained variables obtained from PCA and correlation analysis.

From the residual correlation matrix the $p - k$ eigenvalues are calculated. The McCabe techniques retains the set of k variables satisfying one of the four proposed criteria $M_1 - M_4$:

$$M_1 = \min_{\{k\}}\left(\sum_{m=1}^{p-k} \ell_{\{k\},m}\right) \qquad M_2 = \min_{\{k\}}\left(\prod_{m=1}^{p-k} \ell_{\{k\},m}\right)$$

$$M_3 = \min_{\{k\}}\left(\sum_{m=1}^{p-k} \ell^2_{\{k\},m}\right) \qquad M_4 = \max_{\{k\}}\left(\sum_{m=1}^{\min(k,q)} R^2_{\mathrm{CCA}}\right)$$

where $\ell_{\{k\},m}$ are loadings of the mth eigenvector calculated on retaining the set of k variables. $R^2{}_{\mathrm{CCA}}$ are the square canonical correlation coefficients of the two matrices $\mathbf{S_R}$ and $\mathbf{S_D}$ and the sum runs over the minimum between k (retained variables) and q (deleted variables).

- ***K* correlation analysis**

A method consisting in the iterative procedure of the elimination of one variable at a time, based on the → *multivariate K correlation index* [Todeschini *et al.*, 1998]. All the variables are removed one at a time and the K multivariate correlation $K_{p/j}$ ($j = 1, p$) of the set of $p - 1$ variables is calculated. The jth variable associated with the minimum $K_{p/j}$ correlation is removed from the set of p variables (i.e. the variable which is maximally correlated with all the others), and the procedure is repeated on the remaining variables. The elimination procedure ends when the minimum $K_{p/j}$ is greater than the correlation K_p of the whole set of the remaining variables or when a standardized correlation value, called the ***K* inflation factor** (*KIF*), is less than 0.6 – 0.5.

Variable Selection (*VS*) (: *Variable Subset Selection, VSS*)

The aim of variable subset selection is to reach optimal model complexity in predicting a response variable by a reduced set of independent variables [Hocking, 1976; Miller, 1990; Wikel and Dow, 1993; Tetko *et al.*, 1996; Topliss and Edwards, 1979]. Regression (and classification) models based on optimal subsets of a few predictor variables have several advantages. In fact, simple models show more stable statistical properties, can be more easily interpreted, and give higher predictive power. On the other hand, the major drawback of the variable subset selection procedures consists in the possibility of selecting model variables having → *chance correlation*. Thus, to avoid chance correlation, particular attention must be paid to → *validation techniques*.

Within the framework of the variable subset selection techniques, the general linear regression model can be expressed as:

$$\hat{y}_i = b_0 + \sum_{j=1}^{p} b_j \cdot \mathrm{I}_j^* \cdot x_{ij}$$

where b_j are the regression coefficients, x_{ij} the value of the jth variable for the ith object of the → *data set*, and I_j^* is a binary variable indicating the presence of the jth variable ($\mathrm{I}_j^* = 1$) or its absence ($\mathrm{I}_j^* = 0$) in the final model. The set of all the binary variables is a p-dimensional vector $\mathbf{I}^*$ usually obtained by validation. The actual model dimension k ($k \leq p$) is the sum of all the entries equal to 1 in the $\mathbf{I}^*$ vector.

The methods for variable subset selection can be divided into two main categories: methods which work on the original p variables $\mathbf{x}_1$, $\mathbf{x}_2$, ..., $\mathbf{x}_p$, *direct variable subset selection*, where the best binary vector $\mathbf{I}^*$ is evaluated by considering the relationships among the $\mathbf{x}_j$ variables and a response variable $\mathbf{y}$, i.e.

$$\mathbf{I} = f(\mathbf{x}_1,\ \mathbf{x}_2,\ \ldots,\ \mathbf{x}_p;\ \mathbf{y})$$

and methods which work on linear combinations $\mathbf{t}_1$, $\mathbf{t}_2$, ..., $\mathbf{t}_M$ of the original variables, *indirect variable subset selection*, where the best binary vector $\mathbf{I}^*$ is evaluated by considering the weights of the linear combinations $\mathbf{t}_m$ and the response variables $\mathbf{y}$, i.e.

$$\mathbf{I} = f\left[(w_{11},\ w_{21},\ \ldots,\ w_{p1}),\ (w_{12},\ w_{22},\ \ldots,\ w_{p2}),\ \ldots,\ (w_{1M},\ w_{2M},\ \ldots,\ w_{pM});\ \mathbf{y}\right]$$

where M is the number of significant linear combinations and w_{jm} the weights of the jth variable in the mth component. These new variables can be obtained by Principal Component Analysis (PCA), Partial Least Square regression (PLS), or other related techniques.

The most common methods for variable selection are listed below.

- **all possible models**

The simplest method based on an exhaustive examination of all the possible models of k variables (i.e. the model size) obtained by a set of p variables, where k is a parameter between 1 and a user-defined value (the maximum theoretical value is limited by the number n of objects in the data set). The best model or models is or are evaluated by validation or by parameters depending on the models degrees of freedom.

As the procedure consists in the evaluation of the quality of all the models with one variable (i.e. p univariate models), of all the models with two variables [i.e. $p \times (p-1)$ bivariate models], up to all the possible models with k variables, the greatest disadvantage of this method is the extraordinary increase in the required computer time when p and k are quite large. In fact, the total number t of models is given by the relationship:

$$t = \sum_{k=1}^{L} \frac{p!}{k! \cdot (p-k)!} \leq 2^p - 1$$

where L is the maximum user-defined model size. For example, with 32 total variables ($p = 32$) and $1 \leq k \leq 8$ ($L = 8$), $t = 1 \times 10^7$. The maximum number of possible models is $2^p - 1 = 4\,294\,967\,295 \approx 4 \times 10^9$. The main advantage of this method is the exhaustive search for the best model in the model space.

- **response-variable correlation cutoff**

Variable selection is performed by checking the squared multiple correlation coefficient R^2 or the corresponding cross-validated quantity Q^2 from univariate regression models $\mathbf{y} = b_0 + b_1\,\mathbf{x}_j$, the selection being made separately for each jth variable of the p variables $\mathbf{x}_1$, $\mathbf{x}_2$, ..., $\mathbf{x}_p$.

Variables showing R^2 or Q^2 values lower than a cutoff value, i.e. largely uncorrelated with the $\mathbf{y}$ response, are definitively excluded from the searching of the best models. Acceptable cutoff values for R^2 are usually between 0.05 and 0.01 and for Q^2 less than zero.

This procedure can be used to perform a preliminary screening of the variables, while others are used on the remaining variables, e.g. genetic algorithm variable subset selection or searching for all subset models. Otherwise, the procedure can be iteratively extended to bivariate models, retaining the variables of the best L models (for example, $L = 10$), and then to higher dimensional models, always retaining the best subset of variables appearing in the previously selected models. A disadvantage of this procedure is that a variable with a low correlation with the response, but correlated with residuals and thus able to give a contribution in the final model, is excluded.

- **stepwise regression methods** (SWR)

Groups of commonly used regression methods are proposed to evaluate only a small number of subsets by either adding or deleting variables one at a time according to a specific criterion [Draper and Smith, 1998].

Forward selection (FS-SWR) is a technique starting with no variables in the model and adding one variable at a time until either all variables are entered or until a stopping criterion is satisfied.

The variable considered for inclusion at any step is the one yielding the largest single degree of freedom F-ratio among the variables eligible for inclusion, and this value is larger than a fixed value F_{in}. At each step, the jth variable is added to a k-size model if

$$F_j = \max_j \left(\frac{RSS_k - RSS_{k+j}}{s^2_{k+j}} \right) > F_{in}$$

when RSS is the *residual sum of squares*. The subscript $k + j$ refers to quantities computed when the jth variable is added to the current k variables already in the model.

Backward elimination (BE-SWR) is a stepwise technique starting with a model in which all the variables are included and then deleting one variable at a time. At any step, the variable with the smallest F-ratio is eliminated if this F-ratio does not exceed a specified value F_{out}. At each step, the jth variable is eliminated from a k-size model if

$$F_j = \min_j \left(\frac{RSS_{k-j} - RSS_k}{s^2_k} \right) < F_{\text{out}}$$

The subscript $k - j$ refers to quantities computed when the jth variable is excluded from the current k variables in the model. Obviously, when the total number of variables is greater than the number of objects, this technique cannot be applied in this form.

The most popular stepwise technique combines the two previous approaches (FW and BE) and is called *Elimination-Selection* (ES-SWR) [Efroymson, 1960]. It is basically a forward selection, but at each step (when the number of model variables is greater than two) the possibility of deleting a variable as in the BE approach is considered.

The two major drawbacks of the stepwise procedures are that none of them ensure that the "best" subset of a given size is found and, perhaps more critical, it is not uncommon that the first variable included in FS becomes unnecessary in the presence of other variables.

In order to avoid some drawbacks of the stepwise approaches, the *i-fold stepwise variable selection method* was recently proposed [Lucic *et al.*, 1999b]. This technique is based on descriptor orthogonalization and, at each subsequent step, adds the set of the best i descriptors.

- **Genetic Algorithm – Variable Subset Selection** (GA-VSS)

Variable selection is performed by using *Genetic Algorithms* (GA), based on the evolution of a population of models. In genetic algorithm terminology, the binary vector **I** is called a *chromosome*, which is a p-dimensional vector where each position (a *gene*) corresponds to a variable (1 if included in the model, 0 otherwise). Each chromosome represents a model with a subset of variables.

Once the statistical parameter to optimize is defined (e.g., maximizing Q^2 by a leave-one-out validation procedure), along with the model population size P (for example, P = 100) and the maximum number L of allowed variables in a model (for example, $L = 5$); the minimum number of allowed variables is usually assumed equal to one. Moreover, a *cross-over probability* p_C (usually high, for example, $p_C > 0.9$) and a *mutation probability* p_M (usually small, for example, $p_M < 0.1$) must also be defined by the user.

Once the leading parameters are defined, the genetic algorithm evolution starts based on three main steps:

1. Random initialization of the population

The model population is initially built by random models with a number of variables between 1 and L, and the models are ordered with respect to the selected statistical parameter – the quality of the model – (the best model is in first place, the worst model at position P);

2. Crossover step

From the population, pairs of models are selected (randomly or with a probability proportional to their quality). Then, for each pair of models the common characteristics are preserved (i.e. variables excluded in both models remain excluded, variables included in both models remain included). For variables included in one model and excluded from the other, a random number is tried and compared with the crossover probability p_C: if the random number is lower than the cross-over probability, the excluded variable is included in the model and vice versa. Finally, the statistical parameter for the new model is calculated: if the parameter value is better than the worst value in the population, the model is included in the population, in the place corresponding to its rank; otherwise, it is no longer considered. This procedure is repeated for several pairs (for example 100 times).

3. Mutation step

For each model present in the population (i.e. each chromosome), p random numbers are tried, and one at a time each is compared with the defined mutation probability p_M: each gene remains unchanged if the corresponding random number exceeds the mutation probability, otherwise, it is changed from zero to one or vice versa. Low values of p_M allow only a few mutations, thus obtaining new chromosomes not too different from the generating chromosome.

Once the mutated model is obtained, the statistical parameter for the model is calculated: if the parameter value is better than the worst value in the population, the model is included in the population, in the place corresponding to its rank; otherwise, it is no longer considered.

This procedure is repeated for all the chromosomes (i.e. P times).

4. Stop conditions

The second and third steps are repeated until some stop condition is encountered (e.g., a user-defined maximum number of iterations) or the process is arbitrarily ended.

An important characteristic of the GA-VSS method is that a single model is not necessarily obtained, but the result usually is a population of acceptable models; this characteristic, sometimes considered a disadvantage, provides an opportunity to make an evaluation of the relationships with the response from different points of view. A theoretical disadvantage is that the absolute best model can not be automatically found.

📖 [Goldberg, 1989; Leardi *et al.*, 1992; Leardi, 1994; Luke, 1994; Mestres and Scuseria, 1995; Judson, 1996; Devillers, 1996a; Hopfinger and Patel, 1996; Kemsley, 1998]

• **Genetic Function Approximation** (GFA)

This is variable subset selection algorithm [Rogers and Hopfinger, 1994; Rogers, 1995] that combines genetic algorithms [Holland, 1975] with Friedman's Multivariate Adaptive Regression Splines (MARS) algorithm [Friedman, 1988]. MARS uses splines as basis functions to partition data space as it builds its regression models. The searching for spline-based regression models is improved using a genetic algorithm rather than the original incremental approach. A GFA algorithm is derived from the G/SPLINES algorithm [Rogers, 1992; Rogers, 1991]; it automatically selects which variables are to be used in its basis functions and determines the appropriate number of basis functions.

The initial models are generated by randomly selecting a number of variables from the training set, building basis functions from these variables using user-specified basis function types, and then building the models from random sequences of these basis functions. Improved models are then constructed by performing the genetic crossover operation to recombine the terms of the better performing models.

Model performances are evaluated by using *Friedman's lack-of-fit function* (LOF) [Friedman, 1988], defined as:

$$\mathrm{LOF} = \frac{LSE}{\left(1 - \frac{c+d\cdot p}{n}\right)^2}$$

where LSE is the least-squares error, c the number of basis functions, p the number of variables, n the number of objects, and d a user-defined smoothing parameter.

Using the information contained in the binary vector $\mathbf{I}^*$ collecting the selected variables, it is possible to calculate the regression coefficients by ordinary least squares regression.

📖 [Shi *et al.*, 1998]

• **MUtation and SElection Uncover Models** (MUSEUM)
A variable selection approach based on genetic algorithms, and, more specifically, on an evolutionary algorithm, mutation being the preferred genetic operation [Kubinyi, 1994a; Kubinyi, 1994b; Kubinyi, 1996].

After a random generation of a model population, random mutations of one or a few variables are tried. If better models are obtained after a fixed number of trials, this procedure is repeated on the new generation of models; otherwise random mutations of several variables becomes allowed. Also in this case, if better models are obtained after a fixed number of trials, the first procedure is repeated on the new generation of models; otherwise systematic addition and elimination of the variables of the population models is performed. If better models are obtained, the procedure restarts from the first step; otherwise, if all the variables are checked the procedure ends and the variables of the final population models are checked for their statistical significance and eventually eliminated.

MUSEUM approach evaluates the quality of the models maximizing the *fitness function FIT* defined as:

$$FIT = \frac{R^2 \cdot (n - k - 1)}{(n + k^2) \cdot (1 - R^2)}$$

where n is the number of objects, k the number of model variables, and R^2 the usual explained variance in fitting.

• **sequential search**
A variable selection method proposed to decrease the huge number of models that must be evaluated by all subset model searching.

The statistical parameter to optimize (e.g., maximizing Q^2 by a leave-one-out validation procedure) must be selected and the maximum number L of allowed variables in a model (e.g., $L = 5$). The minimum number of allowed variables is usually taken equal to one. For each size, a small group of models is selected by a preliminary evaluation of the statistical parameter to optimize.

For each model, the sequential search is based on the following procedure:

- Each variable present in the model is excluded, one at a time, and is replaced by all the excluded variables, one at a time. For each replacement, the value of the statistical parameter to optimize is calculated and stored. If k is the number of variables in the model ($1 \leq k \leq L$) and p the total number of available variables, this step involves $k \times (p - k)$ model calculations.
- After all the substitutions are performed, if the best optimized parameter is better than the previous value, the old model is substituted by the best obtained model. In this case, the procedure is iteratively repeated for the new model. Otherwise, if the best optimized parameter is worse than the previous value, the procedure ends and a new model among the initially selected models is taken into consideration.

- **Cluster Significance Analysis** (CSA)

A method proposed for determining which molecular descriptors of a set of compounds are associated with a biological response. The active compounds are expected to be similar to each other with respect to these relevant descriptors and so will cluster together in the space defined by the corresponding descriptors.

This approach, originally proposed for binary response variables [McFarland and Gans, 1986], was extended to the quantitative biological responses **y**, scaled between zero and one [Rose and Wood, 1998] and then called **Generalized Cluster Significance Analysis** (GCSA).

Let **X** be a data matrix of n rows (i.e. the compounds) and p columns (i.e. the descriptors) and **y** the vector of the n biological responses.

The mean squared distance MSD_j was proposed to measure the tightness of the cluster of active compounds with respect to each jth molecular descriptor:

$$MSD_j = \frac{\sum_{s=1}^{n-1} \sum_{t=s+1}^{n} y_s \cdot y_t \cdot \left(x_{sj} - x_{tj}\right)^2}{n \cdot (n-1)}$$

where n is the number of compounds, y_s and y_t the biological responses of compounds s and t, x_{sj} and x_{tj} the jth descriptor values of the two compounds. A small MSD value indicates that the considered descriptor has a good capability to separate compounds with different biological activities.

The MSD calculated as above is proportional to that calculated as:

$$MSD_j = \sum_{i=1}^{n} y_i \cdot \left(x_{ij} - \bar{x}_j^W\right)^2$$

where the weighted mean is calculated as:

$$\bar{x}_j^W = \frac{\sum_{i=1}^{n} y_i \cdot x_{ij}}{\sum_{i=1}^{n} y_i}$$

To reach a statistical evaluation of the clustering capability of each descriptor, a test for significance is performed using a random permutation of the responses and using the permuted values to recalculate MSD values; this calculation is repeated N times (e.g. N = 100 000). Then, for any given descriptor, the number c_j of times giving a value less than or equal to MSD_j is used to obtain the significance level ('p-value') and the standard error s of this estimate:

$$\mathrm{p}_j = \frac{c_j}{\mathrm{N}} \qquad s_j = \sqrt{\frac{\mathrm{p}_j \cdot \left(1 - \mathrm{p}_j\right)}{\mathrm{N}}}$$

The best descriptor is chosen based on the minimum p-value.

If some descriptors are being considered together, the corresponding *MSD* random values are added together, as are the corresponding actual *MSD* values, before the count is taken.

Therefore, the selection of the best subset model can be performed by forward stepwise selection starting from the variable with the lowest p-value (the current model); next each of the variables not yet included in the current model is added to it in turn, producing a set of candidates with corresponding p-values. The candidate model with the lowest p-value is selected and the process is repeated on the new current model.

Moreover, a further implementation was proposed by calculating the conditional probability of candidate models.

📖 [McFarland and Gans, 1990a; McFarland and Gans, 1990b; Ordorica *et al.*, 1993; McFarland and Gans, 1994; McFarland and Gans, 1995]

A number of variable selection techniques were also suggested for the Partial Least Squares (PLS) regression method [Lindgren *et al.*, 1994]. The different strategies for **PLS-based variable selection** are usually based on a rotation of the standard solution by a manipulation of the PLS weight vector **w** or of the regression coefficient vector **b** of the PLS closed form.

Intermediate Least Squares regression (ILS) is an extension of the Partial Least Squares (PLS) algorithm where the optimal variable subset model is calculated as intermediate to PLS and stepwise regression, by two parameters whose values are estimated by cross-validation [Frank, 1987]. The first parameter is the number of optimal latent variables and the second is the number of elements in the weight vector **w** set to zero. This last parameter (ALIM) controls the number of selected variables by acting on the weight vector of each mth latent variable as the following:

$$\max_j\left(w_{jm}\right) \quad \rightarrow \quad w_{jm} = 1 \qquad j = 1, p$$

i.e. setting to one the p – ALIM largest weights, and setting to zero the remaining ALIM weights (thus excluding the corresponding variables). By modifying the number of elements to be set to zero in each weight vector, a whole range of models can be calculated between PLS models (ALIM = 0) and stepwise models (ALIM = p – 1).

Recent interesting developments of PLS-based variable selection are **Interactive Variable Selection for PLS** (IVS-PLS) [Lindgren *et al.*, 1994; Lindgren *et al.*, 1995], where two methods for manipulation of the **w** vector, called *inside-out* and *outside-in* thresholding, are iteratively used, and **Selective PLS**, proposed as a tool for multivariate design and able to separate the variables into a small number of orthogonal groups [Kettaneh-Wold *et al.*, 1994]. Another method, called **Uninformative Variable Elimination by PLS** (UVE-PLS), was proposed based on an analysis of the variance of the regression coefficients obtained by PLS on autoscaled or centred data [Centner *et al.*, 1996]. The fitness to enter the jth variable in the model is evaluated by the function:

$$c_j = \frac{|b_j|}{s(b_j)}$$

where b and $s(b)$ are the regression coefficients and their standard deviations; these last are estimated by the variability of the coefficients in the leave-one-out procedure.

Only the variables with a *c* value greater than a cutoff value are retained in the model. The cutoff value of *c* is estimated in such a way as to exclude from the model all the random variables added to the original variables. The random generated variables range between 0 and 10^{-10}, thus preserving the coefficient variability but negligibly influencing the model.

Other techniques for variable selection were proposed within the framework of the grid-based techniques (often for → *CoMFA* approach).

Among these methods, **Generating Optimal Linear PLS Estimations** (GOLPE) is a variable selection method for selecting by → *experimental design* a limited number of → *interaction energy values*, aimed at obtaining the best predictive PLS models.

The GOLPE procedure begins with the calculation of the PLS model on all variables, and this is followed by variable preselection according to a D-optimal design in the loading space. The D-optimal design works on the original variables described by their loadings in the principal component space, the dimensionality of which is selected according to the usual cross-validation criteria for PLS. The number of variables to be selected is user specified, but it is suggested to keep not less than a half of the variables each time, in an iterative manner [Baroni *et al.*, 1993b; Baroni *et al.*, 1993c]. Several variable subsets showing D-optimality are tried, and the subset with the best prediction ability is retained. Variable effect and its significance on the prediction ability of the PLS model is checked. Moreover, to avoid chance correlation, dummy variables are also introduced in the design matrix. For this kind of constrained problem, D-optimal designs are more efficient than fractional factorial designs (FFD designs), as initially proposed for variable selection [Baroni *et al.*, 1992].

Various **grid region selection methods** have recently been proposed with the aim of significantly reducing the number of field variables and their mutual correlation.

Cross-validated R^2 *guided region selection* (Q^2-GRS) divides the grid box into several small boxes, and CoMFA is separately performed for each box [Cho and Tropsha, 1995; Tropsha and Cho, 1998]. The box associated with Q^2 values greater than a specified threshold value is selected for further analysis. The *genetic algorithm-based PLS* (GA-PLS) has been proposed as a statistical tool able to select field variable combinations by the genetic algorithm using cross-validated R^2 value of the PLS model [Leardi *et al.*, 1992; Leardi, 1994; Hasegawa *et al.*, 1997; Hasegawa and Funatsu, 1998; Kimura *et al.*, 1998]. In this case, only a small number of significant variables are extracted. Furthermore, the *GA-based region selection method* (GA-RGS) uses domains of variables instead of single-field variables [Norinder, 1996; Pastor *et al.*, 1997; Hasegawa *et al.*, 1999b]; *GOLPE-Guided region selection* uses the Voronoi polyhedra for the region representation [Cruciani *et al.*, 1997; Cruciani *et al.*, 1998].

📖 [Allen, 1971] [McCabe, 1975] [Marengo *et al.*, 1992] [Cruciani *et al.*, 1993] [Cruciani and Clementi, 1994] [Cruciani and Watson, 1994] [Greco *et al.*, 1994b] [Greco *et al.*, 1994a] [Lindgren, 1994] [Sutter *et al.*, 1995a] [Sutter and Jurs, 1995b] [Ertepinar *et al.*, 1995] [Piazza *et al.*, 1995] [Lindgren *et al.*, 1996] [Leardi, 1996] [Wold *et al.*, 1996] [Duprat *et al.*, 1998] [Kemsley, 1998] [Kovalishyn *et al.*, 1998] [Leardi and Lupiañez Gonzalez, 1998] [Waller and Bradley, 1999] [Zheng and Tropsha, 2000]

variable subset selection : *variable selection*

variation → **distance matrix**

VEA indices → **eigenvalue-based descriptors** (⊙ eigenvalues of the adjacency matrix)

VED indices → **eigenvalue-based descriptors** (⊙ eigenvalues of the distance matrix)

Verloop parameters : *Sterimol parameters*

Verloop Sterimol parameters : *Sterimol parameters*

vertex adjacency matrix : *adjacency matrix*

vertex centric indices → **centric indices**

vertex chromatic decomposition : *chromatic decomposition*

vertex chromatic information index : *chromatic information index* → **chromatic decomposition**

vertex chromatic number : *chromatic number* → **chromatic decomposition**

vertex complexity → **topological information indices**

vertex coordinate → **indices of neighbourhood symmetry**

vertex cycle matrix → **cycle matrices**

vertex cyclic degree → **cycle matrices** (⊙ vertex cycle matrix)

vertex degree (δ)

The vertex degree of the ith atom δ_i is the count of its σ electrons in the → *H-depleted molecular graph*, i.e. the number of adjacent atoms to the ith atom. This quantity is a local vertex invariant, simply calculated as the sum of the entries a_{ij} in the ith row of the → *adjacency matrix* **A** or as the sum ${}^1f_i$ of the entries equal to one in the ith row of the → *distance matrix* **D**:

$$\delta_i = \sum_{j=1}^{A} a_{ij} = {}^1f_i = \left[\mathbf{A}^2\right]_{ii}$$

where the last term represents the diagonal element of the square of the adjacency matrix [Trinajstic, 1992].

It describes the role of the atom in terms of its connectedness and the count of σ-electrons excluding hydrogen atoms; it is a **σ-electron descriptor**, i.e.

$$\delta_i = \sigma_i - h_i$$

where σ_i and h_i are the number of σ electrons and hydrogens bonded to the ith atom [Kier and Hall, 1986].

The → *vertex degree matrix* of a graph is a diagonal matrix ($A \times A$) whose diagonal elements are the vertex degrees δ_i.

An analogous quantity, called the **bond vertex degree** δ_i^b, is calculated from the → *atom connectivity matrix* **C** as:

$$\delta_i^b = \sum_{j=1}^{A} a_{ij}^b$$

which also accounts for the multiple bonds [Kier and Hall, 1986].

In order to take into account all valence electrons of the ith atom, the vertex degree is replaced by the **valence vertex degree** δ_i^v (also called the **vertex valence**) defined as:

$$\delta_i^v = Z_i^v - h_i = \sigma_i + \pi_i + n_i - h_i$$

where Z_i^v is the number of valence electrons (σ electrons, π electrons and lone pair electrons n) of the ith atom and h_i is the number of hydrogen atoms bonded to it [Kier and Hall, 1986]. This definition holds for atoms of the 2nd principal quantum level (C, N, O, F). For atoms of higher principal quantum levels (P, S, Cl, Br, I), Kier and Hall proposed to account for both valence and non-valence electrons, as the following:

$$\delta_i^v = (Z_i^v - h_i)/(Z_i - Z_i^v - 1)$$

where Z_i is the total number of electrons of the ith atom, i.e. its atomic number. δ_i^v encodes the electronic identity of the atom in terms of both valence electron and core electron counts; it is a **valence electron descriptor**. It is useful to characterize heteroatoms and carbon atoms involved in multiple bonds.

The number of vertices with vertex degree equal to g is called the **vertex degree count** ^gF; therefore to each graph $\mathcal{G}$ the vector

$$\langle {}^1F; {}^2F; {}^3F; {}^4F \rangle$$

can be associated, provided the maximum vertex degree equals four.

Two molecular descriptors PR1 and PR2, called **ramification pair indices**, were proposed based on the vertex degree count of the third order and defined as the number of pairs of vertices with a vertex degree equal to three at a topological distance equal to one and two, respectively [Rios-Santamarina *et al.*, 1998].

A simple **ramification index** was also proposed for acyclic graphs:

$$r = \sum\nolimits_{\delta_i > 2} (\delta_i - 2) = {}^3F + 2 \cdot {}^4F$$

where the sum runs over all the vertex degrees greater than two [Araujo and De La Peña, 1998]. This index is quite similar to the → *quadratic index* and was previously used by Pitzer [Pitzer and Scott, 1941].

It was observed that the difference:

$$\delta_i^v - \delta_i = \pi_i + n_i$$

provides a quantitative measure of the potential of the atom for intermolecular interaction and reaction, the count being of π and n lone pair electrons of the atoms. A significant correlation with the → *Mulliken electronegativity* χ_{MU} was found as:

$$\chi_{MU} = 2.05 \cdot (\delta_i^v - \delta_i) + 6.99$$

Therefore, the **Kier-Hall electronegativity** was proposed based on the difference between valence and simple vertex degree, as:

$$\chi_{KH} \equiv KHE = \frac{\delta_i^v - \delta_i}{L_i^2}$$

where the square of the principal quantum number L_i of the ith atom is used to account for the increase in the screening effect due to inner electrons of higher row elements [Kier and Hall, 1981]. Since electronegativity for C_{sp3} is equal to zero, this scale can also be thought of as relative electronegativity with respect to C_{sp3} . Electronegativity for the hydrogen atom is assumed equal to –0.2; *KHE* is related to Mulliken electronegativity by the relationship:

$$\chi_{MU} = 7.99 \cdot \frac{(\delta_i^v - \delta_i)}{L_i^2} + 7.07$$

where δ^v in this equation is defined as:

$$\delta_i^v = Z_i^v - h_i$$

for all the atoms, i.e. it is the simple valence vertex degree defined above because the principal quantum number has already been considered.

In analogy with the → *total adjacency index* A_V, which is the sum of all the vertex degrees in the molecular graph, three other simple molecular descriptors are defined [Pogliani, 1992a]:

$$D^v = \sum_{i=1}^{A} \delta_i^v \qquad D^b = \sum_{i=1}^{A} \delta_i^b \qquad D^Z = \sum_{i=1}^{A} \delta_i^Z = \sum_{i=1}^{A} \frac{Z_i^v}{L_i}$$

where δ^v is the valence vertex degree, δ^b the bond vertex degree and δ^Z the **Z-delta number**; Z_i^v is the number of valence electrons and L_i the principal quantum number. The third index D^Z is sometimes called the **Pogliani index** [Pogliani, 1996c; Pogliani, 1997b].

The **simple topological index** S [Narumi, 1987] is a molecular descriptor related to → *molecular branching* proposed as the product of the vertex degrees δ:

$$S = \prod_{i=1}^{A} \delta_i$$

where A is the number of atoms. Other related molecular descriptors have been proposed:

$$A = \frac{\sum_{i=1}^{A} \delta_i}{A} = \frac{2 \cdot B}{A} \qquad H = \frac{A}{\sum_{i=1}^{A} 1/\delta_i} \qquad G = \left(\prod_{i=1}^{A} \delta_i \right)^{1/A} = S^{1/A}$$

where A is the **arithmetic topological index**, H the **harmonic topological index**, and G the **geometric topological index**. B is the number of molecule bonds; obviously, the arithmetic topological index is the same for all isomers [Narumi, 1987]. The simple topological index S is quite similar to the → *total connectivity index* χ_T.

Among these molecular descriptors the following simple relationship holds:

$$A \geq G \geq H$$

A **modified valence vertex degree** was also proposed [Hu and Xu, 1997] to distinguish the heteroatoms and defined as:

$$\delta_i' = \delta_i \cdot \sqrt{Z_i}$$

where Z_i is the atomic number of the considered atom. Another modification of the vertex degree has been recently proposed in terms of Mulliken population analysis [Li *et al.*, 2000].

The most popular molecular descriptors based on the vertex degree are the → *connectivity indices*, → *Zagreb indices*, → *Schultz molecular topological index*, and → *degree distance of the graph*.

vertex degree count → **vertex degree**

vertex degree distance → **Schultz molecular topological index**

vertex degree matrix → **Laplacian matrix**

vertex distance code → **distance matrix**

vertex distance complexity → **topological information indices**

vertex distance counts → **distance matrix**

vertex distance degree → **distance matrix**

vertex distance matrix : *distance matrix*

vertex distance sum : *vertex distance degree* → **distance matrix**

vertex eccentricity : *atom eccentricity* → **distance matrix**

vertex information layer index → **information layer index**

vertex orbital information content → **orbital information indices**

vertex path code → **path counts**

vertex path eccentricity : *atom detour eccentricity* → **detour matrix**

vertex path sum : *path degree* → **path counts**

vertex structural code → **self-returning walk counts**

vertex topological indices : *local vertex invariants*

vertex topological state : *topological state* → **topological state matrix**

vertex valence : *valence vertex degree* → **vertex degree**

vertices → **graph**

***V* index** → **topological information indices** (⊙ relative vertex distance complexity)

VLOGP model : *Gombar hydrophobic model* → **lipophilicity descriptors**

volume descriptors (V)

→ *Steric descriptors* and/or → *size descriptors* representing the volume of a molecule. The volume of a molecule can be derived from experimental observation such as the volume of the unit cell in crystals or the molar volume of a solution or from theoretical calculations. In fact, analytical and numerical approaches have been proposed for the calculation of molecular volume where the measure depends directly on the definition of → *molecular surface*; → *van der Waals volume* and → *solvent-excluded volume* are two volume descriptors based on van der Waals surface and solvent-accessible surface, respectively.

A list of other common volume descriptors is given below.

• **molar volume** ($\bar{V}$)

An experimentally measurable quantity defined as:

$$\bar{V} = \frac{MW}{\varrho}$$

where MW is the → *molecular weight* and ϱ the density of the liquid. It is related to → *molar refractivity* via → *refractive index*.

The **molecular volume** V is defined as the volume of the region within which a molecule is constrained by its neighbours. It is calculated from the experimental density ϱ of the liquid as:

$$V = \frac{MW}{\varrho \cdot N_A} = \frac{\bar{V}}{N_A}$$

where MW is the molecular weight and N_A is the Avogadro number [Meyer, 1985a]. The molecular volume as well as the molar volume characterize the bulk compound, comprising both the intrinsic molecular volume V_{VDW} and the volume of the empty "packing" space between molecules, sometimes called **free molecular volume** V_f, i.e. $V = V_{VDW} + V_f$.

Molar volume can also be calculated by additive fragment methods [Elbro *et al.*, 1991; Schotte, 1992] such as the fragment method of LeBas [Reid *et al.*, 1988]. Moreover, it is frequently used as a measure of the → *cavity term* in → *linear solvation energy relationships*.

🕮 [Immirzi and Perini, 1977; Horvath, 1992; van Haelst *et al.*, 1997]

• **Mc Gowan's characteristic volume** (V_X)
A steric descriptor defined as the sum of atomic volume parameters for all atoms in the molecule:

$$V_X = \sum_a w_a - 6.56 \cdot A$$

where w_a are the Mc Gowan volume parameters (Table V-1) and A the number of atoms [Abraham and Mc Gowan, 1987].

Table V-1. Mc Gowan atomic parameters.

Atom	w_a	*Atom*	w_a
H	8.71	S	22.91
C	16.35	F	10.48
N	14.39	Cl	20.95
O	12.43	Br	26.21
P	24.87	I	34.53

• **Corey-Pauling-Koltun volume** (V_{CPK}) (: *CPK volume*)
It is a measure of molecular volume obtained by immersing CPK models in a liquid.

• **geometric volume** (V_{geom})
Defined as the volume of the solid geometric shape of the molecule assuming the atoms as point masses. All the atoms in the molecule are interconnected in such a way that several regular and irregular tetrahedra are formed, their volumes being respectively computable analytically and numerically. Therefore, the geometric volume is obtained by subtracting the common volume from the sum of the volumes of the constituent tetrahedra [Bhattacharjee *et al.*, 1991; Bhattacharjee *et al.*, 1992; Bhattacharjee and Dasgupta, 1994; Bhattacharjee, 1994].

📖 [Edward, 1982b] [Meyer, 1985b] [Meyer, 1986c] [Meyer, 1988a] [Meyer, 1989] [Jaworska and Schultz, 1993] [Dubois and Loukianoff, 1993] [Barratt *et al.*, 1994b] [Connolly, 1994] [van Haelst *et al.*, 1997]

volume profiles → **molecular profiles**

volume-to-surface profiles → **molecular profiles**

Voronoi binding site models → **distance geometry**

Voronoi Field Analsyis (VFA)

Among the → *grid-based QSAR techniques*, Voronoi Field Analysis was proposed with the aim of reducing the very large number of potential → *interaction energy values* assigned to the grid points of → *molecular interaction fields* [Chuman *et al.*, 1998]. Interaction energy values are in fact assigned to each of the **Voronoi polyhedra** into which the superimposed molecular space is decomposed using an approach similar to that used in the → *Voronoi binding site models.*

The molecules in the data set are superimposed taking into account their conformational flexibility, and the total volume of the superimposed molecular space is calculated, after expansion with a 4.0 Å thick shell outside the surface. The total volume is divided into Voronoi polyhedra, each including an atom as reference point. The outer boundaries of the most outer subspaces are not bisecting planes between two reference points.

The reference points are assigned by selecting as the template the simplest molecule or the unsubstituted one, and the position in the template of the atoms including the hydrogens are automatically defined as reference points. As the second step, the largest compound in terms of number of atoms is selected and the atomic positions of this compound are compared with the previous ones: new reference points are defined if no reference point within 1.0 Å of each atom of this molecule is found. The remaining molecules are then selected in order of decreasing size, and atomic positions are compared with reference points as in the previous step. As the final step, a Voronoi polyhedron is assigned to each reference point with its own molecular space; the *Voronoi polyhedron* is a region delimited by a set of planes, each of which bisects as well as being perpendicular to the line connecting the reference point with each of the neighbouring reference points of the adjacent regions. In other words, each polyhedron is a set of points closer to the reference point than to any other.

After the decomposition of the superimposed molecular space into Voronoi polyhedra, a → *grid* exactly containing the expanded molecular surface is defined, with grid points spaced at 0.3 Å, and potential energy values are calculated at each of the lattice points located inside the surface. The steric and electrostatic potential energy values calculated at each kth grid point are transformed into the corresponding Voronoi potential energy values VE_g by summing all the contributions of the grid points belonging to the gth Voronoy polyhedron VP_g:

$$VE_g^T = \sum_k E_k^T \qquad k \in VP_g$$

where the superscript "T" denotes any kind of potential energy value (steric, electrostatic, ...).

[Aurenhammer, 1991]

Voronoi polyhedra → **Voronoi field analysis**

VRA indices → **eigenvalue-based descriptors** (⊙ eigenvalues of the adjacency matrix)

VRD indices → **eigenvalue-based descriptors** (⊙ eigenvalues of the distance matrix)

W

walk → graph

walk connectivity indices → walk counts

walk counts

Walk counts are atomic and molecular descriptors obtained from an → *H-depleted molecular graph* $\mathcal{G}$, based on the graph → *walk* [Rücker and Rücker, 1993; Rücker and Rücker, 1994].

Let **A** be the → *adjacency matrix* of a graph $\mathcal{G}$ and $\mathbf{A^k}$ its kth → *power matrix* (k = 1, 2, ..., A – 1) whose elements are denoted by $a_{ij}^{(k)}$ and A is the number of atoms. The $a_{ij}^{(k)}$ element of the $\mathbf{A^k}$ matrix corresponds to the number of walks of length k (*walk count*) from vertex v_i to vertex v_j. The diagonal entry $a_{ii}^{(k)}$ is the number of self-returning walks for the ith vertex, i.e. the random walks starting and ending at the ith vertex. Therefore, the **atomic walk count** of order k for the ith atom (also called kth order **walk degree** ^kW_i), denoted by $awc_i^{(k)}$, is calculated as the sum of the ith row entries of the kth power adjacency matrix $\mathbf{A^k}$:

$$awc_i^{(k)} \equiv {}^kW_i = \mathcal{R}_i\left(\mathbf{A^k}\right) = \sum_{j=1}^{A} a_{ij}^{(k)}$$

where $\mathcal{R}_i$ is the → *row sum operator*. The atomic walk count $awc_i^{(k)}$ is the total number of walks of length k starting from vertex v_i. → *Self-returning walk counts* are a particular case of walk counts with several important properties.

For k = 1, the matrix $\mathbf{A^1}$ is simply the adjacency matrix, and therefore the atomic walk count coincides with the → *vertex degree* δ_i, i.e.:

$$awc_i^{(1)} = \sum_{j=1}^{A} a_{ij} = {}^1P_i = \delta_i$$

where ^1P_i is the → *atomic path count* of first order.

The atomic walk count is a measure of something like "involvedness" or centrality of the atom in the graph, i.e. a measure of the complexity of the vertex environment. Moreover, the atomic walk count is the → *extended connectivity* defined by Morgan [Razinger, 1982; Rücker and Rücker, 1993; Figueras, 1993].

The atomic walk count can also be evaluated by iterative summation of the walk degrees over all δ_i neighbours [Diudea *et al.*, 1994]:

$$awc_i^{(k+1)} \equiv {}^{k+1}W_i = \left[{}^{k+1}\mathbf{W_C}\right]_{ii} = \sum_{j=1}^{\delta_i} [\mathbf{C}]_{ij} \cdot \left[{}^{k}\mathbf{W_C}\right]_{jj}; \qquad \left[{}^{0}\mathbf{W_C}\right]_{jj} = 1$$

$$\left[{}^{k+1}\mathbf{W_C}\right]_{ij} = \left[{}^{k}\mathbf{W_C}\right]_{ij} = [\mathbf{C}]_{ij}$$

where $^k\mathbf{W_C}$ is a matrix having kth order walk degrees of atoms in the main diagonal, the → *conventional bond order* for each entry corresponding to bonded atoms, and zero otherwise. In fact, it is derived from the → *atom connectivity matrix* **C**. Therefore, the walk degrees obtained in such a way also allow for bond multiplicity in the molecular graph.

The atomic walk counts are the elements of the → *walk-sequence matrix* **SW**.

Atom ID	Walk length k 0	1	2	.....	A–1	Atomic walk count sums
1	1	$awc_1^{(1)}$	$awc_1^{(2)}$	.....	$awc_1^{(A-1)}$	$awcs_1$
2	1	$awc_2^{(1)}$	$awc_2^{(2)}$	.....	$awc_2^{(A-1)}$	$awcs_2$
.....	.....	.....	.....	.....	.....	.....
.....	.....	.....	.....	.....	.....	.....
A	1	$awc_A^{(1)}$	$awc_A^{(2)}$	.....	$awc_A^{(A-1)}$	$awcs_A$
Molecular walk counts	*A*	$mwc^{(1)}$	$mwc^{(2)}$	.....	$mwc^{(A-1)}$	*TWC*

The **walk count atomic code** of each *i*th atom is the ordered sequence of atomic walk counts of increasing length:

$$\left\langle awc_i^{(1)}, awc_i^{(2)}, \ldots, awc_i^{(A-1)} \right\rangle$$

Walks starting from two different vertices v_i and v_j are called **equipotent walks** if their walk count atomic codes are the same. Moreover, if the vertices v_i and v_j are nonequivalent, equipotent walks are called **unusual walks** and the corresponding vertices **unusual vertices** [Randic *et al.*, 1983b].

Moreover, the **atomic walk count sum** of the *i*th atom is the sum of all walks of any length starting from the *i*th atom:

$$awcs_i = \sum_{k=1}^{A-1} awc_i^{(k)}$$

Arranging in increasing order the *A* *awcs* indices, a new molecular vector descriptor, called the **ordered walk count molecular code**, is obtained:

$$\left\langle awcs_{s(1)}, awcs_{s(2)}, \ldots, awcs_{s(A)} \right\rangle$$

where $s(i)$ is an index for ordered sequence. The *awcs* ranking is related to the increasing complexity of the atom molecular environment.

The **molecular walk count** $mwc^{(k)}$ of length *k* (also called the **walk number** or **graph walk count**, $GWC^{(k)}$) is the total number of walks of the *k*th length in the molecular graph and, for any *k* different from zero, it is calculated as the half-sum of all atomic walk counts of the same *k*th length, i.e. as the column half-sum of the **SW** matrix:

$$mwc^{(k)} = \frac{1}{2} \cdot C_k(\mathbf{SW}) = \frac{1}{2} \cdot \sum_{i=1}^{A} awc_i^{(k)} = \frac{1}{2} \cdot \sum_{i=1}^{A} \sum_{j=1}^{A} a_{ij}^{(k)}$$

where C_k is the → *column sum operator* and $a_{ij}^{(k)}$ are the elements of the *k*th power of the adjacency matrix **A**. The molecular walk count of zero order $mwc^{(0)}$ is simply equal to the number *A* of graph vertices.

The molecular walk count is related to → *molecular branching* and size, and in general to the → *molecular complexity* of the graph. In fact, it was found that $mwc^{(k)}$ is directly related to the → *Lovasz-Pelikan index*, i.e. the largest eigenvalue of the adjacency matrix [Cvetkovic and Gutman, 1977]

A **normalized atomic walk count** of the ith atom can be defined from the previous indices by:

$$\overline{awc}_i^{(k)} = \frac{awc_i^{(k)}}{mwc^{(k)}}$$

and can be considered the weight of the ith vertex.

The **total walk count** TWC is the total number of walks of any length in the graph and is calculated as:

$$TWC = \sum_{k=0}^{A-1} mwc^{(k)} = A + \frac{1}{2} \cdot \sum_{k=1}^{A-1} \sum_{i=1}^{A} awc_i^{(k)} = A + \frac{1}{2} \cdot \sum_{i=1}^{A} awcs_i$$

The total walk count is a measure of the molecular complexity, increasing both with increasing size and branching.

The **walk connectivity indices** are obtained from walk degrees by a Randic-type formula [Razinger, 1986]:

$$\chi^W = \sum_b \left(awcs_i \cdot awcs_j\right)_b^{-1/2} \qquad {}^k\chi^W = \sum_b \left(awc_i^{(k)} \cdot awc_j^{(k)}\right)_b^{-1/2}$$

where the sums run over all bonds in the H-depleted molecular graph, and the subscripts i and j refer to vertices incident to the considered bond; the first index χ^W is calculated from the atomic walk count sum *awcs*, i.e. considering all walks of any length from the atom, while the second index ${}^k\chi^W$ is calculated considering only the walks of length k from each atom ($awc^{(k)}$); the atomic count of walks of maximum length D was considered, D being the → *topological diameter*.

→ *Local vertex invariants* proposed to account for both heteroatoms and multiple bonds are the **electronegativity-weighted walk degrees** ${}^kW_E$, defined as [Diudea *et al.*, *1994]*:

$${}^kW_{E,i} = {}^kW_i \cdot {}^kt_i$$

where ${}^kt_i$ is a weighting factor accounting for heteroatoms and multiple bonds derived by a recursive formula as:

$${}^{k+1}t_i = \left[\prod_{j=1}^{\delta_i} \left({}^kt_j\right)^{\pi_{ij}^*}\right]^{1/\sum_j \pi_{ij}^*}$$

where the product runs over the weights of the first neighbours of the ith atom each raised to the conventional bond order π_{ij}^*; vertex weights at the first step ($k = 1$) are valence group carbon-related electronegativities (Table W-1) [Diudea *et al.*, 1996a].

Table W-1. Valence group electronegativities used for the calculation of the electronegativity-weighted walk degrees.

Atom/Hybrid	*t*	*Atom/Hybrid*	*t*	*Atom/Hybrid*	*t*
>C<	1.0000	$-CHBr_2$	1.0672	–H	0.9175
>C=	1.0747	$-CHBr_2$	0.9914	–N<	1.2234
–C≡	1.1476	$-CF_3$	1.3260	=N–	1.3147
=C=	1.1581	$-CCl_3$	1.1932	≡N	1.5288
>CH–	0.9716	$-CBr_3$	1.1266	–NH–	1.1021
=CH–	1.0441	$-CI_3$	1.0088	=NH	1.2474
≡CH	1.2142	>C=O	1.2397	$-NH_2$	1.0644
$-CH_2-$	0.9622	–CH=O	1.1596	$-NHCH_3$	1.0379
$=CH_2$	1.0891	–COOH	1.2220	$-N(CH_3)_2$	1.0292

Atom/Hybrid	*t*	*Atom/Hybrid*	*t*	*Atom/Hybrid*	*t*
$-CH_3$	0.9575	–O–	1.4634	–C≡N	1.2377
$-CH_2F$	1.0674	=O	1.6564	–P<	0.8988
$-CH_2Cl$	1.0305	–OH	1.2325	=P–	0.9658
$-CH_2Br$	1.0110	$-OCH_3$	1.1248	–PH–	0.9124
$-CH_2I$	0.9744	–S–	1.1064	$-PH_2$	0.9170
$-CH_2OH$	1.0228	=S	1.2523	–F	1.6514
$-CH_2SH$	0.9804	$-SCH_3$	1.0073	–Cl	1.3717
$-CHF_2$	1.1897	–NO	1.4063	–Br	1.2447
$-CHCl_2$	1.1089	$-NO_2$	1.4861	–I	1.0262

📖 [Randic, 1980c] [Gao and Hosoya, 1988] [Kunz, 1989] [Shalabi, 1991] [Randic, 1995c] [Rücker and Rücker, 2000]

walk count atomic code → **walk counts**

walk degree : *atomic walk count* → **walk counts**

walk degree layer matrix → **layer matrices**

walk diagonal matrix → **walk matrices**

walk length → **graph**

walk matrices (: *random walk matrices*)

The **walk diagonal matrix** $^{\mathbf{k}}\mathbf{W_M}$ of kth order based on an **M** square $A \times A$ matrix is a diagonal matrix whose diagonal elements are the kth order **weighted walk degrees** $^{k}W_{M,\,i}$, i.e. the sum of the weights (the property collected in matrix **M**) of all walks of length k starting from the ith vertex to any other vertex in the graph, directly calculated as:

$$\left[{}^{\mathbf{k}}\mathbf{W_M}\right]_{ii} = {}^{k}W_{M,i} = \sum_{j=1}^{A} \left[\mathbf{M^k}\right]_{ij}$$

where A is the total number of vertices in the graph and $\mathbf{M^k}$ is the kth power of the matrix **M** [Diudea, 1996a; Diudea and Randic, 1997]. An alternative algorithm to obtain the weighted walk degrees was proposed by Diudea and called the $^{k}W_M$ *algorithm*. This is an algorithm based on the iterative sum of vertex contributions over all first neighbours of the considered ith vertex, defined as:

$$\left[{}^{0}\mathbf{W_M}\right]_{ii} = {}^{0}W_{M,i} = 1 \qquad i = 1, \ldots, A$$

$$\left[{}^{\mathbf{k+1}}\mathbf{W_M}\right]_{ii} = {}^{k+1}W_{M,i} = \sum_{j=1}^{\delta_i} [\mathbf{M}]_{ij} \cdot {}^{k}W_{M,j}$$

where the sum in the second expression runs over the δ_i first neighbours of the ith vertex, i.e. the vertex degree.

If **M** = **A**, **A** being the → *adjacency matrix*, then the weighted walk degree of the ith atom $^{k}W_{M,\,i}$ is just the → *atomic walk count* of length k of the ith atom, i.e. the number of all walks of length k starting from the ith vertex to any other vertex in the graph.

The **weighted walk numbers** are molecular descriptors obtained as the half-sum of all weighted walk degrees of all atoms:

$$^kW_M = \frac{1}{2} \cdot \sum_{i=1}^{A} {^kW_{M,i}} = \frac{1}{2} \cdot \sum_{i=1}^{A} \sum_{j=1}^{A} \left[\mathbf{M^k}\right]_{ij}$$

In the case of $\mathbf{M} = \mathbf{A}$, the global walk number ^kW_A is just the → *molecular walk count* of kth order $mwc^{(k)}$, i.e. the total number of length k walks in the graph.

Based on weighted walk degrees of different orders calculated by the algorithm above defined, the **walk matrix** $\mathbf{W}_{(\mathbf{M_1}, \mathbf{M_2}, \mathbf{M_3})}$ is an asymmetric square $A \times A$ matrix whose i-j entry is defined as:

$$\left[\mathbf{W}_{(\mathbf{M_1}, \mathbf{M_2}, \mathbf{M_3})}\right]_{ij} = {^{[\mathbf{M_2}]_{ij}}} W_{M_1, i} \cdot [\mathbf{M_3}]_{ij}$$

where $\mathbf{M_1}$, $\mathbf{M_2}$ and $\mathbf{M_3}$ are any square $A \times A$ matrices; $W_{M_1,i}$ is the weighted walk degree of the ith vertex, based on the property collected in matrix $\mathbf{M_1}$; $[\mathbf{M_2}]_{ij}$ is the ijth entry of the $\mathbf{M_2}$ matrix giving the length of the walk and the ijth entry $[\mathbf{M_3}]_{ij}$ of the third matrix is used as a weighting factor. The diagonal entries are usually equal to zero.

Summing the entries in each row, → *local vertex invariants* representing a sum of weighted walk degrees of different orders are obtained:

$$W_{M_1, M_2, M_3, i} = \sum_{j=1}^{A} \left({^{[\mathbf{M_2}]_{ij}}} W_{M_1, i} \cdot [\mathbf{M_3}]_{ij} \right)$$

Global weighted walk numbers W_{M_1,M_2,M_3} are obtained applying the → *Wiener operator* $\mathcal{W}$ to the walk matrix as:

$$W_{M_1, M_2, M_3} = \mathcal{W}\left(\mathbf{W}_{(\mathbf{M_1}, \mathbf{M_2}, \mathbf{M_3})}\right) = \frac{1}{2} \cdot \sum_{i=1}^{A} \sum_{j=1}^{A} \left({^{[\mathbf{M_2}]_{ij}}} W_{M_1, i} \cdot [\mathbf{M_3}]_{ij} \right)$$

Appropriate combinations of $\mathbf{M_1}$, $\mathbf{M_2}$ and $\mathbf{M_3}$ give various unsymmetric walk matrices and hence various local and molecular descriptors.

For example, the combination $(\mathbf{A}, \mathbf{1}, \mathbf{1})$, where $\mathbf{A}$ is the adjacency matrix and $\mathbf{1}$ represents a matrix whose off-diagonal elements are all equal to one, gives a walk matrix where in each ith row the vertex degree of the considered ith vertex is repeated in all the columns, the vertex degree being the row sum of the adjacency matrix and coincident with the 1st order atomic walk count awc_i^1. If the matrix $\mathbf{A}$ is substituted by a general matrix $\mathbf{M_1}$, the corresponding row sums of $\mathbf{M_1}$ are collected in the rows of the walk matrix.

Next, considering the combination $(\mathbf{M_1}, \mathbf{1}, \mathbf{M_3})$, the corresponding walk matrix is obtained simply by the relationship:

$$\mathbf{W}_{(\mathbf{M_1}, \mathbf{1}, \mathbf{M_3})} = \mathbf{W}_{(\mathbf{M_1}, \mathbf{1}, \mathbf{1})} \bullet \mathbf{M_3}$$

where the symbol • represents the → *Hadamard matrix product*. Moreover, the global walk number $W_{M_1,1,M_3}$ is equal to the half-sum of the entries of the matrix product $\mathbf{M_1}$ $\mathbf{M_3}$:

$$W_{M_1, 1, M_3} = \frac{1}{2} \cdot \sum_{i=1}^{A} \sum_{j=1}^{A} \left[\mathbf{W}_{(\mathbf{M_1}, \mathbf{1}, \mathbf{M_3})}\right]_{ij} = \frac{1}{2} \cdot \sum_{i=1}^{A} \sum_{j=1}^{A} [\mathbf{M_1} \cdot \mathbf{M_3}]_{ij} = \frac{1}{2} \cdot S(\mathbf{M_1} \cdot \mathbf{M_3})$$

where S is the → *total sum operator*.

This relationship was demonstrated recalling that the sum of the entries of the matrix product $\mathbf{M_1}$ $\mathbf{M_3}$ is equal to the product of the $\mathbf{M_1}$ column sum vector for the $\mathbf{M_3}$ row sum vector:

$$S(\mathbf{M_1 M_3}) = \mathbf{u^T M_1 M_3 u} = \mathbf{cs^T}(\mathbf{M_1})\, \mathbf{rs}(\mathbf{M_3})$$

where **u** is a column unit vector, **cs** and **rs** the → *column sum vector* and the → *row sum vector*, respectively.

The global walk number obtained by the particular walk matrix defined by (**A**, **1**, **D**), where **D** is the → *distance matrix*, coincides with half of the Schultz → *S index*.

Moreover, if also $\mathbf{M_1} = \mathbf{M_3}$, the global walk number $W_{M_1,1,M_1}$ represents the half-sum of the square $\mathbf{M_1}$ matrix. If $\mathbf{M_2}$ is a matrix having all non-diagonal entries equal to an integer n, the global walk number ${}^{n+1}W_{M_1,n,M_1}$ of order n+1 coincides with the walk number ${}^{n+1}W_{M_1}$.

An interesting walk matrix is $\mathbf{W}_{(\mathbf{A},\mathbf{D},\mathbf{1})}$ where $\mathbf{M_1}$ is the adjacency matrix **A**, $\mathbf{M_2}$ is the distance matrix **D** and $\mathbf{M_3}$ is a matrix whose elements equal one. In this matrix the atomic walk counts of increasing orders are collected where the → *topological distance* between each pair of vertices gives the order.

The **reciprocal walk matrix** $\mathbf{W}^{-1}_{(\mathbf{M_1},\mathbf{M_2},\mathbf{M_3})}$ is a square $A \times A$ matrix whose entries are the reciprocals of the corresponding entries of the walk matrix $\mathbf{W}_{(\mathbf{M_1},\mathbf{M_2},\mathbf{M_3})}$:

$$\left[\mathbf{W}^{-1}_{(\mathbf{M_1},\mathbf{M_2},\mathbf{M_3})}\right]_{ij} = \left[\mathbf{W}_{(\mathbf{M_1},\mathbf{M_2},\mathbf{M_3})}\right]^{-1}_{ij}$$

The diagonal entries are always zero. → *Harary indices* are derived from these matrices by applying the Wiener operator.

A **restricted random walk matrix RRW** was also proposed [Randic, 1995c] as an $A \times A$ dimensional square unsymmetric matrix that enumerates restricted (i.e. selected) random walks over a molecular graph $\mathcal{G}$. The *i-j* entry of the matrix is the probability of a random walk starting at vertex v_i and ending at vertex v_j of length equal to the topological distance d_{ij} between the considered vertices:

$$[\mathbf{RRW}]_{ij} = \begin{cases} \dfrac{\left[\mathbf{A}^{\mathbf{d}_{ij}}\right]_{ij}}{{}^{d_{ij}}W_i} & \text{for } i \neq j \\ 0 & \text{for } i = j \end{cases}$$

where ${}^{d_{ij}}W_i$ is the d_{ij}-order atomic walk count of the ith vertex, i.e. the number of walks of length d_{ij} starting at vertex v_i (the elements of the walk matrix $\mathbf{W}_{(\mathbf{A},\mathbf{D},\mathbf{1})}$); $\left[\mathbf{A}^{\mathbf{d}_{ij}}\right]_{ij}$ is the number of all walks of length d_{ij} starting from vertex v_i and ending at vertex v_j; it is equal to one for any pair of vertices in acyclic graphs, while it can be greater than one for a pair of vertices in cyclic graphs; $A^{d_{ij}}$ is the power, equal to the distance d_{ij}, of the adjacency matrix **A**. Therefore, in order to build the **RRW** matrix, the powers of the adjacency matrix **A** are used to calculate walk counts of increasing length, and the distance matrix **D** is used to impose the restrictions on the lengths of walks.

For acyclic graphs, the restricted random walk matrix is simply the reciprocal walk matrix where $\mathbf{M_1} = \mathbf{A}$, $\mathbf{M_2} = \mathbf{D}$, and $\mathbf{M_3} = \mathbf{1}$:

$$\mathbf{RRW} = \mathbf{W}^{-1}_{(\mathbf{A},\mathbf{D},\mathbf{1})}$$

Several matrix invariants can be calculated such as the → *eigenvalues*, the → *determinant* and the coefficients of the → *characteristic polynomial*. Moreover, → *weighted path counts* of mth order are obtained by summing all matrix entries corresponding to vertices separated by a distance of length m. The 1st order weighted path count ${}^1R$ is always equal to the number A of vertices; the 2nd order weighted path count ${}^2R$ is a connection additive index which correlates well with some → *physico-chemical properties*. This sequence is a weighted → *molecular path code*

$$\langle {}^1R, {}^2R, \ldots, {}^LR \rangle$$

where L is equal to $A - 1$, A being the number of vertices. Summing all path numbers from 1 order to $A - 1$ order, the corresponding → *molecular ID number* is obtained. It is called the **restricted walk ID number** (*RWID*) and is defined as:

$$RWID = \sum_{m=1}^{A-1} {}^{m}R = A + \sum_{m=2}^{A-1} {}^{m}R$$

[Diudea, 1996b] [Diudea *et al.*, 1996b] [Diudea and Randic, 1997] [Diudea and Gutman, 1998]

walk matrix → **walk matrices**

walk number : *molecular walk count* → **walk counts**

walk-sequence matrix → **sequence matrices**

WDEN index : *WHIM density* → **WHIM descriptors** (⊙ global WHIM descriptors)

weighted adjacency matrices → **weighted matrices**

weighted atomic self-returning walk counts → **self-returning walk counts**

weighted combination of *COSV* and *ISDFP* → **molecular shape analysis**

weighted detour matrix → **detour matrix**

weighted distance matrices → **weighted matrices**

weighted graph → **graph**

weighted holistic invariant molecular descriptors : *WHIM descriptors*

weighted ID number → **ID numbers**

weighted information indices by volume → **indices of neighbourhood symmetry**

weighted matrices

Matrices derived from vertex- and edge-weighted graphs representing molecules containing heteroatoms and/or multiple bonds.

Weighted adjacency matrices are square symmetric → *sparse matrices* derived from the → *H-depleted molecular graph* whose diagonal elements are not necessarily equal to zero and the off-diagonal elements can be any real positive numbers.

While the vertex → *adjacency matrix* **A** and → *edge adjacency matrix* **E** contain information only about vertex-edge connectivity in the graph, weighted adjacency matrices allow the distinguishing of different bonds and atoms in a molecule.

When the adjacency between a pair of atoms *i-j* is represented by a bond weight (for example, → *bond order*, force constant, ionic character, → *dipole moment*, → *bond distances* or its inverse) and/or each atom is represented by some → *atomic properties* such as atomic numbers, several weighted-vertex adjacency matrices called **atom connectivity matrices ACM** can be defined [Spialter, 1963; Spialter, 1964a; Spialter, 1964b] and their entries are:

$$[\mathbf{A}^{\mathrm{w}}]_{ij} = \begin{cases} w_{ii}^{v} & \text{if } i = j \\ w_{ij}^{e} & \text{if } (i,j) \in \mathcal{E}(\mathcal{G}) \\ 0 & \text{if } (i,j) \notin \mathcal{E}(\mathcal{G}) \end{cases}$$

where w_{ii}^{v} depends on the chemical nature of the ith atom and is usually equal to zero for carbon atom; w_{ij}^{e} depends on the chemical nature of the ith and jth bonded atoms as well as on the bond multiplicity [Ivanciuc *et al.*, 1997a]. $\mathcal{E}(\mathcal{G})$ is the set of the graph edges.

The most well-known is the **atom connectivity matrix C** obtained from multigraphs using the → *conventional bond order* π^* to represent the adjacency between a pair of vertices:

$$[\mathbf{C}]_{ij} = \begin{cases} \pi^*_{ij} & \text{if } (i,j) \in \mathcal{E}(\mathcal{G}) \\ 0 & \text{otherwise} \end{cases}$$

Another important connectivity matrix is obtained weighting each bond *i-j* by the → *edge connectivity* $(\delta_i \cdot \delta_j)^{-1/2}$, δ being the → *vertex degree* of the atoms. This matrix is known as the **Kier-Hall connectivity matrix** or **sparse χ matrix**:

$$[{}^1\boldsymbol{\chi}]_{ij} = \begin{cases} 1/\sqrt{\delta_i \cdot \delta_j} & \text{if } (i,j) \in \mathcal{E}(\mathcal{G}) \\ 0 & \text{otherwise} \end{cases}$$

Analogous connectivity matrices are obtained using → *bond vertex degree* or → *valence vertex degree*. Moreover, from the powers of the sparse χ matrix, → $^k\alpha$ *descriptors* were defined. Related to the sparse χ matrix are also the → *extended adjacency matrices*.

The → *augmented adjacency matrix* is an example of a weighted adjacency matrix where only diagonal elements are replaced by empirical parameters representing heteroatoms. The → *Burden matrix* is another interesting weighted adjacency matrix.

Analogously weighted-edge adjacency matrices are derived for weighted graphs assigning to each edge a weight that can be a bond order or a bond distance. Examples are the → *bond order-weighted edge adjacency matrix* $^\pi\mathbf{E}$ and the → *bond distance-weighted edge adjacency matrix* $^\mathbf{b}\mathbf{E}$. Moreover, the **sparse χ-E matrix** of the graph $\mathcal{G}$ is the sparse χ matrix of the → *line graph* of $\mathcal{G}$, i.e.

$$[{}^1\boldsymbol{\chi} - \mathbf{E}]_{ij} = \begin{cases} 1/\sqrt{\varepsilon_i \cdot \varepsilon_j} & \text{if } (i,j) \in C(\mathcal{G}) \\ 0 & \text{otherwise} \end{cases}$$

where ε is the → *edge degree* [Ivanciuc *et al.*, 1997] and C the set of connections in the graph.

Weighted distance matrices are a generalization of the distance matrix for heteroatom molecular systems with possible multiple bonds. They are square symmetric $A \times A$ matrices derived from the H-depleted molecular graph obtained by weighting the shortest path (i.e. → *geodesic*) between any pair of vertices v_i and v_j. When more than one shortest path exists between a pair of vertices, some rules must be adopted in order to calculate the corresponding matrix element. For example, average values from all the shortest paths can be calculated or the clockwise direction path can be chosen. Another way to obtain weighted distance matrices is to take, as matrix entries, the minimum sum of the edge weights along the path between the considered vertices, which is not necessarily the shortest possible path between them. The diagonal entries of these weighted matrices can be chosen to be different from zero in order to encode information about the chemical nature of the molecule atoms.

The **χ matrix** [Hall, 1990] is the most popular weighted distance matrix, based on the path contributions arising in constructing → *connectivity indices*. Using the vertex degree δ of the atoms, each path p_{ij} between the vertices v_i and v_j is weighted by the **path connectivity** defined as:

$$w_{ij} = (\delta_i \cdot \delta_k \cdot \ \ldots \ \cdot \delta_j)^{-1/2}$$

where δ_i, δ_k, ..., δ_j are the vertex degrees of the atoms belonging to the path p_{ij}.

Using the valence vertex degree δ^v or the bond vertex degree δ^b instead of the simple vertex degree, the corresponding path weights are defined as:

$$w_{ij} = \left(\delta_i^v \cdot \delta_k^v \cdot \ \ldots \ \cdot \delta_j^v\right)^{-1/2} \qquad w_{ij} = \left(\delta_i^b \cdot \delta_k^b \cdot \ \ldots \ \cdot \delta_j^b\right)^{-1/2}$$

Therefore, based on path connectivity the $\boldsymbol{\chi}$ matrix is defined as:

$$[\boldsymbol{\chi}]_{ij} = \begin{cases} w_{ij} & \text{if } i \neq j \\ 0 & \text{if } i = j \end{cases}$$

where w_{ij} is the path connectivity of the shortest path between vertices v_i and v_j. Diagonal entries are zero assuming that there is no path bonding to the atom itself. This matrix was originally proposed as the sparse χ matrix defined above, where only non-zero elements correspond to bonds.

For acyclic graphs, connectivity indices for path subgraphs can be calculated by the → *Wiener operator* applied to the product of the χ matrix and the → *binary sparse matrix* $^{\mathbf{m}}\mathbf{B}$, whose elements are all equal to zero except for those corresponding to the shortest paths *i-j* of length *m* which are equal to one:

$$^{m}\chi_p = \mathcal{W}(\chi \bullet {}^{\mathbf{m}}\mathbf{B}) = \frac{1}{2}\sum_{i=1}^{A}\sum_{j=1}^{A}[\chi \bullet {}^{\mathbf{m}}\mathbf{B}]_{ij}$$

where $\mathcal{W}$ is the Wiener operator and the symbol • indicates the → *Hadamard matrix product*. Moreover, other novel invariants can be derived from the χ matrix such as the largest positive eigenvalue, the spectrum, and row sums for any graphs. For example, the → *characteristic root index* is the sum of the positive eigenvalues of the χ matrix.

Other weighted distance matrices can be calculated by weighting each path by different atomic or bond properties of the elements belonging to the considered shortest path. For example, the **multigraph distance matrix** $^{*}\mathbf{D}$ is a weighted distance matrix where the distance from vertex v_i to vertex v_j is obtained by counting the edges in the shortest path between them, where each edge counts as the inverse of the conventional bond order π^*, i.e. the → *relative topological distance*, and therefore contributes $1 / \pi^*$ to the overall path length [Balaban, 1982; Balaban, 1983a]. The row sum of the matrix elements is called the **multigraph distance degree** $^{*}\sigma$ and is defined as:

$$^{*}\sigma_i = \mathcal{R}(^{*}\mathbf{D}) = \sum_{j=1}^{A}[^{*}\mathbf{D}]_{ij}$$

where $\mathcal{R}$ is the → *row sum operator* and A the → *atom number*. It is a local vertex invariant taking edge multiplicity into account. → *Balaban modified distance connectivity indices* are calculated from the multigraph distance degree. The **chemical distance matrix** is an analogous weighted distance matrix whose entries are distances between any pair of vertices calculated by summing the → *chemical distance* between all pairs of adjacent vertices in the considered path [Balaban *et al.*, 1993a].

The **Barysz distance matrix** $\mathbf{D^Z}$ is a weighted distance matrix accounting simultaneously for the presence of heteroatoms and multiple bonds in the molecule; it is defined as [Barysz *et al.*, 1983a]:

$$[\mathbf{D^Z}]_{ij} = \begin{cases} 1 - \dfrac{Z_C}{Z_i} & \text{if } i = j \\ \displaystyle\sum_{b=1}^{d_{ij}}\left(\frac{1}{\pi_b^*}\cdot\frac{Z_C^2}{Z_{b(1)}\cdot Z_{b(2)}}\right) & \text{if } i \neq j \end{cases}$$

where Z_C is the atomic number of the carbon atom, Z_i the atomic number of the *i*th atom, π^* is the conventional bond order, the sum runs over all d_{ij} bonds involved in the shortest path between vertices v_i and v_j, d_{ij} being the topological distance, and the subscripts b(1) and b(2) represent the two vertices incident to the considered *b* bond.

Note that diagonal elements are atomic weights based on atomic numbers and the quantity in parenthesis is a bond weight based on conventional bond order and atomic numbers of the vertices incident to the considered *b* bond; some values of these atomic and bond parameters are collected in Tables W-2 and W-3.

The **electronegativity-weighted distance matrix $\mathbf{D}^X$** was proposed in analogy with the Barysz distance matrix replacing the atomic number by a relative → *atom electronegativity* [Ivanciuc *et al.*, 1998a]. The atomic electronegativity values S_i are derived from electronegativities χ_S recalculated by Sanderson on the Pauling scale with F, Na, and H atoms having values equal to 4.00, 0.56, and 2.592, respectively, by using a biparametric equation:

$$S_i = 1.1032 - 0.0204 \cdot Z_i + 0.4121 \cdot L_i$$
$$X_i = 0.4196 - 0.0078 \cdot Z_i + 0.1567 \cdot L_i$$

where Z are atomic numbers and L principal quantum numbers, respectively. X_i values are the relative atomic electronegativity to the carbon atom, calculated by dividing the S_i values by the calculated value for carbon $S_C = 2.629$, thus obtaining $X_C = 1$. The weighted matrix $\mathbf{D}^X$ is defined as:

$$[\mathbf{D}^X]_{ij} = \begin{cases} 1 - \dfrac{X_C}{X_i} & \text{if } i = j \\ \sum_{b=1}^{d_{ij}} \left(\dfrac{1}{\pi_b^*} \cdot \dfrac{X_C^2}{X_{b(1)} \cdot X_{b(2)}} \right) & \text{if } i \neq j \end{cases}$$

where π^* is the conventional bond order, the sum runs over all d_{ij} bonds involved in the shortest path between vertices v_i and v_j, d_{ij} being the topological distance, and the subscripts b(1) and b(2) represent the two vertices incident to the considered *b* bond.

Moreover, the **covalent radius-weighted distance matrix $\mathbf{D}^Y$** was also defined [Ivanciuc *et al.*, 1998a] as:

$$[\mathbf{D}^Y]_{ij} = \begin{cases} 1 - \dfrac{Y_C}{Y_i} & \text{if } i = j \\ \sum_{b=1}^{d_{ij}} \left(\dfrac{1}{\pi_b^*} \cdot \dfrac{Y_C^2}{Y_{b(1)} \cdot Y_{b(2)}} \right) & \text{if } i \neq j \end{cases}$$

where Y_i are the relative covalent radii to the carbon atom, recalculated by using a similar biparametric equation based on atomic numbers Z and principal quantum numbers L:

$$Y_i = 1.1191 + 0.0160 \cdot Z_i - 0.0537 \cdot L_i$$

The atomic radii Y_i values are relative to a carbon covalent radius equal to 87.126 picometers ($Y_C = 1$). Relative atomic electronegativities X_i and covalent radii Y_i were also used as atomic weights to account for heteroatoms in the definition of the Balaban modified distance connectivity indices, J^X and J^Y, respectively.

Several common → *topological indices* can be derived from these weighted matrices as a generalization for heteroatom systems including also multiple bonds. The **Barysz index** J_{het} [Barysz *et al.*, 1983a] is a generalization of the → *Balaban distance connectivity index* calculated by applying the → *Ivanciuc-Balaban operator* to the Barysz distance matrix. Moreover, the → *Wiener index* and the → *Schultz molecular topological index* were calculated from the Barysz distance matrix and the Barysz distance plus adjacency matrix, respectively [Nikolic *et al.*, 1993b].

Taking into account only the pairs of bonded atoms, weighted adjacency matrices are analogously defined. Moreover, atomic and bond weights as defined above on atomic numbers, relative electronegativities, relative covalent radii and conventional bond orders constitute a general weighting scheme for chemical graphs which could be used to calculate various molecular descriptors different from those derived from distance or adjacency matrices.

Table W-2. Atomic weights derived from the atomic number Z, relative covalent radius Y, and relative electronegativity X, for some atoms.

Atom	Z	*Z-weight*	X	*X-weight*	Y	*Y-weight*
B	5	−0.200	0.851	−0.175	1.038	0.037
C	6	0.000	1.000	0.000	1.000	0.000
N	7	0.143	1.149	0.130	0.963	−0.038
O	8	0.250	1.297	0.229	0.925	−0.081
Si	14	0.571	0.937	−0.067	1.128	0.113
P	15	0.600	1.086	0.079	1.091	0.083
S	16	0.625	1.235	0.190	1.053	0.050
F	9	0.333	1.446	0.308	0.887	−0.127
Cl	17	0.647	1.384	0.277	1.015	0.015
Br	35	0.829	1.244	0.196	1.303	0.233
I	53	0.887	1.103	0.093	1.591	0.371

Table W-3. Z, X, and Y bond weights for some bonds.

Bond	*Z-weight*	*X-weight*	*Y-weight*	*Bond*	*Z-weight*	*X-weight*	*Y-weight*
C − C	1.000	1.000	1.000	C ≈ S	0.250	0.540	0.633
C = C	0.500	0.500	0.500	C − F	0.667	0.692	1.127
C ≡ C	0.333	0.333	0.333	C − Cl	0.353	0.723	0.985
C ≈ C	0.667	0.667	0.667	C − Br	0.171	0.804	0.767
C − N	0.857	0.870	1.038	C − I	0.113	0.907	0.629
C = N	0.429	0.435	0.519	N − N	0.735	0.757	1.078
C ≡ N	0.286	0.290	0.346	N = N	0.367	0.379	0.539
C ≈ N	0.571	0.580	0.692	N ≈ N	0.490	0.505	0.719
C − O	0.750	0.771	1.081	N − O	0.643	0.671	1.123
C = O	0.375	0.386	0.541	N = O	0.321	0.336	0.561
C − Si	0.429	1.067	0.887	N ≈ O	0.423	0.447	0.748
C − P	0.400	0.921	0.917	O − S	0.281	0.624	1.027
C − S	0.375	0.810	0.950	O = S	0.141	0.312	0.513
C = S	0.188	0.405	0.475				

Also **Schultz weighted distance matrices** were proposed, weighting both vertices and edges in the H-depleted molecular graph to account for heteroatoms and bond multiplicity [Schultz *et al.*, 1994]. Three different graph weights were defined which allow the calculation of different weighted matrices and the corresponding graph invariants.

The vertex weight w_a is based on atomic numbers:

$$w_a = 1 + Z_a - Z_C = 1 + Z_a - 6$$

where Z_a is the atomic number of the ath vertex and Z_C is the atomic number of carbon atom. Note that the vertex weight is equal to unity for all carbon atoms and is defined for characterizing heteroatoms; the heavier the heteroatoms the greater is the weight value (Table W-4).

Table W-4. Schultz's vertex weights for the most common atoms.

Atom	*w*	*Atom*	*w*
Li	–2	F	4
Be	–1	P	10
B	0	S	11
C	1	Cl	12
N	2	Br	30
O	3	I	48

The vertex weights are collected into a diagonal matrix $\mathbf{W}^{\mathbf{V}}$ which is used as a pre-multiplier of the distance matrix **D** in order to give a vertex weighted distance matrix $\mathbf{D}^{\mathbf{V}}$ as:

$$\mathbf{W}^{\mathbf{V}} \cdot \mathbf{D} = \mathbf{D}^{\mathbf{V}}$$

The edge weight w_b is defined as:

$$w_b = \pi_b^* + Z_h - Z_C \quad \text{and} \quad w_b = \pi_b^* + Z_{hi} + Z_{hj} - 2 \cdot Z_C$$

where π_b^* is the conventional bond order of the bth bond, Z_h is the atomic number of the heteroatoms attached to the bth bond and Z_C is the atomic number of the carbon atom; the two expressions refer to the case of one heteroatom and two heteroatoms in the bond, respectively.

When no heteroatoms are present in the bond, the weight simply reduces to the conventional bond order. Note that in the case of elements lying to the left of carbon in the second quantum level of the periodic table the edge weights can be zero or negative.

The edge weighted distance matrix $\mathbf{D}^{\mathbf{E}}$ is obtained by summing the weights of all edges involved in the shortest path between two considered vertices. An edge-vertex weighted distance matrix $\mathbf{D}^{\mathbf{EV}}$ is derived by multiplying the diagonal matrix $\mathbf{W}^{\mathbf{V}}$ of vertex weights by the edge weighted distance matrix:

$$\mathbf{W}^{\mathbf{V}} \cdot \mathbf{D}^{\mathbf{E}} = \mathbf{D}^{\mathbf{EV}}$$

The third weight is related to vertex valence and takes into account the number of bonds incident to a vertex and its pairs of unshared electrons. Each present electron pair counts one, each missing electron pair contributes –1 to the total valence of heteroatoms, and in the case of free radicals the unique electron present in the outer valence shell contributes half a bond to the total valence. The valence weights are collected into a diagonal matrix $\mathbf{W}^{\boldsymbol{\delta}}$ used to give a valence weighted distance matrix $\mathbf{D}^{\boldsymbol{\delta}}$:

$$\mathbf{W}^{\boldsymbol{\delta}} \cdot \mathbf{D} = \mathbf{D}^{\boldsymbol{\delta}}$$

and the edge-vertex-valence weighted distance matrix:

$$\mathbf{W}^{\boldsymbol{\delta}} \cdot \mathbf{D}^{\mathbf{EV}} = \mathbf{D}^{\mathbf{EV}\boldsymbol{\delta}}$$

Analogously, edge-valence and vertex-valence weighted distance matrices can be calculated.

The most common topological indices can be derived from the above defined weighted distance matrices, e.g. the Schultz indices were calculated as well as the → *MTI′ index*, → *determinant*, → *permanent*, → *product of row sums* and the → *hafnian*.

From the vertex-valence weighted distance matrix $\mathbf{D}^{V\delta}$ a **geometric modification number** GM was also derived in order to account for the geometric isomerism of compounds [Schultz *et al.*, 1995]. **Geometric factors** GF equal to +1 are assigned to vertices with *Z* (*cis*) geometry and –1 to vertices with *E* (*trans*) geometry, all other vertices in the graph are assigned geometric factors equal to zero; for large molecules, the GF values of the various paired centres of geometric isomerism are algebraically summed to arrive at correct GF values for each vertex. The geometric factors are collected into a diagonal matrix $\mathbf{W}^{GF}$. Then, this matrix is used as a premultiplier of the sum of the vertex-valence weighted distance matrix $\mathbf{D}^{V\delta}$ with its transposed matrix $(\mathbf{D}^{V\delta})^T$:

$$\mathbf{W}^{GF} \cdot [\mathbf{D}^{V\delta} + (\mathbf{D}^{V\delta})^T]$$

where the matrix sum $\mathbf{D}^{V\delta} + (\mathbf{D}^{V\delta})^T$ is performed to obtain a symmetrical matrix and to incorporate as much information as possible into rows representing the geometric centres of the molecule. The row sums of the final matrix relative to geometric centres give the geometric modification number GM of the molecule. The GM numbers for two isomers (*Z* and *E*) will have opposite signs. The derived GM values can be incorporated (i.e. added) into any given topological index to achieve discrimination among geometric isomers.

The same technique was proposed to discriminate among optical isomers, adding to topological indices a **chiral modification number** CM that is calculated in the same way as the geometric modification number GM, the only difference being to assign **chiral factors** CF of +1 to chiral stereocenters with *R* configuration and CF factors of –1 to stereocenters with *S* configuration; all other vertices in the graph are assigned factor values of zero.

Moreover, **rotamer factors** RF of +1 were proposed to be assigned to single bonded vertices with C geometry and –1 to vertices with T geometry. The **rotamer modification number** RMN based on rotamer factors and the vertex-valence weighted distance matrix allows discrimination among conformational isomers when it is incorporated into the topological indices. It is calculated in the same way as the geometric modification number GM and chiral modification number CM [Schultz *et al.*, 1996].

🕮 [Ivanciuc *et al.*, 2000a]

weighted path counts → **path counts**

weighted self-returning walk counts → **self-returning walk counts**

weighted walk degrees → **walk matrices**

weighted walk numbers → **walk matrices**

WHIM descriptors (: *Weighted Holistic Invariant Molecular descriptors*)

WHIM descriptors are molecular descriptors based on → *statistical indices* calculated on the projections of the atoms along principal axes [Todeschini *et al.*, 1994; Todeschini and Gramatica, 1997a].

WHIM descriptors are built in such a way as to capture relevant molecular 3D information regarding molecular size, shape, symmetry and atom distribution with respect to invariant reference frames. The algorithm consists in performing a → *Princi-*

pal Components Analysis on the centred → *Cartesian coordinates* of a molecule (centred → *molecular matrix*) by using a weighted covariance matrix obtained from different *weighting schemes* for the atoms:

$$s_{jk} = \frac{\sum_{i=1}^{A} w_i (q_{ij} - \bar{q}_j)(q_{ik} - \bar{q}_k)}{\sum_{i=1}^{A} w_i}$$

where s_{jk} is the weighted covariance between the jth and kth atomic coordinates, A is the number of atoms, w_i the weight of the ith atom, q_{ij} and q_{ik} represent the jth and kth coordinate (j, k = x, y, z) of the ith atom respectively, and $\bar{q}$ the corresponding average value.

Six different weighting schemes are proposed: (1) the unweighted case u ($w_i = 1 \quad i = 1, n$, where A is the number of atoms for each compound), (2) atomic mass m, (3) the → *van der Waals volume* v, (4) the Sanderson → *atomic electronegativity* e, (5) the → *atomic polarizability* p and (6) the → *electrotopological state indices* of Kier and Hall S. All the weights are also scaled with respect to the carbon atom, and their values are shown in Table W-5; moreover, as all the weights must be positive, the electrotopological indices are scaled thus:

$$S'_i = S_i + 7 \qquad S'_i > 0$$

In this case only the non-hydrogen atoms are considered and the atomic electrotopological charge of each atom is dependent on its atom neighbour.

Depending on the kind of weighting scheme, different covariance matrices and different principal axes (i.e principal components $\mathbf{t}_m$) are obtained. For example, using atomic masses as the weighting scheme, the directions of the three principal axes are the directions of the → *principal inertia axes*. Thus, the WHIM approach can be viewed as a generalization searching for the principal axes with respect to a defined atomic property (the weighting scheme).

Table W-5. Mass, van der Waals volume, Sanderson electronegativity, and polarizability (atomic original and scaled values).

	Atomic Mass		*vdW Volume*		*Electronegativity*		*Polarizability*	
ID	*m*	*W/M(C)*	*v*	*W/V(C)*	*e*	*W/E(C)*	*p*	*W/P(C)*
H	1.01	0.084	6.709	0.299	2.592	0.944	0.667	0.379
B	10.81	0.900	17.875	0.796	2.275	0.828	3.030	1.722
C	12.01	1.000	22.449	1.000	2.746	1.000	1.760	1.000
N	14.01	1.166	15.599	0.695	3.194	1.163	1.100	0.625
O	16.00	1.332	11.494	0.512	3.654	1.331	0.802	0.456
F	19.00	1.582	9.203	0.410	4.000	1.457	0.557	0.316
Al	26.98	2.246	36.511	1.626	1.714	0.624	6.800	3.864
Si	28.09	2.339	31.976	1.424	2.138	0.779	5.380	3.057
P	30.97	2.579	26.522	1.181	2.515	0.916	3.630	2.063
S	32.07	2.670	24.429	1.088	2.957	1.077	2.900	1.648
Cl	35.45	2.952	23.228	1.035	3.475	1.265	2.180	1.239
Fe	55.85	4.650	41.052	1.829	2.000	0.728	8.400	4.773
Co	58.93	4.907	35.041	1.561	2.000	0.728	7.500	4.261

	Atomic Mass		vdW Volume		Electronegativity		Polarizability	
ID	*m*	*W/M(C)*	*v*	*W/V(C)*	*e*	*W/E(C)*	*p*	*W/P(C)*
Ni	58.69	4.887	17.157	0.764	2.000	0.728	6.800	3.864
Cu	63.55	5.291	11.494	0.512	2.033	0.740	6.100	3.466
Zn	65.39	5.445	38.351	1.708	2.223	0.810	7.100	4.034
Br	79.90	6.653	31.059	1.384	3.219	1.172	3.050	1.733
Sn	118.71	9.884	45.830	2.042	2.298	0.837	7.700	4.375
I	126.90	10.566	38.792	1.728	2.778	1.012	5.350	3.040

For each weighting scheme, a set of statistical indices is calculated on the atoms projected onto each principal component $\mathbf{t}_m$ (m = 1,2,3), i.e. the scores, as described below.

The invariance to translation of the calculated parameters is due to the centering of the atomic coordinates and the invariance to rotation is due to the uniqueness of the PCA solution.

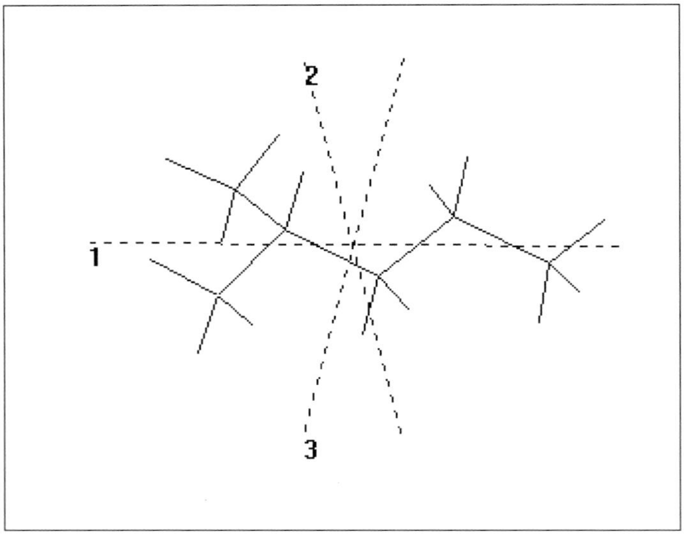

Figure W-1. Principal axes calculated for 2-methylpentane, using mass weights.

- **directional WHIM descriptors**

These are molecular descriptors calculated as univariate statistical indices on the scores of each individual principal component $\mathbf{t}_m$ (m = 1,2,3).

The first group of descriptors consists of the **directional WHIM size** (or **d-WSIZ indices**) descriptors defined as the → *eigenvalues* λ_1, λ_2 and λ_3 of the weighted covariance matrix of the molecule atomic coordinates; they account for the molecular size along each principal direction.

The second group consists of the **directional WHIM shape** (or **d-WSHA indices**) descriptors ϑ_1, ϑ_2 and ϑ_3, calculated as eigenvalue ratios and related to molecular shape:

$$\vartheta_m = \frac{\lambda_m}{\sum_m \lambda_m} \qquad m = 1, 2, 3$$

Because of the closure condition ($\vartheta_1 + \vartheta_2 + \vartheta_3 = 1$), only two of these are independent.

The third group of descriptors consists of the **directional WHIM symmetry** (or **d-WSYM indices**) descriptors γ_1, γ_2 and γ_3, calculated as → *mean information content* on the symmetry along each component with respect to the centre of the scores:

$$\gamma'_m = -\left[\frac{n_s}{n} \cdot \log_2 \frac{n_s}{n} + n_a \cdot \left(\frac{1}{n} \cdot \log_2 \frac{1}{n}\right)\right] \qquad \gamma_m = \frac{1}{1 + \gamma'_m} \qquad 0 < \gamma_m \leq 1$$

n_s, n_a, and n are the number of central symmetric atoms (along the mth component), the number of unsymmetric atoms and the total number of atoms of the molecule, respectively.

Finally, the fourth group of descriptors consists of the inverse of the *kurtosis* κ_1, κ_2 and κ_3, calculated from the fourth order moments of the scores $\mathbf{t}_m$, which are related to the atom distribution and density around the origin and along the principal axes:

$$\kappa_m = \frac{\sum_i t_i^4}{\lambda^2 \cdot n} \qquad \eta_m = \frac{1}{\kappa_m} \qquad m = 1, 2, 3$$

To avoid problems related to infinite (or very high) kurtosis values, obtained when along a principal axis all the atoms are projected in the centre (or near the centre, i.e. leptokurtic distribution), the inverse of the kurtosis is used.

Low values of the kurtosis are obtained when the data points (i.e. the atom projections) assume opposite values ($-t$ and t) with respect to the centre of the scores. When an increasing number of data values are within the extreme values t along a principal axis, the kurtosis value increases (i.e., $\kappa = 1.8$ for a uniform distribution of points, $\kappa = 3.0$ for a normal distribution). When the kurtosis value tends to infinity the corresponding η value tends to zero.

Thus, the group of descriptors η_m can be related to the quantity of unfilled space per projected atom and have been called **directional WHIM density** (or **d-WDEN indices**, or *WHIM emptiness*) descriptors: the greater the η_m values, the greater is the projected unfilled space. The η_m descriptors are used in place of the kurtosis descriptors κ_m (previously proposed).

Therefore, for each weighting scheme w, 11 molecular directional WHIM descriptors (ϑ_3 is excluded) are obtained:

$$\langle \lambda_1 \quad \lambda_2 \quad \lambda_3 \quad \vartheta_1 \quad \vartheta_2 \quad \gamma_1 \quad \gamma_2 \quad \gamma_3 \quad \eta_1 \quad \eta_2 \quad \eta_3 \rangle_w$$

thus resulting in a total of 66 directional WHIM descriptors. For planar compounds, λ_3, γ_3, and η_3 are always equal to zero.

- **global WHIM descriptors**

These are molecular descriptors directly calculated as a combination of the directional WHIM descriptors, thus contemporarily accounting for the variation of molecular properties along with the three principal directions in the molecule. For non-directional WHIM descriptors any information individually related to each principal axis disappears and the description is related only to a global view of the molecule.

In many cases, size descriptors can play, in modelling, a significant role independently of the measured directions, allowing simpler models. Thus, in view of the importance of this quantity, a group of descriptors of the total dimension of a molecule is considered in three different ways, based on the three eigenvalues defined above:

$$\mathrm{T} = \lambda_1 + \lambda_2 + \lambda_3$$
$$\mathrm{A} = \lambda_1\lambda_2 + \lambda_1\lambda_3 + \lambda_2\lambda_3$$
$$\mathrm{V} = \prod_{m=1}^{3}(1+\lambda_m) - 1 = \mathrm{T} + \mathrm{A} + \lambda_1\lambda_2\lambda_3$$

where T and A are, respectively, related to linear and quadratic contributions to the total molecular size. V is the complete expansion, also containing the third order term. These molecular descriptors are called **WHIM size** (or **WSIZ index**) descriptors.

The shape of the molecule is represented by the **WHIM shape** (or **WSHA index**) defined as:

$$K = \frac{\sum_m \left| \frac{\lambda_m}{\sum_m \lambda_m} - \frac{1}{3} \right|}{4/3} \qquad m = 1,2,3 \quad 0 \leq K \leq 1$$

The term 4/3 is the maximum value of the numerator term and is used to scale K between 0 and 1. This expression also has a more general meaning, K being the → *multivariate K correlation index* used to evaluate the global correlation in data [Todeschini, 1997].

For example, for an ideal straight molecule both λ_2 and λ_3 are equal to zero and $K = 1$; for an ideal spherical molecule all three eigenvalues are equal to 1/3 and $K = 0$. For all planar molecules the third eigenvalue λ_3 is 0, there being no variance out of the molecular plane, and K ranges between 0.5 and 1, depending on the molecule linearity.

The K shape term definitely replaces the acentric factor, it being more general than the previously proposed *acentric factor* ω.

The total molecular symmetry is accounted for by the **WHIM symmetry** (or **WSYM index**) defined as:

$$\mathrm{G} = (\gamma_1 \cdot \gamma_2 \cdot \gamma_3)^{1/3}$$

G is the geometric mean of the directional symmetries and equals 1 when the molecule shows a central symmetry along each axis, and tends to 0 when there is a loss of symmetry along at least one axis. Different symmetry values are obtained only when unitary, mass and electrotopological weights are used; for this reason, only three kinds of symmetry parameters are retained: Gu, Gm and Gs.

The total density of the atoms within a molecule is accounted for by the **WHIM density** (or **WDEN index**) defined by the following expression:

$$\mathrm{D} = \eta_1 + \eta_2 + \eta_3$$

The molecular descriptors defined above are generalized molecular properties within each weighting scheme. The global WHIMs are five for each of the six proposed weighting schemes w:

$$\langle \mathrm{T},\ \mathrm{A},\ \mathrm{V},\ K,\ \mathrm{D} \rangle_w$$

plus the symmetry indices Gu, Gm and Gs, giving a total number of 33 descriptors.

WHIM descriptors have been used to model toxicological indices [Todeschini *et al.*, 1996a; Todeschini *et al.*, 1996b; Todeschini *et al.*, 1997b; Shapiro and Guggenheim, 1998b], several physico-chemical properties of polychlorobiphenyls [Gramatica *et al.*, 1998] and polycyclic aromatic hydrocarbons [Todeschini *et al.*, 1995], hydroxyl radical rate constants [Gramatica *et al.*, 1999a], and soil sorption partition coefficients [Gramatica *et al.*, 2000].

📖 [Todeschini and Gramatica, 1997b] [Todeschini and Gramatica, 1997c] [Matter, 1997] [Todeschini and Gramatica, 1998] [Gramatica *et al.*, 1998] [Baumann, 1999] [Gramatica *et al.*, 1999b]

WHIM density → **WHIM descriptors** (⊙ global WHIM descriptors)

WHIM shape → **WHIM descriptors** (⊙ global WHIM descriptors)

WHIM size → **WHIM descriptors** (⊙ global WHIM descriptors)

WHIM symmetry → **WHIM descriptors** (⊙ global WHIM descriptors)

Wiener delta matrix → **Wiener matrix**

Wiener index (*W*) (*: Wiener number*)

One of the first → *topological indices*, introduced by H. Wiener in 1947 as the sum over all bonds of the product of the number of vertices on each side of the bond [Wiener, 1947c; Wiener, 1947a; Wiener, 1947b] and called at the beginning *path number*. Usually, the Wiener index is defined and calculated as the sum of all topological distances in the → *H-depleted molecular graph* [Hosoya, 1971].

It is obtained from the → *distance matrix* **D**:

$$W = \mathcal{W}(\mathbf{D}) = \frac{1}{2}\sum_{i=1}^{A}\sum_{j=1}^{A} d_{ij} = \sum_{k=1}^{D} {}^{k}f \cdot k$$

where $\mathcal{W}$ is the → *Wiener operator*, A is the number of atoms, d_{ij} is the → *topological distance*, D is the → *topological diameter* of the graph and ${}^{k}f$ the number of distances in the graph equal to k, i.e. the kth order → *graph distance count*. The total number of distances is $A \times (A-1)/2$ and the factor 1/2 avoids counting the distances twice.

The **mean Wiener index is defined as:**

$$\bar{W} = \frac{2 \cdot W}{A \cdot (A-1)}$$

The **compactness** is defined as twice the mean Wiener index, i.e. $2\bar{W}$. The smaller the Wiener number, the larger is the compactness of the molecule.

For acyclic graphs the Wiener index, according to its original definition, can also be obtained from the → *Wiener matrix* **W** by summing all of the entries for paths *i-j* of length one ${}^{1}p_{ij}$ (i.e. bonds b) above the main diagonal:

$$W = \sum\nolimits_{{}^{1}p_{ij}} w_{ij} = \frac{1}{2} \cdot \sum_{i=1}^{A}\sum_{j=1}^{A} [\mathbf{W} \bullet \mathbf{A}]_{ij} = \sum_{b=1}^{B} N_{i,b} \cdot N_{j,b}$$

where w_{ij} are the entries of the Wiener matrix and $\mathbf{W} \bullet \mathbf{A}$ is a 1st order → *sparse Wiener matrix* obtained by the → *Hadamard matrix product* of the Wiener matrix **W** and the → *adjacency matrix* **A**. $N_{i,b}$ and $N_{j,b}$ are the numbers of vertices on each side of the bond b, including vertices i and j, respectively. B is the total number of graph edges.

For acyclic graphs it can also be obtained from the 1st-order sparse matrices of the → *symmetric Cluj-distance matrix* ^{1}**CJD,** → *symmetric Cluj-detour matrix* ^{1}**CJΔ** and → *symmetric Szeged matrix* ^{1}**SZ** by applying the → *Wiener operator* or from the corresponding sparse unsymmetric matrices by applying the → *Wiener orthogonal operator.*

According to the original definition, the Wiener index can be calculated by summing the bond contributions w_b^*:

$$W = \sum_{b=1}^{B} w_b^* = \sum_{b=1}^{B}\left(\sum_{i<j} \frac{{}^{\min}P_{ij}^{b}}{{}^{\min}P_{ij}}\right)$$

where B is the number of bonds, ${}^{\min}P_{ij}$ is the number of shortest paths between vertices v_i and v_j, and ${}^{\min}P_{ij}^{b}$ is the number of those shortest paths between vertices v_i and v_j which contain the bond b [Lukovits, 1990b; Lukovits, 1992; Lukovits, 1995c; Juvan and Mohar, 1995]. The first sum is performed for all bonds and for each b bond all pairs of vertices i and j are considered. Contrary to the original Wiener definition, this formula holds for any graphs. For acyclic graphs, the bond contribution w_b^* coincides with the product of the number of vertices on each side of the considered bond, i.e. the number of external paths including the considered bond:

$$w_b^* = \sum_{i<j} \frac{{}^{\min}P_{ij}^{b}}{{}^{\min}P_{ij}} = N_{i,b} \cdot N_{j,b}$$

Moreover, this formula allows the study of the effect of different types of bonds on the global behaviour of the molecule; this led to the splitting of the Wiener index into **partial Wiener indices** related to single, double, triple and aromatic bonds [Lukovits, 1990b]:

$$W = W_S + W_D + W_T + W_A = \sum_{b \in S} w_b^* + \sum_{b \in D} w_b^* + \sum_{b \in T} w_b^* + \sum_{b \in A} w_b^*$$

where W_S is obtained by adding the contributions of all single bonds, W_D by adding the contributions of all double bonds, W_T by adding the contributions of all triple bonds, and W_A by adding the contributions of all aromatic bonds. These partial Wiener indices can be used separately as → *multiple bond descriptors*.

Example : 2-methylpentane

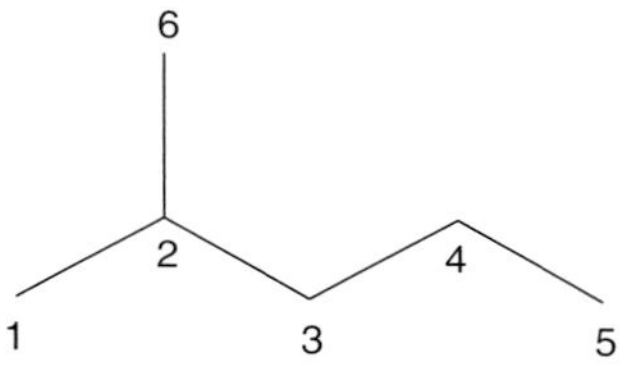

Vertex pair (i, j)	N_i	N_j	$N_i \cdot N_j$
(1, 2)	1	5	5
(2, 3)	3	3	9
(2, 6)	5	1	5
(3, 4)	4	2	8
(4, 5)	5	1	5

Wiener definition

$$W = N_1 \cdot N_2 + N_2 \cdot N_3 + N_2 \cdot N_6 + N_3 \cdot N_4 + N_4 \cdot N_5 = $$
$$= 5 + 9 + 5 + 8 + 5 = 32$$

Distance matrix **D**

Atom	**1**	**2**	**3**	**4**	**5**	**6**
1	0	1	2	3	4	2
2	1	0	1	2	3	1
3	2	1	0	1	2	2
4	3	2	1	0	1	3
5	4	3	2	1	0	4
6	2	1	2	3	4	0

Hosoya definition

$$W = \frac{1}{2} \cdot \sum_{i=1}^{6} \sum_{j=1}^{6} d_{ij} = 32$$

Box W-1.

The same method based on the bond contribution w_b^* for the Wiener index calculation was proposed by Pisanski-Zerovnik [Pisanski and Zerovnik, 1994] together with a more general edge weight w_b based on paths, defined as:

$$w_b^* = \sum_{i<j} \frac{P_{ij}^b}{P_{ij}}$$

where P_{ij} is the number of paths of any length between vertices v_i and v_j and P_{ij}^b the number of those paths which include the b bond; the sum runs over all pairs of vertices in the graph. By summing all the edge weights w_b a global index, called **Pisanski-Zerovnik index** Ω, is obtained:

$$\Omega = \sum_{b=1}^{B} w_b$$

where B is the number of bonds. This index coincides with the Wiener index for any acyclic graph; moreover, the relation $\Omega \geq W$ holds for any graph.

The Wiener index is closely related to the → *Altenburg polynomial* of a graph, in fact graphs with the same Altenburg polynomials always have the same Wiener numbers.

The Wiener index increases with the number of atoms (i.e. with the molecular size) and, for a constant number of atoms, reaches a maximum for linear structure and a minimum for the most branched and cyclic structures. For this reason it was suggested as a measure of → *molecular branching*. It is insensitive to atom type. Moreover, the Wiener index seems to be related to the molecular surface area, thus reflecting molecular compactness and, in some way, the intermolecular forces [Gutman and Körtévlyesi, 1995].

For all connected graphs, the Wiener index is between

$$\frac{A(A-1)}{2} \leq W \leq \frac{A(A^2-1)}{6}$$

where the lower limit refers to a linear → *path graph* and the upper to a → *complete graph*.

Sometimes a **normalized Wiener index** can be found, defined as:

$$W_N = \frac{W}{A^2} \quad \text{or} \quad W_N = \frac{W}{(A+1)^2}$$

where A is the number of atoms.

The **root mean square Wiener index** is defined as:

$$W_{RMS} = \frac{1}{\sqrt{A \cdot (A-1)}} \cdot \left(\sum_{i=1}^{A} \sum_{j=1}^{A} d_{ij}^2 \right)^{1/2}$$

where the sums are performed on the square topological distances. The cube root of the Wiener index was also suggested as a measure of the mean distance among molecule atoms [Platt, 1952; Morales and Araujo, 1993].

The Wiener index is twice the → *Rouvray index*, i.e.

$$I_{ROUV} = 2 \cdot W$$

Other topological indices related to the Wiener index are the → *Kirchhoff number*, the → *quasi-Wiener index*, the → *detour index*, the → *expanded Wiener number*, and the → *all-path Wiener index*.

Moreover, a number of topological indices have been expressed as functions of the Wiener index or as its extensions [Zhu *et al.*, 1996b]; examples are → *detour/Wiener index*, → *hyper-Wiener index*, → *Harary index*, → *total information content on the distance magnitude, mean information content on the distance magnitude*, and → *mean information content on the distance degree magnitude*.

Example : cyclohexane

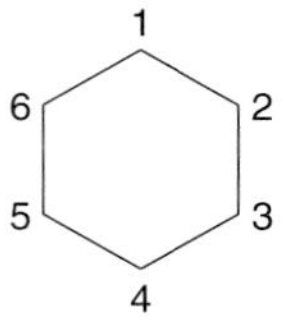

Wiener index bond contributions

Vertex pair *(i, j)*	*Paths*	$^{\min}P_{ij}^{b} / {}^{\min}P_{ij}$					
		(1, 2)	(2, 3)	(3, 4)	(4, 5)	(5, 6)	(6, 1)
(1, 2)	1, 2 1, 6, 5, 4, 3, 2	1	–	–	–	–	–
(1, 3)	1, 2, 3 1, 6, 5, 4, 3	1	1	–	–	–	–
(1, 4)	1, 2, 3, 4 1, 6, 5, 4	0.5	0.5	0.5	0.5	0.5	0.5
(1, 5)	1, 2, 3, 4, 5 1, 6, 5	–	–	–	–	1	1
(1, 6)	1, 2, 3, 4, 5, 6 1, 6	–	–	–	–	–	1
(2, 3)	2, 3 2, 1, 6, 5, 4, 3	–	1	–	–	–	–
(2, 4)	2, 3, 4 2, 1, 6, 5, 4	–	1	1	–	–	–
(2, 5)	2, 3, 4, 5 2, 1, 6, 5	0.5	0.5	0.5	0.5	0.5	0.5
(2, 6)	2, 3, 4, 5, 6 2, 1, 6	1	–	–	–	–	1
(3, 4)	3, 4 3, 2, 1, 6, 5, 4	–	–	1	–	–	–
(3, 5)	3, 4, 5 3, 2, 1, 6, 5	–	–	1	1	–	–
(3, 6)	3, 4, 5, 6 3, 2, 1, 6	0.5	0.5	0.5	0.5	0.5	0.5
(4, 5)	4, 5 4, 3, 2, 1, 6, 5	–	–	–	1	–	–
(4, 6)	4, 5, 6 4, 3, 2, 1, 6	–	–	–	1	1	–
(5, 6)	5, 6 5, 4, 3, 2, 1, 6	–	–	–	–	1	–
	w^*_b =	4.5	4.5	4.5	4.5	4.5	4.5

Wiener index :

$$W = \sum_{b=1}^{6} w_b^* = \sum_{b=1}^{6} \left(\sum_{i<j} \frac{^{\min}P_{ij}^{b}}{^{\min}P_{ij}} \right) = 6 \cdot 4.5 = 27$$

Pisanski-Zerovnik index :

$$\Omega = \sum_{b=1}^{6} w_b = \sum_{b=1}^{6} \left(\sum_{i<j} \frac{P_{ij}^{b}}{P_{ij}} \right) = \sum_{b=1}^{6} \left(\sum_{i<j} 0.5 \right) = 6 \cdot 7.5 = 45$$

Box W-2.

A generalization of the Wiener index to account for heteroatoms and multiple bonds was proposed based on the → *Barysz distance matrix*.

A Wiener-type index, called **3D-Wiener index**, is derived from the → *geometry matrix* G as:

$$^{3D}W_H = \mathcal{W}(G) = \frac{1}{2} \cdot \sum_{i=1}^{A} \sum_{j=1}^{A} r_{ij}$$

where r_{ij} is the interatomic distance between the ith and jth atoms [Mekenyan *et al.*, 1986a; Mekenyan *et al.*, 1986b; Bogdanov *et al.*, 1989; Bogdanov *et al.*, 1990]. This index is obviously more discriminating the 2D-Wiener index and shows different values for different molecular conformations, the largest values corresponding to the most extended conformations, the smallest to the most compact conformations. Therefore, it is considered among → *shape descriptors* since it decreases with increasing sphericity of a structure [Nikolic *et al.*, 1991]. The 3D-Wiener index ^{3D}W can also be calculated by excluding hydrogen atoms [Basak *et al.*, 1999].

Another decomposition of the Wiener index was proposed to define topological substituent indices representing the effects of substituents on a parent molecule and the substituent interactions [Lukovits, 1988]. Let M be the parent structure with two substituents denoted by A and B (Figure W-2), the Wiener index of the molecule can be decomposed thus:

$$W = W_M + S_A + S_B + S_{AB}$$

where W_M is the Wiener index of the parent structure, i.e. the sum of all distances in the parent structure. S_A and S_B are the topological substituent indices defined as:

$$S_A = W_A + n_A \cdot s_a + n_A \cdot n_M + n_M \cdot s_c \quad \text{and} \quad S_B = W_B + n_B \cdot s_b + n_B \cdot n_M + n_M \cdot s_d$$

where W_A and W_B are the Wiener indices of the two substituents; n_A, n_B, and n_M denote the number of atoms in the substituents A and B and in the parent structure M; note that the terms $n_A \cdot n_M$ and $n_B \cdot n_M$ represent the number of times the bond linking A and M as well as B and M has to be traversed; *a* and *b* denote the substitution sites, i.e. the link vertex, in the parent structure for the substituents A and B, respectively, and *c* and *d* denote the link vertices in the substituent A and B, respectively.

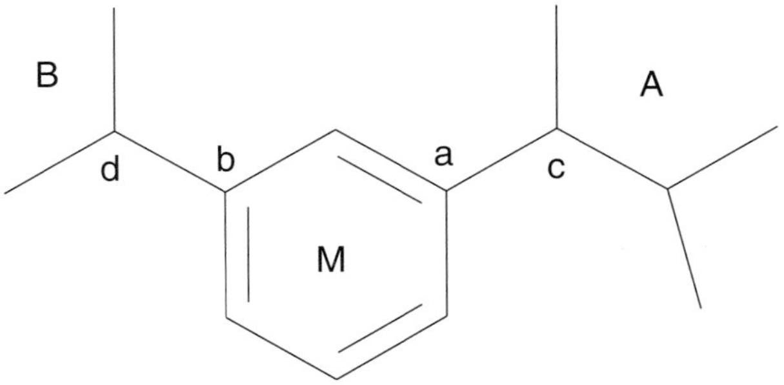

Figure W-2. Disubstituted benzene derivative.

The term *s* is the **connectedness index** characterizing the connectedness of a substitution site defined as [Seybold, 1983a; Seybold, 1983b]:

$$s_k = \sum_{i \in M} d_{ik}$$

where the sum runs over all atoms of the parent molecule and *k* denotes a generic substitution site in M, d_{ik} is the topological distance of the ith atom from the substitution site; the same formula holds for the connectedness of the link vertex in a substituent.

The topological substituent index S_{AB} represents the interaction between the substituents A and B defined as:

$$S_{AB} = n_A \cdot s_d + n_B \cdot n_c + n_A \cdot n_B \cdot (2 + d_{ab})$$

where d_{ab} is the topological distance between the two substitution sites in the parent structure.

The topological substituent indices S_A, S_B, S_{AB} have been used to model the biological activity of series of congeneric compounds; in the models, the Wiener index W_M of the parent structure can be neglected as it represents the constant term due to the main bulk of the molecules. These substituent indices, unlike the → *substituent constants*, can be calculated easily; moreover, they depend on the site at which the substitution takes place and the interaction index S_{AB} overcomes the additivity scheme of the substituent constants.

[Wiener, 1948a] [Wiener, 1948b] [Rouvray, 1976] [Mekenyan *et al.*, 1983] [Balaban *et al.*, 1983] [Canfield *et al.*, 1985] [Rouvray, 1986a] [Gutman and Polansky, 1986b] [Polanski and Bonchev, 1986] [Gutman, 1987b] [Rouvray, 1987] [Polanski and Bonchev, 1987] [Bonchev *et al.*, 1987] [Mohar and Pisanski, 1988] [Senn, 1988] [Polansky *et al.*, 1989] [Polanski and Bonchev, 1990] [Gutman and Rouvray, 1990] [Gutman *et al.*, 1990] [Lukovits, 1990a] [Tratch *et al.*, 1990a] [Graovac and Pisanski, 1991] [Gutman and Soltés, 1991] [Lukovits, 1991] [Mihalic *et al.*, 1992b] [Gutman *et al.*, 1993b] [Gutman *et al.*, 1993a] [Gutman, 1993a] [Gutman, 1993b] [Mendiratta and Madan, 1994] [Gutman, 1994a] [Gutman *et al.*, 1994b] [Lukovits and Gutman, 1994a] [Diudea, 1995b] [Nikolic *et al.*, 1995] [Dobrynin *et al.*, 1995] [Klavzar *et al.*, 1996] [Sagan *et al.*, 1996] [Chepoi and Klavzar, 1997] [Gutman *et al.*, 1997b] [Brendlé and Papirer, 1997] [Dobrynin, 1997a] [Dobrynin, 1997b] [Gutman, 1997] [Pavlovic and Gutman, 1997] [Gutman and Potgieter, 1997] [Dobrynin, 1998a] [Dobrynin, 1998b] [Diudea and Gutman, 1998] [Lukovits, 1998b] [Dobrynin, 1999] [Dobrynin and Gutman, 1999] [Gutman *et al.*, 2000]

Wiener matrix (W)

According to the original definition of the → *Wiener index W* [Wiener, 1947c], the matrix **W** is derived from the → *H-depleted molecular graph* $\mathcal{G}$, based on counts of external paths with respect to a considered path *i-j*, and defined only for acyclic graphs [Randic, 1993c]. The off-diagonal w_{ij} entry corresponds to the number of external paths in the graph $\mathcal{G}$ which contain the path from vertex v_i to vertex v_j calculated as the product of the number of vertices on each side of the path *i-j*, namely N_i and N_j, including both vertices *i* and *j*:

$$w_{ij} = N_i \cdot N_j$$

where

$$N_i = \left|\{a|a \in \mathcal{V}(\mathcal{G});\ d_{ia} < d_{ja};\ p_{ia} \cap p_{ij} = \{i\}\}\right|$$

$$N_j = \left|\{a|a \in \mathcal{V}(\mathcal{G});\ d_{ja} < d_{ia};\ p_{ja} \cap p_{ij} = \{j\}\}\right|$$

By definition, the diagonal entries are equal to zero.

The *i*th row sum of the Wiener matrix is called the **Wiener matrix degree** ϱ_i:

$$\varrho_i = \mathcal{R}_i(\mathbf{W}) = \sum_{j=1}^{A} w_{ij}$$

where $\mathcal{R}_i$ is the → *row sum operator.*

Among the properties of the Wiener matrix, it can be noted that (a) entries corresponding to paths between → *terminal vertices* are necessarily equal to 1; (b) rows of the matrix corresponding to terminal vertices all have entries less than A (the number of vertices); (c) an entry with value $(A - 1)$ shows a terminal bond, because only terminal bonds divide the vertices into $1 + (A - 1)$ partition; (d) terminal vertices have a smaller row sum (Wiener matrix degree) than the centrally located ones associated with larger row sums.

The Wiener matrix for acyclic graphs can also be obtained by the → *Cluj-distance matrix* $\mathbf{CJD_U}$ as:

$$\mathbf{W} = \mathbf{CJD_U} \bullet \mathbf{CJD_U^T} = \mathbf{CJD} \qquad w_{ij} = [\mathbf{CJD_U}]_{ij}[\mathbf{CJD_U}]_{ji} = [\mathbf{CJD}]_{ij}$$

where **CJD** is the → *symmetric Cluj-distance matrix* and the symbol • indicates the → *Hadamard matrix product*.

The **sparse Wiener matrix** of mth *order* $^{\mathbf{m}}\mathbf{W}$ is derived from the Wiener matrix setting to zero all entries except those corresponding to mth order paths $^{m}p_{ij}$. This matrix can be calculated by the Hadamard matrix product of the Wiener matrix and the → *binary sparse matrix* $^{\mathbf{m}}\mathbf{B}$ whose elements corresponding to mth order paths are equal to one, or else zero:

$$^{\mathbf{m}}\mathbf{W} = \mathbf{W} \bullet {}^{\mathbf{m}}\mathbf{B}$$

Several **Wiener matrix invariants** [Randic *et al.*, 1994a] can be calculated.

The Wiener index W can be obtained from the Wiener matrix **W** by summing all entries for paths i-j of length one above the main diagonal, i.e. applying the Wiener operator to the 1st order sparse Wiener matrix $^{\mathbf{1}}\mathbf{W}$. **Higher-order Wiener numbers** ^{m}W are analogously defined for mth-order paths by summing, above the main diagonal, all entries corresponding to paths of the considered length m. Thus, a sequence of Wiener numbers of different order is obtained $(^{1}W, {}^{2}W, {}^{3}W, \ldots, {}^{D}W)$, the sequence length D being determined by the maximum length of the paths in the graph, i.e. the → *topological diameter* [Randic *et al.*, 1993a]:

$$^{m}W = \sum\nolimits_{^{m}p_{ij}} w_{ij} = \frac{1}{2} \cdot \sum_{i=1}^{A} \sum_{j=1}^{A} [^{\mathbf{m}}\mathbf{W}]_{ij}$$

where $^{\mathbf{m}}\mathbf{W}$ is the sparse Wiener matrix of mth order.

The **hyper-Wiener index** WW is defined as:

$$WW = \mathcal{W}(\mathbf{W}) = \frac{1}{2} \cdot \sum_{i=1}^{A} \sum_{j=1}^{A} w_{ij} = \sum_{m=1}^{D} {}^{m}W$$

where $\mathcal{W}$ is the → *Wiener operator* applied to the Wiener matrix [Randic, 1993c]. The hyper-Wiener index can be considered among the → *ID numbers*; it measures the "expansiveness" of a graph, weighting more expansive graphs even more so than does the Wiener index [Klein *et al.*, 1995].

For acyclic graphs, the hyper-Wiener index is also obtained by applying the Wiener operator to the symmetric Cluj-distance matrix **CJD** and the → *Wiener orthogonal operator* to the Cluj-distance matrix $\mathbf{CJD_U}$.

Example : 2-methylpentane

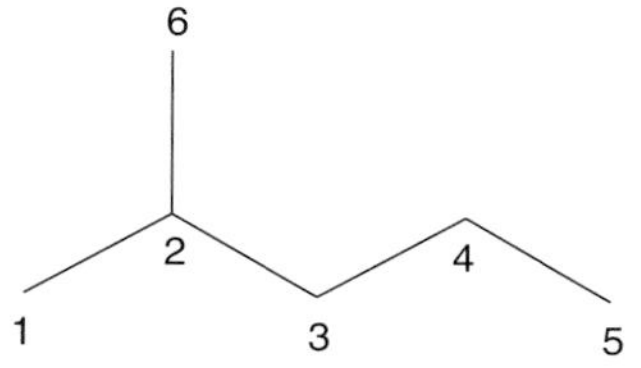

Wiener matrix **W**

Atom	1	2	3	4	5	6	ρi
1	0	5	3	2	1	1	12
2	5	0	9	6	3	5	28
3	3	9	0	8	4	3	27
4	2	6	8	0	5	2	23
5	1	3	4	5	0	1	14
6	1	5	3	2	1	0	12

$WW = 58$

$W \equiv {}^1W = 32 \quad {}^2W = 17 \quad {}^3W = 7 \quad {}^4W = 2$

$JJ^1 = 2 \cdot (12 \cdot 28)^{-1/2} + (28 \cdot 27)^{-1/2} + (27 \cdot 23)^{-1/2} + (23 \cdot 14)^{-1/2} = 0.2413$

$JJ^2 = 2 \cdot (12 \cdot 28 \cdot 27)^{-1/2} + (28 \cdot 27 \cdot 23)^{-1/2} + (27 \cdot 23 \cdot 14)^{-1/2} = 0.0393$

Box W-3.

As the Wiener matrix is defined only for acyclic graphs, the hyper-Wiener index is restricted to that kind of graph too [Lukovits, 1994; Lukovits, 1995b; Lukovits, 1995a; Gutman *et al.*, 1997a]; however, some extensions of the hyper-Wiener index to cycle-containing structures have been proposed [Randic *et al.*, 1993a; Lukovits and Linert, 1994b]. The → *hyper-distance-path index* D_P has been proposed [Klein *et al.*, 1995] as a general formula to calculate the hyper-Wiener index WW for any graph in terms of topological distances d_{ij}:

$$WW \equiv D_P = \frac{1}{2} \cdot \sum_{i=1}^{A} \sum_{j=1}^{A} \frac{d_{ij} \cdot (d_{ij}+1)}{2} = \frac{1}{4} \cdot \sum_{i=1}^{A} \sum_{j=1}^{A} \left(d_{ij}^2 + d_{ij}\right) = \frac{1}{2} \cdot (D_2 + W)$$

$$= \frac{1}{2} \cdot \left[tr\left(\mathbf{D}^2\right)/2 + W\right]$$

where D_2 is the **unnormalized second moment of distances**, i.e. the sum of the squared distances in the graph, and W is the Wiener index. The index D_2 was demonstrated to be equal to half the trace of the distance matrix **D** raised to the second power [Diudea, 1996a; Diudea and Gutman, 1998]; moreover, it is closely related to the Balaban → *mean square distance index*. If mth order → *distance distribution moments* are used instead of the second moment of distances, a whole sequence of hyper-Wiener-type indices is obtained.

The **resistance distance hyper-Wiener index** R' has been proposed [Klein *et al.*, 1995] based on the same general formula as that for the hyper-Wiener index, where the topological distance d_{ij} is replaced by the → *resistance distance* Ω_{ij}:

$$R' = \frac{1}{2} \cdot \sum_{i=1}^{A} \sum_{j=1}^{A} \frac{\Omega_{ij} \cdot (\Omega_{ij}+1)}{2} = \frac{1}{4} \cdot \sum_{i=1}^{A} \sum_{j=1}^{A} \left(\Omega_{ij}^2 + \Omega_{ij}\right)$$

It should be noted that the hyper-distance-path index and the resistance distance hyper-Wiener index are equivalent to the hyper-Wiener index as defined by Randic for acyclic graphs, but they can also be applied to any cycle-containing connected graphs.

In analogy with the Kier-Hall → *connectivity indices* χ and the → *Balaban distance connectivity index* J, **JJ indices** [Randic *et al.*, 1994a] are derived from the Wiener matrix based on the Wiener matrix degrees ϱ_i:

$$^{1}\mathrm{JJ} = \sum_b \left(\varrho_i \cdot \varrho_j\right)_b^{-1/2} \qquad ^{2}\mathrm{JJ} = \sum_{k=1}^{^2P} \left(\varrho_i \cdot \varrho_l \cdot \varrho_j\right)_k^{-1/2}$$

$$^{m}\mathrm{JJ} = \sum_{k=1}^{K} \left(\prod_{a=1}^{n} \varrho_a\right)_k^{-1/2}$$

where k runs over all of the mth order molecular subgraphs constituted by n atoms ($n = m + 1$ for acyclic graphs); K is the total number of mth order subgraphs, and in the case of path subgraphs equals the mth order path count ^{m}P. These indices would have high discriminating power.

The eigenvalues of the Wiener matrix are among → *eigenvalue-based descriptors.*

The **reciprocal Wiener matrix** $\mathbf{W}^{-1}$ is a matrix whose elements are the reciprocal of the corresponding Wiener matrix elements:

$$\left[\mathbf{W}^{-1}\right]_{ij} = [\mathbf{W}]_{ij}^{-1}$$

All elements equal to zero are left unchanged in the reciprocal matrix.

→ *Hyper-Harary indices* are defined for this matrix and → *Harary indices* for the corresponding 1st order sparse matrix $^{1}\mathbf{W}^{-1}$ applying the Wiener operator. Moreover, the **Wiener delta matrix** $\mathbf{W}_{\Delta}$ was also proposed as $\mathbf{W}_{\Delta} = \mathbf{W} - {}^{1}\mathbf{W}$ [Ivanciuc *et al.*, 1997].

📖 [Linert *et al.*, 1995a] [Linert *et al.*, 1995b] [Diudea and Pârv, 1995] [Diudea, 1996b]

Wiener matrix degree → **Wiener matrix**

Wiener matrix eigenvalues → **eigenvalue-based descriptors**

Wiener matrix invariants → **Wiener matrix**

Wiener matrix leading eigenvalue : *1st Wiener matrix eigenvalue* → **eigenvalue-based descriptors** (⊙ Wiener matrix eigenvalues)

Wiener number : *Wiener index*

Wiener operator → **algebraic operators**

Wiener orthogonal operator → **algebraic operators**

Wiener polarity number : *polarity number* → **distance matrix**

Wiener-type indices

→ *Molecular descriptors* calculated by applying the → *Wiener operator* to symmetric matrices derived from molecular graphs or the → *Wiener orthogonal operator* to unsymmetric matrices.

These can be considered as variants of the well-known → *Wiener index*. For some, the term "Wiener-type indices" is restricted to those variants of the Wiener index that coincide with the index itself for acyclic structures but yield different values for cycle-containing structures [Lukovits and Linert, 1998].

Examples of Wiener-type indices are → *detour index*, → *Kirchhoff number* or → *quasi-Wiener index*, → *Szeged index*, → *all-path Wiener index*, → *edge Wiener index*, → *hyper-Wiener index*, → *Cluj-distance index*, → *hyper-Cluj-distance index*, → *Cluj-detour index*, → *hyper-Cluj-detour index*, → *Harary indices*, → *hyper-Harary indices*, → *detour/Wiener index*, and → *hyper-distance-path index*.

Wilcox resonance energy → **resonance indices**

WSHA index : *WHIM shape* → **WHIM descriptors** (⊙ global WHIM descriptors)

WSIZ index : *WHIM size* → **WHIM descriptors** (⊙ global WHIM descriptors)

WSYM index : *WHIM symmetry* → **WHIM descriptors** (⊙ global WHIM descriptors)

WTPT index → **path counts**

***W′/W* index** → **bond order indices** (⊙ graphical bond order)

***WW′/WW* index** → **bond order indices** (⊙ graphical bond order)

X

***X* index** → **topological information indices** (⊙ extended local information on distances)

Xu index

It is a topological molecular descriptor based on the → *adjacency matrix* and → *distance matrix*; it is defined as:

$$Xu = \sqrt{A} \cdot \log L_i = \sqrt{A} \cdot \log \frac{\sum_{i=1}^{A} \delta_i \cdot \sigma_i^2}{\sum_{i=1}^{A} \delta_i \cdot \sigma_i}$$

where A is the number of atoms and L represent the valence average topological distance calculated by → *vertex degree* δ and → *vertex distance degree* σ of all the atoms [Ren, 1999]. It was proposed as a particularly high discriminant molecular descriptor accounting for molecular size and branching.

Y

***Y* index** → **topological information indices** (⊙ mean extended local information on distances)

Y polarity scale → **linear solvation energy relationships** (⊙ dipolarity / polarizability term)

y-randomization test : *y-scrambling* → **validation techniques**

y-scrambling → **validation techniques**

Yukawa-Tsuno equation → **electronic substituent constants** (⊙ resonance electronic constants)

Z

Zagreb indices (M_n)

→ *Topological indices* based on the → *vertex degree* δ of the atoms in the → *H-depleted molecular graph* [Gutman *et al.*, 1975] and defined as:

$$M_1 = \sum_a \delta_a^2 = \sum_g g^2 \cdot {}^gF \qquad M_2 = \sum_b (\delta_i \cdot \delta_j)_b$$

where a runs over the A atoms of the molecule and b over all of the B bonds of the molecule. δ_i and δ_j refer to the vertex degrees of the atoms incident to the considered bond; ${}^gF$ is the → *vertex degree count*, i.e. the number of vertices with the same vertex degree, g is the value of the considered vertex degree. For isomeric series, the Zagreb indices are related to the → *molecular branching*.

The two Zagreb indices are strictly related to zero-order ${}^0\chi$ and first-order ${}^1\chi$ → *connectivity indices*, respectively. The **1st Zagreb index** M_1 (also called **Gutman index**) is also related to the → *Platt number* F and the → *connection number* N_2 by the relationship:

$$M_1 = \mathrm{F} + 2(A-1) = 2 \cdot (\mathrm{N}_2 + A - 1)$$

where A is the number of atoms. The **2nd Zagreb index** M_2 is a part of the → *Schultz molecular topological index*.

By normalization (i.e. imposing a lower bound equal to zero for linear graphs) of the M_1 index the **quadratic index**, denoted by Q and also called **normalized quadratic index**, was defined as [Balaban, 1979]:

$$Q = \frac{\sum_g (g^2 - 2g) \cdot {}^gF + 2}{2} = 3 - 2 \cdot A + \frac{M_1}{2}$$

where g are the different vertex degree values and ${}^gF$ is the vertex degree count. It can be demonstrated that the quadratic index is equal to the difference between the M_1 index of the considered graph and the M_1 index of the corresponding → *linear graph*. Moreover, for acyclic molecules, it can be simply calculated by:

$$Q = 3 \cdot {}^4F + {}^3F$$

where ${}^3F$ and ${}^4F$ are the vertex degree counts of three and four orders, respectively.

The **binormalized quadratic index** Q' is obtained by binormalization of M_1, i.e. imposing a lower bound equal to zero for linear graphs and an upper bound equal to one for star graphs [Balaban, 1979]. In practice, it is calculated by dividing the quadratic index Q by its respective value for the → *star graph*:

$$Q' = \frac{2 \cdot Q}{(A-2) \cdot (A-3)}$$

where A is the number of graph vertices.

The binormalized quadratic index was proposed, together with the → *binormalized centric index* C', to provide information on the topological shape of trees, i.e. the trade-off between linear and star graphs.

Z-counting polynomial → **Hosoya Z index**

Z-delta number → **vertex degree**

Z index : *Hosoya Z index*

^{m}Z numbers → **Hosoya Z matrix**

Z-matrix → **molecular geometry**

Z polarity scale → **linear solvation energy relationships** (⊙ dipolarity / polarizability term)

Z′/Z index → **Hosoya Z matrix**

Greek Alphabet Entries

α A (alpha)

$^k\alpha$ descriptors

Molecular descriptors derived from the powers of the → *sparse* χ *matrix*, by summing all the elements above the main diagonal:

$$^k\alpha = \sum_{i=1}^{A-1} \sum_{j=i+1}^{A} [\boldsymbol{\chi}^{\mathbf{k}}]_{ij}$$

where $\boldsymbol{\chi}^{\mathbf{k}}$ is the kth → *power matrix* (k = 1, 2, 3, ...) of the sparse χ matrix; A is the number of atoms [Randic, 1992d].

The first term $^1\alpha$ coincides with the → *Randic connectivity index* χ; the remaining values show monotonically smaller values. In acyclic graphs, the even powers of $\boldsymbol{\chi}^{\mathbf{k}}$ matrices necessarily make zero contributions as it is not possible to reach an adjacent vertex by an even number of steps.

α scale → **linear solvation energy relationships** (⊙ hydrogen-bond parameters)

α_m scale → **linear solvation energy relationships** (⊙ hydrogen-bond parameters)

α_2^H scale → **linear solvation energy relationships** (⊙ hydrogen-bond parameters)

β B (beta)

β scale → **linear solvation energy relationships** (⊙ hydrogen-bond parameters)

β_m scale → **linear solvation energy relationships** (⊙ hydrogen-bond parameters)

β_2^H scale → **linear solvation energy relationships** (⊙ hydrogen-bond parameters)

λ Λ (lambda)

λ matrix → **layer matrices** (⊙ cardinality layer matrix)

$\lambda\lambda_1$ branching index → **eigenvalue-based descriptors**

π Π (pi)

π-inductive effect → **electronic substituent constants**

π* polarity scale → **linear solvation energy relationships** (⊙ dipolarity / polarizability term)

σ Σ (sigma)

σ-electron descriptor → **vertex degree**

σ electronic constant : *electronic substituent constants*

σ* electronic constant : *Taft σ* constant* → **electronic substituent constants** (⊙ inductive electronic constants)

σ_r electronic constants → **electronic substituent constants** (⊙ field / resonance effect separation)

σ-inductive effect → **electronic substituent constants**

σ_α^* radical substituent constants → **electronic substituent constants** (⊙ field / resonance effect separation)

τ T (tau)

τ matrix → **sequence matrices**

χ X (chi)

χ matrix → **weighted matrices**

χ′/χ index → **bond order indices** (⊙ graphical bond order)

Numerical Entries

0D-descriptors → **molecular descriptors**

1D-descriptors → **molecular descriptors**

1st Wiener matrix eigenvalue → **eigenvalue-based descriptors** (⊙ Wiener matrix eigenvalues)

1st Zagreb index → **Zagreb indices**

1st order sparse matrix → **algebraic operators** (⊙ sparse matrices)

2D-descriptors → **molecular descriptors**

2D-QSAR → **structure/response correlations**

2nd Zagreb index → **Zagreb indices**

3D-Balaban index → **Balaban distance connectivity indices**

3D-connectivity indices → **connectivity indices**

3D-descriptors → **molecular descriptors**

3D-double invariants → **double invariants**

3D-molecule representation of structures based on electron diffraction : *3D-MoRSE descriptors*

3D-MoRSE code → **3D-MoRSE descriptors**

3D-MoRSE descriptors (: *3D-MOlecule Representation of Structures based on Electron diffraction, MoRSE descriptors*)

These descriptors are based on the idea of obtaining information from the 3D atomic coordinates by the transform used in electron diffraction studies for preparing theoretical scattering curves [Soltzberg and Wilkins, 1977]. A generalized scattering function, called the **molecular transform**, can be used as the functional basis for deriving, from a known molecular structure, the specific analytic relationship of both X-ray and electron diffraction. The general molecular transform is:

$$G(\mathbf{s}) = \sum_{i=1}^{A} f_i \cdot \exp(2\pi i \cdot \mathbf{r}_i \cdot \mathbf{s})$$

where $\mathbf{s}$ represents the scattering in various directions by a collection of A atoms located at points $\mathbf{r}_i$; f_i is a form factor taking into account the direction dependence of scattering from a spherical body of finite size. The scattering value s measures the scattering angle as:

$$s = 4\pi \cdot \sin(\vartheta/2)/\lambda$$

where ϑ is the scattering angle and λ the wavelength of the electron beam.

Usually, the above equation is used in a modified form as suggested in 1931 by Wierl [Wierl, 1931]. On substituting the form factors by the atomic property w, considering the molecule to be rigid and setting the instrumental constant equal to one, the following expression is obtained:

$$I(s) = \sum_{i=1}^{A-1} \sum_{j=i+1}^{A} w_i \cdot w_j \cdot \frac{\sin(s \cdot r_{ij})}{s \cdot r_{ij}}$$

where $I(s)$ is the scattered electron intensity, w is an atomic property, chosen as the atomic number, r_{ij} are the → *interatomic distances* between the ith and jth atoms, respectively, and A is the number of atoms.

Soltzberg & Wilkins introduced a number of simplifications in order to obtain a binary code. Only the zero crossing of the $I(s)$ curve, i.e. the s values at which $I(s) = 0$ in the range 1 – 31 $Å^{-1}$ are considered. The s range is then divided into 100 equal intervals, each described by a binary variable equal to 1 if the interval contains a zero crossing, 0 otherwise. Thus, a **3D-MoRSE code** consisting of a 100-dimensional binary vector is obtained.

Gasteiger *et al.* [Schuur and Gasteiger, 1996; Schuur and Gasteiger, 1997] returned to the initial $I(s)$ curve and maintained the explicit form of the curve. For atomic weights w, various physico-chemical properties such as atomic mass, partial atomic charges, and atomic polarizability were considered. In order to obtain → *uniform length descriptors*, the intensity distribution $I(s)$ is made discrete, calculating its value at a sequence of evenly distributed values of e.g., 32 or 64 values in the range 1 – 31 $Å^{-1}$. Clearly, the more values chosen, the finer becomes the resolution in the representation of the molecule.

→ *RDF descriptors* have recently been proposed based on a radial distribution function with some characteristics in common with the $I(s)$ function used to obtain the 3D-MoRSE descriptors.

📖 [Schuur *et al.*, 1996] [Gasteiger *et al.*, 1996] [Selzer *et al.*, 1996] [Gasteiger *et al.*, 1997] [Baumann, 1999]

3D-MTI′ index → **Schultz molecular topological index**

3D principal properties → **principal component analysis**

3D-QSAR → **structure/response correlations**

3D-Schultz index → **Schultz molecular topological index**

3D-Wiener index → **Wiener index**

4D-descriptors → **molecular descriptors**

4D-QSAR : *dynamic QSAR* → **structure/response correlations**

Appendix A. Greek Alphabet

lower	*upper*	*pronounce*	*lower*	*upper*	*pronounce*
α	Α	alpha	ν	Ν	nu
β	Β	beta	ξ	Ξ	xi
γ	Γ	gamma	ο	Ο	omicron
δ	Δ	delta	π	Π	pi
ε	Ε	epsilon	ϱ, ρ	Ρ	rho
ζ	Ζ	zeta	σ	Σ	sigma
η	Η	eta	τ	Τ	tau
θ, ϑ	Θ	theta	υ	Υ	upsilon
ι	Ι	iota	ϕ	Φ	phi
ϰ, κ	Κ	kappa	χ	Χ	chi
λ	Λ	lambda	φ	Ψ	psi
μ	Μ	mu	ω	Ω	omega

Appendix B. Symbols of Molecular Descriptors

The most common elementary symbols used for molecular descriptors are reported in the following tables; the first table collects normal symbols, the second Greek symbols. Symbols denoting molecular descriptors by more complex combinations of elementary symbols are not reported in the tables.

Symbol	*Definition*
A	arithmetic topological index
A	WHIM size
A	number of atoms
$\mathrm{A_E}$	total edge adjacency index
A_{F}	similarity score
$\mathrm{A_V}$	total adjacency index
B	Balaban centric index
B	number of bonds
$\mathrm{B_i}$	Sterimol B parameters
C	Carbò similarity index
C	cyclomatic numer
C	normal centric index
C'	binormalized centric index
C^{+}	cyclicity
$\mathrm{C_\infty}$	characteristic ratio
D	WHIM density
D	topological diameter
D	local dipole index
D_{m}	distance distribution moments
D_{P}	hyper-distance path index
$\mathrm{D^Z}$	Pogliani index
${}^{\mathrm{m}}D$	molecular profiles
E	molecule energies, interaction energies
E	exponential similarity index
$\mathrm{E_s}$	Taft steric constant
$\mathrm{E_s^c}$	Hancock steric constant
$\mathrm{E_s^0}$	Palm steric constant
$\mathrm{E_s'}$	Dubois steric constant
$\mathrm{E_s^F}$	Fujita steric constant
$\mathrm{E_s^{BW}}$	Bowden-Wooldridge steric constant
F	Platt number
$\mathrm{F^-}$	electrophilic frontier electron density index
$\mathrm{F^+}$	nucleophilic frontier electron density index
G	geometric topological index

Symbol	*Definition*
G	gravitational indices
G	WHIM symmetry
H	Hodgkin similarity index
H	harmonic topological index
H	Harary index
H_D	graph distance complexity
I	indicator variables
I	total information content indices
$\bar{I}$	mean information content indices
I_2	graph-theoretical shape coefficient
I_3	geometrical shape coefficient
J	Balaban distance connectivity index
${}^{3D}J$	3D-Balaban index
J_t	J_t index
J_{het}	Barysz index
k'	capacity factor
K	Kekulé number
K	WHIM shape
l	Kuhn length
L	linear similarity index
L	Sterimol length parameter
L_p	persistence length
M_n	Zagreb indices
N_2	connection number
N_6	six-position number
O	ovality index
p	polarity number
P	polarizability
P	polarization
${}^{m}P$	path counts
Q	total absolute atomic charge
Q	quadrupole moment
Q	polarity index
Q	quadratic index
Q'	binormalized quadratic index
Q_H	polar hydrogen factor
r_{ee}	end-to-end distance
R	span
R	Spearmann rank correlation index
R	topological radius
R	Bowden-Young steric constant

Symbol	*Definition*
R'	resistance distance hyper-Wiener index
R_G	radius of gyration
S	simple topological index
S	Meyer-Richards similarity index
S	total softness
S	S index
S^-	electrophilic superdelocalizability
S^+	nucleophilic superdelocalizability
S_{AFF}	Jenkins steric parameter
S_I	inertial shape factor
S_f	substituent front strain
S_X	Austel branching index
T	WHIM size
T	Tanimoto coefficient
T^E	topological electronic descriptors
U	octupole moment
U	U index
V	V index
V	molecular volume
V	WHIM size
$\overline{V}$	molar volume
V_X	Mc Gowan's characteristic volume
V_{CPX}	Corey-Pauling-Koltun volume
V_{GEOM}	geometric volume
w	detour index
W	Wiener index
W^*	quasi-Wiener index
W^{AP}	all-path Wiener index
^kW_M	weighted walk numbers
W_N	normalized Wiener index
W_{RMS}	root mean square Wiener index
$\tilde{W}$	expanded Wiener number
^{3D}W	3D-Wiener index
X	X index
Y	Y index
Z	Hosoya Z index
Z_m	generalized Hosoya indices

Symbol	*Definition*
α	polarizability
α_1	solvent HBD acidity
α_2	solute HBD acidity
β_1	solvent HBA basicity
β_2	solute HBA basicity
β^2	anisotropy of the polarizability
γ_n	directional WHIM symmetry
γ	cyclicity index
δ_H	Hildebrand solubility parameter
${}^1\Delta$	submolecular polarity parameter
ε	molecular eccentricity
ε	edge connectivity index
ε	electropy index
η	Kaliszan shape parameter
η	absolute hardness
η_n	directional WHIM density
ϑ_n	directional WHIM shape
ϕ	folding degree index
Φ	Kier molecular flexibility index
Φ_{BD}	bond flexibility index
κ	Kier shape descriptors
λ_n	directional WHIM size
Λ	interactive polar parameter
λ_1^{LP}	Lovasz-Pelikan eigenvalue
λ_1^{W}	Wiener matrix eigenvalues
μ	dipole moment
ν_x	Charton steric constant
Ξ	Kier steric descriptor
π	Hansch-Fujita hydrophobic substituent constants
π^-	Norrington lipophilic constant
π^*	solvent polarity index
Π	local polarity index
Π^*	dipolarity/polarizability term
σ	electronic substituent constants
σ	Merrifield-Simmons index
τ	total topological state
χ	electronegativity
${}^m\chi$	connectivity indices
χ_T	total structure connectivity index
Ω	Pisanski-Zerovnik index
Ω	electron charge density connectivity index

Symbol	*Definition*
Ω_A	asphericity
Ω_S	spherosity index
ξ^C	eccentric connectivity index

Appendix C. Software

Some packages explicitly related to the calculation of the molecular descriptors for QSAR/QSPR are collected below, in alphabetic order.

General programs for computational quantum-chemistry, molecular modelling and log*P* calculations are not explicitly considered in this list. An extended list of computational chemistry programs can be found at the WebSite http://www.netsci.org/Resources/Software/.

ADAPT	Prof. P.C. Jurs, PennState University, University Park, PA 16802, USA
Description:	A QSAR toolkit with descriptor generation (topological, geometrical, electronic, and physicochemical descriptors), variable selection, regression and artificial neural network modelling.
Reference:	[Jurs *et al.*, 1979]
WebSite:	http://zeus.chem.psu.edu/

ASP	Oxford Molecular Ltd., Oxford Science Park, Oxford OX4 4GA, UK
Description:	Calculates a quantitative measure of molecular similarity based on molecule alignment, shape and electronic properties. Within TSAR 3D package.
WebSite:	http://www.oxmol.com/

$CERIUS^2$	Molecular Simulations Inc. – 9685 Scranton Road, San Diego, CA 92121-7352, USA
Description:	C2-Descriptors+ provides a range of generic descriptors, describing topological, electronic, and structural features.
WebSite:	http://www.msi.com/life/products/cerius2/modules/descriptor.html

CODESSA	Semichem Inc. – 7204 Mullen, Shawnee, KS 66216, USA
Description:	Calculation of several topological, geometrical, constitutional, thermodynamic, electrostatic, and quantum-chemical descriptors, including tools for regression modelling and variable selection.
Reference:	[Katritzky *et al.*, 1995]
WebSite:	http://www.semichem.com/

DRAGON	Prof. R. Todeschini – distributed by Talete srl, via Pisani 13, 20124 Milano, Italy
Description:	Calculation of several sets of molecular descriptors from molecular geometries (topological, geometrical, WHIM, 3D-MoRSE, molecular profiles, etc.).
WebSite:	http://www.disat.unimib.it/chm/

GRIN/GRID	Molecular Discovery Ltd. – West Way House, Elms Parade, Oxford OX2 9LL, UK
Description:	Calculates the GRID empirical force field at grid points. Last release: V.11 – 1993
Reference:	[Goodford, 1985]

HQSAR	Tripos Inc. – 1699 South Hanley Rd., St.Louis, MO 63144-2913, USA
Description:	A part of the SYBYL environment providing hologram descriptors.
WebSite:	http://www.tripos.com/

HYBOT-PLUS	Prof. O. Raevsky – Russian Academy of Science, IPAC.
Description:	Calculation of hydrogen bond and free energy factors.
Reference:	[Raevsky, 1997]
WebSite:	http://www.ipac.ac.ru/qsar/index.htm

HYPERCHEM 6	Hypercube, Inc. – 1115 NW 4th Street, Gainsville, FL 32601, USA
Description:	Calculation of optimized geometries with several computational methods, also providing total surface area, molecular volume, molar refractivity, log P, polarizability and atomic charges. Last release: 6
WebSite:	http://www.hyper.com/

MOLCONN-Z	Prof. L.H. Hall – 2 Davis Street, Quincy, MA 02170, USA
Description:	Successor of MOLCONN-X, MOLCONN-Z calculates the most well-known topological descriptors, including electrotopological and orthogonalized indices. Last release: 3.0
WebSite:	http:// www.eslc.vabiotech.com/molconn/manuals/310s/preface1.html

MULTICASE	Multicase Inc. – PO 22517, Beachwood, OH 44122, USA
Description:	Prediction of biological activities by substructure descriptors.
Reference:	[Klopman, 1992]
WebSite:	http://www.multicase.com/

OASIS	Prof. O. Mekenyan – Bourgas University, 8010 Bourgas, Bulgaria
Description:	Calculation of steric, electronic, and hydrophobic descriptors.
Reference:	[Mekenyan *et al.*, 1990a]
WebSite:	http://omega.btu.bg/~omekenya/

PETRA	Molecular Networks GmbH – Langemarckplatz 1, D-91054 Erlangen (Germany)
Description:	Empirical methods for the calculation of charges and bond energies for use in QSAR.
Reference:	[Gasteiger, 1988; Löw and Saller, 1988]

POLLY	Prof. S. Basak – Minnesota University of Duluth, 5013 Miller Trunk Highway, Duluth, MN 55811, USA
Description:	Calculation of topological connectivity indices. Last release: 2.3
Reference:	[Basak *et al.*, 1988a]

SciQSAR 2D	SciVision – 200 Wheeler Road, Burlington, MA 01803, USA
Description:	Calculation of several topological molecular descriptors (connectivity, shape, electrotopological descriptors).
WebSite:	http://www.scivision.com/

SYBYL/QSAR	Tripos Inc. – 1699 South Hanley Rd., St.Louis, MO 63144-2913, USA
Description:	SYBYL module for the calculation of EVA descriptors, CoMFA and CoMSIA fields, also including several QSAR tools. Last release: 6.1
WebSite:	http://www.tripos.com/

TSAR	Oxford Molecular Ltd., Oxford Science Park, Oxford OX4 4GA, UK
Description:	Statistical and database functions with molecular and substituent property calculations. Within TSAR 3D package.
Reference:	http://www.oxmol.com/

Bibliography

The symbol [R] at the end of a reference denotes a publication with a large list of references. The bibliography covers a period between 1858 and 2000, collects 3300 references, for about 3000 authors and 250 periodicals.

Aboushaaban, R.R., Alkhamees, H.A., Abouauda, H.S. and Simonelli, A.P. (1996). Atom Level Electrotopological State Indexes in QSAR Designing and Testing Antithyroid Agents. *Pharm.Res.*, *13*, 129–136.

Abraham, D.J. and Leo, A. (1987). *Proteins: Struct.Funct.Genetics*, *2*, 130–152.

Abraham, D.J. and Kellogg, G.E. (1993). Hydrophobic Fields. In *3D QSAR in Drug Design. Theory, Methods and Applications.* (Kubinyi, H., ed.), ESCOM, Leiden (The Netherlands), pp. 506–520.

Abraham, M.H., Grellier, P.L. and McGill, R.A. (1987). Determination of Olive Oil-Gas and Hexadecane-Gas Partition Coefficients, and Calculation of the Corresponding Olive Oil-Water and Hexadecane-Water Partition Coefficients. *J.Chem.Soc.Perkin Trans.2*, 797–803.

Abraham, M.H., Grellier, P.L., Hamerton, I., McGill, R.A., Prior, D.V. and Whiting, G.S. (1988). Solvation of Gaseous Non-Electrolytes. *Faraday Discuss.Chem.Soc.*, *85*, 107–115.

Abraham, M.H., Duce, P.P., Prior, D.V., Barratt, D.G., Morris, J.J. and Taylor, P.J. (1989a). Hydrogen Bonding. Part 9. Solute Proton Donor and Proton Acceptor Scales for Use in Drug Design. *J.Chem.Soc.Perkin Trans.2*, 1355–1375.

Abraham, M.H., Grellier, P.L., Prior, D.V., Duce, P.P., Morris, J.J. and Taylor, P.J. (1989b). Hydrogen Bonding. Part 7. A Scale of Solute Hydrogen-Bond Acidity Based on log *K* Values for Complexation in Tetrachloromethane. *J.Chem.Soc.Perkin Trans.2*, 699–711.

Abraham, M.H., Grellier, P.L., Prior, D.V., Morris, J.J. and Taylor, P.J. (1990). Hydrogen Bonding. Part 10. A Scale of Solute Hydrogen-Bond Basicity Using log *K* Values for Complexation in Tetrachorometane. *J.Chem.Soc.Perkin Trans.2*, 521–529.

Abraham, M.H., Whiting, G.S., Alarie, Y., Morris, J.J., Taylor, P.J., Doherty, R.M., Taft, R.W. and Nielsen, G.D. (1990a). Hydrogen Bonding. Part 12. A New QSAR for Upper Respiratory Tract Irritation by Airborne Chemicals in Mice. *Quant.Struct.-Act.Relat.*, *9*, 6–10.

Abraham, M.H., Whiting, G.S., Doherty, R.M. and Shuely, W.J. (1990b). Hydrogen Bonding. Part 13. A New Method for the Characterization of GLC Stationary Phases – The Laffort Data Set. *J.Chem.Soc.Perkin Trans.2*, 1451–1460.

Abraham, M.H., Lieb, W.R. and Franks, N.P. (1991). Role of Hydrogen Bonding in General Anesthesia. *J.Pharm.Sci.*, *80*, 719–724.

Abraham, M.H., Whiting, G.S., Doherty, R.M. and Shuely, W.J. (1991a). Hydrogen Bonding. XVI. A New Solute Solvation Parameter, π_2^H, from Gas Chromatographic Data. *J.Chromat.*, *587*, 213–228.

Abraham, M.H., Whiting, G.S., Doherty, R.M. and Shuely, W.J. (1991b). Hydrogen Bonding. XVII. The Characterisation of 24 Gas-Liquid Chromatographic Stationary Phases Studied by Poole and Co-Workers, Including Molten Salts, and Evaluation of Solute-Stationary Phase Interactions. *J.Chromat.*, *587*, 229–236.

Abraham, M.H. and Whiting, G.S. (1992). Hydrogen Bonding. XXI. Solvation Parameters for Alkylaromatic Hydrocarbons from Gas-Liquid Chromatographic Data. *J.Chromat.*, *594*, 229–241.

Abraham, M.H. (1993a). Hydrogen Bonding. Part 31. Construction of a Scale of Solute Effective or Summation Hydrogen Bond Basicity. *J.Phys.Org.Chem.*, *6*, 660–684.

Abraham, M.H. (1993b). Scales of Solute Hydrogen-Bonding: Their Construction and Application to Physicochemical and Biochemical Processes. *Chem.Soc.Rev.*, *22*, 73–83.

Abraham, M.H., Andonian-Haftvan, J., Whiting, G.S., Leo, A. and Taft, R.S. (1994). Hydrogen Bonding. Part 34. The Factors that Influence the Solubility of Gases and Vapours in Water at 298 K, and a New Method for its Determination. *J.Chem.Soc.Perkin Trans.2*, 1777–1791.

Abraham, M.H., Chadha, H.S., Dixon, J.P. and Leo, A.J. (1994a). Hydrogen Bonding.39. The Partition of Solutes Between Water and Various Alcohols. *J.Phys.Org.Chem.*, *7*, 712–716.

Abraham, M.H., Chadha, H.S. and Mitchell, R.C. (1994b). Hydrogen Bonding. 33. Factors that Influence the Distribution of Solutes Between Blood and Brain. *J.Pharm.Sci.*, *83*, 1257–1268.

Abraham, M.H., Chadha, H.S. and Mitchell, R.C. (1995). Hydrogen Bonding. Part 36. Determination of Blood Brain Distribution Using Octanol-Water Partition Coefficients. *Drug Design & Discovery*, *13*, 123–131.

Abraham, M.H. and Rafols, C. (1995). Factors that Influence Tadpole Narcosis. An LFER Analysis. *J.Chem.Soc.Perkin Trans.2*, 1843–1851.

Abraham, M.H., Andonian-Haftvan, J., Cometto Muniz, J.E. and Cain, W.S. (1996). An Analysis of Nasal Irritation Thresholds Using a New Solvation Equation. *Fund.Appl.Toxicol.*, *31*, 71–76.

Abraham, R.J. and Smith, P.E. (1988). Charge Calculations in Molecular Mechanics IV: A General Method for Conjugated Systems. *J.Comput.Chem.*, *9*, 288–297.

Adamson, G.W., Lynch, M.F. and Town, W.G. (1970). Analysis of Structural Characteristics of Chemical Compounds in a Large Computer-based File. Part II. Atom-Centred Fragments. *J.Chem.Soc.*, *(C)*, 3702–3706.

Affolter, C., Baumann, K., Clerc, J.T., Schriber, H. and Pretsch, E. (1997). Automatic Interpretation of Infrared Spectra. *Mikrochim.Acta*, *14*, 143–147.

Agarwal, A., Pearson, P.P., Taylor, E.W., Li, H.B., Dahlgren, T., Herslof, M., Yang, Y.H., Lambert, G., Nelson, D.L., Regan, J.W. and Martin, A.R. (1993). Three Dimensional Quantitative Structure-Activity Relationships of 5-HT Receptor Binding Data for Tetrahydropyridinylindole Derivatives. A Comparison of the Hansch and CoMFA Methods. *J.Med.Chem.*, *36*, 4006–4014.

Agarwal, K.K. (1998). An Algorithm for Computing the Automorphism Group of Organic Structures with Stereochemistry and a Measure of its Efficiency. *J.Chem.Inf.Comput.Sci.*, *38*, 402–404.

Agrafiotis, D.K. (1997). On the Use of Information Theory for Assessing Molecular Diversity. *J.Chem.Inf.Comput.Sci.*, *37*, 576–580.

Agrafiotis, D.K. and Lobanov, V.S. (1999). An Efficient Implementation of Distance-Based Diversity Measures Based on *k-d* Trees. *J.Chem.Inf.Comput.Sci.*, *39*, 51–58.

Ahmad, P., Fyfe, C.A. and Mellors, A. (1975). *Biochem.Pharmacol.*, *24*, 1103–1109.

Aihara, J. (1976). A Generalized Total π-Energy Index for a Conjugated Hydrocarbon. *J.Org. Chem.*, *41*, 2488–2490.

Aihara, J. (1977a). Aromatic Sextets and Aromaticity in Benzenoid Hydrocarbons. *Bull.Chem.-Soc.Jap.*, *50*, 2010–2012.

Aihara, J. (1977b). Resonance Energies of Benzenoid Hydrocarbons. *J.Am.Chem.Soc.*, *99*, 2048–2053.

Aihara, J. (1978). Resonance Energies of Nonbenzenoid Hydrocarbons. *Bull.Chem.Soc.Jap.*, *51*, 3540–3543.

Akagi, T., Mitani, S., Komyoji, T. and Nagatani, K. (1995). Quantitative Structure-Activity Relationships of Fluazinam and Related Fungicidal N-Phenylpyridinamines Preventive Activity Against Botrytis Cinerea. *J.Pestic.Sci.*, *20*, 279–290.

Albuquerque, M.G., Hopfinger, A.J., Barreiro, E.J. and de Alencastro, R.B. (1998). Four-Dimensional Quantitative Structure-Activity Relationship Analysis of a Series of Interphenylene 7-Oxabicycloheptane Oxazole Thromboxane A_2 Receptor Antagonists. *J.Chem.Inf.Comput.Sci.*, *38*, 925–938.

Alfrey, T. and Price, C.C. (1947). *J.Polym.Sci.*, *2*, 101.

Allen, D.M. (1971). Mean Square Error of Prediction as a Criterion for Selecting Variables. *Technometrics*, *13*, 469–475.

Allen, F.H., Bath, P.A. and Willett, P. (1995). Angular Spectroscopy: Rapid Visualization of Three-Dimensional Substructure Dissimilarity Using Valence Angle or Torsional Descriptors. *J.Chem.Inf.Comput.Sci.*, *35*, 261–271.

Allerhand, A. and Schleyer, P.V.R. (1963a). Nitriles and Isonitriles as Proton Acceptors in Hydrogen Bonding: Correlation of $\Delta\nu_{OH}$ with Acceptor Structure. *J.Am.Chem.Soc.*, *85*, 866–870.

Allerhand, A. and Schleyer, P.V.R. (1963b). Solvent Effects in Infrared Spectroscopic Studies of Hydrogen Bonding. *J.Am.Chem.Soc.*, *85*, 371–380.

Allred, A. and Rochow, E.G. (1958). A Scale of Electronegativity Based on Electronic Forces. *Journal of Inorganic Nuclear Chemistry*, *5*, 264–268.

Allred, A. and Rochow, E.G. (1961). Electronegativity Values from Thermochemical Data. *Journal of Inorganic Nuclear Chemistry*, *17*, 215–221.

Altenburg, K. (1961). Zur Berechnung des Radius Verweighter Molekule. *Kolloid Zeitschr.*, *178*, 112.

Altenburg, K. (1980). Eine Bemerkung zu dem Randicschen "Molekularen Bindungs-Index". *Z.Phys.Chemie*, *261*, 389–393.

Altomare, C., Cellamare, S., Carotti, A. and Ferappi, M. (1993). Linear Solvation Energy Relationships in Reversed-Phase Liquid Chromatography. Examination in Deltabond C_8 as Stationary Phase for Measuring Lipophilicity Parameters. *Quant.Struct.-Act.Relat.*, *12*, 261–268.

Alunni, S., Clementi, S., Edlund, U., Johnels, D., Hellberg, S., Sjöström, M. and Wold, S. (1983). Multivariate Data Analysis of Substituent Descriptors. *Acta Chem.Scand.*, *B 37*, 47–53.

Amat, L., Robert, D., Besalù, E. and Carbó-Dorca, R. (1998). Molecular Quantum Similarity Measures Tuned 3D QSAR: An Antitumoral Family Validation Study. *J.Chem.Inf.Comput.Sci.*, *38*, 624–631.

Amic, D., Davidovic-Amic, D., Juric, A., Lucic, B. and Trinajstic, N. (1995a). Structure-Activity Correlation of Flavone Derivatives for Inhibition of cAMP phosphodiestarase. *J.Chem.Inf.Comput.Sci.*, *35*, 1034–1038.

Amic, D., Davidovic-Amic, D. and Trinajstic, N. (1995b). Calculation of Retention Times of Anthocyanins with Orthogonalized Topological Indices. *J.Chem.Inf.Comput.Sci.*, *35*, 136–139.

Amic, D. and Trinajstic, N. (1995c). On the Detour Matrix. *Croat.Chem.Acta*, *68*, 53–62.

Amic, D., Davidovic-Amic, D., Beslo, D., Lucic, B. and Trinajstic, N. (1997). The Use of the Ordered Orthogonalized Multivariate Linear Regression in a Structure-Activity Study of Coumarin and Flavonoid Derivatives as Inhibitors of Aldose Reductase. *J.Chem.Inf.Comput.Sci.*, *37*, 581–586.

Amic, D., Beslo, D., Lucic, B., Nikolic, S. and Trinajstic, N. (1998a). The Vertex-Connectivity Index Revisited. *J.Chem.Inf.Comput.Sci.*, *38*, 819–822.

Amic, D., Davidovic-Amic, D., Beslo, D., Lucic, B. and Trinajstic, N. (1998b). QSAR of Flavylium Salts as Inhibitors of Xanthine Oxidase. *J.Chem.Inf.Comput.Sci.*, *38*, 815–818.

Amic, D., Davidovic-Amic, D., Beslo, D., Lucic, B. and Trinajstic, N. (2000). Prediction of p*K* Values, Half-Lives, and Electronic Spectra of Flavylium Salts from Molecular Structure. *J.Chem.Inf.Comput.Sci.*, *40*, 967–973.

Amidon, G.L., Yalkowsky, S.H., Anik, S.T. and Valvani, S.C. (1975). Solubility of Nonelectrolytes in Polar Solvents. V. Estimation of the Solubility of Aliphatic Monofunctional Compounds in Water Using a Molecular Surface Area Approach. *J.Phys.Chem.*, *79*, 2239–2246.

Amoore, J.E. (1964). *Ann.N.Y.Acad.Sci.*, *116*, 457.

Andersson, P., Haglund, P., Rappe, C. and Tysklind, M. (1996). Ultraviolet Absorption Characteristic and Calculated Semi-empirical Parameters as Chemical Descriptors in Multivariate Modelling of Polychlorinated Biphenyls. *J.Chemom.*, *10*, 171–185.

Andre, V., Boissart, C., Sichel, F., Gauduchon, P., Le, T.J., Lancelot, J.C., Mercier, C., Chemtob, S., Raoult, E. and Tallec, A. (1997). Mutagenicity of Nitro- and Amino-Substituted Carbazoles in Salmonella typhimurium. III. Methylated Derivatives of 9H-carbazole. *Mut.Res.*, *389*, 247–260.

Andrea, T.A. and Kalayeh, H. (1991). Applications of Neural Networks in Quantitative Structure-Activity Relationships of Dihydropholate Reductase Inhibitors. *J.Med.Chem.*, *34*, 2824–2836.

Andrea, T.A. (1995). Novel Structure-Activity Insights from Neural Network Models. *ACS Symp.Ser.*, *606*, 282–287.

Andrews, P.R., Craik, D.J. and Martin, J.L. (1984). Functional Group Contributions to Drug-Receptor Interactions. *J.Med.Chem.*, *27*, 1648–1657.

Andrews, P.R. (1993). Drug-Receptor Interactions. In *3D QSAR in Drug Design. Theory, Methods and Applications.* (Kubinyi, H., ed.), ESCOM, Leiden (The Netherlands), pp. 13–40.

Anker, L.S., Jurs, P.C. and Edwards, P.A. (1990). Quantitative Structure-Retention Relationship Studies of Odor-Active Aliphatic Compounds with Oxygen-Containing Functional Groups. *Anal.Chem.*, *62*, 2676–2684.

Anzali, S., Barnickel, G., Krug, M., Sadowski, J., Wagener, M., Gasteiger, J. and Polanski, J. (1996). The Comparison of Geometric and Electronic Properties of Molecular Surfaces by Neural Networks: Application to the Analysis of Corticosteroid Binding Globulin Activity of Steroids. *J.Comput.Aid.Molec.Des.*, *10*, 521–534.

Anzali, S., Barnickel, G., Krug, M., Wagener, M. and Gasteiger, J. (1997). Kohonen Neural Network: A Novel Approach to Search for Biosisosteric Groups. In *Computer-Assisted Lead Finding and Optimization.* (van de Waterbeemd, H., Testa, B. and Folkers, G., eds.), Wiley-VCH, Weinheim (Germany), pp. 95–106.

Anzali, S., Gasteiger, J., Holzgrabe, U., Polanski, J., Sadowski, J., Teckentrup, A. and Wagener, M. (1998). The Use of Self-Organizing Neural Networks in Drug Design. In *3D QSAR in Drug Design – Vol. 2* (Kubinyi, H., Folkers, G. and Martin, Y.C., eds.), Kluwer/ESCOM, Dordrecht (The Netherlands), pp. 273–299.

Aoyama, T., Suzuki, Y. and Ichikawa, H. (1990a). Neural Networks Applied to Quantitative Structure-Activity Relationships Analysis. *J.Med.Chem.*, *33*, 2583–2590.

Aoyama, T., Suzuki, Y. and Ichikawa, H. (1990b). Neural Networks Applied to Structure-Activity Relationships. *J.Med.Chem.*, *33*, 905–908.

Aprison, M.H., Galvez-Ruano, E. and Lipkowitz, K.B. (1995). On a Molecular Comparison of Strong and Weak Antagonists at the Glycinergic Receptor. *J.Neur.Res.*, *41*, 259–269.

Åqvist, J. and Tapia, O. (1987). Surface Fractality as a Guide for Studying Protein – Protein Interactions. *J.Mol.Graphics*, *5*, 30–34.

Araujo, O. and Morales, D.A. (1996a). A Theorem About the Algebraic Structure Underlying Orthogonal Graph Invariants. *J.Chem.Inf.Comput.Sci.*, *36*, 1051–1053.

Araujo, O. and Morales, D.A. (1996b). An Alternative Approach to Orthogonal Graph Theoretical Invariants. *Chem.Phys.Lett.*, *257*, 393–396.

Araujo, O. and De La Peña, J.A. (1998). Some Bounds for the Connectivity Index of a Chemical Graph. *J.Chem.Inf.Comput.Sci.*, *38*, 827–831.

Araujo, O. and Morales, D.A. (1998). Properties of New Orthogonal Graph Theoretical Invariants in Structure-Property Correlations. *J.Chem.Inf.Comput.Sci.*, *38*, 1031–1037.

Arcos, J.C. (1987). Structure-Activity Relationships: Criteria for Predicting Carcinogenic Activity of Chemical Compounds. *Environ.Sci.Technol.*, *21*, 743–745.

Argese, E., Bettiol, C., Giurin, G. and Miana, P. (1999). Quantitative Structure-Activity Relationships for the Toxicity of Chlorophenols to Mammalian Submitochondrial Particles. *Chemosphere*, *38*, 2281–2292.

Ariëns, E.J. (1979). *Drug Design – Vol. VII.* Academic Press, New York (NY).

Ariëns, E.J. (1992). QSAR Conceptions and Misconceptions. *Quant.Struct.-Act.Relat.*, *11*, 190–194.

Armstrong, D.R., Perkins, P.G. and Stewart, J.J.P. (1973). Bond Indices and Valency. *J.C.S.Dalton Trans.*, 838–840.

Arnaud, L., Taillandier, G., Kaouadji, M., Ravanel, P. and Tissut, M. (1994). Photosynthesis Inhibition by Phenylureas: A QSAR Approach. *Ecotox.Environ.Safe.*, *28*, 121–133.

Arnold, S.F. (1990). *Mathematical Statistics.* Prentice-Hall, Englewood Cliffs (NJ), 636 pp.

Arteca, G.A. and Mezey, P.G. (1987). A Method for the Characterization of Molecular Conformations. *Int.J.Quantum Chem.Quant.Biol.Symp.*, *14*, 133–147.

Arteca, G.A., Jammal, V.B. and Mezey, P.G. (1988a). Shape Group Studies of Molecular Similarity and Regioselectivity in Chemical Reactions. *J.Comput.Chem.*, *9*, 608–619.

Arteca, G.A., Jammal, V.B., Mezey, P.G., Yadav, J.S., Hermsmeier, M.A. and Gund, T.M. (1988b). Shape Group Studies of Molecular Similarity: Relative Shapes of Van der Waals and Electrostatic Potential Surfaces of Nicotinic Agonists. *J.Mol.Graphics*, *6*, 45–53.

Arteca, G.A. and Mezey, P.G. (1988a). Molecular Conformations and Molecular Shape: A Discrete Characterization of Continua of van der Waals Surfaces. *Int.J.Quant.Chem.*, *34*, 517–526.

Arteca, G.A. and Mezey, P.G. (1988b). Shape Characterization of Some Molecular Model Surfaces. *J.Comput.Chem.*, *9*, 554–563.

Arteca, G.A. and Mezey, P.G. (1989). Shape Group Theory of van der Waals Surfaces. *J.Math.-Chem.*, *3*, 43.

Arteca, G.A. and Mezey, P.G. (1990). A Method for the Characterization of Folding in Protein Ribbon Models. *J.Mol.Graphics*, *8*, 66.

Arteca, G.A. (1991). Molecular Shape Descriptors. In *Reviews in Computational Chemistry – Vol. 9* (Lipkowitz, K.B. and Boyd, D., eds.), VCH Publishers, New York (NY), pp. 191–253. [R]

Arteca, G.A. (1996). Different Molecular Size Scaling Regimes for Inner and Outer Regions of Proteins. *Phys.Rev.A*, *54*, 3044–3047.

Arteca, G.A. (1999). Path-Integral Calculation of the Mean Number of Overcrossings in an Entangled Polymer Network. *J.Chem.Inf.Comput.Sci.*, *39*, 550–557.

Artemi, C. and Balaban, A.T. (1987). Mathematical Modeling of Polymers. Part II. Irriducible Sequences in *n*-ary Copolymers. *MATCH (Comm.Math.Comp.Chem.)*, *22*, 33–66.

Ash, J. E., Warr, W. A., and Willett, P. eds. (1991). *Chemical Structure Systems.* Ellis Horwood, Chichester (UK).

Atkinson, A.C. (1985). *Plots, Transformations, and Regression*. Clarendon Press, Oxford (UK), 282 pp.

Atkinson, R. (1987). A Structure-Activity Relationships for the Estimation of Rate Constants for the Gas-Phase Reactions of OH Radicals with Organic Compounds. *Int.J.Chem.Kinet.*, *19*, 799–828.

Atkinson, R. (1988). Estimation of Gas-Phase Hydroxyl Radical Rate Constants for Organic Compounds. *Environ.Toxicol.Chem.*, *7*, 435–442.

Attias, R. and Dubois, J.-E. (1990). Substructure Systems: Concepts and Classifications. *J. Chem.Inf.Comput.Sci.*, *30*, 2–7.

Attias, R. and Petitjean, M. (1993). Statistical Analysis of Atom Topological Neighborhoods and Multivariate Representations of a Large Chemical File. *J.Chem.Inf.Comput.Sci.*, *33*, 649–656.

Audry, E., Dubost, J.P., Colleter, J.C. and Dallet, Ph. (1986). Une Nouvelle Approche des Relations Structure-Activité: le "Potenciel de Lipophilie Moléculaire". *Eur.J.Med.Chem.*, *21*, 71–72.

Audry, E., Dallet, Ph., Langlois, M.H., Colleter, J.C. and Dubost, J.P. (1989). Quantitative Structure Affinity Relationships in a Series of α_2 Adrenergic Amines Using the Molecular Lipophilicity Potential. *Proceeding of Clinical Biological Research*, *291*, 63.

Audry, E., Dubost, J.P., Langlois, M.H., Croizet, F., Braquet, P., Dallet, Ph. and Colleter, J.C. (1992). Use of Molecular Lipophilicity Potential in QSAR. In *QSAR Design of Bioactive Compounds* (Kuchar, M., ed.), Prous Science, Barcelona (SP), pp. 249–268.

Aurenhammer, F. (1991). Voronoi Diagrams – A Survey of Fundamental Geometric Data Structure. *ACM Comp.Serv.*, *23*, 345–405.

Austel, V., Kutter, E. and Kalbfleisch, W. (1979). A New Easily Accessible Steric Parameter for Structure-Activity Relationships. *Arzneim.Forsch.*, *29*, 585–587.

Austel, V. (1983). Features and Problems of Practical Drug Design. In *Steric Effects in Drug Design (Topics in Current Chemistry, Vol. 114)* (Charton, M. and Motoc, I., eds.), Springer-Verlag, Berlin (Germany), pp. 7–19.

Austel, V. (1995). Experimental Design in Synthesis Planning and Structure-Property Correlations. Experimental Design. In *Chemometrics Methods in Molecular Design – Vol. 2* (van de Waterbeemd, H., ed.), VCH Publishers, New York (NY), pp. 49–62.

Avbelj, F. and Hadzi, D. (1985). *Mol.Pharm.*, *27*, 466–470.

Azzaoui, K., Lafosse, M., Lazar, S., Thiery, V. and Morinallory, L. (1995a). Separation of Benzodioxinic Isomers in LC. A Molecular Modeling Approach for the Choice of the Stationary Phase. *J.Liquid Chromat.*, *18*, 3021–3034.

Azzaoui, K. and Morinallory, L. (1995b). Quantitative Structure-Retention Relationships for the Investigation of the Retention Mechanism in High Performance Liquid Chromatography Using Apolar Eluent with a Very Low Content of Polar Modifiers. *Chromatographia*, *40*, 690–696.

Babic, D., Graovac, A. and Gutman, I. (1991). On a Resonance Energy Model Based on Expansion in Terms of Acyclic Moments: Exact Results. *Theor.Chim.Acta*, *79*, 403–411.

Babic, D., Balaban, A.T. and Klein, D.J. (1995). Nomenclature and Coding of Fullerenes. *J.Chem.Inf.Comput.Sci.*, *35*, 515–526.

Babic, D., Brinkmann, G. and Dress, A. (1997). Topological Resonance Energy of Fullerenes. *J.Chem.Inf.Comput.Sci.*, *37*, 920–923.

Bahnick, D.A. and Doucette, W.J. (1988). Use of Molecular Connectivity Indices to Estimate Soil Sorption for Organic Chemicals. *Chemosphere*, *17*, 1703–1715.

Baird, N.C. (1969). Calculation of "Dewar" Resonance Energies in Conjugated Organic Molecules. *Can.J.Chem.*, *47*, 3535–3538.

Baird, N.C. (1971). Dewar Resonance Energy. *J.Chem.Educ.*, *48*, 509–514.

Baker, F.W., Parish, R.C. and Stock, L.M. (1967). Dissociation Constants of Bicyclo[2.2.2]oct-2-ene-1-carboxylic Acids, Dibenzobicyclo[2.2.2]octa-2,5-diene-1-carboxylic Acids, and Cubanecarboxylic Acids. *J.Am.Chem.Soc.*, *89*, 5677–5685.

Baker, G.A. (1966). Unknown. *J.Math.Phys.*, *7*, 2238.

Baker, J.K. (1979). Estimation of High Pressure Liquid Chromatographic Retention Indices. *Anal.Chem.*, *51*, 1693–1697.

Bakken, G.A. and Jurs, P.C. (1999a). Prediction of Hydroxyl Radical Rate Constants from Molecular Structure. *J.Chem.Inf.Comput.Sci.*, *39*, 1064–1075.

Bakken, G.A. and Jurs, P.C. (1999b). Prediction of Methyl Radical Addition Rate Constants from Molecular Structure. *J.Chem.Inf.Comput.Sci.*, *39*, 508–514.

Balaban, A.T. and Harary, F. (1968). Chemical Graphs. V. Enumeration and Proposed Nomenclature of Benzenoid Catacondensed Polycyclic Aromatic Hydrocarbons. *Tetrahedron*, *24*, 2505–2516.

Balaban, A.T. (1969). Chemical Graphs – VII Proposed Nomenclature of Branched *Cata*-Condensed Benzenoid Polycyclic Hydrocarbons. *Tetrahedron*, *25*, 2949–2956.

Balaban, A.T. (1971). Chemical Graphs. 12. Configurations of Annulenes. *Tetrahedron*, *27*, 6115–6131.

Balaban, A.T. and Harary, F. (1971). The Characteristic Polynomial does not Uniquely Determine the Topology of a Molecule. *J.Chem.Doc.*, *11*, 258–259.

Balaban, A. T. ed. (1976a). *Chemical Applications of Graph Theory.* Academic Press, New York (NY), 390 pp.

Balaban, A.T. (1976b). Chemical Graphs. XXVI. Codes for Configurations of Conjugated Polyenes. *Rev.Roum.Chim.*, *21*, 1045–1047.

Balaban, A.T. (1976c). Chemical Graphs. XXVII. Enumeration and Codification of Staggered Conformations of Alkanes. *Rev.Roum.Chim.*, *21*, 1049–1071.

Balaban, A.T. (1976d). Enumeration of Cyclic Graphs. In *Chemical Applications of Graph Theory* (Balaban, A.T., ed.), Academic Press, London (UK), pp. 63–105.

Balaban, A.T. and Harary, F. (1976). Early History of the Interplay Between Graph Theory and Chemistry. In *Chemical Applications of Graph Theory* (Balaban, A.T., ed.), Academic Press, London (UK), pp. 1–4.

Balaban, A.T. (1978). Chemical Graphs. XXXII. Constitutional and Steric Isomers of Substituted Cycloalkanes. *Croat.Chem.Acta*, *51*, 35–42.

Balaban, A.T. (1979). Chemical Graphs. XXXIV. Five New Topological Indices for the Branching of Tree-Like Graphs. *Theor.Chim.Acta*, *53*, 355–375.

Balaban, A.T. and Motoc, I. (1979). Chemical Graphs. XXXVI. Correlations Between Octane Numbers and Topological Indices of Alkanes. *MATCH (Comm.Math.Comp.Chem.)*, *5*, 197–218.

Balaban, A.T. (1980). Chemical Graphs. XXXVIII. Synthon Graphs. *MATCH (Comm.Math.Comp.Chem.)*, *8*, 159–192.

Balaban, A.T., Barasch, M. and Marcus, S. (1980a). Computer Program for the Recognition of Standard Isoprenoid Structures. *MATCH (Comm.Math.Comp.Chem.)*, *8*, 215–268.

Balaban, A.T., Barasch, M. and Marcus, S. (1980b). Picture Grammars in Chemistry. Generation of Acyclic Isoprenoid Structures. *MATCH (Comm.Math.Comp.Chem.)*, *8*, 193–213.

Balaban, A.T., Chiriac, A., Motoc, I. and Simon, Z. (1980c). *Steric Fit in Quantitative Structure-Activity Relations.* Springer-Verlag, Berlin (Germany), pp. 51–77.

Balaban, A.T. and Rouvray, D.E. (1980). Graph-Theoretical Analysis of the Bonding Topology in Polyhedral Organic Cations. *Tetrahedron*, *36*, 1851–1855.

Balaban, A.T. (1982). Highly Discriminating Distance-Based Topological Index. *Chem.Phys.Lett.*, *89*, 399–404.

Balaban, A.T. (1983a). Topological Indices Based on Topological Distances in Molecular Graphs. *Pure & Appl.Chem.*, *55*, 199–206.

Balaban, A.T. (1983b). Topological Indices: What They Are and What They Can Do. In *Proc. 2nd Balkan Chemistry Days*, Varna.

Balaban, A.T., Motoc, I., Bonchev, D. and Mekenyan, O. (1983). Topological Indices for Structure-Activity Correlations. In *Steric Effects in Drug Design (Topics in Current Chemistry, Vol. 114)* (Charton, M. and Motoc, I., eds.), Springer-Verlag, Berlin (Germany), pp. 21–55.

Balaban, A.T. and Quintar, L.B. (1983). The Smallest Graphs, Trees and 4-Trees with Degenerate Topological Index J. *MATCH (Comm.Math.Comp.Chem.)*, *14*, 213–233.

Balaban, A.T. (1984). Numerical and Non-numerical Methods in Chemistry: Present and Future. *Sigsam Bull.*, *18*, 29–30.

Balaban, A.T. and Filip, P.A. (1984). Computer Program for Topological Index J (Average Distance Sum Connectivity). *MATCH (Comm.Math.Comp.Chem.)*, *16*, 163–190.

Balaban, A.T. (1985a). Applications of Graph Theory in Chemistry. *J.Chem.Inf.Comput.Sci.*, *25*, 334–343.

Balaban, A.T. (1985b). Graph Theory and Theoretical Chemistry. *J.Mol.Struct.(Theochem)*, *120*, 117–142.

Balaban, A.T. (1985c). Symbolic Computation and Chemistry. In *EUROCAL-85* (Buchberger, B., ed.), Springer, Berlin (Germany), pp. 68–79.

Balaban, A.T., Biermann, D. and Schmidt, W. (1985c). Dualist Graph Approach for Correlating Diels-Alder Reactivities of Polycyclic Aromatic Hydrocarbons. *Nouv.J.Chim.*, *9*, 443–449.

Balaban, A.T., Filip, P.A. and Balaban, T.-S. (1985d). Computer Program for Finding all Possible Cycles in Graphs. *J.Comput.Chem.*, *6*, 316–329.

Balaban, A.T., Ionescu-Pallas, N. and Balaban, T.-S. (1985e). Asymptotic Values of Topological Indices J and J′ (Average Distance Sum Connectivities) for Infinite Acyclic and Cyclic Graphs. *MATCH (Comm.Math.Comp.Chem.)*, *17*, 121–146.

Balaban, A.T., Mekenyan, O. and Bonchev, D. (1985a). Unique Description of Chemical Structures Based on Hierarchically Ordered Extended Connectivities (HOC Procedures). I. Algorithms for Finding Graph Orbits and Canonical Numbering of Atoms. *J.Comput.Chem.*, *6*, 538–551.

Balaban, A.T., Mekenyan, O. and Bonchev, D. (1985b). Unique Description of Chemical Structures Based on Hierarchically Ordered Extended Connectivities (HOC Procedures). III. Topological, Chemical, and Stereochemical Coding of Molecular Structure. *J.Comput.Chem.*, *6*, 562–569.

Balaban, A.T. and Tomescu, T. (1985). Chemical Graphs. XLI. Numbers of Conjugated Circuits and Kekulé Structures for Zigzag Catafusenes and (j, k)-hexes; Generalized Fibonacci Numbers. *MATCH (Comm.Math.Comp.Chem.)*, *17*, 91–120.

Balaban, A.T. (1986a). Chemical Graphs. Part 48. Topological Index J for Heteroatom-Containing Molecules Taking into Account Periodicities of Element Properties. *MATCH (Comm.Math.Comp.Chem.)*, *21*, 115–122.

Balaban, A.T. (1986b). Symmetry in Chemical Structures and Reactions. In *Symmetry Unifying Human Understanding* (Hargittai, I., ed.), Pergamon Press, New York (NY), pp. 999–1020.

Balaban, A.T. (1987a). Numerical Modelling of Chemical Structures: Local Graph Invariants and Topological Indices. In *Graph Theory and Topology in Chemistry* (King, R.B. and Rouvray, D.H., eds.), Elsevier, Amsterdam (The Netherlands), pp. 159–176.

Balaban, A.T. (1987b). *Phys.Theor.Chem.*, *51*, 159.

Balaban, A.T. and Artemi, C. (1987). Mathematical Modeling of Polymers. Part I. Enumeration of Non Redundant (Irreducible) Repeating Sequences in Stereoregular Polymers, Elastomers, or in Binary Copolymers. *MATCH (Comm.Math.Comp.Chem.)*, *22*, 3–32.

Balaban, A.T., Brunvoll, J., Cioslowski, J., Cyvin, B.N., Syvin, S.J., Gutman, I., Wenchen, H., Wenjie, H., Knop, J.V., Kovacevic, M., Müller, W.R., Szymanski, K., Tosic, R. and Trinajstic, N. (1987a). Enumeration of Benzenoid and Coronoid Hydrocarbons. *Z.Naturforsch.*, *42C*, 863–870.

Balaban, A.T., Niculescu-Duvaz, I. and Simon, Z. (1987b). Topological Aspects in QSAR for Biologically Active Molecules. *Acta Pharm.Jugosl.*, *37*, 7–36.

Balaban, A.T. (1988a). Chemical Graphs. Part 49. Open Problems in the Area of Condensed Polycyclic Benzenoids: Topological Stereoisomers of Coronoids and Congeners. *Rev.Roum.-Chim.*, *33*, 699–707.

Balaban, A.T. (1988b). Modeling the Chemical Structures Through Graphs & of These Graphs Through Topological Indices. *Rev.Chimie (Bucarest)*, *39*, 1026–1031.

Balaban, A.T. (1988c). Topological Indices and their Uses: a New Approach for the Coding of Alkanes. *J.Mol.Struct.(Theochem)*, *165*, 243–253.

Balaban, A.T., Kennedy, J.W. and Quintas, L.V. (1988). The Number of Alkanes Having n Carbons and a Longest Chain of Length d. An Application of a Theorem of Polya. *J.Chem.Educ.*, *65*, 304–313.

Balaban, A.T. and Tomescu, T. (1988). Alternating 6-Cycles in Perfect Matchings of Graphs Representing Condensed Benzenoid Hydrocarbons. In *Application of Graphs in Chemistry and Physics* (Kennedy, J.W. and Quintas, L.V., eds.), North-Holland, Amsterdam (The Netherlands), pp. 5–16.

Balaban, A.T. and Artemi, C. (1989). Chemical Graphs. Part 51. Enumeration of Nonbranched Catafusenes According to the Numbers of Benzenoid Rings in the Catafusene and Its Longest Linearly Condensed Portion. *Polycyclic Aromatic Compounds*, *1*, 171–189.

Balaban, A.T. and Brocas, J. (1989). Modes of Rearrangements and Reaction Graphs for XeF_6. *J.Mol.Struct.(Theochem)*, *185*, 139–153.

Balaban, A.T. and Ivanciuc, O. (1989). FORTRAN-77 Computer Program for Calculating Topological Index J for Molecules Containing Heteroatoms. In *MATH / CHEM / COMP 1988* (Graovac, A., ed.), Elsevier, Amsterdam (The Netherlands), pp. 193–211.

Balaban, A.T., Catana, C., Dawson, M. and Niculescu-Duvaz, I. (1990a). Applications of Weighted Topological Index J for QSAR of Carcinogenesis Inhibitors (Retinoic Acid Derivatives). *Rev.Roum.Chim.*, *35*, 997–1003.

Balaban, A.T., Ciubotariu, D. and Ivanciuc, O. (1990b). Design of Topological Indices. Part 2. Distance Measure Connectivity Indices. *MATCH (Comm.Math.Comp.Chem.)*, *25*, 41–70.

Balaban, A.T. and Feroiu, V. (1990). Correlation Between Structure and Critical Data of Vapor Pressures of Alkanes by Means of Topological Indices. *Rep.Mol.Theory*, *1*, 133–139.

Balaban, A.T. (1991). Enumeration of Isomers. In *Chemical Graph Theory. Introduction and Fundamentals* (Bonchev, D. and Rouvray, D.E., eds.), Abacus Press/Gordon & Breach, New York (NY), pp. 177–234.

Balaban, A.T. and Balaban, T.-S. (1991). New Vertex Invariants and Topological Indices of Chemical Graphs Based on Information on Distances. *J.Math.Chem.*, *8*, 383–397.

Balaban, A.T., Ciubotariu, D. and Medeleanu, M. (1991). Topological Indices and Real Vertex Invariants Based on Graph Eigenvalues or Eigenvectors. *J.Chem.Inf.Comput.Sci.*, *31*, 517–523.

Balaban, A.T. (1992). Using Real Numbers as Vertex Invariants for Third-Generation Topological Indexes. *J.Chem.Inf.Comput.Sci.*, *32*, 23–28.

Balaban, A.T. and Balaban, T.-S. (1992). Correlation Using Topological Indices Based on Real Graph Invariants. *J.Chim.Phys.*, *89*, 1735–1745.

Balaban, A.T., Joshi, N., Kier, L.B. and Hall, L.H. (1992). Correlations Between Chemical Structure and Normal Boiling Points of Halogenated Alkanes C_1–C_4. *J.Chem.Inf.Comput.Sci.*, *32*, 233–237.

Balaban, A.T., Kier, L.B. and Joshi, N. (1992a). Correlations Between Chemical Structure and Normal Boiling Points of Acyclic Ethers, Peroxides, Acetals, and their Sulfur Analogues. *J.Chem.Inf.Comput.Sci.*, *32*, 237–244.

Balaban, A.T., Kier, L.B. and Joshi, N. (1992b). Structure-Property Analysis of Octane Numbers for Hydrocarbons (Alkanes, Cycloalkanes, Alkenes). *MATCH (Comm.Math.Comp.Chem.)*, *28*, 13–27.

Balaban, A.T. (1993a). Benzenoid Catafusenes: Perfect Matchings, Isomerization, Automerization. *Pure & Appl.Chem.*, *65*, 1–9.

Balaban, A.T. (1993b). Confessions and Reflections of a Graph-Theoretical Chemist. *MATCH (Comm.Math.Comp.Chem.)*, *29*, 3–17.

Balaban, A.T. (1993c). Lowering the Intra- and Intermolecular Degeneracy of Topological Invariants. *Croat.Chem.Acta*, *66*, 447–458.

Balaban, A.T. (1993d). Prediction of Physical Properties from Chemical Structures. In *Recent Advances in Chemical Information* (Collier, H., ed.), Royal Society of Chemistry, Cambridge (UK), pp. 301–317.

Balaban, A.T. (1993e). Solved and Unsolved Problems in Chemical Graph Theory. *Ann.Disc. Math.*, *55*, 109–126.

Balaban, A.T., Bonchev, D. and Seitz, W.A. (1993a). Topological/Chemical Distances and Graph Centers in Molecular Graphs with Multiple Bonds. *J.Mol.Struct.(Theochem)*, *280*, 253–260.

Balaban, A.T. and Catana, C. (1993). Search for Nondegenerate Real Vertex Invariants and Derived Topological Indexes. *J.Comput.Chem.*, *14*, 155–160.

Balaban, A.T. and Diudea, M.V. (1993). Real Number Vertex Invariants: Regressive Distance Sums and Related Topological Indices. *J.Chem.Inf.Comput.Sci.*, *33*, 421–428.

Balaban, A.T., Liu, X., Cyvin, S.J. and Klein, D.J. (1993b). Benzenoids with Maximum Kekule Structure Counts for Given Numbers of Hexagons. *J.Chem.Inf.Comput.Sci.*, *33*, 429–436.

Balaban, A.T. (1994a). Local vs. Global (*i.e.* Atomic versus Molecular) Numerical Modeling of Molecular Graphs. *J.Chem.Inf.Comput.Sci.*, *34*, 398–402.

Balaban, A.T. (1994b). Reaction Graphs. In *Graph Theoretical Approaches to Chemical Reactivity* (Bonchev, D. and Mekenyan, O., eds.), Kluwer, Dordrecht (The Netherlands), pp. 137–180.

Balaban, A.T. (1994c). Real-Number Local (Atomic) Invariants and Global (Molecular) Topological Indices. *Rev.Roum.Chim.*, *39*, 245–257.

Balaban, A.T., Basak, S.C., Colburn, T. and Grunwald, G.D. (1994a). Correlation Between Structure and Normal Boiling Points of Haloalkanes C_1-C_4 Using Neural Networks. *J.Chem.Inf. Comput.Sci.*, *34*, 1118–1121.

Balaban, A.T., Bertelsen, S. and Basak, S.C. (1994b). New Centric Topological Indexes for Acyclic Molecules (Trees) and Substituents (Rooted Trees), and Coding of Rooted Trees. *MATCH (Comm.Math.Comp.Chem.)*, *30*, 55–72.

Balaban, A.T. and Catana, C. (1994). New Topological Indices for Substituents (Molecular Fragments). *SAR & QSAR Environ.Res.*, *2*, 1–16.

Balaban, A.T. (1995a). Chemical Graphs: Looking Back and a Glimpsing Ahead. *J.Chem.Inf. Comput.Sci.*, *35*, 339–350.

Balaban, A.T. (1995b). Local (Atomic) and Global (Molecular) Graph-Theoretical Descriptors. *SAR & QSAR Environ.Res.*, *3*, 81–95.

Balaban, A.T., Liu, X., Klein, D.J., Babic, D., Schmalz, T.G., Seitz, W.A. and Randic, M. (1995). Graph Invariants for Fullerenes. *J.Chem.Inf.Comput.Sci.*, *35*, 396–404.

Balaban, A.T. (1997a). From Chemical Graphs to 3D Molecular Modeling. In *From Chemical Topology to Three-Dimensional Geometry* (Balaban, A.T., ed.), Plenum Press, New York (NY), pp. 1–24.

Balaban, A.T. (1997b). From Chemical Topology to 3D Geometry. *J.Chem.Inf.Comput.Sci.*, *37*, 645–650.

Balaban, A. T. ed. (1997). *From Chemical Topology to Three-Dimensional Geometry.* Plenum Press, New York (NY), 420 pp.

Balaban, A.T., Ivanciuc, O. and Babic, D. (1997). Correlation Between Energies of Proper Fullerenes and their Topological Invariants – Part 1 – Fullerenes with Abutting Pentagons. *Fullerene Sci.Technol.*, *5*, 1479–1506.

Balaban, A.T. (1998). Topological and Stereochemical Molecular Descriptors for Databases Useful in QSAR, Similarity/Dissimilarity and Drug Design. *SAR & QSAR Environ.Res.*, *8*, 1–21.

Balaban, A.T. and Artemi, C. (1998). Mathematical Modeling of Polymers. 3. Enumeration and Generation of Repeating Irriducible Sequences in Linear Bi-, Ter-, Quater-, and Quinquenary Copolymers and in Stereoregular Homopolymers. *Macromol.Chem.*, *189*, 863–870.

Balaban, A.T., Mills, D. and Basak, S.C. (1999). Correlation Between Structure and Normal Boiling Points of Acyclic Carbonyl Compounds. *J.Chem.Inf.Comput.Sci.*, *39*, 758–764.

Balaban, A.T. and Ivanciuc, O. (2000). Historical Development of Topological Indices. In *Topological Indices and Related Descriptors in QSAR and QSPR* (Devillers, J. and Balaban, A.T., eds.), Gordon & Breach, Amsterdam (The Netherlands), pp. 21–57.

Balaban, T.-S., Filip, P. and Ivanciuc, O. (1992). Computer Generation of Acyclic Graphs Based on Local Vertex Invariants and Topological Indices. Derived Canonical Labelling and Coding of Trees and Alkanes. *J.Math.Chem.*, *11*, 79–105.

Balasubramanian, K., Kaufmann, J.J., Koski, W.S. and Balaban, A.T. (1980). Graph Theoretical Characterization and Computer Generation of Certain Carcinogenic Benzenoid Hydrocarbons and Identification of Bay Regions. *J.Comput.Chem.*, *1*, 149–157.

Balasubramanian, K. (1982). Spectra of Chemical Trees. *Int.J.Quant.Chem.*, *21*, 581–590.

Balasubramanian, K. and Randic, M. (1982). The Characteristic Polynomials of Structures with Pending Bonds. *Theor.Chim.Acta*, *61*, 307–323.

Balasubramanian, K. (1984a). Computer Generation of the Characteristic Polynomial of Chemical Graph. *J.Comput.Chem.*, *5*, 387–394.

Balasubramanian, K. (1984b). The Use of Frame's Method for the Characteristic Polynomials of Chemical Graphs. *Theor.Chim.Acta*, *65*, 49–58.

Balasubramanian, K. (1990). Geometry-Dependent Characteristic Polynomials of Molecular Structures. *Chem.Phys.Lett.*, *169*, 224–228.

Balasubramanian, K. (1991). Graph Theory and the PPP Method. *J.Math.Chem.*, *7*, 353–362.

Balasubramanian, K. (1994). Integration of Graph Theory and Quantum Chemistry for Structure-Activity Relationships. *SAR & QSAR Environ.Res.*, *2*, 59–77.

Balasubramanian, K. (1995a). Computer Generation of Nuclear Equivalence Classes Based on the Three-Dimensional Molecular Structure. *J.Chem.Inf.Comput.Sci.*, *35*, 243–250.

Balasubramanian, K. (1995b). Geometry-Dependent Connectivity Indices for the Characterization of Molecular Structures. *Chem.Phys.Lett.*, *235*, 580–586.

Balasubramanian, K. and Basak, S.C. (1998). Characterization of Isospectral Graphs Using Graph Invariants and Derived Orthogonal Parameters. *J.Chem.Inf.Comput.Sci.*, *38*, 367–373.

Balàz, S., Wiese, M., Chi, H.-L. and Seydel, J.K. (1990). Subcellular Pharmacokinetics and Quantitative Structure/Time/Activity Relationships. *Anal.Chim.Acta*, *235*, 195–207.

Ballistreri, A., Bottino, A., Musumarra, G., Fioravanti, R., Biava, M., Porretta, G.C., Simonetti, N. and Villa, A. (1996). Design, Synthesis and Antimycotic Activity of (N-Heteroaryl)Aryemethanamines. *J.Phys.Org.Chem.*, *9*, 61–65.

Balogh, T. and Naray-Szabo, G. (1993). Application of the Average Molecular Electrostatic Field in Quantitative Structure-Activity Relationships. *Croat.Chem.Acta*, *66*, 129–140.

Bangov, I.P. (1988). Use of the ^{13}C-NMR Chemical Shift/Charge Density Linear Relationship for Recognition and Ranking of Chemical Structures. *Anal.Chim.Acta*, *209*, 29.

Bangov, I.P. (1990). Computer-Assisted Structure Generation from a Gross Formula. 3. Alleviation of the Combinatorial Problem. *J.Chem.Inf.Comput.Sci.*, *30*, 277–289.

Bangov, I.P. (1992). Toward the Solution of the Isomorphism Problem in Generation of Chemical Graphs: Generation of Benzenoid Hydrocarbons. *J.Chem.Inf.Comput.Sci.*, *32*, 167–173.

Barbe, J. (1983). Convenient Relations for the Estimation of Bond Ionicity in A-B Type Compounds. *J.Chem.Educ.*, *60*, 640–642.

Barlocco, D., Fanelli, F., Cignarella, G., Villa, S., Cattabeni, F., Balduini, W., Cimino, M. and De Benedetti, P.G. (1996). 2-(Substituted)amino-2,8-diazaspiro[4,5]decan-1,3-diones as Potential Muscarinic Agonists: Synthesis, Modeling and Binding Studies. *Drug Design & Discovery*, *14*, 129–143.

Barlow, T.W. (1995). Self-Organizing Maps and Molecular Similarity. *J.Mol.Graphics*, *13*, 24–27.

Barnard, J.M. (1993). Substructure Searching Methods: Old and New. *J.Chem.Inf.Comput.Sci.*, *33*, 532–538.

Barnard, J.M. (1994). Third International Conference on Chemical Structures. *J.Chem.Inf.Comput.Sci.*, *34*, 1–2.

Baroni, M., Clementi, S., Cruciani, G., Costantino, G., Riganelli, D. and Oberrauch, E. (1992). Predictive Ability of Regression Models. Part II: Selection of the Best Predictive PLS Model. *J.Chemom.*, *6*, 347–356.

Baroni, M., Clementi, S., Cruciani, G., Kettaneh-Wold, N. and Wold, S. (1993a). D-Optimal Designs in QSAR. *Quant.Struct.-Act.Relat.*, *12*, 225–231.

Baroni, M., Costantino, G., Cruciani, G., Riganelli, D., Valigi, R. and Clementi, S. (1993b). Generating Optimal Linear PLS Estimations (GOLPE): An Advanced Chemometric Tool for Handling 3D-QSAR Problems. *Quant.Struct.-Act.Relat.*, *12*, 9–20.

Baroni, M., Costantino, G., Cruciani, G., Riganelli, D., Valigi, R. and Clementi, S. (1993c). GOLPE: An Advanced Chemometric Tool for 3D QSAR Problems. In *Trends in QSAR and Molecular Modelling 92* (Wermuth, C.G., ed.), ESCOM, Leiden (The Netherlands), pp. 256–259.

Barratt, M.D. (1994a). A Quantitative Structure-Activity Relationship (QSAR) for Prediction of α-2μ-Globulin Nephropathy. *Quant.Struct.-Act.Relat.*, *13*, 275–280.

Barratt, M.D., Basketter, D.A. and Roberts, D.W. (1994b). Skin Sensitization Structure-Activity Relationships for Phenyl Benzoates. *Toxicol.Vitro*, *8*, 823–826.

Bartell, L.S. (1963). Resonance Energies from Pauling Bond Orders. *J.Phys.Chem.*, *67*, 1865–1868.

Bartell, L.S. (1964). Semiquantitative Theory of Resonance Using Pauling Bond Orders. *Tetrahedron*, *20*, 139–153.

Barysz, M., Jashari, G., Lall, R.S., Srivastava, A.K. and Trinajstic, N. (1983a). On the Distance Matrix of Molecules Containing Heteroatoms. In *Chemical Applications of Topology and Graph Theory* (King, R.B., ed.), Elsevier, Amsterdam (The Netherlands), pp. 222–230.

Barysz, M., Trinajstic, N. and Knop, J.V. (1983b). On the Similarity of Chemical Structures. *Int.J.Quantum Chem.Quant.Chem.Symp.*, *17*, 441–451.

Barysz, M. and Trinajstic, N. (1984). A Novel Approach to the Characterization of Chemical Structures. *Int.J.Quantum Chem.Quant.Chem.Symp.*, *18*, 661–673.

Barysz, M., Knop, J.V., Pajakovic, S. and Trinajstic, N. (1985). Characterization of Branching. *Pol.J.Chem.*, *59*, 405–432.

Barysz, M., Bonchev, D. and Mekenyan, O. (1986a). Graph-Centre, Self-Returning Walks, and Critical Pressure of Alkanes. *MATCH (Comm.Math.Comp.Chem.)*, *20*, 125–140.

Barysz, M., Plavsic, D. and Trinajstic, N. (1986b). A Note on Topological Indices. *Math.Chem. (Mulcheim/Ruhr)*, *19*, 89–116.

Basak, S.C., Roy, A.B. and Ghosh, J.J. (1980). Study of the Structure-Function Relationship of Pharmacological and Toxicological Agents Using Information Theory. In *Proceedings of the Second International Conference on Mathematical Modelling* (Avula, X.J.R., Bellman, R., Luke, Y.L. and Rigler, A.K., eds.), University of Missouri, Rolla (MS), pp. 851–856.

Basak, S.C., Gieschen, D.P., Magnuson, V.R. and Harriss, D.K. (1982). Structure-Activity Relationships and Pharmacokinetics: A Comparative Study of Hydrophobicity, van der Waals' Volume and Topological Parameters. *Med.Sci.Res.*, *10*, 619–620.

Basak, S.C., Gieschen, D.P., Harriss, D.K. and Magnuson, V.R. (1983). Physico-Chemical and Topological Correlates of the Enzymatic Acetyltransfer Reaction. *J.Pharm.Sci.*, *72*, 934–937.

Basak, S.C. and Magnuson, V.R. (1983). Molecular Topology and Narcosis. A Quantitative Structure-Activity Relationship (QSAR) Study of Alcohols Using Complementary Information Content (CIC). *Arzneim.Forsch.*, *33*, 501–503.

Basak, S.C., Gieschen, D.P. and Magnuson, V.R. (1984a). A Quantitative Correlation of the LC_{50} Values of Esters in *Pimephales promelas* Using Physicochemical and Topological Parameters. *Environ.Toxicol.Chem.*, *3*, 191–199.

Basak, S.C., Harriss, D.K. and Magnuson, V.R. (1984b). Comparative Study of Lipophilicity *versus* Topological Molecular Descriptors in Biological Correlations. *J.Pharm.Sci.*, *73*, 429–437.

Basak, S.C., Monsrud, L.J., Rosen, M.E., Frane, C.M. and Magnuson, V.R. (1986a). A Comparative Study of Lipophilicity and Topological Indices in Biological Correlation. *Acta Pharm. Jugosl.*, *36*, 81–95.

Basak, S.C., Rosen, M.E. and Magnuson, V.R. (1986b). Molecular Topology and Mutagenicity: A QSAR Study of Nitrosoamines. *Med.Sci.Res.*, *14*, 848–849.

Basak, S.C. (1987). Use of Molecular Complexity Indices in Predictive Pharmacology and Toxicology: a QSAR Approach. *Med.Sci.Res.*, *15*, 605–609.

Basak, S.C., Frane, C.M., Rosen, M.E. and Magnuson, V.R. (1987a). Molecular Topology and Acute Toxicity: A QSAR Study of Monoketones. *Med.Sci.Res.*, *15*, 887–888.

Basak, S.C., Magnuson, V.R. and Veith, G.D. (1987b). Topological Indices: Their Nature, Mutual Relatedness, and Applications. In *Mathematical Modelling in Science and Technology* (Avula, X.J.R., Leitmann, G., Mote, C.D.Jr. and Rodin, E.Y., eds.), Pergamon Press, Oxford (UK), pp. 300–305.

Basak, S.C. (1988). Binding of Barbiturates to Cytochrome P_{450}: A QSAR Study Using log P and Topological Indices. *Med.Sci.Res.*, *16*, 281–282.

Basak, S. C., Harriss, D. K., and Magnuson, V. R. (1988a). *POLLY 2.3*. Minnesota University of Duluth.

Basak, S.C., Magnuson, V.R., Niemi, G.J. and Regal, R.R. (1988b). Determining Structural Similarity of Chemicals Using Graph-Theoretic Indices. *Disc.Appl.Math.*, *19*, 17–44.

Basak, S.C. (1990). A Nonempirical Approach to Predicting Molecular Properties Using Graph-Theoretic Invariants. In *Practical Applications of Quantitative Structure-Activity Relationships (QSAR) in Environmental Chemistry and Toxicology* (Karcher, W. and Devillers, J., eds.), Kluwer, Dordrecht (The Netherlands), pp. 83–103.

Basak, S.C., Niemi, G.J. and Veith, G.D. (1990a). A Graph Theoretic Approach to Predicting Molecular Properties. *Math.Comput.Modelling*, *14*, 511–516.

Basak, S.C., Niemi, G.J. and Veith, G.D. (1990b). Optimal Characterization of Structure for Prediction of Properties. *J.Math.Chem.*, *4*, 185–205.

Basak, S.C., Niemi, G.J. and Veith, G.D. (1990c). Recent Developments in the Characterization of Chemical Structure Using Graph-Theoretic Indices. In *Computational Chemical Graph Theory and Combinatorics* (Rouvray, D.H., ed.), Nova, New York (NY), pp. 235–277.

Basak, S.C., Niemi, G.J. and Veith, G.D. (1991). Predicting Properties of Molecules Using Graph Invariants. *J.Math.Chem.*, *7*, 243–272.

Basak, S.C. and Grunwald, G.D. (1993). Use of Graph Invariants, Volume and Total Surface Area in Predicting Boiling Point of Alkanes. *Math.Modelling and Sci.Computing*, *2*, 735–740.

Basak, S.C., Bertelsen, S. and Grunwald, G.D. (1994). Application of Graph Theoretical Parameters in Quantifying Molecular Similarity and Structure-Activity Relationships. *J.Chem.Inf. Comput.Sci.*, *34*, 270–276.

Basak, S.C. and Grunwald, G.D. (1994). Molecular Similarity and Risk Assessment: Analog Selection and Property Estimation Using Graph Invariants. *SAR & QSAR Environ.Res.*, *2*, 289–307.

Basak, S.C., Bertelsen, S. and Grunwald, G.D. (1995). Use of Graph Theoretic Parameters in Risk Assessment of Chemicals. *Toxicol.Lett.*, *79*, 239–250.

Basak, S.C. and Grunwald, G.D. (1995a). Estimation of Lipophilicity from Structural Similarity. *New J.Chem.*, *19*, 231–237.

Basak, S.C. and Grunwald, G.D. (1995b). Molecular Similarity and Estimation of Molecular Properties. *J.Chem.Inf.Comput.Sci.*, *35*, 366–372.

Basak, S.C. and Grunwald, G.D. (1995c). Predicting Mutagenicity of Chemicals Using Topological and Quantum Chemical Parameters: A Similarity Based Study. *Chemosphere*, *31*, 2529–2546.

Basak, S.C. and Grunwald, G.D. (1995d). Tolerance Space and Molecular Similarity. *SAR & QSAR Environ.Res.*, *3*, 265–277.

Basak, S.C., Gute, B.D. and Drewes, L.R. (1996). Predicting Blood-Brain Transport of Drugs: A Computational Approach. *Pharm.Res.*, *13*, 775–778.

Basak, S.C., Gute, B.D. and Grunwald, G.D. (1996a). A Comparative Study of Topological and Geometrical Parameters in Estimating Normal Boiling Point and Octanol/Water Partition Coefficient. *J.Chem.Inf.Comput.Sci.*, *36*, 1054–1060.

Basak, S.C., Gute, B.D. and Grunwald, G.D. (1996b). Estimation of the Normal Boiling Points of Haloalkanes Using Molecular Similarity. *Croat.Chem.Acta*, *69*, 1159–1173.

Basak, S.C., Grunwald, G.D. and Niemi, G.J. (1997a). Use of Graph-Theoretic and Geometrical Molecular Descriptors in Structure-Activity Relationships. In *From Chemical Topology to Three-Dimensional Geometry* (Balaban, A.T., ed.), Plenum Press, New York (NY), pp. 73–116.

Basak, S.C. and Gute, B.D. (1997b). Characterization of Molecular Structures Using Topological Indices. *SAR & QSAR Environ.Res.*, *7*, 1–21.

Basak, S.C., Gute, B.D. and Grunwald, G.D. (1997c). Use of Topostructural, Topochemical, and Geometric Parameters in the Prediction of Vapor Pressure: A Hierarchical QSAR Approach. *J.Chem.Inf.Comput.Sci.*, *37*, 651–655.

Basak, S.C., Gute, B.D. and Grunwald, G.D. (1998). The Relative Effectiveness of Topological, Geometrical and Quantum Chemical Parameters in Estimating Mutagenicity of Chemicals. In *QSAR in Environmental Sciences – Vol. VII* (Chen, F. and *et al.*, eds.), SETAC Press, Pensacola (FL), pp. 245–261.

Basak, S.C., Gute, B.D. and Ghatak, S. (1999). Prediction of Complement-Inhibitory Activity of Benzamidines Using Topological and Geometric Parameters. *J.Chem.Inf.Comput.Sci.*, *39*, 255–260.

Basak, S.C. (2000). Information Theoretic Indices of Neighborhood Complexity and Their Applications. In *Topological Indices and Related Descriptors in QSAR and QSPR* (Devillers, J. and Balaban, A.T., eds.), Gordon & Breach, Amsterdam (The Netherlands), pp. 563–593.

Basilevsky, A. (1994). *Statistical Factor Analysis and Related Methods.* Wiley, New York (NY), 738 pp.

Baskin, I.I., Gordeeva, E.V., Devdariani, R.O., Zefirov, N.S., Palyulin, V.A. and Stankevitch, I.V. (1989). Solving the Inverse Problem of Structure-Property Relations for the Case of Topological Indexes. *Dokl.Akad.Nauk.SSSR*, *307*, 613–617.

Baskin, I.I., Skvortsova, M.I., Stankevitch, I.V. and Zefirov, N.S. (1995). On the Basis of Invariants of Labeled Molecular Graphs. *J.Chem.Inf.Comput.Sci.*, *35*, 527–531.

Baskin, I.I., Palyulin, V.A. and Zefirov, N.S. (1997). A Neural Device for Searching Direct Correlations Between Structures and Properties of Chemical Compounds. *J.Chem.Inf.Comput.Sci.*, *37*, 715–721.

Bate-Smith, E.C. and Westall, R.G. (1950). Chromatographic Behaviour and Chemical Structure. I. Some Naturally Occurring Phenolic Substances. *Biochim.Biophys.Acta*, *4*, 427–440.

Bath, P.A., Poirrette, A.R., Willett, P. and Allen, F.H. (1994). Similarity Searching in Files of Three-Dimensional Chemical Structures: Comparison of Fragment-Based Measures of Shape Similarity. *J.Chem.Inf.Comput.Sci.*, *34*, 141–147.

Bath, P.A., Poirrette, A.R., Willett, P. and Allen, F.H. (1995). The Extent of the Relationship Between the Graph-Theoretical and the Geometrical Shape Coefficients of Chemical Compounds. *J.Chem.Inf.Comput.Sci.*, *35*, 714–716.

Batsanov, S.S. and Durakov, V.I. (1960). *Zhur.Strukt.Khim.*, *1*, 353.

Battershell, C., Malhotra, D. and Hopfinger, A.J. (1981). Inhibition of Dihydrofolate Reductase: Structure-Activity Correlations of Quinazolines Based upon Molecular Shape Analysis. *J.Med.Chem.*, *24*, 812–818.

Bauerschmidt, S. and Gasteiger, J. (1997). Overcoming the Limitations of a Connection Table Description: A Universal Representation of Chemical Species. *J.Chem.Inf.Comput.Sci.*, *37*, 705–714.

Bauknecht, H., Zell, A., Bayer, H., Levi, P., Wagener, M., Sadowski, J. and Gasteiger, J. (1996). Locating Biologically Active Compounds in Medium-Sized Heterogeneous Datasets by Topological Autocorrelation Vectors: Dopamine and Benzodiazepine Agonists. *J.Chem.Inf.Comput.Sci.*, *36*, 1205–1213.

Baum, E.J. (1998). *Chemical Property Estimation: Theory and Application.* Lewis Publishers, Boca Raton (FL).

Baumann, K., Affolter, C., Pretsch, E. and Clerc, J.T. (1997a). Numerical Structure Representation and IR Spectra Prediction. *Mikrochim.Acta*, *14*, 275–276.

Baumann, K. and Clerc, J.T. (1997b). Computer-Assisted IR Spectra Prediction – Linked Similarity Searches for Structures and Spectra. *Anal.Chim.Acta*, *348*, 327–343.

Baumann, K. (1999). Uniform-Length Molecular Descriptors for Quantitative Structure-Property Relationships (QSPR) and Quantitative Structure-Activity Relationships (QSAR): Classification Studies and Similarity Searching. *TRAC*, *18*, 36–46.

Baumer, L., Sala, G. and Sello, G. (1989). Residual Charges on Atoms in Organic Structures: Molecules Containing Charged and Backdonating Atoms. *Tetrahedron Comput.Methodol.*, *2*, 105–118.

Bawden, D. (1983). Computerized Chemical Structure-Handling Techniques in Structure-Activity Studies and Molecular Property Prediction. *J.Chem.Inf.Comput.Sci.*, *23*, 14–22.

Bawden, D. (1990). Applications of Two-Dimensional Chemical Similarity Measures to Database Analysis and Querying. In *Concepts and Applications of Molecular Similarity* (Johnson, M.A. and Maggiora, G.M., eds.), Wiley, New York (NY), pp. 65–76.

Baxter, S.J., Jenkins, H.D.B. and Samuel, C.J. (1996). A Novel Computational Approach to the Estimation of Steric Parameters. III. Extension to Aliphatic Amines and Application to the Adrenergic Blocking Activity of β-Haloalkylamine. *Tetrahedron Lett.*, *37*, 4617–4620.

Bayada, D.M., Hemersma, H. and van Geerestein, V.J. (1999). Molecular Diversity and Representativity in Chemical Databases. *J.Chem.Inf.Comput.Sci.*, *39*, 1–10.

Bazylak, G. (1994). Differentiation of Alkanolamines Properties by Multivariate Analysis of Database Founded by Their Molecular Parameters and Chromatographic Measurements Results. *Chem.Anal.*, *39*, 295–308.

Bearden, A.P. and Schultz, T.W. (1997). Structure-Activity Relationships for *Pimephales* and *Tetrahymena*: A Mechanism of Action Approach. *Environ.Toxicol.Chem.*, *16*, 1311–1317.

Beck, B., Glen, R.C. and Clark, T. (1996). The Inhibition of Alpha Chymotrypsin Predicted Using Theoretically Derived Molecular Properties. *J.Mol.Graphics*, *14*, 130.

Beck, B., Horn, A., Carpenter, J.E. and Clark, T. (1998). Enhanced 3D-Databases: A Fully Electrostatic Database of AM1-Optimized Structures. *J.Chem.Inf.Comput.Sci.*, *38*, 1214–1217.

Beckaus, H.-D. (1978). S_f Parameters – A Measure of the Front Strain of Alkyl Groups. *Angew. Chem.Int.Ed.Engl.*, *17*, 593–594.

Belvisi, L., Bravi, G., Scolastico, C., Vulpetti, A., Salimbeni, A. and Todeschini, R. (1994). A 3D QSAR Approach to the Search for Geometrical Similarity in a Series of Nonpeptide Angiotensin II Receptor Anatagonists. *J.Comput.Aid.Molec.Des.*, *8*, 211–220.

Belvisi, L., Bravi, G., Catalano, G., Mabilia, M., Salimbeni, A. and Scolastico, C. (1996). A 3D QSAR CoMFA Study of Non-Peptide Angiotensin II Receptor Antagonists. *J.Comput.Aid. Molec.Des.*, *10*, 567–582.

Bemis, G.W. and Kuntz, I.D. (1992). A Fast and Efficient Method for 2D and 3D Molecular Shape Description. *J.Comput.Aid.Molec.Des.*, *6*, 607–628.

Benigni, R., Andreoli, C. and Giuliani, A. (1989). Interrelationships among Carcinogenicity, Mutagenicity, Acute Toxicity, and Chemical Structure in a Genotoxicity Data Base. *J.Toxicol.Env.Health*, *27*, 1–20.

Benigni, R. (1991). QSAR Prediction of Rodent Carcinogenicity for a Set of Chemicals Currently Bioassayed by the US National Toxicology Program. *Mutagenesis*, *6*, 423–425.

Benigni, R. and Giuliani, A. (1991). What Kind of Statistics for QSAR Research? *Quant.Struct. Act.Relat.*, *10*, 99–100.

Benigni, R., Andreoli, C., Conti, L., Tafani, P., Cotta Ramusino, M., Carere, A. and Crebelli, R. (1993). Quantitative Structure-Activity Relationship Models Correctly Predict the Toxic and Aneuploidizing Properties of 6-Halogenated Methanes in Aspergillus nidulans. *Mutagenesis*, *8*, 301–305.

Benigni, R. and Giuliani, A. (1993). Analysis of Distance Matrices for Studying Data Structures and Separating Classes. *Quant.Struct.-Act.Relat.*, *12*, 397–401.

Benigni, R. (1994). EVE, a Distance Based Approach for Discriminating Nonlinearly Separable Groups. *Quant.Struct.-Act.Relat.*, *13*, 406–411.

Benigni, R., Andreoli, C. and Giuliani, A. (1994). QSAR Models for Both Mutagenic Potency and Activity Application to Nitroarenes and Aromatic Amines. *Environ.Mol.Mutagen.*, *24*, 208–219.

Benigni, R. and Giuliani, A. (1994). Quantitative Structure-Activity Relationship (QSAR) Studies in Genetic Toxicology. Mathematical Models and the Biological Activity Term of the Relationship. *Mut.Res.*, *306*, 181–186.

Benigni, R., Cotta Ramusino, M., Giorgi, F. and Gallo, G. (1995). Molecular Similarity Matrices and Quantitative Structure-Activity Relationships: A Case Study with Methodological Implications. *J.Med.Chem.*, *38*, 629–635.

Benigni, R., Gallo, G., Giorgi, F. and Giuliani, A. (1999). On the Equivalence Between Different Descriptions of Molecules: Value for Computational Approaches. *J.Chem.Inf.Comput.Sci.*, *39*, 575–578.

Benigni, R., Passerini, L., Livingstone, D.J. and Johnson, M.A. (1999a). Infrared Spectra Information and their Correlation with QSAR Descriptors. *J.Chem.Inf.Comput.Sci.*, *39*, 558–562.

Benigni, R., Passerini, L., Pino, A. and Giuliani, A. (1999b). The Information Content of the Eigenvalues from Modified Adjacency Matrices: Large Scale and Small Scale Correlations. *Quant.Struct.-Act.Relat.*, *18*, 449–455.

Bentley, T.W. and von Schleyer, P.R. (1977). Medium Effects on the Rates and Mechanisms of Solvolytic Reactions. *Adv.Phys.Org.Chem.*, *14*, 1–67.

Berger, B.M. and Wolfe, N.L. (1996). Hydrolysis and Biodegradation of Sulfonylurea Herbicides in Aqueous Buffers and Anaerobic Water-Sediment Systems. Assessing Fate Pathways Using Molecular Descriptors. *Environ.Toxicol.Chem.*, *15*, 1500–1507.

Bergmann, D. and Hinze, J. (1996). Electronegativity and Molecular Properties. *Angew.Chem. Int.Ed.Engl.*, *35*, 150–163.

Bermudez, C.I., Daza, E.E. and Andrade, E. (1999). Characterization and Comparison of Escherichia coli Transfer RNAs by Graph Theory Based on Secondary Structure. *J.Theor. Biol.*, *197*, 193–205.

Bernard, P., Golbraikh, A., Kireev, D.B., Chrétien, J.R. and Rozhkova, N. (1998). Comparison of Chemical Databases: Analysis of Molecular Diversity with Self Organizing Maps (SOM). *Analusis*, *26*, 333–341.

Bernejo, J., Canga, J.S., Gayol, O.M. and Guillen, M.D. (1984). Utilization of Physicochemical Properties and Structural Parameters for Calculating Retention Indices of Alkylbenzenes. *J.Chromatogr.Sci.*, *22*, 252.

Bersohn, M. (1983). A Fast Algorithm for Calculation of the Distance Matrix. *J.Comput.Chem.*, *4*, 110–113.

Bersohn, M. (1987). A Matrix Method for Partitioning the Atoms of a Molecule into Equivalence Classes. *Computers Chem.*, *11*, 67–72.

Berson, J.A., Hamlet, Z. and Mueller, W.A. (1962). The Correlation of Solvent Effects on the Stereoselectivities of Diels-Alder Reactions by Means of Linear Free Energy Relationships. A New Empirical Measure of Solvent Polarity. *J.Am.Chem.Soc.*, *84*, 297–304.

Bersuker, I.B. and Dimoglo, A.S. (1991a). The Electron-Topological Approach to the QSAR Problem. In *Reviews in Computational Chemistry – Vol. 2* (Lipkowitz, K.B. and Boyd, D., eds.), VCH Publishers, New York (NY), pp. 423–460. [R]

Bersuker, I.B., Dimoglo, A.S., Gorbachov, M.Yu. and Vlad, P.F. (1991b). Origin of Musk Fragrance Activity: The Electron-Topological Approach. *New J.Chem.*, *15*, 307–320.

Bersuker, I.B., Dimoglo, A.S., Gorbachov, M.Yu., Vlad, P.F. and Pesaro, M. (1991c). Origin of Musk Fragrance Activity: The Electron Topological Approach. *Nouv.J.Chim.*, *15*, 371.

Berthelot, M., Laurence, C., Lucon, M. and Rossignol, C. (1996). Partition Coefficients and Intramolecular Hydrogen Bonding. 2. The Influence of Partition Solvents on the Intramolecular Hydrogen Bond Stability of Salicyclic Acid Derivatives. *J.Phys.Org.Chem.*, *9*, 626–630.

Bertz, S.H. (1981). The First General Index of Molecular Complexity. *J.Am.Chem.Soc.*, *103*, 3599–3601.

Bertz, S.H. (1983a). A Mathematical Model of Molecular Complexity. In *Chemical Applications of Topology and Graph Theory* (King, R.B., ed.), Elsevier, Amsterdam (The Netherlands), pp. 206–221.

Bertz, S.H. (1983b). On the Complexity of Graphs and Molecules. *Bull.Math.Biol.*, *45*, 849–855.

Bertz, S.H. and Herndon, W.C. (1986). Similarity of Graphs and Molecules. In *Artificial Intelligence Applications in Chemistry*pp. 169–175.

Bertz, S.H. (1988). Branching in Graphs and Molecules. *Disc.Appl.Math.*, *19*, 65–83.

Besalù, E., Carbó, R., Mestres, J. and Solà, M. (1995). Foundations and Recent Developments on Molecular Quantum Similarity. *Top.Curr.Chem.*, *173*, 31–62.

Bhattacharjee, S., Rao, A.S. and Dasgupta, P. (1991). A New Index for Molecular Property Correlation in Halomethanes. *Computers Chem.*, *15*, 319–322.

Bhattacharjee, S., Basak, S.C. and Dasgupta, P. (1992). Molecular Property Correlation in Haloethanes with Geometric Volume. *Computers Chem.*, *16*, 223–228.

Bhattacharjee, S. (1994). Haloethanes, Geometric Volume and Atomic Contribution Method. *Computers Chem.*, *18*, 419–429.

Bhattacharjee, S. and Dasgupta, P. (1994). Molecular Property Correlation in Alkanes with Geometric Volume. *Computers Chem.*, *18*, 61–71.

Bienfait, B. (1994). Applications of High-Resolution Self-Organizing Maps to Retrosynthetic and QSAR Analysis. *J.Chem.Inf.Comput.Sci.*, *34*, 890–898.

Biggs, A.I. and Robinson, R.A. (1961). The Ionisation Constants of Some Substituted Anilines and Phenols: A Test of the Hammett Relation. *J.Chem.Soc.*, 388–393.

Bijloo, G.J. and Rekker, R.F. (1984a). Some Critical Remarks Concerning the Inductive Parameter σ_I. Part III: Parametrization of the Ortho Effect in Benzoic Acids and Phenols. *Quant.Struct.-Act.Relat.*, *3*, 91–96.

Bijloo, G.J. and Rekker, R.F. (1984b). *Quant.Struct.-Act.Relat.*, *3*, 111–115.

Bindal, M.C., Singh, P. and Gupta, S.P. (1980). Quantitative Correlation of Anesthetic Potencies of Halogenated Hydrocarbons with Boiling Point and Molecular Connectivity. *Arzneim.Forsch.*, *30*, 234.

Binsch, G. and Heilbronner, E. (1968). Double Bond Fixation in Non-Alternant π-Electron Systems. *Tetrahedron*, *24*, 1215.

Bintein, S. and Devillers, J. (1994). QSAR for Organic Chemical Sorption in Soils and Sediments. *Chemosphere*, *28*, 1171–1188.

Bird, C.W. (1985). A New Aromaticity Index and its Application to Five-Membered Ring Heterocycles. *Tetrahedron*, *41*, 1409–1414.

Bird, C.W. (1986). The Application of a New Aromaticity Index to Six-Membered Ring Heterocycles. *Tetrahedron*, *42*, 89–92.

Bjorsvik, H.R. and Priebe, H. (1995). Multivariate Data Analysis of Molecular Descriptors Estimated by Use of Semiempirical Quantum Chemistry Methods. Principal Properties for Synthetic Screening of 2-Chloromethyloxirane and Analogous bis-Alkylating C-3 Moieties. *Acta Chem.Scand.*, *49*, 446–456.

Bjorsvik, H.R., Hansen, U.M. and Carlson, R. (1997). Principal Properties of Monodentate Phosphorus Ligands. Predictive Model for the Carbonyl Absorption Frequencies in Ni(Co)(3)l Complexes. *Acta Chem.Scand.*, *51*, 733–741.

Blaney, F.E., Naylor, D. and Woods, J. (1993). Mambas: A Real Time Graphics Environment for QSAR. *J.Mol.Graphics*, *11*, 157–165.

Blickenstaff, R.T., Reddy, S., Witt, R. and Lipkowitz, K.B. (1994). Potential Radioprotective Agents–IV. Schiff Bases. *Bioorg.Med.Chem.*, *2*, 1363–1366.

Blin, N., Federici, C., Koscielniak, T. and Strosberg, A.D. (1995). Predictive Quantitative Structure-Activity Relationships (QSAR) Analysis of Beta 3-Adrenergic Ligands. *Drug Design & Discovery*, *12*, 297–311.

Blum, D.J., Suffet, I.H. and Duguet, J.P. (1993). Estimating the Activated Carbon Adsorption of Organic Chemicals in Water. *Crit.Rev.Environ.Sci.Technol.*, *23*, 121–136.

Blum, D.J., Suffet, I.H. and Duguet, J.P. (1994). Quantitative Structure-Activity Relationship Using Molecular Connectivity for the Activated Carbon Adsorption of Organic Chemicals in Water. *Water Res.*, *28*, 687–699.

Blurock, E.S. (1998). Use of Atomic and Bond Parameters in a Spectral Representation of a Molecule for Physical Property Determination. *J.Chem.Inf.Comput.Sci.*, *38*, 1111–1118.

Bocek, K., Kopecky, J. and Krivucova, M. (1964). *Experientia*, *20*, 667–668.

Bodor, N., Gabanyi, Z. and Wong, C.-K. (1989). A New Method for the Estimation of Partition Coefficient. *J.Am.Chem.Soc.*, *111*, 3783–3786.

Bodor, N., Harget, A. and Huang, M.-J. (1991). Neural Network Studies. 1. Estimation of the Aqueous Solubility of Organic Compounds. *J.Am.Chem.Soc.*, *113*, 9480–9483.

Bodor, N. and Huang, M.-J. (1992a). A New Method for the Estimation of the Aqueous Solubility of Organic Compounds. *J.Pharm.Sci.*, *81*, 954–960.

Bodor, N. and Huang, M.-J. (1992b). An Extended Version of a Novel Method for the Estimation of Partition Coefficients. *J.Pharm.Sci.*, *81*, 272–281.

Bodor, N., Huang, M.-J. and Harget, A. (1992c). Neural Network Studies. 4. An Extended Study of the Aqueous Solubility of Organic Compounds. *Int.J.Quantum Chem.Quant.Chem.Symp.*, *26*, 853–867.

Bodor, N., Huang, M.-J. and Harget, A. (1994). Neural Network Studies. Part 3. Prediction of Partition Coefficients. *J.Mol.Struct.(Theochem)*, *309*, 259–266.

Bodor, N., Buchwald, P. and Huang, M.-J. (1998). Computer-Assisted Design of New Drugs Based on Retrometabolic Concepts. *SAR & QSAR Environ.Res.*, *8*, 41–92.

Boecklen, W.J. and Niemi, G.J. (1994). Multivariate Association of Graph-Theoretic Variables and Physicochemical Properties. *SAR & QSAR Environ.Res.*, *2*, 79–87.

Boethling, R.S. and Sabljic, A. (1989). Screening-Level Model for Aerobic Biodegradability Based on a Survey of Expert Knowledge. *Environ.Sci.Technol.*, *23*, 672–679.

Bogdanov, B., Nikolic, S. and Trinajstic, N. (1989). On the Three-Dimensional Wiener Number. *J.Math.Chem.*, *3*, 299–309.

Bogdanov, B., Nikolic, S. and Trinajstic, N. (1990). On the Three-Dimensional Wiener Number. A Comment. *J.Math.Chem.*, *5*, 305–306.

Bögel, H., Dettmann, J. and Randic, M. (1997). Why is the Topological Approach in Chemistry so Successful? *Croat.Chem.Acta*, *70*, 827–840.

Boggia, R., Forina, M., Fossa, P. and Mosti, L. (1997). Chemometric Study and Validation Strategies in the Structure-Activity Relationship of New Cardiotonic Agents. *Quant.Struct.-Act. Relat.*, *16*, 201–213.

Bojarski, J. and Ekiert, L. (1982). Relationship Between Molecular Connectivity Indices of Barbiturates and Chromatographic Parameters. *Chromatographia*, *15*, 172.

Bojarski, J. and Ekiert, L. (1983). Evaluation of Modified Valence Molecular Connectivity Index for Correlation of Chromatographic Parameters. *J.Liquid Chromat.*, *6*, 73.

Bonaccorsi, R., Scrocco, E. and Tomasi, J. (1970). Molecular SCF Calculations for the Ground State of Some Three-Membered Ring Molecules: (CH2)3, (CH2)2NH, (CH2)2NH2+, (CH2)2O, (CH2)2S, (CH2)2CH2, and N2CH2. *J.Chem.Phys.*, *52*, 5270.

Bonchev, D. and Peev, T. (1973). A Theoretic-Information Study of Chemical Elements. Mean Information Content of a Chemical Element. *God.Vissh khim.- Technol.Inst., Burgas, Bulg.*, *10*, 561–574.

Bonchev, D., Kamenska, V. and Tashkova, C. (1976a). Equations for the Elements in the Periodic Table, Based on Information Theory. *MATCH (Comm.Math.Comp.Chem.)*, *2*, 117–122.

Bonchev, D., Kamenski, D. and Kamenska, V. (1976b). Symmetry and Information Content of Chemical Structures. *Bull.Math.Biol.*, *38*, 119–133.

Bonchev, D., Peev, T. and Rousseva, B. (1976c). Information Study of Atomic Nuclei, Information for Proton-Neutron Composition. *MATCH (Comm.Math.Comp.Chem.)*, *2*, 123–137.

Bonchev, D. and Trinajstic, N. (1977). Information Theory, Distance Matrix, and Molecular Branching. *J.Chem.Phys.*, *67*, 4517–4533.

Bonchev, D. and Kamenska, V. (1978). Information Theory in Describing the Electronic Structures of Atoms. *Croat.Chem.Acta*, *51*, 19–27.

Bonchev, D. and Trinajstic, N. (1978). On Topological Characterization of Molecular Branching. *Int.J.Quantum Chem.Quant.Chem.Symp.*, *12*, 293–303.

Bonchev, D., Knop, J.V. and Trinajstic, N. (1979a). Mathematical Models of Branching. *MATCH (Comm.Math.Comp.Chem.)*, *6*, 21–47.

Bonchev, D., Mekenyan, O., Knop, J.V. and Trinajstic, N. (1979b). On Characterization of Monocyclic Structures. *Croat.Chem.Acta*, *52*, 361–367.

Bonchev, D., Balaban, A.T. and Mekenyan, O. (1980a). Generalization of the Graph Center Concept and Derived Topological Centric Indexes. *J.Chem.Inf.Comput.Sci.*, *20*, 106–113.

Bonchev, D., Mekenyan, O. and Fritsche, H. (1980b). An Approach to the Topological Modeling of Crystal Growth. *J.Cryst.Growth*, *49*, 90–96.

Bonchev, D., Mekenyan, O. and Trinajstic, N. (1980c). Topological Characterization of Cyclic Structures. *Int.J.Quant.Chem.*, *17*, 845–893.

Bonchev, D. and Balaban, A.T. (1981). Topological Centric Coding and Nomenclature of Polycyclic Hydrocarbons. I. Condensed Benzenoid Systems (Polyhexes, Fusenes). *J.Chem.Inf.Comput.Sci.*, *21*, 223–229.

Bonchev, D., Balaban, A.T. and Randic, M. (1981a). The Graph Center Concept for Polycyclic Graphs. *Int.J.Quant.Chem.*, *19*, 61–82.

Bonchev, D., Mekenyan, O. and Trinajstic, N. (1981b). Isomer Discrimination by Topological Information Approach. *J.Comput.Chem.*, *2*, 127–148.

Bonchev, D. (1983). *Information Theoretic Indices for Characterization of Chemical Structures.* Research Studies Press, Chichester (UK).

Bonchev, D. and Mekenyan, O. (1983). Comparability Graphs and Electronic Spectra of Condensed Benzenoid Hydrocarbons. *Chem.Phys.Lett.*, *98*, 134–138.

Bonchev, D., Mekenyan, O. and Balaban, A.T. (1985). Unique Description of Chemical Structures Based on Hierarchically Ordered Extended Connectivities (HOC Procedures). IV. Recognition of Graph Isomorphism and Graph Symmetries. *MATCH (Comm.Math.Comp. Chem.)*, *18*, 83–89.

Bonchev, D., Mekenyan, O. and Balaban, A.T. (1986). Algorithms for Coding Chemical Compounds. In *Mathematics and Computational Concepts in Chemistry* (Trinajstic, N., ed.), Ellis Horwood, Chichester (UK), pp. 34–47.

Bonchev, D., Gutman, I. and Polanski, J. (1987). Parity of the Distance Numbers and Wiener Numbers of Bipartite Graphs. *MATCH (Comm.Math.Comp.Chem.)*, *22*, 209–214.

Bonchev, D. and Polansky, O.E. (1987). On the Topological Complexity of Chemical Systems. In *Graph Theory and Topology in Chemistry* (King, R.B. and Rouvray, D.H., eds.), Elsevier, Amsterdam (The Netherlands), pp. 126–158.

Bonchev, D. (1989). The Concept for the Centre of a Chemical Structure and Its Applications. *J.Mol.Struct.(Theochem)*, *185*, 155–168.

Bonchev, D., Mekenyan, O. and Balaban, A.T. (1989). Iterative Procedure for the Generalized Graph Center in Polycyclic Graphs. *J.Chem.Inf.Comput.Sci.*, *29*, 91–97.

Bonchev, D., Kamenska, V. and Mekenyan, O. (1990). Comparability Graphs and Molecular Properties: IV. Generalizations and Applications. *J.Math.Chem.*, *5*, 43–72.

Bonchev, D. and Rouvray, D. H. eds. (1991). *Chemical Graph Theory: Introduction and Fundamentals.* Abacus Press/Gordon & Breach, New York (NY), 288 pp.

Bonchev, D. and Kier, L.B. (1992). Topological Atomic Indices and the Electronic Charges in Alkanes. *J.Math.Chem.*, *9*, 75–85.

Bonchev, D., Mekenyan, O. and Kamenska, V. (1992). A Topological Approach to the Modeling of Polymer Properties (The TEMPO Method). *J.Math.Chem.*, *11*, 107–132.

Bonchev, D. and Rouvray, D. E. eds. (1992). *Chemical Graph Theory: Reactivity and Kinetics.* Gordon & Breach, New York (NY).

Bonchev, D. and Balaban, A.T. (1993). Central Vertices versus Central Rings in Polycyclic Systems. *J.Math.Chem.*, *14*, 287–304.

Bonchev, D., Kier, L.B. and Mekenyan, O. (1993a). Self-Returning Walks and Fractional Electronic Charges of Atoms in Molecules. *Int.J.Quant.Chem.*, *46*, 635–649.

Bonchev, D., Liu, X. and Klein, D.J. (1993b). Weighted Self-Returning Walks for Structure-Property Correlations. *Croat.Chem.Acta*, *66*, 141–150.

Bonchev, D., Mountain, C.F., Seitz, W.A. and Balaban, A.T. (1993c). Modeling the Anticancer Action of Some Retinoid Compounds by Making Use of the OASIS Method. *J.Med.Chem.*, *36*, 1562–1569.

Bonchev, D., Balaban, A.T., Liu, X. and Klein, D.J. (1994a). Molecular Cyclicity and Centricity of Polycyclic Graphs. I. Cyclicity Based on Resistance Distances or Reciprocal Distances. *Int.J.Quant.Chem.*, *50*, 1–20.

Bonchev, D., Seitz, W.A., Mountain, C.F. and Balaban, A.T. (1994b). Modeling the Anticarcinogenic Action of Retinoids by Making Use of the OASIS Method. 3. Inhibition of the Induction of Ornithine Decarboxylase by Arotinoids. *J.Med.Chem.*, *37*, 2300–2307. [R]

Bonchev, D. (1995). Topological Order in Molecules. 1. Molecular Branching Revisited. *J.Mol. Struct.(Theochem)*, *336*, 137–156.

Bonchev, D. and Gordeeva, E.V. (1995). Topological Atomic Charges, Valencies, and Bond Orders. *J.Chem.Inf.Comput.Sci.*, *35*, 383–395.

Bonchev, D. and Seitz, W.A. (1995). Relative Atomic Moments as Squared Principal Eigenvector Coefficients. *J.Chem.Inf.Comput.Sci.*, *35*, 237–242.

Bonchev, D. and Seitz, W.A. (1996). The Concept of Complexity in Chemistry. In *Concepts in Chemistry: Contemporary Challenge* (Rouvray, D.H., ed.), Research Studies Press, Taunton (UK), pp. 353–381.

Bonchev, D. (1997). Novel Indices for the Topological Complexity of Molecules. *SAR & QSAR Environ.Res.*, *7*, 23–43.

Bonchev, D. (2000). Overall Connectivity and Topological Complexity: A New Tool for QSPR/ QSAR. In *Topological Indices and Related Descriptors in QSAR and QSPR* (Devillers, J. and Balaban, A.T., eds.), Gordon & Breach, Amsterdam (The Netherlands), pp. 361–401.

Bondi, A. (1964). van der Waals Volumes and Radii. *J.Phys.Chem.*, *68*, 441–451.

Bonelli, D., Cechetti, V., Clementi, S., Cruciani, G., Fravolini, A. and Savino, A.F. (1991). The Antibacterial Activity of Quinolones Against Escherichia Coli: A Chemometric Study. *Quant.-Struct.-Act.Relat.*, *10*, 333–343.

Bonjean, M.-C. and Luu Duc, C. (1978). Connectivitè Moleculaire: Rélation dans une Série de Barbituriques. *Eur.J.Med.Chem.*, *13*, 73–76.

Boobbyer, D.N.A., Goodford, P.J., McWhinnie, P.M. and Wade, R.C. (1989). New Hydrogen-Bond Potentials for Use in Determining Energetically Favorable Binding Sites on Molecules of Known Structure. *J.Med.Chem.*, *32*, 1083–1094.

Booth, T.D. and Wainer, I.W. (1996a). Investigation of the Enantioselective Separations of alpha-alkylarylcarboxylic Acids on an Amylose tris(3,5-dimethylphenylcarbamate) Chiral Stationary Phase Using Quantitative Structure-Enantioselective Retention Relationships. Identification of a Conformationally Driven Chiral Recognition Mechanism. *J.Chromat.*, *737A*, 157–169.

Booth, T.D. and Wainer, I.W. (1996b). Mechanistic Investigation into the Enantioselective Separation of Mexiletine and Related Compounds, Chromatographed on an Amylose tris(3,5-dimethylphenylcarbamate) Chiral Stationary Phase. *J.Chromat.*, *741A*, 205–211.

Bordi, F., Mor, M., Morini, G., Plazzi, P.V., Silva, C., Vitali, T. and Caretta, A. (1994). QSAR Study on H-3-Receptor Affinity of Benzothiazole Derivatives of Thioperamide. *Il Farmaco*, *49*, 153–166.

Bordwell, F.G. and Cooper, G.D. (1952). Conjugative Effects of Methylsulfonyl and Methylthio Groupings. *J.Am.Chem.Soc.*, *74*, 1058.

Borisova, N.P. (1969). *Zhur.Strukt.Khim.*, *10*, 1053.

Borodina, Y., Filimonov, D. and Poroikov, V. (1998). Computer-Aided Estimation of Synthetic Compounds Similarity with Endogenous Bioregulations. *Quant.Struct.-Act.Relat.*, *17*, 459–464.

Borth, D.M. and McKay, R.J. (1985). A Difficulty Information Approach to Substituent Selection in QSAR Studies. *Technometrics*, *27*, 25–35.

Borth, D.M. (1996). Optimal Experimental Designs for (Possibly) Censored Data. *Chemom. Intell.Lab.Syst.*, *32*, 25–35.

Bosnjak, N., Aldler, N., Peric, M. and Trinajstic, N. (1987). On the Structural Origin of Chromatographic Retention Data: Alkanes and Cycloalkanes.pp. 103–122.

Böttcher, C.J.F., van Belle, O.C., Bordewijk, P. and Rip, A. (1973). *Theory of Electric Polarization – Vol. 1*. Elsevier, Amsterdam (The Netherlands), 378 pp.

Boudreau, R.J. and Efange, S.M. (1992). Computer-Aided Radiopharmaceutical Design. *Invest. Radiol.*, *27*, 653–658.

Boulu, L.G. and Crippen, G.M. (1989). Voronoi Binding Site Models: Calculation of Binding Modes and Influence of Drug Binding Data Accuracy. *J.Comput.Chem.*, *10*, 673–682.

Boulu, L.G., Crippen, G.M., Barton, H.A., Kwon, H. and Marletta, M.A. (1990). Voronoi Binding Site Model of a Polycyclic Aromatic Hydrocarbon Binding Protein. *J.Med.Chem.*, *33*, 771–775.

Bowden, K. and Young, R.C. (1970). Structure-Activity Relations. I. A Series of Antagonists of Acetylcholine and Histamine at the Postganglionic Receptors. *J.Med.Chem.*, *13*, 225–230.

Bowden, K. and Wooldridge, K.R.H. (1973). *Biochem.Pharmacol.*, *22*, 1015.

Bowden, K. (1990). Electronic Effects in Drug Design. In *Quantitative Drug Design. Vol. 4* (Ramsden, C.A., ed.), Pergamon Press, Oxford (UK), pp. 205–238.

Box, G.E.P., Hunter, W.G. and Hunter, J.S. (1978). *Statistics for Experimenters.* Wiley, New York (NY).

Boyd, J.C., Millership, J.S. and Woolfson, A.D. (1982). The Relationship Between Molecular Connectivity and Partition Coefficients. *J.Pharm.Pharmacol.*, *34*, 364–366.

Boyd, R.J. and Markus, G.E. (1981). Electronegativities of the Elements from a Nonempirical Electrostatic Model. *J.Chem.Phys.*, *75*, 5385–5388.

Bradbury, S.P., Henry, T.R., Niemi, G.J., Carlson, R.W. and Snarski, V.M. (1989). Use of Respiratory-Cardiovalscular Responses of Rainbow Trout (*Salmo gairdneri*) in Identifying Acute Toxicity Syndromes in Fish: Part 3. Polar Narcotics. *Environ.Toxicol.Chem.*, *8*, 247–261.

Bradbury, S.P. and Lipnick, R.L. (1990). Introduction: Structural Properties for Determining Mechanisms of Toxic Action. *Environ.Health Persp.*, *87*, 181–182.

Bradbury, S.P. (1994). Predicting Modes of Toxic Action from Chemical Structure: An Overview. *SAR & QSAR Environ.Res.*, *2*, 89–104.

Bradbury, S.P. (1995). Quantitative Structure-Activity Relationships and Ecological Risk Assessment. An Overview of Predictive Aquatic Toxicology Research. *Toxicol.Lett.*, *79*, 229–237.

Bradbury, S.P., Mekenyan, O., Veith, G.D. and Zaharieva, N. (1995). SAR Modeling of Futile Metabolism: One-Electron Reduction of Quinones, Phenols and Nitrobenzenes. *SAR & QSAR Environ.Res.*, *4*, 109–124.

Bradbury, S.P., Mekenyan, O. and Ankley, G.T. (1996). Quantitative Structure-Activity Relationships for Polychlorinated Hydroxybiphenyl Estrogen Receptor Binding Affinity. An Assessment of Conformer Flexibility. *Environ.Toxicol.Chem.*, *15*, 1945–1954.

Branch, G.E. and Calvin, M. (1941). *The Theory of Organic Chemistry.* Prentice-Hall Inc., New York (NY).

Bratsch, S.G. (1985). A Group Electronegativity Method with Pauling Units. *J.Chem.Educ.*, *62*, 101–103.

Bravi, G., Gancia, E., Mascagni, P., Pegna, M., Todeschini, R. and Zaliani, A. (1997). MS-WHIM, New 3D Theoretical Descriptors Derived from Molecular Surface Properties: a Comparative 3D-QSAR Study in a Series of Steroids. *J.Comput.Aid.Molec.Des.*, *11*, 79–92.

Bray, P.J. and Barnes, R.G. (1957). Estimates of Hammett's Sigma Values from Quadrupole Resonance Studies. *J.Chem.Phys.*, *27*, 551–560.

Brekke, T. (1989). Prediction of Physical Properties of Hydrocarbons Mixtures by Partial-Least-Squares Calibration of Carbon-13 Nuclear Magnetic Resonance Data. *Anal.Chim.Acta*, *223*, 123–134.

Brendlé, E. and Papirer, E. (1997). A New Topological Index for Molecular Probes Used in Inverse Gas Chromatography.2. Application for the Evaluation of the Solid Surface Specific Interaction Potential. *J.Colloid Interf.Sci.*, *194*, 217–224.

Brereton, R.G. (1990). *Chemometrics.* Ellis Horwood, Chichester (UK), 308 pp.

Breyer, E.D., Strasters, J.K. and Khaledi, M.G. (1991). Quantitative Retention-Biological Activity Relationship Study by Micellar Liquid Chromatography. *Anal.Chem.*, *63*, 828–833.

Brickmann, J. (1997). Linguistic Variables in the Molecular Recognition Problem. In *Fuzzy Logic in Chemistry* (Rouvray, D.H., ed.), Academic Press, New York (NY), pp. 225–247.

Briem, H. and Kuntz, I.D. (1996). Molecular Similarity Based on DOCK-Generated Fingerprints. *J.Med.Chem.*, *39*, 3401–3408.

Briens, F., Bureau, R., Rault, S. and Robba, M. (1995). Applicability of CoMFA in Ecotoxicology: A Critical Study on Chlorophenols. *Ecotox.Environ.Safe.*, *31*, 37–48.

Briggs, G.G. (1981). Theoretical and Experimental Relationships Between Soil Adsorption, Octanol-Water Partition Coefficients, Water Solubilities, Bioconcentration Factors, and the Parachor. *J.Agr.Food Chem.*, *29*, 1050–1059.

Briggs, J.M., Marrone, T.J. and McCammon, J.A. (1996). Computational Science: New Horizons and Relevance to Pharmaceutical Design. *Trends Cardiovasc.Med.*, *6*, 198–204.

Brillouin, L. (1962). *Science and Information Theory.* Academic Press (2nd ed.), New York (NY).

Brinck, T., Murray, J.S. and Politzer, P. (1993). Octanol/Water Partition Coefficients Expressed in Terms of Solute Molecular Surface Areas and Electrostatic Potentials. *J.Org.Chem.*, *58*, 7070–7073.

Brint, A.T. and Willett, P. (1987a). Algorithms for the Identification of Three-Dimensional Maximal Common Substructures. *J.Chem.Inf.Comput.Sci.*, *27*, 152–158.

Brint, A.T. and Willett, P. (1987b). Identifying 3-D Maximal Common Substructures Using Transputer Networks. *J.Mol.Graphics*, *5*, 200–207.

Brooker, L.G.S., Craig, A.C., Heseltine, D.W., Jenkins, P.W. and Lincoln, L.L. (1965). Color and Constitution. XIII. Merocyanines as Solvent Property Indicators. *J.Am.Chem.Soc.*, *87*, 2443–2450.

Broto, P., Moreau, G. and Vandycke, C. (1984a). Molecular Structures: Perception, Autocorrelation Descriptor and SAR Studies. Autocorrelation Descriptor. *Eur.J.Med.Chem.*, *19*, 66–70.

Broto, P., Moreau, G. and Vandycke, C. (1984b). Molecular Structures: Perception, Autocorrelation Descriptor and SAR Studies. System of Atomic Contributions for the Calculation of the *n*-Octane/Water Partition Coefficients. *Eur.J.Med.Chem.*, *19*, 71–78.

Broto, P., Moreau, G. and Vandycke, C. (1984c). Molecular Structures: Perception, Autocorrelation Descriptor and SAR Studies. Use of the Autocorrelation Descriptors in the QSAR Study of Two Non-Narcotic Analgesic Series. *Eur.J.Med.Chem.*, *19*, 79–84.

Broto, P. and Devillers, J. (1990). Autocorrelation of Properties Distributed on Molecular Graphs. In *Practical Applications of Quantitative Structure-Activity Relationships (QSAR) in Environmental Chemistry and Toxicology.* (Karcher, W. and Devillers, J., eds.), Kluwer, Dordrecht (The Netherlands), pp. 105–127.

Broughton, H.B., Green, S.M. and Rzepa, H.S. (1992). Rank Correlation of AM1 and PM3 Derived Molecular Electrostatic Potentials (RACEL) with Hammett σ_p-Parameters. *J.Chem.Soc.Chem.Comm.*, 37–39.

Brown, D. and Flagg, E. (1981). *J.Environ.Qual.*, *10*, 382.

Brown, H.C. (1959). The Chemistry of Molecular Shapes. *J.Chem.Educ.*, *36*, 424.

Brown, H.C. and Okamoto, Y. (1958a). Electrophilic Substituent Constants. *J.Am.Chem.Soc.*, *80*, 4979–4987.

Brown, H.C., Okamoto, Y. and Inukai, T. (1958b). Rates of Solvolysis of the *m*- and *p*-Phenyl-, *m*- and *p*-Methylthio-, and *m*- and *p*-Trimethylsilylphenyldimethylcarbinyl Chlorides. Steric Inhibition of Resonance as a Factor in Electrophilic Substituent Constants. *J.Am.Chem.Soc.*, *80*, 4964–4968.

Brown, R.D. and Martin, Y.C. (1996). Use of Structure-Activity Data to Compare Structure-Based Clustering Methods and Descriptors for Use in Compound Selection. *J.Chem.Inf.Comput.Sci.*, *36*, 572–584.

Brown, R.D. (1997). Descriptors for Diversity Analysis. *Persp.Drug Disc.Des.*, *7/8*, 31–49.

Brown, R.D. and Martin, Y.C. (1997). The Information Content of 2D and 3D Structural Descriptors Relevant to Ligand-Receptor Binding. *J.Chem.Inf.Comput.Sci.*, *37*, 1–9.

Brown, R.D. and Martin, Y.C. (1998). An Evaluation of Structural Descriptors and Clustering Methods for Use in Diversity Selection. *SAR & QSAR Environ.Res.*, *8*, 23–39.

Brownlee, R.T.C., Katritzky, A.R. and Topsom, R.D. (1965). Direct Infrared Determination of the Resonance Interaction in Monosubstituted Benzenes. *J.Am.Chem.Soc.*, *87*, 3260–3261.

Brownlee, R.T.C., Katritzky, A.R. and Topsom, R.D. (1966). Distortions of the π-Electron System in Monosubstituted Benzenes. *J.Am.Chem.Soc.*, *88*, 1413–1419.

Brüggemann, R., Altschuh, J. and Matthies, M. (1990). QSAR for Estimating Physicochemical Data. In *Practical Applications of Quantitative Structure-Activity Relationships (QSAR) in Environmental Chemistry and Toxicology* (Karcher, W. and Devillers, J., eds.), Kluwer, Dordrecht (The Netherlands), pp. 197–212.

Brüggemann, R. and Bartel, H.-G. (1999). A Theoretical Concept to Rank Environmentally Significant Chemicals. *J.Chem.Inf.Comput.Sci.*, *39*, 211–217.

Brunoblanch, L. and Estiu, G.L. (1995). Quantum Chemistry in QSAR Anticonvulsivant Activity of Vpa Derivatives. *Int.J.Quant.Chem.*, 39–49.

Brusseau, M.L. (1993). Using QSAR to Evaluate Phenomenological Models for Sorption of Organic Compounds by Soil. *Environ.Toxicol.Chem.*, *12*, 1835–1846.

Buckingham, A.D. (1959). Molecular Quadrupole Moments. *Quarterly Reviews of the Chemical Society*, *13*, 183–214.

Buckingham, A.D. (1967). Permanent and Induced Molecular Moments and Long-Range Intermolecular Forces. In *Advances in Chemical Physics. Vol. 12* (Hirschfelder, J.O., ed.), Interscience Publishers – Wiley, New York (NY), pp. 107.

Buckingham, R.A. (1938). The Classical Equation of State of Gaseous He, Ne and Ar. *Proc. Roy.Soc.London A*, *168*, 264–283.

Buckley, F. and Harary, F. (1990). *Distance Matrix in Graphs.* Addison-Wesley, Redwood City (CA).

Buda, A.B., Auf der Heyde, T. and Mislow, K. (1992). On Quantifying Chirality. *Angew.Chem. Int.Ed.Engl.*, *31*, 989–1007.

Burden, F.R. (1989). Molecular Identification Number for Substructure Searches. *J.Chem.Inf. Comput.Sci.*, *29*, 225–227.

Burden, F.R. (1996). Using Artificial Neural Networks to Predict Biological Activity of Molecules from Simple Molecular Structural Considerations. *Quant.Struct.-Act.Relat.*, *15*, 7–11.

Burden, F.R. (1997). A Chemically Intuitive Molecular Index Based on the Eigenvalues of a Modified Adjacency Matrix. *Quant.Struct.-Act.Relat.*, *16*, 309–314.

Burden, F.R., Brereton, R.G. and Walsh, P.T. (1997). A Comparison of Cross-Validation and non-Cross-Validation Techniques: Application to Polycyclic Aromatic Hydrocarbons Electronic Absorption Spectra. *The Analyst*, *122*, 1015–1022.

Burden, F.R. (1998). Holographic Neural Networks as Nonlinear Discriminants for Chemical Applications. *J.Chem.Inf.Comput.Sci.*, *38*, 47–53.

Burden, F.R. and Winkler, D.A. (1999a). New QSAR Methods Applied to Structure-Activity Mapping and Combinatorial Chemistry. *J.Chem.Inf.Comput.Sci.*, *39*, 236–242.

Burden, F.R. and Winkler, D.A. (1999b). Robust QSAR Models Using Bayesian Regularized Neural Networks. *J.Med.Chem.*, *42*, 3183–3187.

Burke, B.J. and Hopfinger, A.J. (1993). Advances in Molecular Shape Analysis. In *3D QSAR in Drug Design. Theory, Methods and Applications.* (Kubinyi, H., ed.), ESCOM, Leiden (The Netherlands), pp. 276–306.

Burt, C. and Richards, W.G. (1990a). Molecular Similarity: The Introduction of Flexible Fitting. *J.Comput.Aid.Molec.Des.*, *4*, 231–238.

Burt, C., Richards, W.G. and Huxley, P. (1990b). The Application of Molecular Similarity Calculations. *J.Comput.Chem.*, *11*, 1139–1146.

Buydens, L. and Massart, D.L. (1981). Prediction of Gas Chromatographic Retention Indices from Linear Free Energy and Topological Parameters. *Anal.Chem.*, *53*, 1990–1993.

Buydens, L., Coomans, D., Vanbelle, M., Massart, D.L. and Vanden Driessche, R. (1983a). Comparative Study of Topological and Linear Free Energy-Related Parameters for the Prediction of Gas Chromatographic Retention Indices. *J.Pharm.Sci.*, *72*, 1327.

Buydens, L., Massart, D.L. and Geerlings, P. (1983b). Prediction of Gas Chromatographic Retention Indexes with Topological, Physicochemical, and Quantum Chemical Parameters. *Anal. Chem.*, *55*, 738–744.

Buydens, L., Massart, D.L. and Geerlings, P. (1985). Relationship Between Gas Chromatographic Behaviour and Topological, Physicochemical and Quantum Chemically Calculated Charge Parameters for Neuroleptica. *J.Chromatogr.Sci.*, *23*, 304–307.

Bytautas, L. and Klein, D.J. (1998). Chemical Combinatorics for Alkane-Isomer Enumeration and More. *J.Chem.Inf.Comput.Sci.*, *38*, 1063–1078. [R]

Bytautas, L. and Klein, D.J. (1999). Alkane Isomer Combinatorics: Stereostructure Enumeration and Graph-Invariant and Molecular-Property Distributions. *J.Chem.Inf.Comput.Sci.*, *39*, 803–818.

Cabala, R., Svobodova, J., Feltl, L. and Tichy, M. (1992). Direct Determination of Partition Coefficients of Volatile Liquids Between Oil and Gas by Gas Chromatography and Its Use in QSAR Analysis. *Chromatographia*, *34*, 601–606.

Caianiello, E.R. (1953). On the Quantum Field Theory. I. Explicit Solution of Dyson's Equation in Electrodynamics Without Use of Feynman Graphs. *Nuovo Cimento*, *10*, 1634–1652.

Caianiello, E.R. (1956). Proprietà di Pfaffiani e Hafniani. *Recerca, Napoli*, *7*, 25–31.

Caliendo, G., Greco, G., Novellino, E., Perissutti, E. and Santagada, V. (1994). Combined Use of Factorial Design and Comparative Molecular Field Analysis (CoMFA): A Case Study. *Quant. Struct.-Act.Relat.*, *13*, 249–261.

Caliendo, G., Fattorusso, C., Greco, G., Novellino, E., Perissutti, E. and Santagada, V. (1995). Shape-Dependent Effects in a Series of Aromatic Nitro Compounds Acting as Mutagenic Agents on S. Typhimurium TA98. *SAR & QSAR Environ.Res.*, *4*, 21–27.

Calixto, F. and Raso, A. (1982). Retention Index, Connectivity Index and van der Waals Volume of Alkanes. *Chromatographia*, *15*, 521.

Cambon, B. and Devillers, J. (1993). New Trends in Structure-Biodegradability Relationships. *Quant.Struct.-Act.Relat.*, *12*, 49–56.

Camilleri, P., Watts, S.A. and Boraston, J.A. (1988). A Surface Area Approach to Determination of Partition Coefficients. *J.Chem.Soc.Perkin Trans.2*, 1699–1707.

Camilleri, P., Livingstone, D.J., Murphy, J.A. and Manallack, D.T. (1993). Chiral Chromatography and Multivariate Quantitative Structure-Property Relationships of Benzimidazole Sulphoxides. *J.Comput.Aid.Molec.Des.*, *7*, 61–69.

Cammarata, A. and Yau, S.J. (1970). Predictability of Correlations Between *in Vitro* Tetracycline Potencies and Substituent Indices. *J.Med.Chem.*, *13*, 93–97.

Cammarata, A. (1972). Interrelationship of the Regression Models Used for Structure-Activity Analyses. *J.Med.Chem.*, *15*, 573–577.

Cammarata, A. and Bustard, T.M. (1974). Reinvestigation of a "Nonadditive" Quantitative Structure-Activity Relationship. *J.Med.Chem.*, *17*, 981–985.

Cammarata, A. (1979). Molecular Topology and Aqueous Solubility of Aliphatic Alcohols. *J.Pharm.Sci.*, *68*, 839–842.

Campbell, J.L. and Johnson, K.E. (1993). Abductive Networks Generalization, Pattern Recognition, and Prediction of Chemical Behavior. *Can.J.Chem.*, *71*, 1800–1804.

Canfield, E.R., Robinson, R.W. and Rouvray, D.H. (1985). Determination of the Wiener Molecular Branching Index for the General Tree. *J.Comput.Chem.*, *6*, 598–609.

Cao, C.Z. (1996). Distance-Edge Topological Index for Research on Structure-Property Relationships of Alkanes. *Acta Chim.Sin.*, *54*, 533–538.

Caporossi, G. and Hansen, P. (1998). Enumeration of Polyhex Hydrocarbons to $h = 21$. *J.Chem. Inf.Comput.Sci.*, *38*, 610–619.

Caporossi, G., Gutman, I. and Hansen, P. (1999). Variable Neighborhood Search for Extremal Graphs. IV: Chemical Trees with Extremal Connectivity Index. *Computers Chem.*, *23*, 469–477.

Carabédian, M. and Dubois, J.-E. (1998). Large Virtual Enhancement of a ^{13}C NMR Database. A Structural Crowing Extrapolation Method Enabling Spectral Data Transfer. *J.Chem.Inf. Comput.Sci.*, *38*, 100–107.

Carbó-Dorca, R. and Besalù, E. (1996). Extending Molecular Similarity to Energy Surfaces: Boltzmann Similarity Measures and Indices. *J.Math.Chem.*, *20*, 247–261.

Carbó, R., Leyda, L. and Arnau, M. (1980). How Similar is a Molecule to Another? An Electron Density Measure of Similarity Between Two Molecular Structures. *Int.J.Quant.Chem.*, *17*, 1185–1189.

Carbó, R. and Domingo, L. (1987). LCAO-MO Similarity Measures and Taxonomy. *Int.J.Quant.Chem.*, *32*, 517–545.

Carbó, R. and Calabuig, B. (1990). Molecular Similarity and Quantum Chemistry. In *Concepts and Applications of Molecular Similarity* (Johnson, M.A. and Maggiora, G.M., eds.), Wiley, New York (NY), pp. 147–171.

Carbó, R. and Calabuig, B. (1992a). Molecular Quantum Similarity Measures and *N*-Dimensional Representation of Quantum Objects. I. Theoretical Foundations. *Int.J.Quant.Chem.*, *42*, 1681–1693.

Carbó, R. and Calabuig, B. (1992b). Molecular Quantum Similarity Measures and *N*-Dimensional Representation of Quantum Objects. II. Practical Applications. *Int.J.Quant.Chem.*, *42*, 1695–1709.

Carbó, R. and Calabuig, B. (1992c). Quantum Similarity. In *Structure, Interactions and Reactivity* (Fraga, S., ed.), Elsevier, Amsterdam (The Netherlands).

Carbó, R. and Calabuig, B. (1992d). Quantum Similarity Measures, Molecular Cloud Description, and Structure-Properties Relationships. *J.Chem.Inf.Comput.Sci.*, *32*, 600–606.

Carbó, R., Besalù, E., Amat, L. and Fradera, X. (1995). Quantum Molecular Similarity Measures (QMSM) as a Natural Way Leading towards a Theoretical Foundation of Quantitative Structure-Properties Relationships (QSPR). *J.Math.Chem.*, *18*, 237–246.

Carbó, R., Besalù, E., Amat, L. and Fradera, X. (1996). On Quantum Molecular Similarity Measures (QMSM) and Indices (QMSI). *J.Math.Chem.*, *19*, 47–56.

Cardozo, M.G., Iimura, Y., Sugimoto, H., Yamanishi, Y. and Hopfinger, A.J. (1992). QSAR Analyses of the Substituted Indanone and Benzylpiperidine Rings of a Series of Indanone Benzylpiperidine Inhibitors of Acetylcholinesterase. *J.Med.Chem.*, *35*, 584–589.

Carhart, R.E. (1977). Canonical Numbering and Constitutional Symmetry. *J.Chem.Inf.Comput.Sci.*, *17*, 113.

Carhart, R.E. (1978). *J.Chem.Inf.Comput.Sci.*, *18*, 108.

Carhart, R.E., Smith, D.H. and Venkataraghavan, R. (1985). Atom Pairs as Molecular Features in Structure-Activity Studies: Definition and Applications. *J.Chem.Inf.Comput.Sci.*, *25*, 64–73.

Carlson, R. (1992). *Design and Optimization in Organic Synthesis.* Elsevier, Amsterdam (The Netherlands), 536 pp.

Carlton, T.S. (1998). Correlation of Boiling Points with Molecular Structure for Chlorofluoroethanes. *J.Chem.Inf.Comput.Sci.*, *38*, 158–164.

Caron, G., Carrupt, P.-A., Testa, B., Ermondi, G. and Gasco, A. (1996). Insight into the Lipophilicity of the Aromatic N-Oxide Moiety. *Pharm.Res.*, *13*, 1186–1190.

Caron, G., Gaillard, P., Carrupt, P.-A. and Testa, B. (1997). 34. Lipophilicity Behavior of Model and Medicinal Compounds Containing a Sulfide, Sulfoxide, or Sulfone Moiety. *Helv.Chim.Acta*, *80*, 449–462.

Carrieri, A., Altomare, C., Barreca, M.L., Contento, A., Carotti, A. and Hansch, C. (1994). Papain Catalyzed Hydrolysis of Aryl Esters: A Comparison of the Hansch, Docking and CoMFA Methods. *Il Farmaco*, *49*, 573–585.

Carroll, F.I., Mascarella, S.W., Kuzemko, M.A., Gao, Y.G., Abraham, P., Lewin, A.H., Boja, J.W. and Kuhar, M.J. (1994). Synthesis, Ligand Binding, and QSAR (CoMFA and Classical) Study of 3 Beta-(3′-Substituted Phenyl), 3 Beta-(4′-Substituted Phenyl), and 3 Beta (3′,4′-Disubstituted Phenyl)Tropane-2 Beta-Carboxylic Acid Methyl Esters. *J.Med.Chem.*, *37*, 2865–2873.

Carrupt, P.-A., Testa, B. and Gaillard, P. (1997). Computational Approaches to Lipophilicity: Methods and Applications. In *Reviews in Computational Chemistry, Vol. 11* (Lipkowitz, K.B. and Boyd, D., eds.), Wiley-VCH, New York (NY), pp. 241–315.

Carter, P.G. (1949). An Empirical Equation for the Resonance Energy of Polycyclic Aromatic Hydrocarbons. *Trans.Faraday Soc.*, *45*, 597–602.

Cartier, A. and Rivail, J.-L. (1987). Electronic Descriptors in Quantitative Structure-Activity Relationships. *Chemom.Intell.Lab.Syst.*, *1*, 335–347.

Caruso, L., Musumarra, G. and Katritzky, A.R. (1993). "Classical" and "Magnetic" Aromaticities as New Descriptors for Heteroaromatics in QSAR. Part 3 [1]. Principal Properties for Heteroaromatics. *Quant.Struct.-Act.Relat.*, *12*, 146–151.

Casaban-Ros, E., Antón-Fos, G.M., Gálvez, J., Duart, M.J. and Garcìa-Domenech, R. (1999). Search for New Antihistaminic Compounds by Molecular Connectivity. *Quant.Struct.-Act.Relat.*, *18*, 35–42.

Cash, G.G. (1995a). A Fast Computer Algorithm for Finding the Permanent of Adjacency Matrices. *J.Math.Chem.*, *18*, 115–119.

Cash, G.G. (1995b). Correlation of Physicochemical Properties of Alkylphenols with their Graph-Theoretical ε Parameter. *Chemosphere*, *31*, 4307–4315.

Cash, G.G. (1995c). Heats of Formation of Polyhex Polycyclic Aromatic Hydrocarbons from their Adjacency Matrices. *J.Chem.Inf.Comput.Sci.*, *35*, 815–818.

Cash, G.G. (1995d). Prediction of Inhibitory Potencies of Arenesulfonamides Toward Carbonic Anhydrase Using Easily Calculated Molecular Connectivity Indexes. *Struct.Chem.*, *6*, 157–160.

Cash, G.G. (1998). A Simple Means of Computing the Kekulé Structure Count for Toroidal Polyhex Fullerenes. *J.Chem.Inf.Comput.Sci.*, *38*, 58–61.

Cayley, E. (1874). On the Mathematical Theory of Isomers. *Philosophical Magazine*, *67*, 444.

Centner, V., Massart, D.L., de Noord, O.E., De Jong, S., Vandeginste, B.G.M. and Sterna, C. (1996). Elimination of Uniformative Variables for Multivariate Calibration. *Anal.Chem.*, *68*, 3851–3858.

Cercos-del-Pozo, R.A., Perez-Gimenez, F., Salebert-Salvador, M.T. and García-March, F.J. (2000). Discrimination and Molecular Design of New Theoretical Hypolipaemic Agents Using the Molecular Connectivity Functions. *J.Chem.Inf.Comput.Sci.*, *40*, 178–184.

Chan, O., Lam, T.-K. and Merris, R. (1997). Wiener Number as an Immanent of the Laplacian of Molecular Graphs. *J.Chem.Inf.Comput.Sci.*, *37*, 762–765.

Chan, O., Gutman, I., Lam, T.-K. and Merris, R. (1998). Algebraic Connections Between Topological Indices. *J.Chem.Inf.Comput.Sci.*, *38*, 62–65.

Chapman, D. (1996). The Measurement of Molecular Diversity: A Three-Dimensional Approach. *J.Comput.Aid.Molec.Des.*, *10*, 501–512.

Chapman, N. B. and Shorter, J. eds. (1978). *Correlation Analysis in Chemistry.* Plenum Press, New York (NY), 546 pp.

Charton, M. (1963). The Estimation of Hammett Substituent Constants. *J.Org.Chem.*, *28*, 3121–3124.

Charton, M. (1964). Definition of "Inductive" Substituent Constants. *J.Org.Chem.*, *29*, 1222–1227.

Charton, M. (1969). The Nature of the *ortho* Effetc. II. Composition of the Taft Steric Parameters. *J.Am.Chem.Soc.*, *91*, 615–620.

Charton, M. (1971). The Quantitative Treatment of the Ortho Effect. *Prog.Phys.Org.Chem.*, *8*, 235–317.

Charton, M. (1975). Steric Effects. I. Esterification and Acid-Catalyzed Hydrolysis of Esters. *J.Am.Chem.Soc.*, *97*, 1552–1556.

Charton, M. (1976). Steric Effects. 7. Additional ν Constants. *J.Org.Chem.*, *41*, 2217–2220.

Charton, M. (1978a). Applications of Linear Free Energy Relationships to Polycyclic Arenes and to Heterocyclic Compounds. In *Correlation Analysis in Chemistry* (Chapman, N.B. and Shorter, J., eds.), Plenum Press, New York (NY), pp. 175–268.

Charton, M. (1978b). Steric Effects. 13. Composition of the Steric Parameter as a Function of Alkyl Branching. *J.Org.Chem.*, *43*, 3995–4001.

Charton, M. and Charton, B.I. (1978). Steric Effects. 12. Substituents at Phosphorus. *J.Org.Chem.*, *43*, 2383–2386.

Charton, M. (1981). Electrical Effect Substituent Constants for Correlation Analysis. *Prog.Phys.Org.Chem.*, *13*, 119–251.

Charton, M. and Charton, B.I. (1982). The Structural Dependence of Amino Acid Hydrophobicity Parameter. *J.Theor.Biol.*, *99*, 629–644.

Charton, M. (1983). The Upsilon Steric Parameter X: Definition and Determination. In *Steric Effects in Drug Design (Topics in Current Chemistry, Vol. 114)* (Charton, M. and Motoc, I., eds.), Springer-Verlag, Berlin (Germany), pp. 57–91.

Charton, M. (1984). The Validity of the Revised *F* and *R* Electrical Effect Substituent Parameters. *J.Org.Chem.*, *49*, 1997–2001.

Charton, M. (1987). A General Treatment of Electrical Effects. *Prog.Phys.Org.Chem.*, *16*, 287–315.

Charton, M. (1990). The Quantitative Description of Amino Acid, Peptide, and Protein Properties and Bioactivities. *Prog.Phys.Org.Chem.*, *18*, 163–284.

Chastrette, M., Zakarya, D. and El Mouaffek, A. (1986). Relations Structure-Odeur dans une Famille des Muscs Benzénique Nitrés. *Eur.J.Med.Chem.*, *21*, 505–510.

Chastrette, M., Zakarya, D. and Peyraud, J.F. (1994). Structure-Musk Odour Relationships for Indan and Tetralins Using Neural Network. On the Contribution of Descriptors to Classification. *Eur.J.Med.Chem.*, *29*, 343–348.

Chau, P.-L. and Dean, P.M. (1987). Molecular Recognition: 3D Surface Structure Comparison by Gnomonic Projection. *J.Mol.Graphics*, *5*, 97–100.

Chaumat, E., Chamel, A., Taillandier, G. and Tissut, M. (1992). Quantitative Relationships Between Structure and Penetration of Phenylurea Herbicides Through Isolated Plant Cuticles. *Chemosphere*, *24*, 189–200.

Chen, H., Zhou, J. and Xie, G. (1998). PARM: A Genetic Evolved Algorithm to Predict Bioactivity. *J.Chem.Inf.Comput.Sci.*, *38*, 243–250.

Chen, J., Feng, L., Liao, Y., Wang, L. and Hu, H. (1996). Using AM1 Hamiltonian in Quantitative Structure-Properties Relationship Studies of Alkyl(1-Phenylsulfonyl)cycloalkane-Carboxylates. *Chemosphere*, *33*, 537–546.

Chen, Q., Wu, C., Maxwell, D., Krudy, G.A., Dixon, R.A.F. and You, T.J. (1999). A 3D QSAR Analysis of *in vitro* Binding Affinity and Selectivity of 3-Isoxaxazolylsulfonylaminothiophenes as Endothelin Receptor Antagonists. *Quant.Struct.-Act.Relat.*, *18*, 124–133.

Chen, X., Rusinko III, A. and Young, S.S. (1998). Recursive Partitioning Analysis of a Large Structure-Activity Data Set Using Three-Dimensional Descriptors. *J.Chem.Inf.Comput.Sci.*, *38*, 1054–1062.

Cheng, C., Maggiora, G.M., Lajiness, M.S. and Johnson, M.A. (1996). Four Association Constants for Relating Molecular Similarity Measures. *J.Chem.Inf.Comput.Sci.*, *36*, 909–915.

Chenzhong, C. and Zhiliang, L. (1998). Molecular Polarizability. 1. Relationship to Water Solubility of Alkanes and Alcohols. *J.Chem.Inf.Comput.Sci.*, *38*, 1–7.

Chepoi, V. (1996). On Distances in Benzenoid Systems. *J.Chem.Inf.Comput.Sci.*, *36*, 1169–1172.

Chepoi, V. and Klavzar, S. (1997). The Wiener Index and the Szeged Index of Benzenoid Systems in Linear Time. *J.Chem.Inf.Comput.Sci.*, *37*, 752–755.

Cherkasov, A.R., Galkin, V.I. and Cherkasov, R.A. (1998a). A New Approach to the Theoretical Estimation of Inductive Constants. *J.Phys.Org.Chem.*, *11*, 437–447.

Cherkasov, A.R. and Jonsson, M. (1998b). Substituent Effects on Thermochemical Properties of Free Radicals. New Substituent Scales for C-Centered Radicals. *J.Chem.Inf.Comput.Sci.*, *38*, 1151–1156.

Cherqaoui, D., Villemin, D., Mesbah, A., Cense, J.-M. and Kvasnicka, V. (1994). Use of a Neural Network to Determine the Normal Boiling Points of Acyclic Ethers, Peroxides, Acetals and their Sulfur Analogues. *J.Chem.Soc.Faraday Trans.*, *90*, 2015–2019.

Chester, N.A., Famini, G.R., Haley, M.V., Kurnas, C.W., Sterling, P.A. and Wilson, L.Y. (1996). Aquatic Toxicity of Chemical Agent Simulants as Determined by Quantitative Structure-Activity Relationships and Acute Bioassays. *ACS Symp.Ser.*, *643*, 191–204.

Chiorboli, C., Piazza, R., Carassiti, V., Passerini, L. and Tosato, M.L. (1993a). Application of Chemometrics to the Screening of Hazardous Substances. Part II. Advances in the Multivariate Characterization and Reactivity Modelling of Haloalkanes. *Chemom.Intell.Lab.Syst.*, *19*, 331–336.

Chiorboli, C., Piazza, R., Carassiti, V., Passerini, L. and Tosato, M.L. (1993b). Modelling of Ionization Potential of Halogenated Aliphatics. *Quant.Struct.-Act.Relat.*, *12*, 38–43.

Chiorboli, C., Piazza, R., Tosato, M.L. and Carassiti, V. (1993c). Atmospheric Chemistry: Rate Constants of the Gas-Phase Reactions Between Haloalkanes of Environmental Interest and Hydroxyl Radicals. *Coordination Chemistry Reviews*, *125*, 241–250.

Chiorboli, C., Piazza, R., Carassiti, V., Passerini, L., Pino, A., Tosato, M.L. and Riganelli, D. (1996). A Model for the Tropospheric Persistency of Hydrohalo Alkanes. *Gazz.Chim.It.*, *126*, 685–694.

Cho, S.J. and Tropsha, A. (1995). Cross-Validated R^2-Guided Region Selection for Comparative Molecular Field Analysis: A Simple Method to Achieve Consistent Results. *J.Med.Chem.*, *38*, 1060–1066.

Cho, S.J., Garsia, M.L.S., Bier, J. and Tropsha, A. (1996). Structure-Based Alignment and Comparative Molecular Field Analysis of Acetylcholinesterase Inhibitors. *J.Med.Chem.*, *39*, 5064–5071.

Cho, S.J., Zheng, W. and Tropsha, A. (1998). Rational Combinatorial Library Design. 2. Rational Design of Targeted Combinatorial Peptide Libraries Using Chemical Similarity Probe and the Inverse QSAR Approaches. *J.Chem.Inf.Comput.Sci.*, *38*, 259–268.

Chou, J.T. and Jurs, P.C. (1979). Computer-Assisted Computation of Partial Coefficients from Molecular Structures Using Fragment Constants. *J.Chem.Inf.Comput.Sci.*, *19*, 172–178.

Chuman, H., Ito, A., Saishoji, T. and Kumazawa, S. (1995). QSARs and 3 Dimensional Shape Studies of Fungicidal Azolylmethylcyclopentanols Molecular Design of Novel Fungicides Metconazole and Ipconazole. *ACS Symp.Ser.*, *606*, 171–185.

Chuman, H., Karasawa, M. and Fujita, T. (1998). A Novel 3-Dimensional QSAR Procedure – Voronoi Field Analysis. *Quant.Struct.-Act.Relat.*, *17*, 313–326.

Chung, F.R.K. (1988). The Average Distance and the Independence Number. *J.Serb.Chem.Soc.*, *12*, 229–235.

Ciubotariu, D., Deretey, E., Oprea, T.I., Sulea, T., Simon, Z., Kurunczi, L. and Chiriac, A. (1993). Multiconformational Minimal Steric Difference. Structure-Acetylcholinesterase Hydrolysis Rates Relations for Acetic Acid Esters. *Quant.Struct.-Act.Relat.*, *12*, 367–372.

Clare, B.W. (1994). Frontier Orbital Energies in Quantitative Structure-Activity Relationships: A Comparison of Quantum Chemical Methods. *Theor.Chim.Acta*, *87*, 415–430.

Clare, B.W. and Supuran, C.T. (1994). Carbonic Anhydrase Activators. 3. Structure-Activity Correlations of a Series of Isozyme II Activators. *J.Pharm.Sci.*, *83*, 768–773.

Clare, B.W. (1995a). Charge Transfer Complexes and Frontier Orbital Energies in QSAR. A Congeneric Series of Electron Acceptors. *J.Mol.Struct.(Theochem)*, *337*, 139–150.

Clare, B.W. (1995b). Structure-Activity Correlations for Psychotomimetics. 3. Tryptamines. *Aust.J.Chem.*, *48*, 1385–1400.

Clare, B.W. (1995c). The Relationship of Charge Transfer Complexes to Frontier Orbital Energies in QSAR. *J.Mol.Struct.(Theochem)*, *331*, 63–78.

Clark, D.E., Willett, P. and Kenny, P. (1992). Pharmacophoric Pattern Matching in Files of Three-Dimensional Chemical Structures: Use of Smoothed-Bounded Distance Matrices for the Representation and Searching of Conformationally-Flexible Molecules. *J.Mol.Graphics*, *10*, 194–204.

Clark, D.E., Willett, P. and Kenny, P. (1993). Pharmacophoric Pattern Matching in Files of Three-Dimensional Chemical Structures: Implementation of Flexible Searching. *J.Mol. Graphics*, *11*, 146–156.

Clark, D.E. and Murray, C.W. (1995). PRO_LIGAND: An Approach to *de Novo* Molecular Design. 5. Tools for the Analysis of Generated Structures. *J.Chem.Inf.Comput.Sci.*, *35*, 914–923.

Clark, M., Cramer III, R.D. and Van Opdenbosch, N. (1989). Validation of the General Purpose Tripos 5.2 Force Field. *J.Comput.Chem.*, *10*, 982–1012.

Clark, M. and Cramer III, R.D. (1993). The Probability of Chance Correlation Using Partial Least Squares (PLS). *Quant.Struct.-Act.Relat.*, *12*, 137–145.

Clark, R.D., Parlow, J.J., Brannigan, L.H., Schnur, D.M. and Duewer, D.L. (1995). Applications of Scaled Rank Sum Statistics in Herbicide QSAR. *ACS Symp.Ser.*, *606*, 264–281.

Clark, R.D. and Langton, W.J. (1998). Balancing Representativeness Against Diversity Using Optimizable *K*-Dissimilarity and Hierarchical Clustering. *J.Chem.Inf.Comput.Sci.*, *38*, 1079–1086.

Clementi, S., Cruciani, G., Baroni, M. and Costantino, G. (1993a). Series Design. In *3D QSAR in Drug Design. Theory, Methods and Applications.* (Kubinyi, H., ed.), ESCOM, Leiden (The Netherlands), pp. 567–582.

Clementi, S., Cruciani, G., Riganelli, D., Valigi, R., Costantino, G., Baroni, M. and Wold, S. (1993b). Autocorrelation as a Tool for a Congruent Description of Molecules in 3D QSAR Studies. *Pharm.Pharmacol.Lett.*, *3*, 5–8.

Clementi, S. (1995). Data Analyses or Problem Formulations. *J.Chemom.*, *9*, 226–228.

Clementi, S., Cruciani, G., Fifi, P., Riganelli, D., Valigi, R. and Musumarra, G. (1996). New Set of Principal Properties for Heteroaromatics Obtained by GRID. *Quant.Struct.-Act.Relat.*, *15*, 108–120.

Clerc, J.T. and Terkovics, A.L. (1990). Versatile Topological Structure Descriptor for Quantitative Structure/Property Studies. *Anal.Chim.Acta*, *235*, 93–102.

Clifford, A.F. (1959). The Electronegativity of Groups. *J.Phys.Chem.*, *63*, 1227–1231.

Cnubben, N.H., Peelen, S., Borst, J.W., Vervoort, J., Veeger, C. and Rietjens, I.M. (1994). Molecular Orbital Based Quantitative Structure-Activity Relationship for the Cytochrome P450 Catalyzed 4 Hydroxylation of Halogenated Anilines. *Chem.Res.Toxicol.*, *7*, 590–598.

Coats, E.A. (1998). The CoMFA Steroids as a Benchmark Dataset for Development of 3D QSAR Methods. In *3D QSAR in Drug Design – Vol. 3* (Kubinyi, H., Folkers, G. and Martin, Y.C., eds.), Kluwer/ESCOM, Dordrecht (The Netherlands), pp. 199–213.

Cocchi, M., Menziani, M.C. and De Benedetti, P.G. (1992). Theoretical versus Empirical Molecular Descriptors in Monosubstituted Benzenes. *Chemom.Intell.Lab.Syst.*, *14*, 209–224.

Cocchi, M. and Johansson, E. (1993). Amino Acids Characterization by GRID and Multivariate Data Analysis. *Quant.Struct.-Act.Relat.*, *12*, 1–8.

Cocchi, M., Menziani, M.C., Fanelli, F. and De Benedetti, P.G. (1995). Theoretical Quantitative Structure-Activity Relationship Analysis of Congeneric and Noncongeneric Alpha(1) Adrenoceptor Antagonists: A Chemometric Study. *J.Mol.Struct.(Theochem)*, *331*, 79–93.

Cocchi, M., De Benedetti, P.G., Seeber, R., Tassi, L. and Ulrici, A. (1999). Development of Quantitative Structure-Property Relationships Using Calculated Descriptors for the Prediction of the Physicochemical Properties (n_D, ϱ, bp, ε, η) of a Series of Organic Solvents. *J. Chem.Inf.Comput.Sci.*, *39*, 1190–1203.

Coccini,T., Giannoni, L., Karcher, W., Manzo, L., and Roi, R. eds. (1992). *Quantitative Structure/Activity Relationships (QSAR) in Toxicology.* Joint Research Centre – EEC, Brussel (Belgium), 90 pp.

Cohen, J.L., Lee, W. and Lien, E.J. (1974). Dependence of Toxicity on Molecular Structure: Group Theory Analysis. *J.Pharm.Sci.*, *63*, 1068–1072.

Cohen, L.A. and Jones, W.M. (1963). A Study of Free Energy Relationships in Hindered Phenols. Linear Dependence for Solvation Effects in Ionization. *J.Am.Chem.Soc.*, *85*, 3397–3402.

Cohen, N. and Benson, S.W. (1987). Empirical Correlations for Rate Coefficients of Reactions of OH with Haloalkanes. *J.Phys.Chem.*, *91*, 171–175.

Collantes, E.R. and Dunn III, W.J. (1995). Amino Acid Side Chain Descriptors for Quantitative Structure-Activity Relationship Studies of Peptide Analogues. *J.Med.Chem.*, *38*, 2705–2713.

Collantes, E.R., Tong, W. and Welsh, W.J. (1996). Use of Moment of Inertia in Comparative Molecular Field Analysis to Model Chromatographic Retention of Nonpolar Solutes. *Anal.Chem.*, *68*, 2038–2043.

Combes, R.D. and Judson, P.N. (1995). The Use of Artificial Intelligence Systems for Predicting Toxicity. *Pestic.Sci.*, *45*, 179–194.

Compadre, R.L.L., Byrd, C. and Compadre, C.M. (1998). Comparative QSAR and 3-D-QSAR Analysis of the Mutagenicity of Nitroaromatic Compounds. In *Comparative QSAR* (Devillers, J., ed.), Taylor & Francis, Washington (DC), pp. 111–136. [R]

Cone, M.M., Venkataraghavan, R. and McLafferty, F.W. (1977). Molecular Structure Comparison Program for the Identification of Maximal Common Substructures. *J.Am.Chem.Soc.*, *99*, 7668–7671.

Connolly, M.L. (1983a). Analytical Molecular Surface Calculation. *J.Appl.Cryst.*, *16*, 548–558.

Connolly, M.L. (1983b). Solvent-Accessible Surfaces of Proteins and Nucleic Acids. *Science*, *221*, 709–713.

Connolly, M.L. (1985). Computation of Molecular Volume. *J.Am.Chem.Soc.*, *107*, 1118–1124.

Connolly, M.L. (1994). Adjoint Join Volumes. *J.Math.Chem.*, *15*, 339–352.

Constans, P. and Carbó, R. (1995). Atomic Shell Approximation: Electron Density Fitting Algorithm Restricting Coefficients to Positive Values. *J.Chem.Inf.Comput.Sci.*, *35*, 1046–1053.

Convard, T., Dubost, J.P., Le Solleu, H. and Kummer, E. (1994). SmilogP: A Program for a Fast Evaluation of Theoretical Log*P* from the Smiles Code of a Molecule. *Quant.Struct.-Act.Relat.*, *13*, 34–37.

Corbella, R., Rodriguez, M.A., Sanchez, M.J. and Montelongo, F.G. (1995). Correlations Between Gas Chromatographic Retention Data of Polycyclic Aromatic Hydrocarbons and Several Molecular Descriptors. *Chromatographia*, *40*, 532–538.

Cosentino, U., Moro, G., Quintero, M.G., Giraldo, E., Rizzi, C.A., Schiavi, G.B. and Turconi, M. (1993). The Role of Electronic and Conformational Properties in the Activity of 5-HT(3) Receptor Antagonists. *J.Mol.Struct.(Theochem)*, *105*, 275–291.

Coulson, C.A. (1939). The Electronic Structure of Some Polyenes and Aromatic Molecules. VII. Bonds of Fractional Order by the Molecular Orbital method. *Proc.Roy.Soc.London A*, *169*, 413–428.

Coulson, C.A. (1946). *Trans.Faraday Soc.*, *42*, 265.

Coulson, C.A. (1960). Present State of Molecular Structure Calculations. *Rev.Mod.Phys.*, *32*, 170.

Craig, P.N. (1984). *Drug Inf.J.*, *18*, 123–130.

Craig, P.N. (1990). Substructural Analysis and Compound Selection. In *Quantitative Drug Design. Vol. 4* (Ramsden, C.A., ed.), Pergamon Press, Oxford (UK), pp. 645–666.

Craik, D.J. and Brownlee, R.T.C. (1983). Substituent Effects on Chemical Shifts in the Sidechains of Aromatic Systems. *Prog.Phys.Org.Chem.*, *14*, 1–73.

Cramer III, R.D., Redl, G. and Berkoff, C.E. (1974). Substructural Analysis. A Novel Approach to the Problem of Drug Design. *J.Med.Chem.*, *17*, 533–535.

Cramer III, R.D. (1980a). BC(DEF) Parameters. 1. The Intrinsic Dimensionality of Intermolecular Interactions in the Liquid State. *J.Am.Chem.Soc.*, *102*, 1837–1849.

Cramer III, R.D. (1980b). BC(DEF) Parameters. 2. An Empirical Structure-Based Scheme for the Prediction of Some Physical Properties. *J.Am.Chem.Soc.*, *102*, 1849–1859.

Cramer III, R.D. (1983a). BC(DEF) Coordinates. 3. Their Acquisition from Physical Property Data. *Quant.Struct.-Act.Relat.*, *2*, 7–12.

Cramer III, R.D. (1983b). BC(DEF) Coordinates. 4. Correlations with General Anesthesia, Nerve Blockade, and Erythrocyte Stabilization. *Quant.Struct.-Act.Relat.*, *2*, 13–19.

Cramer III, R.D., Bunce, J.D., Patterson, D.E. and Frank, I.E. (1988a). Crossvalidation, Bootstrapping and Partial Least Squares Compared with Multiple Regression in Conventional QSAR Studies. *Quant.Struct.-Act.Relat.*, *7*, 18–25.

Cramer III, R.D., Patterson, D.E. and Bunce, J.D. (1988b). Comparative Molecular Field Analysis (CoMFA). 1. Effect of Shape on Binding of Steroids to Carrier Proteins. *J.Am.Chem.Soc.*, *110*, 5959–5967.

Cramer III, R.D. (1993). Partial Least Squares (PLS): Its Strengths and Limitations. *Persp.Drug Disc.Des.*, *1*, 269–278.

Cramer III, R.D., DePriest, S.A., Patterson, D.E. and Hecht, P. (1993). The Developing Practice of Comparative Molecular Field Analysis. In *3D QSAR in Drug Design. Theory, Methods and Applications.* (Kubinyi, H., ed.), ESCOM, Leiden (The Netherlands), pp. 443–485.

Cramer III, R.D., Clark, R.D., Patterson, D.E. and Ferguson, A.M. (1996). Bioisosterism as a Molecular Diversity Descriptor: Steric Fields of Single "Topomeric" Conformers. *J.Med. Chem.*, *39*, 3060–3069.

Cramer III, R.D., Patterson, D.E., Clark, R.D., Soltanshahi, F. and Lawless, M.S. (1998). Visual Compound Libraries: A New Approach to Decision Making in Molecular Discovery Research. *J.Chem.Inf.Comput.Sci.*, *38*, 1010–1023.

Cramer, C.J., Famini, G.R. and Lowrey, A.H. (1993). Use of Calculated Quantum Chemical Properties as Surrogates for Solvatochromic Parameters in Structure-Activity Relationships. *Acc.Chem.Res.*, *26*, 599–605.

Cramer, C.J. (1995). Continuum Solvation Models: Classical and Quantum Mechanical Implementations. In *Reviews in Computational Chemistry – Vol. 6* (Lipkowitz, K.B. and Boyd, D.B., eds.), VCH Publishers, New York (NY), pp. 1–72.

Crebelli, R., Andreoli, C., Carere, A., Conti, G., Conti, L., Ramusino, M.C. and Benigni, R. (1992). The Induction of Mitotic Chromosome Malsegregation in Aspergillus Nidulans. Quantitative Structure-Activity Relationship (QSAR) Analysis with Chlorinated Aliphatic Hydrocarbons. *Mut.Res.*, *266*, 117–134.

Crebelli, R., Andreoli, C., Carere, A., Conti, L., Crochi, B., Cotta Ramusino, M. and Benigni, R. (1995). Toxicology of Halogenated Aliphatic Hydrocarbons Structural and Molecular Determinants for the Disturbance of Chromosome Segregation and the Induction of Lipid Peroxidation. *Chem.-Biol.Inter.*, *98*, 113–129.

Crippen, G.M. (1977). A Novel Approach to Calculation of Conformation. *J.Comput.Phys.*, *24*, 96–107.

Crippen, G.M. (1978). *J.Comput.Phys.*, *26*, 449.

Crippen, G.M. (1979). Distance Geometry Approach to Rationalizing Binding Data. *J.Med. Chem.*, *22*, 988–997.

Crippen, G.M. (1980). Quantitative Structure-Activity Relationships by Distance Geometry: Systematic Analysis of Dihydrofolate Reductase Inhibitors. *J.Med.Chem.*, *23*, 599–606.

Crippen, G.M. (1981). *Distance Geometry and Conformational Calculations.* Research Studies Press, Letchworth (UK), 58 pp.

Crippen, G.M. (1987). Voronoi Binding Site Models. *J.Comput.Chem.*, *8*, 943–955.

Crippen, G.M. (1991). Chemical Distance Geometry: Current Realization and Future Projection. *J.Math.Chem.*, *6*, 307–324.

Cristante, M., Selves, J.L., Grassy, G. and Colin, J.P. (1993). Structure-Activity Relationship Study on Paraffin Inhibitors for Crude Oils (Inipar Model II). *Anal.Chim.Acta*, *274*, 303–316.

Croizet, F., Langlois, M.H., Dubost, J.P., Braquet, P., Audry, E., Dallet, Ph. and Colleter, J.C. (1990). Lipophilicity Force Field Profile: An Expressive Visualization of the Lipophilicity Molecular Potential Gradient. *J.Mol.Graphics*, *8*, 153–155.

Cronin, M.T.D., Dearden, J.C. and Dobbs, A.J. (1991). QSAR Studies of Comparative Toxicity in Aquatic Organisms. *Sci.Total Environ.*, *109/110*, 431–439.

Cronin, M.T.D. (1992). Molecular Descriptors of QSAR. In *Quantitative Structure/Activity Relationships (QSAR) in Toxicology* (Coccini, T., Giannoni, L., Karcher, W., Manzo, L. and Roi, R., eds.), Joint Research Centre – EEC, Brussel (Belgium), pp. 43–54.

Cronin, M.T.D., Basketter, D.A. and York, M. (1994). A Quantitative Structure-Activity Relationship (QSAR) Investigation of a Draize Eye Irritation Database. *Toxicol.Vitro*, *8*, 21–28.

Cronin, M.T.D. and Dearden, J.C. (1995a). QSAR in Toxicology. 1. Prediction of Aquatic Toxicity. *Quant.Struct.-Act.Relat.*, *14*, 1–7. [R]

Cronin, M.T.D. and Dearden, J.C. (1995b). QSAR in Toxicology. 2. Prediction of Acute Mammalian Toxicity and Interspecies Correlations. *Quant.Struct.-Act.Relat.*, *14*, 117–120.

Cronin, M.T.D. and Dearden, J.C. (1995c). QSAR in Toxicology. 3. Prediction of Chronic Toxicity. *Quant.Struct.-Act.Relat.*, *14*, 329–334.

Cronin, M.T.D. and Dearden, J.C. (1995d). QSAR in Toxicology. 4. Prediction of Non-Lethal Mammalian Toxicological End Points, and Expert Systems for Toxicity Prediction. *Quant. Struct.-Act.Relat.*, *14*, 518–523. [R]

Cronin, M.T.D. (1996). Quantitative Structure-Activity Relationship (QSAR): Analysis of the Acute Sublethal Neurotoxicity of Solvents. *Toxicol.Vitro*, *10*, 103–110.

Cronin, M.T.D. and Schultz, T.W. (1996). Structure-Toxicity Relationships for Phenols to Tetrahymena Pyriformis. *Chemosphere*, *32*, 1453–1468.

Cruciani, G., Baroni, M., Clementi, S., Costantino, G., Riganelli, D. and Skagerberg, B. (1992). Predictive Ability of Regression Models. Part I: Standard Deviation of Prediction Errors (SDEP). *J.Chemom.*, *6*, 335–346.

Cruciani, G., Clementi, S. and Baroni, M. (1993). Variable Selection in PLS Analysis. In *3D QSAR in Drug Design. Theory, Methods and Applications.* (Kubinyi, H., ed.), ESCOM, Leiden (The Netherlands), pp. 551–564.

Cruciani, G. and Clementi, S. (1994). GOLPE: Philosophy and Applications in 3D QSAR. In *Advanced Computer-Assisted Techniques in Drug Discovery* (van de Waterbeemd, H., ed.), VCH Publishers, New York (NY), pp. 61–88.

Cruciani, G. and Watson, K.A. (1994). Comparative Molecular Field Analysis Using GRID Force-Field and GOLPE Variable Selection Methods in a Study of Inhibitors of Glycogen Phosphorylase b. *J.Med.Chem.*, *37*, 2589–2601.

Cruciani, G., Pastor, M. and Clementi, S. (1997). Region Selection in 3D-QSAR. In *Computer-Assisted Lead Finding and Optimization* (van de Waterbeemd, H., Testa, B. and Folkers, G., eds.), Wiley-VCH, Weinheim (Germany), pp. 381–395.

Cruciani, G., Clementi, S. and Pastor, M. (1998). GOLPE-Guided Region Selection. In *3D QSAR in Drug Design – Vol. 3* (Kubinyi, H., Folkers, G. and Martin, Y.C., eds.), Kluwer/ESCOM, Dordrecht (The Netherlands), pp. 71–86.

Crum-Brown, A. (1864). On the Theory of Isomeric Compounds. *Trans.Roy.Soc.Edinburgh*, *23*, 707–719.

Crum-Brown, A. and Fraser, T.R. (1868a). *Trans.Roy.Soc.Edinburgh*, *25*, 151–203.

Crum-Brown, A. and Fraser, T.R. (1868b). *Trans.Roy.Soc.Edinburgh*, *25*, 693–739.

Cuadras, C.M. (1989). Distancias Estadísticas. *Estadistica Española*, *30*, 295–378.

Cummins, D.J. and Andrews, C.W. (1995). Iteratively Reweighted Partial Least Squares: A Performance Analysis by Monte Carlo Simulation. *J.Chemom.*, *9*, 489–507.

Cvetkovic, D.M. and Gutman, I. (1977). Note on Branching. *Croat.Chem.Acta*, *49*, 115–121.

Cvetkovic, D.M. and Gutman, I. (1985). The Computer System GRAPH: A Useful Tool in Chemical Graph Theory. *J.Comput.Chem.*, *7*, 640–644.

Cvetkovic, D.M. and Fowler, P.W. (1999). A Group-Theoretical Bound for the Number of Main Eigenvalues of a Graph. *J.Chem.Inf.Comput.Sci.*, *39*, 638–641.

Cyranski, M. and Krygowski, T.M. (1996). Separation of the Energetic and Geometric Contributions to Aromaticity. 3. Analysis of the Aromatic Character of Benzene Rings in Their Various Topological and Chemical Environments in the Substituted Benzene Derivates. *J.Chem.Inf. Comput.Sci.*, *36*, 1142–1145.

Cyvin, B.N., Brunvoll, J., Cyvin, S.J. and Gutman, I. (1988). All-Benzenoid Systems: Enumeration and Classification of Benzenoid Hydrocarbons. *MATCH (Comm.Math.Comp.Chem.)*, *23*, 163–174.

Da, Y.-Z., Ito, K. and Fujiwara, H. (1992). Energy Aspects of Oil/Water Partition Leading to the Novel Hydrophobic Parameters for the Analysis of Quantitative Structure-Activity Relationships. *J.Med.Chem.*, *35*, 3382–3387.

Da, Y.-Z., Yanagi, J., Tanaka, K. and Fujiwara, H. (1993). Thermochemical Aspects of Partition Quantitative Structure-Activity Relationships of Benzyldimethylalkylammonium Chlorides. *Chem.Pharm.Bull.*, *41*, 227–230.

Dancoff, S.M. and Quastler, H. (1953). *Essays on the Use of Information Theory in Biology.* University of Illinois, Urbana (ILL).

Dang, P. and Madan, A.K. (1994). Structure-Activity Study on Anticonvulsant (Thio) Hydantoins Using Molecular Connectivity Indices. *J.Chem.Inf.Comput.Sci.*, *34*, 1162–1166.

Dannenfelser, R.M., Surendran, N. and Yalkowsky, S.H. (1993). Molecular Symmetry and Related Properties. *SAR & QSAR Environ.Res.*, *1*, 273–292.

Dash, S.C. and Behera, G.B. (1980). A New Steric Parameter to Explain *ortho*-Substituent Effect. *Indian J.Chem.*, *19A*, 541–543.

Davis, A.M., Gensmantel, N.P., Johansson, E. and Marriott, D.P. (1994). The Use of the Grid Program in the 3-D QSAR Analysis of a Series of Calcium Channel Agonists. *J.Med.Chem.*, *37*, 963–972.

Davis, L. ed. (1991). *Handbook of Genetic Algorithms.* Van Nostrand Reinhold, New York (NY).

De Benedetti, P.G. (1992). Electrostatics in Quantitative Structure-Activity Relationship Analysis. *J.Mol.Struct.(Theochem)*, *88*, 231–248.

De Benedetti, P.G., Cocchi, M., Menziani, M.C. and Fanelli, F. (1993). Theoretical Quantitative Structure-Activity Analysis and Pharmacophore Modeling of Selective Noncongeneric Alpha 1a-Adrenergic Antagonists. *J.Mol.Struct.(Theochem)*, *99*, 283–290.

De Benedetti, P.G., Menziani, M.C., Cocchi, M. and Fanelli, F. (1995). Prototropic Molecular Forms and Theoretical Descriptors in QSAR Analysis. *J.Mol.Struct.(Theochem)*, *333*, 1–17.

de Bruijn, J. and Hermens, J.L. (1990). Relationships Between Octanol/Water Partition Coefficients and Total Molecular Surface Area and Total Molecular Volume of Hydrophobic Organic Chemicals. *Quant.Struct.-Act.Relat.*, *9*, 11–21.

de Bruijn, J. and Hermens, J.L. (1993). Inhibition of Acetylcholinesterase and Acute Toxicity of Organophosphorous Compounds to Fish: A Preliminary Structure-Activity Analysis. *Aquat. Toxicol.*, *24*, 257–274.

De Castro, L.F.P. and Reissmann, S. (1995). QSAR in Bradykinin Antagonists. Inhibition of the Bradykinin Induced Contraction of the Isolated Rat Uterus and Guinea Pig Ileum. *Quant.-Struct.-Act.Relat.*, *14*, 249–257.

de Gregorio, C., Kier, L.B. and Hall, L.H. (1998). QSAR Modeling with the Electrotopological State Indices: Corticosteroids. *J.Comput.Aid.Molec.Des.*, *12*, 557–561.

De Julián-Ortiz, V., Garcìa-Domenech, R., Gálvez, J., Soler-Roca, R., García-March, F.J. and Antón-Fos, G.M. (1996). Use of Topological Descriptors in Chromatographic Chiral Separations. *J.Chromat.*, *719A*, 37–44.

De La Guardia, M., Carrión, J.L. and Galdú, M.V. (1988). The Use of Topological Models in Analytical Chemistry. *J.Chemom.*, *3*, 193–207.

De Maria, P., Fini, A. and Hall, F.M. (1973). Thermodynamic Acid Dissociation Constants of Aromatic Thiols. *J.Chem.Soc.Perkin Trans.2*, 1969–1971.

Dean, P.M. (1987). *Molecular Foundations of Drug-Receptor Interaction*. Cambridge University Press, Cambridge (UK), 381 pp.

Dean, P.M. and Callow, P. (1987). Molecular Recognition: Identification of Local Minima for Matching in Rotational 3-Space by Cluster Analysis. *J.Mol.Graphics*, *5*, 159–164.

Dean, P.M. and Chau, P.-L. (1987). Molecular Recognition: Optimized Searching through Rotational 3-Space for Pattern Matches on Molecular Surfaces. *J.Mol.Graphics*, *5*, 152–158.

Dean, P.M., Callow, P. and Chau, P.-L. (1988). Molecular Recognition: Blind-Searching for Regions of Strong Structural Match on the Surfaces of Two Dissimilar Molecules. *J.Mol. Graphics*, *6*, 28–34.

Dean, P.M. (1990). Molecular Recognition: The Measurement and Search for Molecular Similarity in Ligand-Receptor Interaction. In *Concepts and Applications of Molecular Similarity* (Johnson, M.A. and Maggiora, G.M., eds.), Wiley, New York (NY), pp. 211–238.

Dean, P.M. (1993). Molecular Similarity. In *3D QSAR in Drug Design. Theory, Methods and Applications.* (Kubinyi, H., ed.), ESCOM, Leiden (The Netherlands), pp. 150–172.

Dean, P.M. and Perkins, T.D.J. (1993). Searching for Molecular Similarity Between Flexible Molecules. In *Trends in QSAR and Molecular Modelling 92* (Wermuth, C.G., ed.), ESCOM, Leiden (The Netherlands), pp. 207–215.

Dean, P. M. ed. (1995). *Molecular Similarity in Drug Design*. Chapman & Hall, London (UK).

Dearden, J. C. ed. (1983). *Quantitative Approaches to Drug Design*. Elsevier, Amsterdam (The Netherlands).

Dearden, J.C. (1985). Partitioning and Lipophilicity in Quantitative Structure-Activity Relationships. *Environ.Health Persp.*, *61*, 203–228.

Dearden, J.C. (1990). Physico-Chemical Descriptors. In *Practical Applications of Quantitative Structure-Activity Relationships (QSAR) in Environmental Chemistry and Toxicology.* (Karcher, W. and Devillers, J., eds.), Kluwer, Dordrecht (The Netherlands), pp. 25–59. [R]

Dearden, J.C. (1991a). The QSAR Prediction of Melting Point, a Property of Environmental Relevance. *Sci.Total Environ.*, *109/110*, 59–68.

Dearden, J.C., Bradburne, S.J.A. and Abraham, M.H. (1991b). The Nature of Molar Refractivity. In *QSAR: Rational Approaches to the Design of Bioactive Compounds* (Silipo, C. and Vittoria, A., eds.), Elsevier, Amsterdam (The Netherlands), pp. 143–150.

Dearden, J.C., Cronin, M.T.D. and Dobbs, A.J. (1995a). Quantitative Structure-Activity Relationships as a Tool to Assess the Comparative Toxicity of Organic Chemicals. *Chemosphere*, *31*, 2521–2528.

Dearden, J.C., Cronin, M.T.D., Schultz, T.W. and Lin, D.T. (1995b). QSAR Study of the Toxicity of Nitrobenzenes to *Tetrahymena pyriformis*. *Quant.Struct.-Act.Relat.*, *14*, 427–432.

Dearden, J.C. and Ghafourian, T. (1995). Investigation of Calculated Hydrogen Bonding Parameters for QSAR. In *QSAR and Molecular Modelling: Cocepts, Computational Tools and Biological Applications* (Sanz, F., Giraldo, J. and Manaut, F., eds.), Prous Science, Barcelona (Spain), pp. 117–119.

Dearden, J.C., Cronin, M.T.D. and Wee, D. (1997). Prediction of Hydrogen Bond Donor Ability Using New Quantum Chemical Parameters. *J.Pharm.Pharmacol.*, *49, Suppl. 4*, 110–110.

Dearden, J.C. and Ghafourian, T. (1999). Hydrogen Bonding Parameters for QSAR: Comparison of Indicator Variables, Hydrogen Bond Counts, Molecular Orbital and Other Parameters. *J.Chem.Inf.Comput.Sci.*, *39*, 231–235.

Debnath, A.K., Compadre, R.L.L., Debnath, G., Shusterman, A.J. and Hansch, C. (1991). Structure-Activity Relationship of Mutagenic Aromatic and Heteroaromatic Nitro Compounds. Correlation with Molecular Orbital Energies and Hydrophobicity. *J.Med.Chem.*, *34*, 787–797.

Debnath, A.K., Compadre, R.L.L. and Hansch, C. (1992a). Mutagenicity of Quinolines in Salmonella typhimurium TA100. A QSAR Study Based on Hydrophobicity and Molecular Orbital Determinants. *Mut.Res.*, *280*, 55–65.

Debnath, A.K., Compadre, R.L.L., Shusterman, A.J. and Hansch, C. (1992b). Quantitative Structure-Activity Relationship Investigation of the Role of Hydrophobicity in Regulating Mutagenicity in the Ames Test: 2. Mutagenicity of Aromatic and Heteroaromatic Nitro Compounds in Salmonella typhimurium TA100. *Environ.Mol.Mutagen.*, *19*, 53–70.

Debnath, A.K., Debnath, G., Shusterman, A.J. and Hansch, C. (1992c). A QSAR Investigation of the Role of Hydrophobicity in Regulating Mutagenicity in the Ames Test: 1. Mutagenicity of Aromatic and Heteroaromatic Amines in Salmonella typhimurium TA98 and TA100. *Environ.Mol.Mutagen.*, *19*, 37–52.

Debnath, A.K. and Hansch, C. (1992d). Structure-Activity Relationship of Genotoxic Polycyclic Aromatic Nitro-Compounds. Further Evidence for the Importance of Hydrophobicity and Molecular Orbital Energies in Genetic Toxicity. *Environ.Mol.Mutagen.*, *20*, 140–144.

Debnath, A.K., Hansch, C., Kim, K.H. and Martin, Y.C. (1993). Mechanistic Interpretation of the Genotoxicity of Nitrofurans (Antibacterial Agents) Using Quantitative Structure-Activity Relationships and Comparative Molecular Field Analysis. *J.Med.Chem.*, *36*, 1007–1016.

Debnath, A.K., Jiang, S., Strick, N., Lin, K., Haberfield, P. and Neurath, A.R. (1994a). Three Dimensional Structure-Activity Analysis of a Series of Porphyrin Derivatives with Anti HIV-1 Activity Targeted to the V3 Loop of the gp120 Envelope Glycoprotein of the Human Immunodeficiency Virus Type 1. *J.Med.Chem.*, *37*, 1099–1108.

Debnath, A.K., Shusterman, A.J., Compadre, R.L.L. and Hansch, C. (1994b). The Importance of the Hydrophobic Interaction in the Mutagenicity of Organic Compounds. *Mut.Res.*, *305*, 63–72.

Debnath, A.K. (1998). Comparative Molecular Field Analysis (CoMFA) of a Series of Symmetrical Bis-Benzamide Cyclic Urea Derivatives as HIV-1 Protease Inhibitors. *J.Chem.Inf.Comput.Sci.*, *38*, 761–767.

Del Re, G. (1958). A Simple MO-LCAO Method for the Calculation of Charge Distribution in Saturated Organic Molecules. *J.Chem.Soc.*, 4031–4040.

Delaney, J.S., Mullaley, A., Mullier, G.W., Sexton, G.J., Taylor, R. and Viner, R.C. (1993). Rapid Construction of Data Tables for Quantitative Structure-Activity Relationship Studies. *J. Chem.Inf.Comput.Sci.*, *33*, 174–178.

Demeter, D.A., Weintraub, H.J.R. and Knittel, J.J. (1998). The Local Minima Method (LMM) of Pharmacophore Determination: A Protocol for Predicting the Bioactive Conformation of Small, Conformationally Flexible Molecules. *J.Chem.Inf.Comput.Sci.*, *38*, 1125–1136.

Demirev, P.A., Dyulgerov, A.S. and Bangov, I.P. (1991). CTI: A Novel Charge-Related Topological Index with Low Degeneracy. *J.Math.Chem.*, *8*, 367–382.

Denny, W.A., Cain, B.F., Atwell, C., Hansch, C., Panthananickal, A. and Leo, A. (1982). Potential Antitumor Agents. 36. Quantitative Relationships Between Antitumor Activity, Toxicity and Structure for the General Class of 9-Anilinoacridine Antitumor Agents. *J.Med.Chem.*, *25*, 276–315.

DePriest, S.A., Mayer, D., Naylor, C.B. and Marshall, G.R. (1993). 3D QSAR of Angiotensin Converting Enzyme and Thermolysin Inhibitors. A Comparison of CoMFA Models Based on Deduced and Experimentally Determined Active Site Geometries. *J.Am.Chem.Soc.*, *115*, 5372–5384.

DeTar, D.F. and Tenpas, C.J. (1976). Theoretical Calculation of Steric Effects in Ester Hydrolysis. *J.Am.Chem.Soc.*, *98*, 7903–7908.

DeTar, D.F. and Delahunty, C. (1983). Ester Aminolysis: New Reaction Series for the Quantitative Measurement of Steric Effects. *J.Am.Chem.Soc.*, *105*, 2734–2739.

Devillers, J., Chambon, P., Zakarya, D. and Chastrette, M. (1986). Quantitative Structure-Activity Relations of the Lethal Effects of 38 Halogenated Compounds against Lepomis Macrochirus. *C.R.Acad.Sci.,Paris*, *303*, 613–616.

Devillers, J. and Lipnick, R.L. (1990). Practical Applications of Regression Analysis in Environmental QSAR Studies. In *Practical Applications of Quantitative Structure-Activity Relationships (QSAR) in Environmental Chemistry and Toxicology* (Karcher, W. and Devillers, J., eds.), Kluwer, Dordrecht (The Netherlands), pp. 129–144.

Devillers, J. and Karcher, W. eds. (1991). *Applied Multivariate Analysis in SAR and Environmental Studies.* Kluwer Academic Publishers for the European Communities, Dordrecht (The Netherlands), 530 pp.

Devillers, J. ed. (1996). *Genetic Algorithms in Molecular Modeling. Principles of QSAR and Drug Design. Vol. 1.* Academic Press, London (UK), 327 pp.

Devillers, J. (1996a). Genetic Algorithms in Computer-Aided Molecular Design. In *Genetic Algorithms in Molecular Modeling. Principles of QSAR and Drug Design. Vol. 1* (Devillers, J., ed.), Academic Press, London (UK), pp. 131–157.

Devillers, J. ed. (1998). *Comparative QSAR.* Taylor & Francis, Washington (DC), 371 pp. [R]

Devillers, J., Domine, D., Bintein, S. and Karcher, W. (1998). Comparison of Fish Bioconcentration Models. In *Comparative QSAR* (Devillers, J., ed.), Taylor & Francis, Washington (DC), pp. 1–50. [R]

Devillers, J. (2000). Autocorrelation Descriptors for Modeling (Eco)Toxicological Endpoints. In *Topological Indices and Related Descriptors in QSAR and QSPR* (Devillers, J. and Balaban, A.T., eds.), Gordon & Breach, Amsterdam (The Netherlands), pp. 595–612.

Devillers, J. and Balaban, A. T. eds. (2000). *Topological Indices and Related Descriptors in QSAR and QSPR.* Gordon & Breach, Amsterdam (The Netherlands), 824 pp.

Dewar, M.J.S. and Longuet-Higgins, H.C. (1952). The Correspondence Between the Resonance and Molecular Orbital Theories. *Proc.Roy.Soc.London A*, *214*, 482–493.

Dewar, M.J.S. and Grisdale, P.J. (1962a). Substituent Effects. I. Introduction. *J.Am.Chem.Soc.*, *84*, 3539–3541.

Dewar, M.J.S. and Grisdale, P.J. (1962b). Substituent Effects. IV. A Quantitative Theory. *J.Am.Chem.Soc.*, *84*, 3548–3553.

Dewar, M.J.S. and Gleicher, G.J. (1965). Ground States of Conjugated Molecules. III. Classical Polyenes. *J.Am.Chem.Soc.*, *87*, 692–696.

Dewar, M.J.S. (1969a). *The Molecular Orbital Theory of Organic Chemistry.* McGraw-Hill, New York (NY).

Dewar, M.J.S. and de Llano, C. (1969b). Ground States of Conjugated Molecules. XI. Improved Treatment of Hydrocarbons. *J.Am.Chem.Soc.*, *91*, 789–795.

Dewar, M.J.S., Harget, A.J. and Trinajstic, N. (1969). Ground States of Conjugated Molecules. XV. Bond Localization and Resonance Energies in Compounds Containing Nitrogen or Oxygen. *J.Am.Chem.Soc.*, *91*, 6321–6325.

Dewar, M.J.S. and Harget, A.J. (1970a). Ground States of Conjugated Molecules. XVI. Treatment of Hydrocarbons by l.c.a.o. s.c.f. m.o. *Proc.Roy.Soc.London A*, *315*, 443–455.

Dewar, M.J.S. and Harget, A.J. (1970b). Ground States of Conjugated Molecules. XVII. The l.c.a.o. s.c.f. m.o. treatment of Compounds Containing Nitrogen and Oxygen. *Proc.Roy.Soc.London A*, *315*, 457–464.

Dewar, M.J.S., Haselbach, E. and Worley, S.D. (1970). Calculated and Observed Ionization Potentials of Unsaturated Polycyclic Hydrocarbons; Calculated Heats of Formation by Several Semiempirical s.c.f. m.o. Methods. *Proc.Roy.Soc.London A*, *315*, 431–442.

Dewar, M.J.S. and Trinajstic, N. (1970). Resonance Energies of Some Compounds Containing Nitrogen or Oxygen. *Theor.Chim.Acta*, *17*, 235–238.

Dewar, M.J.S., Golden, R. and Harris, J.M. (1971a). Substituent Effects. X. An Improved Treatment (FMMF) of Substituent Effects. *J.Am.Chem.Soc.*, *93*, 4187–4195.

Dewar, M.J.S., Kohn, M.C. and Trinajstic, N. (1971b). Cyclobutadiene and Diphenylcyclobutadiene. *J.Am.Chem.Soc.*, *93*, 3437–3440.

Di Paolo, T., Kier, L.B. and Hall, L.H. (1977). Molecular Connectivity and Structure-Activity Relationship of General Anesthetics. *Mol.Pharm.*, *13*, 31–37.

Di Paolo, T. (1978a). Molecular Connectivity in Quantitative Structure-Activity Relationship Study of Anesthetic and Toxic Activity of Aliphatic Hydrocarbons, Ethers, and Ketones. *J.Pharm.Sci.*, *67*, 566–568.

Di Paolo, T. (1978b). Structure-Activity Relationships of Anesthetic Ethers Using Molecular Connectivity. *J.Pharm.Sci.*, *67*, 564–565.

Di Paolo, T., Kier, L.B. and Hall, L.H. (1979). Molecular Connectivity Study of Halocarbon Anesthetics. *J.Pharm.Sci.*, *68*, 39–42.

Diaconis, P. and Efron, B. (1983). Computer Intensive Methods in Statistics. *Sci.Am.*, 96–108.

Dias, J.R. (1987a). A Periodic Table for Polycyclic Aromatic Hydrocarbons. Part X. On the Characteristic Polynomial and Other Structural Invariants. *J.Mol.Struct.(Theochem)*, *149*, 213–241.

Dias, J.R. (1987b). Facile Calculations of Select Eigenvalues and the Characteristic Polynomial of Small Molecular Graphs Containing Heteroatoms. *Can.J.Chem.*, *65*, 734–739.

Dias, J.R. (1992). Algebraic Structure Count. *J.Math.Chem.*, *9*, 253–260.

Dias, J.R. (1993). *Molecular Orbital Calculations Using Chemical Graph Theory*. Springer Verlag, Berlin (Germany).

Dias, J.R. (1999). Directed toward the Development of a Unified Structure Theory of Polycyclic Conjugated Hydrocarbons: the Aufbau Principle in Structure/Similairty Studies. *J.Chem.Inf. Comput.Sci.*, *39*, 197–203.

Dietz, A. (1995). Yet Another Representation of Molecular Structure. *J.Chem.Inf.Comput.Sci.*, *35*, 787–802.

Dimitrov, S.D. and Mekenyan, O. (1997). Dynamic QSAR: Least Squares Fits with Multiple Predictors. *Chemom.Intell.Lab.Syst.*, *39*, 1–9.

Dimoglo, A.S., Bersuker, I.B. and Gorbachov, M.Yu. (1988). Electron-Topological Study of SAR of Various Inhibitors of α-chymotrypsin. *Khim-Farm.Zh.*, *22*, 1355–1361.

Dimov, N. and Papazova, D. (1978). Calculation of the Retention Indices of C_5-C_9 Cycloalkanes on Squalene. *J.Chromat.*, *148*, 11.

Dimov, N. and Papazova, D. (1979). Correlation Equations for Prediction of Gas Chromatographic Separation of Hydrocarbons on Squalene. *Chromatographia*, *12*, 720.

Dimov, N., Osman, A., Mekenyan, O. and Papazova, D. (1994). Selection of Molecular Descriptors Used in Quantitative Structure Gas Chromatographic Retention Relationships. 1. Application to Alkylbenzenes and Naphthalenes. *Anal.Chim.Acta*, *298*, 303–317.

Dimov, N. and Osman, A. (1996). Selection of Molecular Descriptors Used in Quantitative Structure-Gas Chromatographic Retention Relationships. 2. Isoalkanes and Alkenes. *Anal.Chim.Acta*, *323*, 15–25.

Dimroth, K., Reichardt, C., Seipman, T. and Bohlman, F. (1963). *Justus Liebig Ann.Chem.*, *661*, 1.

Diudea, M.V. and Pârv, B. (1988). Molecolar Topology. 3. A New Centric Connectivity Index (CCI). *MATCH (Comm.Math.Comp.Chem.)*, *23*, 65–87.

Diudea, M.V. and Silaghi-Dumitrescu, I. (1989). Molecular Topology. I. Valence Group Electronegativity as a Vertex Discriminator. *Rev.Roum.Chim.*, *34*, 1175–1182.

Diudea, M.V. and Bal, L. (1990). Recursive Relationships for Computing Y Indices in Some Particular Graphs. *Studia Univ.Babes-Bolyai*, *35*, 17–28.

Diudea, M.V. and Kacso, I.E. (1991). Composition Rules for Some Topological Indices. *MATCH (Comm.Math.Comp.Chem.)*, *26*, 255–269.

Diudea, M.V., Minailiuc, O.M. and Balaban, A.T. (1991). Molecular Topology. IV. Regressive Vertex Degrees (New Graph Invariants) and Derived Topological Indices. *J.Comput.Chem.*, *12*, 527–535.

Diudea, M.V., Horvath, D., Kacso, I.E., Minailiuc, O.M. and Pârv, B. (1992a). Molecular Topology. VIII: Centricities in Molecular Graphs. The MOLCEN Algorithm. *J.Math.Chem.*, *11*, 259–270.

Diudea, M.V., Kacso, I.E. and Minailiuc, O.M. (1992b). Y Indices in Homogeneous Dendrimers. *MATCH (Comm.Math.Comp.Chem.)*, *28*, 61–99.

Diudea, M.V. (1994). Molecular Topology. 16. Layer Matrices in Molecular Graphs. *J.Chem.Inf. Comput.Sci.*, *34*, 1064–1071.

Diudea, M.V., Topan, M. and Graovac, A. (1994). Molecular Topology. 17. Layer Matrices of Walk Degrees. *J.Chem.Inf.Comput.Sci.*, *34*, 1072–1078.

Diudea, M.V. (1995a). Molecular Topology 23. Novel Schultz Analogue Indices. *MATCH (Comm.Math.Comp.Chem.)*, *32*, 85–103.

Diudea, M.V. (1995b). Molecular Topology. 21. Wiener Index of Dendrimers. *MATCH (Comm. Math.Comp.Chem.)*, *32*, 71–83.

Diudea, M.V., Horvath, D. and Bonchev, D. (1995a). Molecular Topology. 14. MOLORD Algorithm and Real Number Subgraph Invariants. *Croat.Chem.Acta*, *68*, 131–148.

Diudea, M.V., Horvath, D. and Graovac, A. (1995b). Molecular Topology. 15. 3D Distance Matrices and Related Topological Indices. *J.Chem.Inf.Comput.Sci.*, *35*, 129–135.

Diudea, M.V. and Pârv, B. (1995). Molecular Topology. 25. Hyper-Wiener Index of Dendrimers. *J.Chem.Inf.Comput.Sci.*, *35*, 1015–1018.

Diudea, M.V. (1996a). Walk Numbers ${}^{e}W_M$: Wiener-Type Numbers of Higher Rank. *J.Chem.Inf. Comput.Sci.*, *36*, 535–540.

Diudea, M.V. (1996b). Wiener and Hyper-Wiener Numbers in a Single Matrix. *J.Chem.Inf.Comput.Sci.*, *36*, 833–836.

Diudea, M.V., Kacso, I.E. and Topan, M.I. (1996a). Molecular Topology.18. A QSPR/QSAR Study by Using New Valence Group Carbon-Related Electronegativities. *Rev.Roum.Chim.*, *41*, 141–157.

Diudea, M.V., Minailiuc, O.M. and Katona, G. (1996b). Molecular Topology. 22. Novel Connectivity Descriptors Based on Walk Degrees. *Croat.Chem.Acta*, *69*, 857–871.

Diudea, M.V. and Pop, C.M. (1996). Molecular Topology. 27. A Schultz-Type Index Based on the Wiener Matrix. *Indian J.Chem.*, *35A*, 257–261.

Diudea, M.V. (1997a). Cluj Matrix Invariants. *J.Chem.Inf.Comput.Sci.*, *37*, 300–305.

Diudea, M.V. (1997b). Cluj Matrix, CJ_U: Source of Various Graph Descriptors. *MATCH (Comm.Math.Comp.Chem.)*, *35*, 169–183.

Diudea, M.V. (1997c). Indices of Reciprocal Properties or Harary Indices. *J.Chem.Inf.Comput.Sci.*, *37*, 292–299.

Diudea, M.V. (1997d). Unsymmetric Matrix CJ_U: Source of Various Graph Invariants. *MATCH (Comm.Math.Comp.Chem.)*, *35*, 169–183.

Diudea, M.V., Ivanciuc, O., Nikolic, S. and Trinajstic, N. (1997c). Matrices of Reciprocal Distance, Polynomials and Derived Numbers. *MATCH (Comm.Math.Comp.Chem.)*, *35*, 309–318.

Diudea, M.V., Katona, G. and Pârv, B. (1997d). Delta Number, D_Δ, of Dendrimers. *Croat. Chem.Acta*, *70*, 509–517.

Diudea, M.V., Minailiuc, O.M. and Katona, G. (1997a). Molecular Topology. 26. SP Indices: Novel Connectivity Descriptors. *Rev.Roum.Chim.*, *42*, 239–249.

Diudea, M.V., Minailiuc, O.M., Katona, G. and Gutman, I. (1997b). Szeged Matrix and Related Numbers. *MATCH (Comm.Math.Comp.Chem.)*, *35*, 129–143.

Diudea, M.V., Pârv, B. and Gutman, I. (1997e). Detour-Cluj Matrix and Derived Invariants. *J.Chem.Inf.Comput.Sci.*, *37*, 1101–1108.

Diudea, M.V., Pârv, B. and Topan, M. (1997f). Derived Szeged and Cluj Indices. *J.Serb.Chem. Soc.*, *62*, 267–276.

Diudea, M.V. and Randic, M. (1997). Matrix Operator, $W(M_1,M_2,M_3)$, and Schultz-Type Indices. *J.Chem.Inf.Comput.Sci.*, *37*, 1095–1100.

Diudea, M.V. and Gutman, I. (1998). Wiener-Type Topological Indices. *Croat.Chem.Acta*, *71*, 21–51.

Diudea, M.V., Katona, G., Lukovits, I. and Trinajstic, N. (1998). Detour and Cluj-Detour Indices. *Croat.Chem.Acta*, *71*, 459–471.

Diudea, M.V. and Randic, M. (1998). Unsymmetric Square Matrices and Schultz-Type Numbers. *J.Chem.Inf.Comput.Sci., in press.*

Dixon, J.S. and Villar, H.O. (1998). Bioactive Diversity and Screening Library Selection via Affinity Fingerprinting. *J.Chem.Inf.Comput.Sci.*, *38*, 1192–1203.

Dixon, S.L. and Jurs, P.C. (1992). Atomic Charge Calculations for Quantitative Structure-Property Relationships. *J.Comput.Chem.*, *13*, 492–504.

Dixon, S.L. and Jurs, P.C. (1993). Estimation of pK_a for Organic Oxyacids Using Calculated Atomic Charges. *J.Comput.Chem.*, *14*, 1460–1467.

Dixon, S.L. and Villar, H.O. (1999). Investigation of Classification Methods for the Prediction of Activity in Diverse Chemical Libraries. *J.Comput.Aid.Molec.Des.*, *13*, 533–545.

Dobrynin, A.A. (1993). Degeneracy of Some Matrix Graph Invariants. *J.Math.Chem.*, *14*, 175–184.

Dobrynin, A.A. and Gutman, I. (1994). On a Graph Invariant Related to the Sum of All Distances in a Graph. *Publ.Inst.Math.(Beograd)*, *56*, 18–22.

Dobrynin, A.A. and Kochetova, A.A. (1994). Degree Distance of a Graph: A Degree Analogue of the Wiener Index. *J.Chem.Inf.Comput.Sci.*, *34*, 1082–1086.

Dobrynin, A.A. (1995). Solving a Problem Connected with Distances in Graphs. *Graph Theory Notes New York*, *28*, 21–23.

Dobrynin, A.A., Gutman, I. and Dömötör, G. (1995). A Wiener-Type Graph Invariant for Some Bipartite Graphs. *Appl.Math.Lett.*, *8*, 57–62.

Dobrynin, A.A. and Gutman, I. (1996). On the Szeged Index of Unbranched Catacondensed Benzenoid Molecules. *Croat.Chem.Acta*, *69*, 845–856.

Dobrynin, A.A. (1997a). A New Formula for the Calculation of the Wiener Index of Hexagonal Chains. *MATCH (Comm.Math.Comp.Chem.)*, *35*, 75–90.

Dobrynin, A.A. (1997b). Congruence Relations for the Wiener Index of Hexagonal Chains. *J.Chem.Inf.Comput.Sci.*, *37*, 1109–1110.

Dobrynin, A.A. (1998a). Formula for Calculating the Wiener Index of Catacondensed Benzenoid Graphs. *J.Chem.Inf.Comput.Sci.*, *38*, 811–814.

Dobrynin, A.A. (1998b). New Congruence Relations for the Wiener Index of Cata-Condensed Benzenoid Graphs. *J.Chem.Inf.Comput.Sci.*, *38*, 405–409.

Dobrynin, A.A. (1999). A Simple Formula for the Calculation of the Wiener Index of Hexagonal Chains. *Computers Chem.*, *23*, 43–48.

Dobrynin, A.A. and Gutman, I. (1999). The Average Wiener Index of Trees and Chemical Trees. *J.Chem.Inf.Comput.Sci.*, *39*, 679–683.

Doherty, P.J., Hoes, R.M., Robbat Jr., A. and White, C.M. (1984). Relationship Between Gas Chromatographic Retention Indices and Molecular Connectivities of Nitrated Polycyclic Aromatic Hydrocarbons. *Anal.Chem.*, *56*, 2697–2701.

Doichinova, I.A., Natcheva, R.N. and Mihailova, D.N. (1994). QSAR Studies of 8-Substituted Xanthines as Adenosine Receptor Antagonists. *Eur.J.Med.Chem.*, *29*, 133–138.

Domine, D., Devillers, J., Wienke, D. and Buydens, L. (1996). Test Series Selection from Nonlinear Neural Mapping. *Quant.Struct.-Act.Relat.*, *15*, 395–402.

Donovan, W.H. and Famini, G.R. (1996). Using Theoretical Descriptions in Structure Activity Relationships: Retention Indices of Sulfur Vescicants and Related Compounds. *J.Chem.Soc. Perkin Trans.2*, 83–89.

Dosmorov, S.V. (1982). Generation of Homogeneous Reaction Mechanism. *Kinet.Catal.*

Doucet, J.P., Panaye, A. and Dubois, J.-E. (1983). Topological Correlations of Carbon-13 Chemical Shifts by Perturbation on a Focus: DARC-PULFO Method. Attenuation and Inversion of α-Methyl Substituent Effects. *J.Org.Chem.*, *48*, 3174–3182.

Doucet, J.P. and Panaye, A. (1998). 3D Structural Information: From Property Prediction to Substructure Recognition with Neural Networks. *SAR & QSAR Environ.Res.*, *8*, 249–272.

Dowdy, D.L., McKone, T.E. and Hsieh, D.P. (1996). Prediction of Chemical Biotransfer of Organic Chemicals from Cattle Diet into Beef and Milk Using the Molecular Connectivity Index. *Environ.Sci.Technol.*, *30*, 984–989.

Doweyko, A.M. (1991). The Hypothetical Active Site Lattice-in Vitro and in Vivo Explorations Using a Three-Dimensional QSAR Technique. *J.Math.Chem.*, *7*, 273–285.

Doweyko, A.M. and Mattes, W.B. (1992). An Application of 3D QSAR to the Analysis of the Sequence Specificity of DNA Alkylation by Uracil Mustard. *Biochemistry, 31*, 9388–9392.

Downs, G.M. and Willett, P. (1999). Similarity Searching in Databases of Chemical Structures. *Rev.Comput.Chem., 7*, 1–66.

Doyle, J.K. and Garver, J.E. (1977). Mean Distance in a Graph. *Disc.Math., 17*, 147–154.

Draber, W. (1996). Sterimol and Its Role in Drug Research. *Z.Naturforsch., 51C*, 1–7.

Draper, N. and Smith, H. (1998). *Applied Regression Analysis.* Wiley, NewYork (NY).

Drefahl, A. and Reinhard, M. (1993). Similarity-Based Search and Evaluation of Environmentally Relevant Properties for Organic Compounds in Combination with the Group Contribution Approach. *J.Chem.Inf.Comput.Sci., 33*, 886–895.

Dross, K., Rekker, R.F., de Vries, G. and Mannhold, R. (1998). The Lipophilic Behaviour of Organic Compounds: 3. The Search for Interconnections Between Reversed/Phase Chromatographic Data and log P_{oct} Values. *Quant.Struct.-Act.Relat., 17*, 549–557.

Duboc, C. (1978). The Correlation Analysis of Nucleophilicity. In *Correlation Analysis in Chemistry* (Chapman, N.B. and Shorter, J., eds.), Plenum Press, New York (NY), pp. 313–355.

Dubois, J.-E., Laurent, D. and Viellard, H. (1966). Système de documentation et d'automatization des recherches des corrèlations (DARC). Principes gènèraux. *C.R.Acad.Sci.,Paris, 263 – Ser.C*, 764–767.

Dubois, J.-E., Laurent, D. and Viellard, H. (1967). Système DARC. Principes des recherches de corrélations et équations gènèrale de topo-information. *C.R.Acad.Sci.,Paris, 264 – Ser.C*, 1019–1022.

Dubois, J.-E., Laurent, D. and Aranda, A. (1973a). Relations Structure Chimique-Activitè Biologique par DARC/PELCO. *J.Chim.Phys., 70*, 1608–1615.

Dubois, J.-E., Laurent, D. and Aranda, A. (1973b). *J.Chim.Phys., 70*, 1616–1624.

Dubois, J.-E., Laurent, D., Panaye, A. and Sobel, Y. (1975a). Hyperstructures formelles d'antériorité. *C.R.Acad.Sci.,Paris, 281 – Ser.C*, 687–690.

Dubois, J.-E., Laurent, D., Panaye, A. and Sobel, Y. (1975b). Système DARC: concept d'hyperstructure formelle. *C.R.Acad.Sci.,Paris, 280 – Ser.C*, 851–854.

Dubois, J.-E. (1976). Ordered Chromatic Graph and Limited Environment Concept. In *Chemical Applications of Graph Theory* (Balaban, A.T., ed.), Academic Press, New York (NY), pp. 333–370.

Dubois, J.-E., Laurent, D., Bost, P., Chambaud, S. and Mercier, C. (1976). Système DARC. Méthode DARC/PELCO. Stratégies de Recherche de Corrélations Appliquées à une Population d'Adamantanamines Antigrippales. *Eur.J.Med.Chem., 11*, 225–236.

Dubois, J.-E., Mercier, C. and Sobel, Y. (1979). Thèorie des graphes chimiques: mèthode Darc/Pelco. Profèrence des corrélations de topologie-information et analyse de fiabilité. *C.R.Acad.Sci.,Paris, 289 – Ser.C*, 89–92.

Dubois, J.-E., Chrétien, J.R., Soják, L. and Rijks, J.A. (1980a). Topological Analysis of the Behaviour of Linear Alkanes up to Tetradecenes in Gas-Liquid Chromatography on Squalene. *J.Chromat., 194*, 121–134.

Dubois, J.-E., MacPhee, J.A. and Panaye, A. (1980b). Steric Effects. III. Composition of the E'_S Parameter. Variation of Alkyl Steric Effects with Substitution. Role of Conformation in Determining Sterically Active and Inactive Sites. *Tetrahedron, 36*, 919–928.

Dubois, J.-E., Sicouri, G., Sobel, Y. and Picchiottino, R. (1984). Système DARC: Opérateurs localizés et co-structures de l'invariant d'une réaction. *C.R.Acad.Sci.,Paris, 298 – Ser.II*, 525–530.

Dubois, J.-E. and Sobel, Y. (1985). DARC System for Documentation and Artificial Intelligence in Chemistry. *J.Chem.Inf.Comput.Sci., 25*, 326–335.

Dubois, J.-E., Mercier, C. and Panaye, A. (1986). DARC Topological System and Computer Aided Design. *Acta Pharm.Jugosl., 36*, 135–169.

Dubois, J.-E., Panaye, A. and Attias, R. (1987). DARC System: Notions of Defined and Generic Substructures. Filiation and Coding of FREL Substructure (SS) Classes. *J.Chem.Inf.Comput.Sci., 27*, 74–82.

Dubois, J.-E., Loukianoff, M. and Mercier, C. (1992). Topology and the Quest for Structural Knowledge. *J.Chim.Phys., 89*, 1493–1506.

Dubois, J.-E. and Loukianoff, M. (1993). DARC "Logic Method" for Molal Volume Prediction. *SAR & QSAR Environ.Res.*, *1*, 63–75.

Dubost, J.P. (1993). 2D and 3D Lipophilicity Parameters in QSAR. In *Trends in QSAR and Molecular Modelling 92* (Wermuth, C.G., ed.), ESCOM, Leiden (The Netherlands), pp. 93–100.

Duewer, D.L. (1990). The Free-Wilson Paradigm *Redux*: Significance of the Free-Wilson Coefficients, Insignificance of Coefficient 'Uncertainties' and Statistical Sins. *J.Chemom.*, *4*, 299–321.

Duffy, J.C., Dearden, J.C. and Rostron, C. (1996). A QSAR Study of Antiinflammatory N-Arylanthranilic Acids. *J.Pharm.Pharmacol.*, *48*, 883–886.

Dunbar Jr, J.B. (1997). Cluster-Based Selection. *Persp.Drug Disc.Des.*, *7/8*, 51–63.

Dunn III, W.J. (1977). Molar Refractivity as an Independent Variable in Quantitative Structure-Activity Studies. *Eur.J.Med.Chem.*, *12*, 109–112.

Dunn III, W.J. and Wold, S. (1978). A Structure-Carcinogenicity Study of 4-Nitroquinoline 1-Oxides Using the SIMCA Method of Pattern Recognition. *J.Med.Chem.*, *21*, 1001.

Dunn III, W.J., Wold, S. and Martin, Y.C. (1978). Structure-Activity Study of β-Adrenergic Agents Using the SIMCA Method of Pattern Recognition. *J.Med.Chem.*, *21*, 922–930.

Dunn III, W.J. and Wold, S. (1980). Structure-Activity Analyzed by Pattern Recognition:The Asymmetric Case. *J.Med.Chem.*, *23*, 595–599.

Dunn III, W.J., Wold, S., Edlund, U. and Hellberg, S. (1984). Multivariate Structure-Activity Relationships Between Data from a Battery of Biological Tests and an Ensemble of Structure Descriptors: The *PLS* Method. *Quant.Struct.-Act.Relat.*, *3*, 131–137.

Dunn III, W.J., Koehler, M.G. and Grigoras, S. (1987). The Role of Solvent-Accessible Surface Area in Determining Partition Coefficients. *J.Med.Chem.*, *30*, 1121–1126.

Dunn III, W.J. and Wold, S. (1990). Pattern Recognition Techniques in Drug Design. In *Quantitative Drug Design. Vol. 4* (Ramsden, C.A., ed.), Pergamon Press, Oxford (UK), pp. 691–714.

Dunn III, W.J. and Rogers, D. (1996). Genetic Partial Least Squares in QSAR. In *Genetic Algorithms in Molecular Modeling. Principles of QSAR and Drug Design. Vol. 1* (Devillers, J., ed.), Academic Press, London (UK), pp. 109–130.

Dunnivant, F.M., Elzerman, A.W., Jurs, P.C. and Hasan, M.N. (1992). Quantitative Structure-Property Relationships for Aqueous Solubilities and Henry's Law Constants of Polychlorinated Biphenyls. *Environ.Sci.Technol.*, *26*, 1567–1573.

Duperray, B., Chastrette, M., Makabeth, M.C. and Pacheco, H. (1976a). Analyse Comparative de Corrélations Hansch, Free-Wilson et Darc-Pelco pour une Famille de Bactéricides: des Phénols Halogénés. *Eur.J.Med.Chem.*, *11*, 323–336.

Duperray, B., Chastrette, M., Makabeth, M.C. and Pacheco, H. (1976b). Analyse de l'Activité Bactéricide de Populations d'Alcools Aliphatiques et de β-Naphtols suivant les Méthodes de Hansch et Darc-Pelco: Effet d'Allongement de Chaîne. *Eur.J.Med.Chem.*, *11*, 433–437.

Duprat, A.F., Huynh, T. and Dreyfus, G. (1998). Toward a Principled Methodology for Neural Network Design and Performance Evaluation in QSAR. Application to the Prediction of LogP. *J.Chem.Inf.Comput.Sci.*, *38*, 586–594.

Durand, P.J., Pasari, R., Baker, J.W. and Tsai, C. (1999). An Efficient Algorithm for Similarity Analysis of Molecules. *Internet Journal of Chemistry, 2 – Article 17.*

Durst, G.L. (1998). Comparative Molecular-Field Analysis (CoMFA) of Herbicidal Protoporphyrinogen Oxidase-Inhibitors Using Standard Steric and Electrostatic Fields and an Alternative LUMO Field. *Quant.Struct.-Act.Relat.*, *17*, 419–426.

Dziembowska, T. (1994). Intramolecular Hydrogen Bonding. *Pol.J.Chem.*, *68*, 1455–1489.

Edward, J.T. (1982a). Correlation of Alkane Solubilities in Water with Connectivity Index. *Can.J.Chem.*, *60*, 2573–2578.

Edward, J.T. (1982b). The Relation of Physical Properties of Alkanes to Connectivity Indices: A Molecular Explanation. *Can.J.Chem.*, *60*, 480–485.

Efron, B. (1982). *The Jackknife, the Bootstrap and Other Resampling Planes.* Society for Industrial and Applied Mathematics, Philadelphia (PA), 92 pp.

Efron, B. (1983). Estimating the Error Rate of a Prediction Rule: Improvement on Cross-Validation. *Journal of American Statistical Association*, *78*, 316–331.

Efron, B. (1987). Better Bootstrap Confidence Intervals. *Journal of American Statistical Association*, *82*, 171–200.

Efroymson, M.A. (1960). Multiple Regression Analysis. In *Mathematical Methods for Digital Computers* (Ralston, A. and Wilf, H.S., eds.), Wiley, New York (NY).

Egolf, L.M. and Jurs, P.C. (1992). Estimation of Autoignition Temperatures of Hydrocarbons, Alcohols and Esters from Molecular Structure. *Ind.Eng.Chem.Res.*, *31*, 1798–1807.

Egolf, L.M. and Jurs, P.C. (1993a). Prediction of Boiling Points of Organic Heterocyclic Compounds Using Regression and Neural Network Techniques. *J.Chem.Inf.Comput.Sci.*, *33*, 616–625.

Egolf, L.M. and Jurs, P.C. (1993b). Quantitative Structure-Retention and Structure-Odor Intensity Relationships for a Diverse Group of Odor-Active Compounds. *Anal.Chem.*, *65*, 3119–3126.

Egolf, L.M., Wessel, M.D. and Jurs, P.C. (1994). Prediction of Boiling Points and Critical Temperatures of Industrially Important Organic Compounds from Molecular Structure. *J.Chem.Inf.Comput.Sci.*, *34*, 947–956.

Ehrenson, S., Brownlee, R.T.C. and Taft, R.W. (1973). *Prog.Phys.Org.Chem.*, *10*, 1.

Eisenberg, D. and McLachlan, A.D. (1986). Solvation Energy in Protein Folding and Binding. *Nature*, *319*, 199–203.

Ekins, S., Bravi, G., Ring, B.J., Gillespie, T.A., Gillespie, J.S., Vandenbranden, M., Wrighton, S.A. and Wikel, J.H. (1999). Three-Dimensional Quantitative Structure Activity Relationship Analyses of Substrates for CYP2B6. *Journal of Pharmacology & Experimental Therapeutics*, *288*, 21–29.

El Tayar, N., Tsai, R.-S., Testa, B., Carrupt, P.-A., Hansch, C. and Leo, A. (1991a). Percutaneous Penetration of Drugs: A Quantitative Structure-Permeability Relationship Study. *J.Pharm.Sci.*, *80*, 744–749.

El Tayar, N., Tsai, R.-S., Testa, B., Carrupt, P.-A. and Leo, A. (1991b). Partitioning of Solutes in Different Solvent Systems: The Contribution of Hydrogen-Bonding Capacity and Polarity. *J.Pharm.Sci.*, *80*, 590–598.

El Tayar, N., Testa, B. and Carrupt, P.-A. (1992a). Polar Intermolecular Interactions Encoded in Partition Coefficients: An Indirect Estimation of Hydrogen-Bond Parameters of Polyfunctional Solutes. *J.Phys.Chem.*, *96*, 1455–1459.

El Tayar, N., Tsai, R.-S., Carrupt, P.-A. and Testa, B. (1992b). Octan-1-ol-Water Partition Coefficients of Zwitterionic α-Amino Acids. Determination by Centrifugal Partition Chromatography and Factorization into Steric/Hydrophobic and Polar Components. *J.Chem.Soc.Perkin Trans.2*, 79–84.

El Tayar, N. and Testa, B. (1993). Polar Intermolecular Interactions Encoded in Partition Coefficients and their Interest in QSAR. In *Trends in QSAR and Molecular Modelling 92* (Wermuth, C.G., ed.), ESCOM, Leiden (The Netherlands), pp. 101–108.

Elass, A., Vergoten, G., Legrand, D., Mazurier, J., Elassrochard, E. and Spik, G. (1996a). Processes Underlying Interactions of Human Lactoferrin with the Jurkat Human Lymphoblastic T-Cell Line Receptor. 1. Quantitative Structure-Affinity Relationships Studies. *Quant.Struct.-Act.Relat.*, *15*, 94–101.

Elass, A., Vergoten, G., Legrand, D., Mazurier, J., Elassrochard, E. and Spik, G. (1996b). Processes Underlying Interactions of Human Lactoferrin with the Jurkat Human Lymphoblastic T-Cell Line Receptor. 2. Comparative Molecular Field Analysis. *Quant.Struct.-Act.Relat.*, *15*, 102–107.

Elbro, H.S., Fredeslund, A. and Rasmussen, P. (1991). Group Contribution Method for the Prediction of Liquid Densities as Function of Temperature for Solvents, Oligomers, and Polymers. *Ind.Eng.Chem.Res.*, *30*, 2576–2582.

Eldred, D.V. and Jurs, P.C. (1999a). Prediction of Acute Mammalian Toxicity of Organophosphorus Pesticide Compounds from Molecular Structure. *Chem.Res.Toxicol.*, *10*, 75–99.

Eldred, D.V., Weikel, C.L., Jurs, P.C. and Kaiser, K.L.E. (1999b). Prediction of Fathead minnow Acute Toxicity of Organic Compounds from Molecular Structure. *Chem.Res.Toxicol.*, *12*, 670–678.

Elk, S.B. and Gutman, I. (1994). Further Properties Derivable from the Matula Numbers of an Alkane. *J.Chem.Inf.Comput.Sci.*, *44*, 331–334.

Elk, S.B. (1995). Expansion of Matula Numbers to Heteroatoms and to Ring Compounds. *J.Chem. Inf.Comput.Sci.*, *35*, 233–236.
Engelhardt, H.L. and Jurs, P.C. (1997). Prediction of Supercritical Carbon Dioxide Solubility of Organic Compounds from Molecular Structure. *J.Chem.Inf.Comput.Sci.*, *37*, 478–484.
Entringer, R.C., Jackson, D.E. and Snyder, D.A. (1976). Distance in Graphs. *Czech.Math.J.*, *26*, 283–296.
Eriksson, L., Jonsson, J., Sjöström, M., Wikström, C. and Wold, S. (1988). Multivariate Derivation of Descriptive Scales for Monosaccharides. *Acta Chem.Scand.*, *B42*, 504–514.
Eriksson, L., Jonsson, J., Hellberg, S., Lindgren, F., Skagerberg, B., Sjöström, M., Wold, S. and Berglind, R. (1990). *Environ.Toxicol.Chem.*, *9*, 1339–1351.
Eriksson, L., Jonsson, J. and Berglind, R. (1993a). External Validation of a QSAR for the Acute Toxicity of Halogenated Aliphatic Hydrocarbons. *Environ.Toxicol.Chem.*, *12*, 1185–1191.
Eriksson, L., Sandström, B.E., Sjöström, M., Tysklind, M. and Wold, S. (1993b). Modelling the Cytotoxicity of Halogenated Aliphatic Hydrocarbons. Quantitative Structure-Activity Relationships for the IC_{50} to Human HeLa Cells. *Quant.Struct.-Act.Relat.*, *12*, 124–131.
Eriksson, L., Berglind, R. and Sjöström, M. (1994a). A Multivariate Quantitative Structure-Activity Relationship for Corrosive Carboxylic Acids. *Chemom.Intell.Lab.Syst.*, *23*, 235–245.
Eriksson, L., Verhaar, H.J. and Hermens, J.L. (1994b). Multivariate Characterization and Modeling of the Chemical Reactivity of Epoxides. *Environ.Toxicol.Chem.*, *13*, 683–691.
Eriksson, L., Hermens, J.L., Johansson, E., Verhaar, H.J. and Wold, S. (1995). Multivariate Analysis of Aquatic Toxicity Data with PLS. *Aquatic Science*, *57*, 217–241.
Eriksson, L. and Johansson, E. (1996). Multivariate Design and Modeling in QSAR. *Chemom. Intell.Lab.Syst.*, *34*, 1–19.
Eriksson, L., Johansson, E., Müller, M. and Wold, S. (1997). Cluster-Based Design in Environmental QSAR. *Quant.Struct.-Act.Relat.*, *16*, 383–390.
Ertepinar, H., Gok, Y., Geban, O. and Ozden, S. (1995). A QSAR Study of the Biological Activities of Some Benzimidazoles and Imidazopyridines Against Bacillus Subtilis. *Eur.J.Med. Chem.*, *30*, 171–175.
Ertl, P. (1997). Simple Quantum-Chemical Parameters as an Alternative to the Hammett Sigma-Constant in QSAR Studies. *Quant.Struct.-Act.Relat.*, *16*, 377–382.
Esaki, T. (1980). *J.Pharmacobiodyn.*, *3*, 562–576.
Essam, J.W., Kennedy, J.W. and Gordon, M. (1977). The Graph-Like State of Matter. Part 8. – LCGI Schemes and the Statistical Analysis of Experimental Data. *J.Chem.Soc.Faraday Trans II*, *73*, 1289–1307.
Estrada, E. and Montero, L.A. (1993). Bond Order Weighted Graphs in Molecules as Structure-Property Indices. *Mol.Eng.*, *2*, 363–373.
Estrada, E. (1995a). Edge Adjacency Relationships and a Novel Topological Index Related to Molecular Volume. *J.Chem.Inf.Comput.Sci.*, *35*, 31–33.
Estrada, E. (1995b). Edge Adjacency Relationships in Molecular Graphs Containing Heteroatoms: A New Topological Index Related to Molar Volume. *J.Chem.Inf.Comput.Sci.*, *35*, 701–707.
Estrada, E. (1995c). Graph Theoretical Invariant of Randic Revisited. *J.Chem.Inf.Comput.Sci.*, *35*, 1022–1025.
Estrada, E. (1995d). Three-Dimensional Molecular Descriptors Based on Electron Charge Density Weighted Graphs. *J.Chem.Inf.Comput.Sci.*, *35*, 708–713.
Estrada, E. (1996). Spectral Moments of the Edge Adjacency Matrix of Molecular Graphs. 1. Definition and Applications to the Prediction of Physical Properties of Alkanes. *J.Chem.Inf. Comput.Sci.*, *36*, 844–849.
Estrada, E. and Gutman, I. (1996). A Topological Index Based on Distances of Edges of Molecular Graphs. *J.Chem.Inf.Comput.Sci.*, *36*, 850–853.
Estrada, E. and Ramirez, A. (1996). Edge Adjacency Relationships and Molecular Topographic Descriptors. Definition and QSAR Applications. *J.Chem.Inf.Comput.Sci.*, *36*, 837–843.
Estrada, E. (1997). Spectral Moments of the Edge-Adjacency Matrix of Molecular Graphs. 2. Molecules Containing Heteroatoms and QSAR Applications. *J.Chem.Inf.Comput.Sci.*, *37*, 320–328.

Estrada, E. and Rodriguez, L. (1997). Matrix Algebraic Manipulation of Molecular Graphs. 2. Harary- and MTI-Like Molecular Descriptors. *MATCH (Comm.Math.Comp.Chem.)*, *35*, 157–167.

Estrada, E., Rodriguez, L. and Gutièrrez, A. (1997). Matrix Algebraic Manipulation of Molecular Graphs. 1. Distance and Vertex-Adjacency Matrices. *MATCH (Comm.Math.Comp.Chem.)*, *35*, 145–156.

Estrada, E. (1998a). Modelling the Diamagnetic Susceptibility of Organic Compounds by a Sub-Structural Graph-Theoretical Approach. *J.Chem.Soc.Faraday Trans.*, *94*, 1407–1410.

Estrada, E. (1998b). Spectral Moments of the Edge Adjacency Matrix in Molecular Graphs. 3. Molecules Containing Cycles. *J.Chem.Inf.Comput.Sci.*, *38*, 23–27.

Estrada, E., Guevara, N. and Gutman, I. (1998a). Extension of Edge Connectivity Index. Relationships to Line Graph Indices and QSPR Applications. *J.Chem.Inf.Comput.Sci.*, *38*, 428–431.

Estrada, E., Guevara, N., Gutman, I. and Rodriguez, L. (1998b). Molecular Connectivity Indices of Iterated Line Graphs. A New Source of Descriptors for QSPR and QSAR Studies. *SAR & QSAR Environ.Res.*, *9*, 229–240.

Estrada, E., Peña, A. and Garcìa-Domenech, R. (1998c). Designing Sedative/Hypnotic Compounds from a Novel Substructural Graph-Theoretical Approach. *J.Comput.Aid.Molec.Des.*, *12*, 583–595.

Estrada, E. (1999). Generalized Spectral Moments of the Iterated Line Graph Sequence. A Novel Approach to QSPR Studies. *J.Chem.Inf.Comput.Sci.*, *39*, 90–95.

Estrada, E. (2000). Edge-Connectivity Indices in QSPR/QSAR Studies. 2. Accounting for Long-Range Bond Contributions. *J.Chem.Inf.Comput.Sci.*, *40*, 1042–1048.

Estrada, E. and Rodriguez, L. (2000). Edge-Connectivity Indices in QSPR/QSAR Studies. 1. Comparison with Other Topological Indices in QSPR Studies. *J.Chem.Inf.Comput.Sci.*, *40*, 1037–1041.

Evans, L.A., Lynch, M.F. and Willett, P. (1978). Structural Search Codes for Online Compound Registration. *J.Chem.Inf.Comput.Sci.*, *18*, 146–149.

Ewing, D.F. (1978). Correlation of *nmr* Chemical Shifts with Hammett σ Values and Analogous Parameters. In *Correlation Analysis in Chemistry* (Chapman, N.B. and Shorter, J., eds.), Plenum Press, New York (NY), pp. 357–396.

Exner, O. (1973). *Prog.Phys.Org.Chem.*, *10*, 411.

Exner, O. (1975). *Dipole Moments in Organic Chemistry.* George Thieme Publishers, Stuttgart (Germany), 156 pp.

Exner, O. (1978). A Critical Compilation of Substituent Constants. In *Correlation Analysis in Chemistry* (Chapman, N.B. and Shorter, J., eds.), Plenum Press, New York (NY), pp. 439–540.

Eyring, H., Walter, J. and Kimball, G. (1944). *Quantum Chemistry.* Wiley, New York (NY).

Fabian, W.M., Timofei, S. and Kurunczi, L. (1995). Comparative Molecular Field Analysis (CoMFA), Semiempirical (AM1) Molecular Orbital and Multiconformational Minimal Steric Difference (MTD) Calculations of Anthraquinone Dye Fiber Affinities. *J.Mol.Struct.(Theochem)*, *340*, 73–81.

Fabic-Petrac, I., Jerman-Blazic, B. and Batagelj, V. (1991). Study of Computation, Relatedness and Activity Prediction of Topological Indices. *J.Math.Chem.*, *8*, 121–134.

Famini, G.R., Kassel, R.J., King, J.W. and Wilson, L.Y. (1991). Using Theoretical Descriptors in Quantitative Structure-Activity Relationships: Comparison with the Molecular Transform. *Quant.Struct.-Act.Relat.*, *10*, 344–349.

Famini, G.R., Ashman, W.P., Mickiewicz, A.P. and Wilson, L.Y. (1992). Using Theoretical Descriptors in Quantitative-Structure-Activity Relationships: Opian Receptor Activity of Fentanyl Compounds. *Quant.Struct.-Act.Relat.*, *11*, 162–170.

Famini, G.R. and Penski, C.A. (1992). Using Theoretical Descriptors in Quantitative Structure-Activity Relationships: Some Physicochemical Properties. *J.Phys.Org.Chem.*, *5*, 395–408.

Famini, G.R., Marquez, B.C. and Wilson, L.Y. (1993). Using Theoretical Descriptors in Quantitative Structure-Activity Relationships: Gas Phase Acidity. *J.Chem.Soc.Perkin Trans.2*, 773–782.

Famini, G.R. and Wilson, L.Y. (1993). Using Theoretical Descriptors in Structure-Activity Relationships Solubility in Supercritical CO2. *J.Phys.Org.Chem.*, *6*, 539–544.

Famini, G.R. and Wilson, L.Y. (1994a). Using Theoretical Descriptors in Linear Solvation Energy Relationships. *Theor.Comput.Chem.*, *1*, 213–241.

Famini, G.R. and Wilson, L.Y. (1994b). Using Theoretical in Quantitative Structure-Property Relationships: 3-Carboxy-Benzisoazole Decarboxylation Kinetics. *J.Chem.Soc.Perkin Trans.2*, 1641–1650.

Famini, G.R., Loumbev, V.P., Frykman, E.K. and Wilson, L.Y. (1998). Using Theoretical Descriptors in a Correlation Analysis of Adenosine Activity. *Quant.Struct.-Act.Relat.*, *17*, 558–564.

Fan, W., El Tayar, N., Testa, B. and Kier, L.B. (1990). Water-Dragging Effect: A New Experimental Hydration Parameter Related to Hydrogen-Bond-Donor Acidity. *J.Phys.Chem.*, *94*, 4764–4766.

Fan, W., Tsai, R.-S., El Tayar, N., Carrupt, P.-A. and Testa, B. (1994). Solute-Water Interactions in the Organic Phase of a Biphasic System. 2. Effects of Organic Phase and Temperature on the "Water-Dragging" Effect. *J.Phys.Chem.*, *98*, 329–333.

Fanelli, F., Menziani, M.C., Carotti, A. and De Benedetti, P.G. (1993). Theoretical Quantitative Structure-Activity Analysis of Quinuclidine-Based Muscarinic Cholinergic Receptor Ligands. *J.Mol.Struct.(Theochem)*, *102*, 63–71.

Fauchère, J.L. and Pliska, V. (1983). Hydrophobic Parameters of Amino Acids Side Chain from the Partitioning of N-Acetyl-Amino-Acid Amides. *Eur.J.Med.Chem.*, *4*, 369–375.

Fauchère, J.L., Quarendon, P. and Kaetterer, L. (1988). Estimating and Representing Hydrophobicity Potential. *J.Mol.Graphics*, *6*, 203–206.

Faulon, J.-L. (1998). Isomorphism, Automorphism Partitioning, and Canonical Labeling Can Be Solved in Polynomial-Time for Molecular Graphs. *J.Chem.Inf.Comput.Sci.*, *38*, 432–444.

Feng, L., Han, S., Wang, L., Wang, Z. and Zhang, Z. (1996a). Determination and Estimation of Partitioning Properties for Phenylthio-Carboxylates. *Chemosphere*, *32*, 353–360.

Feng, L., Wang, L.S., Zhao, Y.H. and Song, B. (1996b). Effects of Substituted Anilines and Phenols on Root Elongation of Cabbage Seed. *Chemosphere*, *32*, 1575–1583.

Ferguson, A.M., Heritage, T.W., Jonathon, P., Pack, S.E., Phillips, L., Rogan, J. and Snaith, P.J. (1997). EVA: A New Theoretically Based Molecular Descriptor for Use in QSAR\QSPR Analysis. *J.Comput.Aid.Molec.Des.*, *11*, 143–152.

Ferguson, J. (1939). The Use of Chemical Potentials as Indices of Toxicity. *Proc.Roy.Soc.London B*, *127*, 387–404.

Ferreira, R. (1963a). Principle of Electronegativity Equalization. Part 1. – Bond Moments and Force Constants. *Trans.Faraday Soc.*, *59*, 1064–1074.

Ferreira, R. (1963b). Principle of Electronegativity Equalization. Part 2. – Bond Dissociation Energies. *Trans.Faraday Soc.*, *59*, 1075–1079.

Figeys, H.P. (1970). Quantum-Chemical Studies on teh Aromaticity of Conjugated Systems – II. Aromatic and Anti-Aromatic Annulenes: the $(4n+2)$-π Electron Rule. *Tetrahedron*, *26*, 5225–5234.

Figueras, J. (1992). Automorphism and Equivalence Classes. *J.Chem.Inf.Comput.Sci.*, *32*, 153–157.

Figueras, J. (1993). Morgan Revisited. *J.Chem.Inf.Comput.Sci.*, *33*, 717–718.

Filimonov, D., Poroikov, V., Borodina, Y. and Gloriozova, T. (1999). Chemical Similarity Assessment Through Multilevel Neighborhoods of Atoms: Definition and Comparison with Other Descriptors. *J.Chem.Inf.Comput.Sci.*, *39*, 666–670.

Filip, P.A., Balaban, T.-S. and Balaban, A.T. (1987). A New Approach for Devising Local Graph Invariants: Derived Topological Indices with Low Degeneracy and Good Correlation Ability. *J.Math.Chem.*, *1*, 61–83.

Finizio, A., Sicbaldi, F. and Vighi, M. (1995). Evaluation of Molecular Connectivity Indices as a Predictive Method of log Kow for Different Classes of Chemicals. *SAR & QSAR Environ.Res.*, *3*, 71–80.

Fisher, M.E. (1966). *J.Combinatorial Theory*, *1*, 105.

Fisher, S.W., Lydy, M.J., Barger, J. and Landrum, P.F. (1993). Quantitative Structure-Activity Relationships for Predicting the Toxicity of Pesticides in Aquatic Systems with Sediment. *Environ.Toxicol.Chem.*, *12*, 1307–1318.

Fleming, I. (1990). *Frontier Orbitals and Organic Chemical Reactions.* Wiley, New York (NY).

Floersheim, P., Nozulak, J. and Weber, H.P. (1993). Experience with Comparative Molecular Field Analysis. In *Trends in QSAR and Molecular Modelling 92* (Wermuth, C.G., ed.), ESCOM, Leiden (The Netherlands), pp. 227–232.

Flory, P.J. (1969). *Statistical Mechanics of Chain Molecules.* Wiley-Interscience, New York (NY).

Flower, D.R. (1998). On the Properties of Bit String-Based Measures of Chemical Similarity. *J.Chem.Inf.Comput.Sci.*, *38*, 379–386.

Folkers, G., Merz, A. and Rognan, D. (1993a). CoMFA as a Tool for Active Site Modelling. In *Trends in QSAR and Molecular Modelling 92* (Wermuth, C.G., ed.), ESCOM, Leiden (The Netherlands), pp. 233–244.

Folkers, G., Merz, A. and Rognan, D. (1993b). CoMFA: Scope and Limitations. In *3D QSAR in Drug Design. Theory, Methods and Applications.* (Kubinyi, H., ed.), ESCOM, Leiden (The Netherlands), pp. 583–618.

Ford, G.P., Katritzky, A.R. and Topsom, R.D. (1978). Substituents Effects in Olefinic Systems. In *Correlation Analysis in Chemistry* (Chapman, N.B. and Shorter, J., eds.), Plenum Press, New York (NY), pp. 269–311.

Forina, M., Boggia, R., Mosti, L. and Fossa, P. (1997). Zupan's Descriptors in QSAR Applied to the Study of a New Class of Cardiotonic Agents. *Il Farmaco*, *52*, 411–419.

Forsyth, D.A. (1973). Semiempirical Models for Substituent Effects in Electrophilic Aromatic Substitution and Side-Chain Reactions. *J.Am.Chem.Soc.*, *95*, 3594–3603.

Foster, R., Hyde, R.M. and Livingstone, D.J. (1978). Substituent Constant for Drug Design Studies Based on Properties of Organic Electron Donor-Acceptor Complexes. *J.Pharm.Sci.*, *67*, 1310–1313.

Fradera, X., Amat, L., Besalù, E. and Carbó, R. (1997). Application of Molecular Quantum Similarity to QSAR. *Quant.Struct.-Act.Relat.*, *16*, 25–32.

Frank, I.E. (1987). Intermediate Least Squares Regression Method. *Chemom.Intell.Lab.Syst.*, *1*, 233–242.

Frank, I.E. and Friedman, J.H. (1989). Classification: Oldtimers and Newcomers. *J.Chemom.*, *3*, 463–475.

Frank, I.E. and Friedman, J.H. (1993). A Statistical View of Some Chemometrics Regression Tools. *Technometrics*, *35*, 109–135.

Frank, I.E. and Todeschini, R. (1994). *The Data Analysis Handbook*. Elsevier, Amsterdam (The Netherlands), 366 pp.

Franke, R. (1984). *Theoretical Drug Design Methods.* Elsevier, Amsterdam (The Netherlands).

Franke, R., Hübel, S. and Streich, W.J. (1985). Substructural QSAR Approaches and Topological Pharmacophores. *Environ.Health Persp.*, *61*, 239–255.

Franke, R. and Streich, W.J. (1985a). Topological Pharmacophores. New Methods and their Application to a Set of Antimalarials. Part 2: Results from LOGANA. *Quant.Struct.-Act.Relat.*, *4*, 51–63.

Franke, R. and Streich, W.J. (1985b). Topological Pharmacophores. New Methods and their Application to a Set of Antimalarials. Part 3: Results from LOCON. *Quant.Struct.-Act.Relat.*, *4*, 63–69.

Franke, R. and Buschauer, A. (1992). Quantitative Structure-Activity Relationships in Histamine H-2-Agonists Related to Impromidine and Arpromidine. *Eur.J.Med.Chem.*, *27*, 443–448.

Franke, R. and Buschauer, A. (1993). Interaction Terms in Free-Wilson Analysis: A QSAR of Histamine H_2-Agonists. In *Trends in QSAR and Molecular Modelling 92* (Wermuth, C.G., ed.), ESCOM, Leiden (The Netherlands), pp. 160–162.

Franke, R., Gruska, A. and Presber, W. (1994). Combined Factor and QSAR Analysis for Antibacterial and Pharmacokinetic Data from Parallel Biological Measurements. *Pharmazie*, *49*, 600–605.

Franke, R., Rose, S.V., Hyde, R.M. and Gruska, A. (1995). The Use of Indicator Variables in QSARs of Chiral Compounds. In *QSAR and Molecular Modelling: Cocepts, Computational Tools and Biological Applications* (Sanz, F., Giraldo, J. and Manaut, F., eds.), Prous Science, Barcelona (Spain), pp. 113–116.

Fratev, F., Bonchev, D. and Enchev, V. (1980). A Theoretical Information Approach to Ring and Total Aromaticity in Ground and Excited States. *Croat.Chem.Acta*, *53*, 545–554.

Free, S.M. and Wilson, J.W. (1964). A Mathematical Contribution to Structure-Activity Studies. *J.Med.Chem.*, *7*, 395–399.

Freeland, R.G., Funk, S.A., O'Korn, L.J. and Wilson, G.A. (1979). The Chemical Abstract Service Chemical Registry System. II. Augmented Connectivity Molecular Formula. *J.Chem.Inf. Comput.Sci.*, *19*, 94–98.

Friedman, J. H. (1988). Multivariate Adaptive Regression Splines. Technical Report N.102, Stanford (CA), Laboratory of Computational Statistics – Dept. of Statistics.

Fuchs, R., Abraham, M.H., Kamlet, M.J. and Taft, R.W. (1982). *J.Phys.Org.Chem.*, *2*, 559.

Fujita, S. (1996). The Sphericity Concept for an Orbit of Bonds. Formulation of Chirogenic Sites in a Homospheric Orbit and of Bond-Differentiating Chiral Reactions with Applications to C_{60}-Adducts. *J.Chem.Inf.Comput.Sci.*, *36*, 270–285.

Fujita, T., Iwasa, J. and Hansch, C. (1964). A New Substituent Constant, π, Derived from Partition Coefficients. *J.Am.Chem.Soc.*, *86*, 5175–5180.

Fujita, T. and Ban, T. (1971). Structure-Activity Study of Phenethylamines as Substrates of Biosynthetic Enzymes of Sympathetic Transmitters. *J.Med.Chem.*, *14*, 148–152.

Fujita, T., Takayama, C. and Nakajima, M. (1973). The Nature and Composition of Taft-Hancock Steric Constants. *J.Org.Chem.*, *38*, 1623–1630.

Fujita, T. and Nishioka, T. (1976). The Analysis of the Ortho Effect. *Prog.Phys.Org.Chem.*, *12*, 49–89.

Fujita, T., Nishioka, T. and Nakajima, M. (1977). Hydrogen-Bonding Parameter and Its Significance in Quantitative Structure-Activity Studies. *J.Med.Chem.*, *20*, 1071–1081.

Fujita, T. (1978). Steric Effects in Quantitative Structure-Activity Relationships. *Pure & Appl. Chem.*, *50*, 987–994.

Fujita, T. (1981). The Ortho Effect in Quantitative Structure-Activity Correlations. *Anal.Chim. Acta*, *133*, 667–676.

Fujita, T. (1983). Substituent Effects in the Partition Coefficient of Disubstituted Benzenes: Bidirectional Hammett-Type Relationships. *Prog.Phys.Org.Chem.*, *14*, 75–113.

Fujita, T. and Iwamura, H. (1983). Applications of Various Steric Constants to Quantitative Analysis of Structure-Activity Relationships. In *Steric Effects in Drug Design (Topics in Current Chemistry, Vol. 114)* (Charton, M. and Motoc, I., eds.), Springer-Verlag, Berlin (Germany), pp. 119–157.

Fujita, T. (1990). The Extrathermodynamic Approach to Drug Design. In *Quantitative Drug Design. Vol. 4* (Ramsden, C.A., ed.), Pergamon Press, Oxford (UK), pp. 497–560.

Fujita, T. (1995). Quantitative Structure-Activity Analysis and Database Aided Bioisosteric Structural Transformation Procedure as Methodologies of Agrochemical Design. *ACS Symp. Ser.*, *606*, 13–34.

Fujita, T. (1997). Recent Success Stories Leading to Commecializable Bioactive Compounds with the Aid of Traditional QSAR Procedures. *Quant.Struct.-Act.Relat.*, *16*, 107–112.

Fujiwara, H., Da, Y.-Z. and Ito, K. (1991). The Energy Aspect of Oil Water Partition and Its Application to the Analysis of Quantitative Structure-Activity Relationships of Aliphatic Alcohols in the Liposome Water Partition System. *Bull.Chem.Soc.Jap.*, *64*, 3707–3712.

Fukui, K., Yonezawa, Y. and Shingu, H. (1954). Theory of Substitution in Conjugated Molecules. *Bull.Chem.Soc.Jap.*, *27*, 423–427.

Fukunaga, J.Y., Hansch, C. and Steller, E.E. (1976). Inhibition of Dihydrofolate Reductase. Structure-Activity Correlations of Quinazolines. *J.Med.Chem.*, *19*, 605–611.

Fuller, F.B. (1971). The Writhing Number of a Space Curve. *Proc.Nat.Acad.Sci.USA*, *68*, 815.

Furay, V.J. and Smith, S. (1995). Toxicity and QSAR of Chlorobenzenes in Two Species of Benthic flatfish, Flounder (Platichthys flesus L) and Sole (Solea solea L). *Bull.Environ.Contam.Toxicol.*, *54*, 36–42.

Furet, P., Sele, A. and Cohen, N.C. (1988). 3D Molecular Lipophilicity Potential Profiles: A New Tool in Molecular Modeling. *J.Mol.Graphics*, *6*, 182–189.

Gabanyi, Z., Surjan, P.R. and Naray-Szabo, G. (1982). *Eur.J.Med.Chem.*, *17*, 307.

Gago, F., Pastor, M., Perez-Butragueño, J., Lopez, R., Alvarez-Builla, J. and Elguero, J. (1994). Hydrophobicity of Heterocycles Determination of the Pi Values of Substituents on N Phenylpyrazoles. *Quant.Struct.-Act.Relat.*, *13*, 165–171.

Gaillard, P., Carrupt, P.-A. and Testa, B. (1994a). The Conformation-Dependent Lipophilicity of Morphine Glucuronides as Calculated from their Molecular Lipophilicity Potential. *Bioorg.Med.Chem.Lett.*, *4*, 737–742.

Gaillard, P., Carrupt, P.-A., Testa, B. and Boudon, A. (1994b). Molecular Lipophilicity Potential, a Tool in 3D QSAR: Method and Applications. *J.Comput.Aid.Molec.Des.*, *8*, 83–96.

Gakh, A.A., Gakh, E.G., Sumpter, B.G. and Noid, D.W. (1994). Neural Network-Graph Theory Approach to the Prediction of the Physical Properties of Organic Compounds. *J.Chem.Inf.Comput.Sci.*, *34*, 832–839.

Galanakis, D., Calder, J.A., Ganellin, C.R., Owen, C.S. and Dunn, P.M. (1995). Synthesis and Quantitative Structure-Activity Relationships of Dequalinium Analogs as K+ Channel Blockers Investigation into the Role of the Substituent at Position 4 of the Quinoline Ring. *J.Med.Chem.*, *38*, 3536–3546.

Gallo, R. (1983). Treatment of Steric Effects. *Prog.Phys.Org.Chem.*, *14*, 115–163.

Gálvez, J., Garcìa-Domenech, R., De Julián-Ortiz, V. and Soler, R. (1994a). Topological Approach to Analgesia. *J.Chem.Inf.Comput.Sci.*, *34*, 1198–1203.

Gálvez, J., Garcìa, R., Salabert, M.T. and Soler, R. (1994b). Charge Indexes. New Topological Descriptors. *J.Chem.Inf.Comput.Sci.*, *34*, 520–525.

Gálvez, J., Garcia, A.E., De Julián-Ortiz, V. and Soler, R. (1995a). A Topological Approach to Drug Design. In *QSAR and Molecular Modelling: Cocepts, Computational Tools and Biological Applications* (Sanz, F., Giraldo, J. and Manaut, F., eds.), Prous Science, Barcelona (Spain), pp. 163–166.

Gálvez, J., Garcìa-Domenech, R., De Julián-Ortiz, V. and Soler, R. (1995b). Topological Approach to Drug Design. *J.Chem.Inf.Comput.Sci.*, *35*, 272–284.

Gálvez, J., Garcìa-Domenech, R., de Gregorio Alapont, C., De Julián-Ortiz, V. and Popa, L. (1996a). Pharmacological Distribution Diagrams: A Tool for *de Novo* Drug Design. *J.Mol.Graphics*, *14*, 272–276.

Gálvez, J., Gomez-Lechón, M.J., Garcìa-Domenech, R. and Castell, J.V. (1996b). New Cytostatic Agents Obtained by Molecular Topology. *Bioorg.Med.Chem.Lett.*, *6*, 2301–2306.

Gamper, A.M., Winger, R.H., Liedl, K.R., Sotriffer, C.A., Varga, J.M., Kroemer, R.T. and Rode, B.M. (1996). Comparative Molecular Field Analysis of Haptens Docked to the Multispecific Antibody IgE(Lb4). *J.Med.Chem.*, *39*, 3882–3888.

Gange, D.M., Donovan, S., Lopata, R.J. and Henegar, K. (1995). The QSAR of Insecticidal Uncouplers. *ACS Symp.Ser.*, *606*, 199–212.

Gantchev, T.G., Ali, H. and Vanlier, J.E. (1994). Quantitative Structure-Activity Relationships Comparative Molecular Field Analysis (QSAR/CoMFA) for Receptor Binding Properties of Halogenated Estradiol Derivatives. *J.Med.Chem.*, *37*, 4164–4176.

Gao, C., Govind, R. and Tabak, H.H. (1992). Application of the Group Contribution Method for Predicting the Toxicity of Organic Chemicals. *Environ.Toxicol.Chem.*, *11*, 631–636.

Gao, H., Williams, C., Labute, P. and Bajorath, J. (1999). Binary Quantitative Structure-Activity Relationship (QSAR) Analysis of Estrogen Receptor Ligands. *J.Chem.Inf.Comput.Sci.*, *39*, 164–168.

Gao, Y. and Hosoya, H. (1988). Topological Index and Thermodynamics Properties. IV. Size Dependency of the Structure-Activity Correlation of Alkanes. *Bull.Chem.Soc.Jap.*, *61*, 3093–3102.

Garcia, E., Lopezdecerain, A., Martinez Merino, V. and Monge, A. (1992). Quantitative Structure Mutagenic Activity Relationships of Triazino Indole Derivatives. *Mut.Res.*, *268*, 1–9.

García-March, F.J., Antón-Fos, G.M., Perez-Gimenez, F., Salebert-Salvador, M.T., Cercos-del-Pozo, R.A. and De Julián-Ortiz, V. (1996). Prediction of Chromatographic Properties for a Group of Natural Phenolic Derivatives by Molecular Topology. *J.Chromat.*, *719A*, 45–51.

Garcìa-Domenech, R., Gálvez, J., Moliner, R. and García-March, F.J. (1991). Prediction and Interpretation of Some Pharmacological Properties of Cephalosporins Using Molecular Connectivity. *Drug Invest.*, *3*, 344–350.

Garcìa-Domenech, R., García-March, F.J., Soler, R., Gálvez, J., Antón-Fos, G.M. and De Julián-Ortiz, V. (1996). New Analgesics Designed by Molecular Topology. *Quant.Struct.-Act.Relat.*, *15*, 201–207.

Garcìa-Domenech, R., de Gregorio Alapont, C., De Julián-Ortiz, V., Gálvez, J. and Popa, L. (1997). Molecular Connectivity to Find β-Blockers with Low Toxicity. *Bioorg.Med.Chem.Lett.*, *7*, 567–572.

Garcìa-Domenech, R. and De Julián-Ortiz, V. (1998). Antimicrobial Activity Characterization in a Heterogeneous Group of Compounds. *J.Chem.Inf.Comput.Sci.*, *38*, 445–449.

Gardner, R.J. (1980). Correlation of Bitterness Thresholds of Amino Acids and Peptides with Molecular Connectivity. *J.Sci.Food Agric.*, *31*, 23.

Gargas, M.L. and Seybold, P.G. (1988). Modeling the Tissue Solubilities and Metabolic Rate Constant (Vmax) of Halogenated Methanes, Ethanes, and Ethylenes. *Toxicol.Lett.*, *43*, 235–256.

Garrone, A., Marengo, E., Fornatto, E. and Gasco, A. (1992). A Study on pK_a(app) and Partition Coefficient of Substituted Benzoic Acids in SDS Anionic Micellar System. *Quant.Struct.-Act.Relat.*, *11*, 171–175.

Gasteiger, J. and Marsili, M. (1980). Iterative Partial Equalization of Orbital Electronegativity: A Rapid Access to Atomic Charges. *Tetrahedron*, *36*, 3219–3228.

Gasteiger, J., Hutchings, M.G., Christoph, B., Gann, L., Hiller, C., Löw, P., Marsili, M., Saller, H. and Yuki, K. (1987). A New Treatment of Chemical Reactivity: Development of EROS, an Expert System for Reaction Prediction and Synthesis Design. *Top.Curr.Chem.*, *137*, 19–73.

Gasteiger, J. (1988). Empirical Methods for the Calculation of Physicochemical Data of Organic Compounds. In *Physical Property Prediction in Organic Chemistry* (Jochum, C., Hicks, M.G. and Sunkel, J., eds.), Springer-Verlag, Berlin (Germany), pp. 119–138.

Gasteiger, J., Röse, P. and Saller, H. (1988). Multidimensional Explorations into Chemical Reactivity: The Reactivity Space. *J.Mol.Graphics*, *6*, 87.

Gasteiger, J., Li, X., Rudolph, C.J., Sadowski, J. and Zupan, J. (1994a). Representation of Molecular Electrostatic Potentials by Topological Feature Maps. *J.Am.Chem.Soc.*, *116*, 4608–4620.

Gasteiger, J., Li, X. and Uschold, A. (1994b). The Beauty of Molecular Surfaces as Revealed by Self-Organizing Neural Networks. *J.Mol.Graphics*, *12*, 90–97.

Gasteiger, J. ed. (1996). *Software Development in Chemistry – Vol. 10*. Fachgruppe Chemie-Information-Computer (CIC), Frankfurt am Mein (Germany), 434 pp.

Gasteiger, J., Sadowski, J., Schuur, J., Selzer, P., Steinhauer, L. and Steinhauer, V. (1996). Chemical Information in 3D Space. *J.Chem.Inf.Comput.Sci.*, *36*, 1030–1037.

Gasteiger, J., Schuur, J., Selzer, P., Steinhauer, L. and Steinhauer, V. (1997). Finding the 3D Structure of a Molecule in its IR Spectrum. *Fresen.J.Anal.Chem.*, *359*, 50–55.

Gasteiger, J. (1998). Making the Computer Understand Chemistry. *Internet Journal of Chemistry*, *1 – Article 33*.

Gautzsch, R. and Zinn, P. (1992a). List Operations on Chemical Graphs. 1. Basic List Structures and Operations. *J.Chem.Inf.Comput.Sci.*, *32*, 541–550.

Gautzsch, R. and Zinn, P. (1992b). List Operations on Chemical Graphs. 2. Combining Basic List Operations. *J.Chem.Inf.Comput.Sci.*, *32*, 551–555.

Gautzsch, R. and Zinn, P. (1994). List Operations on Chemical Graphs. 5. Implementation of Breadth-First Molecular Path Generation and Application in the Estimation of Retention Index Data and Boiling Points. *J.Chem.Inf.Comput.Sci.*, *34*, 791–800.

Gautzsch, R. and Zinn, P. (1996). Use of Incremental Models to Estimate the Retention Indexes of Aromatic Compounds. *Chromatographia*, *43*, 163–176.

Gavezzotti, A. (1983). The Calculation of Molecular Volumes and the use of Volume Analysis in the Investigation of Structured Media and Solid-State Organic Reactivity. *J.Am.Chem.Soc.*, *105*, 5220–5225.

Gawlik, B.M., Sotiriou, N., Feicht, E.A., Schulte-Hostede, S. and Kettrup, A. (1997). Alternatives for the Determination of the Soil Adsorption Coefficient, Koc, of Non-Ionic Organic Compounds – A Review. *Chemosphere*, *34*, 2525–2551.

Geary, R.C. (1954). The Contiguity Ratio and Statistical Mapping. *Incorp.Statist.*, *5*, 115–145.

Geerlings, P., De Proft, F. and Martin, J.M.L. (1996a). Density-Functional Theory Concepts and Techniques for Studying Molecular Charge Distributions and Related Properties. In *Recent Developments and Applications of Modern Density Functional Theory* (Seminario, J.M., ed.), Elsevier, Amsterdam (The Netherlands), pp. 773–809. [R]

Geerlings, P., Langenaeker, W., De Proft, F. and Baeten, A. (1996b). Molecular Electrostatic Potentials vs. DFT Descriptors of Reactivity. In *Molecular Electrostatic Potentials: Concepts and Applications* (Murray, J.S. and Sen, K., eds.), Elsevier, Amsterdam (The Netherlands), pp. 587–617. [R]

Geladi, P. and Kowalski, B.R. (1986). Partial Least Squares Regression: A Tutorial. *Anal.Chim. Acta*, *185*, 1–17.

Geladi, P. and Tosato, M.L. (1990). Multivariate Latent Variable Projection Methods: SIMCA and PLS. In *Practical Applications of Quantitative Structure-Activity Relationships (QSAR) in Environmental Chemistry and Toxicology* (Karcher, W. and Devillers, J., eds.), Kluwer, Dordrecht (The Netherlands), pp. 145–152.

Gelius, U. (1974). Binding Energies and Chemical Shifts in ESCA. *Phys.Scr.*, *9*, 133–147.

Gerstl, Z. and Helling, C.S. (1987). Evaluation of Molecular Connectivity as a Predictive Method for the Adsorption of Pesticides by Soils. *J.Environ.Sci.Health*, *B22*, 55–69.

Ghose, A.K. and Crippen, G.M. (1982). Quantitative Structure-Activity Relationship by Distance Geometry: Quinazolines as Dihydrofolate Reductase Inhibitors. *J.Med.Chem.*, *25*, 892–899.

Ghose, A.K. and Crippen, G.M. (1983). Combined Distance Geometry Analysis of Dihydrofolate Reductase Inhibition by Quinazolines and Triazines. *J.Med.Chem.*, *26*, 996–1010.

Ghose, A.K. and Crippen, G.M. (1984). General Distance Geometry Three-Dimensional Receptor Model for Diverse Dihydrofolate Reductase Inhibitors. *J.Med.Chem.*, *27*, 901–914.

Ghose, A.K. and Crippen, G.M. (1985a). Geometrically Feasible Binding Modes of a Flexible Ligand Molecule at the Receptor Site. *J.Comput.Chem.*, *6*, 350–359.

Ghose, A.K. and Crippen, G.M. (1985b). Use of Physicochemical Parameters in Distance Geometry and Related Three-Dimensional Quantitative Structure-Activity Relationships: A Demonstration Using *Escherichia coli* Dihydrofolate Reductase Inhibitors. *J.Med.Chem.*, *28*, 333–346.

Ghose, A.K. and Crippen, G.M. (1986). Atomic Physicochemical Parameters for Three-Dimensional-Structure-Directed Quantitative Structure-Activity Relationships. I. Partition Coefficients as a Measure of Hydrophobicity. *J.Comput.Chem.*, *7*, 565–577.

Ghose, A.K. and Crippen, G.M. (1987). Atomic Physicochemical Parameters for Three-Dimensional-Structure-Directed Quantitative Structure-Activity Relationships. 2. Modeling Dispersive and Hydrophobic Interactions. *J.Chem.Inf.Comput.Sci.*, *27*, 21–35.

Ghose, A.K., Pritchett, A. and Crippen, G.M. (1988). Atomic Physicochemical Parameters for Three Dimensional Structure Directed Quantitative Structure-Activity Relationships III: Modeling Hydrophobic Interactions. *J.Comput.Chem.*, *9*, 80–90.

Ghose, A.K. and Crippen, G.M. (1990). The Distance Geometry Approach to Modeling Receptor Sites. In *Quantitative Drug Design. Vol. 4* (Ramsden, C.A., ed.), Pergamon Press, Oxford (UK), pp. 715–733.

Ghose, A.K., Logan, M.E., Treasurywala, A.M., Wang, H., Wahl, R.C., Tomczuk, B.E., Gowravaram, M.R., Jaeger, E.P. and Wendoloski, J.J. (1995). Determination of Pharmacophoric Geometry for Collagenase Inhibitors Using a Novel Computational Method and Its Verification Using Molecular Dynamics, NMR, and X-Ray Crystallography. *J.Am.Chem.Soc.*, *117*, 4671–4682.

Ghose, A.K. and Wendoloski, J.J. (1998). Pharmacophore Modellin: Methods, Experimental Verification and Applications. In *3D QSAR in Drug Design – Vol. 2* (Kubinyi, H., Folkers, G. and Martin, Y.C., eds.), Kluwer/ESCOM, Dordrecht (The Netherlands), pp. 253–271.

Ghoshal, N., Mukhopadhayay, S.N., Ghoshal, T.K. and Achari, B. (1993). Quantitative Structure-Activity Relationship Studies Using Artificial Neural Networks. *Indian J.Chem.*, *32B*, 1045–1050.

Gibson, S., McGuire, R. and Rees, D.C. (1996). Principal Components Describing Biological Activities and Molecular Diversity of Heterocyclic Aromatic Ring Fragments. *J.Med.Chem.*, *39*, 4065–4072.

Giese, B. and Beckaus, H.-D. (1978). Front Strain of π and σ Radicals. *Angew.Chem.Int.Ed.Engl.*, *17*, 594–595.

Gilat, G. (1994). On Quantifying Chirality – Obstacles and Problems toward Unification. *J.Math.Chem.*, *15*, 197–205.

Gillet, V.J., Willett, P. and Bradshaw, J. (1998). Identification of Biological Activity Profiles Using Substructural Analysis and Genetic Algorithms. *J.Chem.Inf.Comput.Sci.*, *38*, 165–179.

Gillet, V.J., Willett, P., Bradshaw, J. and Green, D.V.S. (1999). Selecting Combinatorial Libraries to Optimize Diversity and Physical Properties. *J.Chem.Inf.Comput.Sci.*, *39*, 169–177.

Gini, G., Lorenzini, M., Benfenati, E., Grasso, P. and Bruschi, M. (1999). Predictive Carcinogenicity: A Model for Aromatic Compounds, with Nitrogen-Containing Substituents, Based on Molecular Descriptors Using an Artificial Neural Network. *J.Chem.Inf.Comput.Sci.*, *39*, 1076–1080.

Ginn, C.M., Turner, D.B., Willett, P., Ferguson, A.M. and Heritage, T.W. (1997). Similarity Searching in Files of Three-Dimensional Chemical Structures: Evaluation of the EVA Descriptor and Combination of Rankings Using Data Fusion. *J.Chem.Inf.Comput.Sci.*, *37*, 23–27.

Gladstone, J.H. and Dale, T.P. (1858). *Philosophical Transaction of the Royal Society (London)*, *A148*, 887–894.

Glennon, R.A. and Kier, L.B. (1978). LSD Analogs as Serotonin Antagonists: A Molecular Connectivity SAR Analysis. *Eur.J.Med.Chem.*, *13*, 219.

Glennon, R.A., Kier, L.B. and Shulgin, A.T. (1979). Molecular Connectivity Analysis of Hallucinogenic Mescaline Analogs. *J.Pharm.Sci.*, *68*, 906.

Godden, J.W., Xue, L. and Bajorath, J. (2000). Combinatorial Preferences Affect Molecular Similarity/Diversity Calculations Using Binary Fingerprints and Tanimoto Coefficients. *J.Chem. Inf.Comput.Sci.*, *40*, 126–134.

Godfrey, M. (1978). Theoretical Models for Interpreting Linear Correlations in Organic Chemistry. In *Correlation Analysis in Chemistry* (Chapman, N.B. and Shorter, J., eds.), Plenum Press, New York (NY), pp. 85–117.

Golberg, L. ed. (1983). *Structure-Activity Correlation as a Predictive Tool in Toxicology. Fundamentals, Methods, and Applications.* Hemisphere, Washington (DC).

Goldberg, D.E. (1989). *Genetic Algorithms in Search, Optimization and Machine Learning.* Addison-Wesley, Massachusetts (MA).

Golender, V.E., Drboglav, V.V. and Rozenblit, A.B. (1981). *J.Chem.Inf.Comput.Sci.*, *21*, 196.

Goll, E.S. and Jurs, P.C. (1999a). Prediction of the Normal Boiling Points of Organic Compounds from Molecular Strucrures with a Computational Neural Network Model. *J.Chem.Inf.Comput.Sci.*, *39*, 974–983.

Goll, E.S. and Jurs, P.C. (1999b). Prediction of Vapor Pressures of Hydrocarbons and Halohydrocarbons from Molecular Structure with a Computational Neural Network Model. *J.Chem.Inf. Comput.Sci.*, *39*, 1081–1089.

Golub, G.H. and van Loan, C.F. (1983). *Matrix Computations.* John Hopkins University Press, Baltimore (MD).

Gombar, V.K. and Jain, D.V.S. (1987a). Quantification of Molecular Shape & its Correlation with Physicochemical Properties. *Indian J.Chem.*, *26A*, 554–555.

Gombar, V.K., Kumar, A. and Murthy, M.S. (1987b). Quantitative Structure-Activity Relationships: Part IX – A Modified Connectivity Index as Structure Quantifier. *Indian J.Chem.*, *26B*, 1168–1170.

Gombar, V.K. and Enslein, K. (1990). Quantitative Structure-Activity Relationship (QSAR) Studies Using Electronic Descriptors Calculated from Topological and Molecular Orbital (MO) Methods. *Quant.Struct.-Act.Relat.*, *9*, 321–325.

Gombar, V.K., Borgstedt, H.H., Enslein, K., Hart, J.B. and Blake, B.W. (1991). A QSAR Model of Teratogenesis. *Quant.Struct.-Act.Relat.*, *10*, 306–332.

Gombar, V.K., Enslein, K. and Blake, B.W. (1995). Assessment of Developmental Toxicity Potential of Chemicals by Quantitative Structure-Toxicity Relationship Models. *Chemosphere*, *31*, 2499–2510.

Gombar, V.K. and Enslein, K. (1996). Assessment of *n*-Octanol/Water Partition Coefficient: When Is the Assessment Reliable? *J.Chem.Inf.Comput.Sci.*, *36*, 1127–1134.

Gombar, V.K. (1999). Reliable Assessment of Log p of Compounds of Pharmaceutical Relevance. *SAR & QSAR Environ.Res.*, *10*.

Good, A.C. (1992). The Calculation of Molecular Similarity: Alternative Formulas, Data Manipulation and Graphical Display. *J.Mol.Graphics*, *10*, 144–151.

Good, A.C., Hodgkin, E.E. and Richards, W.G. (1992). Utilization of Gaussian Functions for the Rapid Evaluation of Molecular Similarity. *J.Chem.Inf.Comput.Sci.*, *32*, 188–191.

Good, A.C., Peterson, S.J. and Richards, W.G. (1993a). QSAR's from Similarity Matrices. Technique Validation and Application in the Comparison of Different Similarity Evaluation Methods. *J.Med.Chem.*, *36*, 2929–2937.

Good, A.C. and Richards, W.G. (1993). Rapid Evaluation of Shape Similarity Using Gaussian Functions. *J.Chem.Inf.Comput.Sci.*, *33*, 112–116.

Good, A.C., So, S.-S. and Richards, W.G. (1993b). Structure-Activity Relationships from Molecular Similarity Matrices. *J.Med.Chem.*, *36*, 433–438.

Good, A.C. (1995a). 3D Molecular Similarity Indices and their Application in QSAR Studies. In *Molecular Similarity in Drug Design* (Dean, P.M., ed.), Chapman & Hall, London (UK), pp. 24–56.

Good, A.C., Ewing, T.J.A., Gschwend, D.A. and Kuntz, I.D. (1995b). New Molecular Shape Descriptors: Application in Database Screening. *J.Comput.Aid.Molec.Des.*, *9*, 1–12.

Good, A.C. and Kuntz, I.D. (1995). Investigating the Extension of Pairwise Distance Pharmacophore Measures to Triplet-Based Descriptors. *J.Comput.Aid.Molec.Des.*, *9*, 373–379.

Good, A.C. and Richards, W.G. (1998). Explicit Calculation of 3D Molecular Similarity. In *3D QSAR in Drug Design – Vol. 2* (Kubinyi, H., Folkers, G. and Martin, Y.C., eds.), Kluwer/ESCOM, Dordrecht (The Netherlands), pp. 321–338. [R]

Goodford, P.J. (1985). A Computational Procedure for Determining Energetically Favorable Binding Sites on Biologically Important Macromolecules. *J.Med.Chem.*, *28*, 849–857.

Goodford, P.J. (1995). The Properties of Force Fields. In *QSAR and Molecular Modelling: Cocepts, Computational Tools and Biological Applications* (Sanz, F., Giraldo, J. and Manaut, F., eds.), Prous Science, Barcelona (Spain), pp. 199–205.

Goodford, P.J. (1996). Multivariate Characterization of Molecules for QSAR Analysis. *J.Chemom.*, *10*, 107–117.

Gopinathan, M.S. and Jug, K. (1983a). Valency. I. A Quantum Chemical Definition and Properties. *Theor.Chim.Acta*, *63*, 497–509.

Gopinathan, M.S. and Jug, K. (1983b). Valency. II. Applications to Molecules with First-Row Atoms. *Theor.Chim.Acta*, *63*, 511–527.

Gordeeva, E.V., Molchanova, M.S. and Zefirov, N.S. (1990). General Methodology and Computer Program for the Exhaustive Restoring of Chemical Structures by Molecular Connectivity Indices. Solution of the Inverse Problem in QSAR/QSPR. *Tetrahedron Comput.Methodol.*, *3*, 389–415.

Gordeeva, E.V., Katritzky, A.R., Shcherbukhim, V.V. and Zefirov, N.S. (1993). Rapid Conversion of Molecular Graphs to Three-Dimensional Using the MOLGEO Program. *J.Chem.Inf.Comput.Sci.*, *33*, 102–111.

Gordon, M. and Scantlebury, G.R. (1964). Non-Random Polycondensation: Statistical Theory of the Substitution Effect. *Trans.Faraday Soc.*, *60*, 604–621.

Gordon, M. and Kennedy, J.W. (1973). The Graph-Like State of Matter. Part 2. LCGI Schemes for the Thermodynamics of Alkanes and the Theory of Inductive Inference. *J.Chem.Soc.Faraday Trans II*, *69*, 484–504.

Gordon, M.C. (1980). Quantitative Structure-Activity Relationships by Distance Geometry: Systematic Analysis of Dihydrofolate Reductase Inhibitors. *J.Math.Chem.*, *23*, 599–606.

Gordy, W. (1946). A New Method of Determining Electronegativity from Other Atomic Properties. *Physical Review*, *69*, 604–607.

Gordy, W. (1951). Interpretation of Nuclear Quadrupole Couplings in Molecules. *J.Chem.Phys.*, *19*, 792–793.

Gough, J. and Hall, F.M. (1999a). Modeling Antileukemic Activity of Carboquinones with Electrotopological State and Chi Indices. *J.Chem.Inf.Comput.Sci.*, *39*, 356–361.

Gough, J. and Hall, L.H. (1999b). Modeling the Toxicity of Amide Herbicides Using the Electrotopological State. *Environ.Toxicol.Chem.*, *18*, 1069–1075.

Gough, K.M., Belohorcova, K. and Kaiser, K.L. (1994). Quantitative Structure-Activity Relationships (QSARs) of Photobacterium Phosphoreum Toxicity of Nitrobenzene Derivatives. *Sci.Total Environ.*, *142*, 179–190.

Govers, H., Rupert, C. and Aiking, H. (1984). Quantitative Structure-Activity Relationships for Polycyclic Aromatic Hydrocarbons: Correlation Between Molecular Connectivity, Physicochemical Properties, Bioconcentration and Toxicity in *Daphnia pulex*. *Chemosphere*, *13*, 227–236.

Govers, H.A.J. (1990). Prediction of Environmental Behaviour and Effects of Polycyclic Aromatic Hydrocarbons by PAR and QSAR. In *Practical Applications of Quantitative Structure-Activity Relationships (QSAR) in Environmental Chemistry and Toxicology* (Karcher, W. and Devillers, J., eds.), Kluwer, Dordrecht (The Netherlands), pp. 411–432.

Gozalbes, R., Gálvez, J., Garcìa-Domenech, R. and Derouin, F. (1999). Molecular Search of New Active Drugs against *Toxoplasma gondII*. *SAR & QSAR Environ.Res.*, *10*, 47–60.

Graham, C., Gealy, R., Macina, O.T., Karol, M.H. and Rosenkranz, H.S. (1996). QSAR for Allergic Contact Dermatitis. *Quant.Struct.-Act.Relat.*, *15*, 224–229.

Gramatica, P., Navas, N. and Todeschini, R. (1998). 3D-Modelling and Prediction by WHIM Descriptors. Part 9. Chromatographic Relative Retention Time and Physico-Chemical Properties of Polychlorinated Biphenyls (PCBs). *Chemom.Intell.Lab.Syst.*, *40* , 53–63.

Gramatica, P., Consonni, V. and Todeschini, R. (1999a). QSAR Study on the Tropospheric Degradation of Organic Compounds. *Chemosphere*, *38*, 1371–1378.

Gramatica, P., Navas, N. and Todeschini, R. (1999b). Classification of Organic Solvents and Modelling of Their Physico-Chemical Properties by Chemometric Methods Using Different Sets of Molecular Descriptors. *TRAC*, *18*, 461–471.

Gramatica, P., Corradi, M. and Consonni, V. (2000). Modelling and Prediction of Soil Sorption Coefficients of Non-Ionic Organic Pesticides by Molecular Descriptors. *Chemosphere*, *41*, 763–777.

Grant, J.A. and Pickup, B.T. (1995). Gaussian Description of Molecular Shape. *J.Phys.Chem.*, *99*, 3503–3510.

Grant, J.A., Gallardo, M.A. and Pickup, B.T. (1996). A Fast Method of Molecular Shape Comparison: A Simple Application of a Gaussian Description of Molecular Shape. *J.Comput.Chem.*, *17*, 1653–1666.

Graovac, A., Gutman, I. and Trinajstic, N. (1977). *Topological Approach to the Chemistry of Conjugated Molecules*. Springer-Verlag, Berlin (Germany), 123 pp.

Graovac, A. and Gutman, I. (1978). The Determinant of the Adjacency Matrix of the Graph of a Conjugated Molecule. *Croat.Chem.Acta*, *51*, 133–140.

Graovac, A. and Pisanski, T. (1991). On the Wiener Index of a Graph. *J.Math.Chem.*, *8*, 53–62.

Grassy, G. and Lahana, R. (1993). Statistical Analysis and Shape Recognition: Applications to MD Simulations, Conformational Analysis and Structure-Activity Relationships. In *Trends in QSAR and Molecular Modelling 92* (Wermuth, C.G., ed.), ESCOM, Leiden (The Netherlands), pp. 216–219.

Grassy, G., Trape, P., Bompart, J., Calas, B. and Auzou, G. (1995). Variable Mapping of Structure-Activity Relationships. Application to 17-Spirolactone Derivatives with Mineralocorticoid Activity. *J.Mol.Graphics*, *13*, 356–367.

Grdadolnik, S.G. and Mierke, D.F. (1997). Structural Characterization of the Molecular Dimer of the Peptide Antibiotic Vancomycin by Distance Geometry in Four Spatial Dimensions. *J.Chem. Inf.Comput.Sci.*, *37*, 1044–1047.

Greco, G., Novellino, E., Silipo, C. and Vittoria, A. (1991). Comparative Molecular Field Analysis on a Set of Muscarinic Agonists. *Quant.Struct.-Act.Relat.*, *10*, 289–299.

Greco, G., Novellino, E., Pellecchia, M., Silipo, C. and Vittoria, A. (1993). Use of the Hydrophobic Substituent Constant in a Comparative Molecular Field Analysis (CoMFA) on a Set of Anilides Inhibiting the Hill Reaction. *SAR & QSAR Environ.Res.*, *1*, 301–334.

Greco, G., Novellino, E., Fiorini, I., Nacci, V., Campiani, G., Ciani, S.M., Garofalo, A., Bernasconi, P. and Mennini, T. (1994). A Comparative Molecular Field Analysis Model for 6-Arylpyrrolo(2,1-d)(1,5)Benzothiazepines Binding Selectively to the Mitochondrial Benzodiazepine Receptor. *J.Med.Chem.*, *37*, 4100–4108.

Greco, G., Novellino, E., Pellecchia, M., Silipo, C. and Vittoria, A. (1994a). Effects of Variable Sampling on CoMFA Coefficient Contour Maps. *J.Mol.Graphics*, *12*, 67–68.

Greco, G., Novellino, E., Pellecchia, M., Silipo, C. and Vittoria, A. (1994b). Effects of Variable Selection on CoMFA Coefficient Contour Maps in a Set of Triazines Inhibiting DHFR. *J.Comput.Aid.Molec.Des.*, *8*, 97–112.

Greco, G., Novellino, E. and Martin, Y.C. (1997). Approaches to Three-Dimensional Quantitative Structure-Activity Relationships. In *Reviews in Computational Chemistry – Vol. 11* (Lipkowitz, K.B. and Boyd, D.B., eds.), Wiley-VCH, New York (NY), pp. 183–240.

Green, A.L. (1956). A Simple Approximation to the Resonance Energies of Aromatic Molecules. *J.Chem.Soc.*, 1886–1888.

Green, S.M. and Marshall, G.R. (1995). 3D-QSAR: A Current Perspective. *Trends Pharmacol.Sci*, *16*, 285–291.

Grigoras, S. (1990). A Structural Approach to Calculate Physical Properties of Pure Organic Substances: The Critical Temperature, Critical Volume and Related Properties. *J.Comput.Chem.*, *11*, 493–510.

Grigorov, M., Weber, J., Tronchet, J.M.J., Jefford, C.W., Milhous, W.K. and Maric, D. (1997). A QSAR Study of the Antimalarial Activity of Some Synthetic 1,2,4-Trioxanes. *J.Chem.Inf.Comput.Sci.*, *37*, 124–130.

Grob, C.A. and Schlageter, M.G. (1976). 31. The Derivation of Inductive Substituent Constants from pK_a Values of 4-Substituted Quinuclidines. Polar Effects. Part I. *Helv.Chim.Acta*, *59*, 264–276.

Grob, C.A. (1985). 95. Inductive Charge Dispersal in Quinuclidinium Ions. *Helv.Chim.Acta*, *68*, 882–886.

Grossman, S.C. (1985). Chemical Ordering of Molecules: A Graph Theoretical Approach to Structure-Property Studies. *Int.J.Quant.Chem.*, *28*, 1–16.

Grossman, S.C., Jerman-Blazic Dzonova, B. and Randic, M. (1985). A Graph Theoretical Approach to Quantitative Structure-Activity Relationship. *Int.J.Quantum Chem.Quant.Biol. Symp.*, *12*, 123–139.

Grüber, C. and Buss, V. (1989). Quantum-Mechanically Calculated Properties for the Development of Quantitative Structure-Activity Relationships (QSARs). pK_a-Values of Phenols and Aromatic and Aliphatic Carboxylic Acids. *Chemosphere*, *19*, 1595–1609.

Grunenberg, J. and Herges, R. (1995). Prediction of Chromatographic Retention Values (R_M) and Partition Coefficients ($\log P_{oct}$) Using a Combination of Semiempirical Self-Consistent Reaction Field Calculations and Neural Networks. *J.Chem.Inf.Comput.Sci.*, *35*, 905–911.

Grunwald, E. and Winstein, S. (1948). The Correlation of Solvolysis Rates. *J.Am.Chem.Soc.*, *70*, 846–854.

Guo, X. and Zhang, F. (1993). An Efficient Algorithm for Generating all Kekulé Patterns of a Generalized Benzenoid System. *J.Math.Chem.*, *12*, 163–172.

Guo, X., Randic, M. and Klein, D.J. (1996). Analytical Expressions for the Count of LM-Conjugated Circuits of Benzenoid Hydrocarbons. *Int.J.Quant.Chem.*, *60*, 943–958.

Guo, X. and Randic, M. (1999). Trees with the Same Topological Index *JJ. SAR & QSAR Environ.Res.*, *10*.

Gupta, S., Singh, M. and Madan, A.K. (1999). Superpendentic Index: A Novel Topological Descriptor for Predicting Biological Activity. *J.Chem.Inf.Comput.Sci.*, *39*, 272–277.

Gupta, S.P., Singh, P. and Bindal, M.C. (1983). QSAR Studies on Hallucinogens. *Chem.Rev.*, *83*, 633–649.

Gupta, S.P. and Sharma, M.K. (1986). Molecular Connectivity in Hückel's Molecular Orbital Theory – II. Parametrization of Resonance Integral. *MATCH (Comm.Math.Comp.Chem.)*, *21*, 123–132.

Gussio, R., Pattabiraman, N., Zaharevitz, D.W., Kellogg, G.E., Topol, I.A., Rice, W.G., Schaeffer, C.A., Erickson, J.W. and Burt, S.K. (1996). All Atom Models for the Nonnucleoside Binding Site of HIV-1 Reverse Transcriptase Complexed with Inhibitors. A 3D QSAR Approach. *J.Med.Chem.*, *39*, 1645–1650.

Güsten, H., Klasinc, L. and Dubravko, M. (1984). Prediction of the Abiotic Degradability of Organic Compounds in the Troposphere. *Journal of Atmospheric Chemistry*, *2*, 83–93.

Güsten, H., Medven, Z., Sekusak, S. and Sabljic, A. (1995). Predicting Tropospheric Degradation of Chemicals: from Estimation to Computation. *SAR & QSAR Environ.Res.*, *4*, 197–209.

Gute, B.D. and Basak, S.C. (1997). Predicting Acute Toxicity of Benzene Derivatives Using Theoretical Molecular Descriptors: A Hierarchical QSAR Approach. *SAR & QSAR Environ.Res.*, *7*, 117–131.

Gute, B.D., Grunwald, G.D. and Basak, S.C. (1999). Prediction of the Dermal Penetration of Polycyclic Aromatic Hydrocarbons (PAHs): A Hierarchical QSAR Approach. *SAR & QSAR Environ.Res.*, *10*, 1–15.

Gutman, I. and Trinajstic, N. (1972). Graph Theory and Molecular Orbitals. Total π-Electron Energy of Alternant Hydrocarbons. *Chem.Phys.Lett.*, *17*, 535–538.

Gutman, I. and Trinajstic, N. (1973a). Graph Theory and Molecular Orbitals. *Top.Curr.Chem.*, *42*, 49–93.

Gutman, I. and Trinajstic, N. (1973b). Graph Theory and Molecular Orbitals. VIII. Kekulé Structure and Permutations. *Croat.Chem.Acta*, *45*, 539–545.

Gutman, I. (1974). Some Topological Properties of Benzenoid Systems. *Croat.Chem.Acta*, *46*, 209–215.

Gutman, I., Ruscic, B., Trinajstic, N. and Wilcox Jr, C.F. (1975). Graph Theory and Molecular Orbitals. XII. Acyclic Polyenes. *J.Chem.Phys.*, *62*, 3399–3405.

Gutman, I. (1976). Empirical Parameters for Donor and Acceptor Properties of Solvents. *Electrochim.Acta*, *21*, 661–670.

Gutman, I., Milun, M. and Trinajstic, N. (1977). Graph Theory and Molecular Orbitals. 19. Nonparametric Resonance Energies of Arbitrary Conjugated Systems. *J.Am.Chem.Soc.*, *99*, 1692–1704. [R]

Gutman, I. and Randic, M. (1977). Algebraic Characterization of Skeletal Branching. *Chem.-Phys. Lett.*, *47*, 15–19.

Gutman, I. (1978). Topological Formulas for Free-Valency Index. *Croat.Chem.Acta*, *51*, 29–33.

Gutman, I., Bosanac, S. and Trinajstic, N. (1978). Graph Theory and Molecular Orbitals. XX. Local and Long Range Contributions to Bond Order. *Croat.Chem.Acta*, *51*, 293–298.

Gutman, I. (1986). Topological Properties of Benzenoid Systems. XLVIII. Two Contradictory Formulas for Total π-Electron Energy and Their Reconciliation. *MATCH (Comm.Math.-Comp.Chem.)*, *21*, 317–324.

Gutman, I. and El-Basil, S. (1986). Fibonacci Graphs. *MATCH (Comm.Math.Comp.Chem.)*, *20*, 81–94.

Gutman, I. and Polansky, O. (1986a). A Regularity for the Boiling Points of Alkanes and Its Mathematical Modeling. *Z.Phys.Chemie*, *267*, 1152–1158.

Gutman, I. and Polansky, O. (1986b). Wiener Numbers of Polyacenes and Related Benzenoid Molecules. *MATCH (Comm.Math.Comp.Chem.)*, *20*, 115–123.

Gutman, I. (1987a). Acyclic Conjugated Molecules, Trees and their Energies. *J.Math.Chem.*, *1*, 123–144.

Gutman, I. (1987b). Wiener Numbers of Benzenoid Hydrocarbons: Two Theorems. *Chem.Phys. Lett.*, *136*, 134–136.

Gutman, I. (1988). On the Hosoya Index of Very Large Molecules. *MATCH (Comm.Math.-Comp.Chem.)*, *23*, 95–104.

Gutman, I. and Cyvin, S.J. (1988). All-Benzenoid Systems: Topological Properties of Benzenoid Systems. LVII. *MATCH (Comm.Math.Comp.Chem.)*, *23*, 175–178.

Gutman, I., Kennedy, J.W. and Quintas, L.V. (1990). Wiener Numbers of Random Benzenoid Chains. *Chem.Phys.Lett.*, *173*, 403–408.

Gutman, I. and Rouvray, D.H. (1990). A New Theorem for the Wiener Molecular Branching Index of Trees with Perfect Matchings. *Computers Chem.*, *14*, 29–32.

Gutman, I. (1991). Polynomials in Graph Theory. In *Chemical Graph Theory. Introduction and Fundamentals* (Bonchev, D. and Rouvray, D.H., eds.), Abacus Press/Gordon & Breach, New York (NY), pp. 133–176.

Gutman, I. and Soltés, L. (1991). The Range of the Wiener Index and Its Mean Isomer Degeneracy. *Z.Naturforsch.*, *46A*, 865–868.

Gutman, I. (1992a). Remark on the Moment Expansion of Total π-Electron Energy. *Theor.Chim. Acta*, *83*, 313–318.

Gutman, I. (1992b). Some Analytical Properties of the Independence and Matching Polynomials. *MATCH (Comm.Math.Comp.Chem.)*, *28*, 139–150.

Gutman, I., Hosoya, H., Urakovic, G. and Ristic, L. (1992). Two Variants of the Topological Index and the Relations Between Them. *Bull.Chem.Soc.Jap.*, *65*, 14–18.

Gutman, I. (1993a). A New Method for the Calculation of the Wiener Number of Acyclic Molecules. *J.Mol.Struct.(Theochem)*, *285*, 137–142.

Gutman, I. (1993b). Calculating the Wiener Number: The Doyle-Graver Method. *J.Serb.Chem. Soc.*, *58*, 745–750.

Gutman, I., Luo, Y.L. and Lee, S.L. (1993a). The Mean Isomer Degeneracy of the Wiener Index. *J.Chin.Chem.Soc.*, *40*, 195–198.

Gutman, I., Yeh, Y.N., Lee, S.L. and Luo, Y.L. (1993b). Some Recent Results in the Theory of the Wiener Number. *Indian J.Chem.*, *32A*, 651–661.

Gutman, I. (1994a). Formula for the Wiener Number of Trees and Its Extension to Graphs Containing Cycles. *Graph Theory Notes New York*, *27*, 9–15.

Gutman, I. (1994b). Selected Properties of the Schultz Molecular Topological Index. *J.Chem.Inf. Comput.Sci.*, *34*, 1087–1089.

Gutman, I., Lee, S.L., Chu, C.H. and Luo, Y.-R. (1994a). Chemical Applications of the Laplacian Spectrum of Molecular Graphs: Studies of the Wiener Number. *Indian J.Chem.*, *33A*, 603–608.

Gutman, I., Yeh, Y.N., Lee, S.L. and Chen, J.C. (1994b). Wiener Numbers of Dendrimers. *MATCH (Comm.Math.Comp.Chem.)*, *30*, 103–115.

Gutman, I., Bonchev, D., Seitz, W.A. and Gordeeva, E.V. (1995). Complementing the Proof of the Limit of Relative Atomic Moments. *J.Chem.Inf.Comput.Sci.*, *35*, 894–895.

Gutman, I. and Klavzar, S. (1995). An Algorithm for the Calculation of the Szeged Index of Benzenoid Hydrocarbons. *J.Chem.Inf.Comput.Sci.*, *35*, 1011–1014.

Gutman, I. and Körtévlyesi, T. (1995). *Z.Naturforsch.*, *50A*, 669–671.

Gutman, I. and Estrada, E. (1996). Topological Indices Based on Line Graph of the Molecular Graph. *J.Chem.Inf.Comput.Sci.*, *36*, 541–543.

Gutman, I. and Mohar, B. (1996). The Quasi-Wiener and the Kirchhoff Indices Coincide. *J.Chem.Inf.Comput.Sci.*, *36*, 982–985.

Gutman, I. (1997). A Property of the Wiener Number and Its Modifications. *Indian J.Chem.*, *36A*, 128–132.

Gutman, I. and Klavzar, S. (1997). Bounds for the Schultz Molecular Topological Index of Benzenoid Systems in Terms of the Wiener Index. *J.Chem.Inf.Comput.Sci.*, *37*, 741–744.

Gutman, I., Linert, W., Lukovits, I. and Dobrynin, A.A. (1997a). Trees with External Hyper-Wiener Index: Mathematical Basis and Chemical Applications. *J.Chem.Inf.Comput.Sci.*, *37*, 349–354.

Gutman, I., Markovic, S., Popovic, L., Spalevic, Z. and Pavlovic, L. (1997b). The Relation Between the Wiener Indices of Phenylenes and their Hexagonal Squeezes. *J.Serb.Chem.Soc.*, *62*, 207–210.

Gutman, I. and Potgieter, J.H. (1997). Wiener Index and Intermolecular Forces. *J.Serb.Chem. Soc.*, *62*, 185–192.

Gutman, I. and Diudea, M.V. (1998a). Defining Cluj Matrices and Cluj Invariants. *MATCH (Comm.Math.Comp.Chem.)*, *36*.

Gutman, I. and Dobrynin, A.A. (1998b). The Szeged Index – A Success Story. *Graph Theory Notes New York*, *34*, 37–44.
Gutman, I., Linert, W., Lukovits, I. and Tomovic, Z. (2000). The Multiplicative Version of the Wiener Index. *J.Chem.Inf.Comput.Sci.*, *40*, 113–116.
Gutmann, V. (1978). *The Donor-Acceptor Approach to Molecular Interactions.* Plenum Press, New York (NY).
Gutowsky, H.S., McCall, D.W., McGarvey, B.R. and Meyer, L.H. (1952). Electron Distribution in Molecules. I. F^{19} Nuclear Magnetic Shielding and Substituent Effects in Some Benzene Derivatives. *J.Am.Chem.Soc.*, *74*, 4809–4817.
Hadaruga, D.I., Muresan, S., Bologa, C., Chiriac, A., Simon, Z., Cofar, L. and Naray-Szabo, G. (1999). QSAR for Cycloaliphatic Alcohols with Qualitatively Defined Sandalwood Odour Characteristics. *Quant.Struct.-Act.Relat.*, *18*, 253–261.
Hadjipavloulitina, D. and Hansch, C. (1994). Quantitative Structure-Activity Relationships of the Benzodiazepines: A Review and Reevaluation. *Chem.Rev.*, *94*, 1483–1505.
Hadzi, D. and Jerman-Blazic, B. eds. (1987). *QSAR in Drug Design and Toxicology.* Elsevier, Amsterdam (The Netherlands).
Hadzi, D., Kidric, J., Koller, J. and Mavri, J. (1990). The Role of Hydrogen Bonding in Drug-Receptor Interactions. *J.Mol.Struct.*, *237*, 139–150.
Haeberlein, M. and Brinck, T. (1997). Prediction of Water-Octanol Partition Coefficients Using Theoretical Descriptors Derived from the Molecular Surface Area and the Electrostatic Potential. *J.Chem.Soc.Perkin Trans.2*, 289–294.
Hagadone, T.R. (1992). Molecular Substructure Similarity Searching: Efficient Retrieval in Two-Dimensional Structure Databases. *J.Chem.Inf.Comput.Sci.*, *32*, 515–521.
Hahn, M. (1995). Receptor Surface Models. 1. Definition and Construction. *J.Med.Chem.*, *38*, 2080–2090.
Hahn, M. and Rogers, D. (1995). Receptor Surface Models. 2. Application to Quantitative Structure-Activity Relationships Studies. *J.Med.Chem.*, *38*, 2091–2102.
Hahn, M. (1997). Three-Dimensional Shape-Based Searching of Conformationally Flexible Compounds. *J.Chem.Inf.Comput.Sci.*, *37*, 80–86.
Hahn, M. and Rogers, D. (1998). Receptor Surface Models. In *3D QSAR in Drug Design – Vol. 3* (Kubinyi, H., Folkers, G. and Martin, Y.C., eds.), Kluwer/ESCOM, Dordrecht (The Netherlands), pp. 117–133.
Hall, G.G. (1955). The Bond Orders of Alternant Hydrocarbons Molecules. *Proc.Roy.Soc.London A*, *229*, 251–259.
Hall, G.G. (1957). The Bond Orders of Some Conjugated Molecules. *Trans.Faraday Soc.*, *53*, 573–581.
Hall, G.G. (1981). Eigenvalues of Molecular Graphs. *Bull.Inst.Math.Appl.*, *17*, 70–72.
Hall, G.G. (1986). The Evaluation of Moments for Polycyclic Hydrocarbons. *Theor.Chim.Acta*, *70*, 323–332.
Hall, G.G. (1992). Eigenvalue Distributions for the Graphs of Alternant Hydrocarbons. *J.Chem.Inf.Comput.Sci.*, *32*, 11–13.
Hall, G.G. (1993). Eigenvalue Distributions in Alternant Hydrocarbons. *J.Math.Chem.*, *13*, 191–203.
Hall, L.H., Kier, L.B. and Murray, W.J. (1975). Molecular Connectivity. II. Relationship to Water Solubility and Boiling Point. *J.Pharm.Sci.*, *64*, 1974–1977.
Hall, L.H. and Kier, L.B. (1977a). A Molecular Connectivity Study of Electron Density in Alkanes. *Tetrahedron*, *33*, 1953–1957.
Hall, L.H. and Kier, L.B. (1977b). Structure-Activity Studies Using Valence Molecular Connectivity. *J.Pharm.Sci.*, *66*, 642–644.
Hall, L.H. and Kier, L.B. (1978a). A Comparative Analysis of Molecular Connectivity, Hansch, Free-Wilson and Darc-Pelco Methods in the SAR of Halogenated Phenols. *Eur.J.Med.Chem.*, *13*, 89–92.
Hall, L.H. and Kier, L.B. (1978b). Molecular Connectivity and Substructure Analysis. *J.Pharm.Sci.*, *67*, 1743–1747.

Hall, L.H. and Kier, L.B. (1981). The Relation of Molecular Connectivity to Molecular Volume and Biological Activity. *Eur.J.Med.Chem.*, *16*, 399–407.

Hall, L.H. and Kier, L.B. (1984). Molecular Connectivity of Phenols and their Toxicity to Fish. *Bull.Environ.Contam.Toxicol.*, *32*, 354–362.

Hall, L.H. and Kier, L.B. (1986). Molecular Connectivity and Total Response Surface Optimization. *J.Mol.Struct.(Theochem)*, *134*, 309–316.

Hall, L.H. and Aaserud, D. (1989). Structure-Activity Models for Molar Refraction of Alkylsilanes Based on Molecular Connectivity. *Quant.Struct.-Act.Relat.*, *8*, 296–304.

Hall, L.H., Maynard, E.L. and Kier, L.B. (1989). QSAR Investigation of Benzene Toxicity to *Fathead minnow* Using Molecular Connectivity. *Environ.Toxicol.Chem.*, *8*, 783.

Hall, L.H. (1990). Computational Aspects of Molecular Connectivity and its Role in Structure-Property Modeling. In *Computational Chemical Graph Theory* (Rouvray, D.H., ed.), Nova Press, New York (NY), pp. 202–233.

Hall, L.H. and Kier, L.B. (1990). Determination of Topological Equivalence in Molecular Graphs from the Topological State. *Quant.Struct.-Act.Relat.*, *9*, 115–131.

Hall, L. H. (1991). *MolConn-X: A Program for Molecular Topology Analysis.* Hall Associates Consulting, Quincy (MA).

Hall, L.H. and Kier, L.B. (1991). The Molecular Connectivity Chi Indexes and Kappa Shape Indexes in Structure-Property Modeling. In *Reviews in Computational Chemistry – Vol. 2* (Lipkowitz, K.B. and Boyd, D.B., eds.), VCH Publishers, New York (NY), pp. 367–422. [R]

Hall, L.H., Mohney, B. and Kier, L.B. (1991a). The Electrotopological State: An Atom Index for QSAR. *Quant.Struct.-Act.Relat.*, *10*, 43–48.

Hall, L.H., Mohney, B. and Kier, L.B. (1991b). The Electrotopological State: Structure Information at the Atomic Level for Molecular Graphs. *J.Chem.Inf.Comput.Sci.*, *31*, 76–82.

Hall, L.H. and Kier, L.B. (1992a). Binding of Salicylamides: QSAR Analysis with Electrotopological State Indexes. *Med.Chem.Res.*, *2*, 497–502.

Hall, L.H. and Kier, L.B. (1992b). Enumeration, Topological Indexes and Molecular Properties in Alkanes. (Patei, S. and Rapoport, Z., eds.), Wiley, Chichester (UK), pp. 186–213.

Hall, L.H., Dailey, R.S. and Kier, L.B. (1993a). Design of Molecules from Quantitative Structure-Activity Relationship Models. 3. Role of Higher Order Path Counts: Path 3. *J.Chem.Inf. Comput.Sci.*, *33*, 598–603.

Hall, L.H., Kier, L.B. and Frazer, J.W. (1993b). Design of Molecules from Quantitative Structure-Activity Relationship Models. 2. Derivation and Proof of Information Transfer Relating Equations. *J.Chem.Inf.Comput.Sci.*, *33*, 148–152.

Hall, L.H., Mohney, B. and Kier, L.B. (1993c). Comparison of Electrotopological Sate Indexes with Molecular Orbital Parameters: Inhibition of MAO by Hydrazides. *Quant.Struct.-Act. Relat.*, *12*, 44–48.

Hall, L.H. (1995). Experimental Design in Synthesis Planning and Structure-Property Correlations. Total Response Surface Optimization. In *Chemometrics Methods in Molecular Design – Vol. 2* (van de Waterbeemd, H., ed.), VCH Publishers, New York (NY), pp. 91–102.

Hall, L.H. and Kier, L.B. (1995a). Electrotopological State Indices for Atom Types: A Novel Combination of Electronic, Topological, and Valence State Information. *J.Chem.Inf.Comput.Sci.*, *35*, 1039–1045.

Hall, L.H., Kier, L.B. and Brown, B.B. (1995b). Molecular Similarity Based on Novel Atom-Type Electrotopological State Indices. *J.Chem.Inf.Comput.Sci.*, *35*, 1074–1080.

Hall, L.H. and Story, C.T. (1996). Boiling Point and Critical Temperature of a Heterogeneous Data Set: QSAR with Atom Type Electrotopological State Indices Using Artificial Neural Networks. *J.Chem.Inf.Comput.Sci.*, *36*, 1004–1014.

Hall, L.H. and Story, C.T. (1997a). Boiling Point of a Set of Alkanes, Alcohols and Chloroalkanes: QSAR with Atom Type Electrotopological State Indices Using Artificial Neural Networks. *SAR & QSAR Environ.Res.*, *6*, 139–161.

Hall, L.H. and Vaughn, T.A. (1997b). QSAR of Phenol Toxicity Using E-State and Kappa Shape Indices. *Med.Chem.Res.*, *7*, 407–416.

Halova, J., Strouf, O., Zak, P., Sochozova, A., Uchida, N., Yuzuri, T., Sakakibara, K. and Hirota, M. (1998). QSAR of Catechol Analogs Against Malignant Melanoma Using Fingerprint Descriptors. *Quant.Struct.-Act.Relat.*, *17*, 37–39.

Ham, N.S. (1958a). Mobile Bond Orders in the Resonance and Molecular Orbital Theories. *J.Chem.Phys.*, *29*, 1229–1231.

Ham, N.S. and Ruedenberg, K. (1958b). Energy Levels, Atom Populations, Bond Populations in the LCAO MO Model and in the FE MO Model. A Quantitative Analysis. *J.Chem.Phys.*, *29*, 1199–1214.

Ham, N.S. and Ruedenberg, K. (1958c). Mobile Bond Orders in Conjugated Systems. *J.Chem. Phys.*, *29*, 1215–1229.

Hamasaki, T., Masumoto, H., Sato, T., Nagase, H., Kito, H. and Yoshioka, Y. (1995). Estimation of the Hemolytic Effects of Various Organotin Compounds by Structure-Activity Relationships. *Appl.Organometal.Chem.*, *9*, 95–104.

Hamerton, I., Howlin, B.J. and Larwood, V. (1995). Development of Quantitative Structure-Property Relationships for Poly(Arylene Ether)s. *J.Mol.Graphics*, *13*, 14–17.

Hammett, L.P. (1935). Reaction Rates and Indicator Acidities. *Chem.Rev.*, *17*, 67–79.

Hammett, L.P. (1937). The Effect of Structure upon the Reactions of Organic Compounds. Benzene Derivatives. *J.Am.Chem.Soc.*, *59*, 96–103.

Hammett, L.P. (1938). Linear Free Energy Relationships in Rate and Equilibrium Phenomena. *Trans.Faraday Soc.*, *34*, 156–165.

Hammett, L.P. (1940). *Physical Organic Chemistry.* McGraw-Hill, New York (NY).

Hammett, L.P. (1970). *Physical Organic Chemistry: Reaction Rates, Equilibria and Mechanism.* McGraw-Hill, New York (NY).

Han, C.R. (1990). The Calculation of the Ionic Group Electronegativities and Neutral Group Electronegativities. *Acta Chim.Sin.*, *48*, 627–631.

Hancock, C.K. and Falls, C.P. (1961a). A Hammett-Taft Polar-Steric Equation for the Saponification Rates of *m*- and *p*-Substituted Alkyl Benzoates. *J.Am.Chem.Soc.*, *83*, 4214–4216.

Hancock, C.K., Meyers, E.A. and Yager, B.J. (1961b). Quantitative Separation of Hyperconjugation Effects from Steric Substituent Constants. *J.Am.Chem.Soc.*, *83*, 4211–4213.

Hand, D.J. (1981). *Discrimination and Classification.* Wiley, Chichester (UK).

Hand, D.J. (1997). *Construction and Assessment of Classification Rules.* Wiley, Chichester (UK), 214 pp.

Handschuh, S., Wagener, M. and Gasteiger, J. (1998). Superimposition of Three-Dimensional Chemical Structures Allowing for Conformational Flexibility by a Hybrid Method. *J.Chem. Inf.Comput.Sci.*, *38*, 220–232.

Hannay, N.B. and Smyth, C.P. (1946). The Dipole Moment of Hydrogen Fluoride and the Ionic Character of Bonds. *J.Am.Chem.Soc.*, *68*, 171–173.

Hannongbua, S., Lawtrakul, L. and Limtrakul, J. (1996a). Structure-Activity Correlation Study of HIV-1 Inhibitors: Electronic and Molecular Parameters. *J.Comput.Aid.Molec.Des.*, *10*, 145–152.

Hannongbua, S., Lawtrakul, L., Sotriffer, C.A. and Rode, B.M. (1996b). Comparative Molecular Field Analysis of HIV-1 Reverse Transcriptase Inhibitors in the Class of 1((2-Hydroxyethoxy)-Methyl)-6(Phenylthio)Thymine. *Quant.Struct.-Act.Relat.*, *15*, 389–394.

Hansch, C., Maloney, P.P., Fujita, T. and Muir, R.M. (1962). Correlation of Biological Activity of Phenoxyacetic Acids with Hammett Substituent Constants and Partition Coefficients. *Nature*, *194*, 178–180.

Hansch, C., Muir, R.M., Fujita, T., Maloney, P.P., Geiger, F. and Streich, M. (1963). The Correlation of Biological Activity of Plant Growth Regulators and Chloromycetin Derivatives with Hammett Constants and Partition Coefficients. *J.Am.Chem.Soc.*, *85*, 2817–2824.

Hansch, C. and Fujita, T. (1964). $\rho\sigma\pi$ Analysis. A Method for the Correlation of Biological Activity and Chemical Structure. *J.Am.Chem.Soc.*, *86*, 1616–1626.

Hansch, C., Deutsch, E.W. and Smith, R.N. (1965). The Use of Substituent Constants and Regression Analysis in the Study of Enzymatic Reaction Mechanisms. *J.Am.Chem.Soc.*, *87*, 2738–2742.

Hansch, C. and Anderson, S.M. (1967). Structure-Activity Relation in Barbiturates and Its Similarity to That in Other Narcotics. *J.Math.Chem.*, *10*, 745–753.

Hansch, C. and Lien, E.J. (1968). Analysis of the Structure-Activity Relationship in the Adrenergic Blocking Activity of β-Haloalkylamines. *Biochem.Pharmacol.*, *17*, 709–720.

Hansch, C., Quinlan, J.E. and Lawrence, G.L. (1968). The LFER Between Partition Coefficients and the Aqueous Solubility of Organic Liquids. *J.Org.Chem.*, *33*, 347–350.

Hansch, C. (1969). Quantitative Approach to Biochemical Structure-Activity Relationships. *Acc.Chem.Res.*, *2*, 232–239.

Hansch, C., Kutter, E. and Leo, A. (1969). Homolytic Constants in the Correlation of Chloramphenicol Structure with Activity. *J.Med.Chem.*, *12*, 746–749.

Hansch, C. (1970). Steric Parameters in Structure-Activity Correlations. Cholinesterase Inhibitors. *J.Org.Chem.*, *35*, 620–621.

Hansch, C. (1971). Quantitative Structure-Activity Relationships in Drug Design. In *Drug Design, Vol. 1* (Ariëns, E.J., ed.), Academic Press, New York (NY), pp. 271–342.

Hansch, C. and Lien, E.J. (1971). Structure-Activity Relationships in Antifungal Agents. A Survey. *J.Med.Chem.*, *14*, 653–670.

Hansch, C. and Dunn III, W.J. (1972). Linear Relationships Between Lipophilic Character and Biological Activity of Drugs. *J.Pharm.Sci.*, *61*, 1–19.

Hansch, C., Leo, A. and Nikaitani, D. (1972). On the Additive-Constitutive Character of Partition Coefficients. *J.Org.Chem.*, *37*, 3090–3092.

Hansch, C. and Clayton, J.M. (1973). Lipophilic Character and Biological Activity of Drugs II: The Parabolic Case. *J.Pharm.Sci.*, *62*, 1–21.

Hansch, C., Leo, A., Unger, S.H., Kim, K.H., Nikaitani, D. and Lien, E.J. (1973a). "Aromatic" Substituent Constants for Structure-Activity Correlations. *J.Med.Chem.*, *16*, 1207–1216.

Hansch, C., Unger, S.H. and Forsythe, A.B. (1973b). Strategy in Drug Design. Cluster Analysis as an Aid in the Selection of Substituents. *J.Med.Chem.*, *16*, 1217–1222.

Hansch, C. and Yoshimoto, M. (1974). Structure-Activity Relationships in Immunochemistry. 2. Inhibition of Complement by Benzamidines. *J.Med.Chem.*, *17*, 1160–1167.

Hansch, C. and Calef, D.F. (1976). Structure-Activity Relationships in Papain-Ligand Interactions. *J.Org.Chem.*, *41*, 1240–1243.

Hansch, C., Rockwell, S.D., Leo, A. and Steller, E.E. (1977). Substituent Constants for Correlation Analysis. *J.Med.Chem.*, *20*, 304–306.

Hansch, C. (1978). Recent Advances in Biochemical QSAR. In *Correlation Analysis in Chemistry* (Chapman, N.B. and Shorter, J., eds.), Plenum Press, New York (NY), pp. 397–438.

Hansch, C. and Leo, A. (1979). *Substituent Constants for Correlation Analysis in Chemistry and Biology*. Wiley, New York (NY), 352 pp.

Hansch, C., Leo, A. and Taft, R.W. (1991). A Survey of Hammett Substituent Constants and Resonance and Field Parameters. *Chem.Rev.*, *91*, 165–195.

Hansch, C. and Zhang, L.T. (1992). QSAR of HIV Inhibitors. *Bioorg.Med.Chem.Lett.*, *2*, 1165–1169.

Hansch, C. (1993). Quantitative Structure-Activity Relationships and the Unnamed Science. *Acc.Chem.Res.*, *2*, 147–153.

Hansch, C. (1995a). Comparative QSAR Understanding Hydrophobic Interactions. *ACS Symp. Ser.*, *606*, 254–262.

Hansch, C. (1995b). Comparative Quantitative Structure-Activity Relationship Insect Versus Vertebrate Cholinesterase. *ACS Symp.Ser.*, *589*, 281–291.

Hansch, C. and Fujita, T. (1995). Status of QSAR at the End of the Twentieth Century. In *Classical and 3D-QSAR in Agrochemistry (ACS Symposium Series 606)* (Hansch, C. and Fujita, T., eds.), American Chemical Society, Washington (DC), pp. 1–11.

Hansch, C., Hoekman, D., Leo, A., Zhang, L.T. and Li, P. (1995a). The Expanding Role of Quantitative Structure-Activity Relationships (QSAR) in Toxicology. *Toxicol.Lett.*, *79*, 45–53.

Hansch, C. and Leo, A. (1995). *Exploring QSAR. Fundamentals and Applications in Chemistry and Biology*. American Chemical Society, Washington (DC).

Hansch, C., Leo, A. and Hoekman, D. (1995b). *Exploring QSAR. Hydrophobic, Electronic, and Steric Constants*. American Chemical Society, Washington (DC), 557 pp.

Hansch, C., Telzer, B.R. and Zhang, L.T. (1995c). Comparative QSAR in Toxicology: Examples from Teratology and Cancer Chemotherapy of Aniline Mustards. *Crit.Rev.Toxicol.*, *25*, 67–89.

Hansch, C., Hoekman, D. and Gao, H. (1996). Comparative QSAR Toward a Deeper Understanding of Chemicobiological Interactions. *Chem.Rev.*, *96*, 1045–1075.

Hansch, C. and Gao, H. (1997). Comparative QSAR: Radical Reactions of Benzene Derivatives in Chemistry and Biology. *Chem.Rev.*, *97*, 2995–3060.

Hansch, C., Gao, H. and Hoekman, D. (1998). A Generalized Approach to Comparative QSAR. In *Comparative QSAR* (Devillers, J., ed.), Taylor & Francis, Washington (DC), pp. 285–368. [R]

Hansen, P.J. and Jurs, P.C. (1987). Prediction of Olefin Boiling Points from Molecular Structure. *Anal.Chem.*, *59*, 2322–2327.

Hansen, P.J. and Jurs, P.C. (1988a). Chemical Applications of Graph Theory. Part I. Fundamentals and Topological Indices. *J.Chem.Educ.*, *65*, 574–580.

Hansen, P.J. and Jurs, P.C. (1988b). Chemical Applications of Graph Theory. Part II. Isomer Enumeration. *J.Chem.Educ.*, *65*, 661–664.

Hanser, T., Jauffret, P. and Kaufmann, G. (1996). A New Algorithm for Exhaustive Ring Perception in a Molecular Graph. *J.Chem.Inf.Comput.Sci.*, *36*, 1146–1152.

Hansson, G. and Ahnoff, M. (1994). Chromatographic Separation of Amide Diastereomers Correlation with Molecular Descriptors. *J.Chromat.*, *666A*, 505–517.

Harada, A., Hanzawa, M., Saito, J. and Hashimoto, K. (1992). Quantitative Analysis of Structure-Toxicity Relationships of Substituted Anilines by Use of BALB/3T3 Cells. *Environ.Toxicol.Chem.*, *11*, 973–980.

Harary, F. (1969a). *Graph Theory.* Addison-Wesley, Reading (MA).

Harary, F. (1969b). *Proof Techniques in Graph Theory.* Academic Press, San Diego (CA).

Harris, N.V., Smith, C. and Bowden, K. (1992). Antifolate and Antibacterial Activities of 6-Substituted 2,4-Diaminoquinazolines. *Eur.J.Med.Chem.*, *27*, 7–18.

Harrison, A.G., Kebarle, P. and Lossing, F.P. (1961). Free Radicals by Mass Spectrometry. XXI. The Ionization Potentials of Some *meta* and *para* Substituted Benzyl Radicals. *J.Am.Chem. Soc.*, *83*, 777–780.

Hartley, R.V.L. (1928). Transmission of Information. *Bell Syst.Tech.J.*, *7*, 535–563.

Hasegawa, K., Deushi, T., Yaegashi, O., Miyashita, Y. and Sasaki, S. (1995a). Artificial Neural Network Studies in Quantitative Structure-Activity Relationships of Antifungal Azoxy Compounds. *Eur.J.Med.Chem.*, *30*, 569–574.

Hasegawa, K., Shigyou, H. and Sonoki, H. (1995b). Free-Wilson Discriminant Analysis of Antiarrhythmic Phenyl-pyridines Using PLS. *Quant.Struct.-Act.Relat.*, *14*, 344–347.

Hasegawa, K., Kimura, T., Miyashita, Y. and Funatsu, K. (1996a). Nonlinear Partial Least Squares Modeling of Phenyl Alkylamines with the Monoamine Oxidase Inhibitory Activities. *J.Chem.Inf.Comput.Sci.*, *36*, 1025–1029.

Hasegawa, K., Yokoo, N., Watanabe, K., Hirata, M., Miyashita, Y. and Sasaki, S. (1996b). Multivariate Free-Wilson Analysis of Alpha Chymotrypsin Inhibitors Using Pls. *Chemom.Intell. Lab.Syst.*, *33*, 63–69.

Hasegawa, K., Miyashita, Y. and Funatsu, K. (1997). GA Strategy for Variable Selection in QSAR Studies: GA-Based PLS Analysis of Calcium Channel Antagonists. *J.Chem.Inf.Comput.Sci.*, *37*, 329–334.

Hasegawa, K. and Funatsu, K. (1998). GA Strategy for Variable Selection in QSAR Studies: GAPLS and D-Optimal Designs for Predictive QSAR Model. *J.Mol.Struct.(Theochem)*, *425*, 255–262.

Hasegawa, K., Arakawa, M. and Funatsu, K. (1999a). 3D-QSAR Study of Insecticidal Neonicotinoid Compounds Based on 3-Way Partial Least Squares Model. *Chemom.Intell.Lab.Syst.*, *47*, 33–40.

Hasegawa, K., Kimura, T. and Funatsu, K. (1999b). GA Strategy for Variable Selection in QSAR Studies: Application of GA-Based Region Selection to a 3D-QSAR Study of Acetylcholinesterase Inhibitors. *J.Chem.Inf.Comput.Sci.*, *39*, 112–120.

Hassan, M., Bielawski, J.P., Hempel, J.C. and Waldman, M. (1996). Optimization and Visualization of Molecular Diversity of Combinatorial Libraries. *Mol.Div.*, *2*, 64–74.

Hatch, F.T., Colvin, M.E. and Seidl, E.T. (1996). Structural and Quantum Chemical Factors Affecting Mutagenic Potency of Aminoimidazo Azaarenes. *Environ.Mol.Mutagen.*, *27*, 314–330.

Hatrìk, S. and Zahradnìk, P. (1996). Neural Network Approach to the Prediction of the Toxicity of Benzothiazolium Salts from Molecular Structure. *J.Chem.Inf.Comput.Sci.*, *36*, 992–995.

Hays, S.J., Rice, M.J., Ortwine, D.F., Johnson, G., Schwarz, R.D., Boyd, D.K., Copeland, L.F., Vartanian, M.G. and Boxer, P.A. (1994). Substituted 2-Benzothiazolamine as Sodium Flux Inhibitors Quantitative Structure-Activity Relationships and Anticonvulsant Activity. *J.Pharm.Sci.*, *83*, 1425–1432.

Headley, A.D., Starnes, S.D., Wilson, L.Y. and Famini, G.R. (1994). Analysis of Solute/Solvent Interactions for the Acidity of Acetic Acids by Theoretical Descriptors. *J.Org.Chem.*, *59*, 8040–8046.

Headley, A.D., Starnes, S.D., Cheung, E.T. and Malone, P.L. (1995). Solvation Effects on the Relative Basicity of Propylamines. *J.Phys.Org.Chem.*, *8*, 26–30.

Heinzer, V.E.F. and Yunes, R.A. (1996). Using Topological Indices in the Prediction of Gas Chromatographic Retention Indices of Linear Alkylbenzene Isomers. *J.Chromat.*, *719A*, 462–467.

Hemken, H.G. and Lehmann, P.A. (1992). The Use of Computerized Molecular Structure Scanning and Principal Component Analysis to Calculate Molecular Descriptors for QSAR. *Quant.Struct.-Act.Relat.*, *11*, 332–338.

Hemmer, M.C., Steinhauer, V. and Gasteiger, J. (1999). Deriving the 3D Structure of Organic Molecules from Their Infrared Spectra. *Vibrat.Spect.*, *19*, 151–164.

Hendrickson, J.B., Huang, P. and Toczko, A.G. (1987). Molecular Complexity: A Simplified Formula Adapted to Individual Atoms. *J.Chem.Inf.Comput.Sci.*, *27*, 63–67.

Hendriks, M.M.W.B., de Boer, J.H., Smilde, A.K. and Doornbos, D.A. (1992). Multicriteria Decision Making. *Chemom.Intell.Lab.Syst.*, *16*, 175–191.

Henrie II, R.N., Plummer, M.J., Smith, S.E., Yeager, W.H. and Witkowski, D.A. (1993). Discovery and Optimization of a PSI Electron-Accepting 1,2,4-Benzotriazine Herbicide. *Quant. Struct.-Act.Relat.*, *12*, 27–37.

Henry, D.R. and Block, J.H. (1979). Classification of Drugs by Discriminant Analysis Using Fragment Molecular Connectivity Values. *J.Med.Chem.*, *22*, 465–472.

Henry, D.R. and Block, J.H. (1980a). Pattern Recognition of Steroids Using Fragment Molecular Connectivity. *J.Pharm.Sci.*, *69*, 1030.

Henry, D.R. and Block, J.H. (1980b). Steroid Classification by Discriminant Analysis Using Fragment Molecular Connectivity. *Eur.J.Med.Chem.*, *15*, 133.

Henry, D.R., Jurs, P.C. and Denny, W.A. (1982). Structure-Antitumor Activity Relationships of 9-Anilinoacridines Using Pattern Recognition. *J.Math.Chem.*, *25*, 899–908.

Herdan, J., Balaban, A.T., Stoica, G., Simon, Z., Mracec, M. and Niculescuduvaz, I. (1991). Compounds with Potential Cancer Preventing Activity. 1. Synthesis, Physicochemical Properties and Quantum Chemical Indexes of Some Phenolic and Aminophenolic Antioxidants. *Rev.Roum.Chim.*, *36*, 1147–1160.

Heritage, T.W., Ferguson, A.M., Turner, D.B. and Willett, P. (1998). EVA: A Novel Theoretical Descriptor for QSAR Studies. In *3D QSAR in Drug Design – Vol. 2* (Kubinyi, H., Folkers, G. and Martin, Y.C., eds.), Kluwer/ESCOM, Dordrecht (The Netherlands), pp. 381–398.

Hermann, A. and Zinn, P. (1995). List Operations on Chemical Graphs. 6. Comparative Study of Combinatorial Topological Indexes of the Hosoya Type. *J.Chem.Inf.Comput.Sci.*, *35*, 551–560.

Hermann, R.B. (1972). Theory of Hydrophobic Bonding. II. Correlation of Hydrocarbon Solubility in Water with Solvent Cavity Surface Area. *J.Phys.Chem.*, *76*, 2754–2759.

Hermens, J.L. (1995a). Prediction of Environmental Toxicity Based on Structure-Activity Relationships Using Mechanistic Information. *Sci.Total Environ.*, *171*, 235–242.

Hermens, J.L. and Verhaar, H.J. (1995b). QSARs in Environmental Toxicology and Chemistry: Recent Developments. *ACS Symp.Ser.*, *606*, 130–140.

Herndon, W.C. (1973a). Enumeration of Resonance Structures. *Tetrahedron*, *29*, 3–12.

Herndon, W.C. (1973b). Resonance Energies of Aromatic Hydrocarbons. A Quantitative Test of Resonance Theory. *J.Am.Chem.Soc.*, *95*, 2404–2406.

Herndon, W.C. (1974a). Isospectral Molecules. *Tetrahedron Lett.*, *8*, 671–674.

Herndon, W.C. (1974b). Resonance Theory and the Enumeration of Kekulé Structures. *J.Chem. Educ.*, *51*, 10–15.

Herndon, W.C. and Ellzey Jr, M.L. (1974). Resonance Theory. V. Resonance Energies of Benzenoid and Nonbenzenoid π Systems. *J.Am.Chem.Soc.*, *96*, 6631–6642.

Herndon, W.C. and Bertz, S.H. (1987). Linear Notations and Molecular Graph Similarity. *J.Comput.Chem.*, *8*, 367–374.

Herndon, W.C. (1988). Graph Codes and a Definition of Graph Similarity. *Comp.Math.Applic.*, *15*, 303–309.

Hess Jr., B.A. and Schaad, L.J. (1971a). Hückel Molecular Orbital π Resonance Energies. A New Approach. *J.Am.Chem.Soc.*, *93*, 305–310.

Hess Jr., B.A. and Schaad, L.J. (1971b). Hückel Molecular Orbital π Resonance Energies. The Benzenoid Hydrocarbons. *J.Am.Chem.Soc.*, *93*, 2413–2416.

Hess Jr., B.A., Schaad, L.J. and Holyoke Jr., C.W. (1972). On the Aromaticity of Heterocycles Containing the Amine Nitrogen or the Ether Oxygen. *Tetrahedron*, *28*, 3657–3667.

Hess Jr., B.A. and Schaad, L.J. (1973). Hückel Molecular Orbital π-Resonance Energies. Heterocycles Containing Divalent Sulfur. *J.Am.Chem.Soc.*, *95*, 3907–3912.

Hess Jr., B.A., Schaad, L.J. and Holyoke Jr., C.W. (1975). The Aromaticity of Heterocycles Containing the Imine Hydrogen. *Tetrahedron*, *31*, 295–298.

Hetnarski, B. and O'Brien, R.D. (1973). Charge Transfer in Cholinesterase Inhibition. Role of the Conjugation Between Carbamyl and Aryl Groups of Aromatic Carbamates. *Biochemistry*, *12*, 3883–3887.

Hetnarski, B. and O'Brien, R.D. (1975). The Charge-Transfer Constant. A New Substituent Constant for Structure-Activity Relationships. *J.Med.Chem.*, *18*, 29–33.

Hicks, M.G. and Jochum, C. (1990). Substructure Search Systems for Large Chemical Databases. *Anal.Chim.Acta*, *235*, 87–92.

Higuchi, T. and Davis, S.S. (1970). Thermodynamic Analysis of Structure-Activity Relationships of Drugs: Prediction of Optimal Structure. *J.Pharm.Sci.*, *59*, 1376–1383.

Hilal, S.H., Karichoff, S.W. and Carreira, L.A. (1995). A Rigorous Test for SPARC's Chemical Reactivity Models: Estimation of More Than 4300 Ionization pKas. *Quant.Struct.-Act.Relat.*, *14*, 348–355.

Hildebrand, J.H. and Scott, R.L. (1950). *Solubility of Nonelectrolytes*. Rheinhold Publ. Co, New York (NY).

Hill, T.L. (1948). Steric Effects. I. Van der Waals Potential Energy Curves. *J.Chem.Phys.*, *16*, 399–404.

Hine, J. (1962). *Physical Organic Chemistry*. McGraw-Hill Inc., New York (NY).

Hine, J. and Mookerjee, P.K. (1975). The Intrinsic Hydrophilic Character of Organic Compounds. Correlations in Terms of Structural Contributions. *J.Org.Chem.*, *40*, 292–298.

Hinze, J. and Jaffé, H.H. (1962). Electronegativity. I. Orbital Electronegativity of Neutral Atoms. *J.Am.Chem.Soc.*, *84*, 540–546.

Hinze, J. and Jaffé, H.H. (1963a). Electronegativity. III. Orbital Electronegativities and Electron Affinities of Transition Metals. *Can.J.Chem.*, *41*, 1315–1328.

Hinze, J. and Jaffé, H.H. (1963b). Electronegativity. IV. Orbital Electronegativities of the Neutral Atoms of the Periods Three A and Four A and of Positive Ions of Periods One and Two. *J.Phys.Chem.*, *67*, 1501-

Hinze, J., Whitehead, M.A. and Jaffé, H.H. (1963). Electronegativity. II. Bond and Orbital Electronegativites. *J.Am.Chem.Soc.*, *85*, 148–154.

Hinze, J. and Welz, U. (1996). Broad Smiles. In *Software Development in Chemistry – Vol. 10* (Gasteiger, J., ed.), Fachgruppe Chemie-Information-Computer (CIC), Frankfurt am Main (Germany), pp. 59–65.

Hirono, S., Qian, L. and Moriguchi, I. (1991). High Correlation Between Hydrophobic Free Energy and Molecular Surface Area Characterized by Electrostatic Potential. *Chem.Pharm. Bull.*, *39*, 3106–3109.

Hirono, S., Nakagome, I., Hirano, H., Yoshii, F. and Moriguchi, I. (1994). Noncongeneric Structure Pharmacokinetic Property Correlation Studies Using Fuzzy Adaptive Least Squares Volume of Distribution. *Bio.Pharm.Bull.*, *17*, 686–690.

Hirst, J.D. (1996). Nonlinear Quantitative Structure-Activity Relationship for the Inhibition of Dihydrofolate Reductase by Pyrimidines. *J.Med.Chem.*, *39*, 3526–3532.

Hocart, S.J., Reddy, V., Murphy, W.A. and Coy, D.H. (1995). Three-Dimensional Quantitative Structure-Activity Relationships of Somatostatin Analogs. 1. Comparative Molecular Field Analysis of Growth Hormone Release Inhibiting Potencies. *J.Med.Chem.*, *38*, 1974–1989.

Hocking, R.R. (1976). The Analysis and Selection of Variables in Linear Regression. *Biometrics*, *32*, 1–49. [R]

Hodes, L., Hazard, G.F., Geran, R.I. and Richman, S. (1977). A Statistical-Heuristic Method for Automated Selection of Drugs for Screening. *J.Med.Chem.*, *20*, 469–475.

Hodes, L. (1981a). Computer-Aided Selection of Compounds for Antitumor Screening: Validation of a Statistical-Heuristic Method. *J.Chem.Inf.Comput.Sci.*, *21*, 128–132.

Hodes, L. (1981b). Selection of Molecular Fragment Features for Structure-Activity Studies in Antitumor Screening. *J.Chem.Inf.Comput.Sci.*, *21*, 132–136.

Hodgkin, E.E. and Richards, W.G. (1987). Molecular Similarity Based on Electrostatic Potential and Electric Field. *Int.J.Quantum Chem.Quant.Biol.Symp.*, *14*, 105–110.

Hoefnagel, A.J., Oosterbeek, W. and Wepster, B.M. (1984). Substituent Effects. 10. Critique of the "Improved Evaluation of Field and Resonance Effects" Proposed by Swain et al. *J.Org.Chem.*, *49*, 1993–1997.

Hoerl, A.E. and Kennard, R.W. (1970). Ridge Regression: Biased Estimation for Non-Orthogonal Problems. *Technometrics*, *12*, 55.

Holland, J. (1975). *Adaptation in Artificial and Natural Systems*. University of Michigan Press, Ann Arbor (MI).

Hollebone, B.R. and Brownlee, L.J. (1995). A Thermodynamic QSAR Analysis of the Polysubstrate Monooxygenase Responses to Xenobiotic Chemicals. *Toxicol.Lett.*, *79*, 157–168.

Holliday, J.D., Ranade, S.S. and Willett, P. (1995). A Fast Algorithm for Selecting Sets of Dissimilar Molecules from Large Chemical Databases. *Quant.Struct.-Act.Relat.*, *14*, 501–506.

Höltje, H.-D. and Kier, L.B. (1974). A Theoretical Approach to Structure-Activity Relationships of Chloramphenicol and Congeners. *J.Med.Chem.*, *17*, 814–819.

Höltje, H.-D. (1975). Theoretische Untersuchungen zu Struktur-Wirkungsbeziehungen bei Monoachinoxidase-Hemmern der Cyclopropylamin-Reihe. *Arch.Pharm.(Weinheim Ger.)*, *308*, 438–444.

Höltje, H.-D. (1976). Theoretische Untersuchungen zu Struktur-Wirkungsbeziehungen von antihypertensiv wirkenden Benzothiadiazin-1,1-dioxiden. *Arch.Pharm.(Weinheim Ger.)*, *309*, 480–485.

Höltje, H.-D. and Vogelgesang, L. (1979). Theoretische Untersuchung zur Hemmung der Noradrenalin-Rückresorption durch Phenylethylaminanaloge Verbindungen. *Arch.Pharm.(Weinheim Ger.)*, *312*, 578–586.

Höltje, H.-D. (1982). Theoretische Untersuchungen zu Struktur-Wirkungsbeziehungen von ringsubstituierten Verapamil-Derivaten. *Arch.Pharm.(Weinheim Ger.)*, *315*, 317–323.

Höltje, H.-D. and Tintelnot, M. (1984). Theoretical Investigations on Interactions Between Pharmacon Molecules and Receptor Models. V: Construction of a Model for the Ribosomal Binding Site of Cholaramphenicol. *Quant.Struct.-Act.Relat.*, *3*, 6–9.

Höltje, H.-D., Baranowski, P., Spengler, J.P. and Schunack, W. (1985). Ein Bindungsstellenmodel für H_2-Antagonisten vom 4-Pyrimidinon-Typ. *Arch.Pharm.(Weinheim Ger.)*, *318*, 542–548.

Höltje, H.-D., Anzali, S., Dall, N. and Höltje, M. (1993). Binding Site Models. In *3D QSAR in Drug Design. Theory, Methods and Applications* (Kubinyi, H., ed.), ESCOM, Leiden (Germany), pp. 320–354.

Holtz, H.D. and Stock, L.M. (1964). Dissociation Constants for 4-Substituted Bicyclo[2.2.2]octane-1-carboxylic Acids. Empirical and Theoretical Analysis. *J.Am.Chem.Soc.*, *86*, 5188–5194.

Holzgrabe, U. and Hopfinger, A.J. (1996). Conformational Analysis, Molecular Shape Comparison, and Pharmacophore Identification of Different Allosteric Modulators of Muscarinic Receptors. *J.Chem.Inf.Comput.Sci.*, *36*, 1018–1024.

Hong, H., Wang, L. and Han, S. (1996). Prediction Adsorption Coefficients (Koc) for Aromatic Compounds by HPLC Retention Factors (K′). *Chemosphere*, *32*, 343–351.

Hopfinger, A.J. and Battershell, R.D. (1976). Application of SCAP to Drug Design. 1. Prediction of Octanol-Water Partition Coefficients Using Solvent-Dependent Conformational Analyses. *J.Med.Chem.*, *19*, 569–573.

Hopfinger, A.J. (1980). A QSAR Investigation of Dihydrofolate Reductase Inhibition by Baker Triazines Based upon Molecular Shape Analysis. *J.Am.Chem.Soc.*, *102*, 7196–7206.

Hopfinger, A.J. (1981). Inhibition of Dihydrofolate Reductase: Structure-Activity Correlations of 2,4-Diamino-5-benzylpyrimidines Based upon Molecular Shape Analysis. *J.Med.Chem.*, *24*, 818–822.

Hopfinger, A.J. and Potenzone Jr., R. (1982). Ames Test and Antitumor Activity of 1-(*x*-Phenyl)-3.3-dialkyltriazines, Quantitative Structure-Analysis Studies Based Upon Molecular Shape Analysis. *Mol.Pharm.*, *21*, 187–195.

Hopfinger, A.J. (1983). Theory and Application of Molecular Potential Energy Fields in Molecular Shape Analysis: A Quantitative Structure-Activity Relationship Study of 2,4-Diamino-5-benzylpyrimidines as Dihydrofolate Reductase Inhibitors. *J.Med.Chem.*, *26*, 990–996.

Hopfinger, A.J. (1984). A QSAR Study of the Ames Mutagenicity of 1-(X-Phenyl)-3,3-dialkyltriazenes Using Molecular Potential Energy Fields and Molecular Shape Analysis. *Quant. Struct.-Act.Relat.*, *3*, 1–5.

Hopfinger, A.J., Compadre, R.L.L., Koehler, M.G., Emery, S. and Seydel, J.K. (1987). An Extended QSAR Analysis of Some 4-Aminodiphenylsulfone Antibacterial Agents Using Molecular Modeling and LFE-Relationships. *Quant.Struct.-Act.Relat.*, *6*, 111–117.

Hopfinger, A.J. and Burke, B.J. (1990). Molecular Shape Analysis: A Formalism to Quantitatively Establish Spatial Molecular Similarity. In *Concepts and Applications of Molecular Similarity* (Johnson, M.A. and Maggiora, G.M., eds.), Wiley, New York (NY), pp. 173–209.

Hopfinger, A.J. and Kawakami, Y. (1992). QSAR Analysis of a Set of Benzothiopyranoindazole Anticancer Analogs Based upon Their DNA Intercalation Properties as Determined by Molecular Dynamics Simulation. *Anti-Cancer Drug Des.*, *7*, 203–217.

Hopfinger, A.J. and Patel, H.C. (1996). Application of Genetic Algorithms to the General QSAR Problem and to Guiding Molecular Diversity Experiments. In *Genetic Algorithms in Molecular Modeling. Principles of QSAR and Drug Design. Vol. 1* (Devillers, J., ed.), Academic Press, London (UK).

Hopkinson, A.C. (1969). Unimolecular and Bimolecular Mechanisms in the Acid-catalyzed Hydrolysis of Methyl Esters of Aliphatic Monocarboxyilic Acids in Aqueous Sulphuric Acid. *J.Chem.Soc.*, *(B)*, 861–863.

Horvath, A.L. (1988). Estimate Properties of Organic Compounds: Simple Polynomial Equations Relate the Properties of Organic Compounds to their Chemical Structure. *Chem.Eng.*, *95*, 155–158.

Horvath, A.L. (1992). *Molecular Design*. Elsevier, Amsterdam (The Netherlands), 1490 pp. [R]

Horwell, D.C., Howson, W., Higginbottom, M., Naylor, D., Ratcliffe, G.S. and Williams, S. (1995). Quantitative Structure-Activity Relationships (QSARs) of N-Terminus Fragments of NK1 Tachykinin Antagonists: A Comparison of Classical QSARs and Three-Dimensional QSARs from Similarity Matrices. *J.Med.Chem.*, *38*, 4454–4462.

Horwitz, J.P., Massova, I., Wiese, T.E., Wozniak, A.J., Corbett, T.H., Seboltleopold, J.S., Capps, D.B. and Leopold, W.R. (1993). Comparative Molecular Field Analysis of in Vitro Growth Inhibition of L1210 and HCT-8 Cells by Some Pyrazoloacridines. *J.Med.Chem.*, *36*, 3511–3516.

Horwitz, J.P., Massova, I., Wiese, T.E., Besler, B.H. and Corbett, T.H. (1994). Comparative Molecular Field Analysis of the Antitumor Activity of 9H-Thioxanthen-9-One Derivatives Against Pancreatic Ductal Carcinoma 03. *J.Med.Chem.*, *37*, 781–786.

Höskuldsson, A. (1988). PLS Regression Methods. *J.Chemom.*, *2*, 211–228.

Hosoya, H. (1971). Topological Index. A Newly Proposed Quantity Characterizing the Topological Nature of Structural Isomers of Saturated Hydrocarbons. *Bull.Chem.Soc.Jap.*, *44*, 2332–2339.

Hosoya, H. (1972a). Graphical Enumeration of the Coefficients of the Secular Polynomials of the Hückel Molecular Orbitals. *Theor.Chim.Acta*, *25*, 215–222.

Hosoya, H. (1972b). Topological Index and Thermodynamics Properties. I. Empirical Rules on the Boiling Point of Saturated Hydrocarbons. *Bull.Chem.Soc.Jap.*, *45*, 3415–3421.

Hosoya, H. (1972c). Topological Index as a Sorting Device for Coding Chemical Structures. *J.Chem.Doc.*, *12*, 181–183.

Hosoya, H., Hosoi, K. and Gutman, I. (1975a). A Topological Index for the Total π-Electron Energy. Proof of a Generalized Hückel Rule for an Arbitrary Network. *Theor.Chim.Acta*, *38*, 37–47.

Hosoya, H. and Murakami, M. (1975b). Topological Index as Applied to π-Electronic Systems. II. Topolgical Bond Order. *Bull.Chem.Soc.Jap.*, *48*, 3512–3517.

Hosoya, H. and Ohkami, N. (1983). Operator Technique for Obtaining the Recursion Formulas of Characteristic and Matching Polynomials as Applied to Polyhex Graphs. *J.Comput.Chem.*, *4*, 585–593.

Hosoya, H. (1986). Topological Index as a Common Tool for Quantum Chemistry, Statistical Mechanics, and Graph Theory. In *Mathematics and Computational Concepts in Chemistry* (Trinajstic, N., ed.), Ellis Horwood, Chichester (UK), pp. 110–123.

Hosoya, H. (1988). On Some Counting Polynomials in Chemistry. *Disc.Appl.Math.*, *19*, 239–257.

Hosoya, H. (1991). Factorization and Recursion of the Matching and Characteristic Polynomials of Periodic Polymer Networks. *J.Math.Chem.*, *7*, 289–305.

Hosoya, H. (1994). Topological Twin Graphs. Smallest Pair of Isospectral Polyhedral Graphs with Eight Vertices. *J.Chem.Inf.Comput.Sci.*, *34*, 428–431.

Hosoya, H., Gotoh, M., Murakami, M. and Ikeda, S. (1999). Topological Index and Thermodynamic Properties. 5. How Can We Explain the Topological Dependency of Thermodynamic Properties of Alkanes with the Topology of Graphs? *J.Chem.Inf.Comput.Sci.*, *39*, 192–196.

Hu, C.-Y. and Xu, L. (1994). On Hall and Kier's Topological State and Total Topological Index. *J.Chem.Inf.Comput.Sci.*, *34*, 1251–1258.

Hu, C.-Y. and Xu, L. (1996). On Highly Discriminating Molecular Topological Index. *J.Chem.Inf.Comput.Sci.*, *36*, 82–90.

Hu, C.-Y. and Xu, L. (1997). Developing Molecular Identification Numbers by an All-Paths Method. *J.Chem.Inf.Comput.Sci.*, *37*, 311–315.

Hu, Q.H., Wang, X.J. and Brusseau, M.L. (1995). Quantitative Structure-Activity Relationships for Evaluating the Influence of Sorbate Structure on Sorption of Organic Compounds by Soil. *Environ.Toxicol.Chem.*, *14*, 1133–1140.

Huang, M.-J. and Bodor, N. (1994). Quantitative Structure-Inhibitory Activity Relationships of Substituted Phenols on Bacillus Subtilis Spore Germination. *Int.J.Quant.Chem.*, 181–185.

Huang, Q.G., Kong, L. and Wang, L. (1996). Applications of Frontier Molecular Orbital Energies in QSAR Studies. *Bull.Environ.Contam.Toxicol.*, *56*, 758–765.

Huang, Q.G., Song, W.L. and Wang, L.S. (1997). Quantitative Relationship Between the Physiochemical Characteristics as well as Genotoxicity of Organic Pollutants and Molecular Autocorrelation Topological Descriptors. *Chemosphere*, *35*, 2849–2855.

Hübel, S., Rösner, T. and Franke, R. (1980). The Evaluation of Topological Pharmacophores by Heuristic Approaches. *Pharmazie*, *35*, 424–433.

Hückel, E. (1930). Zur Quantentheorie der Doppelbindung. *Zeitschrift für Physik*, *60*, 423–456.

Hückel, E. (1932). Quantentheoretische Beiträge zum Problem der aromatischen und ungesättingten Verbindungen. III. *Zeitschrift für Physik*, *76*, 628–648.

Huggins, M. (1956). Densities and Optical Properties of Organic Compounds in the Liquid State. VI. The Refractive Indices of Paraffin Hydrocarbons and Some of Their Derivatives. *Bull. Chem.Soc.Jap.*, *29*, 336–339.

Huheey, J.E. (1965). The Electronegativity of Groups. *J.Phys.Chem.*, *69*, 3284–3291.

Huheey, J.E. (1966). The Electronegativity of Multiply-Bonded Groups. *J.Phys.Chem.*, *70*, 2086–2092.

Huibers, P.D.T., Lobanov, V.S., Katritzky, A.R., Shah, D.O. and Karelson, M. (1996). Prediction of Critical Micelle Concentration Using a Quantitative Structure-Property Relationship Approach. 1. Nonionic Surfactants. *Langmuir*, *12*, 1462–1470.

Huibers, P.D.T., Lobanov, V.S., Katritzky, A.R., Shah, D.O. and Karelson, M. (1997). Prediction of Critical Micelle Concentration Using a Quantitative Structure-Property Relationship Approach. *J.Colloid Interf.Sci.*, *187*, 113–120.

Huibers, P.D.T. and Katritzky, A.R. (1998). Correlation of the Aqueous Solubility of Hydrocarbons and Halogenated Hydrocarbons with Molecular Structure. *J.Chem.Inf.Comput.Sci.*, *38*, 283–292.

Hunt, P.A. (1999). QSAR Using 2D Descriptors and TRIPOS' SIMCA. *J.Comput.Aid.Molec.Des.*, *13*, 453–467.

Hunter LaFemina, D. and Jurs, P.C. (1985). A Numerical Index for Characterizing Data Set Separation. *J.Chem.Inf.Comput.Sci.*, *25*, 386–388.

Hurst, T. and Heritage, T. W. (1997). HQSAR. A Highly Predictive QSAR Technique Based on Molecular Holograms. 213th ACS National Meeting, San Francisco (CA).

Huuskonen, J., Salo, M. and Taskinen, J. (1998). Acqueous Solubility Prediction of Drugs Based on Molecular Topology and Neural Network Modeling. *J.Chem.Inf.Comput.Sci.*, *38*, 450–456.

Hyde, R.M. and Livingstone, D.J. (1988). Perspectives in QSAR: Computer Chemistry and Pattern Recognition. *J.Comput.Aid.Molec.Des.*, *2*, 145–155. [R]

Iczkowski, R.P. and Margrave, J.L. (1961). Electronegativity. *J.Am.Chem.Soc.*, *83*, 3547–3551.

Idoux, J.P., Hwang, P.T.R. and Hancock, C.K. (1973). Study of the Alkaline Hydrolysis and Nuclear Magnetic Resonance Spectra of Some Thiol Esters. *J.Org.Chem.*, *38*, 4239–4243.

Idoux, J.P., Scandrett, J.M. and Sikorski, J.A. (1977). Conformational Influence of Nonacyl Groups on Acyl Group Properties in N-Monosubstituted Amides and in Other Carboxylic Acid Derivatives: a 7-Position Proximity Effect. *J.Am.Chem.Soc.*, *99*, 4577–4583.

Ihlenfeldt, W.D. and Gasteiger, J. (1994). Hash Codes for the Identification and Classification of Molecular Structure Elements. *J.Comput.Chem.*, *15*, 793–813.

Ikemoto, Y., Motoba, K., Suzuki, T. and Uchida, M. (1992). Quantitative Structure-Activity Relationships of Nonspecific and Specific Toxicants in Several Organism Species. *Environ. Toxicol.Chem.*, *11*, 931–939.

Immirzi, A. and Perini, B. (1977). Prediction of Density in Organic Crystals. *Acta Cryst.*, *33A*, 216–218.

Inamoto, N. and Masuda, S. (1977). Substituent Effects on C-13 Chemical Shifts in Aliphatic and Aromatic Series. Proposal of New Inductive Substituent Parameter (ι; iota) and the Application. *Tetrahedron Lett.*, *18*, 3287–3290.

Inamoto, N., Masuda, S., Tori, K. and Yoshimura, Y. (1978). Effects of Fixed Substituents upon Substituent Chemical Shifts of the C-1 Atom in m- and p-Disubstituted Benzenes. Correlation with Inductive Substituent Parameter (ι). *Tetrahedron Lett.*, *19*, 4547–4550.

Inamoto, N. and Masuda, S. (1982). Revised Method for Calculation of Group Electronegativites. *Chem.Lett.*, 1003–1007.

Isogai, Y. and Itoh, T. (1984). Fractal Analysis of Tertiary Structure. *J.Phys.Soc.Japan*, *53*, 2162.

IUPAC Recommendations. (1997). Glossary of Terms in Computational Drug Design. *Pure & Appl.Chem.*, *69*, 1137–1152.

IUPAC Recommendations. (1998). Glossary of Terms Used in Medicinal Chemistry. *Pure & Appl.Chem.*, *70*, 1129–1143.

Ivanciuc, O. (1989). Design on Topological Indices. 1. Definition of a Vertex Topological Index in the Case of 4-Trees. *Rev.Roum.Chim.*, *34*, 1361–1368.

Ivanciuc, O. and Balaban, A.T. (1992a). Nonisomorphic Graphs with Identical Atomic Counts of Self-Returning Walks: Isocodal Graphs. *J.Math.Chem.*, *11*, 155–167.

Ivanciuc, O. and Balaban, A.T. (1992b). Recurrence Relationships for the Computation of Kekulé Structures. *J.Math.Chem.*, *11*, 169–177.

Ivanciuc, O., Balaban, T.-S., Filip, P. and Balaban, A.T. (1992). Design of Topological Indices. Part 7. Analytical Formulae for Local Vertex Invariants of Linear and Monocyclic Molecular Graphs. *MATCH (Comm.Math.Comp.Chem.)*, *28*, 151–164.

Ivanciuc, O. (1993). Chemical Graph Polynomials. Part 3. The Laplacian Polynomial of Molecular Graphs. *Rev.Roum.Chim.*, *38*, 1499–1508.

Ivanciuc, O., Balaban, T.-S. and Balaban, A.T. (1993a). Chemical Graphs with Degenerate Topological Indices Based on Information on Distances. *J.Math.Chem.*, *14*, 21–33.

Ivanciuc, O., Balaban, T.-S. and Balaban, A.T. (1993b). Design of Topological Indices. Part 4. Reciprocal Distance Matrix, Related Local Vertex Invariants and Topological Indices. *J.Math. Chem.*, *12*, 309–318.

Ivanciuc, O. and Balaban, A.T. (1994a). Design of Topological Indices. Part 5. Precision and Error in Computing Graph Theoretic Invariants for Molecules Containing Heteroatoms and Multiple Bonds. *MATCH (Comm.Math.Comp.Chem.)*, *30*, 117–139.

Ivanciuc, O. and Balaban, A.T. (1994b). Design of Topological Indices. Part 8. Path Matrices and Derived Molecular Graph Invariants. *MATCH (Comm.Math.Comp.Chem.)*, *30*, 141–152.

Ivanciuc, O. (1996). Artificial Neural Networks Applications. 2. Using Theoretical Descriptors of Molecular Structure in Quantitative Structure-Activity Relationships Analysis of the Inhibition of Dihydrofolate Reductase. *Rev.Roum.Chim.*, *41*, 645–652.

Ivanciuc, O. and Balaban, A.T. (1996a). Characterization of Chemical Structures by the Atomic Counts of Self-Returning Walks: On the Construction of Isocodal Graphs. *Croat.Chem.Acta*, *69*, 63–74.

Ivanciuc, O. and Balaban, A.T. (1996b). Design of Topological Indices. Part 3. New Identification Numbers for Chemical Structures: MINID and MINSID. *Croat.Chem.Acta*, *69*, 9–16.

Ivanciuc, O. and Balaban, A.T. (1996c). Design of Topological Indices. Part 6. A New Topological Parameter for the Steric Effect of Alkyl Substituents. *Croat.Chem.Acta*, *69*, 75–83.

Ivanciuc, O., Ivanciuc, T. and Diudea, M.V. (1997a). Molecular Graph Matrices and Derived Structural Descriptors. *SAR & QSAR Environ.Res.*, *7*, 63–87.

Ivanciuc, O., Laidboeur, T. and Cabrol-Bass, D. (1997b). Degeneracy of Topologic Distance Descriptors for Cubic Molecular Graphs: Examples of Small Fullerenes. *J.Chem.Inf.Comput.Sci.*, *37*, 485–488.

Ivanciuc, O., Diudea, M.V. and Khadikar, P.V. (1998). New Topological Matrices and their Polynomials. *Indian J.Chem.*, *37A*, 574–585.

Ivanciuc, O., Ivanciuc, T. and Balaban, A.T. (1998a). Design of Topological Indices. Part 10. Parameters Based on Electronegativity and Covalent Radius for the Computation of Molecular Graph Descriptors for Heteroatom-Containing Molecules. *J.Chem.Inf.Comput.Sci.*, *38*, 395–401.

Ivanciuc, O., Ivanciuc, T. and Balaban, A.T. (1998b). Quantitative Structure-Property Relationship Study of Normal Boiling Points for Halogen-/ Oxygen-/ Sulfur-Containing Organic Compounds Using the CODESSA Program. *Tetrahedron*, *54*, 9129–9142.

Ivanciuc, O., Balaban, A.T. and Babic, D. (1999a). Correlation Between Energies of Proper Fullerenes and their Topological Invariants. Part II. Fullerenes with Isolated Pentagons. *Fullerene Sci.Technol.*, *in press*.

Ivanciuc, O., Ivanciuc, T., Filip, P.A. and Cabrol-Bass, D. (1999b). Estimation of the Liquid Viscosity of Organic Compounds with a Quantitative Structure-Property Model. *J.Chem.Inf.Comput.Sci.*, *39*, 515–524.

Ivanciuc, O. and Balaban, A.T. (2000). The Graph Description of Chemical Structures. In *Topological Indices and Related Descriptors in QSAR and QSPR* (Devillers, J. and Balaban, A.T., eds.), Gordon & Breach, Amsterdam (The Netherlands), pp. 59–167.

Ivanciuc, O. and Ivanciuc, T. (2000). Matrices and Structural Descriptors Computed from Molecular Graphs Distances. In *Topological Indices and Related Descriptors in QSAR and QSPR* (Devillers, J. and Balaban, A.T., eds.), Gordon & Breach, Amsterdam (The Netherlands), pp. 221–277.

Ivanciuc, O. and Devillers, J. (2000). Algorithms and Software for the Computation of Topological Indices and Structure-property Models. In *Topological Indices and Related Descriptors in QSAR and QSPR* (Devillers, J. and Balaban, A.T., eds.), Gordon & Breach, Amsterdam (The Netherlands), pp. 779–804.

Ivanciuc, O., Ivanciuc, T. and Balaban, A.T. (2000a). Vertex- and Edge-Weighted Molecular Graphs and Derived Structural Descriptors. In *Topological Indices and Related Descriptors in QSAR and QSPR* (Devillers, J. and Balaban, A.T., eds.), Gordon & Breach, Amsterdam (The Netherlands), pp. 169–220.

Ivanciuc, O., Ivanciuc, T., Cabrol-Bass, D. and Balaban, A.T. (2000b). Evaluation in Quantitative Structure-Property Relationship Models of Structural Descriptors Derived from Information-Theory Operators. *J.Chem.Inf.Comput.Sci.*, *40*, 631–643.

Ivanciuc, O., Taraviras, S.L. and Cabrol-Bass, D. (2000c). Quasi-Orthogonal Basis Sets of Molecular Graph Descriptors as a Chemically Diversity Measure. *J.Chem.Inf.Comput.Sci.*, *40*, 126–134.

Ivanov, J., Karabunarliev, S. and Mekenyan, O. (1994). A System for Exhaustive 3D Molecular Design Proceeding from Molecular Topology. *J.Chem.Inf.Comput.Sci.*, *34*, 234–243.

Ivanov, J. and Schüürmann, G. (1999). Simple Algorithms for Determining the Molecular Symmetry. *J.Chem.Inf.Comput.Sci.*, *39*, 728–737.

Ivanusevic, M., Nikolic, S. and Trinajstic, N. (1991). A QSAR Study of Antidotal Activity of H-Oximes. *Rev.Roum.Chim.*, *36*, 389–398.

Iwase, K., Komatsu, K., Hirono, S., Nakagawa, S. and Moriguchi, I. (1985). Estimation of Hydrophobicity Based on the Solvent-Accessible Surface Area of Molecules. *Chem.Pharm.Bull.*, *33*, 2114–2121.

Jackel, H. and Nendza, M. (1994). Reactive Substructures in the Prediction of Aquatic Toxicity Data. *Aquat.Toxicol.*, *29*, 305–314.

Jackson, J.E. (1991). *A User's Guide to Principal Components.* Wiley, NewYork (NY).

Jaffé, H.H. (1953). A Reexamination of the Hammett Equation. *Chem.Rev.*, *53*, 191–261.

Jafvert, C.T., Chu, W. and Vanhoof, P.L. (1995). A Quantitative Structure-Activity Relationship for Solubilization of Nonpolar Compounds by Nonionic Surfactant Micelles. *ACS Symp.Ser.*, *594*, 24–37.

Jain, A.N., Dietterich, T.G., Lathrop, R.H., Chapman, D., Critchlow, R.E., Bauer, B.E., Webster, T.A. and Lozano-Perez, T. (1994a). Compass: A Shape-Based Machine Learning Tool for Drug Design. *J.Comput.Aid.Molec.Des.*, *8*, 635.

Jain, A.N., Koile, K. and Chapman, D. (1994b). Compass: Predicting Biological Activities from Molecular Surface Properties. Performance Comparisons on a Steroid Benchmark. *J.Med.Chem.*, *37*, 2315–2327.

Jain, A.N., Harris, N.L. and Park, J.Y. (1995). Quantitative Binding Site Model Generation: Compass Applied to Multiple Chemotypes Targeting the 5-HT_{1A} Receptor. *J.Med.Chem.*, *38*, 1295–1308.

Jalali-Heravi, M. and Parastar, F. (1999). Computer Modeling of the Rate of Glycine Conjugation of Some Benzoic Acid Derivatives: A QSAR Study. *Quant.Struct.-Act.Relat.*, *18*, 134–138.

Janini, G.M., Johnston, K. and Zielinski Jr, W.L. (1975). Use of a Nematic Liquid Crystal for Gas-Liquid Chromatographic Separation of Polyaromatic Hydrocarbons. *Anal.Chem.*, *47*, 670–674.

Jaworska, J.S. and Schultz, T.W. (1993). Quantitative Relationships of Structure-Activity and Volume Fraction for Selected Nonpolar and Polar Narcotic Chemicals. *SAR & QSAR Environ.Res.*, *1*, 3–19.

Jenkins, H.D.B., Kelly, E.J. and Samuel, C.J. (1994). A Novel Computational Approach to the Estimation of Steric Parameters Application to the Menschutkin Reaction. *Tetrahedron Lett.*, *35*, 6543–6546.

Jenkins, H.D.B., Samuel, C.J. and Stafford, J.E. (1995). A Novel Computational Approach to the Estimation of Steric Parameters.2. Extension to Thiazoles. *Tetrahedron Lett.*, *36*, 6159–6162.

Jerman-Blazic Dzonova, B. and Trinajstic, N. (1982). Computer-Aided Enumeration and Generation of the Kekulé Structures in Conjugated Hydrocarbons. *Computers Chem.*, *6*, 121–132.

Jerman-Blazic, B., Fabic-Petrac, I. and Randic, M. (1989). Evaluation of the Molecular Similarity and Property Prediction for QSAR Purposes. *Chemom.Intell.Lab.Syst.*, *6*, 49–63.

Jiang, H.L., Chen, K.X., Wang, H.W., Tang, Y., Chen, J.Z. and Ji, R.Y. (1994). 3D QSAR Study on Ether and Ester Analogs of Artemisinin with Comparative Molecular Field Analysis. *Acta Pharmacol.Sin.*, *15*, 481.

Jiang, Y., Tang, A.-C. and Hoffmann, R.D. (1984). Evaluation of Moments and Their Application in Hückel Molecular Orbital Theory. *Theor.Chim.Acta*, *66*, 183–192.

Jiang, Y. and Zhang, H. (1989). Stability and Reactivity Based on Moment Analysis. *Theor.Chim.Acta*, *75*, 279–297.

Jiang, Y. and Zhang, H. (1990). Aromaticities and Reactivities Based on Energy Partitioning. *Pure & Appl.Chem.*, *62*, 451–456.

Jiang, Y. and Zhu, H. (1994). Evaluation of Level Pattern Indices. *J.Chem.Inf.Comput.Sci.*, *34*, 377–380.

Jiang, Y., Qian, X. and Shao, Y. (1995). The Evaluation of Moments for Benzenoid Hydrocarbons. *Theor.Chim.Acta*, *90*, 135–144.

Joao, H.C., Devreese, K., Pauwels, R., Declercq, E., Henson, G.W. and Bridger, G.J. (1995). Quantitative Structural-Activity Relationship Study of bis-Tetraazacyclic Compounds. A Novel Series of HIV-1 and HIV-2 Inhibitors. *J.Med.Chem.*, *38*, 3865–3873.

Jochum, C. and Gasteiger, J. (1977). Canonical Numbering and Constitutional Symmetry. *J.Chem.Inf.Comput.Sci.*, *17*, 113–117.

Jochum, C., Hicks, M. G., and Sunkel, J. eds. (1988). *Physical Property Prediction in Organic Chemistry.* Springer-Verlag, Berlin (Germany), 554 pp.

John, P.E., Mallion, R.B. and Gutman, I. (1998). An Algorithm for Counting Spanning Trees in Labeled Molecular Graphs Homeomorphic to Cata-Condensed Systems. *J.Chem.Inf.Comput.Sci.*, *38*, 108–112.

Johnson, M.A. (1989). A Review and Examination of Mathematical Spaces Underlying Molecular Similarity Analysis. *J.Math.Chem.*, *3*, 117–145.

Johnson, M.A., Gifford, E. and Tsai, C. (1990a). Similarity Concepts in Modeling Chemical Transformation Pathways. In *Concepts and Applications of Molecular Similarity* (Johnson, M.A. and Maggiora, G.M., eds.), Wiley, New York (NY), pp. 289–320.

Johnson, M. A. and Maggiora, G. M. eds. (1990). *Concepts and Applications of Molecular Similarity.* Wiley, New York (NY), 393 pp.

Johnson, M.A., Basak, S.C. and Maggiora, G.M. (1998). Characterization of Molecular Similarity Methods for Property Prediction. *Math.Comput.Modelling*, *11*, 630–635.

Johnson, R.A. and Wichern, D.W. (1992). *Applied Multivariate Analysis.* Prentice-Hall, Englewood Cliffs (NJ), 642 pp.

Johnson, S.R. and Jurs, P.C. (1999). Prediction of the Clearing Temperatures of a Series of Liquid Crystals from Molecular Structure. *Chemistry of Materials*, *11*, 1007–1023.

Johnston, N., Sadler, R., Shaw, G.R. and Connell, D.W. (1993). Environmental Modification of PAH Composition in Coal Tar Containing Samples. *Chemosphere*, *27*, 1151–1158.

Jolles, G. and Woolridge, K. R. H. eds. (1984). *Drug Design: Fact or Fantasy?* Academic Press, London (UK).

Jolliffe, I.T. (1972). Discarding Variables in a Principal Component Analysis. I. Artificial Data. *Applied Statistics*, *21*, 160–173.

Jolliffe, I.T. (1973). Discarding Variables in a Principal Component Analysis. II. Real Data. *Applied Statistics*, *22*, 21–31.

Jolliffe, I.T. (1986). *Principal Component Analysis.* Springer-Verlag, New York (NY).

Jonathan, P., McCarthy, W.V. and Roberts, A.M. (1996). Discriminant Analysis with Singular Covariance Matrices. A Method Incorporating Cross-Validation and Efficient Randomized Permutation Tests. *J.Chemom.*, *10*, 189–213.

Jonsson, J., Eriksson, L., Hellberg, S. and Wold, S. (1989). *Quant.Struct.-Act.Relat.*, *8*, 204.

Jordan, S.N., Leach, A.R. and Bradshaw, J. (1995). The Application of Neural Networks in Conformational Analysis. 1. Prediction of Minimum and Maximum Interatomic Distances. *J.Chem.Inf.Comput.Sci.*, *35*, 640–650.

Joshi, R.K., Meister, Th., Scapozza, L. and Ha, T.-K. (1993). Development of New Molecular Descriptors Using Conformational Energies from Quantum Calculations and Their Application in QSAR Analysis. In *Trends in QSAR and Molecular Modelling 92* (Wermuth, C.G., ed.), ESCOM, Leiden (The Netherlands), pp. 362–363.

Joshi, R.K., Meister, Th., Scapozza, L. and Ha, T.-K. (1994). A New Quantum Chemical Approach in QSAR-Analysis. Parametrisation of Conformational Energies into Molecular Descriptors Jmn (Steric) and Jsn (Electronic). *Arzneim.Forsch.*, *44*, 779–790.

Judson, P.N. (1992a). QSAR and Expert Systems in the Prediction of Biological Activity. *Pestic.Sci.*, *36*, 155–160.

Judson, P.N. (1992b). Structural Similarity Searching Using Descriptors Developed for Structure-Activity Relationship Studies. *J.Chem.Inf.Comput.Sci.*, *32*, 657–663.

Judson, R. (1996). Genetic Algorithms and Their Use in Chemistry. In *Reviews in Computational Chmistry – Vol. 10* (Lipkowitz, K.B. and Boyd, D., eds.), Wiley – VCH, New York (NY), pp. 1–73.

Jug, K. (1983). A Bond Order Approach to Ring Current and Aromaticity. *J.Org.Chem.*, *48*, 1344.

Jug, K. (1984). Bond Order as a Tool for Molecular Structure and Reactivity. *Croat.Chem.Acta*, *57*, 941–953.

Junghans, M. and Pretsch, E. (1997). Estimation of Partition Coefficients of Organic Compounds. Local Database Modeling with Uniform-Length Structure Descriptors. *Fresen.J.Anal.Chem.*, *359*, 88–92.

Juric, A., Gagro, M., Nikolic, S. and Trinajstic, N. (1992). Molecular Topological Index: An Application in the QSAR Study of Toxicity of Alcohols. *J.Math.Chem.*, *11*, 179–186.

Juric, A., Nikolic, S. and Trinajstic, N. (1997). Topological Resonance Energies of Thienopyrimidines. *Croat.Chem.Acta*, *70*, 841–846.

Jurs, P.C., Chou, J.T. and Yuan, M. (1979). Computer-Assisted Structure-Activity Studies of Chemical Carcinogens. A Heterogeneous Data Set. *J.Med.Chem.*, *22*, 476–483.

Jurs, P.C., Hasan, M.N., Henry, D.R., Stouch, T.R. and Whalen-Pedersen, E.K. (1983). Computer-Assisted Studies of Molecular Structure and Carcinogenic Activity. *Fund.Appl.Toxicol.*, *3*, 343–349.

Jurs, P.C., Stouch, T.R., Czerwinski, M. and Narvaeez, J.N. (1985). Computer-Assisted Studies of Molecular Structure-Biological Activity Relationships. *J.Chem.Inf.Comput.Sci.*, *25*, 296–308.

Jurs, P.C. (1986). *Computer Software Applications in Chemistry.* Wiley, New York (NY).

Jurs, P.C., Hasan, M.N., Hansen, P.J. and Rohrbaugh, R.H. (1988). Prediction of Physicochemical Properties of Organic Compounds from Molecular Structure. In *Physical Property Prediction in Organic Chemistry* (Jochum, C., Hicks, M.G. and Sunkel, J., eds.), Springer-Verlag, Berlin (Germany), pp. 209–233.

Jurs, P.C. and Lawson, R.G. (1991). Analysis of Chemical Structure-Biological Activity Relationships Using Clustering Methods. *Chemom.Intell.Lab.Syst.*, *10*, 81–83.

Jurs, P.C., Dixon, J.S. and Egolf, L.M. (1995). Representations of Molecules. In *Chemometrics Methods in Molecular Design – Vol. 2* (van de Waterbeemd, H., ed.), VCH Publishers, New York (NY), pp. 15–38.

Juvan, M. and Mohar, B. (1995). Bond Contributions to the Wiener Index. *J.Chem.Inf.Comput.Sci.*, *35*, 217–219.

Kaiser, K.L.E. and Niculescu, S.P. (1999). Using Probabilistic Neural Networks to Model the Toxicity of Chemicals to the Fathead Minnow (*Pimephales promelas*): A Study Based on 865 Compounds. *Chemosphere*, *38*, 3237–3245.

Kaliszan, R. (1977). Correlation Between the Retention Indices and the Connectivity Indices of Alcohols and Methyl Esters with Complex Cyclic Structure. *Chromatographia*, *10*, 529.

Kaliszan, R. and Foks, H. (1977). The Relationship Between R_M Values and the Connectivity Indices for Pyrazine Carbothioamide Derivatives. *Chromatographia*, *10*, 346–349.

Kaliszan, R. and Lamparczyk, H. (1978). A Relationship Between the Connectivity Indices and Retention Indices of Polycyclic Aromatic Hydrocarbons. *J.Chromatogr.Sci.*, *16*, 246–248.

Kaliszan, R. (1979). The Relationship Between the Connectivity Indices and the Thermodynamic Parameters Describing the Interaction of Fatt Acid Methyl Esters with Polar and Nonpolar Stationary Phases. *Chromatographia*, *12*, 171.

Kaliszan, R., Lamparczyk, H. and Radecki, A. (1979). A Relationship Between Repression of Dimethylnitrosamine-Demethylase by Polycyclic Aromatic Hydrocarbons and their Shape. *Biochem.Pharmacol.*, *28*, 123–125.

Kaliszan, R. (1981). Chromatography in Studies of Quantitative Structure-Activity Relationships. *J.Chromat.*, *220*, 71–83.

Kaliszan, R., Osmialowski, K., Tomellini, S.A., Hsu, S.-H., Fazio, S.D. and Hartwick, R.A. (1985). Non-Empirical Descriptors of Sub-Molecular Polarity and Dispersive Interactions in Reversed-Phase HPLC. *Chromatographia*, *20*, 705–708.

Kaliszan, R. (1986). Quantitative Relationship Between Molecular Structure and Chromatographic Retention. Implication in Physical, Analytical, and Medicinal Chemistry. *CRC Crit.Rev. Anal.Chem*, *16*, 323–383.

Kaliszan, R. (1987). *Quantitative Structure-Chromatographic Retention Relationships*. Wiley, NewYork (NY), 304 pp.

Kaliszan, R. (1992). Quantitative Structure-Retention Relationships. *Anal.Chem.*, *64*, 619A-631A.

Kaliszan, R., Kaliszan, A., Noctor, T.A., Purcell, W.P. and Wainer, I.W. (1992a). Mechanism of Retention of Benzodiazepines in Affinity, Reversed Phase and Adsorption High Performance Liquid Chromatography in View of Quantitative Structure-Retention Relationships. *J.Chromat.*, *609*, 69–81.

Kaliszan, R., Noctor, T.A. and Wainer, I.W. (1992b). Quantitative Structure-Enantioselective Retention Relationships for the Chromatography of 1,4 Benzodiazepines on a Human Serum Albumin Based HPLC Chiral Stationary Phase An Approach to the Computational Prediction of Retention and Enantioselectivity. *Chromatographia*, *33*, 546–550.

Kaliszan, R. (1993). Quantitative Structure Retention Relationships Applied to Reversed Phase High Performance Liquid Chromatography. *J.Chromat.*, *656A*, 417–435.

Kaliszan, R., Kaliszan, A. and Wainer, I.W. (1993). Deactivated Hydrocarbonaceous Silica and Immobilized Artificial Membrane Stationary Phases in High Performance Liquid Chromatographic Determination of Hydrophobicities of Organic Bases Relationship to Log P and Clogp. *J.Pharmaceut.Biomed.Anal.*, *11*, 505–511.

Kaliszan, R., Nasal, A. and Bucinski, A. (1994). Chromatographic Hydrophobicity Parameter Determined on an Immobilized Artificial Membrane Column Relationships to Standard Measures of Hydrophobicity and Bioactivity. *Eur.J.Med.Chem.*, *29*, 163–170.

Kalivas, J.H. (1995). *Adaption of Simulated Annealing to Chemical Optimization Problems*. Elsevier, Amsterdam (The netherlands), 473 pp.

Kamenska, V., Mekenyan, O., Sterev, A. and Nedjalkova, Z. (1996). Application of the Dynamic Quantitative Structure-Activity Relationship Method for Modeling Antibacterial Activity of Quinolone Derivatives. *Arzneim.Forsch.*, *46*, 423–428.

Kaminski, J.J. (1994). Computer Assisted Drug Design and Selection. *Adv.Drug Deliv.Rev.*, *14*, 331–337.

Kamlet, M.J., Abboud, J.-L.M. and Taft, R.W. (1977). The Solvatochromic Comparison Method. 6. The π^* Scale of Solvent Polarities. *J.Am.Chem.Soc.*, *99*, 6027–6038.

Kamlet, M.J., Jones, M.E., Taft, R.W. and Abboud, J.-L.M. (1979a). Linear Solvation Energy Relationships. Part 2. Correlations of Electronic Spectral Data for Aniline Indicators with Solvent π^* and β Values. *J.Chem.Soc.Perkin Trans.2*, 342–348.

Kamlet, M.J. and Taft, R.W. (1979b). Linear Solvation Energy Relationships. Part 1. Solvent Polarity-Polarizability Effects on Infrared Spectra. *J.Chem.Soc.Perkin Trans.2*, 337–341.

Kamlet, M.J. and Taft, R.W. (1979c). Linear Solvation Energy Relationships. Part 3. Some Reinterpretations of Solvent Effects Based on Correlations with Solvent π^* and α Values. *J.Chem. Soc.Perkin Trans.2*, 349–356.

Kamlet, M.J., Abboud, J.-L.M. and Taft, R.W. (1981a). An Examination of Linear Solvation Energy Relationships. *Prog.Phys.Org.Chem.*, *13*, 485–630.

Kamlet, M.J., Carr, P.W., Taft, R.W. and Abraham, M.H. (1981b). Linear Solvation Energy Relationships. 13. Relationships Between the Hildebrand Solubility Parameter, δ_H,and the Solvatochromic Parameter, π^*. *J.Am.Chem.Soc.*, *103*, 6062–6066.

Kamlet, M.J., Abboud, J.-L.M., Abraham, M.H. and Taft, R.W. (1983). Linear Solvation Energy Relationships. 23. A Comprehensive Collection of the Solvatochromic Parameters, π^*, α, and β and Some Methods for Simplifying the Generalized Solvatochromic Equation. *J.Org.Chem.*, *48*, 2877–2887.

Kamlet, M.J., Abraham, M.H., Doherty, R.M. and Taft, R.W. (1984). Solubility Properties in Polymers and Biological Media. 4. Correlations of Octanol/Water Partition Coefficients with Solvatochromic Parameters. *J.Am.Chem.Soc.*, *106*, 464–466.

Kamlet, M.J., Doherty, P.J., Veith, G.D., Taft, R.W. and Abraham, M.H. (1986a). Solubility Properties in Polymers and Biological Media. 7. An Analysis of Toxicant Properties that Influence Inhibition of Bioluminescence in *Photobacterium phosphoreum* (the Microtox Test). *Environ. Sci.Technol.*, *20*, 690–695.

Kamlet, M.J., Doherty, R.M., Abboud, J.-L.M., Abraham, M.H. and Taft, R.S. (1986b). Solubility. A *New* Look. *Chemtech*, *16*, 566–576.

Kamlet, M.J., Doherty, R.M., Abboud, J.-L.M., Abraham, M.H. and Taft, R.W. (1986c). Linear Solvation Energy Relationships. 36. Molecular Properties Governing Solubilities of Organic Nonelectrolytes in Water. *J.Pharm.Sci.*, *75*, 338–349.

Kamlet, M.J., Doherty, P.J., Taft, R.W., Abraham, M.H., Veith, G.D. and Abraham, D.J. (1987a). Solubility Properties in Polymers and Biological Media. 8. An Analysis of the Factors that Influence Toxicities of Organic Nonelectrolytes to the Golden Orfe Fish (*Leuciscus idus melanotus*). *Environ.Sci.Technol.*, *21*, 149–155.

Kamlet, M.J., Doherty, R.M., Abraham, M.H., Marcus, Y. and Taft, R.W. (1987b). Linear Solvation Energy Relationships. 41. Important Differences Between Aqueous Solubility Relationships for Aliphatic and Aromatic Solutes. *J.Phys.Chem.*, *91*, 1996–2004.

Kamlet, M.J., Doherty, R.M., Fiserova-Bergerova, V., Carr, P.W., Abraham, M.H. and Taft, R.W. (1987c). Solubility Properties in Polymers and Biological Media. 9. Prediction of Solubility and Partition of Organic Nonelectrolytes in Blood and Tissues from Solvatochromic Parameters. *J.Pharm.Sci.*, *76*, 14–17.

Kamlet, M.J., Doherty, R.M., Abraham, M.H., Marcus, Y. and Taft, R.W. (1988a). Linear Solvation Energy Relationships. 46. An Improved Equation for Correlation and Prediction of Octanol/Water Partition Coefficients of Organic Nonelectrolytes (Including Strong Hydrogen Bond Donor Solutes). *J.Phys.Chem.*, *92*, 5244–5255.

Kamlet, M.J., Doherty, R.M., Abraham, M.H. and Taft, R.W. (1988b). *Quant.Struct.-Act.Relat.*, *7*, 71.

Kamlet, M.J., Doherty, R.M., Carr, P.W., Mackay, D., Abraham, M.H. and Taft, R.W. (1988c). Linear Solvation Energy Relationships. 44. Parameter Estimation Rules that Allow Accurate Prediction of Octanol/Water Partition Coefficients and Other Solubility and Toxicity Properties of Polychlorinated Biphenyls and Polycyclic Aromatic Hydrocarbons. *Environ.Sci.Technol.*, *22*, 503–509.

Kang, Y.K. and Jhon, M.S. (1982). Additivity of Atomic Static Polarizabilities and Dispersion Coefficients. *Theor.Chim.Acta*, *61*, 41–48.

Kantola, A., Villar, H.O. and Loew, G.H. (1991). Atom Based Parametrization for a Conformationally Dependent Hydrophobic Index. *J.Comput.Chem.*, *12*, 681–689.

Karabunarliev, S., Mekenyan, O., Karcher, W., Russom, C.L. and Bradbury, S.P. (1996a). Quantum Chemical Descriptors for Estimating the Acute Toxicity of Electrophiles to the Fathead Minnow (Pimephales Promelas). An Analysis Based on Molecular Mechanisms. *Quant.Struct.-Act.Relat.*, *15*, 302–310.

Karabunarliev, S., Mekenyan, O., Karcher, W., Russom, C.L. and Bradbury, S.P. (1996b). Quantum Chemical Descriptors for Estimating the Acute Toxicity of Substituted Benzenes to the Guppy (Poecilia Reticulata) and Fathead minnow (Pimephales promelas). *Quant.Struct.-Act.Relat.*, *15*, 311–320.

Karcher, W. and Devillers, J. eds. (1990a). *Practical Application of Quantitative Structure-Activity Relationships (QSAR) in Environmental Chemistry and Toxicology*. Kluwer Academic Publishers for the European Communities, Dordrecht (The Netherlands), 475 pp.

Karcher, W. and Devillers, J. (1990b). SAR and QSAR in Environmental Chemistry and Toxicology: Scientific Tool or Wishful Thinking? In *Practical Applications of Quantitative Structure-Activity Relationships (QSAR) in Environmental Chemistry and Toxicology* (Karcher, W. and Devillers, J., eds.), Kluwer, Dordrecht (The Netherlands), pp. 1–12.

Karcher, W. and Karabunarliev, S. (1996). The Use of Computer Based Structure-Activity Relationships in the Risk Assessment of Industrial Chemicals. *J.Chem.Inf.Comput.Sci.*, *36*, 672–677.

Karelson, M., Lobanov, V.S. and Katritzky, A.R. (1996). Quantum-Chemical Descriptors in QSAR/QSPR Studies. *Chem.Rev.*, *96*, 1027–1043.

Karelson, M. and Perkson, A. (1999). QSPR Prediction of Densities of Organic Liquids. *Computers Chem.*, *23*, 49–59.

Karelson, M. (2000). *Molecular Descriptors in QSAR/QSPR*. Wiley-Interscience, New York (NY), 430 pp.

Kasai, K., Umeyama, H. and Tomonaga, A. (1988). The Study of Partition Coefficients. The Prediction of Log *P* Value Based on Molecular Structure. *Bull.Chem.Soc.Jap.*, *61*, 2701–2706.

Kato, Y., Itai, A. and Iitaka, Y. (1987). A Novel Method for Superimposing Molecules and Receptor Mapping. *Tetrahedron*, *43*, 5229–5236.

Kato, Y., Inoue, A., Yamada, M., Tomioka, N. and Itai, A. (1992). Automatic Superposition of Drug Molecules Based on their Common Receptor Site. *J.Comput.Aid.Molec.Des.*, *6*, 475–486.

Katritzky, A.R. and Gordeeva, E.V. (1993). Traditional Topological Indices *vs* Electronic, Geometrical, and Combined Molecular Descriptors in QSAR/QSPR Research. *J.Chem.Inf.Comput.Sci.*, *33*, 835–857.

Katritzky, A.R., Ignatchenko, E.S., Barcock, R.A., Lobanov, V.S. and Karelson, M. (1994). Prediction of Gas Chromatographic Retention Times and Response Factors Using a General Quantitative Structure-Property Relationship Treatment. *Anal.Chem.*, *66*, 1799–1807.

Katritzky, A. R., Lobanov, V. S., and Karelson, M. (1995a). *CODESSA. Training Manual*. Gainsville (FL).

Katritzky, A.R., Lobanov, V.S. and Karelson, M. (1995b). QSPR: The Correlation and Quantitative Prediction of Chemical and Physical Properties from Structure. *Chem.Soc.Rev.*, *24*, 279–287.

Katritzky, A.R., Lobanov, V.S., Karelson, M., Murugan, R., Grendze, M.P. and Toomey, J.E.Jr. (1996a). Comprehensive Descriptors for Structural and Statistical Analysis. 1. Correlations Between Structure and Physical Properties of Substituted Pyridines. *Rev.Roum.Chim.*, *41*, 851–867.

Katritzky, A.R., Mu, L. and Karelson, M. (1996b). A QSPR Study of the Solubility of Gases and Vapors in Water. *J.Chem.Inf.Comput.Sci.*, *36*, 1162–1168.

Katritzky, A.R., Mu, L., Lobanov, V.S. and Karelson, M. (1996c). Correlation of Boiling Points with Molecular Structure. 1. A Training Set of 298 Diverse Organics and a Test Set of 9 Simple Inorganics. *J.Phys.Chem.*, *100*, 10400–10407.

Katritzky, A.R., Maran, U., Karelson, M. and Lobanov, V.S. (1997a). Prediction of Melting Points for the Substituted Benzenes: A QSPR Approach. *J.Chem.Inf.Comput.Sci.*, *37*, 913–919.

Katritzky, A.R., Mu, L. and Karelson, M. (1997b). QSPR Treatment of the Unified Nonspecific Solvent Polarity Scale. *J.Chem.Inf.Comput.Sci.*, *37*, 756–761.

Katritzky, A.R., Lobanov, V.S. and Karelson, M. (1998d). Normal Boiling Points for Organic Compounds: Correlation and Prediction by a Quantitative Structure-Property Relationship. *J.Chem.Inf.Comput.Sci.*, *38*, 28–41.

Katritzky, A.R., Mu, L. and Karelson, M. (1998e). Relationships of Critical Temperatures to Calculated Molecular Properties. *J.Chem.Inf.Comput.Sci.*, *38*, 293–299.

Katritzky, A.R., Sild, S. and Karelson, M. (1998a). Correlation and Prediction of the Refractive Indices of Polymers by QSPR. *J.Chem.Inf.Comput.Sci.*, *38*, 1171–1176.

Katritzky, A.R., Sild, S. and Karelson, M. (1998b). General Quantitative Structure-Property Relationship Treatment of the Refractive Index of Organic Compounds. *J.Chem.Inf.Comput.Sci.*, *38*, 840–844.

Katritzky, A.R., Sild, S., Lobanov, V.S. and Karelson, M. (1998c). Quantitative Structure-Property Relationship (QSPR) Correlation of Glass Transition Temperatures of High Molecular Weight Polymers. *J.Chem.Inf.Comput.Sci.*, *38*, 300–304.

Katritzky, A.R., Wang, Y., Sild, S., Tamm, T. and Karelson, M. (1998f). QSPR Studies on Vapor Pressure, Aqueous Solubility, and Prediction of Water-Air Partition Coefficients. *J.Chem.Inf.Comput.Sci.*, *38*, 720–725.

Katritzky, A.R., Tamm, T., Wang, Y. and Karelson, M. (1999a). A Unified Treatment of Solvent Properties. *J.Chem.Inf.Comput.Sci.*, *39*, 692–698.

Katritzky, A.R., Tamm, T., Wang, Y. and Karelson, M. (1999b). QSPR Treatment of Solvent Scales. *J.Chem.Inf.Comput.Sci.*, *39*, 684–691.

Kawashima, Y., Ogawa, T., Kato, M., Nakazato, A., Tsuchida, K., Hatayama, K., Hirono, S. and Moriguchi, I. (1993). Structure-Activity Study of Antihypertensive 1,4-Dihydropyridine Derivatives Having Nitrooxyalkyl Moieties at the 3 and 5 Positions. *Chem.Pharm.Bull.*, *41*, 1060–1065.

Kawashima, Y., Yamada, Y., Asaka, T., Misawa, Y., Kashimura, M., Morimoto, S., Ono, T., Nagate, T., Hatayama, K., Hirono, S. and Moriguchi, I. (1994). Structure-Activity Relationship Study of 6-O-Methylerythromycin-9-O-Substituted Oxime Derivatives. *Chem.Pharm.Bull.*, *42*, 1088–1095.

Kawashima, Y., Sato, M., Yamamoto, S., Shimazaki, Y., Chiba, Y., Satake, M., Iwata, C. and Hatayama, K. (1995). Structure-Activity Relationship Study of TXA(2) Receptor Antagonists 4-(2-(4-Substituted Phenylsulfonylamino)Ethylthio)Phenoxyacetic Acids and Related Compounds. *Chem.Pharm.Bull.*, *43*, 1132–1136.

Kearsley, S.K. and Smith, G.M. (1990). An Alternative Method for the Alignment of Molecular Structures: Maximizing Electrostatic and Steric Overlap. *Tetrahedron Comput.Methodol.*, *3*, 615–633.

Kearsley, S.K., Sallamack, S., Fluder, E.M., Andose, J.D., Mosley, R.T. and Sheridan, R.P. (1996). Chemical Similarity Using Physicochemical Property Descriptors. *J.Chem.Inf.Comput.Sci.*, *36*, 118–127.

Kekulé, A. (1865). *Bull.Soc.Chim.Fr.*, *3*, 98.

Kellogg, G.E., Semus, S.F. and Abraham, D.J. (1991). *HINT*: A New Method of Empirical Hydrophobic Field Calculation for CoMFA. *J.Comput.Aid.Molec.Des.*, *5*, 545–552.

Kellogg, G.E. and Abraham, D.J. (1992a). KEY, LOCK, and LOCKSMITH Complementary Hydropathic Map Predictions of Drug Structure from a Known Receptor Structure of Known Drugs. *J.Mol.Graphics*, *10*, 212.

Kellogg, G.E., Joshi, G.S. and Abraham, D.J. (1992b). New Tools for Modeling and Understanding Hydrophobicity and Hydrophobic Interactions. *Med.Chem.Res.*, *1*, 444–453.

Kellogg, G.E., Kier, L.B., Gaillard, P. and Hall, L.H. (1996). E-State Fields: Applications to 3D QSAR. *J.Comput.Aid.Molec.Des.*, *10*, 513–520.

Kellogg, G.E. (1997). Finding Optimum Field Models for 3-D CoMFA. *Med.Chem.Res.*, *7*, 417–427.

Kemsley, E.K. (1998). A Genetic Algorithm (GA) Approach to the Calculation of Canonical Variates (CVs). *TRAC*, *17*, 24–34.

Kennward, O. and Jones, P.G. (1978). Common Stereochemical Features in Antiepleptic Drugs. A Reinvestigation. *J.Pharm.Pharmacol.*, *30*, 815–817.

Ketelaar, J.A.A. (1958). *Chemical Constitution. An Introduction to the Theory of the Chemical Bond*. Elsevier, Amsterdam (The Netherlands), 448 pp.

Kettaneh-Wold, N., MacGregor, J., Dayal, B. and Wold, S. (1994). Multivariate Design of Process Experiments (M-DOPE). *Chemom.Intell.Lab.Syst.*, *23*, 39–50.

Khadikar, P.V., Deshpande, N.V., Kale, P.P., Dobrynin, A.A., Gutman, I. and Dömötör, G. (1995). The Szeged Index and an Analogy with the Wiener Index. *J.Chem.Inf.Comput.Sci.*, *35*, 547–550.

Kiang, Y.-S. and Tang, A.-C. (1986). A Graphical Evaluation of Characteristic Polynomials of Hückel Trees. *Int.J.Quant.Chem.*, *29*, 229–240.

Kidera, A., Konisci, Y., Ooi, T. and Scheraga, H.A. (1985a). Relation Between Sequence Similarity and Structural Similarity in Proteins. Role of Important Properties of Amino Acids. *J.Prot.Chem.*, *4*, 265–297.

Kidera, A., Konisci, Y., Ooi, T. and Scheraga, H.A. (1985b). Statistical Analysis of the Physical Properties of the 20 Naturally Occurring Amino Acids. *J.Prot.Chem.*, *4*, 23–55.

Kier, L.B. (1971). *Molecular Orbital Theory in Drug Research*. Academic Press, New York (NY).

Kier, L.B., Hall, L.H., Murray, W.J. and Randic, M. (1975a). Molecular Connectivity I: Relationship to Nonspecific Local Anesthesia. *J.Pharm.Sci.*, *64*, 1971–1974.

Kier, L.B., Murray, W.J. and Hall, L.H. (1975b). Molecular Connectivity. 4. Relationships to Biological Activities. *J.Med.Chem.*, *18*, 1272–1274.

Kier, L.B. and Hall, L.H. (1976a). *Molecular Connectivity in Chemistry and Drug Research*. Academic Press, New York (NY), 257 pp.

Kier, L.B. and Hall, L.H. (1976b). Molecular Connectivity VII: Specific Treatment of Heteroatoms. *J.Pharm.Sci.*, *65*, 1806–1809.

Kier, L.B., Murray, W.J., Randic, M. and Hall, L.H. (1976a). Molecular Connectivity Concept Applied to Density. *J.Pharm.Sci.*, *65*, 1226–1230.

Kier, L.B., Murray, W.J., Randic, M. and Hall, L.H. (1976b). Molecular Connectivity V: Connectivity Series Concept Applied to Density. *J.Pharm.Sci.*, *65*, 1226–1230.

Kier, L.B., Di Paolo, T. and Hall, L.H. (1977). Structure-Activity Studies on Odor Molecules Using Molecular Connectivity. *J.Theor.Biol.*, *67*, 585–595.

Kier, L.B. and Hall, L.H. (1977a). Structure-Activity Studies on Hallucinogenic Amphetamines Using Molecular Connectivity. *J.Med.Chem.*, *20*, 1631–1636.

Kier, L.B. and Hall, L.H. (1977b). The Nature of Structure-Acitivity Relationships and their Relation to Molecular Connectivity. *Eur.J.Med.Chem.*, *12*, 307–312.

Kier, L.B. and Glennon, R.A. (1978). Psychotomimetic Phenalkylamines as Serotonin Antagonists: A SAR Analysis. *Life Science*, *22*, 1589.

Kier, L.B. and Hall, L.H. (1978). A Molecular Connectivity Study of Muscarinic Receptor Affinity of Acetylcholine Antagonists. *J.Pharm.Sci.*, *67*, 1408–1412.

Kier, L.B., Simons, R.J. and Hall, L.H. (1978). Structure-Activity Studies on Mutagenicity of Nitrosoamines Using Molecular Connectivity. *J.Pharm.Sci.*, *67*, 725–726.

Kier, L.B. and Hall, L.H. (1979). Molecular Connectivity Analyses of Structure Influencing Chromatographic Retention Indexes. *J.Pharm.Sci.*, *68*, 120–122.

Kier, L.B. (1980a). Molecular Structure Influencing Either a Sweet or Bitter Taste Among Aldoximes. *J.Pharm.Sci.*, *69*, 416.

Kier, L.B. (1980b). Structural Information from Molecular Connectivity ${}^4\chi_{pc}$ Index. *J.Pharm.Sci.*, *69*, 1034–1039.

Kier, L.B. (1980c). Use of Molecular Negentropy to Encode Structure Governing Biological Activity. *J.Pharm.Sci.*, *69*, 807–810.

Kier, L.B. and Hall, L.H. (1981). Derivation and Significance of Valence Molecular Connectivity. *J.Pharm.Sci.*, *70*, 583–589.

Kier, L.B. and Hall, L.H. (1983a). Estimation of Substituent Group Electronic Influence from Molecular Connectivity Delta Values. *Quant.Struct.-Act.Relat.*, *2*, 163.

Kier, L.B. and Hall, L.H. (1983b). General Definition of Valence Delta-Values for Molecular Connectivity. *J.Pharm.Sci.*, *72*, 1170–1173.

Kier, L.B. and Hall, L.H. (1983c). Structural Information and Flexibility Index from the Molecular Connectivity ${}^3\chi_p$ Index. *Quant.Struct.-Act.Relat.*, *2*, 55–59.

Kier, L.B. (1985). A Shape Index from Molecular Graphs. *Quant.Struct.-Act.Relat.*, *4*, 109–116.

Kier, L.B. (1986a). Distinguishing Atom Differences in a Molecular Graph Shape Index. *Quant. Struct.-Act.Relat.*, *5*, 7–12.

Kier, L.B. (1986b). Indexes of Molecular Shape from Chemical Graphs. *Acta Pharm.Jugosl.*, *36*, 171–188.

Kier, L.B. (1986c). Shape Indexes of Orders One and Three from Molecular Graphs. *Quant. Struct.-Act.Relat.*, *5*, 1–7.

Kier, L.B. and Hall, L.H. (1986). *Molecular Connectivity in Structure-Activity Analysis.* Research Studies Press – Wiley, Chichester (UK), 262 pp.

Kier, L.B. (1987a). A Structure Based Approach to Molecular Shape. In *QSAR in Drug Design and Toxicology* (Hadzi, D. and Jerman-Blazic, B., eds.), Elsevier, Amsterdam (The Netherlands).

Kier, L.B. (1987b). Inclusion of Symmetry as a Shape Attribute in Kappa Index Analysis. *Quant. Struct.-Act.Relat.*, *6*, 8–12.

Kier, L.B. (1987c). Indexes of Molecular Shape from Chemical Graph. *Med.Res.Rev.*, *7*, 417–440.

Kier, L.B. (1987d). The Substituent Steric Effect Index Based on the Molecular Graph. *Quant. Struct.-Act.Relat.*, *6*, 117–122.

Kier, L.B. (1989). An Index of Molecular Flexibility from Kappa Shape Attributes. *Quant. Struct.-Act.Relat.*, *8*, 221–224.

Kier, L.B. (1990). Indexes of Molecular Shape from Chemical Graphs. In *Computational Chemical Graph Theory* (Rouvray, D.H., ed.), Nova Science Publishers, New York (NY), pp. 151–174.

Kier, L.B. and Hall, L.H. (1990a). An Electrotopological-State Index for Atoms in Molecules. *Pharm.Res.*, *7*, 801–807.

Kier, L.B. and Hall, L.H. (1990b). The Molecular Connectivity of Non-Sigma Electrons. *Rep.Mol.Theory*, *1*, 121–125.

Kier, L.B. and Hall, L.H. (1991). A Differential Molecular Connectivity Index. *Quant.Struct.-Act.Relat.*, *10*, 134–140.

Kier, L.B., Hall, L.H. and Frazer, J.W. (1991). An Index of Electrotopological State for Atoms in Molecules. *J.Math.Chem.*, *7*, 229–241.

Kier, L.B. and Hall, L.H. (1992a). An Atom-Centered Index for Drug QSAR Models. In *Advances in Drug Design* (Testa, B., ed.), Academic Press, New York (NY).

Kier, L.B. and Hall, L.H. (1992b). An Index of Atom Electrotopological State. In *QSAR in Design of Bioactive compounds, A Telesymposium* (Biaggi, A., ed.), Prous Science, Barcelona (SP).

Kier, L.B. and Hall, L.H. (1992c). Atom Description in QSAR Models: Development and Use of an Atom Level Index. *Adv.Drug Res.*, *22*, 1–38.

Kier, L.B. and Hall, L.H. (1993a). An Atom Centered Index for Drug QSAR Models. *Adv.Drug Res.*, *22*, 1–38.

Kier, L.B. and Hall, L.H. (1993b). The Generation of Molecular Structures for a Graph Based QSAR Equation. *Quant.Struct.-Act.Relat.*, *12*, 383–388.

Kier, L.B., Hall, L.H. and Frazer, J.W. (1993). Design of Molecules from Quantitative Structure-Activity Relationship Models. 1. Information Transfer Between Path and Vertex Degree Counts. *J.Chem.Inf.Comput.Sci.*, *33*, 143–147.

Kier, L.B. (1995). Atom-Level Descriptors for QSAR Analyses. In *Chemometrics Methods in Molecular Design – Vol. 2* (van de Waterbeemd, H., ed.), VCH Publishers, New York (NY), pp. 39–47.

Kier, L.B. and Hall, L.H. (1995). A QSAR Model of the OH Radical Reaction with CFCs. *SAR & QSAR Environ.Res.*, *3*, 97–100.

Kier, L.B. and Testa, B. (1995). Complexity and Emergence in Drug Research. *Adv.Drug Res.*, *26*, 1–43.

Kier, L.B. (1997). Kappa Shape Indices for Similarity Analysis. *Med.Chem.Res.*, *7*, 394–406.

Kier, L.B. and Hall, L.H. (1997a). Quantitative Information Analysis: The New Center of Gravity in Medicinal Chemistry. *Med.Chem.Res.*, *7*, 335–339.

Kier, L.B. and Hall, L.H. (1997b). The E-State as an Extended Free Valence. *J.Chem.Inf.Comput.Sci.*, *37*, 548–552.

Kier, L.B. and Hall, L.H. (1999). *Molecular Structure Description. The Electrotopological State.* Academic Press, London (UK), 246 pp.

Kier, L.B. and Hall, L.H. (2000). Intermolecular Accessibility: The Meaning of Molecular Connectivity. *J.Chem.Inf.Comput.Sci.*, *40*, 792–795.

Kim, K.H. and Martin, Y.C. (1991a). Direct Prediction of Dissociation Constants (pK_a's) of Clonidine-Like Imidazolines, 2-Substituted Imidazoles, and 1-Methyl-2-Substituted-Imidazoles from 3D Structures Using a Comparative Molecular Field Analysis (CoMFA) Approach. *J.Med.Chem.*, *34*, 2056–2060.

Kim, K.H. and Martin, Y.C. (1991b). Direct Prediction of Linear Free Energy Substituent Effects from 3D Structures Using Comparative Molecular Field Analysis. 1. Electronic Effects of Substituted Benzoic Acids. *J.Org.Chem.*, *56*, 2723–2729.

Kim, K.H. and Martin, Y.C. (1991c). Evaluation of Electrostatic and Steric Descriptors for 3D-QSAR: the H^+ and CH_3 Probes Using Comparative Molecular Field Analysis (CoMFA) and the Modified Partial Least Squares Method. In *QSAR: Rational Approaches to the Design of Bioactive Compounds* (Silipo, C. and Vittoria, A., eds.), Elsevier, Amsterdam (The Netherlands), pp. 151–154.

Kim, K.H. (1992a). 3D Quantitative Structure-Activity Relationships Description of Electronic Effects Directly from 3D Structures Using a Grid Comparative Molecular Field Analysis (CoMFA) Approach. *Quant.Struct.-Act.Relat.*, *11*, 127–134.

Kim, K.H. (1992b). 3D Quantitative Structure-Activity Relationships Investigation of Steric Effects with Descriptors Directly from 3D Structures Using a Comparative Molecular Field Analysis (CoMFA) Approach. *Quant.Struct.-Act.Relat.*, *11*, 453–460.

Kim, K.H. (1992c). 3D Quantitative Structure-Activity Relationships Nonlinear Dependence Described Directly from 3D Structures Using a Comparative Molecular Field Analysis (CoMFA) Approach. *Quant.Struct.-Act.Relat.*, *11*, 309–317.

Kim, K.H. (1993a). 3D-Quantitative Structure-Activity Relationships: Describing Hydrophobic Interactions Directly from 3D Structures Using a Comparative Molecular Field Analysis (CoMFA) Approach. *Quant.Struct.-Act.Relat.*, *12*, 232–238.

Kim, K.H. (1993b). Comparison of Classical and 3D QSAR. In *3D QSAR in Drug Design. Theory, Methods and Applications.* (Kubinyi, H., ed.), ESCOM, Leiden (The Netherlands), pp. 619–642.

Kim, K.H. (1993c). Nonlinear Dependence in Comparative Molecular Field Analysis. *J.Comput. Aid.Molec.Des.*, *7*, 71–82.

Kim, K.H. (1993d). Separation of Electronic, Hydrophobic, and Steric Effects in 3D-Quantitative Structure-Activity Relationships with Descriptors Directly from 3D Structures Using a Comparative Molecular Field Analysis (CoMFA) Approach. *Current Topics in Medicinal Chemistry*, *1*, 453–467.

Kim, K.H. (1993e). Use of Indicator Variable in Comparative Molecular Field Analysis. *Med.-Chem.Res.*, *3*, 257–267.

Kim, K.H. (1993f). Use of the Hydrogen-Bond Potential Function in Comparative Molecular Field Analysis (CoMFA): An Extension of CoMFA. In *Trends in QSAR and Molecular Modelling 92* (Wermuth, C.G., ed.), ESCOM, Leiden (The Netherlands), pp. 245–251.

Kim, K.H., Greco, G., Novellino, E., Silipo, C. and Vittoria, A. (1993). Use of the Hydrogen Bond Potential Function in a Comparative Molecular Field Analysis (CoMFA) on a Set of Benzodiazepines. *J.Comput.Aid.Molec.Des.*, *7*, 263–280.

Kim, K.H. (1995a). Comparative Molecular Field Analysis (CoMFA). (Dean, P.M., ed.), Chapman & Hall, London (UK), pp. 291–331.

Kim, K.H. (1995b). Comparison of Classical QSAR and Comparative Molecular Field Analysis: Toward Lateral Validations. In *Classical and Three-Dimensional QSAR in Agrochemistry* (Hansch, C. and Fujita, T., eds.), American Chemical Society, Washington (DC), pp. 302–317.

Kim, K.H. (1995c). Description of the Reversed-Phase High-Performance Liquid Chromatography (RP-HPLC) Capacity Factors and Octanol-Water Partition Coefficients of 2-Pyrazine and 2-Pyridine Analogues Directly from the Three-Dimensional Structures Using Comparative Molecular Field Analysis (CoMFA) Approach. *Quant.Struct.-Act.Relat.*, *14*, 8–18.

Kim, K.H. and Kim, D.H. (1995). Description of Hydrophobicity Parameters of a Mixed Set from their Three-Dimensional Structures. *Bioorg.Med.Chem.*, *3*, 1389–1396.

Kim, K.H. (1998a). List of CoMFA References, 1993–1997. In *3D QSAR in Drug Design – Vol. 3* (Kubinyi, H., Folkers, G. and Martin, Y.C., eds.), Kluwer/ESCOM, Dordrecht (The Netherlands), pp. 317–338. [R]

Kim, K.H., Greco, G. and Novellino, E. (1998b). A Critical Review of Recent CoMFA Applications. In *3D QSAR in Drug Design – Vol. 3* (Kubinyi, H., Folkers, G. and Martin, Y.C., eds.), Kluwer/ESCOM, Dordrecht (The Netherlands), pp. 257–315. [R]

Kimura, T., Miyashita, Y., Funatsu, K. and Sasaki, S.I. (1996). Quantitative Structure-Activity Relationships of the Synthetic Substrates for Elastase Enzyme Using Nonlinear Partial Least Squares Regression. *J.Chem.Inf.Comput.Sci.*, *36*, 185–189.

Kimura, T., Hasegawa, K. and Funatsu, K. (1998). GA Strategy for Variable Selection in QSAR Studies: GA-Based Region Selection for CoMFA Modeling. *J.Chem.Inf.Comput.Sci.*, *38*, 276–282.

King, R. B. ed. (1983). *Chemical Applications of Topology and Graph Theory*. Elsevier, Amsterdam (The Netherlands).

King, R.B. (1991). Experimental Tests of Chirality Algebra. *J.Math.Chem.*, *7*, 69–84.

King, R.D. and Srinivasan, A. (1996). Prediction of Rodent Carcinogenicity Bioassays from Molecular Structure Using Inductive Logic Programming. *Environ.Health Persp.*, *104*, 1031–1040.

Kirby, E.C. (1994). Sensitivity of Topological Indices to Methyl Group Branching in Octanes and Azulenes, or What does a Topological Index Index? *J.Chem.Inf.Comput.Sci.*, *34*, 1030–1035.

Kireev, D.B., Fetisov, V.I. and Zefirov, N.S. (1994). Approximate Molecular Electrostatic Potential Computations. Applications to Quantitative Structure-Activity Relationships. *J.Mol. Struct.(Theochem)*, *110*, 143–150.

Kireev, D.B. (1995). ChemNet: A Novel Neural Network Based Method for Graph/Property Mapping. *J.Chem.Inf.Comput.Sci.*, *35*, 175–180.

Kireev, D.B., Chrétien, J.R. and Raevsky, O.A. (1995). Molecular Modeling and Quantitative Structure-Activity Studies of Anti HIV-1 2-Heteroarylquinoline-4-Amines. *Eur.J.Med.Chem.*, *30*, 395–402.

Kireev, D.B., Chrétien, J.R., Bernard, P. and Ros, F. (1998). Application of Kohonen Neural Networks in Classification of Biologically Active Compounds. *SAR & QSAR Environ.Res.*, *8*, 93–107.

Kirkwood, J.J. and Westheimer, F.H. (1938). The Electrostatic Influence of Substituents on the Dissociation Constants of Organic Acids. I. *J.Chem.Phys.*, *6*, 506–512.

Kitaigorodsky, A.I. (1973). *Molecular Crystals and Molecules.* Academic Press, New York (NY), 553 pp.

Klamt, A. (1996). Estimation of Gas-Phase Hydroxyl Radical Rate Constants of Oxygenated Compounds Based on Molecular Orbital Calculations. *Chemosphere*, *32*, 717–726.

Klappa, S.A. and Long, G.R. (1992). Computer Assisted Determination of the Biological Activity of Polychlorinated Biphenyls Using Gas Chromatographic Retention Indexes as Molecular Descriptors. *Anal.Chim.Acta*, *259*, 89–93.

Klavzar, S. and Gutman, I. (1996). A Comparison of the Schultz Molecular Topological Index with the Wiener Index. *J.Chem.Inf.Comput.Sci.*, *36*, 1001–1003.

Klavzar, S., Rajapaxi, A. and Gutman, I. (1996). On the Szeged and the Wiener Index of Graphs. *Appl.Math.Lett.*, *67*, 45–49.

Klawun, C. and Wilkins, C.L. (1996a). Joint Neural Network Interpretation of Infrared and Mass Spectra. *J.Chem.Inf.Comput.Sci.*, *36*, 249–257.

Klawun, C. and Wilkins, C.L. (1996b). Optimization of Functional Group Prediction from Infrared Spectra Using Neural Networks. *J.Chem.Inf.Comput.Sci.*, *36*, 69–81.

Klebe, G. (1993). Structural Alignment of Molecules. In *3D QSAR in Drug Design. Theory, Methods and Applications.* (Kubinyi, H., ed.), ESCOM, Leiden (The Netherlands), pp. 173–199.

Klebe, G. and Abraham, U. (1993). On the Prediction of Binding Properties of Drug Molecules by Comparative Molecular Field Analysis. *J.Med.Chem.*, *36*, 70–80.

Klebe, G., Abraham, U. and Mietzner, T. (1994a). Molecular Similarity Indices in a Comparative Analysis (CoMSIA) of Drug Molecules to Correlate and Predict their Biological Activity. *J.Med.Chem.*, *37*, 4130–4146.

Klebe, G., Mietzner, T. and Weber, F. (1994b). Different Approaches toward an Automatic Alignment of Drug Molecules: Application to Sterol Mimics, Thrombin and Thermolysin Inhibitors. *J.Comput.Aid.Molec.Des.*, *8*, 751–778.

Klebe, G. (1998). Comparative Molecular Similarity Indices Analysis: CoMSIA. In *3D QSAR in Drug Design – Vol. 3* (Kubinyi, H., Folkers, G. and Martin, Y.C., eds.), Kluwer/ESCOM, Dordrecht (The Netherlands), pp. 87–104.

Klebe, G. and Abraham, U. (1999). Comparative Molecular Similarity Index Analysis (CoMSIA) to Study Hydrogen-Bonding Properties and to Score Combinatorial Libraries. *J.Comput.Aid. Molec.Des.*, *13*, 1–10.

Klein, D.J. (1986). Chemical Graph-Theoretic Cluster Expansion. *Int.J.Quantum Chem.Quant. Chem.Symp.*, *69*, 701–712.

Klein, D.J., Mihalic, Z., Plavsic, D. and Trinajstic, N. (1992). Molecular Topological Index: A Relation with the Wiener Index. *J.Chem.Inf.Comput.Sci.*, *32*, 304–305.

Klein, D.J. and Randic, M. (1993). Resistance Distance. *J.Math.Chem.*, *12*, 81–95.

Klein, D.J. (1995). Similarity and Dissimilarity in Posets. *J.Math.Chem.*, *18*, 321–348.

Klein, D.J., Lukovits, I. and Gutman, I. (1995). On the Definition of the Hyper-Wiener Index for Cycle-Containing Structures. *J.Chem.Inf.Comput.Sci.*, *35*, 50–52.

Klein, D.J. (1997a). Graph Geometry, Graph Metrics, & Wiener. *MATCH (Comm.Math.Comp. Chem.)*, *35*, 7–27.

Klein, D.J. and Babic, D. (1997b). Partial Orderings in Chemistry. *J.Chem.Inf.Comput.Sci.*, *37*, 656–671.

Klein, D.J., Randic, M., Babic, D., Lucic, B., Nikolic, S. and Trinajstic, N. (1997). Hierarchical Orthogonalization of Descriptors. *Int.J.Quant.Chem.*, *63*, 215–222.

Klein, D.J. and Zhu, H.Y. (1998). Distances and Volumina for Graphs. *J.Math.Chem.*, *23*, 179–195.

Klein, D.J. and Gutman, I. (1999). Wiener-Number-Related Sequences. *J.Chem.Inf.Comput.Sci.*, *39*, 534–536.

Klein, D.J., Schmalz, T.G. and Bytautas, L. (1999). Chemical Sub-Structural Cluster Expansions for Molecular Properties. *SAR & QSAR Environ.Res.*, *10*, 131–156.

Klir, G.J. and Folger, T.A. (1988). *Fuzzy Sets, Uncertainty, and Information*. Prentice-Hall, Englewood Cliffs (NJ), 356 pp.

Klopman, G. and Iroff, L. (1981). Calculation of Partition Coefficients by the Charge Density Method. *J.Comput.Chem.*, *2*, 157–160.

Klopman, G. (1984). Artificial Intelligence Approach to Structure-Activities Studies. Computer Automated Structure Evaluation of Biological Activity of Organic Molecules. *J.Am.Chem. Soc.*, *106*, 7315–7321.

Klopman, G., Namboodiri, K. and Schochet, M. (1985). Simple Method of Computing the Partition Coefficient. *J.Comput.Chem.*, *6*, 28–38.

Klopman, G. and Buyukbingol, E. (1988). An Artificial Intelligence Approach to the Study of the Structural Moieties Relevant to Drug-Receptor Interactions in Aldose Reductase Inhibitors. *Mol.Pharm.*, *34*, 852–862.

Klopman, G. and Raychaudhury, C. (1988). A Novel Approach to the Use of Graph Theory in Structure-Activity Relationship Studies. Application to the Qualitative Evaluation of Mutagenicity in a Series of Nonfused Ring Aromatic Compounds. *J.Comput.Chem.*, *9*, 232–243.

Klopman, G., Raychaudhury, C. and Henderson, R.V. (1988). A New Approach to Structure-Activity Using Distance Information Content of Graph Vertices: A Study with Phenylalkylamines. *Math.Comput.Modelling*, *11*, 635–640.

Klopman, G. and Raychaudhury, C. (1990). Vertex Indices of Molecular Graphs in Structure-Activity Relationships: A Study of the Convulsant-Anticonvulsant Activity of Barbiturates and the Carcinogenicity of Unsubstituted Polycyclic Aromatic Hydrocarbons. *J.Chem.Inf. Comput.Sci.*, *30*, 12–19.

Klopman, G. and Henderson, R.V. (1991). A Graph Theory-Based "Expert System" Methodology for Structure-Activity Studies. *J.Math.Chem.*, *7*, 187–216.

Klopman, G. and Wang, S. (1991). A Computer Automated Structure Evaluation (CASE) Approach to Calculation of Partition Coefficient. *J.Comput.Chem.*, *12*, 1025–1032.

Klopman, G. (1992). MULTICASE. 1. A Hierarchical Computer Automated Structure Evaluation Program. *Quant.Struct.-Act.Relat.*, *11*, 176–184.

Klopman, G. and Ptchelintsev, D. (1992). Application of the Computer Automated Structure Evaluation Methodology to a QSAR Study of Chemoreception Aromatic Musky Odorants. *J.Agr.Food Chem.*, *40*, 2244–2251.

Klopman, G., Wang, S. and Balthasar, D.M. (1992). Estimation of Aqueous Solubility of Organic Molecules by the Group Contribution Approach. Application to the Study of Biodegradation. *J.Chem.Inf.Comput.Sci.*, *32*, 474–482.

Klopman, G., Li, J.Y., Wang, S. and Dimayuga, M. (1994). Computer Automated log *P* Calculations Based on an Extended Group Contribution Approach. *J.Chem.Inf.Comput.Sci.*, *34*, 752–781.

Klopman, G. and Rosenkranz, H.S. (1994). Approaches to SAR in Carcinogenesis and Mutagenesis Prediction of Carcinogenicity/Mutagenicity Using Multi-Case. *Mut.Res.*, *305*, 33–46.

Klopman, G. (1998). The MultiCASE Program II. Baseline Activity Identification Algorithm (BAIA). *J.Chem.Inf.Comput.Sci.*, *38*, 78–81.

Knop, J.V., Müller, W.R., Jericevic, Z. and Trinajstic, N. (1981). Computer Enumeration and Generation of Trees and Rooted Trees. *J.Chem.Inf.Comput.Sci.*, *21*, 91–99.

Knop, J.V., Müller, W.R., Szymanski, K. and Trinajstic, N. (1991). On the Determinant of the Adjacency-Plus-Distance Matrix as the Topological Index for Characterizing Alkanes. *J.Chem. Inf.Comput.Sci.*, *31*, 83–84.

Koch, R. (1982). Molecular Connectivity and Acute Toxicity of Environmental Pollutants. *Chemosphere*, *11*, 925–931.

Koehler, M.G., Grigoras, S. and Dunn III, W.J. (1988). The Relationship Between Chemical Structure and the Logarithm of the Partition Coefficient. *Quant.Struct.-Act.Relat.*, *7*, 150–159.

Konstantinova, E.V. (1996). The Discrimination Ability of Some Topological and Information Distance Indices for Graphs of Unbranched Hexagonal Systems. *J.Chem.Inf.Comput.Sci.*, *36*, 54–57.

Kopecky, J., Bocek, K. and Vlachovà, D. (1965). Chemical Structure and Biological Activity on *m*- and *p*-Disubstituted Derivatives of Benzene. *Nature*, *207*, 981–981.

Koppel, I.A. and Paju, A.I. (1974). *Reacts.Sposobnost.Org.Soedin.*, *11*, 137.

Körner, W. (1874). Studi sull'Isomeria delle Così Dette Sostanze Aromatiche a Sei Atomi di Carbonio. *Gazz.Chim.It.*, *4*.

Kosower, E.M. (1958a). The Effect of Solvent on Spectra. I. A New Empirical Measure of Solvent Polarity: Z-Values. *J.Am.Chem.Soc.*, *80*, 3253–3260.

Kosower, E.M. (1958b). The Effect of Solvent on Spectra. II. Correlation of Spectral Absorption Data with Z-Values. *J.Am.Chem.Soc.*, *80*, 3261–3267.

Kourounakis, A. and Bodor, N. (1995). Quantitative Structure-Activity Relationships of Catechol Derivatives on Nerve Growth Factor Secretion in L-M Cells. *Pharm.Res.*, *12*, 1199–1204.

Kovalishyn, V.V., Tetko, I.V., Luik, A.I., Kholodovych, V.V., Villa, A.E.P. and Livingstone, D.J. (1998). Neural Network Studies. 3. Variable Selection in the Cascade-Correlation Learning Architecture. *J.Chem.Inf.Comput.Sci.*, *38*, 651–659.

Kováts, E. (1968). Zu Fragen der Polarität. *Chimia*, *22*, 459.

Kraak, M.S., Wijnands, P., Govers, H.A.J., Admiraal, W.A. and Devoogt, P. (1997). Structural-Based Differences in Ecotoxicity of Benzoquinoline Isomers to the Zebra mussel (Dreissena polymorpha). *Environ.Toxicol.Chem.*, *16*, 2158–2163.

Kränz, H., Vill, V. and Meyer, B. (1996). Prediction of Material Properties from Chemical Structures. The Clearing Temperature of Nematic Liquid Crystal Derived from their Chemical Structures by Artificial Neural Networks. *J.Chem.Inf.Comput.Sci.*, *36*, 1173–1177.

Krivka, P., Jericevic, Z. and Trinajstic, N. (1985). On the Computation of Characteristic Polynomial of a Chemical Graph. *Int.J.Quantum Chem.Quant.Chem.Symp.*, *19*, 129–147.

Kroemer, R.T., Ettmayer, P. and Hecht, P. (1995). 3D-Quantitative Structure-Activity Relationships of Human Immunodeficiency Virus Type-1 Proteinase Inhibitors: Comparative Molecular Field Analysis of 2-Heterosubstituted Statine Derivatives – Implications for the Design of Novel Inhibitors. *J.Med.Chem.*, *38*, 4917–4928.

Kroemer, R.T., Hecht, P. and Liedl, K.R. (1996). Different Electrostatic Descriptors in Comparative Molecular Field Analysis: A Comparison of Molecular Electrostatic and Coulomb Potentials. *J.Comput.Chem.*, *17*, 1296–1308.

Kruszewski, J. and Krygowski, T.M. (1972). Definition of Aromaticity basing on the Harmonic Oscillator Model. *Tetrahedron Lett.*, *36*, 3839–3842.

Krygowski, T.M., Wrona, P.K., Zielkowska, U. and Reichardt, C. (1985). Empirical Parameters of the Lewis Acidity and Basicity for Aqueous Binary Solvent Mixtures. *Tetrahedron*, *41*, 4519–4527.

Krygowski, T.M. (1993). Crystallographic Studies of Inter- and Itramolecular Interactions Reflected in Aromatic Character of π -Electron Systems. *J.Chem.Inf.Comput.Sci.*, *33*, 70–78.

Krygowski, T.M. and Ciesielski, A. (1995a). Local Aromatic Character of C_{60} and C_{70} and their Derivatives. *J.Chem.Inf.Comput.Sci.*, *35*, 1001–1003.

Krygowski, T.M., Ciesielski, A., Bird, C.W. and Kotschy, A. (1995b). Aromatic Character of the Benzene Ring Present in Various Topological Environments in Benzenoid Hydrocarbons. Nonequivalence of Indices of Aromaticity. *J.Chem.Inf.Comput.Sci.*, *35*, 203–210.

Krygowski, T.M., Cyranski, M., Ciesielski, A., Swirska, B. and Leszczynski, P. (1996). Separation of the Energetic and Geometric Contributions to Aromaticity. 2. Analysis of the Aromatic Character of Benzene Rings in their Various Topological Environments in the Benzenoid Hydrocarbons. Crystal and Molecular Structure of Coronene. *J.Chem.Inf.Comput.Sci.*, *36*, 1135–1141.

Krzanowski, W.J. (1988). *Principles of Multivariate Analysis.* Oxford Univ. Press, New York (NY), 564 pp.

Krzyzaniak, J.F., Myrdal, P.B., Simamora, P. and Yalkowsky, S.H. (1995). Boiling Point and Melting Point Prediction for Aliphatic, Non-Hydrogen-Bonding Compounds. *Ind.Eng.Chem.Res.*, *34*, 2530–2535.

Kuanar, M., Kuanar, S.K. and Mishra, B.K. (1999). Correlation of Critical Micelle Concentration of Nonionic Surfactants with Molecular Descriptors. *Indian J.Chem.*, *38*, 113–118.

Kubinyi, H. (1976a). Quantitative Structure-Activity Relationships. 2. A Mixed Approach, Based on Hansch and Free-Wilson Analysis. *J.Med.Chem.*, *19*, 587–600.

Kubinyi, H. (1976b). Quantitative Structure-Activity Relationships. IV. Non-Linear Dependence of Biological Activity on Hydrophobic Character: A New Model. *Arzneim.Forsch.*, *26*, 1991–1997.

Kubinyi, H. and Kehrhahn, O.M. (1976). Quantitative Structure-Activity Relationships. 1. The Modified Free-Wilson Approach. *J.Med.Chem.*, *19*, 578–586.

Kubinyi, H. (1977). Quantitative Structure-Activity Relationships. 7. The Bilinear Model, a New Model for Nonlinear Dependence of Biological Activity on Hydrophobic Character. *J.Med. Chem.*, *20*, 625–629.

Kubinyi, H. (1979). Lipophilicity and Biological Activity. Drug Transport and Drug Distribution in Model Systems and Biological Systems. *Arzneim.Forsch.*, *29*, 1067–1080.

Kubinyi, H. (1988a). Current Problems in Quantitative Structure-Activity Relationships. In *Physical Property Prediction in Organic Chemistry* (Jochum, C., Hicks, M.G. and Sunkel, J., eds.), Springer-Verlag, Berlin (Germany), pp. 235–247.

Kubinyi, H. (1988b). Free Wilson Analysis. Theory, Applications and its Relationship to Hansch Analsyis. *Quant.Struct.-Act.Relat.*, *7*, 121–133. [R]

Kubinyi, H. (1990). The Free-Wilson Method and its Relationship to the Extrathermodynamic Approach. In *Quantitative Drug Design. Vol. 4* (Ramsden, C.A., ed.), Pergamon Press, Oxford (UK), pp. 589–643.

Kubinyi, H. ed. (1993a). *3D QSAR in Drug Design. Theory, Methods, and Applications.* ESCOM, Leiden (The Netherlands), 760 pp.

Kubinyi, H. (1993b). *QSAR: Hansch Analysis and Related Approaches.* VCH Publishers, Weinheim (Germany), 240 pp.

Kubinyi, H. (1994a). Variable Selection in QSAR Studies. I. An Evolutionary Algorithm. *Quant. Struct.-Act.Relat.*, *13*, 285–294.

Kubinyi, H. (1994b). Variable Selection in QSAR Studies. II. A Highly Efficient Combination of Systematic Search and Evolution. *Quant.Struct.-Act.Relat.*, *13*, 393–401.

Kubinyi, H. (1995). From Lipophilicity to 3D QSAR – The Fascination of Computer-Aided Drug Design. In *QSAR and Molecular Modelling: Cocepts, Computational Tools and Biological Applications* (Sanz, F., Giraldo, J. and Manaut, F., eds.), Prous Science, Barcelona (Spain), pp. 2–16.

Kubinyi, H. (1996). Evolutionary Variable Selection in Regression and PLS Analyses. *J.Chemom.*, *10*, 119–133.

Kubinyi, H. (1997). A General View on Similarity and QSAR Studies. In *Computer-Assisted Lead Finding and Optimization* (van de Waterbeemd, H., Testa, B. and Folkers, G., eds.), Wiley-VCH, Weinheim (Germany), pp. 9–28.

Kubinyi, H. (1998a). Similarity and Dissimilarity: A Medicinal Chemist's View. In *3D QSAR in Drug Design – Vol. 2* (Kubinyi, H., Folkers, G. and Martin, Y.C., eds.), Kluwer/ESCOM, Dordrecht (The Netherlands), pp. 226–252.

Kubinyi, H., Folkers, G., and Martin, Y. C. eds. (1998b). *3D QSAR in Drug Design – Vol. 2.* Kluwer/ESCOM, Dordrecht (The Netherlands), 416 pp.

Kubinyi, H., Folkers, G., and Martin, Y. C. eds. (1998c). *3D QSAR in Drug Design – Vol. 3.* Kluwer/ESCOM, Dordrecht (The Netherlands), 352 pp.

Kubinyi, H., Hamprecht, F.A. and Mietzner, T. (1998d). Three-Dimensional Quantitative Similarity-Activity Relationships (3D-QSiAR) from SEAL Similarity Matrices. *J.Med.Chem.*, *41*, 2553–2564.

Kumskov, M.I., Ponomareva, L.A. and Zakharova, M.V. (1995). New Approach to the Solution of QSAR Problem on Organic Compounds. *Zh.Obshch.Khim.*, *65*, 285–286.

Kunz, M. (1986). Entropy and Information Indices of Star Forests. *Collect.Czech.Chem.Comm.*, *51*, 1856–1863.

Kunz, M. (1989). Path and Walk Matrices of Trees. *Collect.Czech.Chem.Comm.*, *54*, 2148–2155.

Kunz, M. (1990). Molecular Connectivity Indices Revisited. *Collect.Czech.Chem.Comm.*, *55*, 630–633.

Kunz, M. (1993). On Topological and Geometrical Distance Matrices. *J.Math.Chem.*, *13*, 145–151.

Kunz, M. (1994). Distance Matrices Yielding Angles Between Arcs of the Graphs. *J.Chem.Inf. Comput.Sci.*, *34*, 957–959.

Kunz, M. and Radl, Z. (1998). Distributions of Distances in Information Strings. *J.Chem.Inf. Comput.Sci.*, *38*, 374–378.

Kupchik, E.J. (1985). *Quant.Struct.-Act.Relat.*, *4*, 132.

Kupchik, E.J. (1986). *Quant.Struct.-Act.Relat.*, *5*, 95.

Kupchik, E.J. (1988). *Quant.Struct.-Act.Relat.*, *7*, 57.

Kupchik, E.J. (1989). General Treatment of Heteroatoms with the Randic Molecular Connectivity Index. *Quant.Struct.-Act.Relat.*, *8*, 98–103.

Kutter, E. and Hansch, C. (1969). Steric Parameters in Drug Design. Monoamine Oxidase Inhibitors and Antihistamines. *J.Med.Chem.*, *12*, 647–652.

Kutulya, L.A., Kuz'min, V.E., Stel'makh, I.B., Handrimailova, T.V. and Shtifanyuk, P.P. (1992). Quantitative Aspects of Chirality. III. Description of the Influence of the Structure of Chiral Compounds on Their Twisting Power in the Nematic Mesophase by Means of the Dissymmetry Function. *J.Phys.Org.Chem.*, *5*, 308–316.

Kuz'min, V.E., Stel'makh, I.B., Bekker, M.B. and Pozigun, D.V. (1992a). Quantitative Aspects of Chirality. I. Method of Dissymmetry Function. *J.Phys.Org.Chem.*, *5*, 295–298.

Kuz'min, V.E., Stel'makh, I.B., Yudanova, I.V., Pozigun, D.V. and Bekker, M.B. (1992b). Quantitative Aspects of Chirality. II. Analysis of Dissymmetry Function Behaviour with Different Changes in the Structure of the Model Systems. *J.Phys.Org.Chem.*, *5*, 299–307.

Kuz'min, V.E., Novikova, N.S., Sidelnikova, T.A. and Triguh, L.P. (1994). Topological Analysis of the Structure-Mesomorphus Property Relationship. *J.Struct.Chem.*, *35*, 471–477.

Kuz'min, V.E., Trigub, L.P., Shapiro, Y.E., Mazurov, A.A., Pozigun, V.V., Gorbatyuk, V.Y. and Andronati, S.A. (1995). Shape Parameters of Peptide Molecules as Descriptors for Solving QSAR Problems. *J.Struct.Chem.Eng.Transl.*, *36*, 465–473.

Kvasnicka, V. and Pospichal, J. (1990). Canonical Indexing and Constructive Enumeration of Molecular Graphs. *J.Chem.Inf.Comput.Sci.*, *30*, 99–105.

Kvasnicka, V., Sklenak, S. and Pospichal, J. (1993). Application of High Order Neural Networks in Chemistry. *Theor.Chim.Acta*, *86*, 257–267.

Kvasnicka, V. and Pospichal, J. (1995). Simple Construction of Embedding Frequencies of Trees and Rooted Trees. *J.Chem.Inf.Comput.Sci.*, *35*, 121–128.

Kyngas, J. and Valjakka, J. (1996). Evolutionary Neural Networks in Quantitative Structure-Activity Relationships of Dihydrofolate Reductase Inhibitors. *Quant.Struct.-Act.Relat.*, *15*, 296–301.

Labanowski, J.K., Motoc, I. and Dammkoehler, R.A. (1991). The Physical Meaning of Topological Indices. *Computers Chem.*, *15*, 47–53.

Laidboeur, T., Cabrol-Bass, D. and Ivanciuc, O. (1997). Determination of Topo-Geometrical Equivalence Classes of Atoms. *J.Chem.Inf.Comput.Sci.*, *37*, 87–91.

Lajiness, M.S. (1990). Molecular Similarity-Based Methods for Selecting Compounds for Screening. In *Computational Chemical Graph Theory* (Rouvray, D.E., ed.), Nova Science Publishers, New York (NY), pp. 299–316.

Lall, R.S. (1981a). Topology and Physical Properties of Acyclic Compounds. *Curr.Sci.*, *50*, 846–849.

Lall, R.S. (1981b). Topology and Physical Properties of n-Alkanes. *Curr.Sci.*, *50*, 668–670.

Lall, R.S. (1981c). Topology of Chemical Reactions. I. The Fragmentation of Hydrocarbons. II. Periocyclic Reactions. *MATCH (Comm.Math.Comp.Chem.)*, *12*, 87–107.

Lall, R.S. (1984). Topology of Oxy-Organic Compounds. *Curr.Sci.*, *53*, 642–643.

Lambert, F.L. (1966). Polarography of Organic Halogen Compounds. III. Quantitative Correlation of the Half-Wave Potentials of Alkyl Bromides with Taft Polar and Steric Constants. *J.Org.Chem.*, *31*, 4184–4188.

Lang, P.Z., Ma, X.F., Lu, G.H., Wang, Y. and Bian, Y. (1996). QSAR for the Acute Toxicity of Nitroaromatics to the Carp (Cyprinus carpio). *Chemosphere*, *32*, 1547–1552.

Langer, T. (1994). Molecular Similarity Determination of Heteroaromatics Using CoMFA and Multivariate Data Analysis. *Quant.Struct.-Act.Relat.*, *13*, 402–405.

Langer, T. and Hoffmann, R.D. (1998a). New Principal Components Derived Parameters Describing Molecular Diversity of Heteroaromatic Residues. *Quant.Struct.-Act.Relat.*, *17*, 211–223.

Langer, T. and Hoffmann, R.D. (1998b). On the Use of Chemical Function-Based Alignments as Input for 3D-QSAR. *J.Chem.Inf.Comput.Sci.*, *38*, 325–330.

Langlois, M., Bremont, B., Rousselle, D. and Gaudy, F. (1993a). Structural Analysis by the Comparative Molecular Field Analysis Method of the Affinity of Beta Adrenoceptor Blocking Agents for 5-Ht1A and 5-Ht1B Receptors. *Eur.J.Pharmacol.*, *244*, 77–87.

Langlois, M.H., Audry, E., Croizet, F., Dallet, Ph., Carpy, A. and Dubost, J.P. (1993b). Topological Lipophilicity Potential: A New Tool for a Fast Evaluation of Lipophilicity Distribution on a Molecular Graph. In *Trends in QSAR and Molecular Modelling 92* (Wermuth, C.G., ed.), ESCOM, Leiden (The Netherlands), pp. 354–355.

Lanteri, S. (1992). Full Validation Procedures for Feature Selection in Classification and Regression Problems. *Chemom.Intell.Lab.Syst.*, *15*, 159–169.

Lassau, C. and Jungers, J.-C. (1968). N/ 397. – L'Influence du Solvant sur la Réaction Chimique. La Quaternation des Amines Tertiaires par l'Iodure de Méthyle. *Bull.Soc.Chim.Fr.*, *7*, 2678–2685.

Laurence, C., Berthelot, M., Lucon, M., Helbert, M., Morris, D.G. and Gal, J.-F. (1984). The Influence of Solvent on the Inductive Order of Substituents from Infrared Measurements on 4-Substituted Camphors: a New Model of Inductive Effects. *J.Chem.Soc.Perkin Trans.2*, 705–710.

Lavenhar, S.R. and Maczka, C.A. (1985). Structure-Activity Considerations in Risk Assessment: a Simulation Study. *Toxicology & Industrial Health*, *1*, 249–259.

Lawson, R.G. and Jurs, P.C. (1990). New Index for Clustering Tendency and Its Application to Chemical Problems. *J.Chem.Inf.Comput.Sci.*, *30*, 36–41.

Le, T.D. and Weers, J.G. (1995). QSPR and GCA Models for Predicting the Normal Boiling Points of Fluorocarbons. *J.Phys.Chem.*, *99*, 6739–6747.

Leach, A.R. (1996). *Molecular Modelling. Principles and Applications.* Longman, Singapore, 596 pp.

Leahy, D.E. (1986). Intrinsic Molecular Volume as a Measure of the Cavity Term in Linear Solvation Energy Relationships: Octanol-Water Partition Coefficients and Aqueous Solubilities. *J.Pharm.Sci.*, *75*, 629–636.

Leahy, D.E., de Meere, A.L.J., Wait, A.R., Taylor, P.J., Tomenson, J.A. and Tomlinson, E. (1989). *Int.J.Pharm.*, *50*, 117–132.

Leahy, D.E., Morris, J.J., Taylor, P.J. and Wait, A.R. (1992a). Model Solvent Systems for QSAR. 3. An LSER Analysis of the Critical Quartet: New Light on Hydrogen Bond Strength and Directionality. *J.Chem.Soc.Perkin Trans.2*, 705–722.

Leahy, D.E., Morris, J.J., Taylor, P.J. and Wait, A.R. (1992b). Model Solvent Systems for QSAR. Part 2. Fragment Values ('*f*-Values') for the 'Critical Quartet'. *J.Chem.Soc.Perkin Trans.2*, 723–731.

Leahy, D.E., Morris, J.J., Taylor, P.J. and Wait, A.R. (1994). Model Solvent Systems for QSAR. 4. The Hydrogen Bond Acceptor Behavior of Heterocycles. *J.Phys.Org.Chem.*, *7*, 743–750.

Leardi, R., Boggia, R. and Terrile, M. (1992). Genetic Algorithms as a Strategy for Feature Selection. *J.Chemom.*, *6*, 267–281.

Leardi, R. (1994). Application of Genetic Algorithms to Feature Selection Under Full Validation Conditions and to Outlier Detection. *J.Chemom.*, *8*, 65–79.

Leardi, R. (1996). Genetic Algorithms in Feature Selection. In *Genetic Algorithms in Molecular Modeling. Principles of QSAR and Drug Design. Vol. 1* (Devillers, J., ed.), Academic Press, London (UK), pp. 67–86.

Leardi, R. and Lupiañez Gonzalez, A. (1998). Genetic Algorithms Applied to Feature Selection in PLS Regression: How and When to Use Them. *Chemom.Intell.Lab.Syst.*, *41*, 195–207.

Lee, B. and Richards, F.M. (1971). The Interpretation of Protein Structure: Estimation of Static Accessibility. *J.Mol.Biol.*, *55*, 379–400.

Lee, D.W., Kim, M.K., Kim, I.W., Park, J.H. and No, K.T. (1996). Studies on the Chromatographic Behaviors of Pd(ii)-alpha-isonitroso-beta-diketone Imine Chelates in Reversed-Phase Liquid Chromatography Using Molecular Descriptors. *Bull.Kor.Chem.Soc.*, *17*, 1158–1161.

Lee, K.W. and Kim, H.J. (1994). Quantitative Structure-Activity Relationship (QSAR) Study by Use of Theoretical Descriptors Quinolone and Naphthyridine. *Bull.Kor.Chem.Soc.*, *15*, 1070–1079.

Lee, K.W., Kwon, S.Y., Hwang, S., Lee, J.U. and Kim, H.J. (1996). Quantitative Structure-Activity Relationships (QSAR) Study on C-7 Substituted Quinolone. *Bull.Kor.Chem.Soc.*, *17*, 147–152.

Lee, S.K., Park, Y.H., Yoon, C.J. and Lee, D.W. (1998). Investigation of Relationships Between Retention Behavior and Molecular Descriptors of Quinolones in PRP-1 Column. *Journal of Microcolumn Separations*, *10*, 133–139.

Lee, S.L. and Yeh, Y.N. (1993). On Eigenvalues and Eigenvectors of Graphs. *J.Math.Chem.*, *12*, 121–135.

Leegwater, D.C. (1989). QSAR-Analysis of Acute Toxicity of Industrial Pollutants to the Guppy Using Molecular Connectivity Indices. *Aquat.Toxicol.*, *15*, 157–168.

Legendre, P. and Legendre, L. (1998). *Numerical Ecology.* Elsevier, Amsterdam (The Netherlands), 854 pp.

Lehtonen, P. (1987). Molecular Connectivity Indices in the Prediction of the Retention of Oxygen-Containing Amines in Reversed-Phase Liquid Chromatography. *J.Chromat.*, *398*, 143–151.

Leicester, S.E., Finney, J.L. and Bywater, R.P. (1988). Description of Molecular Surface Shape Using Fourier Descriptors. *J.Mol.Graphics*, *6*, 104–108.

Leicester, S.E., Finney, J.L. and Bywater, R.P. (1994a). A Quantitative Representation of Molecular Surface Shape. I: Theory and Development of the Method. *J.Math.Chem.*, *16*, 315–341.

Leicester, S.E., Finney, J.L. and Bywater, R.P. (1994b). A Quantitative Representation of Molecular Surface Shape. II: Protein Classification Using Fourier Shape Descriptors and Classical Scaling. *J.Math.Chem.*, *16*, 343–365.

Lekishvili, G. (1997). On the Characterization of Molecular Stereostructure: 1. Cis-Trans Isomerism. *J.Chem.Inf.Comput.Sci.*, *37*, 924–928.

Lennard-Jones, J.E. (1924). On the Determination of Molecular Fields. 11. The Equation of State of a Gas. *Proc.Roy.Soc.London A*, *106*, 463–477.

Lennard-Jones, J.E. (1929). The Electronic Structure of Some Diatomic Molecules. *Trans.Faraday Soc.*, *25*, 668–686.

Leo, A. and Hansch, C. (1971a). Linear Free-Energy Relationships Between Partitioning Solvent Systems. *J.Org.Chem.*, *36*, 1539–1544.

Leo, A., Hansch, C. and Elkins, D. (1971b). Partition Coefficients and their Uses. *Chem.Rev.*, *71*, 525–616.

Leo, A., Jow, P.Y.C., Silipo, C. and Hansch, C. (1975). Calculation of Hydrophobic Constant (Log *P*) from π and f Constants. *J.Med.Chem.*, *18*, 865–868.

Leo, A., Hansch, C. and Jow, P.Y.C. (1976). Dependence of Hydrophobicity of Apolar Molecules on their Molecular Volume. *J.Med.Chem.*, *19*, 611–615.

Leo, A. (1987). Some Advantages of Calculating Octanol-Water Partition Coefficients. *J.Pharm.Sci.*, *76*, 166–168.

Leo, A. (1990). Methods of Calculating Partition Coefficients. In *Quantitative Drug Design. Vol. 4* (Ramsden, C.A., ed.), Pergamon Press, Oxford (UK), pp. 295–319.

Leo, A. (1991). Hydrophobic Parameter: Measurement and Calculation. *Methods Enzymol.*, *202*, 544–591.

Leo, A. (1993). Calculating Log P_{oct} from Structures. *Chem.Rev.*, *93*, 1281–1306.

Lepovic, M. and Gutman, I. (1998). A Collective Property of Trees and Chemical Trees. *J.Chem.Inf.Comput.Sci.*, *38*, 823–826.

Lewis, D.F., Ioannides, C. and Parke, D.V. (1995a). A Quantitative Structure-Activity Relationship Study on a Series of 10 Parasubstituted Toluenes Binding to Cytochrome P4502B4 (Cyp2B4), and Their Hydroxylation Rates. *Biochem.Pharmacol.*, *50*, 619–625.

Lewis, D.F. and Parke, D.V. (1995b). The Genotoxicity of Benzanthracenes: A Quantitative Structure-Activity Study. *Mut.Res.*, *328*, 207–214.

Lewis, D.F.V. (1989). The Calculation of Molar Polarizabilities by the CNDO/2 Method: Correlation with the Hydrophobic Parameter, Log P. *J.Comput.Chem.*, *10*, 145–151.

Lewis, E.S. and Johnson, M.D. (1959). The Substituent Constants of the Diazonium Ion Group. *J.Am.Chem.Soc.*, *81*, 2070–2072.

Lewis, G.N. (1916). The Atom and the Molecule. *J.Am.Chem.Soc.*, *38*, 762–785.

Lewis, G.N. (1923). *Valence and the Structure of Atoms and Molecules.* New York (NY).

Lewis, R.A., Mason, J.S. and McLay, I.M. (1997). Similarity Measures for Rational Set Selection and Analysis of Combinatorial Libraries: The Diverse Property-Derived (DPD) Approach. *J.Chem.Inf.Comput.Sci.*, *37*, 599–614.

Li, L.-F. and You, X.-Z. (1993a). A Topological Index and Its Application. Part 3. Estimations of the Enthalpies of Formation of Mixed Halogen-Substituted Methanes, Silanes and Boron Mixed Halides. *Thermochim.Acta*, *225*, 85–96.

Li, L.-F. and You, X.-Z. (1993b). Molecular Topological Index and Its Application. 1. On the Chemical Shifts of ^{95}Mo NMR and ^{119}Sn Mossbauer Spectroscopy. *Chin.Sci.Bull.*, *38*, 421–425.

Li, L.-F., Zhang, Y. and You, X.-Z. (1995). Molecular Topological Index and Its Application. 4. Relationships with the Diamagnetic Susceptibilities of Alkyl-IVA Group Organometallic Halides. *J.Chem.Inf.Comput.Sci.*, *35*, 697–700.

Li, W.-Y., Guo, Z.-R. and Lien, E.J. (1984). Examination of the Interrelationship Between Aliphatic Group Dipole Moment and Polar Substituent Constants. *J.Pharm.Sci.*, *73*, 553–558.

Li, X., Yu, Q. and Zhu, L. (2000). A Novel Quantum-Topology Index. *J.Chem.Inf.Comput.Sci.*, *40*, 399–402.

Liang, C. and Gallagher, D.A. (1998). QSPR Prediction of Vapor Pressure from Solely Theoretically-Derived Descriptors. *J.Chem.Inf.Comput.Sci.*, *38*, 321–324.

Lias, S.G., Liebman, J.F. and Levin, R.D. (1984). *Journal of Physical Chemistry Reference Data*, *13*, 695.

Lien, E.J., Liao, R.C.H. and Shinouda, H.G. (1979). Quantitative Structure-Activity Relationships and Dipole Moments of Anticonvulsants and CNS Depressants. *J.Pharm.Sci.*, *68*, 463–465.

Lien, E.J., Guo, Z.-R., Li, R.-L. and Su, C.-T. (1982). Use of Dipole Moment as a Parameter in Drug-Receptor Interaction and Quantitative Structure-Activity Relationship Studies. *J.Pharm.Sci.*, *71*, 641–655.

Lien, E.J. and Gao, H. (1995). QSAR Analysis of Skin Permeability of Various Drugs in Man as Compared to in Vivo and in Vitro Studies in Rodents. *Pharm.Res.*, *12*, 583–587.

Liljefors, T. (1998). Progress in Force-Field Calculations of Molecular Interaction Fields and Intermolecular Interactions. In *3D QSAR in Drug Design – Vol. 2* (Kubinyi, H., Folkers, G. and Martin, Y.C., eds.), Kluwer/ESCOM, Dordrecht (The Netherlands), pp. 3–17.

Lin, S.-K. (1996). Correlation of Entropy with Similarity and Symmetry. *J.Chem.Inf.Comput.Sci.*, *36*, 367–376.

Lindgren, F. (1994). Third Generation PLS. Some Elements and Applications. *Ph.D.Thesis*, Umeå University, Umeå, Sweden.

Lindgren, F., Geladi, P., Rännar, S. and Wold, S. (1994). Interactive Variable Selection (IVS) for PLS. Part I: Theory and Algorithms. *J.Chemom.*, *8*, 349–363.

Lindgren, F., Geladi, P., Berglund, A., Sjöström, M. and Wold, S. (1995). Interactive Variable Selection (IVS) for PLS. Part II: Chemical Applications. *J.Chemom.*, *9*, 331–342.

Lindgren, F., Hansen, B., Karcher, W., Sjöström, M. and Eriksson, L. (1996). Model Validation by Permutation Tests: Applications to Variable Selection. *J.Chemom.*, *10*, 521–532.

Lindgren, F. and Rännar, S. (1998). Alternative Partial Least-Squares (PLS) Algorithms. In *3D QSAR in Drug Design – Vol. 3* (Kubinyi, H., Folkers, G. and Martin, Y.C., eds.), Kluwer/ ESCOM, Dordrecht (The Netherlands), pp. 105–113.

Linert, W., Kleestorfer, K., Renz, F. and Lukovits, I. (1995a). Description of Cyclic and Branched-Acyclic Hydrocarbons by Variants of the Hyper-Wiener Index. *J.Mol.Struct.(Theochem)*, *337*, 121–127.

Linert, W., Renz, F., Kleestorfer, K. and Lukovits, I. (1995b). An Algorithm for the Computation of the Hyper-Wiener Index for the Characterization and Discrimination of Branched Acyclic Molecules. *Computers Chem.*, *19*, 395–401.

Linert, W. and Lukovits, I. (1997). Formulas for the Hyper-Wiener and Hyper-Detour Indices of Fosed Bicyclic Structures. *MATCH (Comm.Math.Comp.Chem.)*, *35*.

Lipkowitz, K. B. and Boyd, D. eds. (1990–1999). *Reviews in Computational Chemistry.* Wiley-VCH, New York (NY), Volumes 1–13.

Lipkowitz, K. B. and Boyd, D. eds. (1997). *Reviews in Computational Chemistry. Vol. 11.* Wiley-VCH, New York (NY), 431 pp.

Lipkus, A.H. (1997). A Ring-Imbedding Index and Its Use in Substructure Searching. *J.Chem. Inf.Comput.Sci.*, *37*, 92–97.

Lipkus, A.H. (1999). Mining a Large Database for Peptidomimetic Ring Structures Using a Topological Index. *J.Chem.Inf.Comput.Sci.*, *39*, 582–586.

Lipnick, R.L. (1991). Outliers: Their Origin and Use in the Classification of Molecular Mechanisms of Toxicity. *Sci.Total Environ.*, *109/110*, 131–153.

Lister, D.G., Macdonald, J.N. and Owen, N.L. (1978). *Internal Rotation and Inversion.* Academic Press, London (UK).

Liu, D.X., Jiang, H.L., Chen, K.X. and Ji, R.Y. (1998). A New Approach to Design Virtual Combinatorial Library with Genetic Algorithm Based on 3D Grid Property. *J.Chem.Inf.Comput.Sci.*, *38*, 233–242.

Liu, J. and Qian, C. (1995). Hydrophobic Coefficients of s-Triazine and Phenylurea Herbicides. *Chemosphere*, *31*, 3951–3959.

Liu, L. and Guo, Q.-X. (1999). Wavelet Neural Network and Its Application to the Inclusion of β-Cyclodextrin with Benzene Derivatives. *J.Chem.Inf.Comput.Sci.*, *39*, 133–138.

Liu, Q., Hirono, S. and Moriguchi, I. (1992a). Application of Functional Link Net in QSAR. 1. QSAR for Activity Data Given by Continuous Variate. *Quant.Struct.-Act.Relat.*, *11*, 135–141.

Liu, Q., Hirono, S. and Moriguchi, I. (1992b). Application of Functional Link Net in QSAR. 2. QSAR for Activity Data Given by Ratings. *Quant.Struct.-Act.Relat.*, *11*, 318–324.

Liu, S., Zhang, R., Liu, M. and Hu, Z. (1997). Neural Network-Topological Indices Approach to the Prediction of Properties of Alkene. *J.Chem.Inf.Comput.Sci.*, *37*, 1146–1151.

Liu, S., Cao, C. and Li, Z. (1998). Approach to Estimation and Prediction for Normal Boiling Point (NBP) of Alkanes Based on a Novel Molecular Distance-Edge (MDE) Vector, λ. *J.Chem.Inf.Comput.Sci.*, *38*, 387–394.

Liu, S., Liu, H., Xia, Z., Chenzhong, C. and Li, Z. (1999). Molecular Distance-Edge Vector (μ): An Extension from Alkanes to Alcohols. *J.Chem.Inf.Comput.Sci.*, *39*, 951–957.

Liu, X. and Klein, D.J. (1991). The Graph Isomorphism Problem. *J.Comput.Chem.*, *12*, 1243–1251.

Livingstone, D.J., Hyde, R.M. and Foster, R. (1979). Further Study of an Organic Electron-Donor-Acceptor Related Substituent Constant. *Eur.J.Med.Chem.*, *14*, 393–397.

Livingstone, D.J., Evans, D.A. and Saunders, M.R. (1992). Investigation of a Charge Transfer Substituent Constant Using Computational Chemistry and Pattern Recognition Techniques. *J.Chem.Soc.Perkin Trans.2*, 1545–1550.

Livingstone, D.J. and Salt, D.W. (1992). Regression Analysis for QSAR Using Neural Networks. *Bioorg.Med.Chem.Lett.*, *2*, 213–218.

Liwo, A., Tarnowska, M., Grzonka, Z. and Tempczyk, A. (1992). Modified Free-Wilson Method for the Analysis of Biological Activity Data. *Computers Chem.*, *16*, 1–9.

Llorente, B., Rivero, N., Carrasco, R. and Martinez, R.S. (1994). A QSAR Study of Quinolones Based on Electrotopological State Index for Atoms. *Quant.Struct.-Act.Relat.*, *13*, 419–425.

Lloyd, D. (1996). What is Aromaticity? *J.Chem.Inf.Comput.Sci.*, *36*, 442–447.

Lobato, M., Amat, L., Besalù, E. and Carbó-Dorca, R. (1997). Structure-Activity Relationships of a Steroid Family Using Quantum Similarity Measures and Topological Quantum Similarity Indexes. *Quant.Struct.-Act.Relat.*, *16*, 465–472.

Loew, G.H. and Burt, S.K. (1990). Quantum Mechanics and the Modeling of Drug Properties. In *Quantitative Drug Design. Vol. 4* (Ramsden, C.A., ed.), Pergamon Press, Oxford (UK), pp. 105–123.

Lohninger, H. (1993). Evaluation of Neural Networks Based on Radial Basis Functions and Their Application to the Prediction of Boiling Points from Structural Parameters. *J.Chem.Inf. Comput.Sci.*, *33*, 736–744.

Lohninger, H. (1994). Estimation of Soil Partition Coefficients of Pesticides from their Chemical Structure. *Chemosphere*, *29*, 1611–1626.

Lombardo, F., Blake, J.F. and Curatolo, W.J. (1996). Computation of Brain-Blood Partitioning of Organic Solutes via Free Energy Calculations. *J.Med.Chem.*, *39*, 4750–4755.

Lorentz, H.A. (1880). *Wied.Ann.Phys.*, *9*, 641–665.

Lorentz, L.V. (1880). *Wied.Ann.Phys.*, *11*, 70–103.

Lovasz, L. and Pelikan, J. (1973). On the Eigenvalue of Trees. *Period.Math.Hung.*, *3*, 175–182.

Löw, P. and Saller, H. (1988). PETRA: Software Package for the Calculation of Electronic and Thermochemical Properties of Organic Molecules. In *Physical Property Prediction in Organic Chemistry* (Jochum, C., Hicks, M.G. and Sunkel, J., eds.), Springer-Verlag, Berlin (Germany), pp. 539–543.

Löwdin, P.-Q. (1970). On the Orthogonality Problem. *Adv.Quant.Chem.*, *5*, 185–199.

Lowe, J.P. (1978). *Quantum Chemistry.* Academic Press, New York (NY), 600 pp.

Lowis, D. R. (1997). HQSAR. A New, Highly Predictive QSAR Technique. Tripos Technical Note, *Internet Communication*, http://www.tripos.com

Lowrey, A.H., Cramer, C.J., Urban, J.J. and Famini, G.R. (1995a). Quantum Chemical Descriptors for Linear Solvation Energy Relationships. *Computers Chem.*, *19*, 209–215.

Lowrey, A.H. and Famini, G.R. (1995b). Using Theoretical Descriptors in Quantitative Structure-Activity Relationships HPLC Capacity Factors for Energetic Materials. *Struct.Chem.*, *6*, 357–365.

Lucic, B., Nikolic, S., Trinajstic, N. and Juretic, D. (1995a). The Structure-Property Models can be Improved Using the Orthogonalized Descriptors. *J.Chem.Inf.Comput.Sci.*, *35*, 532–538.

Lucic, B., Nikolic, S., Trinajstic, N., Juretic, D. and Juric, A. (1995b). A Novel QSPR Approach to Physicochemical Properties of the α-Amino Acids. *Croat.Chem.Acta*, *68*, 435–450.

Lucic, B., Nikolic, S., Trinajstic, N., Juric, A. and Mihalic, Z. (1995c). A Structure-Property Study of the Solubility of Aliphatic Alcohols in Water. *Croat.Chem.Acta*, *68*, 417–434.

Lucic, B. and Trinajstic, N. (1997). New Developments in QSPR/QSAR Modeling Based on Topological Indices. *SAR & QSAR Environ.Res.*, *7*, 45–62.

Lucic, B. and Trinajstic, N. (1999a). Multivariate Regression Outperforms Several Robust Architectures of Neural Networks in QSAR Modeling. *J.Chem.Inf.Comput.Sci.*, *39*, 121–132.

Lucic, B., Trinajstic, N., Sild, S., Karelson, M. and Katritzky, A.R. (1999b). A New Efficient Approach for Variable Selection Based on Multiregression: Prediction of Gas Chromatographic Retention Times and Response Factors. *J.Chem.Inf.Comput.Sci.*, *39*, 610–621.

Luco, J.M., Sosa, M.E., Cesco, J.C., Tonn, C.E. and Giordano, O.S. (1994). Molecular Connectivity and Hydrophobicity in the Study of Antifeedant Activity of Clerodane Diterpenoids. *Pestic.Sci.*, *41*, 1–6.

Luco, J.M., Yamin, L.J. and Ferretti, H.F. (1995). Molecular Topology and Quantum Chemical Descriptors in the Study of Reversed-Phase Liquid Chromatography. Hydrogen-Bonding Behavior of Chalcones and Flavanones. *J.Pharm.Sci.*, *84*, 903–908.

Luco, J.M. and Ferretti, H.F. (1997). QSAR Based on Multiple Linear Regression and PLS Methods for the Anti-HIV Activity of a Large Group of HEPT Derivatives. *J.Chem.Inf.Comput.Sci.*, *37*, 392–401.

Luco, J.M. (1999). Prediction of the Brain-Blood Distribution of a Large Set of Drugs from Structurally Derived Descriptors Using Partial Least-Squares (PLS) Modeling. *J.Chem.Inf.Comput.Sci.*, *39*, 396–404.

Luisi, P. (1977). Molecular Conformational Rigidity: An Approach to Quantification. *Naturwissenschaften*, *64*, 569–574.

Luke, B.T. (1994). Evolutionary Programming Applied to the Development of Quantitative Structure-Activity Relationships and Quantitative Structure-Property Relationships. *J.Chem.Inf.Comput.Sci.*, *34*, 1279–1287.

Lukovits, I. and Lopata, A. (1980). Decomposition of Pharmacological Activity Indices into Mutually Independent Components Using Principal Component Analysis. *J.Med.Chem.*, *23*, 449–459.

Lukovits, I. (1983). Quantitative Structure-Activity Relationships Employing Independent Quantum Chemical Indices. *J.Med.Chem.*, *26*, 1104–1109.

Lukovits, I. (1988). Decomposition of the Wiener Topological Index. Application to Drug-Receptor Interactions. *J.Chem.Soc.Perkin Trans.2*, 1667–1671.

Lukovits, I. (1990a). The Generalized Wiener Index for Molecules Containing Double Bonds and the Partition Coefficients. *Rep.Mol.Theory*, *1*, 127–131.

Lukovits, I. (1990b). Wiener Indices and Partition Coefficients of Unsaturated Hydrocarbons. *Quant.Struct.-Act.Relat.*, *9*, 227–231.

Lukovits, I. (1991). General Formulas for the Wiener Index. *J.Chem.Inf.Comput.Sci.*, *31*, 503–507.

Lukovits, I. (1992). Correlation Between Components of the Wiener Index and Partition Coefficients of Hydrocarbons. *Int.J.Quantum Chem.Quant.Biol.Symp.*, *19*, 217–223.

Lukovits, I. (1994). Formulas for the Hyper-Wiener Index of Trees. *J.Chem.Inf.Comput.Sci.*, *34*, 1079–1081.

Lukovits, I. and Gutman, I. (1994a). Edge-Decomposition of the Wiener Number. *MATCH (Comm.Math.Comp.Chem.)*, *31*, 133–144.

Lukovits, I. and Linert, W. (1994b). A Novel Definition of the Hyper-Wiener Index for Cycles. *J.Chem.Inf.Comput.Sci.*, *34*, 899–902.

Lukovits, I. (1995a). A Formula for the Hyper-Wiener Index. In *QSAR and Molecular Modelling: Cocepts, Computational Tools and Biological Applications* (Sanz, F., Giraldo, J. and Manaut, F., eds.), Prous Science, Barcelona (Spain), pp. 53–54.

Lukovits, I. (1995b). A Note on a Formula for the Hyper-Wiener Index of Some Trees. *Computers Chem.*, *19*, 27–31.

Lukovits, I. (1995c). An Algorithm for Computation of Bond Contributions of the Wiener Index. *Croat.Chem.Acta*, *68*, 99–103.

Lukovits, I. (1996a). Indicators for Atoms Included in Cycles. *J.Chem.Inf.Comput.Sci.*, *36*, 65–68.

Lukovits, I. (1996b). The Detour Index. *Croat.Chem.Acta*, *69*, 873–882.

Lukovits, I., Palfi, K., Bako, I. and Kalman, E. (1997a). LKP Model of the Inhibition Mechanism of Thiourea Compounds. *Corrosion*, *53*, 915–919.

Lukovits, I. and Razinger, M. (1997b). On Calculation of the Detour Index. *J.Chem.Inf.Comput.Sci.*, *37*, 283–286.

Lukovits, I. (1998a). An All-Path Version of the Wiener Index. *J.Chem.Inf.Comput.Sci.*, *38*, 125–129.

Lukovits, I. (1998b). Wiener Index: Formulas for Non-Homeomorphic Graphs. *Croat.Chem.Acta*, *71*, 449–458.

Lukovits, I. and Linert, W. (1998). Polarity-Numbers of Cycle-Containing Structures. *J.Chem.Inf.Comput.Sci.*, *38*, 715–719.

Lukovits, I. (1999). Isomer Generation: Syntactic Rules for Detection of Isomorphism. *J.Chem.Inf.Comput.Sci.*, *39*, 563–568.

Luo, Y.-R. and Benson, S.W. (1990). New Electronegativity Scale. 11. Comparison with Other Scales in Correlating Heats of Formation. *J.Phys.Chem.*, *94*, 914–917.

Luo, Z., Wang, R. and Lai, L. (1996). RASSE: A New Method for Structure-Based Drug Design. *J.Chem.Inf.Comput.Sci.*, *36*, 1187–1194.

Lyman, W.J., Reehl, W.F. and Rosenblatt, D.H. (1982). *Handbook of Chemical Property Estimation Methods.* McGraw-Hill, NewYork (NY).

Mabilia, M., Pearlstein, R.A. and Hopfinger, A.J. (1985). Molecular Shape Analysis and Energetics-Based Intermolecular Modelling of Benzylpyrimidine Dihydropholate Reductase Inhibitors. *Eur.J.Med.Chem.*, *20*, 163–174.

Maciel, G.E. and Natterstad, J.J. (1965). Study of ^{13}C Chemical Shifts in Substituted Benzenes. *J.Chem.Phys.*, *42*, 2427–2435.

Mackay, A.L. (1975). On Rearranging the Connectivity Matrix of a Graph – Comments. *J.Chem. Phys.*, *62*, 308–309.

MacPhee, J.A., Panaye, A. and Dubois, J.-E. (1978a). Operational Definition of the Taft Steric Parameter. An Homogeneous Scale for Alkyl groups – Experimental Extension to Highly Hindered Groups. *Tetrahedron Lett.*, *34*, 3293–3296.

MacPhee, J.A., Panaye, A. and Dubois, J.-E. (1978b). Steric Effects – I. A Critical Examination of the Taft Steric Parameter – E_S. Definition of a Revised, Broader and Homogeneous Scale. Extension to Highly Congested Alkyl Groups. *Tetrahedron*, *34*, 3553–3562.

Magee, P.S. (1990). A New Approach to Active-Site Binding Analysis. Inhibitors of Acetylcholinesterase. *Quant.Struct.-Act.Relat.*, *9*, 202–215.

Magee, P.S. (1991). Positional Analysis of Binding Events. In *QSAR: Rational Approaches to the Design of Bioactive Compounds* (Silipo, C. and Vittoria, A., eds.), Elsevier, Amsterdam (The Netherlands), pp. 549–552.

Magee, P.S. (1998). Some Novel Approaches to Modeling Transdermal Penetration and Reactivity with Epidermal Proteins. In *Comparative QSAR* (Devillers, J., ed.), Taylor & Francis, Washington (DC), pp. 137–168. [R]

Mager, P.P., Mager, H. and Barth, A. (1979). *Sci.Pharm.*, *47*, 265–297.

Mager, P.P. (1980). *Sci.Pharm.*, *48*, 117–126.

Mager, P.P. (1995a). A Rigorous QSAR Analysis. *J.Chemom.*, *9*, 232–236.

Mager, P.P. (1995b). Diagnostics Statistics in QSAR. *J.Chemom.*, *9*, 211–221.

Mager, P.P. (1996). A Random Number Experiment to Simulate Resample Model Evaluations. *J.Chemom.*, *10*, 221–240.

Maggiora, G.M. and Johnson, M.A. (1990). Introduction to Similarity in Chemistry. In *Concepts and Applications of Molecular Similarity* (Johnson, M.A. and Maggiora, G.M., eds.), Wiley, New York (NY), pp. 1–13.

Maggiora, G.M., Elrod, D.W. and Trenary, R.G. (1992). Computational Neural Networks as Model Free Mapping Devices. *J.Chem.Inf.Comput.Sci.*, *32*, 732–741.

Magnuson, V.R., Harriss, D.K. and Basak, S.C. (1983). Topological Indices Based on Neighborhood Symmetry: Chemical and Biological Applications. In *Studies in Physical and Theoretical Chemistry* (King, R.B., ed.), Elsevier, Amsterdam (The Netherlands), pp. 178–191.

Maier, B.J. (1992). Wiener and Randic Topological Indices for Graphs. *J.Chem.Inf.Comput.Sci.*, *32*, 87–90.

Makovskaya, V., Dean, J.R., Tomlinson, W.R. and Comber, M. (1995). Octanol-Water Partition Coefficients of Substituted Phenols and Their Correlation with Molecular Descriptors. *Anal. Chim.Acta*, *315*, 193–200.

Malone, J.G. (1933). The Electronic Moment as a Measure of the Ionic Nature of Covalent Bonds. *J.Chem.Phys.*, *1*, 197–199.

Manallack, D.T. and Livingstone, D.J. (1993). The Use of Neural Networks for Data Analysis in QSAR: Chance Effects. In *Trends in QSAR and Molecular Modelling 92* (Wermuth, C.G., ed.), ESCOM, Leiden (The Netherlands), pp. 128–131.

Manallack, D.T., Ellis, D.D. and Livingstone, D.J. (1994). Analysis of Linear and Nonlinear QSAR Data Using Neural Networks. *J.Med.Chem.*, *37*, 3758–3767.

Manallack, D.T. and Livingstone, D.J. (1994). Limitations of Functional Link Nets as Applied to QSAR Data Analysis. *Quant.Struct.-Act.Relat.*, *13*, 18–21.

Manallack, D.T. and Livingstone, D.J. (1995). Relating Biological Activity to Chemical Structure Using Neural Networks. *Pestic.Sci.*, *45*, 167–170.

Manaut, F., Sanz, F., José, J. and Milesi, M. (1991). Automatic Search for Maximum Similarity Between Molecular Electrostatic Potential Distributions. *J.Comput.Aid.Molec.Des.*, *5*, 371–380.

Mandelbrot, B.B. (1982). *The Fractal Geometry of Nature*. Freeman, San Francisco (CA).

Mandloi, M., Sikarwar, A., Sapre, N.S., Kamarkar, S. and Khadikar, P.V. (2000). A Comparative QSAR Study Using Wiener, Szeged, and Molecular Connectivity Indices. *J.Chem.Inf.Comput.Sci.*, *40*, 57–62.

Mann, G. (1967). Conformation and Physical Data of Alkanes and Cycloalkanes. *Tetrahedron*, *23*, 3375–3392.

Mannhold, R. and Dross, K. (1996). Calculation Procedures for Molecular Lipophilicity: A Comparative Study. *Quant.Struct.-Act.Relat.*, *15*, 403–409.

Mannhold, R., Cruciani, G., Dross, K. and Rekker, R.F. (1998a). Multivariate Analysis of Experimental and Computational Descriptors of Molecular Lipophilicity. *J.Comput.Aid.Molec.Des.*, *12*, 573–581.

Mannhold, R., Rekker, R.F., Dross, K., Bijloo, G.J. and de Vries, G. (1998b). The Lipophilic Behaviour of Organic Compounds: 1. An Updating of the Hydrophobic Fragmental Constant Approach. *Quant.Struct.-Act.Relat.*, *17*, 517–536. [R]

Maran, U., Karelson, M. and Katritzky, A.R. (1999). A Comprehensive QSAR Treatment of the Genotoxicity of Heteroaromatics and Aromatic Amines. *Quant.Struct.-Act.Relat.*, *18*, 3–10. [R]

Mardia, K.V., Kent, J.T. and Bibby, J.M. (1988). *Multivariate Analysis*. Academic Press, London (UK), 522 pp.

Marengo, E., Carpignano, R., Savarino, P. and Viscardi, G. (1992). Comparative Study of Different Structural Descriptors and Variable Selection Approaches Using Partial Least Squares in Quantitative Structure-Activity Relationships. *Chemom.Intell.Lab.Syst.*, *14*, 225–233.

Maria, P.-C., Gal, J.-F., de Franceschi, J. and Fargin, E. (1987). Chemometrics of the Solvent Basicity: Multivariate Analysis of the Basicity Scales Relevant to Nonprotogenic Solvents. *J.Am.Chem.Soc.*, *109*, 483–492.

Markovic, S. and Gutman, I. (1991). Dependence of Spectral Moments of Benzenoid Hydrocarbons on Molecular Structure. *J.Mol.Struct.(Theochem)*, *235*, 81–87.

Markovic, S., Gutman, I. and Bancevic, Z. (1995). Correlation Between Wiener and Quasi-Wiener Indices in Benzenoid Hydrocarbons. *J.Serb.Chem.Soc.*, *60*, 633–636.

Markovic, S. and Stajkovic, A. (1997). The Evaluation of Spectral Moments for Molecular Graphs of Phenylenes. *Theor.Chim.Acta*, *96*, 256–260.

Markovic, S. (1999). Tenth Spectral Moment for Molecular Graphs of Phenylenes. *J.Chem.Inf. Comput.Sci.*, *39*, 654–658.

Markovic, S. and Gutman, I. (1999). Spectral Moments of the Edge Adjacency Matrix in Molecular Graphs. Benzenoid Hydrocarbons. *J.Chem.Inf.Comput.Sci.*, *39*, 289–293.

Markowki, W., Dzido, T. and Wawrzynowicz, T. (1978). Correlation Between Chromatographic Parameters and Connectivity Index in Liquid-Solid Chromatography. *Pol.J.Chem.*, *52*, 2063.

Marriott, S. and Topsom, R.D. (1982). Theoretical Studies of the Inductive Effect. – 3. A Theoretical Scale of Field Parameters. *Tetrahedron Lett.*, *23*, 1485–1488.

Marriott, S., Reynolds, W.F., Taft, R.W. and Topsom, R.D. (1984). Substituent Electronegativity Parameters. *J.Org.Chem.*, *49*, 959–965.

Marshall, G.R., Barry, C.D., Bosshard, H.E., Dammkoehler, R.A. and Dunn, D.A. (1979). The Conformational Parameter in Drug Design: The Active Analog Approach. In *Computer-Assisted Drug Design* (Olson, E.C. and Christoffersen, R.E., eds.), American Chemical Society, Washington (DC), pp. 205–226.

Marshall, G.R. and Cramer III, R.D. (1988). Three-Dimensional Structure-Activity Relationships. *Trends Pharmacol.Sci*, *9*, 285–289.

Marshall, G.R. (1993). Binding-Site Modeling of Unknown Receptors. In *3D QSAR in Drug Design. Theory, Methods and Applications.* (Kubinyi, H., ed.), ESCOM, Leiden (The Netherlands), pp. 80–116.

Marsili, M. and Gasteiger, J. (1980). π Charge Distribution from Molecular Topology and π Orbital Electronegativity. *Croat.Chem.Acta*, *53*, 601–614.

Marsili, M. (1988). Computation of Volumes and Surface Areas of Organic Compounds. In *Physical Property Prediction in Organic Chemistry* (Jochum, C., Hicks, M.G. and Sunkel, J., eds.), Springer-Verlag, Berlin (Germany), pp. 249–254.

Marsili, M. and Saller, H. (1993). Analogs A Computer Program for the Design of Multivariate Sets of Analog Compounds. *J.Chem.Inf.Comput.Sci.*, *33*, 266–269.

Martens, H. and Naes, T. (1989). *Multivariate Calibration*. Wiley, Chichester (UK).

Martin, Y.C. and Lynn, K.R. (1971). Quantitative Structure-Activity Relationships in Leucomycin and Lincomycin Antibiotics. *J.Med.Chem.*, *14*, 1162–1166.

Martin, Y.C. (1978). *Quantitative Drug Design. A Critical Introduction*. Marcel Dekker, New York (NY), 425 pp.

Martin, Y.C. (1979). Advances in the Methodology of Quantitative Drug Design. In *Drug Design, Vol. VIII* (Ariëns, E.J., ed.), Academic Press, New York (NY), pp. 1–72.

Martin, Y.C., Danaher, E.B., May, C.S. and Weininger, D. (1988). MENTHOR, a Database System for the Storage and Retrieval of Three-Dimensional Molecular Structures and Associated Data Searchable by Substructural, Biologic, Physical, or Geometric Properties. *J.Comput.Aid. Molec.Des.*, *2*, 15–29.

Martin, Y.C., Lin, C.T. and Wu, J. (1993). Application of CoMFA to D1 Dopaminergic Agonists: A Case Study. In *3D QSAR in Drug Design. Theory, Methods and Applications.* (Kubinyi, H., ed.), ESCOM, Leiden (The Netherlands), pp. 643–659.

Martin, Y.C., Lin, C.T., Hetti, C. and DeLazzer, J. (1995). PLS Analysis of Distance Matrices to Detect Nonlinear Relationships Between Biological Potency and Molecular Properties. *J.Med.Chem.*, *38*, 3009–3015.

Martin, Y.C. (1998). 3D QSAR: Current State Scope, and Limitations. In *3D QSAR in Drug Design – Vol. 3* (Kubinyi, H., Folkers, G. and Martin, Y.C., eds.), Kluwer/ESCOM, Dordrecht (The Netherlands), pp. 3–23.

Mason, J.S. and Pickett, S.D. (1997). Partition-Based Selection. *Persp.Drug Disc.Des.*, *7/8*, 85–114.

Massart, D.L. and Kaufman, L. (1983). *The Interpretation of Analytical Chemical Data by the Use of Cluster Analysis*. Wiley, NewYork (NY).

Massart, D.L., Vandeginste, B.G.M., Buydens, L., De Jong, S., Lewi, P.J. and Smeyers-Verbeke, J. (1997). *Handbook of Chemometrics and Qualimetrics. Part A*. Elsevier, Amsterdam (The Netherlands), 868 pp.

Massart, D.L., Vandeginste, B.G.M., Buydens, L., De Jong, S., Lewi, P.J. and Smeyers-Verbeke, J. (1998). *Handbook of Chemometrics and Qualimetrics. Part B*. Elsevier, Amsterdam (The Netherlands), 714 pp.

Mastryukova, T.A. and Kabachnik, M.I. (1971). Correlation Constants in the Chemistry of Organophosphorus Compounds. *J.Org.Chem.*, *36*, 1201–1205.

Masuda, T., Nakamura, K., Jikihara, T., Kasuya, F., Igarashi, K., Fukui, M., Takagi, T. and Fujiwara, H. (1996). 3D Quantitative Structure-Activity Relationships for Hydrophobic Interactions. Comparative Molecular Field Analysis (CoMFA) Including Molecular Lipophilicity Potentials as Applied to the Glycine Conjugation of Aromatic as well as Aliphatic Carboxylic Acids. *Quant.Struct.-Act.Relat.*, *15*, 194–200.

Matter, H. (1997). Selecting Optimally Diverse Compounds from Structure Databases. A Validation Study of Two-Dimensional and Three-Dimensional Molecular Descriptors. *J.Med.Chem.*, *40*, 1219–1229.

Matter, H. and Pötter, T. (1999). Comparing 3D Pharmacophore Triplets and 2D Fingerprints for Selecting Diverse Compound Subsets. *J.Chem.Inf.Comput.Sci.*, *39*, 1211–1225.

Mayer, D., Naylor, C.B., Motoc, I. and Marshall, G.R. (1987). A Unique Geometry of the Active Site of Angiotensin-Converting Enzyme Consistent with Structure-Activity Studies. *J.Comput. Aid.Molec.Des.*, *1*, 3–16.

Mayer, J.M., van de Waterbeemd, H. and Testa, B. (1982). A Comparison Between the Hydrophobic Fragmental Methods of Rekker and Leo. *Eur.J.Med.Chem.*, *17*, 17.

Mayer, V., Gutman, I. and Gerger, W. (1975). *Monatsh.Chem.*, *106*, 1235.

McCabe, G.P. (1975). Computations for Variable Selection in Discriminant Analysis. *Technometrics*, *17*, 103–109.

McCabe, G.P. (1984). Principal Variables. *Technometrics*, *26*, 137–144.
McClellan, A.L. (1963). *Tables of Experimental Dipole Moments*. Freeman, San Francisco (CA).
McClelland, B.J. (1971). Properties of the Latent Roots of a Matrix: The Estimation of π-Electron Energies. *J.Chem.Phys.*, *54*, 640–643.
McClelland, B.J. (1974). Graphical Method for Factorizing Secular Determinants of Hückel Molecular Orbital Theory. *J.Chem.Soc.Faraday Trans II*, *70*, 1453–1456.
McClelland, B.J. (1982). Eigenvalues of the Topological Matrix. *J.Chem.Soc.Faraday Trans II*, *78*, 911–916.
McDaniel, D.H. and Brown, H.C. (1955). A Quantitative Approach to the Ortho Effects of Halogen Substituents in Aromatic Systems. *J.Am.Chem.Soc.*, *77*, 3756–3763.
McDaniel, D.H. and Brown, H.C. (1958). An Extended Table of Hammett Substituent Constants Based on the Ionization of Substituted Benzoic Acids. *J.Org.Chem.*, *23*, 420–427.
McDaniel, D.H. and Yingst, A. (1964). The Use of Basicity and Oxidative Coupling Potential to Obtain Group Electronegativity. *J.Am.Chem.Soc.*, *86*, 1334–1336.
McFarland, J.W. (1970). On the Parabolic Relationship Between Drug Potency and Hydrophobicity. *J.Med.Chem.*, *13*, 1192–1196.
McFarland, J.W. and Gans, D.J. (1986). On the Significance of Clusters in the Graphical Display of Structure-Activity Data. *J.Med.Chem.*, *29*, 505–514.
McFarland, J.W. and Gans, D.J. (1990a). Cluster Significance Analysis: A New QSAR Tool for Asymmeric Data Sets. *Drug Inf.J.*, *24*, 705–711.
McFarland, J.W. and Gans, D.J. (1990b). Linear Discriminant Analysis and Cluster Significance Analysis. In *Quantitative Drug Design. Vol. 4* (Ramsden, C.A., ed.), Pergamon Press, Oxford (UK), pp. 667–689.
McFarland, J.W. and Gans, D.J. (1994). On Identifying Likely Determinants of Biological Activity in High Dimensional QSAR Problems. *Quant.Struct.-Act.Relat.*, *13*, 11–17.
McFarland, J.W. and Gans, D.J. (1995). Multivariate Data Analysis of Chemical and Biological Data. Cluster Significance Analysis. In *Chemometrics Methods in Molecular Design – Vol. 2* (van de Waterbeemd, H., ed.), VCH Publishers, New York (NY), pp. 295–308.
McGregor, M.J. and Pallai, P.V. (1997). Clustering of Large Databases of Compounds: Using the MDL "Keys" as Structural Descriptors. *J.Chem.Inf.Comput.Sci.*, *37*, 443–448.
McGregor, M.J. and Muskal, S.M. (1999). Pharmacophore Fingerprint. 1. Application to QSAR and Focused Library Design. *J.Chem.Inf.Comput.Sci.*, *39*, 569–574.
McGregor, T.R. (1979). Connectivity Parameters as Predictors of Retention in Gas Chromatography. *J.Chromatogr.Sci.*, *17*, 314.
McHughes, M.C. and Poshusta, R. (1990). Graph-Theoretic Cluster Expansions. Thermochemical Properties of Alkanes. *J.Math.Chem.*, *4*, 227–249.
McKay, B.D. (1977). On the Spectral Characterization of Trees. *Ars Combinatoria*, *3*, 219–232.
McKone, T.E. (1993). The Precision of QSAR Methods for Estimating Intermedia Transfer Factors in Exposure Assessments. *SAR & QSAR Environ.Res.*, *1*, 41–51.
Medeleanu, M. and Balaban, A.T. (1998). Real-Number Vertex Invariants and Schultz-type Indices Based on Eigenvectors of Adjacency and Distance Matrices. *J.Chem.Inf.Comput.Sci.*, *38*, 1038–1047.
Medven, Z., Güsten, H. and Sabljic, A. (1996). Comparative QSAR Study on Hydroxyl Radical Reactivity with Unsaturated Hydrocarbons: PLS versus MLR. *J.Chemom.*, *10*, 135–147.
Mekenyan, O., Bonchev, D. and Trinajstic, N. (1980). Chemical Graph Theory: Modeling the Thermodynamic Properties of Molecules. *Int.J.Quant.Chem.*, *28*, 369–380.
Mekenyan, O., Bonchev, D. and Trinajstic, N. (1981). Algebraic Characterization of Bridged Polycyclic Compounds. *Int.J.Quant.Chem.*, *19*, 929–955.
Mekenyan, O., Bonchev, D. and Trinajstic, N. (1983). Structural Complexity and Molecular Properties of Cyclic Systems with Acyclic Branches. *Croat.Chem.Acta*, *56*, 237–261.
Mekenyan, O., Bonchev, D. and Balaban, A.T. (1984a). Hierarchically Ordered Extended Connectivities. Reflection in the ^{1}H NMR Chemical Shifts of Condensed Benzenoid Hydrocarbons. *Chem.Phys.Lett.*, *109*, 85–88.

Mekenyan, O., Bonchev, D. and Balaban, A.T. (1984b). Unique Description of Chemical Structures Based on Hierarchically Ordered Extended Connectivities (HOC Procedures). V. New Topological Indices, Ordering of Graphs, and Recognition of Graph Similarity. *J.Comput.Chem.*, *5*, 629–639.

Mekenyan, O., Balaban, A.T. and Bonchev, D. (1985a). Unique Description of Chemical Structures Based on Hierarchically Ordered Extended Connectivities (HOC Procedures). VII. Condensed Benzenoid Hydrocarbons and their ^{1}H NMR Chemical Shifts. *J.Magn.Reson.*, *63*, 1–13.

Mekenyan, O., Bonchev, D. and Balaban, A.T. (1985b). Unique Description of Chemical Structures Based on Hierarchically Ordered Extended Connectivities (HOC Procedures). II. Mathematical Proofs for the HOC Algorithm. *J.Comput.Chem.*, *6*, 552–561.

Mekenyan, O. and Bonchev, D. (1986). OASIS Method for Predicting Biological Activity of Chemical Compounds. *Acta Pharm.Jugosl.*, *36*, 225–237.

Mekenyan, O., Bonchev, D., Trinajstic, N. and Peitchev, D. (1986). Modelling the Interaction of Small Organic Molecules with Biomacromolecules. II. A Generalized Concept for Biological Interactions. *Arzneim.Forsch.*, *36*, 421–424.

Mekenyan, O., Peitchev, D., Bonchev, D., Trinajstic, N. and Bangov, I.P. (1986a). Modelling the Interaction of Small Organic Molecules with Biomacromolecules. I. Interaction of Substituted Pyridines wiyh anti-3-azopyridine Antibody. *Arzneim.Forsch.*, *36*, 176–183.

Mekenyan, O., Peitchev, D., Bonchev, D., Trinajstic, N. and Dimitrova, J. (1986b). Modelling the Interaction of Small Organic Molecules with Biomacromolecules. III. Interaction of Benzoates with anti-p-(p′-axophenylazo)-benzoate Antibody. *Arzneim.Forsch.*, *36*, 629–635.

Mekenyan, O., Bonchev, D., Sabljic, A. and Trinajstic, N. (1987). Applications of Topological Indices to QSAR. The Use of the Balaban Index and the Electropy Index for Correlations with Toxicity of Ethers on Mice. *Acta Pharm.Jugosl.*, *37*, 75–86.

Mekenyan, O., Bonchev, D. and Balaban, A.T. (1988a). Topological Indices for Molecular Fragments and New Graph Invariants. *J.Math.Chem.*, *2*, 347–375.

Mekenyan, O., Bonchev, D. and Entchev, V. (1988b). *Quant.Struct.-Act.Relat.*, *7*, 240–244.

Mekenyan, O., Karabunarliev, S. and Bonchev, D. (1990a). The Microcomputer OASIS System for Predicting the Biological Activity of Chemical Compounds. *Computers Chem.*, *14*, 193–200.

Mekenyan, O., Karabunarliev, S. and Bonchev, D. (1990b). The OASIS Concept for Predicting Biological Activity of Chemical Compounds. *J.Math.Chem.*, *4*, 207–215.

Mekenyan, O., Mercier, C., Bonchev, D. and Dubois, J.-E. (1993a). Comparative Study of DARC/PELCO and OASIS Methods. 2. Modeling PNMT Inhibitory Potency of Benzylamines and Amphetamines. *Eur.J.Med.Chem.*, *28*, 811–819.

Mekenyan, O. and Veith, G.D. (1993). Relationships Between Descriptors for Hydrophobicity and Soft Electrophilicity in Predicting Toxicity. *SAR & QSAR Environ.Res.*, *1*, 335–344.

Mekenyan, O., Veith, G.D., Bradbury, S.P. and Russom, C.L. (1993b). Structure-Toxicity Relationships for α, β-Unsaturated Alcohols in Fish. *Quant.Struct.-Act.Relat.*, *12*, 132–136.

Mekenyan, O. and Basak, S.C. (1994). Topological Indices and Chemical Reactivity. In *Graph Theoretic Approaches to Chemical Reactivity* (Bonchev, D. and Mekenyan, O., eds.), T.H.E. Publisher, (The Netherlands).

Mekenyan, O., Ivanov, J., Veith, G.D. and Bradbury, S.P. (1994). Dynamic QSAR: A New Search for Active Conformations and Significant Stereoelectronic Indices. *Quant.Struct.-Act.Relat.*, *13*, 302–307.

Mekenyan, O. and Veith, G.D. (1994). The Electronic Factor in QSAR: MO-Parameters, Competing Interactions, Reactivity and Toxicity. *SAR & QSAR Environ.Res.*, *2*, 129–143.

Mekenyan, O., Veith, G.D., Call, D.J. and Ankley, G.T. (1996). A QSAR Evaluation of Ah Receptor Binding of Halogenated Aromatic Xenobiotics. *Environ.Health Persp.*, *104*, 1302–1310.

Mekenyan, O. and Veith, G.D. (1997). 3D Molecular Design: Searching for Active Conformers in QSAR. In *From Chemical Topology to Three-Dimensional Geometry* (Balaban, A.T., ed.), Plenum Press, New York (NY), pp. 43–71.

Melkova, Z. (1984). Utilization of the Index of Molecular Connectivity in the Study of Antitumor Activity of a Group of Benzo(c)fluorene Derivatives. *Cesk.Farm.*, *33*, 107.

Meloun, M., Militky, J. and Forina, M. (1994). *Chemometrics for Analytical Chemistry.* Ellis Horwood, Bodmin (UK), 400 pp.

Menard, P.R., Lewis, R.A. and Mason, J.S. (1998a). Rational Screening Set Design and Compound Selection: Cascaded Clustering. *J.Chem.Inf.Comput.Sci.*, *38*, 497–505.

Menard, P.R., Mason, J.S., Morize, I. and Bauerschmidt, S. (1998b). Chemistry Space metrics in Diversity Analysis, Library Design, and Compound Selection. *J.Chem.Inf.Comput.Sci.*, *38*, 1204–1213.

Mendiratta, S. and Madan, A.K. (1994). Structure-Activity Study on Antiviral 5-Vinylpyrimidine Nucleoside Analogs Using Wiener's Topological Index. *J.Chem.Inf.Comput.Sci.*, *34*, 867–871.

Menziani, M.C. and De Benedetti, P.G. (1992). Molecular Mechanics and Quantum Chemical QSAR Analysis in Carbonic Anhydrase Heterocyclic Sulfonamide Interactions. *Struct.Chem.*, *3*, 215–219.

Mercier, C. and Dubois, J.-E. (1979). Comparison of Molecular Connectivity and DARC/PELCO Methods: Performance in Antimicrobial, Halogenated Phenol QSARs. *Eur.J.Med.Chem.*, *14*, 415–423.

Mercier, C., Troullier, G. and Dubois, J.-E. (1990). DARC Computer Aided Design in Anticholinergic Research. *Quant.Struct.-Act.Relat.*, *9*, 88–93.

Mercier, C., Mekenyan, O., Dubois, J.-E. and Bonchev, D. (1991). DARC/PELCO and OASIS Methods I. Methodological Comparison. Modeling Purine pKa and Antitumor Activity. *Eur. J.Med.Chem.*, *26*, 575–592.

Merrifield, R.E. and Simmons, H.E. (1980). The Structures of Molecular Topological Spaces. *Theor.Chim.Acta*, *55*, 55–75.

Merrifield, R.E. and Simmons, H.E. (1998). *Topological Methods in Chemistry.* Wiley, New York (NY), 233 pp.

Merschsundermann, V., Rosenkranz, H.S. and Klopman, G. (1994). The Structural Basis of the Genotoxicity of Nitroarenofurans and Related Compounds. *Mut.Res.*, *304*, 271–284.

Mestres, J. and Scuseria, G.E. (1995). Genetic Algorithms: A Robust Scheme for Geometry Optimizations and Global Minimum Structure Problems. *J.Comput.Chem.*, *16*, 729–742.

Meurice, N., Leherte, L. and Vercauteren, D.P. (1998). Comparison of Benzodiazepine-Like Compounds Using Topological Analysis and Genetic Algorithms. *SAR & QSAR Environ.Res.*, *8*, 195–232.

Meyer, A.M. and Richards, W.G. (1991). Similarity of Molecular Shape. *J.Comput.Aid. Molec.Des.*, *5*, 426–439.

Meyer, A.Y. (1985a). Molecular Mechanics and Molecular Shape. Part 1. van der Waals Descriptors of Simple Molecules. *J.Chem.Soc.Perkin Trans.2*, 1161–1169.

Meyer, A.Y. (1985b). Molecular Mechanics and Molecular Shape. Part II. Beyond the van der Waals Descriptors of Shape. *J.Mol.Struct.(Theochem)*, *124*, 93–106.

Meyer, A.Y. (1986a). Molecular Mechanics and Molecular Shape. III. Surface Area and Cross-Sectional Areas of Organic Molecules. *J.Comput.Chem.*, *7*, 144–152.

Meyer, A.Y. (1986b). Molecular Mechanics and Molecular Shape. Part 4. Size, Shape, and Steric Parameters. *J.Chem.Soc.Perkin Trans.2*, 1567–1572.

Meyer, A.Y. (1986c). The Size of Molecules. *Chem.Soc.Rev.*, *15*, 449–474.

Meyer, A.Y. (1988a). Molecular Mechanics and Molecular Shape. Part VI. The Response of Simple Molecules to Bimolecular Association. *J.Mol.Struct.(Theochem)*, *179*, 83–98.

Meyer, A.Y. (1988b). Molecular Mechanics and Molecular Shape. V. On the Computation of the Bare Surface Area of Molecules. *J.Comput.Chem.*, *9*, 18–24.

Meyer, A.Y. (1989). Molecular Mechanics and Molecular Shape. Part VII. Structural Factors in the Estimation of Solvation Energies. *J.Mol.Struct.(Theochem)*, *195*, 147–158.

Meyer, H. (1899). *Arch.Exper.Pathol.Pharmakol.*, *42*, 109–118.

Meylan, W.M., Howard, P.H. and Boethling, R.S. (1992). Molecular Topology/Fragment Contribution Method for Predicting Soil Sorption Coefficients. *Environ.Sci.Technol.*, *26*, 1560–1567.

Meylan, W.M. and Howard, P.H. (1995). Atom/Fragment Contribution Method for Estimating Octanol-Water Partition Coefficients. *J.Pharm.Sci.*, *84*, 83–92.

Mezey, P.G. (1985). Group Theory of Electrostatic Potentials: A Tool for Quantum Chemical Drug Design. *Int.J.Quantum Chem.Quant.Biol.Symp.*, *12*, 113–122.

Mezey, P.G. (1987a). Group Theory of Shapes of Asymmetric Biomolecules. *Int.J.Quantum Chem.Quant.Biol.Symp.*, *14*, 127–132.

Mezey, P.G. (1987b). The Shape of Molecular Charge Distributions: Group Theory without Symmetry. *J.Comput.Chem.*, *8*, 462–469.

Mezey, P.G. (1988a). Global and Local Relative Convexity and Oriented Relative Convexity: Application to Molecular Shapes in External Fields. *J.Math.Chem.*, *2*, 325.

Mezey, P.G. (1988b). Graphical Shapes: Seeing Graphs of Chemical Curves and Molecular Surfaces. *J.Math.Chem.*, *2*, 377.

Mezey, P.G. (1988c). Shape Group Studies of Molecular Similarity: Shape Groups and Shape Graphs of Molecular Contour Surfaces. *J.Math.Chem.*, *2*, 299.

Mezey, P.G. (1989). The Topology of Molecular Surfaces and Shape Graphs. In *Computational Chemical Graph Theory and Combinatorics* (Rouvray, D.H., ed.), Nova Publications, New York (NY).

Mezey, P.G. (1990a). A Global Approach to Molecular Symmetry: Theorems on Symmetry Relations Between Ground- and Excited-State Configurations. *J.Am.Chem.Soc.*, *112*, 3791–3802.

Mezey, P.G. (1990b). Molecular Point Symmetry and the Phase of the Electronic Wave Function: Tools for the Prediction of Critical Points of Potential Energy Surfaces. *Int.J.Quant.Chem.*, *38*, 699–711.

Mezey, P.G. (1990c). Three-Dimensional Topological Aspects of Molecular Similarity. In *Concepts and Applications of Molecular Similarity* (Johnson, M.A. and Maggiora, G.M., eds.), Wiley, New York (NY), pp. 321–368.

Mezey, P. G. ed. (1991a). *Mathematical Modeling in Chemistry.* VCH, Weinheim (Germany).

Mezey, P.G. (1991b). Molecular Surfaces. In *Reviews in Computational Chemistry – Vol. 11* (Lipkowitz, K.B. and Boyd, D., eds.), Wiley – VCH, New York (NY), pp. 265–294.

Mezey, P.G. (1991c). The Degree of Similarity of Three-Dimensional Bodies: Application to Molecular Shape Analysis. *J.Math.Chem.*, *7*, 39–49.

Mezey, P.G. (1992). Shape-Similarity Measures for Molecular Bodies: A Three-Dimensional Topological Approach to Quantitative Shape-Activity Relations. *J.Chem.Inf.Comput.Sci.*, *32*, 650–656.

Mezey, P.G. (1993a). Dynamic Shape Analysis of Molecules in Restricted Domains of a Configuration Space. *J.Math.Chem.*, *13*, 59–70.

Mezey, P.G. (1993b). New Rules on Potential Surface Topology and Critical Point Search. *J.Math.Chem.*, *14*, 79–90.

Mezey, P.G. (1993c). *Shape in Chemistry: An Introduction to Molecular Shape and Topology.* VCH Publishers, New York (NY).

Mezey, P.G. (1993d). Topological Shape Analysis of Chain Molecules: An Application of the GSTE Principle. *J.Math.Chem.*, *12*, 365–374.

Mezey, P.G. (1994). Iterated Similarity Sequences and Shape ID Numbers for Molecules. *J.Chem.Inf.Comput.Sci.*, *34*, 244–247.

Mezey, P.G. (1996). Theorems on Molecular Shape-Similarity Descriptors: External T-Plasters and Interior T-Aggregates. *J.Chem.Inf.Comput.Sci.*, *36*, 1076–1081.

Mezey, P.G. (1997a). Descriptors of Molecular Shape 3D. In *From Chemical Topology to Three-Dimensional Geometry* (Balaban, A.T., ed.), Plenum Press, New York (NY), pp. 25–42.

Mezey, P.G. (1997b). Fuzzy Measures of Molecular Shape and Size. In *Fuzzy Logic in Chemistry* (Rouvray, D.H., ed.), Academic Press, New York (NY), pp. 139–223.

Mezey, P.G. (1999). Holographic Electron Density Shape Theorem and its Role in Drug Design and Toxicological Risk Assessment. *J.Chem.Inf.Comput.Sci.*, *39*, 224–230.

Michotte, Y. and Massart, D.L. (1977). Molecular Connectivity and Retention Indexes. *J.Pharm.Sci.*, *66*, 1630–1632.

Miertus, S., Scrocco, E. and Tomasi, J. (1981). Electrostatic Interaction of a Solute with a Continuum. A Direct Utilization of Ab Initio Molecular Potentials for the Prevision of Solvent Effects. *Chem.Phys.*, *55*, 117–129.

Migliavacca, E., Carrupt, P.-A. and Testa, B. (1997). 116. Theoretical Parameters to Characterize Antioxidants. Part 1. The Case of Vitamin E and Analogs. *Helv.Chim.Acta*, *80*, 1613–1626.

Migliavacca, E., Ancerewicz, J., Carrupt, P.-A. and Testa, B. (1998). Theoretical Parameters to Characterize Antioxidants. Part 2. The Case of Melatonin and Carvedilol. *Helv.Chim.Acta*, *81*, 1337–1348.

Mihalic, Z. and Trinajstic, N. (1991). The Algebraic Modelling of Chemical Structures: On the Development of Three-Dimensional Molecular Descriptors. *J.Mol.Struct.(Theochem)*, *232*, 65–78.

Mihalic, Z., Nikolic, S. and Trinajstic, N. (1992a). Comparative Study of Molecular Descriptors Derived from the Distance Matrix. *J.Chem.Inf.Comput.Sci.*, *32*, 28–37.

Mihalic, Z. and Trinajstic, N. (1992). A Graph-Theoretical Approach to Structure-Property Relationships. *J.Chem.Educ.*, *69*, 701–712.

Mihalic, Z., Veljan, D., Amic, D., Nikolic, S., Plavsic, D. and Trinajstic, N. (1992b). The Distance Matrix in Chemistry. *J.Math.Chem.*, *11*, 223–258.

Miller, A.J. (1990). *Subset Selection in Regression*. Chapman & Hall, London (UK), 230 pp.

Miller, K.J. and Savchik, J.A. (1979). A New Empirical Method to Calculate Average Molecular Polarizabilities. *J.Am.Chem.Soc.*, *101*, 7206–7213.

Miller, K.J. (1990a). Additivity Methods in Molecular Polarizability. *J.Am.Chem.Soc.*, *112*, 8533–8542.

Miller, K.J. (1990b). Calculation of the Molecular Polarizability Tensor. *J.Am.Chem.Soc.*, *112*, 8543–8551.

Millership, J.S. and Woolfson, A.D. (1978). The Relation Between Molecular Connectivity and Gas Chromatographic Retention Data. *J.Pharm.Pharmacol.*, *30*, 483–485.

Millership, J.S. and Woolfson, A.D. (1979). A Study of the Relationship Between Gas Chromatographic Retention Parameters and Molecular Connectivity. *J.Pharm.Pharmacol.*, *31*, 44P-44P.

Millership, J.S. and Woolfson, A.D. (1980). Molecular Connectivity and Gas Chromatographic Retention Parameters. *J.Pharm.Pharmacol.*, *32*, 610–614.

Milne, G.W. A. (1997). Mathematics as a Basis for Chemistry. *J.Chem.Inf.Comput.Sci.*, *37*, 639–644.

Minailiuc, O.M., Katona, G., Diudea, M.V., Strunje, M., Graovac, A. and Gutman, I. (1998). Szeged Fragmental Indices. *Croat.Chem.Acta*, *71*, 473–488.

Minoli, D. (1976). Teorie Combinatorie. – Combinatorial Graph Complexity. *Atti dell'Accademia Nazionale dei Lincei – Rendiconti*, *59*, 651–661.

Mishra, R.K. and Patra, S.M. (1998). Numerical Determination of the Kekulé Structure Count of Some Symmetrical Polycyclic Aromatic Hydrocarbons and their Relationship with π-Electronic Energy (A Computational Approach). *J.Chem.Inf.Comput.Sci.*, *38*, 113–124.

Mislow, K. (1997). Fuzzy Restrictions and Inherent Uncertainties in Chirality Studies. In *Fuzzy Logic in Chemistry* (Rouvray, D.H., ed.), Academic Press, New York (NY), pp. 65–90.

Mitchell, B.E. and Jurs, P.C. (1997). Prediction of Autoignition Temperatures of Organic Compounds from Molecular Structure. *J.Chem.Inf.Comput.Sci.*, *37*, 538–547.

Mitchell, B.E. and Jurs, P.C. (1998a). Prediction of Aqueous Solubility of Organic Compounds from Molecular Structure. *J.Chem.Inf.Comput.Sci.*, *38*, 489–496.

Mitchell, B.E. and Jurs, P.C. (1998b). Prediction of Infinite Dilution Activity Coefficients of Organic Compounds in Aqueous Solution from Molecular Structure. *J.Chem.Inf.Comput.Sci.*, *38*, 200–209.

Mitchell, J.B.O., Alex, A. and Snarey, M. (1999). SATIS: Atom Typing from Chemical Connectivity. *J.Chem.Inf.Comput.Sci.*, *39*, 751–757.

Miyashita, Y., Okuyama, T., Ohsako, H. and Sasaki, S. (1989). Graph Theoretical Approach to Carbon-13 Chemical Shift Sum in Alkanes. *J.Am.Chem.Soc.*, *111*, 3469–3470.

Miyashita, Y., Ohsako, H., Takayama, C. and Sasaki, S. (1992). Multivariate Structure-Activity Relationships Analysis of Fungicidal and Herbicidal Thiocarbamates Using Partial Least Squares Method. *Quant.Struct.-Act.Relat.*, *11*, 17–22.

Miyashita, Y., Li, Z.L. and Sasaki, S. (1993). Chemical Pattern Recognition and Multivariate Analysis for QSAR Studies. *TRAC*, *12*, 50–60.

Modica, M., Santagati, M., Russo, F., Parotti, L., Degioia, L., Selvaggini, C., Salmona, M. and Mennini, T. (1997). [[(Arylpiperazinyl)alkyl]thio]thieno[2,3-d]pyrimidinone Derivatives as High-Affinity, Selective 5-HT1a Receptor Ligands. *J.Med.Chem.*, *40*, 574–585.

Mohar, B. and Pisanski, T. (1988). How to Compute the Wiener Index of a Graph. *J.Math.Chem.*, *2*, 267–277.

Mohar, B. (1989a). Laplacian Matrices of Graphs. *Stud.Phys.Theor.Chem.*, *63*, 1–8.

Mohar, B. (1989b). In *MATH/CHEM/COMP 1988* (Graovac, A., ed.), Elsevier, Amsterdam (The Netherlands), pp. 1–8.

Mohar, B. (1991a). Eigenvalues, Diameter, and Mean Distance in Graphs. *Graphs Comb.*, *7*, 53–64.

Mohar, B. (1991b). The Laplacian Spectrum of Graphs. In *Graph Theory, Combinatorics, and Applications* (Alavi, Y., Chartrand, C. and Ollermann, O.R., eds.), Wiley, New York (NY), pp. 871–898.

Mohar, B., Babic, D. and Trinajstic, N. (1993). A Novel Definition of the Wiener Index for Trees. *J.Chem.Inf.Comput.Sci.*, *33*, 153–154.

Mokrosz, J.L. (1989). Topological Indices in Correlation Analysis. Part 1. Comparison of Molecular Shape with Molecular Connectivity for Some Hydrocarbons. *Quant.Struct.-Act.Relat.*, *8*, 305–309.

Molchanova, M.S. and Zefirov, N.S. (1998). Irredundant Generation of Isomeric Molecular Structures with Some Known Fragments. *J.Chem.Inf.Comput.Sci.*, *38*, 8–22.

Moliner, R., García, F., Gálvez, J., Garcìa-Domenech, R. and Serrano, C. (1991). Nuevos Indices Topológicos en Connnectividad Molecular. Su Aplicación a Algunas Propriedades Fisicoquimicas de un Grupo de Hydrocarburos Alifáticos. *An.R.Acad.Farm.*, *57*, 287–298.

Montanari, C.A., Tute, M.S., Beezer, A.E. and Mitchell, J.C. (1996). Determination of Receptor-Bound Drug Conformations by QSAR Using Flexible Fitting to Derive a Molecular Similarity Index. *J.Comput.Aid.Molec.Des.*, *10*, 67–73.

Morales, D.A. and Araujo, O. (1993). On the Search for the Best Correlation Between Graph Theoretical Invariants and Physicochemical Properties. *J.Math.Chem.*, *13*, 95–106.

Moran, P.A.P. (1950). Notes on Continuous Stochastic Phenomena. *Biometrika*, *37*, 17–23.

Moreau, G. and Broto, P. (1980a). Autocorrelation of Molecular Structures, Application to SAR Studies. *Nouv.J.Chim.*, *4*, 757–764.

Moreau, G. and Broto, P. (1980b). The Autocorrelation of a Topological Structure: A New Molecular Descriptor. *Nouv.J.Chim.*, *4*, 359–360.

Moreau, G. (1997). Atomic Chirality, a Quantitative Measure of the Chirality of the Environment of an Atom. *J.Chem.Inf.Comput.Sci.*, *37*, 929–938.

Morgan, H.L. (1965). The Generation of a Unique Machine Description for Chemical Structures – A Technique Developed at Chemical Abstracts Service. *J.Chem.Doc.*, *5*, 107–113.

Moriguchi, I. (1975). Quantitative Structure-Activity Studies. I. Parameters Relating to Hydrophobicity. *Chem.Pharm.Bull.*, *23*, 247–257.

Moriguchi, I., Kanada, Y. and Komatsu, K. (1976). van der Waals Volume and the Related Parameters for Hydrophobicity in Structure-Activity Studies. *Chem.Pharm.Bull.*, *24*, 1799–1806.

Moriguchi, I. and Kanada, Y. (1977). Quantitative Structure-Activity Studies. Part III. Use of van der Waals Volume in Structure-Activity Studies. *Chem.Pharm.Bull.*, *25*, 926–935.

Moriguchi, I., Hirono, S., Liu, Q. and Nakagome, I. (1992a). Fuzzy Adaptive Least Squares and Its Application to Structure-Activity Studies. *Quant.Struct.-Act.Relat.*, *11*, 325–331.

Moriguchi, I., Hirono, S., Liu, Q., Nakagome, I. and Matsushita, Y. (1992b). Simple Method of Calculating Octanol/Water Partition Coefficient. *Chem.Pharm.Bull.*, *40*, 127–130.

Moriguchi, I., Hirono, S., Nakagome, I. and Hirano, H. (1994). Comparison of Reliability of Log *P* Values for Drugs Calculated by Several Methods. *Chem.Pharm.Bull.*, *42*, 976–978.

Morikawa, T. and Balaban, A.T. (1992). Topological Formulas and Upper/Lower Bounds in Chemical Polygonals Graphs, Particularly in Benzenoid Polyhexes. *MATCH (Comm.Math. Comp.Chem.)*, *28*, 235–247.

Morovitz, H. (1955). Some Order- Disorder- Considerations in Living Systems. *Bull.Math.Biophys.*, *17*, 81–86.

Mõssner, S.G., Lopez de Alda, M.J., Sander, L.C., Lee, M.L. and Wise, S.A. (1999). Gas Chromatographic Retention Behavior of Polycyclic Aromatic Sulfur Heterocyclic Compounds, (dibenzothiophene, naphtho[b]thiophenes, benzo[b]naphthothiophenes and alkyl-substituted derivatives) on Stationary Phases of Different Selectivity. *J.Chromat.*, *841*, 207–228.

Motoc, I., Holban, S., Vancea, R. and Simon, Z. (1977). Minimal Steric Difference Calculated as Nonoverlapping Volumes. Correlations with Enzymatic Hydrolyses of Ribonucleosides. *Studia Biophys.*, *66*, 75–78.

Motoc, I. and Balaban, A.T. (1981). Topological Indices: Intercorrelations, Physical Meaning, Correlational Ability. *Rev.Roum.Chim.*, *26*, 593–600.

Motoc, I. and Dragomir, O. (1981). Molecular Interactions in Biological Systems. Steric Interactions. The SIBIS Algorithm. *Math.Chem.*, *12*, 117–126.

Motoc, I. and Balaban, A.T. (1982a). Testing the Geometrical Meaning of *Taft*-Type Steric Constants. *Rev.Roum.Chim.*, *27*, 735–739.

Motoc, I., Balaban, A.T., Mekenyan, O. and Bonchev, D. (1982b). Topological Indices: Inter-Relations and Composition. *MATCH (Comm.Math.Comp.Chem.)*, *13*, 364–404.

Motoc, I. (1983). Molecular Shape Descriptors. In *Steric Effects in Drug Design (Topics in Current Chemistry, Vol. 114)* (Charton, M. and Motoc, I., eds.), Springer-Verlag, Berlin (Germany), pp. 93–105.

Motoc, I. (1984a). Biological Receptor Maps. 2. Steric Maps of Benzoate Antibody and Carbonic Anhydrase. *Quant.Struct.-Act.Relat.*, *3*, 47–51.

Motoc, I. (1984b). Biological Receptor Maps. I. Steric Maps. The SIBIS Method. *Quant.Struct.-Act.Relat.*, *3*, 43–47.

Motoc, I. and Marshall, G.R. (1985). Van der Waals Volume Fragmental Constants. *Chem.Phys. Lett.*, *116*, 415–419.

Mouvier, G. and Dubois, J.-E. (1968). N/ 224. – Réactivité des Composés Éthyléniques: Réaction de Bromation. XVIII. – Applications des Relations Linéaires D'énergie Libre au Cas des Alcènes. *Bull.Soc.Chim.Fr.*, *4*, 1441–1445.

Mowshowitz, A. (1968a). Entropy and the Complexity of Graphs. I. An Index of the Relative Complexity of a Graph. *Bull.Math.Biophys.*, *30*, 175–204.

Mowshowitz, A. (1968b). Entropy and the Complexity of Graphs. II. The Information Content of Digraphs and Infinite Graphs. *Bull.Math.Biophys.*, *30*, 225–240.

Mowshowitz, A. (1968c). Entropy and the Complexity of Graphs. III. Graphs with Prescribed Information Content. *Bull.Math.Biophys.*, *30*, 387–414.

Mowshowitz, A. (1968d). Entropy and the Complexity of Graphs: IV. Entropy Measures and Graphical Structure. *Bull.Math.Biophys.*, *30*, 533–546.

Mracec, M., Kurunczi, L., Nusser, T., Simon, Z. and Naray-Szabo, G. (1996). QSAR Study with Steric (MTD), Electronic and Hydrophobicity Parameters on Psychotomimetic Phenylalkylamines. *J.Mol.Struct.(Theochem)*, *367*, 139–149.

Mracec, M., Muresan, S., Simon, Z. and Naray-Szabo, G. (1997). QSARs with Orthogonal Descriptors on Psychotomimetic Phenylalkylamines. *Quant.Struct.-Act.Relat.*, *16*, 459–464.

Mullay, J. (1984). Atomic and Group Electronegativities. *J.Am.Chem.Soc.*, *106*, 5842–5847.

Mullay, J. (1985). Calculation of Group Electronegativity. *J.Am.Chem.Soc.*, *107*, 7271–7275.

Müller, M. and Klein, W. (1991). Estimating Atmospheric Degradation Processes by SARs. *Sci. Total Environ.*, *109/110*, 261–273.

Müller, M. and Kördel, W. (1996). Comparison of Screening Methods for the Estimation of Adsorption Coefficients on Soil. *Chemosphere*, *32*, 2493–2504.

Müller, M. (1997). Quantum Chemical Modelling of Soil Sorption Coefficients: Multiple Linear Regression Models. *Chemosphere*, *35*, 365–377.

Müller, W.R., Szymanski, K., Knop, J.V. and Trinajstic, N. (1987). An Algorithm for Construction of the Molecular Distance Matrix. *J.Comput.Chem.*, *8*, 170–173.

Müller, W.R., Szymanski, K., Knop, J.V., Nikolic, S. and Trinajstic, N. (1990a). On the Enumeration and Generation of Polyhex Hydrocarbons. *J.Comput.Chem.*, *11*, 223–235.

Müller, W.R., Szymanski, K., Knop, J.V. and Trinajstic, N. (1990b). Molecular Topological Index. *J.Chem.Inf.Comput.Sci.*, *30*, 160–163.

Müller, W.R., Szymanski, K., Knop, J.V., Mihalic, Z. and Trinajstic, N. (1993). The Walk ID Number Revisited. *J.Chem.Inf.Comput.Sci.*, *33*, 231–233.

Müller, W.R., Szymanski, K., Knop, J.V., Mihalic, Z. and Trinajstic, N. (1995). Note on Isocodal Graphs. *J.Chem.Inf.Comput.Sci.*, *35*, 871–873.

Mulliken, R.S. (1928a). *Physical Review*, *32*, 186.

Mulliken, R.S. (1928b). *Physical Review*, *32*, 761.
Mulliken, R.S. (1934). A New Electroaffinity Scale, together with Data on Valence States and an Ionization Potential and Electron Affinities. *J.Chem.Phys.*, *2*, 782–793.
Mulliken, R.S. (1935). *J.Chem.Phys.*, *3*, 573.
Mulliken, R.S. (1955a). Electronic Population Analysis on LCAO-MO Molecular Wave Functions. I. *J.Chem.Phys.*, *23*, 1833–1840.
Mulliken, R.S. (1955b). Electronic Population Analysis on LCAO-MO Molecular Wave Functions. II. Overlap Populations, Bond Orders, and Covalent Bond Energies. *J.Chem.Phys.*, *23*, 1841–1846.
Muresan, S., Bologa, C., Mracec, M., Chiriac, A., Jastorff, B., Simon, Z. and Naray-Szabo, G. (1995). Comparative QSAR Study with Electronic and Steric Parameters for cAMP Derivatives with Large Substituents in Position 2, Position 6 and Position 8. *J.Mol.Struct.(Theochem)*, *342*, 161–171.
Murray, J.S. and Politzer, P. (1991). Correlations Between the Solvent Hydrogen-Bond-Donating Parameter α and the Calculated Molecular Surface Electrostatic Potential. *J.Org.Chem.*, *56*, 6715–6717.
Murray, J.S., Ranganathan, S. and Politzer, P. (1991). Correlations Between the Solvent Hydrogen Bond Acceptor Parameter β and the Calculated Molecular Electrostatic Potential. *J.Org. Chem.*, *56*, 3734–3737.
Murray, J.S., Brinck, T. and Politzer, P. (1993). Partition Coefficients of Nitroaromatics Expressed in Terms of their Molecular Surface Areas and Electrostatic Potentials. *J.Phys.Chem.*, *97*, 13807–13809.
Murray, J.S., Lane, P., Brinck, T., Paulsen, K., Grice, M.E. and Politzer, P. (1993a). Relationships of Critical Constants and Boiling Points to Computed Molecular Surface Properties. *J.Phys.Chem.*, *97*, 9369–9373.
Murray, J.S., Lane, P., Brinck, T. and Politzer, P. (1993b). Relationships Between Computed Molecular Properties and Solute-Solvent Interactions in Supercritical Solutions. *J.Phys.Chem.*, *97*, 5144–5148.
Murray, J.S., Brinck, T., Lane, P., Paulsen, K. and Politzer, P. (1994). Statistically-Based Interaction Indices Derived from Molecular Surface Electrostatic Potentials: A General Interaction Properties Function (GIPF). *J.Mol.Struct.(Theochem)*, *307*, 55–64.
Murray, J.S., Gagarin, S.G. and Politzer, P. (1995). Representation of C_{60} Solubilities in Terms of Computed Molecular Surface Electrostatic Potentials and Areas. *J.Phys.Chem.*, *99*, 12081–12083.
Murray, J.S., Brinck, T. and Politzer, P. (1996). Relationships of Molecular Surface Electrostatic Potentials to Some Macroscopic Properties. *Chem.Phys.*, *204*, 289–299.
Murray, J.S., Lane, P. and Politzer, P. (1998). Effects of Strongly Electron-Attracting Components on Molecular Surface Electrostatic Potentials: Application to Predicting Impact Sensitivities of Energetic Molecules. *Mol.Phys.*, *93*, 187–194.
Murray, J.S. and Politzer, P. (1998). Statistical Analysis of the Molecular Surface Electrostatic Potential: An Approach to Describing Noncovalent Interactions in Condensed Phases. *J.Mol. Struct.(Theochem)*, *425*, 107–114.
Murray, J.S., Abu-Awwad, F. and Politzer, P. (1999). Prediction of Aqueous Solvation Free Energies from Properties of Solute Molecular Surface Electrostatic Potentials. *Journal of Physical Chemistry A*, *103*, 1853–1856.
Murray, W.J., Hall, L.H. and Kier, L.B. (1975). Molecular Connectivity III: Relationship to Partition Coefficients. *J.Pharm.Sci.*, *64*, 1978–1981.
Murray, W.J., Kier, L.B. and Hall, L.H. (1976). Molecular Connectivity. 6. Examination of the Parabolic Relationship Between Molecular Connectivity and Biological Activity. *J.Med. Chem.*, *19*, 573–578.
Murray, W.J. (1977). Molecular Connectivity and Steric Parameters. *J.Pharm.Sci.*, *66*, 1352–1354.
Murrell, J.N. and Harget, A.J. (1972). *Semi-Empirical Self-Consistent-Field Molecular Orbital Theory of Molecules*. Wiley-Interscience, London (UK).

Murugan, R., Grendze, M.P., Toomey, J.E.Jr., Katritzky, A.R., Karelson, M., Lobanov, V.S. and Rachwal, P. (1994). Predicting Physical Properties from Molecular Structure. *Chemtech*, *24*, 17–23.

Myers, R.H. (1986). *Classical and Modern Regression with Applications*. Duxbury Press, Boston (MA).

Myrdal, P., Ward, G.H., Simamora, P. and Yalkowsky, S.H. (1993). AQUAFAC: Aqueous Functional Group Acitivity Coefficients. *SAR & QSAR Environ.Res.*, *1*, 53–61.

Nagy, P.J., Tokarski, J. and Hopfinger, A.J. (1994). Molecular Shape and QSAR Analyses of a Famly of Substituted Dichlorodiphenyl Aromatase Inhibitors. *J.Chem.Inf.Comput.Sci.*, *34*, 1190–1197.

Nakayama, A., Hagiwara, K., Hashimoto, S. and Shimoda, S. (1993). QSAR of Fungicidal Delta(3)-1,2,4-Thiadiazolines Reactivity Activity Correlation of SH Inhibitors. *Quant.Struct.-Act.Relat.*, *12*, 251–255.

Nakayama, S., Shigezumi, S. and Yoshida, M. (1988). Method for Clustering Proteins by Use of All Possible Pairs of Amino Acids as Structural Descriptors. *J.Chem.Inf.Comput.Sci.*, *28*, 72–78.

Naray-Szabo, G. and Balogh, T. (1993). The Average Molecular Electrostatic Field as a QSAR Descriptor. 4. Hydrophobicity Scales for Amino Acid Residues Alpha. *J.Mol.Struct.(Theochem)*, *103*, 243–248.

Narumi, H. and Hosoya, H. (1980). Topological Index and Thermodynamics Properties. II. Analysis of Topological Factors on the Absolute Entropy of Acyclic Saturated Hydrocarbons. *Bull. Chem.Soc.Jap.*, *53*, 1228–1237.

Narumi, H. and Hosoya, H. (1985). Topological Index and Thermodynamics Properties. III. Classification of Various Topological Aspects of Properties of Acyclic Saturated Hydrocarbons. *Bull.Chem.Soc.Jap.*, *58*, 1778–1786.

Narumi, H. (1987). New Topological Indices for Finite and Infinite Systems. *MATCH (Comm. Math.Comp.Chem.)*, *22*, 195–207.

Navajas, C., Poso, A., Tuppurainen, K. and Gynther, J. (1996). Comparative Molecular Field Analysis (CoMFA) of MX Compounds Using Different Semiempirical Methods. LUMO Field and Its Correlation with Mutagenic Activity. *Quant.Struct.-Act.Relat.*, *15*, 189–193.

Needham, D.E., Wei, I.C. and Seybold, P.G. (1988). Molecular Modeling of the Physical Properties of the Alkanes. *J.Am.Chem.Soc.*, *110*, 4186–4194.

Nefati, H., Diawara, B. and Legendre, J.J. (1993). Predicting the Impact Sensitivity of Explosive Molecules Using Neuromimetic Networks. *SAR & QSAR Environ.Res.*, *1*, 131–136.

Nefati, H., Cense, J.-M. and Legendre, J.J. (1996). Prediction of the Impact Sensitivity by Neural Networks. *J.Chem.Inf.Comput.Sci.*, *36*, 804–810.

Nelson, T.M. and Jurs, P.C. (1994). Prediction of Aqueous Solubility of Organic Compounds. *J.Chem.Inf.Comput.Sci.*, *34*, 601–609.

Nemba, R.M. and Balaban, A.T. (1998). Algorithm for the Direct Enumeration of Chiral and Achiral Skeletons of a Homosubstituted Derivative of a Monocyclic Cycloalkane with a Large and Factorizable Ring Size *n*. *J.Chem.Inf.Comput.Sci.*, *38*, 1145–1150.

Nevalainen, T. and Kolehmainen, E. (1994). New QSAR Models for Polyhalogenated Aromatics. *Environ.Toxicol.Chem.*, *13*, 1699–1706.

Newman, M.S. (1950). Some Observations Concerning Steric Factors. *J.Am.Chem.Soc.*, *72*, 4783–4786.

Nguyen-Cong, V. and Rode, B.M. (1996a). Quantum Pharmacological Analysis of Structure-Activity Relationships for Mefloquine Antimalarial Drugs Using Optimal Transformations. *J.Chem.Inf.Comput.Sci.*, *36*, 114–117.

Nguyen-Cong, V., Vandang, G. and Rode, B.M. (1996b). Using Multivariate Adaptive Regression Splines to QSAR Studies of Dihydroartemisinin Derivatives. *Eur.J.Med.Chem.*, *31*, 797–803.

Nicklaus, M.C., Milne, G.W.A. and Burke, T.R. (1992). QSAR of Conformationally Flexible Molecules: Comparative Molecular Field Analysis of Protein-Tyrosine Kinase Inhibitors. *J.Comput.Aid.Molec.Des.*, *6*, 487–504.

Niemi, G.J., Basak, S.C., Veith, G.D. and Grunwald, G.D. (1992). Prediction of Octanol/Water Partition Coefficient (K_{OW}) with Algorithmically Derived Variables. *Environ.Toxicol.Chem.*, *11*, 893–900.

Nikolic, S., Trinajstic, N., Mihalic, Z. and Carter, S. (1991). On the Geometric-Distance Matrix and the Corresponding Structural Invariants of Molecular Systems. *Chem.Phys.Lett.*, *179*, 21–28.

Nikolic, S., Plavsic, D. and Trinajstic, N. (1992). On the Z-Counting Polynomial for Edge-Weighted Graphs. *J.Math.Chem.*, *9*, 381–387.

Nikolic, S., Medicsaric, M. and Matijevicsosa, J. (1993a). A QSAR Study of 3-(Phthalimidoalkyl)-Pyrazolin-5-Ones. *Croat.Chem.Acta*, *66*, 151–160.

Nikolic, S., Trinajstic, N. and Mihalic, Z. (1993b). Molecular Topological Index: An Extension to Heterosystems. *J.Math.Chem.*, *12*, 251–264.

Nikolic, S., Trinajstic, N. and Mihalic, Z. (1995). The Wiener Index: Development and Applications. *Croat.Chem.Acta*, *68*, 105–129. [R]

Nikolic, S., Trinajstic, N., Juric, A. and Mihalic, Z. (1996a). The Detour Matrix and the Detour Index of Weighted Graphs. *Croat.Chem.Acta*, *69*, 1577–1591.

Nikolic, S., Trinajstic, N., Juric, A., Mihalic, Z. and Krilov, G. (1996b). Complexity of Some Interesting (Chemical) Graphs. *Croat.Chem.Acta*, *69*, 883–897.

Nikolic, S. and Trinajstic, N. (1998a). Modeling the Aqueous Solubility of Aliphatic Alcohols. *SAR & QSAR Environ.Res.*, *9*, 117–126.

Nikolic, S., Trinajstic, N. and Baucic, I. (1998b). Comparison Between the Vertex- and Edge-Connectivity Indices for Benzenoid Hydrocarbons. *J.Chem.Inf.Comput.Sci.*, *38*, 42–46.

Nilakantan, R., Bauman, N., Dixon, J.S. and Venkataraghavan, R. (1987). Topological Torsion: A New Molecular Descriptor for SAR Applications. Comparison with Other Descriptors. *J.Chem.Inf.Comput.Sci.*, *27*, 82–85.

Nilakantan, R., Bauman, N. and Venkataraghavan, R. (1993). New Method for Rapid Characterization of Molecular Shapes: Applications in Drug Design. *J.Chem.Inf.Comput.Sci.*, *33*, 79–85.

Nirmalakhandan, N.N. and Speece, R.E. (1988a). Prediction of Aqueous Solubility of Organic Chemicals Based on Molecular Structure. *Environ.Sci.Technol.*, *22*, 328–338.

Nirmalakhandan, N.N. and Speece, R.E. (1988b). Structure-Activity Relationships. *Environ.Sci. Technol.*, *22*, 606–615.

Nirmalakhandan, N.N. and Speece, R.E. (1989a). Prediction of Aqueous Solubility of Organic Chemicals Based on Molecular Structure. 2. Application to PNAa, PCBs, PCDDs, etc. *Environ.Sci.Technol.*, *23*, 708–713.

Nirmalakhandan, N.N. and Speece, R.E. (1989b). QSAR Model for Predicting Henry's Constant. *Environ.Sci.Technol.*, *22*, 1349–1357.

Nirmalakhandan, N.N. and Speece, R.E. (1993). Prediction of Activated Carbon Adsorption Capacities for Organic Vapors Using Quantitative Structure-Activity Relationship Methods. *Environ.Sci.Technol.*, *27*, 1512–1516.

Nirmalakhandan, N.N., Sun, B., Arulgnanendran, V.J., Mohsin, M., Wang, X.H., Prakash, J. and Hall, N. (1994). Analyzing and Modeling Toxicity of Mixtures of Organic Chemicals to Microorganisms. *Water Sci.Technol.*, *30*, 87–96.

No, K.T., Grant, J.A., Jhon, M.S. and Scheraga, H.A. (1990a). Determination of Net Atomic Charges Using a Modified Partial Equalization of Orbital Electronegativity Method. 2. Application to Ionic and Aromatic Molecules as Models for Polypeptides. *J.Phys.Chem.*, *94*, 4740–4746.

No, K.T., Grant, J.A. and Scheraga, H.A. (1990b). Determination of Net Atomic Charges Using a Modified Partial Equalization of Orbital Electronegativity Method. 1. Application to Neutral Molecules as Models for Polypeptides. *J.Phys.Chem.*, *94*, 4732–4739.

No, K.T., Cho, K.H., Jhon, M.S. and Scheraga, H.A. (1993). An Empirical Method to Calculate Average Molecular Polarizablities from the Dependence of Effective Atomic Polarizabilties on Net Atomic Charge. *J.Am.Chem.Soc.*, *115*, 2005–2014.

Nord, L.I., Fransson, D. and Jacobsson, S.P. (1998). Prediction of Liquid Chromatographic Retention Times of Steroids by Three-Dimensional Structure Descriptors and Partial Least Squares Modeling. *Chemom.Intell.Lab.Syst.*, *44*, 257–269.

Norinder, U. (1991). Theoretical Amino Acid Descriptors. Application to Bradykinin Potentiating Peptides. *Peptides*, *12*, 1223–1227.

Norinder, U. (1992). Experimental Design Based Quantitative Structure Toxicity Relationship of Some Local Anesthetics Using the Pls Method. *J.Appl.Toxicol.*, *12*, 143–147.

Norinder, U. and Hogberg, T. (1992). PLS Based Quantitative Structure-Activity Relationship for Substituted Benzamides of Clebopride Type. Application of Experimental Design in Drug Design. *Acta Chem.Scand.*, *46*, 363–366.

Norinder, U. (1993). Multivariate Free-Wilson Analysis of Some N-Alkylmorphinan-6-one Opioids Using PLS. *Quant.Struct.-Act.Relat.*, *12*, 119–123.

Norinder, U. (1994). Theoretical Descriptors of Nucleic Acid Bases. Application to DNA Promotor Sequences. *Quant.Struct.-Act.Relat.*, *13*, 295–301.

Norinder, U., Florvall, L. and Ross, S.B. (1994). A PLS Quantitative Structure-Activity Relationship Study of Some Monoamine Oxidase Inhibitors of the Phenyl Alkylamine Type. *Eur. J.Med.Chem.*, *29*, 191–195.

Norinder, U. (1996). Single and Domain Mode Variable Selection in 3D QSAR Applications. *J.Chemom.*, *10*, 95–105.

Norinder, U., Österberg, T. and Artursson, P. (1997). Theoretical Calculation and Prediction of CACO-2 Cell Permeability Using MolSurf Parametrization and PLS Statistics. *Pharm.Res.*, *14*, 1786–1791.

Norinder, U. (1998). Recent Progress in CoMFA Methodology and Related Techniques. In *3D QSAR in Drug Design – Vol. 3* (Kubinyi, H., Folkers, G. and Martin, Y.C., eds.), Kluwer/ESCOM, Dordrecht (The Netherlands), pp. 25–39.

Norinder, U., Sjöberg, P. and Österberg, T. (1998). Theoretical Calculation and Prediction of Brain-Blood Partitioning of Organic Solutes Using MolSurf Parametrization and PLS Statistics. *J.Pharm.Sci.*, *87*, 952–959.

Norinder, U., Österberg, T. and Artursson, P. (1999). Theoretical Calculation and Prediction of Intestinal Absorption of Drugs in Humans Using MolSurf Parametrization and PLS Statistics. *Eur.J.Pharm.Sci.*, *8*, 49–56.

Norrington, F.E., Hyde, R.M., Williams, S.G. and Wootton, R. (1975). Physicochemical-Activity Relations in Practice. 1. A Rational and Self-Consistent Data Bank. *J.Med.Chem.*, *18*, 604–607.

Nouwen, J., Lindgren, F., Hansen, B. and Karcher, W. (1996). Fast Screening of Large Databases Using Clustering and PCA Based on Structure Fragments. *J.Chemom.*, *10*, 385–398.

Nouwen, J., Lindgren, F., Hansen, B. and Karcher, W. (1997). Classification of Environmentally Occurring Chemicals Using Structural Fragments and PLS Discriminant Analysis. *Environ. Sci.Technol.*, *31*, 2313–2318.

Novic, M. and Zupan, J. (1996). A New General Approach and Uniform Structure Representation. In *Software Development in Chemistry – Vol. 10* (Gasteiger, J., ed.), Fachgruppe Chemie-Information-Computer (CIC), Frankfurt am Main (Germany), pp. 47–58.

Novic, M., Nikolovska-Coleska, Z. and Solmajer, T. (1997). Quantitative Structure-Activity Relationship of Flavonoid $p56^{lck}$ Protein Tyrosine Kinase Inhibitors. A Neural Network Approach. *J.Chem.Inf.Comput.Sci.*, *37*, 990–998.

Nusser, T., Balogh, T. and Naray-Szabo, G. (1993). The Average Molecular Electrostatic Field as a QSAR Descriptor. 5. Hydrophobicity Indexes for Small Molecules. *J.Mol.Struct.*, *297*, 127–132.

Nys, G.G. and Rekker, R.F. (1973). Statistical Analysis of a Series of Partition Coefficients with Special Reference to the Predictability of Folding of Drug Molecules. The Introduction of Hydrophobic Fragmental Constants (*f* Values). *Eur.J.Med.Chem.*, *8*, 521–535.

Nys, G.G. and Rekker, R.F. (1974). The Concept of Hydrophobic Fragmental Constants (*f*-Values). II. Extension of its Applicability to the Calculation of Lipophilicities of Aromatic and Heteroaromatic Structures. *Eur.J.Med.Chem.*, *9*, 361–375.

Oberrauch, E. and Mazzanti, V. (1990). Partial-Least-Squares Models for the Octane Number of Alkanes Based on Subgraph Descriptors. *Anal.Chim.Acta*, *235*, 177–188.

Okamoto, Y. and Brown, H.C. (1958a). Rates of Solvolysis of Phenyldimethylcarbinyl Chlorides Containing Substituents ($-NMe_3^+$, $-CO_2$) Bearing a Charge. *J.Am.Chem.Soc.*, *80*, 4976–4979.

Okamoto, Y., Inukai, T. and Brown, H.C. (1958b). Rates of Solvolysis of Phenyldimethylcarbinyl Chlorides Containing Meta Directing Substituents. *J.Am.Chem.Soc.*, *80*, 4969–4972.

Okamoto, Y., Inukai, T. and Brown, H.C. (1958c). Rates of Solvolysis of Phenyldimethylcarbinyl Chlorides in Methyl, Ethyl and Isopropyl Alcohols. Influence of the Solvent on the Value of the Electrophilic Substituent Constant. *J.Am.Chem.Soc.*, *80*, 4972–4976.

Okey, R.W. and Stensel, H.D. (1996a). A QSAR-Based Biodegradability Model. A QSBR. *Water Res.*, *30*, 2206–2214.

Okey, R.W., Stensel, H.D. and Martis, M.C. (1996b). Modeling Nitrification Inhibition. *Water Sci.Technol.*, *33*, 101–107.

Okouchi, S. and Saegusa, H. (1989). Prediction of Soil Sorption Coefficients of Hydrophobic Organic Pollutants by Adsorbability Index. *Bull.Chem.Soc.Jap.*, *62*, 922–924.

Onicescu, O. (1966). Energie informationelle. *C.R.Acad.Sci.,Paris, 263 – Ser.A*, 841–842.

Oprea, T.I., Ciubotariu, D., Sulea, T. and Simon, Z. (1993). Comparison of the Minimal Steric Difference (MTD) and Comparative Molecular Field Analysis (CoMFA) Methods for Analysis of Binding of Steroids to Carrier Proteins. *Quant.Struct.-Act.Relat.*, *12*, 21–26.

Oprea, T.I. and Garcia, A.E. (1996). Three-Dimensional Quantitative Structure-Activity Relationships of Steroid Aromatase Inhibitors. *J.Comput.Aid.Molec.Des.*, *10*, 186–200.

Oprea, T.I. and Waller, C.L. (1997). Theoretical and Practical Aspects of Three-Dimensional Quantitative Structure-Activity Relationships. In *Reviews in Computational Chemistry – Vol. 11* (Lipkowitz, K.B. and Boyd, D., eds.), Wiley-VCH, New York (NY), pp. 127–182.

Ordorica, M.A., Velazquez, M.L., Ordorica, J.G., Escobar, J.L. and Lehmann, P.A. (1993). A Principal Component and Cluster Significance Analysis of the Antiparasitic Potency of Praziquantel and Some Analogs. *Quant.Struct.-Act.Relat.*, *12*, 246–250.

Ortiz, A.R., Pisabarro, M.T., Gago, F. and Wade, R.C. (1995). Prediction of Drug Binding Affinities by Comparative Binding Energy Analysis. *J.Med.Chem.*, *38*, 2681–2691.

Osmialowski, K., Halkiewicz, J., Radecki, A. and Kaliszan, R. (1985). Quantum Chemical Parameters in Correlation Analysis of Gas-Liquid Chromatographic Retention Indices of Amines. *J.Chromat.*, *346*, 53–60.

Osmialowski, K., Halkiewicz, J. and Kaliszan, R. (1986). Quantum Chemical Parameters in Correlation Analysis of Gas-Liquid Chromatographic Retention Indices of Amines. II. Topological Electronic Index. *J.Chromat.*, *361*, 63–69.

Osmialowski, K. and Kaliszan, R. (1991). Studies of Performance of Graph Theoretical Indices in QSAR Analysis. *Quant.Struct.-Act.Relat.*, *10*, 125–134.

Osten, D.W. (1988). Selection of Optimal Regression Models via Cross-Validation. *J.Chemom.*, *2*, 39.

Oth, J.F.M. and Gilles, J.-M. (1968). Mobilite Conformationnelle et Isomerie de Valence Rapide Reversible dans le [16] Annulene. *Tetrahedron Lett.*, *1968*, 6259–6264.

Otsuji, Y., Kubo, M. and Imoto, E. (1960). Reactivities of Heterocyclic Compounds. IX. A Method to Systematize the Reactivities of the Substituents in Aromatic and Heteroaromatic Compounds. *Chem.Abs.*, *54*, 24796–24796.

Ouyang, Z., Yuan, S., Brandt, J. and Zheng, C. (1999). An Effective Topological Symmetry Perception and Unique Numbering Algorithm. *J.Chem.Inf.Comput.Sci.*, *39*, 299–303.

Overton, E. (1901). *Studien über die Narkose zugleich ein Beitrag zur allgemeinen Pharmakologie.* Verlag Gustav Fischer, Jena (Germany), 141 pp.

Overton, E. (1991). *Studies on Narcosis (English translation).* Chapman & Hall, London (UK).

Oxford Molecular Ltd. (1999). *TSAR – Reference manual.* The Magdalen Centre, Oxford Science Park, Sandford-on-Thames, Oxford (UK).

Pagliara, A., Khamis, E., Trinh, A., Carrupt, P.-A., Tsai, R.-S. and Testa, B. (1995). Structural Properties Governing Retention Mechanisms on RP-HPLC Stationary Phases Used for Lipophilicity Measurements. *J.Liquid Chromat.*, *18*, 1721–1745.

Pagliara, A., Caron, G., Lisa, G., Fan, W., Gaillard, P., Carrupt, P.-A., Testa, B. and Abraham, M.H. (1997a). Solvatochromic Analysis of di-*n*-butyl Ether/Water Partition Coefficients as Compared to Other Solvent Systems. *J.Chem.Soc.Perkin Trans.2*, 2639–2643.

Pagliara, A., Carrupt, P.-A., Caron, G., Gaillard, P. and Testa, B. (1997b). Lipophilicity Profiles of Ampholytes. *Chem.Rev.*, *97*, 3385–3400.

Palm, K., Luthman, K., Ungell, A.-L., Strandlund, G. and Artursson, P. (1996). Correlation of Drug Absorption with Molecular Surface Properties. *J.Pharm.Sci.*, *85*, 32–39.

Palm, K., Luthman, K., Ungell, A.-L., Strandlund, G., Beigi, F., Lundahl, P. and Artursson, P. (1998). Evaluation of Dynamic Polar Molecular Surface Area as Predictor of Drug Absorption: Comparison with Other Computational and Experimental Predictors. *J.Med.Chem.*, *41*, 5382–5392.

Palm, V.A. (1972). *Fundamentals of the Quantitative Theory of Organic Reactions.* Khimiya, Leningrad (Rus).

Palyulin, V.A., Baskin, I.I., Petelin, D.E. and Zefirov, N.S. (1995). Novel Descriptors of Molecular Structure in QSAR and QSPR Studies. In *QSAR and Molecular Modelling: Cocepts, Computational Tools and Biological Applications* (Sanz, F., Giraldo, J. and Manaut, F., eds.), Prous Science, Barcelona (Spain), pp. 51–52.

Palyulin, V.A., Radchenko, E.V. and Zefirov, N.S. (2000). Molecular Field Topology Analysis Method in QSAR Studies of Organic Compounds. *J.Chem.Inf.Comput.Sci.*, *40*, 659–667.

Panaye, A., MacPhee, J.A. and Dubois, J.-E. (1980). Steric Effects. II. Relationship Between Topology and Steric Parameter E'_S – Topology as a Tool for the Correlation and Prediction of Steric Effects. *Tetrahedron*, *36*, 759–768.

Papadopoulos, M.C. and Dean, P.M. (1991). Molecular Structure Matching by Simulated Annealing. IV. Classification of Atom Correspondences in Sets of Dissimilar Molecules. *J.Comput.Aid.Molec.Des.*, *5*, 119–133.

Pariser, R. and Parr, R.G. (1953a). A Semi-Empirical Theory of the Electronic Spectra and Electronic Structure of Complex Unsaturated Molecules. I. *J.Chem.Phys.*, *21*, 466–471.

Pariser, R. and Parr, R.G. (1953b). A Semi-Empirical Theory of the Electronic Spectra and Electronic Structure of Complex Unsaturated Molecules. II. *J.Chem.Phys.*, *21*, 767–776.

Parr, R.G. and Pearson, R.G. (1983). Absolute Hardness: Companion Parameter to Absolute Electronegativity. *J.Am.Chem.Soc.*, *105*, 7512–7516.

Parr, R.G. and Yang, W. (1989). *Density-Functional Theory of Atoms and Molecules.* Oxford Science Publications, New York (NY), 334 pp.

Pascual-Ahuir, J.L. and Silla, E. (1990). GEPOL: An Improved Description of Molecular Surfaces. I. Building the Spherical Surface Set. *J.Comput.Chem.*, *11*, 1047–1060.

Pastor, M. and Alvarez-Builla, J. (1991). The Edisfar Programs Rational Drug Series Design. *Quant.Struct.-Act.Relat.*, *10*, 350–358.

Pastor, M. and Alvarez-Builla, J. (1994). New Developments of Edisfar Programs Experimental Design in QSAR Practice. *J.Chem.Inf.Comput.Sci.*, *34*, 570–575.

Pastor, M., Cruciani, G. and Clementi, S. (1997). Smart Region Definition: A New Way to Improve the Predictive Ability and Interpretability of Three-Dimensional Quantitative Structure-Activity Relationships. *J.Med.Chem.*, *40*, 1455–1464.

Patil, G.S., Bora, M. and Dutta, N.N. (1995). Empirical Correlations for Prediction of Permeability of Gases Liquids Through Polymers. *J.Memb.Sci.*, *101*, 145–152.

Patterson, D.E., Cramer III, R.D., Ferguson, A.M., Clark, R.D. and Weinberger, L.E. (1996). Neighborhood Behavior: A Useful Concept for Validation of "Molecular Diversity" Descriptors. *J.Med.Chem.*, *39*, 3049–3059.

Pauling, L. (1932). The Additivity of the Energies of Normal Covalent Bonds. *Proc.Nat.Acad. Sci.USA*, *14*, 414–416.

Pauling, L. and Wilson, E.B. (1935). *Introduction to Quantum Mechanics.* McGraw-Hill, New York (NY).

Pauling, L. (1939). *The Nature of the Chemical Bond.* Cornell University Press, Ithaca (NY).

Pauling, L. and Pressman, D. (1945). The Serological Properties of Simple Substances. IX. Hapten Inhibitiion of Precipitation of Antisera Homologous to the *o*-, *m*-, and *p*-Azophenylarsonic Acid Groups. *J.Am.Chem.Soc.*, *67*, 1003–1012.

Pavani, R. and Ranghino, G. (1982). A Method to Compute the Volume of a Molecule. *Computers Chem.*, *6*, 133–135.

Pavlikova, M., Lacko, I., Devinsky, F. and Mlynarcik, D. (1995). Quantitative Relationships Between Structure, Aggregation Properties and Antimicrobial Activity of Quaternary Ammonium Bolaamphiphiles. *Collect.Czech.Chem.Comm.*, *60*, 1213–1228.

Pavlovic, L. and Gutman, I. (1997). Wiener Numbers of Phenylenes: An Exact Result. *J.Chem. Inf.Comput.Sci.*, *37*, 355–358.

Payares, P., Diaz, D., Olivero, J., Vivas, R. and Gomez, I. (1997). Prediction of the Gas Chromatographic Relative Retention Times of Flavonoids from Molecular Structure. *J.Chromat.*, *771*, 213–219.

Pearlman, R.S. (1980). Molecular Surface Areas and Volumes and their Use in Structure/Activity Relationships. In *Physical Chemical Properties of Drugs* (Yalkowsky, S.H., Sinkula, A.A. and Valvani, S.C., eds.), Marcel Dekker, New York (NY), pp. 321–347.

Pearlman, R.S. (1993). 3D Molecular Structures: Generation and Use in 3D Searching. In *3D QSAR in Drug Design. Theory, Methods and Applications.* (Kubinyi, H., ed.), ESCOM, Leiden (The Netherlands), pp. 41–79.

Pearlman, R.S. and Smith, K.M. (1998). Novel Software Tools for Chemical Diversity. In *3D QSAR in Drug Design – Vol. 2* (Kubinyi, H., Folkers, G. and Martin, Y.C., eds.), Kluwer/ ESCOM, Dordrecht (The Netherlands), pp. 339–353.

Pearlman, R. S. (1999). Novel Software Tools for Addressing Chemical Diversity., *Internet Communication*, http://www.netsci.org/Science/Combichem/feature08.html

Pearlman, R.S. and Smith, K.M. (1999). Metric Validation and the Receptor-Relevant Subspace Concept. *J.Chem.Inf.Comput.Sci.*, *39*, 28–35.

Peijnenburg, W.J., Thart, M.J., Denhollander, H.A., Vandemeent, D., Verboom, H.H. and Wolfe, N.L. (1992a). QSARs for Predicting Reductive Transformation Rate Constants of Halogenated Aromatic Hydrocarbons in Anoxic Sediment Systems. *Environ.Toxicol.Chem.*, *11*, 301–314.

Peijnenburg, W.J., Debeer, K.G., Dehaan, M.W., Denhollander, H.A., Stegeman, M.H. and Verboom, H. (1992b). Development of a Structure-Reactivity Relationship for the Photohydrolysis of Substituted Aromatic Halides. *Environ.Sci.Technol.*, *26*, 2116–2121.

Peijnenburg, W.J., Debeer, K.G., Denhollander, H.A., Stegeman, M.H. and Verboom, H. (1993). Kinetics, Products, Mechanisms and QSARs for the Hydrolytic Transformation of Aromatic Nitriles in Anaerobic Sediment Slurries. *Environ.Toxicol.Chem.*, *12*, 1149–1161.

Pepperrell, C.A. and Willett, P. (1991). Techniques for the Calculation of Three-Dimensional Structural Similarity Using Inter-Atomic Distances. *J.Comput.Aid.Molec.Des.*, *5*, 455–474.

Pepperrell, C.A. (1994). *Three-Dimensional Chemical Similarity Searching.* Research Studies Press – Wiley, Taunton (UK), 304 pp.

Perez-Gimenez, F., Antón-Fos, G.M., García-March, F.J., Salebert-Salvador, M.T., Cercos-del-Pozo, R.A. and Jaenoltra, J. (1995). Prediction of Chromatographic Parameters for Some Anilines by Molecular Connectivity. *Chromatographia*, *41*, 167–174.

Perrin, D.D., Dempsey, B. and Serjeant, E.P. (1981). *pKa Prediction for Organic Acids and Bases.* Chapman & Hall, London (UK).

Petelin, D.Y., Palyulin, V.A. and Zefirov, N.S. (1992). Topological Indexes Based on Weights of the Molecular Graph Vertices for Investigations in the Field of QSAR and QSPR. *Dokl. Akad.Nauk.SSSR*, *324*, 1019–1022.

Petitjean, M. and Dubois, J.-E. (1990). Topological Statistics on a Large Structural File. *J.Chem. Inf.Comput.Sci.*, *30*, 332–343.

Petitjean, M. (1992). Applications of the Radius-Diameter Diagram to the Classification of Topological and Geometrical Shapes of Chemical Compounds. *J.Chem.Inf.Comput.Sci.*, *32*, 331–337.

Petitjean, M. (1996). Three-Dimensional Pattern Recognition from Molecular Distance Minimization. *J.Chem.Inf.Comput.Sci.*, *36*, 1038–1049.

Piazza, R., Pino, A., Marchini, S., Passerini, L., Chiorboli, C. and Tosato, M.L. (1995). Modelling Physico-Chemical Properties of Halogenated Benzenes: QSAR Optimization through Variables Selection. *SAR & QSAR Environ.Res.*, *4*, 59–71.

Pickett, S.D., Luttmann, C., Guerin, V., Laoui, A. and James, E. (1998). DIVSEL and COMPLIB – Strategies for the Design and Comparison of Combinatorial Libraries Using Pharmacophoric Descriptors. *J.Chem.Inf.Comput.Sci.*, *38*, 144–150.

Piggott, J.R. and Withers, S.J. (1993). Modern Statistics and Quantitative Structure-Activity Relationships in Flavor. *ACS Symp.Ser.*, *528*, 100–108.

Pimentel, G.C. and McClellan, A.L. (1960). *The Hydrogen Bond*. Freeman, San Francisco (CA), 575 pp.

Pisanski, T. and Zerovnik, J. (1994). Weights on Edges of Chemical Graphs Determined by Paths. *J.Chem.Inf.Comput.Sci.*, *34*, 395–397.

Pitzer, K.S. (1940). The Vibration Frequencies and Thermodynamic Functions of Long Chain Hydrocarbons. *J.Chem.Phys.*, *8*, 711–720.

Pitzer, K.S. and Scott, D.W. (1941). The Thermodynamics of Branched-Chain Paraffins. The Heat Capacity, Heat of Fusion and Vaporization, and Entropy of 2,3,4-Trimethylpentane. *J.Am.Chem.Soc.*, *63*, 2419–2422.

Pitzer, K.S. (1955a). The Volumetric and Thermodynamic Properties of Fluids. I. Theoretical Basis and Virial Coefficients. *J.Am.Chem.Soc.*, *77*, 3427–3433.

Pitzer, K.S., Lippmann, D.Z., Curl, R.F., Huggins, C.M. and Peterson, D.E. (1955b). The Volumetric and Thermodynamic Properties of Fluids. II. Compressibility Factor, Vapor Pressure and Entropy of Vaporization. *J.Am.Chem.Soc.*, *77*, 3433–3440.

Pizarro Millán, C., Forina, M., Casolino, C. and Leardi, R. (1998). Extraction of Representative Subsets by Potential Functions Method and Genetic Algorithms. *Chemom.Intell.Lab.Syst.*, *40*, 33–52.

Platt, D.E. and Silverman, B.D. (1996). Registration, Orientation and Similarity of Molecular Electrostatic Potentials through Multipole Matching. *J.Comput.Chem.*, *17*, 358–366.

Platt, J.R. (1947). Influence of Neighbor Bonds on Additive Bond Properties in Paraffins. *J.Chem.Phys.*, *15*, 419–420.

Platt, J.R. (1952). Prediction of Isomeric Differences in Paraffin Properties. *J.Phys.Chem.*, *56*, 328–336.

Platt, J.R. (1954). The Box Model and Electron Densities in Conjugated Systems. *J.Chem.Phys.*, *22*, 1448–1455.

Platts, J.A., Butina, D., Abraham, M.H. and Hersey, A. (1999). Estimation of Molecular Linear Free Energy Relation Descriptors Using a Group Contribution Approach. *J.Chem.Inf.Comput.Sci.*, *39*, 835–845.

Platts, J.A., Abraham, M.H., Butina, D. and Hersey, A. (2000). Estimation of Molecular Linear Free Energy Relationship Descriptors by a Group Contribution Approach. 2. Prediction of Partition Coefficients. *J.Chem.Inf.Comput.Sci.*, *40*, 71–80.

Plavsic, D., Nikolic, S., Trinajstic, N. and Klein, D.J. (1993a). Relation Between the Wiener Index and the Schultz Index for Several Classes of Chemical Graphs. *Croat.Chem.Acta*, *66*, 345–353.

Plavsic, D., Nikolic, S., Trinajstic, N. and Mihalic, Z. (1993b). On the Harary Index for the Characterization of Chemical Graphs. *J.Math.Chem.*, *12*, 235–250.

Plavsic, D., Soskic, M., Landeka, I., Gutman, I. and Graovac, A. (1996a). On the Relation Between the Path Numbers ^1Z, ^2Z and the Hosoya Z Index. *J.Chem.Inf.Comput.Sci.*, *36*, 1118–1122.

Plavsic, D., Soskic, M., Landeka, I. and Trinajstic, N. (1996b). On the Relation Between the P′/P Index and the Wiener Number. *J.Chem.Inf.Comput.Sci.*, *36*, 1123–1126.

Plavsic, D., Soskic, M., Dakovic, Z., Gutman, I. and Graovac, A. (1997). Extension of the Z Matrix to Cycle-Containing and Edge-Weighted Molecular Graphs. *J.Chem.Inf.Comput.Sci.*, *37*, 529–534.

Plavsic, D., Soskic, M. and Lers, N. (1998). On the Calculation of the Molecular Descriptor χ'/χ. *J.Chem.Inf.Comput.Sci.*, *38*, 889–892.

Plavsic, D. (1999). On the Definition and Calculation of the Molecular Descriptor R'/R. *Chem. Phys.Lett.*, *304*, 111–116.

Pleiss, M.A. and Grunewald, G.L. (1983). An Extension of the *f*-Fragment Method for the Calculation of Hydrophobic Constants (Log P) of Conformationally Defined Systems. *J.Med.Chem.*, *26*, 1760–1764.

Plesnik, J. (1984). On the Sum of All Distances in a Graph or Digraph. *J.Graph Theory*, *8*, 1–21.

Plummer, E.L. (1995). Successful Application of the QSAR Paradigm in Discovery Programs. *ACS Symp.Ser.*, *606*, 240–253.

Pogliani, L. (1992a). Molecular Connectivity Model for Determination of Isoelectric Point of Amino Acids. *J.Pharm.Sci.*, *81*, 334–336.

Pogliani, L. (1992b). Molecular Connectivity: Treatment of Electronic Structure of Amino Acids. *J.Pharm.Sci.*, *81*, 967–969.

Pogliani, L. (1993a). Molecular Connectivity Model for Determination of T_1 Relaxation Times of α-Carbons of Amino Acids and Cyclic Dipeptides. *Computers Chem.*, *17*, 283–286.

Pogliani, L. (1993b). Molecular Connectivity Model for Determination of Physicochemical Properties of α-Amino Acids. *J.Phys.Chem.*, *97*, 6731–6736.

Pogliani, L. (1994a). Molecular Connectivity Descriptors of the Physicochemical Properties of the α-Amino Acids. *J.Phys.Chem.*, *98*, 1494–1499.

Pogliani, L. (1994b). On a Graph Theoretical Characterization of Cis/Trans Isomers. *J.Chem.Inf. Comput.Sci.*, *34*, 801–804.

Pogliani, L. (1994c). Structure Property Relationships of Amino Acids and Some Dipeptides. *Amino Acids*, *6*, 141–153.

Pogliani, L. (1995a). Modeling the Solubility and Activity of Amino Acids with the LCCI Method. *Amino Acids*, *9*, 217–228.

Pogliani, L. (1995b). Molecular Modeling by Linear Combinations of Connectivity Indexes. *J.Phys.Chem.*, *99*, 925–937.

Pogliani, L. (1996a). A Strategy for Molecular Modeling of a Physicochemical Property Using a Linear Combination of Connectivity Indexes. *Croat.Chem.Acta*, *69*, 95–109.

Pogliani, L. (1996b). Modeling Purines and Pyrimidines with the Linear Combination of Connectivity Indices-Molecular Connectivity "LCCI-MC" Method. *J.Chem.Inf.Comput.Sci.*, *36*, 1082–1091.

Pogliani, L. (1996c). Modeling with Special Descriptors Derived from a Medium-Sized Set of Connectivity Indices. *J.Phys.Chem.*, *100*, 18065–18077.

Pogliani, L. (1997a). Modeling Biochemicals with Leading Molecular Connectivity Terms. *Med. Chem.Res.*, *7*, 380–393.

Pogliani, L. (1997b). Modeling Enthalpy and Hydration Properties of Inorganic Compounds. *Croat.Chem.Acta*, *70*, 803–817.

Pogliani, L. (1997c). Modeling Properties of Biochemical Compounds with Connectivity Terms. *Amino Acids*, *13*, 237–255.

Pogliani, L. (1999a). Modeling Properties with Higher-Level Molecular Connectivity Descriptors. *J.Chem.Inf.Comput.Sci.*, *39*, 104–111.

Pogliani, L. (1999b). Modeling with Semiempirical Molecular Connectivity Terms. *J.Phys.Chem.*, *103*, 1598–1610.

Pogliani, L. (1999c). Properties of Molecular Connectivity Terms and Physicochemical Properties. *J.Mol.Struct.(Theochem)*, *466*, 1–19.

Pogliani, L. (2000). The Concept of Graph Mass in Molecular Graph Theory. A Case in Data Reduction Analysis. In *QSAR Studies by Molecular Descriptors* (Diudea, M.V., ed.), Nova Science Publishers, New York (NY), in press.

Polanski, J. and Rouvray, D.E. (1976a). Graph-Theoretical Treatment of Aromatic Hydrocarbons. I: The Formal Graph-Theoretical Description. *MATCH (Comm.Math.Comp.Chem.)*, *2*, 63–90.

Polanski, J. and Rouvray, D.E. (1976b). Graph-Theoretical Treatment of Aromatic Hydrocarbons. II: The Analysis of All-Benzenoid Systems. *MATCH (Comm.Math.Comp.Chem.)*, *2*, 91–109.

Polanski, J. and Bonchev, D. (1986). The Wiener Number of Graphs. I. General Theory and Changes Due to Graph Operations. *MATCH (Comm.Math.Comp.Chem.)*, *21*, 133–186.

Polanski, J. and Bonchev, D. (1987). The Minimum Distance Number of Trees. *MATCH (Comm. Math.Comp.Chem.)*, *21*, 314–344.

Polanski, J. and Bonchev, D. (1990). Theory of the Wiener Number of Graphs. II. Transfer Graphs and Some of Their Metric Properties. *MATCH (Comm.Math.Comp.Chem.)*, *25*, 3–39.

Polanski, J. (1997). The Receptor-Like Neural Network for Modeling Corticosteroid and Testosterone Binding Globulins. *J.Chem.Inf.Comput.Sci.*, *37*, 553–561.

Polanski, J., Gasteiger, J., Wagener, M. and Sadowski, J. (1998). The Comparison of Molecular Surfaces by Neural Networks and its Applications to Quantitative Structure Activity Studies. *Quant.Struct.-Act.Relat.*, *17*, 27–36.

Polansky, O.E. and Derflinger, G. (1963). Über den Zusammenhang von Bindungslängen und Elektronegativitäten. *Theor.Chim.Acta*, *1*, 308–315.

Polansky, O.E., Randic, M. and Hosoya, H. (1989). Transfer Matrix Approach to the Wiener Number of Catacondensed Benzenoids. *MATCH (Comm.Math.Comp.Chem.)*, *24*, 3–28.

Polansky, O.E. (1991). Elements of Graph Theory for Chemists. In *Chemical Graph Theory. Introduction and Fundamentals* (Bonchev, D. and Rouvray, D.H., eds.), Abacus Press/Gordon & Breach, New York (NY), pp. 42–96.

Politzer, P. (1987). A Relationship Between the Charge Capacity and the Hardness of Neutral Atoms and Groups. *J.Chem.Phys.*, *86*, 1072–1073.

Politzer, P., Lane, P., Murray, J.S. and Brinck, T. (1992). Investigation of Relationships Between Solute Molecule Surface Electrostatic Potentials and Solubilities in Supercritical Fluids. *J.Phys.Chem.*, *96*, 7938–7943.

Politzer, P., Murray, J.S., Lane, P. and Brinck, T. (1993). Relationships Between Solute Molecular Properties and Solubility in Supercritical CO_2. *J.Phys.Chem.*, *97*, 729–732.

Politzer, P., Murray, J.S. and Flodmark, P. (1996). Relationship Between Measured Diffusion Coefficients and Calculated Molecular Surface Properties. *J.Phys.Chem.*, *100*, 5538–5540.

Politzer, P., Murray, J.S., Grice, M.E., DeSalvo, M. and Miller, E. (1997). Calculation of Heats of Sublimation and Solid Phase Heats of Formation. *Mol.Phys.*, *91*, 923–928.

Politzer, P. and Murray, J.S. (1998). Relationships Between Lattice Energies and Surface Electrostatic Potentials and Areas of Anions. *Journal of Physical Chemistry A*, *102*, 1018–1020.

Politzer, P., Murray, J.S. and Abu-Awwad, F. (2000). Prediction of Solvation Free Energies from Computed Properties of Solute Molecular Surfaces. *Int.J.Quant.Chem.*, *76*, 643–647.

Pompe, M., Razinger, M., Novic, M. and Veber, M. (1997). Modelling of Gas Chromatographic Retention Indices Using Counterpropagation Neural Networks. *Anal.Chim.Acta*, *348*, 215–221.

Pompe, M. and Novic, M. (1999). Prediction of Gas-Chromatographic Retention Indices Using Topological Descriptors. *J.Chem.Inf.Comput.Sci.*, *39*, 59–67.

Ponec, R. and Strand, M. (1990). A Novel Approach to the Characterization of Molecular Similarity. The 2nd Order Similarity Index. *Collect.Czech.Chem.Comm.*, *55*, 896–902.

Ponec, R., Amat, L. and Carbó-Dorca, R. (1999). Molecular Basis of Quantitative Structure-Properties Relationships (QSPR): A Quantum Similarity Approach. *J.Comput.Aid.Molec. Des.*, *13*, 259–270.

Pople, J.A. (1953). Electron Interaction in Unsaturated Hydrocarbons. *Trans.Faraday Soc.*, *49*, 1375–1385.

Popoviciu, V., Holban, S., Badilescu, I.I. and Simon, Z. (1978). *Studia Biophys.*, *69*, 75–76.

Poshusta, R. and McHughes, M.C. (1989). Embedding Frequencies of Trees. *J.Math.Chem.*, *3*, 193–215.

Poso, A., Tuppurainen, K., Ruuskanen, J. and Gynther, J. (1993). Binding of Some Dioxins and Dibenzofurans to the Ah Receptor. A QSAR Model Based on Comparative Molecular Field Analysis (CoMFA). *J.Mol.Struct.(Theochem)*, *101*, 259–264.

Poso, A., Tuppurainen, K. and Gynther, J. (1994). Modeling of Molecular Mutagenicity with Comparative Molecular Field Analysis (CoMFA): Structural and Electronic Properties of MX Compounds Related to TA100 Mutagenicity. *J.Mol.Struct.(Theochem)*, *110*, 255–260.

Potts, R.O. and Guy, R.H. (1995). A Predictive Algorithm for Skin Permeability: The Effects of Molecular Size and Hydrogen Bond Activity. *Pharm.Res.*, *12*, 1628–1663.

Pratesi, P., Caliendo, G., Silipo, C. and Vittoria, A. (1992). A QSAR Approach to the Study of Structural Requirements of Muscarinic Receptor Ligands. 2. Antagonists. *Quant.Struct.-Act.Relat.*, *11*, 151–161.

Primas, H. (1981). *Chemistry, Quantum Mechanics and Reductionism*. Springer-Verlag, Berlin (Germany), 452 pp.

Pritchard, H.O. and Skinner, H.A. (1955). The Concept of Electronegativity. *Chem.Rev.*, *55*, 745.

Pritchard, H.O. (1963). Equalization of Electronegativity. *J.Am.Chem.Soc.*, *85*, 1876–1876.

Purcell, W.P., Bass, G.E. and Clayton, J.M. (1973). *Strategy of Drug Design. A Molecular Guide to Biological Activity*. Wiley, New York (NY).

Purdy, R. (1996). A Mechanism Mediated Model for Carcinogenicity Model Content and Prediction of the Outcome of Rodent Carcinogenicity Bioassays Currently Being Conducted on 25 Organic Chemicals. *Environ.Health Persp.*, *104*, 1085–1094.

Puri, R.D., Mirgal, S.V., Ramaa, C.S. and Kulkarni, V.M. (1996). Chromatographically Derived Hydrophobicity Parameters in QSAR Analysis of Diarylsulphone Analogs. *Indian J.Chem.*, *35B*, 1271–1274.

Pussemier, L., De Borger, R., Cloos, P. and Van Bladel, R. (1989). Relation Between the Molecular Structure and the Adsorption of Arylcarbamate, Phenylurea and Anilide Pesticides in Soil and Model Organic Adsorbents. *Chemosphere*, *18*, 1871–1882.

Qian, L., Hirono, S., Matsushita, Y. and Moriguchi, I. (1992). QSARs Based on Fuzzy Adaptive Least Squares Analysis for the Aquatic Toxicity of Organic Chemicals. *Environ.Toxicol. Chem.*, *11*, 953–959.

Quayle, O.R. (1953). The Parachors of Organic Compounds. An Interpretation and Catalogue. *Chem.Rev.*, *53*, 439–589.

Quintas, L.V. and Slater, P.J. (1981). Pairs of Non-Isomorphic Graphs Having the Same Path Degree Sequence. *MATCH (Comm.Math.Comp.Chem.)*, *12*, 75–86.

Rabinowitz, J.R. and Little, S.B. (1991). Prediction of the Reactivities of Cyclopenta-Polynuclear Aromatic Hydrocarbons by Quantum Mechanical Methods. *Xenobiotica*, *21*, 263–275.

Radecki, A., Lamparczyk, H. and Kaliszan, R. (1979). A Relationship Between the Retention Indices on Nematic and Isotropic Phases and the Shape of Polycyclcic Aromatic Hydrocarbons. *Chromatographia*, *12*, 595–599.

Raevsky, O.A., Grigor'ev, V.J., Kireev, D.B. and Zefirov, N.S. (1992a). Complete Thermodynamic Description of H Bonding in the Framework of Multiplicative Approach. *Quant.Struct.-Act.Relat.*, *11*, 49–63.

Raevsky, O.A., Grigor'ev, V.J., Kireev, D.B. and Zefirov, N.S. (1992b). Correlation Analysis and H Bond Ability in Framework of QSAR. *J.Chim.Phys.Phys-Chim.Biol.*, *89*, 1747–1753.

Raevsky, O.A., Grigor'ev, V.J. and Mednikova, E. (1993). QSAR H-Bonding Descriptions. In *Trends in QSAR and Molecular Modelling 92* (Wermuth, C.G., ed.), ESCOM, Leiden (The Netherlands), pp. 116–119.

Raevsky, O.A, Sapegin, A. and Zefirov, N. (1994). The QSAR Discriminant-Regression Model. *Quant.Struct.-Act.Relat.*, *13*, 412–418.

Raevsky, O.A, Dolmatova, L., Grigor'ev, V.J., Lisyansky, I. and Bondarev, S. (1995). Molecular Recognition Descriptors in QSAR. In *QSAR and Molecular Modelling: Cocepts, Computational Tools and Biological Applications* (Sanz, F., Giraldo, J. and Manaut, F., eds.), Prous Science, Barcelona (Spain), pp. 241–245.

Raevsky, O.A. (1997). Hydrogen Bond Strength Estimation by Means of the HYBOT Program Package. In *Computer-Assisted Lead Finding and Optimization.* (van de Waterbeemd, H., Testa, B. and Folkers, G., eds.), Wiley-VCH, Weinheim (Germany), pp. 367–378.

Ralev, N., Karabunarliev, S., Mekenyan, O., Bonchev, D. and Balaban, A.T. (1985). Unique Description of Chemical Structures Based on Hierarchically Ordered Extended Connectivities (HOC Procedures). VIII. General Principles for Computer Implementation. *J.Comput.Chem.*, *6*, 587–591.

Ramsden, C. A. ed. (1990). *Quantitative Drug Design. Vol. 4.* Pergamon Press, Oxford (UK), 766 pp.

Randic, M. (1974). On the Recognition of Identical Graphs Representing Molecular Topology. *J.Chem.Phys.*, *60*, 3920–3928.

Randic, M. (1975a). Graph Theoretical Approach to Local and Overall Aromaticity of Benzenoid Hydrocarbons. *Tetrahedron*, *31*, 1477–1481.

Randic, M. (1975b). On Characterization of Molecular Branching. *J.Am.Chem.Soc.*, *97*, 6609–6615.

Randic, M. (1975c). On Rearrangement of the Connectivity Matrix of a Graph. *J.Chem.Phys.*, *62*, 309–310.

Randic, M. (1975d). On Unique Numbering of Atoms and Unique Codes for Molecular Graphs. *J.Chem.Inf.Comput.Sci.*, *15*, 105–108.

Randic, M., Trinajstic, N. and Zivkovic, T. (1976). Molecular Graphs Having Identical Spectra. *J.Chem.Soc.Faraday Trans II*, *72*, 244–256.

Randic, M. (1977). On Canonical Numbering of Atoms in a Molecule and Graph Isomorphism. *J.Chem.Inf.Comput.Sci.*, *17*, 171–180.

Randic, M. (1978a). Fragment Search in Acyclic Structures. *J.Chem.Inf.Comput.Sci.*, *18*, 101–107.

Randic, M. (1978b). The Structural Origin of Chromatographic Retention Data. *J.Chromat.*, *161*, 1–14.

Randic, M. (1979). Characterization of Atoms, Molecules, and Classes of Molecules Based on Paths Enumeration. *MATCH (Comm.Math.Comp.Chem.)*, *7*, 5–64.

Randic, M., Brissey, G.M., Spencer, R.B. and Wilkins, C.L. (1979). Search for All Self-Avoiding Paths for Molecular Graphs. *Computers Chem.*, *3*, 5–13.

Randic, M. and Wilkins, C.L. (1979a). Graph-Based Fragment Search in Polycyclic Structures. *J.Chem.Inf.Comput.Sci.*, *19*, 23–31.

Randic, M. and Wilkins, C.L. (1979b). Graph-Theoretical Approach to Recognition of Structural Similarity in Molecules. *J.Chem.Inf.Comput.Sci.*, *19*, 31–37.

Randic, M. and Wilkins, C.L. (1979c). Graph Theoretical Ordering of Structures as a Basis for Systematic Searches for Regularities in Molecular Data. *J.Phys.Chem.*, *83*, 1525–1540.

Randic, M. and Wilkins, C.L. (1979d). Graph Theoretical Study of Structural Similarity in Benzomorphans. *Int.J.Quantum Chem.Quant.Biol.Symp.*, *6*, 55–71.

Randic, M. and Wilkins, C.L. (1979e). On a Graph Theoretical Basis for Ordering of Structures. *Chem.Phys.Lett.*, *63*, 332–336.

Randic, M. (1980a). Chemical Shift Sums. *J.Magn.Reson.*, *39*, 431–436.

Randic, M. (1980b). Graphical Enumeration of Conformations of Chains. *Int.J.Quantum Chem. Quant.Biol.Symp.*, *7*, 187–197.

Randic, M. (1980c). Random Walks and Their Diagnostic Value for Characterization of Atomic Environment. *J.Comput.Chem.*, *1*, 386–399.

Randic, M., Brissey, G.M., Spencer, R.B. and Wilkins, C.L. (1980). Use of Self-Avoiding Paths for Characterization of Molecular Graphs with Multiple Bonds. *Computers Chem.*, *4*, 27–43.

Randic, M. and Wilkins, C.L. (1980). A Procedure for Characterization of the Rings of Molecule. *J.Chem.Inf.Comput.Sci.*, *20*, 36–46.

Randic, M., Brissey, G.M. and Wilkins, C.L. (1981). Computer Perception of Topological Symmetry via Canonical Numbering of Atoms. *J.Chem.Inf.Comput.Sci.*, *21*, 52–59.

Randic, M. (1982). On Evaluation of the Characteristic Polynomial for Large Molecules. *J.Comput.Chem.*, *3*, 421–435.

Randic, M. and Woodworth, W.L. (1982). Characterization of Acyclic Graphs by Successive Dissection. *MATCH (Comm.Math.Comp.Chem.)*, *13*, 291–313.

Randic, M. (1983). On Alternative Form of the Characteristic Polynomial and the Problem of Graph Recognition. *Theor.Chim.Acta*, *62*, 485–498.

Randic, M., Kraus, G.A. and Jerman-Blazic Dzonova, B. (1983a). Ordering of Graphs as an Approach to Structure-Activity Studies. In *Chemical Applications of Topology and Graph Theory* (King, R.B., ed.), Elsevier, Amsterdam (The Netherlands), pp. 18–22.

Randic, M., Woodworth, W.L. and Graovac, A. (1983b). Unusual Random Walks. *Int.J.Quant.Chem.*, *24*, 435–452.

Randic, M. (1984a). Nonempirical Approach to Structure-Activity Studies. *Int.J.Quantum Chem.Quant.Biol.Symp.*, *11*, 137–153.

Randic, M. (1984b). On Molecular Identification Numbers. *J.Chem.Inf.Comput.Sci.*, *24*, 164–175.

Randic, M. (1986a). Compact Molecular Codes. *J.Chem.Inf.Comput.Sci.*, *26*, 136–148.

Randic, M. (1986b). Molecular ID Numbers: By Design. *J.Chem.Inf.Comput.Sci.*, *26*, 134–136.

Randic, M., Oakland, D.O. and Klein, D.J. (1986). Symmetry Properties of Chemical Graphs. IX. The Valence Tautomerism in the P_7^{3-} Skeleton. *J.Comput.Chem.*, *7*, 35–54.

Randic, M., Jerman-Blazic, B., Rouvray, D.H., Seybold, P.G. and Grossman, S.C. (1987). The Search for Active Substructures in Structure-Activity Studies. *Int.J.Quantum Chem.Quant. Biol.Symp.*, *14*, 245–260.

Randic, M. (1988a). Molecular Topographic Descriptors. *Stud.Phys.Theor.Chem.*, *54*, 101–108.

Randic, M. (1988b). On Characterization of Three-Dimensional Structures. *Int.J.Quantum Chem.Quant.Biol.Symp.*, *15*, 201–208.

Randic, M. (1988c). Ring ID Numbers. *J.Chem.Inf.Comput.Sci.*, *28*, 142–147.

Randic, M., Hansen, P.J. and Jurs, P.C. (1988a). Search for Useful Graph Theoretical Invariants of Molecular Structure. *J.Chem.Inf.Comput.Sci.*, *28*, 60–68.

Randic, M., Jericevic, Z., Sabljic, A. and Trinajstic, N. (1988b). On the Molecular Connectivity and π-Electronic Energy in Polycyclic Hydrocarbons. *Acta Phys.Pol.*, *74*, 317–330.

Randic, M. and Trinajstic, N. (1988). Composition as a Method for Data Reduction: Application to Carbon-13 NMR Chemical Shifts. *Theor.Chim.Acta*, *73*, 233–246.

Randic, M. (1989). Aromaticity in Polycyclic Conjugated Hydrocarbons Dianions. *J.Mol.Struct. (Theochem)*, *185*, 249–274.

Randic, M. and Jurs, P.C. (1989). On a Fragment Approach to Structure-Activity Correlations. *Quant.Struct.-Act.Relat.*, *8*, 39–48.

Randic, M. (1990a). Design of Molecules with Desired Properties. A Molecular Similarity Approach to Property Optimization. In *Concepts and Applications of Molecular Similarity* (Johnson, M.A. and Maggiora, G.M., eds.), Wiley, New York (NY), pp. 77–145.

Randic, M. (1990b). The Nature of the Chemical Structure. *J.Math.Chem.*, *4*, 157–184.

Randic, M., Jerman-Blazic, B. and Trinajstic, N. (1990). Development of 3-Dimensional Molecular Descriptors. *Computers Chem.*, *14*, 237–246.

Randic, M. (1991a). Correlation of Enthalpy of Octanes with Orthogonal Connectivity Indices. *J.Mol.Struct.(Theochem)*, *233*, 45–59.

Randic, M. (1991b). Generalized Molecular Descriptors. *J.Math.Chem.*, *7*, 155–168.

Randic, M. (1991c). Novel Graph Theoretical Approach to Heteroatoms in Quantitative Structure-Activity Relationships. *Chemom.Intell.Lab.Syst.*, *10*, 213–227.

Randic, M. (1991d). On Computation of Optimal Parameters for Multivariate Analysis of Structure-Property Relationship. *J.Comput.Chem.*, *12*, 970–980.

Randic, M. (1991e). Orthogonal Molecular Descriptors. *New J.Chem.*, *15*, 517–525.

Randic, M. (1991f). Resolution of Ambiguities in Structure-Property Studies by Use of Orthogonal Descriptors. *J.Chem.Inf.Comput.Sci.*, *31*, 311–320.

Randic, M. (1991g). Search for Optimal Molecular Descriptors. *Croat.Chem.Acta*, *64*, 43–54.

Randic, M. (1992a). Chemical Structure. What is "She"? *J.Chem.Educ.*, *69*, 713–718.

Randic, M. (1992b). In Search of Structural Invariants. *J.Math.Chem.*, *9*, 97–146.

Randic, M. (1992c). Representation of Molecular Graphs by Basic Graphs. *J.Chem.Inf.Comput.Sci.*, *32*, 57–69.

Randic, M. (1992d). Similarity Based on Extended Basis Descriptors. *J.Chem.Inf.Comput.Sci.*, *32*, 686–692.

Randic, M. (1993a). Comparative Regression Analysis. Regressions Based on a Single Descriptor. *Croat.Chem.Acta*, *66*, 289–312.

Randic, M. (1993b). Fitting of Nonlinear Regressions by Orthogonalized Power Series. *J.Comput.Chem.*, *14*, 363–370.

Randic, M. (1993c). Novel Molecular Descriptor for Structure-Property Studies. *Chem.Phys. Lett.*, *211*, 478–483.

Randic, M., Guo, X., Oxley, T. and Krishnapriyan, H. (1993a). Wiener Matrix: Source of Novel Graph Invariants. *J.Chem.Inf.Comput.Sci.*, *33*, 709–716.

Randic, M., Mihalic, Z., Nikolic, S. and Trinajstic, N. (1993b). Graph-Theoretical Correlations – Artifacts or Facts? *Croat.Chem.Acta*, *66*, 411–434.

Randic, M. and Seybold, P.G. (1993). Molecular Shape as a Critical Factor in Structure-Property-Activity Studies. *SAR & QSAR Environ.Res.*, *1*, 77–85.

Randic, M. and Trinajstic, N. (1993a). In Search for Graph Invariants of Chemical Interest. *J.Mol.Struct.*, *300*, 551–571.

Randic, M. and Trinajstic, N. (1993b). Viewpoint 4 – Comparative Structure-Property Studies: the Connectivity Basis. *J.Mol.Struct.(Theochem)*, *284*, 209–221.

Randic, M. (1994a). Curve-Fitting Paradox. *Int.J.Quantum Chem.Quant.Biol.Symp.*, *21*, 215–225.

Randic, M. (1994b). Hosoya Matrix – A Source of New Molecular Descriptors. *Croat.Chem. Acta*, *67*, 415–429.

Randic, M., Guo, X., Oxley, T., Krishnapriyan, H. and Naylor, L. (1994a). Wiener Matrix Invariants. *J.Chem.Inf.Comput.Sci.*, *34*, 361–367.
Randic, M., Kleiner, A.F. and DeAlba, L.M. (1994b). Distance/Distance Matrices. *J.Chem.Inf. Comput.Sci.*, *34*, 277–286.
Randic, M., Mihalic, Z., Nikolic, S. and Trinajstic, N. (1994c). Graphical Bond Orders: Novel Structural Descriptors. *J.Chem.Inf.Comput.Sci.*, *34*, 403–409.
Randic, M. and Trinajstic, N. (1994). Isomeric Variations in Alkanes: Boiling Points of Nonanes. *New J.Chem.*, *18*, 179–189.
Randic, M. (1995a). Molecular Profiles. Novel Geometry-Dependent Molecular Descriptors. *New J.Chem.*, *19*, 781–791.
Randic, M. (1995b). Molecular Shape Profiles. *J.Chem.Inf.Comput.Sci.*, *35*, 373–382.
Randic, M. (1995c). Restricted Random Walks on Graphs. *Theor.Chim.Acta*, *92*, 97–106.
Randic, M. and Razinger, M. (1995a). Molecular Topographic Indices. *J.Chem.Inf.Comput.Sci.*, *35*, 140–147.
Randic, M. and Razinger, M. (1995b). On Characterization of Molecular Shapes. *J.Chem.Inf. Comput.Sci.*, *35*, 594–606.
Randic, M. (1996a). Molecular Bonding Profiles. *J.Math.Chem.*, *19*, 375–392.
Randic, M. (1996b). Orthosimilarity. *J.Chem.Inf.Comput.Sci.*, *36*, 1092–1097.
Randic, M. (1996c). Quantitative Structure-Property Relationship – Boiling Points of Planar Benzenoids. *New J.Chem.*, *20*, 1001–1009.
Randic, M., Klein, D.J., El-Basil, S. and Calkins, P. (1996a). Resonance in Large Benzenoid Hydrocarbons. *Croat.Chem.Acta*, *69*, 1639–1660.
Randic, M. and Krilov, G. (1996). Bond Profiles for Cuboctahedron and Twist Cuboctahedron. *Int.J.Quantum Chem.Quant.Biol.Symp.*, *23*, 127–139.
Randic, M. and Mezey, P.G. (1996). Palindromic Perimeter Codes and Chirality Properties of Polyhexes. *J.Chem.Inf.Comput.Sci.*, *36*, 1183–1186.
Randic, M., Morales, D.A. and Araujo, O. (1996b). Higher-Order Fibonacci Numbers. *J.Math. Chem.*, *20*, 79–94.
Randić, M. and Razinger, M. (1996). Molecular Shapes and Chirality. *J.Chem.Inf.Comput.Sci.*, *36*, 429–441.
Randic, M. (1997a). Linear Combinations of Path Numbers as Molecular descriptors. *New J.Chem.*, *21*, 945–951.
Randic, M. (1997b). On Characterization of Chemical Structure. *J.Chem.Inf.Comput.Sci.*, *37*, 672–687.
Randic, M. (1997c). On Characterization of Cyclic Structures. *J.Chem.Inf.Comput.Sci.*, *37*, 1063–1071.
Randic, M. (1997d). On Molecular Branching. *Acta Chim.Sloven.*, *44*, 57–77.
Randic, M. (1997e). Resonance in Catacondensed Benzenoid Hydrocarbons. *Int.J.Quant.Chem.*, *63*, 585–600.
Randic, M. and DeAlba, L.M. (1997). Dense Graphs and Sparse Matrices. *J.Chem.Inf.Comput.Sci.*, *37*, 1078–1081.
Randic, M. and Krilov, G. (1997a). Characterization of 3-D Sequences of Proteins. *Chem.Phys. Lett.*, *272*, 115–119.
Randic, M. and Krilov, G. (1997b). On Characterization of Molecular Surfaces. *Int.J.Quant.Chem.*, *65*, 1065–1076.
Randic, M., Müller, W.R., Knop, J.V. and Trinajstic, N. (1997a). The Characteristic Polynomial as a Structure Discriminator. *J.Chem.Inf.Comput.Sci.*, *37*, 1072–1077.
Randic, M., Plavsic, D. and Razinger, M. (1997b). Double Invariants. *MATCH (Comm.Math. Comp.Chem.)*, *35*, 243–259.
Randic, M. and Razinger, M. (1997). On Characterization of 3D Molecular Structure. In *From Chemical Topology to Three-Dimensional Geometry* (Balaban, A.T., ed.), Plenum Press, New York (NY), pp. 159–236.
Randic, M. (1998a). On Characterization of Molecular Attributes. *Acta Chim.Sloven.*, *45*, 239–252.

Randic, M. (1998b). On Structural Ordering and Branching of Acyclic Saturated Hydrocarbons. *J.Math.Chem.*, *24*, 345–358.

Randic, M. (1998c). Topological Indices. In *Encyclopedia of Computational Chemistry* (von Schleyer, P.R., ed.), Wiley, London (UK).

Randic, M., DeAlba, L.M. and Harris, F.E. (1998a). Graphs with the Same Detour Matrix. *Croat.Chem.Acta*, *71*, 53–68.

Randic, M. and Dobrowolski, J.Cz. (1998b). Optimal Molecular Connectivity Descriptors for Nitrogen-Containing Molecules. *Int.J.Quant.Chem.*, *70*, 1209–1215.

Randic, M., El-Basil, S., Nikolic, S. and Trinajstic, N. (1998c). Clar Polynomials of Large Benzenoid Systems. *J.Chem.Inf.Comput.Sci.*, *38*, 563–574.

Randic, M. (1999a). On Characterization of Shape of Molecular Graphs. *Journal of Molecular Modeling*, *in press*.

Randic, M. and Basak, S.C. (1999b). Multiple Regression Analysis with Optimal Molecular Descriptors. *SAR & QSAR Environ.Res.*, *in press*.

Randic, M. and Basak, S.C. (1999c). Optimal Molecular Descriptors Based on Weighted Path Numbers. *J.Chem.Inf.Comput.Sci.*, *39*, 261–266.

Randic, M. and Guo, X. (1999). Giant Benzenoid Hydrocarbons. Superphenalene Resonance Energy. *New J.Chem.*, *23*, 251–260.

Randic, M. and Krilov, G. (1999). On a Characterization of the Folding of Proteins. *Int.J.Quant.Chem.*, *75*, 1017–1026.

Randic, M. and Pompe, M. (1999). On Characterization of the CC Double Bond in Alkenes. *SAR & QSAR Environ.Res.*, *10*.

Rarey, M. and Dixon, J.S. (1998). Feature Trees: A New Molecular Similarity Measure Based on Tree Matching. *J.Comput.Aid.Molec.Des.*, *12*, 471–490.

Rashevsky, N. (1955). Life, Information Theory and Topology. *Bull.Math.Biophys.*, *17*, 229–235.

Rashevsky, N. (1960). Life, Information Theory, Probability, and Physics. *Bull.Math.Biophys.*, *22*, 351–364.

Rastelli, G., Costantino, L. and Albasini, A. (1995). Theoretical and Experimental Study of Flavones as Inhibitors of Xanthine Oxidase. *Eur.J.Med.Chem.*, *30*, 141–146.

Ravanel, P., Taillander, G., Tissut, M. and Benoit-Guyod, J.L. (1985). Effects of Chlorophenols on Isolated Plant Mitochondria Activities: A QSAR Study. *Ecotox.Environ.Safe.*, *9*, 300–320.

Rawlings, J.O. (1988). *Applied Regression Analysis.* Wadsworth & Brooks/Cole, Pacific Grove (CA).

Ray, S.K., Basak, S.C., Raychaudhury, C., Roy, A.B. and Ghosh, J.J. (1981). Quantitative Structure-Activity Relationship Studies of Bioactive Molecules Using Structural Information Indices. *Indian J.Chem.*, *20B*, 894–897.

Ray, S.K., Basak, S.C., Raychaudhury, C., Roy, A.B. and Ghosh, J.J. (1982). A Quantitative Structure-Activity Relationship Study of N-Alkylnorketobemidones and Triazinones Using Structural Information Content. *Arzneim.Forsch.*, *32*, 322–325.

Ray, S.K., Basak, S.C., Raychaudhury, C., Roy, A.B. and Ghosh, J.J. (1983). The Utility of Information Content, Structural Information Content, Hydrophobicity and van der Waals Volume in the Design of Barbiturates and Tumor Inhibitory Triazenes. *Arzneim.Forsch.*, *33*, 352–356.

Ray, S.K., Gupta, D.K., Basak, S.C., Raychaudhury, C., Roy, A.B. and Ghosh, J.J. (1985). Weighted Information Indices & Anxioselective Anxiolytic Drug Design. *Indian J.Chem.*, *24B*, 1149–1153.

Raychaudhury, C., Ray, S.K., Ghosh, J.J., Roy, A.B. and Basak, S.C. (1984). Discrimination of Isomeric Structures Using Information Theoretic Topological Indices. *J.Comput.Chem.*, *5*, 581–588.

Raychaudhury, C. and Klopman, G. (1990). *Bull.Soc.Chim.Belg.*, *99*, 255.

Raychaudhury, C., Banerjee, A., Bag, P. and Roy, S. (1999). Topological Shape and Size of Peptides: Identification of Potential Allele Specific Helper T Cell Antigenic Sites. *J.Chem.Inf.Comput.Sci.*, *39*, 248–254.

Raychaudhury, C. and Nandy, A. (1999). Indexing Scheme and Similarity Measures for Macromlecular Sequences. *J.Chem.Inf.Comput.Sci.*, *39*, 243–247.

Razinger, M. (1982). Extended Connectivity in Chemical Graphs. *Theor.Chim.Acta*, *61*, 581–586.

Razinger, M., Chrétien, J.R. and Dubois, J.-E. (1985). Structural Selectivity of Topological Indexes in Alkane Series. *J.Chem.Inf.Comput.Sci.*, *25*, 23–27.

Razinger, M. (1986). Discrimination and Ordering of Chemical Structures by the Number of Walks. *Theor.Chim.Acta*, *70*, 365–378.

Read, R.C. and Corneil, D.G. (1977). The Graph Isomorphism Desease. *J.Serb.Chem.Soc.*, *1*, 339–363.

Reddy, K.N. and Locke, M.A. (1994a). Prediction of Soil Sorption of Herbicides Using Semi-Empirical Molecular Properties. *Weed Science*, *42*, 453–461.

Reddy, K.N. and Locke, M.A. (1994b). Relationships Between Molecular Properties and log P and Soil Sorption (Koc) of Substitued Phenylureas: QSAR models. *Chemosphere*, *28*, 1929–1941.

Reddy, K.N. and Locke, M.A. (1996). Molecular Properties as Descriptors of Octanol/Water Partition Coefficients of Herbicides. *Water Air Soil Pollution*, *86*, 389–405.

Reddy, K.N., Dayan, F.E. and Duke, S.O. (1998). QSAR Analysis of Protoporphyrinogen Oxidase Inhibitors. In *Comparative QSAR* (Devillers, J., ed.), Taylor & Francis, Washington (DC), pp. 197–233. [R]

Reichardt, C. (1965). Empirical Parameters of the Polarity of Solvents. *Angew.Chem.Int.Ed. Engl.*, *4*, 29–39.

Reichardt, C. and Dimroth, K. (1968). *Fortschr.Chem.Forsch.*, *11*, 1.

Reichardt, C. (1990). *Solvents and Solvent Effects in Organic Chemistry.* VCH, New York (NY).

Reid, R.C., Prausnitz, J.M. and Poling, B.E. (1988). *The Properties of Gases and Liquids.* McGraw-Hill, New York (NY).

Reinhard, M. and Drefahl, A. (1999). *Handbook for Estimating Physicochemical Properties of Organic Compounds.* Wiley, New York (NY), 228 pp.

Rekker, R.F. (1977a). *The Hydrophobic Fragment Constant.* Elsevier, Amsterdam (The Netherlands).

Rekker, R.F. (1977b). *The Hydrophobic Fragmental Constant. Its Derivation and Applications. A Means of Characterizing Membrane Systems.* Elsevier, Amsterdam (The Netherlands), 390 pp.

Rekker, R.F. and De Kort, H.M. (1979). The Hydrophobic Fragmental Constant: An Extension to a 1000 Data Point Set. *Eur.J.Med.Chem.*, *14*, 479–488.

Rekker, R.F. (1992). The History of Drug Research From Overton to Hansch. *Quant.Struct.-Act.Relat.*, *11*, 195–199.

Rekker, R.F. and Mannhold, R. (1992). *Calculation of Drug Lipophilicity. The Hydrophobic Fragmental Constant Approach.* VCH Publishers, Weinheim (Germany).

Rekker, R.F. and de Vries, G. (1993). A Basic Confrontation of Rekker's Revised Σf-System with HPLC Retention Data Obtained on a Mixed Series of Aliphatic and Aromatic Hydrocarbons. In *Trends in QSAR and Molecular Modelling 92* (Wermuth, C.G., ed.), ESCOM, Leiden (The Netherlands), pp. 132–136.

Rekker, R.F., ter Laak, A.M. and Mannhold, R. (1993). On the Reliability of Calculated Log P-values: Rekker, Hansch/Leo and Suzuki Approach. *Quant.Struct.-Act.Relat.*, *12*, 152–157.

Rekker, R.F., Mannhold, R., Bijloo, G.J., de Vries, G. and Dross, K. (1998). The Lipophilic Behaviour of Organic Compounds: 2. The Development of an Aliphatic Hydrocarbon/Water Fragmental System via Interconnection with Octanol/Water Partitioning Data. *Quant.Struct.-Act.Relat.*, *17*, 537–548.

Ren, B. (1999). A New Topological Index for QSPR of Alkanes. *J.Chem.Inf.Comput.Sci.*, *39*, 139–143.

Reynolds, C.A., Burt, C. and Richards, W.G. (1992a). A Linear Molecular Similarity Index. *Quant.Struct.-Act.Relat.*, *11*, 34–35.

Reynolds, C.A., Essex, J.W. and Richards, W.G. (1992b). Atomic Charges for Variable Molecular Conformations. *J.Am.Chem.Soc.*, *114*, 9075–9079.

Reynolds, C.H. (1995). Estimating Lipophilicity Using the GB/SA Continuum Solvation Model: A Direct Method for Computing Partition Coefficients. *J.Chem.Inf.Comput.Sci.*, *35*, 738–742.

Reynolds, C.H., Druker, R. and Pfahler, L.B. (1998). Lead Discovery Using Stochastic Cluster Analysis (SCA): A New Method for Clustering Structurally Similar Compounds. *J.Chem.Inf. Comput.Sci.*, *38*, 305–312.

Reynolds, W.F. (1983). Polar Substituent Effects. *Prog.Phys.Org.Chem.*, *14*, 165–203.

Reynolds, W.F. and Topsom, R.D. (1984). Field and Resonance Substituent Constants for Aromatic Derivatives: Limitations of Swain's Revised *F* and *R* Constants for Predicting Aromatic Substituent Effects. *J.Org.Chem.*, *49*, 1989–1992.

Rhyu, K.-B., Patel, H.C. and Hopfinger, A.J. (1995). A 3D-QSAR Study of Anticoccidial Triazines Using Molecular Shape Analysis. *J.Chem.Inf.Comput.Sci.*, *35*, 771–778.

Richard, A.J. and Kier, L.B. (1980). SAR Analysis of Hydrazide Monoamine Oxidase Inhibitors Using Molecular Connectivity. *J.Pharm.Sci.*, *69*, 124.

Richard, A.M. (1991). Quantitative Comparison of Molecular Electrostatic Potentials for Structure-Activity Studies. *J.Comput.Chem.*, *12*, 959–969.

Richard, A.M. and Hunter, E.S. (1996). Quantitative Structure-Activity Relationships for the Developmental Toxicity of Haloacetic Acids in Mammalian Whole Embryo Culture. *Teratology*, *53*, 352–360.

Richards, F.M. (1977). Areas, Volumes, Packing, and Protein Structures. *Ann.Rev.Biophys.Bioeng.*, *6*, 151–176.

Richards, W.G. and Hodgkin, E.E. (1988). Molecular Similarity. *Chemistry in Britain*, 1141–1144.

Richards, W.G. (1993). Molecular Similarity. In *Trends in QSAR and Molecular Modelling 92* (Wermuth, C.G., ed.), ESCOM, Leiden (The Netherlands), pp. 203–206.

Richards, W.G. (1995). Molecular Similarity and Dissimilarity. In *Modelling of Biomolecular Structures and Mechanisms* (Pullman, A., Jortner, J. and Pullman, B., eds.), Kluwer, Dordrecht (The Netherlands), pp. 365–369.

Richet, M.C. (1893). Noté sur la Rapport entre la Toxicité et les Propriétés Physiques des Corps. *Compt.Rend.Soc.Biol.(Paris)*, *45*, 775–776.

Ridings, J.E., Manallack, D.T., Saunders, M.R., Baldwin, J.A. and Livingstone, D.J. (1992). Multivariate Quantitative Structure-Toxicity Relationships in a Series of Dopamine Mimetics. *Toxicology*, *76*, 209–217.

Rios-Santamarina, I., Garcìa-Domenech, R., Gálvez, J., Cortijo, J., Santamaria, P. and Marcillo, E. (1998). New Bronchodilators Selected by Molecular Topology. *Bioorg.Med.Chem.Lett.*, *8*, 477–482.

Robbat Jr., A., Corso, N.P., Doherty, P.J. and Marshall, D. (1986a). Multivariate Relationships Between Gas Chromatographic Retention Index and Molecular Connectivity of Mononitrated Polycyclic Aromatic Hydrocarbons. *Anal.Chem.*, *58*, 2072–2077.

Robbat Jr., A., Corso, N.P., Doherty, P.J. and Wolf, M.H. (1986b). Gas Chromatographic Chemiluminescent Detection and Evaluation of Predictive Models for Identifying Nitrated Polycyclic Aromatic Hydrocarbons in a Diesel Fuel Particulate Extract. *Anal.Chem.*, *58*, 2078–2084.

Robert, D. and Carbó-Dorca, R. (1998a). A Formal Comparison Between Molecular Quantum Similarity Measures and Indices. *J.Chem.Inf.Comput.Sci.*, *38*, 469–475.

Robert, D. and Carbó-Dorca, R. (1998b). Analyzing the Triple Density Molecular Quantum Similarity Measures with the INDSCAL Model. *J.Chem.Inf.Comput.Sci.*, *38*, 620–623.

Robert, D., Amat, L. and Carbó-Dorca, R. (1999). Three-Dimensional Quantitative Structure-Activity Relationships from Tuned Molecular Quantum Similarity Measures: Prediction of the Corticosteroid-Binding Globulin Binding Affinity for a Steroid Family. *J.Chem.Inf.Comput.Sci.*, *39*, 333–344. [R]

Roberts, J.D. and Moreland, W.T. (1953). Electrical Effects of Substituent Groups in Saturated Systems. Reactivities of 4-Substituted Bicyclo[2.2.2]octane-1-carboxylic Acids. *J.Am.Chem.Soc.*, *75*, 2167–2173.

Robinson, D.D., Barlow, T.W. and Richards, W.G. (1997a). Reduced Dimensional Representations of Molecular Structure. *J.Chem.Inf.Comput.Sci.*, *37*, 939–942.

Robinson, D.D., Barlow, T.W. and Richards, W.G. (1997b). The Utilization of Reduced Dimensional Representations of Molecular Structure for Rapid Molecular Similarity Calculations. *J.Chem.Inf.Comput.Sci.*, *37*, 943–950.

Robinson, D.D., Lyne, P.D. and Richards, W.G. (1999). Alignment of 3D-Structures by the Method of 2D-Projections. *J.Chem.Inf.Comput.Sci.*, *39*, 594–600.

Rodriguez Delgado, M.A., Sanchez, M.J., Gonzalez, V. and Garcìa-Montelongo, F. (1993). Correlations Between Retention Data of Polycyclic Aromatic Hydrocarbons in Micellar Liquid Chromatography and Several Molecular Descriptors. *Fresen.J.Anal.Chem.*, *345*, 748–752.

Rogers, D. (1991). G/SPLINES: A Hybrid of Friedman's Multivariate Adaptive Regression Splines (MARS) Algorithm with Holland's Genetic Algorithm. In *The Proceedings of the Fourth International Conference on Genetic Algorithms*, San Diego (CA).

Rogers, D. (1992). Data Analysis Using G/SPLINES. In *Advances in Neural Processing Systems 4*Kaufmann, San Mateo (CA).

Rogers, D. and Hopfinger, A.J. (1994). Application of Genetic Function Approximation to Quantitative Structure-Activity Relationships and Quantitative Structure-Property Relationships. *J.Chem.Inf.Comput.Sci.*, *34*, 854–866.

Rogers, D. (1995). Genetic Function Approximation: A Genetic Approach to Building Quantitative Structure-Activity Relationship Models. In *QSAR and Molecular Modelling: Cocepts, Computational Tools and Biological Applications* (Sanz, F., Giraldo, J. and Manaut, F., eds.), Prous Science, Barcelona (Spain), pp. 420–426.

Rohrbaugh, R.H. and Jurs, P.C. (1985). Prediction of Gas Chromatographic Retention Indexes of Selected Olefins. *Anal.Chem.*, *57*, 2770–2773.

Rohrbaugh, R.H. and Jurs, P.C. (1986). Prediction of Gas Chromatographic Retention Indexes of Polycyclic Aromatic Compounds and Nitrated Polycyclic Aromatic Compounds. *Anal. Chem.*, *58*, 1210–1212.

Rohrbaugh, R.H. and Jurs, P.C. (1987a). Descriptions of Molecular Shape Applied in Studies of Structure/Activity and Structure/Property Relationships. *Anal.Chim.Acta*, *199*, 99–109.

Rohrbaugh, R.H. and Jurs, P.C. (1987b). Molecular Shape and the Prediction of High-Performance Liquid Chromatographic Retention Indexes of Polycyclic Aromatic Hydrocarbons. *Anal.Chem.*, *59*, 1048–1054.

Romanowska, K. (1992). The Application of the Graph Theoretical Method in the QSAR Scheme Possibilities and Limits. *Int.J.Quant.Chem.*, *43*, 175–195.

Rorije, E., Van Wezel, M.C. and Peijnenburg, W.J.G.M. (1995). On the Use of Backpropagation Neural Networks in Modeling Environmental Degradation. *SAR & QSAR Environ.Res.*, *4*, 219–235.

Rorije, E. and Peijnenburg, W.J.G.M. (1996). QSARs for Oxidation of Phenols in the Aqueous Environment, Suitable for Risk Assessment. *J.Chemom.*, *10*, 79–93.

Rose, V.S., Wood, J. and MacFie, H.J.H. (1991). Single Class Discrimination Using Principal Component Analysis (Scd PCA). *Quant.Struct.-Act.Relat.*, *10*, 359–368.

Rose, V.S., Wood, J. and MacFie, H.J.H. (1992). Generalized Single Class Discrimination (GSCD). A New Method for the Analsyis of Embedded Structure-Activity Relationships. *Quant.Struct.-Act.Relat.*, *11*, 492–504.

Rose, V.S. and Wood, J. (1998). Generalized Cluster Significance Analysis and Stepwise Cluster Significance Analysis with Conditional Probabilities. *Quant.Struct.-Act.Relat.*, *17*, 348–356.

Rosen, R. (1990). An Approach to Molecular Similarity. In *Concepts and Applications of Molecular Similarity* (Johnson, M.A. and Maggiora, G.M., eds.), Wiley, New York (NY), pp. 369–382.

Roussel, C., Piras, P. and Heitmann, I. (1997). An Approach to Discriminating 25 Commercial Chiral Stationary Phases from Structural Data Sets Extracted from a Molecular Database. *Biomedical Chromatography*, *11*, 311–316.

Rouvray, D.E. (1986). The Role of the Topological Distance Matrix in Chemistry. In *Mathematical and Computational Concepts in Chemistry* (Trinajstic, N., ed.), Ellis Horwood, Chichester (UK), pp. 295–306.

Rouvray, D.E. (1988). Novel Applications of Topological Indices. *J.Mol.Struct.(Theochem)*, *165*, 9–20.

Rouvray, D.H. (1973). The Search for Useful Topological Indices in Chemistry. *Am.Sci.*, *61*, 729–735.

Rouvray, D.H. (1975). The Value of Topological Indices in Chemistry. *MATCH (Comm.Math. Comp.Chem.)*, *1*, 125–134.

Rouvray, D.H. (1976). The Topological Matrix in Quantum Chemistry. In *Chemical Applications of Graph Theory* (Balaban, A.T., ed.), Academic Press, New York (NY), pp. 175–222.

Rouvray, D.H. and Balaban, A.T. (1979). Chemical Applications of Graph Theory. In *Applications of Graph Theory* (Wilson, R.J. and Beineke, L.W., eds.), Academic Press, London (UK), pp. 177–221.

Rouvray, D.H. (1986a). Predicting Chemistry from Topology. *Sci.Am.*, *24*, 40–47.

Rouvray, D.H. (1986b). The Prediction of Biological Activity Using Molecular Connectivity Indices. *Acta Pharm.Jugosl.*, *36*, 239–252.

Rouvray, D.H. and Pandey, R.B. (1986). The Fractal Nature, Graph Invariants, and Physicochemical Properties of Normal Alkanes. *J.Chem.Phys.*, *85*, 2286–2290.

Rouvray, D.H. (1987). The Modeling of Chemical Phenomena Using Topological Indices. *J.Comput.Chem.*, *8*, 470–480.

Rouvray, D.H. (1988). The Challenge of Characterizing Branching in Molecular Species. *Disc. Appl.Math.*, *19*, 317–338.

Rouvray, D.H. and El-Basil, S. (1988). Novel Applications of Topological Indices. Part 4. Correlation of Arene Absorption Spectra with the Randic Molecular Connectivity Index. *J.Mol.-Struct.(Theochem)*, *165*, 9–20.

Rouvray, D.H. (1989). The Limits of Applicability of Topological Indices. *J.Mol.Struct.(Theochem)*, *185*, 187–201.

Rouvray, D. H. ed. (1990a). *Computational Chemical Graph Theory.* Nova Press, New York (NY).

Rouvray, D.H. (1990b). The Evolution of the Concept of Molecular Similarity. In *Concepts and Applications of Molecular Similarity* (Johnson, M.A. and Maggiora, G.M., eds.), Wiley, New York (NY), pp. 15–42.

Rouvray, D.H. (1991). The Origins of Chemical Graph Theory. In *Chemical Graph Theory. Introduction and Fundamentals* (Bonchev, D. and Rouvray, D.H., eds.), Abacus Press/Gordon & Breach, New York (NY), pp. 1–39.

Rouvray, D.H. and Kumazaki, H. (1991). Prediction of Molecular Flexibility in Halogenated Alkanes via Fractal Dimensionality. *J.Math.Chem.*, *7*, 169–185.

Rouvray, D.H. ed. (1997). *Fuzzy Logic in Chemistry.* Academic Press, New York (NY), 364 pp.

Rovero, P., Riganelli, D., Fruci, D., Vigano, S., Pegoraro, S., Revoltella, R., Greco, G., Butler, R., Clementi, S. and Tanigaki, N. (1994). The Importance of Secondary Anchor Residue Motifs of HLA Class I Proteins: A Chemometric Approach. *Mol.Immunol.*, *31*, 549–554.

Rowberg, K.A., Even, M., Martin, E. and Hopfinger, A.J. (1994). QSAR and Molecular Shape Analyses of Three Series of 1-(Phenylcarbamoyl)-2-Pyrazoline Insecticides. *J.Agr.Food Chem.*, *42*, 374–380.

Roy, A.B., Raychaudhury, C., Ghosh, A., Ray, S.K. and Basak, S.C. (1983). Information-Theoretic Topological Indices of a Molecule and Their Applications in QSAR. In *Quantitative Approaches to Drug Design* (Dearden, J.C., ed.), Elsevier, Amstedam (The Netherlands), pp. 75–76.

Roy, A.B., Basak, S.C., Harriss, D.K. and Magnuson, V.R. (1984). Neighborhood Complexities and Symmetry of Chemical Graphs and their Biological Applications. In *Mathematical Modelling in Science and Technology* (Avula, X.J.R., Kalman, R.E., Liapis, A.I. and Rodin, E.Y., eds.), Pergamon Press, New York (NY), pp. 745–750.

Roy, T.A., Krueger, A.J., Mackerer, C.R., Neil, W., Arroyo, A.M. and Yang, J.J. (1998). SAR Models for Estimating the Percutaneous Absorption of Polynuclear Aromatic Hydrocarbons. *SAR & QSAR Environ.Res.*, *9*, 171–185.

Rücker, C. and Rücker, G. (1992). Understanding the Properties of Isospectral Points and Pairs in Graphs: The Concept of Orthogonal Relation. *J.Math.Chem.*, *9*, 207–238.

Rücker, C. and Rücker, G. (1994). Mathematical Relation Between Extended Connectivity and Eigenvector Coefficients. *J.Chem.Inf.Comput.Sci.*, *34*, 534–538.

Rücker, G. and Rücker, C. (1990). Computer Perception of Constitutional (Topological) Symmetry: TOPSYM, a Fast Algorithm for Partitioning Atoms and Pairwise Relations among Atoms into Equivalence Classes. *J.Chem.Inf.Comput.Sci.*, *30*, 187–191.

Rücker, G. and Rücker, C. (1991a). Isocodal and Isospectral Points, Edges, and Pairs in Graphs and How to Cope with Them in Computerized Symmetry Recognition. *J.Chem.Inf.Comput.Sci.*, *31*, 422–427.

Rücker, G. and Rücker, C. (1991b). On Using the Adjacency Matrix Power Method for Perception of Symmetry and for Isomorphism Testing of Highly Intricate Graphs. *J.Chem.Inf.Comput.Sci.*, *31*, 123–126.

Rücker, G. and Rücker, C. (1993). Counts of All Walks as Atomic and Molecular Descriptors. *J.Chem.Inf.Comput.Sci.*, *33*, 683–695.

Rücker, G. and Rücker, C. (1998). Symmetry-Aided Computation of the Detour Matrix and the Detour Index. *J.Chem.Inf.Comput.Sci.*, *38*, 710–714.

Rücker, G. and Rücker, C. (1999). On Topological Indices, Boiling Points, and Cycloalkanes. *J.Chem.Inf.Comput.Sci.*, *39*, 788–802.

Rücker, G. and Rücker, C. (2000). Walk Counts, Labyrinthicity, and Complexity of Acyclic and Cyclic Graphs and Molecules. *J.Chem.Inf.Comput.Sci.*, *40*, 99–106.

Ruedenberg, K. (1958). Theorem on the Mobile Bond Orders of Alternant Conjugated Systems. *J.Chem.Phys.*, *29*, 1232–1233.

Rum, G. and Herndon, W.C. (1991). Molecular Similarity Concepts. 5. Analysis of Steroid-Protein Binding Constants. *J.Am.Chem.Soc.*, *113*, 9055–9060.

Russell, C.J., Dixon, S.L. and Jurs, P.C. (1992). Computer-Assisted Study of the Relationship Between Molecular Structure and Henry's Law Constant. *Anal.Chem.*, *64*, 1350–1355.

Russom, C.L., Bradbury, S.P., Broderius, S.J., Hemmermeister, D.E. and Drummond, R.A. (1997). Predicting Modes of Toxic Action from Chemical Structure: Acute Toxicity in the Fathed minnow (*Pimephales promelas*). *Environ.Toxicol.Chem.*, *16*, 948–967.

Ryan, T.P. (1997). *Modern Regression Methods.* Wiley, NewYork (NY), 516 pp.

Sabljic, A. and Trinajstic, N. (1981). Quantitative Structure-Activity Relationships: The Role of Topological Indices. *Acta Pharm.Jugosl.*, *31*, 189–214.

Sabljic, A. and Protic, M. (1982). Relationship Between Molecular Connectivity Indices and Soil Sorption Coefficients of Polycyclic Aromatic Hydrocarbons. *Bull.Environ.Contam.Toxicol.*, *28*, 162–165.

Sabljic, A. (1983). Quantitative Structure-Toxicity Relationship of Chlorinated Compounds: A Molecular Connectivity Investigation. *Bull.Environ.Contam.Toxicol.*, *30*, 80–83.

Sabljic, A. and Protic-Sabljic, M. (1983). Quantitative Structure-Activity Study on the Mechanism of Inhibition of Microsomal *p*-Hydroxylation of Aniline by Alcohols. *Mol.Pharm.*, *23*, 213–218.

Sabljic, A. (1984). Predictions of the Nature and Strength of Soil Sorption of Organic Pollutants by Molecular Topology. *J.Agr.Food Chem.*, *32*, 243–246.

Sabljic, A. and Protic, M. (1984). Molecular Connectivity: A Novel Method for Prediction of Bioconcentration Factor of Hazardous Chemicals. *Chem.-Biol.Inter.*, *42*, 301–310.

Sabljic, A. (1985). Calculation of Retention Indices by Molecular Topology. Chlorinated Benzenes. *J.Chromat.*, *319*, 1–8.

Sabljic, A. (1987). On the Prediction of Soil Sorption Coefficients of Organic Pollutants from Molecular Structures: Application of Molecular Connectivity Model. *Environ.Sci.Technol.*, *21*, 358–366.

Sabljic, A. (1988). Application of Molecular Topology for the Estimation of Physical Data for Environmental Chemicals. In *Physical Property Prediction in Organic Chemistry* (Jochum, C., Hicks, M.G. and Sunkel, J., eds.), Springer-Verlag, Berlin (Germany), pp. 335–348.

Sabljic, A. (1989). Quantitative Modeling of Soil Sorption for Xenobiotic Chemicals. *Environ. Health Persp.*, *83*, 179–190.

Sabljic, A. (1990). Topological Indices and Environmental Chemistry. In *Practical Applications of Quantitative Structure-Activity Relationships (QSAR) in Environmental Chemistry and Toxicology.* (Karcher, W. and Devillers, J., eds.), Kluwer, Dordrecht (The Netherlands), pp. 61–82.

Sabljic, A. and Güsten, H. (1990a). Predicting the Night-Time NO_3 Radical Reactivity in the Troposphere. *Atmospheric Environment*, *1*, 73–78.

Sabljic, A., Güsten, H., Schönherr, J. and Riederer, M. (1990b). Modeling Plant Uptake of Airborne Organic Chemicals. 1. Plant Cuticle/Water Partitioning and Molecular Connectivity. *Environ.Sci.Technol.*, *24*, 1321–1326.

Sabljic, A. (1991). Chemical Topology and Ecotoxicology. *Sci.Total Environ.*, *109/110*, 197–220.

Sabljic, A. and Piver, W.T. (1992). Quantitative Modeling of Environmental Fate and Impact of Commercial Chemicals. *Environ.Toxicol.Chem.*, *11*, 961–972.

Sabljic, A., Gusten, H., Hermens, J.L. and Opperhuizen, A. (1993). Modeling Octanol/Water Partition Coefficients by Molecular Topology Chlorinated Benzenes and Biphenyls. *Environ.Sci. Technol.*, *27*, 1394–1402.

Sabljic, A. and Horvatic, D. (1993). GRAPH III: A Computer Program for Calculating Molecular Connectivity Indices on Microcomputers. *J.Chem.Inf.Comput.Sci.*, *33*, 292–295.

Sabljic, A., Güsten, H., Verhaar, H.J. and Hermens, J.L. (1995). QSAR Modeling of Soil Sorption. Improvements and Systematics of log Koc vs. log Kow Correlations. *Chemosphere*, *31*, 4489–4514.

Saçan, M.T. and Inel, Y. (1993). Prediction of Aqueous Solubility of PCBs Related to Molecular Structure. *Tr.J.Chemistry*, *17*, 188–195.

Saçan, M.T. and Inel, Y. (1995). Application of the Characteristic Root Index Model to the Estimation of N-Octanol Water Partition Coefficients. Polychlorinated Biphenyls. *Chemosphere*, *30*, 39–50.

Saçan, M.T. and Balcioglu, I.A. (1996). Prediction of Soil Sorption Coefficient of Organic Pollutants by the Characteristic Root Index Model. *Chemosphere*, *32*, 1993–2001.

Sadowski, J. and Gasteiger, J. (1993). From Atoms and Bonds to Three-Dimensional Atomic Coordinates: Automatic Model Builders. *Chem.Rev.*, *93*, 2567–2581.

Sadowski, J., Gasteiger, J. and Klebe, G. (1994). Comparison of Automatic Three-Dimensional Model Builders Using 639 X-ray Structures. *J.Chem.Inf.Comput.Sci.*, *34*, 1000–1008.

Sadowski, J., Wagener, M. and Gasteiger, J. (1995). Assessing Similarity and Diversity of Combinatorial Libraries by Spatial Autocorrelation Functions and Neural Networks. *Angew.Chem. Int.Ed.Engl.*, *34*, 2674–2677.

Sagan, B.E., Yeh, Y.N. and Zhang, P. (1996). The Wiener Polynomial of a Graph. *Int.J.Quant. Chem.*, *60*, 959–969.

Sakhartova, O.V. and Shatz, V.D. (1984). Vybor uslovii elyuirovaniya v obrashchenno-phazovoi khromatographii. Priblizhennaya apriornaya otsenka uderzhivaniya poliphunktsionalnykh kislorodsoderzhashchikh soedinemii. *Zh.Anal.Khim.*, *39*, 1496.

Salem, L. (1966). *Molecular Orbital Theory of Conjugated Systems.* Benjamin, New York (NY).

Salo, M., Sarna, S. and Vuorela, H. (1994). Statistical Evaluation of Molecular Descriptors and Quantitative Structure-Property Relationship Studies of Retinoids. *J.Pharm.Biomed.Anal.*, *12*, 867–874.

Salo, M., Siren, H., Volin, P., Wiedmer, S. and Vuorela, H. (1996). Structure Retention Relationships of Steroid Hormones in Reversed Phase Liquid Chromatography and Micellar Electrokinetic Capillary Chromatography. *J.Chromat.*, *728A*, 83–88.

Salt, D.W., Yildiz, N., Livingstone, D.J. and Tinsley, C.J. (1992). The Use of Artificial Neural Networks in QSAR. *Pestic.Sci.*, *36*, 161–170.

Salter, G.J. and Kell, D.B. (1995). Solvent Selection for Whole Cell Biotransformations in Organic Media. *Crit.Rev.Biotechnol.*, *15*, 139–177. [R]

Salvador, J.M., Hernandez, A., Beltram, A., Duran, R. and Mactutis, A. (1998). Fast Partial-Differential Synthesis of the Matching Polynomial of C_{72-100}. *J.Chem.Inf.Comput.Sci.*, *38*, 1105–1110.

Samata, A.K., Ray, S.K., Basak, S.C. and Bose, S.K. (1982). Molecular Connectivity and Antifungal Activity. *Arzneim.Forsch.*, *32*, 1515.

Sanderson, R.T. (1951). An Interpretation of Bond Lengths and a Classification of Bonds. *Science*, *114*, 670–672.

Sanderson, R.T. (1952). Electronegativity. I. Orbital Electronegativity of Neutral Atoms. *J.Chem.Educ.*, *29*, 540–546.

Sanderson, R.T. (1954). Electronegativities in Inorganic Chemistry. III. *J.Chem.Educ.*, *31*, 238.

Sanderson, R.T. (1955). Relation of Stability to Pauling Electronegativities. *J.Chem.Phys.*, *23*, 2467–2468.

Sanderson, R.T. (1971). *Chemical Bonds and Bond Energy.* Academic Press, New York (NY).

Sanderson, R.T. (1983). *Polar Covalence.* Academic Press, New York (NY).

Sanderson, R.T. (1988). Principles of Electronegativity. Part I. General Nature. *J.Chem.Educ.*, *65*, 112.

Sanz, F., Giraldo, J., and Manaut, F. eds. (1995). *QSAR and Molecular Modelling: Concepts, Computational Tools and Biological Applications.* Prous Science, Barcelona (Spain), 688 pp.

Sarkar, R., Roy, A.B. and Sarkar, P.K. (1978). Topological Information Content of Genetic Molecules. I. *Math.Biosci.*, *39*, 299–312.

Sasaki, Y., Takagi, T., Kawaki, H. and Iwata, A. (1980). Novel Substituent Entropy Constant σ_{s0} represents the Molecular Connectivity χ and its related Indices. *Chem.Pharm.Bull.*, *31*, 330.

Sasaki, Y., Takagi, T., Yamazato, Y., Iwata, A. and Kawaki, H. (1981). Utility of the Substituent Entropy Constants σ_{so} in the Studies of Quantitative Structure-Activity Relationships. *Chem. Pharm.Bull.*, *29*, 3073–3075.

Sasaki, Y., Kubodera, H., Matuszaki, T. and Umeyama, H. (1991). Prediction of Octanol/Water Partition Coefficients Using Parameters Derived from Molecular Structures. *J.Pharmacobiodyn.*, *14*, 207–214.

Sasaki, Y., Takagi, T. and Kawaki, H. (1992). On the Estimation of the Quantitative Structure-Activity Relationships Descriptor Sigma(S)Circle for Aliphatic Compound. *Chem.Pharm. Bull.*, *40*, 565–569.

Sasaki, Y., Takagi, T. and Kawaki, H. (1993). Rational Estimation of the QSAR (Quantitative Structure-Activity Relationships) Descriptors Sigma (S-Degrees), and Their Applications for Medicinals Now Available. *Chem.Pharm.Bull.*, *41*, 415–423.

Satoh, H., Sacher, O., Nakata, T., Chen, L., Gasteiger, J. and Funatsu, K. (1998). Classification of Organic Reactions: Similarity of Reactions Based on Changes in the Electronic Features of Oxygen Atoms at the Reaction Sites. *J.Chem.Inf.Comput.Sci.*, *38*, 210–219.

Sawada, M., Tsuno, Y. and Yukawa, Y. (1972). The Substituent Effect. II. Normal Substituent Constants for Polynuclear Aryls from the Hydrolysis of Arylcarbinyl Benzoates. *Bull.Chem. Soc.Jap.*, *45*, 1206–1209.

Saxena, A.K. (1995a). Physicochemical Significance of Topological Parameters, Connectivity Indices and Information Content. Part 1: Correlation Studies in the Sets with Aromatic and Aliphatic Substituents. *Quant.Struct.-Act.Relat.*, *14*, 31–38.

Saxena, A.K. (1995b). Physicochemical Significance of Topological Parameters, Connectivity Indices and Information Content. Part 2: Correlation Studies with Molar Refractivity and Lipophylicity. *Quant.Struct.-Act.Relat.*, *14*, 142–150.

Schaad, L.J. and Hess Jr., B.A. (1972). Hückel Molecular Orbital π Resonance Energies. The Question of the σ Structure. *J.Am.Chem.Soc.*, *94*, 3068–3074.

Schaad, L.J. and Hess Jr., B.A. (1977). Theory of Linear Equations as Applied to Quantitative Structure-Activity Correlations. *J.Med.Chem.*, *20*, 619–625.

Schaad, L.J., Hess Jr., B.A., Purcell, W.P., Cammarata, A., Franke, R. and Kubinyi, H. (1981). Compatibility of the Free-Wilson and Hansch Quantitative Structure-Activity Relations. *J.Med.Chem.*, *24*, 900–901.

Schmalz, T.G., Zivkovic, T. and Klein, D.J. (1987). Cluster Expansion of the Hückel Molecular Energy of Acyclic: Applications to PI Resonance Theory. *Stud.Phys.Theor.Chem.*, *54*, 173–190.

Schmalz, T.G., Klein, D.J. and Sandleback, B.L. (1992). Chemical Graph-Theoretical Cluster Expansion and Diamagnetic Susceptibility. *J.Chem.Inf.Comput.Sci.*, *32*, 54–57.

Schneider, H.-J., Rüdiger, V. and Raevsky, O.A. (1993). The Incremental Description of Host-Guest Complexes: Free Energy Increments Derived from Hydrogen Bonds Applied to Crown Ethers and Cryptands. *J.Org.Chem.*, *58*, 3648–3653.

Schnur, D. (1999). Design and Diversity Analysis of Large Combinatorial Libraries Using Cell-Based Methods. *J.Chem.Inf.Comput.Sci.*, *39*, 36–45.

Schomaker, V. and Stevenson, D.P. (1941). Some Revisions of the Covalent Radii and the Additivity Rule for the Lengths of Partially Ionic Single Covalent Bonds. *J.Am.Chem.Soc.*, *63*, 37–40.

Schotte, W. (1992). Prediction of the Molar Volume at the Normal Boiling Point. *Chem.Eng.J.*, *48*, 167–172.

Schramke, J.A., Murphy, S.F., Doucette, W.J. and Hintze, W.D. (1999). Prediction of Aqueous Diffusion Coefficients for Organic Compounds at 25/C. *Chemosphere*, *38*, 2381–2406.

Schultz, H.P. (1989). Topological Organic Chemistry. 1. Graph Theory and Topological Indices of Alkanes. *J.Chem.Inf.Comput.Sci.*, *29*, 227–228.

Schultz, H.P., Schultz, E.B. and Schultz, T.P. (1990). Topological Organic Chemistry. 2. Graph Theory, Matrix Determinants and Eigenvalues, and Topological Indices of Alkanes. *J.Chem. Inf.Comput.Sci.*, *30*, 27–29.

Schultz, H.P. and Schultz, T.P. (1991). Topological Organic Chemistry. 3. Graph Theory, Binary and Decimal Adjacency Matrices, and Topological Indices of Alkanes. *J.Chem.Inf.Comput.Sci.*, *31*, 144–147.

Schultz, H.P., Schultz, E.B. and Schultz, T.P. (1992). Topological Organic Chemistry. 4. Graph Theory, Matrix Permanents, and Topological Indices of Alkanes. *J.Chem.Inf.Comput.Sci.*, *32*, 69–72.

Schultz, H.P. and Schultz, T.P. (1992). Topological Organic Chemistry. 5. Graph Theory, Matrix Hafnians and Pfaffians, and Topological Indices of Alkanes. *J.Chem.Inf.Comput.Sci.*, *32*, 364–368.

Schultz, H.P., Schultz, E.B. and Schultz, T.P. (1993). Topological Organic Chemistry. 7. Graph Theory and Molecular Topological Indices of Unsaturated and Aromatic Hydrocarbons. *J.Chem.Inf.Comput.Sci.*, *33*, 863–867.

Schultz, H.P. and Schultz, T.P. (1993). Topological Organic Chemistry. 6. Graph Theory and Molecular Topological Indices of Cycloalkanes. *J.Chem.Inf.Comput.Sci.*, *33*, 240–244.

Schultz, H.P., Schultz, E.B. and Schultz, T.P. (1994). Topological Organic Chemistry. 8. Graph Theory and Topological Indices of Heteronuclear Systems. *J.Chem.Inf.Comput.Sci.*, *34*, 1151–1157.

Schultz, H.P., Schultz, E.B. and Schultz, T.P. (1995). Topological Organic Chemistry. 9. Graph Theory and Molecular Topological Indices of Stereoisomeric Organic Compounds. *J.Chem. Inf.Comput.Sci.*, *35*, 864–870.

Schultz, H.P., Schultz, E.B. and Schultz, T.P. (1996). Topological Organic Chemistry. 10. Graph Theory and Topological Indices of Conformational Isomers. *J.Chem.Inf.Comput.Sci.*, *36*, 996–1000.

Schultz, H.P. and Schultz, T.P. (1998). Topological Organic Chemistry. 11. Graph Theory and Reciprocal Schultz-Type Molecular Topological Indices of Alkanes and Cycloalkanes. *J.Chem. Inf.Comput.Sci.*, *38*, 853–857.

Schultz, H.P. and Schultz, T.P. (2000). Topological Organic Chemistry. 12. Whole-Molecule Schultz Topological Indices of Alkanes. *J.Chem.Inf.Comput.Sci.*, *40*, 107–112.

Schultz, T.W., Kier, L.B. and Hall, L.H. (1982). Structure-Toxicity Relationships of Selected Nitrogenous Heterocyclic Compounds. III. Relations Using Molecular Connectivity. *Bull.Environ.Contam.Toxicol.*, *28*, 373.

Schultz, T.W. and Moulton, B.A. (1985). Structure-Activity Relationships of Selected Pyridines. I. Substituent Constant Analysis. *Ecotox.Environ.Safe.*, *10*, 97–111.

Schultz, T.W. (1987a). The Use of the Ionization Constant (pK_a) in Selecting Models of Toxicity in Phenols. *Ecotox.Environ.Safe.*, *14*, 178–183.

Schultz, T.W. and Cajina-Quezada, M. (1987b). Structure-Activity Relationships for Mono Alkylated or Halogenated Phenols. *Toxicol.Lett.*, *37*, 121–130.

Schultz, T.W., Lin, D.T., Wilke, T.S. and Arnold, L.M. (1990). Quantitative Structure-Activity Relationships for the *Tetrahymena Pyriformis* Population Growth Endpoint: A Mechanism of Action Approach. In *Practical Applications of Quantitative Structure-Activity Relationships (QSAR) in Environmental Chemistry and Toxicology* (Karcher, W. and Devillers, J., eds.), Kluwer, Dordrecht (The Netherlands), pp. 241–262.

Schultz, T.W., Sinks, G.D. and Bearden, A.P. (1998). QSAR in Aquatic Toxicology: A Mechanism of Action Approach Comparing Toxic Potency to *Pimephales promelas*, *Tetrahymena pyriformis*, and *Vibrio fischeri*. In *Comparative QSAR* (Devillers, J., ed.), Taylor & Francis, Washington (DC), pp. 51–109. [R]

Schultz, T.W. and Cronin, M.T.D. (1999). Response-Surface Analyses for Toxicity to *Tetrahymena pyriformis*: Reactive Carbonyl-Containing Aliphatic Chemicals. *J.Chem.Inf.Comput.Sci.*, *39*, 304–309.

Schuur, J. and Gasteiger, J. (1996). 3D-MoRSE Code – A New Method for Coding the 3D Structure of Molecules. In *Software Development in Chemistry – Vol. 10* (Gasteiger, J., ed.), Fachgruppe Chemie-Information-Computer (CIC), Frankfurt am Main (Germany), pp. 67–80.

Schuur, J., Selzer, P. and Gasteiger, J. (1996). The Coding of the Three-Dimensional Structure of Molecules by Molecular Transforms and Its Application to Structure-Spectra Correlations and Studies of Biological Activity. *J.Chem.Inf.Comput.Sci.*, *36*, 334–344.

Schuur, J. and Gasteiger, J. (1997). Infrared Spectra Simulation of Substituted Benzene Derivatives on the Basis of a 3D Structure Representation. *Anal.Chem.*, *69*, 2398–2405.

Schüürmann, G. (1990). Quantitative Structure-Property Relationships for the Polarizability, Solvatochromic Parameters and Lipophilicity. *Quant.Struct.-Act.Relat.*, *9*, 326–333.

Schüürmann, G. (1995). Quantum Chemical Approach to Estimate Physicochemical Compound Properties Application to Substituted Benzenes. *Environ.Toxicol.Chem.*, *14*, 2067–2076.

Schüürmann, G. (1996). Modelling pK_a of Carboxylic Acids and Chlorinated Phenols. *Quant. Struct.-Act.Relat.*, *15*, 121–132.

Schüürmann, G., Somashekar, R.K. and Kristen, U. (1996). Structure-Activity Relationships for Chlorophenol and Nitrophenol Toxicity in the Pollen Tube Growth Test. *Environ.Toxicol. Chem.*, *15*, 1702–1708.

Schüürmann, G., Segner, H. and Jung, K. (1997). Multivariate Mode-of-Action Analysis of Acute Toxicity of Phenols. *Aquat.Toxicol.*, *38*, 277–296.

Schweitzer, R.C. and Morris, J.B. (1999). The Development of a Quantitative Structure Property Relationship (QSPR) for the Prediction of Dielectric Constants Using Neural Networks. *Anal. Chim.Acta*, *384*, 285–303.

Scsibrany, H. and Varmuza, K. (1992a). Common Substructures in Groups of Compounds Exhibiting Similar Mass Spectra. *Fresen.J.Anal.Chem.*, *344*, 220–222.

Scsibrany, H. and Varmuza, K. (1992b). Topological Similarity of Molecules Based on Maximum Common Substructures. In *Software Development in Chemistry – Proceedings of the 7th CIC-Workshop "Computers in Chemistry"* (Ziessow, D., ed.), Berlin/Gosen (Germany).

Seel, M., Turner, D.B. and Willett, P. (1999). Effect of Parameter Variations on the Effectiveness of HQSAR Analyses. *Quant.Struct.-Act.Relat.*, *18*, 245–252.

Seibel, G.L. and Kollman, P.A. (1990). Molecular Mechanics and the Modeling of Drug Structures. In *Quantitative Drug Design. Vol. 4* (Ramsden, C.A., ed.), Pergamon Press, Oxford (UK), pp. 125–138.

Seiler, P. (1974). Interconversion of Lipophilicities from Hydrocarbon/Water Systems into Octanol/Water System. *Eur.J.Med.Chem.*, *9*, 473–479.

Sekusak, S. and Sabljic, A. (1992). Soil Sorption and Chemical Topology. *J.Math.Chem.*, *11*, 271–280.

Sekusak, S. and Sabljic, A. (1993). Calculation of Retention Indices by Molecular Topology. III. Chlorinated Dibenzodioxins. *J.Chromat.*, *628*, 69–79.

Selassie, C.D. and Klein, T.E. (1998). Comparative Quantitative Structure Activity Relationships (QSAR) of the Inhibition of Dihydrofolate Reductase. In *Comparative QSAR* (Devillers, J., ed.), Taylor & Francis, Washington (DC), pp. 235–284. [R]

Sello, G. (1998). Similarity Measures: Is it Possible to Compare Dissimilar Structures? *J.Chem. Inf.Comput.Sci.*, *38*, 691–701.

Selzer, P., Schuur, J. and Gasteiger, J. (1996). Simulation of IR Spectra with Neural Networks Using the 3D-MoRSE Code. In *Software Development in Chemistry – Vol. 10* (Gasteiger, J., ed.), Fachgruppe Chemie-Information-Computer (CIC), Frankfurt am Main (Germany), pp. 293–302.

Senn, P. (1988). The Computation of the Distance Matrix and the Wiener Index for Graphs of Arbitrary Complexity with Weighted Vertices and Edges. *Computers Chem.*, *12*, 219–227.

Serilevy, A., Salter, R., West, S. and Richards, W.G. (1994a). Shape Similarity as a Single Independent Variable in QSAR. *Eur.J.Med.Chem.*, *29*, 687–694.

Serilevy, A., West, S. and Richards, W.G. (1994b). Molecular Similarity, Quantitative Chirality, and QSAR for Chiral Drugs. *J.Med.Chem.*, *37*, 1727–1732.

Serrano, J.L., Marcos, M., Melendez, E., Albano, C., Wold, S. and Elguero, J. (1985). Classification of Mesogenic Benzalazines by Multivariate Data Analysis. *Acta Chem.Scand.*, *B 39*, 329–341.

Seybold, P.G. (1983a). Topological Influences on the Carcinogenicity of Aromatic Hydrocarbons. I. The Bay Region Geometry. *Int.J.Quantum Chem.Quant.Biol.Symp.*, *10*, 95–101.

Seybold, P.G. (1983b). Topological Influences on the Carcinogenicity of Aromatic Hydrocarbons. II. Substituent Effects. *Int.J.Quantum Chem.Quant.Biol.Symp.*, *10*, 103–108.

Seybold, P.G., May, M. and Bagal, U.A. (1987). Molecular Structure-Property Relationships. *J.Chem.Educ.*, *64*, 575–581.

Seydel, J. K. ed. (1985). *QSAR and Strategies in the Design of Bioactive Compounds.* VCH, Weinheim (Germany).

Shalabi, A.S. (1991). Random Walks: Computations and Applications to Chemistry. *J.Chem.Inf. Comput.Sci.*, *31*, 483–491.

Shankar Raman, V. and Maranas, C.D. (1998). Optimization in Product Design with Properties Correlated with Topological Indices. *Computers Chem.Eng.*, *22*, 747–763.

Shannon, C. (1948). A Mathematical Theory of Communication. *Bell Syst.Tech.J.*, *27*, 379–423.

Shannon, C. and Weaver, W. (1949). *The Mathematical Theory of Communication.* University of Illinois Press, Urbana (ILL).

Shapiro, S. and Guggenheim, B. (1998a). Inhibition of Oral Bacteria by Phenolic-Compounds. Part 1. QSAR Analysis Using Molecular Connectivity. *Quant.Struct.-Act.Relat.*, *17*, 327–337.

Shapiro, S. and Guggenheim, B. (1998b). Inhibition of Oral Bacteria by Phenolic-Compounds. Part 2. Correlations with Molecular Descriptors. *Quant.Struct.-Act.Relat.*, *17*, 338–347.

Sharma, V., Goswami, R. and Madan, A.K. (1997). Eccentric Connectivity Index: A Novel Highly Discriminating Topological Descriptor for Structure-Property and Structure-Activity Studies. *J.Chem.Inf.Comput.Sci.*, *37*, 273–282.

Shatz, V.D., Sakhartova, O.V., Brivkalne, L.A. and Belikov, V.A. (1984). Vybor uslovii elyuirovaniya v obrashchenno-phazovoi khromatographii. Indeks svyazyvaemosti i uderzhivanie uglevodorodov i prosteishikh kislorodsoderzhashchikh soedinemii. *Zh.Anal.Khim.*, *39*, 94.

Shelley, C.A. and Munk, M.E. (1977). *J.Chem.Inf.Comput.Sci.*, *17*, 110.

Shemetulskis, N.E., Dunbar Jr, J.B., Dunbar, B.W., Moreland, D.W. and Humblet, C. (1995). Enhancing the Diversity of a Corporate Database Using Chemical Database Clustering and Analysis. *J.Comput.Aid.Molec.Des.*, *9*, 407–416.

Sheridan, R.P., Nilakantan, R., Dixon, J.S. and Venkataraghavan, R. (1986). The Ensemble Approach to Distance Geometry: Application to the Nicotinic Pharnacophore. *J.Med.Chem.*, *29*, 899–906.

Sheridan, R.P. and Venkataraghavan, R. (1987). New Methods in Computer-Aided Drug Design. *Acc.Chem.Res.*, *20*, 322.

Sheridan, R.P., Nilakantan, R., Rusinko III, A., Bauman, N., Haraki, K. and Venkataraghavan, R. (1989). 3DSEARCH: A System for Three-Dimensional Structure Searching. *J.Chem.Inf.Comput.Sci.*, *29*, 255–260.

Sheridan, R.P., Miller, M.D., Underwood, D.J. and Kearsley, S.K. (1996). Chemical Similarity Using Geometric Atom Pair Descriptors. *J.Chem.Inf.Comput.Sci.*, *36*, 128–136.

Sheridan, R.P. and Miller, M.D. (1998). A Method for Visualizing Recurrent Topological Substructures in Sets of Active Molecules. *J.Chem.Inf.Comput.Sci.*, *38*, 915–924.

Shi, L.M., Fan, Y., Myers, T.G., O'Connor, P.M., Paull, K.D., Friend, S.H. and Weinstein, J.N. (1998). Mining the NCI Anticancer Drug Discovery Databases: Genetic Function Approximation for the QSAR Study of Anticancer Ellipticine Analogues. *J.Chem.Inf.Comput.Sci.*, *38*, 189–199.

Shorter, J. (1978). Multiparameter Extensions of the Hammett Equation. In *Correlation Analysis in Chemistry* (Chapman, N.B. and Shorter, J., eds.), Plenum Press, New York (NY), pp. 119–173.

Shpilkin, S.A., Smolenskii, E.A. and Zefirov, N.S. (1996). Topological Structure of the Configuration Space and the Separation of Spin and Spatial Variables for *N*-Electron Systems. *J.Chem. Inf.Comput.Sci.*, *36*, 409–412.

Shusterman, A.J. (1992). Predicting Chemical Mutagenicity by Using Quantitative Structure-Activity Relationships. *ACS Symp.Ser.*, *484*, 181–190.

Siegel, S. and Komarmy, J.M. (1960). Quantitative Relationships in the Reactions of *trans*-4-X-Cyclohexanecarboxylic Acids and their Methyl Esters. *J.Am.Chem.Soc.*, *82*, 2547–2553.

Silipo, C. and Hansch, C. (1975). Correlation Analysis. Its Application to the Structure-Activity Relationship of Triazines Inhibiting Dihydrofolate Reductase. *J.Am.Chem.Soc.*, *97*, 6849–6861.

Silipo, C. and Vittoria, A. (1990). Three-Dimensional Structure of Drugs. In *Quantitative Drug Design. Vol. 4* (Ramsden, C.A., ed.), Pergamon Press, Oxford (UK), pp. 153–204.

Silipo, C. and Vittoria, A. eds. (1991). *QSAR: Rational Approaches to the Design of Bioactive Compounds.* Elsevier, Amsterdam (The Netherlands), 576 pp.

Silla, E., Tunon, I. and Pascual-Ahuir, J.L. (1991). GEPOL: An Improved Description of Molecular Surfaces. II. Computing the Molecular Area and Volume. *J.Comput.Chem.*, *12*, 1077–1088.

Silverman, B.D. and Platt, D.E. (1996). Comparative Molecular Moment Analysis (CoMMA): 3D-QSAR without Molecular Superposition. *J.Med.Chem.*, *39*, 2129–2140.

Silverman, B.D., Pitman, M.C., Platt, D.E. and Rigoutsos, I. (1998a). Molecular Moment Similarity Between Clozapine and Substituted [(4-Phenylpiperazinyl)-methyl] benzamides:Selective Dopamine D4 Agonist. *J.Comput.Aid.Molec.Des.*, *12*, 525–532.

Silverman, B.D., Platt, D.E., Pitman, M. and Rigoutsos, I. (1998b). Comparative Molecular Moment Analysis (CoMMA). In *3D QSAR in Drug Design – Vol. 3* (Kubinyi, H., Folkers, G. and Martin, Y.C., eds.), Kluwer/ESCOM, Dordrecht (The Netherlands), pp. 183–198.

Silverman, B.D., Pitman, M.C., Platt, D.E. and Rigoutsos, I. (1999). Molecular Moment Similarity Between Several Nucleoside Analogs of Thymidine and Thymidine. *Journal of Biomolecular Structure and Dynamics*, *16*, 1169–1175.

Simmons, K.A., Dixson, J.A., Halling, B.P., Plummer, E.L., Plummer, M.J., Tymonko, J.M., Schmidt, R.J., Wyle, M.J., Webster, C.A., Baver, W.A., Witkowski, D.A., Peters, G.R. and Gravelle, W.D. (1992). Synthesis and Activity Optimization of Herbicidal Substituted 4-Aryl-1,2,4-Triazole-5(1H)-Thiones. *J.Agr.Food Chem.*, *40*, 297–305.

Simon, Z., Dragomir, N., Plauchitiu, M.G., Holban, S., Glatt, H. and Kerek, F. (1973). Receptor Site Mapping for Cardiotoxic Aglicones by the Minimal Steric Difference Method. *Eur.J.Med. Chem.*, *15*, 521–527.

Simon, Z. and Szabadai, Z. (1973). Minimal Steric Difference-Parameter and the Importance of Steric Fit for Quantitative Structure-Activity Correlations. *Studia Biophys.*, *39*, 123–132.

Simon, Z. (1974). Specific Interactions. Intermolecular Forces, Steric Requirements, and Molecular Size. *Angew.Chem.Int.Ed.Engl.*, *13*, 719–727.

Simon, Z. and Szabadai, Z. (1975). *Studia Biophys.*, *51*, 49–57.

Simon, Z., Chiriac, A., Motoc, I., Holban, S., Ciubotariu, D. and Szabadai, Z. (1976a). *Studia Biophys.*, *55*, 217–226.

Simon, Z., Holban, S., Motoc, I., Mracec, M., Chiriac, A., Kerek, F., Ciubotariu, D., Szabadai, Z., Pop, R.D. and Schwartz, I. (1976b). *Studia Biophys.*, *59*, 181–197.

Simon, Z., Badilescu, I.I. and Racovitan, T. (1977). Mapping of Dihydrofolate Reductase Receptor Site by Correlations with Minimal Topological (Steric) Difference. *J.Theor.Biol.*, *66*, 485–495.

Simon, Z., Chiriac, A., Holban, S., Ciubotariu, D. and Mihalas, G.I. (1984). *Minimum Steric Difference. The MTD Method for QSAR Studies.* Research Studies Press, Letchworth (UK), 174 pp.

Simon, Z., Balaban, A.T., Ciubotariu, D. and Balaban, T.-S. (1985a). QSAR for Carcinogenesis by Polycyclic Aromatic Hydrocarbons and Derivatives in Terms of Delocalization Energy, Minimal Sterical Differences and Topological Indices. *Rev.Roum.Chim.*, *30*, 985–1000.

Simon, Z., Ciubotariu, D. and Balaban, A.T. (1985b). Reactivity and Stereochemical Parameters in QSAR for Carcinogenic Polycyclic Hydrocarbon Derivates. In *QSAR and Strategies in the Design in Bioactive Compounds* (Seydel, J.K., ed.), VCH Verlagsgesellschaft, Weinheim (Germany), pp. 370–373.

Simon, Z. and Bohl, M. (1992). Structure-Activity Relations in Gestagenic Steroids by the MTD Method. The Case of Hard Molecules and Soft Receptors. *Quant.Struct.-Act.Relat.*, *11*, 23–28.

Simon, Z. (1993). MTD and Hyperstructure Approaches. In *3D QSAR in Drug Design. Theory, Methods and Applications.* (Kubinyi, H., ed.), ESCOM, Leiden (The Netherlands), pp. 307–319.

Singer, J.A. and Purcell, W.P. (1967). Relationships Among Current Quantitative Structure-Activity Models. *J.Med.Chem.*, *10*, 1000–1002.

Singh, N., Gupta, R.L. and Roy, N.K. (1996). Synthesis and Quantitative Structure-Activity Relationships of Aryl-2-Chloroethyl-Methyl Phosphate Fungicides. *Indian J.Chem.*, *35B*, 697–702.

Singh, P., Ojha, T.N., Sharma, R.C. and Tiwari, K.S. (1993). Quantitative Structure-Activity Relationship Study of Benzodiazepine Receptor Ligands. 3. *Indian J.Chem.*, *32B*, 555–561.

Singh, P., Ojha, T.N., Tiwari, S. and Sharma, R.C. (1996). Fujita-Ban and Hansch Analyses of A(1)-Adenosine Receptor Binding and A(2)-Adenosine Receptor Binding Affinities of Some 4-Amino(1,2,4)Triazolo(4,3-a)-Quinoxalines. *Indian J.Chem.*, *35B*, 929–934.

Singh, V.K., Tewari, V.P., Gupta, D.K. and Srivastava, A.K. (1984). Calculation of Heat of Formation: – Molecular Connectivity and IOC-ω Technique, A Comparative Study. *Tetrahedron*, *40*, 2859–2863.

Sixt, S., Altschuh, J. and Brüggemann, R. (1995). Quantitative Structure-Toxicity Relationships for 80 Chlorinated Compounds Using Quantum Chemical Descriptors. *Chemosphere*, *30*, 2397–2414.

Sixt, S., Altschuh, J. and Brüggemann, R. (1996). Estimation of pk_a for Organic Oxyacids Using Semiempirical Quantum Chemical Methods. In *Software Development in Chemistry – Vol. 10* (Gasteiger, J., ed.), Fachgruppe Chemie-Information-Computer (CIC), Frankfurt am Main (Germany), pp. 147–153.

Sjöberg, P., Murray, J.S. and Brinck, T. (1990). Average Local Ionization Energies on the Molecular Surfaces of Aromatic Systems as Guides to Chemical Reactivity. *Can.J.Chem.*, *68*, 1440–1443.

Sjöberg, P. (1997). MOLSURF – A Generator of Chemical Descriptors for QSAR. In *Computer-Assisted Lead Finding and Optimization.* (van de Waterbeemd, H., Testa, B. and Folkers, G., eds.), Wiley-VCH, Weinheim (Germany), pp. 81–92.

Sjöström, M. and Wold, S. (1976). *Chem.Scripta*, *9*, 200.

Sjöström, M. and Eriksson, L. (1995). Experimental Design in Synthesis Planning and Structure-Property Correlations. Applications of Statistical Experimental Design and PLS Modeling in QSAR. In *Chemometrics Methods in Molecular Design – Vol. 2* (van de Waterbeemd, H., ed.), VCH Publishers, New York (NY), pp. 63–90.

Skagerberg, B., Bonelli, D., Clementi, S., Cruciani, G. and Ebert, C. (1989). Principal Properties for Aromatic Substituents. A Multivariate Approach for Design in QSAR. *Quant.Struct.-Act.Relat.*, *8*, 32–38.

Skorobogatov, V.A. and Dobrynin, A.A. (1988). Metric Analysis of Graphs. *MATCH (Comm. Math.Comp.Chem.)*, *23*, 105–151.

Skvortsova, M.I., Baskin, I.I., Slovokhotova, O.L., Palyulin, V.A. and Zefirov, N.S. (1993). Inverse Problem in QSAR/QSPR Studies for the Case of Topological Indices Characterizing Molecular Shape (Kier Indices). *J.Chem.Inf.Comput.Sci.*, *33*, 630–634.

Skvortsova, M.I., Baskin, I.I., Stankevitch, I.V., Palyulin, V.A. and Zefirov, N.S. (1998). Molecular Similarity. 1. Analytical Description of the Set of Graph Similarity Measures. *J.Chem.Inf. Comput.Sci.*, *38*, 785–790.

Slater, J.C. and Kirkwood, J.G. (1931). *Physical Review*, *37*, 682.

Slater, J.M. and Paynter, J. (1994). Prediction of Gas Sensor Response Using Basic Molecular Parameters. *The Analyst*, *119*, 191–195.

Small, P.A. (1953). *J.Appl.Chem.*, *3*, 71–80.

Smeeks, F.C. and Jurs, P.C. (1990). Prediction of Boiling Points of Alcohols from Molecular Structure. *Anal.Chim.Acta*, *233*, 111–119.

Smith, C., Payne, V., Doolittle, D.J., Debnath, A.K., Lawlor, T. and Hansch, C. (1992). Mutagenic Activity of a Series of Synthetic and Naturally Occurring Heterocyclic Amines in Salmonella. *Mut.Res.*, *279*, 61–73.

Smith, D.H. and Carhart, R.E. (1976). Structural Isomerism of Mono- and Sesquiterpenoid Skeletons. *Tetrahedron*, *32*, 2513–2519.

Smith, E.G. and Baker, P.A. (1975). *The Wiswesser Line-Formula Chemical Notation (WLN)*. Chemical Information Management, Cherry Hill (NJ).

Smith, R.N., Hansch, C. and Ames, M.M. (1975). Selection of a Reference Partitioning System for Drug Design Work. *J.Pharm.Sci.*, *64*, 599–606.

Smolenskii, E.A. (1964). Application of the Theory of Graphs to Calculations of the Additive Structural Properties of Hydrocarbons. *Russ.J.Phys.Chem.*, *38*, 700–702.

Sneath, P.H.A. and Sokal, R.R. (1973). *Numerical Taxonomy*. Freeman, San Francisco (CA).

So, S.-S. and Richards, W.G. (1992). Application of Neural Networks: Quantitative Structure-Activity Relationships of the Derivatives of 2,4-Diamino-5-(Substituted-Benzyl) Pyrimidines as DHFR Inhibitors. *J.Med.Chem.*, *35*, 3201–3207.

So, S.-S. and Karplus, M. (1996a). Evolutionary Optimization in Quantitative Structure-Activity Relationship: An Application of Genetic Neural Networks. *J.Med.Chem.*, *39*, 1521–1530.

So, S.-S. and Karplus, M. (1996b). Genetic Neural Networks for Quantitative Structure-Activity Relationships: Improvements and Application of Benzodiazepine Affinity for Benzodiazepine/$GABA_A$ Receptors. *J.Med.Chem.*, *39*, 5246–5256.

So, S.-S. and Karplus, M. (1997a). Three-Dimensional Quantitative Structure-Activity Relationships from Molecular Similarity Matrices and Genetic Neural Networks. 1. Method and Validations. *J.Med.Chem.*, *40*, 4347–4359.

So, S.-S. and Karplus, M. (1997b). Three-Dimensional Quantitative Structure-Activity Relationships from Molecular Similarity Matrices and Genetic Neural Networks. 2. Applications. *J.Med.Chem.*, *40*, 4360–4371.

So, S.-S. and Karplus, M. (1999). A Comparative Study of Ligand-Receptor Complex Binding Affinity Prediction Methods Based on Glycogen Phosphorylase Inhibitors. *J.Comput.Aid. Molec.Des.*, *13*, 243–258.

Soltzberg, L.J. and Wilkins, C.L. (1977). Molecular Transforms: a Potential Tool for Structure-Activity Studies. *J.Am.Chem.Soc.*, *99*, 439–443.

Son, S.H., Han, C.K., Ahn, S.K., Yoon, J.H. and No, K.T. (1999). Development of Three-Dimensional Descriptors Represented by Tensors: Free Energy of Hydration Density Tensor. *J.Chem.Inf.Comput.Sci.*, *39*, 601–609.

Soskic, M. and Sabljic, A. (1993). Herbicidal Selectivity of (E)-3-(2,4 Dichlorophenoxy)Acrylates QSAR Study with Molecular Connectivity Indexes. *Pestic.Sci.*, *39*, 245–250.

Soskic, M., Klaic, B., Magnus, V. and Sabljic, A. (1995). Quantitative Structure-Activity Relationships for N-(Indol-3-Ylacetyl)Amino Acids Used as Sources of Auxin in Plant Tissue Culture. *Plant Growth Regul.*, *16*, 141–152.

Soskic, M. and Sabljic, A. (1995). QSAR Study of 4-Hydroxypyridine Derivatives as Inhibitors of the Hill Reaction. *Pestic.Sci.*, *45*, 133–141.

Soskic, M., Plavsic, D. and Trinajstic, N. (1996a). 2-Difluoromethylthio-4,6-bis-(monoalkylamino)-1,3,5-triazines a Inhibitor of Hill Reaction: A QSAR Study with Orthogonalized Descriptors. *J.Chem.Inf.Comput.Sci.*, *36*, 146–150.

Soskic, M., Plavsic, D. and Trinajstic, N. (1996b). Link Between Orthogonal and Standard Multiple Linear Regression Models. *J.Chem.Inf.Comput.Sci.*, *36*, 829–832.

Sotomatsu, T. and Fujita, T. (1989). The Steric Effect of Ortho Substituents on the Acidic Hydrolysis of Benzamides. *J.Org.Chem.*, *54*, 4443–4448.

Spialter, L. (1963). The Atom Connectivity Matrix (ACM) and its Characteristic Polynomial (ACMCP): A New Computer-Oriented Chemical Nomenclature. *J.Am.Chem.Soc.*, *85*, 2012–2013.

Spialter, L. (1964a). The Atom Connectivity Matrix (ACM) and its Characteristic Polynomial (ACMCP). *J.Chem.Doc.*, *4*, 261–269.

Spialter, L. (1964b). The Atom Connectivity Matrix Characteristic Polynomial (ACMCP) and its Physico-Geometric (Topological) Significance. *J.Chem.Doc.*, *4*, 269–274.

Spillane, W.J., McGlinchey, G., Muricheartaigh, I.O. and Benson, G.A. (1983). Structure-Activity Studies on Sulfamate Sweeteners. 3. Structure-Taste Relationships for Heterosulfamates. *J.Pharm.Sci.*, *72*, 934.

Srivastava, S., Richardson, W.W., Bradley, M.P. and Crippen, G.M. (1993). Three-Dimensional Receptor Modeling Using Distance Geometry and Voronoi Polyhedra. In *3D QSAR in Drug Design. Theory, Methods and Applications.* (Kubinyi, H., ed.), ESCOM, Leiden (The Netherlands), pp. 409–430.

Srivastava, A., Mishra, N. and Khan, A.A. (1995). QSAR on Some Antileukemic 1′-Substituted 9-Anilinoacridines with Tau Index τ and Electronic Parameters. *Indian J.Chem.*, *34B*, 364–366.

Stankevitch, I.V., Skvortsova, M.I. and Zefirov, N.S. (1995). On a Quantum Chemical Interpretation of Molecular Connectivity Indices for Conjugated Hydrocarbons. *J.Mol.Struct.(Theochem)*, *342*, 173–179.

Stankevitch, M.I., Stankevitch, I.V. and Zefirov, N.S. (1988). Topological Indices in Organic Chemistry. *Russian Chemical Review*, *57*, 191–208.

Stanley, H. E. and Ostrowsky, N. eds. (1990). *Correlations and Connectivity: Geometric Aspects of Physics, Chemistry, and Biology.* Kluwer, Dordrecht (The Netherlands).

Stanton, D.T. and Jurs, P.C. (1990). Development and Use of Charged Partial Surface Area Structural Descriptors in Computer-Assisted Quantitative Structure-Property Relationship Studies. *Anal.Chem.*, *62*, 2323–2329.

Stanton, D.T., Jurs, P.C. and Hicks, M.G. (1991). Computer-Assisted Prediction of Normal Boiling Points of Furans, Tetrahydrofurans, and Thiophenes. *J.Chem.Inf.Comput.Sci.*, *31*, 301–310.

Stanton, D.T., Egolf, L.M., Jurs, P.C. and Hicks, M.G. (1992). Computer-Assisted Prediction of Normal Boiling Points of Pyrans and Pyrroles. *J.Chem.Inf.Comput.Sci.*, *32*, 306–316.

Stanton, D.T. and Jurs, P.C. (1992). Computer-Assisted Study of the Relationship Between Molecular Structure and Surface Tension of Organic Compounds. *J.Chem.Inf.Comput.Sci.*, *32*, 109–115.

Stanton, D.T., Murray, W.J. and Jurs, P.C. (1993). Comparison of QSAR and Molecular Similarity Approaches for a Structure-Activity Relationship of DHFR Inhibitors. *Quant.Struct.-Act.Relat.*, *12*, 239–245.

Stanton, D.T. (1999). Evaluation and Use of BCUT Descriptors in QSAR and QSPR Studies. *J.Chem.Inf.Comput.Sci.*, *39*, 11–20.

Stanton, D.T., Morris, T.W., Roychoudhury, S. and Parker, C.N. (1999). Application of Nearest-Neighbor and Cluster Analyses in Pharmaceutical Lead Discovery. *J.Chem.Inf.Comput.Sci.*, *39*, 21–27.

Stegeman, M.H., Peijnenburg, W.J. and Verboom, H. (1993). A Quantitative Structure-Activity Relationship for the Direct Photohydrolysis of Meta Substituted Halobenzene Derivatives in Water. *Chemosphere*, *26*, 837–849.

Stein, N. (1995). New Perspectives in Computer-Assisted Formal Synthesis Design – Treatment of Delocalized Electrons. *J.Chem.Inf.Comput.Sci.*, *35*, 305–309.

Stein, T.M., Gordon, S.H. and Greene, R.V. (1999). Amino Acids as Plasticizers – II. Use of Quantitative Structure-Property Relationships to Predict the Behavior of Monoammonium-monocarboxylate Plasticizers in Starch-Glycerol Blends. *Carbohydr.Polym.*, *39*, 7–16.

Sternberg, M.J., King, R.D., Lewis, R.A. and Muggleton, S. (1994). Application of Machine Learning to Structural Molecular Biology. *Phil.Trans.Roy.Soc.(London) B*, *344*, 365–371.

Stewart, J.P. (1990). MOPAC: A Semiempirical Molecular Orbital Program. *J.Comput.Aid. Molec.Des.*, *4*, 1–105.

Steyaert, G., Lisa, G., Gaillard, P., Boss, G., Reymond, F., Girault, H.H., Carrupt, P.-A. and Testa, B. (1997). Intermolecular Forces Expressed in 1,2-Dichloroethane-Water Partition Coefficients. A Solvatochromic Analysis. *J.Chem.Soc.Faraday Trans.*, *93*, 401–406.

Still, W.C., Tempczyk, A., Hawley, R.C. and Hendrickson, T. (1990). Semi-Analytical Treatment of Solvation for Molecular Mechanics and Dynamics. *J.Am.Chem.Soc.*, *112*, 6127–6129.

Stoklosa, H.J. (1973). Computer Program for Calculation of Charge Distributions in Molecules. *J.Chem.Educ.*, *50*, 290–290.

Stone, M. (1974). *J.R.Stat.Soc., B 36*, 111–133.

Stouch, T.R. and Jurs, P.C. (1986). A Simple Method for the Represantation, Quantification, and Comparison of the Volumes and Shapes of Chemical Compounds. *J.Chem.Inf.Comput.Sci., 26*, 4–12.

Streich, W.J., Dove, S. and Franke, R. (1980). On the Rotational Selection of Test Series. 1. Principal Component Method Combined with Multidimensional Mapping. *J.Med.Chem., 23*, 1452–1456.

Streich, W.J. and Franke, R. (1985). Topological Pharmacophores. New Methods and their Applications to a Set of Antimalarials. Part 1. The Methods LOGANA and LOCON. *Quant.Struct.-Act.Relat., 4*, 13–18.

Streitweiser, A.Jr. (1961). *Molecular Orbital Theory for Organic Chemists.* Wiley, New York (NY).

Stuer-Lauridsen, F. and Pedersen, F. (1997). On the Influence of the Polarity Index of Organic Matter in Predicting Environmental Sorption of Chemicals. *Chemosphere, 35*, 761–773.

Stuper, A.J. and Jurs, P.C. (1978). Structure-Activity Studies of Barbiturates Using Pattern Recognition Techniques. *J.Pharm.Sci., 67*, 745–751.

Sudgen, S. (1924). The Variation of Surface Tension with Temperature and some Related Functions. *J.Chem.Soc., 125*, 32–41.

Sulea, T., Kurunczi, L. and Simon, Z. (1995). Dioxin-Type Activity for Polyhalogenated Arylic Derivatives. A QSAR Model Based on MTD-Method. *SAR & QSAR Environ.Res., 3*, 37–61.

Sulea, T., Oprea, T.I., Muresan, S. and Chan, S.L. (1997). A Different Method for Steric Field Evaluation in CoMFA Improves Model Robustness. *J.Chem.Inf.Comput.Sci., 37*, 1162–1170.

Sulea, T., Kurunczi, L., Oprea, T.I. and Simon, Z. (1998). MTD-ADJ: A Multiconformational Minimal Topologic Difference for Determining Bioactive Conformers Using Adjusted Biological Activities. *J.Comput.Aid.Molec.Des., 12*, 133–146.

Sulea, T. and Purisima, E.O. (1999). Desolvation Free Energy Field Derived from Boundary Element Continuum Dielectric Calculations. *Quant.Struct.-Act.Relat., 18*, 154–158.

Sundaram, A. and Venkatasubramanian, V. (1998). Parametric Sensitivity and search-Space Characterization Studies of genetic Algorithms for Computer-Aided Polymer Design. *J.Chem.Inf.Comput.Sci., 38*, 1177–1191.

Susarla, S., Masunaga, S. and Yonezawa, Y. (1996). Kinetics of Halogen Substituted Aniline Transformation in Anaerobic Estuarine Sediment. *Water Sci.Technol., 34*, 37–43.

Sutter, J.M., Dixon, J.S. and Jurs, P.C. (1995a). Automated Descriptor Selection for Quantitative Structure-Activity Relationships Using Generalized Simulated Annealing. *J.Chem.Inf.Comput.Sci., 35*, 77–84.

Sutter, J.M. and Jurs, P.C. (1995b). Selection of Molecular Descriptors for Quantitative Structure-Activity Relationships. In *Adaption of Simulated Annealing to Chemical Optimization Problems* (Kalivas, J.H., ed.), Elsevier, Amsterdam (The Netherlands), pp. 111–132.

Sutter, J.M. and Jurs, P.C. (1996). Prediction of Aqueous Solubility for a Diverse Set of Heteroatom-Containing Organic Compounds Using a Quantitative Structure-Property Relationship. *J.Chem.Inf.Comput.Sci., 36*, 100–107.

Sutter, J.M., Peterson, T.A. and Jurs, P.C. (1997). Prediction of Gas Chromatographic Retention Indices of Alkylbenzene. *Anal.Chim.Acta, 342*, 113–122.

Suzuki, T. and Kudo, Y. (1990). Automated Log *P* Estimation Based on Combined Additive Modeling Methods. *J.Comput.Aid.Molec.Des., 4*, 155–198.

Suzuki, T. (1991). Development of an Automatic Estimation System for Both the Partition Coefficient and Aqueous Solubility. *J.Comput.Aid.Molec.Des., 5*, 149–166.

Suzuki, T., Ohtaguchi, K. and Koide, K. (1992a). Application of Principal Components Analysis to Calculate Henry's Constant from Molecular Structure. *Computers Chem., 16*, 41–52.

Suzuki, T., Ohtaguchi, K. and Koide, K. (1992b). Computer-Aided Prediction of Solubilities of Organic Compounds in Water. *J.Chem.Eng.Jpn., 25*, 729–734.

Suzuki, T., Ohtaguchi, K. and Koide, K. (1992c). Correlation Between Solubilities in Water and Molecular Descriptors of Hydrocarbons. *J.Chem.Eng.Jpn., 25*, 434–438.

Suzuki, T., Ohtaguchi, K. and Koide, K. (1996). Computer-Assisted Approach to Develop a New Prediction Method of Liquid Viscosity of Organic Compounds. *Computers Chem.Eng.*, *20*, 161–173.

Svozil, D., Sevcik, J.G. and Kvasnicka, V. (1997). Neural Network Prediction of the Solvatochromic Polarity/Polarizability Parameter π_2^H. *J.Chem.Inf.Comput.Sci.*, *37*, 338–342.

Swaan, P.W., Koops, B.C., Moret, E.E. and Tukker, J.J. (1998). Mapping the Binding Site of the Small Intestinal Peptide Carrier (PepT1) Using Comparative Molecular Field Analysis. *Recept.*, *6*, 189.

Swain, C.G. and Lupton Jr., E.C. (1968). Field and Resonance Components of Substituent Effects. *J.Am.Chem.Soc.*, *90*, 4328–4337.

Swain, C.G., Unger, S.H., Rosenquist, N.R. and Swain, M.S. (1983). Substituent Effects on Chemical Reactivity. Improved Evaluation of Field and Resonance Components. *J.Am.Chem. Soc.*, *105*, 492–502.

Swain, C.G. (1984). Substituent and Solvent Effects on Chemical Reactivity. *J.Org.Chem.*, *49*, 2005–2010.

Swinborne-Sheldrake, R., Herndon, W.C. and Gutman, I. (1975). Kekulé Structures and Resonance Energies of Benzenoid Hydrocarbons. *Tetrahedron Lett.*, *10*, 755–758.

Szász, Gy., Papp, O., Vámos, J., Hankó-Novák, K. and Kier, L.B. (1983). Relationships Between Molecular Connectivity Indices, Partition Coefficients and Chromatographic Parameters. *J.Chromat.*, *269*, 91–95.

Szymanski, K., Müller, W.R., Knop, J.V. and Trinajstic, N. (1985). On Randic's Molecular Identification Numbers. *J.Chem.Inf.Comput.Sci.*, *25*, 413–415.

Szymanski, K., Müller, W.R., Knop, J.V. and Trinajstic, N. (1986a). Molecular ID Numbers. *Croat.Chem.Acta*, *59*, 719–723.

Szymanski, K., Müller, W.R., Knop, J.V. and Trinajstic, N. (1986b). On the Identification Numbers for Chemical Structures. *Int.J.Quantum Chem.Quant.Chem.Symp.*, *20*, 173–183.

Taft, R.W. (1952). Polar and Steric Substituent Constants for Aliphatic and *o*-Benzoate Groups from Rates of Esterification and Hydrolysis of Esters. *J.Am.Chem.Soc.*, *74*, 3120–3128.

Taft, R.W. (1953a). Linear Steric Energy Relationships. *J.Am.Chem.Soc.*, *75*, 4538–4539.

Taft, R.W. (1953b). The General Nature of the Proportionality of Polar Effects of Substituent Groups in Organic Chemistry. *J.Am.Chem.Soc.*, *75*, 4231–4238.

Taft, R.W. (1953c). The Separation of Relative Free Energies of Activation to Three Basic Contributing Factors and the Relationship of These to Structure. *J.Am.Chem.Soc.*, *75*, 4534–4537.

Taft, R.W. (1956). Separation of Polar, Steric, and Resonance Effects in Reactivity. In *Steric Effects in Organic Chemistry* (Newman, M.S., ed.), Wiley, New York (NY), pp. 556–675.

Taft, R.W. and Lewis, I.C. (1958). The General Applicability of a Fixed Scale of Inductive Effects. II. Inductive Effects of Dipolar Substituents in the Reactivities of *m*- and *p*-Substituted Derivatives of Benzene. *J.Am.Chem.Soc.*, *80*, 2436–2443.

Taft, R.W., Ehrenson, S., Lewis, I.C. and Glick, R.E. (1959). Evaluation of Resonance Effects on Reactivity by Application of the Linear Inductive Energy Relationship. VI. Concerning the Effects of Polarization and Conjugation on the Mesomeric Order. *J.Am.Chem.Soc.*, *81*, 5352–5361.

Taft, R.W. and Lewis, I.C. (1959). Evaluation of Resonance Effects on Reactivity by Application of the Linear Inductive Energy Relationship. V. Concerning a σ_R Scale of Resonance Effects. *J.Am.Chem.Soc.*, *81*, 5343–5352.

Taft, R.W. (1960). Sigma Values from Reactivities. *J.Phys.Chem.*, *64*, 1805–1815.

Taft, R.W., Price, E., Fox, I.R., Lewis, I.C., Andersen, K.K. and Davis, G.T. (1963a). Fluorine Nuclear Magnetic Resonance Shielding in *meta*-Substituted Fluorobenzenes. The Effect of Solvent on the Inductive Order. *J.Am.Chem.Soc.*, *85*, 709–724.

Taft, R.W., Price, E., Fox, I.R., Lewis, I.C., Andersen, K.K. and Davis, G.T. (1963b). Fluorine Nuclear Magnetic Resonance Shielding in *p*-Substituted Fluorobenzenes. The Influence of Structure and Solvent on Resonance Effects. *J.Am.Chem.Soc.*, *85*, 3146–3156.

Taft, R.W. and Grob, C.A. (1974). Concerning the Separation of Polar and Resonance Effects in the Ionization of 4-Substituted Pyridinium Ions. *J.Am.Chem.Soc.*, *96*, 1236–1238.

Taft, R.W. and Kamlet, M.J. (1976). The Solvatochromic Comparison Method. 2. The α-Scale of Solvent Hydrogen-Bond Donor (HBD) Acidities. *J.Am.Chem.Soc.*, *98*, 2886–2894.

Taft, R.W. and Kamlet, M.J. (1979). Linear Solvation Energy Relationships. Part 4. Correlations with and Limitations of the α Scale of Solvent Hydrogen Bond Donor Acidities. *J.Chem.Soc. Perkin Trans.2*, 1723–1729.

Taft, R.W. (1983). Protonic Acidities and Basicities in the Gas Phase and in Solution: Substituent and Solvent Effects. *Prog.Phys.Org.Chem.*, *14*, 247–350.

Taft, R.W., Abboud, J.-L.M. and Kamlet, M.J. (1984). Linear Solvation Energy Relationships. 28. An Analysis of Swain's Solvent "Acity" and "Basity" Scales. *J.Org.Chem.*, *49*, 2001–2005.

Taft, R.W., Abboud, J.-L.M., Kamlet, M.J. and Abraham, M.H. (1985a). Linear Solvation Energy Relations. *J.Solut.Chem.*, *14*, 153–186.

Taft, R.W., Abraham, M.H., Famini, G.R., Doherty, R.M. and Kamlet, M.J. (1985b). Solubility Properties in Polymers and Biological Media 5: An Analysis of the Physicochemical Properties which Influence Octanol-Water Partition Coefficients of Aliphatic and Aromatic Solutes. *J.Pharm.Sci.*, *74*, 807–814.

Taft, R.W. and Topsom, R.D. (1987). The Nature and Analysis of Substituent Electronic Effects. *Prog.Phys.Org.Chem.*, *16*, 1–84.

Taillander, G., Domard, M. and Boucherle, A. (1983). QSAR et Séries Aromatiques: Propositions de Paramètres Stériques. *Il Farmaco*, *38*, 473–487.

Takagi, T., Tange, K., Jikihara, T., Onozawa, N., Sakashita, K., Fujiwara, H. and Sasaki, Y. (1993). Revision of the Dual Scaling Method for Successive Categories Data and Its Application to Quantitative Structure-Activity Relationships (QSAR). *Bull.Chem.Soc.Jap.*, *66*, 3606–3612.

Takahashi, Y., Miashita, Y., Tanaka, Y., Hayasaka, H., Abe, H. and Sasaki, S. (1985). Discriminative Structural Analysis Using Pattern Recognition Techniques in the Structure-Taste Problem of Perillartines. *J.Pharm.Sci.*, *73*, 737–741.

Takahashi, Y., Sukekawa, M. and Sasaki, S.I. (1992). Automatic Identification Molecular Similarity Using Reduced-Graph Representation of Chemical Structure. *J.Chem.Inf.Comput.Sci.*, *32*, 639–643.

Takeuchi, K., Kuroda, C. and Ishida, M. (1990). Prolog-Based Functional Group Perception and Calculation of 1-Octanol/Water Partition Coefficients Using Rekker's Fragment Method. *J.Chem.Inf.Comput.Sci.*, *30*, 22–26.

Takihi, N., Rosenkranz, H.S., Klopman, G. and Mattison, D.R. (1994). Structural Determinants of Developmental Toxicity. *Risk Analysis*, *14*, 649–657.

Tanaka, A., Nakamura, K., Nakanishi, I. and Fujiwara, H. (1994). A Novel and Useful Descriptor for Hydrophobicity, Partition Coefficient Micellar-Water, and Its Application to a QSAR Study of Antiplatelet Agents. *J.Med.Chem.*, *37*, 4563–4566.

Tanaka, A. and Fujiwara, H. (1996). Quantitative Structure-Activity Relationship Study of Fibrinogen Inhibitors, ((4-(4-Amidinophenoxy)Butanoyl)Aspartyl)Valine (FK633) Derivatives, Using a Novel Hydrophobic Descriptor. *J.Med.Chem.*, *39*, 5017–5020.

Tanford, C. (1957). The Location of Electrostatic Charges in Kirkwood's Model of Organic Ions. *J.Am.Chem.Soc.*, *79*, 5348–5352.

Tanford, C. (1961). *Physical Chemistry of Macromolecules*. Wiley, New York (NY).

Tanford, C. (1973). *The Hydrophobic Effect*. Wiley, New York (NY).

Tang, Y., Jiang, H.L., Chen, K.X. and Ji, R.Y. (1996). QSAR Study of Artemisinin (Qinghaosu) Derivatives Using Neural Network Method. *Indian J.Chem.*, *35B*, 325–332.

Tao, S. and Lu, X. (1999a). Estimation of Organic Carbon Normalized Sorption Coefficient (K_{OC}) for Soils by Topological Indices and Polarity Factors. *Chemosphere*, *39*, 2019–2034.

Tao, S., Piao, H., Dawson, R., Lu, X. and Hu, H. (1999b). Estimation of Organic Carbon Normalized Sorption Coefficient (K_{OC}) for Soils Using the Fragment Constant Method. *Environ.Sci. Technol.*, *33*, 2719–2725.

Taylor, P.J. (1990). Hydrophobic Properties of Drugs. In *Quantitative Drug Design. Vol. 4* (Ramsden, C.A., ed.), Pergamon Press, Oxford (UK), pp. 241–294.

ter Laak, A.M., Tsai, R.-S., Donné-Op den Kelder, G.M., Carrupt, P.-A., Testa, B. and Timmerman, H. (1994). Lipophilicity and Hydrogen-Bonding Capacity of H_1-Antihistaminic Agents in Relation to their Central Sedative Side-Effects. *Eur.J.Pharm.Sci.*, *2*, 373–384.

Testa, B. and Purcell, W.P. (1978). A QSAR Study of Sulfonamide Binding to Carbonic Anhydrase as Test of Steric Models. *Eur.J.Med.Chem.*, *13*, 509–514.

Testa, B. and Seiler, P. (1981). Steric and Lipophobic Components of the Hydrophobic Fragmental Constant. *Arzneim.Forsch.*, *31*, 1053–1058.

Testa, B. and Kier, L.B. (1991). The Concept of Molecular Structure in Structure-Activity Relationship Studies and Drug Design. *Med.Res.Rev.*, *11*, 35–48.

Testa, B., El Tayar, N., Altomare, C., Carrupt, P.-A., Tsai, R.-S. and Carotti, A. (1993). The Hydrogen Bonding of Drugs: Its Experimental Determination and Role in Pharmacokinetics and Pharmacodynamics. (Angeli, P., Gulini, U. and Quaglia, W., eds.), Elsevier, Amsterdam (The Netherlands), pp. 61–72.

Testa, B., Carrupt, P.-A., Gaillard, P., Billois, F. and Weber, P. (1996). Lipophilicity in Molecular Modeling. *Pharm.Res.*, *13*, 335–343.

Testa, B., Kier, L.B. and Carrupt, P.-A. (1997). A Systems Approach to Molecular Structure, Intermolecular Recognition, and Emergence-Dissolvence in Medicinal Research. *Med.Res. Rev.*, *17*, 303–326.

Testa, B., Raynaud, I. and Kier, L.B. (1999). What Differentiates Free Amino Acids and Aminoacyl Residues? An Exploration of Conformational and Lipophilicity Spaces. *Helv.Chim. Acta*, *82*, 657–665.

Tetko, I.V., Livingstone, D.J. and Luik, A.I. (1995). Neural Network Studies. 1. Comparison of Overfitting and Overtraining. *J.Chem.Inf.Comput.Sci.*, *35*, 826–833.

Tetko, I.V., Villa, A.E.P. and Livingstone, D.J. (1996). Neural Network Studies. 2. Variable Selection. *J.Chem.Inf.Comput.Sci.*, *26*, 794–803.

Tetko, I.V. (1998). Application of a Pruning Algorithm to Optimize Artificial Neural Networks for Pharmaceutical Fingerprinting. *J.Chem.Inf.Comput.Sci.*, *38*, 660–668.

Tetteh, J., Suzuki, T., Metcalfe, E. and Howells, S. (1999). Quantitative Structure-Property Relationships for the Estimation of Boiling Point and Flash Point Using a Radial Basis Function Neural Network. *J.Chem.Inf.Comput.Sci.*, *39*, 491–507.

Thangavel, P. and Venuvanalingam, P. (1993). Algorithm for the Computation of Molecular Distance Matrix and Distance Polynomial of Chemical Graphs on Parallel Computers. *J.Chem. Inf.Comput.Sci.*, *33*, 412–414.

Thibaut, U. (1993). Applications of CoMFA and Related 3D QSAR Approaches. In *3D QSAR in Drug Design. Theory, Methods and Applications.* (Kubinyi, H., ed.), ESCOM, Leiden (The Netherlands), pp. 661–696. [R]

Thibaut, U., Folkers, G., Klebe, G., Kubinyi, H., Merz, A. and Rognan, D. (1994). Recommendations for CoMFA Studies and 3D-QSAR Publications. *Quant.Struct.-Act.Relat.*, *13*, 1–3.

Thinh, T.P. and Trong, T.K. (1976). Estimation of Standard Heats of Formation, ΔH_T^f, Standard Entropies of Formation, ΔS_T^f,Standard Free Energies of Formation, ΔG_T^f, and Absolute Entropies, ΔS_T of Hydrocarbons from Group Contributions: An Accurate Approach. *Can.J. Chem.Eng.*, *54*, 344.

Thomas, E.R. and Eckert, C.A. (1984). Prediction of Limiting Activity Coefficients by a Modified Separation of Cohesive Energy Density Model and UNIFAC. *I&EC.Process Des.Dev.*, *23*, 194–209.

Thomsen, M., Rasmussen, A.G. and Carlsen, L. (1999). SAR/QSAR Approaches to Solubility, Partitioning and Sorption of Phthalates. *Chemosphere*, *38*, 2613–2624.

Thull, U., Kneubuhler, S., Gaillard, P., Carrupt, P.-A., Testa, B., Altomare, C., Carotti, A., Jenner, P. and McNaught, K.S. (1995). Inhibition of Monoamine Oxidase by Isoquinoline Derivatives Qualitative and 3D Quantitative Structure-Activity Relationships. *Biochem.Pharmacol.*, *50*, 869–877.

Timofei, S., Kurunczi, L., Schmidt, W. and Simon, Z. (1996). Lipophilicity in Dye Cellulose Fiber Binding. *Dyes & Pigments*, *32*, 25–42.

Timofei, S. and Fabian, W.M.F. (1998). Comparative Molecular Field Analysis of Heterocyclic Monoazo Dye-Fiber Affinities. *J.Chem.Inf.Comput.Sci.*, *38*, 1218–1222.

Tinker, J. (1981). Relating Mutagenicity to Chemical Structure. *J.Chem.Inf.Comput.Sci.*, *21*, 3–7.

Todeschini, R., Cazar, R. and Collina, E. (1992). The Chemical Meaning of Topological Indices. *Chemom.Intell.Lab.Syst.*, *15*, 51–59.

Todeschini, R., Lasagni, M. and Marengo, E. (1994). New Molecular Descriptors for 2D- and 3D-Structures. Theory. *J.Chemom.*, *8*, 263–273.

Todeschini, R., Gramatica, P., Marengo, E. and Provenzani, R. (1995). Weighted Holistic Invariant Molecular Descriptors. Part 2. Theory Development and Applications on Modeling Physico-Chemical Properties of PolyAromatic Hydrocarbons (PAH). *Chemom.Intell.Lab.Syst.*, *27*, 221–229.

Todeschini, R., Bettiol, C., Giurin, G., Gramatica, P., Miana, P. and Argese, E. (1996a). Modeling and Prediction by Using WHIM Descriptors in QSAR Studies. Submitochondrial Particles (SMP) as Toxicity Biosensors of Chlorophenols. *Chemosphere*, *33*, 71–79.

Todeschini, R., Vighi, M., Provenzani, R., Finizio, A. and Gramatica, P. (1996b). Modeling and Prediction by Using WHIM Descriptors in QSAR Studies: Toxicity of Heterogeneous Chemicals on Daphnia Magna. *Chemosphere*, *32*, 1527–1545.

Todeschini, R. (1997). Data Correlation, Number of Significant Principal Components and Shape of Molecules. The K Correlation Index. *Anal.Chim.Acta*, *348*, 419–430.

Todeschini, R. and Gramatica, P. (1997a). 3D-Modelling and Prediction by WHIM Descriptors. Part 5. Theory Development and Chemical Meaning of WHIM Descriptors. *Quant.Struct.-Act.Relat.*, *16*, 113–119.

Todeschini, R. and Gramatica, P. (1997b). 3D-Modelling and Prediction by WHIM Descriptors. Part 6. Application of WHIM Descriptors in QSAR Studies. *Quant.Struct.-Act.Relat.*, *16*, 120–125.

Todeschini, R. and Gramatica, P. (1997c). The WHIM Theory: New 3D Molecular Descriptors for QSAR in Environmental Modelling. *SAR & QSAR Environ.Res.*, *7*, 89–115.

Todeschini, R., Moro, G., Boggia, R., Bonati, L., Cosentino, U., Lasagni, M. and Pitea, D. (1997a). Modeling and Prediction of Molecular Properties. Theory of Grid-Weighted Holistic Invariant Molecular (G-WHIM) Descriptors. *Chemom.Intell.Lab.Syst.*, *36*, 65–73.

Todeschini, R., Vighi, M., Finizio, A. and Gramatica, P. (1997b). 3D-Modelling and Prediction by WHIM Descriptors. Part 8. Toxicity and Physico-Chemical Properties of Environmental Priority Chemicals by 2D-TI and 3D-WHIM Descriptors. *SAR & QSAR Environ.Res.*, *7* , 173–193.

Todeschini, R., Consonni, V. and Maiocchi, A. (1998). The K Correlation Index: Theory Development and its Applications in Chemometrics. *Chemom.Intell.Lab.Syst.*, *46*, 13–29.

Todeschini, R. and Gramatica, P. (1998). New 3D Molecular Descriptors: The WHIM Theory and QSAR Applications. In *3D QSAR in Drug Design – Vol. 2* (Kubinyi, H., Folkers, G. and Martin, Y.C., eds.), Kluwer/ESCOM, Dordrecht (The Netherlands), pp. 355–380.

Todeschini, R., Consonni, V., Galvagni, D. and Gramatica, P. (1999). A New Molecular Structure Representation: Spectral Weighted Molecular (SWM) Signals for Studies of Molecular Similarity. *Quimica Analitica*, *18*, 41–47.

Tokarski, J. and Hopfinger, A.J. (1994). Three-Dimensional Molecular Shape Analysis – Quantitative Structure-Activity Relationship of a Series of Cholecystokinin-A Receptor Antagonists. *J.Med.Chem.*, *37*, 3639–3654.

Tominaga, Y. and Fujiwara, I. (1997a). Data Structure Comparison Using Fractal Analysis. *Chemom.Intell.Lab.Syst.*, *39*, 187–193.

Tominaga, Y. and Fujiwara, I. (1997b). Novel 3D Descriptors Using Excluded Volume: Application to 3D Quantitative Structure-Activity Relationships. *J.Chem.Inf.Comput.Sci.*, *37*, 1158–1161.

Tominaga, Y. (1998a). Data Structure Comparison Using Box Counting Analysis. *J.Chem.Inf.Comput.Sci.*, *38*, 867–875.

Tominaga, Y. (1998b). Novel 3D Descriptors Using Excluded Volume 2: Application to Drug Classification. *J.Chem.Inf.Comput.Sci.*, *38*, 1157–1160.

Tominaga, Y. (1999). Comparative Study of Class Data Analysis with PCA-LDA, SIMCA, PLS, ANNs, and k-NN. *Chemom.Intell.Lab.Syst.*, *49*, 105–115.

Tompe, P., Clementis, G., Petnehazy, I., Jaszay, Z.M. and Toke, L. (1995). Quantitative Structure Electrochemistry Relationships of Alpha,Beta Unsaturated Ketones. *Anal.Chim.Acta*, *305*, 295–303.

Tong, W., Collantes, E.R., Chen, Y. and Welsh, W.J. (1996). A Comparative Molecular Field Analysis Study of N-Benzylpiperidines as Acetylcholinesterase Inhibitors. *J.Med.Chem.*, *39*, 380–387.

Tong, W., Lowis, D.R., Perkins, R., Chen, Y., Welsh, W.J., Goddette, D.W., Heritage, T.W. and Sheehan, D.M. (1998). Evaluation of Quantitative Structure-Activity Relationship Method for Large-Scale Prediction of Chemicals Binding to the Estrogen Receptor. *J.Chem.Inf.Comput.Sci.*, *38*, 669–677.

Topliss, J.G. and Shapiro, E.L. (1975). Quantitative Structure-Activity Relationships in the Δ^6-6-Substituted Progesterone Series. A Reappraisal. *J.Med.Chem.*, *18*, 621–623.

Topliss, J.G. and Edwards, R.P. (1979). Chance Factors in Studies of Quantitative Structure-Activity Relationships. *J.Med.Chem.*, *22*, 1238–1244.

Topliss, J. G. ed. (1983). *Quantitative Structure-Activity Relationships of Drugs*. Academic Press, New York (NY).

Topliss, J.G. (1993). Some Observation on Classical QSAR. *Persp.Drug Disc.Des.*, *1*, 253–268.

Topsom, R.D. (1976). The Nature and Analysis of Substituent Electronic Effects. *Prog.Phys. Org.Chem.*, *12*, 1–20.

Topsom, R.D. (1987a). Electronic Substituent Effects in Molecular Spectroscopy. *Prog.Phys. Org.Chem.*, *16*, 193–235.

Topsom, R.D. (1987b). Some Theoretical Studies of Electronic Substituent Effects in Organic Chemistry. *Prog.Phys.Org.Chem.*, *16*, 125–191.

Topsom, R.D. (1987c). Substituent Effects on Ground-State Molecular Structures and Charge Distributions. *Prog.Phys.Org.Chem.*, *16*, 85–124.

Tosato, M.L., Chiorboli, C., Eriksson, L. and Jonsson, J. (1991). Multivariate Modelling of the Rate Constant of the Gas-Phase Reaction of Haloalkanes with the Hydroxyl Radical. *Sci.Total Environ.*, *190/110*, 307–325.

Tosato, M.L., Piazza, R., Chiorboli, C., Passerini, L., Pino, A., Cruciani, G. and Clementi, S. (1992). Application of Chemometrics to the Screening of Hazardous Chemicals. *Chemom. Intell.Lab.Syst.*, *16*, 155–167.

Trapani, G., Carotti, A., Franco, M., Latrofa, A., Genchi, G. and Liso, G. (1993). Structure Affinity Relationships of Some Alkoxycarbonyl 2H-Pyrimido or Alkoxycarbonyl-4H-Pyrimido-(2,1-b)Benzothiazol-2-One or 4-One Benzodiazepine Receptor Ligands. *Eur.J.Med.Chem.*, *28*, 13–21.

Tratch, S.S., Devdariani, R.O. and Zefirov, N.S. (1990a). Combinatorial Models and Algorithms in Chemistry. Topological-Configurational Analogs of the Wiener Index. *Zh.Org.Khim.*, *26*, 921–932.

Tratch, S.S., Stankevitch, I.V. and Zefirov, N.S. (1990b). Combinatorial Models and Algorithms in Chemistry. The Expanded Wiener Number – A Novel Topological Index. *J.Comput.Chem.*, *11*, 899–908.

Tratch, S.S., Lomova, O.A., Sukhachev, D.V., Palyulin, V.A. and Zefirov, N.S. (1992). Generation of Molecular Graphs for QSAR Studies: An Approach Based on Acyclic Fragment Combinations. *J.Chem.Inf.Comput.Sci.*, *32*, 130–139.

Tratnyek, P.G. and Holgné, J. (1991). Oxidation of Substituted Phenols in the Environment: A QSAR Analysis of Rate Constants for Reaction with Singlet Oxygen. *Environ.Sci.Technol.*, *25*, 1596–1604.

Trinajstic, N., Jericevic, Z., Knop, J.V., Müller, W.R. and Szymanski, K. (1983). Computer Generation of Isomeric Structures. *Pure & Appl.Chem.*, *55*, 379–390.

Trinajstic, N., Klein, D.J. and Randic, M. (1986a). On Some Solved and Unsolved Problems of Chemical Graph Theory. *Int.J.Quantum Chem.Quant.Chem.Symp.*, *20*, 699–742.

Trinajstic, N., Randic, M. and Klein, D.J. (1986b). On the Quantitative Structure-Activity Relationship in Drug Research. *Acta Pharm.Jugosl.*, *36*, 267–279.

Trinajstic, N. (1988). The Characteristic Polynomial of a Chemical Graph. *J.Math.Chem.*, *2*, 197–215.

Trinajstic, N. (1991). Graph Theory and Molecular Orbitals. In *Chemical Graph Theory. Introduction and Fundamentals* (Bonchev, D. and Rouvray, D.H., eds.), Abacus Press/Gordon & Breach, New York (NY), pp. 235–279.

Trinajstic, N. (1992). *Chemical Graph Theory.* CRC Press, Boca Raton (FL), 322 pp.

Trinajstic, N., Babic, D., Nikolic, S., Plavsic, D., Amic, D. and Mihalic, Z. (1994). The Laplacian Matrix in Chemistry. *J.Chem.Inf.Comput.Sci.*, *34*, 368–376.

Trinajstic, N., Nikolic, S., Lucic, B. and Amic, D. (1996). On QSAR Modeling. *Acta Pharm. Jugosl.*, *46*, 249–263.

Trinajstic, N., Nikolic, S., Lucic, B., Amic, D. and Mihalic, Z. (1997). The Detour Matrix in Chemistry. *J.Chem.Inf.Comput.Sci.*, *37*, 631–638.

Tripos Inc. (1997). *HQSAR Software.*

Tropsha, A. and Cho, S.J. (1998). Cross-Validated R2 Guided Region Selection for CoMFA Studies. In *3D QSAR in Drug Design – Vol. 3* (Kubinyi, H., Folkers, G. and Martin, Y.C., eds.), Kluwer/ESCOM, Dordrecht (The Netherlands), pp. 57–69.

Trucco, E. (1956a). A Note on the Information Content of Graphs. *Bull.Math.Biophys.*, *18*, 129–135.

Trucco, E. (1956b). On the Information Content of Graphs: Compound Symbols; Different States for Each Point. *Bull.Math.Biophys.*, *18*, 237–253.

Tsai, R.-S., Testa, B., El Tayar, N. and Carrupt, P.-A. (1991). Structure-Lipophilicity Relationships of Zwitterionic Amino Acids. *J.Chem.Soc.Perkin Trans.2*, 1802.

Tsai, R.-S., Fan, W., El Tayar, N., Carrupt, P.-A., Testa, B. and Kier, L.B. (1993). Solute-Water Interactions in the Organic Phase of a Biphasic System. 1. Structural Influence of Organic Solutes on the "Water-Dragging" Effect. *J.Am.Chem.Soc.*, *115*, 9632–9639.

Tsai, R.-S., Carrupt, P.-A. and Testa, B. (1995). Measurement of Partition Coefficient Using Centrifugal Partition Chromatography. Method Development and Application to the Determination of Solute Properties. In *Modern Countercurrent Chromatography* (Conway, W.D. and Petroski, R.J., eds.), American Chemical Society, New York (NY), pp. 143–154.

Tsantilikakoulidou, A. and Kier, L.B. (1992a). A Quantitative Structure-Activity Relationship (QSAR) Study of Alkylpyrazine Odor Modalities. *Pharm.Res.*, *9*, 1321–1323.

Tsantilikakoulidou, A., Kier, L.B. and Joshi, N. (1992b). The Use of Electrotopological State Indexes in QSAR Studies. *J.Chim.Phys.Phys-Chim.Biol.*, *89*, 1729–1733.

Tuppurainen, K., Lotjonen, S., Laatikainen, R. and Vartiainen, T. (1992). Structural and Electronic Properties of MX Compounds Related to TA100 Mutagenicity: A Semiempirical Molecular Orbital QSAR Study. *Mut.Res.*, *266*, 181–188.

Tuppurainen, K. and Lotjonen, S. (1993). On the Mutagenicity of MX Compounds. *Mut.Res.*, *287*, 235–241.

Tuppurainen, K. (1994). QSAR Approach to Molecular Mutagenicity. A Survey and a Case Study: MX Compounds. *J.Mol.Struct.(Theochem)*, *112*, 49–56.

Tuppurainen, K. (1999a). EEVA (Electronic Eigenvalue): A New QSAR/QSPR Descriptor for Electronic Substituent Effects Based on Molecular Orbital Energies. *SAR & QSAR Environ.Res.*, *10*, 39–46.

Tuppurainen, K. (1999b). Frontier Orbital Energies, Hydrophobicity and Steric Factors as Physical QSAR Descriptors of Molecular Mutagenicity. A Review with a Case Study: MX Compounds. *Chemosphere*, *38*, 3015–3030.

Turner, D.B., Willett, P., Ferguson, A.M. and Heritage, T.W. (1995). Similarity Searching in Files of Three-Dimensional Structures: Evaluation of Similarity Coefficients and Standardization Methods for Field-Based Similarity Searching. *SAR & QSAR Environ.Res.*, *3*, 101–130.

Turner, D.B., Willett, P., Ferguson, A.M. and Heritage, T.W. (1997). Evaluation of a Novel Infrared Range Vibration-Based Descriptor (EVA) for QSAR Studies. 1. General Application. *J.Comput.Aid.Molec.Des.*, *11*, 409–422.

Turner, D.B., Costello, C.L. and Jurs, P.C. (1998). Prediction of Critical Temperatures and Pressures of Industrially Important Organic Compounds from Molecular Structure. *J.Chem.Inf. Comput.Sci.*, *38*, 639–645.

Turner, D.B., Willett, P., Ferguson, A.M. and Heritage, T.W. (1999). Evaluation of a Novel Molecular Vibration-Based Descriptor (EVA) for QSAR Studies: 2. Model Validation Using a Benchmark Steroid Dataset. *J.Comput.Aid.Molec.Des.*, *13*, 271–296.

Turro, N.J. (1986). Geometric and Topological Thinking in Organic Chemistry. *Angew.Chem.Int. Ed.Engl.*, *25*, 882–901.

Tute, M.S. (1990). History and Objectives of Quantitative Drug Design. In *Quantitative Drug Design – Vol. 4* (Ramsden, C.A., ed.), Pergamon Press, Oxford (UK), pp. 1–31.

Tversky, A. (1977). Features of Similarity. *Psychol.Rev.*, *84*, 327–352.

Tysklind, M., Lundgren, K., Rappe, C., Eriksson, L., Jonsson, J., Sjöström, M. and Ahlborg, U.G. (1992). Multivariate Characterization and Modeling of Polychlorinated Dibenzo Para Dioxins and Dibenzofurans. *Environ.Sci.Technol.*, *26*, 1023–1030.

Tysklind, M., Lundgren, K., Rappe, C., Eriksson, L., Jonsson, J. and Sjöström, M. (1993). Multivariate Quantitative Structure-Activity Relationships for Polychlorinated Dibenzo-p-Dioxins and Dibenzofurans. *Environ.Toxicol.Chem.*, *12*, 659–672.

Tysklind, M., Tillitt, D., Eriksson, L., Lundgren, K. and Rappe, C. (1994). A Toxic Equivalency Factor Scale for Polychlorinated Dibenzofurans. *Fund.Appl.Toxicol.*, *22*, 277–285.

Ugi, I., Wochner, M., Fontain, E., Bauer, J., Gruber, B. and Karl, R. (1990). Chemical Similarity, Chemical Distance, and Computer-Assisted Formalized Reasoning by Analogy. In *Concepts and Applications of Molecular Similarity* (Johnson, M.A. and Maggiora, G.M., eds.), Wiley, New York (NY), pp. 239–288.

Unger, S.H. and Hansch, C. (1973). On Model Building in Structure-Activity Relationships. A Reexamination of Adrenergic Blocking Activity of β-Halo-β-arylalkylamines. *J.Med.Chem.*, *16*, 745–749.

Unger, S.H. and Hansch, C. (1976). Quantitative Models of Steric Effects. *Prog.Phys.Org.Chem.*, *12*, 91–118.

Unger, S.H., Cook, J.R. and Hollenberg, J.S. (1978). Simple Procedure for Determining Octanol-Aqueous Partition, Distribution, and Ionization Coefficients by Reversed-Phase High-Pressure Liquid Chromatography. *J.Pharm.Sci.*, *67*, 1364–1367.

Unger, S.H., Cheung, P.S., Chiang, G.H. and Cook, J.R. (1986). In *Partition Coefficient Determination and Estimation* (Dunn III, W.J., Block, J.H. and Hansch, C., eds.), Pergamon, New York (NY), pp. 69–81.

Urrestarazu Ramos, E., Vaes, W., Verhaar, H.J. and Hermens, J.L. (1998). Quantitative Structure-Activity Relationships for the Aquatic Toxicity of Polar and Nonpolar Narcotic Pollutants. *J.Chem.Inf.Comput.Sci.*, *38*, 845–852.

Valko, K. and Slegel, P. (1992). Chromatographic Separation and Molecular Modeling of Triazines with Respect to Their Inhibition of the Growth of L1210/R71 Cells. *J.Chromat.*, *592*, 59–63.

Vallat, P., Fan, W., El Tayar, N., Carrupt, P.-A. and Testa, B. (1992). Solvatochromic Analysis of the Retention Mechanism of Two Novel Stationary Phases Used for Measuring Lipophilicity by RP-HPLC. *J.Liquid Chromat.*, *15*, 2133–2151.

Vallat, P., Gaillard, P., Carrupt, P.-A., Tsai, R.-S. and Testa, B. (1995). 37. Structure-Lipophilicity and Structure-Polarity Relationships of Amino Acids and Peptides. *Helv.Chim.Acta*, *78*, 471–485.

van Aalten, D.M.F., Bywater, R., Findlay, J.B.C., Hendlich, M., Hooft, R.W.W. and Vriend, G. (1996). PRODRG, a Program for Generating Molecular Topologies and Unique Molecular Descriptors from Coordinates of Small Molecules. *J.Comput.Aid.Molec.Des.*, *10*, 255–262.

van Bekkum, H., Verkade, P.E. and Wepster, B.M. (1959). *Rec.Trav.Chim.*, *78*, 815.

van de Waterbeemd, H. and Testa, B. (1983). *Int.J.Pharm.*, *14*, 29–41.

van de Waterbeemd, H. (1986). *Hydrophobicity of Organic Compounds, Vol. 1*. Compudrug, Budapest (HUN).

van de Waterbeemd, H. and Testa, B. (1987). The Parametrization of Lipophilicity and Other Structural Properties in Drug Design. In *Advances in Drug Research – Vol. 16* (Testa, B., ed.), Academic Press, London (UK), pp. 85–225.

van de Waterbeemd, H., El Tayar, N., Carrupt, P.-A. and Testa, B. (1989). Pattern Recognition Study of QSAR Substituent Descriptors. *J.Comput.Aid.Molec.Des.*, *3*, 111–132.

van de Waterbeemd, H. (1992). The History of Drug Research: From Hansch to the Present. *Quant.Struct.-Act.Relat.*, *11*, 200–204.

van de Waterbeemd, H. (1993). Recent Progress in QSAR-Technology. *Drug Design & Discovery*, *9*, 277–285. [R]

van de Waterbeemd, H., Carrupt, P.-A., Testa, B. and Kier, L.B. (1993a). Multivariate Data Modeling of New Steric, Topological and CoMFA-Derived Substituent Parameters. In *Trends in QSAR and Molecular Modelling 92* (Wermuth, C.G., ed.), ESCOM, Leiden (The Netherlands), pp. 69–75.

van de Waterbeemd, H., Clementi, S., Costantino, G., Carrupt, P.-A. and Testa, B. (1993b). CoMFA-Derived Substituent Descriptors for Structure-Property Correlations. In *3D QSAR in Drug Design. Theory, Methods and Applications.* (Kubinyi, H., ed.), ESCOM, Leiden (The Netherlands), pp. 697–707.

van de Waterbeemd, H. ed. (1994). *Chemometric Methods in Molecular Design*. VCH, Weinheim (Germany).

van de Waterbeemd, H. ed. (1995). *Advanced Computer-Assisted Techniques in Drug Discovery.* VCH Publishers, Weinheim (Germany), 359 pp.

van de Waterbeemd, H., Costantino, G., Clementi, S., Cruciani, G. and Valigi, R. (1995). Experimental Design in Synthesis Planning and Structure-Property Correlations. Disjoint Principal Properties of Organic Substituents. In *Chemometric Methods in Molecular Design – Vol. 2* (van de Waterbeemd, H., ed.), VCH Publishers, New York (NY), pp. 103–112.

van de Waterbeemd, H. ed. (1996). *Structure-Property Correlations in Drug Research*. Academic Press, R.G.Landes Co., Austin (TX).

van de Waterbeemd, H., Camenisch, G., Folkers, G. and Raevsky, O.A. (1996). Estimation of CACO-2 Cell Permeability Using Calculated Molecular Descriptors. *Quant.Struct.-Act.Relat.*, *15*, 480–490.

van de Waterbeemd, H., Carter, R.E., Grassy, G., Kubinyi, H., Martin, Y.C., Tute, M.S., Willett, P., Haasnoot, C.G., Kier, L.B., Muller, K., Rose, S.V., Weber, J., Wibley, K.S., Wold, S., Boyd, D.B., Clark, D.E., Dehaen, C., Heindel, N.D., Kratochvil, P., Kutscher, B., Lewis, R.A., Mabilia, M., Metanomski, W.V., Polymeropoulos, E.E. and Tollenaere, J.P. (1997a). Glossary of Terms Used in Computational Drug Design. *Pure & Appl.Chem.*, *69*, 1137–1152.

van de Waterbeemd, H., Testa, B., and Folkers, G. eds. (1997b). *Computer-Assisted Lead Finding and Optimization*. Wiley-VCH, Weinheim (Germany), 554 pp.

van Haelst, A.G., Paulus, R.H.W.L. and Govers, H.A.J. (1997). Calculation of Molecular Volumes of Tetrachlorobenzyltoluenes. *SAR & QSAR Environ.Res.*, *6*, 205–214.

Vansteen, B.J., Vanwijngaarden, I., Tulp, M.T. and Soudijn, W. (1994). Structure Affinity Relationship Studies on 5HT(1A) Receptor Ligands. 2. Heterobicyclic Phenylpiperazines with N4-Aralkyl Substituents. *J.Med.Chem.*, *37*, 2761–2773.

Vanvlaardingen, P.A., Steinhoff, W.J., Devoogt, P. and Admiraal, W.A. (1996). Property-Toxicity Relationships of Azaarenes to the Green Alga Scenedesmus Acuminatus. *Environ.Toxicol. Chem.*, *15*, 2035–2042.

Varkony, T.H., Shiloach, Y. and Smith, D.H. (1979). Computer-Assisted Examination of Chemical Compounds for Structural Similarities. *J.Chem.Inf.Comput.Sci.*, *19*, 104–111.

Varmuza, K., Penchev, P.N. and Scsibrany, H. (1998). Maximum Common Substructures of Organic Compounds Exhibiting Similar Infrared Spectra. *J.Chem.Inf.Comput.Sci.*, *38*, 420–427.

Varmuza, K. and Scsibrany, H. (2000). Substructure Isomorphism Matrix. *J.Chem.Inf.Comput.Sci.*, *40*, 308–313.

Veith, G.D. and Mekenyan, O. (1993). A QSAR Approach for Estimating the Aquatic Toxicity of Soft Electrophiles (QSAR for Soft Electrophiles). *Quant.Struct.-Act.Relat.*, *12*, 349–356.

Vendrame, R., Braga, R.S., Takahata, Y. and Galvão, D.S. (1999). Structure-Activity Relationship Studies of Carcinogenic Activity of Polycyclic Aromatic Hydrocarbons Using Molecular Descriptors with Principal Component Analysis and Neural Network Methods. *J.Chem.Inf. Comput.Sci.*, *39*, 1094–1104.

Venturelli, P., Menziani, M.C., Cocchi, M., Fanelli, F. and De Benedetti, P.G. (1992). Molecular Modeling and Quantitative Structure-Activity Relationship Analysis Using Theoretical Descriptors of 1,4-Benzodioxan (WB-4101) Related Compounds Alpha-1-Adrenergic Antagonists. *J.Mol.Struct.(Theochem)*, *95*, 327–340.

Verhaar, H.J., Vanleeuwen, C.J. and Hermens, J.L. (1992). Classifying Environmental Pollutants. 1. Structure-Activity Relationships for Prediction of Aquatic Toxicity. *Chemosphere*, *25*, 471–491.

Verhaar, H.J., Eriksson, L., Sjöström, M., Schüürmann, G., Seinen, W. and Hermens, J.L. (1994). Modeling the Toxicity of Organophosphates: A Comparison of the Multiple Linear Regression and PLS Regression Methods. *Quant.Struct.-Act.Relat.*, *13*, 133–143.

Verhaar, H.J., Urrestarazu Ramos, E. and Hermens, J.L. (1996). Classifying Environmental Pollutants. 2: Separation of Class 1 (Baseline Toxicity) and Class 2 ('Polar Narcosis') Type Compounds Based on Chemical Descriptors. *J.Chemom.*, *10*, 149–162.

Verloop, A. (1972). The Use of Linear Free Energy Parameters and Other Experimental Constants in Structure-Activity Studies. In *Drug Design, Vol. III* (Ariëns, E.J., ed.), Academic Press, New York (NY), pp. 133–187.

Verloop, A., Hoogenstraaten, W. and Tipker, J. (1976). Development and Application of New Steric Substituent Parameters in Drug Design. In *Drug Design, Vol. 7* (Ariëns, E.J., ed.), Academic Press, New York (NY), pp. 165–207.

Verloop, A. (1985). In *QSAR and Strategies in the Design of Bioactive Compounds* (Seydel, J.K., ed.), VCH Publishers, Weinheim (Germany), pp. 98–104.

Verloop, A. (1987). *The STERIMOL Approach to Drug Design*. Marcel Dekker, New York (NY).

Villemin, D., Cherqaoui, D. and Cense, J.-M. (1993). Neural Networks Studies. Quantitative Structure-Activity Relationship of Mutagenic Aromatic Nitro Compounds. *J.Chim.Phys.Phys-Chim.Biol.*, *90*, 1505–1519.

Villemin, D., Cherqaoui, D. and Mesbah, A. (1994). Predicting Carcinogenicity of Polycyclic Aromatic Hydrocarbons from Backpropagation Neural Network. *J.Chem.Inf.Comput.Sci.*, *34*, 1288–1293.

Vinogradov, S.N. and Linnell, R.H. (1971). *Hydrogen Bonding*. Van Nostrand Reinhold, New York (NY).

Violon, D. (1999). Multiple Regression Analysis of Octanol/Water Partition Coefficients of Non-Ionic Monomeric Radiographic Contrast Media with the Combination of Three Molecular Descriptors. *Brit.J.Radiol.*, *72*, 44–47.

Viswanadhan, V.N., Ghose, A.K., Revankar, G.R. and Robins, R.K. (1989). Atomic Physicochemical Parameters for Three Dimensional Structure Directed Quantitative Structure-Activity Relationships. 4. Additional Parameters for Hydrophobic and Dispersive Interactions and their Application for an Automated Superposition of Certain Naturally Occurring Nucleoside Antibiotics. *J.Chem.Inf.Comput.Sci.*, *29*, 163–172.

Viswanadhan, V.N., Reddy, M.R., Bacquet, R.J. and Erion, M.D. (1993). Assessment of Methods Used for Predicting Lipophilicity: Application to Nucleosides and Nucleoside Bases. *J.Comput.Chem.*, *14*, 1019–1026.

Viswanadhan, V.N., Ghose, A.K., Singh, U.C. and Wendoloski, J.J. (1999). Prediction of Solvation Free Energies of Small Organic Molecules: Additive-Constitutive Models Based on Molecular Fingerprints and Atomic Constants. *J.Chem.Inf.Comput.Sci.*, *39*, 405–412.

Voelkel, A. (1994). Structural Descriptors in Organic Chemistry – New Topological Parameter Based on Electrotopological State of Graph Vertices. *Computers Chem.*, *18*, 1–4.

Vogel, A.I. (1948). Physical Properties and Chemical Constitution. Part XXIII. Miscellaneous Compounds. Investigation of the so-called Co-ordinate or Dative Link in Esters of Oxy-Acids and in Nitro-Paraffins by Molecular Refractivity Determinations. Atomic, Structural, and Group Parachors and Refractivities. *J.Chem.Soc.*, 1833–1855.

Vogel, A.I., Cresswell, W.T., Jeffery, G.H. and Leicester, J. (1951). Calculation of the Refractive Indices of Liquid Organic Compounds: Bond Molecular Refraction. *Chem.& Ind.*, *5*, 376–376.

Voiculetz, N., Balaban, A.T., Niculescu-Duvaz, I. and Simon, Z. (1990). *Modeling of Cancer Genesis and Prevention*. CRC Press, Boca Raton (FL).

Volkenstein, M.V. (1963). *Configurational Statistics of Polymeric Chains*. Wiley-Interscience, New York (NY).

von der Lieth, C.-W., Stumpf-Nothof, K. and Prior, U. (1996). A Bond Flexibility Index Derived from the Constitution of Molecules. *J.Chem.Inf.Comput.Sci.*, *36*, 711–716.

Voronoi, G.F. (1908). *Reine Angew.Math.*, *134*, 198.

Vracko, M. (1997). A Study of Structure-Carcinogenic Potency Relationship with Artificial Neural Networks. The Using of Descriptors Related to Geometrical and Elecronic Structures. *J.Chem.Inf.Comput.Sci.*, *37*, 1037–1043.

Wade, R.C. (1993a). Molecular Interaction Fields. In *3D QSAR in Drug Design. Theory, Methods and Applications.* (Kubinyi, H., ed.), ESCOM, Leiden (The Netherlands), pp. 486–502.

Wade, R.C., Clark, K.J. and Goodford, P.J. (1993b). Further Development of Hydrogen Bond Functions for Use in Determining Energetically Favorable Binding Sites on Molecule of Known Structure. 1. Ligand Probe Groups with the Ability to Form Two Hydrogen Bonds. *J.Med.Chem.*, *36*, 140–147.

Wade, R.C. and Goodford, P.J. (1993c). Further Development of Hydrogen Bond Functions for Use in Determining Energetically Favorable Binding Sites on Molecule of Known Structure. 2. Ligand Probe Groups with the Ability to Form More Than Two Hydrogen Bonds. *J.Med.Chem.*, *36*, 148–156.

Wagener, M., Sadowski, J. and Gasteiger, J. (1995). Autocorrelation of Molecular Surface Properties for Modeling *Corticosteroid Binding Globulin* and Cytosolic *Ah* Receptor Activity by Neural Networks. *J.Am.Chem.Soc.*, *117*, 7769–7775.

Wagner, G.C., Colvin, J.T., Allen, J.P. and Stapleton, H.J. (1985). Fractal Models of Protein Structures, Dynamics, and Magnetic Relaxation. *J.Am.Chem.Soc.*, *107*, 5589–5594.

Walker, P.D., Maggiora, G.M., Johnson, M.A., Petke, J.D. and Mezey, P.G. (1995). Shape Group Analysis of Molecular Similarity: Shape Similarity of Six-Membered Aromatic Ring Systems. *J.Chem.Inf.Comput.Sci.*, *35*, 568–578.

Waller, C.L. and McKinney, J.D. (1992). Comparative Molecular Field Analysis of Polyhalogenated Dibenzo-p-Dioxins, Dibenzofurans, and Biphenyls. *J.Med.Chem.*, *35*, 3660–3666.

Waller, C.L. and Marshall, G.R. (1993). Three Dimensional Quantitative Structure-Activity Relationship of Angiotesin Converting Enzyme and Thermolysin Inhibitors. 2. A Comparison of CoMFA Models Incorporating Molecular Orbital Fields and Desolvation Free Energies Based on Active Analog and Complementary Receptor Field Alignment Rules. *J.Med.Chem.*, *36*, 2390–2403.

Waller, C.L., Oprea, T.I., Giolitti, A. and Marshall, G.R. (1993). Three Dimensional QSAR of Human Immunodeficiency Virus (I) Protease Inhibitors. 1. A CoMFA Study Employing Experimentally Determined Alignment Rules. *J.Med.Chem.*, *36*, 4152–4160.

Waller, C.L. (1994). A Three-Dimensional Technique for the Calculation of Octanol-Water Partition Coefficients. *Quant.Struct.-Act.Relat.*, *13*, 172–176.

Waller, C.L., Wyrick, S.D., Kemp, W.E., Park, H.M. and Smith, F.T. (1994). Conformational Analysis, Molecular Modeling, and Quantitative Structure-Activity Relationship Studies of Agents for the Inhibition of Astrocytic Chloride Transport. *Pharm.Res.*, *11*, 47–53.

Waller, C.L., Minor, D.L. and McKinney, J.D. (1995). Using Three-Dimensional Quantitative Structure-Activity Relationships to Examine Estrogen Receptor Binding Affinities of Polychlorinated Hydroxybiphenyls. *Environ.Health Persp.*, *103*, 702–707.

Waller, C.L., Evans, M.V. and McKinney, J.D. (1996a). Modeling the Cytochrome P450 Mediated Metabolism of Chlorinated Volatile Organic Compounds. *Drug.Metab.Disposition*, *24*, 203–210.

Waller, C.L. and Kellogg, G.E. (1996b). Adding Chemical Information to CoMFA Models with Alternative 3D QSAR Fields. *Netscience*, *2(1)*.

Waller, C.L. and Bradley, M.P. (1999). Development and Validation of a Novel Variable Selection Technique with Application to Multidimensional Quantitative Structure-Activity Relationship Studies. *J.Chem.Inf.Comput.Sci.*, *39*, 345–355.

Walsh, D.B. and Claxton, L.D. (1987). Computer-Assisted Structure-Activity Relationships of Nitrogenous Cyclic Compounds Tested in Salmonella Assays for Mutagenicity. *Mut.Res.*, *182*, 55–64.

Walters, D.E. and Hopfinger, A.J. (1986). Case Studies of the Application of Molecular Shape Analysis to Elucidate Drug Action. *J.Mol.Struct.(Theochem)*, *134*, 317–323.

Walters, D.E. and Hinds, R.M. (1994). Genetically Evolved Receptor Models: A Computational Approach to Construction of Receptor Models. *J.Med.Chem.*, *37*, 2527–2537.

Walters, D.E. (1998). Genetically Evolved Receptor Models (GERM) as a 3D QSAR Tool. In *3D QSAR in Drug Design – Vol. 3* (Kubinyi, H., Folkers, G. and Martin, Y.C., eds.), Kluwer/ ESCOM, Dordrecht (The Netherlands), pp. 159–166.

Walther, D. (1974). *J.Prakt.Chem.*, *316*, 604.

Wang, C.-X., Shi, Y.-Y. and Huang, F.-H. (1990). Fractal Study of Tertiary Structure of Proteins. *Phys.Rev.A*, *41*, 7043.

Wang, S.M., Milne, G.W.A. and Klopman, G. (1994). Graph Theory and Group Contributions in the Estimation of Boiling Points. *J.Chem.Inf.Comput.Sci.*, *34*, 1242–1250.

Wang, S.M. and Milne, G.W.A. (1993). Applications of Computers to Toxicological Research. *Chem.Res.Toxicol.*, *6*, 748–753.

Wang, T. and Zhou, J. (1998). 3DFS: A New 3D Flexible Searching System for Use in Drug Design. *J.Chem.Inf.Comput.Sci.*, *38*, 71–77.

Wang, X.Z. and Chen, B.H. (1998). Clustering of Infrared Spectra of Lubricating Base Oils Using Adaptive Resonance Theory. *J.Chem.Inf.Comput.Sci.*, *38*, 457–462.

Warne, M.S., Connell, D.W., Hawker, D.W. and Schüürmann, G. (1989a). Prediction of the Toxicity of Mixtures of Shale Oil Components. *Ecotox.Environ.Safe.*, *18*, 121–128.

Warne, M.S., Connell, D.W., Hawker, D.W. and Schüürmann, G. (1989b). Quantitative Structure-Activity Relationships for the Toxicity of Selected Shale Oil Components to Mixed Marine Bacteria. *Ecotox.Environ.Safe.*, *17*, 133–148.

Warne, M.S., Boyd, M.A., Meharg, E.M., Osborn, D., Killham, D., Lindon, J.C. and Nicholson, J.K. (1999a). Quantitative Structure-Toxicity Relationships for Halobenzenes in Two Species of Bioluminescent Bacteria, *Pseudomonas fluorescens* and *Vibrio fischeri*, Using an Atom-Centered Semi-Empirical Molecular-Orbital Based Model. *SAR & QSAR Environ.Res.*, *10*, 17–38.

Warne, M.S., Osborn, D., Lindon, J.C. and Nicholson, J.K. (1999b). Quantitative Structure-Toxicity Relationships for Halogenated Substituted-Benzenes to *Vibrio fischeri*, Using Atom-Based Semi-Empirical Molecular-Orbital Descriptors. *Chemosphere*, *38*, 3357–3382.

Warthen, J.D., Schmidt, W.F., Cunningham, R.T., Demilo, A.B. and Fritz, G.L. (1993). Quantitative Structure-Activity Relationships (QSAR) of Trimedlure Isomers. *J.Chem.Ecol.*, *19*, 1323–1335.

Wayner, D.D.M. and Arnold, D.R. (1984). Substituent Effects on Benzyl Radical Hyperfine Coupling Constants. Part 2. The Effect of Sulphur Substituents. *Can.J.Chem.*, *62*, 1164–1168.

Wehrens, R., Pretsch, E. and Buydens, L. (1998). Quality Criteria of Genetic Algorithms for Structure Optimization. *J.Chem.Inf.Comput.Sci.*, *38*, 151–157.

Weiner, M.L. and Weiner, P.H. (1973). A Study of Structure-Activity Relationships of a Series of Dyphenylaminopropanols by Factor Analysis. *J.Med.Chem.*, *16*, 655–661.

Weininger, D. (1988). SMILES, a Chemical Language and Information System. 1. Introduction to Methodology and Encoding Rules. *J.Chem.Inf.Comput.Sci.*, *28*, 31–36.

Weininger, D., Weininger, A. and Weininger, J.L. (1989). SMILES. 2. Algorithm for Generation of Unique SMILES Notation. *J.Chem.Inf.Comput.Sci.*, *29*, 97–101.

Weininger, D. (1990). SMILES. 3. DEPICT: Graphical Depiction of Chemical Structures. *J.Chem.Inf.Comput.Sci.*, *30*, 273–283.

Weininger, D. and Weininger, J.L. (1990). Chemical Structures and Computers. In *Quantitative Drug Design. Vol. 4* (Ramsden, C.A., ed.), Pergamon Press, Oxford (UK), pp. 59–82.

Weininger, S.J. (1984). The Molecular Structure Conundrum: Can Classical Chemistry be reduced to Quantum Chemistry? *J.Chem.Educ.*, *61*, 939–944.

Wells, M.J.M., Clark, C.R. and Patterson, R.M. (1981). Correlation of Reversed-Phase Capacity Factors for Barbiturates with Biological Activities, Partition Coefficients, and Molecular Connectivity Indices. *J.Chromatogr.Sci.*, *19*, 573.

Wells, M.J.M., Clark, C.R. and Patterson, R.M. (1982). Investigation of *N*-Alkylbenzamides by Reversed-Phase Liquid Chromatography. III. Correlation of Chromatographic Parameters with Molecular Connectivity Indices for C_1 to C_5 *N*-alkylbenzamides. *J.Chromat.*, *235*, 61–74.

Wells, P.R. (1968a). *Linear Free Energy Relationships*. Academic Press, New York (NY).

Wells, P.R. (1968b). *Prog.Phys.Org.Chem.*, *6*, 111.

Wentang, C., Ying, Z. and Feibai, Y. (1993). New Computer Representation for Chemical Structures: Two-Level Compact Connectivity Tables. *J.Chem.Inf.Comput.Sci.*, *33*, 604–608.

Wermuth, C.G. ed. (1993). *Trends in QSAR and Molecular Modelling 92*. ESCOM, Leiden (The Netherlands), 595 pp.

Wessel, M.D. and Jurs, P.C. (1994). Prediction of Reduced Ion Mobility Constants from Structural Information Using Multiple Linear Regression Analysis and Computational Neural Networks. *Anal.Chem.*, *66*, 2480–2487.

Wessel, M.D. and Jurs, P.C. (1995a). Prediction of Normal Boiling Points for a Diverse Set of Industrially Important Organic Compounds from Molecular Structure. *J.Chem.Inf.Comput.Sci.*, *35*, 841–850.

Wessel, M.D. and Jurs, P.C. (1995b). Prediction of Normal Boiling Points of Hydrocarbons from Molecular Structure. *J.Chem.Inf.Comput.Sci.*, *35*, 68–76.

Wessel, M.D., Sutter, J.M. and Jurs, P.C. (1996). Prediction of Reduced Ion Mobility Constants of Organic Compounds from Molecular Structure. *Anal.Chem.*, *68*, 4237–4243.

Wessel, M.D., Jurs, P.C., Tolan, J.W. and Muskal, S.M. (1998). Prediction of Human Intestinal Absorption of Drug Compounds from Molecular Structure. *J.Chem.Inf.Comput.Sci.*, *38*, 726–735.

Westheimer, F.H. and Kirkwood, J.J. (1938). The Electrostatic Influence of Substituents on the Dissociation Constants of Organic Acids. II. *J.Chem.Phys.*, *6*, 513–517.

Wheland, G.W. (1955). *Resonance in Organic Chemistry*. Wiley, New York (Ny).

White, J.H. (1969). Self-Linking and the Gauss Integral in Higher Dimension. *Am.J.Math.*, *91*, 693.

Whitley, D.C. (1998). van der Waals Surface Graphs and the Shape of Small Rings. *J.Chem.Inf.Comput.Sci.*, *38*, 906–914.

Wiberg, K.B. (1968). Application of the Pople-Santry-Segal CNDO Method to the Cyclopropylcarbinyl and Cyclobutyl Cation and to Bicyclobutane. *Tetrahedron*, *24*, 1083–1096.

Wieland, T. (1996). The Use of Structure Generators in Predictive Pharmacology and Toxicology. *Arzneim.Forsch.*, *46*, 223–227.

Wiener, H. (1947a). Correlation of Heat of Isomerization, and Differences in Heats of Vaporization of Isomers, among the Paraffin Hydrocarbons. *J.Am.Chem.Soc.*, *69*, 2636–2638.

Wiener, H. (1947b). Influence of Interatomic Forces on Paraffin Properties. *J.Chem.Phys.*, *15*, 766–766.

Wiener, H. (1947c). Structural Determination of Paraffin Boiling Points. *J.Am.Chem.Soc.*, *69*, 17–20.

Wiener, H. (1948a). Relationship of Physical Properties of Isomeric Alkanes to Molecular Structure Surface Tension, Specific Dispersion and Critical Solution Temperature in Aniline. *J.Phys.Colloid Chem.*, *52*, 1082–1089.

Wiener, H. (1948b). Vapour Pressure-Temperature Relations among the Branched Paraffin Hydrocarbons. *J.Phys.Chem.*, *52*, 425–430.

Wierl, K. (1931). *Ann.Phys.(Leipzig)*, *8*, 521–564.

Wiese, M. (1993). The Hypothetical Active-Site Lattice. In *3D QSAR in Drug Design. Theory, Methods and Applications*. (Kubinyi, H., ed.), ESCOM, Leiden (The Netherlands), pp. 431–442.

Wikel, J.H. and Dow, E.R. (1993). The Use of Neural Networks for Variable Selection in QSAR. *Bioorg.Med.Chem.Lett.*, *3*, 645–651.

Wikel, J.H., Sofia, M.J., Saussy, D.L. and Bemis, K.G. (1994). QSAR Study of Ortho Phenylphenol Leukotriene B4 Receptor Antagonists. *Bioorg.Med.Chem.Lett.*, *4*, 795–800.

Wilcox Jr, C.F. (1968). Solubility of Molecules Containing (4n)-Rings. *Tetrahedron Lett.*, *7*, 795–800.

Wilcox Jr, C.F. (1969). Stability of Molecules Containing Nonalternant Rings. *J.Am.Chem.Soc.*, *91*, 2732–2736.

Wilding, W.V. and Rowley, R.L. (1986). A Four Parameter Corresponding States Method for the Prediction of Thermodynamic Properties of Polar and Nonpolar Fluids. *Int.J.Thermophys.*, *7*, 525–539.

Wildman, S.A. and Crippen, G.M. (1999). Prediction of Physicochemical Parameters by Atomic Contributions. *J.Chem.Inf.Comput.Sci.*, *39*, 868–873.

Wilkerson, W.W., Galbraith, W., Gansbrangs, K., Grubb, M., Hewes, W.E., Jaffee, B., Kenney, J.P., Kerr, J. and Wong, N. (1994). Antiinflammatory 4,5 Diarylpyrroles Synthesis and QSAR. *J.Med.Chem.*, *37*, 988–998.

Wilkerson, W.W. (1995a). A Quantitative Structure-Activity Relationship Analysis of a Series of 2′-(2,4 Difluorophenoxy)-4′-Substituted Methanesulfonilides. *Eur.J.Med.Chem.*, *30*, 191–197.

Wilkerson, W.W., Copeland, R.A., Covington, M. and Trzaskos, J.M. (1995b). Antiinflammatory 4,5-Diarylpyrroles. 2. Activity as a Function of Cyclooxygenase-2 Inhibition. *J.Med.Chem.*, *38*, 3895–3901.

Wilkins, C.L. and Randic, M. (1980). A Graph Theoretical Approach to Structure-Property and Structure-Activity Correlations. *Theor.Chim.Acta*, *58*, 69–71.

Wilkins, C.L., Randic, M., Schuster, S.M., Markin, R.S., Steiner, S. and Dorgan, L. (1981). A Graph-Theoretic Approach to Quantitative Structure-Activity/Reactivity Studies. *Anal.Chim.Acta*, *133*, 637–645.

Willett, P. and Winterman, V.A. (1986). A Comparison of Some Measures for the Determination of Inter-Molecular Structural Similarity. *Quant.Struct.-Act.Relat.*, *5*, 18–25.

Willett, P. (1987). *Similarity and Clustering in Chemical Information Systems.* Research Studies Press, Letchworth (UK), 254 pp.

Willett, P. (1988). Ranking and Clustering of Chemical Structure Databases. In *Physical Property Prediction in Organic Chemistry* (Jochum, C., Hicks, M.G. and Sunkel, J., eds.), Springer-Verlag, Berlin (Germany), pp. 191–207.

Willett, P. (1990). Algorithms for the Calculation of Similarity in Chemical Structure Databases. In *Concepts and Applications of Molecular Similarity* (Johnson, M.A. and Maggiora, G.M., eds.), Wiley, New York (NY), pp. 43–63.

Willett, P. (1991). *Three-Dimensional Chemical Structure Handling.* Research Studies Press – Wiley, Taunton (UK).

Willett, P. (1997). Computational Tools for the Analysis of Molecular Diversity. *Persp.Drug Disc.Des.*, *7/8*, 1–11.

Willett, P. (1998). Chemical Similarity Searching. *J.Chem.Inf.Comput.Sci.*, *38*, 983–996.

Wilson, L.Y. and Famini, G.R. (1991). Using Theoretical Descriptors in Quantitative Structure-Activity Relationships: Some Toxicological Indices. *J.Med.Chem.*, *34*, 1668–1674.

Winberg, N. and Mislow, K. (1995). A Unification of Chirality Measures. *J.Math.Chem.*, *17*, 35–53.

Winiwarter, S., Bonham, N.M., Ax, F., Hallberg, A., Lennernäs, H. and Karlén, A. (1998). Correlation of Human Jejunal Permeability (in Vivo) of Drugs with Experimentally and Theoretically Derived Parameters. A Multivariate Data Analysis Approach. *J.Med.Chem.*, *41*, 4939–4949.

Winkler, D.A. and Burden, F.R. (1998a). Holographic QSAR of Benzodiazepines. *Quant.Struct.-Act.Relat.*, *17*, 224–231.

Winkler, D.A., Burden, F.R. and Watkins, A.J.R. (1998b). Atomistic Topological Indices Applied to Benzodiazepines Using Various Regression Methods. *Quant.Struct.-Act.Relat.*, *17*, 14–19.

Wipke, W.T. and Dyott, T.M. (1974). Stereochemically Unique Naming Algorithm. *J.Am.Chem. Soc.*, *96*, 4834–4842.

Wirth, K. (1986). Coding of Relational Descriptions of Molecular Structures. *J.Chem.Inf.Comput.Sci.*, *26*, 242–249.

Wise, S.A., Bonnett, W.J., Guenther, F.R. and May, W.E. (1981). A Relationship Between Reversed-phase C_{18} Liquid Chromatographic Retention and the Shape of Polycyclic Aromatic Hydrocarbons. *J.Chromatogr.Sci.*, *19*, 457–465.

Wold, S. (1978). Cross-Validatory Estimation of the Number of Components in Factor and Principal Components Models. *Technometrics*, *20*, 397–405.
Wold, S. and Sjöström, M. (1978). Linear Free Energy Relationships as Tools for Investigating Chemical Similarity. Theory and Practice. In *Correlation Analysis in Chemistry* (Chapman, N.B. and Shorter, J., eds.), Plenum Press, New York (NY), pp. 1–54.
Wold, S. and Dunn III, W.J. (1983). Multivariate Quantitative Structure-Activity Relationships (QSAR): Conditions for their Applicability. *J.Chem.Inf.Comput.Sci.*, *23*, 6–13.
Wold, S., Ruhe, A., Wold, H. and Dunn III, W.J. (1984). The Collinearity Problem in Linear Regression. The Partial Least Squares (PLS) Approach to Generalized Inverses. *SIAM J.Sci. Stat.Comput.*, *5*, 735–743.
Wold, S. (1991). Validation of QSAR's. *Quant.Struct.-Act.Relat.*, *10*, 191–193.
Wold, S., Johansson, E. and Cocchi, M. (1993). PLS – Partial Least Squares Projection of Latent Structures. In *3D QSAR in Drug Design. Theory, Methods and Applications.* (Kubinyi, H., ed.), ESCOM, Leiden (The Netherlands), pp. 523–550.
Wold, S. (1994). PLS for Multivariate Linear Modeling. In *Chemometric Methods in Molecular Design. Vol. 2* (van de Waterbeemd, H., ed.), VCH Publishers, Weinheim (Germany), pp. 195–218.
Wold, S. and Eriksson, L. (1995). Statistical Validation of QSAR Results. Validation Tools. In *Chemometrics Methods in Molecular Design – Vol. 2* (van de Waterbeemd, H., ed.), VCH Publishers, New York (NY), pp. 309–318.
Wold, S., Kettaneh-Wold, N. and Tjessem, K. (1996). Hierarchical Multiblock PLS and PC Models for Easier Interpretation and as an Alternative to Variable Selection. *J.Chemom.*, *10*, 463–482.
Woolley, R.G. (1976). On the Description of High-Resolution Experiments in Molecular Physics. *Chem.Phys.Lett.*, *44*, 73–75.
Woolley, R.G. (1978a). Further Remarks on Molecular Structure in Quantum Theory. *Chem. Phys.Lett.*, *55*, 443.
Woolley, R.G. (1978b). Must a Molecule Have a Shape? *J.Am.Chem.Soc.*, *100*, 1073–1078.
Wootton, R., Cranfield, R., Sheppey, G.C. and Goodford, P.J. (1975). Physicochemical-Activity Relationships in Practice. 2. Rational Selection of Benzenoid Substituents. *J.Med.Chem.*, *18*, 607–613.
Worth, A.P. and Cronin, M.T.D. (1999). Embedded Cluster Modelling – A Novel Method for Analysing Embedded Data Sets. *Quant.Struct.-Act.Relat.*, *18*, 229–235.
Xiao, Y., Qiao, Y., Zhang, J., Lin, S. and Zhang, W. (1997). A Method for Substructure Search by Atom-Centered Multilayer Code. *J.Chem.Inf.Comput.Sci.*, *37*, 701–704.
Xie, Q., Sun, H., Xie, G. and Zhou, J. (1995). An Iterative Method for Calculation of Group Electronegativities. *J.Chem.Inf.Comput.Sci.*, *35*, 106–109.
Xu, L. (1992). Molecular Toplogical Index a_N and its Extension. *J.Serb.Chem.Soc.*, *57*, 485–495.
Xu, L., Wang, H.-Y. and Su, Q. (1992a). A Newly Proposed Molecular Topological Index for the Discrimination of *Cis/Trans* Isomers and for the Studies of QSAR/QSPR. *Computers Chem.*, *16*, 187–194.
Xu, L., Wang, H.-Y. and Su, Q. (1992b). Correlation Analysis in Structure and Chromatographic Data of Organophosphorus Compounds by GAI. *Computers Chem.*, *16*, 195–199.
Xu, L., Ball, J., Dixon, S.L. and Jurs, P.C. (1994). Quantitative Structure-Activity Relationships for Toxicity of Phenols Using Regressions Analysis and Computational Networks. *Environ. Toxicol.Chem.*, *13*, 841–851.
Xu, L., Yao, Y.Y. and Wang, H.-Y. (1995). New Topological Index and Prediction of Phase Transfer Energy for Protonated Amines and Tetraalkylamines Ions. *J.Chem.Inf.Comput.Sci.*, *35*, 45–49.
Xue, L., Godden, J., Gao, H. and Bajorath, J. (1999). Identification of a Preferred Set of Molecular Descriptors for Compound Classification Based on Principal Component Analysis. *J.Chem. Inf.Comput.Sci.*, *39*, 699–704.
Yalkowsky, S.H. and Valvani, S.C. (1979). Solubilities and Partitioning 2. Relationships Between Aqueous Solubilities, Partition Coefficients, and Molecular Surface Areas of Rigid Aromatic Hydrocarbons. *Journal of Chemical and Engineering Data*, *24*, 127–129.

Yalkowsky, S.H., Dannenfelser, R.M., Myrdal, P. and Simamora, P. (1994a). Unified Physical Property Estimation Relationships (UPPER). *Chemosphere*, *28*, 1657–1673.

Yalkowsky, S.H., Myrdal, P., Dannenfelser, R.M. and Simamora, P. (1994b). UPPER II: Calculation of Physical Properties of the Chlorobenzenes. *Chemosphere*, *28*, 1675–1688.

Yamagami, C., Kawase, K. and Fujita, T. (1999). Hydrophobicity Parameters Determined by Reversed-Phase Liquid Chromatography. XIII A new Hydrogen-accepting Scale of Monosubstituted (Di)azines for the Relationship Between Retention Factor and Octanol-Water Partition Coefficient. *Quant.Struct.-Act.Relat.*, *18*, 26–34.

Yamamoto, Y. and Otsu, T. (1967). Effects of Substituents in Radical Reactions: Extension of the Hammett Equation. *Chem.& Ind.*, 787–789.

Yang, G., Lien, E.J. and Guo, Z. (1986). Physical Factors Contributing to Hydrophobic Constant π. *Quant.Struct.-Act.Relat.*, *5*, 12–18.

Yang, J.A. and Kiang, Y.-S. (1983). *Acta Chim.Sin.*, *41*, 884.

Yang, S.Y., Bumgarner, J.G., Kruk, L.F. and Khaledi, M.G. (1996). Quantitative Structure-Activity Relationships Studies with Micellar Electrokinetic Chromatography Influence of Surfactant Type and Mixed Micelles on Estimation of Hydrophobicity and Bioavailability. *J.Chromat.*, *721A*, 323–335.

Yang, W. and Mortier, W.J. (1986). The Use of Global and Local Molecular Parameters for the Analysis of the Gas-Phase Basicity of Amines. *J.Am.Chem.Soc.*, *108*, 5708–5711.

Yang, Y.-Q., Xu, L. and Hu, C.-Y. (1994). Extended Adjacency Matrix Indices and their Applications. *J.Chem.Inf.Comput.Sci.*, *34*, 1140–1145.

Yao, Y.Y., Xu, L., Yang, Y.-Q. and Yuan, X.S. (1993a). Study on Structure-Activity Relationships of Organic Compounds: Three New Topological Indices and their Applications. *J.Chem.Inf. Comput.Sci.*, *33*, 590–594.

Yao, Y.Y., Xu, L. and Yuan, X.S. (1993b). A New Topological Index for Research on Structure-Property Relationships of Alkane. *Acta Chim.Sin.*, *51*, 463–469.

Yi-Qui, Y., Xu, L. and Hu, C.-Y. (1994). *J.Chem.Inf.Comput.Sci.*, *34*, 1140–1145.

Yoneda, Y. (1979). An Estimation of the Thermodynamic Properties of Organic Compounds in the Ideal Gas State. I. Acyclic Compounds and Cyclic Compounds with a Ring of Cyclopentane, Cyclohexane, Benzene or Naphthalene. *Bull.Chem.Soc.Jap.*, *52*, 1297–1314.

Yoshida, F. and Topliss, J.G. (1996). Unified Model for the Corneal Permeability of Related and Diverse Compounds with Respect to Their Physicochemical Properties. *J.Pharm.Sci.*, *85*, 819–823.

Young, S.S. and Hawkins, D.M. (1995). Analysis of a 2^9 Full Factorial Chemical Library. *J.Med. Chem.*, *38*, 2784–2788.

Young, S.S., Profeta, S., Unwalla, R.J. and Kosh, J.W. (1997). Exploratory Analysis of Chemical Structure, Bacterial Mutagenicity and Rodent Tumorigenicity. *Chemom.Intell.Lab.Syst.*, *37*, 115–124.

Young, S.S. and Hawkins, D.M. (1998). Using Recursive Partitioning to Analyze a Large SAR Data Set. *SAR & QSAR Environ.Res.*, *8*, 183–193.

Yukawa, Y. and Tsuno, Y. (1959). *Bull.Chem.Soc.Jap.*, *32*, 971.

Yukawa, Y., Tsuno, Y. and Sawada, M. (1966). *Bull.Chem.Soc.Jap.*, *39*, 2274.

Yukawa, Y., Tsuno, Y. and Sawada, M. (1972a). The Substituent Effect. I. Normal Substituent Constants from the Hydrolysis of Substituted-benzyl Benzoates. *Bull.Chem.Soc.Jap.*, *45*, 1198–1205.

Yukawa, Y., Tsuno, Y. and Sawada, M. (1972b). The Substituent Effect. III. The Basicities of Polynuclear Aryl Methyl Ketones. *Bull.Chem.Soc.Jap.*, *45*, 1210–1216.

Zabrodsky, H. and Avnir, D. (1995). Continuous Symmetry Measures. 4. Chirality. *J.Am.Chem. Soc.*, *117*, 462–473.

Zakarya, D., Belkhadir, M. and Fkih-Tetouani, S. (1993a). Quantitative Structure-Biodegradability Relationships (QSBRs) Using Modified Autocorrelation Method (MAM). *SAR & QSAR Environ.Res.*, *1*, 21–27.

Zakarya, D., Tiyal, F. and Chastrette, M. (1993b). Use of the Multifunctional Autocorrelation Method to Estimate Molar Volumes of Alkanes and Oxygenated Compounds Comparison Between Components of Autocorrelation Vectors and Topological Indexes. *J.Phys.Org.Chem.*, *6*, 574–582.

Zakarya, D. and Chastrette, M. (1998). Contribution of Structure-Odor Relationships to the Elucidation of the Origin of Musk Fragrance Activity. In *Comparative QSAR* (Devillers, J., ed.), Taylor & Francis, Washington (DC), pp. 169–195. [R]

Zaliani, A. and Gancia, E. (1999). MS-WHIM Scores for Amino Acids: A New 3D-Description for Peptide QSAR and QSPR Studies. *J.Chem.Inf.Comput.Sci.*, *39*, 525–533.

Zar, J.H. (1984). *Biostatistical Analysis*. Prentice-Hall, Englewood Cliffs (NJ), 718 pp.

Zefirov, N.S., Kirpichenok, M.A., Izmailov, F.F. and Trofimov, M.I. (1987). Scheme for the Calculation of the Electronegativities of Atoms in a Molecule in the Framework of Sanderson's Principle. *Dokl.Akad.Nauk.SSSR*, *296*, 883–887.

Zefirov, N.S., Palyulin, V.A. and Radchenko, E.V. (1991). Problem of Generation of Structures with Specified Properties. Solution of the Inverse Problem for Balaban Centric Index. *Dokl. Akad.Nauk.SSSR*, *316*, 921–924.

Zefirov, N.S., Palyulin, V.A., Skvortsova, M.I. and Baskin, I.I. (1995). Inverse Problems in QSAR. In *QSAR and Molecular Modelling: Cocepts, Computational Tools and Biological Applications* (Sanz, F., Giraldo, J. and Manaut, F., eds.), Prous Science, Barcelona (Spain), pp. 40–41.

Zefirov, N.S., Palyulin, V.A. and Radchenko, E.V. (1997a). Molecular Field Topology Analysis in Studies of Quantitative Structure-Activity Relationships for Organic Compounds. *Dokl.Akad. Nauk.SSSR*, *352*, 23–26.

Zefirov, N.S. and Tratch, S.S. (1997b). Some Notes on Randic-Razinger's Approach to Characterization of Molecular Shape. *J.Chem.Inf.Comput.Sci.*, *37*, 900–912.

Zerovnik, J. (1996). Computing the Szeged Index. *Croat.Chem.Acta*, *69*, 837–843.

Zerovnik, J. (1999). Szeged Index of Symmetric Graphs. *J.Chem.Inf.Comput.Sci.*, *39*, 77–80.

Zetzsch, C. (1982). Predicting the Abiotic Degradability of Organic Chemicals in the Atmosphere by OH Using Structure Reactivity Relations. Chemicals in the Environment, Copenhagen (The Netherlands), 302–312.

Zhang, R., Liu, S., Liu, M. and Hu, Z. (1997). Neural Network-Molecular Descriptors Approach to the Prediction of Properties of Alkenes. *Computers Chem.*, *21*, 335–341.

Zhang, Y. (1982a). Electronegativities of Elements in Valence States and their Applications. 1. Electronegativities of Elements in Valence States. *Inorg.Chem.*, *21*, 3886–3889.

Zhang, Y. (1982b). Electronegativities of Elements in Valence States and their Applications. 2. A Scale of Strengths of Lewis Acids. *Inorg.Chem.*, *21*, 3889–3893.

Zheng, W., Cho, S.J. and Tropsha, A. (1998). Rational Combinatorial Library Design. 1. Focus-2D: A New Approach to the Design of Targeted Combinatorial Chemical Libraries. *J.Chem. Inf.Comput.Sci.*, *38*, 251–258.

Zheng, W. and Tropsha, A. (2000). Novel Variable Selection Quantitative Structure-Property Relationship Approach Based on the *k*-Nearest-Neighbor Principle. *J.Chem.Inf.Comput.Sci.*, *40*, 185–194.

Zhou, J., Xie, Q., Sun, D., Xie, G., Cao, L. and Xu, L. (1993). Structure-Activity Relationships on Pesticides: A Development in Methodology and Its Sofware System. *J.Chem.Inf.Comput.Sci.*, *33*, 310–319.

Zhu, H.Y. and Klein, D.J. (1996a). Graph-Geometric Invariants for Molecular Structures. *J.Chem.Inf.Comput.Sci.*, *36*, 1067–1075.

Zhu, H.Y., Klein, D.J. and Lukovits, I. (1996b). Extensions of the Wiener Number. *J.Chem.Inf. Comput.Sci.*, *36*, 420–428.

Zivkovic, T., Trinajstic, N. and Randic, M. (1975). On Conjugated Molecules with Identical Topological Spectra. *Mol.Phys.*, *30*, 517–532.

Zivkovic, T. (1990). On the Evaluation of the Characteristic Polynomial of a Chemical Graph. *J.Comput.Chem.*, *11*, 217–222.

Zupan, J. and Novic, M. (1997). General Type of a Uniform and Reversible Representation of Chemical Structures. *Anal.Chim.Acta*, *348*, 409–418.

Zupan, J. and Gasteiger, J. (1999). *Neural Networks for Chemistry and Drug Design*. Wiley-VCH Publishers, Weinheim (Germany).

Zupan, J., Vracko, M. and Novic, M. (2000). New Uniform and Reversible Representation of 3-D Chemical Structures. *Acta Chim.Sloven.*, *47*, 19–37.

Zweerszeilmaker, W.M., Horbach, G.J. and Witkamp, R.F. (1997). Differential Inhibitory Effects of Phenytoin, Diclofenac, Phenylbutazone and a Series of Sulfonamides on Hepatic Cytochrome P4502c Activity in Vitro, and Correlation with Some Molecular Descriptors in the Dwarf Goat (caprus hircus aegagrus). *Xenobiotica*, *27*, 769–780.